I0055834

ASTROPARTICLE, PARTICLE, SPACE PHYSICS AND DETECTORS FOR PHYSICS APPLICATIONS

Proceedings of the 13th ICATPP Conference

Astroparticle, Particle, Space Physics, Radiation Interaction, Detectors and Medical Physics Applications

The book series is dedicated to an up-to-date coverage of investigations, physics requirements, survey of technologies and performance of detectors employed — or to be employed — in fundamental and particle physics experiments at accelerators, underground laboratories, submarine facilities, and in space environment, including Earth magnetosphere and heliosphere, for astroparticle and cosmic-ray physics experiments and astrophysics research, special applications like medical imaging, radiotherapy, simulation codes for dose estimates in radiotherapy, biological effects, radiation effects on devices, synchrotron radiation and new advanced detectors.

---

*Published**

Vol. 2 Astroparticle, Particle and Space Physics, Detectors and Medical Physics Applications
Proceedings of the 8th Conference
edited by M. Barone, E. Borchi, C. Leroy, P.-G. Rancoita, P.-L. Riboni & R. Ruchti

Vol. 3 Astroparticle, Particle and Space Physics, Detectors and Medical Physics Applications
Proceedings of the 9th Conference
edited by M. Barone, E. Borchi, A. Gaddi, C. Leroy, L. Price, P.-G. Rancoita & R. Ruchti

Vol. 4 Astroparticle, Particle and Space Physics, Detectors and Medical Physics Applications
Proceedings of the 10th Conference
edited by M. Barone, A. Gaddi, C. Leroy, L. Price, P.-G. Rancoita & R. Ruchti

Vol. 5 Astroparticle, Particle and Space Physics, Detectors and Medical Physics Applications
Proceedings of the 11th Conference
edited by C. Leroy, P.-G. Rancoita, M. Barone, A. Gaddi, L. Price & R. Ruchti

Vol. 6 Cosmic Rays for Particle and Astroparticle Physics
Proceedings of the 12th ICATPP Conference
edited by S. Giani, C. Leroy & P.-G. Rancoita

Vol. 7 Astroparticle, Particle, Space Physics and Detectors for Physics Applications
Proceedings of the 13th ICATPP Conference
edited by S. Giani, C. Leroy, L. Price, P.-G. Rancoita & R. Ruchti

*The complete list of the published volumes in the series can be found at http://worldscibooks.wspc.com.sg/series/apsprdma_series.shtml.

ASTROPARTICLE, PARTICLE, SPACE PHYSICS AND DETECTORS FOR PHYSICS APPLICATIONS

Proceedings of the 13th ICATPP Conference

Villa Olmo, Como, Italy 3–7 October 2011

Editors

S Giani
CERN, Switzerland

C Leroy
Université de Montréal, Canada

L Price
Argonne National Laboratory, USA

P-G Rancoita
INFN Milano-Bicocca, Italy

R Ruchti
University of Notre Dame, USA

NEW JERSEY • LONDON • SINGAPORE • BEIJING • SHANGHAI • HONG KONG • TAIPEI • CHENNAI

Published by

World Scientific Publishing Co. Pte. Ltd.
5 Toh Tuck Link, Singapore 596224
USA office: 27 Warren Street, Suite 401-402, Hackensack, NJ 07601
UK office: 57 Shelton Street, Covent Garden, London WC2H 9HE

British Library Cataloguing-in-Publication Data
A catalogue record for this book is available from the British Library.

Astroparticle, Particle, Space Physics, Radiation Interaction, Detectors and Medical Physics Applications — Vol. 7
ASTROPARTICLE, PARTICLE, SPACE PHYSICS AND DETECTORS FOR PHYSICS APPLICATIONS
Proceedings of the 13th ICATPP Conference

ISBN-13 978-981-4405-06-5
ISBN-10 981-4405-06-X

Printed in Singapore.

ORGANIZING COMMITTEES

Organizing Committee:

S. Giani	CERN
C. Leroy	University of Montreal
L. Price	ANL
P.G. Rancoita	INFN Milano-Bicocca (chairman)
R. Ruchti	University of Notre Dame

Local Organizers:

M. Falasconi	Centro Volta, Conference Secretariat
F. Mercalli	Director of Centro Volta

Session Organizers and International Advisory Committee:

D. Abbaneo	CERN
S. Baccaro	ENEA-Rome
A. Bellerive	Carleton University, Ottawa
P. Binko	ISDC and University of Geneva
A. Capone	INFN and University "La Sapienza", Rome
M. Cirelli	CNRS IPhT Saclay and CERN
F. Donato	INFN and Univ. of Torino
F. Favata	HQ ESA
S. Gabici	Laboratoire APC, CNRS and University of Paris 7
M. Gervasi	INFN and Univ. of Milano-Bicocca
A. Ibarra	TUM Munich
F. Jansen	DLR Bremen
K. Kudela	Slovak Academy of Sciences, Kosice
E. Nappi	INFN-Bari

J. Pinfold	University of Alberta
S. Pospisil	CTU in Prague
M.S. Potgieter	North-West University, Potchefstroom
J. Seguinot	College de France, Paris
A. Strong	Max Planck Institut, Garching

Scientific Organization Assistence:

S. Della Torre	INFN Milano-Bicocca and University of Insubria
D. Grandi	INFN Milano-Bicocca
M. Tacconi	INFN and Univ. of Milano-Bicocca

Plenary Session Organizers:

A. Bellerive	(a) Dark Matter Searches and Man-Made Neutrinos (b) Underground and Underwater Experiments
M. Cirelli and S. Gabici	Production of Cosmic Rays from Exotic Matter and Astrophysical Sources
F. Donato, K. Kudela and M.S. Potgieter	Cosmic Rays production and propagation in the Galaxy and Heliosphere
F. Favata	Space Experiments
E. Nappi and J. Seguinot	Advanced Detectors and Particle Identification
J. Pinfold	Experimental Observations of Cosmic Rays (a) in Space and (b) at Ground
L. Price and R. Ruchti	High Energy Physics Experiments
R. Ruchti	Synergy in High Energy and Cosmic Rays Physics

Parallel Session Organizers:

D. Abbaneo	Tracker and Applications
S. Baccaro	Devices and Material in Radiation
S. Giani	Software Applications for High Energy Physics and Space Environment
A. Bellerive	(a) Dark Matter Searches and (b) Underground and Underwater Experiments, Reactor and Accelerator neutrinos

M. Cirelli and S. Gabici	Production of Cosmic Rays from Exotic Matter and Astrophysical Sources
F. Favata	Space Experiments
S. Giani and C. Leroy	Poster Session
F. Donato, K. Kudela and M.S. Potgieter	Cosmic Rays production and propagation in the Galaxy and Heliosphere
C. Leroy	Calorimetry and Applications
E. Nappi and J. Seguinot	Advanced Detectors and Particle Identification
J. Pinfold	Experimental Observations of Cosmic Rays (a) in Space and (b) at Ground
J. Pinfold and R. Ruchti	Broader Impacts Activities in High Energy Physics and Astrophysics
S. Pospisil	Position Sensitive Detectors & Tracking Devices
L. Price and R. Ruchti	High Energy Physics Experiments

The Article Committee:

S. Giani	CERN
C. Leroy	University of Montreal
P.G. Rancoita	INFN Milano-Bicocca

Conference Secretariat:

M. Falasconi	Centro di Cultura Scientifica A. Volta, Como
N. Tansini	Centro di Cultura Scientifica A. Volta, Como
G. Fontanive	Centro di Cultura Scientifica A. Volta, Como

PREFACE

The exploration of the subnuclear world is done through increasingly complex experiments covering a wide range of energy and performed in a large variety of environments going from particle accelerators, underground detectors up to satellites and space laboratory. The achievement of these research programs calls for novel techniques, new materials and new instrumentation to be used in detectors, often of large scale. Therefore, particle physics is at the forefront of technological advance and also leads to many applications. Among these, are imaging applications in medical and biological fields reported in the session on position sensitive detectors and tracking devices of the conference. The International Conference on Advanced Technology and Particle Physics is held every two years (the 12th Edition particularly dedicated to Cosmic Rays for Particle and Astroparticle Physics was exceptionally organized in October 2010). The Conference held during the week 3-7 October 2011 at the "Centro di Cultura Scientifica A. Volta" was the 13th Edition and again welcomed a large participation. There were about 250 participants representing more than 150 institutions from 23 countries. The participants were experienced researchers but also graduate students and recent postdoctoral fellows, students receiving financial support from the Conference organization.

The conference allows a regular review of the advances made in all technological aspects of the experiments at various stages, data taking, upgrade or in preparation. The open and flexible format of the Conference is conducive to fruitful exchanges of points of view among participants that permit the measure of the progresses made and indicate research directions. This year many contributions were reporting first results obtained by the LHC experiments and progress done in preparation of these experiments upgrades, including plan for the LHC machine upgrade. Progresses and results from space experiments, in particular AMS-02 (set into operation on the International Space Station on 19 May 2011), were also reported.

Plenary and parallel sessions covered:

1. Advanced Detectors and Particle Identification (organized by E. Nappi and J. Seguinot),
2. Dark Matter Searches, Underground and Underwater Experiments, Reactor and Accelerator neutrinos (organized by A. Bellerive),
3. Space Experiments (organized by F. Favata),
4. Cosmic Rays Production and Propagation in the Galaxy and Heliosphere (organized by F. Donato, K. Kudela and M.S. Potgieter),
5. Calorimetry and Applications (organized by C. Leroy),
6. Software Applications for High Energy Physics and Space Environment (organized by S. Giani),
7. Production of Cosmic Rays from Exotic Matter and Astrophysical Sources (organized by M. Cirelli and S. Gabici),
8. Experimental Observations of Cosmic Rays in Space and at Ground (organized by J. Pinfold),
9. High Energy Physics Experiments (organized by L. Price and R. Ruchti),
10. Position Sensitive Detectors and Tracking Devices (organized by S. Pospisil),
11. Broader Impacts Activities in High Energy Physics and Astrophysics (organized by J. Pinfold and R. Ruchti),
12. Synergy in High Energy and Cosmic Rays Physics (organized by R. Ruchti),
13. Devices and Material in Radiation (organized by S. Baccaro) and
14. Tracker and Applications (organized by D. Abbaneo).

A poster exhibition was organized by S. Giani and C. Leroy. The posters were divided into three general categories as follows: i) Astroparticle, Underground Experiments, Space Physics and Cosmic Rays, ii) High-Energy Physics Experiments, Trackers, Calorimetry, Software and Data Systems and iii) Advanced Detectors and Medical Physics Applications. The poster papers accepted for publications have been included in the appropriate section of the Proceedings, as function of their topic. From the quality of the material of the posters, it is clear that a lot of efforts went into their preparation making the posters a truly important part of this Conference.

The Article Committee was set to follow the article submission, review and publication in the Conference Proceedings.

We would like to thank the staff of the Centro A. Volta for the excellent support provided to the Conference organization at Villa Olmo. In particular, we would like to extend our appreciation and thanks to the Secretariat of Centro di Cultura Scientifica A. Volta for their help and efficiency with the organization of the Conference and its running. The help of Mauro Tacconi, from INFN and Milano Bicocca University, in the preparation of the Conference proceedings is gratefully acknowledged.

The organizers would like to thank the strong support of INFN, Centro Volta, UTEF-CTU in Prague and the Milano-Bicocca University which made the conference possible.

Finally, we would like to thank the speakers for the high quality of their contributions and the participants for their enthusiasm in attending the Conference and contributing to the discussions.

Article and Organizing Committee

Simone Giani
Claude Leroy
Pier-Giorgio Rancoita

Organizing Committee

Larry Price
Randal Ruchti

December 2011

CONTENTS

Space Experiments and Cosmic Rays Observations

L. Baldini	Science highlights from the Fermi Observatory
M. Bertaina	The JEM-EUSO Mission
P. Bobik	Estimation of JEM-EUSO experiment observation efficiency
E. Bottacini	Combining the SWIFT/BAT and the INTEGRAL/ISGRI observations
T. Bretz	A status report: FACT — a fact!
W.J. Burger	Aiglon: A Magnetic Spectrometer for Geophysics
P. Cristofari	Multi-TeV gamma rays from supernova remnants
G. Decerprit	Constraints on the origin of ultra-high-energy cosmic rays from cosmogenic photons and neutrinos
D. D'Urso	The AMS-02 Silicon Tracker: Status and Performances
N. Giglietto	Lunar gamma ray emission as observed by FERMI
T.G. Guzik	The CALorimeter Electron Telescope (CALET) for High-Energy Astroparticle Physics on the International Space Station
D. Haas	LOFT — The Large Observatory For X-ray Timing — scientific goals and technical realization
C.C. Hsu	Studying Cosmic Ray Composition Around the Knee Region with the ANTARES Telescope
H. Kellermann	Absolute Measurement of the Reflectivity and the Point Spread Function of the MAGIC Telescopes
F. Kislat	Status of the IceTop air shower array at the South Pole
L. Knötig	Light Sensor Candidates for the Cherenkov Telescope Array
G. Marsella	Recent results from the ARGO-YBJ experiment

N.E. Mavromatos	Recent Results from Indirect and Direct Dark Matter Searches — Theoretical Scenarios
A. Mennella	The microwave sky after one year of Planck operations
E. Mocchiutti	The PAMELA Experiment: Five Years of Cosmic Rays Investigation
C. Monte	Observation of blazars with the high energy SED peak in the Fermi-LAT band
Y.F. Novoseltsev	EAS spectrum vs the total number of high energy muons and the mass composition of primary cosmic rays in the energy region 10^{15}-10^{16} eV
G. Pareschi	Status of the Cerenkov Telescope Array (CTA)
M. Pearce	PoGOLite: a balloon-borne X-ray polarimeter
S. Rainó	Fermi-LAT analysis of Fermi, Planck, Swift and radio selected samples of AGN
R. Rando	Fermi results in cosmic ray physics
H. Schoorlemmer	Cosmic rays detected with the Auger Engineering Radio Array
V.G. Sinitsyna	Cosmic Ray Acceleration in Historical Supernova Remnants in our Galaxy
V.Y. Sinitsyna	Search for neutrino decay at SHALON
O. Tibolla	Recent Results from the MAGIC Telescopes
V. Togo	Search for massive penetrating particles in the Cosmic Radiation
N. Tomassetti	AMS Observations of Light Cosmic Ray Isotopes and Implications for their Production in the Galaxy

Science highlights from the *Fermi* Observatory

Luca Baldini for the *Fermi*-LAT collaboration
INFN-Sezione di Pisa, Largo B. Pontecorvo, 3
Pisa, 56127, Italy
** E-mail:luca.baldini@pi.infn.it*

Successfully launched in June 2008, the Fermi Gamma-ray Space Telescope, formerly named GLAST, has been observing the high-energy gamma-ray sky with unprecedented sensitivity for more than three years, opening a new observational window on a wide variety of astrophysical objects. This paper is a short overview of the main science highlights, not including those aspects of the broad *Fermi* science menu which are covered in separate contributions presented at this conference (most notably Cosmic-ray studies, Galactic sources and Active Galactic Nuclei).

Keywords: Gamma-ray astrophysics, Gamma-Ray Bursts, Dark Matter.

1. Introduction

Designed to survey the gamma-ray sky in the broad energy range from 20 MeV to more than 300 GeV, with the additional capabilities of studying transient phenomena at lower energies, the *Fermi* γ*-ray Space Telescope* is the reference space-borne gamma-ray observatory of this decade.

Fermi carries two instruments on-board: the Gamma-ray Burst Monitor (GBM)[1] and the Large Area Telescope (LAT).[2] The GBM, sensitive in the energy range between 8 keV and 40 MeV, is designed to observe the full unocculted sky with rough directional capabilities (at the level of one to a few degrees) for the study of transient sources, particularly Gamma-Ray Bursts (GRBs). The LAT is a pair conversion telescope for photons above 20 MeV up to a few hundreds of GeV.

1.1. *The Large Area Telescope*

The LAT is a 4×4 array of identical towers, each one made by a tracker-converter module (hereafter tracker) and a calorimeter module. A segmented anti-coincidence detector (ACD) covers the tracker array and a programmable trigger and data acquisition system completes the instrument. Though owing most of the basic design to its predecessors—particularly the Energetic Gamma-Ray Experiment Telescope (EGRET)[3] on-board the *CGRO* mission—the LAT exploits modern particle-physics detector technology, which allows for a breakthrough leap in the instrument performance.

Each tracker module features 16 tungsten layers, promoting the conversion of γ-rays into e^+/e^- pairs, and 18 x-y pairs of single-sided, 228 μm silicon-strip detector planes—for a total of 1.5 radiation lengths of material on-axis. The silicon-sensor technology allows precise tracking (with no detector-induced dead time and no use of consumables) and the capability to self-trigger.

Each calorimeter module consists of 96 CsI(Tl) crystals, arranged in a hodoscopic configuration (for a total depth of $\sim$ 8.6 radiation lengths on axis). The calorimeter provides an intrinsically three-dimensional image of the shower development, which is crucial both for the energy reconstruction (especially at high-energy, where a significant part of the shower can leak out of the back of the instrument) and for background rejection.

The anti-coincidence detector, a set of plastic scintillators surrounding the tracker, is the first defense of the LAT against the overwhelming background due to charged CRs. In order to limit the "self-veto" effect—due to the back-splash of secondaries from high-energy particles hitting the calorimeter—it is segmented in 89 tiles providing spatial information that can be correlated with the signal from the tracker and the calorimeter.

The design, construction and operation of such a complex detector is a fascinating subject on its own and the interested readers can refer to[2] and references therein for further details. The LAT largely surpasses the previous generations of γ-ray telescopes in terms of effective area ($\sim$ 8000 cm^2 on axis above 1 GeV), energy range (from 20 MeV to more that 300 GeV), instrumental dead time (typically 26.5 μs), angular resolution (ranging from a few degrees to $\sim 0.1^\circ$ depending on energy) and field of view ($\sim$ 2.4 sr at 1 GeV). It has the ability to observe 20% of the sky at any time which, in the nominal scanning mode of operation, enables it to view the entire sky every three hours.

1.2. *Mission status and publicly released data products*

The operation of the instrument through the first three years of the mission was smooth at a level which is probably beyond the more optimistic prelaunch expectations. The LAT has been collecting science data for more than 99% of the time spent outside the South Atlantic Anomaly (SAA)[a]. The remaining tiny fractional down-time accounts for both hardware issues and detector calibrations. We note in passing that during this conference the LAT reached 200 billion event triggers taken on orbit, of which almost 40 billions were down-linked to ground for offline processing. More than 650 million gamma-ray candidates (i.e. events passing the background rejection selection) were made public and distributed to the Community through the Fermi Science Support Center (FSSC)[b].

Over the first three years of mission the LAT collaboration has put a considerable effort toward a better understanding of the instrument and of the environment in which it operates. In addition to that a continuous effort was made to in order to make the advances public as soon as possible. In August 2011 the first new event classification (*Pass 7*) since launch was released, along with the corresponding *Instrument Response Functions.* Compared with the pre-launch (*Pass 6*) classification, it features a greater and more uniform exposure, with a significance enhancement in acceptance below 100 MeV. The first major test case for *Pass 7* was the second LAT catalog, which will be briefly described in section 3.

2. The high-energy γ-ray sky seen by *Fermi*

The high-energy gamma-ray sky is dominated by diffuse emission: more than 70% of the photons detected by the LAT are produced in the interstellar space of our Galaxy by interactions of high-energy cosmic rays with matter and low-energy radiation fields. An additional diffuse component with an almost-isotropic distribution—and therefore thought to be extragalactic in origin—accounts for another significant fraction of the LAT photon sample. The rest consists of various different types of point-like or extended sources: Active Galactic Nuclei (AGN) and normal galaxies, pulsars and their relativistic wind nebulae, globular clusters, binary systems, shock-waves remaining from supernova explosions and nearby solar-system bodies like the Sun and the Moon.

[a]The SAA is the region, extending over much of South America, where the inner van Allen radiation belts lie closest to the earth surface. Fermi crosses the SAA several times a day (spending $\sim$ 13% of the time inside) and the LAT does not acquire science data during the passages.
[b]The FSSC is available at http://fermi.gsfc.nasa.gov/ssc

3. The Second *Fermi*-LAT catalog

The Second *Fermi*-LAT catalog[4] (2FGL)—representing the successor to the LAT Bright Source List[5] and the First *Fermi*-LAT catalog[6] (1FGL)—is the deepest catalog ever produced in the energy band between 100 MeV and 100 GeV. Compared to the 1FGL, it features several significant improvements: it is based on data from 24 (vs. 11) months of observation and makes use of the new *Pass 7* event selection (described in section 1.2) and of a higher resolution of the diffuse Galactic and isotropic emission. In addition to that, suitable spatial templates are used for 12 specific extended sources and spectral shapes different from simple power-laws are used in the spectral analysis.

The 2FGL catalog contains 1873 sources, whose sky-distribution is shown in figure 1. 127 of them are *firmly identified*, based either on periodic variability (e.g. pulsars) or on spatial morphology or on correlated variability. In addition to that 1170 are *reliably associated* with sources known at other wavelengths, while 576 (i.e. 31% of the total number of entries in the catalog) are still *unassociated*. It is interesting to note that 352 sources previously listed in the 1FGL, for various reasons, do not have a counterpart in the 2FGL.

A characteristic feature of the gamma-ray sky is its variability. The temporal variability is characterized for each source in the 2FGL via light curves in monthly time bins.

In addition to transients known for decades like GRBs (not included in 2FGL, see section 3.1) and flares from AGN, the LAT observed unexpected phenomena like the variability of the gamma-ray emission from the Crab nebula[7] and gamma-rays from the recurrent nova V407 Cyg.

3.1. *Other catalogs*

The *Fermi*-LAT collaboration is releasing to the Community additional, specific catalogs, on different timescales. The second catalog of Active Galactic Nuclei,[8] a follow up on the first AGN catalog,[9] was prepared at about the same time of the 2FGL. Similarly the second pulsar catalog, following up on the one released in 2010,[10] is currently in preparation. Finally, the first GRB catalog is in an advanced stage and should see the light soon.

4. Gamma-Ray Bursts

Gamma-Ray Burst (GRB) are the most violent explosions in the Universe. The isotropic energy emission, that can be estimated based on the measured

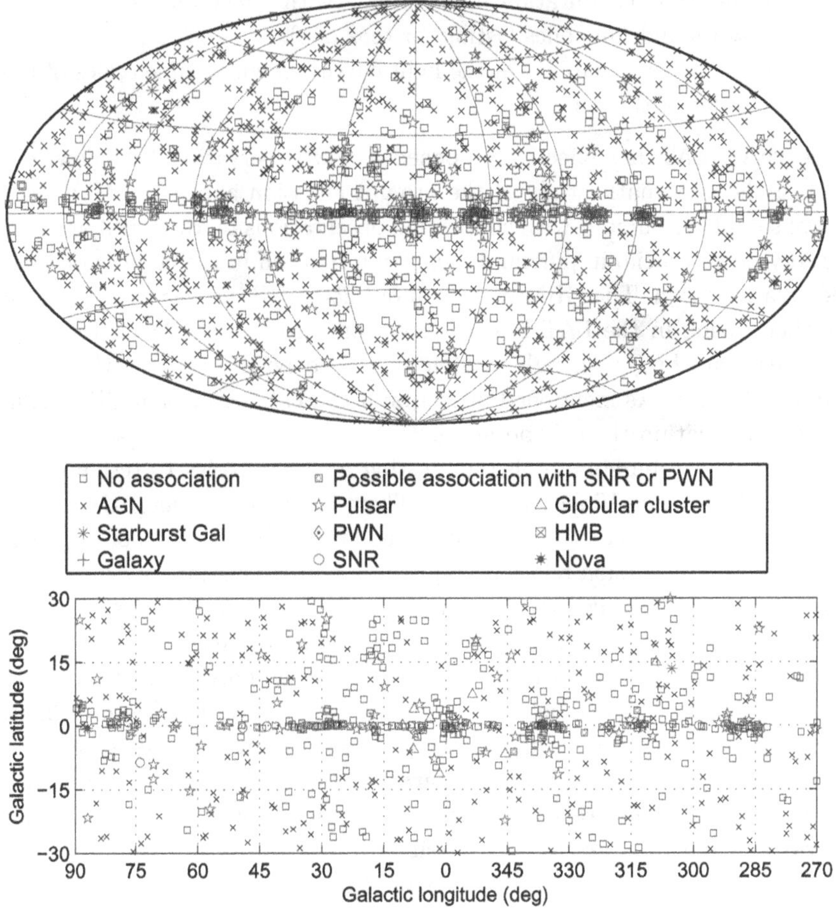

Fig. 1. 2FGL sky map (top) and close-up on the inner Galaxy (bottom) showing sources by source classes.[4]

fluence, can be as high as 10^{54} erg for the brightest bursts, with a huge energy release in keV–MeV gamma-rays. This naturally suggests a narrow beaming of the emission—still, a beaming factor of the order of 10^{-3} implies a considerable energy budget of 10^{51} erg.

GRBs have been extensively studied at relatively low energies over the last two decades, the most notable observed features being:

- the cosmological origin (they are isotropically distributed in the sky and have been observed up to a redshift $z \approx 9$);

- the bimodal duration distribution (with *short* bursts lasting for ≈ 1 s and *long* ones lasting for tens of s);
- the rapid variability of the light curves, down to time scales of the order of the ms.

The Fermi LAT has effectively opened a new observational window, enabling the systematic study of the prompt emission above 100 MeV on a large sample of bursts—with supporting observations by the GBM at lower energy being crucial in putting the new information in the context of what is already known. The GBM detects about 250 GRB per year, roughly half of them being in the LAT field of view; some 10% of the latter (or ≈ 10 per year) are bright enough at high energy to be detected by the Large Area Telescope. As mentioned in section 3.1, a catalog of such GRBs is in production within the collaboration.

Gamma-ray bursts are also interesting as laboratories to test possible Lorentz Invariance Violations (LIV). There are indeed several theoretical frameworks that predict (or can accommodate) a modification of the standard photon dispersion relation, implying a non trivial energy dependence of the speed of light in vacuum. The experimental signature of such a scenario would be the fact that two photons of different energies, emitted simultaneously, will travel with different velocities and arrive to the observer with some relative time delay. Even a tiny variation in the speed of light, when accumulated over cosmological distances, can in principle be revealed at energies much lower than the characteristics scale $m_{\rm QG}c^2$. Indeed, the short duration, rapid variability and cosmological distance make GRBs for perfect candidates to constraint possible LIV.

The most stringent limit on possible LIV comes from a 31 GeV photon emitted 0.829 s after trigger time for GRB 090510,[11] a Gamma-Ray Burst with a measured redshift of $z = 0.903 \pm 0.003$. Sure enough, we don't know *exactly* when this particular photon was emitted, relative to the much more abundant low-energy gamma-rays, so that this arrival time cannot be readily interpreted as a *time delay*. However, if we assume that the photon itself was not emitted before the beginning of the precursor of the burst (30 ms before the GBM trigger), that gives us all the necessary ingredients to set a robust lower limit $m_{\rm QG} > 1.19 m_{\rm P}$ in the linear subluminal case. In fact this is the most stringent limit of its kind available so far, and, effectively, it strongly disfavors any scenario characterized by a linear, subluminal LIV. The assumption on which it is based is very reasonable and somewhat supported by the fact that in none of the bursts detected by the Fermi LAT high-energy photons were detected before the onset of the low-energy emission—on the contrary, in many cases there was evidence for a spectral evolution with a consequent delay of the high-energy emission onset. The reader is referred to[11] for further details.

5. Indirect Dark Matter searches

One of the major open issues in our understanding of the Universe is the existence of an extremely-weakly interacting form of matter, the Dark Matter (DM), supported by a wide range of observations including large scale structures, the cosmic microwave background and the isotopic abundances resulting from the primordial nucleosynthesis. Complementary to *direct* searches being carried out in underground facilities and at accelerators, the *indirect* search for DM is one of the main items in the broad *Fermi* Science menu.

The word *indirect* denotes here the search for signatures of Weakly Interactive Massive Particle (WIMP) annihilation or decay processes through the final products (gamma-rays, electrons and positrons, antiprotons) of such processes. Among many other ground-based and space-borne instruments, the LAT plays a prominent role in this search through a variety of distinct search targets: gamma-ray lines, Galactic and isotropic diffuse gamma-ray emission, dwarf satellites, CR electrons and positrons.

5.1. *Gamma-ray lines*

The quest for a possible narrow line in the diffuse γ-ray emission arises naturally since photons can be produced in two-body DM particle annihilations $\chi\chi \to \gamma X$ or decays $\chi \to \gamma X$. Since, in most scenarios, DM particles are electrically neutral (and therefore do not couple directly to photons) such processes do not occur at tree level and the branching ratios are expected to be strongly suppressed. Despite this, a photon line, if present, is relatively easy to identify and distinguish from the standard astrophysical sources of gamma rays—whose flux is dominant in most situations. Therefore, this discovery channel features a distinctive experimental signature that, if observed, would incontrovertibly indicate new physics at work.

The detector response to a monochromatic line is obviously not a monochromatic line and the effect of the finite energy resolution cannot be ignored. Nonetheless the response can be modeled by means of Monte Carlo simulations and verified with tests at accelerators so that it can be effectively folded into the procedure and used to asses the statistical significance of a possible line component in the measured count spectra.

No significant evidence of gamma-ray line(s) has been found in the first

11 months of data, between 30 and 200 GeV[12] and work is ongoing to extend the energy range of the analysis and include more data. The detailed discussion of the upper limits obtained and their relevance in the context of specific DM models is beyond the scope of this brief overview.

5.2. *DM signals from the Galactic halo*

The Milky Way halo and the Galactic center are obvious candidates for indirect dark matter searches in gamma-rays: given the large DM content, a large annihilation signal can be potentially expected. The main challenge is presented by the strong gamma-ray foreground comprising the Galactic diffuse emission. Indeed, the detailed modeling of this foreground is currently the main limiting factor for DM searches in this channel.

In the case of the Galactic center a firm assessment of the conventional astrophysical signals is furthermore complicated by the problem of source confusion, due to the limited angular resolution of the instrument, and of pile-up along the line of sight.

5.3. *Dwarf galaxies*

Dwarf satellites of the Milky Way are among the cleanest targets for indirect dark matter searches in gamma-rays. They are systems with a very large mass/luminosity ratio (i.e. systems which are largely DM dominated). The LAT detected no significant emission from any of such systems and the upper limits on the γ-ray flux allowed us to put very stringent constraints on the parameter space of well motivated WIMP models.[13]

A combined likelihood analysis of the 10 most promising dwarf galaxies, based on 24 months of data and pushing the limits below the thermal WIMP cross section for low DM masses (below a few tens of GeV), has been recently presented.[14]

6. Conclusions

Fermi turned three years in orbit on June, 2011, and it is definitely living up to its expectations in terms of scientific results delivered to the community. The mission is planned to continue at least two more years (likely more) with many remaining opportunities for discoveries. All the Fermi gamma-ray data (along with the science analysis software tools developed and used by the Collaboration) are publicly available through the Fermi Science Support Center.

Acknowledgments

The *Fermi* LAT Collaboration acknowledges support from a number of agencies and institutes for both development and the operation of the LAT as well as scientific data analysis. These include NASA and DOE in the United States, CEA/Irfu and IN2P3/CNRS in France, ASI and INFN in Italy, MEXT, KEK, and JAXA in Japan, and the K. A. Wallenberg Foundation, the Swedish Research Council and the National Space Board in Sweden. Additional support from INAF in Italy and CNES in France for science analysis during the operations phase is also gratefully acknowledged.

References

1. Meegan, C. et al., **662**, 469 (2003).
2. Atwood, W. B. et al., *Astrophys. J.* **697**, 1071 (2009).
3. Thompson, D. et al., *Astrophys. J. Suppl. Series* **86**, 629 (1993).
4. The Fermi-LAT Collaboration, *ArXiv e-prints 1108.1435* (2011).
5. Abdo, A. A. et al., *Astrophys. J. Suppl. Series* **183**, 46 (2009).
6. Abdo, A. A. et al., *Astrophys. J. Suppl. Series* **188**, 405 (2010).
7. Abdo, A. A. et al., *Science* **331**, 739 (2011).
8. The Fermi-LAT collaboration, *ArXiv e-prints 1108.1420* (2011).
9. Abdo, A. A. et al., *Astrophys. J.* **715**, 429 (2010).
10. Abdo, A. A. et al., *Astrophys. J. Suppl. Series* **187**, 460 (2010).
11. Abdo, A. A. et al., *Nature* **462**, 331 (2009).
12. Abdo, A. A. et al., *Physical Review Letters* **104**, 091302 (2010).
13. Abdo, A. A. et al., *Astrophys. J.* **712**, 147 (2010).
14. Ackermann, M. et al., *ArXiv e-prints 1108.3546* (2011).

THE JEM-EUSO MISSION

M. BERTAINA* for the JEM-EUSO Collaboration

Physics Department, University of Torino & INFN Torino
Via P. Giuria 1, Torino, 10125, Italy
**E-mail: bertaina@to.infn.it*

The main scientific goals, the scientific and observational requirements, and the expected performances of the Extreme Universe Space Observatory (EUSO) onboard the Japanese Experiment Module of the International Space Station are being described. Designed as the first mission to explore the Ultra High Energy Universe from space, JEM-EUSO will monitor, nighttime, the earth's atmosphere to record the UV (300-400 nm) tracks generated by the Extensive Air Showers (EAS) produced by UHE primaries propagating in the atmosphere.

Keywords: cosmic rays; JEM-EUSO; space observation.

1. Introduction

EUSO is the first space mission devoted to the exploration of the Universe through the detection of Ultra High Energy Cosmic Rays (UHECR) and neutrinos with energy E > 100 EeV. Being selected by the European Space Agency (ESA) as a mission attached to the Columbus module of the International Space Staton (ISS), after successfully completing the ESA Phase-A in 2004, it could not proceed to Phase-B. In 2006 the Japanese team redefined the mission as an observatory attached to the *Kibo* module of the Japanese Experiment Module, and started a renewed phase-A study as JEM-EUSO.[1]

JEM-EUSO is intended to address basic problems of fundamental physics and high energy astrophysics by investigating the nature and origin of UHECR. The corresponding jump in statistics due to the by-far larger exposure than presently ground-based running experiment, will clarify the origin (sources) of UHECR and, possibly, the particles mechanisms operating at energies well beyond those achievable by man-made accelerators. Furthemore, the spectrum of scientific goals of JEM-EUSO includes also as explorative objectives, the detection of high energy gamma rays and neutrinos, the study of cosmic magnetic fields, and testing relativity and quantum gravity effects at extreme energies. In parallel, along the mission, JEM-EUSO will sistematically survey atmospheric phenomena over the Earth Surface.

In the JEM-EUSO concept, the Earth's atmosphere is a giant detector. UHECRs collide with atmospheric nuclei and produce EAS. JEM-EUSO observes the fluorescence light emitted by the nitrogen molecules excited by the EAS charged particles and the reflected signal at ground of the Cherenkov emission associated with the shower development. Viewing from the ISS orbit, the $\pm 30°$ Field-of-View (FoV) of the telescope corresponds to an observational area at ground larger than 1.9×10^5 km^2.

The threshold energy of the detector is around 3×10^{19} eV. Increase in exposure and energy threshold is realized by inclining the telecope from nadir to tilted mode, to extend the range of observation up to 10^{21} eV. JEM-EUSO will be launched by JAXA H2B rocket and conveyed by H-II Transfer Vehicle on ISS and attached to the Exposure Facility of JEM.

2. The main scientific objectives

The Cosmic Radiation can be considered as the Particle channel complementing the Electromagnetic one of conventional astronomy. Given current uncertainties,[2,3] the expected number of events that will be detected by JEM-EUSO in 3 years mission (nadir mode) will be between 500 and 800 with energy above 5×10^{19} eV. Such number of events makes possible the following targets: a) identification of sources by high-statistics arrival direction analysis; b) measurement of the energy spectra from individual sources to constrain acceleration or emission mechanisms.

The photo-pion production of UHE proton interaction with the Cosmic Microwave Background (CMB) strongly suppresses the UHECR spectrum above $\sim 4 \times 10^{19}$ eV (the GZK-cutoff[4]) effectively setting a horizon at nominally $\sim$100 Mpc. The same applies for nuclei due to photo-disentegration. The result is that that UHECR sample the very local Universe, where the Large Scale Mass Distribution (LSMD) is inhmogeneous, therefore, the source distribution should emerge from the UHECR flux and arrival direction. Such analysis is being attempted by current operating observatories, however, the results are ambiguous. JEM-EUSO has the great advantage of significantly increase the exposure with a very low declination dependence, which is a combination of two essential parameters to improve significantly this kind of analysis.

If several sources are found with at least a dozen of observed events, the ob-

served differences in spectral features among those sources, combined with a multi-wavelength approach, will provide direct clues on the identity of the sources and the acceleration mechanisms involved. In fact, the spectrum of a source located around 5 Mpc should manifest a very low GZK cut-off effect, while a similar source around 50 Mpc should show a much steeper energy spectrum.[5]
The pattern of the energy dependent distorsions of the sources point spread functions as a result of the Galactic Magnetic Field, over the celestial sphere can be used to infer the large scale structure of the Galactic Magnetic Field itself.

3. The JEM-EUSO telescope

The JEM-EUSO telescope[6] is an extremely-fast, highly-pixelized, large-aperture and large-FoV digital camera, working in near-UV wavelength range (300-400 nm) with single photon counting capability (see fig. 1). The telescope mainly consists of four parts: collecting optics,[7] focal surface detector (FS),[8] electronics[9] and structure.[10]
The optics is made by two curved double sided Fresnel lenses with 2.65 m external diameter, a precision middle lens and a pupil. The UV photons are focused onto the FS with an angular resolution of ~0.07°. The FS detector converts the incident photons to electric pulses, which are counted by the electronics in 2.5 μs Gate Time Units (GTU). When a signal pattern of an EAS is found, the trigger[11] is issued and the intensity of the signal in the triggered and surrounding pixels is sent to the ground operation center. The list of the main parameters of JEM-EUSO telescope is reported in table 1.

Table 1. Parameters of JEM-EUSO telescope.

Field of View	$\pm 30°$
Observational area	$> 1.9 \times 10^5$ km^2
Optical bandwith	$330 \div 400$ nm
Focal Surface Area	4.5 m^2
Number of pixels	3.2×10^5
Pixel size	2.9 mm
Pixel size at ground	~ 550 mm
Spatial resolution	0.07°
Event time sampling	2.5 μs
Duty cycle $\times$ cloud impact	~14%

The intensity of the fluorescence and Cherenkov light from EAS de-

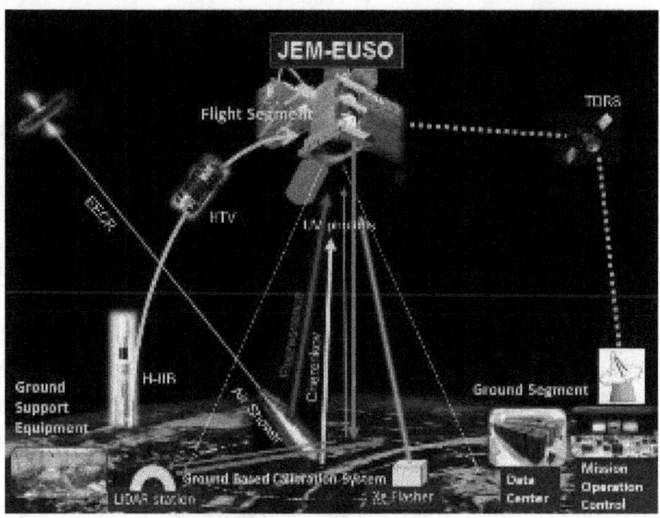

Fig. 1. Principle of the JEM-EUSO telescope to detect UHECRs.

pends on the transparency of the atmosphere, the cloud coverage and the height of cloud top. For that reasons, JEM-EUSO will account also for an Atmospheric Monitoring (AM) system[12] to estimate as precisely as possible the sky conditions and estimate the effective observing time with high accuracy. AM will consist of an infrared camera, a LIDAR system and by the slow data mode of the telescope itself.

4. The technique and the Expected Performance

The main advantages of JEM-EUSO compared to any existing or planned ground-based experiment are the significant increase of aperture and the full-sky coverage with an almost uniform exposure. Moreover, as the EAS maximum develops for most zenith angles at altitudes higher than 3-5 km from ground, the measurements will be possible even in cloudy sky conditions. Compared to ground-based detectors, the duty cycle will be, therefore, limited in primis by the moon phase, while the cloud impact will be less important than for ground-based observations.

The continuous development of the detector since the ESA time has considerably improved the performance of the instrument. The main contributions have to be ascribed to the optics (with ~1.5 better throughput and ~1.5 better focusing capability), to the photo-detector (with ~1.6 higher detection efficiency), to the better geometrical layout of the focal surface to

reduce the dead space, and to the improved performance of the electronics which allows exploitation of more complex trigger algorithms.
One of the key elements to estimate the performance of JEM-EUSO is the evaluation ot its exposure. This can be factorized in three main contributions: the trigger aperture, the observational duty cycle and the cloud impact. The observational duty cycle, meant as the fraction of time in which EAS observation is not hampered by the brightness of the sky has been evaluated by analyzing the measurements of the Tatiana satellite and rescaling them to the ISS orbit[13] and it accounts for ~20%.
The peculiarity of the observation from space is the possibility of observing CR also in some cloudy conditions (i.e. if the shower maximum is above the cloud top), which is tipically not the case for ground-based telescopes. In order to quantify the effective observational time, a study on the distribution of clouds as a function of altitude, optical depth and geographical location has been performed using different meteorological data sets.[14] Then, showers have been simulated according to such distributions, trigger efficiency curves estimated and by further requiring showers with good quality, the cloud impact has been evaluated.[15] The result indicates that the average fraction of the observational time where the measurement will not be hampered by atmospheric factors is ~70%. This number convoluted with the 20% duty cycle observational time, provides a final 14% multiplication factor to be applied to the aperture to determine the exposure.
The last parameter needed to estimate the aperture and the exposure is the trigger efficiency which depends on the average night glow background. In the following, the nominal value of 500 $\mathrm{ph/m^2/ns/sr}$[16] has been assumed. Figure 2 shows the full aperture, and annual exposure of JEM-EUSO in nadir mode for the full FoV of the detector together with different quality cuts.[17] 80-90% aperture is already reached at energies $\sim$2-3$\times10^{19}$ eV when the foot print of the shower is located in the central part of the FoV (R<125 km from nadir) and with zenith angles $\theta > 60^\circ$, and it slightly increases at $\sim 5\times10^{19}$ eV if the entire FoV is considered. In the most stringent conditions JEM-EUSO has an annual exposure equivalent to Auger (~7000 $\mathrm{km^2}$ sr yr) while it reaches ~60000 $\mathrm{km^2}$ sr yr at 10^{20} eV, 9 times Auger equivalent. The above conditions are well enough to have a sufficient range of overlapping energies (at least 1 order of magnitude in energies, starting from 2-3$\times$ 10^{19} eV) with the ground-based experiments to cross-check systematics and performance. At high energies it will be able to accumulate statistics at a pace per year of about 1 order or magnitude higher than currently existing ground based detectors. Preliminary results on the tilted

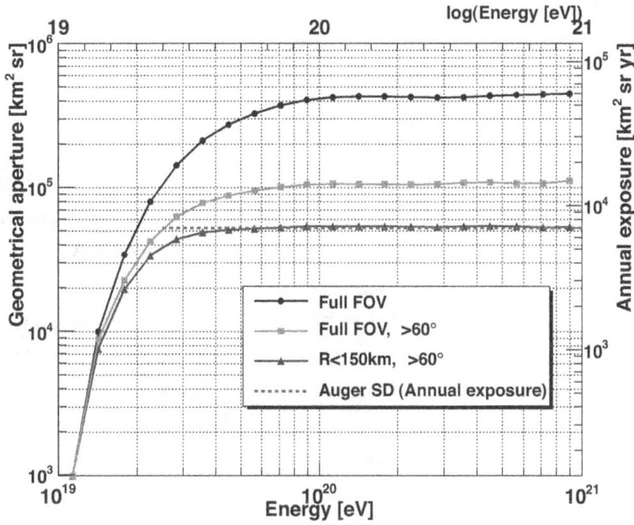

Fig. 2. Aperture and annual exposure of JEM-EUSO for different quality cuts.

mode at 35° indicate that the statistics could be increased by another factor of 2-3 at the highest energies. Such exposures properly combined would guarantee the detection of more than 500 events in 3-5 years mission at energy E > 5×10^{19} eV, assuming an Auger-like energy spectrum.[2]
JEM-EUSO performance[18] is estimated using ESAF,[19] a software for the simulation of space based UHECR detectors. Regarding the energy reconstruction, at the current status of development of the instrument and of the reconstruction algorithms, proton showers with zenith angle $\theta > 60°$ are reconstructed in clear-sky conditions with a typical energy resolution $\Delta E/E$ of about 25% (20%) at energies around 4×10^{19} (10^{20}) eV. The energy resolution slightly worsen for more vertical showers where it accounts for ~30% around 10^{20} eV. This result indicates that the JEM-EUSO perfomance is sound already below the range of energies (E >5×10^{19}) where it is expected to give scientific results, thus confirming a meaningful overlapping with ground based experiments in a sufficient wide energy range. Regarding the angular direction analysis, the current simulations indicate that showers of energy E $\sim 7\times10^{19}$ and zenith angle $\theta > 60°$ can be reconstructed with a 68% separation angle less than 2.5°. Concerning the X_{max} uncertainty, studies are still on going, however, the preliminary results indicate that around 10^{20} eV, the $\sigma_{X_{max}}$ is better than 70 g/cm^2.

5. Conclusions

Many new technological items have been developed to realize the mission since the old ESA-EUSO time. The study is now successfully in progress. The performance results indicate that the instrument satisfies the main scientific requirements needed to meet the scienfic goals of the mission. In the forthcoming two years a couple of tests on a prototype of the telescope made by a system of fresnel lenses and one PDM fully equipped are being planned. The first one will be performed at the Telescope Array site where such a prototype will be deployed for calibration purposes. In parallel, a similar one will fly on stratospheric balloons for engineering purposes such as the verification of the trigger electronics and the measurement of the nightglow background in conditions as close as possible to what is expected from ISS, and possibly to observe the first EAS from space.

6. Acknowledgments

This work was partially supported by the Italian Ministry of Foreign Affairs, General Direction for the Cultural Promotion and Cooperation.

References

1. Y. Takahashi, *New Journal of Physics* **11**, 065009 (2009); T. Ebisuzaki et al. *Nucl. Phys. B (Proc. Suppl.)* **175-176**, 237 (2008).
2. J. Abraham et al., *Science* **318(5852)**, 938 (2007); J. Abraham et al., *Astrop. Phys.* **29(3)**, 188 (2008); P. Abreu et al., *Astrop. Phys.* **34(5)**, 314 (2010).
3. R.U. Abbasi et al., *Astrop. Phys.* **30(4)**, 175 (2008).
4. K. Greisen, *Phys. Lett.* **16**, 148 (1966); G.T. Zatsepin & V.A. Kuz'min *JETP Phys. Lett.* **4**, 78 (1966).
5. G. Medina Tanco et al. *Proc. 32nd ICRC* **#0956**, (2011).
6. F. Kajino et al. *NIM A* **623** 422 (2010).
7. A. Zuccaro Marchi et al. *Proc. 32nd ICRC* **#0852**, (2011).
8. Y. Kawasaki et al. *Proc. 32nd ICRC* **#0472**, (2011).
9. M. Casolino et al. *Proc. 32nd ICRC* **#1219**, (2011).
10. M. Ricci et al. *Proc. 32nd ICRC* **#0335**, (2011).
11. O. Catalano et al. *Proc. 31nd ICRC* **#0326**, (2009).
12. A. Neronov et al. *Proc. 32nd ICRC* **#0301**, (2011).
13. P. Bobik et al. *Proc. 32nd ICRC* **#0886**, (2011).
14. F. Garino et al. *Proc. 32nd ICRC* **#0398**, (2011).
15. L. Saez Cano et al. *Proc. 32nd ICRC* **#1034**, (2011).
16. G. La Rosa et al. *Astrop. & Space Science* **276**, 219 (2001).
17. K. Shinozaki et al. *Proc. 32nd ICRC* **#0979**, (2011).
18. A. Santangelo et al. *Proc. 32nd ICRC* **#0991**, (2011).
19. C. Berat et al. *Astrop. Phys.* **33(4)**, 22 (2010).

ESTIMATION OF JEM-EUSO EXPERIMENT OBSERVATION EFFICIENCY[1]

P. BOBIK*, K. KUDELA, B. PASTIRCAK
Institute of Experimental Physics, Slovak Academy of Sciences
Kosice, 04001, Slovakia
**E-mail: bobik@saske.sk*

G. GARIPOV, B. KHRENOV, P. KLIMOV, V. MOROZENKO
Skobeltsyn Institute of Nuclear Physics, Moscow State University, Russia

M. BERTAINA
Dipartimento di Fisica Generale, Università di Torino, Italy

A. SANTANGELO
Institute fuer Astronomie und Astrophysik Kepler Center for Astro and Particle Physics
Eberhard Karls University, Tuebingen, Germany

K. SHINOZAKI
RIKEN Advanced Science Institute, Japan

J. URBAR
Department of Surface and Plasma Science, Charles University, Czech Republic

JEM-EUSO experiment will search for UHECR by monitoring UV light produced in their interaction with atmosphere from International Space Station. We have estimated an operational duty cycle for JEM-EUSO experiment along the ISS trajectory by the analytical evaluation of possible UV light sources on the Earth nightside. Main sources are UV moon light and UV background intensities created by nightglow and stars. Effect of artificial sources of UV light in populated areas is also estimated.

[1]This work was supported by Slovak Academy of Sciences MVTS JEM-EUSO

1. Introduction

1.1. *JEM-EUSO observation efficiency*

The JEM-EUSO experiment [1,2] will search for UV light produced in interactions of ultra high energy cosmic rays (UHECR) with atmosphere on the Earth's night side. We estimated operational duty cycle for JEM-EUSO detector on low earth orbit previously from Universitetski Tatiana satellite measurements [3,4,5] and from simulations based on moonlight intensity evaluation along ISS trajectory [6]. In second approach ISS trajectory was traced with one minute time-steps. The moonlight was estimated [7,8] from the Moon position and phase at evaluated ISS positions. The operational duty cycle was evaluated as a time during the night when UV intensity from reflected moon light was less than the selected allowed value. Because JEM-EUSO deal with reflected (not direct) moon light, presence of Moon over the horizon does not necessarily mean that we cannot measure showers. At maximum reflected moon light is roughly 30 times higher than moonless UV background over the oceans. Let us also note that because of the orbital position of JEM-EUSO detector, we can also partly measure with clouds in the observed FOV [9]. Previous approach from [8] did not take into account another sources of UV light on the Earth night side i.e. nightglow, zodiacal light, integrated faint star light and artificial lights. In this article we present simulation counting with these sources.

1.2. *DMSP satellite program*

We use Defense Meteorological Satellite Program [10] database annual averages of cloud free moonless light intensities on the earth night side for estimation of artificial lights influence to JEM-EUSO operational efficiency. Data in 30 arcseconds grid on surface describe light pollution of cities mainly in visible range (350–2000 nm in 63 levels scale). We assume UV intensity proportional to visible and estimate UV intensity over oceans in DSMP data to be equivalent to intensity 500 UV photons/(m^2 sr ns). Intensity over oceans in DSMP data is described by values 2 (31.8% from used data set) and 3 (50.9% from data set). We set value 2.62 to be so called oceanequivalent i.e. UV intensity estimated for cloud free and moonless conditions over oceans. To find city position we take data with value 3 times higher than oceanequivalent intensity i.e. 7.8, level 8 and higher in DMSP data.

Let us note that UV light spectrum produced by different cities differ significantly. Many kinds of lamps are used over the world and no one of them significantly dominates [11]. For example Chicago, Tokyo and Hong Kong images [12] in visible part of spectrum has different colors. Orange color of Chicago and Hong Kong is probably sign of domination of sodium lamps in the city, green light of Tokyo is due to metal halide lamps. Both lamps have different spectrum in UV [11], sodium lamp do not emit in UV. Sodium lamps and mercury lamps are mainly used for the street lighting. From the fact that some cities will be in UV less visible than in DSMP data we conclude that used DSMP data can be used for conservative estimation of city light effect for JEM-EUSO measurements.

2. Method and Results

2.1. *JEM-EUSO duty cycle simulation for moonlight and UV background*

We use ISS trajectory provided by NASA SSCweb [13]. For every position of ISS during the period from 2005 till 2007 (period selected to have simulation comparable with estimation based on Universitetsky Tatiana measurements) we have evaluated a position of the Sun (solar zenith angle) and Moon (Moon phase and lunar zenith angle) and calculated the reflected UV moonlight intensity I_{Moon} (θ, α) at low orbit in the range 300–400 nm [10]. For the night defined by solar zenith angle higher than 109.18° we have evaluated the duty cycle for a set of moonlight induced background values. We add a nominal oceanequivalent background intensity I_{BG} to every point along ISS trajectory to add to model influence of UV background created by nightglow, zodiacal light and integrated faint star light. Total UV intensity is evaluated as

$$I = I_{SUN} + I_{MOON} + I_{BG} \tag{1}$$

I_{SUN} is equal zero, because in the operational duty cycle only points on the night side (i.e. where solar zenith angle is higher than defined value) are counted, I_{BG} is set to 0 and 500 UV photons/(m^2 sr ns) (discussion about possible another values of I_{BG} is in the 2.3 part of the article).

2.2. *City lights influence*

JEM-EUSO detector field of view (FOV hereafter) in nadir mode is 140 000 km^2 on the earth (value for 400 km orbit, depending on altitude of ISS [14]). We start with conservative approach to estimate effect of city lights to operational efficiency of experiment. In this approach we refuse from operational duty cycle measurements in detector PDMs where any city light with intensity over level 3 times higher than oceanequivalent background (i.e. 1500 ph/(m^2 sr ns)) appear in the PDM projection on the Earth surface. Let us note that this means that we exclude any PDM measurements where even one small city (resolution 30 arcsec in DSMP data give ~1 km resolution on Earth) was found in the PDM projection on Earth [1,15]. For every selected point of ISS trajectory (1 minute steps) all 137 PDMs projection on Earth was scanned for city appearance. If part of PDMs was city free, we count them in the operational efficiency of the experiment. The result compared to evaluation based just on moonlight effect is presented in Figure 1. For allowed background 1500 ph/(m^2 ns sr) we get as city lights effect reduction of detector operational efficiency by 2%, from 21.43% for simulation counted only with moon light to 19.43% for simulation counted with moon light and city light. When UV oceanequivalent background (500 ph/(m^2 ns sr)) is taken into account, effect to duty cycle is 2.92% (from 21.43% to 18.51%). Every PDM contain 2304 PMT pixels (all JEM-EUSO detector has 315648 pixels). If only 2 of them will see city (two JEM-EUSO pixel are roughly one DMSP pixel) we conclude all PDMs to be blind.

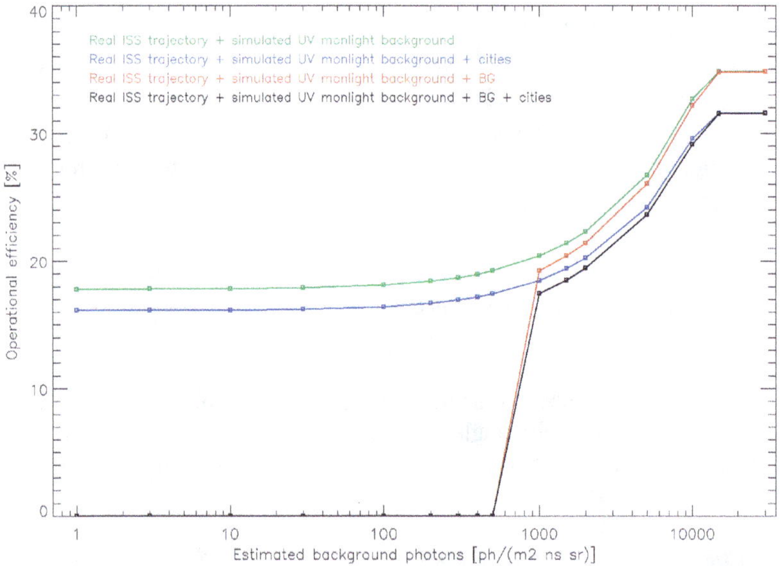

Figure 1. Operational duty cycle evaluated along real ISS trajectory in years 2005 till 2007 with simulated moonlight (green line), moonlight together with oceanequivalent UV background (red), moonlight together with oceanequivalent UV background (blue) and all sources i.e. moon, oceanequivalent background and cities together (black).

To summarize previous statements, at present stage, 1 bright pixel in the PDM is blinding the entire PDM. If the 1st trigger level could work at EC level (9 elementary cells in PDM), we could gain ~1% (from 18.51% back to 19%) in operational duty cycle.

2.3. *Higher background and Summary*

Intensity of UV background during moonless and cloudless night (from nightglow, zodiacal light and integrated faint star light) is still open question. However 500 ph/(m^2 ns sr) is to date the most expected value at ISS orbit. We made estimation for different values I_{BG} from 300 till 700 ph/(m^2 ns sr). Effect to operational efficiency with moon light and city lights counted together with UV background I_{BG} is presented in Table 1. For increasing or decreasing value of oceanequivalent background by 100 ph/(m^2 ns sr) is operational efficiency affected approximately by 0.2%.

Table 1. Oceanequivalent background influence on operational efficiency

I_{BG} [ph/(m^2 ns sr)]	Operational efficiency [%]
300	18.90
400	18.70
500	18.51
600	18.31
700	18.11

To summarize all effects taken in account in evaluation of operational efficiency see Table 2.

Table 2. Summary of all effects, I_{BG} in last column is 500 ph/(m^2 ns sr)

$I_{Allowed}$ [ph/(m^2 ns sr)]	I_{SUN} > 109.18°	I_{MOON} only [%]	Cities only [%]	I_{SUN} + I_{MOON} [%]	I_{SUN} + I_{BG} + I_{MOON} [%]	I_{SUN} + I_{BG} + I_{MOON} + Cities [%]
1		50.00	90.14	17.83	0.00	0.00
10		50.11	90.14	17.85	0.00	0.00
100		51.14	90.18	18.14	0.00	0.00
300		53.45	90.18	18.72	0.00	0.00
500		55.92	90.26	19.25	0.00	0.00
1000	34.84	62.06	90.26	20.41	19.25	17.46
1500		**68.08**	**91.06**	**21.43**	**20.41**	**18.51**
5000		89.73	95.97	26.73	26.07	23.61
10000		97.85	98.81	32.69	32.20	29.15
15000		99.99	100.00	34.83	34.80	31.55
30000		100.00	100.00	34.84	34.84	31.58

Acknowledgments

This work was supported by Slovak Academy of Sciences MVTS JEM-EUSO.

References

1. Y. Takahashi, the JEM-EUSO Collaboration, *New Journal of Physics* **11**, 065009 (2009).
2. M. Casolino et al., Astrophys. Space Sci. Trans. 7 (2011) 477.
3. G. K. Garipov, B. A. Khrenov, M. I. Panasyuk, V. I. Tulupov, A. V. Shirokov, I. V. Yashin, H. Salazar, *Astroparticle Phys.* **24**, 400x (2005).
4. G. K. Garipov, M. I. Panasyuk, V. I. Tulupov, B. A. Khrenov, A. V. Shirokov, I. V. Yashin, H. Salazar, *Journal of Experimental and Theoretical Physics Letters* **82**, 185 (2005).
5. V. A. Sadovnichy et. al., *Cosmic Research* **45**, 273 (2007).
6. C. Berat, D. Lebrun, F. Montanet, J. Adams, *Proceedings of the 28th ICRC*, 927 (2003).
7. F. Montanet, *EUSO-SIM-REP-009-1.2* (2004).
8. P. Bobik, G. Garipov, B. Khrenov, P. Klimov, V. Morozenko, K. Shinozaki, M. Bertaina, A. Santangelo, K. Kudela, B. Pastircak, J. Urbar, for the JEM-EUSO collaboration, *Proceedings of the 32th ICRC*, HE1.4, 886 (2011).
9. G. Sáez Cano, J. A. Morales de los Ríos, K. Shinozaki, F. Fenu, H. Prieto, L. del Peral, N. Pacheco Gómez, J. Hernández, A. Santangelo, M. D. Rodríguez Frías, for the JEM-EUSO Collaboration, *Proceedings of the 32th ICRC*, HE1.4, 1034 (2011).
10. NASA NOAA, http://www.ngdc.noaa.gov/dmsp/
11. Ch. D. Elvidge, D. M. Keith, B. T. Tuttle and K. E. Baugh, Sensors, **10**, 3961 (2010).
12. D. Pettit, NASA Ask Magazine, **38**, 34 (2010).
13. NASA SSCweb, http://sscweb.gsfc.nasa.gov/cgi-bin/Locator.cgi
14. K. Shinozaki, M. E. Bertaina, S. Biktemerova, P. Bobik, F. Fenu, A. Guzman, K. Higashide, G. Medina Tanco, T. Mernik, J. A. Morales de los Rios Pappa, D. Naumov, M. D. Rodriguez - Frias, G. Saéz Cáno, A. Santangelo, on behalf of JEM-EUSO Collaboration, *Proceedings of the 32th ICRC*, HE1.3, 979 (2011).
15. F. Kajino, T. Ebisuzaki, H. Mase, K. Tsuno, Y. Takizawa, Y. Kawasaki, K. Shinozaki, H. Ohmori, S. Wada, M. Inoue, N. Sakaki, J. Adams, M. Christl, R. Young, C. Ferguson, M. Bonamente, A. Santangelo, M. Teshima, E. Parizot, P. Gorodetzky, O. Catalano, P. Picozza, M. Casolino, M. Bertaina, M. Panasyuk, B.A. Khrenov, I. H. Park, A. Neronov, G. Medina-Tanco, D. Rodriguez-Frias, J. Szabelski, P. Bobik, R. Tsenov, International Symposium on the Recent Progress of Ultra-High Energy Cosmic Ray Observation, AIP Conference Proceedings, 1367, 197, (2011).

COMBINING THE *SWIFT*/BAT AND THE INTEGRAL/ISGRI OBSERVATIONS

EUGENIO BOTTACINI
W. W. Hansen Experimental Physics Laboratory and Kavli Institute for Particle Astrophysics and Cosmology, Stanford University, 452 Lomita Mall Stanford, Stanford 94305, California, USA
E-mail: eugenio.bottacini@stanford.edu

MARCO AJELLO
SLAC National Laboratory and Kavli Institute for Particle Astrophysics and Cosmology, 2575 Sand Hill Road Menlo Park 94025, California, USA

Current surveys of Active Galactic Nuclei (AGN) find only a very small fraction of AGN contributing to the Cosmic X-ray Background CXB at energies above 15 keV. Roughly 99% of the CXB is so far unresolved. In this work we address the question of the unresolved component of the CXB with the combined surveys of INTEGRAL and *Swift*. These two currently flying X-ray missions perform independent surveys at energies above 15 keV. Our approach is to perform the independent surveys and merge them in order to enhance the exposure time and reduce the systematic uncertainties. We do this with resampling techniques. As a result we obtain a new survey over a wide sky area of 6200 deg^2 that is a factor ~4 more sensitive than the survey of *Swift* or INTEGRAL alone. Our sample comprises more than 100 AGN. We use the extragalactic source sample to resolve the CXB by more than a factor 2 compared to current parent surveys.

1. Coded mask detectors and current X-ray surveys

Coded-mask instruments start to dominate when X-ray photons are too energetic to be focused with gazing mirrors. Currently flying missions are able to focus X-rays up to 10 keV. At energies above 15 keV the INTEGRAL Soft Gamma-Ray Imager (ISGRI: [1]) and the Burst Alert Telescope (BAT: [2]) coded-mask detectors represent both a major improvement for the imaging of the sky at hard X-ray energies. ISGRI and BAT are flying on board the INTEGRAL [3] and the *Swift* [4] satellites respectively. These two coded-mask detectors are at the forefront of this technology.

The major advantage is that coded-mask instruments have a large field of view and they are therefore very suited for surveys. The ongoing hard X-ray surveys of BAT [5,6] and ISGRI [7,8] continuously shed light onto the properties of the AGN, blazars, Gamma Ray Bursts, Galaxy Clusters and many other astrophysical objects. The drawback of the coded-mask technique is that, by design, ~50% of photons are blocked by the mask and the instrument itself is background dominated, limiting these instruments in sensitivity. This limit becomes evident when determining the contribution of point-like sources to the Cosmic X-ray Background (CXB). The spectrum of the CXB shows a peak ~30 keV [9], where the resolved fraction of point-like sources is a mere ~1%. While at lower energies (~1 keV) focusing X-ray telescopes (e.g. Chandra, XMM-Newton) resolve 100% of the CXB. As a consequence, the sources contributing to the peak of the CXB are undetected.

2. Combining the ISGRI and the BAT observations

In [7] the authors compare the exposure of the BAT detected sources in the ISGRI instrument that are detected and un-detected by ISGRI. They conclude that the none-detection by ISGRI is just due to the low exposure of the sources in the instrument. The lower exposure of these sources in the ISGRI instrument is mainly due to the different pointing strategies of the two satellites. BAT is quasi-randomly pointing the sky (and thus covering uniformly the whole sky) while INTEGRAL performs predetermined pointing spending a considerable amount of time observing the Galactic center.

A natural way of increasing the exposure on the hard X-ray sky is obtained by combining the BAT and the ISGRI observations. Indeed, it is possible to merge the independently produced sky survey images by BAT and ISGRI to obtain a much deeper X-ray survey image. We first perform the two independent surveys of BAT and ISGRI in the energy range 18 – 55 keV. The BAT survey is performed following the recipe as in [5]. ISGRI data are processed using the official INTEGRAL Off-line Science Analysis (OSA) software [10] accounting for bright sources and their variability and time-dependent background models. The mosaic images are resampled using a bilinear interpolation obtaining new aligned images of the same pixel size. The newly obtained images are cross-calibrated and finally merged. The advantage of this method is not only to reduce the statistical noise due to increase exposure time, but also the heavy systematic errors that affect coded-mask detectors [11] are reduced being the systematic uncertainties of the two instruments not correlated.

We have applied this technique to an extragalactic sky area of 6200 deg^2. To study the quality of the final mosaic image we investigate the pixel significance distribution (see Figure 1) that is the ratio of the intensities and their associated errors of the mosaic image. The significance histogram follows a

Gaussian distribution with zero mean and unitary variance. The errors of the final mosaic are computed propagating the errors of both BAT and ISGRI surveys considering that the covariance term between the both surveys is zero being their associated errors uncorrelated. This survey is complete to a flux of $\sim 1.2 \times 10^{-11}$ erg cm^{-2} s^{-1} and reaches a flux limit down to $\sim 3 \times 10^{-12}$ erg cm^{-2} s^{-1}. Setting the source detection threshold to 4.8 sigma we detect more than 100 AGN, galaxy clusters, galaxies, and Galactic sources.

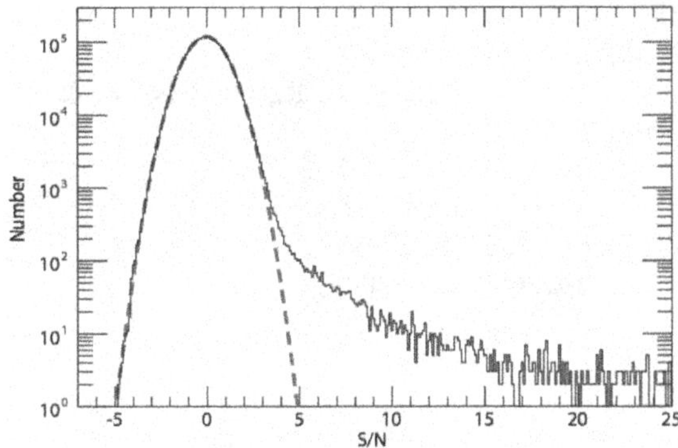

Figure 1. The distribution of the pixel significances of the merged mosaic image of 6200 deg^2. The red dashed line is a fitted Gaussian distribution with dispersion $\sigma = 1.0$. There are no wings in the distribution. The noise is under control. Positive significances above the value 4.8 represent real detected sources.

3. Conclusions

We have combined the BAT and ISGRI surveys in the energy range 18 – 55 keV. This allows one to greatly enhance the exposure on the hard X-ray sky. In turn statistical and systematic uncertainties are well behaved. This allows sampling fluxes to a limit of $\sim 3 \times 10^{-12}$ erg cm^{-2} s^{-1}. As a result we detect more than 100 AGN over a sky area of 6200 deg^2. This permits resolving the CXB by more than a factor 2 compared to the sample in [12].

Acknowledgments

We thank the INTEGRAL and the *Swift* team for the observations and the support. E.B. acknowledges support through SAO grant GO1-12144X.

References

1. Lebrun, F. et al., *A&A*, **411**, L141 (2003)
2. Barthelmy, S. D. et al., *Space Science Reviews*, **120**, 143 (2005)
3. Winkler, C. et al., *A&A*, **411**, L1 (2003)
4. Gehrels, N. et al., *ApJ*, **611**, 1005 (2004)
5. Ajello, M. et al., *ApJ*, **678**, 102 (2008)
6. Cusumano, G., *American Institute of Physics Conference Series*, Vol. 1126, American Institute of Physics Conference Series, ed. J. Rodriguez & P. Ferrando, **104–107** (2009)
7. Bird, A.J. et al., *ApJS*, **186**, 1 (2010)
8. Beckmann, V., Soldi, S., Shrader, C. R., Gehrels, N., & Produit, N., *ApJ*, **652**, 126 (2006)
9. Türler, M. et al., *A&A*, **512**, A49 (2010)
10. Courvoisier, T. J.-L. et al, *A&A*, **411**, L53 (2003)
11. Skinner, G., *ApOpt*, **47**, 2739 (2008)
12. Ajello, M. et al., *ApJ*, **699**, 603 (2009)

A status report: FACT – a fact!

T. Bretz[1] (thomas.bretz@epfl.ch)

H. Anderhub[2], M. Backes[3], A. Biland[2], A. Boller[2], I. Braun[2], V. Commichau[2], L. Djambazov[2], D. Dorner[4], C. Farnier[4], A. Gendotti[2], O. Grimm[2], H.P. von Gunten[2], D. Hildebrand[2], U. Horisberger[2], B. Huber[2a], K.-S. Kim[2b], J.-H. Köhne[3], T. Krähenbühl[2], B. Krumm[3], M. Lee[2b], J.-P. Lenain[4], E. Lorenz[2c], W. Lustermann[2], E. Lyard[4], K. Mannheim[5], M. Meharga[4], D. Neise[3], F. Nessi-Tedaldi[2], A.-K. Overkemping[3], F. Pauss[2], D. Renker[2d], W. Rhode[3], M. Ribordy[1], R. Rohlfs[4], U. Röser[2], J.-P. Stucki[2], J. Schneider[2], J. Thaele[3], O. Tibolla[5], G. Viertel[2], P. Vogler[2], R. Walter[4], K. Warda[3], Q. Weitzel[2]

[1] *École Polytechnique Fédérale de Lausanne, CH-1015 Lausanne*
[2] *ETH Zurich, Institute for Particle Physics, Schafmattstrasse 20, CH-8093 Zürich*
[3] *TU Dortmund, Experimental Physics 5, Otto-Hahn-Str. 4, D-44221 Dortmund*
[4] *ISDC, University of Geneva, Chemin d'Ecogia 16, CH-1290 Versoix*
[5] *Universität Würzburg, Emil-Fischer-Str. 31, D-97074 Würzburg*

[a] *also at University of Zurich, CH-8057 Zurich*
[b] *also at Kyungpook National University, 702-701 Daegu (Korea)*
[c] *also at Max-Planck-Institut für Physik, D-80805 München*
[d] *also at Technische Universität München, D-85748 Garching*

In the past years, the second generation of imaging air-Cherenkov telescopes has proven its power detecting weak sources with high sensitivity and low energy threshold. The goal to further improve the sensitivity and lower the energy threshold requires a robust and highly efficient sensor technology. A promising detector technology are silicon based photon detectors, namely Geiger-mode avalanche photo-diodes (G-APDs). They promise robustness and easy manageability compared photo-multiplier tubes so far in use.
To prove the applicability of this technology for Cherenkov telescopes, one of the former HEGRA telescopes was revived and equipped with a camera using G-APDs as photo sensors. Since G-APDs are comparably small, solid light guides are used to significantly increase the light collection area of each sensor. With this technologies, the First G-APD Cherenkov Telescopes (FACT) promises an increase in sensitivity and decrease in energy threshold, compared with a classical photo-multiplier based camera.

Keywords: Geiger-mode Avalanche Photodiode; G-APD; Solid light-guides; Imaging atmospheric Cherenkov Telescope; IACT; FACT

1. Introduction

Currently, the Cherenkov Telescope Array (CTA), a new generation of Cherenkov telesopes (CT), is under development. One of the goals of the project is to gain high sensitivies at energies above 1 TeV. Since in this energy region the observations of Cherenkov telescopes are usually flux limited, many small telescopes should be build and distributed over a large area to increase the collection area.

For this a technology is needed which allows to build small and inexpensive CTs and operate them with reasonable maintanace costs. Using Geiger-mode avalanche photo diodes (G-APDs) instead of Photo-multiplier tubes (PMTs) for photo detection seems to be a very promising, inexpensive and robust technology.

The camera of the First G-APD Cherenkov Telescope (FACT) is build to prove that G-APDs are an inexpensive and easy to handle alternative to PMTs, and to gain experience in operating such a device. Finally, it will serve as a monitoring telescope, dedicated to the observation of the brightest known blazars.[1]

2. Overview

The telecope is located on the Observatorio Roque de los Muchachos at the Canary island La Palma. For the installation of the camera, the former HEGRA CT3 telescope mount was refurbished. A new drive system, similar to one of the close-by 17 m MAGIC telescopes, has been installed but scaled down in power.[2] Additonally, the old disc-like mirrors have been replaced by re-coated hexagonal mirrors yielding a total effective reflective area of $9.5\,\mathrm{m}^2$ and improved reflectivity.

The camera consists of 1440 channels, each equipped with a G-APD and a solid light-concentrator. After pre-amplification, nine channels are summed at a time to build the trigger signal which is discriminated and or'ed together for the final trigger decision. The time jitter of the trigger signal, distributed to the readout boards, is well below 1 ns.

For the readout, the DRS4 analog ring-buffer is used to buffer the analog signal until the trigger signal arrives. Finally, the signal is digitized by a 25 MHz ADC. Storage of the signal in the DRS can be adjusted between 800 MHz and 5 GHz. The anticipated sampling frequency is 2 GHz.

The electronics is build on eighty boards, fourty pre-amplifier with trigger mezzanine boards and fourty readout boards plugged into four crates. Each board contains the readout chain of 36 channels. Except the G-APD bias-power supply, all electronics is integrated into the camera.

Communication and data trasmission is done over fourty Ethernet connections routed through two Ethernet switches and connected to four Ethernet ports in the data acquisition PC.

For details on the electronics see Vogler.[3]

2.1. *Usage of G-APDs*

G-APDs have a couple of advantages compared to classical Photo-multiplier tubes (PMTs). The main advantages are their robustness and the need of relatively low voltages below 100 V.

As typical disadvantages of G-APDs, usually their small size, high dark count rate, their high afterpulsing probability, their internal optical crosstalk and the temperature dependance of their properties are mentioned. For CTs they do not apply as discussed hereafter.

Physical size G-APDs used in FACT have a sensitive area of $9\,\mathrm{mm}^2$ which is small compared to typical PMTs. In general, CTs use light guides to fill the insensitive area between the PMTs, usually not optimized for the best light compression ratio. If they are optimized for best compression, they can significantly increase the sensitive area of G-APDs by factors of 10 to 17. Their size is still small enough that they do not suffer much from transmission losses. Additionally, they reduce Fresnel losses if glued to the sealing window and the G-APD. Due to total reflection, reflection losses become negligible.

For details on solid light concentrators see Huber.[4]

Dark count rate At typical operation temperature, around and below room temperature, the dark count rate is in the order of 5 MHz. Compared to the typical count rates for photons from the diffuse night-sky background, which, in our case, are in the order of 50 MHz, this is negligible.

Optical crosstalk Typical crosstalk probabilities in the order of 15% to 20% give a quite high uncertainty on the single photon level. However, in Cherenkov astronomy signals are usually in the order of tens to hundreds of photons. Therefore, crosstalk just increases the average signal and slightly its fluctuation. These additional fluctuations are still small compared to the intrinsic fluctuations of the shower development.

Afterpulses The afterpulse probability Since afterpulses of G-APDs are in the general case smaller than the inducing signal and their temporal distribution is exponentically decreasing, i.e. they are incoherent, they cannot induce fake triggers. Since afterpulses come very close to the inducing pulse they mainly prolongate the signal and increase the fluctuations of the falling edge. For Cherenkov astronomy this is of no importance.

Temperature dependence Photo detection efficiency, afterpulse probability, optical crosstalk and gain depend on the temperature due to a change of the breakdown resistor, i.e. a change of the so called over-voltage. Changing the applied voltage, this effect can be compensated. Only the dark count rate, originating in thermal noise, depends exclusively on the temperature. But if temperatures are kept well below 35°C, it stays below an acceptable limit. To achive this, a heat isolator is used to shield the sensors from the waste heat of the electronics. Apart from that, normal convection is enough to keep the sensor temperatures around the ambient temperature, which does usally not exceed 30°C at night at La Palma.

To maintain the over-voltage a feedback-system is used. For this the camera is flashed with a temperature stabilized light pulser. By changing the bias-voltage such that its measured average signal amplitude is kept stable, the gain and hence the over-voltage can be kept stable. The constant gain also ensures a homogenous trigger response.

For more information on the feedback system and the calibration see Krähenbühl.[5]

3. Status

All electronics has been build and tested. After continous full system tests, ongoing for several weeks, the camera has recently been shipped to La Palma and was installed Oct 3rd 2011. First system test on La Palma were successfull. Three pixels are known to show large noise and two are known to be dead.

A single photo-electron spectrum could be extracted from all working channels. Both light-pulsers, the internal and external one show reasonable and stable signals.

The camera control and readout is working properly and stable. The total transfer rate through the four Ethernet cards of the PC reach is about 350 MB/s, which is about 100 times faster than needed for regular data-taking. This excess will allow the application of a software trigger whihc is expected to further lower the energy threshold of the telescope.

For more details on existing measurements and test results see Biland.[6]

Fig. 1. Amplitude of the external light-pulser with a cover in front of the camera.

4. Conclusion

All lab tast performed were successfull. All parameters are within their specifications. Stability test did not show major problems. The camera has successfully been installed and first simple system test showed no problem. In the coming weeks the camera will be comissioned.

After comissioning is finished, standard operation will be started. The FACT collaboration aims for an immediate publicizing of the data. This can be an important potential for the development of future software projects.

From the current experience of the FACT project it can already be concluded that G-APDs are an alternative for future Cherenkov telescopes.

Acknowledgment The important contributions from ETH Zurich grant ETH-10.08-2 as well as the funding of novel photo-sensor research by the german BMBF Verundforschung are gratefully acknowledged. We also thank the Instituto de Astrofisica de Canarias allowing us to operate the telescope at the Observatorio Roque de los Muchachos in La Palma, and the Max-Planck-Institut für Physik for providing us with the mount of the former HEGRA CT3 telescope.

References

1. Bretz, T., Backes, M., Braun, I., et al. 2008, AIPC, 1085, 850
2. Bretz, T., Dorner, D., Wagner, R. M., & Sawallisch, P. 2009, APh, 31, 92
3. Patrick Vogler et al., in proceedings of 32nd ICRC 2011
4. Ben Huber et al., in proceedings of 32nd ICRC 2011
5. Thomas Krähenbühl et al., in proceedings of 32nd ICRC 2011
6. Adrina Biland et al., in proceedings of 32nd ICRC 2011

Aiglon: A Magnetic Spectrometer for Geophysics

G. Ambrosi, R. Battiston, W.J. Burger*, C. Fidani and M. Ionica

Dipartimento di Fisica e Sezione INFN Perugia,
Viale A. Pascoli, I-06123 Perugia (PG), Italy
* *corresponding author, email: William.Burger@cern.ch*

G. Castellini

Instituto di Fisica "N. Carrara"
Viale Madonna del Piano 10, I-50019 Sesto Fiorentqino (FI), Italy

C. Guandalini and G. Laurenti

Sezione INFN Bologna,
Viale B. Pichat 6/2, I-40127 Bologna (BO), Italy

S. Lucidi

Laboratorio SERMS, Polo di Terni, Università di Perugia,
Viale Pentima Bassa 21, I-05100 Terni (TR), Italy

A magnetic spectrometer has been designed for the China Seismological Experiment Satellite (CSES). The satellite, which includes instruments to measure the electromagnetic field, will record the field variations and the particle precipitation from the inner radiation belts, phenomena which have been observed in relation to strong earthquakes. Similar in conception to the AMS and PAMELA magnetic spectrometers, Aiglon represents a significant evolution with respect to the space instruments used for previous seismic studies.

Keywords: Charged particle spectrometers, Tracking and position sensitive detectors, Spaceborne and space research instruments, Inner radiation belts

1. Introduction

Earthquakes are accompanied by mechanical, geochemical, hydrological, electromagnetic and seismic precursor phenomena produced by the transformations due to the compression and dilatation of the Earth's crust. In particular, the modification of the local radon concentration, a radioactive gas[a], which increases during dilatation, provides an explanation for the variations observed in the local electrical field. The resulting ionization of the air molecules, coupled with important gas discharges which accompany the pre-shock period, including volatile metallic aerosols (Hg, As, Sb), and the subsequent formation of heavy molecular ions of low mobility, produce an important anomalous vertical electric field. The observed phenomena are the basis of an seismo-ionospheric coupling model which explains the electron plasma density variations in the ionosphere, between altitudes of 140-350 km, observed with ionosondes during the preparation phase of strong earthquakes.[1]

Electromagnetic phenomena of seismic origin have been put forth to explain the short-term variations in the particle fluxes in near-Earth orbit reported by instruments on the Salyut and Mir space stations, the Intercosmos-Bulgaria-1300, Meteor-03 and SAMPEX satellites.[2,3] The sudden increase in the count rates of the energetic electrons ($> 4\,\mathrm{MeV}$) is attributed to the presence of low frequency electromagnetic emission of seismic origin, which precipitates the trapped particles into the loss cone. The evidence is a rather sharp correlation between the observation time of the particle bursts and the beginning of the earthquake active phase, $\sim 4\,\mathrm{h}$ prior to the latter.

DEMETER is the first satellite dedicated to the observation of seismic phenomena. Although, no correlation has been reported between seismic activity and particle precipitation, transmitter-induced precipitation of inner-radiation belt electrons was observed in the very low frequency range (10-25 kHz).[4,5] The energy range of the solid state particle detector, 0.07-1 MeV, is lower than those of the instruments used in the studies quoted above. Since the axis of the field-of-view is oriented orthogonal to the Earth's dipole field throughout its orbit, the device has a limited acceptance for the electron pitch angles near the loss cone.

2. The Aiglon magnetic spectrometer

The Aiglon magnetic spectrometer is designed to detect electrons in the energy range of 5-50 MeV with good energy and angular resolutions, in order to improve the sensitivity of the measurement with respect to the parameters characterizing the motion of the trapped particles, their rigidity and pitch angle. The influence of multiple Coulomb scattering on the rigidity measurement is reduced by the active collimation provided by filter planes

[a] α-emitter, ^{222}Rn half-life 3.8 d

composed of edgeless silicon microstrip detectors.[6] Aiglon, the high energy particle detector (HEPD) on the CSES satellite, will be complemented by a LEPD, composed of silicon strip detectors and a total energy counter, to measure electrons in the energy range $\sim$ 0.1-10 MeV.

The magnetic spectrometer is presented in Fig.1. The 13.8 cm inner diameter, 18.3 cm long NdFeB cylindrical magnet is a Hallbach array with a 522 G field oriented perpendicular to the cylinder axis. The outer and inner silicon planes are composed of $4.1 \times 7.2 \times 0.02\,\text{cm}^3$ silicon microstrip sensors grouped together to form ladders, analogous to the operational subunits of the AMS-02 silicon tracker.[7] The segmented filter planes are composed of $1.0 \times 7.0 \times 0.02\,\text{cm}^3$ edgeless silicon microstrip sensors, each separated by a 1 mm gap. The scintillator planes consist of 4, $7.0 \times 28.0 \times 0.3\,\text{cm}^3$ plastic scintillator bars which provide a measure of the particle time-of-flight and the event trigger. The interior wall of the magnet is lined with plastic scintillators to veto events accompanied by interactions in the material of the magnet.

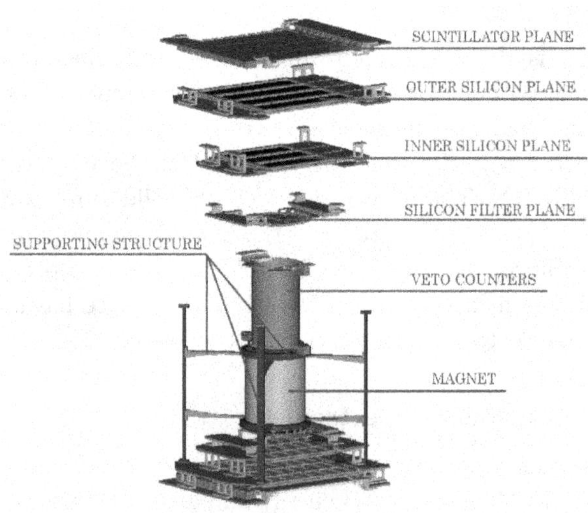

Fig. 1. The principal components of the Aiglon magnetic spectrometer.

3. Principle of operation and performance

The particle rigidity is determined by a measurement of the incident and exit directions in the two pairs of inner/outer silicon ladder planes and the two filter planes. With respect to the distance (5 cm) between the inner ladder and filter planes, the slit width of the filter planes (1 mm) is chosen to limit the contribution of large angle events, indicated by the Gaussian curves in Fig.2, which represent the projected tracker error of the incident particles due to multiple Coulomb scattering. When the particle passes by a slit in the filter, the direction is reconstructed using the closest predicted slit position and the position recorded in the nearest silicon ladder plane, and the rigidity determined with a three-point reconstruction (Fig.2).

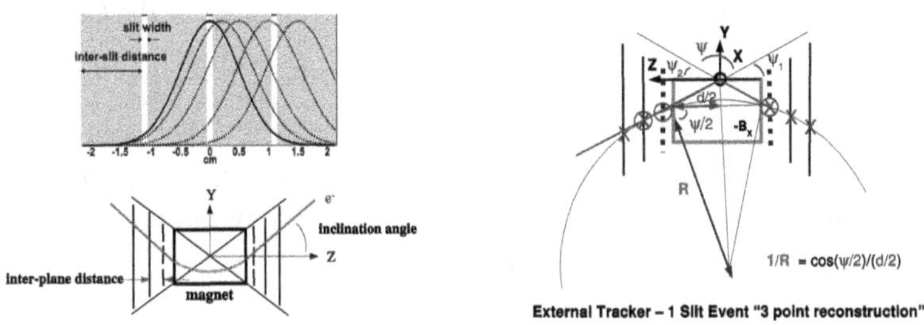

Fig. 2. The slit width, inter-plane and inter-slit distances are defined on the left. For an upstream 1-slit event (right), the deflection angle in the magnetic field is obtained with three points: the closest predicted slit position (circle in the upstream **+z** filter plane), the hit in the nearest ladder plane (circle in the second upstream **+z** ladder plane) and the hit in the second filter plane (circle in the downstream **-z** filter plane). The deflection angle of a downstream 1-slit event is obtained in a similar manner with the roles of the upstream/downstream planes reversed.

The electron rigidity resolution is presented on the left in Fig.3. The difference between the reconstructed and generated rigidity for the downstream slit events are shown in Fig.4. At the lowest momentum, there is a non-negligible contribution from the Gaussian tails (open circles in Fig.3). The width of the projected Gaussian position decreases with increasing momentum and the resolution is eventually limited by the slit width. The inter-plane distance, the slit dimensions and spacing have been chosen to optimize the performance for $\sim 10\,\mathrm{MV}$ electrons.

The resolution has been estimated with a Geant3 simulation[8] over the full acceptance of the spectrometer. The rigidity resolution has been con-

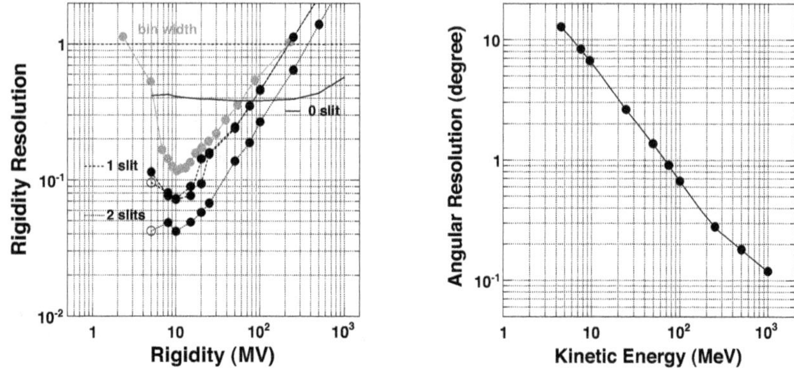

Fig. 3. The electron rigidity resolution (left) for 0, 1 and 2 slit events, and the angular resolution (right). The open circles indicate that the contribution of the Gaussian tails in the reconstructed rigidity distributions exceeds 50%.

voluted with the expected electron energy spectra αE^{-3} between 2 and 200 MeV,[9] to obtain an estimate of the resolution defined in terms of the rigidity bin widths which limit the migration of events between bins to less than 50%, indicated by the gray points in Fig.3.

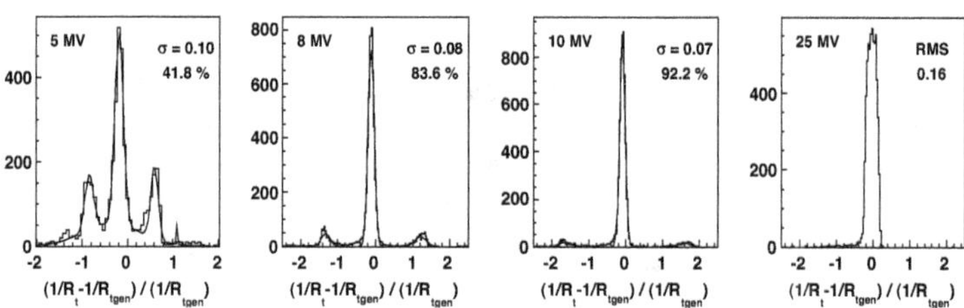

Fig. 4. The difference between the reconstructed and generated transverse rigidity for the downstream slit events. The fitted Gaussian width of the central peak and its relative contribution to the total area of the multiple-Gaussian fit are indicated. The standard deviations of the distributions are quoted for rigidities $\geq$ 25 MV, $\geq$ 20 MV for the upstream slit events.

The corresponding angular resolution is presented on the right in Fig.3. The incident angle is defined by the double-sided silicon sensors of the two upstream ladder planes. The angular resolution is limited by the 3 mm thick time-of-flight scintillators.

Aiglon is compared to existing satellite detectors used for seismic studies in Table 1. The 11-y data base of the MEPED detectors of the NOAA Polar Operational Environmental Satellites (POES) was used in a study to better distinguish seismic effects from the influence of seasonal variations and solar activity.[11]

Table 1. Comparison with existing satellite detectors. The Aiglon geometric factors for electrons (1-slit events) and protons are indicated.

detector	geometric factor $cm^2 sr$	aperture	pitch angle	range electron MeV	range proton MeV
SAMPEX	1.7	58°	0°-90°	1-4	19-28
PET[3]				4-20	28-64
DEMETER	1.2	32°	90°	0.07-2.4	
IDP[4]					
NOAA	0.1	30°	0°-90°	0.03- 2.5	0.03-6.9
MEPED[10]			90°		
AIGLON	3-5 (e), 30 (p)	70°	0°-90°	5-50	30-300

The proton acceptance quoted in Table 1 corresponds to the full geometric acceptance of the magnetic spectrometer. The proton energy is obtained by the time-of-flight measurement. The minimum proton energy threshold is $\sim 30\,\mathrm{MeV}$. The maximum energy for a 3σ separation between the flight times of the electrons and protons varies between 100 and 400 MeV for 300 and 100 ps resolutions for the scintillators. With a 100 (150) ps time resolution, the proton energy resolution would be better than 10 (20)% up to 100 MeV.

4. Power, mass and data rate

Power and mass estimates are based on the existing Tracker Data Reduction (TDR) crates of the AMS-02 silicon tracker. Each crate processes the signals of 24 576 channels with a power consumption of 84 W, including the bias and front-end electronic operating voltages furnished by the Tracker Power Distribution (TPD) unit. The TDR and TPD masses are respectively 14.5 and 8 kg.

The Aiglon electronic design is based on the TDR architecture suitably modified to handle the functionality required for all detector types, as well as higher level functions (trigger processing, event building, operational control and data transfer to the exterior). The total number of readout channels in Aiglon (9 200) represent a fraction (0.375) of a TDR crate which provide preliminary estimates for the power (31.5 W), and mass (13.4 kg), including the 5 kg mass of the permanent magnet. The limits imposed on the satellite for the HEPD are 25 W and 15 kg.

The estimates are conservative. For example, the present-day versions of the front-end electronics of the silicon strip detectors have a power consumption 50% less than those used for AMS. The readout pitch of Aiglon silicon detectors is a factor two larger, while the performance was evaluated with a position resolution (80 μm) a factor of 4 larger than the resolution of AMS-02 silicon tracker. A further optimization is possible.

The average expected total event rate for electrons between 5-500 MeV and protons between 35-500 MeV is 1.6 kHz including the region of the South Atlantic Anomaly (SAA), and 700 Hz outside the SAA. The corresponding 24 h data samples, with an event size of 460 bits, are 64 and 28.3 Gbits respectively. The data rate estimates are based on an extrapolation of the electron and proton fluxes of the MEDPED data of the NOAA 15 satellite, with an energy dependence of E^{-3}.

The Beijing ground station has a bandwidth of 66 Mbits/s. With an expected daily transmission time of 2000 s, 132 Gbits per day can be transmitted. *A priori* the on-line event rate is limited to several kHz by the readout electronics, effectively excluding the region of the SAA. In this case, the event data of HEPD represent 20% of the total daily transmission capacity of the satellite.

The event-by-event operation mode, coupled with energy and angular resolutions, represent an unprecedented degree of sensitivity for the study of the effects of seismic activity on the inner radiation belts.

References

1. S. Pulinets and K. Boyarchuk, in *Ionospheric Precursors of Earthquakes* (Springer, Berlin, 2004).
2. S.Yu. Aleksandrin *et al.*, *Annales Geophysicae* **21** 597 (2003).
3. V. Sgrigna *et al.*, *J. Atmospheric and Solar-Terrestrial Phy.* **67** 1448 (2005).
4. J.A. Sauvaud *et al.*, *Planetary and Space Science* **54** 502 (2006).
5. K.L. Graf *et al.*, *J. Geophysical Research* **144** A07205 (2009).
6. V. Avati *et al.*, *Nucl. Instr. and Method. A* **518** 264 (2004).
7. J. Alcaraz *et al.*, *Nucl. Instr. and Method. A* **593** 376 (2008).
8. R. Brun *et al.*, *CERN Report DD/EE/84-1* (1987).
9. X. Li *et al.*, *Geophyics. Research Letters* **24** 923 (1997).
10. D.S. Evans and M.S. Greer, *National Oceanic and Atmospheric Administration Technical Memorandum 1.4* (2004).
11. C. Fidani *et al.*, *Remote Sensing* **2** 2170 (2010).

MULTI-TeV GAMMA RAYS FROM SUPERNOVA REMNANTS

P. CRISTOFARI* and S. GABICI

Laboratoire APC, CNRS, Universit Paris 7 Denid Diderot, Paris, France
* *E-mail: pierre.cristofari@apc.univ-paris7.fr*

Supernova remnants have been detected in TeV γ-rays, thereby suggesting that they are powerful cosmic ray accelerators. If these objects are the sources of galactic cosmic rays, they should be able to accelerate particles up to the energy of the cosmic ray knee (~4 PeV). However there is sill no clear evidence that they can accelerate protons up to these energies. Future gamma-ray observations in the ~100 TeV might contribute to the solution of this issue. Such photons are indeed produced in proton-proton interactions experienced by PeV cosmic rays. This energy range, to date almost unexplored, will be probed by the next generation of telescopes, such as the Cherenkov Telescope Array. In this paper, we discuss the impact of such future observations for our understanding of particle acceleration at supernova remnant shocks.

1. Introduction

Supernova remnants (SNRs) are believed to accelerate galactic cosmic rays up to the knee (~4 PeV) and beyond via diffusive shock acceleration (see e.g., ref. 1 for a review). Although this is a very popular idea, an unambiguous evidence to proof (or refute) this hypothesis is still missing. The detection of a number of SNRs in TeV gamma rays (see e.g., ref. 2) is encouraging, since gamma rays are expected to be produced in hadronic interactions between the accelerated cosmic rays and the ambient gas swept us by the shock. However, it still does not constitute a conclusive proof since competing leptonic processes, namely, inverse Compton scattering may also explain these observations. For example, the gamma ray emission detected from the historical SNR Tycho has most likely an hadronic origin,[3,4] while the emission from the SNR RX J1713.7-3946[5,6] is most likely leptonic. For most of the other SNRs detected in gamma rays the situation is less clear.

An unambiguous way to discriminate between hadronic and leptonic emission from SNRs would be to perform observations in the multi-TeV range, up to ~ 100 TeV and beyond. This is because at these energies, due to the Klein-Nishina effect, the efficiency of inverse Compton scattering is severely reduced. Thus the interpretation of the emission in this energy region would reduce to the only possible mechanism: decay of neutral pions produced in hadronic CR interactions with the ambient gas. If SNRs indeed are the sources of CRs, they must act, at some stage of their evolution, as CR PeVatrons, and accelerate CRs up to the energy of the knee (~ 4 PeV). In this case, some of them have to emit gamma rays up to hundreds of TeVs, because the photons produced in proton-proton interactions carry an energy which is $\sim 10\%$ of the parent proton energy.[7]

The Cherenkov Telescope Array (CTA) is an initiative to build the most sensitive array operating in the TeV energy domain.[8] The accessible energy range will extend up to 100 TeV and beyond. The aim of this paper is to investigate the role of CTA in detecting SNRs during their PeVatron phase. A preliminary estimate of the number of such objects that might be detected at multi-TeV energies is provided. Such detections would constitute a solid evidence for the fact that SNRs indeed accelerate CRs up to the knee.

2. Gamma ray emission from SNRs

To model the gamma ray emission from a SNR we rely on the simple approach developed in ref. 9. The spectrum of accelerated particles at the shock is assumed to be a power law in momentum: $f_0(p) \propto p^{-\alpha}$ and it is normalized by assuming that the pressure of CRs at the shock is a fraction ξ_{CR} of the shock ram pressure: $P_{CR}^0 = \xi_{CR} \varrho u_{sh}^2$. The evolution of the SNR shock radius and velocity is described by analytic expression for both supernovae of type I and II and for both ejecta-dominated and Sedov phases (see ref. 9 for references). During the ejecta dominated phase the maximum energy of accelerated particles increases roughly linearly with time, while during the Sedov phase, it is assumed to decrease with time according to $p_{max} \propto t^{-\delta}$, with $p_{max} = 4$ PeV (the energy of the knee) and 10 GeV at the beginning and end of the Sedov phase, respectively. The spatial distribution of CRs inside the SNR is computed from the CR transport equation after dropping the diffusion term (i.e. CRs with $p < p_{max}$ are well confined in the SNR), and the gas distribution from the hydrodynamic equations. The gamma ray emission is then calculated by using the analytic expressions given in ref. 10, multiplied by 1.4 and 1.5 to take into account the presence of He in the target material[11] and in the accelerated particles,[12] respectively.

Though very simple, the model described above allows one to predict

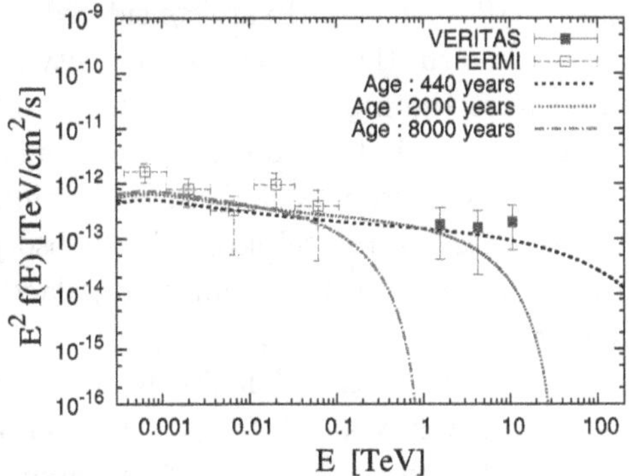

Fig. 1. Gamma ray emission from Tycho. Data are from VERITAS and FERMI-LAT observations. Curves represents the model predictions for different ages of the SNR.

with fair accuracy the emission from SNRs. To illustrate this, we plot in Fig. 1 the gamma ray emission from the historical SNR Tycho. Data from VERITAS[3] and FERMI-LAT[4] are shown as filled and open squares, respectively. Curves represents the model's predictions for a SNR of age 440 yr (the age of Tycho), 2000 yr, and 8000 yr, respectively. Following ref. 4 we adopt a distance of the SNR of 2.78 kpc, an ambient gas density of 0.3 cm^{-3}, and an explosion energy equal to 10^{51} erg. The spectrum of the accelerated CRs is $\alpha = 2.2$ and a shock compression factor equal to 7 is assumed. The CR acceleration efficiency needed to fit the data is $\xi_{CR} = 0.15$, of the same order of efficiencies quoted in other works (see e.g.,ref. 4).

For ages larger than few hundred years, a cutoff appears in the gamma ray spectrum, at an energy which decreases with time. This reflects the fact that particles with energy $p_{max}(t)c$ cannot be confined as the shock speed decreases and thus leave the system (see ref. 13 for a review). This is an important fact since the decrease of the maximum energy determines the time during which an SNR can be visible in the multi-TeV energy range.

We also applied the model to the SNR RX J1713.7-3946 for an ambient gas density of 0.2 cm^{-3} and an explosion energy of 10^{51} erg and obtained, for an efficiency of $\xi_{CR} \approx 0.1$, a flux much smaller than the observed one, in agreement with the leptonic interpretation.

3. Results

We developed a Monte Carlo code to simulate the position and the time of the explosion of supernovae in the Galaxy. We assume that 3 supernovae explode each century, with a spatial distribution given by ref. 14. Following ref. 9 , SNRs of type Ia, IIP, Ib/c, and IIb are considered. Once the position and age of each SNR is known, the ambient gas density at the location of the SNR is computed from the fits to the gas density distribution in the Galaxy provided in ref. 15. It is then possible to calculate the expected hadronic gamma ray emission from each SNR and compare it with the expected sensitivity of a gamma ray telescope. The absorption of gamma ray photons due to pair production in the infrared galactic background is also taken into account.[16]

Being interested in the multi-TeV domain, we computed the number of sources with a gamma ray flux at 50 TeV which is above the expected sensitivity of CTA for 50 hours of observation.[8] We assumed the array to be located at the same site as the H.E.S.S. telescopes. When needed, the sources extension is accounted for, by rescaling the sensitivity by a factor $\propto \vartheta_s/\vartheta_{PSF}$ where ϑ_s is the source size and $\vartheta_{PSF} \approx 4$ arcmin is the instrument angular resolution. The choice of 50 TeV is connected to the fact that at this energy and above, the contribution from inverse Compton scattering is expected to be strongly suppressed due to Klein-Nishina effect and thus the emission has most likely an hardonic origin. Moreover, photons with energies above 50 TeV are produced by CRs with energies above $\approx$ 500 TeV and thus this choice of energy range allows us to study PeVatrons.

Figure 2 shows the results of the simulation. The Galaxy has been simulated 400 times and the histograms represent the fraction of such realizations as a function of the number of SNRs detectable above 50 TeV. The histogram on the left has been produced by assuming that the spectrum of the CRs at the SNR shock is a power law with a hard spectral index: $f_0(p) \propto p^{-3.7}$, while in the right panel a soft spectrum $f_0(p) \propto p^{-4.3}$ has

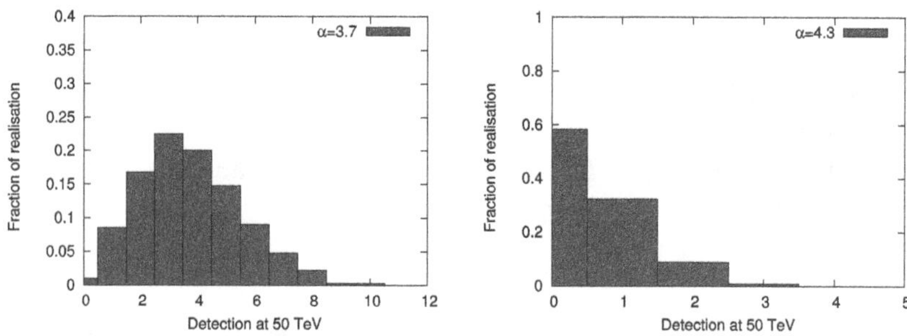

Fig. 2. Fraction of the simulated configurations as a function of the number of detectable SNRs at 50 TeV. In the left (right) panel a power law spectrum with index $\alpha = 3.7$ (4.3) has been assumed for the CRs at the shock.

been adopted. The hard spectrum has been considered to mimic the hardening (with respect to $\alpha = 4$) of the CR spectrum expected for modified shocks (i.e. high acceleration efficiency),[17] while the soft spectrum fits better with CR data and is allowed in some models of shock acceleration.[18] The acceleration efficiency is adjusted to reproduce the measured CR luminosity of the Galaxy and is $\xi = 0.2$ (0.3) for the left (right) panel. The average number of detectable SNR at 50 TeV is ≈ 3.7 and ≈ 0.5 for the hard and soft spectrum, respectively. Even in the most pessimistic case (right panel), the detection of at least one SNR is obtained in $\approx 40\%$ of the realizations. Thus, the detections of PeVatrons in the multi-TeV energy domain seem feasible with instruments of next generation such as CTA. Such detections would provide an evidence for the fact that SNRs can accelerate CRs up to the knee. The results presented here are preliminary. Several effects still needs to be considered, including the spiral structure of the Galaxy.Also, an extension of this studies to lower energies, currently probed by instruments like H.E.S.S., is also needed to check the consistency of our predictions with available data.

SG acknowledges support from the EU [FP7 grant agr. n° 256464]. The work by PC and SG is supported by ANR - Agence Nationale de la Recherche.

References

1. A. M. Hillas, *J. Phys. G* **31**, 95 (2005).
2. F. Aharonian and et al, *Rep. Prog. Phys.* **71**, 096901 (2008).
3. V. A. Acciari and et al, *ApJ Lett.* **730**, L20+ (2011).
4. F. Giordano and et al, *ArXiv e-prints, arXiv:1108.0265* (2011).
5. F. A. Aharonian and et al, *Nature* **432**, 75 (2004).
6. A. A. Abdo and et al, *ApJ* **734**, 28 (2011).
7. S. Gabici and F. A. Aharonian, *ApJ Lett.* **665**, L131 (2007).
8. The CTA Consortium, *ArXiv e-prints, arXiv:1008.3703* (2010).
9. V. S. Ptuskin and V. N. Zirakashvili, *A&A* **429**, 755 (2005).
10. S. R. Kelner and et al, *Phys. Rev. D* **74**, 034018 (2006).
11. M. Mori, *Astroparticle Physics* **31**, 341 (2009).
12. D. Caprioli, P. Blasi and E. Amato, *Astroparticle Physics* **34**, 447 (2011).
13. S. Gabici, *ArXiv e-prints, arXiv:1108.4844* (2011).
14. G. L. Case and D. Bhattacharya, *ApJ* **504**, 761 (1998).
15. T. Shibata, T. Ishikawa and S. Sekiguchi, *ApJ* **727**, 38 (2011).
16. I. V. Moskalenko, T. A. Porter and A. W. Strong, *ApJ Lett.* **640**, L155 (2006).
17. M. A. Malkov and L. O'C Drury, *Rep. Prog. Phys.* **64**, 429 (2001).
18. V. Ptuskin, V. Zirakashvili and E.-S. Seo, *ApJ* **718**, 31 (2010).

Constraints on the origin of ultra-high-energy cosmic rays from cosmogenic photons and neutrinos

Guillaume Decerprit*

DESY, Platanenallee 6, 15738 Zeuthen, Germany
E-mail:guillaume.decerprit@desy.de

We study the production of cosmogenic neutrinos and photons during the propagation of ultra-high-energy cosmic rays (UHECRs). For a wide range of models in cosmological evolution of source luminosity, composition and maximum energy E_{max}, we calculate the expected flux of cosmogenic secondaries and compare their flux to the experimental one measured by Fermi & IceCube. Most of these models yield significant neutrino fluxes for current experiments. Plus, we discuss the possibilities of signing the presence of UHE proton sources either within or outside the cosmic ray horizon using neutrinos or photons observations.

Keywords: Astroparticle physics; Neutrinos; Gamma-rays; Cosmic Rays.

1. Introduction

A multi-messenger approach of the Ultra-High Energy Cosmic Rays (UHECR) question is a promising way to unveil their origin.[1,2] Indeed, the production of UHE neutrinos from UHECR propagation was shown.[3] Because *cosmogenic neutrinos* travel freely, they were considered as interesting probes of UHE phenomena in the universe.[4] Cosmogenic gamma-rays are also produced during the propagation of UHECRs. Unlike neutrinos, these very high-energy gamma-rays,(VHEGR) from 10^{11} to a few 10^{18} eV, interact rapidly and produce electromagnetic cascades. Above 10 EeV, gamma-rays can propagate tens of Mpc. These cosmogenic VHEGR can then probe the UHECR acceleration in the local universe[5] because the cascades pile up below 100 GeV, such that Fermi data can constrain the evolution of sources and thus the neutrino fluxes expected.

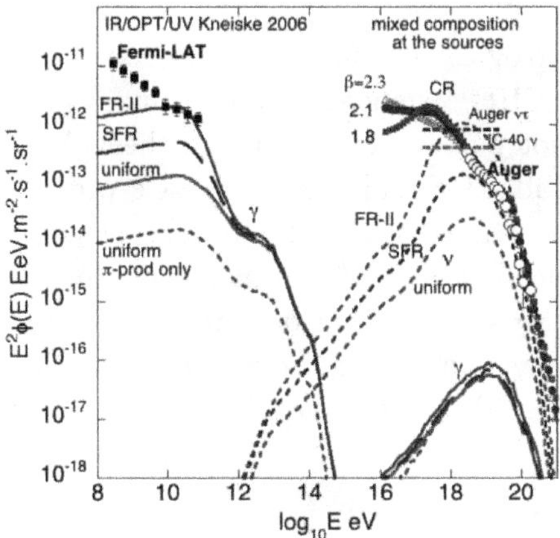

Fig. 1. Cosmic ray (markers), neutrino (dashed lines) and photon (solid lines) spectra for proton-dominated mixed composition and spectral indices $\beta = 2.3$ (uniform), 2.1 (SFR), 1.8 (FR-II) compared to Auger spectrum (open circles) and Fermi diffuse gamma-ray spectrum[7] (black squares). The contribution of the pion mechanism to the photon spectrum is shown (dashed lines). The IceCube limit (IC-40, red dashed line) is shown.[8]

2. Calculation principles

For the propagation of UHECR nuclei we used the code described in,[6] where interactions of protons and nuclei were modeled. The production of secondary photons, pairs and neutrinos and the development of electromagnetic cascades now completes this Monte Carlo framework.[1]

We calculate secondary photons and neutrino fluxes for different models of composition, maximum acceleration energy at the sources, and cosmological evolution of the UHECR luminosity. The cosmic ray spectrum measured by the Pierre Auger Observatory[9] allows us to normalize our UHECR and secondary neutrino and photon fluxes and to choose compatible spectral indices. In most cases, we consider, at the source, $E_{max} = Z \times 10^{20.5}$ eV and perform an exponential cut-off above. Sources are continuously distributed between $z = 8$ and 4 Mpc.

3. Results

Constraints from GeV-TeV gamma-rays

An example of our calculation is shown in Fig. 1 for three hypotheses (different colors) on the source evolution. One can see their strong influence on UHE neutrino fluxes. The strongest evolution models (FR-II) are directly constrained by their diffuse multi-GeV gamma-ray fluxes.

Constraints from UHE gamma-rays

Because the cosmological evolution of the source luminosity does not strongly influence UHE cosmogenic photon fluxes (see above), the Fermi diffuse flux does not put any constraint on their observability. The UHE integrated photon flux for some of the models presented above is shown on Fig. 2. The predictions displayed here are below the experimental upper limits. This means that UHE cosmogenic photons would be difficult to detect. Within 5 years of Auger data-taking, the three most optimistic cases displayed in Fig. 2, corresponding to very high values of E_{max} or hard spectral indices, might be constrained.

Spotting single sources with UHE gamma-rays: Independently of the global composition at the highest energies, the detection of individual sources accelerating protons above 10^{20} eV would help understanding the origin of UHECRs. Cosmic ray observatories are also expected to spot individual sources by observing UHE photons. Their fluxes are strongly influenced by the distribution of local sources. We studied the UHE photon flux of some sources accelerating protons. The example of a source at 20 Mpc and different values of E_{max} is shown in Fig. 3, where we assumed that

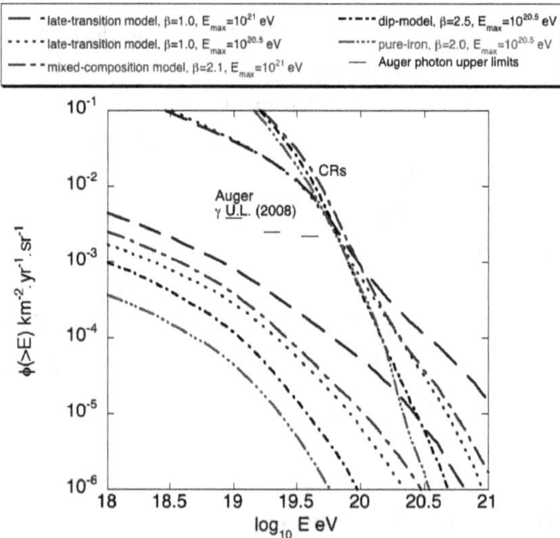

Fig. 2. Integrated fluxes of cosmic rays and photon compared to Auger, for different models assuming a SFR source evolution. The photon fluxes are always below their corresponding cosmic ray fluxes and therefore share the same line style. The legend lies in the panel itself. Upper limits on the UHE photon flux from the Auger.[10]

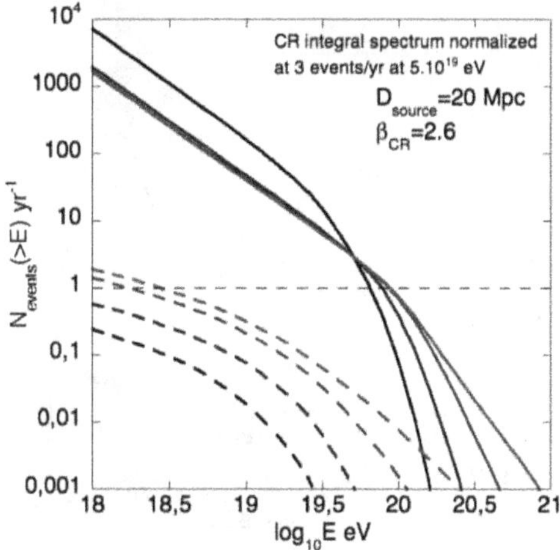

Fig. 3. UHECR and UHE photon-integrated fluxes (dashed) for a source located at 20 Mpc and different values of E_{max} (from lowest to highest photon flux): $10^{19.5}$, 10^{20}, $10^{20.5}$ and 10^{21} eV and $\beta = 2.6$. The integrated flux is normalized assuming that the source provides three cosmic ray events above 5×10^{19} eV.

the source was providing three events per year above 5×10^{19} eV (~10% of the Auger flux). Obviously the rate of expected photons is between ~0.3 and 1 per year above 10^{19} eV. UHE photons are best suited to spot local sources rather than distant ones, due to the cascading.

Constraints from UHE neutrinos

In 3 years, IceCube should provide upper limits down to the level of the predictions for the SFR scenario.[8] The non-observation of UHE neutrinos would then strongly constrain the cosmological evolution of the sources.

For the mixed composition model, not only Fermi, but also IceCube clearly rules out the FR-II scenario and constrains less optimistic scenarios between the FR-II and SFR cases. The cosmogenic neutrino fluxes implied by the diffuse gamma-ray flux are model-dependent, because for a given neutrino flux expectation, different GeV-TeV diffuse gamma fluxes are possible.

Spotting single sources with neutrinos If one assumes that powerful UHE proton sources exist but are rare and outnumbered by less efficient accelerators in the local universe, then detectable diffuse neutrino fluxes are expected. A contribution to the UHECR spectrum as low as, e.g., ~10% at

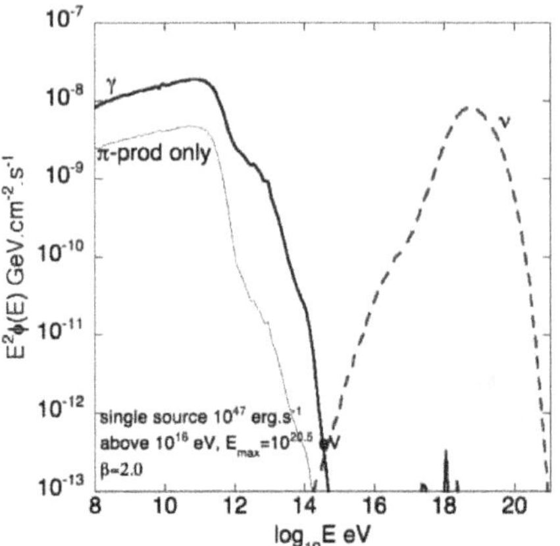

Fig. 4. For a source of protons at 1 Gpc with a luminosity of 10^{47} erg s^{-1}, the photon and neutrino fluxes are shown.

10^{19} eV from sources accelerating protons above 10^{20} eV with a cosmological evolution of type FR-II would produce $\sim$ as many UHE neutrinos as the SFR case shown in Fig. 1. The observation of such a diffuse neutrino flux would represent a signature of the existence of remote UHE proton accelerators. Neutrino and photon fluxes for a 1 Gpc pure-proton source are shown in Fig. 4. It could be detected by a neutrino detector with sensitivity $\sim 10^{-8}$ erg s^{-1} $\geq 10^{17}$ eV. For sources below D$\leq$ 500 Mpc, the luminosity has to be lowered not to overshoot the cosmic ray spectrum, thus giving less optimistic neutrino fluxes.

References

1. G. Decerprit and D. Allard, *Astronomy & Astrophysics* (2011).
2. D. Allard, N. G. Busca, G. Decerprit *et al.*, *Journal of Cosmology and Astro-Particle Physics* **10**, p. 33 (Octtober 2008).
3. V. S. Berezinsky and G. T. Zatsepin, *Phys. Lett. B* **28**, p. 423 (1969).
4. M. Ahlers, L. Anchordoqui and S. Sarkar, *Phys.Rev. D* **79**, p. 083009 (2009).
5. M. Ahlers and J. Salvado (2011).
6. D. Allard, E. Parizot, E. Khan, S. Goriely and A. V. Olinto, *Astron. Astrophys.* **443**, L29 (2005).
7. A. A. A. et al. [Fermi Collaboration], *ApJ* **720**, p. 435 (2010).
8. R. A. et al. [IceCube collaboration], *Phys. Rev. D.* **83(11)**, p. 092003 (2011).
9. A. et al. [Pierre Auger Collaboration], *Phys. Rev. B.* **685**, p. 239 (2010).
10. J. Abraham *et al.*, *Astropart. Phys.* **31**, 399 (2009).

The AMS-02 Silicon Tracker: Status and Performances

D. D'Urso[1] for the AMS-02 Collaboration

Physics Department, University and INFN section of Perugia, Perugia, I-06124, Italy

[1] *E-mail: domenico.durso@pg.infn.it*

www.unipg.it

The Alpha Magnetic Spectrometer (AMS-02) is a space based high energy physics experiment operating on the International Space Station (ISS) since May. AMS-02 will measure the different cosmic radiation components allowing the search of primordial antimatter and dark matter annihilation products. Exploiting a large acceptance and a data taking of at least 10 years, AMS-02 will detect more than 10^{10} charged particles in the GV-TV rigidity range. The tracking device is composed by 2 planes at the ends of the apparatus and 7 layers of silicon sensors in the permanent magnet (0.15T) bore. The measurement of the curvature radius of the charged particles bent trajectories allows the estimation of particle rigidity and charge sign. The tracker is composed by 2264 double-sided silicon sensors (72x41 mm^2, 300 μm thick) assembled in 192 read-out units, for a total of $\approx$ 200.000 read-out channels. The status of the AMS-02 tracker, after these first months of data taking in space, its performances and potentialities will be presented.

Keywords: Silicon Tracker; Cosmic Rays; Anti-Matter; Dark Matter.

1. Introduction

The Alpha Magnetic Spectrometer[1–3] (AMS-02) is a large acceptance ($\approx$ 0.5 m^{2}sr) cosmic ray (CR) space detector which will perform a precise measurement of the galactic cosmic radiation in the GV-TV rigidity range. It was installed on the International Space Station (ISS) on May 19th 2011 and it will operate all along the ISS lifetime.

The main objective of AMS-02 is to search for primordial antimatter looking for the presence of anti-nuclei into the cosmic radiation. It will also search for indirect dark matter signals, measuring spectra of e^+, $\bar{p}$, γ, it will study spectrum and chemical composition of cosmic radiation, it will perform γ-ray astronomy and it will search for exotic signals as strangelets. Redundant measurements of charge and the energy of particles are performed by different subdetectors to minimize systematic errors and to get a high detector sensitivity.

The core of AMS-02 detector is the silicon tracker (ST). It allows to estimate particle rigidity and charge sign via the curvature radius of charged particle bent trajectories, and their charge absolute value measuring the energy deposited by incident particles.

2. The silicon tracker

The AMS-02 ST is composed by 9 layers of silicon double sided sensors, arranged in 192 ladders, disposed on 6 honeycomb carbon fiber planes with diameter of $\sim$ 1 m. At the center of AMS-02 detector is located the "inner tracker", composed by 3 double layer planes, contained within a permanent magnet of 0.15 T, closed on the top by one single layer tracker plane. Two single layer tracker planes are placed respectively at the top of AMS-02 and at the bottom, above the electromagnetic calorimeter. The AMS-02 tracker has been designed to survive/operate in a wide temperature range, between -20/-10 and 40/20 °C. Each ladder is the assemby of a variable

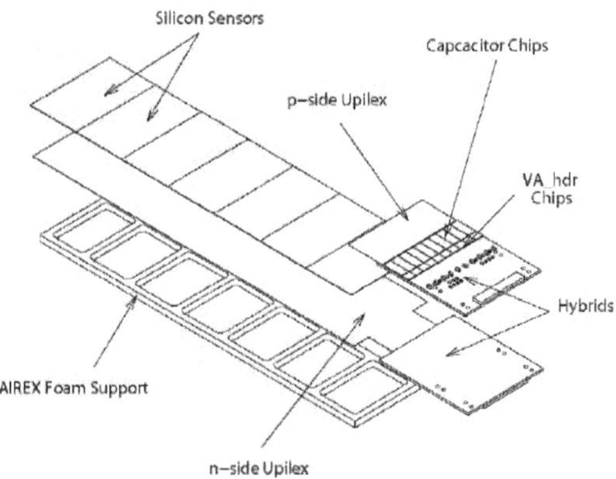

Fig. 1. Main components of a silicon ladder.

number (from 7 to 15) of double sided (p^+-n-n^+) silicon microstrip wafers (28×60 cm^2 $\times$ 0.3 mm^3) supported by a light structure of carbon skinned Airex (see fig. 1) and it is coupled to its read-out chain. The silicon wafer is a high resistivity n-type sensor.[4] On the two surfaces, p^+ and n^+ strips are implanted along orthogonal directions (p-side and n-side).

The sensor design has an implantation (read-out) strip pitches of 27.5 (110) μm on the p-side, the bending direction, and 104 (208) μm on the n-side. The connection between microstrips and the read-out electronis is via micro-wire bonds in the implementation direction on the p-side and via the upilex cable on the n-side. Each ladder has a total of 1024 read-out channles, 640 on the p-side and 384 on the n-side, and 2 front-end electronics, the Hybrids, where signals are processed by low noise high dynamic range (1-100 MIP) VA_hdr chips.[5]

Due to the limited band-width, signals collected by all subdetectors have to be processed on board. Tracker analog signals of the front-end electronics are processed by Tracker Data Reduction boards (TDRs) that digitize them, apply calibration data, remove pedestal and the common noise, search for clusters and perform on line data reduction.

Detector operation requires an active cooling system. A large set of temperature sensors are distributed all along the ST to monitor the detector status and the functionality of the tracker cooling system: on the central plane of the inner tracker, above and below the inner tracker, on the first and last tracker plane.

3. Status and Performances

Since its first activation, the tracker has the expected behavior in terms of signals and noise level.

An on-line monitoring tool has been developed to monitor ST response as a function of time. A web application has been implemented, based on a mysql database, the AMS Monitoring Interface, AMI.[6] It collects data coming from temperature sensors, currents and voltages of crates and electronics, calibration measurements, data size of ST events and so on.

The ST response is monitored performing a detector calibration every 46 minutes, when passing above the equator, measuring pedestal and noise level of each strip and assigning a status bit to each of them to avoid to use noisy or dead strips as cluster seeds. The ST detector calibration response is stable in time, as seen from the monitoring of pedestal and noise levels. In fig. 2 the average noise measured on the y-side as a function of time, for all the layers, is shown. AMS-02 Tracker is also characterized by the uniform ladder response, monitored measuring the signal corresponding to p and He charge for each ladder, and a high ladder efficiency (most of ladders have an efficiency above 90%).

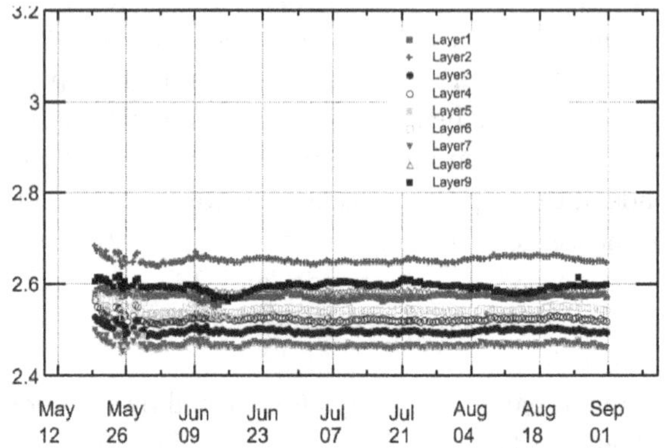

Fig. 2. Average noise measured on the y-side, for all of the layers, as a function of time.

AMS-02 Tracker can reach a rigidity resolution of about 10% at 10 GeV and has a Maximum Detectable Rigidity of about 2 TeV[7] for Z=1. On each layer, it has a spatial resolution of $\sim$ 10 μm, in the bending direction, and of $\sim$ 30 μm in the not bending direction.

4. Conclusions

AMS-02 is collecting data since May 19th: all of the subdetectors and the DAQ are properly working. The AMS-02 ST has the expected behavior with high stability of pedestal and noise levels. Furthermore, it has a high ladder uniformity and efficiency response. Intensive studies are presently on going to understand all the detector systematics and to perform the tracker alignment, crucial to exploit all the AMS-02 detector potentiality.

References

1. J. Alcaraz et al., *Physics Reports* **366**, 331 (2002); K. Lübelsmeyer et al. *Nucl. Instr. Method A* **654**, 639 (2011).
2. The AMS Collaboration, *Proc. 32nd ICRC, Beijing, China* (2011).
3. A. Contin these proceedings.
4. N. Dinu, E. Fiandrini and L. Fanò, *INFN/AE*, 1 (2006).
5. O. Toker et al., *Nucl. Instr. and Method A* **340**, 572 (1994), B. Alpat et al. *Nucl. Instr. and Method A* **446**, 552 (2000).
6. G. Alberti and P. Zuccon, *Journal of Physics: Conf. Series (CHEP-10)* (2011).
7. B. Alpat at al., *Nucl. Instr. and Method A* **613**, 207 (2010).

LUNAR GAMMA RAY EMISSION AS OBSERVED BY FERMI

M. BRIGIDA AND N. GIGLIETTO †

INFN-Bari and Dipartimento Interateneo di Fisica "M.Merlin" dell'Università di Bari
Via Amendola 173 70126 BARI - Italy

We report the Fermi-LAT observations of the lunar emission during the extended period of low solar activity. During this period the CR-induced emission was the brightest. While the Moon was detected by the EGRET instrument on the CGRO with low statistics, Fermi is the only gamma-ray mission capable of detecting the Moon and monitoring it over the full 24th solar cycle. We present the gamma-ray images of the Moon, its spectrum, and flux measurements in comparison with models and previous EGRET results.

1. Introduction

The gamma-ray emission from the Moon is due to the cosmic-ray interactions with Moon surface. The γ-rays are produced from decays of neutral pions and kaons. Cosmic-ray interactions with the lunar surface are well established and the gamma-ray spectrum has been computed almost analytically[1]. More recent calculations have been performed using GEANT and a detailed description of the regolite lunar surface[2] and GEANT4[3] taking into account all the interactions with the specific composition of the lunar rock ([2],[4]). These calculations indicate that the kinematics of the collisions of cosmic rays hitting the lunar surface, produce a secondary particle cascade that develops deep in the rock. A small fraction of low energy secondary pions are directed toward to lunar surface and decay to produce a soft γ-ray spectrum.

The detection of the γ-ray emission from the Moon was made during early analysis of EGRET data [5] and provided an integral flux F(E>100 MeV)=(4.7 $\pm$ 0.7) $\times 10^{-7}$ cm^{-2} s^{-1}; this was confirmed in recent work using a different analysis[6].

With its large effective area the *Fermi*-LAT should be able to explore in detail the features of the lunar emission. In particular, since is well known the anti-correlation of the cosmic ray fluxes with solar activity, it is very important to have the possibility to follow the evolution of the actual 24th solar cycle

† giglietto@ba.infn.it

starting from the recent prolonged solar minimum toward the maximum expected in between 2012-2013.

In this paper we report preliminary results from the analysis of the first months of observations, updating previous Fermi observations of lunar gamma-ray emission[7-9].

2. Data Selection

The LAT data used in this analysis of the quiescent solar emission was collected from August 4, 2008 until February 4, 2011. During the period covered by this analysis the Sun was at the beginning of the 24th Solar cycle, hence in a period of minimum activity. Then, the quiescent solar gamma-ray flux during this period is expected to be at its maximum.

For this analysis we have selected photons with energies E> 100 MeV coming within 20° from the Moon direction and satisfying the Pass 6 Diffuse class screening criteria [6], corresponding to the events with the highest probability to be considered as photons.

Since the Moon is quickly moving across the sky, the data are selected in a moving frame centered on the instantaneous source position, computed using an interface to JPL libraries, taking into account parallax corrections [10].
In order to have a clean sample of photons emitted by the Moon, contributions due to any others sources has been reduced with the following selection:

- zenith cut of 105o to eliminate photons from the Earth's limb;
- the Moon should be at least 30o below or above the galactic plane in order to reduce the diffuse components and avoid the brightest sources on the galactic plane;
- angular separation between Moon and Sun should be more that 20o in order to remove the Sun emission component;
- a minimal angular distance of 20° from any other bright celestial source with flux > 4×10-7 ph cm-2s-1 above 100 MeV as selected from the 1FGL LAT source catalog [11].

All these selections remove around 35% of the data set.

3. Analysis method and results

To study the spectrum of the gamma-ray emission we have used the standard maximum likelihood tool provided with the LAT science tools. This analysis was performed by fitting a "fake" moon data to model the background, similarly to the procedure used in [12] and the Moon data sample with either a simple power law or other functional forms like a logParabola. The observed spectrum is best described by a logParabola function with a Test Statistic (TS) value of 20212.8 and the resulting integral flux above 100 MeV is 1.04 ± 0.02[stat] ± 0.2[sys] x 10^{-6} $cm^{-2}s^{-1}$.

Observation of the lunar flux over the full solar cycle can be used to monito the evolution of the low energy cosmic ray flux (E<10 GeV) avoiding the effects due to the geomagnetic cutoff. Since the significance of the lunar emission is high over the whole data sample, we can divide safely the sample in 2-3 months interval, in order to follow the CR flux evolution. Figure 1, for example show the neutron monitor rates detected on Earth, on the McMurdo Neutron Monitor station*, and exhibits an evident decrease of the rate, connected to the starting solar activity.

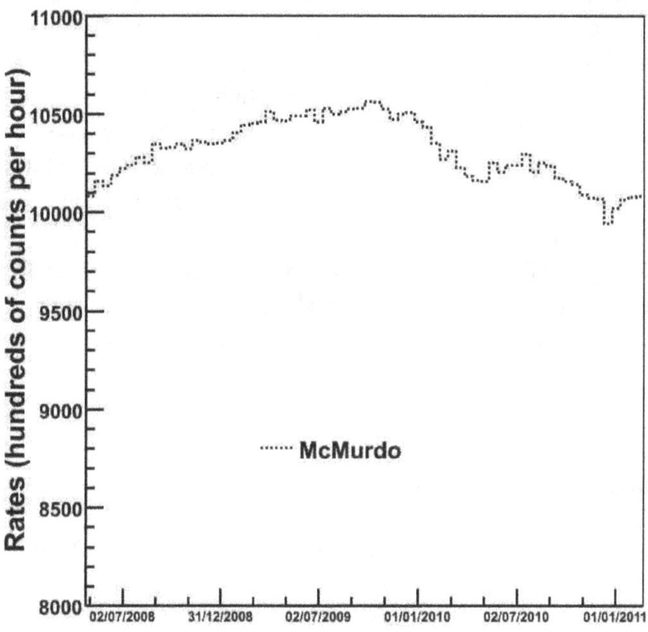

Figure 1. Rates of McMurdo neutron monitor station, measured as hundreds o count per hour, starting from the Fermi-LAT launch to the beginning of 2011.

* operated by Bartol Research Institute, see http://neutronm.bartol.udel.edu

4. Conclusions

The Fermi-LAT observations has reported the lunar emission with high precision and significance. The measured flux is in agreement with the flux estimations computed for the minimum of the solar activity. The high significance of the detection can be used to monitor the solar cycle that modulates the cosmic ray low energy flux, probing it outside the geomagnetic field.

Acknowledgment

The *Fermi* LAT Collaboration acknowledges generous ongoing support from a number of agencies and institutes that have supported both the development and the operation of the LAT as well as scientific data analysis. These include the National Aeronautics and Space Administration and the Department of Energy in the United States, the Commissariat à l'Energie Atomique and the Centre National de la Recherche Scientifique / Institut National de Physique Nucléaire et de Physique des Particules in France, the Agenzia Spaziale Italiana and the Istituto Nazionale di Fisica Nucleare in Italy, the Ministry of Education, Culture, Sports, Science and Technology (MEXT), High Energy Accelerator Research Organization (KEK) and Japan Aerospace Exploration Agency (JAXA) in Japan, and the K. A. Wallenberg Foundation, the Swedish Research Council and the Swedish National Space Board in Sweden.

Additional support for science analysis during the operations phase from the following agencies is also gratefully acknowledged: the Istituto Nazionale di Astrofisica in Italy and the Centre National d'Etudes Spatiales in France.

References

1. D. J. Morris, D. J., J. Geophys. Res., **89**, 10685 (1984).
2. I. V. Moskalenko and T. A. Porter, Astrophys. J. **670** (2007) 1467, [arXiv:0708.2742 [astro-ph]].
3. S. Agostinelli *et al.* [GEANT4 Collaboration], Nucl. Instrum. Meth. A **506** (2003) 250.
4. Moskalenko, I. V., & Porter, T. A., International Cosmic Ray Conference, (2008), **2**, 759 (2008).
5. Thompson, D. J.,Bertsch, D. L., Morris, D. J., & Mukherjee, R., J., Geophys. Res., **102**, 14735 (1997).

6. E. Orlando and A. W. Strong, Astron. Astrophys. **480** (2008) 847, [arXiv:0801.2178 [astro-ph]].
7. N. Giglietto, N., AIP Conf. Proc. **1112** 238 (2009).
8. Brigida, M.,44th Rencontres de Moriond Proceedings (2009), 115.
9. N. Giglietto, International Cosmic Ray Conference Lodz (2009), arXiv:0912.3734 [astro-ph.HE].
10. http://iau-comm4.jpl.nasa.gov/access2ephs.html
11. Abdo, A. A., et al. ApJS, 188, 405 (2010)
12. Abdo, A. A., et al., APJ 734 (2011)116

THE CALORIMETRIC ELECTRON TELESCOPE (CALET) FOR HIGH-ENERGY ASTROPARTICLE PHYSICS ON THE INTERNATIONAL SPACE STATION

T. GREGORY GUZIK,
for the CALET Collaboration
Department of Physics & Astronomy, Louisiana State University
Baton Rouge, LA 70803-4001, U.S.A
email: guzik@phunds.phys.lsu.edu

During a five year mission the CALET space experiment, currently under development by collaborators in Japan, Italy and the United States, will measure the flux of Cosmic Ray electrons (and positrons) from 1 GeV to 20 TeV, gamma rays from 10 GeV to 10 TeV and nuclei with Z=1 to 40 from 10 GeV to 1,000 TeV. These measurements are essential to investigate possible nearby astrophysical sources of high-energy electrons, study the details of galactic particle propagation and search for dark matter signatures. The instrument consists of a module to identify the particle charge, a thin imaging calorimeter (3 radiation lengths) with tungsten plates interleaving scintillating fiber planes, and a thick calorimeter (27 radiation lengths) composed of lead tungstate logs. CALET has the depth, imaging capabilities and energy resolution necessary for excellent separation between hadrons, electrons and gamma rays. The instrument is currently being prepared for launch, during the 2014 time frame, to the International Space Station (ISS) for installation on the Japanese Experiment Module – Exposed Facility (JEM-EF). This paper summarizes the instrument design and performance.

1. Introduction

Matter accelerated to velocities very close to the speed of light and originating external to our solar system has been studied at Earth for close to 100 years, yet these cosmic rays have tenaciously held on to some of their fundamental secrets. In particular, particle acceleration associated with supernova remnant (SNR) shocks appears to be the best explanation for how galactic cosmic rays (GCR) below a few PeV achieve their high energies. Further, evidence that particle acceleration is taking place at SNRs is provided by electron synchrotron and gamma-ray emission measurements, but no *direct detection* of accelerated particles from specific sources has yet been confirmed.

Cosmic ray transport through the galaxy is understood to be a diffusion process, where the GCR hadronic component may traverse the distance equivalent of hundreds of galactic diameters during their lifetime, thereby randomizing their trajectory and losing connection with their original source. High-energy electrons, however, have radiative energy losses that limit their lifetime and, consequently, the distance they can diffuse away from their source. As a result, the highest energy electrons that we see at Earth very likely originate from only a few local (within ~1 kpc of the Earth) sources and, thus, the GCR electron energy spectrum at energies 100 GeV to ~10 TeV should show structure associated with specific sources [1]. Furthermore, electrons and positrons are products of the annihilation of many exotic particles speculated as dark matter candidates and could appear as features in the spectrum.

Recent observations of positrons by PAMELA [2] up to ~150 GeV and electrons plus positrons by ATIC [3], PPB-BETS [4], H.E.S.S. [5] and Fermi-LAT [6] from 300 GeV to 800 GeV have all shown an excess of particles relative to what would be expected from diffusive propagation of cosmic ray electrons from sources uniformly distributed around the galaxy. Explanations for this excess include local SNR or Pulsar Wind Nebulae (PWN) sources as well as the annihilation signature of various dark matter particles [3]. Each of these instruments has a particular, different, limitation that precludes spectral shape details of the excess to be definitively characterized and used to distinguish between the various source models. What is needed is a new space instrument designed to provide high resolution, accurate spectrum measurements for electrons over a broad energy range up to ~20 TeV.

2. The CALorimetric Electron Telescope

The CALET instrument is designed to measure electrons from 1 GeV to 20 TeV and, in addition, protons and nuclei from several 10 GeV to 1000 TeV as well as

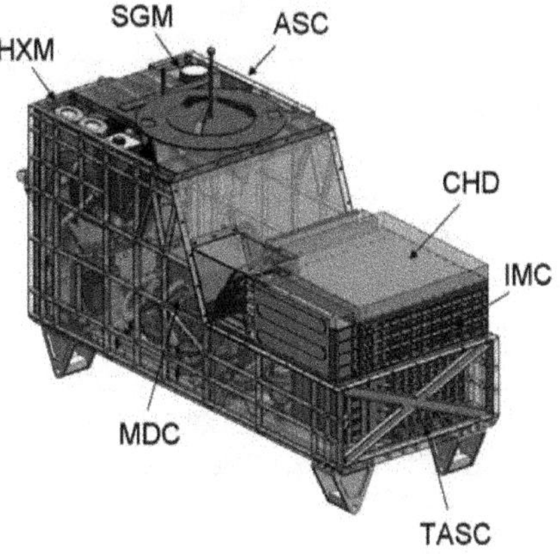

Figure 1: The CALET instrument consisting of the CHD, IMC, TASC and other sub-systems (see text) mounted on the JEM-EF attachment pallet.

high-energy gamma-rays from 10 GeV to 10 TeV. The instrument is being designed and built in Japan, with hardware contributions from Italy and assistance from collaborators in the United States and Italy, in anticipation of a launch on HTV-5 in the 2014 time frame. The CALET pallet (Figure 1), which includes a gamma-ray burst instrument (HXM, SGM) [7], the Advanced Stellar Compass (ASC) for attitude determination, and the Mission Data Controller (MDC), is planned to be attached to the JEM-EF, Port #9 on the ISS to measure cosmic ray spectra in space for five years [8]

Figure 1, as well as the instrument schematic included as Figure 2, show that CALET consists of three primary detectors: the Charge Detector (CHD), the Imaging Calorimeter (IMC) and the Total Absorption Calorimeter (TASC). The instrument weight will be approximately 600 kg, and the detectors will have a total absorber thickness of 30 radiation lengths (X_o), 1.3 proton interaction lengths (λ), and an effective geometrical factor for high-energy electrons of 1,200 cm^2 sr.

The CHD is used to determine the charge of incident particles and is designed to cover the range from Z = 1 to Z ~40 with a charge resolution of ~0.1e for CNO to 0.2e for Fe [9]. This detector is composed of two layers orthogonally arranged to determine the position of the incident particle, with 14 plastic scintillator segments in each layer covering an area of approximately 450 mm by 450 mm. This segmented configuration reduces the number of backscatter particles that hit each scintillator. The CHD readout electronics is very similar to that used for the TASC and a 16-bit ADC is used to cover the required dynamic range.

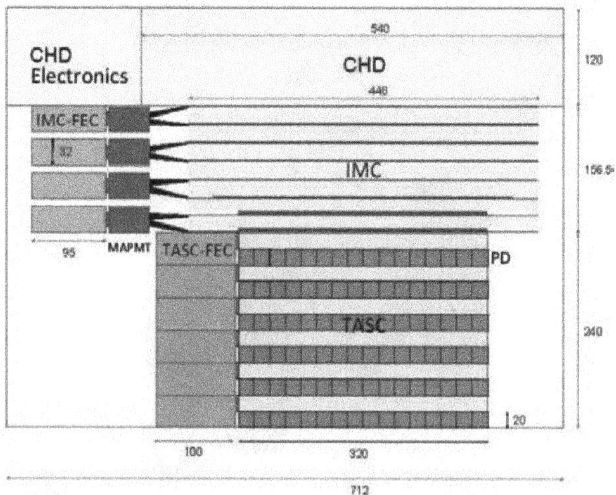

Figure 2: A schematic of the CALET detector and electronics layout.

The IMC has dimensions of about 450 mm by 450 mm and consists of 7 layers of tungsten plates each separated by 2 layers of 1 mm square cross section scintillating fiber (SciFi) belts arranged in the x and y directions with an additional x,y SciFi layer pair at the top. Each of the first five IMC layers is 0.2 X_0 thick and the following 2 layers are 1.0 X_0 thick. This provides the precision to 1) separate the incident particles, 2) determine the shower starting point, and 3) establish incident particle trajectory. The readout for the SciFi layers consists of multianode photomultiplier tubes (MA-PMT), such as the Hamamatsu R5900.

The TASC, with a vertical thickness of 27 X_0, measures the development of the electromagnetic shower to 1) determine the particle total energy and 2) separate electrons and γ-rays from hadrons. The TASC is composed of 12 layers of Lead Tungstate (PWO) "logs" where each log has dimensions of 20 mm (H) 19 mm (W) 326 mm (L) with 16 logs per layer. Alternate layers are orthogonal to each other to provide an x,y coordinate for tracking the shower core. The top layer is used for triggering, and a PMT for readout is attached to the log face. A photodiode and avalanche photodiode package (PD/APD) is used for the readout

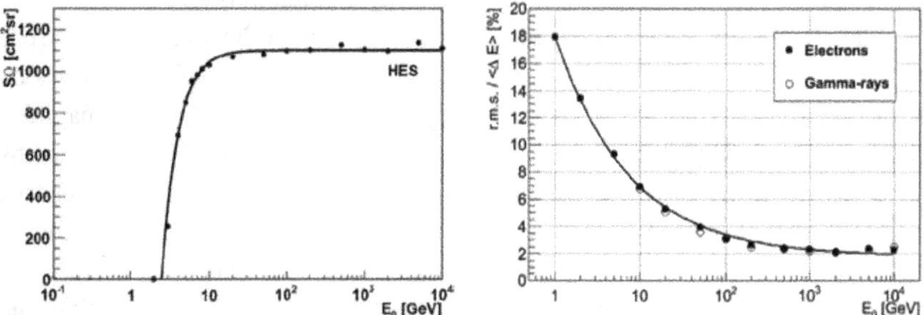

Figure 3: The expected CALET geometrical factor (left) and energy resolution (right) as a function of energy.

of the other layers. The PD and APD have separate high and low gain shaper amplifiers and 16-bit ADCs to cover the expected very high dynamic range.

3. The Expected CALET Performance

The CALET instrument design has been extensively studied to determine performance characteristics using simulations [10] as well as tests of prototype systems at particle accelerators and during balloon flights [9]. The left panel of Figure 3 shows the geometrical factor as a function of energy for the High Energy Shower (HES) trigger mode. The HES trigger becomes 100% efficient for electrons above about 10 GeV. HES will be the primary mission trigger, giving an energy independent geometry factor of 1,200 cm^2sr from ~10 GeV to

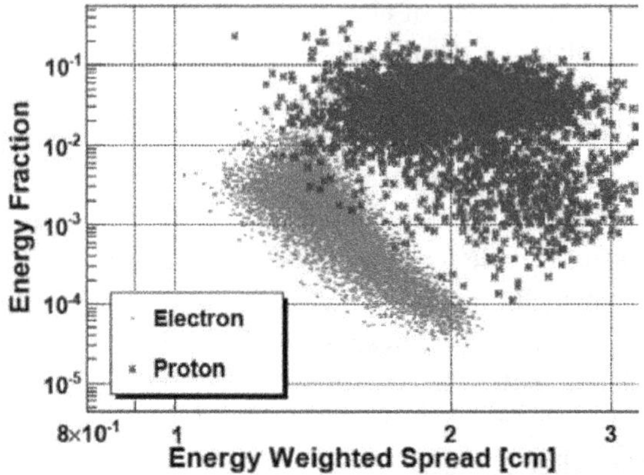

Figure 4: Energy fraction vs lateral spread for electrons and protons in CALET.

>20 TeV. The right panel of Figure 3 shows the expected energy resolution of electrons and gamma-rays as a function of energy. From ~100 GeV to > 20 TeV the energy resolution is <3% due to the very thick (30 X_o) calorimeter. For protons and heavier particles the expected energy resolution is ~30%

The ability for CALET to provide excellent separation between electrons and protons is shown in Figure 4. CALET separates electrons from protons by measuring the difference in shower development. Electron electromagnetic showers are generally narrower and develop earlier relative to those generated by hadrons. In Figure 4 the lateral width of the shower is given as the energy weighted spread on the x-axis and on the y-axis is the fraction of energy deposited in the bottom TASC layer. The simulation included 1.6 x 10^6 protons (blue) distributed as a power law with index -2.7 from 1 TeV to 100 TeV and electrons with 1 TeV incident energy. When cuts are applied 95% of the

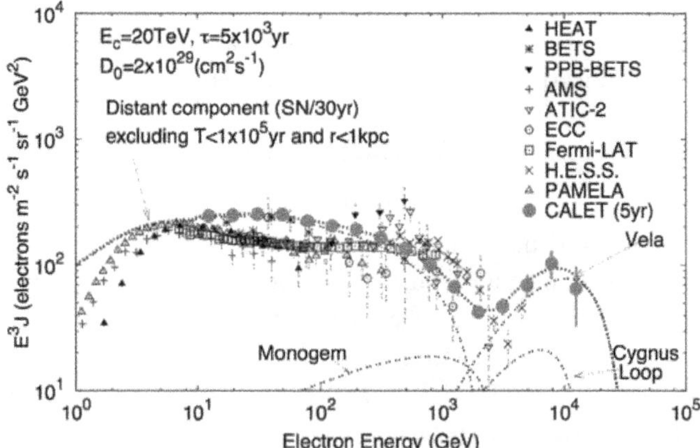

Figure 5: Expected CALET electron energy spectrum for a five year mission.

electrons (red) survive, but only a few protons are selected implying a proton rejection power at 1 TeV of 2.0 x 10^5.

4. Expected Science Results

For a five year mission CALET is expect to record about 1000 electrons with energy > 1 TeV and corresponding larger numbers at lower energy. Figure 5 shows a potential electron energy spectrum determined by CALET along with previous measurements and possible source signatures [11]. CALET will have the resolution and statistics necessary to characterize any features in the electron spectrum up to ~20 TeV.

Further, CALET will be able to extend Fermi-LAT gamma-ray observations to 10 TeV with high precision to search for possible dark matter signatures such as annihilation lines or an excess in the extra-galactic diffuse gamma-ray flux. With the CHD, CALET will be able to extend measurements of the H to Fe energy spectrum to 1000 TeV and elemental abundances up to Z ~40. Accurate measurements of B/C will provide new constraints on cosmic ray propagation and ultra-heavy composition will help distinguish between alternate models of cosmic ray origin.

Acknowledgments

This effort is supported by NASA in the United States, by JAXA in Japan and ASI in Italy.

References

1. Kobayashi, T. et al., *Ap. J.*, **601**, 340-351, (2004)
2. O.Adriani et al., *Phys. Rev. Lett.*, 106: 201101 (2011)
3. J.Chang et al., *Nature*, **456**: 362 (2008)
4. S.Torii et al., *arXiv:0809.0760 [astro-ph]*, (2008)
5. F.Aharonian et al., *Astron. & Astrophys.*, **508:** 561 (2009)
6. M.Ackermann et al., *Phys. Rev. D*, **82**: 092004 (2010)
7. Yamaoka, K. et al., Proc. 32nd Intl. Conf. on Cos Rays, OG2.5, 0839, (2011)
8. Torii, S. et al., Proc. 32nd Intl. Conf. on Cos. Rays, OG1.5, 0615, (2011)
9. Marrocchesi, P.S. et al., *Nucl. Instr. & Meth. A*, doi:10.1016, (2011)
10. Akaike, Y. et al., Proc. 32nd Intl. Conf. on Cos. Rays, OG1.5, 0769, (2011)
11. Yoshida, K. et al., Proc. 32nd Intl. Conf. on Cos. Rays, OG1.5, 0766, (2011)

LOFT – THE LARGE OBSERVATORY FOR X-RAY TIMING – SCIENTIFIC GOALS AND TECHNICAL REALIZATION

D. HAAS*, M. FEROCI**

SRON Netherlands Institute for Space Research, Utrecht, The Netherlands
**E-mail: D.Haas@sron.nl*
*** INAF/IASF Rome, Italy*

LOFT (`http://isdc.unige.ch/loft`), the Large Observatory for X-ray Timing is a candidate M-class mission within the ESA cosmic vision program and will compete for a launch opportunity around 2020. X-ray observations at a high time resolution (5-10 μs) of compact objects (as the Galactic and extra-Galactic neutron stars and black holes) provide a unique tool to investigate strong-field gravity, and give direct access to measurements of black hole masses and spins, and to the equation of state of ultradense matter. For this, good spectral resolution and a large effective area are needed. These scientific goals will be achieved using two instruments on LOFT: The Large Area Detector (LAD) of LOFT achieves an effective area of >10 m^2 (more than an order of magnitude larger than current space-borne X-ray detectors) in the 2-30 keV range (up to 50 keV in expanded mode), yet still fits a conventional platform and small/medium-class launcher. The LAD will be realized using silicon drift detectors (SDDs) similar to the ones used currently at LHC/Alice. The LAD is complemented by a Wide Field Monitor (WFM) which provides important diagnostic information and enhanced sensitivity, spectral and timing resolution to allow for a large parameter space for discovery. The LOFT mission concept as well as the LAD and WFM design will be presented.

Keywords: LOFT; SDD; Silicon Drift Detectors; LAD; WFM.

1. Introduction

The science case for LOFT is presented elsewhere[1] in these proceedings. This article will concentrate on the technical description and current design status of the Large Area Detector (LAD) and the Wide Field Monitor (WFM) aboard LOFT and on the implementation of these instruments within the satellite platform. A summary of the main characteristics of the LAD and WFM is shown in Table 1.

Table 1. Main characteristics of the LAD and WFM.

Instrument Characteristic	LAD	WMF
Detector Type	SDD	SDD
Mass [kg]	762	46
Peak Power [W]	713	45
Operating T [^{o}C]	< -30	< -20
Detector size	18 m^2	1460 cm^2
Energy range [keV]	2-50	2-50
Energy resolution [FWHM]	<260 eV @ 6 keV	<500 eV @ 6 keV
Pixel size	0.97 mm x 35 mm	200 μm
Field of View	40 x 40 arcmin	180° x 90° FWZR
Angular resolution	N/A	<5 arcmin
Typ/Max data rate [kbps]	200/1000	50/90

2. The Large Area Detector (LAD)

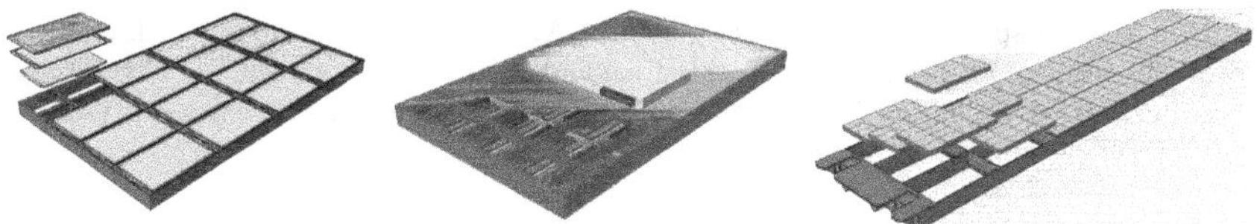

Fig. 1. Left: Front-side view of a Module, showing the mounting of collimator, SDD and the FEE; Center: Back-side view of a Module, showing the radiating surface and the MBEE; Right: A LOFT Detector Panel with assembled Modules and interfaces to the deployment mechanism.

The Large Area Detector (LAD) of LOFT is designed as a classical collimated experiment. A set of 6 Detector Panels are tiled with 2016 Silicon Drift Detectors (SDDs), electrically and mechanically organized in groups of 16, referred to as Modules. Each of the 6 Panels hosts 21 Modules, each one in turn composed of 16 SDDs (see Figure 1).

The SDDs are 450 μm thick and operate in the energy range 2-50 keV. They have strong heritage from the SDDs developed for the Inner Tracking System (ITS) in the ALICE experiment of the Large Hadron Collider (LHC) at CERN.[2,3]

The field of view of the LAD is limited to ~40 arc min by X-ray collimators. These are developed by using the technique of micro-capillary plates, the same used for the micro-channel plates: a 3 mm thick sheet of Lead glass is perforated by a huge number of micro-pores, ~20 μm diameter, ~4-6 μm wall thickness. The stopping power of Pb in the glass over the large number of walls that off-axis photons need to cross is effective in collimating X-rays below 50 keV. In order to accommodate for the internal misalignments of the instrument and for attitude uncertainties, the response of the collimator in the central ~10-15 arc min angle will be flat (flat-top response) to avoid any spurious modulation of the detected source flux.

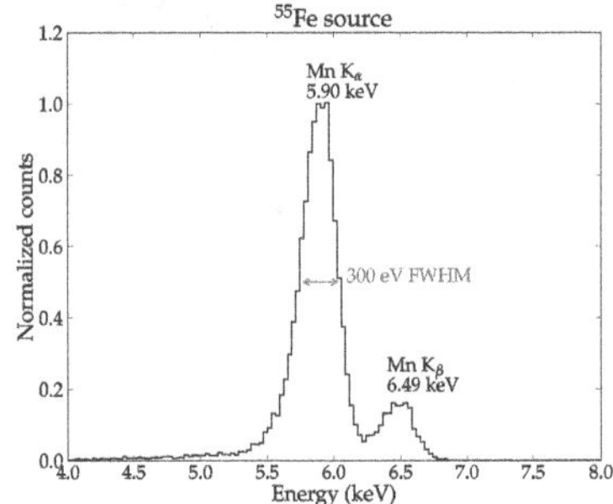

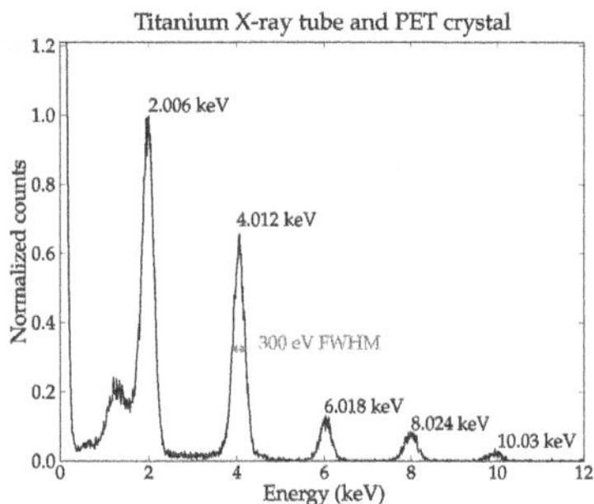

Fig. 2. Energy spectra measured using a spare ALICE detector equipped with discrete read- out electronics, at room temperature. The FWHM energy resolution was measured as 300 eV at 5.9 keV. The minimum line energy is ~1.5 keV (the spurious Al k-fluorescence comes from the detector box).

The energy resolutions of the SDDs has been measured at room temperature with an unoptimized Alice-SDD and are shown in Figure 2. These preliminary results give high confidence, that the desired energy resolution will be achieved at low temperatures and with an optimized design of the detectors.

In addition, preliminary tests for possible effects due to radiation damage have been performed. The chosen SDDs show stable leakage currents (crucial to maintain the energy resolution) at total fluences of $2.3{\cdot}10^8$ protons/cm^2 (worst case scenario for a suboptimal orbit over the full mission lifetime) and start to increase only gradually at even higher fluences. Expected fluences on 'good' orbits are one order of magnitude lower.

The LAD readout electronics follows a modular approach. Each Detector is equipped with its own Front-End Electronics (FEE). The FEE includes a dedicated ASIC (heritage from the StarX32 development). The

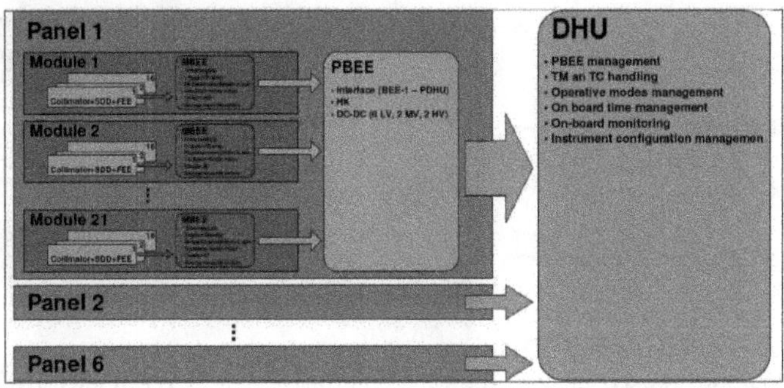

Fig. 3. A block diagram showing the organization and structure of the LAD readout electronics

ASIC has to take care of trigger detection, provide the trigger map of all triggered anodes, perform the A/D conversion and reject background (mainly from cosmic rays). The FEEs of the 16 Detectors in a Module converge into a single Module Back End Electronics (MBEE). One Panel Back-End Electronics (PBEE) for each Detector Panel is in charge of interfacing in parallel the 21 MBEE included in a PBEE, making the Module the basic redundant unit. A block diagram of the LAD electronics is shown in Figure 3.

3. The Wide Field Monitor (WFM)

The LOFT baseline WFM is a coded aperture imaging experiment designed on the heritage of the SuperAGILE experiment,[4] successfully operating in orbit since 2007[5] . With the ~100 μm position resolution provided by its Silicon micro-strip detector, SuperAGILE demonstrated the feasibility of a compact, large-area, light , and low-power high resolution X-ray imager, with steradian-wide field of view. The LOFT WFM applies the same concept but using SDDs, with improvements on the low energy threshold and energy resolution.

The working principle of the WFM is the classical sky encoding by coded masks[6] and is widely used in space borne instruments (e.g. INTEGRAL, RXTE/ASM, Swift/BAT). By using SDDs, with a position resolution <100 μm, a coded mask at <200 mm provides an angular resolution better than 5 arcmin. The coded mask imaging is the most effective technique to observe simultaneously steradian-wide sky regions with arcmin angular resolution.

Each WFM camera is a 1-D coded mask imager. After proper deconvolution is applied to the detector images, a single point-like source will

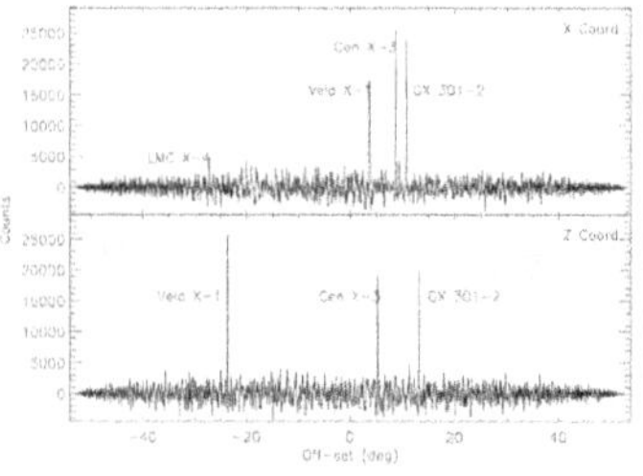

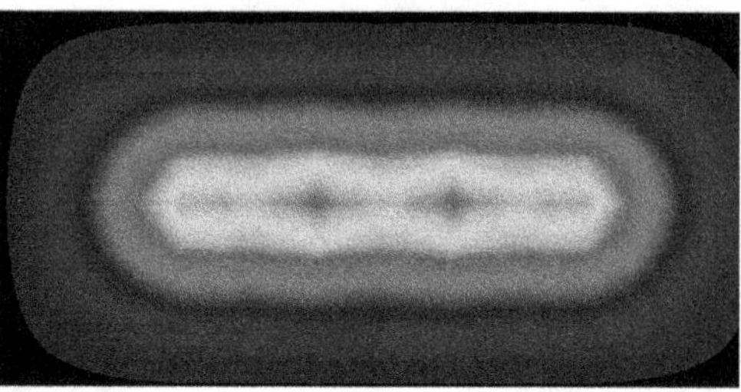

Fig. 4. Left: An example of 1D de-convoluted mask images. The source sky positions are projected onto two 1D images. Right: A qualitative plot of WFM effective area. The long direction covers the sky region accessible to the LAD by a rotation around the Solar panel array axis. The short direction corresponds to the amplitude of the maximum Sun aspect angle for the Solar panel array.

appear as a single peak over a flat background. The position of the peak corresponds to the projection of the sky coordinates onto the WFM reference frame. If more than one source is present in the observed sky region, the image will show a corresponding number of individual peaks, whose amplitude will depend on the intensity of the source and on the exposed detector area at that specific sky location. By observing simultaneously the same sky region with two cameras oriented at 90° to each other (such pair composing one WFM Unit), one can derive the precise 2D position of the sources, by intersecting the two orthogonal 1D projections. In Figure 4 we show an example of two such de-convoluted 1D-images from a real observation of the Vela region taken with the SuperAGILE experiment[7] as well as the expected LOFT WFM effective area.

An alternative design for the WFM has been proposed and it is currently considered as an option, to be subject to trade-off studies over the next few months. It is based on the use of Double Sided Silicon Strip Detectors (DSSD) that offer directly a 2D fine position resolution: the position is derived by a set of orthogonal strips on the two sides of the Si tile, collecting the electrons and holes, respectively. The interception between the up and bottom strips identifies the point of interaction of the photon/particle.

The WFM readout electronics will be very similar to the LAD architecture and is not presented here.

4. The LOFT Satellite

The LOFT payload is an extensive array of X-ray detectors with a total geometric area of $\sim$18 m^2. A preliminary evaluation of the mission has identified a LOFT configuration based on 6 panels, connected by hinges at

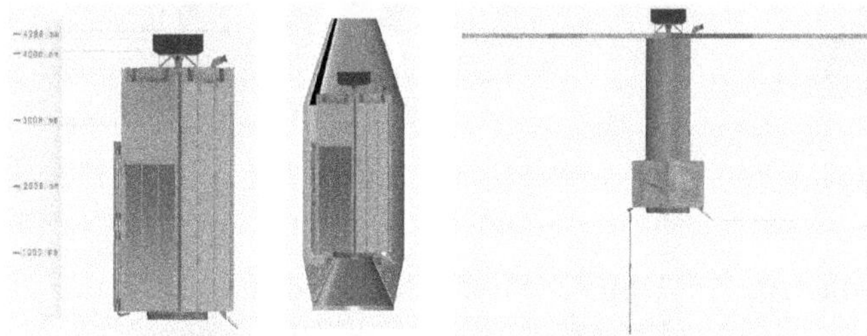

Fig. 5. The proposed baseline configuration for LOFT: folded (left), inside the Vega fairing (center) and deployed (right).

the optical bench located at the top of a tower, in an arrangement similar to that previously used for SAR antenna wings. This arrangement allows the stowing of the satellite inside the launch vehicle fairing, with the Wide Field Monitor (WFM) hosted on top of the tower. The satellite will operate in a low equatorial earth orbit in order to reduce the background and the radiation damage effect of the South Atlantic Anomaly. The science return of the mission has been evaluated assuming a medium-small class and a Vega launcher. The maximum area achievable is a product of both the size of the detector array and the solar array size that can be accommodated, as power is an important limiting factor. Figure 5 shows a preliminary concept compatible with Vega and sized for 12 m^2 of detector array effective area and 15 m^2 of solar array.

5. Outlook

LOFT is a simple mission, relying on solid hardware heritage, offering both breakthrough and observatory science. It is one of the 4 missions selected by ESA as a candidate for the future M3-mission (launched around 2021) and is currently in its assessment phase. Further down selection of the missions by ESA will take place in early 2013.

References

1. L. Stella, Fundamental Physics with LOFT, in *these proceedings.*
2. Vacchi et al., 1991 NIM A306 187
3. Rashevsky et al., 2002 NIM A485 54
4. Feroci et al., 2007 NIM A581 728
5. Feroci et al., 2010 A&A 510 A9
6. Fenimore & Cannon 1978, Appl. Opt 17 337
7. Feroci et al., 2010 A&A 510 A9

Studying Cosmic Ray Composition Around the Knee Region with the ANTARES Telescope

C.C. Hsu

NIKHEF,
Amsterdam, 1025SR, the Netherlands
**E-mail: cchsu@nikhef.nl*

The composition of the cosmic rays in the *knee* region ($\approx 10^4$ TeV/nucleus) of all particle spectrum is considered to be the result of the particle acceleration and propagation from the astrophysical sources. The steeply falling cosmic ray spectrum makes a direct measurement of the composition difficult, but it can be inferred from the measurements of the showers generated by the interaction of the primary cosmic ray with the Earth atmosphere. In particular the characteristics of the muon bundles produced in the showers depend on the primary cosmic ray nature. The ANTARES telescope is situated 2.5 km under the Mediterranean Sea off the coast of Toulon, France. It is taking data in its complete configuration since 2008 with nearly 900 photomultipliers installed on 12 lines. The trigger rate is a few Hz dominated by atmospheric muons. A method using a multiple layered neural network as a classifier was developed to estimate the relative contribution of proton and iron showers from the energy and multiplicity distribution of the muon tracks reaching the detector. The performance of the method estimated from simulation will be discussed.

Keywords: Cosmic Ray, Knee, Composition, Antares

1. Introduction

Although the main goal of ANTARES telescope is to look for high energy neutrinos coming from the deep space, it also provides us opportunities to study cosmic ray physics. One of the important topics is to distinguish the different chemical compositions around knee region of its spectrum. The cosmic ray spectrum is known as a power law with power index about -2.7 up to few PeV. Then the slope changes to -3.1 until the energy around 4×10^{18} eV.[1] The origin of the knee could be generally summarized into either astrophysical origin or particle physics origin. It is still a puzzle and believed to be the key issue to the problem of the origin of galactic cosmic rays.

Most people attributed the knee to the sudden reduction in Galactic trapping efficiency. A popular explanation is that the knee is associated with an upper limit of acceleration energy by galactic supernovae. Another popular scenario is the leakage of particles from the Galaxy, since the Larmor radius of a proton in the galactic magnetic field increases with its energy and finally exceeds the thickness of the galactic disk. Additionally, there is a minority of theorists who proposed that the knee is due to a single, recent and local supernova remnant (SNR) or a rapidly rotating pulsar interacting with radiation from its parent SNR.[2]

If the knee is caused by the maximum energy attained during the acceleration process or it is due to leakage from the Galaxy, the energy spectra for individual elements with charge Z would exhibit a cut-off at an energy $E_c^Z = \mathrm{Z}\times E_c^p$, where E_c^p is the cut-off energy of protons. The sum of the flux of all elements with their individual cut-off makes up the all-particle spectrum. In this picture, the knee is related to the proton cut-off and the steeper spectrum above the knee is a consequence of the subsequent cut-off of heavier elements, resulting in a relatively smooth spectrum above the knee.[3]

2. Analysis Strategies

The ANTARES detector is located at 40 km off the coast of Toulon, France , at a depth of 2475 m in the Mediterranean Sea. It consists of 12 flexible strings, each with a total height of 450 m, separated by about 60 m. Each string carries 75 10-inch Hamamatsu photo multipliers (PMTs) housed in glass spheres, the so-called optical modules (OMs).[4] The OMs are arranged in 25 storeys (three optical modules per storey) separated by 14.5m.

Since ANTARES is deeply buried under the sea, only the muon components from the air shower will survive at detector level. The muons will emit Cherenkov radiation only when passing through the sea water, which can be detected using photomultiplier tubes. To be registered by the ANTARES detector, muons have to travel at least 2.5 km of sea water and still be energetic enough to trigger the detector. The muon bundles properties (such as multiplicity) are strongly related to the primary energy and species of the nuclei. Two useful analysis methods are combined and then applied on MC samples. The first one is a cluster finding algorithm, and the second one is an electromagnetic shower searching algorithm.[5]

In both analysis methods, we rotate our software coordinate system such that the z axis is along the reconstructed muon track. We define the plane which is passing through the detector center and perpendicular to the recon-

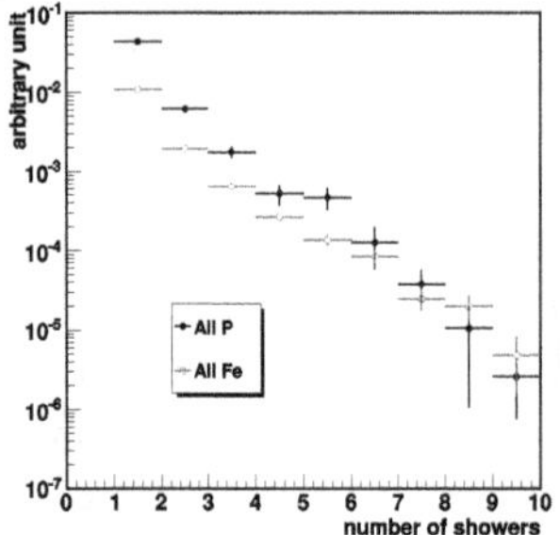

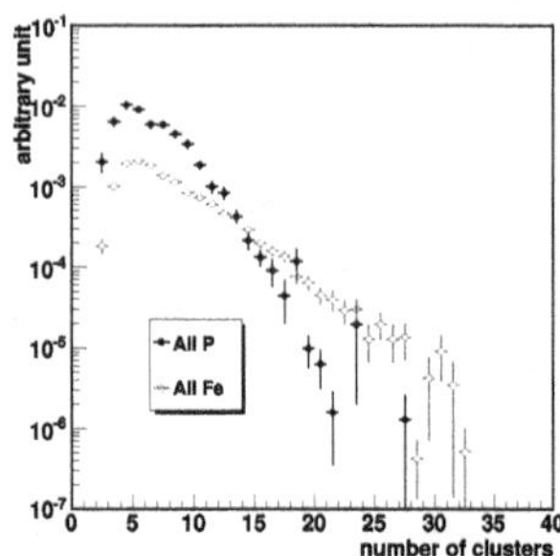

Fig. 1. The relative distribution of N_{shower} (left plot) and $N_{cluster}$ (right plot) parameters after the reconstruction quality cuts assuming that all the particles from cosmic rays are proton (black circles) and iron (empty circles), respectively.

structed track as *detection plane.* The projections of all the hits positions on the detection plane are also calculated, including the time correction information.

In the cluster finding algorithm, the clusters are formed if the hits within the cluster fulfill the following three conditions: (i) $| T_i - T_j | \leq | r_i - r_j| /C/n_g$ where T_i and T_j are the times for any two hit pairs in the hit cluster. r_i and r_j are the associated positions of these two hits. n_g is the index of refraction in the sea water and C is the speed of light. (ii) Any two hits within the cluster should be in the same or neighboring strings or floors. (iii) $| T_i - T_j | \leq |(z_i - z_j)|/C + d \times tan\theta_c /c + T_{ext}$, where T_{ext} are maximum extra time, here we set 20 ns; z_i and z_j are the rotated Z positions of the two hits. θ_c is the Cherenkov angle in the sea water. We parametrized the hit patterns and get useful parameters.

The electromagnetic shower searching method gives us additional information about the muon bundles. The high energy muons suffer from catastrophic energy losses. Once it happens, electromagnetic showers are initiated either by γ or e^+ and e^- pairs. We calculate the origin of the selected hits on the reconstructed axis assuming the hits are from Cherenkov radiation and search for the peaks using the TSpectrum function in the ROOT package. The details of the algorithm can be found in.[5] Five more parameters are obtained by this method. Combining the two algorithms, we have in total 42 parameters. Each parameter has a different discrimination power for the chemical species. To achieve multi-dimensional comparisons, we use the existing package "TMVA" (Toolkit for Multivariate Data Analysis with ROOT) for this analysis.[6] It hosts a large variety of multivariate classification algorithms. Training, testing, performance evaluation and ap-

plication of all available classifiers is carried out simultaneously.
Several methods are implemented inside TMVA packages. In order to cross check the results, three different methods were chosen for this analysis: Multilayer Perception MLP, MLPBNN and TMlpANN. We reduced the numbers of parameters to five in order to optimize the performances. These five parameters are N_{hit}, $N_{cluster}$ and N_{npe}, representing number of hits, clusters, and photoelectrons from cluster-finding algorithm. N_{shower} and N_{base} are the numbers of showers and the number of photoelectrons of the baseline from EM shower-finding algorithm.

3. Analysis on MC samples

A full MC simulation was adopted in this analysis. The air showers induced by the primary nuclei with energy ranging from 1 to 10^5 TeV /nucleon and zenith angle between 0° and 85° are simulated using the CORSIKA (Version 6.2)[7] and the hadronic interaction model QGSJET.01c. All muons reaching the sea level, with energies larger than the threshold energy, are propagated through sea water to the detector. At last, muons are transported through the ANTARES sensitive volume, Cherenkov light is produced and the detector response is simulated. Background noises were added afterwards. The standard ANTARES trigger is simulated. The muon direction and position are reconstructed using a multi-stage fitting procedure, which basically maximizes the likelihood of the observed hit times as a function of the muon direction and position.[8]
We further divide the MC samples into 3 independent subsets for training, testing and evaluating the TMVA analysis. Each event is weighted according to different models of CR composition. The iron component is tagged as signal whereas the protons are tagged as background. The "pure" signal distributions are obtained, if we assume that all the coming cosmic rays are iron. On the other hand, the background samples come from assuming all cosmic rays are protons. The distributions of N_{shower} and $N_{cluster}$ assuming the all particle spectrum are, respectively, proton and iron, are shown in Fig 1. We carefully checked the output of the neural network from test samples in order to avoid the over-training effects.

4. Results and Conclusions

The training and test events fed into the neural network were subjected to a ser ies of cuts. The main quality cut is the so-called Λ cut.[8] Its purpose is to keep good quality on reconstructed events. Our approach is to put differ-

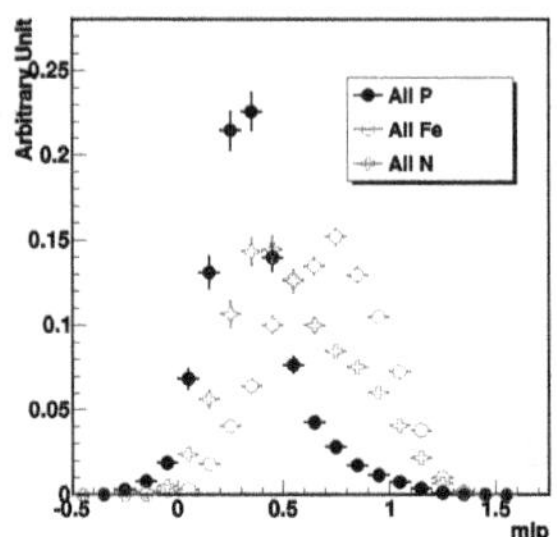

Fig. 2. Output computed by the neural network according to different cosmic ray knee models. Here we take three extreme composition hyphothesis: All particles are protons, nitrogen and irons. The built-up neural network gives different distributions from these three models.

ent weight in each event according to differnt physics models and check the output of different physics models from neural work. An example is shown on Fig 2. From the plot, the differences between three extreme models are clearly seen. In the future, with this method, we are able to compute the compatibilities between different realistic physics models and data based on the neural work. In summary, we developed a method to estimate the ratio of the heavy elements in the triggered cosmic ray events based on the information from the muon tracks and electromagnetic showers in ANTARES detector. To combine all the discrimination powers from established multi-parameters needs the help from multi-variate analysis (neural network). The analysis of ANTARES real data with the goal of deriving ratio between different groups of elements in the cosmic ray spectra is ongoing and will yield results in the near future.

References

1. J. Bluemer, R. Engel and Hoerandel, *J. R. Prog. Part. Nucl. Phys* **63**, 293 (2009).
2. T. Wibig and A. W. Wolfendale, *astro-ph/1012.4562* (2010).
3. J. R. Hoerandel, *Astropart. Phys.* **21**, 241 (2004).
4. M. Ageron *et al.*, *Nucl. Instrum. Meth.* **A656**, 11 (2011).
5. S. Mangano, *Nucl. Instrum. Meth. A* **588**, p. 107 (2008).
6. A. Hocker, P. Speckmayer, J. Stelzer, F. Tegenfeldt and H. Voss Prepared for PHYSTAT-LHC Workshop on Statistical Issues for LHC Physics, Geneva, Switzerland, 27-29 Jun 2007.
7. D. Heck, G. Schatz, T. Thouw, J. Knapp and J. N. Capdevielle FZKA-6019.
8. A. Heijboer, *PhD Thesis* .

Absolute Measurement of the Reflectivity and the Point Spread Function of the MAGIC Telescopes

H. KELLERMANN*, R. MIRZOYAN, C. SCHULTZ†, J. HOSE, M. TESHIMA
Max-Planck-Institiut für Physik, 80805 München, Germany
**E-mail: hkellerm@mpp.mpg.de*

M. GARCZARCZYK
Instituto de Astrofisica de Canarias, La Palma, Islas Canarias

In the past twenty years the ground-based Imaging Atmospheric Cherenkov Telescope (IACT) technique has revolutionized the understanding of cosmic rays. Over 120 sources of cosmic rays of both galactic and extragalactic origin have been observed in the very high energy regime of 50 GeV to 100 TeV. One key parameter of these measurements is the gamma ray flux, which is derived from the detection of gamma ray induced Cherenkov light in the earth's atmosphere. The ratio between the total Cherenkov light hitting the primary IACT mirror and the projection on its camera has a direct impact on the precision of the flux measured. We have further improved an existing method for measuring *in situ* the reflectivity of the mirror of a telescope. The method is based on the simultaneous measurement of the brightness of both, a selected star directly and its image in the focal plane of the telescope. We applied this method to both 17 m diameter MAGIC IACTs operating on the Canary Island of La Palma. In this report we want to present the details of this method as well as results of the reflectivity measurements.

Keywords: focusing; reflectivity; Cherenkov telescope

1. Introduction

The Imaging Atmospheric Cherenkov Telescope (IACT) technique is using an indirect method to observe galactic and extragalactic sources of high energy cosmic rays. These cosmic particles induce flashes of Cherenkov light when impinging the atmosphere of the earth. To detect those faint flashes with adequate efficiency, a large mirror surface is required. Profound knowledge of the properties of the atmosphere acting as calorimeter, of the

†Current address: Dipartimento di Fisica, Università di Padova and INFN

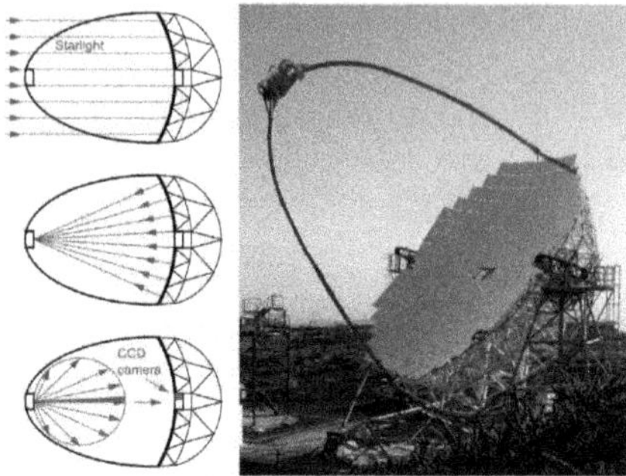

Fig. 1. Left: Schematic drawing of the measurement technique; Right: MAGIC II telescope. Photography: Robert Wagner

optics, and the detector is essential to correctly deduce physical characteristics of the cosmic sources from the Cherenkov light observations. As part of the continuous effort to achieve a better understanding of the MAGIC telescopes' imaging properties and to improve its absolute calibration, the technique to measure the amount of light focused in the focal plane was further improved.[1] The results of these measurements are spot radius dependent fractions of light concentrated in the focal plane over the total amount of incident light collected by the IACT mirrors (referred to as *focused reflectivity* in the following).

2. Technique for measuring the focused reflectivity

Our technique enables the measurement of the focused reflectivity of all MAGIC mirrors, without the need to unmount the mirror panels from the dish. For this purpose the telescope is pointed to a bright star whose image is projected onto a diffusely reflecting material (Spectralon) in the focal plane. We use a highly sensitive CCD camera located in the center of the reflector to simultaneously observe both the star directly and its projection on the Spectralon target (see fig. 1). Taking into account the geometry and the optical properties of the Spectralon reflector, the focused reflectivity of the MAGIC mirrors can be determined using the following formula:[1]

$$R_{\text{foc}} = \frac{\Phi_{\text{indirect}}}{\Phi_{\text{direct}}} \cdot \frac{r^2}{A_{\text{eff}}} \cdot \frac{\Omega_{\text{eff}}}{R_{\text{sp}}} \cdot \frac{1}{\cos(4.24^\circ)^{1.15}} \qquad (1)$$

with

Φ_{direct}	sum of CCD counts in the region of the star
Φ_{indirect}	sum of CCD counts in the region of the reflected image
r	distance between the CCD camera and the diffuse reflector
A_{eff}	total effective area of the mirrors
Ω_{eff}	effective solid angle (derived from radiation characteristics)
R_{sp}	total reflectivity of the Spectralon
$\frac{1}{\cos(4.24°)^{1.15}}$	correction factor for small angular offset of the camera from reflector's center and its deviation from the Lambertian law

Φ_{direct} and Φ_{indirect} are derived from the CCD images (see fig. 2).
The optical properties of the Spectralon reflector[2] have a large impact on

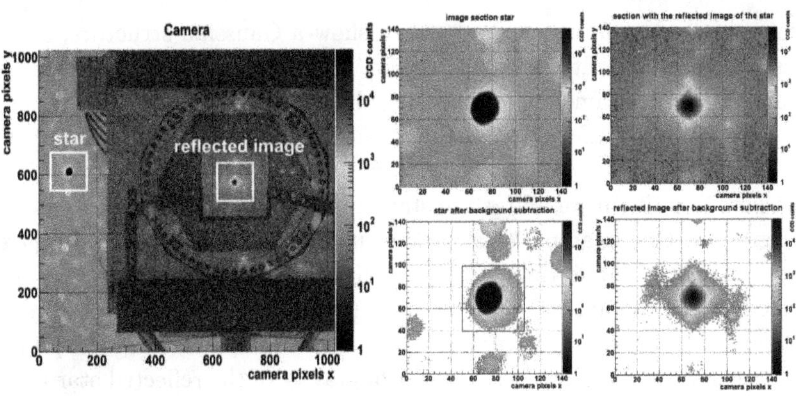

Fig. 2. Left: CCD picture of a star and its image on the diffuse reflector (Spectralon) in front of the MAGIC II telescope camera; Right: Section around the star and its reflected image before (top) and after background subtraction (bottom)

the final results and thus were precisely determined in the laboratory. Samples were illuminated with a laser and angular scans were performed using Si-photodiodes as a detector, while a second diode permanently monitored the beam intensity. This way we could characterize the diffuse Spectralon reflector with high precision. In contrast to the manufacturer's specifications, moderate but yet significant deviations from a Lambertian reflector could be observed (see fig. 3).

For the CCD image analysis we selected a section of 140×140 pixels around each, the star and its reflected image. Since the background is significantly different in both sections, we decided to create two separate intensity histograms in order to remove the background individually. For low CCD

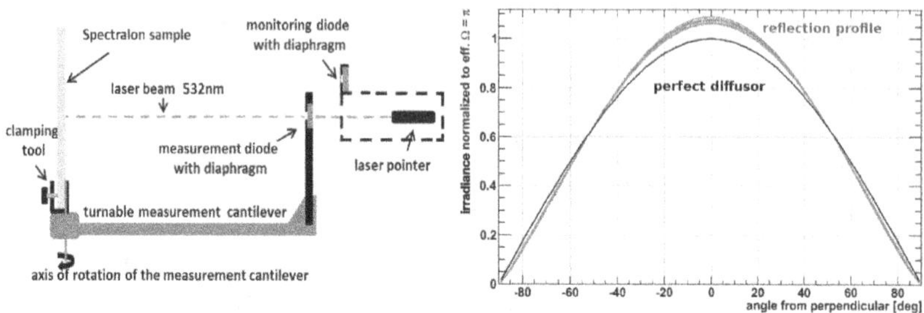

Fig. 3. Left: Setup for measuring the angular scattering intensity profile; Right: Angular reflection characteristic of different Spectralon samples (grey) and a perfect diffuse Lambertian reflector (black)

values the background was expected to show a Gaussian structure, thus a gauss function was fitted to this region. The background was subtracted separately for each image section using the following criteria:

$$E_{\text{CCD}} \leq \mu + 3\sigma \Rightarrow E_{\text{CCD}} = 0$$
$$E_{\text{CCD}} > \mu + 3\sigma \Rightarrow E_{\text{CCD}} = E_{\text{CCD}} - \mu$$

where μ is the mean and σ the standard deviation of the corresponding background. One example of a background subtracted image is displayed in fig. 2 on the lower right side. The light flux from the star Φ_{direct} is defined as the amount of light in the square box in fig. 2 lower left side. A radius dependent reflectivity curve is calculated by summing the photon counts inside circles around the center of gravity of the reflected star image with increasing radius (see fig.4). The following important quantities can be extracted from the curves:

r_{RMS} [mm] RMS radius of the 2-dimensional reflected image
R_{RMS}[%] percentage of light reflected within r_{RMS}
R_{Pixel}[%] light reflected onto the radius of one MAGIC camera pixel (0.05°)
R_{Total}[%] fraction of incident light at a radius of 70 mm

3. Results of the measurements

The table below shows the final results of our measurements for both telescopes and mirror types. The MAGIC I reflector is equipped with 0.25 m^2 all-*Al* mirrors, while the MAGIC II reflector includes 143 all-*Al* and 104 glass mirrors with a size of 1 m^2 each. The results were obtained by averaging over several measurements using different stars and zenith angles in

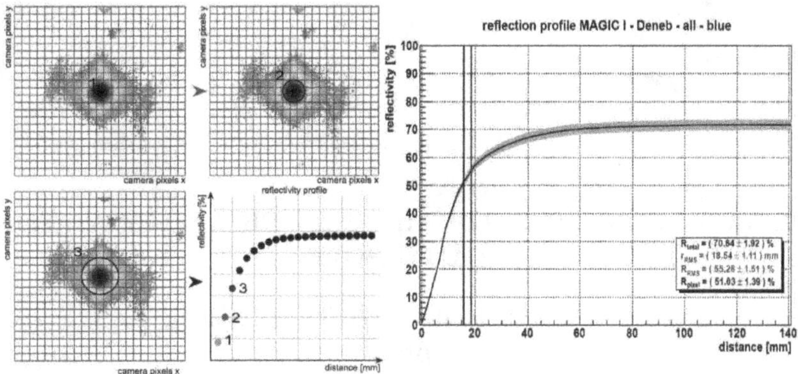

Fig. 4. Left: Example for spot-radius dependent reflectivity profile; Right: Cumulative reflectivity profile

the wavelength regime from 380 to 520 nm. Our measurements show that the absolute amount of reflected light is larger for the glass mirrors, while the *Al* mirrors provide slightly better focusing. The *Al* mirrors of MAGIC I seem to have a better reflectivity than those of MAGIC II but MAGIC II has a better point spread function (PSF).
All results and more detailed information are available online.[3]

telescope/mirror	$\mathbf{r_{RMS}}$ **[mm]**	$\mathbf{R_{RMS}}$ **[%]**	$\mathbf{R_{Pixel}}$ **[%]**	$\mathbf{R_{Total}}$ **[%]**
MAGIC I	19.25 ± 0.61	55.01 ± 0.78	48.91 ± 1.83	71.21 ± 0.72
MAGIC II all	16.06 ± 0.73	55.21 ± 2.11	54.61 ± 1.53	74.68 ± 0.58
MAGIC II *Al*	14.42 ± 0.58	54.28 ± 1.07	55.93 ± 1.46	67.16 ± 0.74
MAGIC II glass	16.22 ± 0.15	56.31 ± 1.64	54.90 ± 2.05	81.36 ± 0.28

4. Acknowledgment

We are grateful to Dr. Adrian Biland for his strong support in optimizing the PSF of the MAGIC Telescopes.

References

1. R. Mirzoyan *et al.*, *Astropart.Phys.* **27**(July 2007).
2. B. Pichler *et al.*, *Nuclear Instruments and Methods in Physics Research* **422**(March 2000).
3. H. Kellermann, *Diploma thesis* (March 2011), `http://wwwmagic.mppmu.mpg.de/publications/theses/HKellermann_dipl.pdf`.

STATUS OF THE ICETOP AIR SHOWER ARRAY AT THE SOUTH POLE

F. KISLAT* for the IceCube Collaboration

DESY

D-15738 Zeuthen, Germany

**E-mail: fabian.kislat@desy.de*

The IceTop air shower array is the surface component of the IceCube Neutrino Observatory at the geographic South Pole. The combination of IceTop and IceCube provides a new and powerful tool to measure cosmic ray composition in the energy range between about 300 TeV and 1 EeV by detecting the electromagnetic component at the surface in coincidence with the muon bundle in the deep underground detector. The paper will give an overview of the current status of the detector and the first physics results will be presented.

Keywords: Cosmic rays; IceTop; Experiment.

1. Introduction

IceTop is an air shower array and the surface component of the IceCube Neutrino Observatory located at the geographic South Pole.[1] It comprises 81 detector stations covering an area of 1 km^2 above the neutrino telescope in the ice. Construction of IceCube and IceTop was completed in the 2010/11 austral summer.[2]

The main purpose of IceTop is the measurement of the cosmic ray energy spectrum and chemical composition in the energy range from 10^{14} to 10^{18} eV. At the so-called "knee" at an energy of about $4 \cdot 10^{15}$ eV the energy spectrum steepens from a spectral index of about -2.7 to about -3.1. Many models[3] predict that this steepening is accompanied by a change of the chemical composition of cosmic rays in the energy range above the knee. A good measurement of the composition and spectrum in this energy range is thus crucial in order to understand the acceleration mechanisms and the propagation of cosmic rays. However, the mass determination is notoriously difficult because measurements are indirect and thus afflicted with large systematic errors. While several experiments have already observed a change in composition above the knee, details of the features remain unclear, reducing discriminative power.

The combination of IceTop and IceCube offers the unique capability to separate the core of high-energy muons from the electromagnetic component of an air shower. This provides a very powerful way of measuring cosmic ray composition in the PeV energy range.

2. IceTop detector

The IceTop stations are located on the surface above IceCube strings, which are instrumented with light detectors at a depth between 1450 m and 2450 m. They are arranged on a triangular grid with a nominal spacing of 125 m. Three stations are located at intermediate positions at the center of the array forming an infill with a smaller spacing in order to reduce the detector threshold. Each station consists of two tanks with a diameter of 1.82 m filled with ice to a height of 90 cm. In each tank, Cherenkov light emitted by relativistic charged particles traversing the ice is recorded by two "Digital Optical Modules" (DOMs).[4]

The DOMs consist of a photomultiplier tube and electronic circuitry for readout and digitization. The two DOMs inside an IceTop tank are operated at two different gains (high gain: $5 \cdot 10^6$; low gain: 10^5) in order to increase the linear dynamic range of the tank. After a trigger, a DOM records and digitizes the PMT signal for about 422 ns with a sampling rate of 300 MSPS. The two tanks of a station are operated in a local coincidence mode requiring both high-gain DOMs to trigger in order to reduce the trigger rate. Additionally, for all DOM triggers a simple charge and time stamp are recorded (soft local coincidence, SLC).

Near-vertical muons with GeV energies leave a distinct peak in the charge distribution recorded by a tank when no local coincidence is required. The position of this peak is referred to as 1 Vertical Equivalent Muon (VEM) and is used to calibrate the tank signals. All signal charges are thus expressed in units of VEM.

3. Air shower reconstruction

The main primary energy sensitive observable in IceTop is the shower size S_{125}. It is determined by fitting the lateral distribution of charges with a custom lateral distribution function.[5] The position of the shower core is determined from the lateral charge distribution in the same fit, whereas the shower direction is reconstructed in a fit of the signal times to a function describing the shower front. With this reconstruction a core position resolution of about 8 m and an angular resolution better than 0.5° was obtained for sufficiently high energies with the 26-station configuration of the detector.[6]

4. Spectrum and composition of cosmic rays

Like all indirect measurements, IceTop has to rely on Monte Carlo simulations in order to relate the measured shower parameters to the properties of the primary particle. Therefore, it is important to exploit several systematically independent composition sensitive observables.

4.1. *Coincident events in IceTop and IceCube*

Due to the interplay of decay and interaction of charged pions during the first few interaction lengths, the multiplicity of high-energy muons is larger for showers initiated by heavier primaries. Thus, measuring the number of TeV muons in IceCube in coincidence with the shower size in IceTop is a very powerful way to measure primary mass.

Figure 1 (left) shows a simulation of the correlation between the parameter K_{70}, which measures the muon energy at a distance of 70 m from the bundle axis in the deep ice, and the electromagnetic component shower size S_{125}. Different primary masses populate different bands in this plot, and the shading indicates the proton content in the simulation: dark grey is 100% iron, light grey is 100% protons. The relation between the $K_{70} - S_{125}$ space and the mass-energy space is non-linear. The measured data are overlayed.

A neural network has been used to derive the cosmic-ray energy spectrum and mass composition from data taken with the 40 stations and 40 strings configuration of IceCube.[7] The resulting energy spectrum is in good agreement with previous results from other experiments and the measured dependence of mean logarithmic mass on energy (Fig. 1, right) can be described with the polygonato model.[8]

4.2. *Muons in IceTop*

The muon energy distribution in an air shower peaks at a few GeV. These muons are produced at a much later stage of shower development than the high-energy muons that can reach IceCube. Distinguishing the signals produced by individual particles in IceTop tanks is not possible. However, because muons in the GeV range always create a signal of about 1 VEM

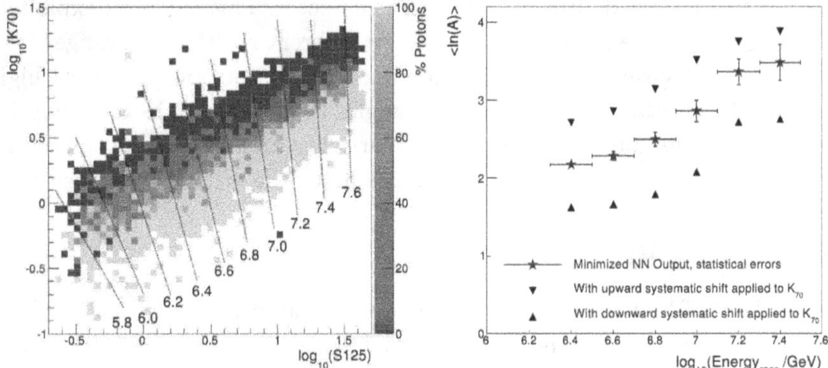

Fig. 1. Left: Muon bundle size as a function of the shower size at the surface for protons and iron. The numbered lines indicated lines of constant primary energy. Right: Mean logarithmic mass determined from IceCube-40 data.

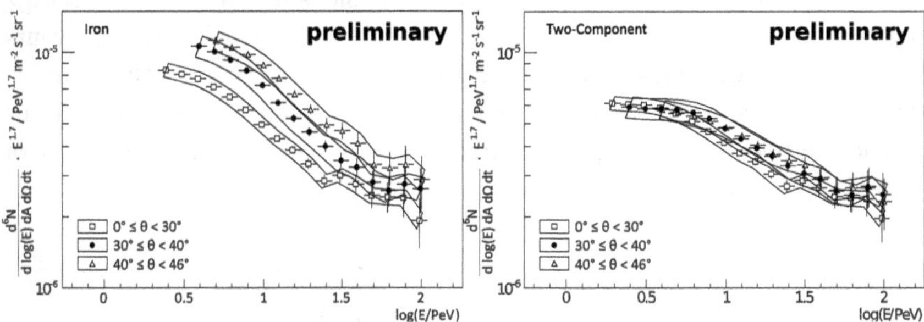

Fig. 2. Unfolded energy spectra in three different zenith angle ranges, assuming pure iron (left) and a mixture of protons and iron (right). Under the assumption of an isotropic cosmic ray flux, experimental data are in disagreement with pure iron below 15 PeV at a 99% confidence level.

they can be detected at the periphery of an air shower where the signal expectation value from the electromagnetic component is much smaller than 1 VEM. The abundance of these muons depends on the primary mass. Thus, they can be exploited to determine the mass composition, which is currently being investigated.

4.3. *Inclined showers*

Showers initiated by heavy primaries develop faster in the atmosphere than those initiated by lighter particles. Studying showers at various zenith angles allows one to sample the average longitudinal development of air showers at several slant depths with the same detector. This has been exploited by determining energy spectra from three different zenith angle ranges (see Fig. 2). Since cosmic rays can be assumed to be isotropic to a very high degree, the spectra in different zenith angles have to agree. In the analysis, this has been used to determine the all-particle cosmic-ray energy spectrum and limit the range of potential assumptions on the primary composition.[6]

5. Other results and conclusions

Since May 2011 IceTop and IceCube are operated in their final configuration with 86 strings and 81 stations. First results on composition and energy spectrum of cosmic rays are available, while further composition-sensitive observables are being investigated. In future, combining several measurements of the spectrum and composition of cosmic rays will allow us to reduce systematic errors and to make conclusions about the physics of cosmic-ray interactions in the atmosphere. Besides its high-energy cosmic-ray physics program IceTop also allows the study of heliospheric events through the variation of individual DOM trigger rates.[9] Furthermore, a search for air showers without muons in IceCube allowed setting a limit on the relative abundance of PeV photons.[10] A study of the anisotropy of cosmic rays is currently in progress.

References

1. A. Achterberg *et al.*, *Astropart. Phys.* **26**, 155 (2006).
2. R. Abbasi *et al.*, The IceTop Air Shower Array: detector overview, physics goals and first results, in *Proc. 32nd ICRC*, (Beijing, China, 2011).
3. J. Hörandel, *Astropart. Phys.* **21**, 241 (2004).
4. R. Abbasi *et al.*, *Nucl. Instrum. Meth.* **A601**, 294 (2009).
5. S. Klepser, Reconstruction of Extensive Air Showers and Measurement of the Cosmic Ray Energy Spectrum in the Range of 1-80 PeV at the South Pole, PhD thesis, Humboldt-Universität zu Berlin, (Berlin, Germany, 2008).
6. F. Kislat, Measurement of the Energy Spectrum of Cosmic Rays with the 26-Station Configuration of the IceTop Detector, PhD thesis, Humboldt-Universität zu Berlin, (Berlin, Germany, 2011).
7. R. Abbasi *et al.*, Cosmic Ray Composition from the 40-string IceCube/IceTop Detectors, in *Proc. 32nd ICRC*, (Beijing, China, 2011).
8. J. Hörandel, *Astropart. Phys.* **19**, 193 (2003).
9. R. Abbasi *et al.*, *Astrophys. J. Lett.* **689**, L65 (2008).
10. R. Abbasi *et al.*, Searching for PeV gamma rays with IceCube, in *Proc. 32nd ICRC*, (Beijing, China, 2011).

Light Sensor Candidates for the Cherenkov Telescope Array

M. L. Knötig*, R. Mirzoyan, M. Kurz, J. Hose, E. Lorenz, T. Schweizer, M. Teshima
Max-Planck-Institut für Physik, 80805 München, Germany
**E-mail: mknoetig@mpp.mpg.de*

P. Buzhan, E. Popova
Moscow Engineering and Physics Institute, 115409 Moscow, Russia

J. Bolmont, J.-P. Tavernet, P. Vincent
LPNHE, Université Pierre et Marie Curie Paris 6, 75252 Paris Cedex 5, France

M. Shayduk
Deutsches Elektronen-Synchrotron (DESY), 15738 Zeuthen, Germany

On behalf of the
CTA Focal Plane Instrumentation WP

We report on the characterization of candidate light sensors for use in the next-generation Imaging Atmospheric Cherenkov Telescope project called Cherenkov Telescope Array, a major astro-particle physics project of about 100 telescopes that is currently in the prototyping phase. Our goal is to develop with the manufacturers the best possible light sensors (highest photon detection efficiency, lowest crosstalk and afterpulsing). The cameras of those telescopes will be based on classical super-bi-alkali Photomultiplier tubes but also Silicon Photomultipliers are candidate light sensors. A full characterisation of selected sensors was done. We are working in close contact with several manufacturers, giving them feedback and suggesting improvements.

Keywords: PMT; SiPM; MPPC; GAPD; Quantum Efficiency; CTA

1. Introduction

The atmosphere is opaque for cosmic rays. But when particles with very high energy (VHE) of at least some tens of GeV hit the atmosphere they produce extended showers of secondary particles. These emit blueish Cherenkov light flashes due to their high speed, exceeding that of light in the atmosphere. The faint flashes can be collected by a telescope and projected onto it's fine pixelised camera. This method is called Imaging Atmospheric Cherenkov Telescope (IACT) technique and it is currently the most successful one with over 120 sources of VHE gamma-rays discovered.[1] The camera consists of hundreds of ultra fast and highly sensitive light sensors, fast enough to follow the development of the shower.[2] All the current telescopes use Photomultiplier Tubes (PMT) — with the exception of the FACT telescope[a].

2. The Cherenkov Telescope Array

The Cherenkov Telescope Array (CTA) is the major project for the next generation ground based VHE gamma-ray astronomy. Current systems of IACTs use at most four telescopes. The plan is to build an array of about 100 telescopes of three different sizes (large ∼23m, middle ∼12m and small ∼6m) that will provide a ten times higher sensitivity compared to current systems. The camera of each telescope will comprise ∼2000 of ultra-fast PMTs which means a total need of ∼ 150000 PMTs.

Table 1. The FPI sensor wish list — a selection of parameters in comparison with a target PMT

Parameter	Range Specification	Hamamatsu R11920-100
Spectral Sensitivity Range	290 - 600 nm	300 - 650 nm
Peak Quantum Efficiency	35%	(35.6±1.7)%
Average QE over Cherenkov Spectrum	> 21%	(22.8±1.0)%
Afterpulsing at 4 ph.e. Threshold	< 0.02%	≃0.03%
Transit Time Spread, single ph.e, FWHM	< 1.3 ns	(1.3±0.1) ns
Collection Efficiency 1.st Dynode	96%	≃93%

3. Photomultiplier Tubes

A development program was started with Hamamatsu K.K. and Electron Tubes Enterprises Ltd and a full characterisation of selected PMT samples was done. This combined measurements of quantum efficiency, afterpulses, single photo electron response, transit time spread and light emission. After two years the development program resulted in the PMT candidate R11920-100 from Hamamatsu which combines the cathode of the Hamamatsu R9420

[a]Their approach is to use Silicon Photomultipliers (SiPM) from Hamamatsu (MPPC)[3]

— providing very high quantum efficiency — and the dynode structure of the Hamamatsu R8619 — providing very low afterpulsing. In Table 1 we show a comparison between some selected parameters from the wish list and the measurement results for three samples of the R11920-100. The quantum efficiency in particular is high and has its maximum at about 340–370 nm with an average peak quantum efficiency of (35.6±1.7)%, as can be seen in Fig. 1.

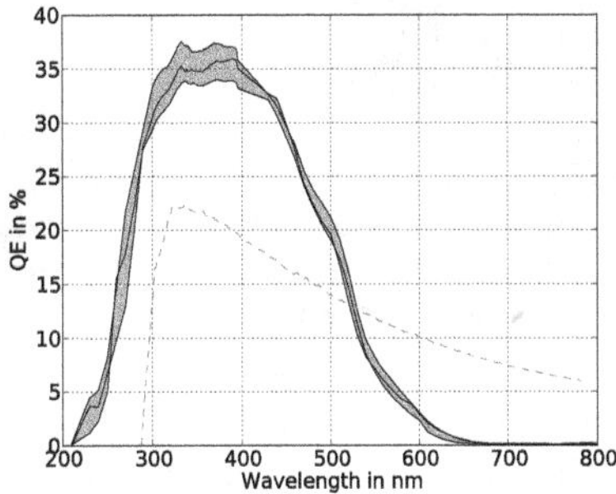

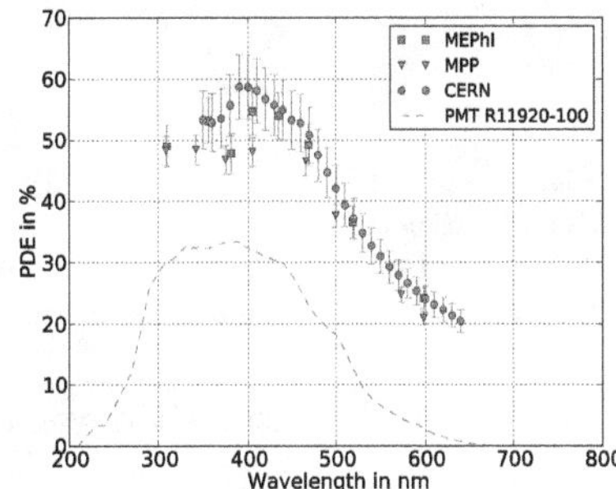

Fig. 1. Left: Quantum efficiency of three R11920-100 samples over wavelength. The dashed line is the differential Cherenkov light spectrum of a 100GeV photon hitting the atmosphere at zenith angle = 0; Right: Photo detection efficiency of the MEPhI SiPM 100b 1x1mm structure compared to the R11920-100 (dashed). Circles: Y. Musienko, CERN; Squares: P. Buzhan, MEPhI; Triangles: M. Knötig, MPP

PMTs show weak emission of light from the dynodes.[4] For a better understanding of the R11920-100, we investigated and measured this(Figure 2). We assume that this light, after many reflections, could arrive at the cathode and produce light-induced afterpulses. These shall become visible only after few tens of nanoseconds after the laser pulse illuminates the cathode. Such information is communicated to manufacturers and should help to improve the PMT, in this case the light-induced afterpulsing. We have a new setup with a fast gated image intensifier of the Hamamatsu C9546 series which will allow us to measure coincidences with a delayed pulse.

Hamamatsu and Electron Tubes have both worked to increase the collection efficiency (CE) of their PMTs. The next generation prototype from Hamamatsu reaches a CE of 96% due to changes in the input window curvature and stabilised cathode-to-first-dynode voltage.

The main specifications (Table 1) of CTA light sensors are close to being

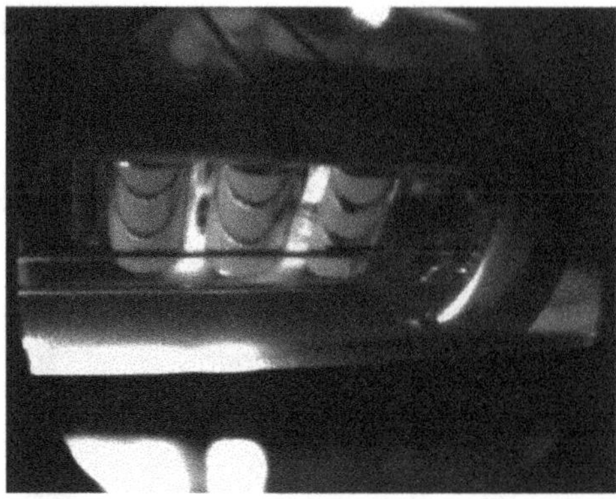

Fig. 2. Overlay of an optical and a light emission picture showing the R11920-100 dynode structure. A pulsed laser is shooting at the cathode on the right at a rate of 20MHz and 440nm. The voltage applied is 1.1kV and the emission is integrated over 1s. One can see the light emission between dynodes two, four and six.

fulfilled by the currently developed PMTs. We are optimistic that they all will be met before the the beginning of the main construction phase.

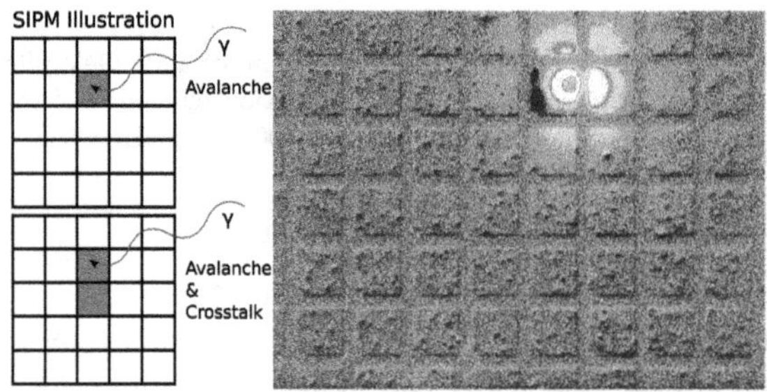

Fig. 3. Left: Illustration of a crosstalk event in a SiPM, artificially increasing the signal; Right: Emission morphology from the SiPM surface, type 100b 1x1mm MEPhI at 38V, the grey spot marks the focused laser

4. Silicon Photomultipliers

A Silicon Photomultiplier, known as SiPM but also as MPPC and G-APD array is a novel solid state photo sensor with a potential of 2-3 times higher photon detection efficiency (PDE) compared to classical PMTs. They consist of a matrix of small avalanche photo diodes with a common anode and are operated in Geiger mode. To become useful for CTA the PDE, the crosstalk between cells (Figure 3) and the dark rate need to improve. In contrast to the PMT's afterpulses, reduction of the crosstalk is still the biggest challenge for the application of SiPM in the CTA project as it reduces the amplitude resolution and requires a higher trigger threshold.[5]

We present the measurements of the recent prototype SiPM from Moscow Engineering and Physics Institute (MEPhI) in cooperation with Excelitas of the p-on-n type. As three independent measurements show it has a high and flat plateau with PDE(400 nm) $\approx$ 50% (Figure 1) and good sensitivity in the ultraviolet region. The crosstalk effect was suppressed by introducing trenches between the cells, a second p-n junction and by ion implantation and is about 5% at 14% relative overvoltage, defined as $\frac{\Delta U}{U_{applied}}$ where $\Delta U = U_{applied} - U_{breakdown}$.

We have developed a new method for imaging the crosstalk morphology. The idea is to shoot with a laser onto the surface of a selected cell and to count the number of photons emitted by the avalanches[6] in the neighbouring cells. Interestingly, as one can see in Fig. 3, the avalanches emit light only locally around the focused laser spot. Our comparison of this method with the classical one shows very good agreement.

5. Conclusions

The PMT development for CTA is in progress and already now has almost achieved the specified targets. During the next two years further optimization of PMTs from both Hamamatsu and Electron Tubes Enterprises is foreseen.

SiPM development is ongoing. We consider their use interesting in many applications when the PDE of SiPM will become 1.5-2 times higher than that of PMTs at comparable costs. New MEPhI SiPMs are getting there with peak PDE $\simeq$ 50% at $P_{crosstalk} \simeq 5\%$. Soon these are going to become a commercial product. In a couple of years this type of sensor could become a serious alternative to the PMT in CTA.

References

1. Tevcat: An online catalog for very high energy gamma-ray astronomy `http://tevcat.uchicago.edu`.
2. A. Ostankov, *NIM A* **471**, 188 (2001).
3. H. Anderhub *et al.*, *NIM A* **628**, 107 (2011).
4. H. R. Krall, *Nuclear Science, IEEE Transactions on* **14**, 455 (1967).
5. P. Buzhan *et al.*, *NIM A* **610**, 131 (2009).
6. R. Mirzoyan, R. Kosyra and H.-G. Moser, *NIM A* **610**, 98 (2009).

RECENT RESULTS FROM THE ARGO-YBJ EXPERIMENT*

G. MARSELLA

Dipartimento di Ing. Dell'Innovazione, Università del Salento and INFN sez. Lecce, via Arnesano
Lecce, LE 73100, Italy
giovanni.marsella@le.infn.it

On behalf of ARGO-YBJ collaboration

The ARGO-YBJ experiment, installed at the Yangbajing Cosmic Ray Laboratory (Tibet, China), at 4300 m a.s.l., is a detector 100x110m^2 large, made by a layer of Resistive Plate Counters (RPCs) consisting of a central carpet with almost full coverage extending over an area of about 5.500 m^2, surrounded by a guard ring with partial coverage. The high space-time granularity, the full-coverage technique and the high altitude location make this detector a unique device for a detailed study of the atmospheric shower characteristics with an energy threshold of a few hundred GeV. The large field of view, the high duty cycle enable the ARGO-YBJ experiment to monitor the sky in a continuous way. A summary of recent results in Gamma-ray Astronomy and Cosmic Ray Physics will be presented and reviewed.

1. Introduction

The ARGO-YBJ experiment (Astrophysical Radiation with Ground-based Observatory at YangBajing) has been designed to study cosmic rays and cosmic gamma-radiation at energy larger than few hundred GeV, by detecting air showers at high altitude with wide-aperture and high duty cycle[1].

ARGO-YBJ is operating in its complete layout since 2007 allowing a complete and detailed three dimensional reconstruction of the shower front with unprecedented spatial and time resolution. The space-time structure of extensive air showers depends on primary mass, energy and arrival direction and on the interaction mechanisms with air nuclei. Measurements of shower parameters with several detection techniques would be required for a detailed knowledge of the shower front. A flat array like ARGO-YBJ can measure the particles arrival times and their densities at ground. The digital readout allows detecting shower secondary particles down to very low density and the high space-time granularity is able to provide a fine sampling of the shower front close to the core. The time profile of the shower front can be reconstructed by the time of fired pads.

* This work is supported by Italian INFN and Chinese CAS.

In this work the recent results concerning the gamma ray observation of the sky and the cosmic ray physics will be presented. Concerning gamma ray astronomy, the long term survey of the brightest sources with their flare activities will be presented together with the last sources detected. Regarding cosmic ray physics, the latest results on the calculation of the proton-air cross section, limits on the antiproton-proton ratio using the moon shadow, composition, cosmic ray anisotropies and some more results will be presented.

1.1. *The Detector*

The apparatus is a single layer detector logically divided into 153 units called clusters (7.64 x 5.72m^2), each made of 12 Resistive Plate Counters (RPCs) operating in streamer mode. Each RPC is read out using 10 pads (62 x 56 cm^2), which are further divided into 8 pick-up strips providing a larger particle counting dynamic range.

The signal coming from all the strips of a given pad are sent to the same channel of a multi-hit TDC. Pads are the time elemental units for measuring the pattern of the shower front with time resolution of about 1.8 ns. The percentage of active area in the central array is 92%. To improve the reconstruction capability, the surrounding area has been partially instrumented with a guard ring of RPCs extending the detector equipped surface up to 100 x 111 m^2.

An analogical readout system has been implemented to enhance the energy range to few hundred TeV and to reconstruct the shower profile close to the core with a very high accuracy and detail, allowing a better characterization of shower structures[2,3].

1.2. *Detector performance*

To check the detector stability many parameters are continuously monitored such as the trigger rate, the particle multiplicity distribution and the angular distribution. The data show a very stable behaviour of the detector along all the data taking periods since the end of November 2007.

To calculate the angular resolution, the pointing accuracy and the energy resolution the Moon shadow has been used[4]. The deficit in the cosmic ray flux from the Moon direction has been observed. The width of deficit allows to calculate the angular resolution, while the position of the peak allows to determine the pointing accuracy. The westward displacement of the shadow allows an absolute calibration of the energy. The Moon is observed with a sensitivity of about 10 standard deviation per month for events with a multiplicity $N_{Hit} > 40$ and zenith angle $\Theta < 50^{\circ}$ corresponding to a proton median energy $E_p \sim 1.8$ TeV. The angular and energy resolutions are in good agreement with the expected values obtained by Monte Carlo analysis. In the energy range

1-30 TeV the estimated energy uncertainty is smaller than 13%. The month analysis shows that the pointing accuracy is stable within 0.1^{o} while the angular resolution is stable at a level of 10%.

2. Gamma Astronomy

One of the main tasks for which ARGO-YBJ has been designed is the sky monitoring of high energy Gamma Ray sources at energy above 300 GeV. In three years of data taking many sources have been observed. The background has been calculated using both the *time swapping*[5] and the *equi-zenith*[6] methods. The two methods give results comparable within the statistical errors. At present no gamma/hadron discrimination is applied. In this work only the most significant sources will be discussed.

2.1. *The Crab Nebula*

The Crab Nebula has been observed for 3.5 years[7]. The data are selected with a particle multiplicity $N_{Hit}>40$, corresponding to a mean energy $E_{\gamma}=1$TeV for a Crab-like spectrum, and a zenith $\Theta<40^{o}$. The transit time is 5.8h per day and the source culminate at a zenith angle $\Theta_{max}=8^{o}$. A signal with a significance of 17 standard deviations has been detected, corresponding to an integrated sensitivity of 0.3 Crab units. To evaluate the energy spectrum a Monte Carlo analysis has been developed in order to reproduce the Crab emission path in the sky and the detector response in terms of the multiplicity in function of the spectrum parameters. The best fitting spectrum obtained is:

$$dN/dE=(3.0\pm0.3)\times10^{-11}\times E^{(-2.59\pm0.09)}\ \text{photons cm}^{-2}\,\text{s}^{-1}\,\text{TeV}^{-1}$$

which is in good agreement with the observations of the other detectors such as HESS[8] and MAGIC[9]. Concerning the energy range sampled, the 84% of the events correspond to showers with an energy $E_{\gamma}>300$GeV while only the 8% correspond to events with an energy $E_{\gamma}>10$TeV. The measured Point Spread Function (PSF) is in good agreement with the Monte Carlo expectation.

In 2010 the Agile satellite reported an unexpected strong flare activity from the Crab at energy above 100 MeV[10]. Fermi-LAT confirmed such result[11], and now at least three strong flare activities have been reported between the years 2009 and 2011[12,13]. ARGO-YBJ data during 3.5 years of data taking, excluding the periods of ten days around the reported flare activities, are compatible with a steady flux emission. The daily significance distribution is Gaussian with a mean value $s=0.31\pm0.03$ and an rms= 0.99 ± 0.02. An excess in the flux has been observed in coincidence with the flare activities in 2010 and in 2011, while no excess is evident in February 2009[14]. During the September

2010 flare 8 days of data across the 3 Fermi-LAT peaks have been integrated and an excess corresponding to 3.2 s.d. has been observed while a value of 0.55 s.d. was expected. The excess is obtained with a multiplicity selection of N_{Hit}>40, corresponding to a mean energy E_{γ}=1TeV. During the flare of April 2011 6 days across the peak detected by AGILE have been integrated with a multiplicity selection of N_{Hit}>100, corresponding to a mean energy E_{γ}=3TeV. An excess of 3.5 s.d. has been detected while a value of 0.62 s.d. was expected.

2.2. *Markarian 421*

Mrk421 is characterized by a strong flaring activity both in X-rays and in TeV γ–rays. ARGO-YBJ since November 2007 is observing the source allowing a long term monitoring in the VHE range[15]. The data were collected by the ARGO-YBJ experiment in the period from November 2007 to December 2010. The total live time is more than 1000 days. A clear signal from Mrk 421 with significance greater than 11 s.d. is observed using events with Npad > 60. With such a significance, a light curve has been determined in order to study the correlation with X-ray flux, and the evolution of the spectral energy distribution (SED). A cumulative light curve has been compared with the data from ROSSI/RXTE[16] and SWIFT[17] satellites in the energy band 2-12 keV and 15-50 keV respectively. A good correlation between X-Ray and TeV data has been found and the active and quiet periods are well distinguished. A detailed multiwavelength analysis has been done during the strong flare activities in 2008 and in 2010. The spectral indexes in X-ray and in TeV bands becomes harder and the relation between X-ray and TeV fluxes is quadratic, indicating evidences that support the SSC acceleration model[18].

2.3. *MGRO sources*

MILAGRO was operating from year 2000 up to 2006. The experiment was sensitive to gamma sources in the range 20-100 TeV and the observation of signals from known sources and the detection of new sources was claimed. ARGO-YBJ observed 13 of the sources detected by MILAGRO with a sensitivity larger than 3 s.d. with the 3,5 years of data collected, but only two of that sources were detected with more than 5 s.d.: MGROJ1908+06 and MGROJ2031+41[19].

The first is a Pulsar Wind Nebula discovered by MILAGRO with about 8 s.d., a flux 80% of the Crab and an angular extension less than 2.6°. The source was confirmed by HESS[20] and VERITAS[21]. In particular HESS indicates an angular extension of the source of 0.34° and a lower flux, not in agreement with MILAGRO observation. ARGO-YBJ detected the sources with a significance of

5.74 s.d.. The extension is $0.50^{\circ}\pm0.35^{\circ}$, in agreement with HESS. The calculated flux is $dN/dE = (2.2 \pm 0.4)\cdot 10^{-13}\cdot (E/TeV)^{-2.3\pm0.3} cm^{-2} sec^{-1} TeV^{-1}$ in agreement with MILAGRO flux. The Disagreement with HESS is puzzling, opening questions on the extension or possible flare activities of the source.

MGROJ2031+41 has been observed with a significance of 6.3 s.d.. The extension results $(0.2+0.4-0.2)^{\circ}$ and the flux $dN/dE \approx E^{-2.8 \pm 0.4}$ corresponding to 30% Crab units. The extension is consistent with MAGIC[22] and HEGRA[23] results, while the flux results to be higher than what observed by MAGIC, HEGRA and WHIPPLE.

Quite intriguing is the case of MGROJ2019+37. This is the most intense source in MILAGRO catalogue. This source have been observed by MILAGRO with a significance of 12.4 s.d., the observed spectrum is $dN/dE = 5.4\ 10^{-12} E^{-1.83}$ $\exp(-E/22.4)\ sec^{-1} cm^{-2} TeV^{-1}$ and the source extension reported is $1.1^{\circ}\pm0.5^{\circ}$. No signal has been detected by ARGO-YBJ and an upper limit to its flux has been calculated.

Higher statistics and gamma/hadron separation algorithms are necessary in order to solve this puzzles and contribute to better define the mechanism of cosmic rays acceleration.

3. Cosmic Rays

Many important results have been reached in the study of Cosmic Rays in the energy range 1-100 TeV. In this work the most relevant results will be presented.

3.1. *Light component spectrum*

Showers recorded by ARGO-YBJ with a number of fired strips in the multiplicity interval 500÷50000 are mainly induced by primaries in the energy range 1÷300 TeV. Requiring quasi-vertical showers ($\Theta < 30^{\circ}$) with core landing inside a fiducial area ($50\times50\ m^2$) and applying a selection criterium based on the particle density, a sample of events mainly induced by proton and helium nuclei has been selected. An unfolding technique based on the Bayesian approach has been applied to the strip multiplicity in order to obtain the differential energy spectrum of the light-component (proton and helium nuclei) in the energy range 5÷250 TeV[24]. The main uncertainty affecting this analysis is due to systematic effects which do not exceed 10%. The ARGO-YBJ data, as shown in Fig.1, well agree with the recent results from the balloon-borne CREAM[25] experiment and imply that the proton spectrum in this energy range is flatter than the lower energy measurements.

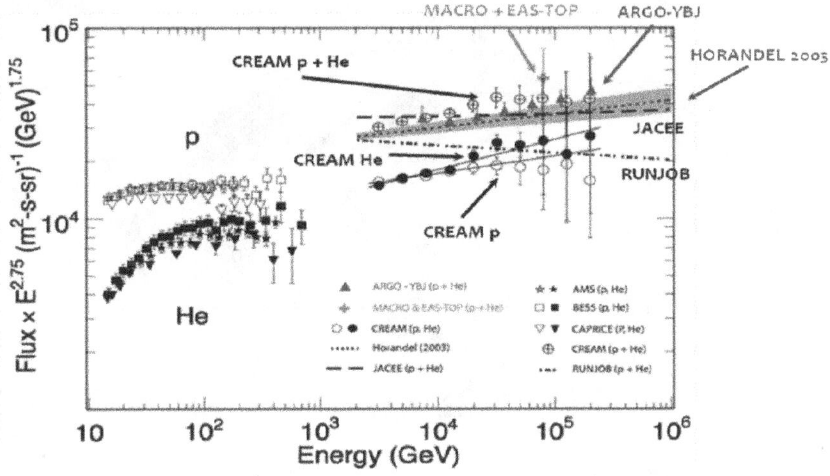

Figure 1. The differential energy spectrum of the light-component (proton and helium) measured by ARGO-YBJ (filled triangles) compared with the proton spectrum and helium spectrum measured by other experiments

3.2. *Large and medium scale anisotropies*

Many experiments have observed two large regions in their sidereal skymap: an excess located at R.A. 65 named "tail-in", and a deficit at R.A. 200 named "loss-cone", such as the TibetIII, with a modal energy of 3 TeV[26], Milagro, with a median energy of 6 TeV[27], and Super-Kamiokande-I[28], with a median energy of 10TeV, and IceCube, in the south pole with a median energy 14 TeV[29]. All these regions are basically consistent with ARGO-YBJ observations[30]. The data have been collected selecting reconstructed events with zenith angle <45°. Standard cuts on χ^2 conical fit have been applied to assure the quality of the directional reconstruction; $1.3\text{x}10^{11}$ events survived the selection criteria. Equi-zenith angle method has been used to evaluate the background as it can eliminate various detecting effects caused by instrumental and environmental variations, such as changes in pressure and temperature.

The intensity skymaps have been divided in seven multiplicity bins. Each 2-dimensional skymap is projected in 1-dimensional function of the RA. The distribution is fitted with a first harmonic function to get amplitude and phase values of large-scale anisotropy. The fitted amplitude as a function of median energy shows an increment with the increase of energy from hundreds GeV to several TeV region to decrease again at tens of TeV. Such an energy spectrum is consistent with an anisotropy due to the possible acceleration of particles arriving from the heliotail direction[31].

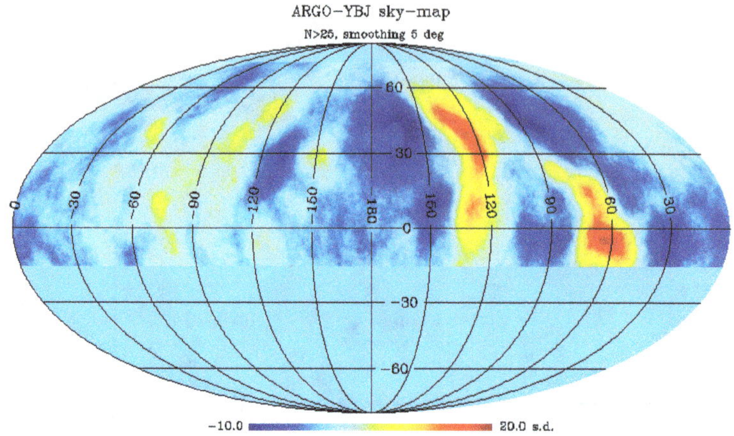

Figure 2. The medium scale Anisotropy in ARGO-YBJ skymap.

A medium scale anisotropy has been detected for the first time by MILAGRO at an energy of about 10 TeV[32]. Two regions have been defined. The first region between RA 66°-76° and Dec 10°-20° has been detected with 15 s.d. of significance corresponding to a fractional excess of $6x10^{-4}$. The second region between RA 117°-141° and Dec 15°-50° has been detected with 12.4 s.d. of significance corresponding to a fractional excess of $4x10^{-4}$. ARGO-YBJ detected the same anisotropy at energy around 1 TeV and with a significance of 16 s.d., as shown in Fig.2. In the first region the fractional excess is 10^{-3} while in the second region is $6x10^{-4}$. The large statistic allow a detailed study of the two regions. Respect to MILAGRO, in ARGO-YBJ analysis some changes in the localization of the two regions have been adopted in order to optimize the selection of the excess. Two spectra have been calculated. They show a harder shape with respect to isotropic Cosmic Ray spectrum with a cut off at 8 TeV in the first region and at 2 TeV in the second region. There is currently no explanation for these local enhancement in Cosmic Ray flux.

3.3. *Antiproton/proton ratio*

Using data on Moon shadow, limits on antiparticle flux can be derived. Protons are deflected towards West, antiprotons are deflected towards East, so 2 symmetric shadows are expected. If the displacement is large and the angular resolution small enough we can distinguish between the 2 shadows and from the deepness determine the relative abundance of antiprotons and protons. If no event deficit on the antimatter side is observed an upper limit on antiproton content can be calculated. An upper limit has been determined with ARGO-YBJ data at 2 energy values at 90% c.l.: 5% at 1.4 TeV and 6% at 5 TeV[33].

3.4. *Sun Shadow*

The sun shadow has been observed during the 3.5 years of data taking reaching a significance of 43 s.d.. From the displacement of the sun shadow form the sun position is possible to measure the Interplanetary Magnetic Field. Well reconstructed events from directions within 6° around the Sun are selected. They must be reconstructed within 150 m from the center of the array with a multiplicity of at least 100 hits. In order to study the spatial distribution of IMF over solar longitudes, the ARGO-YBJ data mentioned above are divided into 12 groups according to the position of the Earth in terms of solar longitudes when the events are recorded. More specifically, events in each group fly along trajectories within a sector of 30° in the ecliptic. In order to compensate for the Earth orbital effect, the synodic Carrington period of 27.3 days and corresponding Carrington longitudes are used to describe the position of the Earth in the sectors. This clearly reveal periodical distribution patterns over solar longitudes, indicating that cosmic rays are deflected differently by IMF in the 12 sectors. The angular position of the shadow can be used as a measure of IMF as a function of the solar longitude. The IMF shows two different behavior in different periods. Anyway, in both periods the measurements are of the same order of amplitude (2.0±0.2nT) and are consistent in the alternating periodical pattern. The ARGO-YBJ permit a shorter scale measurement of the variation of the IMF[34].

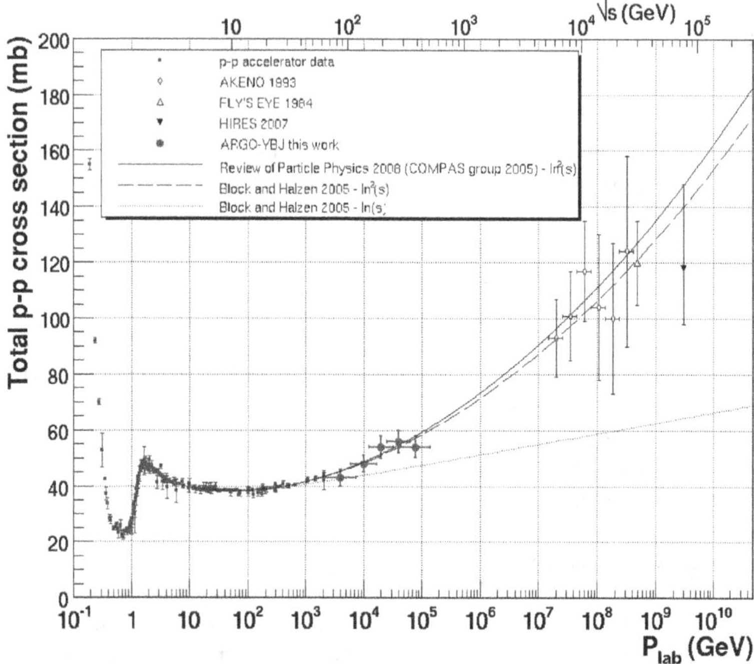

Figure 3. Total p-p cross section obtained by ARGO-YBJ together with the same quantity published by other CR experiments

3.5. *The p-p cross section*

The proton-air cross section measurement analysis is based on the shower flux attenuation for different zenith angles and exploits the detector accuracy in reconstructing the shower properties. For fixed primary energy and shower age, such attenuation is expressed by the absorption length, Λ, connected to the primary interaction length in the atmosphere by the relation: $\Lambda = k \cdot \lambda_{INT}$ where k depends on the shower development in the atmosphere, on its fluctuations and on the detector response. The actual value of k has been evaluated by Monte Carlo simulations. For primary protons, the interaction length is related to the p-air interaction cross-section by: $\sigma_{p\text{-}AIR}$ [mb] = $2.41\text{x}10^4/\ \lambda_{INT}$. The ARGO-YBJ detector location and features, which ensure the capability of reconstructing the detected showers in a very detailed way, has been exploited (see [35] for the analysis details). All the possible systematic sources have been taken into account and evaluated on the basis of a full Monte Carlo simulation. Finally, Glauber theory has been used to infer the total proton-proton cross section from the measured proton-air cross section, in an energy region not yet explored by accelerator experiments (Fig. 3).

4. Conclusions

ARGO-YBJ detector is taking data steadily since 2006 at rate of ~ 3.6 kHz with a duty cycle close to 90 %. The Moon shadow is continuously monitored with a sensitivity of about 10 σ per month, reaching, at the end of 2009, a sensitivity of 55 σ, which allows a good detector calibration of angular resolution, pointing accuracy and energy determination.

The excellent stability of the detector during all the data taking period made possible many observation in Gamma Astronomy and long term monitoring of the sky. In Gamma Ray astronomy after more than 1000 days of sky survey, without any gamma/hadron selection, 4 sources have been detected with significance greater than 5 s.d.. In particular the Crab and Markarian 421 have been monitored constantly, and excesses in their fluxes have been detected in coincidence with X-ray and Gev observations. A first light component spectrum has been measured using a Bayesian unfolding method in an energy range between 5 and 250 TeV. The result is in quite good agreement with CREAM measurements. Large and Medium scale anisotropy in cosmic rays has been observed at a mean energy of about 1 TeV. From the study of the shape of the moon shadow a first limit on antiproton/proton ratio in cosmic rays has been calculated. Finally the total proton-proton cross section has been measured. The result is obtained in an energy region not yet covered by accelerators.

Acknowledgments

This work is supported in China by NSFC (No. 10120130794), the Chinese Ministry of Science and Technology, the Chinese Academy of Sciences, the Key Laboratory of Particle Astrophysics, CAS, and in Italy by the Istituto Nazionale di Fisica Nucleare (INFN) and the Ministero dell'Istruzione, dell'Università e della Ricerca (MIUR).

References

1. The ARGO-YBJ Project, Addendum to the Proposal (1998).
2. Aielli, G. et al., *NIM A*, **562**, 92 (2006).
3. M. Iacovacci et al., *31st ICRC*, Lodz, Poland (2009).
4. B. Bartoli et al., *Phys. Rev. D*, **84**, 022003 (2011)
5. Aielli, G. et al., *ApJ*, **714**, L208, (2010).

6. Amenomori, M., et al., *ApJ* **633**, 1005, (2005).
7. Aielli G. et al., *Nucl. Instr. and Meth.* A (2010), doi: 10.1016/j.nima.2010.08.005
8. Aharonian, F.A., et al., *ApJ* **457**, (2006) 899.
9. Albert, J., et al., *ApJ* **674**, (2008) 1037.
10. Tavani, M., et al., *A&A* **502**, (2009) 995.
11. Buehler, R., et al., *Astron. Telegram* **2861** (2010).
12. Tavani, M. et al., *Science* **331**, (2011) 736.
13. Abdo, A.A. et al., *Science* **331**, (2011) 739.
14. B. Bartoli et al., SUBMITTED to THE ASTROPHYSICAL JOURNAL LETTERS (2011).
15. Bartoli, B. et al., *ApJ*, **734**, (2011) 110.
16. RXTE/ROSSI, http://xte.mit.edu/asmlc/ASM.html
17. SwiftCollaboration, Swift/BATHardX-rayTransientMonitor, /http://heasarc.gsfc.nasa.gov/docs/swift/results/transients.
18. Fossati, G., Buckley, J. H., Bond, I. H., et al., *ApJ*, **677**, (2008) 906.
19. Abdo, A. A., et el., *ApJ*, **664**, (2007) L91.
20. Aharonian F. et al., *A&A*, **499**, (2009) 723.
21. Ward, J. E., *AIP Conf. Proc.* (2008) 1085.
22. Aharonian, F., et al. *ApJL.* **675**, (2008) L25.
23. Aharonian, F. et al., *A.&A.* **431**, (2005) 197.
24. S.Bussino, E. De Marinis and S.M.Mari *Astropart. Phys.* **22** (2004) 81.
25. H. S. Ahn et al. *The Astrophysical Journal Letter* **714** (2010) L89.
26. Amenomori M. et al., *Science*, **314**, (2006) 439.
27. Abdo, A. A. et al. *ApJS* **183**, (2009) L46.
28. Guillian G. et al., *Phys. Rev. D*, **75**, (2007) 6.
29. Abbasi R. et al., , *Proc. 31th ICRC*, Lodz, Poland (2009).
30. Di Sciascio et al., Proc. *TeV Part. Astroph.,* Stockholm, Sweden (2011).
31. Nagashima, K., Fujimoto, K., & Jacklyn, R. M., *J. Geophys. Res.*, **103**, (1998) 17429.
32. Abdo A.A. et al, *Phys.Rev.Lett.***101**, (2008) 221101
33. Di Sciascio G. and Iuppa R. et al., 32nd ICRC, Beijing, China (2011).
34. Aielli, G. et al., *ApJ*, **729**, (2011) 113.
35. Aielli, G. et al., *Phys. Rew. D* **80**, (2009) 092004.

RECENT RESULTS FROM INDIRECT AND DIRECT DARK MATTER SEARCHES: THEORETICAL SCENARIOS

N. E. MAVROMATOS
Theoretical Particle Physics and Cosmology Group, Department of Physics, King's College London, Strand, London, WC2R 2LS, UK
E-mail: Nikolaos.Mavromatos@kcl.ac.uk

In this review, I discuss briefly theoretical scenarios concerning the interpretation of recent results from indirect and direct dark matter searches, with emphasis on the former.

Keywords: Dark Matter; Theory and Phenomenology; Indirect Searches

1. Introduction

There is current evidence from a plethora of astrophysical measurements that the energy budget of our Universe consists of more than 70% of a mysterious Dark Energy component, responsible for its current accelerating expansion, and another 23% of Dark Matter (DM), also of unknown origin. In this article I will concentrate on theoretical interpretations of recent results from indirect DM searches, that is excess of γ-rays and neutrinos from galactic sources or the Sun above the expected cosmic backgrounds, as well as matter-antimatter asymmetries (positron excess) around the Earth observed recently in the cosmic ray (CR) spectrum by PAMELA and confirmed by FERMI.[1]

The structure of this article is as follows: In the next section 2, I review the properties of DM candidates in supersymmetric(SUSY)/supergravity(SUGRA) models, placing the emphasis on the (significant) dependence of the various predictions (in particular with relevance to indirect DM searches) on the specific theoretical model used. This discussion has particular relevance these days, where LHC results seem to disfavour large parts of the parameter space of simplest SUSY models. In sec. 3 I explain how WIMP models (including SUSY ones) can accommodate the results on positron excess in Cosmic Ray spectra. In sec. 4, I discuss sterile neutrinos as DM candidates in non supersymmetric models. In sec. 5, I describe other interesting particle physics candidates of DM, charged or neutral (fermionic), which may be strongly interacting with SM particles (Strongy Interacting Massive Particles (SIMP)). Finally, in section 6, I discuss other ideas on DM, including axions as well as the possibility of having Dark Atoms in non supersymmetric extensions of the SM, or mediation of the interactions of DM with SM particles via the exchange of Z' gauge bosons, pertaining to extra $U(1)'$ gauge groups.

2. Supersymmetry/Supergravity and indirect DM searches

The most extensively studied model so far, from the point of view of supersymmetry searches at colliders and in particular LHC, is the five-parameter Constrained Minimal Supersymmetric Standard Model (CMSSM) (and its minimal Supergravity (mSUGRA) variant) with R-symmetry conservation. The Cold DM candidate in this class of models is the neutralino, which is the Lightest SUSY Particle (LSP) in the spectrum and hence stable. Indirect searches for neutralinos χ are motivated by the fact that neutralino annihiliation in the galaxies produces gamma ray excess in the relevant spectra, and this constitute a signal for this type of DM. Although there are attempts to provide model independent fits to such photon spectra,[2] nevertheless due to the weakness of the signal there is significant sensitivity to the particular theoretical model for DM, as we now discuss. The most studied example are the photon spectra from neutralino χ annihilation at the core of our Galaxy. The total annihilation cross section rates of the mSUGRA or CMSSM have been studied in ref. 3 and the photon spectra from the core of our galaxy due to LSP (χ) annihilation have been estimated in the region of the parameter space of the models that are compatible with the WMAP constraints: (i) the stau $\overline{\tau}_1 - \chi$ co-annihilation strip, (ii) the focus point region (in which χ has an enhanced Higgsino component) and (iii) the funnel at large tanβ, in which the annihilation rate is enhanced by poles of nearby heavy MSSM Higgs bosons. The important point to notice is that, as the relevant calculations show, annihilation attenuates rapidly with decreasing tanβ and increasing $m_{1/2}$. The analysis of ref. 3 has been quite thorough, involving detailed calculations of WMAP compatible branching fractions of $\chi - \chi$ annihilation into SM particle pairs at certain characteristic CMSSM benchmark points. The resulting CMSSM total γ-ray flux as a function of the energy threshold has been computed. The prospects for detection depend crucially on the astrophysical γ-ray background, which has three known components so far: (a) diffuse galactic emission (DGE), from nucleon-nucleon interactions producing π^0 which subsequently decay to gamma rays, and electron bremsstrahlung as it it scattered by a nucleus, (b) Isotropic Extragalactic (possibly) Contributions (IGRB) from a plethora of sources, Active galactic Nuclei (AGN), Galaxy Clusters, Ultra High Energy Cosmic rays, Blazars and Star forming Galaxies, and (c) Resolved Point Sources (RPS), which constitute an important part of photon background from the direction of the Galactic Centre. The current sensitivity of the FERMI satellite data is unfortunately hidden by the above background components, especially if uncertainties in the effective area of the detector are taken into account. The situation will hopefully improve in the next few years, with the reduction of systematic errors. However, as the analysis of ref. 3 demonstrates, the prospects for CMSSM LSP indirect detection are not great. In particular, in the low tanβ case, it will be very difficult to detect a γ ray signal along the co-annihilation strip but the focus point region has better prospects of detection due to the larger annihilations at that region. On the other hand, better prospects seem to characterise the large tanβ case, due to larger annihilation cross section in the co-annihilation, funnel and focus point regions. In general, it will always be more difficult to pin down the CMSSM than other supersymmetric models via searches for energetic photons from astrophysical sources. Hence collider searches are much superior in this respect for falsifying CMSSM, mSUGRA models.

Another set of indirect DM tests is that of neutrinos produced as a result of Capture and Annihilation of LSP in the Sun. The dominant process for the production of neutrinos is the annihilation of the LSP into SM particles, mainly tau pairs, the subsequent decay of which produces neutrinos and muons, and it is the detection of muons that eventually provides the main indirect DM test: $\chi - \chi \to \overline{\tau}\,\tau, \quad \tau \to \mu\,\nu_\mu\,\overline{\nu}_\tau$. Muon detection energy threshold is an important parameter for these tests. The fluxes of neutrinos (and hence muons) have been calculated again for CMSSM in ref. 4, at the same benchmark points as for the γ-ray spectra, with the conclusion that the detectability of CMSSM by the ICE CUBE/DEEP CORE detector is not straightforward, given that the signal above the background depends on the shape of the neutrino spectrum. The detailed mechanism for the production of detectable neutrino and thus muon flux from the Sun is the following: in the beginning we have the Gravitational Capture of LSP from a galaxy by the Sun, then the LSP scatters off a nucleus in the Sun, loses energy and is captured (as it cannot escape Sun's gravitational potential). Further scatterings in the Sun during the LSP's fall towards the solar centre take place, resulting in thermalization (equilibrium situation) at the Solar Centre. The increase of thermalized LSP populations implies an increase in the LSP annihilation rates. There is significant dependence of the calculated LSP annihilation rates on the solar model used as well as the particle physics model (spin dependent and spin independent cross sections). Indirect Searches for LSP DM via annihilations yielding high energy neutrinos (and hence muons) is *not the most promising* route for discovering SUSY, at least within CMSSM. But, as the detailed analysis of refs. 4,5 has indicated, there are models beyond the CMSSM, such as the Non Universal Higgs Mass variants of CMSSM, with sensitivity close to that of ICE CUBE/DEEP CORE. In such models, ICE CUBE/DEEP CORE friendly fluxes there are in regions where the LSP has significant Higgsino component, which implies larger LSP masses as compared to the corresponding CMSSM case along the focus point regions.

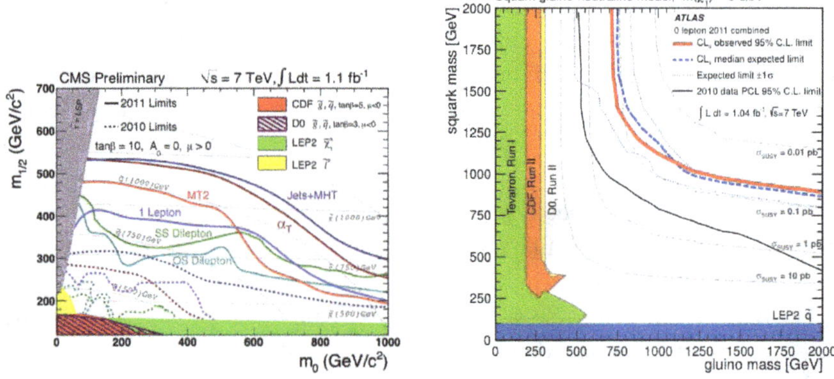

Fig. 1. Current SUSY exclusion limits from the latest LHC results. Left panel : CMS Experiment, Right panel: ATLAS Experiment (ATLAS Collaboration, arXive:1109.6572)

Unfortunately, at present there seem to be stringent exclusion bounds from the LHC experiments (cfr fig. 1) for the CMSSM , mSUGRA models, excluding low SUSY partner masses of the type that constitute interesting regions of the CMSSM and its variants in the above indirect DM searches. Nevertheless, since the current exclusion limits from LHC (cfr fig. 1) pertain to missing transfer energy in interactions involving energetic jets and may be leptons, there are still regions of the parameter space that allow for minimal SUSY extensions of the SM: electroweak production, e.g. gaugino-gaugino production, compressed spectra (low sparticle mass differences, which imply low-momentum jets) and third-generation sparticle production. Further exclusion will require improvement of systematic uncertainties, higher energies and relevant optimisation of analyses. However, if the physical SUSY is realised through other models, then the conclusions may be completely different. Below we shall discuss two such departures from mSUGRA. The first concerns SUSY models with broken R-parity, in which a long lived Gravitino ($\tilde{G}$), with life time longer than the age of our Universe , plays the rôle of LSP. In some interesting variants of this class of models,[6] with bilinear R parity Violation (RPV), neutrino masses are generated in an intrinsically supersymmetric way. The most promising (indirect detection) signal of the gravitino DM are monochromatic gamma-rays as a result of the gravitino decay modes $\tilde{G} \to \nu\gamma$, where ν indicates neutrinos. The model can be constrained LHC data, neutrino oscillations, the WMAP astrophysical constraints on the relevant relic abundance $\Omega_\chi h^2$, γ-ray line searches (via Fermi, EGRET satellites). The allowed gravitino masses are below 1 GeV, with the corresponding life times longer than about 10^{28} sec.

The second simplest class of models beyond the CMSSM, mSUGRA is provided by their coupling to cosmic time dependent scalar fields (dilatons) $\phi(t)$. The motivation for the use of such extensions is that they entail relaxing to zero Dark Energy asymptotically in cosmic time,[7] compatible with the current astrophysical data. Such couplings affect DM thermal species abundances, as they modify the Boltzmann Equation by appropriate source terms dependent on the dilaton cosmic rates, $d\phi/dt$. The presence of the source and the associated corrections to $\Omega_\chi h^2$ may result in an O(10) dilution of the thermal relic abundance of the neutralino LSP in the CMSSM, while the baryon density remains unchanged. This results in more room for supersymmetry being available in the $(m_0\, m_{1/2})$ parameter space of the model,[7] compatible with the WMAP data, and thus heavier partners. The latter feature leads to new LHC signatures, for instance h($\to$ bb) + jets + MET, Z($\to \ell\ell$) + jets + MET and 2τ + jets + MET, (MET= missing transverse energy) are favoured in new regions. Such regions may be probed in the short future by the LHC detector. The fact that in such extensions of the CMSSM heavier partners are allowed, also implies larger annihilation cross section, and thus the above mentioned indirect DM signals via gamma rays and neutrinos from LSP annihilation have better prospects of detection, in comparison with the CMSSM case. However, the coupling of the dilaton to the relevant gravitino terms in the (conformal) supergravity Lagrangian will affect the gravitino decays rates, for instance $\tilde{G} \to \chi + Z^0$, where χ is the neutralino of the CMSSM, thus affecting the DM relic density and therefore implying stronger Big Bang Nucleosynthesis (BBN) constraints. It is therefore important that detailed cosmological studies of such dilaton extended mSUGRA/CMSSM models are performed.[8]

3. PAMELA/FERMI e^+ excess and WIMPs

Before proceeding to other DM candidates and their indirect searches, we should mention that the observed asymmetries between matter and antimatter by PAMELA, which have been confirmed by FERMI, in particular the observed positron excess in the Cosmic Ray (CR) spectra, but the absence of antiproton ($\overline{p}$) excess, can be accommodated[9] within existing models of SUSY neutralino DM, although their most likely explanation may be astrophysical (pulsar emission[1]). Neutralino DM ineractions with SM particles produce charginos (next to lightest) , whose subsequent decay can produce peak in the spectrum of cosmic leptons, yielding a signal analogous to that seen by PAMELA , ATIC and FERMI (peak in the CR positron spectrum). The example studied in ref. 9 considered masses of neutralino of order 110 GeV and chargino of order 250 GeV . Such values are excluded by the current LHC data, and the question arises whether such scenarios survive the full LHC exclusion data, after four years of running.

However, in general, the PAMELA data may be compatible with generic WIMP DM. In particular, it has been argued in ref. 10 that heavy ($m_\chi \gg$ 1GeV) DM annihilation can produce a jet structure which may result in antideuteron ($\overline{d}$) excess that can explain the lack of antriproton peak in the CR spectra, as observed by PAMELA anf FERMI. The result seems pretty robust in the sense that astrophysical uncertainties do not affect significantly the ratio of concentrations $\overline{p}/\overline{d}$. The antideuteron signal is significantly enhanced for DM masses above 1 TeV. Moreover, in models where the Cosmology is modified, e.g. by considering low reheating temperatures of the Universe after inflation in certain quintessence models,[11] the relic density of DM WIMPs is found significantly enhanced compared to standard cosmology. In such models the calculated induced fluxes of e^-, e^+ in CR, produced by LSP annihilation into e^-e^+, $\mu^-\mu^+$, $\tau^+\tau^-$, indicate agreement with the results of PAMELA and FERMI as far as positron peak is concerned. Thus, although pulsar emission seems adequate to explain the current astrophysical data on observed ring of antimatter around the Earth, nevertheless several DM model explanations are also at play. Further astrophysical searches are therefore essential in order to settle this.

In particular, if the pulsar explanation is the natural mechanism, then as we have heard in this meeting,[1] proton asymmetries in the CR spectra should be observed. This will hopefully be settled in the near future.

4. Sterile Neutrinos as DM

The SM CP violation cannot explain the observed matter-antimatter (baryon-antibaryon) asymmetry in the Universe. Several ideas beyond the SM (such as GUT models, Supersymmetry, Extra Dimensions *etc.*) have been proposed in an attempt to resolve this issue. Right-handed super-massive neutrinos may provide extensions of SM with extra CP Violation that can explain the origin of the observed matter-antimatter asymmetry in the Universe. Such a scenario has been proposed in ref. 12 as a non-supersymmetric minimal extension of the SM, called νMSM. The model may have several species N of right handed singlet Majorana neutrinos. The Model with one extra singlet fermion is excluded by the data, while models with 2 or 3 singlet fermions work well in reproducing the Baryon Asymmetry and are consistent with experimental data on neutrino oscillations . In particular, the Model with N=3 works fine, and in fact it allows one of the Majorana fermions to almost decouple from the rest of the SM fields, thus providing a candidate for light (kEV region of mass) sterile neutrino Dark Matter. The other two right-handed neutrinos are degenerate in mass, and in fact much heavier than the third,[12] specifically one has: Mass N_2 (N_3) / (Mass N_1) = O(10^5), with the masses of $N_{2,3} < M_W = O(100)$GeV , with M_W the electroweak symmetry breaking mass scale. The light neu-

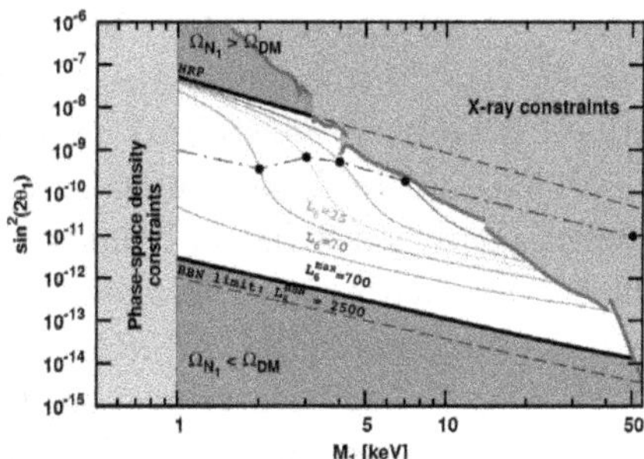

Fig. 2. Astrophysical Constraints for the lightest sterile neutrino mass and mixing angle θ_1 (with SM particles) in the νMSM model (from ref. 12).

trino masses are determined by the sea saw mechanism in the model. The lightest singlet neutrino is not stable, but its life time can be longer than the life time of the Universe, since its coupling with the SM matter can be extremely weak . Under such conditions, its contributions to the mass of light neutrinos are well within experimental errors, and hence there is a consistent oscillation phenomenology of νMSM with light sterile neutrino DM. Taking into account the interactions of the light neutrino with its heavier sterile partners in the νMSM, one may derive detailed constraints on the mass and couplings of the light sterile neutrino DM.[12] The reader should bear in mind that the decaying light sterile neutrino will produce narrow spectral lines in the spectra of DM dominated astrophysical objects, such as halos of galaxies etc. This will constitute a means of its detection. The νMSM model is found consistent with constraints from BBN , structure formation data in the universe and other astrophysical constraints. The allowed mass ranges and mixing angles θ_1 are depicted in fig. 2.[12]

5. Strongly Interacting Massive Particles (SIMPs) as DM candidates

Strongly interacting Massive particle (SIMP) matter may be (part) of DM, although much more severely constrained. A rather old idea[13] for a DM candidate is that the latter consists of Charged Massive Particles (CHAMP). If the whole of DM, as originally assumed,[13] consists of such charged particles, then cosmological compatibilities require them to be heavy, 20 TeV $< M_{\rm Ch} <$ 1000 TeV . Indeed, if of charge + 1, they will result in Superheavy remnants of H isotopes in the Universe. CHAMPs are assumed particle-antiparticle symmetric, so charge -1 anti-CHAMP may bind with ^{4}He nuclei and after BBN. Mostly, however, they bind to protons to behave like superheavy stable neutrons. Such bound states bring severe constraints in their relic populations. Less severe constraints are imposed if CHAMPS constitute only (a small) part of DM: if neutral DM decays (at late eras) to CHAMPs then the above-mentioned stringent bounds may be re-evaluated, for instance it has been estimated[14] that consistently with all current astrophysics constraints, the fraction of CHAMP in the galactic halo may be less than $0.41.4 \times 10^{-2}$. Also, it has been argued recently[15] that Galactic magnetic fields parallel to the disc prevent CHAMPS from entering the disc (hence their non detection on Earth), if their charge q_X and mass are in the range: $10^2(q_X/e)^2 \leq m_X/(TeV) \leq 10^8(q_X/e)^2$. Such CHAMPS exert important influence on the DM density profiles: they interact with ordinary matter via magnetic field mediation and hence affect the visible Universe in the sense that their density profiles depend on the Galaxy: moderate effects appear in large elliptical galaxies and the Milky Way, while there is expulsion of CHAMPS with moderate charge (Coulomb Interactions not important) from spherical Dwarf Galaxies in agreement with observations . Moreover, their DM Annihilation patterns are different from those of Cold Dark Matter (CDM) model: due to the attractive Coulomb potential between X^+ and X^- there is an increased annihilation cross section (relative to CDM models) by a factor c/v (Sommerfield-Sakharov effect) ; after CHAMP becomes non relativistic, the annihilation rate falls off slower than in CDM, their kinetic energies scale as (1 + z) with redshift z, and their present annihilation rate depends on the fraction of X^- bound to baryons.

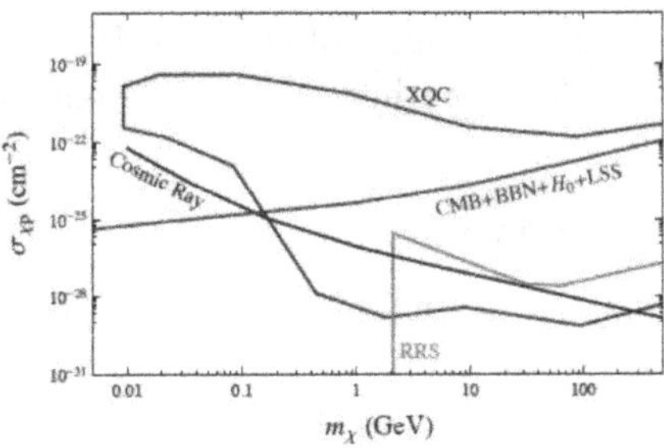

Fig. 3. Current Astrophysical Constraints for neutral fermionic SIMP (from ref.[16]).

There are various experimental searches that impose stringent constraints on charged SIMPs: as already said, the most important of them is associated with the formation of *Bound States SIMP-Nucleons*. Indeed, if these are formed then the associated constraints exclude the models. Hence one should avoid such bound states, e.g. by invoking repulsive forces between SIMPS and nucleons. To this end one may assume that SIMPS are fermionic and electrically neutral, since charged SIMPS of the same charge as nucleons (protons) could affect the Universe neutrality. By assuming scalar field φ mediators, as well as that the scalar force is less than that of two pions, so that bound states due to the scalar mediator do not form, the pertinent part of the interacting Lagrangian is[16] $L_{\rm int} = -g_X\varphi\overline{X}X - g_N\varphi\overline{N}N$, with $g_N g_X < 0, \quad m_X, m_\varphi > 0$. Important information is included in the SIMP(X) annihilation cross section σ_{XX}, which must be less than $\sigma_{XX} \leq 3m_X\mathrm{GeV}^{-3}$ in order not to affect the shape of Galactic halo (the above upper bound is placed by an analysis of merging galaxies data, such as those of the Bullet Cluster[16]). Such upper bounds allow for sufficiently strongly interacting particles. Another important constraint comes from SIMP(X)-nucleon(N) cross section σ_{XN} which must be less than $\sigma_{XN} \leq \frac{4g_N^2}{g_X^2}\frac{m_X}{(1GeV)} \times 10^{-27} cm^2$ so as not to affect the galactic halo shape. The resulting constraints from a plethora of astrophysical measurements, including X-ray quantum calorimetry (XQC) and Cosmic Rays are depicted in fig. 3, indicating that light neutral fermionic SIMPs with masses less than 1 GeV are allowed by the current constraints. An important collider signature of neutral fermionic SIMPS are DM di-jets produced in colliders, such as LHC. Indeed, as discussed in ref. 16, the scattering length of a neutral fermionic SIMP $L_X = L_n \frac{\sigma_{Xp}^{\rm inela}}{\sigma_{np}^{\rm inela}}$, where $n(p)$ indicates neutron (proton), can be smaller than the calorimeter size, so the SIMP can deposit energy in the form of DM jets. If the DM is neutral, then such jets would be trackless, and hence very different from the QCD one.

6. Other Interesting DM Possibilities

Axions are also interesting candidates for DM, with a theoretical motivation, since their presence is associated with a resolution of the strong CP problem in QCD. In this talk, due to lack of time, I will not discuss them in detail. I will simply mention that the axion pseudoscalar field $\tilde{a}$ couples to the electromagnetic U(1) part of a GUT gauge group via terms of the form $\tilde{a}\vec{E} \cdot \vec{B}$, where E and B are the electric and magnetic fields respectively. Such couplings imply that an axion field can be converted to a photon in the presence of an external magnetic field (Primakoff effect), which is the basis for their potential observation. The CAST experiment at CERN[17] has placed the most stringent limits today to the QCD axions, given that no signal over background has been observed in the experiment. The preliminary limit on the axion couplings and masses are $g_a < 2 - 2.5 \times 10^{-10}\text{GeV}^{-1}$ for the mass range 0.39 eV $< m_a < 0.65$ eV. The experiment, in addition to the QCD axions, has also placed bounds on the couplings and compactification radius of Kaluza-Klein axions in extra dimensional theories.

In ref. 18, *Dark Atoms* from stable charged fundamental constituents of matter beyond the Standard Model (new quarks and leptons) have been conjectured to exist. Severe constraints from anomalous isotopes in the Universe imply that only charge -2 object (X^{--}) is allowed, not charge +1 or -1, hence the relevant models are necessarily non supersymmetric. There are bound states of X^{--} with primordial Helium 4 He^{++} to form neutral O-He atoms of warm DM. The non trivial (but unclear) nuclear physics of such bound states has been argued in ref. 18 to provide resolutions to various DM puzzles. The O-He atoms may be responsible for the observed constant and annual modulation of underground detectors (like DAMA, CoGENT), while their decays to stable constituents may also explain the observed PAMELA and FERMI excess events. A clear signature of such models would be the appearance in the matter of DAMA detectors of anomalously heavy (> 1 TeV) Sodium Isotopes.

Another interesting DM scenario has been presented in ref. 19. DM (which, in the model, is a right-handed sneutrino) communicates with SM particles via mediating light particles, e.g. Z' bosons of extra $U(1)'$ groups that appear in extensions of the SM. Such Z' couple to the DM elastic cross sections, and can lead (via DM annihilation) to excess of γ-rays from, say, the Galactic Centre, which are compatible with observations .

Acknowledgement

This work was supported in part by the London Centre for Terauniverse Studies (LCTS), using funding from the European Research Council via the Advanced Investigator Grant 267352.

References

1. See plenary talks by F. Mocchiutti (PAMELA) and R. Rando (FERMI).
2. See, e.g.: A. de la Cruz-Dombriz, V. Gammaldi, [arXiv:1109.5027 [hep-ph]].
3. J. Ellis, K. A. Olive, V. C. Spanos, [arXiv:1106.0768 [hep-ph]].
4. J. Ellis, K. A. Olive, C. Savage, V. C. Spanos, Phys. Rev. **D81**, 085004 (2010).
5. J. Ellis, K. A. Olive, C. Savage, V. C. Spanos, Phys. Rev. **D83**, 085023 (2011).
6. D. Restrepo, M. Taoso, J. W. F. Valle, O. Zapata, [arXiv:1109.0512 [hep-ph]].
7. A. B. Lahanas, N. E. Mavromatos, D. V. Nanopoulos, Phys. Lett. **B649**, 83-90 (2007); B. Dutta *et al.*, Phys. Rev. **D79**, 055002 (2009).
8. N. E. Mavromatos and V. C. Spanos, to appear.
9. A. B. Flanchik, [arXiv:1101.5920 [astro-ph.HE]].
10. M. Kadastik, M. Raidal, A. Strumia, Phys. Lett. **B683**, 248-254 (2010).
11. C. Pallis, Nucl. Phys. **B831**, 217-247 (2010).
12. M. Shaposhnikov, Prog. Theor. Phys. **122**, 185 (2009) and references therein.
13. A. De Rujula, S. L. Glashow, U. Sarid, Nucl. Phys. **B333**, 173 (1990); G. D. Starkman, A. Gould, R. Esmailzadeh, S. Dimopoulos, Phys. Rev. **D41**, 3594 (1990).
14. F. J. Sanchez-Salcedo, E. Martinez-Gomez, J. Magana, JCAP **1002**, 031 (2010).
15. L. Chuzhoy, E. W. Kolb, JCAP **0907**, 014 (2009).
16. Y. Bai, A. Rajaraman, [arXiv:1109.6009 [hep-ph]] and references therein.
17. I. G. Irastorza, *et al.*, J. Phys. Conf. Ser. **309**, 012001 (2011).
18. M. Y. Khlopov, [arXiv:1012.5756 [astro-ph.CO]] and Poster here.
19. M. R. Buckley, D. Hooper, J. L. Rosner, Phys. Lett. **B703**, 343-347 (2011); M. R. Buckley, D. Hooper, T. M. P. Tait, Phys. Lett. **B702**, 216-219 (2011).

The microwave sky after one year of *Planck* operations

Planck Collaboration, presented by A. Mennella*
Dipartimento di Fisica, Università degli Studi di Milano,
Milano, 20133, Italy
**E-mail: aniello.mennella@fisica.unimi.it*
www.fisica.unimi.it

The ESA *Planck* satellite, launched on May 14th, 2009, is the third generation space mission dedicated to the measurement of the Cosmic Microwave Background (CMB), the first light in the Universe. *Planck* observes the full sky in nine frequency bands from 30 to 857 GHz and is designed to measure the CMB anisotropies with an unprecedented combination of sensitivity, angular resolution and control of systematic effects. In this presentation we summarise the *Planck* instruments performance and discuss the main scientific results obtained after one year of operations in the fields of galactic and extragalactic astrophysics.

Keywords: Cosmology; Cosmic Microwave Background; Space experiments.

1. Introduction

The Cosmic Microwave Background (CMB) is constituted by relic photons that were coupled to barionic matter in the hot primordial plasma and travelled in the expanding universe when it became neutral, after ~380000 years after the big bang. Today we detect it as a highly isotropic microwave background at the temperature of $\sim 2.73\,$K, with anisotropies at the level of $\Delta T/T \sim 10^{-5}$. These anisotropies trace the matter density distribution in the universe immediately before matter-radiation decoupling.

Planck, launched on 14 May 2009, is the first European and third generation CMB space mission after COBE and WMAP; the *Planck* instruments are designed to extract all the cosmological information encoded in the CMB temperature anisotropies with an accuracy set by cosmic variance and astrophysical confusion limits. *Planck* will image the sky in nine frequency bands ranging from 30 to 857 GHz, leading to a full-sky map of the CMB temperature fluctuations with signal-to-noise > 10 and angular resolution $< 10'$. In addition, all *Planck* bands between 30 and 353 GHz are sensitive to linear polarisation.

Planck performance is sized to map the CMB anisotropies over the entire angular range dominated by primordial fluctuations. This will lead to accurate estimates of cosmological parameters that describe the geometry, dynamics, and matter-energy content of the universe. The *Planck* polarisation measurements are expected to deliver complementary information on cosmological parameters and to provide a unique probe of the thermal history of the universe in the early phase of structure formation. *Planck* will also test the inflationary paradigm with unprecedented sensitivity through studies of non-Gaussianity and of B-mode polarisation as a signature of primordial gravitational waves.

The wide frequency range of *Planck* is required primarily to ensure accurate discrimination of foreground emissions from the cosmological signal. However, the nine maps also represent a rich data set for galactic and extragalactic astrophysics. In this paper we present an overview of the status of the *Planck* mission and of the preliminary results in the field of galactic and extragalactic astrophysics obtained after one year of operations.

2. The *Planck* satellite and its operations

Planck (see left panel of Fig. 1) is a spinner constituted of two modules:[1] (i) a payload module containing telescope, instruments, a baffle that provides stray-light rejection and radiative cooling, and three conical "V-groove" radiators that thermally decouple the warm and cold satellite modules; (ii) a service module containing the warm satellite and instrument electronics, the solar cells, the cryo-coolers, the main on-board computer, the telecommand receivers and telemetry transmitters, and the attitude control system with its sensors and actuators.

The telescope[2] is a dual-reflector off-axis aplanatic Gregorian telescope with 1.5 m primary projected aperture, pointing at 85° with respect to the spin axis. It focusses the sky radiation on the secondary mirror focal plane, hosting the feed horn antennas of the two instruments: the Low Frequency Instrument (LFI) and the High Frequency Instrument (HFI).

Planck was launched together with the ESA's Herschel observatory on 14 May 2009 (13:12 UT) from the Centre Spatial Guyanais in Kourou (French Guyana) on an Ariane 5 ECA rocket,[3] and was placed in its final Lissajous orbit around the second Lagrangian point of the Earth-Sun system ("L2") after three large manoeuvres. The right panel in Fig. 1 sketches the orbit and provides the basic details of the scanning strategy.

Planck started cooling down radiatively shortly after launch. Approximately two months were necessary to complete the cooldown and reach the nominal temperatures of the *Planck* cryogenic stages:[4] (i) 50 K, reached by passive cooling, for the telescope, baffle and upper V-groove, (ii) 20 K provided by the Hydrogen *Planck* Sorption Cooler, for the LFI focal plane and HFI pre-cooling, (iii) 4.5 K provided by a ^{4}He-JT Stirling cooler, for the HFI focal plane feeds and LFI ref-

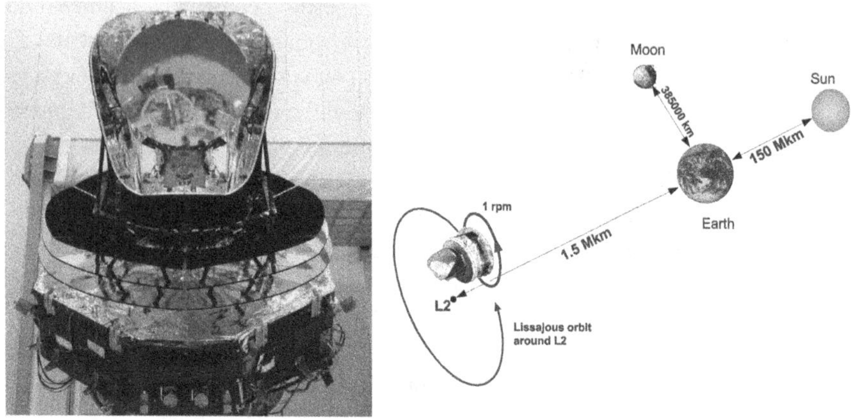

Fig. 1. Left panel: the *Planck* satellite a few days before launch. Right panel: *Planck* orbits around the second Lagrangian point, L2, and scans the sky in near-great circles by spinning at 1 r.p.m. The satellite is repointed approximately hourly by 2′ so that the solar aspect angle is kept constant and the full sky is observed in about six months.

erence loads and (iv) 1.6 K and 0.1 K, provided by a ^{3}He-^{4}He open cycle dilution cooler, for the HFI filters and bolometers, respectively.

During cooldown, several activities were carried out to verify instrument functionality, tune the main parameters and check that scientific performance was comparable with ground measurements.[4,5] At the end of this phase the two instrument were fully tuned and ready for routine operations. On 13 August 2009 *Planck* entered the so-called "First Light Survey" (FLS), a two-week period during which *Planck* operated as if it were in its routine phase. The conclusion was that the *Planck* payload required no further tuning of its instruments, so that the FLS was accepted as a valid part of the first *Planck* survey.

The instruments have continued working smoothly and their parameters have not been changed ever since. The only notable event has been the planned switch-over from the nominal to the redundant Sorption Cooler which happened during August 2010. Nominal operations will continue until the end of the dilution cooler refrigerant, which is foreseen by the end of January 2012. At that time the HFI will become not operative and the LFI will continue scanning the sky for up to another year, depending on the residual duration of the Sorption Cooler lifetime.

3. Instruments and scientific performance

The LFI[6] (left panel of Fig. 2) is an array of 11 microwave coherent pseudo-correlation differential receivers in Ka, Q and V bands. They are based on low noise amplifiers using indium-phosphide high electron mobility transistors, operated at a temperature of ~20 K. The HFI[7] (right panel of Fig. 2) is designed around 52 bolometers operated at 0.1 K and fed by corrugated feedhorns and bandpass filters within a back-to-back conical horn optical waveguide. Twenty of the bolometers are sensitive to total power, and the remaining 32 units are arranged in pairs of orthogonally-oriented polarisation-sensitive bolometers (PSBs).

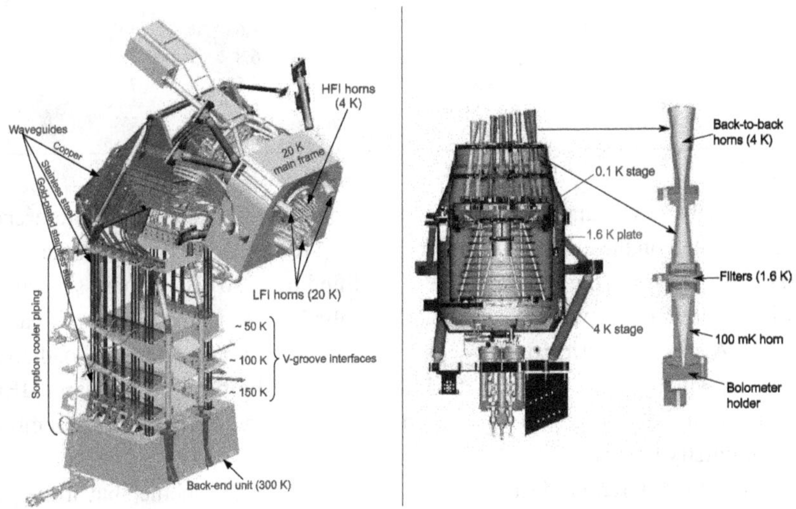

Fig. 2. Left panel: 3-D view of the LFI instrument. Right panel: the HFI focal plane and a schematic of the bolometric optical chain.

The in-flight measured main performance parameters of the *Planck* instruments are summarised in Table 1.

3.1. *The Low Frequency Instrument*

The instrument consists of a ~ 20 K focal plane unit hosting the corrugated feed horns, orthomode transducers (OMTs), and receiver front-end modules (FEMs). A set of 44 composite waveguides interfaced with the three V-groove radiators[4] connects the front-end modules to the warm (~ 300 K) back-end unit (BEU), which contains further radio frequency amplification, detector diodes, and electronics for data acquisition and bias supply.

In each receiver the feed horn is connected to an OMT, which splits the incoming radiation into two perpendicular linear polarisation components that propagate through two independent pseudo-correlation differential radiometers. In each ra-

Table 1. *Planck* performance parameters determined from flight data.

Channel	$N_{detectors}$	ν_{center} [GHz]	Mean Beam FWHM	Mean Beam Ellipticity	White-noise Sensitivity [$\mu K_{RJ}\, s^{1/2}$]	White-noise Sensitivity [$\mu K_{CMB}\, s^{1/2}$]	Calibr. Unc. [%]
30 GHz	4	28.5	32.65	1.38	143.4	146.8	1
44 GHz	6	44.1	27.92	1.26	164.7	173.1	1
70 GHz	12	70.3	13.01	1.27	134.7	152.6	1
100 GHz	8	100	9.37	1.18	17.3	22.6	2
143 GHz	11	143	7.04	1.03	8.6	14.5	2
217 GHz	12	217	4.68	1.14	6.8	20.6	2
353 GHz	12	353	4.43	1.09	5.5	77.3	2
545 GHz	3	545	3.80	1.25	4.9	...	7
857 GHz	3	857	3.67	1.03	2.1	...	7

diometer, the sky signal is continuously compared with a stable 4.5 K reference load mounted on the external shield of the HFI 4 K box. After being summed by a first hybrid coupler, the two signals are amplified by ~ 30 dB. A phase shift alternating at 4096 Hz between 0° and 180° is applied in one of the two amplification chains. A second hybrid coupler separates back the sky and reference load components, which are further amplified, detected and digitised in the warm BEU. After the digital conversion, data are downsampled, requantised and assembled into telemetry packets.

The LFI differential design is similar to the WMAP scheme[8] but it uses an internal blackbody load as a stable reference. The receiver is balanced via a gain modulator factor that provides good suppression of systematic effects and $1/f$ noise knee frequencies of ~ 50 mHz.[5]

Data are continuously calibrated using the dipole modulation in the CMB. In particular our calibration signal is the sum of the solar dipole and the orbital dipole, which is the contribution from *Planck*'s orbital velocity around the Sun. This model allows us to reconstruct the mean value of the calibration constant with an accuracy better than 1%. Variations in the radiometer gain on timescales of the order of few days are traced with an accuracy better than 0.1%.[9]

3.2. *The High Frequency Instrument*

The HFI is a bolometric array designed to reach photon-noise limited sensitivity. Its architecture is based on independent optical chains collecting the light from the telescope and feeding it to bolometric detectors. Sixteen channels (at 100, 143, 217 and 353 GHz) feed pairs of polarisation-sensitive bolometers (PSBs), while the remaining 20 channels (at 143, 217, 353, 545 and 857 GHz) feed spider-web bolometers (SWBs), which are not sensitive to polarisation.

A typical optical chain is shown in the right panel of Fig. 2. At 4 K, the back-to-back horns provide initial geometrical and spectral selection of the radiation, and a first set of filters blocks the highest frequency and most energetic part of the background. In order to ensure proper positioning and cooling, these elements are attached onto three thermal stages (at 4.5 K, 1.4 K and 0.1 K) in a nested arrangement. The six HFI spectral bands cover all frequencies from 84 GHz up to 1 THz via adjacent bands with close to 33% relative bandwidth.

Bolometers consist of (i) an absorber that transforms the incoming radiation into heat; (ii) a thermometer that is thermally linked to the absorber and measures the temperature changes; and (iii) a weak thermal link to a thermal sink, to which the bolometer is attached. In the SWBs,[10,11] the absorbers consist of metallic grids deposited on a Si3 N4 substrate in the shape of a spider web. The absorber of PSBs is a rectangular grid with metallization in one direction.[12] Electrical fields parallel to this direction develop currents and then deposit some power in the grid, while perpendicular electrical fields propagate through the grid without significant interaction. A second PSB perpendicular to the first one absorbs the other polarisation. Such a PSB pair measures two polarisations of radiation collected by the same horns and filtered by the same devices.

The readout electronics consist of 72 channels that perform impedance measurements of 52 bolometers, two blind bolometers and 16 accurate low temperature thermometers. It is split into three boxes: the 50 K JFET box, the 300 K pre-amplifier unit (PAU), which provides an amplification of the low level voltages by a factor of 1000, and the 300 K readout electronic unit (REU), which provides a further variable gain amplification and contains all the interfaces between the analogue and the digital electronics. Digital signal are then processed in the data processing unit (DPU) that downsample and compress data before sending to telemetry.

The primary absolute calibration of HFI is based on extended sources. At low frequencies (353 GHz and below), the orbital dipole and the Solar dipole provide good absolute calibrators. At high frequencies (545 GHz and above), Galactic emission is used.

Bolometer data are affected by by glitches resulting from the interaction of cosmic rays with the bolometers. Because the glitch rate is of typically one per second and many of them are characterised by long decay times (from some tens of ms up to 2 s), simple flagging would lead to an unacceptably large loss of data. Therefore only the initial part of thc glitch is flagged and discarded, while the tail is fit with a model which is subtracted from the TOI. The temporal redundancy of the sky observations is used to separate the glitch signal from the sky signal.

4. The *Planck* microwave sky

The First Light Survey started on 13 August 2009 and was completed on 27 August, yielding maps of a strip of the sky, one for each of *Planck*'s nine frequencies. Each map is a ring about 15 degrees wide, stretching across the full sky. The coloured strip in the left panel of Fig. 3 represents data from the 70 GHz channel (the most insensitive to foregrounds contamination) superimposed to an optical image of the whole sky. The two insets show a relatively foreground-free sky patch for the LFI-70 GHz (left) and HFI-100 GHz (right) *Planck* channels. The CMB anisotropy structure is clearly visible at high galactic latitudes and in the two insets, showing consistency in the data from the two frequency channels.

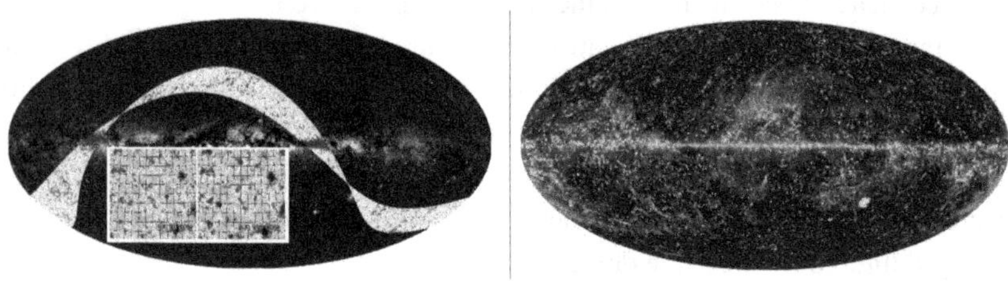

Fig. 3. Left: the *Planck* First Light Survey. The 14 days survey (data from LFI 70 GHz channel) is superimposed to an optical image of the whole sky. Two insets show details of a region dominated by the CMB anisotropies in the 70 and 100 GHz channels, that are less sensitive to foregrounds contamination. Right: the set of *Planck* compact sources (galactic and extragalactic) detected during the first year of operations.

On 11 January 2011 the *Planck* Collaboration released its first set of scientific data to the public: the Early Release Compact Source Catalogue (ERCSC), a list of more than 15000 unresolved and compact sources extracted from the first complete all-sky survey (see right panel of Fig. 3). The ERCSC[13] consists of nine lists of sources, extracted independently from each of *Planck*'s frequency bands, and two lists of sources extracted using multi-band criteria: (i) "Cold Cores," cold and dense locations in the Insterstellar Medium of the Milky Way, and (ii) clusters of galaxies, selected using the spectral signature left on the Cosmic Microwave Background by the Sunyaev-Zeldovich (SZ) effect.

In Figure 4 we show the full set of maps of the astrophysical foregrounds at the nine *Planck* frequencies, where the CMB component has been removed as described in Refs. [9,14]. This set of maps represents the first full-sky view of the microwave sky ever observed by a single experiment in such a wide frequency span.

These nine maps represent a key element in the data processing pipeline that

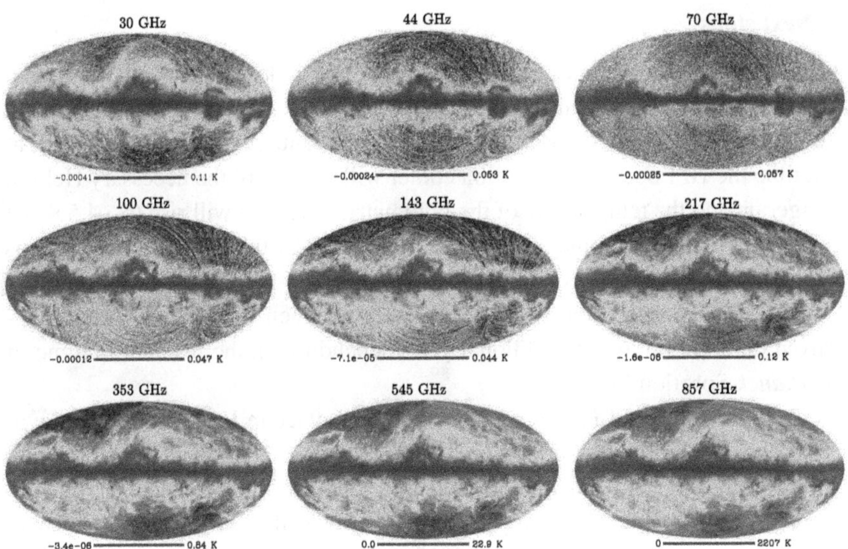

Fig. 4. The nine *Planck* foreground full-sky maps.

extracts the CMB anisotropies separating them from the foreground signals and also provide a wealth of new information shedding light on many open questions in galactic and extragalactic astrophysics. A set of 18 papers have been submitted by the *Planck* team together with the release of the ERCSC discussing preliminary scientific results obtained from the catalogue and from the maps which were used as input for the production of the ERCSC.

Although covering the details of such studies is outside the scope of this paper (see the complete list of references provided in [3]) we outline here the main areas of scientific interest:

- detection of galaxy clusters via the Sunyaev-Zeldovich (SZ) effect and correlation with X-ray follow-up observations (e.g. with the XMM-Newton X-ray observatory);
- statistics of the spectral distribution of radio sources in the microwave region;
- detection and statistics of "Cold Cores", i.e. regions of cold dust gravitationally collapsing in star-forming regions;
- study of the physical properties of the interstellar medium ("dark gas", i.e. gas not spatially correlated with known tracers of neutral and molecular gas, molecular clouds, anomalous emission from interstellar dust which can be interpreted as arising from small spinning grains).

5. Next steps

At the time of writing *Planck* has completed the fourth sky survey and it is entering the fifth. The two instruments have so far maintained their initial performance and are expected to continue observing the sky until January 2012. As expected, at that time the HFI open-cycle dilution cooler will run out of the necessary Helium refrigerant and the temperature of the bolometric detectors will rise to ~4.5 K. At this temperature the bolometers will become essentially blind to the CMB photons and the HFI will end its operations.

An extended operation period is currently foreseen for the LFI, with an approved duration up to one additional year, depending on the residual lifetime of the *Planck* Sorption Cooler.

The next release of *Planck* products will take place in January 2013, and will cover data acquired in the period up to 27 November 2010. At this time the first *Planck* cosmological results will be released together with the necessary data to support them. A third release of products is foreseen after January 2014, to cover the data acquired beyond November 2010 and until the end of *Planck* operations.

6. Conclusions

Planck, the third space mission dedicated to the measurements of the CMB anisotropies, has been successfully scanning the sky for more than two years and has started its fifth sky survey as we write. Its focal plane instruments are the most sensitive microwave detectors ever flown and have confirmed and maintained their predicted performance.

The first public *Planck* data release, on 11 January 2011, has provided the scientific community with a preliminary full-sky catalogue of compact, unresolved sources in a largely unexplored frequency range. These early Planck results have already started shedding new light on several astrophysical issues like detection of new galaxy clusters, the physics of stellar formation in cold gas regions, the nature of "dark gas", etc. The first release of cosmological data and results is foreseen by January 2013, and a second one, based on the complete *Planck* dataset, will happen in early 2014.

After 19 years *Planck* is approaching its end of operations but the data analysis is currently at full speed to provide the scientific community with the highest quality CMB data ever released. The best still has to come.

Acknowledgements

Planck is operated by ESA via its Mission Operations Centre located at ESOC (Darmstadt, Germany). ESA also coordinates scientific operations via the *Planck*

Science Office located at ESAC (Madrid, Spain). Two Consortia, comprising around 50 scientific institutes within Europe, the USA, and Canada, and funded by agencies from the participating countries, developed the scientific instruments LFI and HFI, and continue to operate them via Instrument Operations Teams located in Trieste (Italy) and Orsay (France). The Consortia are led by the Principal Investigators: J.-L. Puget in France for HFI (funded principally by CNES and CNRS/INSU-IN2P3) and N. Mandolesi in Italy for LFI (funded principally via ASI). NASA's US *Planck* Project, based at JPL and involving scientists at many US institutions, contributes significantly to the efforts of these two Consortia. In Finland, the *Planck* LFI 70 GHz work was supported by the Finnish Funding Agency for Technology and Innovation (Tekes). This work was also supported by the Academy of Finland, CSC, and DEISA (EU).

References

1. J. A. Tauber, N. Mandolesi, J. Puget *et al.*, *A&A* **520**, p. A1 (September 2010).
2. J. A. Tauber, H. U. Norgaard-Nielsen, P. A. R. Ade *et al.*, *A&A* **520**, p. A2 (September 2010).
3. Planck Collaboration *et al.*, *A&A* **536**, p. A2 (December 2011).
4. Planck Collaboration *et al.*, *A&A* **536**, p. A3 (December 2011).
5. A. Mennella, M. Bersanelli, R. C. Butler *et al.*, *A&A* **536**, p. A4 (December 2011).
6. M. Bersanelli, N. Mandolesi, R. C. Butler *et al.*, *A&A* **520**, p. A4 (September 2010).
7. J. Lamarre, J. Puget, P. A. R. Ade *et al.*, *A&A* **520**, p. A9 (September 2010).
8. N. Jarosik, C. Barnes, C. L. Bennett *et al.*, *ApJS* **148**, p. 29 (September 2003).
9. A. Zacchei, D. Maino, C. Baccigalupi *et al.*, *A&A* **536**, p. A5 (December 2011).
10. J. J. Bock, D. Chen, P. D. Mauskopf and A. E. Lange, *Space Sci. Rev.* **74**, p. 229 (October 1995).
11. P. D. Mauskopf, J. J. Bock, H. del Castillo, W. L. Holzapfel and A. E. Lange, *Appl. Opt.* **36**, p. 765 (February 1997).
12. W. C. Jones, R. Bhatia, J. J. Bock and A. E. Lange, A Polarization Sensitive Bolometric Receiver for Observations of the Cosmic Microwave Background, in *Society of Photo-Optical Instrumentation Engineers (SPIE) Conference Series*, ed. T. G. P. . J. Zmuidzinas, Society of Photo-Optical Instrumentation Engineers (SPIE) Conference Series, Vol. 4855 February 2003.
13. Planck Collaboration *et al.*, *A&A* **536**, p. A7 (December 2011).
14. Planck HFI Core Team *et al.*, *A&A* **536**, p. A6 (December 2011).

THE PAMELA EXPERIMENT: FIVE YEARS OF COSMIC RAYS INVESTIGATION

E. MOCCHIUTTI*, M. BOEZIO, V. BONVICINI, R. SARKAR, C. PIZZOLOTTO, A. VACCHI, G. ZAMPA and N. ZAMPA

INFN, Sezione di Trieste, Padriciano 99, I–34149 Trieste, Italy
**E-mail: Emiliano.Mocchiutti@ts.infn.it*

R. CARBONE, V. FORMATO and G. JERSE

INFN, Sezione di Trieste, Padriciano 99, I–34149 Trieste, Italy
and University of Trieste, Department of Physics, via A. Valerio 2, I–34127 Trieste, Italy

S. BORISOV, N. DE SIMONE, V. DI FELICE, N. N. NIKONOV, F. PALMA, P. PICOZZA and R. SPARVOLI

INFN, Sezione di Roma "Tor Vergata", Via della Ricerca Scientifica 1, I–00133 Rome, Italy
and University of Rome "Tor Vergata", Department of Physics, Via della Ricerca Scientifica 1, I–00133 Rome, Italy

M. CASOLINO, M. P. DEPASCALE, C. DE SANTIS, V. MALVEZZI, and L. MARCELLI

INFN, Sezione di Roma "Tor Vergata", Via della Ricerca Scientifica 1, I–00133 Rome, Italy

O. ADRIANI, L. BONECHI and P. SPILLANTINI

INFN, Sezione di Firenze, Via Sansone 1, I–50019 Sesto Fiorentino, Florence, Italy
and University of Florence, Department of Physics, Via Sansone 1, I–50019 Sesto Fiorentino, Florence, Italy

M. BONGI, S. BOTTAI, N. MORI, P. PAPINI, S. B. RICCIARINI and E. VANNUCCINI

INFN, Sezione di Firenze, Via Sansone 1, I–50019 Sesto Fiorentino, Florence, Italy

G. C. BARBARINO

INFN, Sezione di Napoli, Via Cintia, I–80126 Naples, Italy
and University of Naples "Federico II", Department of Physics, Via Cintia, I–80126 Naples, Italy

D. CAMPANA, L. CONSIGLIO and G. OSTERIA

INFN, Sezione di Napoli, Via Cintia, I–80126 Naples, Italy

G. A. BAZILEVSKAYA, A. N. KVASHNIN, O. MAKSUMOV and Y. I. STOZHKOV

Lebedev Physical Institute, Leninsky Prospekt 53, RU–119991 Moscow, Russia

R. BELLOTTI, A. BRUNO and A. MONACO

INFN, Sezione di Bari, Via Amendola 173, I–70126 Bari, Italy and University of Bari, Department of Physics, Via Amendola 173, I–70126 Bari, Italy

F. CAFAGNA

INFN, Sezione di Bari, Via Amendola 173, I–70126 Bari, Italy

E. A. BOGOMOLOV, S. Y. KRUTKOV and G. VASILYEV

Ioffe Physical Technical Institute, Polytekhnicheskaya 26, RU–194021 St. Petersburg, Russia

P. CARLSON, M. PEARCE, L. ROSSETTO and J. WU

KTH, Department of Physics, and the Oskar Klein Centre for Cosmoparticle Physics, AlbaNova University Centre, SE–10691 Stockholm, Sweden

G. CASTELLINI

IFAC, Via Madonna del Piano 10, I–50019 Sesto Fiorentino, Florence, Italy

A. M. GALPER, L. GRISHANTSEVA, A. V. KARELIN, S. V. KOLDASHOV, A. LEONOV, V. MALAKHOV, A. G. MAYOROV, V. V. MIKHAILOV, M. RUNTSO, S. A. VORONOV, Y. T. YURKIN and V. G. ZVEREV

Moscow Engineering and Physics Institute, Kashirskoe Shosse 31, RU–11540 Moscow, Russia

M. RICCI

INFN, Laboratori Nazionali di Frascati, Via Enrico Fermi 40, I–00044 Frascati, Italy

W. MENN and M. SIMON

Universität Siegen, D–57068 Siegen, Germany

In five years of data taking in space, the PAMELA experiment collected very interesting data on the charged cosmic radiation over a wide energy range (from 100 MeV up to 1 TeV, depending from the species) with unprecedent statistics. The apparatus comprises a time–of–flight system, a silicon–microstrip magnetic spectrometer, a silicon–tungsten electromagnetic calorimeter, an anticoincidence system, a shower tail counter scintillator and a neutron detector. PAMELA is providing fundamental data not only to search for dark matter signals but also to better understand cosmic–ray propagation models. Main results after five years of data taking will be presented.

Keywords: Cosmic rays; CR composition; energy spectra and interactions; Spaceborne and space research instruments.

1. Introduction

PAMELA is a dedicated satellite borne experiment conceived by the WiZard collaboration to study the anti–particle component of the cosmic radiation. In this work the main results obtained after five years of data taking are reviewed.

2. Physics goals and instrument description

The PAMELA physics goal is the precise measurement of the cosmic ray composition at 1 Astronomical Unit (AU). Its 70 degrees, 350–610 km quasi-polar orbit makes it particularly suited to study items of galactic, heliospheric and trapped nature. PAMELA has been mainly conceived to perform high-precision spectral measurement of antiprotons and positrons and to search for antinuclei, over a wide energy range. Besides the study of cosmic antimatter, the instrument setup and the flight characteristics allow many additional scientific goals to be pursued.[1]

The instrument is installed inside a pressurized container (2 mm aluminum window) attached to the Russian Resurs–DK1 Earth–observation satellite that was launched into Earth orbit by a Soyuz–U rocket on June 15th 2006 from the Baikonur cosmodrome in Kazakhstan. The mission has been extended from the foreseen three years to the actual unlimited acquisition tight to the satellite life time. PAMELA was first switched on June 21st 2006 and it has been collecting data continuously since July 11th 2006 for more than 20 TB of collected data. To date about 1230 days of data have been analyzed, corresponding to more than one billion recorded triggers.

A schematic overview of the PAMELA apparatus is shown in Fig. 1. It comprises the following subdetectors, arranged as shown in figure (from top to bottom): a time–of–flight system (TOF — S1, S2, S3); a magnetic spectrometer; an anticoincidence system (AC — CARD, CAT, CAS); an electromagnetic imaging calorimeter; a shower tail catcher scintillator (S4) and a neutron detector.

The TOF system is made of plastic scintillators. Its timing resolution allows albedo–particle identification and mass discrimination below 1 GeV/c. The TOF provides also a fast signal for triggering the data acquisition.

The central components of PAMELA is a magnetic spectrometer, made by a permanent magnet and a tracking system composed of six planes of double-sided silicon sensors. This device is used to determine the rigidity (momentum divided by charge) and the charge of particles crossing the magnetic cavity. The rigidity measurement is done through the reconstruction of the trajectory based on the impact points on the tracking planes and the resulting determination of the curvature due to the Lorentz force. The direction of bending of the particle (*i.e.* the discrimination of the charge sign) is the key method used to separate

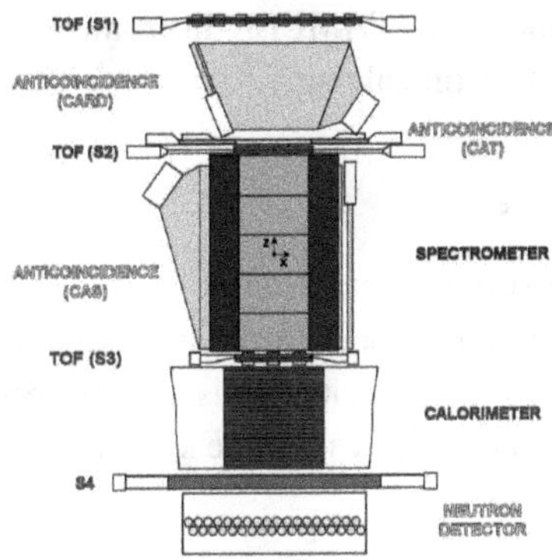

Fig. 1. A schematic view of the PAMELA apparatus. The instruments is ~1.3 m tall and has a mass of 470 kg. The average power consumption is 355 W. Magnetic field lines are oriented parallel to the y direction.

matter from anti-matter. The acceptance of the spectrometer, which also defines the overall acceptance of the PAMELA experiment, is 21.5 cm^2sr and the spatial resolution of the tracking system is better than 4 μm up to a zenith angle of 10°, corresponding to a maximum detectable rigidity exceeding 1 TV.

The spectrometer is surrounded by a plastic scintillator veto shield, aiming to identify false triggers and multiparticle events generated by secondary particles produced in the apparatus.

The main task of the electromagnetic calorimeter (16.3 X_0, 0.6 λ_0), mounted below the spectrometer, is to select positrons and antiprotons from the large background constituted by protons and electrons, respectively. Positrons have to be identified from a background of protons that is about 10^3 times the positrons component at 1 GeV/c, increasing to 5×10^3 at 10 GeV/c. Antiprotons have to be selected from a background of electrons that decreases from 5×10^3 times the antiproton component at 1 GeV/c to less than 10^2 times above 10 GeV/c. This means that PAMELA must be able to separate electrons from hadrons at a level better than 10^5. Much of this rejection power in PAMELA is provided by the calorimeter. Besides the electron-hadron separation, the calorimeter directly measures the energy of electrons and positrons. The high granularity of the calorimeter and the use of silicon strip detectors provide detailed information on the longitudinal and lateral profiles of particles' interactions as well as a measure of the deposited energy.

A plastic scintillator system mounted beneath the calorimeter aids the identification of high-energy electrons and is followed by a neutron detection system for the selection of high-energy electrons which shower in the calorimeter. More technical details about the entire PAMELA instrument and launch preparations can be found in PAMELA technical paper.[2]

3. Particle Measurements

The measurement of most common particles (protons, nuclei and electrons) in the cosmic radiation is easier respect to the anti-particle measurements due to their abundance. Hence the sample selection is usually very efficient and the main effort must be put in the study of any possible systematic uncertainty.

3.1. *Protons and helium nuclei spectra*

Protons and helium nuclei are the most abundant components of the cosmic radiation. Precise measurements of their fluxes are needed to understand the acceleration and subsequent propagation of cosmic rays in the Galaxy. The results reported by PAMELA[3] in the rigidity range 1 GV – 1.2 TV, Fig. 2 on the left, show that the spectral shapes of these two species are different and cannot be well described by a single power law. These data challenge the current paradigm of cosmic-ray acceleration in supernova remnants followed by diffusive propagation in the Galaxy. More complex processes of acceleration and propagation of cosmic rays may be required to explain the spectral structures observed.

3.2. *Negative electron spectrum*

As it will be discussed in Sec. 4.2, the positron fraction rise measured by PAMELA could be due to a very soft electron (e^-) spectrum. It is therefore im-

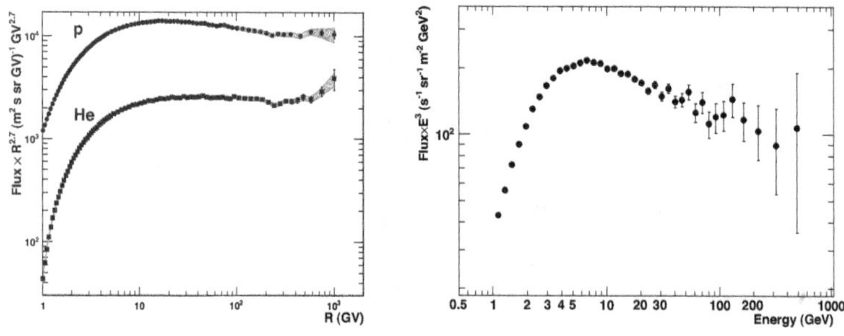

Fig. 2. On the left, proton (circles) and helium nuclei (boxes) flux measured with PAMELA as function of the rigidity; shaded area represent the estimated systematic error. On the right, electron flux as measured with PAMELA.

portant to precisely measure the negative electron spectrum in order to put constraints in the interpretation of the positron fraction rise. Moreover, if a primary positron source exists, it is difficult to explain the generation and acceleration of positrons without generating and accelerating the same amount of electrons. This implies that a negative electron spectrum measurement with high enough statistic and precision should reveal spectral features in the same energy range at which the positron fraction seems to deviate from the expected background.

The PAMELA apparatus is able to separate negative electrons from positrons up to about 600 GeV.[4] Figure 2, on the right, shows the negative electron spectrum as measured by PAMELA. Although the overall results can be easily described by a single power law, a hardening of the spectrum may be present at high energies. This interesting feature seems to be in agreement with the positron fraction measurement.

3.3. *Nuclei*

Light nuclei up to Oxygen are detectable with the dE/dx measured by the scintillator of the TOF system. It is possible to study with high statistics the secondary/primary cosmic ray nuclear and isotopic abundances such as B/C, Be/C, Li/C and ^{3}He/^{4}He. These measurements can constrain existing production and propagation models in the galaxy, providing detailed information on the galactic structure and the various mechanisms involved. The preliminary B/C ratio as function of kinetic energy per nucleon measured by PAMELA is in good agreement with previous measurements.

4. Anti–Particle Measurement

The main task of PAMELA is to measure the antimatter components of the cosmic-ray. At high energy, main sources of background in the antimatter samples result from spillover (protons in the antiproton sample and electrons in the positron sample) and from like-charged particles (electrons in the antiproton sample and protons in the positron sample). Spillover background originates from the wrong determination of the charge sign due to measured deflection uncertainty; its extent is related to the spectrometer performances and its effect is to set a limit to the maximum rigidity up to which the measurement can be extended. The like-charged particle background is related to the capability of the instrument to perform electron–hadron separation.

4.1. *Antiproton to proton ratio*

Electrons in the antiproton sample can be easily rejected by applying conditions on the calorimeter shower topology, while the main source of background originates from spillover protons. In order to reduce the spillover background and accurately measure antiprotons up to the highest possible energy, strict selection criteria were imposed on the quality of the fit. To measure the antiproton–to–proton flux ratio the different calorimeter selection efficiencies for antiprotons and protons were estimated. The difference is due to the momentum dependent interaction cross-sections for the two particles. These efficiencies were studied using both simulated antiprotons and protons, and proton samples selected from the flight data. In this way it was possible to normalize the simulated proton and therefore the antiproton selection efficiency.

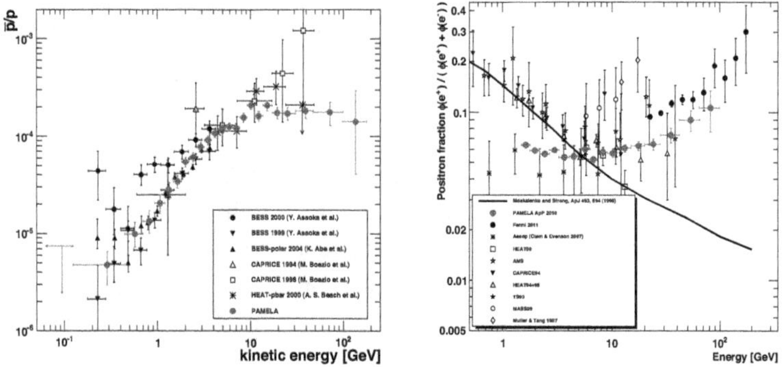

Fig. 3. Left panel: the antiproton-to-proton flux ratio obtained by PAMELA compared with recent measurements. Right panel: the PAMELA positron fraction compared to other experimental results and the standard model prediction for secondary positron production.

The left panel of Fig. 3 shows the antiproton–to–proton flux ratio measured by the PAMELA experiment[5,6] compared with other recent measurements.[7–11] Only statistical errors are shown since the systematic uncertainty is less than a few percent of the signal, which is significantly lower than the statistical uncertainty. The PAMELA data are in excellent agreement with recent data from other experiments, the antiproton–to–proton flux ratio increases smoothly with energy up to about 10 GeV and then levels off. The data follow the trend expected from secondary production calculations and our results are sufficiently precise to place tight constraints on secondary production calculations and contributions from exotic sources.[6]

4.2. *Positron fraction*

Protons are the main source of background in the positron sample and an excellent positron identification is needed to reduce the contamination at a negligible level.

The method used to obtain the published results is the proton background estimation method. This approach consists in keeping a very high selection efficiency and in quantifying the residual proton contamination by the mean of a so-called "spectral analysis".[12] In a conservative approach, the proton distributions needed to estimate the contamination were obtained using the flight calorimeter data without any dependence on simulations or test beam data. Results are shown in the right panel of Fig. 3 where PAMELA data[12] are compared to some recent measurements[13–21] and to the standard model theoretical prediction for secondary positron production.[22] At low energy PAMELA data are lower than most of the other data and this can be interpreted as an observation of charge-sign dependent solar modulation effects. Between about 6 and 10 GeV the PAMELA positron fraction is compatible with other measurements and above 10 GeV it increases significantly with energy. The PAMELA data cannot be described by the standard model of secondary production, black line in Fig. 3, right panel. The secondary production model has its indetermination due to the knowledge of the fluxes of primary particles, of the interaction cross-sections, of the average amount of traversed matter, and of the electron spectrum. A pure secondary component has difficulties in explaining the rise at $E>10$ GeV without using an unrealistic soft electron spectrum and ad hoc tuning of the other parameters,[23] suggesting the existence of other primary sources.[24] Many explanations about the origin of the positron excess have been postulated. These models can be divided in terms of astrophysical sources, like pulsars[25] or the distribution of Supernovæ remnants[26] in the Galaxy, or more speculative ones, like annihilation of new type of dark matter[27] or of the lightest superparticle dark matter.[28]

5. Other Measurements

5.1. *Solar modulation galactic cosmic rays*

Since protons and helium nuclei are detected by PAMELA with very high statistics it is possible to precisely study time variations and transient phenomena during the present 23rd solar minimum.

A long term measurement of the proton, electron and nuclear flux at 1 AU provides information on propagation phenomena occurring in the heliosphere.[29] As already mentioned, the possibility to measure the anti-particle spectra allows also charge dependent solar modulation effects to be studied. The proton flux as measured by PAMELA in different time intervals shows an increasing flux of galactic cosmic rays corresponding to a decreasing solar activity. This effect is in agreement with the increase in the fluxes as measured by neutron monitors at ground.[30]

5.2. *Re–entrant albedo and trapped particles measurements*

Albedo particles are secondary particles produced by cosmic-rays interacting with the Earth's atmosphere that are scattered upward. When these particles lack sufficient energy to leave the Earth's magnetic field they re-enter the atmosphere in the opposite hemisphere but at a similar magnetic latitude and they are called re-entrant albedo particles. The measurement of the composition and spectra of the secondary cosmic rays particles provides a tool for the fine tuning of models used in air shower simulation programs. Due to its orbit PAMELA is able to provide a world map of the primary and re-entrant albedo particles, allowing fine details in the spectra, especially in the sub-cutoff region, to be discerned.[31,32] It is, in fact, possible to observe structures in the spectra in the sub-cutoff region which are also reproduced by simulations.[33]

The 70° orbit of the Resurs–DK1 satellite allows for continuous monitoring of the electron and proton belts. The high energy (>80 MeV) component of Van Allen Belts can be monitored during the time and it is possible to perform a detailed mapping of these regions.[34]

5.3. *Solar energetic particles*

Due to the period of solar minimum few significant solar events with energy high enough to be detectable are expected. The observation of solar energetic particle (SEP) events with a magnetic spectrometer permit several aspects of solar and heliospheric cosmic ray physics to be addressed for the first time. PAMELA detected the December 2006 SEP event and results are going to appear on ApJ.[35]

6. Conclusions

PAMELA is continuously taking data and the mission is planned to continue until the satellite will stay in orbit. The increase in statistics will allow higher energies to be studied. An analysis for positron flux till low energy (down to 100 MeV), and primary cosmic rays nuclei is in progress and will be the topic of future publications.

Acknowledgment

We would like to acknowledge contributions and support from: Italian Space Agency (ASI), Deutsches Zentrum für Luft– und Raumfahrt (DLR), The Swedish National Space Board, Swedish Research Council, Russian Space Agency (Roskosmos, RKA). R. S. wishes to thank the TRIL program of the International Center of Theoretical Physics, Trieste, Italy that partly sponsored his activity.

References

1. P. Picozza *et al.*, *Proceedings of 20th ECRS, Lisbon – Portugal 2006.*
2. P. Picozza *et al.*, Astrophys. J., **27**, 296 (2007).
3. O. Adriani *et al.*, Science, **332**, 6025 (2011).
4. O. Adriani *et al.*, Phys. Rev. Lett., **106**, 201101 (2011).
5. O. Adriani *et al.*, Phys. Rev. Lett., **102**, 051101 (2009).
6. O. Adriani *et al.*, Phys. Rev. Lett., **105**, 121101 (2010).
7. M. Boezio *et al.*, Astrophys. J., **561**, 787 (2001).
8. A. S. Beach *et al.*, Phys. Rev. Lett., **87**, 271101 (2001).
9. M. Boezio *et al.*, Astrophys. J., **487**, 415 (1997).
10. Y. Asaoka *et al.*, Phys. Rev. Lett., **88**, 051101 (2002).
11. K. Abe *et al.*, Phys. Lett. B, **670**, 103 (2008).
12. O. Adriani *et al.*, Astropart. Phys., **34**, 1 (2010).
13. M. Ackermann *et al.*, Astro–Ph/1109.0521 preprint (2011).
14. R. L. Golden *et al.*, Astrophys. J., **436**, 769 (1994).
15. J. Alcaraz *et al.*, Phys. Lett. B, **484**, 10 (2000).
16. J. J. Beatty *et al.*, Phys. Rev. Lett., **93**, 241102 (2004).
17. S. W. Barwick *et al.*, Astrophys. J., **482**, L191 (1997).
18. H. Gast, J. Olzem and S. Schael, *Proceeding of XLI Rencontres De Moriond, Electroweak Interaction And Unified Theories, La Thuile, Italy, 2006* 2006, p. 421.
19. M. Boezio *et al.*, Astrophys. J., **532**, 653 (2000).
20. J. Clem and P. Evenson, *Proceedings of 30th International Cosmic Ray Conference, Merida – Mexico 2006*, Edited By Caballero R., D'olivo J. C., Medina–Tanco G. And Valdés–Galicia J. F. (Universidad Nacional Autónoma De México, Mexico City, Mexico) 2008, Vol. 6. p. 27.
21. D. Müller and K. K. Tang, Astrophys. J., **312**, 183 (1987).
22. I. Moskalenko and A. Strong, Astrophys. J., **493**, 694 (1998).
23. T. Delahaye, F. Donato, N. Fornengo, J. Lavalle, R. Lineros, P. Salati and R. Taillet, Astro–Ph/0809.5268 preprint (2008).
24. P. D. Serpico, Phys. Rev. D, **79**, 021302 (2009).
25. H. Yuksel, M. D. Kistler and T. Stanev, Astro–Ph/0810.2784 preprint (2008).
26. N. J. Shaviv, E. Nakar and T. Piran, Astro–Ph/0902.0376 preprint (2009).
27. D. Hooper and S. Profumo, Phys. Reports, **453**, 29 (2007).
28. P. Grajek *et al.*, Astro–Ph/0812.4555 preprint (2008).
29. N. De Simone *et al.*, Astrophys. Space Sci. Trans., **7**, 425 (2011).
30. For example Http://neutronm.bartol.udel.edu/modplot.html .
31. O. Adriani *et al.*, J. of Geophys. Res., **114/A12**, A12218 (2009).
32. O. Adriani *et al.*, Astrophys. J., **737**, L29 (2011).
33. M. Honda *et al.*, Phys. Rev. D, **70/4**, 043008 (2004).
34. M. Casolino *et al.*, J. Of The Phys. Society Of Japan, **78(A)**, 35 (2009).
35. O. Adriani *et al.*, Astro–Ph/1107.4519 preprint (2011).

Observation of blazars with the high energy SED peak in the Fermi-LAT band

C. Monte*[1], S. Ciprini[2,3], S. Rainó[1] and G. Tosti[4,5]
on behalf of the *Fermi*-LAT collaboration

[1] *Istituto Nazionale di Fisica Nucleare, Sezione di Bari, 70126, Bari, Italy*
[2] *ASI Science Data Center, 00044, Frascati (Rome), Italy*
[3] *INAF Rome Observatory, 00040, Monte Porzio Catone (Rome), Italy*
[4] *Dipartimento di Fisica, Universit degli Studi di Perugia, 06123, Perugia, Italy*
[5] *Istituto Nazionale di Fisica Nucleare, Sezione di Perugia, 06123, Perugia, Italy*
* *Corresponding author e-mail: claudia.monte@ba.infn.it*

After three years of scientific activity with the Large Area Telescope (LAT), the primary instrument onboard Fermi, the second catalog of the active galactic nuclei (2LAC) detected by the Fermi Large Area Telescope is now complete. It is a follow-up of the first LAT AGN catalog, (1LAC). The 2LAC includes a number of analysis refinements and additional association methods which have substantially increased the number of associations over 1LAC. Among the AGN included both in the 1LAC and in the 2LAC, those blazars that have their Spectral Energy Distribution (SED) high-energy peak centered on the Fermi-LAT band (from 20 MeV to 300 GeV) are of particular interest. The brightest of these sources have been analyzed covering a period of 22 months of Fermi LAT gamma-ray data in order to investigate their spectral features in the gamma-ray band and to characterize the temporal evolution of their gamma-ray spectra.

Keywords: Gamma rays astronomical observations, Active Galactic Nuclei

1. Introduction

The second catalog of active galactic nuclei (AGN) detected by the Fermi Large Area Telescope (LAT) in two years of scientific operation has been completed and it is almost published. This catalog, known as 2LAC,[1] includes 1016 gamma-ray sources located at high Galactic latitudes ($|b| > 10\,°$) that have been detected with a test statistic (TS) greater than 25 and associated statistically with AGNs. Because some 2LAC sources are associated with multiple AGNs, a clean sample of LAT sources with unique association has been defined: it includes 885 AGNs, comprising 395 BL Lacertae objects (BL Lacs), 310 flat-spectrum radio quasars (FSRQs), 156 candidate blazars of unknown type, 8 misaligned AGNs, four narrow-line Seyfert 1 (NLS1s), 10 AGNs of other types and 2 starburst galaxies. The 2LAC represents a significant improvement relative to the First LAT AGN Catalog (1LAC[2]), with 52% more associated sources.

2. The selected sample

The blazars included in our sample has been extracted starting from the 1LAC. All blazars detected with TS greater then 100 were considered. A preliminary analysis has been performed in the energy band from 100 MeV to 300 GeV using a deconvolution technique (unfolding method[3]) to reconstruct the source energy spectra from the one year observed data after background subtraction. Looking at the reconstructed spectral-energy distributions (SEDs) obtained in this way, a selection has been made among the blazars, including in our sample only those objects showing their SED high energy peak in the Fermi-LAT energy band. First of all, the sources with the high energy peak greater than 200 MeV have been included; then also sources that have an extrapolated peak between 20 MeV and 200 MeV have been added. The final sample includes 17 sources (12 BL Lacs and 5 FSRQs).

3. Fermi-LAT data analysis

For all the sources in our sample, we analyzed the Fermi-LAT data covering a period of 22 months from August 4, 2008 (the starting date of the science phase of the mission) to June 4, 2010. For each source, only events with energy in the range from 200 MeV to 300 GeV have been selected in a Region of Interest (RoI) of 15° centered around the source itself. Only photons belonging to the so called "Diffuse class"[4] have been retained for the analysis. In order to avoid background contamination from the bright Earth limb, time intervals where the Earth entered in the LAT field of view have been excluded from the data sample. We have also excluded observations in which the source under investigation was viewed at zenith angles larger than 105°, where Earth's albedo gamma-rays increase the background contamination. Each source has been analyzed using a binned maximum likelihood technique, implemented in the *gtlike*[a] analysis tool developed by the LAT team. The model of the region of interest to be used in the Likelihood includes the source of interest, all the sources included

[a] *http://fermi.gsfc.nasa.gov/ssc/data/analysis/documentation/Cicerone/*

in the 1FGL within the 15° RoI, the sources external to the RoI up to a distance of 20° that can contribute to the RoI, the model for the Galactic diffuse background component and the model for the isotropic background[b]. Three different methods of analysis have been used. In the first case, the

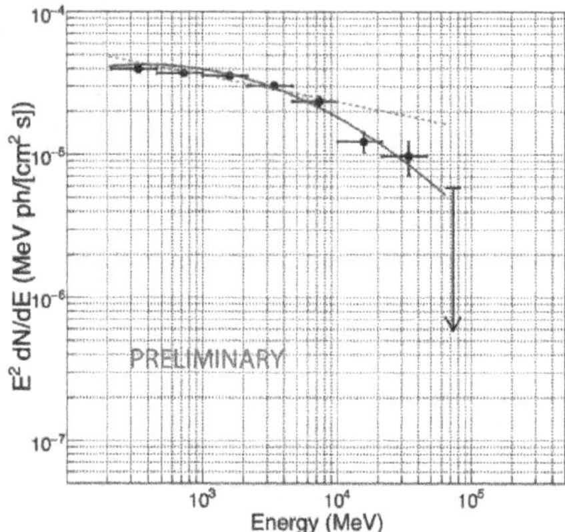

Fig. 1. SED of PKS 0537-441 obtained in the 22 months data analysis. The continuous line and the dotted line represent the results of the gtlike analysis performed in the whole energy band modeling the source under investigation respectively with a Log Parabola and with a Power Law spectrum.

whole energy range has been divided into energy bins and the gtlike tool has has been used to calculate the flux in each band: a Power Law parametric model with Spectral Index fixed at 2, is assumed in each individual energy bin both for the source spectrum and for the background components. In the second and third method of analysis, a maximum likelihood fit is performed in the whole energy band (200 MeV ÷ 300 GeV) modeling the source under investigation respectively with a Power Law (PL) or with a Log Parabola (LP). The results of the analysis with the different methods are shown in figure 1 for the source PKS 0537-441 (1FGL J0538.8-4404): the black filled points show the results of the binned gtlike analysis performed dividing the energy range into 3 bins per decade, the continuous blue line and the dotted red line represent the results of the gtlike analysis performed in the whole energy band modeling the source under investigation respectively with a Log Parabola and with a Power Law spectrum.

---

[b] *http://fermi.gsfc.nasa.gov/ssc/data/access/lat/BackgroundModels.html*

4. Spectral Energy Distributions and Light Curves

The flux light curves for our sources have been generated dividing the total observation period (22 months) in 1 month time bins and applying the maximum likelihood fit implemented in the gtlike tool across the full energy band for each time bin. The 22 months light curve for the PKS 0537-441 is

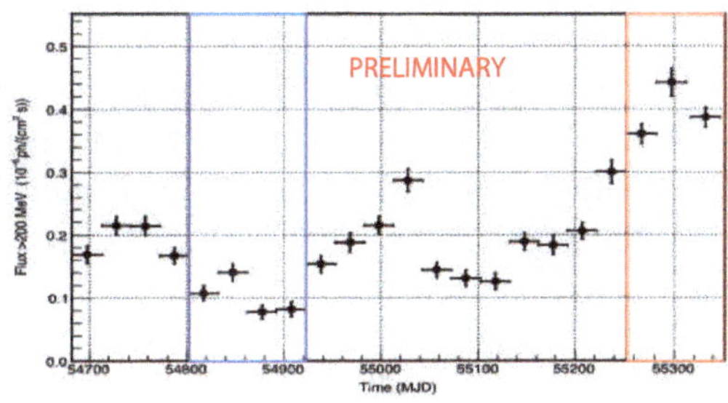

Fig. 2. 22 months flux light curve of PKS 0537-441.

shown in figure 2. Looking at the monthly light curve for each source, the time periods in which the source showed the maximum flux (high state)

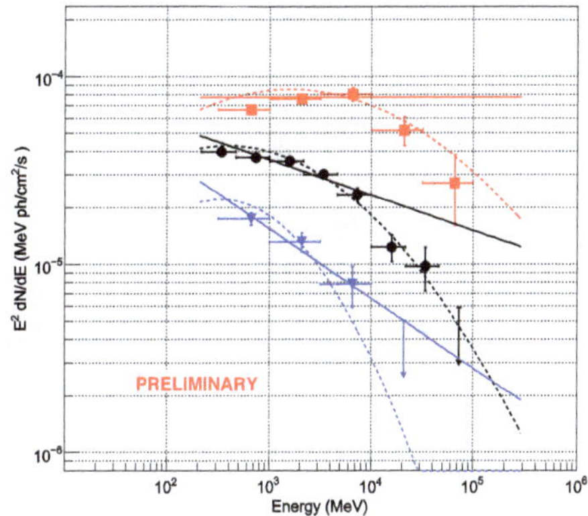

Fig. 3. The SED of the PKS 0537-441 in the 22 months period (black filled circles), in the high state (red filled squares) and in the low state (blue filled triangles). The continuous lines and the the dotted lines represent the results of the gtlike analysis performed in the whole energy band modeling the source under investigation respectively with a Log Parabola and with a Power Law spectrum in the three time periods.

and the minimum flux (low state) have been considered separately and the gtlike analysis has been repeated for each source both in the high and in the low state with the three different techniques used for the analysis in the 22 months period, as described before. The shaded area on the left (blue) and that on the right (red) in figure 2 represent respectively the selected low state and high state for the source PKS 0537-441. The results of the gtlike analysis performed both in high and low state and for comparison in the 22 months period are shown in figure 3 for the source PKS 0537-44: filled points represent the results of the gtlike analysis performed dividing the energy range into 3 bins per decade, while the continuous and the dotted lines represent the results of the gtlike fit obtained modeling the source respectively with a LogParabola and a PowerLaw.

5. Conclusions

A sample of 17 sources has been selected from the 1LAC. A preliminary spectral analysis performed over a period of 22 months seems to show that the Log Parabola function is better than the Power Law function in order to model the shape of the SED high-energy peak in the gamma-ray energy range for most of our sources. The spectral analysis performed in the selected high and low state for each source seems to show that there are some spectral features related to the variability of the sources, that are still under investigation.

Acknowledgments

The *Fermi* LAT Collaboration acknowledges support from a number of agencies and institutes for both development and the operation of the LAT as well as scientific data analysis. These include NASA and DOE in the United States, CEA/Irfu and IN2P3/CNRS in France, ASI and INFN in Italy, MEXT, KEK, and JAXA in Japan, and the K. A. Wallenberg Foundation, the Swedish Research Council and the National Space Board in Sweden. Additional support from INAF in Italy and CNES in France for science analysis during the operations phase is also gratefully acknowledged.

References

1. M. Ackermann al., *Submitted to ApJ*, arXiv:1108.1420v2
2. A. A. Abdo al., *ApJ*, **715**, 429 (2010)
3. M. N. Mazziotta, Proc. of the 31 ICRC, LODZ (2009)
4. W.B. Atwood al., *ApJ* **697**, 1071 (2009)

EAS spectrum vs the total number of high energy muons and the mass composition of primary cosmic rays in the energy region $10^{15} - 10^{16}$ eV

Yu.F. Novoseltsev*, R.V. Novoseltseva and G.M. Vereshkov+

Institute for Nuclear Research of RAS, Moscow, 117312, Russia
**E-mail: novoseltsev@inr.ru*, +*E-mail: gveresh@gmail.com*

The method of a determination of Primary Cosmic Ray mass composition is presented. Data processing is based on the theoretical model representing the EAS spectrum vs the total number of muons as the superposition of the spectra corresponding to different kinds of primary nuclei. The method consists of two stages. At the first stage, the permissible intervals of mass fractions f_i are determined on the base of the EAS spectrum vs the total muon number ($E_\mu \geq$ 235 GeV). At the second stage, the permissible intervals of f_i are narrowed by fitting procedure. Within the framework of three components (protons, helium and heavy nuclei), the mass composition in the region $10^{15} - 10^{16}$ eV has been defined: $f_p = 0.235 \pm 0.02$, $f_{He} = 0.290 \pm 0.02$, $f_H = 0.475 \pm 0.03$.

1. Initial conditions

We present the method of determination of CR mass composition on the base of data on the EAS spectrum vs the total number of high energy muons.

We use the data on high multiplicity muon events ($n_\mu \geq 114$) collected at the Baksan underground scintillation telescope [1]. In [2] the muon multiplicity spectrum (i.e., the number m of muons hitting the facility at unknown position of EAS axis) at $m \geq 20$ was measured at zenith angles $\theta \leq 20^o$. The threshold energy of muons coming from this solid angle is 235 GeV. In papers [3–5] we developed the method of recalculation from the multiplicity spectrum to the EAS spectrum vs the total number of muons, $I(n_\mu)$. With the help of the method we have combined the results reported in [2] and [5,6] and obtained the EAS spectrum vs the total number of muons in the range $75 \leq n_\mu \leq 4000$, which corresponds to the primary energy range of $10^{15} \leq E_N \leq 10^{17}$ eV (Fig.1). It should be clarified that the data at $n_\mu > 2000$ are obtained for the muon threshold energy $E_{th} =$

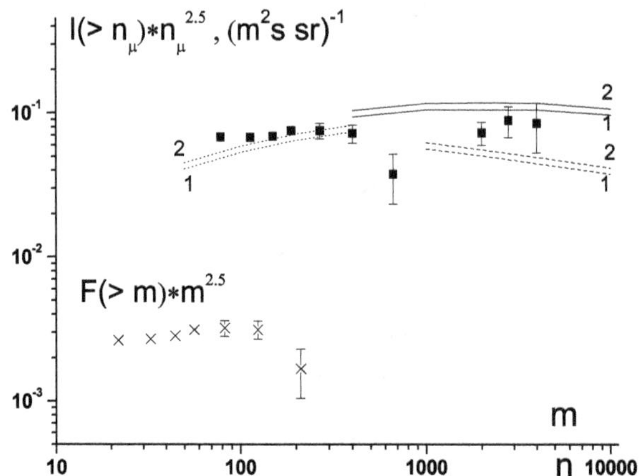

Fig. 1. Squares are the EAS spectrum vs n_μ (experimental data). The muon threshold energy is $E_{th} = 235$ GeV if $n_\mu < 1000$ and $E_{th} = 220$ GeV at $n_\mu > 1000$ [3,5]. Crosses show the muon multiplicity spectrum obtained in [2] (m and $F(m)$ correspond to the multiplicity spectrum). Solid curves are expected fluxes ($E_{th} = 220$ GeV) for the case $E_k = Z \cdot 3 \cdot 10^{15}$ eV, dashed curves – the case $E_k = 3 \cdot 10^{15}$ eV/nucleus. Dotted curves show expected fluxes for the case $E_k = Z \cdot 3 \cdot 10^{15}$ eV at $E_{th} = 235$ GeV.

220 GeV, while the points at $n_\mu < 700$ have $E_{th} = 235$ GeV which is the threshold energy in the experiment [2]. In Fig.1, the expected fluxes are calculated for $E_{th} = 235$ GeV ($n_\mu < 1000$, dotted curves) and $E_{th} = 220$ GeV ($n_\mu > 1000$, solid and dashed curves). Numbers near curves denote the mass composition variants: 1 is the low energy composition (the nuclei fractions in percentage are 39, 24, 13, 13, 11), 2 is the composition (1).

The data at $n_\mu < 700$ can be used for retrieval of information on the CR mass composition in the region $E_N = 10^{15} - 10^{16}$ eV. Let us remark here that the data at $m = 125$ $m = 212$ in [2] are obtained with essential systematic errors: according to our estimates, the values of m in these points are underestimated 4% and 10% respectively [7], therefore we restrict ourselves to the data at $m \leq 82$ ($n_\mu \leq 270$). As initial conditions, we use the mass composition obtained by Swordy [8] with the help of compilation of results of direct measurements at energies $\simeq 100$ TeV per nucleus[a] (A is

[a]in comparison with the composition presented in [8], in (1) the proton fraction is increased by 5% (at the expense of helium nuclei) in accordance with data of [9]

the number of nucleons in a nucleus)

$$\begin{array}{lccccc} & \text{p} & \text{He} & \text{CNO} & \text{Ne-S} & \text{Fe} \\ A & 1 & 4 & 14 & 28 & 56 \\ f,\% & 25 & 31 & 19 & 12 & 13 \end{array} \quad (1)$$

and the proton flux at the energy $E_p = 100$ TeV measured in the JACEE experiment [9].

$$D_p(100\ TeV) = 2.95 \times 10^{-10}\ (m^2 \cdot s \cdot sr \cdot GeV)^{-1} \quad (2)$$

Then the total flux of nuclei with energy of $E = 100$ TeV is equal to $D_{tot} = D_p/0.25 = 11.8 \times 10^{-10}\ (m^2 \cdot s \cdot sr \cdot GeV)^{-1}$, that is in a good agreement with the result obtained at Tibet array [10].

Our goal is to determine the mass composition evolution (from the composition (1)) into the region $E_N = 10^{15} - 10^{16}$ eV on the base of data on the number of high energy muons ($E_\mu \geq 235$ GeV) in EAS. To this end, we will use the measured fluxes of multiple muon events with the multiplicity into differential intervals $n_{\mu i} \leq n_\mu \leq n_{\mu(i+1)}$. At the first stage, we determine the permissible intervals of fractions f_i, which ensure an agreement with experimental data within the limits of one standard deviation . And then, the results are refined with the help of fitting procedure.

To obtain the more certain results we fix the CR energy spectrum, namely we adopt the conservative scenario:
i) the slope change of the spectrum occurs at the same energy per unit charge $E_k(Z) = 3$ PeV$\times Z$,
ii) the spectra of all nuclei kinds have the slope exponents $\gamma_1 = 2.7$ before the "knee" and $\gamma_2 = 3.1$ after the "knee"

$$D_A(E) = I_A E^{-2.7}(1 + E/E_k(Z))^{-0.4}. \quad (3)$$

This scenario is supported by experimental data well enough.

It should be emphasized, we do not attempt to use the data at $n_\mu > 2000$ because the energy spectra of nuclei at $E_N > 10^{16}$ eV are poorly understood.

2. Permissible domains

The flux of events with muon multiplicity $n \geq n_\mu$ produced by nuclei with A nucleons can be written in the form

$$I_A(n \geq n_{\mu i}) = \int_{E_{th}(A)}^{\infty} F_A(E) D_N(E) P_A(E,\ n \geq n_{\mu i}) dE, \quad (4)$$

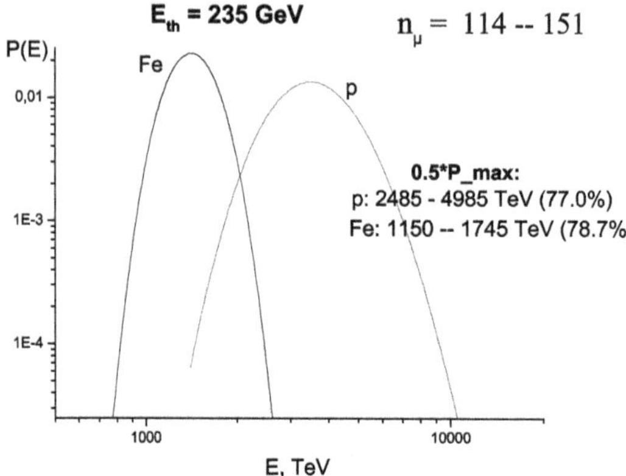

Fig. 2. Energy distributions of protons and iron nuclei making a contribution to the flux of muon events with $114 \leq n_\mu < 151$. (Areas under curves (p and Fe) are equal to 1.) The widths of distributions at half-height and fractions of events in these regions are indicated.

here $F_A(E)$ is the fraction of nuclei with A nucleons with energy E per nucleon, $D_N(E)$ is the total flux of nuclei with the same energy per nucleon, $P_A(E,\ n \geq n_\mu)$ is the probability that the number of muons (with $E_\mu \geq 235$ GeV) in EAS produced by nucleus "A" is $n \geq n_\mu$, $E_{th}(A)$ is the threshold energy of nuclei with A nucleons.

We assume that the multiplicity of muons in EAS is described by the negative binomial distribution $B_A(E,\ n)$, then $P_A(E,\ n \geq n_\mu) = \sum_{n \geq n_\mu} B_A(E,\ n)$.

The flux of events with $n_{\mu i} \leq n_\mu \leq n_{\mu(i+1)}$ has the form

$$J_A(\Delta n_{\mu i}) = I_A(n \geq n_{\mu i}) - I_A(n \geq n_{\mu(i+1)}) = \int\limits_{E^i_{th}(A)}^{\infty} F_A(E) D_N(E) \times$$

$$P_A(E,\ n \geq n_{\mu i}) dE - \int\limits_{E^{i+1}_{th}(A)}^{\infty} F_A(E) D_N(E) P_A(E,\ n \geq n_{\mu i+1}) dE = \overline{F}_A R_{iA}, \tag{5}$$

where the first index of the matrix R_{ij} points out to muon multiplicity ($n_{\mu i}$) and the second one pertains to a nucleus sort. $\overline{F}_A$ is the fraction of nuclei "A" averaged over the energy region which gives the main contribution in the integral (5) (as is seen in Fig.2, the region is rather narrow). Thus we work in the approximation $F_j(E) \simeq \overline{F}_j = const$ and will drop the symbol of averaging hereinafter.

To avoid possible methodical errors, we use only 4 points in the spectrum $I_{tot}(\geq n_\mu) \equiv I(\geq n_\mu)$: at $n_\mu = 114, 151, 189, 268$ (these points have the same exposure time). In table 1, the input data are presented: muon multiplicity intervals Δn_μ, the numbers and fluxes of events in given intervals of n_μ.

Table 1. Integral (I) and finite-difference (J) fluxes of events (EAS) with the given number of muons ($E_\mu \geq 235$ GeV). The flux $J(\Delta n_\mu)$ is defined according to (5).

n_μ	$I(\geq n_\mu) \times 10^7$, $(m^2 \cdot s \cdot sr)^{-1}$	Δn_μ	$N(\Delta n_\mu)$	$J \times 10^7$, $(m^2 \cdot s \cdot sr)^{-1}$
114	4.887 ± 0.209	114 - 151	277	2.419 ± 0.145
151	2.468 ± 0.150	151 - 189	106	0.920 ± 0.089
189	1.548 ± 0.121	189 - 268	98	0.906 ± 0.092
268	0.642 ± 0.079	≥ 268	66	0.642 ± 0.079

We will solve a direct problem and define the region of F_j values which is compatible with equations (couplings)

$$\sum_j R_{ij} \times F_j = J_i, \quad (i = 1, 2, ..., 4) \tag{6}$$

where J_i is the observed flux of events (EAS) with muon multiplicity from i-th interval – $n_{\mu i} \leq n_\mu < n_{\mu(i+1)}$.

Next we will pass to the energy per nucleus and decrease the number of independent variables with the help of relations

$$f_3(E) = f_4(E) = f_5(E), \quad (model\ II) \tag{7}$$

$$or \quad f_3(E) = 1.5 f_4(E), \quad f_4(E) = f_5(E) \quad (model\ I). \tag{8}$$

where $f_j(E)$ is the fraction of nuclei of kind j at the same energy per nucleus.

The relations (7) are fulfilled at low energies ($E_N \sim 100$ GeV) and the relations (8) are valid at $E_N \simeq 100$ TeV (see mass composition (1)). We will find the solution of equations (6) in both cases, and discuss later which variant ((7) or (8)) is more preferable.

Passing from five variables F_j to three variables f_k $(k = 1, 2, 3)$, it is convenient to rewrite the equations (6) as follows (we multiply and divide the j-th term by $A_j^{1.7}$ in each equation):

$$\sum_{j=1}^{3} R3_{ij} \times B_j = J_i, \quad (i = 1, 2, ..., 4) \tag{9}$$

$$where \quad \begin{aligned} & R3_{ij} = R_{ij}/A_j^{1.7}, \quad B_j = F_j \times A_j^{1.7}, \ j = 1, 2 \\ & R3_{i3} = \sum_{j=3}^{5} R_{ij}/A_j^{1.7}, \quad B_3 = \frac{1}{3} \sum_{j=3}^{5} F_j \times A_j^{1.7}. \end{aligned} \tag{10}$$

$$In\ addition \qquad f_k = B_k \times \left[\sum_{j=1}^{5} F_j \times A_j^{1.7} \right]^{-1} . \tag{11}$$

To determine B_j we will use independent pairs of equations (9) (for example for i = 1,2 or i = 2,3 etc.), and for closure of the equation system we use the normalization condition $f_1 + f_2 + 3f_3 = 1$, which (taking into account (9), (11)) can be read so

$$B_1 + B_2 + 3B_3 = \sum_{j=1}^{5} F_j \times A_j^{1.7} \tag{12}$$

Executing all operations mentioned above we have obtained the permissible intervals of fractions f_i which ensure an agreement with experimental data within the limits of one standard deviation[b]:

$$\begin{aligned} Model\ I: \quad & 0.182 \leq f_p \leq 0.250 , \qquad 0.287 \leq f_{He} \leq 0.321 , \\ & 0.203 \leq f_N \leq 0.211 , \qquad 0.135 \leq f_{Si} = f_{Fe} \leq 0.141 . \end{aligned} \tag{13}$$

$$\begin{aligned} Model\ II: \quad & 0.207 \leq f_p \leq 0.271 , \qquad 0.280 \leq f_{He} \leq 0.321 , \\ & 0.150 \leq f_N = f_{Si} = f_{Fe} \leq 0.157 . \end{aligned} \tag{14}$$

3. Fitting procedure

The second stage of data processing consists in narrowing of permissible intervals of f_i. With this end in view, we carry out the simultaneous fit of 4 integral points (see table 1) and 3 finite difference points:

$$\begin{aligned} I_{n_\mu} &\equiv I(\geq n_\mu) , \qquad n_\mu = 114, \quad 151, \quad 189, \quad 268 , \\ J_{n_\mu} &\equiv I(\geq n_\mu) - I(\geq n'_\mu) , \; n_\mu - n'_\mu = 114 - 151, \; 151 - 189, \; 189 - 268 . \end{aligned} \tag{15}$$

We perform fitting of 7 points (15) requiring that:
i) all data (7 points) are satisfied within the limits of one standard deviation,
ii) fitted parameters (f_i) are within permissible intervals.

$$\begin{aligned} ModelI: \quad & f_p = 0.236 \pm 0.003 , \; f_{He} = 0.290 \pm 0.003 , \; f_N = 0.204 \pm 0.001 , \\ & f_{Si} = f_{Fe} = 0.1356 \pm 0.0007 . \end{aligned} \tag{16}$$

---

[b] we have somewhat simplified (and reduced) the presentation; for more details see [11]

The fractions of light and heavy nuclei are

$$f_{light} = f_p + f_{He} = 0.526 \pm 0.005 \ , \qquad f_{heavy} = 0.474 \pm 0.003 \ . \tag{17}$$

$$\begin{aligned} Model\ II: \quad & f_p = 0.240 \pm 0.005 \ , \ f_{He} = 0.299 \pm 0.015 \ , \\ & f_N = f_{Si} = f_{Fe} = 0.1535 \pm 0.004 \end{aligned} \tag{18}$$

$$f_{light} = f_p + f_{He} = 0.539 \pm 0.019 \ , \qquad f_{heavy} = 0.461 \pm 0.010 \ . \tag{19}$$

We estimate the error value about 10% [11]. In this context Model I and Model II lead to the same results

$$f_p = 0.236 \pm 0.020, \ f_{He} = 0.290 \pm 0.020, \ f_H = 0.474 \pm 0.030 \tag{20}$$

The result (20) should be read as the estimation (rather precise) of CR mass composition in the energy region of $10^{15} - 10^{16}$ eV. Thus our analysis points out that CR mass composition become some more heavy in comparison with the one (1).

Acknowledgments

This work was supported by RFBR grant 11-02-12043, the "Neutrino Physics and Neutrino Astrophysics" Program for Basic Research of the Presidium of the Russian Academy of Sciences and the Federal Targeted Program of Ministry of Science and Education of Russian Federation "Research and Development in Priority Fields for the Development of Russia's Science and Technology Complex for 2007-2013", contract no.16.518.11.7072.

References

1. E.N. Alexeyev et al., Proc. 16th ICRC, Kyoto, **10**, 276 (1979)
2. A.V. Voevodsky at al., *Rus.J. Izvestiya of RAS, ser.phys.* **58,N12**, 127 (1994)
3. V.N. Bakatanov, Yu.F. Novoseltsev et al. *Astrop. Phys.* **12**, 19 (1999)
4. Yu.F. Novoseltsev. *Rus.J. Nuclear Phys.* **63**, 1129 (2000)
5. V.N. Bakatanov, Yu.F. Novoseltsev, R.V. Novoseltseva. Proc. 27th ICRC, Hamburg, **1**, 84 (2001)
6. V.N. Bakatanov, Yu.F. Novoseltsev et al. *Astrop. Phys.* **8**, 59 (1997)
7. Yu.F. Novoseltsev. Doctor of Sciences Thesis, Institute for Nuclear Research of RAS, Moscow, 2003
8. Swordy S.P. Proc. 23th ICRC, Calgary, 1993, Rapporteur Papers, p.243
9. Asakimori K. et al. Proc. 24th ICRC, Rome, **2**, 728 (1995)
10. Amenomori M. et al. Proc. 24th ICRC, Rome, **2**, 736 (1995)
11. Yu.F. Novoseltsev, R.V. Novoseltseva, G.M. Vereshkov. arXiv:1108.4245v1

STATUS OF THE CERENKOV TELESCOPE ARRAY (CTA)

GIOVANNI PARESCHI
INAF – Osservatorio Astronomico di Brera, Via E.Bianchi 46
Merate, 23807, Italy
e-mail: giovanni.pareschi@brera.inaf.it

ON BEHALF OF THE CTA COLLABORATION

The Very High Energy band (above a few tens of GeV up to 100 TeV) is the natural domain where the study of the astrophysical sources is tangled with the realm of the particle physics. Several outstanding results were obtained so far by the HESS, MAGIC, and VERITAS Cherenkov arrays, both on Galactic and extra-Galactic sources, such as the investigations of the innermost surroundings of super-massive black-holes or the detailed morphology of several supernova remnants. The forthcoming Cherenkov Telescope Array (CTA), with its innovative approach based on the use of three different sizes of telescopes, should obtain a one-order-of-magnitude improvement with respect to the current Cherenkov telescope performance. In this talk I will review both the science goals and the technological aspects of the CTA, also reporting on the status of the project.

1. Introduction

With the advent of the Imaging Atmospheric Cherenkov Telescopes (IACTs) [1] in late 1980's [2], ground-based observation of TeV gamma-rays came into reality and, since the first source detected at TeV energies in 1989 the number of gamma-ray sources has rapidly grown up to 125 (see the catalogue at the link: http://tevcat.uchicago.edu/), all of them a part a few cases all discovered with Cherenkov telescopes. The technique was first pioneered by the *Whipple* experiment since 1985, leading to the discovery of TeV gamma-rays from the Crab Nebula in 1989. This first result was followed by the discovery of the TeV emission from the first extragalactic source (Mrk 421), showing that acceleration processes are taking part in AGNs too. Since 2003, as the new generation experiments (HESS, MAGIC, CANGAROO and VERITAS) have been started to observe the gamma-ray sky, the number of VHE sources started to rapidly increase. New classes of sources were detected at GeV-TeV energies both galactic (e.g. Galactic Center, Pulsar, Wind Nebulae, Pulsars and Binary Systems) and extragalactic (e.g. Blazars, radio-galaxies, star-forming galaxies) as well as about a dozen of unknown new TeV sources. It showed for the first time that an array of IACTs could be properly used as a real astronomical observatory able to survey a large portion of the sky with a high sensitivity. The recent advances of TeV γ-ray astronomy have shown that the 10 GeV – 100 TeV energy band is crucial to investigating the physics prevailing in extreme conditions found in remote cosmic objects as well as to testing fundamental physics. Nevertheless, after the launch of two gamma-ray dedicated satellites (AGILE and Fermi), the gamma-ray astronomy is now living a sort of *Golden Age* and opening unprecedented opportunities of multiwavelength observations on a very wide energy range. In such an exciting scenario, a new generation of ground-based VHE gamma-ray instruments are needed in order to significantly improve the sensitivity, the operational bandwidth, the field of view and the angular resolution. The international VHE astrophysics community is moving towards such a new generation of Cherenkov experiments and now both in Europe and USA the Agencies and a large consortium of Institutes support the implementation of the Cherenkov Telescope Array (CTA) observatory [3]. CTA is conceived to allow both detection and in-depth study of large samples of known source types, and to explore a wide range of classes of suspected gamma-ray emitters beyond the sensitivity of current instruments. CTA will be a combination of the well proven technology of Cherenkov telescopes (with some tens deployed over a large area) and of new wide-field detectors and optics [4].

1.1. *CTA requirements and goals*

The CTA requirements and aims are reported hereafter:

- **Sensitivity:** CTA will be about a factor 10 more sensitive than any existing instrument (see Fig. 1). In its core energy range, from about 100 GeV to several TeV, CTA will have milliCrab sensitivity in 50 h of data taking, a factor 1000 below the strength of the strongest steady sources of very-high-energy gamma-rays, and a factor 10000 below the highest flux measured in bursts. This large dynamic range will allow the study of weaker and new type sources, reducing the selection bias in the taxonomy of known source types.
- **Angular resolution:** Current instruments are able to resolve extended sources can then be resolved, but they cannot probe the fine structures visible in other wavebands (as e.g. substructures of SNR shock fronts). Selecting a subset of gamma-ray induced cascades detected simultaneously by many telescopes, CTA will reach angular resolutions in the arc-minute range, a factor 5 lower than current instruments, from about 0.1 deg to 0.02 deg.
- **Temporal resolution:** With its large detection area, CTA will resolve flaring and time-variable emission on sub-minute time-scales, currently not accessible with present Cherenkov telescopes.
- **Energy range:** CTA is aiming to cover, with a single facility, almost four orders of magnitude in energy range (see Fig. 1), from a few tens of GeV up to 100 TeV. This will enable to distinguish between key hypotheses such as the electronic or hadronic origin of highest energy gamma-rays. Combined with the FERMI satellite gamma-ray observatory, the two instruments will provide an unprecedented coverage of more than 7 orders of magnitude in energy.

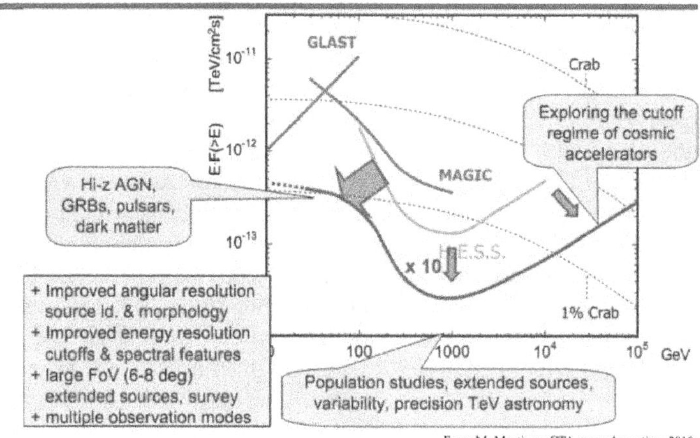

Fig. 1. Flux sensitivity versus photon energy for CTA.

- **Flexibility:** Consisting of a large number of individual telescopes, CTA can be operated in a wide range of configurations, allowing on the one hand the in-depth study of individual objects with unprecedented sensitivity, on the other hand the simultaneous monitoring of tens of objects (relevant for flaring sources) and any combination in between.
- **Survey capability:** Groups of telescopes can point at adjacent fields in the sky, with their fields of view overlapping, providing an increase of sky area surveyed per time unit by an order of magnitude, and for the first time enabling a full-sky survey at high sensitivity. The galactic plane should be completely explored at ~ 0.001 Crab in 250 hours and the full sky at ~ 0.01 Crab in 1 year.
- **Number of sources:** Extrapolating from the intensity distribution of known sources, CTA is expected to enlarge the catalogue of objects detected from currently 125 objects to beyond 1000 objects.
- **Global coverage and integration:** CTA aims to provide global coverage of the sky from two observatory sites: a Southern array for the exploration of the galactic plane and extragalactic sources and one Northern array mainly devoted to the study of extragalactic sources. For both sites a transparent access policy will be adopted and identical tools to extract and analyse data will be used.

CTA, will be, for the first time in this field, operated as a true observatory, open to the entire astrophysics and particle physics community, and providing support for easy access and analysis of data. Data will be made publicly available and, possibly, will be accessible through Virtual Observatory tools. The large amount of data obtained and open to public access will also allow data mining in addition to targeted observation proposals favouring multiwavelength studies.

2. The array configuration

In the atmospheric showers originated by a primary gamma-ray, the Cherenkov light intensity is almost proportional to the gamma-ray energy. In general, large pupil mirrors are needed to trigger low energy γ-rays, while small mirrors are sufficient enough to trigger high energy gamma-rays. Moreover, due to the very low values of the gamma-ray fluxes at high energy, future Cherenkov telescopes must be able to catch events with a core (impact point) very far from the telescope position, reaching, in this way, effective areas of the order of millions of square meters; to trigger far showers, imaged at large off-axis angles, Cherenkov telescopes must be provided by sufficiently large fields of view. Optical dish diameters and field of view are the first parameters to be considered before analyzing other specific aspects as optical design, mirrors structure, focal camera sensors and electronics. Due to the widespread of the spectral band, and to the phenomenological constraints three different specific set of telescopes with different characteristics are envisaged for CTA to cover the energy interval from 10 GeV up to 100 TeV and more:

- **very low energies** (< 200 GeV) need very large optical dishes (>20 m diameter). A few unit of 24 m telescopes (Large System Telescope – LST) with 4-5° FoV, 2000 -3000 pixels and an angular resolution of ~ 0.1° will be implemented.
- **in the extreme energy range** 10-100 TeV, the intensity of the Cherenkov signal allows one to image very far showers with small (< 10 m) optical dishes provided by a sufficient wide angle (~8-10° full). Forthy/Eighty small telescopes of 4-7 m diameter (Small System Telescopes, SSTs) with 8-10° FoV, 1000-2000 pixels and ~ 0.2° - 0.3° angular resolution, will be implemented, but just for the Southern array, since the high energy sources visible from the Earth are located in our Galaxy.
- **intermediate energy band** (0.2–10) TeV will be covered by telescopes with average characteristics for what concerns dish diameter and field of view. Twenty 12 m telescopes in each site, with 6-8° FoV, 2000 pixels and angular resolution of ~ 0.18° will for the array of medium size telescopes(MSTs).

An important technology development program is going on to study the implementation of each of the three kinds of telescopes to be realized and the associated subsystems, with particular reference to the structures [5,6], cameras [7] and mirrors [8]. In parallel Monte Carlo simulations are being carried out in order to verify the telescope performances as a function of the technological solutions and trade-offs [9].

3. Summary

The Cherenkov Telescope Array (CTA) is presently the worldwide project for the next generation of ground-based Cherenkov Telescopes for Very High Energy (VHE) gamma-ray astronomy. The CTA project started in 2010 a three-year EU-funded Preparatory Phase with the goal of being ready to start construction by the end of 2013; the entire international community, and in particular the US groups, are participating in the study. For that phase the vast amount of activities needed is organized in a number of Work Packages whose work is advancing. Prototypes of components, systems and even some telescopes are being constructed, which shall allow the final decisions to be made and construction to begin in 2013, with aim of completing the deployment of both arrays by 2018. CTA will then be the major observatory in VHE gamma-ray astronomy, combining guaranteed astrophysics and physics returns with significant discovery potential.

Acknowledgments

We gratefully acknowledge financial support from the agencies and organizations listed in this page: http://www.cta-observatory.org/?q=node/22.

References

1. J.A. Hinton & W. Hofmann, Ann. Rev. Astron. Astrophys., 47, 523 (2010)
2. T.C. Weekes, AIP Conf. erence Proc., Vol. 1085, pp. 3-17 (2008)
3. W. Hoffman and M. Martinez for the CTA Consortium, Design Concepts for CTA, *arXiv:1008.3703v2 (2009)*
4. M. Martinez, "CTA: where do we stand and where do we go ?," Proc. of 32nd Int. Cosmic Ray Conference, in press (2011)
5. M. Teshima, CTA-LST Large Size Telescope, Proc. of 32nd Int. Cosmic Ray Conference, in press (2011)
6. R. J. White et al., Telescopes for the High Energy Section of the Cherenkov Telescope Array, Proc. of 32nd Int. Cosmic Ray Conference, in press (2011)
7. G. Puehlhofer et al., FlashCam: A concept and design of a fully digital camera for the Cherenkov Telescope Array CTA, Proc. of 32nd Int. Cosmic Ray Conference, Beijing, in press (2011)
8. A. Foerster et al., Mirror Development for CTA, Proc. of 32nd Int. Cosmic Ray Conference, in press (2011)
9. F. Di Pierro et al., Performance studies of the CTA observatory, Proc. of 32nd Int. Cosmic Ray Conference, in press (2011)

PoGOLite - A BALLOON-BORNE X-RAY POLARIMETER

M. PEARCE*

KTH, Department of Physics; and the Oskar Klein Centre, AlbaNova University Centre,
106 91 Stockholm, Sweden
**E-mail: pearce@kth.se*
www.particle.kth.se

For the PoGOLite Collaboration

PoGOLite is a balloon-borne soft gamma-ray (X-ray) polarimeter operating in the 25-80 keV energy band. The polarisation of incoming photons is determined using Compton scattering and photo-absorption events reconstructed in an array of plastic scintillator detector cells surrounded by a BGO side anti-coincidence shield and a polyethylene neutron shield. Observations take place from a stratospheric balloon operating at an altitude of ~40 km. A custom attitude control system keeps the polarimeter field-of-view aligned to targets of interest. The maiden 'pathfinder' flight of PoGOLite took place from the Esrange Space Centre in July 2011.

Keywords: X-ray polarisation; scientific ballooning; Crab pulsar; Cygnus X-1.

1. Introduction

The field of X-ray astronomy was born in 1962 with observations of Scorpius X-1 from an Aerobee sounding rocket. Observations outside of the Earth's atmosphere were key to this endeavor since X-rays are readily absorbed by the Earth's atmosphere. Since then a large number of instruments have been flown on sounding rocket, balloon and satellite platforms. Common to almost all missions is that the incoming photon flux is characterised by its energy, arrival time and direction on the sky. The polarisation of the photon flux is usually not measured directly.

Polarised X-rays are expected from the high-energy processes at work within compact astrophysical objects such as pulsars, accreting black holes and jet-dominated active galaxies.[1] Polarisation is also expected to provide valuable information regarding the processes underlying solar flares[2] and gamma-ray bursts.[3] The polarisation arises naturally for synchrotron radiation in large-scale ordered magnetic fields and for photons propagating through a strong magnetic field. Polarisation can also result from anisotropic Compton scattering. In all cases, the orientation of the polarisation plane and degree of polarisation is a powerful probe of the physical environment around compact astrophysical sources.

Despite the wealth of sources accessible to polarisation measurements and the importance of these measurements, there has been only one successful mission with dedicated instrumentation. The Crab nebula was studied at 2.6 keV and 5.2 keV using a Bragg reflectometer flown on the OSO-8 satellite in 1976.[4] Measurements making inventive use of the IBIS and SPI instruments on-board the Integral satellite have reinvigorated the field of late, with polarisation measurements and limits reported for both the Crab and Cygnus X-1.[5–7] It is noted, however, that the IBIS and SPI instruments were not designed for polarimetric measurements and that their response to polarised radiation was not studied prior to launch.

PoGOLite[8] is a balloon-borne soft gamma-ray (25 - ~100 keV) polarimeter which is optimised for the study of point-like astrophysical objects. It is expected that as low as 10% polarisation can be measured from 200 mCrab sources during a 6 hour long balloon-borne observation at an altitude of approximately 40 km.

2. Measurement principle and polarimeter design

The polarisation of photons with an energy between approximately 25 keV and 1 MeV can be measured using Compton scattering.[9] The differential cross-section for Compton scattering is defined by the Klein-Nishina equation which states that the azimuthal scattering angle, ϕ, of a photon will be modulated as $\cos^2 \phi$ with maxima perpendicular to the polarisation plane of the photon. In PoGOLite, the scattering angle is defined by a Compton scattering followed by a photoelectric absorption in a hexagonal close-packed array of plastic scintillators. A schematic overview of the PoGOLite instrument is shown in figure 1 (left). Compton scattering and photo-absorption events are identified in an array of phoswich detector cells (PDC), made of plastic and BGO scintillators and surrounded by a BGO side anticoincidence shield (SAS). The full-size PoGOLite instrument consists of 217 PDC units. Initially, a reduced volume 'pathfinder' instrument comprising 61 PDC units will be evaluated in flight as shown in figure 2. Each PDC is composed of a thin-walled tube (well) of slow plastic scintillator (fluorescence decay time ~285 ns), a solid rod of fast plastic scintillator (decay time ~2 ns), and a short bismuth germanate oxide (BGO) crystal (decay time ~300 ns), all viewed by one photomultiplier tube (PMT). The wells serve as a charged particle anticoincidence, the fast scintillator rods as active photon detectors, and the bottom BGOs act as a lower anticoincidence. Each well is sheathed in thin layers of tin and lead foils to provide passive collimation.

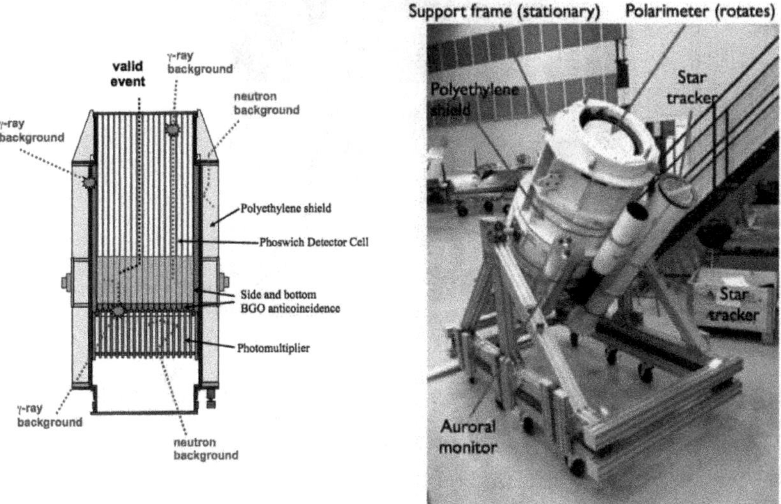

Fig. 1. **Left:** A schematic overview of the PoGOLite polarimeter. The overall length is approximately 1.5 m. The main components are identified along with possible signal and background interactions. The volume below the photomultipliers houses data acquisition electronics, including a fluid-based cooling system. The polarimeter components are located in a pressure vessel assembly which rotates around the viewing axis. **Right:** The polarimeter assembly mounted in a transport wagon, prior to integration with the attitude control system. Two star trackers and an auroral monitor are attached to the body of the polarimeter. Due to the relatively high latitude (67°N) of the pathfinder flight, there is a potential background from polarised X-rays arising from auroral bremsstrahlung at an altitude of 90-100 km.[10]

Gamma-rays entering within the field of view of the instrument (2 degrees × 2 degrees (FWHM), defined by the slow plastic scintillator tubes) will hit one of the fast plastic scintillators and may be Compton scattered, with a probability that depends on the photon energy. The scattered photon may escape, be photo-absorbed in another detector cell, or undergo a second scattering. Electrons resulting from a photo-absorption will deposit

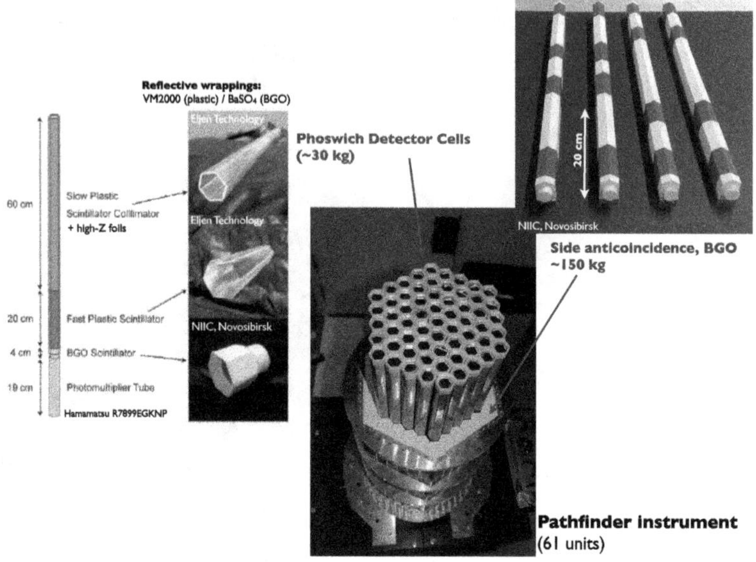

Fig. 2. The pathfinder polarimeter comprises 61 Phoswich Detector Cells (PDCs) surrounded by a segmented BGO anticoincidence shield. The composition of the PDCs is shown (left) as well as the side anticoincidence units (right) which comprise 3 BGO crystals, each of length 20 cm.

their energy in the plastic scintillator and produce a signal at the PMT. A trigger based on the photoelectron energy deposit initiates high speed (37.5 MHz) waveform sampling (15 pre- and 35 post-trigger samples) of PMT outputs from all PDCs. Online vetoes can be selected based on an upper discriminator level to reject cosmic ray events and pulse-shape discrimination to reject off-axis events with a slow scintillator component. The locations of the PDCs in which the Compton scatter and photo-absorption are detected determine the azimuthal scattering angle. The geometry of the PDC arrangement limits the polar scattering angle to approximately (90±30) degrees, roughly orthogonal to the incident direction. Little of the energy of an incident gamma ray photon is lost at the Compton scattering site, while most of the energy is deposited at the photo-absorption site. In spite of the relatively poor energy resolution of plastic scintillator, it is straightforward to differentiate Compton scattering sites from photo-absorption sites. The polarisation plane can be derived from the azimuthal distribution of scattering angles. The degree of polarisation (%) is determined from the ratio of the measured counting rate modulation around the azimuth to that predicted for a 100% polarised beam (from simulations calibrated with experiments at polarised photon beams). The polarimeter assembly is placed inside a pressure vessel system which is in turn housed inside a structure which allows the polarimeter to rotate along the viewing axis. This allows instrument asymmetries to be systematically studied during flight. The fully assembled pathfinder polarimeter is shown in figure 1 (right).

3. Performance studies

The PoGOLite design has been optimised using computer simulations, data collected in the laboratory with radioactive sources[11] and at accelerator facilities using polarised photon beams[12] and particle beams. As an example, studies of the pathfinder polarimeter using ~60 keV photons from a ^{241}Am radioactive source are described here. The response of the polarimeter is quantified in terms of the modulation factor which describes the amplitude of the counting rate asymmetries reconstructed in the polarimeter. The response to an unpolarised beam is shown in figure 3. The modulation curve is constructed for a pair-wise combination of detector cells in order to reduce the effect of alignment systematics. The resulting modulation factor is compatible with zero, which is important given that polarisation is a positive definite measurement in the presence of unknown systematics. A polarised beam was generated by Compton scattering the ^{241}Am beam through 90 degrees in a small plastic block. The resulting scattered beam will be approximately 100% polarised with an energy of ~53 keV. An example modulation curve obtained during this study is also shown in figure 3.

Figure 4 shows the background contributions to a Crab observation by the PoGOLite pathfinder as simulated using Geant4. The minimum ionising particle signature in scintillators arising from charged cosmic-ray backgrounds is readily reduced using a simple pulse height analysis. A background due to neutrons (mostly albedo) generated by cosmic-ray interactions with the Earth's atmosphere is seen to dominate. Neutrons in the energy range 500 keV to 2 MeV are the main component of the false trigger rate, with a flux of 0.7 neutrons $s^{-1}cm^{-2}$ predicted below 10 MeV. For this reason a polyethylene shield (with thickness 15 cm around the scattering scintillators) has been added to the PoGOLite design. This reduces the neutron background by an order of magnitude. In order to measure residual background during flight, a dedicated neutron detector comprising

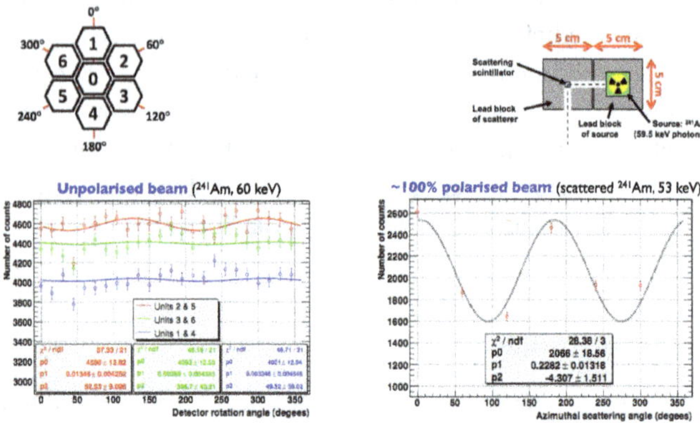

Fig. 3. **Left:** Modulation curves obtained in the laboratory when illuminating the central detector cells of the pathfinder polarimeter shown in figure 2 with unpolarised ~60 keV photons from an ^{241}Am radioactive source. During measurements the polarimeter is rotated around the viewing axis. Opposite pairs of PDCs are considered in order to reduce the effect of alignment systematics. No significant overall modulation is found. **Right:** The modulation curve obtained when illuminating the central detector cells of the pathfinder polarimeter with ~100% polarised ~53 keV photons from an ^{241}Am radioactive source. A clear modulation signal is obtained.

a 5 mm thick $LiCaAlF_6$ crystal with 2% Eu doping complemented with BGO anticoincidence is mounted in the vicinity of the fast scattering scintillators.[13] The ^{6}Li in the crystal has a large neutron capture cross-section (940 barn) and is sensitive to neutrons with energies <10 MeV. The photon background arises from both atmospheric and galactic sources. This background is suppressed due to the narrow field-of-view provided by the PDC design combined with the segmented anticoincidence system.

4. Attitude control system (ACS)

The ACS keeps the viewing axis of the polarimeter aligned to the sidereal motion of observation targets and also compensates for local perturbations such as flight-train torsion and changing stratospheric winds. Pointing to within ~5% of the polarimeter field-of-view (corresponding to ~0.1°) is required to secure a minimum detectable polarisation of better than 10% for a 1 Crab source during the pathfinder flight during a 6 hour long ob-

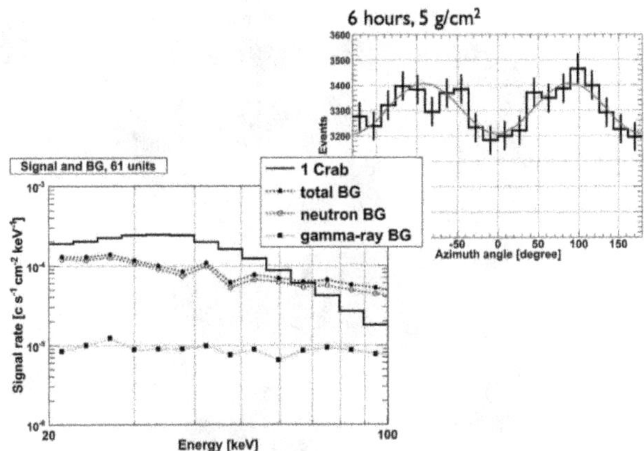

Fig. 4. **Left:** Signal and background energy spectra obtained from Geant4 simulations of the pathfinder polarimeter. The atmospheric neutron background is found to dominate. **Right:** The modulation curve expected from a 6 hour long observation of the Crab with 5 g/cm^2 of residual atmosphere.

servation. The polarimeter is shown mounted in the ACS gimbal assembly in figure 5. Custom torque motors[14] act directly on the polarimeter elevation axis. Azimuthal positioning is achieved with a flywheel assembly which connects to the flight train through a momentum dump motor which allows angular momentum stored in the flywheel to be reset upon saturation. Control signals to the motor systems are generated by a real-time computer system which monitors the attitude sensors, comprising a differential GPS system, a micromechanical accelerator/gyroscope package, angular encoders, an inclinometer and a magnetometer. These primary attitude sensors are augmented by two star trackers. One star tracker is developed from a design successfully used on previous missions[15] and has a field-of-view of 2.6°×1.9°. Ground-based tests[16] have shown that stars down to 10th magnitude can be resolved assuming the background light conditions at 40 km.[17] The second star tracker is a more compact design with in-house designed optics providing a field-of-view of 5.0°×3.7°. Both trackers require a stabilized gondola to operate (provided by the differential GPS system and gyroscopes) and can either operate in a star pattern matching mode (providing relatively slow, but absolute position fixes to the ACS) or can be commanded to lock onto a bright star providing ACS updates at a rate of order 1 Hz. The latter case is foreseen to be the primary operating mode during flight.

Fig. 5. The PoGOLite attitude control system. The polarimeter is mounted in a gimbal assembly where custom direct drive torque motors act on the polarimeter elevation axle. The balloon flight train connects to the gimbal assembly through a combined momentum dump/flywheel system which is used for azimuthal positioning.

5. Gondola and ancillary systems

As shown in figure 6, the polarimeter and ACS are installed in a gondola assembly (designed by SSC Esrange). The upper part of the gondola is connected to the ACS gimbal and also provides a mounting point for the cooling system radiator and pump. The lower section houses batteries, power control electronics and communications equipment. The gondola frame is covered in lightweight honeycomb panels which enhance the structural rigidity and help protect the polarimeter from damage during landing. Two glass-fibre booms are attached to the upper gondola. The booms span 10 m and have a GPS antenna mounted at each end. The antenna separation provides the required baseline for the differential GPS system alone to meet the $\sim 0.1°$ pointing accuracy requirement. Other GPS antennae for Esrange flight systems, Iridium antennae for over-the-horizon communications and magnetometers for the ACS and auroral monitor are also mounted on the booms. Beneath the lower gondola, a four-sided 'skirt' of solar panels (each side comprising 5 panels, with overall dimensions (3.5×1.5) m is mounted along with landing crash pads, ballast hoppers and E-Link (ethernet over radio) communication antennae. The gondola stands approximately 5.2 m tall and has a flight-ready mass of approximately 2 T.

Fig. 6. The complete gondola assembly configured for flight.

6. Pathfinder flight

The PoGOLite pathfinder payload was launched from the Esrange Space Centre at approximately 02:00 (local time) on July 7th 2011. The primary observation goals of the pathfinder mission were the Crab and Cygnus X-1. A flight duration of approximately 5 days was foreseen with multiple target observations combined with background studies. Flight termination was tentatively planned for the vicinity of Victoria Island, Northern Canada. Unfortunately, the flight did not develop as expected. At ~04:30 the balloon altitude leveled out at 35 km and at ~05:20 the altitude started to decrease due to a suspected balloon leak. The decision to terminate the flight was taken soon after. During the short time aloft, it was only possible to make initial function tests, but both the polarimeter and attitude control system were found to operate as expected. For example, during the final stages of the flight it was possible to point the polarimeter at Cygnus X-1 and confirm the pointing solution using guide stars. It was also possible to confirm that scattering photon events could be identified. Since the altitude was very low at this point (33 km and dropping), the registered photons were of atmospheric origin. At ~07:20 the balloon was cut from the PoGO-Lite gondola close to Sweden's highest mountain, Kebnekaise. The gondola touched down by parachute at 08:01 and was subsequently recovered close to the village of Nikkaluokta and returned to Esrange. An investigation into the cause of the balloon failure is currently underway. The gondola suffered relatively minor structure damage and did an excellent job in protecting the scientific payload and ancillary equipment from more extensive damage. The solar cell array hanging under the gondola attracted significant damage and the majority of the panels will need to be replaced. Tests and refurbishment of the payload components is now underway. It is foreseen to re-fly the PoGOLite pathfinder during summer 2012.

References

1. R. Bellazzini et al. (editors), *X-ray polarimetry. A new window on astrophysics*, Cambridge Contemporary Astrophysics, Cambridge University Press (2010).
2. T. Bai and R. Ramaty, Ap. J. 219 (1978) 705. J. Leach and V. Petrosian, Ap. J. 269 (1983) 715.
3. K. Toma et al., Ap. J. 698 (2009) 1042.
4. M.C. Weisskopf et al., Ap. J. 208 (1976) L125. M.C. Weisskopf et al., Ap. J. 220 (1978) L117.
5. M. Forot et al., Ap. J. 688 (2008) L29.
6. A.J. Dean et al., Science 321 (2008) 1183.
7. P. Laurent et al., Science 332 (2011) 438.
8. T. Kamae et al., Astroparticle Physics 30 (2008) 72.
9. F. Lei et al., Space Sci. Rev. 82 (1997) 309.
10. S. Larsson et al., ESA-SP-647, ESAPAC Proceedings, Visby, Sweden (2007) 513.
11. M. Kiss, KTH doctoral thesis (2011). `http://kth.diva-portal.org`.
12. Y. Kanai et al., Nucl. Instr. and Meth. A 570 (2007) 61.
13. H. Takahashi et al., IEEE NSS MIC Conference Record (2010).
14. J-E. Strömberg, Proc. 20th ESA-PAC Symposium (2011).
15. M. Rex et al., Proc. SPIE 6269 (2006) 62693H.
16. C. Marini Bettolo, KTH doctoral thesis (2010). `http://kth.diva-portal.org`.
17. K.L. Dietz et al., Optical Engineering 41(2002) 26.

Fermi-LAT analysis of Fermi, Planck, Swift and radio selected samples of AGN

S. Rainó*[1], F. Gargano[1], C .Monte[1], S. Cutini[2], D. Gasparrini[2]

for the *Fermi*-LAT Collaboration
*E-mail: silvia.raino@ba.infn.it

J. Leon Tavares[3], G. Polenta[2,4]

for the Planck Collaboration

[1]*Istituto Nazionale di Fisica Nucleare - Sezione di Bari, I-70126, Bari, Italy*

[2]*Agenzia Spaziale Italiana (ASI) Science Data Center, I-00044 Frascati (Roma), Italy,*

[3]*Aalto University Metsähovi Radio Observatory, Metsäovintie 114, FIN-02540 Kylml, Finland*

[4]*INAF - Osservatorio Astronomico di Roma, I-00044, Monte Porzio Catone (Roma), Italy*

Blazars are jet-dominated extragalactic objects characterized by the emission of strongly variable non-thermal radiation across the entire electromagnetic spectrum. The use of multi-frequency simultaneous data is essential in order to understand the physical processes that take place in these objects. It is now possible to assemble high-quality multi-frequency simultaneous broadband spectra of large and statistically well-defined samples of radio-loud AGN, thanks to Planck, Fermi and Swift simultaneously on orbit, complemented with other space and ground-based observatories. For this study, we have selected four samples of sources. The first three samples are flux limited in the high energy part of the electromagnetic spectrum: the soft X-ray (0.1-2 keV) sample includes 43 sources from the Rosat All Sky Survey Bright Source Catalog, the hard X-ray (15-150 keV) sample includes 34 sources from the Swift-BAT 54 months source catalog and the gamma-ray sample includes 50 sources from the Fermi-LAT 3 months Bright AGN Source List. The fourth sample is radio flux limited, including 104 bright northern and equatorial radio-loud AGN (most of which have been monitored at Metsahovi Radio observatory for many years) with average radio flux density at 37 GHz greater than 1 Jy. We present the methods applied and the results of the analysis performed using Fermi-LAT data for all sources in the four different samples of AGN.

1. Introduction

Blazars are well-known jet-dominated extragalactic objects characterized by the emission of strongly variable and polarized non-thermal radiation across the entire electromagnetic spectrum, from radio to high energy γ-rays. The strong emission of blazars at all wavelengths makes them the dominant type of extragalactic sources in the radio, μ-wave, and γ-ray bands where the accrection and other thermal emission processes do not produce significant amounts of radiation. For these reasons blazars are hard to find at optical and X-ray frequencies, while dominating the newly explored μ-wave and γ-ray high Galactic latitude sky. The advent of the Fermi[1] and, more recently, the Planck satellite,[2] capable of probing deeply these two last observing windows, combined with the versatility of the Swift observatory,[3] and the observations by a number of ground based observatories, is giving us the unprecedented opportunity to collect multi-frequency data for very large samples of blazars in order to assemble simultaneous broad-band spectra.

2. The samples

In order to explore the blazars' parameters space from different viewpoints we have adopted different criteria to select the list of blazars to be observed simultaneously by Planck, Swift and Fermi. The first three samples of blazars are flux limited in the high energy part of the electromagnetic spectrum. The soft X-ray flux limited sample, including 43 sources, was defined starting from the Rosat All-Sky Survey Bright Source Catalog (1RXS[4]), selecting all the blazars with count rate larger than 0.3 counts/s in the 0.1-2.4 keV energy band, and radio flux density S_{5GHz} >200 mJy. The hard X-ray flux limited sample, including 34 sources, was defined starting from the *Swift*-BAT 54 month source catalog,[5] selecting all blazars with X-ray flux $>10^{-11}$ erg cm^{-2}s^{-1} in the 15-150 keV energy band and radio flux density S_{5GHz} >100 mJy. The γ-ray flux limited sample, including 50 sources, was created starting from the *Fermi*-LAT Bright Source List,[6] selecting all the high galactic ($|\ b\ |>10°$) blazars detected with high significance (TS>100), with a flux cut F(E>100 MeV)$>8\times10^{-8}$ph cm^{-2}s^{-1} and radio flux density S_{5GHz} >1 Jy. The last sample,[7] including 104 sources, is the radio flux density limited sample.[4] It includes all northern and equatorial radio-loud AGN with declination $\geq$-10° that have a measured average radio flux density S_{37GHz} >1 Jy.

3. *Fermi*-LAT data analysis

A period from August 4, 2008 to November 4, 2010 has been analyzed selecting for each source only photons above 100 MeV, belonging to the diffuse class[1] which have a low probability of background contamination. For each source, we selected only photons within a 15° Region of Interest (RoI) centered around the source itself. The data were analyzed with a binned maximum likelihood technique[8] using the analysis software (gtlike) developed by the LAT team. A model accounting for the diffuse emission as well as for the nearby γ-ray sources is included in the fit. For the evaluation of the γ-ray SEDs, the whole energy range from 100 MeV to 300 GeV is divided into 2 equal logarithmically spaced bins per decade. In each energy bin the standard gtlike binned analysis has been applied assuming a power law spectrum for all the point sources in the model, with photon index fixed to 2. The flux of the source in all selected energy bins is evaluated, requiring in each bin a Test Statistics (TS) greater than 10 and the ratio between the flux and its error greater than 0.5. If these conditions are not satisfied, an upper limit (UL) is evaluated in that energy bin. For each source, we have considered three different integration periods:

(1) simultaneous observations, data accumulated during the period of Planck observation of the source;
(2) quasi-simultaneous observations, data integrated over a period of two months centered on the Planck observing period of the source;
(3) 27 month Fermi integration, data integrated over a period of 27 months from August 4, 2008 to November 4, 2010.

4. Spectral Energy Distributions

The plot of radio to γ-ray flux distributions in the Logν-LogF_ν, widely known as a Spectral Energy Distribution (SED), is a powerful method of studying the physics of blazars. Figures 1 and 2 report the SEDs respectively of PKS 1124-186 (from the soft X-ray sample), PKS B1830-210 (from the hard X-ray sample), PKS 1502+106 (from the γ-ray sample) and, finally PKS 1510-089 (from the radio sample). In all plots, red filled points (or UL) show simultaneous multi-frequency data, green points (or UL) show γ-ray data integrated over a period of 2 months centered on the Planck observing period, or ground-based data taken quasi-simultaneously and blue points (or UL) show Fermi data integrated over 27 months; literature or archival data are shown in light gray.

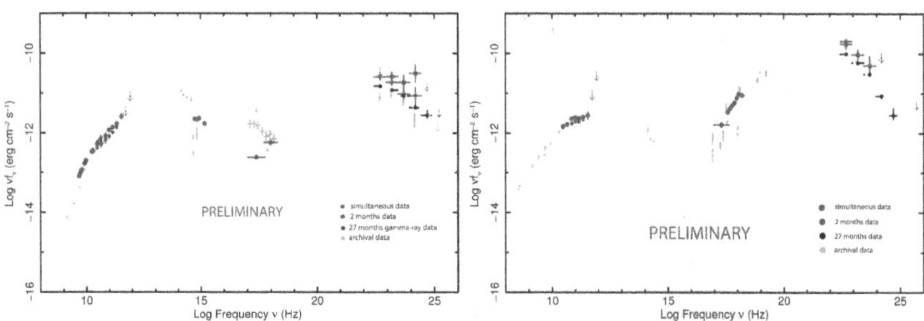

Fig. 1. Left: SED of PKS 1124-186 from the soft X-ray sample. Right: SED of PKS B1830-210 from the hard X-ray sample.

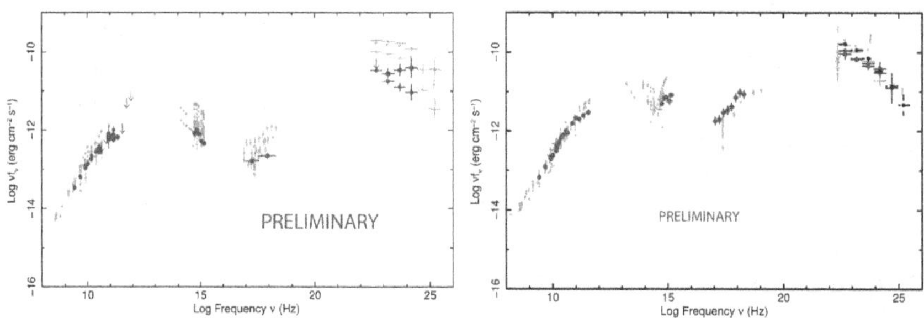

Fig. 2. Left: SED of PKS 1502+106 from the γ-ray sample. Right: SED of PKS 1510-089 from the radio sample.

5. Conclusions

We have collected Planck, Swift, Fermi and ground based simultaneous multi-frequency data for a great number of blazars included in four statistically well defined samples. The acquisition of this unprecedented multifrequency multi-satellite data set was used to build well sampled simultaneous SEDs. The comparison between our simultaneous data with literature archival measurementes shows that SEDs built with non-simultaneous data suffer from uncertainties in the μ-wave region that are relatively modest and generally limited to about a factor 2 while the high energy part of the spectrum is much more affected with uncertainties due to flux variatio up to a factor of ten or more.

References

1. Abdo A. et al., 2009, ApJ, 697,1071
2. Tauber J. et al, 2010, A&A, 520, A1
3. Gehrels N. et al., 2004, ApJ, 611, 1005
4. Voges W. et al., 1999, A&A, 349, 389
5. Cusumano G. et al., 2010, A&A, 524, A64
6. Abdo A.A. et al., 2009, ApJ, 700, 597
7. Planck Coll. et al., 2011, arXiv:1101.2047v1
8. Mattox J. Et al., 1996, ApJ, 461, 396

Fermi results in cosmic ray physics

R. RANDO

Dipartimento di Fisica, Università di Padova,
& INFN, Sezione di Padova,
via Marzolo 8,
Padova, 35131, Italy,
**E-mail: riccardo.rando@pd.infn.it*

on behalf of the Fermi LAT Collaboration

Since the beginning of its operations in Summer 2008 the *Fermi* Gamma-Ray Space Telescope has been monitoring a vast number of high energy astrophysical sources, among which several are commonly thought to be plausible sources of cosmic rays (CRs). In addition to the gamma emission from individual sources like supernova remnants, *Fermi* has been observing γ rays from the interaction of CRs with the galactic medium, while locally it measures with outstanding accuracy the flux of primary electrons in the Earth orbit up to 1 TeV. After three years of operation the collected information gives novel insights to the long standing puzzle of the CR origin.

Keywords: gamma rays, cosmic rays, gamma rays: observations, telescopes.

1. Introduction

In the following sections we describe the *Fermi* Large Area Telescope (LAT) results in the field of CR physics. In Sec. 2 we briefly introduce the instrument; in Sec. 3 we describe direct measurements of the local abundances of CRs; in Sec. 4 we describe some observations of diffuse γ-ray emission produced by CR interactions in the Galaxy, and in Sec. 5 in other galaxies.

Unless otherwise specified, in this text with "electrons" we refer to both electron and positrons, resorting to e^+ and e^- if we need to differentiate.

2. The Fermi Large Area Telescope

The *Fermi* Large Area Telescope (LAT) is a pair-conversion γ-ray telescope, designed to detect photons in the range from $\sim$ 20 MeV up to more than 300 GeV.[1] The LAT tracker includes tungsten (W) foils to increase the probability of a γ ray converting into an electron-positron pair within its volume, and 36 layers of silicon strip detectors to detect the passage of the secondary ionizing particles. Below the tracker, a CsI imaging calorimeter measures the energy deposited by the electromagnetic shower and allows to fully reconstruct the energy of the impinging γ rays. A plastic scintillator AntiCoincidence Detector surrounds the tracker to veto the events caused by the passage of a charged particle.

The LAT onboard *Fermi* is orbiting at 565 km a.s.l. with an inclination of 25.6°, immersed in a dense background of energetic charged particles that need to be recognized and rejected at all levels: at the trigger level, to maximize livetime; at the downlink level, to preserve bandwidth; at the event analysis level, to minimize background counts in science analysis. While all stages were designed to fulfill the main purpose of the LAT, namely the collection of the incoming γ rays by rejecting everything else, for calibration and diagnostic purposes a certain amount of other particles are purposefully allowed into the data stream.

At a very early time it was recognized how the LAT can perform as an extraordinary detector of other energetic particles, especially electrons, whose shower development closely follows that of γ rays. This led to a further tuning of the trigger and filter stages to optimize the downlinked charged-particle samples for science analysis. Notably all triggers with an energy deposition in the calorimeter above 20 GeV are collected (*high-pass threshold*) because above this threshold the impact on the data transmit bandwidth is negligible.

3. Measurement of local cosmic rays

As described in Sec. 2, the LAT is capable of performing direct measurements of CRs present along its orbit. We will describe in the following the LAT measurement of the electron component of the CR flux. The LAT electron analysis has large similarities with the γ-ray analysis, and is thus at a very mature stage. On the other hand, a direct measurement of the hadronic component is complicated, primarily because of the poor LAT energy resolution for hadronic cascades, and it is at present not complete. An early LAT indirect measurement of the proton flux at the Earth through the secondary γ-rayemission reports agreement with expectations from previous observations.[2] Other observations strongly related with the local abundances of CRs can be found in Ref. 3.

3.1. *Direct measurement of cosmic rays electrons*

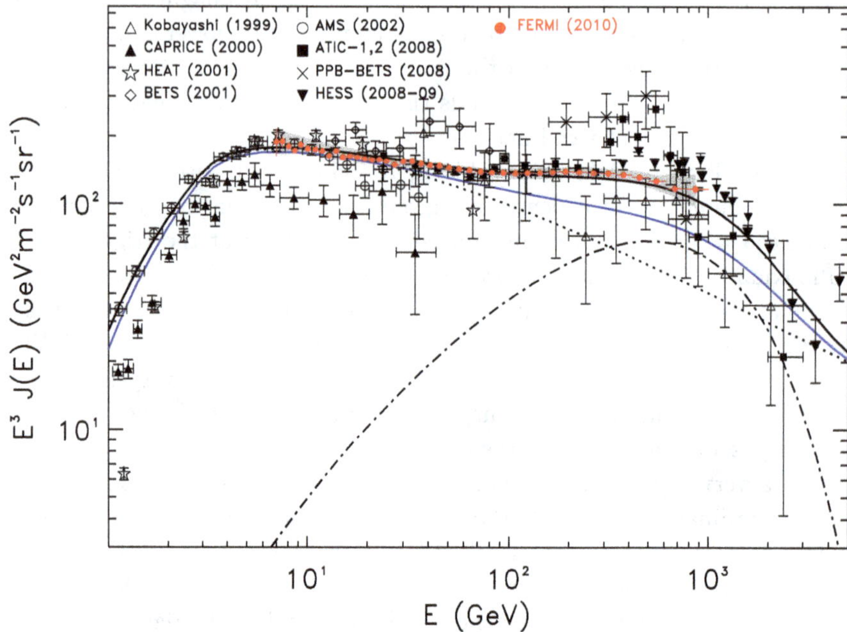

Fig. 1. The $e^{+}+e^{-}$ spectrum measured by the *Fermi* LAT above 7 GeV.[4] full circles: LAT data; blue line: GALPROP model: dotted line: conventional GALPROP model tuned to reproduce the LAT data; dashed line: additional nearby CR source.

As already mentioned, an electromagnetic shower initiated by a CR electron closely matches that of a γ ray. Above the *high-pass* threshold of the onboard filter all triggers are sent to ground, making it relatively easy to model the LAT behavior and to assess the systematics of the analysis. An early report of the LAT observations of primary electrons[5] indicated a harder spectrum than what was assumed by commonly-used CR propagation models. A further effort allowed us to decrease the low energy threshold to 7 GeV by using an additional diagnostic trigger channel and a dedicated analysis to disentangle the effect caused by the modulation due to the Earth's magnetic field;[4] in Fig. 1 the latter LAT measurement is shown.

While all LAT data can be fitted with a simple power law, the spectral shape suggests a hardening just below 100 GeV. To properly render the effect a model is proposed, composed by two components. A conventional

GALPROP model (i.e. not tuned to reproduce LAT γ-ray data) easily reproduces the low energy part; at high energies an additional source of CR electrons is included. Inclusion of an additional source is motivated by the fact that at $\sim$ 100 GeV primary electrons observed at the solar system must be produced nearby (within $\sim$ 1.6 kpc) to survive the strong radiative losses during their propagation. Therefore, at high energy the flux should be dominated by a few, close sources, and the stochasticity of source location and power naturally affect the smoothness of the electron spectrum. Possible nearby sources range from pulsars to local clumps of dark matter; more details can be found in Ref. 4.

3.2. *Cosmic rays electron anisotropy*

In Sec. 3.1 we mentioned that inclusion of a nearby source of CRs would naturally reproduce the spectrum observed by the LAT. Such a close source could cause a measurable amount of anisotropy in the electron flux, and the location of an anisotropy in the skymap could be an indication of the underlying CR accelerator.

The search for a percent level or lower anisotropy in the CR flux is significantly complicated by the LAT behavior: the scanning strategy and the power-down in correspondence of the South Atlantic Anomaly imply some small but noticeable disuniformity in the exposure of different regions of the sky. Even if this is compensated by dividing the flux by the estimated geometric factor, small uncertainties in the LAT efficiency and in its temporal stability would affect the observation of any astrophysical effect. To prevent this, the LAT analysis is based on the construction of an electron skymap under the assumption of no anisotropy (*null hypothesis*) using the data themselves to properly sample the *real* (unknown) geometric factor and its possible variations.

Currently published results[6] are based on $\sim$ 2 years of data taking. No signal is observed; under the assumption of a dipole morphology we probe an anisotropy δ going from $\sim$ 0.004 at lower energies (50 GeV) up to $\sim$ 0.1 at 500 GeV. The uncertainties on the magnitude of a signal caused even by well known pulsars (i.e. Vela) are so large that we currently cannot exclude any scenario.

3.3. *Electrons and Positrons*

PAMELA results concerning the unexpected increase of the positron fraction above a few GeV[7] have generated a lively amount of speculations. The

LAT is not equipped with a magnet to allow charge discrimination, so conventional separation of e^- and e^+ fluxes is not possible. It is well known though that the geomagnetic field can be used to discriminate e^- and e^+: the geomagnetic cutoff and the east-west difference in flux are modulated to a measurable extent.[8]

For our measurement of the e^+ flux between 20 and 200 GeV we employ the Earth shadow.[9] Primary e^- can reach the LAT if their trajectories do not cross the Earth (including the denser regions of the atmosphere); looking at an electron count skymap, the Earth shadow is moved westwards by a significant amount due to the terrestrial magnetic field. The same is true for e^+, and the Earth shadow is moved eastwards. This implies there is a region to the west where we find only primary e^+, and one to the east where we find only primary e^-; around each there is a limb region that has to be excluded being rich with secondary e^- and e^+ produced by CRs interacting in the Earth atmosphere. Of course rejection of the large proton background complicates this picture considerably.

Background rejection, acceptance and systematics are estimated both via Monte Carlo simulations selection and by a template fitting technique using signal and background distributions to separate the different contributions. The outcome of this analysis is a total electron spectrum in excellent agreement with the LAT results described in Sec. 3.1; the positron fraction is found to increase in good agreement with PAMELA results. We note that this is the first time the absolute e^+ spectrum has been measured above 50 GeV, and the first time the e^+ fraction has been measured above 100 GeV.

4. Cosmic rays in our Galaxy

While direct measurement of the primary CRs is possible only locally, indirect observations greatly extend the range accessible to our investigation. The LAT detects the diffuse γ-ray emission caused by CRs interacting with the interstellar medium and radiation field, probing the density of CRs across the entire Galaxy. In addition, the LAT can resolve individual Galactic sources of γ rays, e.g. supernova remnants (SNRs), considered among the most probable sites for CR acceleration; investigation of the properties of these regions sheds a new light on the mechanisms giving birth to high-energy CRs. For more details on galactic LAT observations see Ref. 10 in these proceedings.

4.1. *Diffuse emission from CR interactions*

Of particular interest is the problem of the distributions of CRs in our Galaxy. γ rays are an excellent probe of the CR interactions with the interstellar medium and radiation field, as they are not deflected by the galactic magnetic fields and preserve the spatial structure of the emission regions.

Being the amount of γ rays produced a function of the CR density and of the density of the target medium, since EGRET we are aware of the advantages of decomposing the γ-ray emission in terms of spatial templates of the galactic hydrogen. The amount of the target gas mass at different galactocentric radii can be derived by radio observations, and by assuming that at every location the γ-ray emissivity is proportional to the gas mass and to the CR abundance, the latter can be traced all over the Galaxy. The *Fermi* LAT collaboration has reported on the Galactic γ-ray emission from several regions.[11–14]

Among all the various topics covered, we mention only one related with the distributions of CRs in the outer Galaxy. γ-ray emissivity is expected to decrease beyond the solar circle; on the contrary, our observations indicate it is flat within systematic uncertainties. Possible explanations vary from a CR density above expectations in the outer galaxy, to a thicker diffusion volume for Galactic CRs, or to vast amounts of untraced gas. See e.g. Ref. 14 for a discussion.

4.2. *Galactic CR accelerators*

Among the individual galactic sources resolved by the LAT, SNRs are thought to be the key to the puzzle of the CR origin: in fact, in SNR efficient conversion of the shock kinetic energy into particle energy can reproduce the CR spectrum up to the knee. The LAT can observe the correlated γ ray emission and bring new insights on the processes at work in each site. A long standing issue is whether the γ-ray emission should be attributed to leptonic or hadronic processes; the associated γ ray spectra are advocated to present significant differences in the GeV energy band and can be distinguished by a detector with sufficient performance, such as the LAT. The amount of observations is too ample to be discussed here, so we only describe three different SNRs, indicative of three distinct categories of galactic CRsources in our sample.

The young SNR RX J1713.7-3946 is characterized by bright synchrotron emission in the X band, and is detected at TeV energies; LAT observations show a power-law spectrum, as expected for IC emission.[15] It should be noted that the magnitude of the magnetic field derived from the variation rate in filaments is larger by at least an order of magnitude with respect to the maximum value compatible with the reported IC emission, so under assumption of a leptonic model the average field must be significantly lower that the field in the filaments. On the other hand, a hadronic scenario would require a major revision of current shock-acceleration models.

The young SNR Cas A is in the early Sedov phase, so it already produces CRs at the maximum energies possible in its entire lifetime. It is characterized by synchrotron X emission in filaments, and it is observed at TeV energies. γ emission in the LAT energy range shows a rather flat energy spectrum,[16] to be compared to a clear cutoff at TeV energies. From the variability of the X emission one estimates the magnetic field to be ~ 0.3 mG; a leptonic scenario capable to explain the LAT spectrum would require a magnetic field lower by a factor ~ 3, and still would be at odds with the decrease at TeV energies. On the other hand, a hadronic scenario is in good agreement with γ-ray observations.

W44 is a good example of how a SNR can affect the surrounding medium. W44 is a middle-aged SNR, known to be interacting with a nearby H_2 cloud on the basis of excited CO lines and OH masers. W44 is a bright source for the LAT, with a significance above 60 σ, showing significant spatial extension.[17] A model assuming hadronic emission fits the LAT spectrum reasonably well, while leptonic models assuming bremsstrahlung require, among other ad-hoc assumptions, a high ambient density and a electron-to-proton injection ratio 10 times what is observed for the CR abundances near Earth. Furthermore, the similarity between the γ-ray emission and the infrared morphology, the latter tracing the shocked molecular Hydrogen, supports the inference that the bulk of the emission is due to the interaction of the SNR with the molecular cloud.

For more details on acceleration of CRs in SNR shocks see Ref. 18, in these proceedings.

5. Cosmic rays in other galaxies

About half of the individual LAT sources are Active Galactic Nuclei, in which energetic emission is produced in processes connected with the accretion of mass on the supermassive black hole at the center of the galaxy. In addition to these, LAT observes a few nearby galaxies where emission is associated with interaction of CRs with the interstellar medium, similarly to what happens in our Galaxy.[19–22]

One of the most suggestive results from these observations is shown in

Fig. 2: luminosity in the LAT band (above 100 MeV) scales as a power law with the star formation rate, hinting to a global relation valid over a wide range of galactic properties.

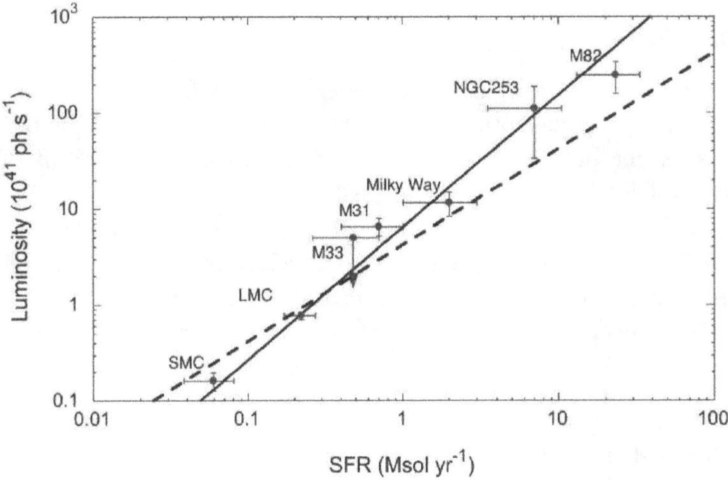

Fig. 2. γ-ray luminosity above 100 MeV versus star formation rates for galaxies in the Local Group and nearby starburst galaxies, as observed by the LAT.[21] Solid line: power law fit, local group only; dashed line: same, with slope fixed at 1.

6. Conclusions

The *Fermi* LAT allows both direct and indirect measurement of CR abundances.

Local measurements of the electron spectrum indicate a possible nearby source, but anisotropy measurements are still inconclusive. The separation of e^+ and e^- spectra confirms the increase in the e^+ fraction above a few GeV observed by PAMELA.

On a galactic scale, *Fermi* LAT γ observations indicate either a larger scale height for the CR propagation, or possibly a flatter distribution of CRs with the galactocentric radius with respect to previous expectations. An additional possibility involves vast amounts of gas beyond what is estimated by the current tracers. Individual sources believed to be the sites of CR acceleration are detected and at times resolved, indicating a variety of scenarios: sites where efficient and inefficient hadronic acceleration occurs are both suggested by the data, and possible sources interacting with the surrounding medium are examined.

On an even wider scale, observation of galaxies in the local group and nearby starbust galaxies suggests an intriguing relation between gamma luminosity and star formation rate, hinting to a possible global relation.

Acknowledgments

The *Fermi* LAT Collaboration acknowledges support from a number of agencies and institutes for both development and the operation of the LAT as well as scientific data analysis. These include NASA and DOE in the United States, CEA/Irfu and IN2P3/CNRS in France, ASI and INFN in Italy, MEXT, KEK, and JAXA in Japan, and the K. A. Wallenberg Foundation, the Swedish Research Council and the National Space Board in Sweden. Additional support from INAF in Italy and CNES in France for science analysis during the operations phase is also gratefully acknowledged.

References

1. W. B. Atwood *et al.*, *Astrophys. J.* **697**, 1071(June 2009).
2. A. A. Abdo *et al.*, *Phys. Rev. D* **80**, 122004(December 2009).
3. M. Brigida, in *ICATPP-13, Como, Italy, Oct 3-7, 2011*, (World Scientific, 2011).
4. M. Ackermann *et al.*, *Phys. Rev. D* **82**, 092004(November 2010).
5. A. A. Abdo *et al.*, *Phys. Rev. Lett.* **102**, 181101(May 2009).
6. M. Ackermann *et al.*, *Phys. Rev. D* **82**, 092003(November 2010).
7. O. Adriani *et al.*, *Nature* **458**, 607(April 2009).
8. D. Mueller and K.-K. Tang, *Astrophys. J.* **312**, 183(January 1987).
9. M. Ackermann *et al.*, *ArXiv e-prints* (September 2011).
10. F. Giordano, in *ICATPP-13, Como, Italy, Oct 3-7, 2011*, (World Scientific, 2011).
11. A. A. Abdo *et al.*, *Astrophys. J.* **703**, 1249(Octtober 2009).
12. A. A. Abdo *et al.*, *Phys. Rev. Lett.* **103**, 251101(December 2009).
13. A. A. Abdo *et al.*, *Astroph. J.* **710**, 133(February 2010).
14. M. Ackermann *et al.*, *Astrophys. J.* **726**, 81(January 2011).
15. A. A. Abdo, *Astrophys. J.* **734**, 28(June 2011).
16. A. A. Abdo *et al.*, *Astrophys. J.* **710**, L92(February 2010).
17. A. A. Abdo *et al.*, *Science* **327**, 1103(February 2010).
18. M. Lemoine-Goumard, in *ICATPP-13, Como, Italy, Oct 3-7, 2011*, (World Scientific, 2011).
19. A. A. Abdo *et al.*, *Astron. & Astrophys.* **523**, A46+(November 2010).
20. A. A. Abdo *et al.*, *Astron. & Astrophys.* **512**, A7+(March 2010).
21. A. A. Abdo *et al.*, *Astron. & Astrophys.* **523**, L2+(November 2010).
22. A. A. Abdo *et al.*, *Astrophys. J. Lett.* **709**, L152(February 2010).

Cosmic rays detected with the Auger Engineering Radio Array

H. Schoorlemmer[1], for the Pierre Auger Collaboration[2]

1) IMAPP, Radboud University Nijmegen
Nijmegen, The Netherlands
**E-mail: h.schoorlemmer@science.ru.nl*

2) Observatorio Pierre Auger, Av. San Martín Norte 304, 5613 Malargüe, Argentina
(Full author list:http://www.auger.org/archive/authors_2011_10.html)

At the Pierre Auger Observatory in Argentina, the first stage of the Auger Engineering Radio Array (AERA) has been deployed. It is located close to the low-energy enhancements of the observatory and currently consists of 24 autonomous radio detector stations. In the coming years, the number of detection stations will grow to 160 units covering almost 20 km^2. Since April of this year AERA is measuring radio emission from cosmic-ray induced air showers. These measurements are confirmed by simultaneous measurements of the particle detectors and the fluorescence telescopes of the observatory. AERA will provide us with new insights on the radio emission mechanisms of air showers with energies above 10^{17} eV.

Keywords: Radio detection, cosmic rays, AERA

1. Introduction

The Pierre Auger Observatory[1] is designed to measure ultra-high energy cosmic rays. It consists of an array of surface detectors (SD) and fluorescence detectors (FD). The SD is an array of 1600 water-Cherenkov particle detectors covering an area of 3000 km^2. The FD array is composed of 24 telescopes placed at four locations overlooking the SD. In addition a denser array of particle detectors (AMIGA[2]) and three fluorescence telescopes with higher elevation of viewing angle (HEAT[3]) have been added to measure lower energy cosmic rays.

Cosmic-ray-induced extensive air showers (EASs) emit radiation as short pulses ($\sim$ 30 ns) that can be detected by antennas in the MHz domain. This was first shown by Jelley[4] in 1965. The radio-detection technique of EASs has been revived by experiments like CODALEMA[5] and LOPES.[6] At the Pierre Auger Observatory, the first stage of a next-generation radio detector has been constructed: the Auger Engineering Radio Array[7](AERA). In the following years this array will be expanded to have a collecting area of about 20 km^2.

Simulation packages for the radio emission, like REAS[8] and MGMR,[9] predict a dependence of the radio signal on the longitudinal development of the EAS in the atmosphere. Therefore, radio detection could be a tool to assess the composition of the cosmic ray primary. The collocation with HEAT, AMIGA, and the FD gives us the opportunity to verify and calibrate the radio measurement. The advantage of radio measurements is that in principle the uptime of AERA can be 100%, while the fluorescence telescope can only be operated maximally 10% of the time. The goal of AERA is to calibrate the radio emission mechanisms and measure the composition of cosmic rays above 10^{17} eV.

2. Auger Engineering Radio Array

AERA will consist of 160 radio detector stations (RDSs), which will be deployed in three stages. Stage 1, consisting of 24 RDSs placed in a regular grid with spacing of 150 m is already deployed and its commissioning started in the beginning of 2011. Stages 2 and 3 will have a detector grid with a larger spacing and will be deployed in the coming years.
In stage 1, each RDS has a dual-polarization log-periodic dipole antenna, oriented to point north-south and east-west. These antennas are constructed to be sensitive to frequencies between 27 and 84 MHz. The signals are amplified and filtered in an analog chain and afterwards digitized. The RDSs are developed to be autonomous stations, obtaining power from solar panels and providing a first level trigger based upon only on the radio signal. After having triggered, the time-stamp of the event is sent over an optical fiber network to the central radio station. If multiple stations have triggered in time coincidence, the full waveform data from the RDSs are collected and written to disk.

3. Self-triggered cosmic-ray events

Although the AERA site is remote, there is still a lot of human interference in the 27-84 MHz bandwidth. One type of interference is due to narrowband transmitters, which can be suppressed digitally in real time. This allows a lower signal threshold to trigger on air showers. Another type of noise sources are short pulses, which mimic a real signal coming from an EAS. The rate of these pulses is so high, that it becomes hard to process them all. Luckily they have some distinct features which are used to discriminate between them and cosmic rays. Most of these pulses originate from certain directions on the horizon, which allows for a directional veto. Another feature is that a (large) fraction of these pulses arrive at the RDSs with a 100 Hz repetition rate, which might be related to the frequency used on the Argentinean power grid. Using the timing of subsequent events, the single-station timing can be used to veto some of these pulses.

After implementing and applying these vetoes, the recorded events are compared offline with events recorded by the SD. In the beginning of April 2011 this resulted in the first self-triggered events that were confirmed by a time coincidence with an SD event. Soon thereafter, the first coincidence between AERA, SD, and FD was measured. This proved the feasibility of super-hybrid measurements of cosmic rays.
In Fig. 1 the first confirmed radio signal from an EAS measured by AERA is shown together with a view of an event measured in 8 RDSs.

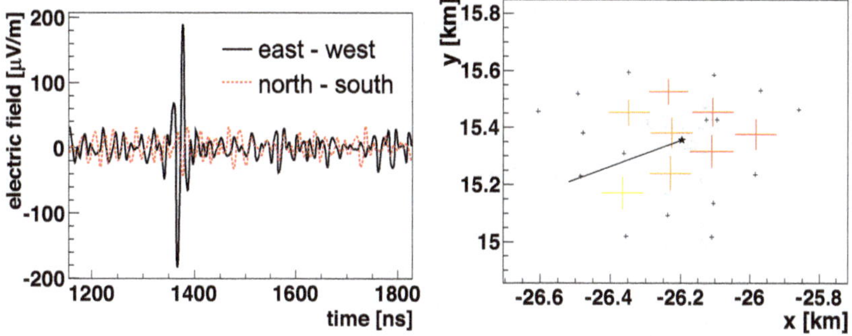

Fig. 1. Left: The first cosmic-ray-induced pulse measured by AERA that was confirmed by the SD. Right: Top view of the array with a reconstructed event. The size of the colored crosses gives an indication of the signal strength in each polarization per station. The color coding indicates the relative timing of the pulses, and the black line is pointing from the barycenter of the radio signals towards the reconstructed direction.

4. First Results

The direction of the shower axis of an EAS can be estimated to first order by fitting a plane to the arrival times of the radio pulses. In Fig. 2 the direction of the events reconstructed by AERA and the SD are compared. On average the angular separation between the two reconstructed directions is 3°. This shows that not only the timing, but also the direction is in agree-

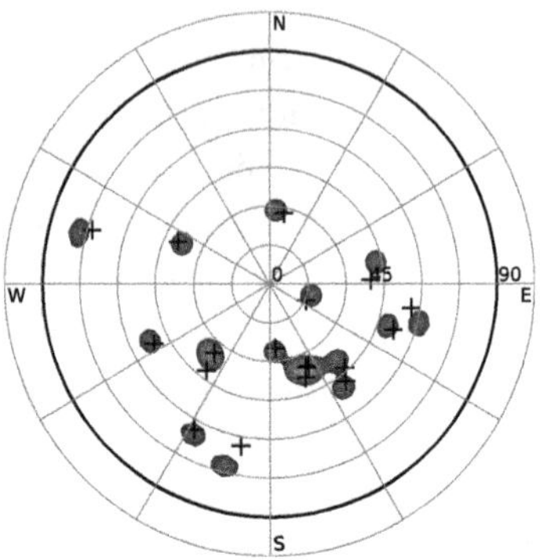

Fig. 2. The distributions of the arrival directions of the first events measured with AERA. Each direction reconstructed by radio is smeared by a Gaussian of 3 degrees. The black crosses indicate the direction of the corresponding SD events.

ment with the corresponding SD measurement. The dominant contribution

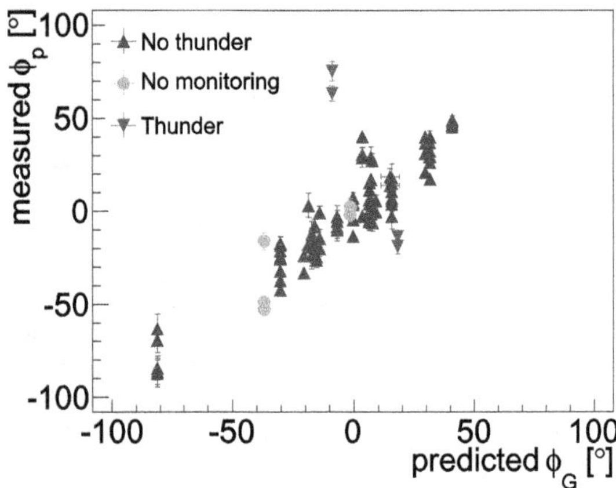

Fig. 3. The angle of polarization predicted from a pure geomagnetic emission mechanism compared the to measured angle of polarization. Green triangles indicate that atmospheric electric field monitoring was on, but there was no indication of thunderstorms. The yellow circles indicate that the monitoring system was not measuring, while the red triangles indicate thunderstorm conditions.

for the observed radio signal is due to radiation caused by the deflections of the charged particles in the geomagnetic field. The orientation of the

resulting electric field is expected to be perpendicular to the shower axis $\hat{a}$ and the direction of the geomagnetic field $\hat{b}$. The expected orientation of the electric field in the horizontal plane can be described by the angle of polarization $\phi_G = \tan^{-1}((\hat{a} \times \hat{b})_{\rm ns}/(\hat{a} \times \hat{b})_{\rm ew})$, in which ns is north-south component and ew is east-west component. The orientation of the magnetic field is known, and the shower axis is obtained from the SD. Figure 3 shows the comparison of ϕ_G to the measured polarization angles (ϕ_p) per individual RDS. From Fig. 3 it is clear that the dominant geomagnetic emission process is confirmed by AERA.
As an example of the importance of atmospheric monitoring for radio, the measurements of variations of the atmospheric electric field are used to indicate thunderstorm conditions. The largest outliers from the general trend in Fig. 3 are measured during thunderstorm conditions.

5. Conclusions and outlook

The first stage of AERA is deployed and taking data. The first milestone has been reached by measuring self-triggered coincidences with the SD and the FD. This proves the feasibility of super-hybrid detection of EASs, and provides the unique opportunity to cross-calibrate the radio signals against well-established techniques.
Directional reconstruction using the radio signals alone is in agreement with the direction reconstruction using the SD information. The polarization of the measured radio signals (except for those measured during thunderstorm conditions) confirms the dominant geomagnetic radiation mechanism.

References

1. The Pierre Auger Collaboration, *Nucl. Instr. and Meth. A* **523**, 50 (2004).
2. F. Sánchez for the Pierre Auger Collaboration, *Proc. 32st ICRC, Bejing,China; arXiv:1107.4807* (2011).
3. T.H. J. Mathes for the Pierre Auger Collaboration, *Proc. 32st ICRC, Bejing,China; arXiv:1107.4807* (2011).
4. J. V. Jelley et al., *Nature* **205**, p. 658 (1965).
5. D. Ardouin et al. (CODALEMA collaboration), *Nucl. Instr. and Meth. A* **555**, 148 (2005).
6. H. Falcke et al.(LOPES collaboration), *Nature* **435**, 313 (2005).
7. J.L. Kelley, for the Pierre Auger Collaboration, *Proc. 32st ICRC, Bejing,China; arXiv:1107.4807* (2011).
8. T. Huege, R. Ulrich and R. Engel, *Astropart. Physics* **30**, 96 (2008).
9. K. D. de Vries, A. M. van den Berg, O. Scholten and K. Werner, *Astroparticle Physics* **34**, 267 (2010).

COSMIC RAY ACCELERATION IN HISTORICAL SUPERNOVA REMNANTS IN OUR GALAXY

V.G. Sinitsyna and V.Y. Sinitsyna
P.N.Lebedev Physical Institute, Russian Academy of Science
Leninsky prospect 53, Moscow, 119991, Russia
E-mail: sinits@sci.lebedev.ru

We present the results of our observations of two types of Galactic supernova remnants with the SHALON mirror Cherenkov telescope: the plerion Crab Nebula, Geminga (probably plerion) and the shell-type supernova remnants Cassiopeia A and Tycho. The experimental data have confirmed the prediction of the theory about the hadronic generation mechanism of very high energy (800 GeV - 100 TeV) γ-rays in Tycho's supernova remnant. The data obtained suggest that the very high energy γ-ray emission in the objects being discussed is different in origin.

Keywords: TeV γ-rays; Tycho's SNR; Cas A; Crab; Geminga; electronic and hadronic Cosmic Ray components.

Introduction

One of the main aims of very high energy (VHE) γ-ray astronomy is to detect and investigate objects in the Galaxy and beyond where cosmic rays are accelerated and to study the generation mechanisms of elementary particles in active astrophysical objects. The hypothesis that supernova remnants (SNRs) are unique candidates for cosmic-ray sources[1–3] has been prevalent from the very outset of cosmic-ray physics. Recent observations of several SNRs in X-rays and TeV γ-rays will help in solving the problem of the origin of cosmic rays and are key to understanding the mechanism of particle acceleration at a propagating shock wave.

Crab Nebula (SN 1054)

The Crab Nebula, most famous supernova remnant, plays an important role in the modern astrophysics. No other space object has such impact on the progress and development of the modern experimental and theoretical astrophysics methods. Since the first detection with ground based telescope the Crab has been observed by the number of independent groups (Fig. 1)

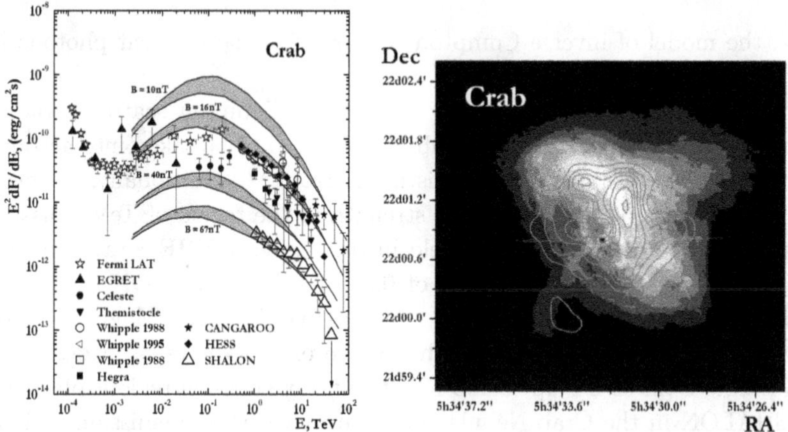

Fig. 1. **left** – The Crab Nebula γ-quantum integral spectrum by SHALON[4–8] in comparison with other experiments: Fermi LAT, EGRET, COS-B, CELESTE, CAT, Asgat, Whipple, Themistocle, HEGRA CT2, CANGAROO, Tibet, CASA-MIA ;[9] **right** – A Chandra X-ray image of Crab Nebula. The central part 200" × 200" of Crab Pulsar Wind Nebula in the energy range 0.2-4 keV. The contour lines show the TeV - structure by SHALON observations.

using different methods of registration of γ-initiated showers [4–10] As in other ranges of the electromagnetic spectrum, the Crab Nebula is a standard source for TeV γ-ray astronomy. Perhaps the most important fact is that this source with a stable flux can be used to calibrate Cherenkov telescopes in both Northern and Southern Hemispheres. However, quite recently, the AGILE[11] and Fermi LAT[12] satellite experiments have reported on a flare exceeding the nominal flux from the Crab Nebula in the energy range from 100 MeV to 2 - 5 GeV by a factor of 4, which was assumed to be absolutely stable and, consequently, was used as a standard candle. No flux increase was detected in the observations of the MAGIC[13] and VERITAS[14] ground-based telescopes in the same period.

The spectrum of γ-rays from the Crab Nebula has been measured in the energy range 0.8 TeV to 30 TeV at the SHALON Observatory by the Atmospheric Cerenkov Technique[4–8] with a statistical significance[15] of 36.1σ. The integral energy spectrum is well described by the single power law $I(>E_\gamma) \propto E_\gamma^{-1.40\pm0.07}$(Fig. 1 left).

Crab Nebula has an extraordinary broad spectrum, attributed to synchrotron radiation of electrons with energies from GeV to PeV. This continuous spectrum appears to terminate near 10^8 eV and photons, produced by relativistic electrons and positrons ($\sim 10^{15}$ eV) via Inverse Compton, form a new component of spectrum in GeV TeV energy range. To make a description of the intensity and spectral shape in the TeV region of > 0.8

TeV, the model of inverse Compton scattering of the ambient photons in the nebula in the Ref.[10] is used.

For the purpose of calculating the Inverse Compton scattered radiation, the complete spectrum of Crab Nebula need to be taken into account to deduce the spectrum of relativistic electrons.[10] Additionally, we need the assuming about magnetic field strength in the region of TeV emission (Figs. 1). The average magnetic field in the region of VHE γ-ray emission is extracted from the comparison of 0.8 – 30 TeV (SHALON data) and X-ray (Chandra data [16]) emission regions (Fig. 1, right). Magnetic fields of representative regions in Chandra image of Crab have been derived[16] and ranges from 62 nT up to 153 nT. The γ-ray emission regions observed by SHALON in the Crab Nebula correlate well with the emission regions of synchrotron photons in the energy range 0.4 - 2.1 keV and the average magnetic field strength in the TeV regions is 67 ± 7 nT.

Finally, the TeV γ-quantum spectrum of Crab by SHALON is generated via Inverse Compton of soft, mainly optical, photons which are produced by relativistic electrons and positrons, in the nebula region around 1.5'(Fig. 1) from the pulsar with specific average magnetic field of about 67 ± 7 nT.

Tycho's SNR (SN 1572)

Tycho Brage supernova remnant has been observed by SHALON telescope of Tien-Shan high-mountain observatory since 1996 (Figs. 2, 3). This object has long been considered as a candidate to cosmic ray hadrons source in Northern Hemisphere.[17–20]

Tycho's SNR has been detected by SHALON at TeV energies[4–8] (in observations of 1996 - 2010 years) with a statistical significance[15] of 17 σ. The integral γ-ray flux above 0.8 TeV was estimated as $(0.52 \pm 0.05) \times$

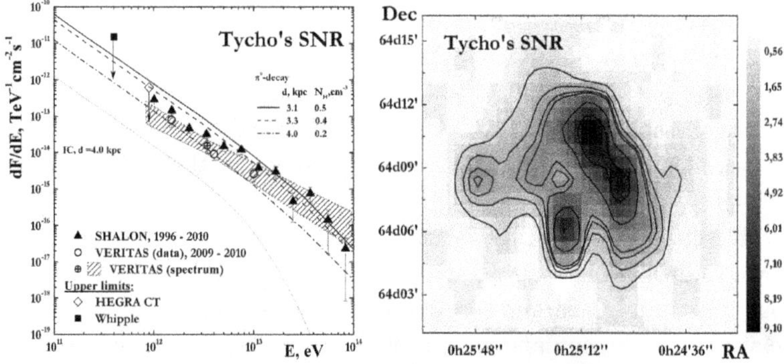

Fig. 2. **left:** The Tycho's SNR γ-ray differential spectrum by SHALON in comparison with VERITAS data[21] and theoretical models;[17,18] **right:** The SHALON image of γ-ray emission from Tycho's SNR.

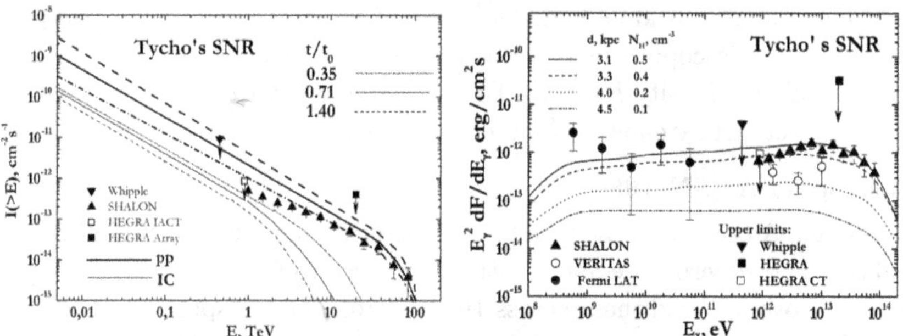

Fig. 3. **left:** The Tycho's SNR γ-ray spectrum by SHALON in comparison with other experiments and calculations: IC emission (thin lines), π° - decay (thick lines).[17] **right:** Spectral energy distribution of the γ-ray emission from Tycho's SNR by SHALON[4–8] in comparison with other experiment data Fermi LAT,[22] Whipple,[23,24] HEGRA,[25] VERITAS[21] and with theoretical predictions.[19,20] All cases have dominant hadronic γ-ray flux.

$10^{-12} cm^{-2} s^{-1}$ (Fig. 3). Figures 2 and 3 show the observational results for the Tycho's SNR. The energy spectrum of Tycho's SNR at $0.8-20$ TeV can be approximated by the power law $F(> E_O) \propto E^{k_\gamma}$ with $k_\gamma = -0.96 \pm 0.06$. Also, the energy spectrum of γ-rays in the observed energy region from 0.8 TeV is well described by the power law with exponential cutoff, $I(> E_\gamma/1TeV) = (0.42 \pm 0.09) \times 10^{-12} \times (E_\gamma/1TeV)^{-0.93 \pm 0.09} exp(-E_\gamma/35TeV)$. The energy spectrum of supernova remnant Tycho's SNR is harder than Crab spectrum.

Recently, Tycho's SNR was also confirmed with VERITAS[21] telescope in observations of 2008 - 2010 years. The high energy γ-ray emission from Tycho'SNR was detected with Fermi LAT[22] in the range 400 MeV - 100 GeV. Figure 2 presents the Tycho's SNR differential spectrum by SHALON[4–8] comparing with VERITAS[21] data and theoretical models.[17,18]

A nonlinear kinetic model of cosmic ray acceleration in supernova remnants is used in [17,18] (Fig. 3, left), to describe the properties of Tycho's SNR. The expected flux of γ-quanta from π°-decay, $F_\gamma \propto E_\gamma^{-1}$, extends up to ~ 30 TeV, while the flux of γ-rays originated from the Inverse Compton scattering has a sharp cutoff above the few TeV, so the detection of γ-rays with energies of ~ 10 to 80 TeV by SHALON (Fig. 3) is an evidence of their hadronic origin.[17,19,20] Figure 3, right presents spectral energy distribution of the γ-ray emission from Tycho's SNR, as a function of γ-ray energy ϵ_γ for a mechanical SN explosion energy of $E_{SN} = 1.2 \times 10^{51}$ erg and four different distances d and corresponding values of the interstellar medium number densities N_H. All cases have dominant hadronic γ-ray flux.[20] The additional information about parameters of Tycho's SNR can be predicted in frame of nonlinear kinetic model [17,19,20] if the TeV γ- quantum spectrum of SHALON telescope is taken into account: a source distance 3.1 - 3.3 kpc and an ambient density N_H $0.5-0.4$ cm^{-3} and the expected π°-decay γ-ray energy spectrum extends up to about 100 TeV.

Cassiopeia A (SN 1680)

Cassiopeia A (Cas A) is the youngest of historical supernova remnant in our Galaxy. The supernova explosion that gave rise to Cas A occurred around 1680. Its overall brightness across the electromagnetic spectrum makes it a unique object for studying high-energy phenomena in SNRs. Cas A was detected in TeV γ rays, first by HEGRA[26] and later confirmed by MAGIC[27] and VERITAS.[28] The high energy γ-ray emission from Cas A was detected with Fermi LAT[29] in the range 500 MeV - 50 GeV.

Cas A was observed with SHALON telescope during the 27 hours of autumn 2010.[8] All observations were made with the standard procedure of SHALON experiment during moonless nights. The γ-ray source associated with the SNR Cassiopeia A was detected above 800 GeV with a statistical significance[15] of 7.1 σ with a γ-quantum flux above 0.8 TeV of $I_{CasA}(> 0.8TeV) = (0.68 \pm 0.13) \times 10^{-12} cm^{-2}s^{-1}$. The γ-ray integral spectrum is presented in Fig. 4, left. It is compatible with a power law with an index $k_{\gamma} = 1.32 \pm 0.18$.

The favored scenarios in which the γ-rays of 500 MeV - 10 TeV energies are emitted in the shell of the SNR like Cas A are considered in.[29,30] The γ-ray emission could be produced by electrons accelerated at the forward shock through relativistic bremsstrahlung (NB) or IC.[29] Alternatively, the GeV γ-ray emission could be produced by accelerated CR hadrons through interaction with the background gas and then π°-decay. Figure 4, right presents spectral energy distribution of the γ-ray emission

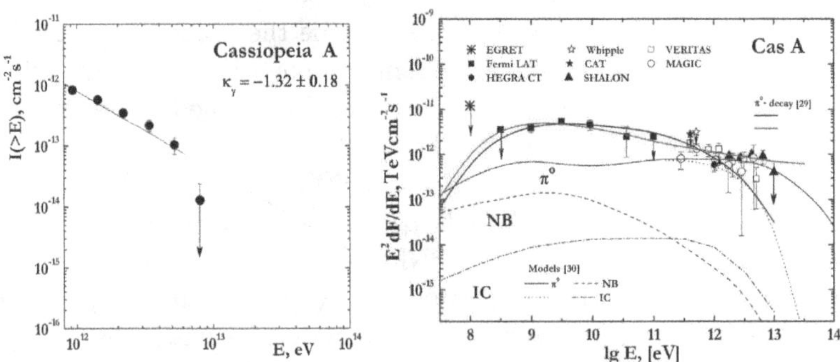

Fig. 4. **left:** The Cas A γ-quantum integral spectrum with power index of $k_{\gamma} = -1.32 \pm 0.18$ by SHALON experiment; **right:** Spectral energy distribution of the γ-ray emission from Cas A.

from Cas A by SHALON in comparison with other experiment data Fermi LAT,[29] HEGRA,[26] MAGIC,[27] VERITAS,[28] EGRET,[31] CAT,[32] Whipple[33] and with theoretical predictions.[29,30] The detection of very high energy γ-ray emission at 5 - 10 TeV and the hard spectrum below 1 TeV would favor the π°-decay origin of the γ-rays in Cas A SNR.

References

1. V. S. Berezinskii, S. V. Bulanov, V. L. Ginzburg, et al., *Astrophysics of Cosmic Rays*, Ed. by V. L. Ginzburg (Nauka, Moscow, 1984) [in Russian].
2. E. G. Berezhko and G. F. Krymskii, *Usp. Fiz. Nauk* **154(1)**, 49 (1988)
3. S. P. Reynolds, *Ann. Rev. Astron. Astrophys.* **46**, 89 (2008).
4. V. G. Sinitsyna, *AIP Conf. Proc.*,**515**, 205; 293 (1999).
5. V. G. Sinitsyna, S. I. Nikolsky, et al., *Izv. Ross. Akad. Nauk Ser. Fiz.*, **66(11)**, 1654; 1661 (2002); *ibid.* **69(3)**, 422 (2005).
6. V. G. Sinitsyna, et al.,*in Proc. of 27th ICRC, Hamburg*, **3**, 2665 (2001); of *29th ICRC, Puna*, **4**,231 (2005); of *30th ICRC, Merida*, **2**, 543 (2007).
7. V. G. Sinitsyna et al., *Nucl. Phys. B (Proc.Suppl.)*, **196**, 437 (2009); *ibid.*, **175-176**, 455 (2008); *ibid.*, **151**, 112 (2006); *ibid.*, **122**, 247, 409 (2003); *ibid.*, **97**, 215, 219 (2001); *ibid.*, **75A**, 352 (1999).
8. V. G. Sinitsyna and V. Yu. Sinitsyna *Astronomy Letters*, **37(9)**, 621 (2011).
9. V. G. Sinitsyna et al. *Cosmic Rays for Particle and Astroparticle Physics* Ed. by S. Giany, C. Leroy and P.-G. Rancoita, **6**, 3 (2011).
10. A. M. Hillas, et al., ApJ, 1998, **503**: 744.
11. M. Tavani et al., *Science* **331**, 736 (2010).
12. A. A. Abdo, et al., *Science* **331**(6018), 739 (2010).
13. M.Mariotti, *Astron. Telegramm* No. 2967 (2010).
14. R. Ong, *Astron. Telegramm* No. 2968 (2010).
15. T.-P. Li and Y.-Q. Ma, *ApJ*, **272**, 317 (1983).
16. F. D. Seward, W. H. Tucker and R. A. Fesen, *ApJ.*, **652**, 1277 (2006).
17. H. J. Völk, et al., in *Proc. 27th ICRC, Hamburg*, **2**, 2469 (2001).
18. E. G. Berezhko et al., *Astrophys. Space Sci.*, **309**, 385 (2007).
19. H. J. Völk, et al., in *Proc. 29th ICRC, Pune*, **3**, 235 (2005).
20. H. J. Völk, E. G. Berezhko, L. T. Ksenofontov *A&A*, **483(2)**, 529 (2008).
21. V. A. Acciari et al. *arXiv:astro-ph/ 1102.3871v1*
22. F. Giordano, et al., *arXiv:astro-ph/ 1108.0265v.*
23. T. C. Weekes, *AIP Conf. Proc.*, **515**, 3 (1999).
24. J. H. Buckley et al., A&A, 1998, **329**: 639.
25. J. Prahl and C.Prosch, in *Proc. 25th ICRC, Durban*, **3**, 217 (1997).
26. F. Aharonian, et al., *A&A*, **370**, 112 (2001).
27. J. Albert et al., *A&A*, **474**, 937 (2007).
28. V. A. Acciari, et al., *ApJ*, **714**, 163 (2010).
29. A.A. Abdo, et al., *ApJ*, **710**, L92 (2010).
30. E. G. Berezhko, G. Pühlhofer, H. J. Völk, *A&A*, **400**, 971 (2003).
31. J. A. Esposito et al., *ApJ.*, **461**, 820 (1996).
32. P. Goret, et al., *Proc. of 26th ICRC*, **3**, 496 (1999).
33. R. W. Lessard, et al., *Proc. of 26th ICRC*, **3**, 488 (1999).

SEARCH FOR NEUTRINO DECAY AT SHALON

V.G. Sinitsyna[1], M. Masip[2], S.I. Nikolsky[1], V.Y. Sinitsyna[1]

P.N.Lebedev Physical Institute, Russian Academy of Science
Leninsky prospect 53, Moscow, 119991, Russia
E-mail: sinits@sci.lebedev.ru
[2]*CAFPE and Departamento de Fosica Teoretica y del Cosmos Universidad de Granada,E-18071 Granada Spain*

The SHALON Cherenkov telescope has recorded over 2×10^6 extensive air showers during the past 17 years. The analysis of the signal at different zenith angles has included observations from the sub-horizontal direction $\Theta = 97°$. This inclination defines an Earth skimming trajectory with 7 km of air and around 1000 km of rock in front of the telescope. During a period of 324 hours of observation, after a cut of shower-like events that may be caused by chaotic sky flashes or reflections on the snow of vertical showers, we have detected 5 air showers of TeV energies. We argue that these events may be caused by the decay of a long-lived penetrating particle entering the atmosphere from the ground and decaying in front of the telescope. We show that this particle can it not be a muon or a tau lepton. As a possible explanation, we discuss two scenarios with an unstable neutrino of mass $m \approx 0.5$ GeV and $c\tau \approx 30$ m. Remarkably, one of these models has been recently proposed to explain an excess of electron-like neutrino events at MiniBooNE.

Keywords: neutrino decay, SHALON atmospheric Cherenkov telescope

Introduction

Cosmic rays may also offer an opportunity to study the properties of elementary particles. The main objective in experiments like IceCube[1] or Auger[2] is to determine a flux of neutrinos or protons as they interact with terrestrial matter. These interactions involve energies not explored so far at particle colliders, so their study should lead us to a better understanding of that physics. In addition, the *size* of the detector and its distance to the interaction point is much larger there than in colliders, which may leave some room for unexpected effects caused by long-lived particles. It could well be that in the near future cosmic rays play in particle physics a complementary role similar to the one played nowadays by cosmology (in

Table 1. The atmosphere depth at different zenith angles

Zenith angle, Θ°	Atmosphere depth, g/cm^2	Number of Cherenkov burst per hour
0°	670	1100 ± 210
72°	2250	7 ± 1.14
76°	3000	1.8 ± 0.5
84°	5950	0.5 ± 0.01

aspects like dark matter, neutrino masses, etc.).

In this paper we describe what we think may be one of such effects. It occurs studying the response of the SHALON atmospheric Cherenkov telescope[3] to air showers from different zenith angles, in a sub-horizontal configuration where the signal from cosmic rays should vanish.

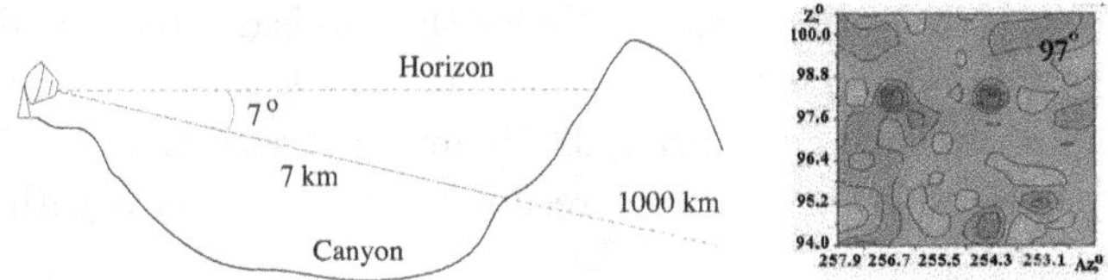

Fig. 1. **left:** Configuration at $\theta = 97^o$. **right:** the typical Cherenkov burst recorded under 97° zenith angle, but have not shower characteristics. The amplitude of gray - scale shower images is proportional to the QDC count.

The SHALON mirror telescope

SHALON is a gamma-ray telescope[4,5] of Tien-Shan high-mountain observatory. It has been operating since 1992.[3,5,6] During this period it has detected gamma-ray signals from well known and also from new sources of different type: Crab Nebula, Tycho's SNR, Geminga, Mkn 421, Mkn 501, NGC 1275, SN2006 gy, 3c454.3 and 1739+522.[6,7]

Observations at large zenith angles have been aimed on study of spectra of the air showers induced by cosmic rays crossing through different atmosphere thickness and events accompanying the passing of EAS and cosmic ray particles near horizon. The observation at large zenith angles 72°, 76°, 84° showed that the efficiency of Cherenkov light detection drops essentially as a zenith angle increases, perhaps because of dissipation and absorption in the atmosphere. So, the comparison of observation results shown that at the zenith angle 84° the number of observed showers is ~ 25 times less than expected by estimation with neglecting by absorption and dissipation of Cherenkov photons in the atmosphere (table 1).

Cherenkov bursts below the horizon

The study of extensive air showers at large zenith angles included observations at the sub-horizontal direction $\theta = 97^o$.[3,6] The configuration of the telescope is depicted in Fig. 1. SHALON Cherenkov mirror telescope is located at 3340 m a.s.l. The mountain range lies in the east direction and is more than 4300 m a.s.l. The mountain range is about 20 km long. The mountain slope has a structure which is irregular on the scales less than typical shower size; it is covered with the forest. The thickness of matter in the telescopic field of view is from 2000 to 800 kms; the viewed mountain slope area is $> 7 \times 10^5\ m^2$. For telescope located about 7 kms away from the mountain slope horizontally, the shadow of mountain is about 7° in elevation. In actual conditions the mirror telescope placement the distance till the opposite slope of the gorge is $\sim$ 7 km or $\sim$ 16.5 radiation units of length, that is quite enough for the development of an electromagnetic cascade till the structure characteristic for the rarefied atmosphere. Observations at 97° zenith angle have been done in cloudless nights in absence of artificial lights and dry air. During 324 hours of observation at 97° zenith angle 323 short-range bursts were recorded. In accordance with the ex-

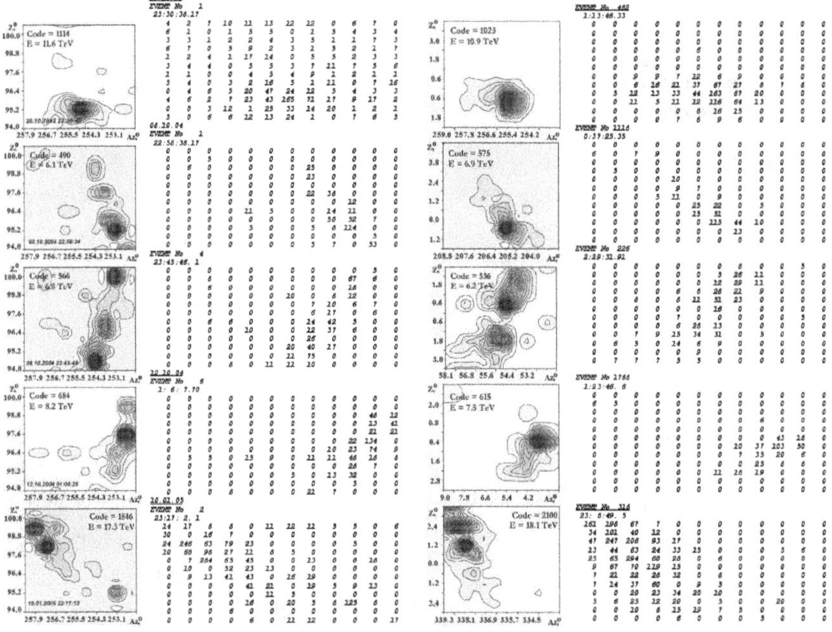

Fig. 2. **left:** Cherenkov Radiation of Extensive Air Showers Observed at 97° Zenith Angles by SHALON; **right:** Cherenkov Radiation of Extensive Air Showers Observed at 0° Zenith Angles by SHALON

isting ideas and estimations[3,8,9] an appearance of electron-photon cascade from upward direction in current SHALON experiment conditions, which are above described, can be connected with passing of weakly interacting particles through rock and earth matter. The identity of upward neutrino initiated showers to cosmic ray ones in frames of concerning experiment has been performed in accordance to the following parameters used in the gamma-ray astronomy. The parameters used to characterize the shower image are image maximum position (x_{max}, y_{max}); length, width; the relation of two previous described Hillas parameters: *Length/Width*; two parameters sensible to the shower shape: *Int*0, *Int*1; The *Int*0 is the ratio of Cherenkov light intensity in pixel with maximum pulse amplitude to the light intensity in the eight surrounding;[6] The *Int*1 is the ratio of Cherenkov light intensity in pixel with maximum pulse amplitude to the light intensity in the in all the pixels except for the nine in the center of the matrix. In addition, the selection criteria we are using (*Int*0, *Int*1, *Length/Width*) are of the relative nature to describe 2-D shower structure, which is also different in current experiment conditions from vertical one by less than 10%. The parameter proportional to the energy of the shower is Code.

The SHALON databank (since 1992) contains a millions of verified showers from vertical cosmic ray observations with their parameters, so the selection of the showers with a set of parameters of any sample can be performed. Reconstruction of shower coming direction using the analysis of shower shape and position of shower maximum (in case of non-gamma shower) is performed with accuracy $< 0.5°$ which is enough to judge on whether it upward shower or near horizontal. Horizontal and down going shower is out of field of view because of narrow-beaming of Cherenkov telescope relative to ice or water neutrino telescopes.

During 324 hours of observations 5 events were detected (Figs. 2 left) which have expected angular characteristics of a light burst of an electron-photon cascade developing within a telescope observation angle. These showers have energy in the range of about $6-17.5$ TeV. These 5 events have form characteristics and parameters similar (within 10% error) to those observed at 0° zenith angle (Figs. 2 right). These cascades look like the usual extensive air showers generated in atmosphere with narrow light shape. The background for this events can be some reflections of cosmic ray EAS in the mountain slope. First of all it could be a reflection of showers initiated by particles born in interaction of very high energy cosmic rays and rock matter nucleons. The energy of detected showers is more than 6 TeV. There is no albedo particles of such high energies. One more source of particles with high transverse energy is jet production. The reflection from snow which can mimed the EAS shape is excluded due to the irregular and woody structure of opposite slope (in addition: there is no snow there till the start of November). The probability of hadronic jet production with energy of observed showers is ten orders of magnitude less than one for detection of shower generated by secondary particles of UHE neutrino interaction.

All other 318 events of detection of short-range light bursts in the atmosphere have not a narrow angle light direction and are chaotically distributed along the whole matrix or its part of a light-receiver (see Fig. 1, left for 97°; no other type of shower image were found among the 318 mentioned pictures). These events may be interpreted as a reflection of a Cherenkov burst from a snow mountain slope or as an ionization luminescence of the atmosphere while an extensive air showers transition within a telescope observation angle.

Earth-skimming neutrino interactions

The flux of sub-horizontal events is around 6×10^{-6} times the flux of TeV cosmic rays reaching the atmosphere. Such a large flux seems to eliminate the possibility that these events are due to neutrino interactions in the air or within the last ≈ 20 cm of rock. The interaction length of a 10 TeV neutrino is $\approx 10^5$ km.[10] This implies that only one out of 10^9 of them will interact to produce such an event. The expected neutrino flux from pion and kaon decays at 10 TeV is a per cent fraction of the primary proton flux, whereas the flux from the prompt decay of charmed hadrons, although uncertain, should be still smaller at these energies.[11] Therefore, the expected number of events from atmospheric neutrino interactions is 10^5 times smaller than the one observed. On the other hand, a flux of primary (non-atmospheric) neutrinos large enough would be inconsistent with observations at neutrino telescopes.

Another possibility that can be readily excluded is the decay in the air of a muon or a tau lepton produced inside the rock. A 10 TeV muon could emerge if it is produced ≈ 1 km inside the rock[10] (one out of 10^5 incident neutrinos will produce a muon there). However, the muon decay length at TeV energies is around 10^4 km, so the probability that it decays in the air in front of the telescope is again too small. The neutrino fluxes required to explain the events from μ decays or from ν interactions are then similar (and excluded). The probability for tau lepton production in the rock and decay in the air is not higher. The tau becomes *long-lived* at $\approx 10^8$ GeV. At 10 TeV it should be produced within the last meter of rock ($c\tau\gamma \approx 0.5$ m), which reduces very much the number of events.

Heavy neutrino decay

Therefore, we have to explore possible explanations based on new physics. The ideal candidate should be a long-lived massive particle, neutral, frequently produced in air showers, and very penetrating: able to cross 1000 km of rock and decay within the 7 km of air in front of the telescope. If this particle has (possibly suppressed) couplings to the W and/or Z bosons, its mass m_h should be larger than m_μ (to decay in the last 7 km of air) and smaller than m_τ (to cross 1000 km of rock without decaying). Notice that if its decay length at 10 TeV is $c\tau\gamma \approx 1000$ km, at GeV energies it will tend to decay far from the detectors in colliders.

An obvious possibility is a sterile neutrino. We take two Weyl spinors n and n^c and add a Dirac mass term together with a Yukawa coupling to the lepton family $L = (\nu_l\ l)$,

$$-\mathcal{L}_\nu = m_n\, nn^c + y_\nu\, h^\dagger L n^c + \text{h.c.} \tag{1}$$

Then the Higgs VEV v induces mixing between n and ν_l:

$$-\mathcal{L}_\nu \supset m_n\, nn^c + m_{EW}\, \nu_l n^c = m_h\, \nu_h n^c\,, \tag{2}$$

where $m_{EW} = y_\nu v/\sqrt{2}$, $m_h = \sqrt{m_n^2 + m_{EW}^2}$, $\nu_h = c_\alpha n + s_\alpha \nu_l$, $s_\alpha = m_{EW}/m_h$, and the orthogonal combination $-s_\alpha n + c_\alpha \nu_l$ remains massless. The mixing implies couplings of ν_h to the W and Z gauge bosons; the first one will appear suppressed by $U_{lh} = s_\alpha$, whereas the flavour-changing (heavy to light) Z coupling will be proportional to $c_\alpha s_\alpha$.

A first ν_h model that we would like to discuss has been recently proposed by Gninenko[12] to explain an anomaly at MiniBooNE.[13] He claims that the excess of electron-like events in the interactions of the $\langle E\rangle \approx 800$ MeV ν_μ beam could be caused by the decay of a heavy neutrino if $m_h \approx 0.5$ GeV, $c\tau_h \leq 30$ m, and $|U_{\mu h}|^2 \approx 10^{-3}$. This explanation requires a large transition magnetic moment,[14] $\mu_{tran} \approx 10^{-10}\mu_B$, which implies a dominant decay mode: $\nu_h \to \gamma\nu$. The final photon would convert into a e^+e^- pair with a small opening angle that would be indistinguishable from an electron in MiniBooNE. At the same time, this dominant decay channel could make the required value of $U_{\mu h}$ consistent with bounds $|U_{\mu h}|^2 \leq 10^{-5}$ from BEBC,[15] CHARM[16] and CHARM2,[17] as these experiments look for decays into final states with charged particles ($\nu_h \to ee\nu, \mu e\nu, \mu\pi$).

It is easy to see that such a particle could have an impact on the SHALON events. At 10 TeV its decay length is $\lambda_h \approx 600$ km. If ν_h is produced in the atmosphere with that energy, the probability that it crosses $\lambda \approx 1000$ km of rock and decays within the $\Delta\lambda \approx 7$ km of air in front of the telescope is

$$p = e^{-\lambda/\lambda_h}\left(1 - e^{-\Delta\lambda/\lambda_h}\right) \approx 0.002 \tag{3}$$

This implies that the atmospheric flux of heavy neutrinos should be a 1/1000 fraction of the TeV flux of primary cosmic rays. This large flux seems difficult to achieve because ν_h is not produced in pion or kaon decays (as $m_h > m_{\pi,K}$), it appears only in a $|U_{\mu h}|^2 \approx 10^{-3}$ fraction of charmed hadron decays into muons.

A slightly more frequent production rate could be expected in a second model, where ν_h has a sizeable component along the tau neutrino. NOMAD[18] has set limits $|U_{\tau h}|^2 \leq 10^{-2}$ from $D_s \to \tau \nu_h$, and then $\nu_h \to \nu_\tau ee$, but they apply only to neutrinos lighter than $m_{D_s} - m_\tau \approx 0.19$ GeV. Cosmological and supernova bounds on $U_{\tau h}$ apply to lighter values of m_h as well.[19] On the other hand, LEP bounds cover just the range $m_h > 3$ GeV[20] (decays in the detector of lighter neutrinos are too rare). Therefore, a possible candidate could have a 0.2–0.4 GeV mass, $|U_{\tau h}|^2 \approx 0.1$, and negligible mixings with the other two families. The dominant decay channels would be into $\nu_\tau \pi^0$ and into $\nu_\tau ee$, $\nu_\tau \mu\mu$. If its decay length at the TeV energies of the sub-horizontal events is around 1000 km, then the probability of decay in the air in front of SHALON is ≈ 0.003. Its production in air showers would be through tau decay; one can expect $|U_{\tau h}|^2 \approx 0.1$ heavy neutrinos from each tau produced in the atmosphere. These tau leptons would mainly come from the prompt decay of charmed D_s mesons, and also from mesons containing a bottom quark. The flux required, a per cent of the TeV proton flux, seems still too large. Notice, however, that there are also large uncertainties in the flux and energy of the sub-horizontal events, or in the tau production rate in the atmosphere by cosmic rays.[11]

Summary and discussion

When a cosmic ray enters the atmosphere it produces an extended air shower with thousands of secondary particles. Obviously, if there is any new physics it will be contained in a fraction of these events. Now, if this *exotic* physics includes a long-lived particle, we think that there is the potential for its discovery in cosmic ray experiments. Generically, to be detectable the particle must *survive* after the rest of the shower has been absorbed by the atmosphere (*e.g.*, a long-lived gluino in horizontal air showers[21]) or the ground (a stau in neutrino telescopes[22]). In particular, a long-lived neutral particle could propagate to the center of a neutrino telescope and start there a contained shower when it decays. However, this event would look indistinguishable from a standard neutrino interaction.

In this paper we discuss several air showers obtained at SHALON in a configuration (see Fig. 1) where the expected number of events is zero. Around 1000 km of rock absorb the atmospheric flux of any standard particles but neutrinos. Neutrino interactions in the rock are frequent, but they are not observable as they *disappear* in just half a meter of soil. A few muons could be produced during the last km and emerge from the rock, but then the probability of muon decay within the 7 km of air in front of the telescope is too small. The crucial difference with a neutrino telescope is that here the probability of a *visible* ν interaction (in the air or the last centimeters of rock) is negligible.

We argue that these events may correspond to the decay of a neutral particle after it is produced in the atmosphere and has crossed 1000 km of rock. We have studied a couple of models where this particle is a heavy neutrino, and have concluded that although the required production rate seems higher than the expected one, due to a number of uncertainties on the flux and the energy of the exotic events or on the production of charmed particles in the atmosphere, none of these possibilities should be excluded.

References

1. A. Achterberg, et al., *Astropart. Phys.* , **26**, 155 (2006).
2. J. Abraham, et al., *Nucl. Instrum. Meth. A*, **523**, 50 (2004).
3. V. G. Sinitsyna , in *Proc. of Toward a Major Atmospheric Cherenkov Detector-II, Calgary, July 17-18 1993*, ed. R. C. Lamb, 91, 1993.
4. S. I. Nikolsky and V. G.Sinitsyna, *VANT., Ser. Tekhn. Fiz. Eksp.* **2**, 30 (1987).
5. V. G. Sinitsyna, *in Proc. of Toward a Major Atmospheric Cerenkov Detector-VI, Snowbird, Utah,* 13-16 Aug 1999.
6. V. G. Sinitsyna et al., *Nucl. Phys. Proc. Suppl.*, **122**, 247 (2003); *ibid.*,**175-176**, 455, 544 (2008); *ibid.*,**196**, 251, 433, 437, 442 (2009).
7. V. Y. Sinitsyna et al., in Proc. of 28th ICRC, Tsukuba, 2369 (2004).
8. L.K. Resvanis, *Nucl.Phys.B (Proc.Suppl.)*, **151**, 279 (2006).
9. D. Fargion, *ApJ*, **570**, 909 (2002).
10. C. Amsler et al., *Phys. Lett. B*, **667**, 1 (2008).
11. C. G. S. Costa *Astropart. Phys.*, **16**, 193 (2001).
12. S. N. Gninenko, *arXiv:0902.3802 [hep-ph]*.
13. A. A. Aguilar-Arevalo et al., *Phys. Rev. Lett.*, **98**, 231801 (2007).
14. R. N. Mohapatra and P. B. Pal: *World Sci. Lect. Notes Phys* bf60, 1 (1998).
15. A. M. Cooper-Sarkar et al., *Phys. Lett. B*, **160**, 207 (1985).
16. F. Bergsma et al. *Phys. Lett. B*, **166**, 473 (1986).
17. P. Vilain et al., *Phys. Lett. B*, **343**, 453 (1995).
18. P. Astier et al., *Phys. Lett. B*, **506**, 27 (2001). *[arXiv:hep-ex/0101041]*.
19. A. D. Dolgov et al., *Nucl. Phys. B*, **590**, 562 (2000).
20. O. Adriani et al., *Phys. Lett. B*, **295**, 371 (1992).
21. J. I. Illana, M. Masip and D. Meloni, *Phys. Rev. D*, **75**, 055002 (2007).
22. M. Ahlers, J. I. Illana, M. Masip and D. Meloni *JCAP*, **0708**, 008 (2007).

Recent Results from the MAGIC Telescopes

O. Tibolla* on behalf of the MAGIC collaboration

ITPA, Universität Würzburg, Campus Hubland Nord,
Emil-Fischer-Str. 31 D-97074 Würzburg, Germany
**E-mail: omar.tibolla@gmail.com ;*
Omar.Tibolla@astro.uni-wuerzburg.de

MAGIC (Major Atmospheric Gamma−ray Imaging Cherenkov Telescope) is a system of two 17 meters Cherenkov telescopes, sensitive to very high energy (VHE; $> 10^{11}$ eV) gamma radiation above an energy threshold of 50 GeV. The first telescope was built in 2004 and operated for five years in standalone mode. A second MAGIC telescope (MAGIC−II), at a distance of 85 meters from the first one, started taking data in July 2009. Together they integrate the MAGIC stereoscopic system. Stereoscopic observations have improved the MAGIC sensitivity and its performance in terms of spectral and angular resolution, especially at low energies.

We report on the status of the telescope system and highlight selected recent results from observations of galactic and extragalactic gamma−ray sources. The variety of sources discussed includes pulsars, galactic binary systems, clusters of galaxies, radio galaxies, quasars, BL Lacertae objects and more.

Keywords: Gamma−ray: instruments; Gamma−ray: observations; Galactic astrophysics; Extragalactic astrophysics; Cosmic Rays

1. MAGIC

The Major Atmospheric Gamma−ray Imaging Cherenkov Telescope (MAGIC) is a system of two 17−meters Atmospheric Cherenkov Telescopes (shown in Fig. 1) located at the *Observatorio del Roque de los Muchachos* in the island of *La Palma*, 2200 meters above sea level. MAGIC-I has been in operation since 2004 and the stereoscopic system has been operation since 2009. MAGIC has an enhanced duty cycle up to $\sim 17\%$ as it is able to operate in presence of moderate moonlight and twilight.

The performances of the MAGIC stereoscopic system are reported in[1] . The low energy threshold of 50 GeV (or 25 GeV with special trigger setup[2]) allows observations of the distant universe and overlaps with the Fermi

Fig. 1. MAGIC Telescopes in Observatorio del Roque de los Muchachos.

satellite; the angular resolution is $\sim 0.1°$ at 100 GeV, down to $\sim 0.05°$ above 1 TeV; the energy resolution is 20% at 100 GeV and goes down to 15% at 1 TeV.

Another very important feature of MAGIC telescopes is their light structure (ultralight carbon fiber frame), that allows fast repositioning (less than 20 seconds for a 180° repositioning), for fast follow-up observation of gamma-ray bursts (GRBs).

In order to achieve an easier maintenance, better sensitivity and performances (in particular for extended sources), on June 15th, the MAGIC telescopes were shut down to perform a major upgrade of the hardware:

- Both telescopes will be equipped with a new 2 GSamples/s readout based on DRS4 chip (linear, low dead time, low noise);
- The camera of MAGIC-I will be upgraded to a clone of the MAGIC-II camera, i.e. from 577 to 1039 pixels to match the camera geometry and the trigger area of MAGIC-I (currently this is planned for 2012);
- Both telescopes will be equipped with *sumtrigger* (threshold ~25 GeV^2) covering the total conventional trigger area (planned for 2012 as well).

The MAGIC scientific program covers different aspects of high energy astrophysics:

- Galactic Objects: Supernovae Remnants (SNRs), Pulsars and Pulsar Wind Nebulae (PWNe).
- Extragalactic Objects: Active Galactic Nulcei (AGNs), starburst galaxies, clusters of galaxies and Gamma-Ray Bursts (GRBs).
- Fundamental physics, such as the origin of Cosmic Rays (CRs; that can of course be studied indirectly, by means of studying SNRs for instance, but also directly, considering the showers initiated by primary CRs), Dark Matter (DM) searches and the possible tests of Lorentz invariance violations.

1.1. *Galactic observations*

Recent MAGIC results on Galactic science are here highlighted in four sections:

- Crab Pulsar Wind Nebula.
- Crab pulsar.
- Extended sources (PWNe and SNRs).
- X-ray binaries (XRBs) and Galactic microquasars.

1.1.1. *Crab nebula*

The Crab nebula is the the prototype of young PWN and it has been considered the "standard candle" of Imaging Atmospheric Cherenkov Telescopes (IACTs) so far; however, in the past year, questions have been raised about its flux constancy, after it was seen flaring in GeV gamma-rays (with both *AGILE* and *Fermi LAT*,[3]), a year-scale longer variability has been observed in X-rays and ARGO-YBJ[4] reported an increase in the TeV gamma-rays flux. MAGIC observed the Crab during the September 2010 flare, finding no indication for variability above 300 GeV[5] , and during the April 2011 flare (shown in Fig. 2 and reported in[6]) and no variability observed in the energy range 700 GeV - 10 TeV. However, given the daily binning of MAGIC data, any shorter term variability cannot be excluded so far.

1.1.2. *Crab pulsar*

Most models for gamma-ray emission from pulsars (such as polar cap, outer or slot gap) predict exponential or super-exponential cut-offs in pulsar spectral energy distribution at a few GeV and this is indeed what *Fermi LAT* has observed in the 100 MeV - 10 GeV energy range (e.g.[7]) for many pulsars. Thanks to its low energy threshold MAGIC has the capabilities to test

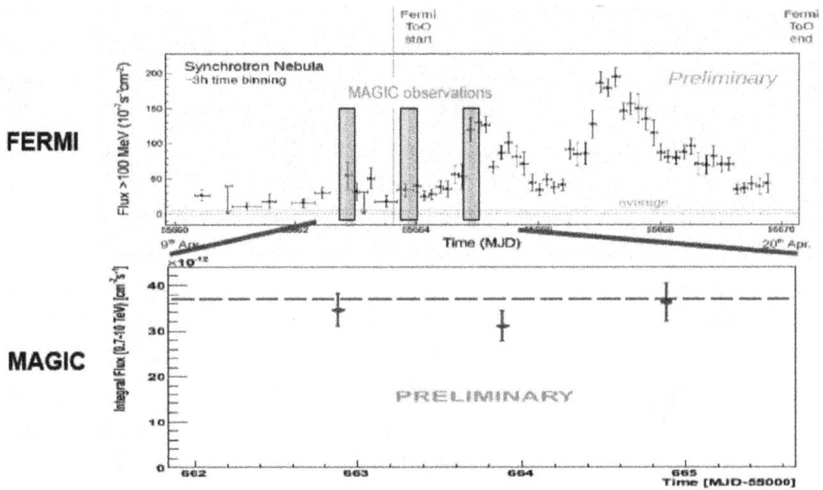

Fig. 2. Crab nebula was observed for 3 nights of data in April 2011: during the third night Fermi LAT was observing a flux above 100 MeV that was 15 times higher than the steady flux.[6]

this trend; the Crab pulsar has been observed for 59 hours (between October 2007 and February 2009) with MAGIC-I, allowing the extraction of detailed phase–resolved spectra between 25 GeV and 100 GeV: the spectra show a power–law behavior and the cut-off extrapolation is ruled out at more than 5 standard deviations[8] .

After fall 2009, the Crab pulsar was observed for 73 hours in stereoscopic mode; the phase–resolved spectra extracted from those observations agree with the ones obtained with MAGIC-I, their simple power law behavior is confirmed and they extend well beyond a cut-off at few GeV energies[9] . Moreover MAGIC data are in good agreement with *Fermi LAT* and VERITAS[10] observations; hence "standard" models above mentioned cannot explain this observed behaviour. Is the Crab pulsar atypical or do other pulsars also have such a VHE power law tail?

1.1.3. *Extended sources*

Thanks to the improved stereoscopic system, MAGIC is more performant also for studying extended sources.

- HESS J1857+026, a VHE unidentified gamma-ray source discovered by H.E.S.S. in 2008 and after suggested to be a PWN pow-

ered by the energetic pulsar PSR J1856+0245, has been detected by MAGIC in 2010, allowing us to investigate its energy dependent morphology[11] . Its spectrum fits well with a power law, consistent with the H.E.S.S. one and its extrapolation; in order to agree with the LAT data a spectral turnover is needed at 10-100 GeV: this could be naturally explained with an Inverse Compton turnover, which would confirm the leptonic nature suggested for this source.
- W51 was detected in GeV gamma-rays energies by *Fermi LAT*[12] and at TeV energies by H.E.S.S.[13] ; its gamma-ray emission is thought to have its origin in the SNR/MC interaction. MAGIC clearly detected W51 in 2010 (more than 8 standard deviations in 31 hours of observation), confirming its extent ($\sim 0.16°$) and the fact that the VHE emission spatially coincides with shocked MC[14] . Its spectral shape would confirm the hadronic origin of the gamma-ray signal as well.

1.1.4. *XRBs and Galactic microquasars*

Two nice examples of binaries have been observed by MAGIC in the last years:

- The high mass X-ray binary system (HXRB) LSI+61 303 consists of a compact object of unknown nature (i.e. either a Neutron Star or a Black Hole) orbiting around a Be star of 13 solar masses, with a period of ~27 days, and it is located at 2 kpc of distance. It was discovered in VHE gamma-rays by MAGIC in 2005 and regularly monitored since then. In 2008, LSI+61 303 faded, however in 2009 we managed to detect it during this low VHE state; and more recently, between Autumn 2010 and Spring 2011, the luminosity of this HXRB was back to the level at which it was first detected[15] ; eventual correlations with the superorbital periodicity (~4.6 years) observed in radio are currently under investigations.
- HESS J0632+057 was discovered by H.E.S.S. in 2007; it was the first point-like unidentified source seen at VHE energies and it is the first binary discovery triggered by VHE observations. It was detected with MAGIC in 2011[16] in coincidence with a high X-ray activity period. Currently this source is monitored with MAGIC, H.E.S.S. and VERITAS and hence is a very nice example of synergy among different IACTs

However there is another class of XRBs, i.e. objects that show an accretion disk around the compact object and jet-like structures orthogonal to it, the so-called Galactic microquasars, monitored by MAGIC and not detected so far: e.g Cygnus X-1 has been observed for more than 100 hours leading to no detection, and, more recently, also upper limits on GRS1915+105 and Scorpius X-1 have been released by our collaboration[17][18] ; the current upper limits put constraints to the gamma-ray luminosity to be a very small fraction of the kinetic luminosty of the jets.

1.2. *Extragalactic observations*

The importance of an improved stereoscopic system reflects also on extragalactic science; in fact in the last 12 months, seven extragalactic objects have been discovered at VHE energies thanks to MAGIC: 4 BL Lac objects (1ES1741+196, 1ES 1215+303, MAGIC J2001+435 and B3 2247+381), 2 radio-galaxies (NGC 1275 and IC 310, visible in Fig. 3) and one Flat Spectrum Radio Quasar (PKS 1221+21).

Another successful strategy in order to discover new extragalactic objects in VHE gamma-rays is represented by the optical trigger, i.e. by monitoring regularly candidate sources by the optical KVA telescope in La Palma (close to MAGIC site) and observing the candidates with MAGIC during their high optical states. This led us to discover several BL Lac objects, such as Mrk180, 1ES1011+496, S5 0716+714, B3 2247+381 and 1ES1215+303.

Another crucial improvement is represented by the complementarity with gamma-ray satellites, that allow us generate much more detailed Spectral Energy Distributions of the sources, by covering in details the High Energy component over 5 decades, and, by means of monitoring these sources also at Radio, optical and X-ray wavelengths, allow us to cover simultaneously more than 17 decades in energy (e.g.[19]).

However, present day the extragalactic VHE science is no longer restricted to BL Lac objects: MAGIC detected successfully also Radio galaxies (such as M87, IC 310 and NGC 1275, e.g.[20]) and FSRQs[21][22] .

1.2.1. *Quasars, the most distant objects*

After the surprising detection by MAGIC, 3C 279 (i.e. the most distant object ever detected at VHE; z=0.536) has been re-observed by MAGIC[21] , confirming that its emission is harder than expected, and showing that the Universe is more transparent to gamma-rays than previously predicted.

This low upper limit to the Extragalactic Background Light (EBL)

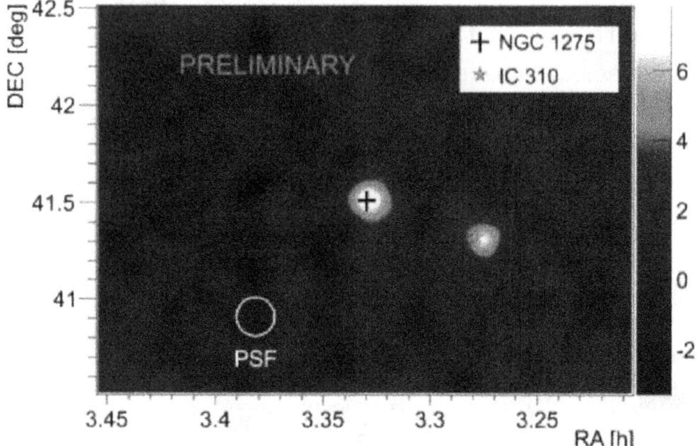

Fig. 3. MAGIC preliminary count map of the Perseus cluster region, showing the detection of the two radio galaxies, IC 310 and NGC 1275, in the same field of view.[20]

has been confirmed with the discovery at VHE of another FSRQ: PKS 1222+21[22] . Observations of PKS 1222+21 and 3C 279 show the same features:

- Emission up to hundreds of GeV.
- Fast variability (e.g in PKS 1222+21 we saw 9 minutes doubling times).
- No signs of intrinsic cut-off.

To reconcile those hard spectra with such a high variability is still a challenge for the theoretical models of photon emission in this type of sources. In fact, in the standard picture[23][24][25] , if gamma-rays are produced outside the Broad Line Region (BLR) by Inverse Compton scattering of dusty torus photons, we can explain explain the smaller-than-expected but it is hard to explain the fast variability; in contrast, if gamma-rays are produced inside the BLR by Inverse Compton scattering of BLR photons, we can explain the variability, but we expect strong absorption and Klein-Nishina suppression (i.e. a cut-off at energies lower than 100 GeV). More complex models are currently under evaluation, such as strong recollimations of the jet, or the presence of blobs or minijets inside the jet, or the so-called two-zone model (i.e. a large emission zone inside the BLR and in addition a small blob outside).

1.2.2. *GRBs*

As mentioned in section 1 MAGIC was especially designed to search for prompt emission of GRBs, thanks to its fast repositioning (less than 20 seconds for 180°) automatically after an alert. We are observing in average ~1 GRB/month and so far none has been detected at VHE energies.

Recently, following a X-ray detection, GRB110328 was observed was observed. After a multiwavelength follow-up, it turned out to be hardly classified as a GRB due to its long-lasting activity. The nature of this source is still uncertain[26] .

1.3. *Fundamental physics*

1.3.1. *Cosmic Rays*

The first MAGIC results on CR electrons spectrum, visible in Fig. 4, are based on 14 hours of data taken in 2009-2010: the $e^{\pm}$ spectrum has been measured in the energy range between 100 GeV and 3 TeV. The energy distribution agrees well with the previous measurements of H.E.S.S. and Fermi LAT (and the peak detectedby ATIC cannot be excluded or confirmed)[27] .

A related initiative (following the experience of the ARTEMIS experiment) consists in probing the e^+/e^- ratio at 300-700 GeV by measuring the shadowing of the CR flux by the Moon. This measurable effect (for 50 hours of observations) has been estimated to be ~4.4% of the Crab flux in the range 300-700 GeV; given its small observability window (i.e. the most favorable observation periods are the spring equinox and the autumn equinox for e^+ and e^- respectively) this measure could be possible with MAGIC integrating data over a few years (for details[28]).

1.3.2. *Dark Matter: indirect searches*

Super-Symmetrical (SUSY) extensions of the standard model foresee the existence of stable, weakly interacting particles (e.g. the lightest neutralino) which could account for part of the dark matter in the Universe. The annihilation of neutralinos may give rise to gamma-rays in the energy range accesible to MAGIC. Such signals have been sought with MAGIC in several targets:

- Galaxy Clusters. MAGIC searches concentrated on the Perseus cluster, that is really challenging for several reasons: (1) for the presence of NGC 1275 and IC 310, (2) since the expected flux

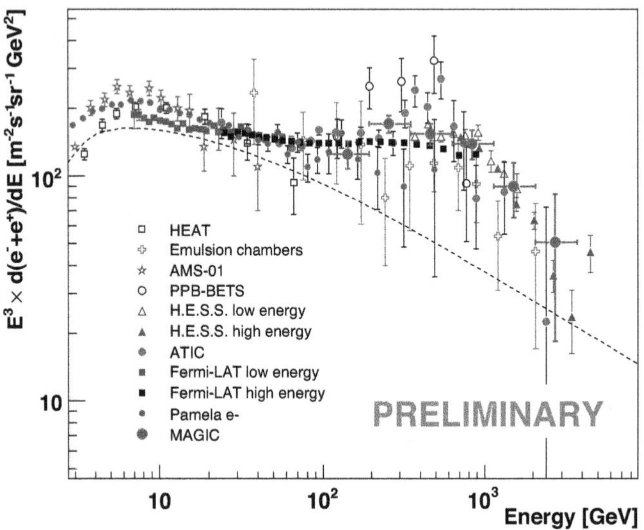

Fig. 4. The $e^{\pm}$ CR spectrum measured by MAGIC is in good agreement with the previous measurements[27] .

much smaller than the one coming from CRs and moreover (3) it could possibly have a very extended DM profile.

- Unidentified Fermi Objects. 1FGL J0338.8+1313 and 1FGL J2347.3+0710 show a hard spectrum in *Fermi LAT* data and they could be DM micro-spikes in the Galactic halo; they have been observed for relatively short exposures ($\sim$10 hours) and lead to no detection so far[29] .
- Dwarf spheroidal galaxies. Several of them were observed in the past (such as Draco and Willman-1) and recently Segue-1, i.e. the most DM dominated object known so far ($M/L > 1000$), has been observed for 29 hours, showing no significant excess[30] .

DM indirect searches leaded to no detections so far and the derived upper limits are still above theoretical expectations.

2. Acknowledgments

The MAGIC Collaboration would like to thank the Instituto de Astrofisica de Canarias for the excellent working conditions at the Observatorio del

Roque de los Muchachos in La Palma. The support of the German BMBF and MPG, the Italian INFN, the Swiss National Fund SNF, and the Spanish MICINN is gratefully acknowledged. This work was also supported by the Marie Curie program, by the CPAN CSD2007-00042 and MultiDark CSD2009-00064 projects of the Spanish Consolider-Ingenio 2010 programme, by grant DO02-353 of the Bulgarian NSF, by the YIP of the Helmholtz Gemeinschaft, by the DFG Cluster of Excellence "Origin and Structure of the Universe", by the DFG Collaborative Research Centers SFB823/C4 and SFB876/C3, and by the Polish MNiSzW grant 745/N-HESS-MAGIC/2010/0.

References

1. The MAGIC collaboration, arXiv:1108.1477 (2011).
2. Rissi et al., IEEE, 56, 3840 (2009).
3. Baldini et al. (Fermi LAT collaboration), these proceedings.
4. Marsella et al. (ARGO YBJ collaboration), these proceedings.
5. Mariotti et al.(MAGIC collaboration), ATel. 2967 (2011).
6. Zanin et al.(MAGIC collaboration), 32nd ICRC (2011).
7. Abdo et al. (Fermi LAT collaboration) ApJS, 187, 460 (2010).
8. The MAGIC collaboration, arXiv:1108.5391 (2011).
9. The MAGIC collaboration, arXiv:1109.6124 (2011).
10. Aliu et al. (VERITAS collaboration), arXiv:1108.3797 (2011).
11. Klepser et al.(MAGIC collaboration), 32nd ICRC (2011).
12. Abdo et al. (Fermi LAT collaboration) ApJ, 706, L1 (2009).
13. Fiasson et al. (H.E.S.S collaboration), 31st ICRC (2009).
14. Carmona et al.(MAGIC collaboration), 32nd ICRC (2011).
15. Jogler et al.(MAGIC collaboration), 32nd ICRC (2011).
16. Mariotti et al.(MAGIC collaboration), ATel. 3161 (2011).
17. The MAGIC collaboration,ApJ, 735, L5 (2011).
18. Rico et al.(MAGIC collaboration), 32nd ICRC (2011).
19. The MAGIC collaboration, arXiv:1106.1589 (2011).
20. Hildebrand et al.(MAGIC collaboration), 32nd ICRC (2011).
21. The MAGIC collaboration, A&A, 530, id.A4 (2011).
22. The MAGIC collaboration, ApJ, 730, L8 (2011).
23. Dermer et al., ApJ, 692, 32 (2009).
24. Ghisellini and Tavecchio, MNRAS, 397, 985 (2009).
25. Sikora et al., ApJ, 704, 38 (2009).
26. Berger et al.(MAGIC collaboration), 32nd ICRC (2011).
27. Borla Tridon et al.(MAGIC collaboration), 32nd ICRC (2011).
28. Colin et al.(MAGIC collaboration), arXiv:0907.1026 (2009).
29. Nieto et al.(MAGIC collaboration), 32nd ICRC (2011).
30. The MAGIC collaboration,JCAP, 06, 035 (2011).

Search for massive penetrating particles in the Cosmic Radiation

V. Togo* for the SLIM Collaboration†

INFN sez. Bologna, v.le Berti Pichat 6/2, I-40127 Bologna, Italy.
**E-mail: Vincent.Togo@bo.infn.it*

Different exotic particles have been suggested to be part of the cosmic radiation flux, such as Magnetic Monopoles, Strange Quark Matter, Q-balls in a very wide mass range, from multi-TeV to 10^{17} GeV and beyond.
The SLIM experiment based on a large array of nuclear track detectors at the Chacaltaya Laboratory (Bolivia) reached the highest sensitivity for Intermediate mass MMs, nuclearites and charged Q-balls. After discussing the experimental procedure, the results obtained by the SLIM experiment are compared with the sensitivity reachable by future searches with large volume neutrino telescopes and space based experiments.

Keywords: Magnetic Monopoles, Strage Quark Matter, Nuclearites, Q-balls.

1. Introduction

Magnetic Monopoles (MMs) were introduced in 1931 by Dirac to explain the quantization of the electric charge[1] : $e\,g = n\hbar c/2 = n \times g_D$, where $n = 1, 2, ...$; $g_D = \hbar c/2e = 68.5\,e$ and e are respectively the Dirac unit and the electron charges.

Pointlike, magnetic monopoles are usually referred to as "classical" or "Dirac" monopoles, they have been extensively searched for at every new accelerator/collider.

So-called "primordial" GUT monopoles are topological "point" defects possibly produced in the Early Universe at the phase transition corresponding to the spontaneous breaking of the Unified Gauge group into subgroups, one of which is U(1)[2,3] . GUT MMs would have masses as large as

†The SLIM Collaboration: S. Balestra, S. Cecchini, M. Cozzi, D. Di Diferdinando, M. Errico, F. Fabbri, G. Giacomelli, R. Giacomelli, M. Giorgini, A. Kumar, S. Manzoor, J. McDonald, G. Mandrioli, S. Marcellini, A. Margiotta, E. Medinaceli, L. Patrizii, J. Pinfold, V. Popa, I.E. Qureshi, O. Saavedra, Z. Sahnoun, G. Sirri, M. Spurio, V. Togo, C. Valieri, A. Velarde, A. Zanini.

$10^{16} - 10^{17}$ GeV$/c^2$. They could be a component of the galactic cold dark matter with typical velocities of $\sim 10^{-3}c$ (c : velocity of light) and can be searched for in the cosmic radiation.

Later phase transitions could have lead to the production of Intermediate Mass Monopoles (IMMs)[4] with masses in the range $10^5 \div 10^{13}$ GeV$/c^2$. These later may be accelerated to very high velocities in one coherent domain of the galactic magnetic field[5–7] .

With its high altitude exposure, the SLIM (Search for LIght magnetic Monopoles) experiment deployed at the Chacaltaya laboratory in Bolivia, was sensitive to downgoing light and intermediate mass monopoles. As a by-product, it also searched for Nuclearites (nuggets of strange quark matter (SQM))[8,9] and Q-balls (supersymmetric states of squarks, sleptons and Higgs fields)[10] , other exotic particles which could also contribute to the galactic dark matter.

In the following we briefly present the apparatus and the experimental procedure. Then, we give the resulting flux upper limits for IMMs, SQM nuggets and Q-balls set by SLIM and compare them with the sensitivity reachable by other searches with neutrino telescopes and space based experiments.

2. Experimental procedure

The SLIM experiment was a large array ($\sim$427 m^2 area) of Nuclear Track Detectors (NTDs) deployed at the Chacaltaya laboratory (5230 m a.s.l.). The detector modules (each 24.5×24.5 cm^2) consisted of stacks composed of three layers of CR39®, three layers of Makrofol DE®, two layers of Lexan and a 1 mm thick aluminium absorber to slow down or stop nuclear recoils. The detectors were exposed to the cosmic radiation for an average of 4.22 years[11] after which they were brought back to the Bologna Laboratory where they were etched in "strong" and "soft" etching conditions, as described in Refs. 12–14, and analyzed.

When a charged particle crosses a nuclear track detector it produces damages along its trajectory forming the so-called latent track. The following chemical etching leads to the formation of etch-pit cones in both front and back faces of the detector foils, provided the etching velocity v_T along the track is larger than the etching velocity of the bulk material, v_B. The response of the detector is given by the etching rate ratio $p = v_T/v_B$ which is dependent on the Restricted Energy Loss (REL) of the incident ion in the medium and thus on $(Z/\beta)^2$ with Z the charge, given in units of e, and β (v/c) the velocity of the incident ion. The calibration with heavy relativistic ion beams (In^{49+}, Pb^{82+} and Fe^{26+}) at accelerator facilities allowed the build of a data set values of (p-1) versus REL.

The strong etching conditions for CR39 were : 8N KOH + 1.5% ethyl alcohol at 75 °C for 30 hours leading to a threshold at $Z/\beta \sim 14$, equivalent to REL $\sim$ 200 MeV g^{-1} cm^2. The soft etching conditions were : 6N NaOH + 1% ethyl alcohol at 70 °C for 40 hours and gave a threshold at the level of $Z/\beta \sim 7$, corresponding to REL $\sim$ 50 MeV g^{-1} cm^2. About 50 m^2 of the SLIM CR39 contained 0.1% of DOP additive in order to reduce the neutron induced background by raising the detector threshold[15] to $Z/\beta \sim 21$ in strong etching conditions corresponding to REL $\sim$ 460 MeV g^{-1} cm^2, and $Z/\beta \sim 13$ for soft etching, corresponding to REL $\sim$ 170 MeV g^{-1} cm^2. Finally, for Makrofol the threshold was at $Z/\beta \sim 50$, corresponding to REL $\sim$ 2.5 GeV g^{-1} cm^2 in 6N KOH + 20% ethyl alcohol at 50 °C for 10 hours etching conditions.

The energy loss of IMMs, SQM and Q-balls is constant in the detector modules thus, the chemical etching of SLIM NTDs was expected to lead to the formation of collinear and equally sized etch-pit cones in both top and bottom faces of each detector sheet. The standard analysis procedure was the following: all uppermost CR39 layers were etched in strong etching conditions, then scanned using a high magnification stereo microscope looking for a twofold coincidence between equally sized front and back cones. The bottom CR39 sheets were softly etched when a track was considered as a possible "candidate" in the uppermost layers (about 10% of the cases). The third CR39 sheet was etched only in few cases, when there was still a possible uncertainty. A few % of Makrofol foils were etched for similar reasons. More details about the experimental analysis procedure of NTDs, are given in Ref. 14 and references therein.

3. Results and Discussion

SLIM was sensitive to MMs with $g = 2g_D$ in the range $4 \times 10^{-5} < \beta < 1$ and $g = g_D$ for $\beta > 10^{-3}$. It was also sensitive to downgoing nuclearites and Q-balls with $\beta > 10^{-4}$. With an exposure of 4.22 years no candidate event was found; this allowed the setting of a 90% C.L. flux upper limit of $\sim 1.3 \cdot 10^{-15}$ cm^{-2} s^{-1} sr^{-1} for downgoing IMMs, Nuclearites and Q-balls.

In Fig.1 are plotted flux upper limits versus mass of IMMs as set by SLIM and other experiments for two different MM velocities. From this plot we see that SLIM was able to set flux upper limits to MMs with masses of several orders of magnitude lower than earlier experiments[16,17] ; more stringent limits may be given by large volume neutrino telescopes.

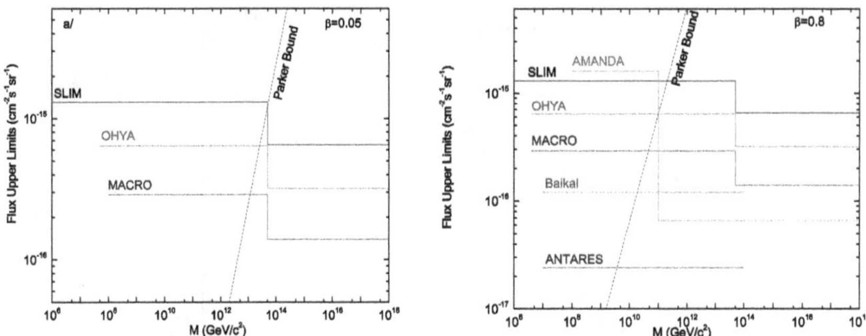

Fig. 1. Experimental 90% CL flux upper limits versus mass for MMs with (a) $\beta = 0.05$, (b) $\beta = 0.8$ at the detector level from different experiments.

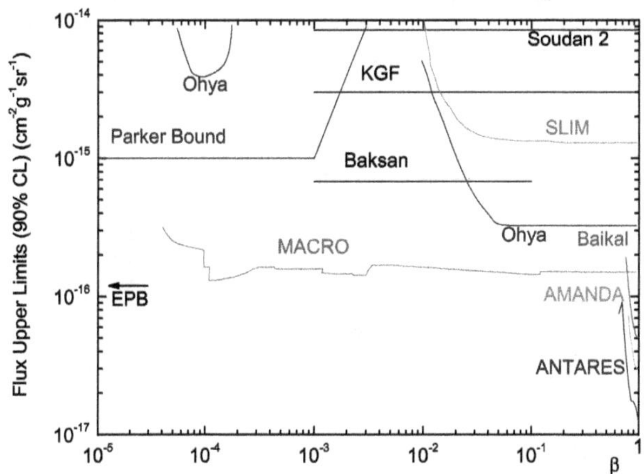

Fig. 2. The 90% CL upper limits vs β for a flux of cosmic GUT monopoles with magnetic charge $g = g_D$.

Searches with underwater and underice neutrino telescopes are sensitive to relativistic MMs which would be detected by their large amount of Cherenkov radiation, $\gtrsim 8300$ times that of muons.

In Fig.2 the 90% CL flux upper limits versus β for MMs with $g = g_D$ as set by the SLIM[14] , MACRO[16] , Ohya[17] , Baksan[18] , Baikal[19] , AMANDA[20] and ANTARES[21] experiments are shown. The Baikal and AMANDA limits were obtained assuming that relativistic MMs would reach the detector from "below", i.e. after crossing the Earth (which is unlikely).

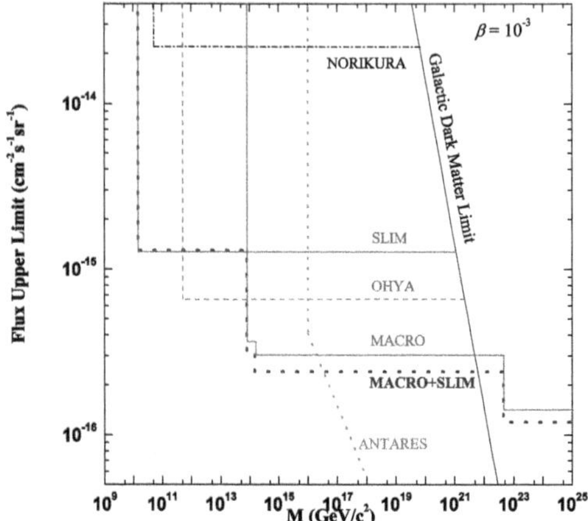

Fig. 3. 90% CL flux upper limits vs mass for intermediate and high mass nuclearites with $\beta = 10^{-3}$ obtained from various searches with NTDs. A combined flux from the MACRO and SLIM experiments and a preliminary result from the ANTARES neutrino telescope[23] are also shown.

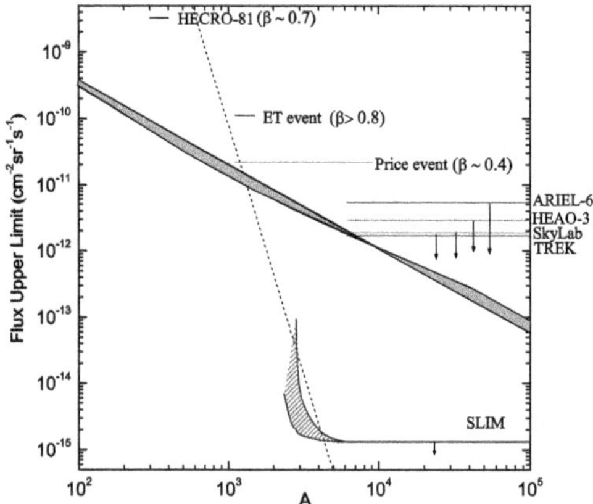

Fig. 4. 90% CL flux upper limits vs mass number A for relativistic strangelets from on-board balloon and space experiments and by the SLIM detector at mountain altitude.[22]

Flux upper limits on nuclearites were obtained as by-products of magnetic monopole searches[22] . In Fig. 3 the most stringent limits for nuclearites with $\beta = 10^{-3}$ are at the level of $1.4 \div 3 \times 10^{-16}$ cm^{-2}s^{-1}sr^{-1}. The galactic dark matter bound was obtained assuming that all dark matter is composed of nuclearites. Similar limits were obtained for charged Q-balls.

Smaller SQM bags with mass $< 10^6 \div 10^7$ GeV/c^2, usually called "strangelets", could also be searched for with SLIM. In Fig. 4 we show the flux upper limits for relativistic strangelets obtained with SLIM and by experiments onboard stratospheric balloons and in space[24–27] . The fluxes of some cosmic ray unusual events are also shown (the three uppermost horizontal lines)[28–30] .

4. Conclusion

The analysis of 427 m^2 of CR39 NTDs of the SLIM experiment exposed for 4.22 years at an altitude of 5230 m a.s.l. did not show any candidate event. This resulted in new upper limits (90% C.L.) on the flux of IMMs at the level of 1.3×10^{-15} cm^{-2}s^{-1}sr^{-1}. These limits could be improved with much larger detectors, in particular by large volume neutrino telescopes[20,21,31] .

As a by-product of MM searches some experiments obtained stringent limits on nuclearites and on Q-balls in the cosmic radiation. Future searches with neutrino telescopes[23] and in space[32,33] may reach improved sensitivities to nuclearites, Q-balls and to strangelets of smaller masses.

References

1. P. A. M. Dirac, *Proc. R. Soc. London* **133**, 60 (1931); *Phys. Rev.* **74**, 817 (1948).
2. G.'t Hooft, *Nucl. Phys. B* **29**, 276 (1974).
3. A. M. Polyakov, *JETP Lett.* **20**, 194 (1974); N.S. Craigie et al., *Theory and Detection of MMs in Gauge Theories*, (World Scientific, Singapore, 1986).
4. G. Lazarides et al., *Phys. Rev. Lett.* **58**, 1707 (1987); T. W. Kephart and Q. Shafi, *Phys. Lett. B* **520**, 313 (2001).
5. P. Bhattacharjee and G. Sigl, Phys. Rept. **327** (2000) 109.
6. T. J. Weiler and T. K. Kephart, *Nucl. Phys. B. Proc. Suppl.* **51**, 218 (1996); T. W. Kephart and T. J. Weiler , *Astropart. Phys.* **4**, 217 (1996).
7. C. O. Escobar and R. A. Vazquez, *Astropart. Phys.* **10**, 197 (1999).
8. E. Witten, *Phys. Rev. D* **30**, 272 (1984).
9. A. De Rujula and S. L. Glashow, *Nature* **312**, 734 (1984); A. De Rujula, *Nucl. Phys. A* **434**, 605 (1985).
10. S. Coleman, *Nucl. Phys. B* **262**, 263 (1985).
    A. Kusenko et al., *Phys. Lett. B* **418**, 46 (1998).
11. D. Bakari et al., hep-ex/0003028.
    S. Cecchini et al., *Il Nuovo Cimento C* **24**, 639 (2001).
12. S. Balestra et al., *Czech. J. Phys.* **56**, A221 (2006), hep-ex/0601019.
    S. Cecchini et al., *Radiat. Meas.* **40**, 405 (2005).
13. S. Balestra et al., *Nucl. Instrum. Meth. B* **254**, 254 (2007).
    S. Manzoor et al., *Radiat. Meas.* **40**, 433 (2005).

S. Cecchini et al., *Radiat. Meas.* **34**, 55 (2001).
G. Giacomelli et al., *Nucl. Instrum. Meth. A* **411**, 41 (1998).

14. S. Balestra et al., *Eur. Phys. J. C* **55**, 57 (2008), arXiv: 0801.4913 [hep-ex].
15. G. Tarlé et al., *Nature* **393**, 556 (1981).
16. M. Ambrosio et al., MACRO Coll., *Eur. Phys. J. C* **25**, 511 (2002); *Phys. Lett. B* **406**, 249 (1997); *Phys. Rev. Lett.* **72**, 608 (1994).
17. S. Orito et al., *Phys. Rev. Lett.* **66**, 1951 (1991).
18. E.N. Alexeyev et al., 21st ICRC **10**, 83 (1990); Yu.F. Novoseltsev et al., *Nucl. Phys. B* **151**, 337 (2006).
19. V. Aynutdinov et al., astro-ph/0507713; R. Wischnewski et al. 30th ICRC, arXiv:0710.3064 [astro-ph].
20. R. Abbasi et al., *Eur. Phys. J. C* **69**, 361 (2010); H. Wissing et al., 30th ICRC **4**, 799 (2007). arXiv:0711.0353 [astro-ph].
21. G. Giacomelli, arXiv:1105.1245 [astro-ph.IM].
22. S. Cecchini et al., *Eur. Phys. J. C* **57**, 525 (2008).
23. G.E. Păvălaş, *AIP Conf. Proc.* **1304**, 454 (2010), arXiv:1010.2071 [astro-ph.HE]. V. Popa, Proc. 32nd ICRC, Beijing (2011).
24. P. H. Fowler et al., *Astrophys. J.* **314**, 739 (1987).
25. W. R. Binn et al., *Astrophys. J.* **347**, 997 (1989).
26. E. K. Shirk and P. B. Price, *Astrophys. J.* **220**, 719 (1978).
27. A. J. Westphal et al., *Nature* **396**, 50 (1998); B. A. Weaver et al., *Nucl. Instrum. Meth. B* **145**, 409 (1998).
28. P. B. Price and M. H. Salamon, *Phys. Rev. Lett.* **56**, 1226 (1986); D. Ghosh and S. Chatterjea, *Europhys. Lett.* **12**, 25 (1990).
29. T. Saito et al., *Phys. Rev. Lett.* **65**, 2094 (1990).
30. M. Ichimura et al., *Il Nuovo Cimento A* **106**, 843 (1993).
31. D. Hardtke et al., *Proc. International Workshop on Exotic Physics with Neutrino Telescopes*, 89 (2006).
32. J. Sandweiss, *J. Phys. G* **30**, S51 (2004).
33. E. Finch, *J. Phys. G* **32**, S251 (2006).

AMS Observations of Light Cosmic Ray Isotopes and Implications for their Production in the Galaxy

Nicola Tomassetti

Perugia University and INFN, I-06122 Perugia, Italy
**E-mail: nicola.tomassetti@pg.infn.it*

Observations of light isotopes in cosmic rays provide information on their origin and propagation in the Galaxy. Using the data collected by AMS-01 in the STS-91 space mission, we report our final results on the isotopic composition of hydrogen and helium between 200 MeV and 1.4 GeV per nucleon. These measurements are in good agreement with the previous data and set new standards of precision. We discuss the role of isotopic composition data in modeling the cosmic ray production, acceleration and diffusive transport in the Galaxy.

Keywords: cosmic rays — isotopic composition — nuclear reactions

1. Introduction

The rare secondary isotopes ^{2}H, ^{3}He and LiBeB are produced by collisions of primary cosmic rays (CRs) such as ^{1}H, ^{4}He or CNO with the interstellar matter (ISM). Secondary to primary ratios such as ^{2}H/^{4}He, ^{3}He/^{4}He or B/C give us information on the propagation of CRs through the ISM. In many CR propagation studies the key parameters are inferred using the B/C ratio and used to predict the secondary production for other rare species ($\bar{p}$, $\bar{d}$, ...) under the implicit assumption that all CRs experience the same propagation histories [1–3]. It is therefore important to test the CR propagation with nuclei of different mass-to-charge ratios. In this work we report the new AMS-01 observations for the ^{2}H/^{4}He and ^{3}He/^{4}He ratios and compare them with propagation calculations. We study how these data are described by the models consistent with the B/C ratio within their astrophysical uncertainties (related to the CR transport parameters) and nuclear uncertainties (intrinsic of the ^{2}H and ^{3}He production rates).

2. Observations

AMS-01 operated on 1998 June in a 10-day space shuttle mission, STS-91, at an altitude of $\sim$380 km. The spectrometer was composed of a permanent magnet, a silicon micro-strip tracker, time-of-flight scintillators, an aerogel Čerenkov detector and anti-coincidence counters [4]. Results on isotopic spectra have been recently published [5] including the ratios ^{2}H/^{4}He, ^{3}He/^{4}He, ^{6}Li/^{7}Li, ^{7}Be/(^{9}Be+^{10}Be) and ^{10}B/^{11}B in the range $0.2-1.4$ GeV of kinetic energy per nucleon. Figure 1 shows the AMS-01 energy spectra of

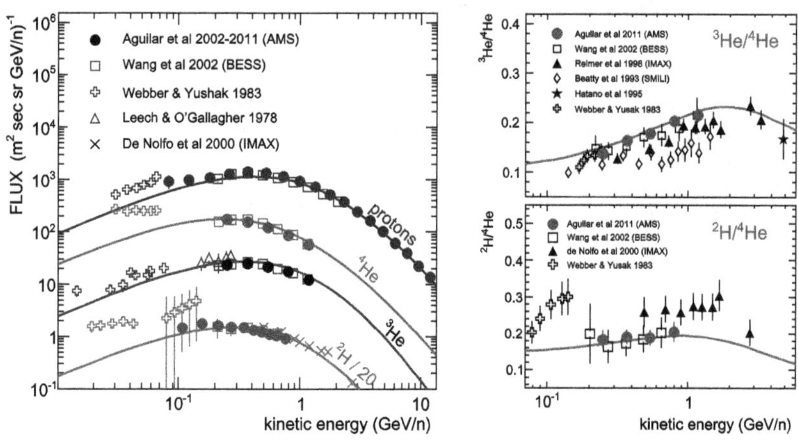

Fig. 1. Left: energy spectra of CR proton, deuteron (divided by 20), ^{3}He and ^{4}He. Right: isotopic ratios ^{2}He/^{4}He and ^{3}He/^{3}He. Calculations are shown together with the experimental data [5–11].

proton, deuteron, helium isotopes, and the ratios ^{2}He/^{4}He and ^{3}He/^{3}He. These data are free from atmospheric background. Other data come from balloon borne experiments [6–11].

3. Cosmic Ray Transport and Interactions

The Galactic CR transport is characterized by diffusion in the turbulent magnetic field, nuclear interactions, decays, energy losses and diffusive reacceleration. We describe the data using `GALPROP-v50.1` which numerically solves the CR propagation equation in a cylindrical diffusive region for given source and matter distributions [3]. We adopt the "conventional model" which finely reproduces the primary CR fluxes and the B/C ratio at intermediate energies under a diffusion-reacceleration scenario. We model the heliospheric propagation under the *force-field* approximation [12], using the parameter $\phi = 500\,\mathrm{MV}$ to characterize the modulation strength for 1998 June. To compute the ^{2}H and ^{3}He production rate in the ISM, we consider the reactions ^{4}He$\to$^{3}He, ^{4}He$\to$^{3}H, ^{4}He$\to$^{2}H, CNO$\to$^{3}He, and $p + p \to \pi$+^{2}H (the ^{3}H isotopes subsequently decay into ^{3}He). Most of the reactions involve the ^{4}He spallation cross sections (CSs) that we have adapted from the parametrizations of [13]. Contributions from of heavier nuclei (CNO or Fe, giving $\sim$5% of the flux) are included using the CS data from the latest `GALPROP-v54` version [14]. For collisions with He targets ($\sim$10% of the ISM) the algorithm of [15] is used. We assume the *straight-*

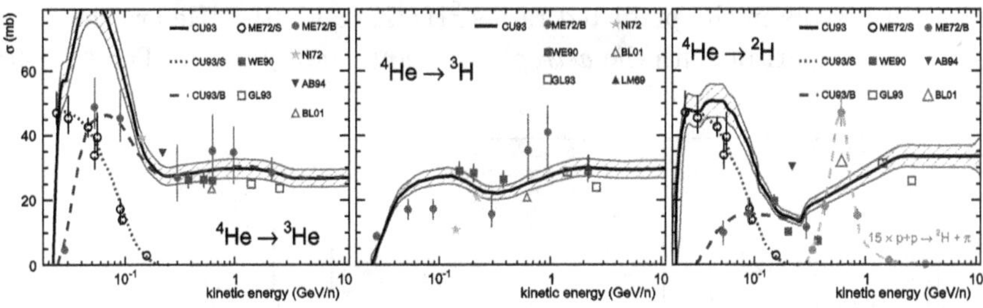

Fig. 2. CS parametrization [13] for the channels ^{4}He$\to$^{3}He, ^{4}He$\to$^{3}H, ^{4}He$\to$^{2}H and $p + p \to \pi$+^{2}H (multiplied by 15). The experimental data are found in [13,16,17].

ahead approximation to link the fragment-progenitor energies E and E' through $\sigma(E, E') \approx \sigma(E)\delta(E - E')$, that is valid at some percent of accuracy when the progenitor is heavier than the fragment [13,18]. For the *p-p* fusion channel the kinetic energy per nucleon is not conserved: we assume $\sigma(E, E') \approx \sigma(E)\delta(E - \xi E')$, where $\xi \approx 4$ is the average inelasticity for the ^{2}H production [16]. This reaction contributes to the ^{2}H flux at $E \lesssim 250\,\mathrm{MeV\,nucleon}^{-1}$. The main CSs are shown in Fig. 2 together with the data. For ^{2}H and ^{3}He the contributions of break-up (B) and stripping (S) reactions are shown separately. Predictions for CR spectra and for the ratios ^{2}H/^{4}He and ^{3}He/^{4}He under this setup are reported in Fig. 1.

4. Model Uncertainties for the ^{2}H and ^{3}He Productions

We consider two classes of uncertainties. The *astrophysical uncertainties* are related to the transport parameters constrained by the B/C data. The relevant parameters are δ (diffusion coefficient spectral index), v_A (Alfvénic speed) and the ratio between D_0 (diffusion coefficient normalization) and L (halo height). We have performed a grid scan in the parameter space $\{\delta, v_A, D_0/L\}$ by running GALPROP several times, while the other inputs (*e.g.* source parameters or ϕ) are kept fixed. In order to derive the astrophysical errors for the $Z \leq 2$ predictions, we select the models compatible with the B/C data within one sigma of uncertainty. Our purpose is estimating the parameter uncertainties rather than determining their values (*e.g.*, as recently done in [19]). The *nuclear uncertainties* on the ^{2}H and ^{3}He calculations are those arising from uncertainties in their production CSs. In order to estimate these uncertainties using the CS data, we re-fit the normalization factors of their parametrizations. The error bands are shown in Fig. 2 for the main production CSs of ^{2}H and ^{3}He. Their corresponding errors in the predicted ratios ^{2}H/^{4}He and ^{3}He/^{4}He are shown in Fig. 3 in comparison with the *astrophysical uncertainty* bands. The AMS-01

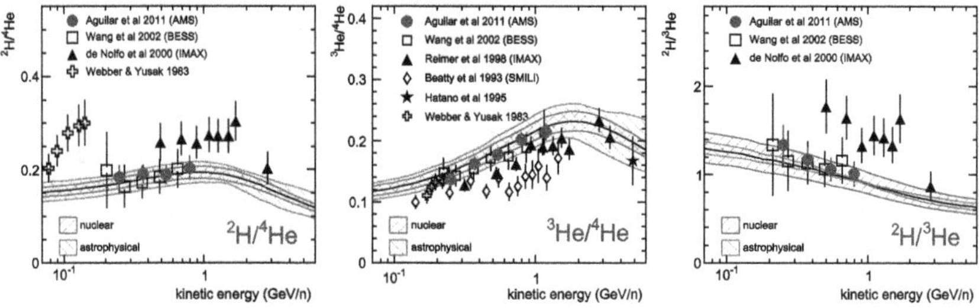

Fig. 3. Astrophysical and nuclear uncertainty bands for the predicted ratios ^{2}H/^{4}He, ^{3}He/^{4}He and ^{2}H/^{3}He in comparison with the experimental data [5–11].

data agree well with the calculations within the *astrophysical uncertainties*, indicating a good consistency with the B/C-based propagation picture. It is also clear that these ratios carry valuable information on the CR transport parameters and can be used to tighten the constraints given by the B/C ratio. On the other hand, the *nuclear uncertainties* represent an intrinsic limitation on the accuracy of the predictions. Unaccounted errors or biases in the CS estimates cause errors on the predicted ratios which, in turn, may lead to a mis-determination of the CR transport parameters. For high precision data upcoming from PAMELA or AMS-02, CS errors may become the dominant source of uncertainty. A strategy to test the model consistency with CR data is given by the comparison with secondary to secondary ratios such as ^{2}H/^{3}He. In fact the ^{2}H and ^{3}He isotopes have similar astrophysical origin, so that their ratio is almost insensitive to the propagation physics and can be used to probe the net effect of nuclear interactions. Thus, a mis-consistency between calculations and ^{2}H/^{3}He data would indicate systematic biases in the CSs that cannot be re-absorbed by a different choice of the propagation parameters. From Fig. 3, the nuclear uncertainty in the ^{2}H/^{3}He ratio is larger than the astrophysical one.

5. Conclusions

We have compared the AMS-01 observations of the ^{2}H/^{4}He and ^{3}He/^{4}He ratios in CRs with model predictions for their production in the ISM. These ratios are well described by propagation models consistent with the B/C ratio, suggesting that He and CNO nuclei experience similar propagation histories. The accuracy of the secondary CR calculations depends on the reliability of the CSs employed. CS parametrizations may be improved using more refined calculations or more precise accelerator data. The use of ratios such as ^{2}H/^{3}He, ^{6}Li/^{7}Li or ^{10}B/^{11}B can represent a possible diagnostic test for the reliability of the calculations: any CR propagation model, once tuned on secondary to primary ratios, must correctly reproduce the secondary to secondary ratios as well. Precision modeling may also require a more refined solar modulation description in place of the simple *force-field*. This aspect can be better inspected by the AMS-02 log-term observations.

References

1. Strong, A. W., et alii, 2007, Ann. Rev. Nucl. & Part. Sci., 57, 285–327
2. Putze, A., et alii, 2010, A&A, 516, 66
3. Trotta, R., et alii, 2011, ApJ, 729, 106–122
4. Aguilar, M., et alii, 2002, Phys. Rep., 366, 331–405 (AMS-01)
5. Aguilar, M., et alii, 2011, ApJ, 746, 105 (AMS-01)
6. de Nolfo G.A. et alii, 2000, in Proc. ACE2000 Symp., AIP 528, 425 (IMAX)
7. Reimer, O., et alii, 1998, ApJ, 496, 490 (IMAX)
8. Ahlen, S. P., et alii, 2000, ApJ, 534, 757–769 (SMILI-II)
9. Wang, J. Z., et alii, 2002, ApJ, 564, 244–259 (BESS)
10. Hatano, Y., et alii, 1995, Phys. Rev. D, 52, 6219–6223
11. Webber, W. R., & Yushak, S. M, 1983, ApJ, 275, 391–404
12. Gleeson, L. J., & Axford, W. I., 1968, ApJ, 154, 1011
13. Cucinotta, M. G., et alii, 1993, NASA–TR 3285
14. Vladimirov, A. E., et alii, 2011, Comp. Phys. Comm., 182–5, 1156–1161
15. Ferrando, M., et alii, 1988, PRC, 37–4, 1490–1502
16. Meyer, J. P., 1972, A&A Suppl., 7, 417–467
17. Blinov, A. V., & Chadeyeva, M. V., 2008, Phys. Part. & Nucl., 39, 526–559
18. Kneller, J. P., et alii, 2003, ApJ, 589, 217–224
19. Coste, B., et alii, 2011, arXiv: astro-ph/1108.4349

Production and Propagation of Cosmic Rays in the Galaxy and Heliosphere

Exploring Particle Acceleration in Gamma-Ray Binaries

V. Bosch-Ramon

Dublin Institute for Advanced Studies,
Fitzwilliam 31, Dublin 2, Ireland
**E-mail: valenti@cp.dias.ie*

F. M. Rieger

Max-Planck-Institut fur Kernphysik,
P.O. Box 103980, 69029 Heidelberg, Germany

Binary systems can be powerful sources of non-thermal emission from radio to gamma rays. When the latter are detected, then these objects are known as gamma-ray binaries. In this work, we explore, in the context of gamma-ray binaries, different acceleration processes to estimate their efficiency: Fermi I, Fermi II, shear acceleration, the converter mechanism, and magnetic reconnection. We find that Fermi I acceleration in a mildly relativistic shock can provide, although marginally, the multi-10 TeV particles required to explain observations. Shear acceleration may be a complementary mechanism, giving particles the final boost to reach such a high energies. Fermi II acceleration may be too slow to account for the observed very high energy photons, but may be suitable to explain extended low-energy emission. The converter mechanism seems to require rather high Lorentz factors but cannot be discarded a priori. Standard relativistic shock acceleration requires a highly turbulent, weakly magnetized downstream medium; magnetic reconnection, by itself possibly insufficient to reach very high energies, could perhaps facilitate such a conditions. Further theoretical developments, and a better source characterization, are needed to pinpoint the dominant acceleration mechanism, which need not be one and the same in all sources.

Keywords: binary systems; non-thermal; gamma rays

1. Introduction

During the last decade, binary systems have turned out to be a new class of gamma-ray sources whose numbers are growing with the increasing sensitivity of the new instrumentation (see Refs. 1–13). Different types of objects pertain to this class, like microquasars, binaries hosting a non-accreting pulsar, massive star binaries, and even symbiotic stars. All these sources share the characteristic of hosting powerful outflows that interact with themselves and their environment, and dissipate their energy partially in the form of non-thermal particles. Given the typical compactness of the emitter, the dynamical and radiation timescales are short, yielding rapidly variable emission that can reach high luminosities, and also very high energies. Particularly interesting in this regard are those gamma-ray binaries that reach energies $\gg$ TeV. In some cases, like LS 5039,[14,15] the emitting particles could be as energetic as ~ 100 TeV, which poses serious constraints on any particle acceleration model that aims at explaining the observed radiation, as noted by Khangulyan et al. (15). Rieger et al. (16) discussed different diffusive acceleration processes for microquasars, and also concluded that any mechanism responsible of the very high-energy emission should run at its limits. In this work, we carry out a semi-quantitative analysis of the requirements and efficiency of diffusive acceleration processes (Fermi I, Fermi II and shear acceleration[17–19]), the converter mechanism and magnetic reconnection (e.g. Refs. 20,21), for gamma-ray binaries in general. Our goal is to take a first look at the problem of extreme acceleration in compact galactic sources, in which dense radiation fields, together with highly supersonic sometimes relativistic bulk motion, shear layers, turbulence, and magnetic fields, are expected.

In Figure 1, a sketch of the general binary scenario discussed here is presented, showing the relevant elements that could play a role in the production of very energetic particles. An interaction structure is formed due to the presence of, for instance, a powerful relativistic outflow from a compact object (e.g. a jet or a pulsar wind) and a strong stellar wind. Two stellar winds could also form a similar though non-relativistic structure. For the case of a jet, it will be more collimated, but jet disruption may also lead to a broadening of the interaction region. Powerful shocks are expected to form at the colliding region: a termination shock in the pulsar, and an asymmetric re-collimation shock when a jet is present. In the jet scenario, internal shocks can also occur. The contact discontinuity between the different flows involved is subject to Rayleigh-Taylor and/or Kelvin-Helmholtz instabilities (neglecting the role of the magnetic field), which will trigger turbulence downstream the flow, as well as mixing from the two different media. Despite of its complexity, the picture can be approximately analyzed (see, e.g., Refs. 22–24), and different acceleration site and mechanism candidates can be proposed. In the presence of strong shocks, diffusive acceleration could be the dominant mechanism. For magnetized, highly turbulent and diluted media Fermi II could be at work, and if strong velocity gradients are present, shear may occur as well. For highly relativistic and radiation or matter dense environments the converter mechanism could play a role, whereas magnetic reconnection could take place under the presence of strong irregular magnetic fields. The modeling of observations tends to favor leptonic models (e.g. Refs. 15,25–29), although hadronic models cannot be discarded (e.g. Ref. 30; see also Ref. 31 and references therein). We will focus here mainly on electron acceleration, strongly affected at the highest energies by synchrotron losses, but some of our conclusions apply to protons as well.

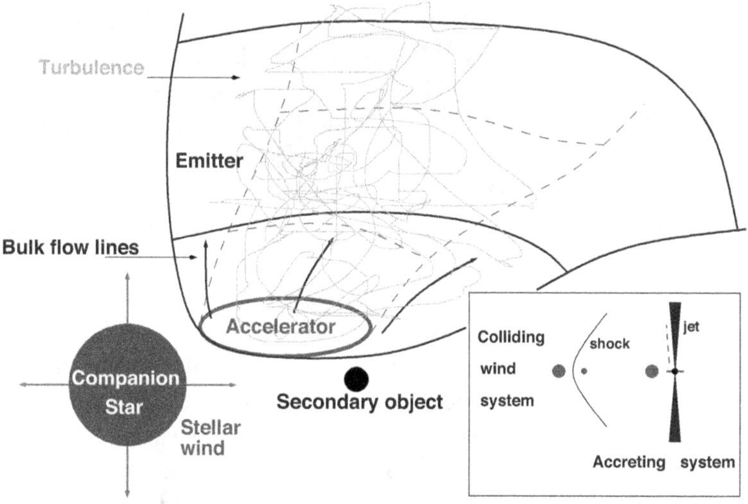

Fig. 1. Sketch of the generic binary scenario.

2. Diffusive acceleration mechanisms

Diffusive shock or Fermi I acceleration takes place through repeated particle bouncing upstream-downstream of a shock front. The particle deflection is mediated by magnetic field irregularities of the background plasma. In each cycle, the particle energy gain is $\Delta E/E \propto (v_{\rm s}/c)$, where $v_{\rm s}$ is the shock velocity. For a mildly relativistic shock and in the Bohm diffusion limit, the acceleration timescale is $t_{\rm acc} \sim 2\pi\,(c/v_{\rm s})^2\,E/qBc \approx 3\,(0.5\,c/v_{\rm s})^2\,E_{\rm TeV}\,B_{\rm G}^{-1}$ s, where $E_{\rm TeV}$ is the particle energy in TeV and $B_{\rm G}$ the magnetic field in Gauss; $v_{\rm s} \sim 0.5\,c$ would be the validity limit of the non-relativistic assumption. For reasonable radiation and matter densities, in a finite size *homogeneous* accelerator, dominant particle/energy losses are diffusion escape, and adiabatic and synchrotron cooling, with typical timescales

$t_{\rm diff} = R^2/2D = 15000\,R_{12}^2\,B_{\rm G}\,E_{\rm TeV}^{-1}$ s, $t_{\rm ad} \sim R/v = 100\,R_{12}/v_{10}$ s, and $t_{\rm sy} = 1/a_{\rm s}B^2E \approx 390\,B_{\rm G}^{-2}\,E_{\rm TeV}^{-1}$ s, where $R = 10^{12}\,R_{12}$ cm and $v = 10^{10}\,v_{10}$ cm s^{-1} are typical lengths and flow velocities in the region, and D the Bohm diffusion coefficient. This yields a maximum energy for Fermi I acceleration of $E_{\rm max} = \min[E_{\rm max}^{\rm diff}, E_{\rm max}^{\rm ad}, E_{\rm max}^{\rm sy}]$, where $E_{\rm max}^{\rm diff} \sim 74\,R_{12}\,B_{\rm G}\,(0.5\,c/v_{\rm s})^{-1}$ TeV, $E_{\rm max}^{\rm ad} \sim 36\,B_{\rm G}\,R_{12}\,v_{10}^{-1}\,(0.5\,c/v_{\rm s})^{-2}$ TeV, and $E_{\rm max}^{\rm sy} \sim 12\,B_{\rm G}^{-1/2}\,(0.5\,c/v_{\rm s})^{-1}$ TeV. These simple estimates already show that for slow shocks (e.g. between two massive star winds), or outside the range $B \sim 0.1 - 1$ G, it is very hard to accelerate particles up to $E \gtrsim 10$ TeV. It is worth noting that particle acceleration in highly relativistic shocks could work,[32] but requires rather specific magnetic field conditions downstream, such as high turbulence and relatively small magnetization. This might be hard to realize in pulsar winds,[33] in which a significant toroidal B-field is usually expected to remain downstream the termination shock (see however below).

In stochastic particle acceleration or Fermi II, the situation is worse than in Fermi I. This process is driven only by stochastic collisions with randomly moving, magnetic field irregularities. It is a second order process, i.e. $\Delta E \propto (v_{\rm A}/c)^2$, where $v_{\rm A}$ is the Alfven speed. Since $t_{\rm acc} \sim (c/v_{\rm A})^2\,E/qBc$, it is required that turbulent energy will be a significant fraction of the plasma energy and that $v_{\rm A} \to c$. It means that B should be around equipartition with the turbulence rest mass energy density. Under such a condition, if $B \sim 1$ G (i.e. optimal for $v_{\rm A} \sim 0.5\,c$), then $n \sim 100$ cm^{-3}, hardly feasible for a compact binary. Downstream of a pulsar wind termination shock, relativistic Alfvenic speed may be achieved, although size and turbulence-energy requirements favor the region behind the pulsar with respect to the star. This however requires negligible stellar wind contamination and adiabatic losses and may be unrealistic (see Ref. 23), so a proper assessment of the Fermi II efficiency here requires a detailed study. In general, stochastic acceleration seems to be more suitable to explain extended and low-energy emission, in regions downstream shocks or rich in instabilities.

Shear acceleration is, like Fermi II, a stochastic process, but relies on an additional global velocity gradient in the flow $\sim \Delta u/\Delta R$. In the mildly relativistic case, $t_{\rm acc} \sim 3(\Delta R)^2/r_{\rm g}c = 300\,B_{\rm G}\,\Delta R_{11}^2\,E_{\rm TeV}^{-1}$ ($\Delta R = 10^{11}\,\Delta R_{11}$ cm; $r_{\rm g}$ is the particle gyroradius), and therefore the acceleration timescale has the same dependence on E as synchrotron. In order to operate then, shear has to overcome synchrotron cooling, implying $B \lesssim \Delta R_{11}^{-2/3}$ G. Shear also requires of an injection process, since otherwise adiabatic or advection (escape) losses will block it, i.e. $E > 3\,\Delta R_{11}\,v_{10}\,B_G(\Delta R/R)_{0.1}$ TeV, where $(\Delta R/R)_{0.1}$ means $\Delta R = R/10$. In principle, Fermi I could be a good injection candidate, since it does no require high B to operate. Fermi II otherwise needs higher B that may render synchrotron cooling dominant over shear acceleration. The shear maximum energy (in the mildly relativistic case) is limited by $r_{\rm g} = \Delta R$, i.e., $E^{sh}_{\rm max} \sim q\,B\,\Delta R \approx 30\,B_{\rm G}\,\Delta R_{11}$ TeV.

3. The converter mechanism and magnetic reconnection

In conventional relativistic ($\Gamma \gg 1$) shock scenarios, charged particles are quickly overtaken by the shock so that they do not have enough time to isotropise in the upstream region. Thus, when caught up by the shock, the shock normal/particle motion angle θ will be $\sim 2/\Gamma$, and the energy gain, $\Delta E/E \sim \Gamma^2\,\theta^2/2$, will be reduced down to ~ 2. However, particles may have time to isotropise upstream if they managed to switch to a neutral state and propagate far from the shock without deflections in the B-field. For instance, an electron of energy E cooling via Klein-Nishina (KN) inverse Compton (IC) can transfer most of its energy to a scattered photon. For pair creation mean free paths $l \sim \Gamma^2\,r_{\rm g}$, the gamma ray will pair create with an ambient photon far enough from the shock to allow the subsequent electron (or positron) to get deflected by $\theta \sim 1$, i.e., $\Delta E/E \sim \Gamma^2/2$. This effectively implies $t_{\rm acc} \sim r_{\rm g}/c$, the electrodynamical limit (e.g. Ref. 34). If otherwise l is too small, then $\theta \to 2/\Gamma$, i.e., the standard case. Once started, the mechanism proceeds efficiently and yields a rather hard electron spectrum until l, roughly $\propto E$, becomes $\gtrsim \Gamma\,R$. After that, the electron spectrum becomes softer. In binary systems with bright UV stars, the converter mechanism could yield hard electron spectra up to $\sim$ TeV for $\Gamma \sim 100$ (see also Ref. 35). This might be the case in microquasar $e^{\pm}$-jets before suffering mass-loading (caveat: gamma-ray absorption), or at the reaccelerated shocked pulsar wind.[36] The energy is limited by $r_{\rm g} \sim \Gamma\,R$, although synchrotron losses can reduce this limit. For a pulsar termination shock in a UV stellar field the process cannot work efficiently, since for $\theta \sim 1$, a distance ahead the shock of $\Gamma^2\,r_{\rm g} \approx 3\times 10^{21}\,(\Gamma/10^6)^2\,E_{\rm TeV}\,B_{\rm G}^{-1}$ cm would be required. As shear acceleration, the converter mechanism requires an injection mechanism. In the leptonic case, E should be enough to pair-create in the ambient photon field ($\sim$ 30 GeV for stellar photons).

Magnetic reconnection is perhaps the less well characterized process among those discussed here. Numerical calculations show that, beside bulk acceleration in the current sheet up to $v_{\rm A}$, non-thermal particles can be also produced.[21] Potentially, the mechanism is fast with $t_{\rm acc} \sim r_{\rm g}/c$ (assuming $\epsilon \sim B$, where ϵ is the current sheet electric field), and particles may reach energies limited by $r_g \sim \Delta R$, where ΔR is the current sheet size. However, unless the current sheet occupies a significant fraction of the source, the process will not yield the required multi-10 TeV particle energies. A possibility could be many reconnection sites, which could be equivalent to the Fermi II acceleration process. An interesting role of magnetic reconnection may be the dissipation of an alternating polarity B-field in the jet base (e.g., Ref. 37), or at the pulsar wind termination (e.g., Ref. 38), thereby possibly allowing further acceleration via a Fermi I-type mechanism in relativistic shocks (e.g., Ref. 39).

4. Conclusions

The present work indicates that for typical conditions expected in gamma-ray binaries, the production of photons with energies $\gtrsim$ 10 TeV indeed requires very efficient acceleration with $t_{acc} < 10 - 100\, r_g/c$. Although less strictly, this conclusion also applies to protons. All this implies strong turbulence, and relatively weak magnetic fields (for electrons), as well as at least mildly relativistic speeds. Fermi I acceleration, although marginal, seems to be the best candidate in mildly relativistic outflows, but for highly relativistic flows the situation is less clear. Shear acceleration and Fermi II could also operate, but the latter is unlikely to help at TeV energies. The converter mechanism and magnetic reconnection cannot be discarded, although they require quite specific conditions. A better source characterization is needed for a proper assessment of the feasibility of all these mechanisms.

5. Acknowledgments

We thank Maxim Barkov, Evgeny Derishev and Dmitry Khangulyan for fruitful discussions. This research has received funding from the European Union Seventh Framework Program (FP7/2007-2013) under grant agreement PIEF-GA-2009-252463. V.B.-R. acknowledges support by the Spanish Ministerio de Ciencia e Innovación (MICINN) under grants AYA2010-21782-C03-01 and FPA2010-22056-C06-02.

References

1. F. Aharonian et al., *A&A* **442**, 1(Octtober 2005).
2. F. Aharonian et al., *Science* **309**, 746(July 2005).
3. J. Albert et al., *Science* **312**, 1771(June 2006).
4. J. Albert et al., *ApJL* **665**, L51(August 2007).
5. A. A. Abdo et al., *ApJL* **701**, L123(August 2009).

6. A. A. Abdo et al., *ApJL* **706**, L56(November 2009).
7. M. Tavani et al., *Nature* **462**, 620(December 2009).
8. M. Tavani et al., *ApJL* **698**, L142(June 2009).
9. A. A. Abdo et al., *Science* **326**, 1512(December 2009).
10. S. Sabatini et al., *ApJL* **712**, L10(March 2010).
11. A. A. Abdo et al., *Science* **329**, 817(August 2010).
12. S. D. Bongiorno, A. D. Falcone and M. Stroh et al., *ApJL* **737**, L11+(August 2011).
13. R. H. D. Corbet, C. C. Cheung and M. Kerr et al., *The Astronomer's Telegram* **3221**, 1(March 2011).
14. F. Aharonian et al., *A&A* **460**, 743(December 2006).
15. D. Khangulyan, F. Aharonian and V. Bosch-Ramon, *MNRAS* **383**, 467(January 2008).
16. F. M. Rieger, V. Bosch-Ramon and P. Duffy, *Ap&ss* **309**, 119(June 2007).
17. L. O. Drury, *Reports on Progress in Physics* **46**, 973(August 1983).
18. E. Fermi, *Physical Review* **75**, 1169(April 1949).
19. F. M. Rieger and P. Duffy, *ApJ* **617**, 155(December 2004).
20. E. V. Derishev, F. A. Aharonian, V. V. Kocharovsky and V. V. Kocharovsky, *Phys. Rev. D* **68**, 043003(August 2003).
21. S. Zenitani and M. Hoshino, *ApJL* **562**, L63(November 2001).
22. M. Perucho and V. Bosch-Ramon, *A&A* **482**, 917(May 2008).
23. V. Bosch-Ramon and M. V. Barkov, in press, *A&A* (2011).
24. A. T. Okazaki, S. Nagataki and T. Naito et al., in press, *PASJ* (2011).
25. V. Bosch-Ramon, J. M. Paredes, G. E. Romero and M. Ribó, *A&A* **459**, L25(November 2006).
26. D. Khangulyan, S. Hnatic, F. Aharonian and S. Bogovalov, *MNRAS* **380**, 320(September 2007).
27. G. Dubus, B. Cerutti and G. Henri, *A&A* **477**, 691(January 2008).
28. A. Sierpowska-Bartosik and D. F. Torres, *Astroparticle Physics* **30**, 239(December 2008).
29. V. Zabalza, J. M. Paredes and V. Bosch-Ramon, *A&A* **527**, A9+(March 2011).
30. G. E. Romero, D. F. Torres, M. M. Kaufman Bernadó and I. F. Mirabel, *A&A* **410**, L1(Octtober 2003).
31. V. Bosch-Ramon and D. Khangulyan, *International Journal of Modern Physics D* **18**, 347 (2009).
32. L. Sironi and A. Spitkovsky, *ApJ* **726**, 75(January 2011).
33. M. Lemoine and G. Pelletier, *MNRAS* **402**, 321(February 2010).
34. F. A. Aharonian, A. A. Belyanin, E. V. Derishev, V. V. Kocharovsky and V. V. Kocharovsky, *Phys. Rev. D* **66**, 023005(July 2002).
35. B. E. Stern and J. Poutanen, *MNRAS* **372**, 1217(November 2006).
36. S. V. Bogovalov, D. V. Khangulyan, A. V. Koldoba, G. V. Ustyugova and F. A. Aharonian, *MNRAS* **387**, 63(June 2008).
37. Y. Lyubarsky, *ApJL* **725**, L234(December 2010).
38. Y. Lyubarsky and M. Liverts, *ApJ* **682**, 1436(August 2008).
39. L. Sironi and A. Spitkovsky, *ArXiv e-prints* (July 2011).

Cosmic Rays propagation in the Heliosphere

P. Bobik[1], M.J. Boschini[2,4], C. Consolandi[2], S. Della Torre[2,5,*], M. Gervasi[2,3], D. Grandi[2], K. Kudela[1], F. Noventa[2,3], S. Pensotti[2,3], P.G. Rancoita[2]

[1] *Institute of Experimental Physics, Kosice (Slovak Republic)*
[2] *Istituto Nazionale di Fisica Nucleare, INFN Milano-Bicocca, Milano (Italy)*
[3] *Department of Physics, University of Milano Bicocca, Milano (Italy)*
[4] *CILEA, Segrate (MI) (Italy)*
[5] *Department of Physics and Maths, University of Insubria, Como (Italy)*
**E-mail: stefano.dellatorre@mib.infn.it*

The cosmic rays modulation inside the heliosphere is well described by a transport equation introduced by Parker in 1958. We used the HelMod Monte Carlo code to reproduce the modulation effect in the inner heliosphere depending from solar activity and solar polarity. In this 2-D MonteCarlo approach we include a general treatment of the diffusion tensor that compound an enhancement of perpendicular diffusion coefficient at high solar latitude and a polar increased magnetic field. In our simulation we considered a heliosphere that changes the modulation parameter with the distance from the Earth, including periods prior to the one we intend to simulate. In this work we furthermore exploited the energy distribution of injected particles to the observed flux. We compared HelMod results with data of BESS-97, AMS-98, BESS-98, BESS-99, BESS-2000, BESS-2002 and PAMELA; this covering a period of 11 years and two solar polarity, these simulations are well in agreement with experimental data.

Keywords: Cosmic rays; Solar Modulation; Monte Carlo; Stochastic Differential Equation

1. Introduction

The *solar modulation* is a cosmic rays flux reduction for energy below 10$\sim$20 GeV, depending on solar activity, particle charge and interplanetary magnetic field (IMF) polarity. It is described by the so called the Parker's equation[1], this is a transport equation including diffusion, convection, magnetic drift and adiabatic energy loss. In this work we present the Heliospheric Modulation (HelMod) Monte Carlo code to integrate the Parker's equation; we first describe the present model, than we show some results from the Monte Carlo approach and, finally, we compare our results with experimental data along the 23th solar cycle.

2. Model Description

HelMod Code uses a Monte Carlo technique to integrate Parker's equation, in a bi-dimensional (radius and co-latitude) approximation, from the boundary of an effective heliosphere (in present model located at 100 AU) down to Earth orbit. It carried on a set of Stochastic Differential Equations (SDEs) fully equivalent to the Parker's equation[2]:

$$\begin{aligned}\Delta r &= \frac{1}{r^2}\frac{\partial}{\partial r}(r^2 K^S_{rr})\Delta t - \frac{\partial}{\partial\mu(\theta)}\left[\frac{K^S_{r\theta}\sqrt{1-\mu^2(\theta)}}{r}\right]\Delta t \\ &+(V_{\rm sw}+v_{{\rm dr},r}+v_{\rm HCS})\Delta t+\omega_r\sqrt{\frac{K^S_{rr}K^S_{\theta\theta}-(K^S_{r\theta})^2}{0.5K^S_{\theta\theta}}\Delta t} \\ &-\omega_{\mu(\theta)}K^S_{r\theta}\sqrt{\frac{2}{K^S_{\theta\theta}}\,\Delta t}\,, \end{aligned} \tag{1}$$

$$\begin{aligned}\Delta\mu(\theta) &= -\frac{1}{r^2}\frac{\partial}{\partial r}\left(rK^S_{\theta r}\sqrt{1-\mu^2(\theta)}\right)\Delta t+\frac{\partial}{\partial\mu(\theta)}\left\{\frac{K^S_{\theta\theta}[1-\mu^2(\theta)]}{r^2}\right\}\Delta t \\ &-\frac{v_{{\rm dr},\theta}\sqrt{1-\mu^2(\theta)}}{r}\Delta t+\omega_{\mu(\theta)}\sqrt{\frac{2K^S_{\theta\theta}[1-\mu^2(\theta)]}{r^2}\Delta t}\,, \end{aligned} \tag{2}$$

$$\Delta T = -\frac{2}{3}\frac{\alpha_{\rm rel}V_{\rm sw}T}{r}\Delta t. \tag{3}$$

here r is the helio-centric distance, $\mu(\theta) = \cos\theta$ is the cosine of solar co-latitude, T is the kinetic energy (per nucleon), t is the time, V_{sw} the Solar Wind (SW) velocity, K^S_{ij} is the i-j component of symmetric part of the diffusion tensor, ω_i is a random number following a Gaussian distribution with a zero mean value and a standard deviation of one (e.g., see Ref. 3) and $\alpha_{\rm rel}=(T+2T_0)/(T+T_0)$[4], where T_0 is particle's rest energy; here the drift velocity is splitted in regular drift (radial drift v_{D_r}, latitudinal drift v_{D_θ}) and neutral sheet drift ($v_{D_{NS}}$), as described in Ref. 5, and is scaled using the tilt angle[6] ($\alpha_{\rm t}$) of the neutral sheet as described in Ref. 7. This set represent the stochastic variation on radius, co-latitude and energy of a quasi-particle object in a simulated effective heliosphere. The symmetric part of the diffusion tensor, for a reference system with the 3rd coordinate along the average magnetic-field, could be simplified in a pure diagonal matrix that contains both the transverse ($K_{\perp\theta}$ and $K_{\perp r}$) and parallel ($K_{||}$) components (e.g., see Ref. 8). It has to be remarked that the general transformations of the symmetric and antisymmetric parts of the diffusion tensor from field-aligned to

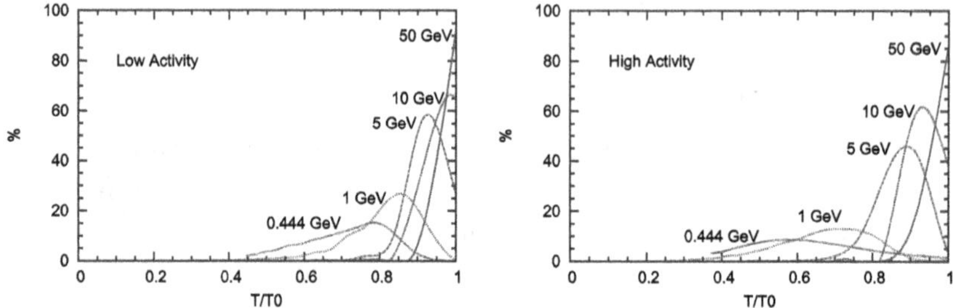

Fig. 1. Energy distribution at Earth for injected energy T_0

heliospheric (spherical) coordinates can be found in Ref. 9. As suggested by Ref. 10 the parallel diffusion coefficient is given by

$$K_{||} = \beta K_0 K_P(P,t) \left[\frac{B_\oplus}{3B}\right] \tag{4}$$

with $\beta = v/c$, v the particle velocity and c the speed of light; the diffusion parameter K_0 accounts for the dependence on the solar activity and, in the present model, is parametrized as function of Smoothed Sunspot Numbers (SSN, see figure 1 of Ref. 11); $B_\oplus$ is the value of IMF at the Earth orbit - typically $\approx 5\,\mathrm{nT}$, but it varies as a function of the time - obtained from Ref. 12[a]; B is the magnitude of the large scale IMF; finally, the term K_P takes into account the dependence on the rigidity P of the GCR particle usually expressed in GV. In the present model K_P is assumed to be equal to the particle rigidity[13] and we tuned different setup for periods approaching the solar maximum and periods not dominated by High Solar Activity, reflecting the change of the magnetic field topology between this periods. For period of low solar activity, we increase gradually the solar wind speed, from the low value on Earth orbit obtained from Ref. 12, to a maximum value of $\simeq 760\,\mathrm{km/s}$[14]; For period approaching the solar maximum we assume a speed - obtained from Ref. 12 - independent of the colatitude. As IMF we modified the Parker model[15] introducing a small latitudinal components as described in Ref. 16. In the present code, the effective heliosphere was subdivided in 15 spherical regions. In each region, the parameters (i.e., SW speed, SSN, $B_\oplus$, α_t) are determined from the solar activity at the time of the solar wind ejection[17]. In this work we use the Local Interstellar Spectrum from Ref. 18. It is suggested, by Ref. 8, an *enhanced* $K_{\perp\theta}$, in order to reproduce the amplitude and rigidity dependence of the latitudinal gradients

[a]OMNIWEB data available at http://omniweb.gsfc.nasa.gov/form/dx1.html

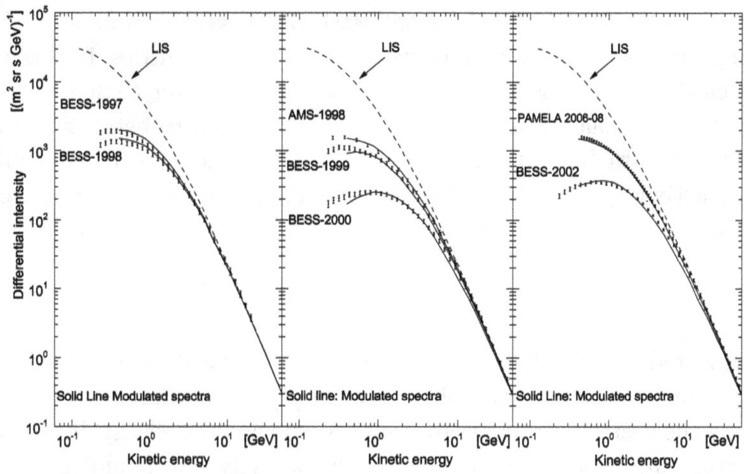

Fig. 2. HelMod results compared with data along the solar cycle 23.

of GCR differential intensities for protons and electrons (see e.g., Ref. 19); in this work we tested this description (with a factor 10 of enhancement in polar regions) compared to a $K_{\perp\theta}$ independent of colatitude.

3. Results

We simulated mono-energetic particles during a low activity period (i.e. 1998) and an high activity period (i.e. 2002); we obtained the average propagation time, from the boundary down to the Earth, and the energy distribution in order to evaluate how propagation process are related to energy loss. In fig. 1 we show the energy distribution (in percentage) at Earth orbit for particle injected with energy T_0 equal to 0.444 ($\sim$ 1 GV), 1, 5, 10 and 50 GeV, for both low and high activity periods. From a comparison between the two plot in fig. 1, combining with the result that particles in high activity period spend inside the heliosphere about time double with respect to low activity period, we conclude that HelMod particles lose more energy in high activity period than in low activity period. This effect is greater at low energy, and vanish with increasing the energy. This agrees with the experimental evidence of a greater modulation effect during solar maximum[20].

We compared the present model results with data collected during cycle 23, including or excluding the polar enhancement of $K_{\perp\theta}$. The dataset we considered in our analysis are: AMS[21] (1998), BESS[20] (1997,1998, 1999,

2000 and 2002) and PAMELA[22] (2006–08), that covers both periods of low and high solar activity, as well as two opposite solar polarities. In this work we focused our analysis on particles with energy greater than 1 GV. We found that the polar enhancement of $K_{\perp\theta}$ is required to better reproduce low activity data (i.e. 1997,1998 and 2006–08), while is no more required for high activity periods (i.e. 1999,2000 and 2002). The best results for each experiment are showed in fig. 2.

4. Conclusion

We presented the HelMod Monte Carlo code for the study of cosmic rays propagation in the inner heliosphere. The Monte Carlo approach allow us to compare the average particles propagation time and the energy distribution at Earth, during both low and high solar activity. We found as in high solar activity period, the time spent by particle is double with respect low activity period, to get the Earth orbit; the energy distribution also shows as particles in high activity period lose more energy than ones in low activity periods; these are consistent with the greater modulation effect observed. We finally compare our results with data along the solar cycle 23. We find that the polar enhancement of $K_{\perp\theta}$ is required during low activity periods, but could neglected during high solar activity.

Acknowledgments

Authors acknowledge the use of NASA/GSFCs Space Physics Data Facilitys OMNIWeb service, and OMNI data.

References

1. E. N. Parker, *Plan. & Space Sci.* **13**, 9 (1965).
2. P. Bobik, M. Boschini, C. Consolandi, S. Della Torre, M. Gervasi, D. Grandi, K. Kudela, S. Pensotti and P. Rancoita, *Astrophis. J.* **743** (2011); available at the web site: http://arxiv.org/abs/1110.4315.
3. W. M. Kruells and A. Achterberg, *Astron. and Astrophys* **286**, 314 (1994).
4. L. J. Gleeson and W. I. Axford, *Astrophis. J. Lett.* **149**, L115 (1967).
5. M. S. Potgieter and H. Moraal, *Astrophis. J.* **294**, 425 (1985).
6. Wilcox solar observatory data available at: http://wso.stanford.edu/tilts.html (2011).
7. R. A. Burger and M. S. Potgieter, *Astrophis. J.* **339**, 501 (1989).
8. M. S. Potgieter, *J. Geophys. Res.* **105**, 18295 (2000).
9. R. A. Burger, T. P. J. Krüger, M. Hitge and N. E. Engelbrecht, *Astrophis. J.* **674**, 511 (2008).

10. M. S. Potgieter and J. A. Le Roux, *Astrophis. J.* **423**, 817 (1994).
11. Bobik, P, Boschini, M.J., Della Torre, S., Gervasi, M., Grandi, D., Kudela, K. and Rancoita, P. G., Proton and antiproton modulation in the heliosphere for different solar conditions and AMS-02 measurements prediction, in *Astroparticle, Particle and Space Physics, Detectors and Medical Physics Applications - Proc. of the 12th ICATPP Conference on Cosmic Rays for Particle and Astroparticle Physics*, ed. C. Leroy, P.-G. Rancoita, M. Barone, A. Gaddi, L. Price, & R. Ruchti 2011.
12. J. H. King and N. E. Papitashvili, *Journal of Geophysical Research (Space Physics)* **110**, A02104(February 2005).
13. J. S. Perko, *Astron. and Astrophys.* **184**, 119 (1987).
14. D. J. McComas, B. L. Barraclough, H. O. Funsten, J. T. Gosling, E. Santiago-Muñoz, R. M. Skoug, B. E. Goldstein, M. Neugebauer, P. Riley and A. Balogh, *J. Geophys. Res.* **105**, 10419 (2000).
15. E. N. Parker, *Astrophis. J.* **128**, 664 (1958).
16. U. Langner, *Ph.D. Thesis, Potchestroom University* (2004).
17. P. Bobik, M. J. Boschini, S. Della Torre, M. Gervasi, D. Grandi, K. Kudela and P. G. Rancoita, Galactic Cosmic Rays Modulation and Prediction for the AMS-02 Mission, in *Astroparticle, Particle and Space Physics, Detectors and Medical Physics Applications*, ed. C. Leroy, P.-G. Rancoita, M. Barone, A. Gaddi, L. Price, & R. Ruchti 2010.
18. R. A. Burger, M. S. Potgieter and B. Heber, *J. Geophys. Res.* **105**, 27447 (2000).
19. M. S. Potgieter, Implications of Enhanched Perpendicular Diffusion in the Heliospheric Modulation of Cosmic Rays, in *25th International Cosmic Ray Conference*, , International Cosmic Ray Conference Vol. 21997.
20. Y. Shikaze *et al.*, *Astrop. Phys.* **28**, 154 (2007).
21. AMS Collaboration, M. Aguilar *et al.*, *Phys. Rep.* **366**, 331 (2002).
22. O. Adriani *et al.*, *Science* **332**, 69 (2011).

Numerical modelling of the heliosphere

S.E.S. FERREIRA*, M.S. POTGIETER, R.D. STRAUSS

Centre for Space Research, North-West University,
Potchefstroom, 2520, South Africa
**E-mail: stefan.ferreira@nwu.ac.za*

K. SCHERER, H. FICHTNER

Institut für Theoretische Physik, Lehrstuhl IV: Weltraum- und Astrophysik,
Ruhr-Universitat Bochum, Bochum, Germany

The interaction of the solar wind with the surrounding interstellar medium (ISM) forms a cavity in the ISM. This cavity is called the heliosphere (influence sphere) of the Sun. Numerical models were developed using hydrodynamic (HD) and magneto-hydrodynamic (MHD) equations to simulate the interaction of the different fluids in the heliosphere. These models showed that the heliosphere is driven by the interaction of the ionized solar wind, which expands away from the Sun at supersonic speeds, and the local interstellar medium (LISM). Also of importance is the effect of neutral particles in the LISM and magnetic fields. These descriptions then provide the background environment in which cosmic ray transport and acceleration can be calculated as they are transported from the heliospheric boundary up to Earth.

Keywords: heliosphere; hydrodynamic; magneto-hydrodynamic; solar wind

1. Introduction

By the early 1970s the hydrodynamic treatment of expanding stellar winds (such as the solar wind) was well established, e.g. [1,2]. The studies of the solar wind-interstellar medium interaction found that it consist of several parts that have to be solved self-consistently: The first interaction to be calculated is that of the ionized particles in the solar wind and local interstellar medium (LISM). Next, the effect of neutral particles in the LISM on this interaction needs to be taken into account. Due to the ionized nature of the solar wind and LISM, both these plasmas carry magnetic fields. Therefore, beside the various mutual interactions between ionized and neutral particles, a solution to the electrodynamic equations describing the interaction between the magnetic fields and the plasma needs to be considered.

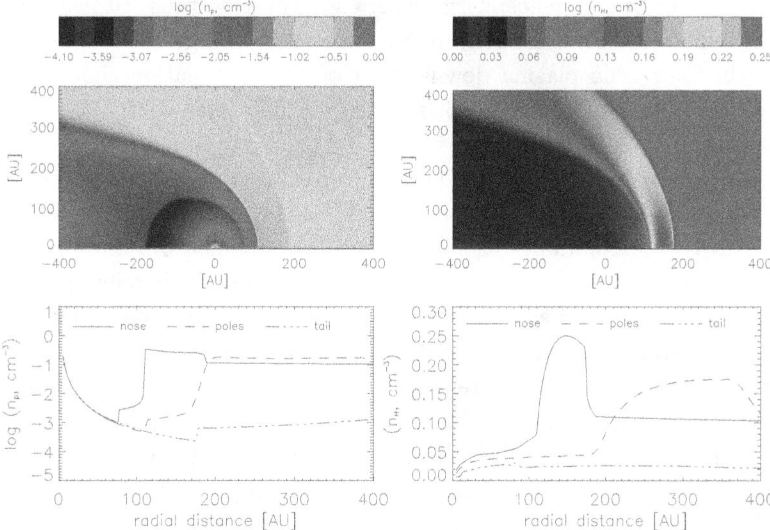

Fig. 1. The heliosphere model. Shown on the left is the solar wind-LISM proton density and on the right the neutral H density as particles per cubic centimeter. The interstellar wind blows from the positive to the negative x-axis, given in AU. The y-axis points along the polar direction also in AU. The top panels show the density as countours in the meridional plane, and the bottom panels the radial profiles in the nose (0°), pole (90°) and tail direction (180°) respectively. From [3].

As computational resources increased, more detailed solutions to the solar wind-LISM interaction could be calculated. Initially, hydrodynamic formulations of the solar wind-LISM interaction were used, e.g. [4–10]. Recently, models including the heliospheric magnetic field were developed by [11,12] also including cosmic rays by found in this region by [13–16]. See recent review by [17] concerning cosmic rays in the heliosphere.

2. The heliosphere

The solar wind is the source of the heliosphere which originates from the solar corona, where the outward force on the particles due to the magnetic field outweighs the gravity. The size and features of the heliosphere are mainly determined by the interaction of the solar wind and the LISM. See e.g. [7,18] for reviews. However, the LISM is partly ionized consisting mainly of protons and hydrogen. This interstellar neutral H undergo charge exchange processes with the LISM plasma and a subpopulation is created. The charge exchange mean free path is sufficiently small in the region of decelerated LISM flow and a wall of neutral H will is formed in the upstream direction of the heliospheric nose, e.g. [19]. Neutral H atoms that cross the heliopause into the heliosphere experience charge exchange with the very hot subsonic plasma downstream of the termination shock and also experience charge exchange with the supersonic solar wind inside the terminationshock. See e.g. [4,9,20].

Another important fluid affecting the heliospheric geometry is the so called pickup ions (PUIs). These ions created out of the interaction of neutral H with the surrounding plasma remove both momentum and energy from the solar wind flow. The solar wind gets decelerated, therefore reducing the ram pressure and subsequently the size of the heliosphere. See e.g. [18,21] for an overview. The initial PUI population is unstable, generating magnetic turbulence which scatters both PUIs and cosmic rays. Furthermore cosmic rays themselves, either galactic or anomalous [22], may also interact with these different fluids when their pressure at the termination shock is sufficient to modify the shock structure and position. See e.g. [23] for modelling results.

In order to compute the ineraction of these different fluids in the hydrodynamic limit, e.g. [9,10], we solve the following set of Euler equations:

$$\frac{\partial}{\partial t}\rho_i + \nabla \cdot (\rho_i \mathbf{u}_i) = Q_{p,i}, \tag{1}$$

$$\frac{\partial}{\partial t}(\rho_i \mathbf{u}_i) + \nabla \cdot (\rho_i \mathbf{u}_i \mathbf{u}_i + P_i \mathbf{I}) = \mathbf{Q}_{m,i}, \tag{2}$$

$$\frac{\partial}{\partial t}(\frac{\rho_i}{2}\mathbf{u}_i^2 + \frac{P_i}{\gamma_i - 1}) + \nabla \cdot (\frac{\rho_i}{2}\mathbf{u}_i^2 \mathbf{u}_i + \frac{\gamma_i \mathbf{u}_i P_i}{\gamma_i - 1}) = Q_{e,i} \tag{3}$$

which describe the balance of mass, momentum, and energy of the protons in the solar wind and LISM (i=p) and the neutral H population (i=H), to calculate the heliospheric geometry and plasma flow. For the PUIs (i=PUI) only the equation for the conservation of mass is taken into account. The fluid quantities are the mass density ρ_i, velocity $\mathbf{u}_i$, and pressure P_i, $\mathbf{I}$ denotes the unity tensor, and γ_i the adiabatic indices of the components. The time is t and Q_i denote the sources related to the interaction between various species.

An example of the heliospheric structure, in terms of number density, is shown in Figure 1. Shown on the left is the solarwind-LISM proton density and on the right the neutral H density. The top panels show the density as countours in the meridional plane, and the bottom panels radial profiles in the nose, poles and tail direction respectively. The inclusion of neutral H in the model changes the structure and size of the heliosphere because of the removal of momentum from the supersonic solar wind by charge-exchange,

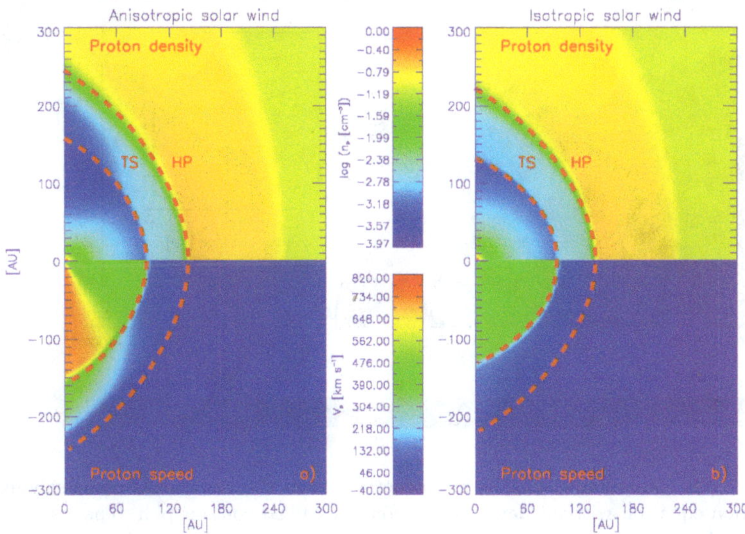

Fig. 2. The heliosphere in the nose regions in terms of solar wind speed (top) and density (bottom) for a isotropic wind (solar maximum) and an anisotropic wind (solar minimum) when it changes from 400 km s^{-1} to 800 km s^{-1}.

e.g. [4,9]. The important heliospheric structures like the TS, heliopause, hydrogen wall and even bow shock are clearly visible.

3. The asymmetric heliosphere

Note that Figure 1 is only applicable for an isotropic solar wind speed. However, for solar minimum conditions there is a distinct transition between the speed of the solar wind at the polar regions (fast) and the speed at the equatorial regions (slow). The fast solar wind has a speed of $\sim$ 800 km s^{-1} and the speed of the slow wind is $\sim$ 400 km s^{-1}. The speed of the solar wind for solar maximum conditions is in the range of 350 - 700 km s^{-1} and is highly variable. The effect of a changing solar wind speed over the poles on the heliospheric geometry are illustrated in Figure 2, which shows the heliosphere in the nose regions in terms of solar wind speed and density for a isotropic wind (solar maximum) and an anisotropic wind (solar minimum). Shown is that the heliosphere is also more elongated in the poleward directions, e.g. [6,9,10], because of the latitudinal variation of the solar wind momentum flux. Another important feature shown is that from a cosmic ray modulation point of view, as solar activity changes the

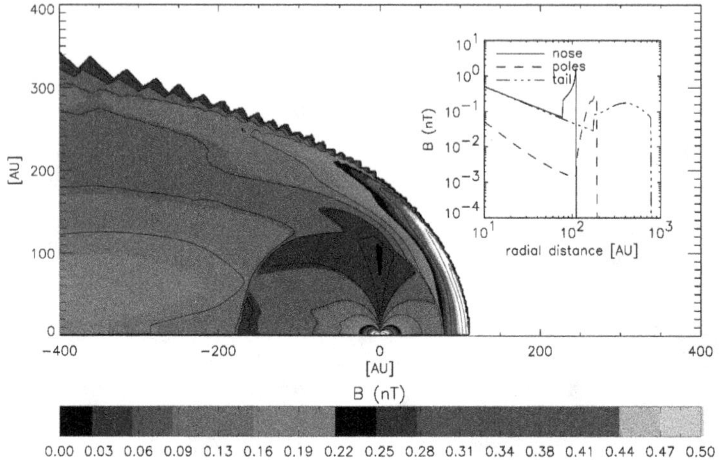

Fig. 3. The computed heliospheric magnetic field in the meridional plane solving the induction equation without any modifications as a filled contour plot. The insert shows the radial profile of the field in the nose (solid), pole (dashed) and tail (dashed-dotted) directions of the heliosphere. From [3].

termination shock moves, especially at the polar and tail regions changing the modulation volume for cosmic rays. See e.g. [10] for dynamic modeling results.

The heliosphere, resulting under the relative motion between the LISM and the Sun has an asymmetric (bullet-shaped) structure. For instance the position of the termination shock in the tail region is related to the position in the nose through the relation found by [24] as $r_{TS,tail} = (2.08 \pm 0.04) r_{TS,nose}$ with the position of the heliopause in the nose related to the postion of the termination shock as $r_{HP} = (1.39 \pm 0.01) r_{TS,nose}$. Recent results from [12,25] indicates that in addition, an interstellar magnetic field in the order of a few μG can distort the heliosphere resulting in a significant north/south asymmetry. It was also shown by [25] that an interstellar field of a strength greater than 4 μG can account for a 10 AU difference in the TS radius. Such an asymmetry seems to be observed during the crossing by the Voyager 1 and Voyager 2 spacecraft. Recently [26,27] also argued from a cosmic ray perspective the existence of such an asymmetry, but see also [28] who argue that the interstellar cosmic ray flux must neither be isotropic nor homogeneous at the heliospheric boundaries.

4. The heliospheric magnetic field

To calculate the heliospheric magnetic field $\mathbf{B}$ in plasma flow with velocity $\mathbf{v}$, Faraday's law

$$\frac{\partial \mathbf{B}}{\partial t} - \nabla \times (\mathbf{v} \times \mathbf{B}) = 0. \tag{4}$$

is solved to obtain the different components with a constraining step added to ensure $\nabla \cdot \mathbf{B} = 0$. Figure 3 shows the computed B in the meridional plane as a filled contour plot. The insert shows the radial profile in the nose (solid), pole (dashed) and tail (dashed-dotted) directions of the heliosphere. As expected for the equatorial regions and inside the TS, $B \propto \frac{1}{r}$, which is more or less the analytic magnetic field described by [29] with some minor deviation due to the effect of PUIs and neutral H. For the poles, $B \propto \frac{1}{r^2}$ for a large part of the inner heliosphere where the field is almost completely radial, while for the outer heliospheric polar regions the azimuthal contribution dominates resulting in $B \propto \frac{1}{r}$.

An important feature in Figure 3 is the existence of a highly amplified field in the heliosheath, where a magnetic wall emerges from the amplification of the B_ϕ component, caused by the flow deceleration. As shown, over the shock, B abruptly increases by a factor corresponding to the compression ratio. Further into the heliosheath a steady increase is computed due to the decelerating plasma. Beyond the heliopause $\mathbf{B}$=0 due to the boundary condition imposed here. For a discussion on cosmic ray transport and acceleration in this heliospheric environment see our other contributions to this volume.

Acknowledgments

S.E.S.F, M.S.P and R.D.S are grateful for partial financial support granted to them by the South African National Research Foundation (NRF) and the Centre for High Performance Computing (CHPC) in Cape Town. H.F and K.S are grateful to the Deutsche Forschungsgemeinschaft for funding the Heliocauses project and to the Bundesministerium für Forschung und Bildung for supporting the South Africa-German collaboration (SUA 08/011)

References

1. E. N. Parker, *Astrophysical Journal* **134**, p. 20 (1961).
2. T. E. Holzer, *Journal of Geophysical Research* **77**, p. 5407 (1972).
3. S. E. S. Ferreira, M. S. Potgieter and K. Scherer, *Astrophysical Journal* **659**, p. 1777 (2007).

4. G. P. Zank, H. L. Pauls, L. L. Williams and D. T. Hall, *Journal of Geophysical Research* **101**, p. 21639 (1996).
5. V. V. Izmodenov, *Astronomy Letters* **23**, p. 221 (1997).
6. H. L. Pauls and G. P. Zank, *Journal of Geophysical Research* **102**, p. 19779 (1997).
7. T. E. Holzer, *Annual Review of Astronomy and Astrophysics* **27**, p. 199 (1989).
8. T. Borrmann and H. Fichtner, *Advances in Space Research* **35**, p. 2091 (2005).
9. H. J. Fahr, T. Kausch and H. Scherer, *Astronomy and Astrophysics* **357**, p. 268 (2000).
10. K. Scherer and S. E. S. Ferreira, *Astrophysics and Space Sciences Transactions* **1**, p. 17 (2005).
11. N. V. Pogorelov, G. P. Zank and T. Ogino, *Astrophysical Journal* **644**, p. 1299 (2006).
12. M. Opher, E. C. Stone and P. C. Liewer, *Astrophysical Journal* **640**, L71 (2006).
13. V. Florinski and J. R. Jokipii, *Astrophysical Journal* **591**, p. 454 (2003).
14. S. E. S. Ferreira and K. Scherer, *Astrophysical Journal* **616**, p. 1215 (2004).
15. U. W. Langner and M. S. Potgieter, *Advances in Space Research* **35**, p. 2084 (2005).
16. B. Ball, M. Zhang, H. Rassoul and T. Linde, *Astrophysical Journal* **634**, p. 1116 (2005).
17. V. Florinski, S. E. S. Ferreira and N. V. Pogorelov, *Space Science Reviews* , 28 (2011).
18. G. P. Zank, *Space Science Reviews* **89**, p. 413 (1999).
19. J. L. Linsky and B. E. Wood, *Astrophysical Journal* **463**, 254 (1996).
20. V. B. Baranov and Y. G. Malama, *Journal of Geophysical Research* **100**, 14755 (1995).
21. J. D. Richardson and E. C. Stone, *Space Science Reviews* **143**, 7 (2009).
22. R. D. Strauss, M. S. Potgieter and S. E. S. Ferreira, *Advances in Space Research* **48**, 65 (2011).
23. J. A. le Roux and H. Fichtner, *Journal of Geophysical Research* **102**, 17365 (1997).
24. H. Müller, P. C. Frisch, V. Florinski and G. P. Zank, *Astrophysical Journal* **647**, p. 1491 (2006).
25. N. V. Pogorelov, J. Heerikhuisen, G. P. Zank, J. J. Mitchell and I. H. Cairns, *Advances in Space Research* **44**, 1337 (2009).
26. R. Manuel, S. E. S. Ferreira and M. S. Potgieter, *Advances in Space Research* **48**, 874 (2011).
27. M. D. Ngobeni and M. S. Potgieter, *Advances in Space Research* **48**, 300 (2011).
28. K. Scherer, H. Fichtner, R. D. Strauss, S. E. S. Ferreira, M. S. Potgieter and H.-J. Fahr, *Astrophysical Journal* **735**, 128 (2011).
29. E. N. Parker, *Astrophysical Journal* **128**, 664 (1958).

CR electrons and positrons: what we have learned in the latest three years and future perspectives

Daniele Gaggero, Dario Grasso

Department of Physics, Pisa University,
Largo B. Pontecorvo 3, 56127 Pisa Italy
**E-mail:* daniele.gaggero@pi.infn.it

After the PAMELA finding of an increasing positron fraction above 10 GeV, the experimental evidence of the presence of a new electron and positron spectral component in the cosmic ray zoo has been recently confirmed by Fermi-LAT. We show as a simple phenomenological model which assumes the presence of an electron and positron extra component peaked at $\sim$ 1 TeV allows a consistent description of all available data sets. We then describe the most relevant astrophysical uncertainties which still prevent to determine $e^{\pm}$ source properties from those data and the perspectives of forthcoming experiments.

Keywords: Proceedings; World Scientific Publishing.

1. Introduction

Recent experimental results raised a wide interest about the origin and the propagation of the leptonic component of the cosmic radiation.

Among the most striking of those results, there is the observation performed by the PAMELA satellite experiment that the positron to electron fraction $e^+/(e^- + e^+)$ rises with energy from 10 up to 100 GeV at least (Adriani *et al.* 2008[1]). This appeared in contrast with the predictions of the standard cosmic ray scenario and could therefore be interpreted as the smoking gun of new physics, unless a very soft electron spectrum was assumed.

The significance of this anomaly increased when the Fermi-LAT space observatory measured the $e^- + e^+$ spectrum in the 7 GeV - 1 TeV energy range with unprecedented accuracy and found it to be compatible with a power-law with index $\gamma(e^{\pm}) = -3.08 \pm 0.05$ (Abdo *et al.* 2009,[2] Ackermann *et al.* 2010[3]); this slope is significantly harder than what estimated on the basis of previous measurements: the hypothesis of a steep spectrum was therefore excluded.

More recently, the same collaboration provided a further, and stronger, evidence of the positron anomaly by providing direct measurement of the absolute e^+ and e^- spectra, and of their fraction, between 20 and 200 GeV using the Earth magnetic field. A steady rising of the positron fraction was observed by this experiment up to that energy in agreement with that found by PAMELA. In the same energy range, the e^- spectrum was fitted with a power-law with index $\gamma(e^-) = -3.19 \pm 0.07$ which is in agreement with what recently measured by PAMELA between 1 and 625 GeV (Adriani *et al.* 2011[4]). Most importantly, Fermi-LAT measured, for the first time, the e^+ spectrum in the 20 - 200 GeV energy interval and showed it is fitted by a power-law with index $\gamma(e^+) = -2.77 \pm 0.14$.

We will show in the following paragraph how all those measurements rule out the standard scenario in which the bulk of electrons reaching the Earth in the GeV - TeV energy range are originated by Supernova Remnants (SNRs) and only a small fraction of secondary positrons and electrons comes from the interaction of CR nuclei with the interstellar medium (ISM). Then we will see how the alternative scenario in which the presence of electron + positron component peaked at $\sim$ 1 TeV is invoked allows a consistent description of all the available data sets. Finally we will discuss to which extent astrophysical and particle physics uncertainties still affect our modeling of cosmic ray leptons origin and propagation and how forthcoming measurements are expected to reduce those uncertainties.

2. The necessity of a primary extra-component

After the release of Fermi-LAT $e^- + e^+$ spectrum, it was clearly pointed out in several papers (see e.g. Grasso *et al.* 2009[5] and Di Bernardo *et al. 2011*[6]) that both Fermi-LAT and PAMELA measurements described in the Introduction are in contrast with a standard single-component scenario in which positrons are the secondary products of the nuclear component of cosmic rays (CRs) interacting with the interstellar medium (ISM).

The main problems encountered by this kind of models can be summarized as follows.

- As explained many times (see e.g. Serpico 2011[7] for a recent review), they cannot reproduce the rising positron-to-electron ratio measured by PAMELA and recently confirmed by Fermi-LAT;
- They are unable to reproduce all the features revealed by Fermi-LAT in the CRE spectrum, in particular the flattening observed at around 20 GeV and the softening at $\sim$ 500 GeV. In fact, if

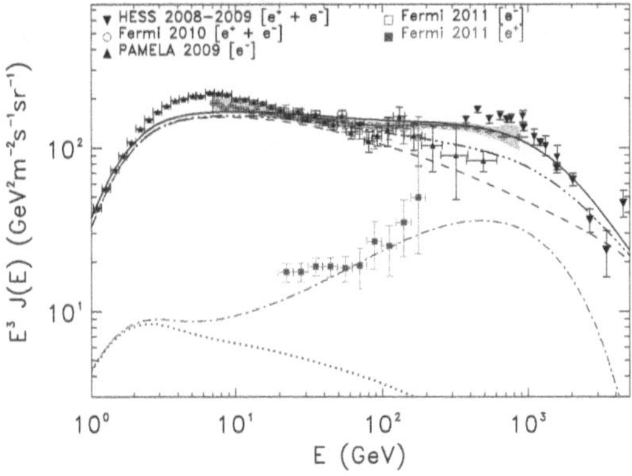

Fig. 1. *Fermi-LAT and PAMELA data on electrons + positrons and electrons are compared to a double component phenomenological model. The absolute positron spectrum is compared to a single and double component phenomenological model.* ***Dotted line:*** *e^+ in single-component scenario.* ***Dot-dashed line:*** *e^+ in double-component scenario.* ***Triple dotted-dashed line, solid line:*** *e^- and e^-+e^+ in double-component scenario.* ***Dashed line:*** *e^- diffuse background in double-component scenario. The Kolmogorov diffusion setup is adopted.*

such models are normalized against data in the 20 - 100 GeV energy range, where systematical and theoretical uncertainties are the smallest, they clearly fail to match CRE Fermi-LAT and PAMELA e^- data outside that range. A different normalization results in even worse fits.

With the release of the e^- and e^+ separate spectra by the Fermi-LAT collaboration the problems with the single component scenario became even worse. In fact, the e^+ spectrum (Fig. 1) is clearly inconsistent with the predictions of a single component scenario computed with `DRAGON` numerical diffusion package (and similar results are obtained with `GALPROP`). Even without considering numerical models, the simple consideration that the reported positron spectral slope is -2.77 ± 0.14 reveals how these data are incompatible with a purely secondary origin from proton spallation on interstellar gas: the source slope should be the same as the proton spectrum, i.e. $\simeq -2.75$ (Adriani *et al.* 2011[8]) and no room is then left for the unavoidable steepening due to energy-dependent diffusion and energy losses.

A double component scenario is the most straightforward solution to these problems.

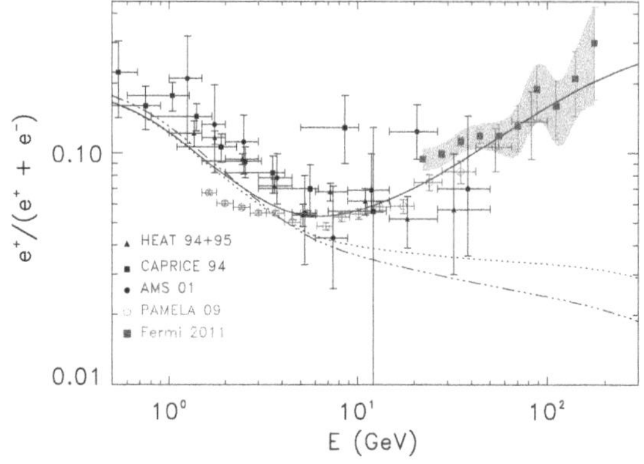

Fig. 2. *Fermi-LAT and PAMELA data on the positron ratio are compared to a single and double component phenomenological model.* ***Dot-dashed line:*** *positron ratio in single-component scenario.* ***Dotted line:*** *positron ratio in double-component scenario due to conventional secondary positron production.* ***Solid line:*** *positron ratio in double-component scenario including extra-component. The progagation setup and modulation potential are the same of Fig. 1. The solar modulation potential is taken* $\Phi = 550$ *MV in all figures of this paper.*

The idea dates back to the pioneering work by F. Aharonian and A. Atoyan 1995[9] and was extensively studied after the release of ATIC and PAMELA data in 2008 (see e.g. Hooper *et al. 2009*[10] and Profumo 2008[11]).

More recently, we contributed to several papers in which it was shown that a consistent interpretation of the $e^+ + e^-$ spectrum measured by Fermi-LAT and the PAMELA positron fraction can be naturally obtained in that framework (Grasso *et al.* 2009,[5] Ackermann *et al.* 2010,[3] Di Bernardo *et al.* 2011[6]).

For example, in Fig. 1 and Fig. 2 we show that the double component model proposed in Ackermann *et al.* 2010[3] reproduces the data mentioned above and also the e^+ and e^- separate spectra, and their ratio, recently released by the Fermi-LAT collaboration and not yet available at the time. The model represented in those figures assumes a propagation setup characterized by a cylindrical diffusive halo with half-thikness of 4 kpc; a diffusion coefficient scaling with rigidity like $\rho^{1/3}$ (corresponding to a Kolmogorov-like diffusion within the quasi-linear approximation) and a relatively strong reacceleration (the Alfvén velocity is $v_A = 30$ kms^{-1}). Solar modulation is treated here as charge independent in the force field approximation by fixing the modulation potential Φ against proton data taken in the same solar phase. In that model, the standard e^- primary component is tuned to fit Fermi-LAT data at low energy in the presence of the extra-component becoming dominant at higher energies; the injection slope for the primary electron component is set to -2.70 above 2 GeV, while under that energy a slope of -1.6 is adopted, in accord with recent constraints from the synchrotron spectra (see Jaffe et al. 2011[12]). The extra component, instead, originates from a primary source of electron+positron pairs; it has an injection spectrum modelled in a simple way as a power-law with index -1.5 plus an exponential cutoff at 1.2 TeV; the spatial distribution of this source is the same as the standard one and the propagation parameters are also the same; the normalization is tuned so that Fermi-LAT and PAMELA data at high energy are matched by the sum of standard + extra component. Both components are computed with **DRAGON** (even if it was checked that the same result can be obtained with **GALPROP**).

An issue remains open about the origin of the discrepancy between the prediction of this, or similar, models and the positron fraction measured by PAMELA below 10 GeV. In the next section we will show as that discrepancy may be interpreted as the consequence of an incorrect choice of the propagation setup and discuss other uncertainties which can affect the electron and positron spectra in that low energy range.

3. LOW ENERGY. Impact of astrophysical uncertainties

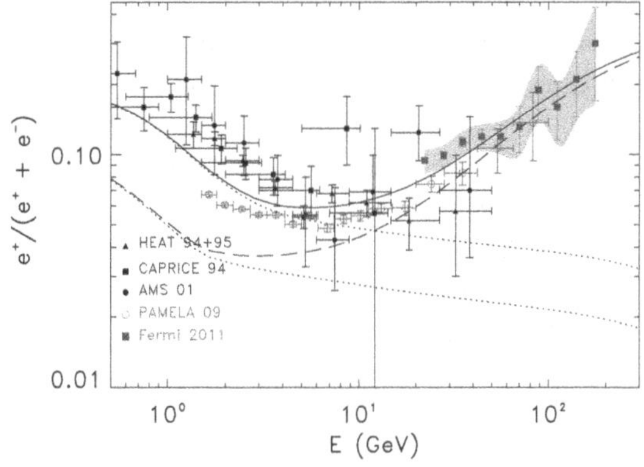

Fig. 3. *Effect of changing the diffusion halo height. Solid line: h = 1 kpc; dashed: h = 10 kpc.*

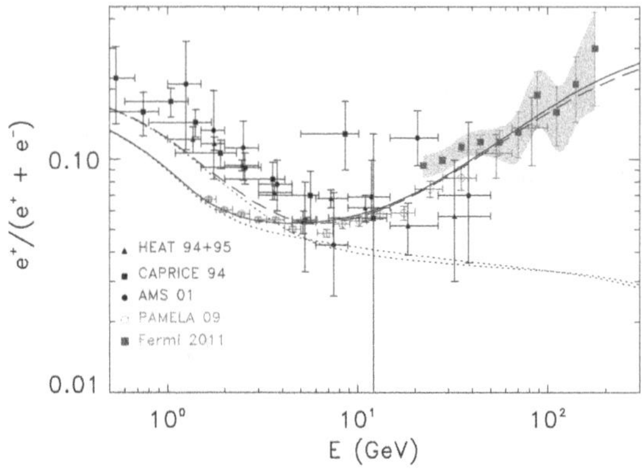

Fig. 4. *Effect of the diffusion setup. Solid line: KRA; dashed line: KOL.*

Cosmic ray electrons and positrons, either belonging to the standard or the extra component, propagate in the Galaxy undergoing several physical processes: diffusion, reacceleration, energy losses. Such complex motion is effectively described by a well-known diffusion-loss equation (Berezinskii *et al.* 1990[13]). In this equation several free parameters are involved: the height of the halo in which the propagation takes place, the normalization and energy dependence of the diffusion coefficient (the latter parametrized by the parameter δ), the Alfvén velocity that influences the effectiveness of reacceleration; moreover, several astrophysical inputs need to be considered: the injection spectrum, the spatial distribution of the source term, the interstellar radiation field, the gas distribution.

The free parameters that appear in the diffusion-loss equation are constrained by some CR observables such as Boron-to-Carbon (B/C) or antiproton-to-proton ratio; different *diffusion setups* exist in the literature, obtained through comparison of experimental data with the prediction of semi-analytical codes (Maurin *et al.* 2001,[14] Donato *et al.* 2004[15]) or numerical packages such as `DRAGON` or `GALPROP` (see e.g. Di Bernardo *et al.* 2010[16] for `DRAGON`-related models and Trotta *et al.* 2011[17] for a `GALPROP`-based analysis).

The uncertainties related to the diffusion model and to the astrophysical inputs were discussed in the latest years in several papers making use of semi-analytic codes (e.g. Delahaye *et al.* 2010[18]). In the following we will briefly analyse the impact of these uncertainties adopting the `DRAGON` code.

One of the most relevant parameter is the halo height. According to the analytical computations by Bulanov and Dogel,[19] while at low energy the electrons (or positrons) are distributed throughout all the diffusion halo, as the energy increases the electrons occupy a smaller and smaller fraction of the halo due to energy losses. This is relevant especially for the secondary positron spectrum. In fact, since their injection power is determined by the CR nuclei density in the Galactic disk, a thicker halo results in a larger dilution of their density in the halo hence a in smaller flux on the Earth. Numerical computations confirm the expectation of this heuristic argument as shown in Fig. 3. From the plot it is also evident that large halo heights are disfavoured by the data.

Even fixing the height of the diffusion halo, the choice of the diffusion setup can also affect the low energy spectra of CR leptons. This is evident from Fig. 4 where we compare the predictions of two different models which both reproduce nuclear CR data:

- a Kraichnan-like diffusion setup with $\delta = 0.5$ and moderate reacceleration (that was pointed out as the preferred one in a `DRAGON`-based maximum likelihood analysis with focus on both B/C and antiproton high energy data[16])
- a Kolmogorov-like diffusion setup with $\delta = 0.33$ and high reacceleration (that was pointed out as the preferred one in a `GALPROP`-based maximum likelihood analysis with focus on B/C data[17])

It is clear from that plot that the Kraichnan-like setups allows a better fit of low-energy positron ratio measured by PAMELA; this consideration, together with several other facts (high reacceleration models do not permit a good fit of antiproton data and cannot reproduce the spectrum of the synchrotron emission of the Galaxy), led us to conclude that models with strong reacceleration are disfavoured.

4. High energy uncertainties and the nature of the extra-component

In the double component scenario discussed in Sec. 2, the positron spectrum above ~ 10 GeV is dominated by the primary extra component. The nature of its source is one of the hottest matter of debate in the CR physics.

Galactic pulsars were suggested as natural source candidates of a primary CR positron component well before PAMELA results (Aharonian and Atoyan, 1995.[9]) More recently, it was noticed that a single, nearby, pulsar (such as Monogem or Geminga) could explain the positrons fraction excess found by PAMELA (Hooper *et al.* 2009[10]).

In the Fermi-LAT era, we showed (Grasso *et al. 2009*[5] and Di Bernardo *et al.* 2010[6]) that also the $e^+ + e^-$ measured by that experiment can consistently be explained in the same terms: if one considers the observed nearby pulsars within 2 kpc and assumes that a relevant fraction of their rotational energy is transferred into $e^+ + e^-$ pairs ($\simeq 30\%$), under reasonable assumpions on the injection spectrum and cutoff it is possible to reproduce all existing data. In the cosmic ray channel, this scenario has two possible testable consequences:

- the detection of a CR electron anisotropy towards the most relevant sources (in our analysis, Monogem and Geminga[10]);
- the presence of some bumpiness in the e^- and e^+ spectra in the TeV region due to the contribution of several pulsars.

Those two signatures are somehow complementary: if a single pulsar give the dominant contribution to the extra component a large anisotropy and a small bumpiness should be expected; if several pulsars contribute the opposite scenario is expected.

So far no positive detection of CRE anisotropy was reported by the Fermi-LAT collaboration, but some stringent upper limits were published. In Di Bernardo *et al.* 2010[6] we showed that the pulsar scenario is still compatible with these upper limits. Also, no evidence of spectral bumpiness has been found so far in the $e^+ + e^-$ spectrum.

It should be noted that several astrophysical uncertainties prevent accurate predictions of the CRE anisotropy and of the spectral bumpiness. For example, unknown irregularities in the local structure of the Galactic magnetic field may distort the angular distribution of the CRE flux due to a nearby pulsar. Furthermore, due to the stochastic nature of the e^- emission of nearby SNRs, the CRE standard component is expected to be subject to fluctuations which may produce anisotropies and spectral bumpiness which may hide those due to pulsars.

The other possible scenario to explain the origin of the extra component is more exotic but very appealing as it invokes DM annihilihation/decay as the origin of the $e^\pm$ extra component. Plenty of papers were published on that subject after the release of PAMELA and Fermi-LAT results (see e.g. He 2009[20] for a review). That scenario, however, present some problems. The most important are the following ones.

- It requires a heavy DM particle mass – O(TeV) – and an annihi-

lation cross section much higher than that predicted by standard cosmology if one assumes that DM is a thermal relic.
- Since no excess was detected for antiprotons, the annihilation/decay channels must include *only leptons* (lepto-philic DM).

Although several DM models which may fulfil those conditions were developed, another issue arises when electroweak corrections are taken into account. Those corrections, in fact, give rise – even in a lepto-philic scenario – to soft electroweak gauge bosons, and hence to antiprotons, at the end of their decay chains (Ciafaloni *et al.* 2011[21]). Since those exotic $\bar{p}$ are produced mainly in the Galactic Center region, the flux reaching the Earth strongly depends on the properties of CR propagation in the Galaxy. As we discussed in Sec. 3, these properties are still subject to strong uncertainties. It was shown in Evoli *et al.* 2011[22] that, accounting for those uncertainties, a scenario in which a heavy DM particle annihilates into muons is still compatible with the antiproton constraints. In the same paper it was also shown that AMS-02 is expected to constrain even more these models since its sensitivity to antiprotons will be much higher.

5. Conclusions and future perspectives

In this contribution we argued as recent experimental data rule out the standard scenario in which CR positrons are produced only by CR spallation onto the ISM and showed as an empirical model which invokes an extra $e^{\pm}$ component fulfils all data sets. We also discussed several uncertainties which still prevent to infer some of the properties of CR electron and positron sources. We argued that at low energy those uncertainties are dominated by our poor knowledge of CR propagation (which prevent an accurate determination of the injection spectrum of the e^{-} standard component) while at high energy the effect of the stochastic nature of astrophysical sources prevails (which makes more difficult to decide between the astrophysical and DM origin of the extra component).

Forthcoming experiments like AMS-02 and CALET are expected to reduce drastically the uncertainties on the propagation parameters by providing more accurate measurements of the spectra of the nuclear components of CR. Fermi-LAT and those experiments are also expected to provide more accurate measurements of the CRE spectrum and anisotropy looking for features which may give a clue of the nature of the extra component.

References

1. O. A. *et al.* [PAMELA collaboration], *Nature* **458**, 607(April 2009).
2. A. A. A. *et al.* [Fermi Collaboration], *Physical Review Letters* **102**, p. 181101(May 2009).
3. M. A. *et al.* [Fermi Collaboration], *Physical Review D* **82**, p. 092004(November 2010).
4. O. A. *et al.* [PAMELA collaboration], *Physical Review Letters* **106**, p. 201101(May 2011).
5. D. Grasso, S. Profumo, A. W. Strong, L. Baldini, R. Bellazzini, E. D. Bloom, J. Bregeon, G. di Bernardo, D. Gaggero, N. Giglietto, T. Kamae, L. Latronico, F. Longo, M. N. Mazziotta, A. A. Moiseev, A. Morselli, J. F. Ormes, M. Pesce-Rollins, M. Pohl, M. Razzano, C. Sgro, G. Spandre and T. E. Stephens, *Astroparticle Physics* **32**, 140(September 2009).
6. G. di Bernardo, C. Evoli, D. Gaggero, D. Grasso, L. Maccione and M. N. Mazziotta, *Astroparticle Physics* **34**, 528(February 2011).
7. P. D. Serpico, *ArXiv e-prints* (August 2011).
8. O. A. *et al.* [PAMELA collaboration], *Science* **332**, p. 69(April 2011).
9. A. M. Atoyan, F. A. Aharonian and H. J. Völk, *Physical Review D* **52**, 3265(September 1995).
10. D. Hooper, P. Blasi and P. Dario Serpico, *Journal of Cosmology and Astroparticle Physics* **1**, p. 25(January 2009).
11. S. Profumo, *ArXiv e-prints* (December 2008).
12. T. R. Jaffe, A. J. Banday, J. P. Leahy, S. Leach and A. W. Strong, *Monthly Notices of the Royal Astronomical Society* **416**, 1152(September 2011).
13. V. S. Berezinskii, S. V. Bulanov, V. A. Dogiel and V. S. Ptuskin, *Astrophysics of cosmic rays* 1990.
14. D. Maurin, F. Donato, R. Taillet and P. Salati, *Astrophysical Journal* **555**, 585(July 2001).
15. F. Donato, N. Fornengo, D. Maurin, P. Salati and R. Taillet, *Physical Review D* **69**, p. 063501(March 2004).
16. G. di Bernardo, C. Evoli, D. Gaggero, D. Grasso and L. Maccione, *Astroparticle Physics* **34**, 274(December 2010).
17. R. Trotta, G. Jóhannesson, I. V. Moskalenko, T. A. Porter, R. Ruiz de Austri and A. W. Strong, *Astrophysical Journal* **729**, p. 106(March 2011).
18. T. Delahaye, J. Lavalle, R. Lineros, F. Donato and N. Fornengo, *Astronomy and Astrophysics* **524**, p. A51(December 2010).
19. S. V. Bulanov and V. A. Dogel, *Astrophysics and Space Science* **29**, 305(August 1974).
20. X.-G. He, *Modern Physics Letters A* **24**, 2139 (2009).
21. P. Ciafaloni, D. Comelli, A. Riotto, F. Sala, A. Strumia and A. Urbano, *Journal of Cosmology and Astroparticle Physics* **3**, p. 19(March 2011).
22. C. Evoli, I. Cholis, D. Grasso, L. Maccione and P. Ullio, *ArXiv e-prints* (August 2011).

HELIOSPHERE DIMENSION AND COSMIC RAY MODULATION

P. Bobik[1], M.J. Boschini[2,4], C. Consolandi[2], S. Della Torre[2,5], M. Gervasi[2,3,*], D. Grandi[2], K. Kudela[1], F. Noventa[2,3], S. Pensotti[2,3], P.G. Rancoita[2], D. Rozza[2,3]

[1] *Institute of Experimental Physics, Kosice (Slovak Republic)*
[2] *Istituto Nazionale di Fisica Nucleare, INFN Milano-Bicocca, Milano (Italy)*
[3] *Department of Physics, University of Milano Bicocca, Milano (Italy)*
[4] *CILEA, Segrate (MI) (Italy)*
[5] *Department of Physics and Maths, University of Insubria, Como (Italy)*
**E-mail: massimo.gervasi@mib.infn.it*

The differential intensities of Cosmic Rays at Earth were calculated using a 2D stochastic Montecarlo diffusion code and compared with observation data. We evaluated the effect of stretched and compressed heliospheres on the Cosmic Ray intensities at the Earth. This was studied introducing a dependence of the diffusion parameter on the heliospherical size. Then, we found that the optimum value of the heliospherical radius better accounting for experimental data. We also found that the obtained values depends on solar activity. Our results are compatible with Voyager observations and with models of heliospherical size modulation.

Keywords: Cosmic rays; Solar Modulation; Monte Carlo simulations

1. The 2D Model of the Heliosphere

HelMod Code[1] solves the bi-dimensional Parker's particle transport equation[2]. A Monte Carlo technique is applied on a set of Stochastic Differential Equations (SDEs) fully equivalent to the Parker's equation[3]. The model takes into account particle drift effects and latitudinal dependence of the solar wind speed and of the Interplanetary Magnetic Field (IMF). It is described in details in Ref. 1. In the model, the IMF from Parker[4] is modified introducing a small latitudinal components as described in Ref. 5. For periods of low solar activity, we take a solar wind speed gradually increasing from the Earth position up to a maximum value near the heliospherical poles ($\simeq 760\,\mathrm{km/s}$)[6]. For periods approaching the solar maximum we assume a solar wind speed independent on the latitude.

The symmetric part of the diffusion tensor, in a reference frame with one axis aligned with the Parker's magnetic-field, is purely diagonal containing transverse ($K_{\perp\theta}$ and $K_{\perp r}$) and parallel ($K_{||}$) components[7]. The diffusion coefficients are given by[8]

$$\begin{aligned} K_{||} &= \beta\, K_0(t)\, K_P(P) \left[\tfrac{B_\oplus}{3B}\right] , \\ K_{\perp r} &= \rho_k\ K_{||} , \\ K_{\perp\theta} &= \iota(\theta)\ \rho_k\ K_{||} , \end{aligned} \tag{1}$$

where $\beta = v/c$, v the particle velocity and c the speed of light; the diffusion parameter K_0 accounts for the dependence on the solar activity; $B_\oplus$ is the measured value of IMF at the Earth position - typically $\approx 5\,\mathrm{nT}$, but changing with time - obtained from Ref. 9; B is the magnitude of the large scale IMF as a function of heliocentric coordinates; finally, the term K_P takes into account the dependence on the rigidity P of the GCR particle usually expressed in GV. In the present model $K_P \approx P$ (e.g., see Ref.[10]). Furthermore $\rho_k = 0.05$ and, as described in Ref. 1,

$$\iota(\theta) = \begin{cases} 10\ , & \text{in the polar regions,} \\ 1\ , & \text{in the equatorial region.} \end{cases} \tag{2}$$

After the transformations from 3D field-aligned into 2D heliospherical coordinates[11], the symmetric components of the diffusion tensor contains both diagonal (K_{rr} and $K_{\theta\theta}$) and off-diagonal terms ($K_{r\theta}$ and $K_{\theta r}$), resulting by a proper combination of $K_{\perp\theta}$, $K_{\perp r}$ and $K_{||}$[1].

2. The Diffusion Parameter

K_0 accounts for the dependence on the solar activity. We estimated K_0 by using the modulation strength ϕ_s, in the framework of the Force Field (FF) approximation[12]. ϕ_s was evaluated starting from Neutron Monitor (NM) counting rates in Ref. 13. We moreover used a practical correlation of K_0 with the level of solar activity in the different solar phases[1]. We used, as solar activity monitor, the Smoothed Sunspot Number (SSN).

This method is sensitive to the modulation of the GCR flux integrated over the full heliosphere, from the outer boundary to the Earth position, down to a lower limit in rigidity of $\sim$ (2-3) GV. This limit is fixed by the sensitivity of the NM network, due to the geomagnetic rigidity cut-off and to the atmospheric yield function. The outer boundary of the heliosphere is located at the position of the Termination Shock. Beyond this limit the model of heliosphere we are using is not more valid. Moreover, the additional modulation occurring in the heliosheat only affects particles with rigidity well below 1 GV[14,15]. The method is sensitive also to the LIS used for the estimation of the modulation strength, but several LIS spectra do not differ each other above this rigidity limit. Finally the diffusion parameter depends on the outer boundary position, as follows from the FF approximation:

$$K_0(t) = \frac{V_{\rm sw}(t)\,(R_{\rm TS} - R_{\rm earth})}{3\phi_{\rm s}(t)} \,. \tag{3}$$

In Ref. 1 the boundary of heliosphere was placed at 100 AU. The solar cavity was split in 15 spherical regions to take into account the time spent by SW to travel outward. In each region of the interplanetary space, the parameters (i.e., SW speed, SSN, $B_\oplus$, tilt angle) are related to the solar activity at the time of the injection of the solar wind diffusing in that region[16]. In this way modulated intensities of protons, down to $\sim$ 400 MeV, were simulated and successfully compared with experimental data covering roughly one solar cycle. We did not find significant differences changing the position of the outer boundary of the heliosphere[1].

3. Heliospherical Size and Diffusion Parameter

In the past years the position of the Termination Shock was estimated through the observations of Voyager 1 and Voyager 2 spacecrafts (see Refs. 17,18 and Table 1). In addition several authors (see Refs. 19,20) suggest that the size of the heliospere should change with the solar activity, following a quasi-periodic feature, roughly anti-correlated with the SSN.

Table 1. Voyager crossings of Termination Shock.

	R_{TS} (AU)	solar latitude (deg)
Voyager 1	94.0	+ 34.3
Voyager 2	83.7	− 27.5

Following these results we evaluated the effect of stretched and compressed heliospheres on the Cosmic Ray intensities at the Earth introducing a dependence of the diffusion parameter on the heliospherical size. We defined a new diffusion parameter K_0^*, introducing the parameter $r(R_{TS}, P)$ sensitive to the position of the Termination Shock:

$$K_0^*(R_{TS}) = r(R_{TS}, P)\ K_0(100\ \text{AU}) \tag{4}$$

$$r(R_{TS}, P) = 1 + f(P)\left[\frac{R_{TS}(\text{AU}) - 100}{99}\right] . \tag{5}$$

$r(R_{TS}, P)$ allows to modify the value of the diffusion parameter adapting it to a different volume of the heliosphere, determined by R_{TS}. $r(R_{TS}, P)$ is fully effective below a rigidity limit P_1. We also defined a transition function $f(P)$:

$$f(P) = \begin{cases} 0\ , & \text{for} \quad P \geq P_2\ , \\ (P_2 - P)/(P_2 - P_1)\ , & \text{for} \quad P_1 < P < P_2\ , \\ 1\ , & \text{for} \quad P \leq P_1\ . \end{cases} \tag{6}$$

For rigidity higher than P_2, the dependence on R_{TS} can be neglected. Here the diffusion parameter is still defined for an heliospherical dimension of 100 AU. The dependence on the heliospherical radius R_{TS} is then effective at rigidity lower than P_2. Using the novel diffusion parameter $K_0^*(R_{TS}, P)$ we simulated the modulated spectra, for different values of R_{TS}, P_1 and P_2, extending the modulated spectra down to a lower rigidity.

4. Results

We compare our simulated spectra with proton data extended down to a kinetic energy of 200 MeV. Here we present results obtained using the following rigidity parameters: $P_2 = P_1 = 1.0$ GV. We used the Local Interstellar Spectrum (LIS) from Ref. 21 and compared it with the LIS form GALPROP[22]. In Fig. 1 the results compared with AMS-01 data[23] are shown, assuming $R_{TS} = 120$ AU, as discussed later on.

We estimated the best value of R_{TS}, looking at the RMS differences (η_{RMS}) with experimental data:

$$\eta_{\text{RMS}} = \sqrt{\frac{\sum_i (\eta_i/\sigma_{\eta,i})^2}{\sum_i 1/\sigma_{\eta,i}^2}}\ , \tag{7}$$

with

$$\eta_i = \frac{f_{\text{sim}}(T_i) - f_{\text{exp}}(T_i)}{f_{\text{exp}}(T_i)}\ , \tag{8}$$

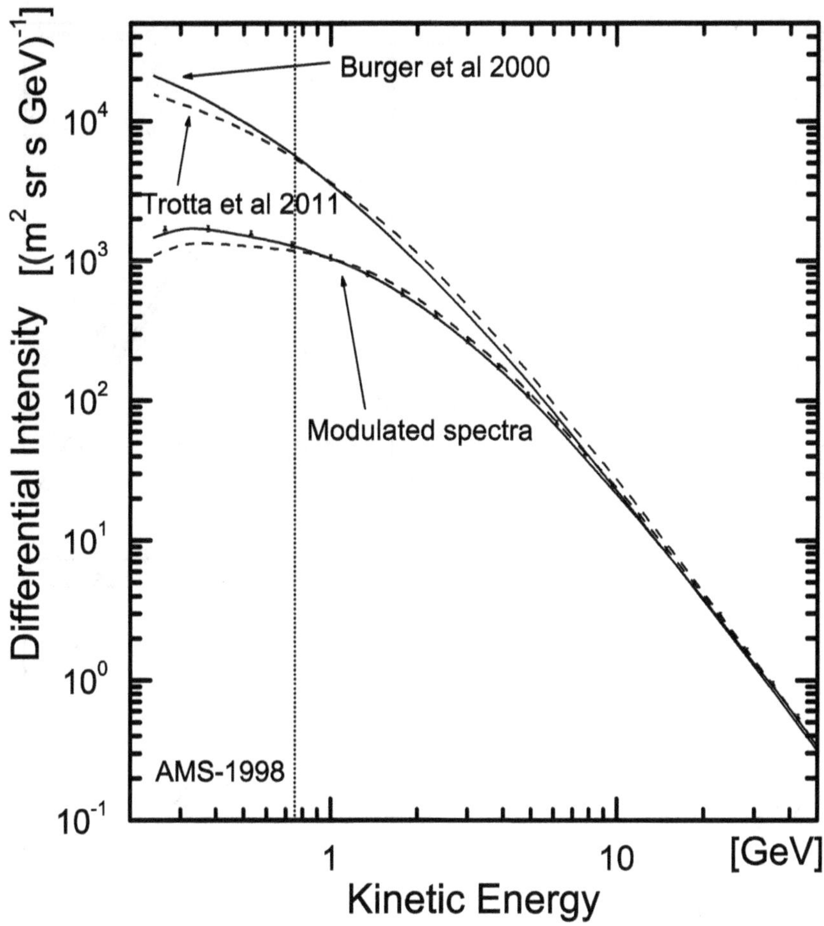

Fig. 1. Modulated proton spectra and comparison with AMS-01 data[23]. LIS are taken form Ref. 21 and GALPROP[22]. The vertical dotted line represents the lower limit of the sensitivity of the NM. Above this limit the two LIS are not significantly different.

where T_i is the average energy of the i-th energy bin of the differential intensity distribution and $\sigma_{\eta,i}$ are the error bars including the experimental and Monte Carlo uncertainties. For each experimental spectrum we got the best values of R_{TS} shown in Table 2 together with the minimum value of $\eta_{\rm RMS}$. Data from BESS flights are given in Ref. 24, data from AMS-01 are given in Ref. 23.

In Fig. 2 modulated spectra, obtained using values of R_{TS} reported in Table 2, are shown in comparison with BESS experimental data. Modulated

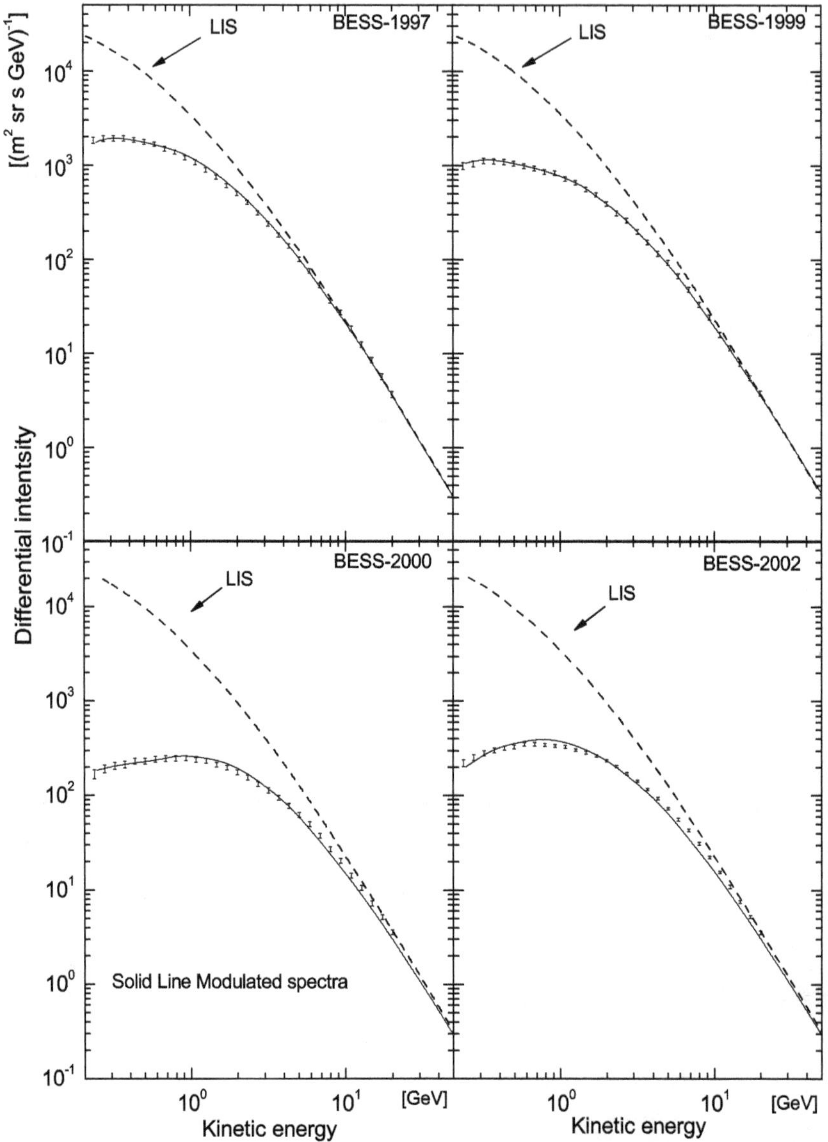

Fig. 2. Modulated proton spectra and comparison with BESS-1997, BESS-1999, BESS-2000, BESS-2002 observing data[24].

spectrum compared with AMS-01 data has been shown in Fig. 1. We did not use data measured by Pamela[25] because published spectra start from

Table 2. Best values of R_{TS}, its minimum and maximum values, and RMS differences between simulations and experimental data.

	R_{TS}^{best} (AU)	R_{TS}^{min} (AU)	R_{TS}^{max} (AU)	$\eta_{\rm RMS}$ (%)
BESS-1997	115	100	130	7.05
AMS-1998	120	110	135	4.86
BESS-1999	120	110	130	3.35
BESS-2000	140	125	150	10.00
BESS-2002	105	95	115	11.78

400 MeV, while our analysis is more sensitive below this limit. In Table 2 we report the interval of values of R_{TS} where $\eta_{\rm RMS}$ does not change by more than $\sim (2-3)$ % from its minimum value, reported in the last column. This variation roughly represents the uncertainty of the computation itself, and it is determined comparing simulations and data at energies larger than $(10-20)$ GeV, i.e. above the region of solar modulation. Results are shown in Fig. 3 in comparison with models[20] and Voyager measurements[17,18].

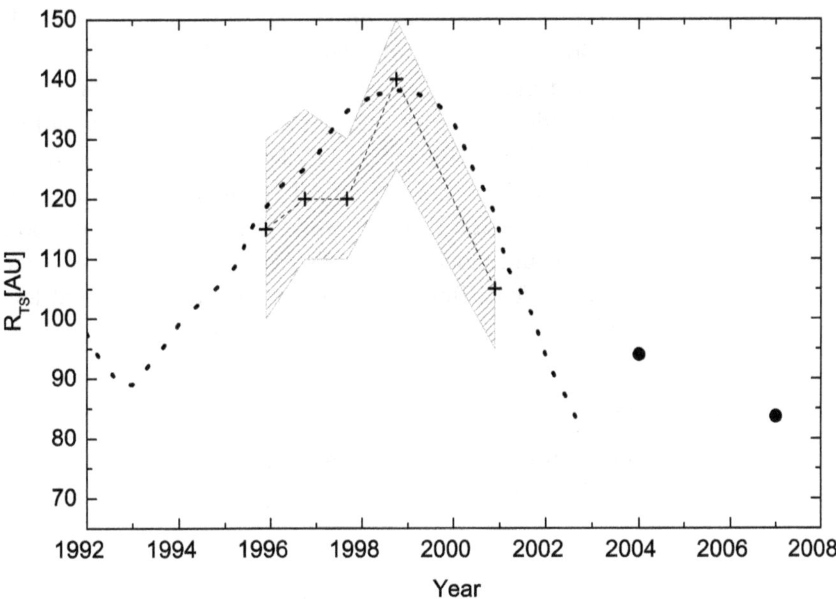

Fig. 3. R_{TS} best value for the several experiments (crosses) in comparison with Voyager data[17,18] (dots) and models[20] (dashed line). The shadow represents the region between the minimum and maximum value, as reported in Table 2.

As shown in Fig. 1 and Fig. 2, modulated spectra succeed to fit observing data, in particular during periods of low solar activity. For more accurate results we need more accurate experimental data. Current error bars are of the order of 5% or even larger. Moreover systematic deviations are present looking at different data sets at energy above (20–30) GV, where spectra are not affected by solar modulation. In addition current observation data, taken on board of stratospheric balloons or space orbiters, may be contaminated at low energy by secondaries produced inside the Earth magnetosphere. A LIS spectrum with a slightly different shape could be also preferred to fit better low energy data. Finally a refinement of our model could be requested, starting with a slightly different values of P_2 and P_1, in order to smooth the ripple present in some spectrum. In the future a model with an aspheric Heliosphere can be also developed.

5. Conclusions

We presented the HelMod 2D Monte Carlo code for the study of Cosmic Rays propagation in the inner heliosphere. Both heliospherical shape and size are supposed to be relevant for the modulation process. We introduced a dependence of the diffusion parameter on the heliospherical size, which accounts for the variation with time and solar activity. We compare modulated spectra with experimental data covering the solar cycle 23. Then we found, for our 2D model, the best value of the heliospherical radius, which changes with time. Most of the solar modulation occurs in the inner heliosphere and differences in the heliospherical radius are effective only at energy below a few hundred MeV. Our results are not in contradiction with Voyager observations and models of TS distance as a function of solar activity. We found that LIS form Ref. 21 fits better observation data at low energy.

Acknowledgments

Authors acknowledge the use of NASA/GSFCs Space Physics Data Facilitys OMNIWeb service, and OMNI data.

References

1. P. Bobik, G. Boella, M.J. Boschini, C. Consolandi, S. D. Torre, M. Gervasi, D. Grandi, K. Kudela, S. Pensotti, P.G. Rancoita, M. Tacconi, *Astrophys. J.*, in press, (2011); arXiv:1110.4315 [astro-ph.SR] 19 Oct 2011.
2. E. N. Parker, *Plan. & Space Sci.* **13**, p. 9 (1965).

3. P. Bobik, M. Boschini, C. Consolandi, S. Della Torre, M. Gervasi, D. Grandi, K. Kudela, S. Pensotti and P.G. Rancoita, *Astrophys. Space Sci. Trans.* **7**, 245 (2011).
4. E. N. Parker, *Astrophis. J.* **128**, 664 (1958).
5. U. Langner, *Ph.D. Thesis, Potchestroom University* (2004).
6. D. J. McComas, B. L. Barraclough, H. O. Funsten, J. T. Gosling, E. Santiago-Muñoz, R. M. Skoug, B. E. Goldstein, M. Neugebauer, P. Riley and A. Balogh, *J. Geophys. Res.* **105**, 10419 (2000).
7. M. S. Potgieter, *J. Geophys. Res.* **105**, 18295 (2000).
8. M. S. Potgieter and J. A. Le Roux, *Astrophis. J.* **423**, p. 817 (1994).
9. Omniweb data available at: http://omniweb.gsfc.nasa.gov/form/dx1.html (2011).
10. J. S. Perko, *Astron. and Astrophys.* **184**, 119 (1987).
11. R. A. Burger, T. P. J. Krüger, M. Hitge and N. E. Engelbrecht, *Astrophis. J.* **674**, 511 (2008).
12. L. J. Gleeson and W. I. Axford, *Astrophis. J. Lett.* **149**, L115 (1967).
13. I.G. Usoskin, et al., *J. Geophys. Res.* **110**, A12108 (2005).
14. Bobik, P., K. Kudela, M. Boschini, D. Grandi, M. Gervasi, P.G. Rancoita, Solar modulation model with reentrant particles, *Adv. Space Res.*, **41** (2008), 339-342, doi:10.1016/j.asr.2007.02.085.13.
15. Scherer, K., Fichtner, H., Strauss, R. D., et al., *Astrophys. J.*, **735**, 128 (2011), doi:10.1088/0004-637X/735/2/128.
16. P. Bobik, M. Boschini, C. Consolandi, S. Della Torre, M. Gervasi, D. Grandi, K. Kudela, S. Pensotti and P.G. Rancoita, Proton and antiproton modulation in the heliosphere for different solar conditions and AMS-02 measurements prediction, in *Astroparticle, Particle and Space Physics, Radiation Interaction, Detectors and Medical Physics Applications* - vol. **6** - *Proc. of the 12th ICATPP Conference on Cosmic Rays for Particle and Astroparticle Physics,* 360-368 (2011), eds. S. Giani, C. Leroy,P.-G. Rancoita, ISBN-13 978-981-4329-02-6.
17. E.C. Stone, A.C. Cummings, F.B. McDonald, B.C. Heikkila, N. Lal and W.R. Webber, *Science* **309**, 2017 (2005).
18. E.C. Stone, A.C. Cummings, F.B. McDonald, B.C. Heikkila, N. Lal and W.R. Webber, *Nature* **454**, 71 (2008).
19. Y.C. Whang, L.F. Burlaga, *Geophys. Res. Lett.* **27**, 1607.
20. Y.C. Whang, L.F. Burlaga, Y.-M. Wang, and N.R. Sheeley Jr., *Geophys. Res. Lett.* **31**, L03805 (2004), doi:10.1029/2003GL018679.
21. R. A. Burger, M. S. Potgieter and B. Heber, *J. Geophys. Res.* **105**, 27447 (2000).
22. R. Trotta, G. Johannesson, I.V. Moskalenko, T.A. Porter, R. Ruiz de Austri, and A.W. Strong, *Astrophis. J.* **729**, 106 (2011), doi:10.1088/0004-637X/729/2/106.
23. AMS Collaboration, M. Aguilar *et al.*, *Phys. Rep.* **366**, 331(2002).
24. Y. Shikaze *et al.*, *Astropart. Phys.* **28**, 154 (2007), doi:10.1016/j.astropartphys.2007.05.001.
25. O. Adriani *et al.*, *Science* **332**, 69 (2011), DOI: 10.1126/science.1199172.

ON QUASI-PERIODIC VARIATIONS OF COSMIC RAYS.

K. KUDELA, P. BOBÍK
IEP SAS, Kosice, Slovakia,kkudela@kosice.upjs.sk

G. BOELLA, M.J. BOSCHINI, C. CONSOLANDI, S. DELLA TORRE, M. GERVASI, D. GRANDI, S. PENSOTTI, P.G. RANCOITA AND M. TACCONI
INFN, Milano-Bicocca, Milano, Italy

Selected results in studies of quasi-periodic variations of cosmic rays obtained from ground based measurements, especially from neutron monitor Climax, are reviewed. At high frequencies the dominant periodicity is the solar diurnal variation and its higher harmonics. The slope of power spectrum density at high frequencies is flatter for solar maxima than for solar minima. Wavelet spectra are useful tool for checking the fine strucure of quasi-periodicities and their temporal behaviour. ~27 day, ~13.5 day and ~9 day quasi-periodicity is far more complex than the diurnal one. For ~27 day variation, two local peaks are temporarily observed, one near 27 and another near 30-31 days with variable contribution to the signal. ~150 days, ~1.7 yr periodicities are seen in the Lomb-Scargle periodogram.

1. Introduction

Quasi-periodic and periodic variations in cosmic ray (CR) intensity observed at Earth are reported already long time ago. The research of solar diurnal wave, being the fixed one, started probably from papers [1-5]. Reports on its higher harmonics, namely on semi-diurnal one, can be found since papers [6,7] and on the tri-diurnal one since [8,9]. In those papers references to earlier studies are mentioned. In the longer quasi-periodicities probably first attention was paid to ~ 11 yr, ~22 yr and ~27 d. The detailed review is in [10,11] and references therein. Here we present first results of the study we recently started, devoted to checking of characteristics of selected quasi-periodicities (q-per) in CR direct measurements on the ground, mainly by the neutron monitor (NM) Climax. The motivation is driven by the problem of relation of CR to the atmospheric processes (started probably from [12] and references therein); by the recent paper reporting results of CLOUD experiment [13]; as well as by the availability of long term series of CR data from various NM and muon detectors.

Also, the description of q-per in CR flux at Earth over long time period, can be one of the tools for testing the theories of CR modulation in the heliosphere (e.g. [14,15] and references therein).

2. Shape of power spectra of CR time series

Using the daily counting rates from Climax NM, skipping the data influenced by the GLEs, interpolating linearly the gaps, and applying the FFT method, the power spectrum plotted in Figure 1 is obtained.

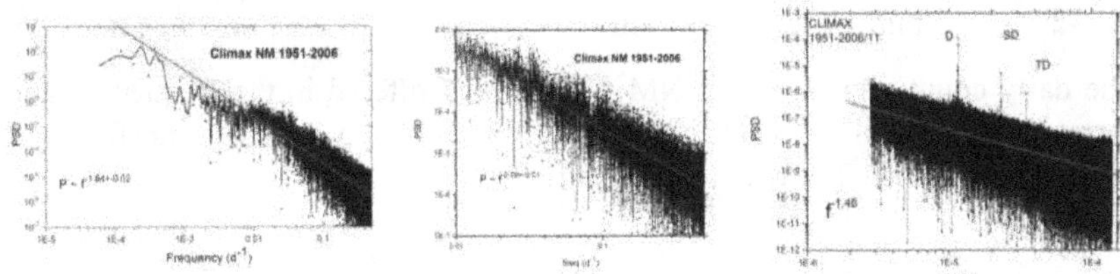

Fig. 1. Power spectrum density (PSD by FFT method) of daily averages (left, middle panels) and of hourly (right panel) count rate of Climax NM. Data downloaded From ftp://ulysses.sr.unh.edu/NeutronMonitor/DailyAverages.1951-.txt. Line (left, middle panels) – cubic b-spline connection.

The plot on the left is similar to [16] for the period until 1996. At higher frequencies (middle panel, $f < 5.8 \times 10^{-6}$ Hz) the slope is consistent with the theory [17] where the authors indicate that "...including the effects of non-field-aligned diffusion, which dominates the power spectrum of NMs at low frequencies ($< 5 \times 10^{-6}$ Hz) and produces a spectrum of f^{-2}" .The right panel clearly indicates the presence of the fixed frequencies of the diurnal, semi-diurnal and tri-diurnal waves. Probably the fourth harmonic is present too. The spectral slope is lower, around -1.5. At $f > 5.10^{-6}$ Hz the theory with field-aligned diffusion is satisfactory for explanation [18]. Solar diurnal anisotropy is resulting from the co-rotational streaming of particles past Earth [19]. Difference in the slopes from NM Lomnický štít (LS) for different phases of solar activity cycle is in Table 1.

Table 1. Slopes (γ) in $f^{-\gamma}$ fit of of PSD at $f > 5.10^{-6}$ Hz based on hourly data LS compared with the slopes of IMF **B** (downloaded from omniweb NASA site):

	2002 max	2009 min
NM LS	-1.52±0.02	-1.08±0.03
IMF B	-1.89±0.02	-1.72±0.02

The flattening of the spectra for solar minimum is apparent. The values of γ for NMs are lower because of presence of diurnal wave and its harmonics. Qualitatively the picture is in agreement with the slopes of IMF measured in different solar cycle phase at large distances [20, Voyager-1]. [21] studied power spectra of CR in range 2 – 500 days and found significant differences in the individual spectra of solar maxima for different cycles. Spectra in range $2.7 \times 10^{-7} - 1.4 \times 10^{-4}$ Hz measured by muon detector examined [22]. The spectra are reported to be flatter and have lower power when the interplanetary magnetic field (IMF) is directed away from the Sun above the current sheet $(A > 0)$ than when the IMF is directed in opposite way $(A < 0)$.

3. ~27, ~13.5 and ~9 day variation

The daily count rate means of NM Climax was filtered in the frequency range corresponding to time scale 25 – 33.3 days and the wavelet transform was applied on the data. The result is in Figure 2.

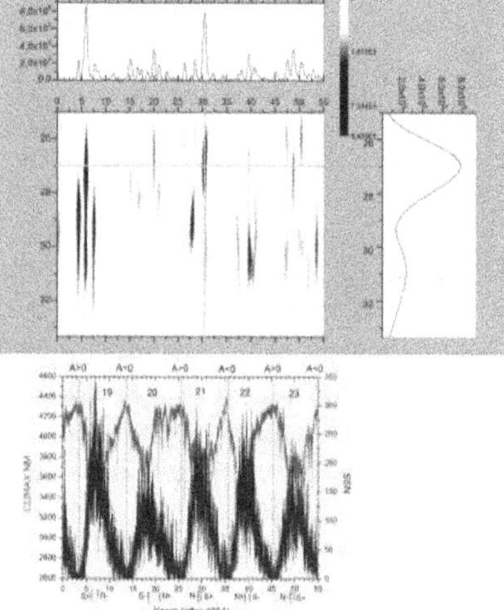

Fig. 2. Middle panel: wavelet spectrum density (Morlet) of Climax NM daily means. The upper panel is cross section of the density at 27 days, right panel is cross section over periods 25 – 33.3 d for the time of solar maxima ~ 1981.

The structure with essential peaks at ~ 27 days and ~ 30-31 days is found at solar maxima, similar to [23] based on data of GCR, EPHIN on SOHO. However, the structure is more complex. Transport models [24] and measurements analysis [25] suggest dependence on solar magnetic field polarity. This feature will be examined by wavelet technique in more detail in continuation of the present study. We examined the vicinity of ~13.5 and ~9 day period. At ~13.5 days we confirm the result [26] indicating the maximum contribution is correlated with

SSN. At ~9 days the results are in [27].

4. ~ 150 day q-per

Reference [28] reported 154 day periodicity in solar X ray and gamma ray flares. Reference [29] reported various periodicities in solar activity time series. There is no explanation for 150-157 day period found in several data sets. 154-day periodicity is found in the near-Earth IMF strength and solar wind speed [30]; also in solar proton events [31]. Reference [32] using Voyager 1 data, in outer heliosphere at ACR indicates that q-per variations are in phase, with O, He having periods ~ 151 days, while protons exhibit a period ~ 146 days. In CR the earlier results can be found e.g. in [33, 34]. Figure 3 indicates possible double structure in Climax data.

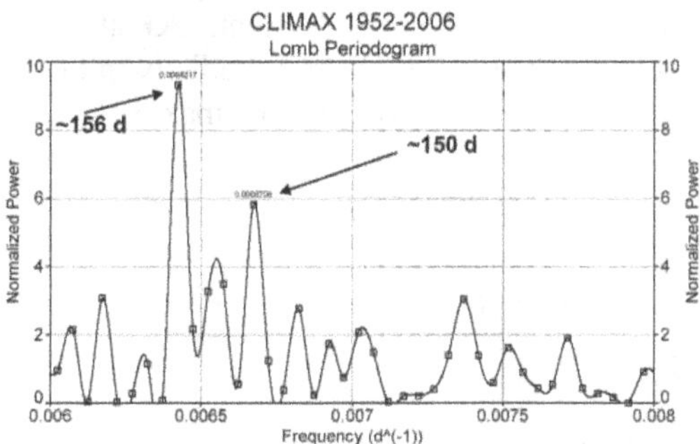

Fig. 3. Lomb-Scargle periodogram of the Climax NM data in the frequency range 7 x 10-8 – 9.3 x 10-8 Hz for the entire interval of the measurements.

5. Q-per ~ 1. 7 year

This q-per was in CR reported first in [35], analyzed by wavelet technique in [36], found in the outer heliosphere in Voyager data [37]. Earlier, using NM data Calgary and Deep River, Reference [38] indicated that a 20m peak occurs, as well as a spectrum instability in the neighborhood of the periods 6 - 18 m. Recently [39] reports that length of the q-2 year periodicity in even and odd numbered cycles differs by ~2 months. In cycles 20 and 22, T = 22–23.5 months, in cycles 21 and 23, T = 20.2–20.8 m. Reference [40] by examining solar magnetic fluxes in the period 1971–1998 found that ~ 1.7 year is the dominant fluctuation for all the types of fluxes analyzed (total, closed, open, low and high latitude open fluxes) and has a strong tendency to appear during the descending phase of solar activity. The periodicities of ~1.3 year and ~1.7 years were seen neither often nor prominently in several solar activity indices [41]. Reference [42] relates a strong 1.68-year oscillation in GCR fluxes to a corresponding oscillation in the open solar magnetic flux and infer CR propagation paths confirming the predictions of theories in which drift is important in modulating CR flux. Reference [43] indicates an interesting approach to the ~1.6 yr variation. The author calculated the solar motion due to the inner (terrestrial) planets (Mercury, Venus, Earth, Mars) for the years 1868–2030 and indicated spectrum of periods shows the dominant period of 1.6 years (V-E) and further periods of 2.13 years (E-Ma), (QBO), 0.91 years (V-Ma), 0.8 years ((V-E)/2) and 6.4 years. Paper [44] reported that starting with Alfven's original suggestion, it is possible to develop a quantitative equilibrium theory for the trapping of CR in the magnetic field of the Sun, where in addition to the effect of scattering in the geomagnetic field there is also taken into account the direct absorption of the CR by five heavenly bodies Mars, Venus, the Earth, the Sun, and the Moon. This may be one of candidates for the link between the result [42] and ~ 1.7 year q-per observed in CR. Results of spectral analysis of surface atmospheric electricity data (42 years of Potential Gradient, PG at Nagycenk, Hungary) showed also ~1.7 year q-per [45]. ~1.7 year periodicity in the PG data is present 1978 – 1990, but absent in 1963 – 1977. The Climax NM shows clearly ~1.7 yr q-per in Figure 4.

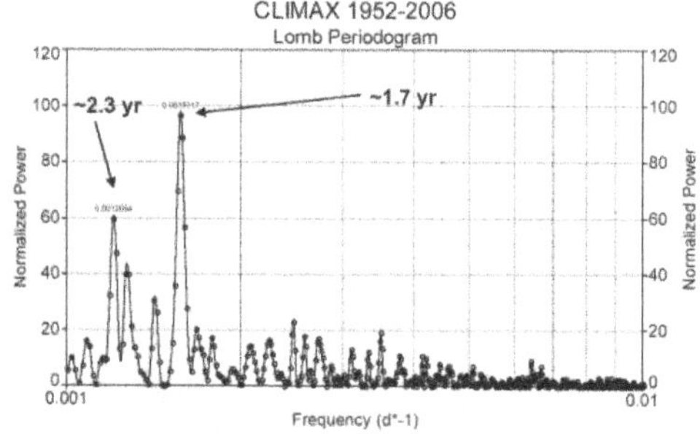

Fig. 4. Lomb-Scargle periodogram of Climax NM daily means indicating the presence of ~ 2.3 yr in addition to ~1.7 yr one.

~2.3 years reported also by [46] in coronal index calculated using Fe XIV 530.3 nm coronal emission line ground-based measurements from the worldwide net of the coronal stations [47]. The wavelet analysis at various cut-offs for NM and for muon detector data is required to clarify whether that q-per has a cut-off in energy and how it is evolved over several tens of years.

6. Above 1.7 year

Reference [48] identified the presence of ~ 5 year variability in CR over epochs of low solar activity. It is desirable to investigate whether a correlated 5-year signal exists in other geophysical and biological records, and if so, it could provide an additional source of data on the characteristics of the sun at times of low solar activity. Paper [49] studied solar wind speed and density of q-per: although the spectrum shows remarkable peaks at the wavelengths 0.5, 0.7, 1.0, 1.3 years, additional significant peaks for solar wind speed are also found. The averages of solar wind ion density showed a periodic variation with three nearly equal peaks at intervals ~ 5.1 ± 0.2 yr. The 9.8, 3.8, and 1.7–2.2 yr periods are the most significant found in interplanetary proton flux at 190 – 440 MeV [50]. Periodogram of the NM Climax for long periodicities is in Figure 5.

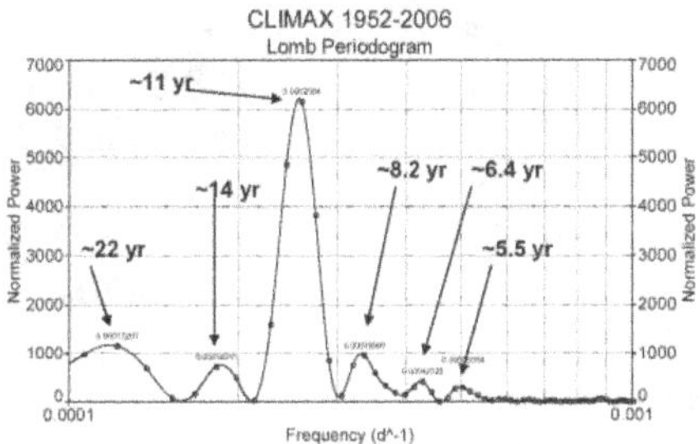

Fig. 5. Lomb-Scargle periodogram of Climax data for periodicities longer than 1000 days. All q-per below ~11 year found here were reported by different methods in the data until 1996 [51].

7. Summary

First step in systematic study of q-per variations of CR by ground measurements was done. For long time series the slope of PSD at $f < 5.8 \times 10^{-6}$ Hz is approx. -2 (in agreement with theory [17]). At $f > 5.10^{-6}$ Hz the spectra is flatter for solar minima than for solar maxima (at least for 1 solar cycle), in qualitative similarity with IMF B. For 1964-2006 the phase of diurnal wave has ~22 year periodicity and amplitude ~11 years (3 NMs with geomagnetic cut-off < 4 GV), amplitude is correlated with magnitude of IMF [52]. Wavelet analysis shows hat ~27 day q-per has complicated character, two maxima (~27, ~30-31 days), with variable contribution in time. ~13.5 day q-per over 20-23 solar cycle is similar to the results of older paper for cycles 18-20 based on ionisation chamber data. ~9 day q-per is observed in selected intervals, correlation with v, B. Probably double structure is in 150 – 156 d q-per. Temporal evolution of ~1.7 yr seen at several NMs is important for checking relations to atmospheric electricity. The wavelet technique is a tool for testing the presence of other periodicities reported e.g. in [53-55] and, if used for the same intervals and applied on time series of CR, solar, geomagnetic and interplanetary activity, may help in discriminating the links of CR to the atmospheric processes.

Acknowledgments

We acknowledge the University of New Hampshire, "National Science Foundation Grant ATM-9912341" for Climax NM data. KK wishes to acknowledge support by the grant agency VEGA, project 2/0081/10.

References

1. S.E. Forbush, Magnetism Atmospheric Elec., 42, 1 (1937).
2. S.F. Singer, Nature 170, 63 – 64 (1952).
3. J.L. Thompson, Phys. Rev. 54, 93–96 (1938).
4. E.A. Brunberg and A. Dattner, Tellus, 6, 1, 78-83 (1954).
5. H.S. Ahluwalia and A.J. Dessler, Planet. Space Sci., 9, 5, 195-210 (1952).
6. H.S. Ahluwalia, Proc. Phys. Soc. 80, 472 (1962).
7. P. Nicolson and V. Sarabhai, Proc. Phys. Soc. 60, 509 (1948).
8. S. Mori, S. Yasue and M. Ichinose, paper MOD-37, Proc. 12th ICRC, Hobart, 2, 666 (1971).
9. H.S. Ahluwalia and S. Singh, Proc. 13th ICRC, Denver, 2, 948 (1973).
10. L.I. Dorman, Variations of galactic cosmic rays, Moscow, MGU Publ. House, pp. 214, in Russian (1975).
11. L.I. Dorman, Cosmic rays in the Earth's atmosphere and underground, Kluwer, pp. 841 (2004).
12. H. Svensmark and E. Friis-Christensen, J. Atmos. Sol. Terr. Phys., 59, 1225–1232 (1997).
13. J. Kirkby, et al., Nature 476, 429–433 (25 August 2011).
14. M.S. Potgieter and H. Moraal, Astrophys. J., 294, 425–440 (1985).
15. J.F. Valdés-Galícia, J.F., Adv. Space Res., 35, 755–767 (2005).
16. H. Mavromichalaki, et al., Ann. Geophys., 21, 1681-1689 (2003).
17. J.R. Jokipii and A.J. Owens, GRL, 1,329 (1974b).
18. J.R. Jokipii and A.J. Owens, JGR, 81, 13, 2094-2096 (1974a).
19. M.L. Duldig, Publ. Astron. Soc. Austr., 18, 12-40 (2010).
20. L.F. Burlaga and N.F. Ness, JGR, 103, A12, 29,719-29,732 (1998).
21. M.A. El-Boriye and S.S. Al-Thoyaib, Solar Phys., 209, 397–407 (2002a).
22. I. Sabbah and M.L. Duldig, Solar Phys., 243, 231–235 (2007).
23. P. Dunzlaff et al., Ann. Geophys., 26, 3127–3138 (2008).
24. A. Gil et al., Adv. Space Res., 35, 687-690 (2005).
25. I.G. Richardson, Space Sci. Rev., 111, 267-376 (2004).
26. O. Filisetti and V. Mussino, Rev. Bras. de Fisica, 12, NP 4 (1982).
27. I. Sabbah and K. Kudela, JGR, 116, A04103, (2011).
28. E. Rieger et al., Nature, 312, 623 (1984).
29. J. Pap, W.K. Tobiska and S.D. Bouwer, Sol. Phys., 129, 165-189 (1990).

30. H.V. Cane, I.G. Richardson and T.T. von Rosenvinge, GRL, 25, 4437 (1998).
31. S. Gabriel, R. Evans, and J. Feynman, Sol. Phys., 128, 415 (1990).
32. M.E. Hill, D.C. Hamilton and S.M. Krimigis, JGR, 106, A5,8315 (2001).
33. K. Kudela et al., Sol. Phys., 266, 173–180 (2010).
34. H. Mavromichalaki et al., Ann. Geophys., 21, 1681-1689 (2003).
35. J.F. Valdes-Galicia, J. F., et al., Sol. Phys., 67, 409 – 417 (1996).
36. K. Kudela et al., Sol. Phys., 205, 165 – 175 (2002).
37. C. Kato et al., JGR, 108, A10, 1367 (2003).
38. K. Kudela, A.G. Ananth and D. Venkatesan, JGR 96, A9, 15,871 (1991).
39. V.P. Okhlopkov, Moscow U. Phys. Bull, 66, 1, 99–103 (2011).
40. B.V. Mendoza, V. M. Velasco and J. F. Valdés-Galicia, Sol. Phys., 233, 319-330 (2006).
41. R.P. Kane, Sol. Phys., 227, 155–175 (2005).
42. A. Rouillard, A. and M.A. Lockwood, Ann. Geophys., 22, 4381–4395 (2004).
43. I. Charvátová, Ann. Geophys., 25, 1227–1232 (2007).
44. E.O. Kane, J.B. Shanley and J.A. Wheeler, Rev. Mod. Phys., 21, 1, 51 (1949).
45. R.G. Harrison and F. Märcz, GRL, 34, L23816, 2007GL031714 (2007).
46. H. Mavromichalaki et al., Adv. Space Res. ,35, 410–415 (2005).
47. M. Rybanský, Bull. Astron. Inst. Czech. 26, 367–377 (1975).
48. K.G. Mccracken, K.G.,Beer, J. and McDonald, F.B., GRL, 29, NO. 24, 2161, 2002.
49. M.A. El-Boriye, Solar Phys., 208, 345–358 (2002b).
50. M. Laurenza et al., JGR, 114, A01103 (2009).
51. H. Mavromichalaki et al., Ann. Geophys., 21, 1681-1689 (2003).
52. K. Kudela et al., Proc. 24th ECRS, Kosice, 374-378 (2008).
53. P. Chowdhury, P. et al., Astrophys. Space Sci., 326, 191-201 (2010).
54. R. Agarwal et al., Proc. 32nd ICRC, Beijing, paper 0132 (2011).
55. N. Zarrouk and R. Bennaceur, Acta Astronautica, 65, 1-2, 262-272 (2009).

Acceleration of cosmic rays at supernova remnant shocks: constraints from gamma-ray observations

M. Lemoine-Goumard

Centre d'Études Nucléaires de Bordeaux Gradignan
Université Bordeaux 1, CNRS/IN2P3
33175 Gradignan, France
E-mail: lemoine@cenbg.in2p3.fr
Funded by contract ERC-StG-259391 from the European Community

In the past few years, gamma-ray astronomy has entered a golden age. At TeV energies, only a handful of sources were known a decade ago, but the current generation of ground-based imaging atmospheric Cherenkov telescopes has increased this number to more than one hundred. At GeV energies, the Fermi Gamma-ray Space Telescope has increased the number of known sources by nearly an order of magnitude in its first 2 years of operation. The recent detection and unprecedented morphological studies of gamma-ray emission from shell-type supernova remnants is of great interest, as these analyses are directly linked to the long standing issue of the origin of the cosmic-rays. However, these detections still do not constitute a conclusive proof that supernova remnants accelerate the bulk of Galactic cosmic-rays, mainly due to the difficulty of disentangling the hadronic and leptonic contributions to the observed gamma-ray emission. In this talk, I will review the most relevant cosmic ray related results of gamma ray astronomy concerning supernova remnants.

Keywords: cosmic-rays; supernova remnants

1. The cosmic-ray mystery

1.1. *The link between cosmic-rays and supernova remnants*

The association between supernova remnants (SNRs) and Galactic cosmic rays (CRs) is very popular since 1934, when Baade and Zwicky argued that this class of astrophysical objects can account for the required CR energetics [1]. Indeed, in order to maintain the cosmic-ray energy density in the Galaxy, about 3 supernovae per century should transform 10 percent of their kinetic energy in cosmic-ray energy. This argument has also been supported by E. Fermi's proposal of a very general mechanism for particle acceleration, which is very efficient if applied at SNR shocks [2]. The extremely interesting point of the diffusive shock acceleration (DSA) mechanism is that it naturally yields power-law spectra for the energy distribution of accelerated particles. However, until recently there were absolutely no observational evidence concerning the acceleration of protons and nuclei in SNRs. Indeed, through their interaction with the interstellar magnetic fields, the charged particles arriving on Earth have lost all directional information and cannot be used to pinpoint the sources. That is why, almost 100 years after their discovery by V. Hess, the origins of the cosmic-rays and their cosmic accelerators remain unknown.

Astronomy with gamma-rays provides a means to study these sources of high energy particles. Indeed, cosmic rays (ionized nuclei of all species, but mostly protons, plus a small fraction of electrons) can interact with ambient matter and photons producing gamma-rays via two different channels. One mechanism invokes the interaction of accelerated protons at supernova remnants shocks with interstellar material generating neutral pions which in turn decay into gamma rays. We call this mechanism the hadronic scenario. A second competing channel exists in the inverse Compton scattering of the photon fields in the surroundings of the SNR by the same relativistic electrons that generate the synchrotron X-ray emission. This is the leptonic scenario. Being of leptonic or hadronic origin, these gamma-rays are not affected while they travel to Earth and can therefore be used to pinpoint the cosmic accelerators in our Galaxy.

1.2. *Gamma-ray experiments*

Two major breakthroughs in gamma-ray astronomy occurred in recently.
Firstly, after more than 20 years of development, the first source of very high energy gamma-rays, the Crab Nebula, was discovered in 1989 by the Whipple telescope. Since this date the technical progresses in this field have led to important scientific results, especially by the Cherenkov telescopes H.E.S.S., VERITAS and MAGIC. These ground-based experiments for gamma-ray astronomy rely on the development of cascades (air-showers) initiated by astrophysical gamma-rays. Such cascades only persist to ground-level above 1 TeV and only produce significant Cherenkov light above a few GeV, setting a fundamental threshold to the range of this technique. Today, more than 120 gamma-ray sources have been detected with high significance, 17 being associated to supernova remnants or molecular clouds.
Second, in space, the Large Area Telescope (LAT) onboard the Fermi satel-

lite has considerably improved our knowledge of the 0.1-100 GeV gamma-ray sky with 1873 objects detected in only two years of observation [3]. It has moved the field from the detection of a small number of sources to the detailed study of several classes of Galactic and extragalactic objects. A complete study of association of the 1873 sources detected show that $\sim 4\%$ of them are associated to supernova remnants [3].
There is no doubt today that supernova remnants can accelerate efficiently particles up to 10^{14} eV. The question is whether these particles are protons or electrons and if they can be accelerated up to the knee of the cosmic-ray spectrum (10^{15} eV).

1.3. *First evidence of efficient particle acceleration in supernova remnants with X-ray satellites*

Accelerated electrons producing gamma-ray emission through inverse Compton scattering also radiate through synchrotron emission when spiraling in a magnetic field. This emission extends from the radio to the X-ray domain. While radio synchrotron emission is observed in most SNRs (in 203 over the 217 observed Galactic SNRs, [4]), X-ray synchrotron emission is observed only in a few remnants up to now. In some of these X-ray detected SNRs, the X-ray synchrotron emission exhibits a filamentary emission just behind the blast wave. One plausible explanation is that the magnetic field is large enough ($\sim 100\,\mu$G) to induce strong radiative losses in the high energy electrons [5, 6]. If the magnetic field is indeed amplified at the limbs, the maximum energy at which particles can be accelerated is much larger there ($>$ 1000 TeV) than outside the limbs (E $\approx$ 25 TeV if B $\approx 10\,\mu$G).
Recently, a discovery of the brightening and decay of X-ray hot spots in the shell of the SNR RX J1713.7-3946 on a one-year timescale has been reported by Uchiyama and collaborators [7]. This rapid variability implies that electron acceleration needs to take place in a strongly magnetized environment, indicating amplification of the magnetic field by a factor of more than 100.
A last evidence of very efficient particle acceleration in supernova remnants is provided by the postshock plasma temperatures observed in SNRs 1E 0102.2-7219 and RCW 86, that are lower than expected for their measured shock velocities [8, 9]. For the first time, by comparing the measured post-shock proton temperature with the one determined using the shock velocity, the authors presented the evidence that $> 50\%$ of the post-shock pressure is produced by cosmic rays.
There are strong indirect arguments confirming that electrons and protons

are accelerated up to at least TeV energies (maybe even PeV) in supernova remnants. A direct signature of accelerated protons is expected through pion decay emission in the GeV-TeV gamma ray range.

2. Detection of supernova remnants in gamma-rays

The sample of supernova remnants detected in gamma-rays is now extremely large: it goes from evolved supernova remnants interacting with molecular clouds (MC) up to young shell-type supernova remnants and historical supernova remnants. The Fermi-LAT even detected one evolved supernova remnants without MC interaction, Cygnus loop. This section will review the main characteristics of the detected SNRs.

2.1. *Supernova remnants interacting with molecular clouds*

The Fermi LAT Collaboration has so far reported the discoveries of five middle aged ($\sim 10^4$ yrs) remnants interacting with molecular clouds: W51C [10], W44 [11], IC 443 [12], W49 [13] and W28 [14]. Apart from W44, they have all been detected in the TeV regime as well. These SNRs are generally much brighter in GeV than in TeV in terms of energy flux (due to a spectral steepening arising at a few GeV), which emphasizes the importance of the GeV observations. The interaction with a molecular cloud provides the target material that allows one to enhance the gamma-ray emission, either through bremsstrahlung by relativistic electrons or by pion-decay gamma-rays produced by high-energy protons. The observed large luminosity of the GeV gamma-ray emission precludes the inverse-Compton scattering off the CMB and interstellar radiation fields as the main emission mechanism since it would require an extremely low density (to suppress the bremsstrahlung and proton-proton interaction), a low magnetic field to enhance the gamma/X-ray flux ratio and an unrealistically large energy injected into protons. In addition, the break in the electron spectrum corresponding to the gamma-ray spectrum directly appears in the radio data leading to a bad modeling of the radio data and therefore disfavours the bremsstrahlung process. A model in which gamma-rays are produced via proton-proton interaction gives the most satisfactory explanation for the GeV gamma-rays observed in SNRs interacting with molecular gas as seen in Figure 1 for the case of W51C.

There are two different types of hadronic scenarios to explain the GeV gamma-ray emission arising from such SNRs: the "Runaway CR" model [15, 16] and the "Crushed Cloud" model [17]. The Runaway CR model considers gamma-ray emission from molecular clouds illuminated by run-away CRs that have escaped from their accelerators, whereas the Crushed Cloud model invokes a shocked molecular cloud into which cosmic-ray particles are adiabatically compressed and accelerated resulting in enhanced synchrotron and pion-decay gamma-ray emissions.

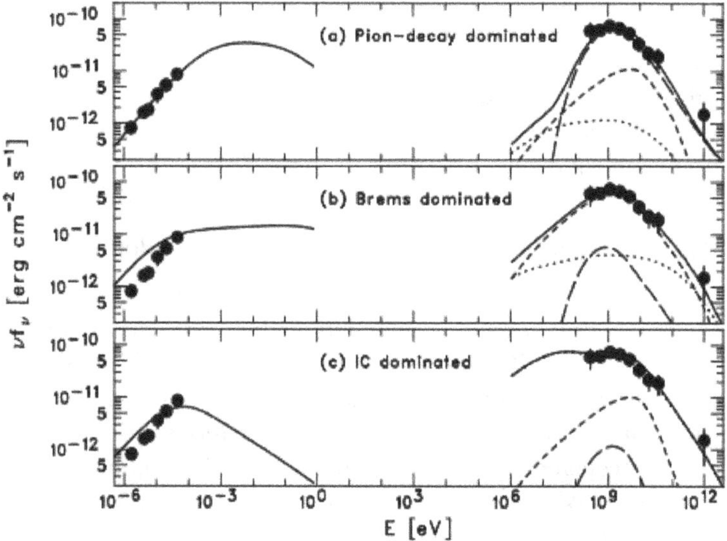

Fig. 1. Different scenarios proposed for the multiwavelength modeling of W51C [10]. The radio emission (from Moon & Koo 1994) is explained by synchrotron radiation, while the gamma-ray emission is modeled by different combinations of pion-decay (long-dashed curve), bremsstrahlung (dashed curve), and IC scattering (dotted curve). The sum of the three component is shown as a solid curve. See [10] for more details.

2.2. *Young shell-type supernova remnants*

Four young shell-like SNRs with clear shell-type morphology resolved in VHE gamma-rays have been detected by H.E.S.S.: RX J1713.7-3946 [18, 19], RX J0852.04622 - also known as Vela Junior - [20], SN 1006 [21] and HESS J1731-347 [22]. A fifth case, RCW 86 [23], might be added to this list although the TeV shell morphology has not yet been clearly proved. Two of them, RX J1713.7-946 [24] and Vela Junior [25], have been detected by Fermi-LAT allowing direct investigation of young shell-type SNRs as sources of cosmic rays. Concerning RX J1713.7-3946, the Fermi-LAT spectrum is well described by a very hard power-law with a photon index of $\Gamma = 1.5 \pm 0.1$ that coincides in normalization with the steeper H.E.S.S.-detected gamma-ray spectrum at higher energies. The GeV measurements with Fermi-LAT do not agree with the expected fluxes around 1 GeV in most hadronic models published so far (e.g., Berezhko & Voelk 2010 [26]) and requires an unrealistically large density of the medium. The agreement with the expected IC spectrum is better (as can be seen in Figure 2) but requires a very low magnetic field of $\sim 10\,\mu$G in comparison to the one measured in the thin filaments by X-ray observations. It is possible to reconcile a high magnetic field with the leptonic model if GeV gamma rays are radiated not only from the filamentary structures seen by Chandra, but also from other regions in the SNR where the magnetic field may be weaker. Similar conclusions are reported for Vela Junior supernova remnant even though in this case the hadronic scenario can not be ruled out. However, being of hadronic or leptonic origin, the GeV-TeV gamma-ray detections imply a low maximal energy for the accelerated particles of $\sim$ 100 TeV, well below the knee of the cosmic-ray spectrum.

2.3. *Historical supernova remnants*

Two historical SNRs have been detected both at GeV and TeV energies: Cassiopeia A (Cas A) [27, 28, 29] and Tycho [30, 31].

Cas A is the remnant of SN 1680. It is the brightest radio source in our Galaxy and its overall brightness across the electromagnetic spectrum makes it a unique laboratory for studying high-energy phenomena in SNRs. A multiwavelength modeling of Cas A does not allow a discrimination between the hadronic and leptonic scenarios. However, regardless of the origin of the observed gamma rays, this modeling implies that the total content of CRs accelerated in Cas A is $\sim(1-2)\times10^{49}$ erg, and the magnetic field amplified at the shock can be constrained as $B \approx 0.12$ mG. Even though Cas A is considered to have entered the Sedov phase, the total amount of CRs accelerated in the remnant constitutes only a minor fraction ($\sim$ 2%) of the total kinetic energy of the supernova, which is well below the $\sim$ 10% commonly used to maintain the cosmic-ray energy density in the Galaxy.

Tycho's SNR (SN 1572) is classified as a Type Ia (thermonuclear explosion of a white dwarf) based on observations of the light-echo spectrum. Thanks to the large amount of data available at various wave bands, this remnant can be considered one of the most promising object where to test the shock acceleration theory and hence the CR – SNR connection. First, using the precise radio and X-ray observations of this SNR, Morlino & Caprioli (2011)

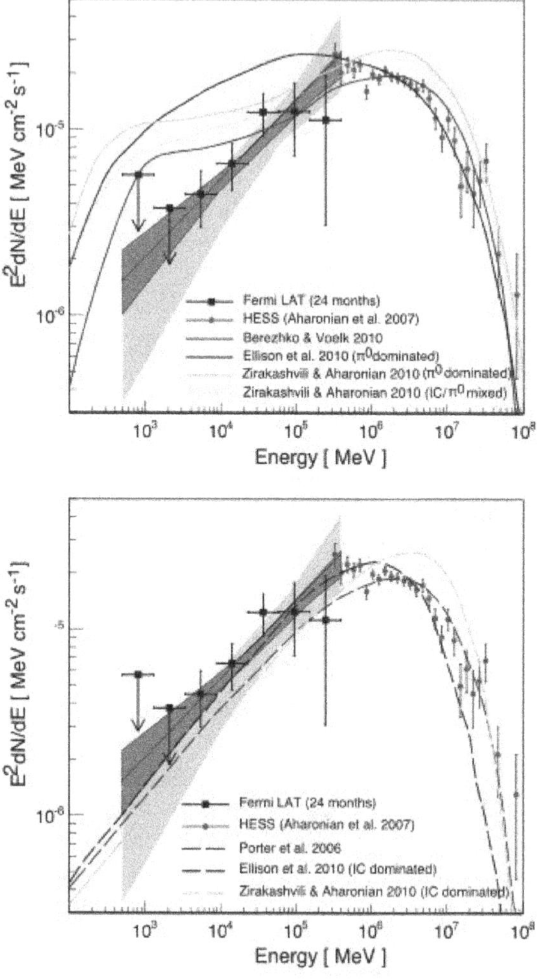

Fig. 2. Energy spectrum of RX J1713.7-3946 in gamma rays. Shown is the Fermi-LAT [24] detected emission in combination with the energy spectrum detected by H.E.S.S. [19]. See [24] for more details.

[32] have shown that the magnetic field at the shock has to be $> 200\mu$G to reproduce the data. Then, using multiwavenlength data, especially the GeV and TeV detections, they could infer that the gamma-ray emission detected from Tycho cannot be of leptonic origin, but has to be due to accelerated protons (this result is consistent with another modeling proposed in 30). These protons are accelerated up to energies as large as $\sim$500 TeV, with

a total energy converted into CRs estimated to be about 12 per cent of the forward shock bulk kinetic energy. This is much more reasonable in the context of acceleration of Galactic cosmic-rays in SNRs.

3. Where are the PeVatrons ?

The recent GeV and TeV detections of supernova remnants confirm the theoretical predictions that supernova remnants can operate as powerful cosmic ray accelerators. However, if these objects are responsible for the bulk of galactic cosmic rays, they should be able to accelerate protons and nuclei at least up to 10^{15} eV and therefore act as PeVatrons. Gabici and Aharonian (2007) [33] have shown that the spectrum of nonthermal particles extends to PeV energies only during a relatively short period of the evolution of the remnant since high energy particles are the first to escape from the supernova remnant shock. For this reason one may expect spectra of secondary gamma-rays extending to energies beyond 10 TeV only from less than 1 kyr old supernova remnants. In this respect, Tycho could be considered as a half-PeVatron at least, since there is no evidence of a cut-off in the VERITAS data. One may wonder how many PeVatrons are expected to be detectable in our Galaxy. A simple estimate has been provided by Gabici and Aharonian (2007): assuming a rate of $\sim$3 supernovae per century in our Galaxy, this directly implies that only a dozen of PeVatrons are present in the Galaxy on average and hence that they are likely to be distant and weak. This emphasizes the importance of TeV observations by the future generation of Cherenkov telescopes such as the Cherenkov Telescope Array (CTA) which will have a better effective area in the energy range already covered but that will also allow the observation up to 100 TeV of sources such as Tycho, therefore constraining the maximal energy at which protons are being accelerated in young SNRs.

Acknowledgements

I thank all the members of the Fermi GALACTIC and HESS SNR-PWN working groups for valuable discussion. I gratefully acknowledge funding from the European Community (contract ERC-StG-259391).

References

1. W. Baade and F. Zwicky, *Proceedings of the National Academy of Science* **20**, 259 (1934).
2. A. R. Bell, *MNRAS* **182**, 147 (1978).

3. A. A. Abdo, M. Ackermann, M. Ajello *et al.*, *ApJS* **submitted** (2011).
4. D. A. Green, *Bulletin of the Astronomical Society of India* **32**, p. 335 (2004).
5. J. M. Vink, J. & Laming, *ApJ* **584**, p. 758 (2003).
6. J. Ballet, *Advances in Space Research* **37**, p. 1902 (2006).
7. Y. Uchiyama *et al.*, *Nature* **449**, 576 (2007).
8. J. P. Hughes, C. E. Rakowski and A. Decourchelle, *ApJ* **543**, L61 (2000).
9. E. A. Helder *et al.*, *Science* **325**, p. 719 (2009).
10. A. A. Abdo, M. Ackermann, M. Ajello *et al.*, *ApJL* **706**, 1 (2009).
11. A. A. Abdo, M. Ackermann, M. Ajello *et al.*, *Science* **327**, 1103 (2010).
12. A. A. Abdo, M. Ackermann, M. Ajello *et al.*, *ApJ* **712**, 459 (2010).
13. A. A. Abdo, M. Ackermann, M. Ajello *et al.*, *ApJ* **722**, 1303 (2010).
14. A. A. Abdo, M. Ackermann, M. Ajello *et al.*, *ApJ* **718**, 348 (2010).
15. F. A. Aharonian and A. M. Atoyan, *A&A* **309**, 917 (1996).
16. Y. Ohira, K. Murase and R. Yamazaki, *MNRAS* **410**, 1577 (2011).
17. Y. Uchiyama, R. Blandford, S. Funk, S. Tajima and T. Tanaka, *ApJL* **723**, 122 (2010).
18. F. Aharonian, A. G. Akhperjanian, K. Aye *et al.*, *Nature* **432**, 75 (2004).
19. F. Aharonian, A. G. Akhperjanian, A. R. Bazer-Bachi *et al.*, *A&A* **437**, L7 (2005).
20. F. Aharonian, A. G. Akhperjanian, A. R. Bazer-Bachi *et al.*, *A&A* **464**, 235 (2007).
21. F. Acero, F. Aharonian, A. G. Akhperjanian *et al.*, *A&A* **512**, A62+ (2010).
22. A. Abramowski, F. Acero, F. Aharonian *et al.*, *A&A* **531**, A81+ (2011).
23. F. Aharonian, A. G. Akhperjanian, U. B. de Almeida *et al.*, *ApJ* **692**, 1500 (2009).
24. A. A. Abdo, M. Ackermann, M. Ajello *et al.*, *ApJ* **734**, p. 28 (2011).
25. T. Tanaka *et al.*, *ApJL* **740**, L51 (2011).
26. E. G. Berezhko and H. J. Voelk, *A&A* **511**, A34+ (2010).
27. A. A. Abdo, M. Ackermann, M. Ajello *et al.*, *ApJL* **710**, p. L92 (2010).
28. J. Albert *et al.*, *A&A* **474**, p. 937 (2007).
29. V. A. Acciari *et al.*, *ApJ* **714**, p. 163 (2010).
30. F. Giordano *et al.*, *ApJ* **submitted** (2011).
31. V. A. Acciari *et al.*, *ApJL* **730**, L20 (2011).
32. G. Morlino and D. Caprioli, *A&A* **accepted** (2011).
33. S. Gabici and F. A. Aharonian, *ApJ* **665**, L131 (2007).

Particle acceleration and magnetic field amplification in supernova remnants

A. Marcowith*

*Laboratoire Univers et Particules de Montpellier, univeristé Montpellier II, CNRS Montpellier, 34095, France *E-mail: Alexandre.Marcowith@univ-montp2.fr*
http://web.lupm.univ-montp2.fr/

This review addresses the issue of cosmic rays acceleration in supernova remnants in connection with the amplification of magnetic fluctuations. Possible scenarios of magnetic field amplification are discussed with a special emphasise on the contribution of instabilities driven by cosmic ray currents as well as the saturation process and the properties of the turbulence. We scan acceleration efficiencies of different class of supernova remnants and there respective contribution to the cosmic ray spectrum. We finally review some aspects of the recent numerical effort in modelling the magnetic field amplification in supernova remnant shocks.

Keywords: Cosmic-Rays - Supernova Remants - Particle acceleration - Magnetic turbulence.

1. Introduction

Diffusive shock acceleration (DSA) is likely the most promising mechanism for producing supra-thermal up to relativistic particles in wide variety of shock waves.[1–4] This process is thought to be efficient enough to accelerate energetic particles to energies above 10^{15} eV at supernova remnant (SNRs) shocks and hence to be at the origin of galactic cosmic rays (GCRs). However, to reach such high energies DSA requires particles in there path around the shock front to be scattered by magnetic fluctuations with amplitude substantially larger than the ambient interstellar ones. Observational supports of strong magnetic field amplification (MFA) have gone through X-ray synchrotron radiation filaments detected from several young SNR.[5–8] These point towards a direct connection between high energy particle production at SNR shocks and the MFA process; i.e. a scenario where the energetic particles accelerated at the shock front self-generate the turbulence required for the DSA process to work.

A critical issue of MFA is the dynamic range of the self-generated turbulence. The most energetic particles are expected to explore scalelengths ahead the shock up to 9 or 10 order of magnitudes larger than the thermal ion Larmor radii that mediate the shock front. Also the magnetic field fluctuations must include sufficient power in the long wavelength limits in order to confine the highest energetic particles around the shock. All these aspects have to be included into a self-consistent modelling of the non-linear DSA. The dynamic issue is a real challenge for both analytical and numerical experiments trying to capture the physics of the DSA.
The problem of turbulence production in connection with CR acceleration and DSA has been investigated with the help of magneto-hydrodynamic (MHD) models since the 1960s first in the case of interstellar medium[9,10] and in SNR shocks.[11] These work investigated the *resonant* current instability where modes are produced with scale lengths comparable with the cosmic ray gyroradius. However[12] did uncovered the case of short-wavelength perturbations produced in a shock precursor where the CR current efficiently generates *non-resonant* modes with wavelengths much shorter than the gyroradii of the CRs. Most of the attempt to account self-consistently to the non-linear back-reaction of turbulence generation over the DSA process did only involved the resonant modes[13] (but see[14]). In addition to current instabilities, a number of non-resonant instabilities have been investigated for DSA. A non-resonant acoustic instability, where the cosmic ray pressure gradient in the shock precursor amplifies compressional disturbances, was investigated by[15] (and references therein). Also a short-scale dynamo process driven by the interactions of the CR pressure gradient and density perturbations in the precursor has been investigated by.[16] These possibilities will be not further discussed in this review focused on current driven instabilities only.

2. Magnetic field instabilities in non-relativistic shocks

The generation of CRs ahead the shock front added to the fact that these energetic particles can extract a substantial (about a few tens percent) fraction of the kinetic energy of the flow produce a large precursor. This precursor is a medium inhomogeneously filled in by energetic particles (identified with CR hereafter) and magnetic fluctuations that can perturbate the background interstellar medium. The gradient of energetic particle is a source of free energy that can drive in different ways plasma instabilities. Also the streaming of CRs with respect to the background medium is another source of free energy that drives current type instabilities. We will focus our discussion on different aspects of the connection between the non-resonant current instability and DSA.

2.1. *Current driven instabilities*

Following[12] the CR current in the shock precursor drives fluctuations at scales shorter than the CR Larmor radius. The modes are purely growing, incompressible propagating along the background magnetic field and have a circular polarization at the opposite to the CR gyromotion. In such disturbances the CR current is basically unmagnetized and is equilibrated by a return current in the background plasma. CRs also drive a resonant instability: here the fastest growing modes are propagating along the background magnetic field and have a circular polarization in the sense of CR gyromotion.[7] The non-resonant fluctuations grow the fastest (see section 3). A multi-dimensional analysis shows that the fastest growing modes are those generated along the background magnetic field independently of the direction the CR current with respect to it. But no instability is found for perpendicular propagating modes.[18]

2.2. *Saturation and turbulence*

The focus of much of the analytical and numerical work on the Bell instability concerns the saturation level and the spectral properties of the instability[12,14,19,20] but no yet clear answer has came out. The instability is possibly quenched by the magnetic tension exerted on the stretched field lines that equilibrate the Lorentz force due to the return current. In that case the power spectrum generated scales as k^{-1}.[12] But non-linear cascading effects can produce a steeper spectrum in k^{-2}.[19] It is of much difficulties to anticipate over the final result and one usually has to rely on numerical experiments (see section 4). However both previous analytical estimates predict the same order of magnitude for the level of magnetic fluctuations at the shock front:

$$B_{sat}/B_\infty \simeq M_{a,\infty} \times (V_{sh}/c)^{1/2} \times (P_{CR}/\rho_\infty V_{sh}^2)^{1/2} \ ,$$

where $B_\infty, M_{a,\infty}, \rho_\infty$ are the magnetic field, the Alfvenic Mach number (ratio of the shock velocity V_{sh} to the ambient Alfvén velocity), and the mass density of the ambient medium respectively. The pressure in CR P_{CR} is a fraction of the kinetic energy density of the flow that may reach 30-50 % in the strongly non-linear cases. The above formula shows that the saturated magnetic field scales as $V_{sh}^{3/2}$, which with the help of X-ray filaments can be tested.[21]

2.3. *High energy particle confinement problem*

A major drawback of the non resonant instability is that it does not permit a good confinement of the highest energies around the shock. Indeed the main scale of the fluctuations are $l_* \sim 2 \times 10^{-3}$ parsec $\ll \mathrm{R_{sh}}$ for a shock velocity of $V_{sh} = 10000$km/s and radius $R_{sh} \geq 1$parsec in the Sedov phase, an ambient density $n_\infty = 1\mathrm{cm}^{-3}$ and magnetic field $B_\infty = 3\mu$Gauss, for maximum CR energies of one PeV.[12] The explanation is simply due to the fact that the CR mean free path in a turbulence with a coherence length shorter than the particle Larmor radius increases rapidly with the particle energy as E^2.[22] This results in a poor confinement performances around the shock and an early escape of the high energy particles in the upstream medium. Possible remedies have been recently put forward (see[23,24]). This issue is still opened in particular considering the consequences on the high energy end of the CR spectrum accelerated at the shock front.

3. Application to supernova remnants

Using the above saturation level it is interesting to evaluate the properties of the interstellar medium that can provide the highest particle energies. The non-resonant instability grows the fastest in medium of high ambient (ionized) density in the case of fast shock waves with velocities of the order of 10 000 km/s.[12] This points towards SN blast waves propagating in dense stellar winds that may be found some classes of SN; especially in the case of type IIb. In the latter class, the shock propagates into a dense slow wind of a red supergiant star as it is the case for Cassiopeia A. Another source SN 1993 J is thought to be rather favorable to this instability to develop and accelerate CRs. With typical ambient densities of $n_{\mathrm{ext}} \geq 10^4\mathrm{cm}^{-4}$ type IIb SN could contribute even to acceleration of protons beyond the knee up to 10^{17} eV.[25,26] It should be however kept in mind that non resonant and resonant modes grow together and that the growth rate of the former scales as V_{sh}^{-3} and drops much faster than the growth rate of the latter. Apart the mechanism of generation of long wavelength perturbations above, the resonant instability hence should dominate late generation of magnetic field and provide longer CR confinement around the shock for all class of SN.[19,20]

4. Numerical approaches

There are different methods to handle the problem of MFA and particle acceleration at SNR shock waves. All have the same drawback already mentioned above that they are limited in dynamic scales. Let us distinguish among methods adapted to the microscopic physics of the shock acceleration process (shock formation, energetic particle injection): particle-in-cell (PIC) and hybrid simulations. Hybrid methods including energetic particles that are more adapted to the mesoscopic scales (diffusive shock acceleration). Finally methods involving magnetohydrodynamic (MHD) are better configured for large scale physics (diffusive shock acceleration and CR escape).

In the case of non resonant instability discussed above if the approximation that the CR current is only weakly affected by the short-scale fluctuations holds, non-linear MHD-type simulations, where the CR current is fixed as an external parameter, can be used.[12,27] These simulations typically show a high saturation level of the short-scale turbulence consistent with the analytical estimates. They also show non-linear spectral energy transfer (i.e. the cascading of turbulence energy) into both larger and smaller scales. However, since long-wavelength magnetic fluctuations on scales comparable to or larger than the CR gyroradius cause a non-negligible response of the CR current, the MHD simulations are limited to short-scale fluctuations only. In a real system, long-wavelength fluctuations will occur and they will induce perturbations perpendicular to the local mean magnetic field (see[28]).

PIC and hybrid plasma simulations have permitted some important breakthroughs. In the linear regime, the simulated growth rates conrm the basic theoretical predictions of Bell. Amplication factors above 10 for the magnetic eld were achieved in some simulations.[29,30] Usually the instability is shown to saturate if CRs get deflected in the amplied eld. Recent PIC and hybrid simulations have start a parametric exploration of the particle acceleration at non-relativistic shock. PIC simulations by[31] show a different behaviour in injection capacities of electrons and protons depending on the shock obliquity (direction of the mean magnetic field with the normal to the shock front). Parallel shock favour proton injection while perpendicular ones favor electron injection. Also hydrid simulations have provided first evidences of an optimum efficiency in the proton acceleration process in the parallel shock configuration. Shocks produce more energetic particles at moderate Alfvén Mach numbers; close to 10, while faster shocks produce higher energetic particles.[32]

But investigation of the highest energetic particles and there back reaction over the fluid flow require the use of MHD and kinetic approaches. Coupled effects of MFA, particle acceleration and hydrodynamic solutions have been obtained in 1D (in space and energy) thank to the semi-analytical method developped by Blasi and collaborators (see[33–35]). One important result of these work is the MFA pumps a part of the energy otherwise injected into CRs and hence decreases the effect of CR backreaction over the magnetised flow. Monte-Carlo methods have started to include non resonant modes either.[14] They show that the turbulent spectrum can be very sensitive to the effect of non-linear cascading. In extreme case of no cascading effect the downstream turbulent spectrum looses its scale invariant shape and is decomposed into several narrow peaks of width $\Delta k < k_{peak}$ that can produce particular signatures in X-ray polarization synchrotron radiation.[36] Finally there are several attempts to describe the high energy particle propagation in the turbulent medium in a self-consistent way including there radiative signatures. Coupled MHD-kinetic simulations where the CR trajectories are reconstructed in a snapshot of a MHD solution of the simulation of the non resonant instability have been developed by.[37] The work[38] did instead assumed given diffusion coefficients and calculated the X- and gamma-ray filaments one can expect downstream shocks of young SNR for instance by the Cerenkov Telescope Array (CTA) system . Both work did not include CR back-reaction yet.

5. Perspectives and conclusions

Magnetic field amplification (MFA) in young SNRs is now established in several objects. MFA is likely connected with the acceleration of energetic particles which self generate the fluctuations necessary for DSA to operate. Several instabilities have been investigated since early 2000. One of most studied involves the effect of the current produced by the charged particles in the upstream medium. Early theoretical estimates found that the instability may produce amplification factors as large as 100. However there are still an important issue concerning the saturation process and the properties of the turbulence. Multiple numerical experiments try to catch the different dynamic of the shock acceleration problem. PIC and hybrid methods show that the defection of CR likely leads to more modest amplificiation levels of ~ 10. PIC provide evidences that protons are likely injected at parallel shocks while electrons are preferentially injected at perpendicular shocks. Hybrid simulations show that in parallel shocks the acceleration efficiency peaks at moderate Alfvenic Mach numbers. However high energy particle dynamics are catched by kinetic-MHD simulations and semi-analytical approaches. There are first evidences that MFA pump a part of the energy originally put in CR in non-linear models. Calculations the particle trajectories are promising to obtain accurate high energy spectra and images that may be provided by the CTA experiment.

References

1. L.O'.C. Drury, 1983, Reports on Progress in Physics 46, 973
2. R.D. Blandford & D.C. Eichler, 1987, Physics Reports, 154, 1
3. F.C. Jones & D.C. Ellison, 1991, SSR, 58, 259
4. M. A. Malkov & L.O'.C. Drury, 2001, Reports on Progress in Physics, 64, 429
5. E.V. Gotthelf et al, 2001, ApJ, 552, L39
6. U. Hwang et al, 2002, ApJ, 581, 1101
7. J. Ballet, 2006, Adv Sp Res, 37, 1902
8. J. Vink, 2008, AIP, 1085, 169
9. D.G. Wentzel, 1974, ARAA, 12, 71
10. C.J. Cesarsky, 1980, ARAA, 18, 289
11. A.R. Bell, 1978, MNRAS, 182, 174
12. A.R. Bell, 2004, MNRAS, 353, 550
13. A.E. Vladimirov, A.M. Bykov & D.C. Ellison, 2008, ApJ, 688, 1084
14. A.E. Vladimirov, A.M. Bykov & D.C. Ellison, 2009, ApJ, 703, L29
15. L.O.'C. Drury & S.A.E.G. Falle, MNRAS, 223, 353
16. A. Beresnyak, T.W. Jones & A. Lazarian, 2009, ApJ, 707, 1541
17. A.R. Bell & S.G. Lucek, 2001, MNRAS, 321, 433
18. A.R. Bell, 2005, MNRAS, 358, 181
19. G. Pelletier, M.Lemoine & A.Marcowith, 2006, A&A, 453, 181
20. A.Marcowith, M.Lemoine & G. Pelletier, 2006, A&A, 453, 193
21. H.J. Voelk, E.G. Berezhko, L.T. Ksenofontov, 2005, A&A, 433, 229
22. E. Parizot, 2004, Nu Phys, 136, 169
23. A.M. Bykov, S.M. Osipov & D.C. Ellison, 2011, MNRAS, 410, 39
24. K. Schure & A.R. Bell, 2011, MNRAS, in press
25. V. Tatischeff, 2009, A&A, 499, 191
26. V.S. Ptuskin, V.N. Zirakashvili & E.S. Seo, 2010, ApJ, 718, 31
27. V.N. Zirakashvili & V.S. Ptuskin, 2008, ApJ, 678, 939
28. A.Bykov, S.M. Osipov & I.N. Toptyghin, Astron Letters, 35, 555
29. L. Gargaté et al, 2010, ApJ, 711, L127
30. M.A. Riquelme & A. Spitkovsky, 2010, ApJ, 717, 1054
31. M.A. Riquelme & A. Spitkovsky, 2011, ApJ, 733, 63
32. L. Gargaté & A. Spitkovsky, 2011, arXiv1107.0762
33. P. Blasi et al, 2007, MNRAS, 375, 1471
34. D. Caprioli et al, 2008, ApJ, 679, L139
35. D. Caprioli et al, 2010, MNRAS, 407, 1773
36. A.M. Bykov et al, 2011, ApJ, 735, L40
37. B. Reville et al, 2008, MNRAS, 386, 509
38. A.Marcowith & F. Casse, 2010, A&A, 515, 90

IMPORTANCE OF CHARGE-SIGN DEPENDENT SOLAR MODULATION FOR PAMELA AND AMS 2 OBSERVATIONS

M. S. POTGIETER
Centre for Space Research, North-West University,
2520 Potchefstroom, South Africa
Email: Marius.Potgieter@nwu.ac.za

Galactic cosmic rays serve as probes of solar activity and for very specific modulation conditions in the heliosphere, particularly when they are protons, anti-protons, electrons and positrons. By observing these particles and their anti-particles over a wide range of energies on various spacecraft, satellites and balloons, a better understanding is gained about the basics of cosmic ray transport and modulation, and various heliospheric phenomena such as the 11 and 22 year modulation cycles, and gradient and curvature drifts. Significant progress is being made in this field, stimulated by observations in the outer heliosphere by the two Voyager spacecraft and in the inner heliosphere by ULYSSES, PAMELA and other space and balloon missions. Because of its contemporary relevance, the basic causes and consequences of charge-sign dependent solar modulation and the 22 year cycle are briefly discussed.

1. Causes of Charge-sign Dependent Solar Modulation

Entering the heliosphere, galactic cosmic rays (GCRs) encounter a turbulent solar wind with an imbedded heliospheric magnetic field (HMF) leading to significant global and temporal changes in their intensity, also with energy, a process known as the solar modulation of GCRs. This process could happen already from where the heliosphere begins to disturb the local interstellar medium, well beyond the heliopause (HP) [1]. Traditionally, however, the HP has been assumed as the location where significant reductions in GCRs commence, and is usually considered to be the true modulation boundary. With the two Voyager spacecraft still operational far inside the (inner) heliosheath, it is now an observational fact that solar modulation of GCRs is still significant as far out as ~120 AU from the Sun [2]. The process is of course strongly dependent on energy. How far away from Earth the solar modulation of GCRs will eventually subside and where the local interstellar medium subsequently will be encountered by the Voyagers is presently debated.

Once inside the heliosphere, surely on the upwind side of the solar wind termination shock, GCRs quickly sense the gradients and curvature of the HMF and subsequently experience gradient and curvature drifts, one of four major solar modulation mechanisms. The HMF also has a wavy current sheet, separating inward directed magnetic field lines in one hemisphere from the outward directed ones in the other hemisphere. This current sheet becomes more wavy with increasing solar activity and forms therefore an important part of particle drifts. Every ~11 years the polarity of the HMF changes during periods of extreme solar activity so that GCRs from then on drift in opposite directions.

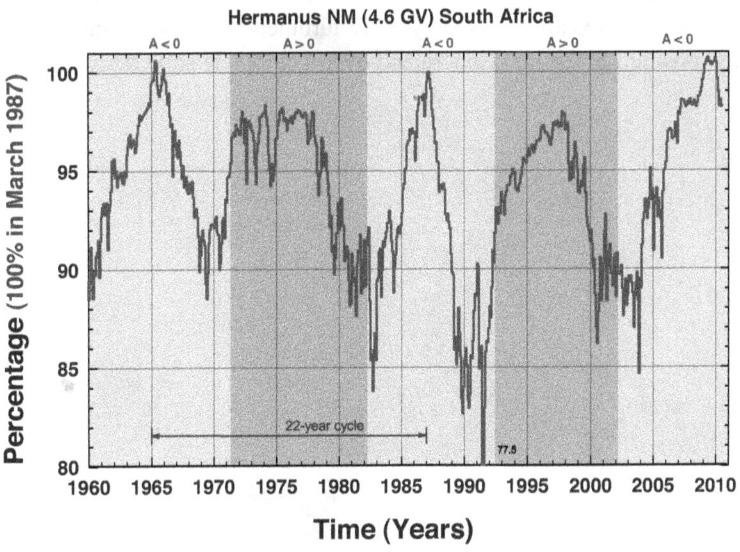

Figure 1. Long-term GCR record observed by the Hermanus Cosmic Ray Monitor in South Africa with the intensity in March 1987 taken as 100%, corrected for atmospheric pressure to eliminate daily and seasonal variations. Notice the 11 year and 22 year cycles: Peaks are formed during the A < 0 solar magnetic field polarity cycles in contrast to A > 0 cycles because of gradient, curvature and current sheet particle drifts that are also responsible for the occurrence of charge-sign dependent modulation inside the heliosphere.

This means that while protons (positrons) will drift towards the inner heliosphere primarily through the northern regions of the heliosphere and then mainly outwards along the wavy current sheet (A > 0 cycle), anti-protons (electrons) will drift inwards mainly along the current sheet and outwards through the polar regions, thus sampling different modulation conditions during the same solar cycle. Apart from GCR modulation becoming charge-sign dependent, current sheet drifts also cause a clear 22-year modulation cycle. This is shown in Figure 1. GCR drift effects are increasing and decreasing with the solar cycle. Apart from charge-sign dependence, several other modulation effects caused by particle drifts in the HMF have been reported, e.g. a 22 year cycle in GCR intensity gradients. See the reviews by [3,4].

2. Evidence of charge-sign dependent solar modulation

Simultaneous measurements of GCR electrons and positrons (or protons and anti-protons) are a crucial test of our present understanding of charge-sign dependent modulation in the heliosphere, in particular how the amount of drifts that they experience are affected by HMF turbulence as a function of decreasing energy, position in the heliosphere and over a complete solar activity cycle. It is

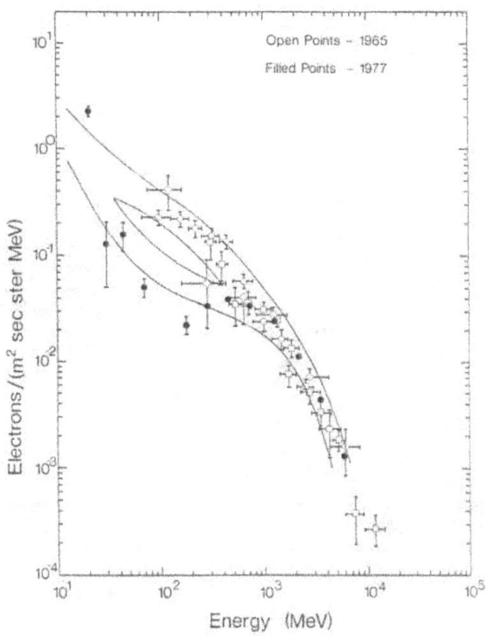

Figure 2. GCR electron observations for two consecutive solar minimum modulation periods in 1965 (open circles) and 1977 (filled circles) compared to the predictions of a modulation model (solid lines) containing gradient, curvature and current sheet drifts, clearly depicting a 22 year cycle [8], see also Figure 14 in [9].

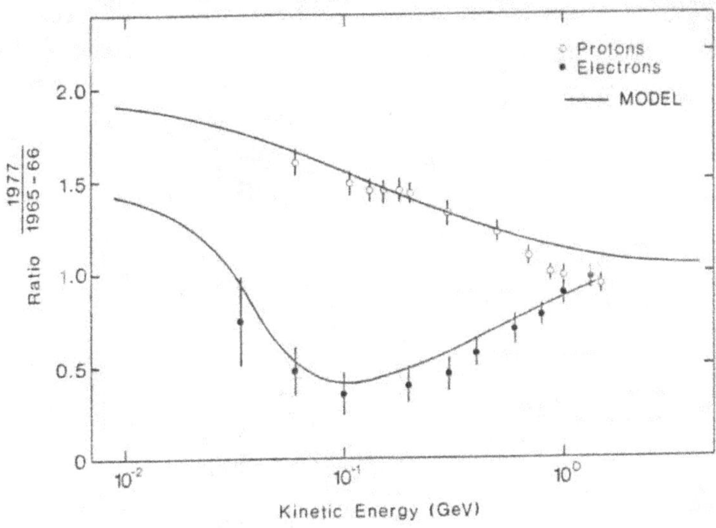

Figure 3. Ratios of proton and electron measurements for 1977 (A > 0) to 1965-66 (A < 0) as a function of energy compared to the predictions made with a drift modulation model illustrating how differently protons behave to electrons during two solar minimum periods with opposite solar magnetic field polarity [8,9].

expected that the effects of drifts on GCRs should be become more evident closer to minimum solar activity.

Observations of GCRs and their anti-particles have been done over the years and are presently been made simultaneously by the PAMELA and the AMS 2 missions [5,6]. Before these simultaneous measurements, charge-sign dependent solar modulation was mostly studied using electrons (as the sum of electrons and positrons) and protons (or helium) of the same rigidity. The first observational evidence of charge-sign dependent solar modulation was reported by Webber [7] and modelled concurrently by Potgieter [8] using a first generation drift model [9]. This is shown in Figure 2. The corresponding charge-sign dependent effect, comparing protons and electrons measurements made during two consecutive solar minimum periods (1965 as A < 0 and 1977 as A > 0), is illustrated in Figure 3. Evidently, electrons behave differently from protons during consecutive solar minimum epochs, exhibiting a 22 year cycle.

Recent model predictions of charge-sign dependent modulation are shown in Figures 4 and 5 respectively for the intensity ratios of electrons, positrons, protons and anti-protons [10,11].

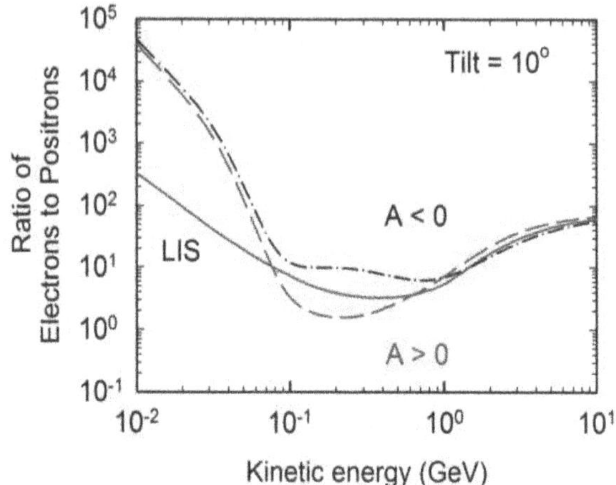

Figure 4. Computed electron to positron ratios at Earth for two polarity cycles (A > 0 e.g. 1997, 2020 and A < 0 e.g. 1985, 2009) of the HMF compared to the ratio of the LIS at the HP (120 AU). Differences above about 80 MeV are caused by gradient, curvature and current sheet drifts in the heliosphere during solar minimum activity [10].

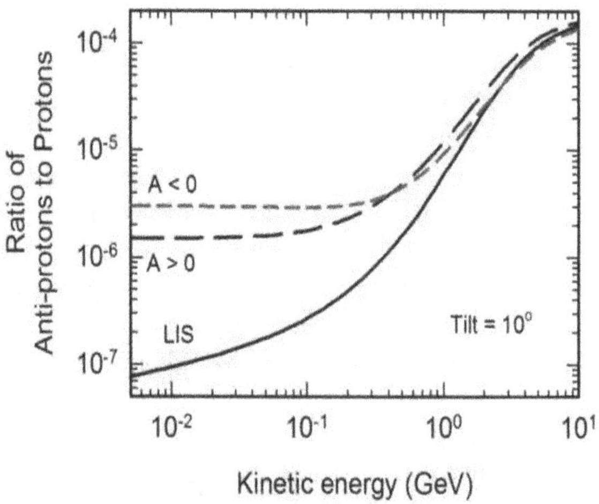

Figure 5. Similar to Figure 4 but for computed ratios of galactic protons to anti-protons, during solar minimum modulation, compared with the ratio of the corresponding LIS at 120 AU [11]. Charge-sign dependent modulation is evident up to ~10 GeV. See also [12,13,14].

3. Conclusions

Charge-sign dependent solar modulation has been described in detail by numerical drift models, with the first observational and computational evidence reported in the early 1980's [7,8]. Until recently, charge-sign dependence had to be studied mostly through comparison of GCR nuclei with the total flux of electrons and positrons [3]. Now it is possible to test gradient and curvature drift theory using simultaneous long-duration space observations of electrons and positrons, also protons and anti-protons, from PAMELA and future observations by AMS 2 [13,14]. The figures presented in this work are examples of these general predictions for the 22 year modulation cycle and the accompanying charge-sign dependent solar modulation.

References

1. K. Scherer, H. Fichtner, R. du T. Strauss, S. E. S. Ferreira, M. S. Potgieter and H.-J. Fahr, *Astrophys. J.* **735**, 128 (2011).
2. E. C. Stone, A. C. Cummings, F. B. McDonald, B. C. Heikkila, N. Lal and W. R. Webber, *Nature* **454**, 71 (2008).
3. B. Heber and M. S. Potgieter, *Space Sci. Rev.* **127**, 117 (2006).
4. M. S. Potgieter, *Space Sci. Rev.* on line (2011).
5. O. Adriani, G. C. Barbarino, G. A. Bazilevskaya, R. Bellotti, et al., *Nature* **458**, 407 (2009).
6. R. Battiston, *Proc. 11th ICATPP*, ed. C. Leroy, P. G. Rancoita, M. Barone, A. Gaddi, L. Price and R. Ruchti, World Scientific, 741 (2010).
7. W. R. Webber, J. C. Kish and D. A. Schrier, *Proc. 18th Int. Cosmic Ray Conf.* **3**, 35 (1983).
8. M. S. Potgieter, Ph.D. thesis, Potchefstroom University, South Africa. (1984).
9. M. S. Potgieter and H. Moraal, *Astrophys. J.* **294**, 425 (1985).
10. U. W. Langner and M. S. Potgieter, *Adv. Space Res.* **34**, 144 (2004).
11. U. W. Langner and M. S. Potgieter, *J. Geophys. Res.* **109**, A01103 (2004).
12. W. R. Webber and M. S. Potgieter, *Astrophys. J.* **344**, 779 (1989).
13. P. Bobik, M. J. Boschini, C. Consolandi, S. Della Torre, M. Gervasi, D. Grandi, K. Kudela, S. Pensotti and P. G. Rancoita, ASTRA, 7, 245 (2011).
14. R. D . Strauss, M. S. Potgieter, M. Boezio, N. De Simone, V. Di Felice, I. Büsching A. Kopp, *Proc. 13th ICATPP*, this volume (2011).

THE HELIOSPHERIC TRANSPORT OF PROTONS AND ANTI-PROTONS: A STOCHASTIC MODELLING APPROACH TO PAMELA OBSERVATIONS

R. D. STRAUSS* and M. S. POTGIETER

Centre for Space Research, North-West University,
Potchefstroom, South Africa
**E-mail: DuToit.Strauss@nwu.ac.za*

M. BOEZIO

INFN, Structure of Trieste and Physics Department of University of Trieste,
Trieste, Italy

N. DE SIMONE and V. DI FELICE

INFN, Structure of Rome "Tor Vergata" and Physics Department of University of Rome "Tor Vergata", Rome, Italy

A. KOPP

Institut für Experimentelle und Angewandte Physik,
Christian-Albrechts-Universität zu Kiel, Germany

I. BÜSCHING

Institut für Theoretische Physik, Lehrstuhl IV: Weltraum- und Astrophysik,
Ruhr-Universität Bochum, Germany

Using a newly developed 5D comic ray modulation model, we study the modulation of galactic protons and anti-protons inside the heliosphere. This is done for different heliospheric magnetic field polarity cycles, which, in combination with drifts, lead to charge-sign dependent cosmic ray transport. Computed energy spectra and intensity ratios for the different cosmic ray populations are shown and discussed. Modelling results are extensively compared to recent observations made by the PAMELA space borne particle detector. Using a stochastic transport approach, we also show pseudo-particle traces, illustrating the principle behind charge-sign dependent modulation.

Keywords: heliosphere; protons; anti-protons; cosmic ray drifts; stochastic differential equations.

1. Introduction

Caused by a combination of dynamic processes driven by the Sun, galactic cosmic ray (CR) intensities decrease from the outer to the inner heliosphere; a process referred to as CR modulation, or specifically, CR solar modulation. In this study, the charge-sign dependent modulation of galactic protons (p) and anti-protons ($\bar{p}$) is modelled by solving the relevant transport equation in five dimensions (three spatial spherical coordinates, energy and time). The resulting intensities at Earth are compared to the recent PAMELA observations [1, 2] to illustrate the effect and extent of charge-sign dependent modulation.

2. Cosmic Ray Transport in the Heliosphere

The transport of CRs through the heliosphere is described by the transport equation (TPE), given by [3] as

$$\frac{\partial f}{\partial t} = -\left(\vec{V}_{sw} + \langle \vec{v}_d \rangle\right) \cdot \nabla f + \nabla \cdot (\mathbf{K} \cdot \nabla f) + \frac{P}{3}\left(\nabla \cdot \vec{V}_{sw}\right) \frac{\partial f}{\partial P}, \quad (1)$$

in terms of the isotropic distribution function f, related to the differential intensity by $j = P^2 f$, where P is the particle rigidity. The different processes incorporated in the TPE are, from left to right, temporal changes, convection, drifts, spatial diffusion and adiabatic energy losses. Although all of these processes are understood, the exact form of the drift and diffusion coefficients remains relatively uncertain. Recent studies have however made significant progress in this regard, e.g. [4, 5, 6, 7].

3. The Stochastic Solver

The TPE, which is a 5D partial differential equation, can be rewritten into a set (of 4) of two dimensional stochastic differential equations (SDE), which are easier to handle numerically. The resulting SDEs can then be solved simultaneously by employing e.g. the Euler numerical scheme. The formalism of this approach is discussed by [8, 9], while heliospheric implementations are discussed by e.g. [10, 11, 12]. The heliospheric model used here is discussed by [13, 14]. Essentially, an initial phase space coordinate is chosen (where e.g. j will be evaluated), $x_i^{t=0}, i \in \{r, \theta, \phi, E\}$. The evolution of the trajectory of this *pseudo-particle* is then calculated backwards in time until a modulation boundary (the heliopause for galactic CRs) is reached. The process is then repeated for N of these particles, with j calculated as the average of these different realizations.

4. Modelling the PAMELA p and $\bar{p}$ Measurements

4.1. *Transport parameters*

The most general form of the pitch-angle averaged CR drift velocity is given by e.g. [6, 15] as

$$\langle \vec{v}_d \rangle = \frac{qv}{3} \nabla \times \lambda_d \mathbf{e}_B, \tag{2}$$

where q is the particle charge, v its speed, $\mathbf{e}_B$ a unit vector directed along the heliospheric magnetic field (HMF) and λ_d the drift scale, defined by

$$\lambda_d = \frac{(\omega\tau)^2}{1+(\omega\tau)^2} r_L, \tag{3}$$

with r_L the Larmor radius and $\omega\tau$ the so-called scattering parameter. Because of the resulting factor $\sim q \cdot \mathbf{e}_B$ in the equation above, the drift velocity depends on both the HMF polarity and the CR charge, leading to charge-sign dependent modulation [16]. In this study, the approach of [15] is followed, whereby $\omega\tau \sim P$. This drift suppression factor causes λ_d to be decreased at low energies. Note that, in the absence of scattering, $\lambda_d = r_L$. Both λ_d and r_L are shown in Panel (a) of Fig. 1 as a function of rigidity.

For the mean free path directed parallel to the mean HMF ($\lambda_{||}$), we adopt the analytical form of [17, 18], which is based on results from quasi-linear theory (QLT, [19]) along with a dynamical model for slab turbulence discussed by [20]. For the mean free paths directed perpendicular to the HMF (in the $r-\phi$ plane labelled by $\lambda_{\perp r}$ and $\lambda_{\perp\theta}$ directed in the θ-direction, completing the right handed coordinate system) we approximate the non-linear guiding centre (NLGC, 4) results of [21] as

$$\lambda_\perp = a\lambda_{||} \cdot \left[\frac{P}{P_0}\right]^{-b}, \tag{4}$$

with $a = 0.01$ (e.g. [22]), $P_0 = 1$ GV and $b = 1/3$ to produce an almost energy dependent form for $\lambda_\perp$ below ~ 5 GV. Note that isotropic perpendicular diffusion is assumed, where $\lambda_\perp = \lambda_{\perp r} = \lambda_{\perp\theta}$.

The solar wind velocity, $\vec{V}_{sw}$, is assumed to be directed radially outwards, with a latitude dependent speed in accordance with the Ulysses observations discussed in [23]. The heliopause, where the local interstellar spectrum (LIS) is specified, is assumed to be located at $r_{HP} = 130$ AU. For protons and anti-protons, we use the LIS of [24] as parametrized by [25]. For all simulations, a constant heliospheric current sheet (HCS) tilt angle of $\alpha = 5°$ is assumed, while an unmodified Parker [26] HMF is implemented, normalized to a magnitude of 4 nT at Earth.

4.2. *Modelling results*

Modelled energy spectra for p and $\bar{p}$ are shown in Panels (b) and (c) of Fig. 1 respectively. For p, the modelled $A < 0$ spectra are compared to the observed PAMELA spectrum of the past solar minimum. The p $A > 0$ results were obtained by using the same transport parameters, only switching the HMF polarity. As a Parker HMF is used, this spectrum should be interpreted as an upper limit for the p flux in the next $A > 0$ solar minimum. The modelled $\bar{p}$ spectrum is also consistent with the PAMELA observations, using exactly the same drift and diffusion coefficients used for p (except of course for the charge). Note that, for both p and $\bar{p}$, the energy spectra at low energies exhibit the well known $j \propto E$ adiabatic range. For $\bar{p}$, the LIS is also quite close to this spectral shape, so that the $\bar{p}$ spectra exhibit the $j \propto E$ limit very quickly. This produce very little difference between the $\bar{p}$ intensities for the two drift cycles, as compared to that of p.

The modelled $\bar{p}/p$ ratio is shown in Panel (d) of Fig. 1 for both HMF cycles at Earth. These results are also compared successfully to PAMELA observations. At very low energies the ratio becomes independent of energy due to the adiabatic effect discussed above. The ratio of $\bar{p}/p$ differs by a factor of $\sim$ 10 between drift cycles at 10 MeV, with the difference decreasing with increasing energy. Beyond $\sim$ 30 GeV the ratio becomes independent of drifts (and also modulation) as a *no modulation* high energy regime is reached where the mean free path becomes equivalent to the size of the heliosphere. Because the modelled $A > 0$ spectrum is an upper limit, the modelled ratio for the $A > 0$ cycle should be considered as a lower limit prediction for the next solar minimum.

Lastly, Fig. 2 shows pseudo-particle traces for protons (blue) and anti-protons (red) in the $A < 0$ (top panel) and $A > 0$ (bottom panel) HMF cycles. The shaded surface shows the HCS. These trajectories show the well-known charge-sign dependent difference is drift directions, with p and $\bar{p}$ sampling different regions of the heliosphere in the same drift cycle. In the $A > 0$ cycle, protons reach Earth by drifting from the polar regions, while they reach Earth from the equatorial (HCS) region in the $A < 0$ cycle. For $\bar{p}$ these drift directions reverse. For further discussions regarding charge-sign dependent modulation, see also the modelling done by [27, 16,

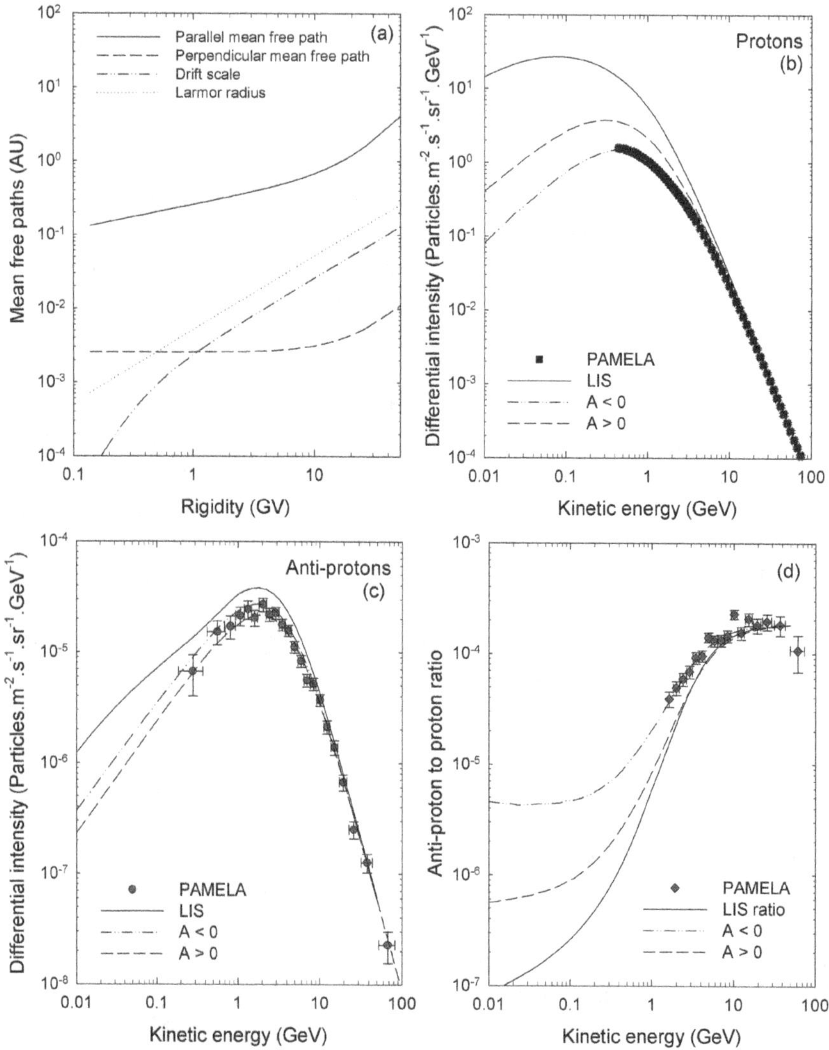

Fig. 1. Panel (a) shows the mean free paths (at Earth in the equatorial plane) used in this study; $\lambda_{||}$ as the solid line, $\lambda_{\perp\theta,r}$ as the dashed line and λ_d by the dashed-dotted line. Panel (b) shows calculated galactic proton energy spectra for the previous $A < 0$ solar minimum (dashed-dotted line), compared to the PAMELA proton spectrum of [1] *The predicted* $A > 0$ spectrum (dashed line) for the next solar minimum in $\sim$ 2021 should be considered as an upper limit. Panel (c) shows similar modelling results, but now for galactic anti-protons, compared to the PAMELA anti-proton spectrum of [2]. Panel (d) shows modelled results for the anti-proton to proton ratio (dashed-dotted line for the past $A < 0$ cycle and the dashed line for the nearing $A > 0$ minimum). The solid line is the ratio of the two LIS, while the PAMELA observations are taken from [2].

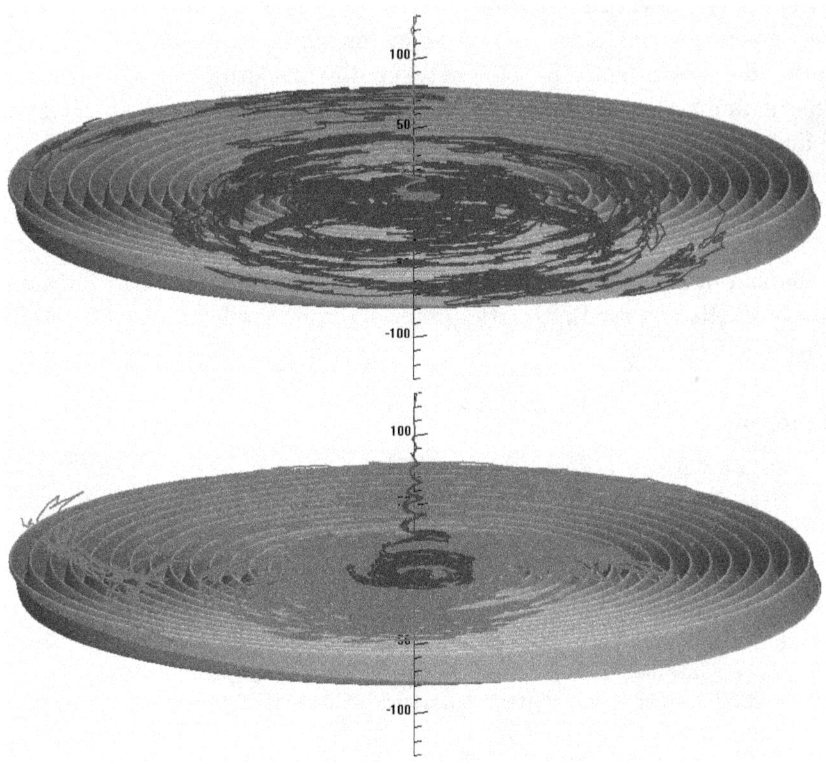

Fig. 2. Pseudo-particle traces of protons (blue) and anti-protons (red) in the $A < 0$ (top panel) and the $A > 0$ (bottom panel) HMF polarity cycles. The grey surface shows the wavy HCS.

28], as well as the $\bar{p}$ and p modelling of [25, 29].

5. Discussion and Conclusions

Using the SDE modelling approach, we presented results from our newly developed 5D CR modulation model. The model was applied to the modulation of galactic protons and anti-protons. Charge-sign dependent modulation, being very evident when studying the modulation of these CR species, was discussed and illustrated. The modelled energy spectra (for both protons and anti-protons) for the $A < 0$ HMF cycle were compared successfully to recent PAMELA observations, while upper limits for the fluxes in the next ($A > 0$) cycle were given. The same was done for the $\bar{p}/p$ ratio. From this ratio, it is evident that drift effects are present up to ~ 10 GeV. Drift effects are however most prominent at low energies, at 10 MeV the $\bar{p}/p$ ratio differs by a factor of ~ 10 between different drift cycles, while these effects diminish at the higher energies. With the upcoming solar minimum p and $\bar{p}$ results from the AMS-2 mission for the $A > 0$ cycle eminent, we predict a value for the $\bar{p}/p$ ratio that is lower than measured in the present $A < 0$ cycle. We conclude by noting that if the modulation conditions during the next solar minimum (in ~ 2021) would be the same as in the $A < 0$ minimum of 2009 (coined an unusual solar minimum with record breaking levels of CRs observed), galactic proton intensities will be even higher due to drift effects.

References

1. O. Adriani et al., Science **332**, 69 (2011).
2. O. Adriani et al., *Phys. Rev. Lett.* **105**, 121101 (2010).
3. E. N. Parker, *Planet. Space Sci.* **13**, 9 (1965).
4. W. H. Matthaeus et al., *Astrophys. J. Lett.* **590**, L53 (2003).
5. A. Shalchi, Nonlinear Cosmic Ray Diffusion Theories, Springer (2009).
6. J. Minnie et al., *Astrophys. J.* **670**, 1149 (2007).
7. R. A. Burger & D. J. Visser, *Astrophys. J.* **725**, 1366 (2010).
8. C. W. Gardiner, Handbook of Stochastic Methods, Springer (1983).
9. P. E. Kloeden & E. Platen, Numerical Solution of Stochastic Differential Equations, Springer (1999).
10. H. Fichtner et al., *Astron. Astrophys.* **314**, 650 (1996).
11. M. Zhang, *Astrophys. J.* **512**, 409 (1999).
12. M. Gervasi et al., *Nuc. Phys.* **78**, 26 (1999).
13. R. D. Strauss et al., *Astrophys. J.* **735**, 83 (2011).
14. R. D. Strauss et al., *J. Geophys. Res.* (submitted) (2011).
15. R. A. Burger et al, *J. Geophys. Res.* **105**, 27447 (2000).
16. M. S. Potgieter & H. Moraal, *Astrophys. J.* **294**, 425 (1985).
17. N. E. Engelbrecht & R. A. Burger, *Adv. Space Res.* **45**, 1015 (2010).
18. R. A. Burger et al., *Astrophys. J.* **674**, 511 (2008).
19. J. R. Jokipii, *Astrophys. J.* **146**, 480 (1966).
20. A. Teufel & R. Schlickeiser, *Astron. Astrophys.* **397**, 15 (2003).
21. A. Shalchi et al., *Astrophys. J.* **604**, 675 (2004).
22. J. Giacalone & J. R. Jokipii, *Astrophys. J.* **520**, 204 (1999).
23. J. L. Phillips et al., *Geophys. Res. Lett.* **22**, 3301 (1995).
24. I. V. Moskalenko et al., *Geophys. Res. Lett.* **22**, 3301 (2002).
25. U. W. Langner & M. S. Potgieter, *Geophys. Res. Lett.* **109**, A01103 (2004).
26. E. N. Parker, *Geophys. Res. Lett.* **128**, 664 (1958).
27. B. Heber & M. S. Potgieter, *Space Sci. Rev.* **127**, 117 (2006).
28. M. S. Potgieter, *Adv. Space Res.* **16**, 191 (1995).
29. U. W. Langner & M. S. Potgieter, *Adv. Space Res.* **34**, 144 (2004).

Dark Matter Searches, Underwater and Underground Experiments

V. Antonelli	Standard Model tests with man-made neutrino beams and liquid Argon detectors
L.I. Bodine	The Focal Plane Detector System for the Karlsruhe Tritium Neutrino Experiment
C. Bronner	Results from T2K and accelerator oscillation experiments
A. Fava	Status report of ICARUS T600
A.D. Ferella	Recent experimental results in direct Dark Matter searches
N. Heidrich	COBRA — Neutrinoless Double Beta Decay Experiment
M.Y. Khlopov	Dark Atoms of Dark Matter and their Stable Charged Constituents
V.Y. Kozlov	The EDELWEISS dark matter search experiment: towards enhanced sensitivity
K. Lang	Neutrino Physics Without Neutrinos: A review of selected world-wide searches for neutrinoless double beta decays
J. Maneira	Combined Analysis of all Three Phases of Solar Neutrino Data from the Sudbury Neutrino Observatory
A. Meregaglia	Search for the θ_{13} mixing angle with the Double Chooz experiment
F. Nova	A search for neutrinoless double beta decay: recent results from the NEMO-3 experiment and plans for SuperNEMO
C.A.B. Oliveira	The NEXT experiment at the LSC
M. Prall	The KATRIN Experiment

S. H. Seo	DM-Ice: A Direct Dark Matter Search at the South Pole
Y. Takeuchi	Recent results from Super-Kamiokande
M. Weber	The Enriched Xenon Observatory (EXO) for double beta decay
U. Wichoski	Status of the PICASSO Experiment
A. Wright	The DarkSide Program at LNGS

Standard Model tests with man-made neutrino beams and liquid Argon detectors

V. Antonelli*

Department of Physics, Milano University,
Via Celoria 16, 22100-Milano, Italy
**E-mail: vito.antonelli@mi.infn.it*

The accelerator neutrino experiments offer new possibilities for elementary particle and astroparticle physics. With the very high intensity man-made neutrino beams provided for experiments like the superbeams (and possibly, in future, β beams) it will be possible, not only to answer open questions about neutrino, but also to perform medium and low energy Standard Model tests and check the theory stability. We performed a deep investigation of the possibilities offered by present and future experiments and here we discuss how to extract a competitive low energy estimate of Weinberg angle by using the beam of T2K experiment and a liquid Argon near detector. We also discuss a possible application of our idea to future experiments under discussion at the moment, like oscillation experiments realized with a neutrino beam produced by CERN-PS and analyzed by means of ICARUS and another liquid Argon near detector.

Keywords: Accelerator neutrino; S. M. tests; LAr TPC; T2K; PS,ICARUS.

1. Neutrino, elementary particle physics and astrophysics

The recent astonishing result obtained by the OPERA collaboration about the velocity of neutrinos in the CERN-Gran Sasso beam, if confirmed, would force us to revise a theory like relativity which up to now passed successfully all the experimental tests and is at the basis of our present way of describing the physical world. This would be for sure the most relevant, but not the first, example of the fundamental role of neutrino in the development of our knowledge of elementary particle physics and astrophysics and in creating a connection between the two. It has been essential in proving the validity of the Standard Model and testing this theory and the results obtained in neutrino physics had also a great impact on astrophysical theories, for example in the improvement of solar models and of our knowledge of the stars, the galaxies and of the Supernovae properties and in the development of consistent cosmological models. The study of high energy neutrinos in conjunction with gamma and X ray astronomy is becoming more and more important in understanding the mechanisms ruling the formation and development of cosmic rays.

The discovery of neutrino mass and oscillation may be considered the first clear hint of the need to go beyond the Standard Model. Nowadays there is a great attention for the LHC results. However, it has been already stressed in literature that it would be essential to study, in parallel to the high energy also the high intensity frontier, that is to look for processes forbidden or strongly disfavored by the theory or investigate special corners of the phase space where the data available are not so many up to now, in order to check the theory stability and potentially find signals of new physics. Neutrino will play, as usual, a relevant role in this research project, mainly thanks to the new experiments that started running recently or are planned for the near future A new era already started, in which the main answers to open questions (starting from the exact mass and mixing patterns) could come from experiments using artificial neutrino beams (from reactors and accelerators) and, at least in some cases, from appearance experiments. In a first stage, started already, the experiments will use conventional neutrino beams, produced by a secondary meson beam. This is the case of T2K (in the recent past), MINOS and of the experiments using the CERN-Gran Sasso beam. They are supposed to confirm and improve the results obtained by the experiments with atmospheric and solar neutrino and by KamLAND and mainly to lower the limit on θ_{13} mixing angle and to find appearance signals. In particular OPERA found already the first ν_τ appearance signal. A particular role is played already by the T2K experiment, whose case will be treated in detail in section (3). The future of accelerator neutrino physics is not completely clear at the moment and it will depend also on the value of θ_{13} angle. After the second phase, with the presence of the superbeams (T2K in its 2nd phase, NOνA that should start working in 2013 in the States and, in case, a superbeam in Europe) that will be characterized by a great increase of neutrino beam luminosity, it could be worthwhile to pass to new generation of experiments (neutrino factories and/or β beams) with neutrinos produced directly by a primary beam decay. All of these experiments are designed to solve some long standing puzzles of neutrino physics, but they will also offer the possibility of using neutrino beams with intensities never reached before to perform precision tests of the Standard Model at energies much lower than the ones deeply investigated by high energy colliders. It could be, for instance, interesting to extract a low energy estimate of Weinberg angle and to compare it with the high energy determinations.

2. Low energy Standard Model tests

For energies close to the nucleon masses, typical of the superbeams, the golden channel would be elastic and quasi elastic contributions to neutrino-nucleon interactions. They are still sizable and do not present the uncertainty associated to parton distributions, that would be the disadvantage of the inelastic contribution. The (quasi) elastic scattering takes place through six different channels (charged and neutral, for neutrino and neutrino) and fixing, for instance, the electric form factors of nucleons one is left with 6 other parameters: the Weinberg angle and 5 hadronic form factors (the magnetic form factors of nucleons, $G_M^{(n,p)}(Q^2)$, the strange magnetic one, $G_M^S(Q^2)$ and the 2 axial and strange axial form factors, $G_A(Q^2)$ and $G_A^S(Q^2)$). Therefore one would have to solve a system of 6 coupled equations with 6 unknown, but we proved that the equation for Weinberg angle can be solved analytically[1] and written in terms of measurable quantities and this suggests the idea that a simultaneous fit of the weak mixing angle and some of the hadronic form factors is feasible.

In the numerical analysis to extract Weinberg angle it will be essential to select a narrow energy region, around 1 GeV, in which the cross section is not too low and the quasi elastic scattering contribution is still significant and a good energy recconstruction will also be important to separate the events in different energy bins and increase the number of information entering the statistical analysis. The most delicate experimental task would be to detect and reconstruct the neutral current signal in elastic events. The kinematic requirements impose quite severe cuts and in particular the low threshold (around 0.766 MeV) for the proton kinetic energy, would be too severe for an hypothetic Cerenkov detector. A liquid Argon time projection chamber (LAr TPC), like the one developed for ICARUS, would, instead, be the ideal detector for our purposes, due to its capalibity of detecting charged particles even at low energies, imposing much less severe kinematic requirements. Assuming a conservative kinematic cut of the momentum of the order $p \geq 300 MeV/c$ about 75% of the events survive. The detection of neutral current events on nucleons remains a very difficult task, even with this kind of detector. The possibility is under further investigation,[2] but, for the time being to be conservative, we assumed in our analysis that only the currents on protons would be detected, reducing the number of experimental observables from 6 to 4.

To prove the possibility of extracting from the data of future experiments a competive low energy estimate of Weinberg angle and to estimate the accuracy reachable in the realistic cases of beams available at T2K and similar future experiments, we a generated a fictitious set of data by using realistic spectrum distributions for neutrino beams, corresponding to specific experiments. Comparing these data with the theoretical results (which depend upon the parameters we are interested in) we performed a simultaneous fit of Weinberg angle and hadronic form factors and evaluated the accuracy of the fit results. We developed the statistical analysis in two different ways, as explained in the next section.

3. The T2K and the PS-CERN ICARUS cases

The T2K[3] long baseline uses a neutrino beam produced at the JPARC protosynchrotron and sent to the SuperKamiokande detector. Its main goal is to measure the value of $sin^2\theta_{13}$ with a nominal sensitivity 20 times better than the present limits. T2K aims also to improve the determination of the atmospheric parameters and possibly to search for sterile neutrinos. We showed that its beam could be very useful also to perform low energy tests of the Standard Model. After a 1st run, a 2nd one started in November 2010, with the aim to reach a sensitivity on $sin^2\theta_{13}$ of the order 0.05. Unfortunately the JPARC area was affected by the recent earthquake and this determined a slowing down of the research project, but the plans of the collaboration are to restart the data taking on December 2011.

In our analysis we assumed to put a LAr TPC near detector 2 off axis with respect to the beam line. In this way, we got a narrow distribution centered around a value between 800 and 900 MeV. We first performed an analysis assuming that the functional expressions of the hadronic form factors are known and we adopted the same expressions (power series) in data generation and in the theoretical fit. We fitted simultaneously the weak mixing angle and a subset of 2 or 3 form factors chosen between the magnetic and strange magnetic ones. The accuracy reached in Weinberg angle determination was satisfactory in all the cases, as shown in table (3).

The simultaneous fit of Weinberg angle and the 3 magnetic form factors is still well compatible with the input value and with uncertainty below 1%. To evaluate the impact on the fit of our partial ignorance of the functional form of the form factors, we repeated the analysis keeping in the fit the same form factor expressions and varying the expressions used in data generation. We found a conservative estimate of the global uncertainty corresponding to simultaneous variation of all the fitted form factors of the order of 0.60%.

Fitted parameters	Results of the fit
$sin^2(\theta_W)$, $G^p_M(Q^2)$, $G^S_M(Q^2)$	$sin^2(\theta_W) = 0.23116 \pm 0.00040$
$sin^2(\theta_W)$, $G^n_M(Q^2)$, $G^S_M(Q^2)$	$sin^2(\theta_W) = 0.23085 \pm 0.00040$
$sin^2(\theta_W)$, $G^p_M(Q^2)$, $G^n_M(Q^2)$	$sin^2(\theta_W) = 0.23145 \pm 0.00178$

Note: Fit of e.w. mixing angle and 2 form factors for an input of $\sin^2(\theta_W = 0.23120)$.

In the 2nd part of our study we dropped any assumptions on form factors and reproduced them with a neural network. It is possible to reproduce in this way up to 2 magnetic form factors still getting a satisfacory fit. The extension to 3 form factors would require the adoption of more complicated algorithms, like genetic algorithms and we decided to defer the investigation of this case to the moment in which we will be in presence of real experimental data. For a more detailed discussion of the results obtained in all the kind of analyses we refer the interested reader to.[2]

A proposal is under discussion to move ICARUS detector to CERN and use it, with a smaller LAr TPC near detector, to analize a series of neutrino anomalies by studying the neutrino beam produced using the CERN Proton Synchrotron (PS). This would provide us with an ideal high intensity beam. The near detector mass would be relatively low: 150 tons (plus additional 500 tons situated at a distance 7 times higher). The analysis would, instead, benefit of a very high intensity neutrino flux, made possible because the near detector is closest to the production point (the distance for T2K was higher, 1000 m) and, therefore, the neutrino flux on the detector is increased roughly of a factor 60 for geometrical reason. Assuming 2.5×10^{20} P.o.T., the flux on the detector would be at least a factor 10 higher than in the T2K case. Hence, it should be possible also in this case to recover a competitive low energy estimate of Weinberg angle and least 2 of the 3 magnetic form factors. An accurate analyis of this opportunity is in progress.

References

1. V. Antonelli, G. Battistoni, P. Ferrario, S. Forte, *Nucl.Phys.Proc.Suppl.* **168**, 192 (2007) and reference therein.
2. V. Antonelli, G. Battistoni, P. Ferrario and S. Forte, work in preparation to appear soon.
3. For T2K description, its runs and status, see Bronner's talk at this conference.

The Focal Plane Detector System for the Karlsruhe Tritium Neutrino Experiment

L. I. BODINE on behalf of the KATRIN Collaboration

Center for Experimental Nuclear Physics and Astrophysics & Department of Physics, University of Washington, Seattle, WA 98195, USA
E-mail: lbodine@uw.edu

The Focal Plane Detector system for the Karlsruhe Tritium Neutrino experiment will detect low-energy electrons that pass through the analyzing spectrometers. The electrons are guided by superconducting solenoid magnets to a multi-pixel silicon PIN-diode array in an extreme high-vacuum environment. Contact between the detector and front-end electronics is made by an array of spring-loaded pogo-pins, providing a reliable, low-background connection that allows detector wafers to be easily replaced. We report on design and commissioning tests performed at the University of Washington prior to final installation.

Keywords: PIN-diode array, neutrino mass, low-background counting

1. Introduction

The Karlsruhe Tritium Neutrino (KATRIN) experiment is designed to make a model-independent measurement of the electron-antineutrino mass by searching for distortion at the endpoint of the T_2-beta-decay spectrum. The experiment is designed to reach a sensitivity of 0.2 eV,[1] an order of magnitude improvement over the current laboratory-based limit.[2]

The Focal Plane Detector (FPD) system will detect low-energy electrons that pass through the KATRIN spectrometers. The system was designed to meet stringent background and electromagnetic constraints. Figure 1 shows the basic layout of the FPD system.

The detector is a custom monolithic PIN-diode array fabricated by Canberra Belgium on a 125-mm diameter, 500-μm thick, n-type silicon wafer. The array consists of 148 44-mm^2 pixels arranged in a dartboard-style pattern. A TiN coating provides ohmic contact for bias and readout.

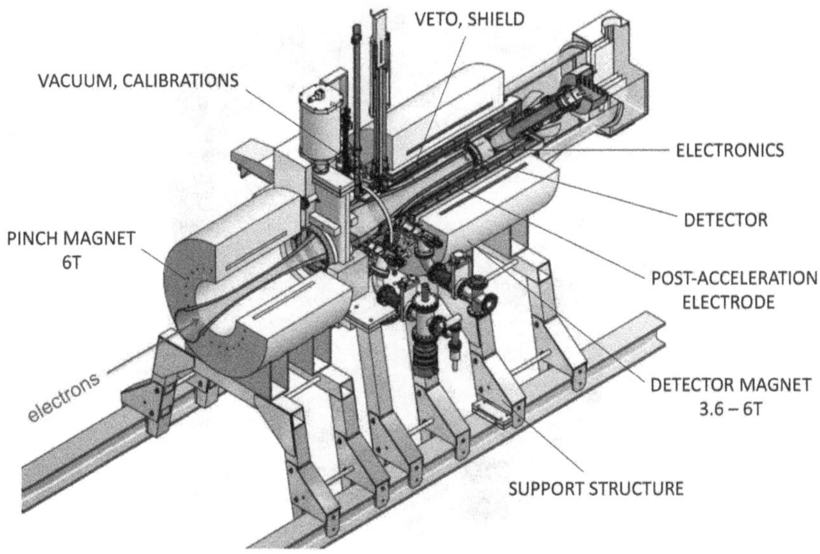

Fig. 1. Cut-away model of the KATRIN Focal Plane Detector system

2. Custom Detector Mount

System specifications necessitated the development of a low-background, extreme-high-vacuum-compatible, easily-replaceable detector mount. Our mount utilizes an array of spring-loaded pogo pins to make electrical contact with the detector. The pins are mounted on a vacuum feedthrough with preamplifier connections in a separate high-vacuum chamber immediately behind the feedthrough. L-shaped pins clamp a copper ring around the edge of the wafer, holding the wafer in contact with the pogo pins. Kapton insulating rings are placed on both sides of the detector to prevent electrical contact with the mount. Figure 2 shows a model of the detector mount.

A prototype mount was constructed to test detector performance under mechanical stress. Pin compressions between 0.38 mm and 1.1 mm were tested. Although the wafer deflects by up to 0.3 mm under pressure from the pins, electrical contact to the detector is unaffected. Gamma spectra from a ^{241}Am source taken with the prototype detector mount demonstrate that energy resolution is unaffected by pin-induced stress, as shown in Fig. 3. Direct leakage-current measurements also showed no dependence on pin compression.[3]

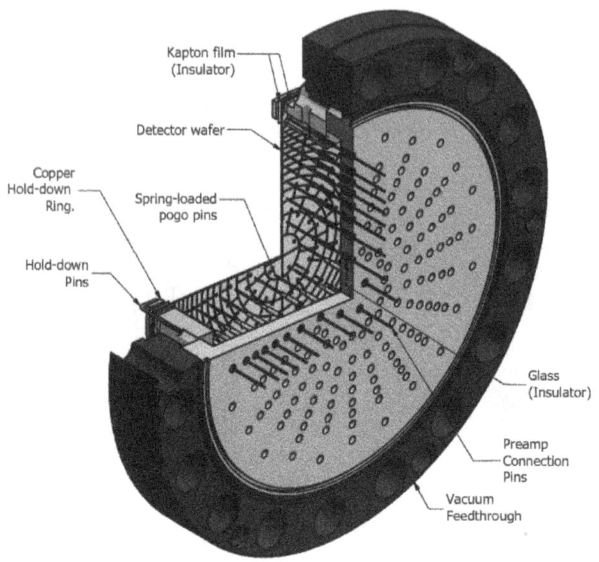

Fig. 2. Model of the KATRIN FPD detector mount

3. Initial Construction and Characterization

The FPD system was constructed and characterized at the University of Washington prior to shipment to Karlsruhe, Germany. ^{241}Am spectra, variable-energy photoelectron spectra, pulsed-LED spectra, and background measurements were taken to characterize its performance.

The system had 131 working channels with dead channels largely attributable to noisy preamplifiers for which there were no replacements at the time . The average resolution of the ^{241}Am 59.5-keV gamma peak was 1.51 ± 0.03 keV while the average resolution of the 18.6-keV electron peak was 1.60 ± 0.02 keV. The RMS non-linearity of the electronics was less than 0.3%.

The detector background goal was 1 mHz/keV in the region of interest. Low-background requirements necessitated the development of Geant4-based background simulation software to aid system design.[4] Because the background was highly sensitive to FPD components, radio-assays of materials were performed prior to construction. The measured system background is in good agreement with the pre-construction simulations as shown in Fig. 4.

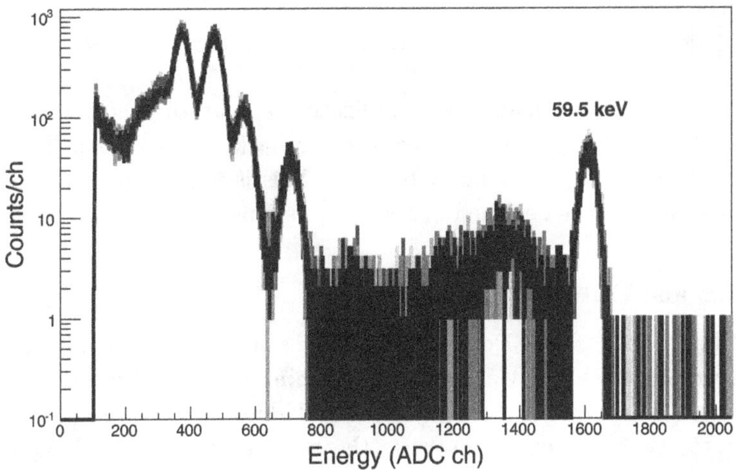

Fig. 3. Single-pixel ^{241}Am spectra for seven pogo-pin array translations between 0.38 mm and 1.1 mm

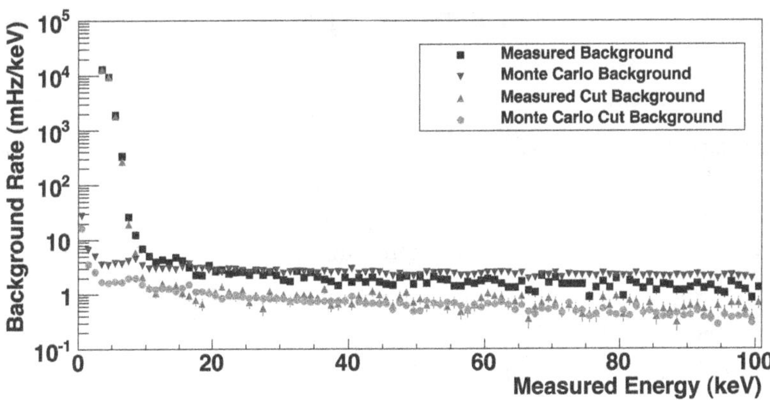

Fig. 4. FPD background spectra before and after multi-pixel and veto cuts. The sharp peak below 10 keV in measured spectra arises from threshold effects.

The overall performance is characterized by a figure of demerit F, such that minimizing F maximizes neutrino mass sensitivity. F is defined in Eq. 1, where $f(E_L, E_U)$ is the detector response function within an energy range bounded by E_L and E_U, $b_{det}(E_L, E_U)$ is the background rate from the FPD system, and b_{spec} is the background rate from the spectrometer.

$$F = \frac{\left(f(E_L, E_U) + \frac{b_{det}(E_L, E_U)}{b_{spec}}\right)^{1/6}}{f(E_L, E_U)^{2/3}} \tag{1}$$

The results of performance tests indicate that, for an energy range of 14.6–20.3 keV and an assumed spectrometer background of 9 mHz, the FPD system has a figure of merit of <1.2. This is within the acceptable range for achieving the designed neutrino-mass sensitivity.

4. Status and Outlook

The KATRIN Focal Plane Detector system was successfully commissioned and characterized at the University of Washington. Prototype detector-mount tests, background measurements and detector-performance tests demonstrate that the system satisfies the requirements of the KATRIN experiment.

The system was shipped to Germany for installation in the detector hall in June 2011. The system is under reconstruction and is expected to be ready for use in commissioning the KATRIN main spectrometer in early 2012. A technical paper covering design and performance is in preparation.

Acknowledgements

Primary support for this research is provided by the US Department of Energy under contract DE-FG-FG02-97ER41020.

The KATRIN collaboration is grateful for the continued support of Marijke Keeters, Mathieu Morrelle and their Canberra Benelux team.

References

1. *KATRIN Design Report 2004*, technical report (2005).
2. K. Nakamura (Particle Data Group), *J. Phys. G: Nucl. Part. Phys.* **37**, p. 075021 (2010).
3. B. A. VanDevender *et al.*, Performance of a TiN-coated monolithic silicon pin-diode array under mechanical stress, submitted to Nuclear Instruments and Methods A, (2011).
4. M. L. Leber, Monte Carlo calculations of the intrinsic detector background for the Karlsruhe Tritium Neutrino Experiment, PhD thesis, University of Washington, (WA, USA, 2010).

Results from T2K and accelerator oscillation experiments

C. BRONNER* on behalf of the T2K collaboration

Laboratoire Leprince Ringuet, Ecole Polytechnique, Palaiseau, 91128, France
**E-mail: bronner@llr.in2p3.fr*
http://llr.in2p3.fr

Various experiments use secondary neutrino beams produced by accelerators to study neutrino oscillations. In this article, we will review oscillation results from a number of those experiments (MINOS, OPERA), and focus more on results from T2K. This long baseline off-axis experiment uses a beam of muon neutrinos produced in J-PARC in Japan to study muon neutrino disappearance in order to measure atmospheric parameters, as well as studying electron neutrino appearance to measure the 13 mixing angle. We will present in particular very recent results of those measurements obtained by MINOS and T2K.

Keywords: Neutrino oscillations, T2K, mixing angles, θ_{13}

1. Neutrino oscillations and neutrino beams

1.1. *Neutrino oscillations*

Neutrinos are fermions which come in three flavors ν_e, ν_μ and ν_τ. Those are the flavor eigenstates ν_α, in which neutrinos are produced through weak interactions. There are also three mass eigenstates ν_i which are the propagation eigenstates. We can assume that the mass eigenstates are different from the flavor eigenstates, and then decompose a flavor eigenstate in the basis of mass eigenstates:

$$|\nu_\alpha\rangle = \sum_i U_{\alpha i}|\nu_i\rangle \tag{1}$$

This defines the mixing matrix U, or Pontecorvo-Maki-Nakagawa-Sakata matrix. Using this expression and some assumptions, we can obtain the probability to detect a neutrino that was produced in a flavor α in a different flavor β:

$$P(\nu_\alpha \to \nu_\beta) = \delta_{\alpha\beta} - 4\sum_i \sum_{j,j<i} \Re\left(U^*_{\alpha i}U_{\beta i}U_{\alpha j}U^*_{\beta j}\right)\sin^2\left(1.27\Delta m^2_{ij}\frac{L}{E}\right)$$
$$+2\sum_i \sum_{j,j<i} \Im\left(U^*_{\alpha i}U_{\beta i}U_{\alpha j}U^*_{\beta j}\right)\sin\left(2.54\Delta m^2_{ij}\frac{L}{E}\right) \tag{2}$$

Where:

- L is the distance in kilometers travelled by neutrinos
- E is their energy in GeV
- $\Delta m^2_{ij} = m^2_i - m^2_j$ is the difference of the squares of masses of mass eigenstates i and j.

This probability can be different from $\delta_{\alpha\beta}$ if the Δm^2_{ij} are non-zero and the mixing matrix satisfies certain conditions. In this case, a neutrino produced in a certain flavor can be detected in a different flavor: this is the phenomenon of neutrino oscillations.

1.2. *Parameters describing neutrino oscillations*

The mixing matrix U can be taken as the product of three rotation matrices:

$$U = \begin{pmatrix} 1 & 0 & 0 \\ 0 & c_{23} & s_{23} \\ 0 & -s_{23} & c_{23} \end{pmatrix} \begin{pmatrix} c_{13} & 0 & s_{13}e^{-i\delta} \\ 0 & 1 & 0 \\ -s_{13}e^{i\delta} & 0 & c_{13} \end{pmatrix} \begin{pmatrix} c_{12} & s_{12} & 0 \\ -s_{12} & c_{12} & 0 \\ 0 & 0 & 1 \end{pmatrix}$$

where $c_{ij} \equiv \cos\theta_{ij}$, $s_{ij} \equiv \sin\theta_{ij}$.

This gives a total of six parameters describing neutrino oscillations:

- three mixing angles θ_{12}, θ_{23} and θ_{13}
- two mass splittings Δm^2_{21} and Δm^2_{23}
- the CP-violation phase δ.

Experiments studying neutrino oscillations aim at measuring those parameters. At the start of T2K, two parameters were still unknown: θ_{13} and δ.[1]

1.3. *Making of a neutrino beam*

Experiments described in this article use conventional neutrino beams, made using a particle accelerator. Such a beam is produced in the following way: first, a particle accelerator is used to accelerate protons. Those protons are then sent to a target, usually made of graphite. The collisions produce secondary particles, in particular pions and kaons. Those hadrons are focused using electromagnetic horns, and sent into a decay volume in which they decay in flight into neutrinos and other particles. This produces a nearly pure ($\approx 99\%$) muon neutrino beam.

Different flavors of neutrinos can be looked for in this muon neutrinos beam:

- appareance of electronic neutrinos ν_e allows one to measure θ_{13}
- disappearance of muonic neutrinos ν_μ allows one to measure atmospheric parameters θ_{23} and Δm^2_{32}
- appearance of tau neutrinos ν_τ for observation of $\nu_\mu \to \nu_\tau$ oscillation

2. Long baseline oscillation experiments

Experiments covered in this article are long baseline experiments, which means that neutrinos travel a long distance (several hundred kilometers) between their production and detection points.

2.1. *T2K*

The Tokai to Kamioka[2] experiment uses a neutrino beam produced at the J-PARC center located at Tokai on the east coast of Japan. The neutrinos, produced from a 30 GeV proton beam, travel 295 kilometers to the far detector Super-Kamiokande. The experiment also includes a group of near detectors, located 280 meters away from the target. A schematic view of the experiment is shown on figure 1.

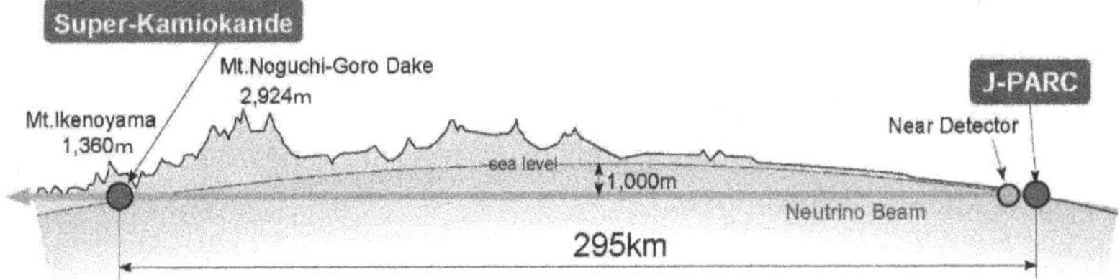

Fig. 1. Overview of the T2K experiment

The particularity of T2K is to be the first experiment to use the off-axis beam technique: the far detector is not located on the axis of the beam, but in a direction making a 2.5° angle with the beam direction. This allows one to select the neutrino energy as can be seen on figure 2: we obtain a narrow band beam, peaked at the maximum of oscillation probability 0.6 GeV. This also reduces the high energy tail of the beam which produces background at the far detector, and reduces the intrinsic ν_e component of the beam to 0.5% at peak energy.

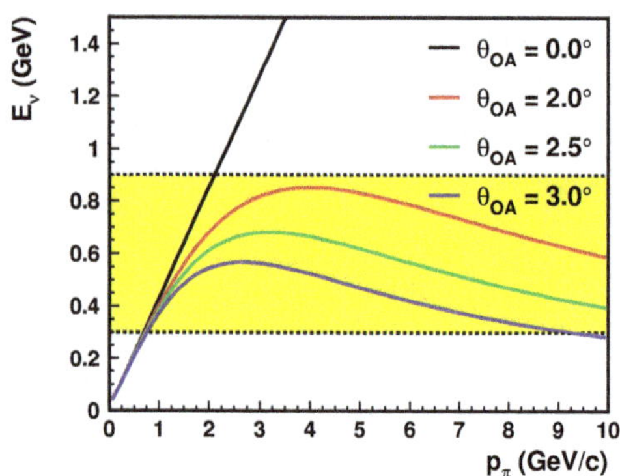

Fig. 2. Energy of produced neutrino as a function of pion momentum, for different off-axis angles. This energy decreases when the off-axis angle increases.

The near detectors are separated into two groups: the INGRID detector is located on the axis of the beam, and monitors its stability (direction and event rate). The ND280 is located off-axis in the direction of the far detector, and is used to measure the fluxes before oscillation of the various components of the neutrino beam, neutrino cross-sections, as well the rate of neutral current π^0 interactions which are a background for ν_e measurements at the far detector.

The main goals of the experiment are to measure θ_{13} through ν_e appearance, and to improve the measurement of atmospheric parameters through ν_μ disappearance measurements.

2.2. *MINOS*

The MINOS experiment uses a beam of neutrinos produced by 120 GeV protons at the NuMI beamline of Fermilab (United States). This is an on-axis experiment, with a far and a near detector of similar design, made of a sandwich of iron and scintillators in a magnetized volume. The far detector is located 735 kilometers away from the neutrinos production point. The experiment measures the atmospheric parameters, and will also look for ν_e appearance to measure θ_{13}.

2.3. *OPERA*

The OPERA experiment looks for ν_τ appearance in a muon neutrino beam produced at the CERN SPS using 400 GeV protons. The OPERA detector is located 732 kilometers away from the target in Gran Sasso, Italy. To be able to identify ν_τ decays, the experiment needs very precise tracking, and the detector uses emulsion techniques for this reason.

3. Electron neutrino appearance

3.1. *T2K analysis overview*

In T2K, the expected number of events at the far detector is computed by simulation, assuming no oscillations. The simulations of flux and interactions are tuned to existing data. In particular, results of the NA61 experiment at CERN which measures hadrons production by collision of a 30 GeV proton beam on a graphite target are used for tuning of the resulting neutrino flux. The near detectors are used to normalize simulation to data. The expected number of events at the far detector is given by the formula:

$$N_{SK}^{exp} = \frac{R_{ND}^{\mu,data}}{R_{ND}^{\mu,MC}} \times N_{SK}^{MC} \tag{3}$$

where the ratio represents the normalization of simulation to data at the near detectors, and N_{SK}^{MC} is the number of events at the far detector predicted by the simulation.

The normalization at the near detectors is done with the inclusive ν_μ charged current interactions measurement. This was done using data from the first year of data taking (2.9×10^{19} POT), and based on particle identification using the Time Projection Chambers. This gives a 90% purity and 38% efficiency for selecting ν_μ charged current interactions. The normalization factor was found to be:

$$\frac{R_{ND}^{\mu,data}}{R_{ND}^{\mu,MC}} = 1.036 \pm 0.028(\text{stat})^{+0.044}_{-0.037}(\text{det. syst}) \pm 0.038(\text{phy. syst})$$

3.2. *Number of events at the far detector*

For ν_e appearance analysis, only the number of events at the far detector is used. The expected number of events at the far detector for $\theta_{13} = 0$, and the associated systematic error obtained by the previous method are detailed in tables 1 and 2.

Table 1. Expected number of events in the T2K far detector

	Number of expected events
Beam ν_e component	0.8
Neutral Current background	0.6
Oscillation from solar term	0.1
Total	1.5

Table 2. Systematic error on expected number of events at far detector

Error source	Systematic error
Neutrino flux	$\pm 8.5\%$
Neutrino cross sections	$\pm 14.0\%$
Near detectors systematics	$^{+5.6\%}_{-5.2\%}$
Far detector systematics	$\pm 14.7\%$
Near detectors statistics	$\pm 2.7\%$
Total	$^{+22.8\%}_{-22.7\%}$

The final expectation is:

$$N_{SK}^{exp} = 1.5 \pm 0.3 \text{ events}$$

Events are selected in the far detector in data from the first two years of data taking by applying fiducial and selection cuts. Six events remain after all cuts. When looking at the position of those events in the detector, it appears that they are clustered at large radius. Comparisons of simulated and data distributions without fiducial cut in the inner detector, as well as in the outer detector were made to look for possible background contamination. The two sets of distribution agree, and no evidence for such contamination were found.

3.3. *Results*

Six events have been observed as compared to an expectation of 1.5 ± 0.3 in the case of null θ_{13}. This excludes the null oscillation hypothesis at 2.5σ level.[3] The Feldman-Cousins unified method[4] was used to draw the confidence interval, assuming $\Delta m^2_{32} = 2.4 \times 10^{-3} eV^2$ and $sin^2 2\theta_{23} = 1$, for both normal and inverted hierarchy (fig 3).

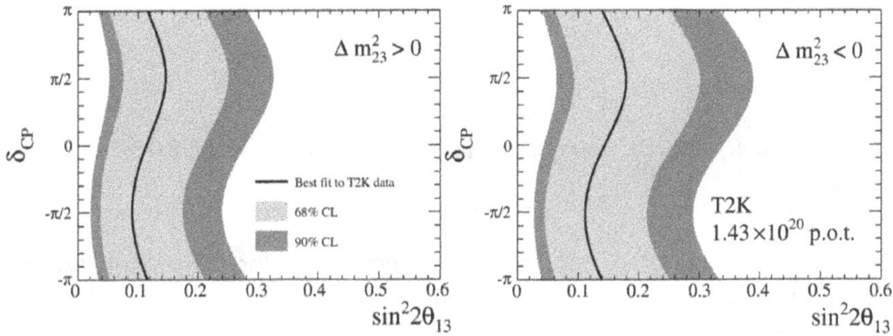

Fig. 3. Confidence intervals for θ_{13} from T2K ν_e appearance results.

3.4. *Comparison with MINOS results*

The MINOS collaboration released their 2011 ν_e appearance search results, using data totalizing 8.2×10^{20} POT.[5] In the case of null θ_{13} the expected number of events was 49.5 ± 2.8 (syst) $\pm$ 7.0 (stat). 62 events were observed in the data, which corresponds to a 1.7σ excess. The 90% confidence level intervals obtained are compatible with a null θ_{13}. The boundaries of those intervals can be compared with those obtained by T2K (fig 4).

There is an overlap between the 90% CL intervals found by the two experiments. It also can be noticed that for the first time T2K exclude a zero θ_{13} at 90% CL, and that in the normal hierarchy, the sensitivity of MINOS is now below the CHOOZ limit.

4. Muon neutrino disappearance

4.1. *T2K results*

The T2K ν_μ disappearance analysis is done is a two flavors oscillation framework. The expected number of events at the far detector in the absence of

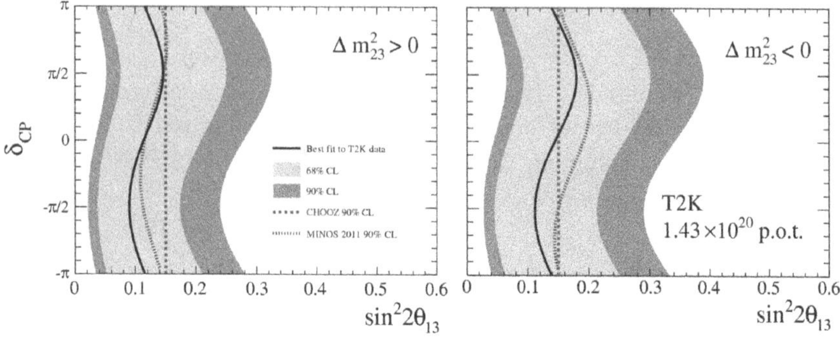

Fig. 4. 90% CL intervals for θ_{13} from T2K, MINOS and CHOOZ

oscillations is computed using the same method as for the ν_e appearance case, and is found to be 104. Events in the far detector are selected in the data by applying fiducial cut and three selection cuts. 31 events passed all cuts, which excludes the non-oscillation hypothesis at 4.4 σ level. A second independant analysis on those data with different energy binning excludes this hypothesis at 4.3 σ level.

The energy spectra of neutrinos observed at the far detector can be fitted to extract the oscillation parameters. This was done using two methods: the first one is a maximum likelihood with systematic errors fit, while the second one is a likelihood ratio without systematic errors fit. The confidence intervals obtained by those two methods can be seen on figure 5, where the solid lines correspond to the first method, and the dashed lines to the second one. The two methods are found to be in good agreement, with differences coming from not using systematic errors fit in the second one.

4.2. *Comparison with results from other experiments*

The atmospheric parameters are also measured by MINOS, as well as by the Super-Kamiokande collaboration using atmospheric neutrinos. 90% CL intervals obtained by those experiments are shown on figure 6.

It appears that the different results are compatible as there is an overlap between the intervals obtained by those experiments. Concerning accelerator neutrinos experiments, the best sensitivity is currently obtained by the MINOS ν_μ disappearance results.

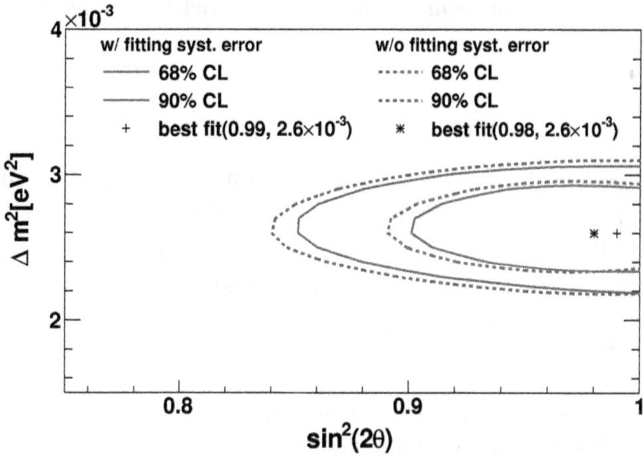

Fig. 5. 90% CL intervals for the atmospheric parameters, from T2K ν_μ disappearance results.

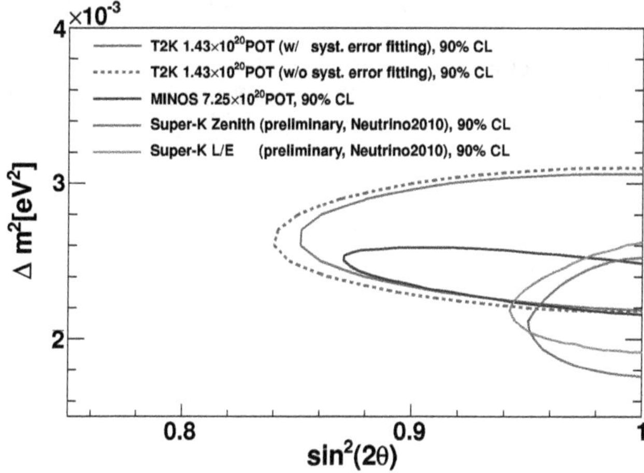

Fig. 6. 90% CL intervals for the atmospheric parameters, from T2K, MINOS and Super-Kamiokande

5. Tau neutrinos appearance

OPERA presented results on the search for tau neutrino appearance, using 92% of the 2008-2009 data sample, corresponding to 4.8×10^{19} POT.[6] 1.65 events were expected, and one candidate event was observed.[7] The probability of this event being a background fluctuation was found to be 15%.

6. Status and prospects

The T2K experiment released results on ν_e appearance and ν_μ disappearance, showing for the first time indication of a non-zero θ_{13} at 90% confidence level. The experiment is currently recovering from the earthquake that hit Japan in March 2011, and plans to restart at the end of 2011.

References

1. M. Mezzetto and T. Schwetz, *J.Phys.G* **37** (2010), doi: 10.1088/0954-3899/37/10/103001, arXiv: 1003.5800.
2. K. Abe *et al.*, *Nucl. Instrum. Methods* (2011), article in press, doi: 10.1016/j.nima.2011.06.067, arXiv:1106.1238 [physics.ins-det].
3. K. Abe *et al.*, *Phys. Rev. Lett* **107** (2011), doi: 10.1103/PhysRevLett.107.041801.
4. G. J. Feldman and R. D. Cousins, *Phys. Rev. D* **57**, 3873(Apr 1998).
5. P. Adamson *et al.*, *Accepted for publication in Phys. Rev. Lett* (2011), arXiv: 1108.0015.
6. The OPERA collaboration, arXiv: 1107.2594.
7. N. Agafonova *et al.*, *Phys. Lett. B* **691**, 138 (2010), arXiv: 1006.1623.

Status report of ICARUS T600

A. FAVA* for the ICARUS Collaboration
I.N.F.N. — Sezione di Padova,
Padova, I-35131, Italy
**E-mail: fava@pd.infn.it*

ICARUS T600 at the INFN-LNGS Gran Sasso Laboratory is the first underground large mass Liquid Argon TPC: exposed to the CERN-CNGS neutrino beam, it has smoothly began taking data since October 2010. Its excellent resolution and 3D imaging allow an unprecedented event visualization quality combined with a good calorimetric reconstruction and the electronic event processing. In addition to the $\nu_\mu \rightarrow \nu_\tau$ oscillation and sterile neutrino search, atmospheric neutrino and matter stability will be studied.

Keywords: Style file; LaTeX; Proceedings; World Scientific Publishing.

1. ICARUS T600 experiment

The ICARUS T600 LAr-TPC detector, presently taking data in Hall B of the INFN Gran Sasso underground National Laboratory (LNGS), is the largest liquid Argon TPC ever built. Its detection technique, based on the collection of scintillation light (5000 γ/mm at 128 nm wavelength) combined with the stereoscopic recording of the ionization signal ($\sim$ 6000 electrons per mm), was first proposed by C. Rubbia in 1977.[1]

1.1. *Detector layout and liquid Argon purity*

The ICARUS T600 detector[2] (see Fig.1) consists of a large cryostat split into two identical, adjacent and independet half-modules, with an overall mass of about 760 tons of ultra-pure liquid Argon at 89 K temperature. Each half-module, with internal dimensions $3.6 \times 3.9 \times 19.6$ m^3, houses two Time Projection Chambers (TPCs) separated by a common cathode. An electric field E_D = 500 V/cm, kept uniform by field shaping electrodes, ensures the coverage of the 1.5 m maximum drift distance in 1 ms. The anode of each TPC is made of three parallel wire planes, 3 mm apart, oriented at 0° and ±60° w.r.t. the horizontal direction: in total 53248 wires are installed.

By appropriate voltage biasing, the first two planes (Induction-1, 2) are transparent to drift electrons and measure them in a non-destructive way, whereas the ionization charge is finally collected by the last one (Collection). The signals coming from each wire are continuously read and digitized at 25 MHz (1 t-sample $\sim$ 400 ns) and recorded in multi-event circular buffers operated with a 3-levels veto able to give different priorities to the trigger sources, thus minimizing DAQ dead-time. Building rate is kept as high as possible ($>$ 0.7 Hz) by splitting data-flow into 4 parallel streams, one per TPC chamber, afterwards combined once consistency has been checked.

Fig. 1. The ICARUS T600 detector in Hall B at the LNGS underground laboratory (left) and a simple sketch of the inner TPCs structure (right).

Scintillation light detection allows one to determine the absolute time of the ionizing events. For this purpose arrays of Photo Multiplier Tubes (PMTs), operating at the LAr cryogenic temperature[3] and made sensible to VUV scintillation light (= 128 nm) by applying a wavelength shifter layer (TPB), are installed behind the wire planes.
In ICARUS T600 detector an elaborate cryogenic plant allows one to reduce and keep at an exceptionally low level the electro-negative impurities, especially water and Oxygen, filtering both LAr and GAr with Oxysorb/Hydrosorb filters. The electron lifetime is monitored by studying the attenuation of the charge signal as a function of the drift time along "clean" through-going muon tracks in Collection view. With the liquid recirculation turned on, the free electron lifetime (τ_e) is constantly above 6 ms in both cryostats (Fig. 2); this corresponds to 0.05 ppb O_2 equivalent impurity concentration, producing 16% charge attenuation at the maximum 1.5 m drift distance.

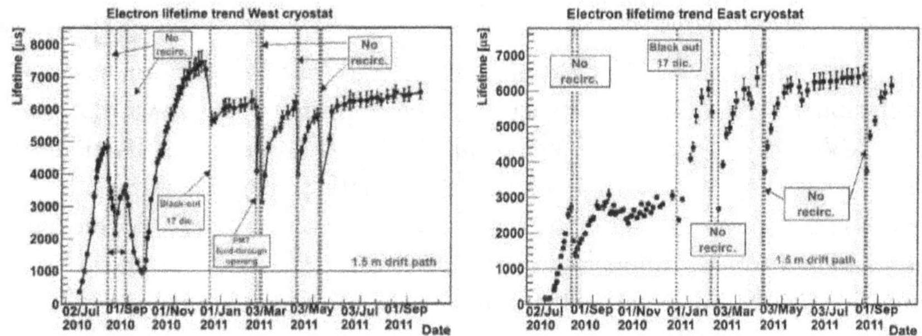

Fig. 2. Free electron lifetime evolution with time in 2010-2011 run, for both cryostats.

1.2. *Trigger strategy*

The main ICARUS T600 trigger system relies on the scintillation light signals, using the analog sum of signal from PMTs with a $\sim$ 85 photo-electron discrimination threshold for each of the four TPC chambers.

(i) For CNGS neutrino events $\sim$ 60 μs gate is open at the predicted proton extraction time, ("early warning" signal sent from CERN to LNGS 80 ms before the extraction); a trigger is generated when the PMT sum signal is present in at least one TPC chamber within the gate. About 80 events/day are recorded (1 mHz rate), well distributed in the 10.5 μs proton spill width.

(ii) For cosmic rays an efficient reduction of the spurious signals, still maximizing the detection of low energy events, is provided by the coincidence of the PMT sum signals of the two adjacent chambers in the same module, relying on the 50% cathode transparency. A trigger rate of $\sim$ 18 mHz per cryostat has been achieved leading to about 100 events/hour on the full T600 (expected: 160 events/hour), out of which only 6% are empty. Recently the PMTs HV biasing system has been upgraded in order to increase the c-ray detector efficiency.

Performance of the trigger system can be pushed forward by using the charge information. A new dedicated DR-slw algorithm, able to online identify hits with 1 board (32 wires) modularity,[4] has been software implemented since May 3rd in an independent 2-levels trigger for CNGS events. Events are collected at 200 mHz rate every time the CNGS gate is opened, and empty events are rejected in real time at 10^4 level. The same algorithm is being hardware implemented to trigger on low energy localized events.

2. Physics programme

The main goal of the ICARUS T600 programme[5] is the search for $\nu_\mu \to \nu_\tau$ oscillation in the CNGS beam, i.e. an almost pure ν_μ beam with $E_\nu \sim$ 17.4 GeV, traveling over 732 km from CERN to Gran Sasso. Particulary attractive is the $\tau \to e\nu\nu$ channel where kinematical selection criteria based on missing transverse momentum allow full rejection with 50% efficiency the associated background. On the same beam the search for sterile neutrinos in LNSD parameter space is also performed, looking for an excess of ν_e CC events. ICARUS T600 is studying also atmospheric neutrinos.
Finally, thanks to the powerful background rejection and its 3×10^{32} nucleons, ICARUS T600 can play a role in proton decay search, in particular in interesting exotic channels not accessible to Čerenkov detectors.

3. 2010–2011 data taking and analysis with CNGS beam

ICARUS-CNGS run started in stable conditions on October 1st, collecting $5.8 \cdot 10^{18}$ pot out of the $8 \cdot 10^{18}$ delivered by CERN up to November 22nd. The CNGS beam restarted on March 19th 2011: $3.9 \cdot 10^{19}$ pot have been collected up to September 30th out of the $4.2 \cdot 10^{19}$ pot delivered, with detector duty cycle in excess of 93%. Data collected in 2010 run have been used for training and tuning the analysis software tools. 169 neutrino events have been identified, into 434 t fiducial volume, in good agreement with the expectations (172) accounting for fiducial volume and DAQ dead-time.

- Momentum of long μ tracks is determined by multiple scattering with a Kalman filter algorithm[6] with $\sim$ 16% resolution. The resulting spectrum is in good agreement with expectations (Fig. 3 left).
- All tracks are fully 3D reconstructed using a polygonal line algorithm[7] and particles (mainly μ, π, K and p) are identified with a neural network approach by studying the event topology and the energy deposition per track length unit as a function of the particle range (dE/dx versus range).
- Electrons, identified by the characteristic e.m. showering, are well separated from π^0 by γ reconstruction, dE/dx signal comparison and π^0 invariant mass measurement.[8] This guarantees 90% efficiency identification of the leading ν_e CC e^-, while almost fully rejecting NC interactions.

Besides the event-by-event approach, the calorimetric energy measurement of the CNGS ν_μCC events has been performed: leptonic and hadronic part have been separately reconstructed, and MC has been corrected for non-containment and non-compensation (Fig. 3 right).

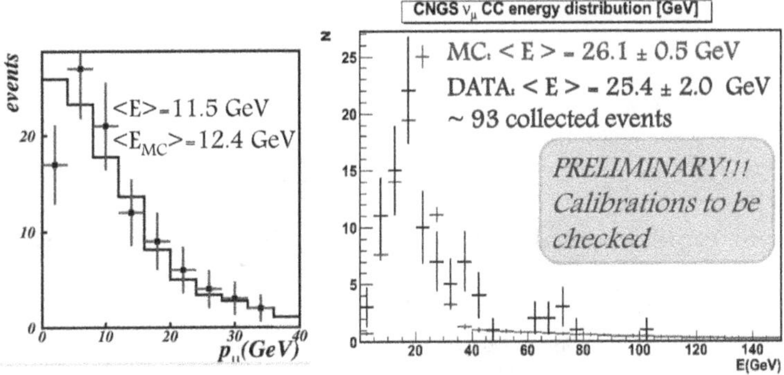

Fig. 3. Muon momentum recontructed by multiple scattering (left) and ν_μCC energy spectrum (right) compared with MC expectations

4. Conclusions

The ICARUS T600 LAr-TPC, installed underground at LNGS, has been succefully collecting CNGS ν_μ events since October 2010, searching for $\nu_\mu \to \nu_\tau$ oscillations and LSND-like ν_e excess, studying atmospheric neutrinos and exploring the nucleon stability in few selected channels.
While the 2011 data taking is smoothly going over, the analysis of the 2010 CNGS events demonstrated unique imaging capability, spatial and calorimetric resolutions of the LAr-TPC technique.

References

1. C. Rubbia, *The Liquid-Argon Time Projection Chamber: A new Concept for Neutrino Detector*, CERN-EP 77-08 (1977).
2. S. Amerio *et al.*, *Nucl. Instr. and Methods A* **527**, 329 (2004).
3. A. Ankowski *et al.*, *Nucl. Instr. and Methods A* **556**, 146 (2006).
4. B. Baibussinov *et al.*, *Journal of Instrumentation* **5**, p. P12006 (2010).
5. F. Arneodo *et al.*, *The ICARUS Experiment, A Second-Generation Proton Decay Experiment and Neutrino Observatory at the Gran Sasso Laboratory*, LNGS-P 28/2001 (2001).
6. A. Ankowski *et al.*, *The European Physical Journal C - Particles and Fields* **48**, 667 (2006), 10.1140/epjc/s10052-006-0051-3.
7. B. Kegl *et al.*, *Pattern Analysis and Machine Intelligence, IEEE Transactions on* **22**, 281 (mar 2000).
8. A. Ankowski *et al.*, *Acta Physica Polonica B* **41**, p. 103 (2010).

Recent experimental results in direct Dark Matter searches

A. D. FERELLA
Physik-Institut, Universität Zürich,
8057Zürich, Switzerland
**E-mail: ferella@physik.uzh.ch*
http://www.physik.unizh.ch

Dark Matter is one of the most intriguing and challenging puzzles of modern astroparticle physics and many experimental efforts are being put into its solution. The best motivated Dark Matter particle is the Weakly Interacting Massive Particle (WIMP). The WIMP direct detection principles are explained, and a review of the leading experiments and their recent results is given.

Keywords: Dark Matter; WIMP.

1. Introduction

The solution of the Dark Matter and Dark Energy puzzles is surely at the main challenge of modern cosmology. Both have been theorized, basing on strong observational evidences, and constitute the main success of the most accredited cosmological models. Yet none of them has been directly detected. In this review the Dark Matter problem will be discussed and the approaches to directly detect it, in the form of a special category of particles, will be presented.

2. Evidence for Dark Matter

The experimental evidence of Dark Matter comes from astronomical and astrophysical observations at different scales and with completely different techniques. From galactic to cosmological scale all evidences strongly suggest that more than 95% of the Universe is made of invisible and unknown types of matter and energy. In this section the observations pointing to the existence of a missing mass in the Universe will be briefly discussed.

Presence of non luminous matter in the galaxies is found in the observation of the so called rotation curves of galaxies, i.e. the graph of circular velocities of stars and gas versus their distance from the galactic center. The milestone in the study of such rotation curves was put by the pioneering work by V. Rubin and W.K. Ford in 1970.[1]

Assuming that Newtonian dynamics is applicable also at such distances, the circular velocity of the stars in a galaxy (and of the interstellar medium) is found to be approximately constant with the radius, while expected to decrease (based on the luminous matter) which points to the presence of a non luminous matter extending far beyond the optical and gaseous disks.

The first experimental evidence of a missing mass in the structures of the Universe came from the observation of the velocity dispersion of galaxies in the Coma cluster (Fritz Zwicky[2] in 1933). By measuring the radial velocities of eight galaxies in the cluster Zwicky found an unexpectedly large velocity dispersion which would imply the existence of some kind of "Dark Matter".

The Inter-Cluster Medium (ICM) is a gas (mostly ionized hydrogen and helium) heated up by the gravitation induced movement and for dispersion greater than 300 Km/s, the gas emits radiation in the X-ray region. Thus, the dynamics of a cluster can be inferred by analyzing its X-ray profile. The first systematic observations of X-ray dynamics of galaxy clusters were made by Forman et al. in 1985.[3]

One of the most convincing "direct proof[s] of Dark Matter" comes from this extragalactic scale. In 2006, D. Clowe et al.[4] combined the observations in the visible (by HST), the X-ray (by Chandra) and the weak lensing of two colliding clusters of galaxies, called the Bullet cluster (1E 0657-558). They found that while the major baryonic component of the two clusters (the ICM, detected with the X-ray) interacts electromagnetically and thus gets slowed down by the collision, the main masses of the two clusters (detected by weak lensing) cross undisturbed each other without interacting. Also the visible objects were not greatly affected by the collision, and most passed right through, given the relatively low density of starts in the clusters.

Another very stringent evidences of the existence of a non-baryonic component of matter in the Universe comes from the precision measurement of the Cosmic Microwave Background (CMB) radiation and especially of its anisotropies.

The CMB was predicted in 1948 by George Gamow,[5] as a relic radiation from ~300,000 years after the Big Bang. The power spectrum of the CMB depends on the value of the "cosmological parameters", i.e. a set of less than ten numbers (depending on the model) which usually describe the matter content of the Universe (baryons, Dark Matter, Dark Energy, neutrinos), its age (hubble parameter), its global geometry (curvature parameter) and the properties of the initial fluctuations (amplitude and spectral index).

From the analysis of the WMAP experiment data,[6] using the flat Λ-CDM model [a] it is found that $\Omega_\chi h^2 = 0.1099 \pm 0.0062$, $\Omega_b h^2 = 0.02273 \pm 0.00062$ and $h = 0.719^{+0.026}_{-0.027}$, where Ω_χ is the Dark Matter density, Ω_b is the density of baryonic matter and h is the Hubble parameter.

A comprehensive picture on the matter content of the Universe at cosmological scale comes from the combination of the results from CMB, Ia supernovae[7] and

[a] we use here the convention of indicating with $\Omega_i \equiv \rho_i/\rho_c$ the density of each matter/energy component of the Universe related to the so called "critical density", $\rho_c = 1.88\, h^2\, 10^{-29}$ g/cm^3, defined as the average total density corresponding to a flat Universe.

the REFLEX galaxy cluster survey,[8] which can be summarized as follows:

$$\Omega_{Tot} = 1.02 \pm 0.02, \quad (1)$$
$$\Omega_{\Lambda} = 0.73 \pm 0.04, \quad (2)$$
$$\Omega_{M} = 0.27 \pm 0.04, \quad (3)$$
$$\Omega_{b} = 0.044 \pm 0.004, \quad (4)$$
$$\Omega_{\chi} = 0.22 \pm 0.04, \quad (5)$$

with Ω_M being the total mass density, $\Omega_M = \Omega_\chi + \Omega_b$.

3. Nature of Dark Matter

Once that the existence of Dark Matter is assessed, we need to define its characteristics and nature. Basing always on experimental evidence and with the help of theoretical prediction we will now try to depict the "identikit" of a Dark Matter particle.

It has already been observed in the previous section that the Dark Matter interacts gravitationally (i.e. is constituted by **massive** particles), is made of **non-baryonic** particles and, being invisible to any radiation sensitive device, such particles have to be also **electrically neutral**.

In the frame of the Big Bang theory, we can assume that in the early stages of the Universe, the Dark Matter particles were in thermal equilibrium with the others. However, in order to provide the present significant abundance and to satisfy the cosmological requirements $\Omega_M \cong 0.3$, they had to have decoupled, before the present time;[9] furthermore they have to be **stable** or at least have a lifetime longer than the age of the Universe.

Dark Matter candidates may be classified as 'hot' (relativistic) or 'cold' (non-relativistic) according to their energy at the time when they de-coupled from the rest of the Universe. The observations on the present Universe suggest a Dark Matter being predominantly cold, i.e. **non-relativistic**. This comes from the fact that the tiny fluctuations in the matter-density of the early Universe have evolved into the large scale structure we see today. Moreover, anisotropies in the cosmic microwave background radiation, cannot be created by the fluctuations in the baryonic matter density alone, and thus Dark Matter is required. However, if Dark Matter were hot it would not be able to assemble in confined region and the Universe structures we observe nowadays would have been much more isotropic.

3.1. *The WIMP miracle*

The evolution of the number density of a particle χ over time of the Universe t follows the Boltzmann equation:

$$\frac{dn_\chi}{dt} + 3Hn_\chi = - < \sigma_A v > (n_\chi^2 - (n_\chi^{eq})^2)$$

H is the Hubble constant, $n_\chi$is the number-density, $< \sigma_A v >$ is the thermally averaged annihilation cross section times velocity of the species χ, and n_χ^{eq} is the number-density in thermal equilibrium, so the term $3Hn_\chi$ gives the expansion of the Universe.

After relatively simple considerations and calculations, we can find that the total density of χ particles (Ω_χ) is:

$$\Omega_\chi = 1.66\ g^{1/2}\frac{T_0^3}{\rho_c m_{Pl} < \sigma_A |v> }.$$

Substituting $T_0 == 2.35 \cdot 10^{-4} eV$ (current Universe temperature), $\rho_c = 1.05 \times 10^4\ h^2\ eV \cdot cm^{-3}$, $m_{Pl} = 1.22 \cdot 10^{28}\ eV$ and $g^{1/2} \sim 1$, we obtain:

$$\Omega_\chi h^2 = \frac{m_\chi n_\chi}{\rho_c} \simeq \frac{3 \cdot 10^{-27} cm^3 s^{-1}}{<\sigma_a v>}$$

Therefore, in the case of Dark Matter particles in order to obtain the measured density (equation 5) we obtain that $< \sigma_a v > \sim 10^{-26} \div 10^{-25} cm^3 s^{-1}$, which is very close to a weak interaction cross section (of the order of $< \sigma_a v > \sim 10^{-25} cm^3 s^{-1}$). For this reason it is believed that a hypothetical "**Weakly Interacting** Massive Particle" (WIMP) could solve the Dark Matter puzzle and this concept defines the so called *WIMP miracle.*

No solution to the Dark Matter problem can be found in the framework of the Standard Model of Particle Physics, but on the other hand WIMPs are predicted in many supersymmetric extensions of this Model, with largely different masses and interaction cross-sections.

4. Direct Detection of WIMPs

If WIMPs exist and are really the dominant constituent of Dark Matter, they must be present also in the Milky Way[10] and, though they only very rarely interact with conventional matter, should nonetheless be detectable in sufficiently sensitive experiments on Earth. The WIMP flux on Earth is of the order of $10^5\ \text{cm}^{-2}\text{s}^{-1}$, large enough to allow the detection of the nuclear recoils caused by their elastic scattering off target nuclei of Earth based detectors.[11] Direct Dark Matter search experiments, indeed, aim to detect the interactions of WIMPs in dedicated low background detectors, by measuring the rate, R, and energy, E_R, of the induced nuclear recoils and possibly, in directional experiments, the direction. Since the WIMP–nucleon relative velocity v is non-relativistic, the recoil energy E_R can be easily expressed in terms of the scattering angle in the center of mass frame, θ:

$$E_R = \frac{|\vec{q}|^2}{2m_N} = \frac{\mu^2_{\chi-N} v^2}{m_N}(1 - cos\theta),$$

where m_N and m_χ are the masses of the target nucleus and of the WIMP, $|\vec{q}| = \sqrt{2m_N E_R}$ is the momentum transfer and $\mu_{\chi-N} = \frac{m_\chi m_N}{m_\chi + m_N}$ is the WIMP-nucleus reduced mass. The differential nuclear recoil rate induced by the WIMPs can be written as:

$$R = \int_{E_{th}}^{\infty} dE_R \frac{\rho_0 \sigma_0}{m_N m_\chi} F^2(E_R) \int_{v_{min}}^{v_{esc}} v f(v) dv.$$

Here E_{th} is the energy threshold of the detector, ρ_0 is the local Dark Matter density, σ_0 is the cross section at zero momentum transfer, $f(v)$ is the WIMP velocity distribution in the galactic halo, v_{min} is the minimum velocity required for the WIMP to generate the recoil energy E_R and v_{esc} is the galactic escape velocity. $F^2(E_R)$ is the nuclear form factor, which accounts for the fact that the de Broglie wavelength associated with the momentum transfer is of the same order as the nuclear dimensions; thus the bigger the nucleus the stronger this effect becomes.

The main astrophysical uncertainties lie in the velocity distribution $f(v)$ (commonly assumed Maxwellian) and in the density ρ_0 (usually assumed equal to 0.3 GeVc^{-2}cm^{-3}. Detecting the direction of the WIMPs would provide a viable solution to the velocity distribution function problem.

If WIMPs are neutralinos, i.e. Majorana fermions, they can have only scalar or axial coupling with quarks, which, in this specific non-relativistic regime, translates into a spin-independent coupling and a coupling between the neutralino spin and the nucleon spin. In the spin-independent case, the full coherence results in a cross section $\sigma_0 \propto A^2$, for a target nucleus of mass number A, while in the spin-dependent case the cross section is dominated by the total net spin of the nucleus. In most cases, the coherent term will dominate because it has the A^2 enhancement. However, neutralinos with dominantly gaugino or higgsino states may only couple through the spin-dependent term.

As a result of the Earth motion relative to the WIMP halo, the event rate is expected to modulate with a period of one year with the maximum on the 2nd of June. To detect this characteristic modulation signature, large masses are required, since the effect is of the order of ~2% with respect to the total event rate. A stronger diurnal direction modulation of the WIMP signal is also expected. The Earth rotation about its axis, oriented at angle with respect to the WIMP "wind", changes the signal direction by 90 degrees every 12 hours, with a resulting 30% modulation on respect to the total rate.

Nuclear recoils induced by WIMPs are detected exploiting the three basic phenomena associated with the energy loss of charged particles in target media: scintillation, ionization and heat. All the detectors used to perform this rare event search are also sensitive to the environmental radiation associated with cosmic rays and radioactivity in construction materials and the environment. At the current limits[12,13] the expected WIMP rate is less than 1 event per kg per year and significant SUSY parameter space exists down to 10^{-3} event per kg per year. Exploring this parameter space requires ton-scale detectors with nearly vanishing backgrounds.

Dark matter search experiments are located in deep-underground sites, to attenuate the cosmic muons flux by a factor 10^5 to 10^7. In addition, the detectors are typically enclosed by thick layers of absorbing materials (lead for γ's and hydrogen-rich compounds for neutrons), in order to reduce also the contribution due to the environmental radioactivity. Moreover shielding and detector components have to be as well selected to have low radioactivity, to allow significant background reduction. Since the mean free path of a high energy γ-ray or a neutron is of the order of centimeters, while the mean free path of a WIMP is of the order of light-years, identification of multi site events constitute a powerful background rejection tool. Finally in many Dark Matter direct search experiments background discrimination mechanisms are used, based on the fact that nuclear recoils (signals) and electron recoils generate (backgrounds) generate different signals in the detectors, due to the different nature.

5. WIMPs direct detection experiments (a selection)

A large variety of underground experiments all over the world aim to the direct detection of the Milky Way halo's WIMPs. Only a small selection of them is presented in this review.

5.1. *DAMA/LIBRA – A possible evidence?*

The DAMA/LIBRA experiment started operation 1990 at Gran Sasso underground laboratory (LNGS). The detector (DAMA) was initially based on nine 9.7 kg highly radio-pure NaI(Tl) scintillators shielded from radioactive background. The collaboration has then upgraded the detector to a sensitive mass of 250 kg of NaI(Tl). This new experiment, called LIBRA, is running since March 2003. The threshold provided for both experiments is 2 keV.

The DAMA experiment belongs to the first generation of Dark Matter direct detection experiments requiring a large detector exposure. Although the NaI(Tl) scintillator provides some discrimination between nuclear recoils and electronic recoils based on pulse shape, the collaboration published its data without any background reduction. The DAMA/LIBRA collaboration combined the 290 kg×year exposure of DAMA with the with the 530 ton×year exposure of LIBRA (total 820 kg×year), confirming in both cases a consistent modulation signal[14] (as reported in Figure 1). The total significance of the signal is 8.2 σ.

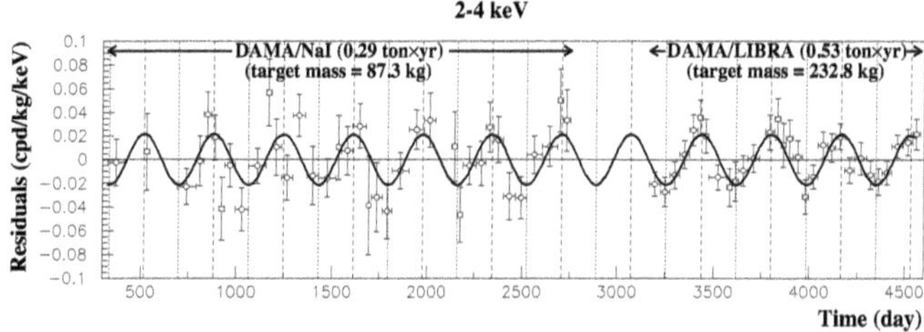

Fig. 1. DAMA annual modulation signal from a model independent fit to the cosine function, showing a period of oscillation of 1.00 ± 0.01 year and offset t_0 equal to 140 ± 22 days.[14]

5.2. *CoGeNT – Hint for light WIMPs?*

The CoGeNT experiment is based on a 440 g, low threshold (~ 0.4 keV) P-type Point Contact (PPC) germanium detector. The PPC technology employed allows effective surface background events rejection, thanks to the good position sensitivity. The detector is installed in the Soudan Underground Laboratory (SUL) and operated from December 2009 to March 2011; it acquired 442 live-days of data for a total exposure of $\sim$ 146 kg$\times$day. Also the CoGeNT collaboration reports an annual modulation signal[15] (see Figure 2) with a statistical significance of 2.8 σ, 16.6$\pm$3.8% modulation amplitude, period 347$\pm$29 days and minimum in Oct. 16$\pm$12d.

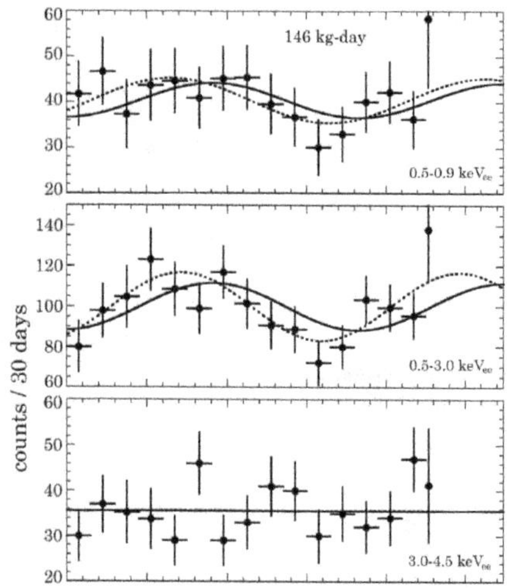

Fig. 2. CoGeNT data:[15] rate versus time in several energy regions. The best-fit modulation is shown as dashed line. The solid line indicates a prediction for a 7 GeV/c^2 WIMP in a galactic halo with Maxwellian velocity distribution. No indication of modulation is observed for the surface background events.

5.3. *CRESST-II – Evidence, but background?*

The CRESST-II (Cryogenic Rare Event Search with Superconducting Thermometers) experiment is also located at the LNGS. The experiment is based on 9 scintillating $CaWO_4$ crystals of cylindrical shape, each with a mass of $\sim$ 300 g. The crystals are operated as cryogenic calorimeters at temperatures of $\sim$ 10 mK. The energy deposited by an interacting particle is mainly converted into phonons, which are then detected with a transition edge sensor (TES). A small fraction of the energy deposited in the crystal goes into scintillation light, which is detected by a cryogenic light detector in order to detect. According to the different phonon to light yields ratio particles can be identified (see Figure 3) and background rejected.

The collaboration recently published[16] the results from the analysis of the data collected between 2009 and 2011 (total net exposure of 730 kg$\times$d), where the region of interest was defined in the energy interval 12 ÷ 40 keV. "Sixty-seven events are found in the acceptance region where a WIMP signal in the form of low energy nuclear recoils would be expected". With a maximum likelihood analysis they found that all the background sources are not sufficient to account for such a big excess of events.

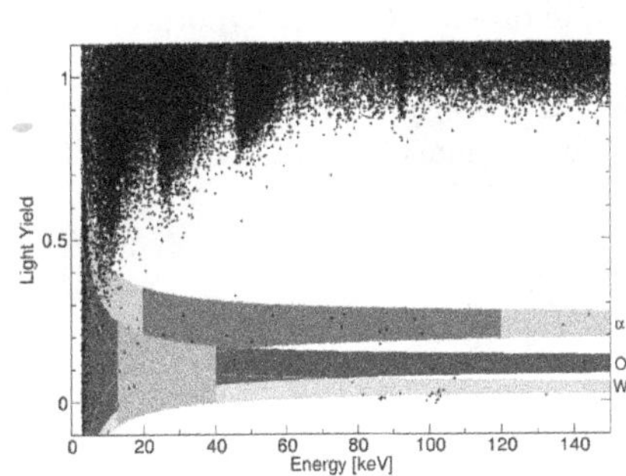

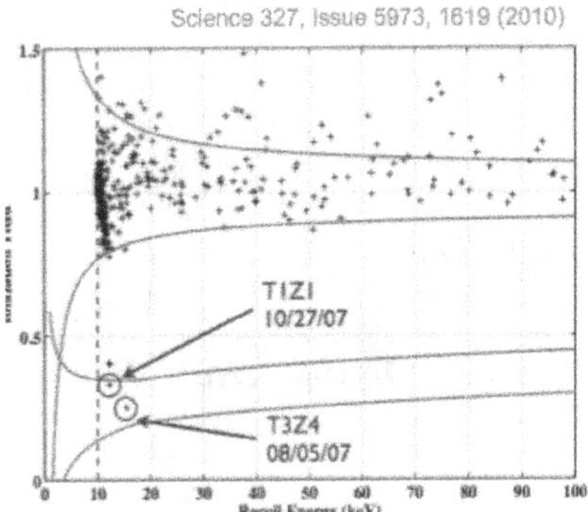

Fig. 3. CRESST-II data of one crystal ("detector module"): light yield versus recoil energy. Electronic recoil events occur at high light yield (~ 1). The shaded areas represent the alpha, oxygen, and tungsten recoil bands, as indicated in the Figure. The acceptance region, the reference region in the α-band, and the events observed are also shown.

Fig. 4. CDMS data: Ionization yield versus recoil energy for all the events passing the analysis cuts as published by Ahmed et al.[12] The solid lines indicate the 2σ electronic and nuclear recoil bands, while the vertical dashed line represents the energy threshold used for the analysis. Highlighted are also the two events found in the signal region.

5.4. *CDMS – No evidence...*

To detect WIMPs interactions, CDMS uses the so called ZIP (Z-dependent Ionization Phonon) detector technology, consisting in disk-shaped germanium (250 g) or silicon (100 g) crystals operated at cryogenic temperatures (~ 50 mK). The final CDMS-II experiment consists of a total of 30 detectors, 19 Germanium and 11 Silicon crystals. The technique allows the simultaneous detection of ionization and phonon signals. By looking at the charge to phonon yield ratio, the CDMS detectors can identify and possibly reject the background events. Electron recoil events from surface contamination result in a reduced ionization signal, leading to the leakage of such events into the nuclear recoil identification region. A phonon timing cut, developed by the collaboration, provides an effective discrimination against these events and the resulting rejection power for the electron recoil background is of the order of 10^6.

The data acquired between July 2007 and September 2008 with the complete detector have been analyzed and the results published in 2010 for a total exposure of 612 kg×day.[12] Figure 4 shows the event distribution in the ionization yield versus recoil energy parameter space. Two events have been found in the signal region. Since the probability for these event to be an upward fluctuation of the background is quite high (23%) the collaboration didn't claim any evidence of Dark Matter particle interaction and set an limit on the WIMP-nucleon spin-independent cross-section of 3.8 $\times 10^{-44}$ cm^2 for a WIMP mass of 70 GeV/c^2.

In 2010 the collaboration published a second paper[17] where they extenced their analysis also to low energy events (2 keV recoil energy threshold), so renouncing to the background discrimination, in order to test the hypothesis of low mass ($\lesssim$ 10 GeV/c^2) WIMPs. This results provide a stronger constraint than the previous one for WIMPs with mass $<$ 9 GeV/c^2 GeV/c^2 and "excludes parameter space associated with possible low-mass WIMP signals from the DAMA/LIBRA and CoGeNT" and CRESST-II experiments.

5.5. *XENON100 – No evidence*

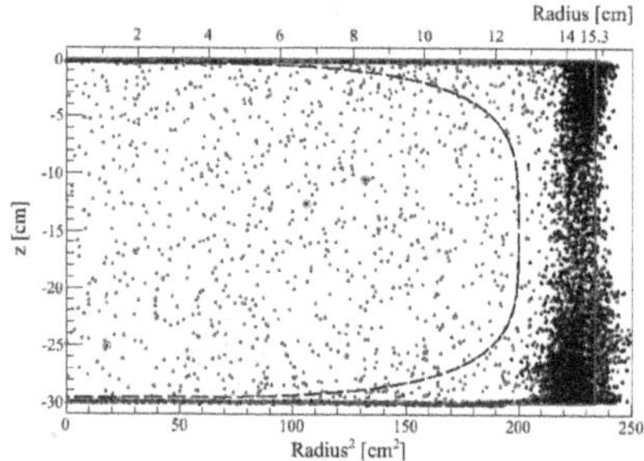

Fig. 5. XENON100 data: 3D distribution of all the events (gray) and of the events below the 99.75% rejection (black dots) in the energy region of interest (8.4–44.6) keV$_{nr}$. The dashed line indicates the boundary of the 48 kg fiducial volume, while the gray line indicate the TPC dimensions. The three circles indicate the three events found in the region of interest.

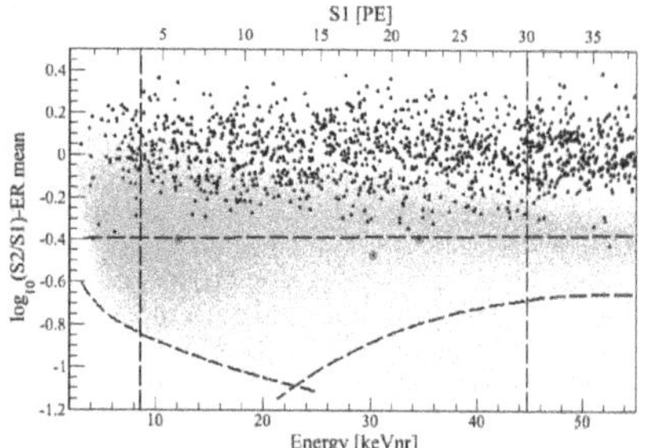

Fig. 6. XENON100 data: flattened $\log_{10}(S2_{bottom}/S1)$ versus nuclear recoil energy for all the events passing all the analysis cuts. The gray points indicate the nuclear recoil event distribution from a ^{241}AmBe calibration run. The dashed lines indicate the energy region selected for the final analysis, the software threshold S2 $>$ 300 p.e, the 99.75 % rejection line and the 3 σ lower bound of the nuclear recoil band.

The XENON100 experiment is the most sensitive of the WIMP direct detection experiment operative to date. This experiment exploits the time projection chamber (TPC) technology based on Liquid Xenon (LXe), with simultaneous detection of the ionization (via proportional scintillation) and the direct scintillation signals. The amplitude and timing of the signals, as well as the 3-dimensional event localization capability, enables these TPCs to effectively reject background.

The XENON collaboration is following a phased approach to the direct detection of WIMPs in liquid Xe, with a series of detectors of increasingly larger mass and lower background. The goal is to realize within the 2014 an experiment with a ton scale fiducial target (XENON1T) to search for WIMPs with almost two orders of magnitude better sensitivity than the current best limit. After the successful results of the first 10 kg scale prototype, XENON10,[18,19] the XENON collaboration has designed and built a second generation experiment with a mass increase of a factor 10 and a background reduction of a factor 100, in order to achieve the sensitivity goal of $\sim$ 50 times better than XENON10.

The XENON100 experiment is installed at LNGS. The TPC sensitive volume is surrounded by an active liquid xenon veto. The total mass of Xe required to fill the detector is 170 kg, of which approximately 65 kg are in the target volume. By looking at the different light to charge yield ratio background events can be rejected. Moreover the double phase technique allows one to further reduce the background by applying fiducial volume cuts (exploiting the efficient self-shielding features of the LXe) and by single scatter selection criteria.

The collaboration has recently published the results of the analysis on a 100.9 live-days data sample acquired between January and June 2010[13] (with a total "acceptance corrected" exposure of 1471 kg$\times$day). Three candidate events were found in the signal region, where 1.8 $\pm$ 0.6 background events were expected, as shown in Figure 5 and 6. Therefore a limit has been put on the WIMP-nucleon spin-independent cross-section of 7.0 $\times 10^{-45}$ cm^2 for a WIMP mass of 50 GeV/c^2, which is the most stringent limit to date. Moreover also the XENON100 results exclude the parameter space associated with the signals of the DAMA/LIBRA, CoGeNT and CRESST-II experiments.

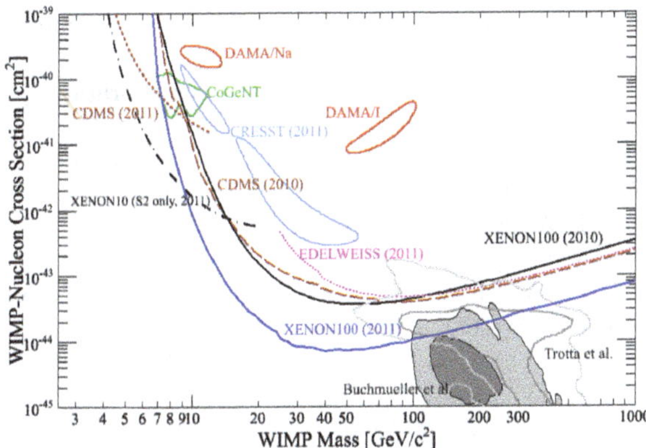

Fig. 7. Spin Independent elastic WIMP-nucleon cross section as a function of the WIMP mass. Exclusion limits from the most sensitive experiments are shown, as well as 90% favored regions from DAMA/LIBRA, CoGeNT and CRESST-II.

6. Summary

In Figure 7 the parameter space associated with the DAMA/LIBRA, CoGeNT and CRESST-II signals and the exclusion limits from the most significant experiments (including the ones presented in this short review). Many more experiments are either taking data, in commissioning runs, in construction phase or being designed with the precise aim of solving the Dark Matter puzzle. The tensions between the experiments claiming and those rejecting light WIMPs will hopefully be solved, or at least relaxed by the upcoming experiments. In order to improve the reliability of the results, CoGeNT is going to start a new run (after a forced shut off due to fire in SUL), CRESST is expected to start a new run with reduced background and XENON100 is taking data with reduced background and lower energy threshold.

It is interesting to notice that in the last five years the sensitivity of the experiments improved by two or three orders of magnitude and in this challenge detectors based on noble liquids seem to have best sensitivity and the most promising, thanks to their easy scalability on respect to cryogenic crystals, for example.

Finally it is important to notice that other WIMP detection mechanisms, like indirect detection and appearance at high energy particle colliders, are already trying to answer the pressing question on the real nature of Dark Matter particles.

References

1. V. C. Rubin, W. K. Ford, Jr., Astrophys. J. **159** (1970) 379-403.
2. F. Zwicky, *Helv. Phys. Acta* **6** (1933) 110.
3. Forman,W., Jones,C., Tucker, W. *Astrophys. J.* **293** (1985) 102.
4. D. Clowe, M. Bradac, A. H. Gonzalez, M. Markevitch, S. W. Randall, C. Jones, D. Zaritsky, Astrophys. J. **648** (2006) L109-L113. [astro-ph/0608407].
5. G. Gamow, Phys. Rev. **74** (1948) 505.
6. J. Dunkley *et al.* [WMAP Collaboration], Astrophys. J. Suppl. **180** (2009) 306-329. [arXiv:0803.0586 [astro-ph]].
7. A. G. Riess *et al.* [Supernova Search Team Collaboration], Astrophys. J. **607** (2004) 665-687. [astro-ph/0402512].
8. P. Schuecker, H. Bohringer, C. A. Collins, L. Guzzo, Astron. Astrophys. **398** (2003) 867-878. [astro-ph/0208251].
9. M.S. Turner, E.W. Kolb, 1990, *the early Universe (Frontiers in Physics)*, (Addisn-Wesley, U.S.);
10. M. R. Merrifield, Astronom. J., **103** (1992) 1552.
11. M.W.Goodman, E.Witten, *Phys. Rev. D* **31** (1985) 3059.
12. Z. Ahmed *et al.* [The CDMS-II Collaboration], Science **327** (2010) 1619-1621. [arXiv:0912.3592 [astro-ph.CO]].
13. E. Aprile *et al.* [XENON100 Collaboration], Phys. Rev. Lett. **107** (2011) 131302. [arXiv:1104.2549 [astro-ph.CO]].
14. R. Bernabei *et al.* [DAMA Collaboration], Eur. Phys. J. **C56** (2008) 333-355. [arXiv:0804.2741 [astro-ph]].
15. C. E. Aalseth, P. S. Barbeau, J. Colaresi, J. I. Collar, J. Diaz Leon, J. E. Fast,

N. Fields, T. W. Hossbach *et al.*, Phys. Rev. Lett. **107** (2011) 141301. [arXiv:1106.0650 [astro-ph.CO]].
16. G. Angloher, M. Bauer, I. Bavykina, A. Bento, C. Bucci, C. Ciemniak, G. Deuter, F. von Feilitzsch *et al.*, [arXiv:1109.0702 [astro-ph.CO]].
17. Z. Ahmed *et al.* [CDMS-II Collaboration], Phys. Rev. Lett. **106** (2011) 131302. [arXiv:1011.2482 [astro-ph.CO]].
18. J. Angle et al. (XENON10 Collaboration), Phys. Rev. Lett. 100, 021303 (2008).
19. J. Angle et al. (XENON10 Collaboration), Phys. Rev. Lett. **101** (2008) 091301.

COBRA – Neutrinoless Double Beta Decay Experiment

Nadine Heidrich*
for the COBRA Collaboration

Institute of Experimental Physics, University of Hamburg,
Luruper Chaussee 149, 22761 Hamburg, Germany
** E-mail: nadine.heidrich@desy.de*
www.cobra-experiment.org

The COBRA experiment is searching for neutrinoless double beta decay using CdZnTe semiconductor detectors. The main focus is on Cd-116, with a decay energy of 2814 keV well above the highest naturally occurring gamma lines. Furthermore, Te-130, with a high natural abundance, and Cd-106, a double β^+ emitter, are under investigation. Advantageous is the possibility to operate the detectors at room temperature. Besides coplanar grid detectors, pixelised detectors are considered. The latter ones would allow for particle discrimination, therefore providing efficient background reduction.
The current status of the experiment is described, including the upgrade of the R&D set–up in spring 2011 at the LNGS underground laboratory, the different detector concepts and the latest half – life limits. Furthermore, studies on the use of liquid scintillator for background suppression and Monte – Carlo simulations are presented.

Keywords: COBRA; Neutrinoless double beta decay; CdZnTe; Pixelated detector; Coplanar grid detector; LNGS

1. The Concept of COBRA

The aim of the next–generation experiment COBRA[1] is to measure the neutrinoless double beta decay ($0\nu\beta\beta$) by using cadmium zinc telluride (CdZnTe) semiconductor detectors.
The concept for a large scale set–up consists of an array of CdZnTe detectors with a total mass of 420 kg enriched in ^{116}Cd up to 90 %. With a background rate of the order of 10^{-3} counts/keV/kg/year, the experiment would be sensitive to a half–life $T_{1/2}$ larger than 10^{26} years, corresponding to a Majorana mass term $m_{\beta\beta}$ smaller than 50 meV. CdZnTe contains

*Speaker

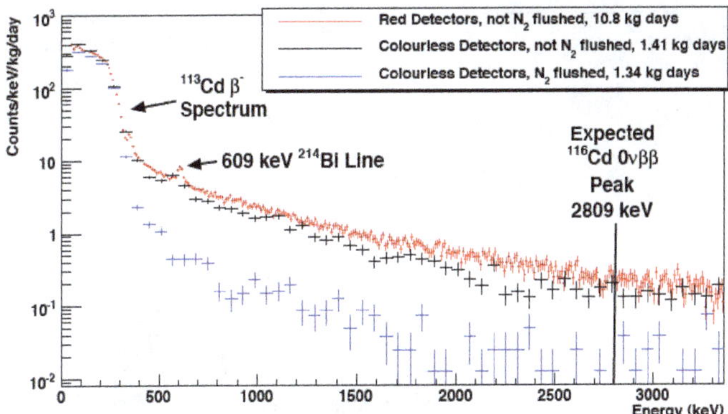

Fig. 1. Results of measurements with CPG detectors at LNGS. Main background sources are the red passivation and radon, which was solved by changing the passivation and a nitrogen flushing.

five double beta emitters, some of them being β^+ emitters. Therefore, the detector material provides its own source, allowing for high masses and a high detection efficiency. In addition, CdZnTe has important advantages. The material is radiopure, can be operated at room temperature and due to the industrial development of CdZnTe detectors, it is a maturing technology.

As ^{116}Cd has a high Q–value (2814 keV), a good matrix element and a large phase space it is the most interesting isotope, next to ^{106}Cd, which also has a high Q–value (2771 keV) and ^{130}Te with a high natural abundance (33.8 %).

2. Detector Technologies

Within the R&D program, two different detector technologies are investigated. These are coplanar grid detectors (CPG) and pixelated detectors.

The CPG technology was delevoped for CdZnTe[2] to counteract the missing hole signal at the cathode of the detector. Due to a different drift velocity the hole signal is mostly lost during data aquisition. This problem can be solved by an anode which is structured into two comb–shaped parts isolated from each other and on a slitly different potential.

With such a readout approach, high masses with only a small number of readout channels are possible. Near to the region of interest energy resolutions better than 2 % FWHM at 2615 keV with cost efficient low resolution detectors were achieved. In Figure 1 background spectra of CPG dectectors

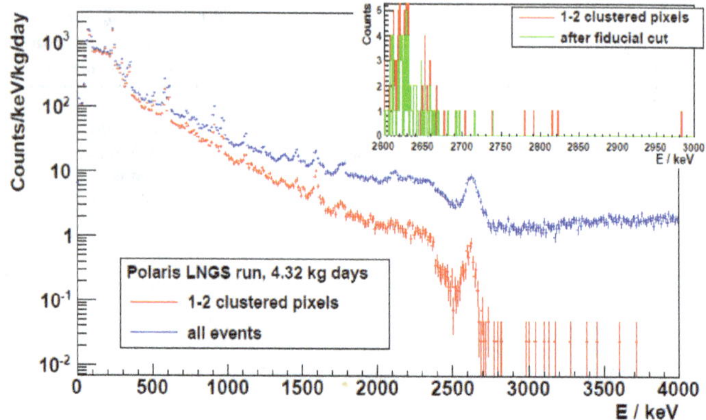

Fig. 2. Results of a measurement with the Polaris detector at LNGS. Zero Events remain in the ROI (2.75 – 2.85 MeV) after cuts to boundary and clustered pixels.

at LNGS are shown. The main background sources are the red passivation of the detectors and radon from the air. Therefore the passivation was changed and the test set–up is flushed with nitrogen. So far a background rate of ~ 5 counts/keV/kg/year was reached and next to six half–life limits above 10^{20} years,[3] seven half–life limits within a factor of the world best limits were measured.

Pixelated CdZnTe detectors add tracking capabilities to the pure energy measurement of CPG detectors. Due to the small–pixel effect,[4] the missing hole signal problem is solved by the pixel technology itself. The three different pixel systems under investigation can be classified in large volume detectors ($2 - 6\,\text{cm}^3$) with a large pixel pitch (~ 1 mm) and thin detectors ($0.3 - 2$ mm) with a small pixel pitch ($\sim 100\,\mu$m).

The Polaris system from the University of Michigan[5] and the WUSTL system from the Washington University of St. Louis refer to the large volume detectors. The Polaris detector has a size of $2 \times 2 \times 1.5\,\text{cm}^3$ and a mass of 36 g. An energy resolution of 0.78 % FWHM at 662 keV has been reached. Due to 11×11 pixels, direct particle identification is not possible. But with cuts to boundary and clustered pixel, the background can be reduced significantly. In Figure 2, low background data taken with the Polaris detector at LNGS is pictured. Zero events in the region of interest ($2.75 - 2.85$ MeV) remain after both cuts.

The WUSTL detector has a size of $2 \times 2 \times 0.5\,\text{cm}^3$ and is available with different pixel systems from 8×8 pixels up to 100×100 pixels. An energy resolution of 1.5 % FWHM at 583 keV has been achieved. As well as with

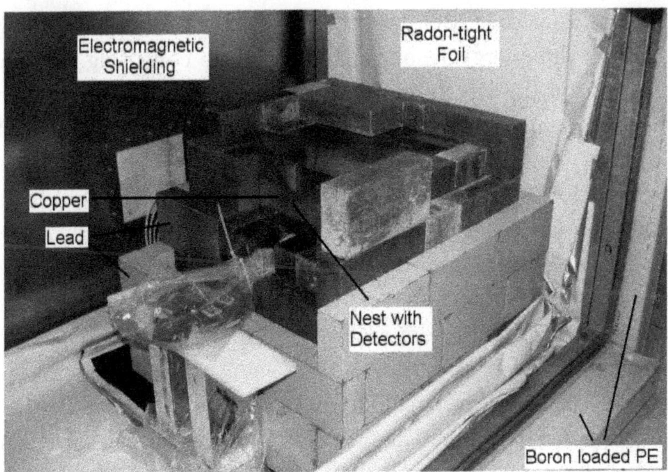

Fig. 3. R&D set–up at LNGS

the Polaris detector particle identification is not an opportunity due to the large pixel pitch.

The third type of pixel detector is the Timepix system from the Medipix3 Collaboration. With a size of $14 \times 14 \times 0.3\,\text{mm}^3$ (Si detector) and $14 \times 14 \times 1\,\text{mm}^3$ (CdTe detector) with 256×256 or 128×128 pixels it belongs to the thin systems with small pixel pitch. The Timepix detector acts like a solid state TPC and therefore offers the possibility of identifying particles via track reconstruction, an unique technique in this field. The background can be reduced essentially, as particles like αs, βs and muons are clearly distinguishable.

3. R&D Activities

The broad R&D program encompasses many different activities, such as a test set–up at LNGS (Figure 3), able to house upto 64 CPG detectors with a size of $1\,\text{cm}^3$. In spring 2011, the test set–up was moved to the former Heidelberg–Moscow hut and the whole passive shielding has been improved. The first layer consists of high purity copper and followed by 2 tons of lead. Ultra low background lead is used for the inner brick layer. A radon–tight foil surrounds the lead, followed by an electromagnetic shielding and 7 cm of boron loaded polyethelyn. The whole set–up is flushed with nitrogen. Since summer 2011 16 detectors are running with a new FADC readout. An upgrade to the 64 array is underway.

The performance of the CPG detectors depends on the quality of the material. Next to the energy resolution, the charge collecting efficiency is a major contributor. To determine these parameters, prior the installation in the test set–up each detector is examined with a highly collimated γ–ray source.
With the new FADC readout pulse shape analysis is possible. Thus the background can be reduced by rejecting unwanted pulses like electronic noise and surface events. Also, the discrimination of single–site and multi–site events is feasible.
Another R&D activity is the operation of unpassivated CdZnTe detectors in liquid scintillator with the aim of decreasing the background, while increasing the detection efficiency. In addition, the scintillator would act as an activ veto. For the determination of the optimal shielding design, Monte–Carlo studies are carried out and the radiopurity of the used materials are measured in a maintained low background facility. Furthermore, crystal growth and the enrichment of ^{116}Cd are being investigated.

4. Conclusion

The COBRA experiment is currently in the R&D phase with a set–up at LNGS. CdZnTe detectors offer innovative and promising possibilities for a low rate experiment. With CPG detectors, seven half–life limits were measured within a factor 3 of the world best limits. Also, six half–life limits above 10^{20} years were achieved. Furthermore, pixel detectors are able to reduce the background significantly by cuts to boundary and clustered pixels, or by direct particle identification. This would be a unique technique in this field. Therefore, COBRA has great large potential as a next generation detector for the search for neutrinoless double beta decay.

References

1. K. Zuber, *Physics Letters B* **519**, 1 (2001).
2. P. N. Luke, *Nuclear Science, IEEE Transactions on* **42**, 207 (1995).
3. J. V. Dawson et al., *Phys. Rev. C* **80**, p. 025502 (2009).
4. H. H. Barrett et al., *Phys. Rev. Lett.* **75**, 156 (1995).
5. F. Zhang et al., *Nuclear Science, IEEE Transactions on* **54**, 843 (2007).

DARK ATOMS OF DARK MATTER AND THEIR STABLE CHARGED CONSTITUENTS

Maxim Yu. Khlopov[1,2,3]

[1]*National Research Nuclear University "Moscow Engineering Physics Institute", 115409 Moscow, Russia*
[2]*Centre for Cosmoparticle Physics "Cosmion" 115409 Moscow, Russia*
[3]*APC laboratory 10, rue Alice Domon et Léonie Duquet 75205 Paris Cedex 13, France*
E-mail: khlopov@apc.univ-paris7.fr

Direct searches for dark matter lead to serious problems for simple models with stable neutral Weakly Interacting Massive Particles (WIMPs) as candidates for dark matter. A possibility is discussed that new stable quarks and charged leptons exist and are hidden from detection, being bound in neutral dark atoms of composite dark matter. Stable -2 charged particles O^{--} are bound with primordial helium in O-helium (OHe) atoms, being specific nuclear interacting form of composite dark matter. The positive results of DAMA experiments can be explained as annual modulation of radiative capture of O-helium by nuclei. In the framework of this approach test of DAMA results in detectors with other chemical content becomes a nontrivial task, while the experimental search of stable charged particles at LHC or in cosmic rays acquires a meaning of direct test for composite dark matter scenario.

1. Introduction

It was shown recently[1,2] that new stable charged particles can exist, if they are hidden in neutral atom-like states. To avoid anomalous isotopes overproduction, stable particles with charge ± 1 (like tera-electrons[3,4]) should be absent, so that stable negatively charged particles should have charge -2 only. This possibility cannot take place in SUSY models but a row of alternative models predict such particles (see Refs. in[2]).

In the asymmetric case, corresponding to excess of -2 charge species, O^{--}, they bind in "dark atoms" with primordial ^4He as soon as it is formed in the Standard Big Bang Nucleosynthesis. Such dark atoms, called O-helium (OHe), are assumed to be the dominant form of the modern dark matter, giving rise to a Warmer than Cold dark matter scenario.[2,5]

Interaction of OHe with nuclei in underground detectors can explain positive results of dark matter searches in DAMA/NaI (see for review[6]) and DAMA/LIBRA[7] experiments by annual modulations of radiative capture of O-helium, resolving the controversy between these results and the results of other experimental groups.

2. Some Features of O-Helium Universe

As soon as primordial helium is formed in the Big bang nucleosynthesis, all free O^{--} are trapped by ^4He in O-helium "atoms" ($^4He^{++}O^{--}$). The radius of Bohr orbit in these "atoms"[1,2] $R_o \sim 1/(Z_O Z_{He} \alpha m_{He}) \approx 2 \cdot 10^{-13}$ cm is nearly equal to the radius of helium nucleus.

Due to nuclear interactions of its helium constituent with nuclei in the cosmic plasma, the O-helium gas is in thermal equilibrium with plasma and radiation on the Radiation Dominance (RD) stage, while the energy and momentum transfer from plasma is effective. The radiation pressure acting on the plasma is then transferred to density fluctuations of the O-helium gas and transforms them in acoustic waves at scales up to the size of the horizon.

At temperature $T < T_{od} \approx 200 S_3^{2/3}$ eV the energy and momentum transfer from baryons to O-helium is not effective[1,2] and O-helium gas decouples from plasma. It starts to dominate in the Universe after $t \sim 10^{12}$ s at $T \leq T_{RM} \approx 1$ eV and O-helium "atoms" play the main dynamical role in the development of gravitational instability, triggering the large scale structure formation. The composite nature of O-helium determines the specifics of the corresponding warmer than cold dark matter scenario.

Being decoupled from baryonic matter, the OHe gas does not follow the formation of baryonic astrophysical objects (stars, planets, molecular clouds...) and forms dark matter halos of galaxies. It can be easily seen that O-helium gas is collisionless for its number density, saturating galactic dark matter. Taking the average density of baryonic matter one can also find that the Galaxy as a whole is transparent for O-helium in spite of its nuclear interaction. Only individual baryonic objects like stars and planets are opaque for it.

3. Radiative Capture of OHe in the Underground Detectors

3.1. *O-helium in the terrestrial matter*

The evident consequence of the O-helium dark matter is its inevitable presence in the terrestrial matter, which appears opaque to O-helium and stores all its in-falling flux.

After they fall down terrestrial surface, the in-falling OHe particles are effectively slowed down due to elastic collisions with matter. Then they drift, sinking down towards the center of the Earth. Near the Earth's surface, the O-helium abundance is determined by the equilibrium between the in-falling and down-drifting fluxes.

At a depth L below the Earth's surface, the drift timescale is $t_{dr} \sim L/V$, where $V \sim 400 S_3 \,\mathrm{cm/s}$ is the drift velocity and $m_o = S_3 \,\mathrm{TeV}$ is the mass of O-helium. It means that the change of the incoming flux, caused by the motion of the Earth along its orbit, should lead at the depth $L \sim 10^5 \,\mathrm{cm}$ to the corresponding change in the equilibrium underground concentration of OHe on the timescale $t_{dr} \approx 2.5 \cdot 10^2 S_3^{-1} \,\mathrm{s}$.

The equilibrium concentration, which is established in the matter of underground detectors at this timescale, is given by

$$n_{oE} = n_{oE}^{(1)} + n_{oE}^{(2)} \cdot sin(\omega(t - t_0)) \tag{1}$$

with $\omega = 2\pi/T$, $T = 1yr$ and t_0 the phase. So, there is a constant concentration and its annual modulation with amplitude $n_{oE}^{(2)}$.

3.2. *Potential of O-helium interaction with nuclei*

The explanation[2] of the results of DAMA/NaI[6] and DAMA/LIBRA[7] experiments is based on the idea that OHe, slowed down in the matter of detector, can form a few keV bound state with nucleus, in which OHe is situated **beyond** the nucleus. Therefore the positive result of these experiments is explained by annual modulation in reaction of radiative capture of OHe

$$A + (^4He^{++}O^{--}) \rightarrow [A(^4He^{++}O^{--})] + \gamma \tag{2}$$

by nuclei in DAMA detector.

The approach of[2] assumes the following picture: OHe is a neutral atom in the ground state, perturbed by Coulomb and nuclear forces of the approaching nucleus. The sign of OHe polarizability changes with the distance: at larger distances Stark-like effect takes place - nuclear Coulomb force polarizes OHe so that nucleus is attracted by the induced dipole moment of OHe, while as soon as the perturbation by nuclear force starts to dominate the nucleus polarizes OHe in the opposite way so that He is situated more close to the nucleus, resulting in the repulsive effect of the helium shell of OHe. When helium is completely merged with the nucleus the interaction is reduced to the oscillatory potential of O^{--} with homogeneously charged merged nucleus with the charge $Z + 2$.

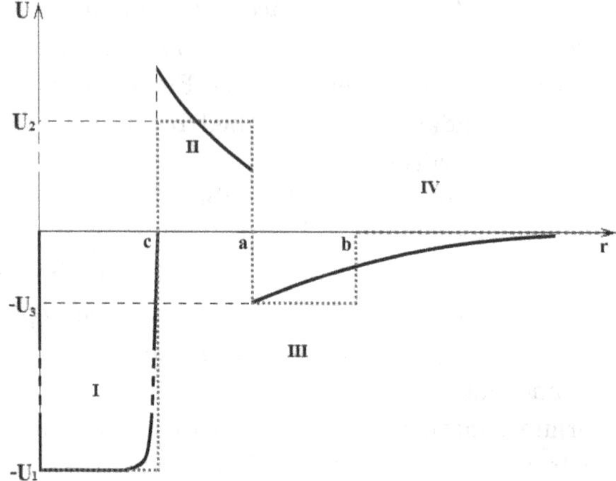

Fig. 1. The approximation of rectangular well for potential of OHe-nucleus system.

To simplify the solution of Schrodinger equation the potential was approximated in[2] by a rectangular potential, presented on Fig. 1.

Solution of Schrodinger equation determines the condition, under which a low-energy OHe-nucleus bound state appears in the region III.

3.3. *Radiative capture of O-helium by sodium*

The rate of radiative capture of OHe by nuclei can be calculated[2] with the use of the analogy with the radiative capture of neutron by proton with the account for: i) absence of M1 transition that follows from conservation of orbital momentum and ii) suppression of E1 transition in the case of OHe. Since OHe is isoscalar, isovector E1 transition can take place in OHe-nucleus system only due to effect of isospin nonconservation, which can be measured by the factor $f = (m_n - m_p)/m_N \approx 1.4 \cdot 10^{-3}$, corresponding to the difference of mass of neutron,m_n, and proton,m_p, relative to the mass of nucleon, m_N. In the result the rate of OHe radiative capture by nucleus with atomic number A and charge Z to the energy level E in the medium with temperature T is given by

$$\sigma v = \frac{f \pi \alpha}{m_p^2} \frac{3}{\sqrt{2}} (\frac{Z}{A})^2 \frac{T}{\sqrt{A m_p E}}. \tag{3}$$

Formation of OHe-nucleus bound system leads to energy release of its binding energy, detected as ionization signal. In the context of our approach the existence of annual modulations of this signal in the range 2-6 keV and absence of such effect at energies above 6 keV means that binding energy of Na-OHe system in DAMA experiment should not exceed 6 keV, being in the range 2-4 keV. The amplitude of annual modulation of ionization signal can reproduce the result of DAMA/NaI and DAMA/LIBRA these experiments for $E_{Na} = 3\,\text{keV}$. The account for energy resolution in DAMA experiments[8] can explain the observed energy distribution of the signal from monochromatic photon (with $E_{Na} = 3\,\text{keV}$) emitted in OHe radiative capture.

At the corresponding nuclear parameters there is no binding of OHe with iodine and thallium.[2]

It should be noted that the results of DAMA experiment exhibit also absence of annual modulations at the energy of MeV-tens MeV. Energy release in this range should take place, if OHe-nucleus system comes to the deep level inside the nucleus. This transition implies tunneling through dipole Coulomb barrier and is suppressed below the experimental limits.

For the chosen range of nuclear parameters, reproducing the results of DAMA/NaI and DAMA/LIBRA, our results[2] indicate that there are no levels in the OHe-nucleus systems for heavy nuclei. In particular, there are no such levels in Xe, what seem to prevent direct comparison with DAMA results in XENON100 experiments. The existence of such level in Ge and the comparison with the results of CDMS and CoGeNT experiments need special study.

4. Conclusions

The results of dark matter search in experiments DAMA/NaI and DAMA/LIBRA can be explained in the framework of our scenario without contradiction with the results of other groups. The proposed explanation is based on the mechanism of low energy binding of OHe with nuclei. Within the uncertainty of nuclear physics parameters there exists a range at which OHe binding energy with sodium is in the interval 2-4 keV. Annual modulation in radiative capture of OHe to this bound state leads to the corresponding energy release observed as an ionization signal in DAMA detector.

With the account for high sensitivity of the numerical results to the values of nuclear parameters and for the approximations, made in the calculations, the presented results can be considered only as an illustration of the possibility to explain puzzles of dark matter search in the framework of composite dark matter scenario. An interesting feature of this explanation is a conclusion that the ionization signal expected in detectors with the content, different from NaI, should be dominantly in the energy range beyond 2-6 keV. Therefore test of results of DAMA/NaI and DAMA/LIBRA experiments by other experimental groups can become a very nontrivial task.

The presented approach sheds new light on the physical nature of dark matter. Specific properties of dark atoms and their constituents are challenging for the experimental search. The development of quantitative description of OHe interaction with matter confronted with the experimental data will provide the complete test of the composite dark matter model. It challenges search for stable double charged particles at accelerators and cosmic rays as direct experimental probe for charged constituents of dark atoms of dark matter.

References

1. M. Yu. Khlopov, *JETP Lett.* **83**, 1 (2006).
2. M. Y. Khlopov, A. G. Mayorov and E. Y. Soldatov, *J. Phys.:* Conf. Ser. **309**, 012013 (2011).
3. S. L. Glashow, arXiv:hep-ph/0504287.
4. D. Fargion and M. Khlopov, arXiv:hep-ph/0507087.
5. M. Y. Khlopov and C. Kouvaris, *Phys. Rev.* D **78**, 065040 (2008)
6. R. Bernabei *et al.*, *Rivista Nuovo Cimento* **26**, 1 (2003)
7. R. Bernabei *et al.* [DAMA Collaboration], *Eur.Phys.J* **C56**, 333 (2008) arXiv:0804.2741 [astro-ph].
8. R. Bernabei *et al.* [DAMA Collaboration], Nucl. Instrum. Meth. A **592** (2008) 297 [arXiv:0804.2738 [astro-ph]].

The EDELWEISS dark matter search experiment: towards enhanced sensitivity

V. Yu. Kozlov* for the EDELWEISS collaboration

Karlsruhe Institute of Technology, Institut für Kernphysik, Postfach 3640, 76021 Karlsruhe, Germany
**E-mail: Valentin.Kozlov@kit.edu*

EDELWEISS-2 is a Ge-bolometer experiment searching for WIMP dark matter and located in the underground laboratory, Laboratoire Souterrain de Modane (LSM, France). The collaboration uses new cryogenic detectors with an improved background rejection (interleaved electrode design and continues further developments. An operation of ten 400-g bolometers at LSM together with an active muon veto shielding has been achieved. Results based on a total effective exposure of 384 kgd obtained in 2009/2010 have been published recently. A cross-section for spin-independent scattering of WIMPs on the nucleon of $4.4 \cdot 10^{-8}$ pb is excluded at 90%CL for a WIMP mass of 85 GeV. This bolometer data and the latest measurements with 800-g detectors are presented. Further plans of the collaboration to reach sensitivity of $5 \cdot 10^{-9}$ pb and for a next generation experiment, EURECA, are discussed.

Keywords: dark matter; bolometer detectors; background; underground physics.

1. The Edelweiss dark matter search experiment

EDELWEISS-2 is a direct dark matter search experiment looking for WIMP[a] candidates. A very low interaction rate of WIMPs with a nuclei of a terrestrial detector is expected, i.e. for the commonly accepted scenario less than 0.01 event per day and per kg of a target material, thus one has to overcome various background issues. This is why such experiments go deep underground and invest a lot of effort into detector developments and background studies. EDELWEISS-2 profits from 4850 m.w.e. of shielding provided by LSM depth. This reduces a neutron flux by four orders of magnitude and the muon flux by six orders of magnitude, down to $\sim$5 muons/m^2/day. Bolometers of high-purity Ge are used in the experiment both as the detec-

[a]Weakly Interacting Massive Particle

tors and the target material. These bolometers are operated at a temperature of 18 mK, and equipped with NTD[b] sensors and aluminum electrodes to simultaneously measure phonon and ionization signals. The ratio of the two signals, so-called Q-value, is different for nuclear and electron recoils with nuclear recoils having Q$\sim$0.3 when normalized to Q=1 for electron recoils. This separation in Q allows a powerful γ/β-background rejection on a per event basis. An additional background, however, arises from so-called surface events, which are being electron recoils have Q-values in the region-of-interest due to an incomplete charge collection near the detector surface. EDELWEISS collaboration developed bolometers with *interdigitized electrode design*[1] (ID) to successfully separate between bulk and surface events. This is achieved by alternately biasing coplanar concentric ring electrodes located on both detector sides. By tuning the electric potentials the electrical field near the surface is modified such that a created charge is accumulated either on electrodes with the highest potential, called *fiducial* electrodes, or in addition on low potential electrodes, called correspondently *veto* electrodes. These detectors show very efficient rejection of near surface events at about $6{\cdot}10^{-5}$ contribution from surface β interactions.[1] A general overview of the set-up is shown in Fig. 1: Dilution refrigerator occupies the central part of the set-up and can host up to 40 kg of detectors. A 20 cm lead shield reduces the external γ-background. Neutrons constitute a prominent background component as they, like WIMPs, produce nuclear recoils and thus may mimic a WIMP signal. Several processes lead to neutron production: muon interaction in vicinity of the set-up, ^{238}U fission in lead shield and the rock, (α,n) reactions. A 50 cm layer of polyethylene is used to moderate neutrons while a muon veto system consisting of 100 m^2 of plastic scintillators serves to tag muons. Complementary studies of neutron background are performed with a Rn-monitor, ^{3}He proportional counters and a counter for muon-induced neutrons based on 1 m^3 of Gd-loaded liquid scintillator.[2] With an upcoming upgrade of the set-up the scientific goal of EDELWEISS experiment is to reach a sensitivity of $5{\cdot}10^{-9}$ pb for the WIMP-nucleon spin-independent (SI) cross-section in 2013.

2. Final results using 4-kg array of ID detectors

Ten 400 g ID-detectors were operated since April 2009 till May 2010 constituting 418 days in total, out of which 325 days have been considered for the WIMP search, 10.1 and 6.4 days were dedicated to gamma and neutron

[b]neutron-transmutation-doped germanium

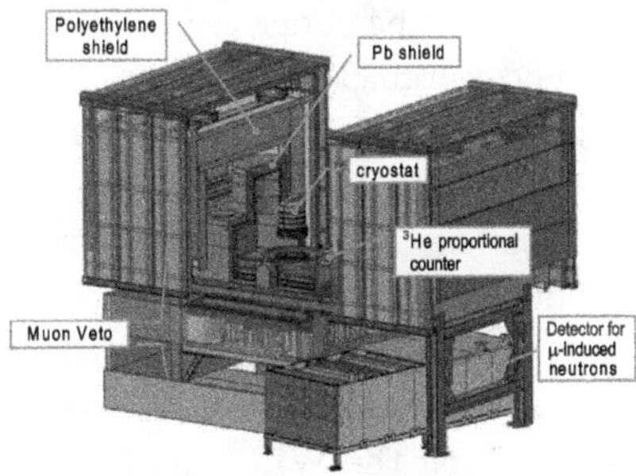

Fig. 1. General layout of the EDELWEISS experimental set-up with additional ^{3}He proportional counter for thermal neutrons and liquid scintillator detector to measure muon-induced neutrons.

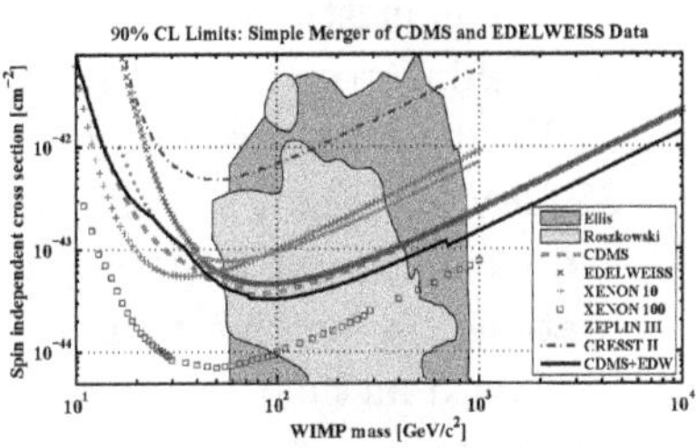

Fig. 2. Final results of the EDELWEISS-2 experiment: the upper limits on the WIMP-nucleon SI cross-section as a function of WIMP mass (the EDELWEISS-2 only data are marked with crosses[3] while the solid black line represents combined CDMS-EDELWEISS analysis[5]). The grey shaded area correspond to theoretical SUSY predictions.

calibrations, respectively. During a physics run, an every day regeneration of the detectors was performed in order to avoid the formation of space charges. The WIMP search data were analyzed offline with 2 independent analysis chains. Here were give a summury while the full analysis procedure and the results are described in detail in Ref. 3. The average energy resolutions for the detectors were ~1.2 keV (FWHM) for the phonon channel and ~0.9 keV (FWHM) for the ionization channel. The energy threshold for the WIMP search was defined a priori as 20 keV, above which the efficiency is independent of energy. The periods of data taking retained for analysis were selected hour-by-hour solely on the basis of the measured baseline resolution of the heat and ionization signals. A reduced χ^2 cut was applied on the pulse fit for heat and fiducial signals of each event. An average fiducial mass of 160±5 g was measured using the low-energy peaks (~10 keV) of ^{65}Zn and ^{68}Ge isotopes which result from the cosmogenic activation of germanium. Coincident events between two bolometers or with a trigger in the muon veto within an appropriate time window are rejected. After all cuts, the effective exposure is 384 kg·d.[3] Five events were found in the nuclear recoil band, four between 20 and 23 keV and one at 172 keV. Background studies to assess the known residual gamma, beta and neutron backgrounds, using calibration data, radioactivity measurements of materials and Monte Carlo studies of the detectors have been carried out, giving 3 events expected from known backgrounds. The five events exhibit good data quality, with signals well above the noise levels of the detectors. The SI upper limit for WIMP-nucleon elastic scattering cross sections has been derived using the standard Yellin prescription,[4] halo model and parameters (see Ref. 3 for details). EDELWEISS-2 reaches an optimum sensitivity of $4.4 \cdot 10^{-8}$ pb for $M_\chi = 85$ GeV/c^2 (Fig. 2).

The use of the same target material allowed the CDMS and EDELWEISS collaborations to combine their results of direct dark matter searches. A straightforward method of combination was chosen for its simplicity before data were exchanged between the experiments. The total data set represents 614 kg·d equivalent exposure.[5] The upper limit on the WIMP-nucleon SI cross-section is derived: a cross section of $3.3 \cdot 10^{-8}$ pb is excluded at 90% C.L for a WIMP mass of 90 GeV/c^2 where this analysis is most sensitive (Fig. 2). At higher WIMP masses the combination improves the individual limits, by a factor 1.6 above 700 GeV/c^2. Further details are found in Ref. 5.

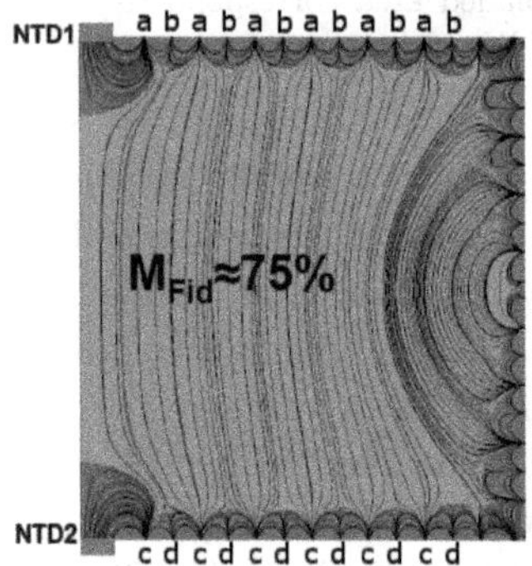

Fig. 3. Calculations performed for the new FID800 detector show the improved fiducial volume of 75%, which is defined by the electrical field lines.

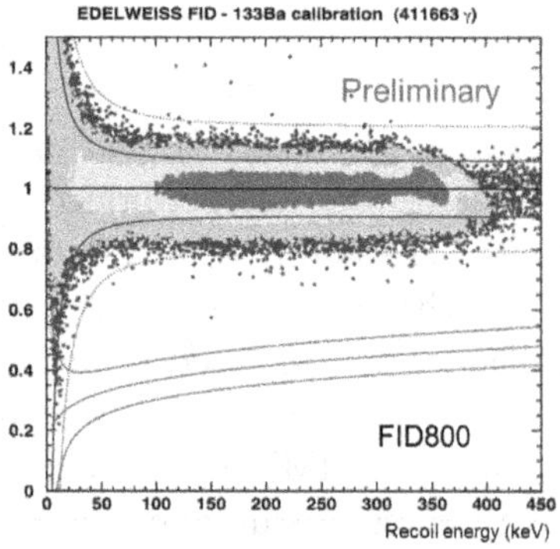

Fig. 4. Ionization yield versus energy after fiducial cuts, obtained with new FID800 detectors using the ^{133}Ba γ-source.

3. Outlook: towards $\sigma_{SI} = 5 \cdot 10^{-9}$ pb sensitivity

To go beyond the present performance requires a number of improvements which lead to the EDELWEISS-3 project. A new generation of detectors with interleaved electrodes covering also the lateral surfaces of the crystal has been developed: the *fully interdigitized* (FID) bolometers (Fig. 3). With twice the mass (800 g) and better volume to surface ratio, FID800 detectors exhibit a fiducial mass of ~600 g, much increased from the ~160 g of ID400 detectors. Further benefit comes from using two, instead of only one, NTD sensors for phonon measurements and new surface treatments to improve further on the rejection efficiency of near surface events.[6] A set of four FID800 detectors have been successfully tested at LSM and, for example, 411663 gammas were detected as fiducial events during a ^{133}Ba calibration while all near surface events were rejected resulting in an empty nuclear recoil band (Fig. 4). Forty of FID800 detectors will be installed in the upgraded EDELWEISS setup. The upgrade implies improved cryogenics, new cabling, installation of additional polyethelyne shield between the lead layer and the cryostat, supplementary muon veto modules, use of the new integrated DAQ and electronics, e.g. implemention of fast ionization channel with 40 MS/s. The goal of the funded EDELWEISS-3 project is with an exposure aim of 3000 kg·d to reach a WIMP-nucleon scattering cross-section sensitivity of $5 \cdot 10^{-9}$ pb. The further development continues towards improvement of read-outs and detectors and will also profit from more cryogenic test stands becoming available in laboratories. The detector R&D on longer term aims to reach a few 100 eV thresholds on both ionization and heat channels while on shorter term use a heat-only detector with the threshold of 2 keV to probe low-mass WIMPs. The ongoing research together with the detailed studies of the background conditions in LSM, in particular, muon-induced neutrons put a good base for a dark matter experiment of next generation, EURECA,[7] a 1-ton cryogenic detector array.

Acknowledgments: The help of the technical staff of the Laboratoire Souterrain de Modane is gratefully acknowledged. This work is supported in part by the Helmholtz Alliance for Astroparticle Physics, by the French Agence Nationale pour la Recherche and the Russian Foundation for Basic Research (grant No. 07-02-00355-a).

References

1. A. Broniatowski et al., *Phys. Lett. B* **681**, 305 (2009) and references there in; *arXiv:0905.0753.*
2. V. Yu. Kozlov et al. *Astropart. Phys.* **34**, 97 (2010); *arXiv:1006.3098.*
3. E. Armengaud et al., *Phys Lett B* **702**, 329 (2011); *arXiv:1103.4070.*
4. S. Yellin *Phys. Rev. D* **66**, 032005 (2002).
5. Z. Ahmed et al., *Phys. Rev. D* **84**, 011102 (2011); *arXiv:1105.3377.*
6. S. Marnieros et al., *AIP Conference Proceedings* **1185**, 635 (American Institute of Physics, Melville, NY, 2009).
7. H. Kraus et al., *Nucl. Phys. B, (Proc. Suppl.)* **173**, 168 (2007).

NEUTRINO PHYSICS WITHOUT NEUTRINOS: A REVIEW OF SELECTED SEARCHES FOR NEUTRINOLESS DOUBLE BETA DECAYS

K. LANG*

Department of Physics, University of Texas at Austin,
Austin, TX 78712, USA
**E-mail: lang@physics.utexas.edu*
http://www.hep.utexas.edu/cpf/lang/index.html

A discovery of neutrino oscillations, proving that neutrinos are massive, has strengthen the motivation for experiments seeking an observation of the neutrinoless double beta decay which provides the only practical way to determine whether neutrinos are Majorana or Dirac particles. A number of ingenious techniques to detect such a transition and suppress an omnipresent natural radioactivity are being pursued world-wide. We briefly review the recent progress, the current status, and near-term prospects of selected experiments.*

Keywords: neutrinos, double beta decay, neutrino mass, Majorana neutrino.

1. Introduction

If neutrinos, now known to have mass, are Majorana particles[1] (i.e., for which a particle and its antiparticle represent the same field) then they are most likely to be detected in processes represented by Feynman diagrams, shown in Fig. 1, in which a neutrino connects two vertices such that the transition changes the lepton number by two units, $\Delta L = 2$. A double beta decay ($\beta\beta$) provides the best practical source for such a reaction. The final state of this process results in two electrons ($\beta^-\beta^-$) or positrons ($\beta^+\beta^+$) whose energy sum is equal to the energy of nuclear transition between (A, Z) and $(A, Z \pm 2)$. This seemingly simple feature provides the main observable for detecting $\beta\beta$ and poses experimental challenges in searches for such transitions.

*Some recent reviews of this field are: W. Rodejohann, arXiv:1106.1334; A.S. Barabash, *Phys. Rev. C* **81**,035501(2010); J.J. Gomez-Cadenas *et al.*, arXiv:1010.5112; F.T. Avignone, S.R Elliott, and J. Engel, *Rev. Mod. Phys.* **80**, 481(2008).

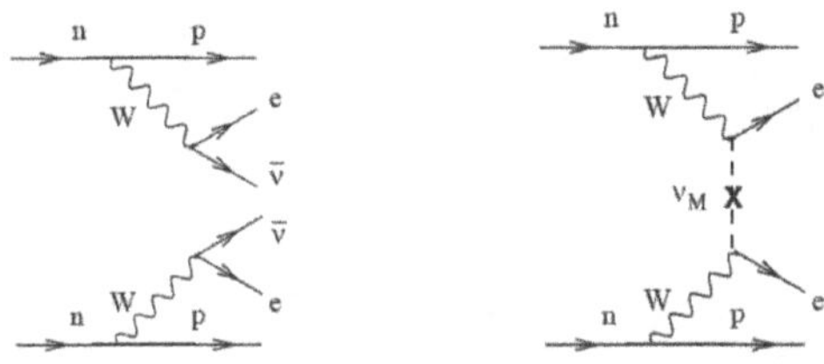

Fig. 1. Feynman diagrams for two-neutrino (left) and neutrinoless double beta decay (right). Figures borrowed from reference [3].

Half-lives for $2\nu\beta\beta$ and $0\nu\beta\beta$, expressed by Eq. 1 and 2, depend on the phase factor, nuclear matrix element, and for $0\nu\beta\beta$ also on the effective neutrino mass:

$$[T^{2\nu}_{1/2}(A,Z)]^{-1} = |M^{2\nu}(A,Z)|^2 G^{2\nu}(Q_{\beta\beta},Z) \tag{1}$$

$$[T^{0\nu}_{1/2}(A,Z)]^{-1} = |M^{0\nu}(A,Z)|^2 G^{0\nu}(Q_{\beta\beta},Z)\langle m_{\beta\beta}\rangle^2, \tag{2}$$

where $M^{2(0)\nu}(A,Z)$ is the nuclear matrix element and $G^{2(0)\nu}(Q_{\beta\beta},Z)$ is a known phase space factor that depends on the transition energy, $Q_{\beta\beta}$, of the process with 2 (0) neutrinos. The "effective" Majorana mass of the electron neutrino, $\langle m_{\beta\beta}\rangle$, is a sum of the mass eigenstates weighted by the squared elements of the PMNS neutrino mixing matrix:[2] $\langle m_{\beta\beta}\rangle = \sum_{i=1,2,3} U^2_{ei} m_i$. Sensitivity of $0\nu\beta\beta$ searches may be gauged by plotting $\langle m_{\beta\beta}\rangle$ versus the lightest mass, as shown on the left of Fig. 2

There are two main experimental goals: observe the neutrinoless double beta decay, which would constitute a major discovery, and then infer the value of $\langle m_{\beta\beta}\rangle$. Unfortunately, the theoretical knowledge of the nuclear matrix elements is limited, as shown in the right plot of Fig. 2, so much progress is needed to reliably connect $T^{0\nu}_{1/2}$ with $\langle m_{\beta\beta}\rangle$. Currently, extraction of the value of $\langle m_{\beta\beta}\rangle$ from half-lives has a range of values reflecting the range of theoretical calculations and uncertainties. Despite of this, $0\nu\beta\beta$ processes may provide the best option to measure the absolute energy scale of the neutrino mass. Neutrino oscillations give differences of neutrino mass squared, while single beta decay searches that use the end-point energy shape (e.g., $^3H \rightarrow {}^3He$, or $^{187}Re \rightarrow {}^{187}Os$)[3,4] may turn out insufficient in sensitivity.

2. Principal challenges in $0\nu\beta\beta$ searches

In addition to mentioned theoretical difficulties, there are three principal intertwined experimental challenges to reach sensitivity in the desired $\langle m_{\beta\beta}\rangle$

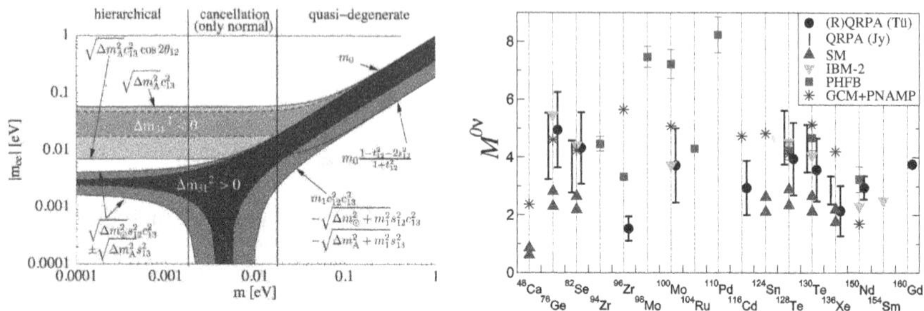

Fig. 2. A summary of theoretical options for $\langle m_{\beta\beta}\rangle$ (from W. Rodejohann, arXiv:1106.1334)[left] and predictions of matrix elements for most common $0\nu\beta\beta$ isotopes (from V. Rodin, presented at TAUP 2011) [right].

range of mili-electron-volts (corresponding to a half-life of about 10^{26} y or higher, depending on an isotope): 1) suppression of backgrounds due to natural radioactivity, 2) achieving detector performance with sufficient energy resolution to identify a final state of $0\nu\beta\beta$, and 3) obtaining sufficiently large exposure (mass×time) requiring high isotopic enrichment. Table 1 displays isotopes with the highest values of $Q_{\beta\beta}$ which are the most practical, thus common, isotopes used in searches for $0\nu\beta\beta$. Eleven isotopes shown in this table are ordered by the $Q_{\beta\beta}$ value.

Table 1. A list of 11 neutrinoless double beta decays with $Q_{\beta\beta}>2$ MeV, ordered from the highest energy. Shown are values of the $0\nu\beta\beta$ energy, the natural abundance of isotope, and names of experiments exploiting the reaction. SuperNEMO's baseline is ^{82}Se but ^{48}Ca and ^{150}Nd are also being considered.

Double beta transition	$Q_{\beta\beta}$ (MeV)	Natural abund. (%)	Experiments
^{48}Ca → ^{48}Ti	4.271	0.187	CANDLES, NEMO-3, (SuperNEMO)
^{150}Nd → ^{150}Sm	3.367	5.6	SNO+, NEMO-3, (SuperNEMO)
^{96}Zr → ^{96}Mo	3.350	2.8	NEMO-3
^{100}Mo → ^{100}Ru	3.034	9.6	NEMO-3
^{82}Se → ^{82}Kr	2.995	9.2	LUCIFER, NEMO-3, SuperNEMO
^{116}Cd → ^{116}Sn	2.802	7.5	COBRA, LUCIFER, NEMO-3
^{130}Te → ^{130}Xe	2.533	34.5	CUORICINO, CUORE, NEMO-3
^{136}Xe → ^{136}Ba	2.479	8.9	EXO, KamLAND-Zen, NEXT
^{124}Sn → ^{124}Te	2.228	5.64	
^{76}Ge → ^{76}Se	2.040	7.8	GERDA, MAJORANA
^{110}Pa → ^{110}Cd	2.013	11.8	

Ranking by difficulty, the most pivotal issue that must be solved by

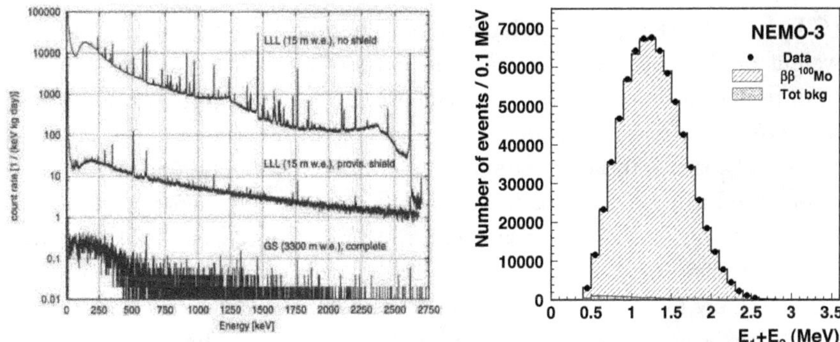

Fig. 3. Left: An example of natural radioactivity spectrum measured with HPGe detectors for three levels two levels of overburden with and without an additional local shielding. Right: Preliminary[5] NEMO-3 $2\nu\beta\beta$ two-electron spectrum of ^{100}Mo $\rightarrow$ ^{100}Ru decays.

any experiment is suppression of backgrounds due to natural radioactivity. Typically, it must be suppressed by many orders of magnitude and there are no obvious choices how to accomplish such a daunting goal. The left plot of Fig. 3 shows three spectra due to natural radioactivity with small and an increasing overburden shielding.

The highest-energy gamma line at 2.614 MeV due to ^{208}Tl is an important mark delineating experimental approaches to search for $0\nu\beta\beta$. Experiments with isotopes for which $0\nu\beta\beta$ transition energy is smaller than the ^{208}Tl line require an exquisite energy resolution to establish a new peak for $0\nu\beta\beta$ in the region with, perhaps well-suppressed, but almost always unavoidable remnants of the natural radioactivity spectrum. Depending on specific materials used for constructing a detector, backgrounds vary and eliminating or suppressing them requires specialized solution. Each experiment must also deal with a tail of an irreducible $2\nu\beta\beta$ energy spectrum, as illustrated by NEMO-3 results, shown on the right of Fig. 3. This preliminary, almost background-free, spectrum has about 700,00 of ^{100}Mo $\rightarrow^{100}$ Ru decays, the largest statistics ever collected in $0\nu\beta\beta$.

The identification of the final state is expected through an observation of a narrow energy peak in the $0\nu\beta\beta$ spectrum. The challenge is to collect a sufficient number of events so that a $0\nu\beta\beta$ transition can be actually discovered. Additionally, the significance of measurement is given by Eq. 3.[6]

$$T^{0\nu}_{1/2}(n_\sigma) = \frac{4.16 \times 10^{26} y}{n_\sigma} \left(\frac{\epsilon \times a}{W}\right) \sqrt{\frac{M \times t}{b \times \Delta E}} \qquad (3)$$

where n_σ is a number of standard deviations for a given confidence level

(C.L), a is an isotopic abundance, ϵ is detection efficiency, W is molecular weight of the isotope, M is the total mass of the source (in kg), t is time of data collection (in years), b is background rate (in counts/$keV \cdot kg \cdot y$), and ΔE is the energy resolution (in keV).

Finally, the isotopic abundance, the cost of enrichment, and radio-purity of enriched isotope must be factored into planning experiments with a significant mass. Only ^{130}Te has a large natural abundance of 34.5%, while the most desirable ^{48}Ca (because of its high $Q_{\beta\beta}$ value of 4.271 MeV) has an extremely low abundance of 0.187% and is extremely difficult to enrich.

There is no obvious optimization how to achieve high sensitivity to $0\nu\beta\beta$. Experiments "bet" on different technological aspects of detectors, usually using their own past experience. Generally, older techniques are enhanced to produce more observables, and essentially all experiments proceed cautiously with a staged approach such that mid-course corrections or improvements can be incorporated in the future.

3. Experimental landscape

There are several experiments in operation, under construction, or are being developed which should be capable of measuring double beta decay lifetimes with unprecedented sensitivities.[6,7] Some employ quite ingenious techniques and the R&D program of new ideas is ongoing. Experiments with isotopes for which $Q_{\beta\beta} < 2.614$ MeV require an exquisite energy resolutions, while those above can relax this requirement but must muster other performance parameters which would allow to eliminate backgrounds and identify the two-electron final state. Below, we will briefly summarize main features of selected current efforts in this field.

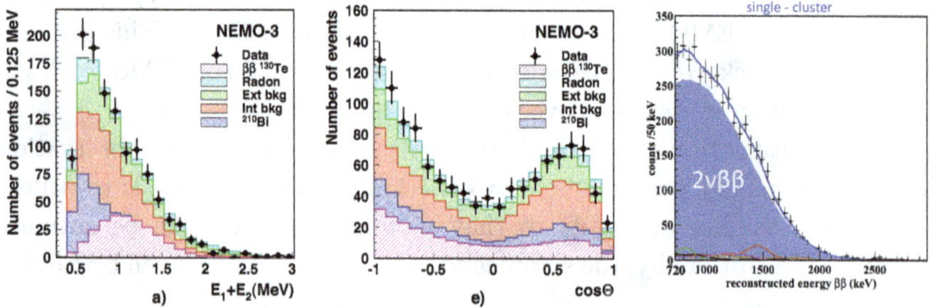

Fig. 4. Left two plots show measured by NEMO-3[8] two-electron energy spectrum and the cosine distribution of an angle between two electrons for candidate event of ^{130}Te $\rightarrow$ ^{130}Xe transition. The right plot shows the first observation of the ^{136}Xe $\rightarrow$ ^{136}Ba.[9]

3.1. *News of the year 2011*

There are two results reported this year (2011), reproduced in Fig. 4, which deserve a special note. The NEMO-3 collaboration reported[8] a first real time measurement of the $2\nu\beta\beta$ decay of ^{130}Te $\rightarrow$ ^{130}Xe with the half-life $T^{2\nu}_{1/2} = [7.0 \pm 0.9(\text{stat}) \pm 1.1(\text{syst})] \times 10^{20}$ y. This result firmly establishes this transition inferred earlier through geochemical analysis. The EXO-200 experiment reported[9] the first observation of $2\nu\beta\beta$ decay ^{136}Xe $\rightarrow$ ^{136}Ba with $T^{2\nu}_{1/2} = [2.11 \pm 0.04(\text{stat}) \pm 0.21(\text{syst})] \times 10^{21}$ y. This value is surprisingly lower than previously reported lower limits for this process.

3.2. *Overview of selected $2\nu\beta\beta$ experiments*

Current generation experiments pursue a staged approach by initially employing a few kilograms of isotopes to reach sensitivity in the range of $200 - 500$ meV energy scale for $\langle m_{\beta\beta} \rangle$ to demonstrate the validity of their technique. Most of these efforts are underway. At the same time most experiments plan much larger scale future incarnations of their detectors aimed at reaching $20 - 50$ meV $\langle m_{\beta\beta} \rangle$. We briefly review selected efforts.

3.2.1. *NEMO-3 and SuperNEMO*

The NEMO-3 experiment[10] ran in the Modane Underground Laboratory in the Fréjus Tunnel since 2003 until 2011. The detector used wire drift chamber tracker and large plastic scintillating blocks to measure the topology, energy, and timing of electrons emerging from thin foils in the center of the tracking volume. NEMO-3 employed seven different isotopes to construct foils, with most notable mass of ^{100}Mo of 6.9 kg and ^{82}Se of 0.93 kg. Data from the entire running period are currently being analyzed but NEMO-3 has published preliminary and already best-to-date results on life-times of all isotopes in their use. For ^{100}Mo and ^{82}Se they obtained:[5] $T^{2\nu}_{1/2} = [7.11 \pm 0.02(\text{stat}) \pm 0.54(\text{syst})] \times 10^{18}$ y for ^{100}Mo, and $T^{2\nu}_{1/2} = [9.6 \pm 0.1(\text{stat}) \pm 1.0(\text{syst})] \times 10^{19}$ y for ^{92}Se. They have also reported[5] preliminary lower limits on $0\nu\beta\beta$ half-lives: $T^{0\nu}_{1/2} > 1.0 \times 10^{24}$ y for ^{100}Mo and $T^{0\nu}_{1/2} > 3.2 \times 10^{23}$ y for ^{82}Se. Higher precision half-lives and more stringent $2\nu\beta\beta$ limits are expected from the full NEMO-3 data set. The lower limit from ^{100}Mo translates to $\langle m_{\beta\beta} \rangle < 310 - 960$ meV, where the spread reflects uncertainties in the nuclear matrix element.

The next generation experiment, SuperNEMO, will employ the technique pioneered by NEMO-3 but will ultimately house about 100 kg of an isotopic source. The baseline isotope is ^{82}Se but this technique has flexibility to use any other source. The collaboration iconsiders ^{48}Ca and ^{150}Nd if sufficient amounts of these isotopes can be enriched, currently viewed as an extremely challenging task of its own. SuperNEMO's goal is to reach half-life sensitivity of about 10^{26} y and thus $20 - 50$ meV for $\langle m_{\beta\beta} \rangle$.

3.2.2. *EXO-200 and EXO*

The EXO-200 experiment is a first phase of the EXO experiment. It is a time projection chamber (TPC) using liquid Xe (LXe) both as the source of nuclear decays and the detection medium.[11] The TPC has the geometry of a cylinder with 40 cm diameter and 44 cm length with the LXe fiducial mass of 63 kg. Each end-cap has 250 large-area avalanche photodiodes that allow for simultaneous readout of ionization and scintillation in the LXe. As mentioned, EXO-200 has reached unprecedented sensitivities based on the exposure of 5.4 kg·y. The "single site" signal spectrum shown in Fig. 4 is clean and allows to project that the experiment will be able to set stringent bounds on $0\nu\beta\beta$ as the analysis and date collection continues.

In the future, the EXO experiment plans to incorporate larger mass and tagging of the final state ^{136}Ba^{++} to eliminate backgrounds. The proposed scheme is challenging and has never been achieved so the collaboration pursues an active R&D which also involves high-pressure Xe instead of LXe as an option for the future detector.

3.2.3. *CUORICINO and CUORE*

CUORICINO operated in years 2003-2008 as an initial phase of the CUORE experiment. Both detectors use TeO_2 crystals operated at low temperature of 8-10 mK as bolometers of energy resolution of 5-7 keV FWHM at 2.6 MeV. In the total exposure of 19.75 kg·y they have set a limit[12] for $^{130}\mathrm{Te} \to {}^{130}\mathrm{Xe}$ of $T^{0\nu}_{1/2} > 2.8 \times 10^{24}$ y thus providing presently the lowest upper bound on $\langle m_{\beta\beta} \rangle < 300 - 700$ meV, where the spread reflects uncertainties in the nuclear matrix element.

CUORE,[13] currently under development, will employ the same technology with about 200 kg of ^{130}Te to reach sensitivity of 1.6×10^{26} y in a five-year run. The first batch of 52 crystals has been installed in the CUORICINO cryostat and will start taking data shortly. More crystals are being produced (a total of 988 are planned) and work on improved shielding for CUORE is ongoing.

3.2.4. *GERDA and MAJORANA*

GERDA[14] and MAJORANA[15] experiments employ high purity germanium (HPGe) diodes which are characterized by good energy resolution of about 2% at 2 MeV. The two collaboration cooperate on their R&D but pursue very different shielding ideas. GERDA for their Phase I has constructed a large tank of water which surrounds a cryogenic system with liquid argon as the immediate shielding medium. As it turns out, ^{42}Ar and ^{40}K are a source of background, as recently determined in preliminary measurements, so various mitigating strategies are pursued, including an additional copper mini-vessel to additionally surround germanium. In their Phase I, they will use about 18 kg of ^{76}Ge while for Phase II the plan calls for 40 kg, and 1 ton for Phase III. The projected sensitivity is 1.5×10^{26} for the first two phases.

MAJORANA has similar goals but their shielding is a more traditional compact high-purity copper and lead. To minimize background due to high energy muon interactions they have commissioned an underground plant for copper electro-forming and have also shown that broad-energy point-contact HPGe diodes provide the best energy resolution and lowest noise. They will have two independent cryostats and will start data taking with 20-30 kg of ^{76}Ge in 2013. For a 1 ton scale experiments the two collaborations plan to merge their effort into a single project.

3.2.5. *KamLAND-Zen and SNO+*

The KamLAND-Zen experiment[16] uses the KamLAND detector with liquid scintillator (dodecane with pseudocumene and PPO) for reactor neutrinos, now in operation for several years, with an addition of a radio-pure nylon mini-balloon filled with scintillator and diluted ^{136}Xe. The balloon is deployed in the middle of the detector and holds about 400 kg of ^{136}Xe diluted in a decane-based scintillator. Due to relatively poor energy resolution, the $0\nu\beta\beta$ signal will have to be identified as a shape change of the energy spectrum near 2.479 MeV which is likely dominated by backgrounds due to ^{214}Bi, ^{40}K, and ^{208}Tl. The experiment is being commissioned and projects to reach about 80 meV for $\langle m_{\beta\beta} \rangle$ sensitivity after two years of running. If successful, they would increase the size of the balloon and ^{136}Xe content to about 1 ton.

SNO+[17] also uses liquid scintillator (linear-alkybenzene or LAB) but plans to dilute $^{nat}NdCl_3$ salt inside the entire detector. The initial phase of the experiment will use 1000 kg of salt which gives about 43 kg of ^{150}Nd. Just as KamLAND-Zen, they will look for a shape change of the energy spectrum but at a lot higher value of 3.367 MeV where the expected background is mostly due to ^{208}Tl. The experiment is under construction, and the existing acrylic vessel is being carefully cleaned of radon daughters to minimize backgrounds. Data taking is to start in 2013 and after three years of data taking they expect to reach $\langle m_{\beta\beta} \rangle$ sensitivity of 175 meV. The next phase of the experiment would use ^{150}NdCl$_3$ salt, but this would require production of a large mass of ^{150}Nd which remains speculative at this time as only the AVLIS[18] technique could be used for large scale enrichment process.

3.2.6. *NEXT, LUCIFER, COBRA*

Among many other ideas for future searches of $0\nu\beta\beta$ decays we introduce three experiments. NEXT will be a high pressure (15 bar) ^{136}Xe TPC where the readout uses scintillation light and very high yield light due to electroluminescence achieved through a high electric field near one of the TPC end-caps. This additional observable provides a powerful tool in eliminating background and improves the energy resolution to about 1%. The design of a 100 kg detector is completed[19] so the detector construction in the Canfranc lab is projected to start soon with data taking commencing in 2014.

Following the idea to augment the energy resolution by additional observables which would help to suppress backgrounds and identify the two-electron final state of $0\nu\beta\beta$, the LUCIFER collaboration is developing scintillating bolometers using $CdWO_4$, $ZnMoO_4$, and ZnSe crystals.[20] In their R&D program they have shown that betas and alphas can be identified using both the heat signal (bolometry) and detecting the scintillation light. They plan to focus on Zn^{82}Se crystals and with about 18 kg of ^{82}Se they project reaching 2.3×10^{26} y sensitivity. The start of the detector construction is proposed for 2014.

The COBRA experiment[21] is developing a high light-yield CZT crystal (ZnCdTe) to study ^{116}Cd, ^{106}Cd and ^{130}Te. This versatile physics program may be significantly helped by readout segmentation, using a Timepix chip, allowing tracking of electrons, alphas, and muons. Design of a full scale detector is expected in 2014.

4. Summary and outlook

As is perhaps clear from this brief overview, there are many experiments pursuing a search for $0\nu\beta\beta$ using different isotopes and ingenious techniques, constantly revised and improved. No experiment has a sure path to success since they all probe unprecedented sensitivities in signal detection and background suppression. Table 2 gives an overview of selected experiments discussed above. This is a dynamic field with a number of other efforts on the way so the future looks very promising and exciting.

Table 2. Main features of the upcoming $0\nu\beta\beta$ experiments and their projected sensitivities which usually include earlier "demonstrating" phases.

Experiment	Main isotope	Mass (kg)	Principle of operation	Start year	Projected sensitivity
SuperNEMO	^{82}Se	100	Tracking and calorimetry	2014	1×10^{26}
EXO-200	^{136}Xe	63	Liquid Xe TPC	2011	6.4×10^{25}
CUORE	^{130}Te	200	Bolometric crystals	2012	1.6×10^{26}
GERDA	^{76}Ge	40	HPGe semiconductor	2012	1.5×10^{26}
MAJORANA	^{76}Ge	30	HPGe semiconductor	2013	5×10^{25}
KamLAND	^{136}Xe	389	Liq. scint. w/ diluted ^{136}Xe	2011	4×10^{26}
SNO+	^{150}Nd	40	Liq. scint. w/ diluted $NdCl_3$	2013	1×10^{24}
NEXT	^{136}Xe	100	Gas TPC w/ electroluminescence	2014	1×10^{26}
LUCIFER	^{116}Cd	18	Scintillating/bolometric crystals	2014	2.3×10^{26}
COBRA	^{116}Cd	420	Pixelized CZT semiconductor	2014	1×10^{26}

References

1. E. Majorana, Nuovo Cim. **14**, 171 (1937).
2. B. Pontecorvo, *Sov. Phys. JETP* **34**, 247 (1958);
   Z. Maki, M. Nakagawa, and S. Sakata, *Prog. Theor. Phys.* **28**, 870 (1962).
3. A. Osipowicz *et al.*, arXiv:0109033.
4. A. Monfardini, *et al.*, Nucl. Instrum. Meth. **A559**, 346-348 (2006).
5. Preliminary NEMO-3 results presented at TAUP 2011.
6. F.T. Avignone, III, S.R. Elliott, J. Engel, Rev. Mod. Phys. **80**,481-516(2008).
7. W. Rodejohann, arXiv:1106.1334[hep-ph].
8. R. Arnold *et al.*, Phys. Rev. Lett. **107**, 062504 (2011).
9. N. Ackerman, *et al.*, arXiv:1108.4193.
10. R. Arnold *et al.*, Nucl. Instrum. Meth. **A536**, 79-122 (2005).
11. H. Drumm *et al.*, Nucl. Instr. Meth. **A 176** (1980) 333.
12. C. Arnaboldi *et al.*, Phys. Rev. **C78**, 035502 (2008).
13. E. Andreotti, *et al.*, JINST **4**, P09003 (2009).
14. S. Schonert *et al.*, Nucl. Phys. Proc. Suppl. **145**, 242-245 (2005).
15. C. E. Aalseth *et al.*, Nucl. Phys. Proc. Suppl. **138**, 217-220 (2005).
16. KamLAND-Zen
17. M. C. Chen *et al.*, [arXiv:0810.3694 [hep-ex]].
18. The AVLIS (Atomic Vapor Laser Isotope Separation) technique was used in the USA, France, and Japan but currently no facility is available for ^{150}Nd.
19. E. Gomez *et al.*, arXiv:1106.3630,
20. C. Nones *et al.*, Nucl. Phys. Proc. Suppl. **217**, 56-58 (2011).
21. K. Zuber, Phys. Lett. B **519**, 1 (2001).

Combined Analysis of all Three Phases of Solar Neutrino Data from the Sudbury Neutrino Observatory

J. Maneira*, on behalf of the SNO Collaboration.

Laboratório de Instrumentação e Física Experimental de Partículas, Av. Elias Garcia 14, 1°, 1000-149 Lisboa, Portugal
**E-mail: maneira@lip.pt*

The Sudbury Neutrino Observatory (SNO) took data between 1999 and 2006 in three phases, with distinct neutron detection methods, in order to measure the ^{8}B solar neutrino flux via the neutral current reaction on deuterium with different systematic uncertainties as well as the electron neutrino flux via the charged current reaction. The results of the three separate phases were already published, as well as the results of a combined analysis of the first two phases.The combination of the full 3-phase data set was recently finalized and submitted for publication. The new results were shown in this communication. The neutron signal identification and calibration in the third phase data was improved, as well the signal extraction and 3-flavor neutrino oscillation analyses. By making full use of the knowledge of correlated systematic uncertainties between phases, the combined 3-phase analysis provided the most accurate solar neutrino flux measurements from SNO, and the most precise constraints on the neutrino oscillation parameters.

1. Introduction

The Sudbury Neutrino Observatory (SNO) was designed to measure the flux of solar ^{8}B neutrinos, in both the electron-neutrino flavor and in all flavors, as proposed by Herb Chen.[1] The SNO detector with heavy water observed ^{8}B neutrinos via neutral current (NC), charged current (CC) and elastic scattering (ES) reactions:

$$(NC) \quad \nu_x + d \rightarrow p + n + \nu_x, \tag{1}$$

$$(CC) \quad \nu_e + d \rightarrow p + p + e^-, \tag{2}$$

$$(ES) \quad \nu_x + e^- \rightarrow \nu_x + e^-. \tag{3}$$

The NC reaction is equally sensitive to all three active neutrino flavors, allowing the determination of the total ^{8}B neutrino flux, Φ_{B}, independently of any specific active neutrino flavor oscillation hypothesis. The rate of CC reactions is only sensitive to ν_es, and comparing this to the NC reaction rate, it was possible to determine the neutrino survival probability as a function of energy, independently of any specific prediction of Φ_B.

As a result of measurements with the SNO detector and other experiments, it is now well-established that neutrinos are massive and that they change flavor through neutrino oscillations. Improving the precision of the SNO measurements is relevant in order to improve the precision of the neutrino oscillation parameters, as well as for Solar Physics. The best possible precision can be achieved with a fully combined analysis of all three phases of SNO. In this communication we present this final combined analysis of all solar neutrino data from the SNO experiment, recently submitted for publication.[2]

2. The SNO experiment

The SNO detector[3] consisted of an inner volume containing heavy water (D_2O) within a 12 m diameter transparent acrylic vessel (AV). Over 7×10^6 kg of H_2O between the rock and the AV shielded the D_2O from external radioactive backgrounds. An array of 9456 photomultiplier tubes (PMTs), installed on an 18 m diameter geodesic structure (PSUP), detected Cherenkov radiation produced in both the D_2O and H_2O, directly by recoil electrons from both the ES and CC reactions and indirectly by γ-rays from the NC reaction. The detector was located in Vale's Creighton mine near Sudbury, Ontario, Canada, with the center of the detector at a depth of 2092 m (5890$\pm$94 meters water equivalent). The heavy water has since been removed and preparations are ongoing for the new SNO+ experiment.[4]

The SNO detector operated in three phases distinguished by how the neutrons from the NC interactions were detected. In Phase I, the detected neutrons captured on deuterons in the D_2O releasing a single 6.25 MeV γ-ray. In Phase II, 2×10^3 kg of NaCl were added to the D_2O, and the neutrons captured predominantly on ^{35}Cl nuclei, which have a much larger neutron capture cross-section than deuterium nuclei and emit multiple gammas providing a more isotropic light distribution. In Phase III, an array of ^{3}He-filled proportional counters, made of high purity nickel, was deployed in the D_2O.[5] Neutrons were detected by capture on ^{3}He. The resulting triton and proton had a total kinetic energy of 0.76 MeV, and the voltage they induced in the counters was recorded as a function of time ("waveform"). The Neutral Current Detection, or NCD, array consisted of 36 strings filled with ^{3}He, and an additional 4 strings filled with ^{4}He, used to study backgrounds.

3. Combined analysis

The combined analysis of the data from all three phases of SNO accounts for any correlations in the systematic uncertainties between phases. The data were split into day and night sets in order to search for matter effects as the neutrinos propagated through the Earth. The data periods used in this analysis, as well as the event cuts to select good candidates were the same as our most recent analyses of data from these phases.[6,7]

The general form of the analysis was a fit to Monte Carlo-derived probability density functions (PDFs) for each of the possible signal and background types (except for backgrounds originating from radioactivity in the PMTs, for which the PDFs were described by an analytical function).

As with previous analyses of SNO data, the following four variables were calculated for each event recorded with the PMT array: the effective electron kinetic energy, T_{eff} ; the cube of the radial position, r, divided by 600 cm (the radius of the AV), $\rho = (r[\text{cm}]/600)^3$; the isotropy of the detected light, β_{14}; and the angle of the reconstructed electron propagation relative to the direction of the Sun, $\cos\theta_{\odot}$. The energy deposited in the gas of a proportional counter, E_{NCD}, was calculated for each event recorded with the NCD array, and the correlated waveform was determined.[8]

3.1. *NCD array analysis*

The NCD array observed neutrons, alphas, and events caused by instrumental backgrounds (their removal is described in Reference[8]). The previous analysis of data from Phase III[7] distinguished between neutron and alpha events by fitting the E_{NCD} spectrum, but the large number of alpha events resulted in large statistical and systematic uncertainties.

In this analysis, the waveform of each event was fitted to libraries of known neutron and alpha waveforms.[9] Calibration with neutron sources provided the neutron waveforms, while data from the ^{4}He-filled strings provided the alpha waveform. Further information based on the kurtosis and skewness of the waveform was also used in the analysis. A cut based on the results of the waveform fits reduced the number of alpha events by more than 98%, while retaining 74.78% of the neutron events.

The analyses of the neutron waveform, as well as the determination of their detection efficiency, rely heavily on calibrations with a ^{24}Na source distributed uniformly throughout the detector.[10] The systematic uncertainty on the neutron/alpha separation was 0.87%, to be added to a 0.64% uncertainty on E_{NCD} neutron PDF, and the larger statistical uncertainty, for a total of 7.1% on the number of neutrons obtained from the fit of the NCD array data.

3.2. *Combined 3-phase fit*

The event variables T_{eff}, ρ, β_{14}, and $\cos\theta_{\odot}$ were used to construct 4-dimensional PDFs for Phases I and II, and 3-dimensional PDFs for Phase III (the β_{14} variable was unnecessary). The combined fit to all phases was performed using the maximum likelihood technique[11] (a Bayesian approach to the fit was also carried out, and used as cross-check[12]). Comparisons of Monte Carlo simulations with calibration data defined the variation range for the systematic uncertainties. Constraints were placed on various "nuisance" parameters associated with these and the rate of background events.

Although there were multiple sets of data in this fit, the result was a single average Φ_{B} for day and night, a ν_e survival probability as a function of neutrino energy, E_ν, during the day, $P^{\text{d}}_{ee}(E_\nu)$, and an asymmetry between the day and night survival probabilities, $A_{ee}(E_\nu)$. $P^{\text{d}}_{ee}(E_\nu)$ was parameterized by

$$P^{\text{d}}_{ee}(E_\nu) = c_0 + c_1(E_\nu[\text{MeV}] - 10) + c_2(E_\nu[\text{MeV}] - 10)^2, \tag{4}$$

where c_0, c_1, and c_2 were the parameters to be fitted. Expanding the function around 10 MeV, which corresponds approximately to the peak in the detectable ^{8}B neutrino spectrum, reduced correlations between c_0, c_1, and c_2. For the same reasons, $A_{ee}(E_\nu)$ was parameterized by

$$A_{ee}(E_\nu) = a_0 + a_1(E_\nu[\text{MeV}] - 10). \tag{5}$$

This approach disentangled the detector response from the fit result as $P^{\text{d}}_{ee}(E_\nu)$ and $A_{ee}(E_\nu)$ were functions of E_ν as opposed to T_{eff}.

4. Results from combined fit to all data

The results of the combined fit to all data using the maximum likelihood technique are shown in Table 1. This result was consistent with but more precise than both the BPS09(GS), $(5.88 \pm 0.65) \times 10^6\,\text{cm}^{-2}\text{s}^{-1}$, and BPS09(AGSS09), $(4.85{\pm}0.58){\times}10^6\,\text{cm}^{-2}\text{s}^{-1}$, solar model predictions[13] The day ν_e survival probability at 10 MeV, c_0, was inconsistent at very high significance with the null hypothesis that there were no neutrino oscillations. The results for parameters c_1 and c_2 are consistent with no spectral distortions, but also with the LMA prediction (within the 1 σ band). The results for parameters a_0 and a_1 are consistent with no day/night distortions of the ν_e survival probability.

Table 1. Results from the maximum likelihood fit. Note that Φ_B is in units of $\times 10^6\,\mathrm{cm}^{-2}\mathrm{s}^{-1}$.

	Best fit	Stat. uncertainty	Syst. uncertainty
Φ_B	5.25	± 0.16	$^{+0.11}_{-0.13}$
c_0	0.317	± 0.016	± 0.009
c_1	0.0039	$^{+0.0065}_{-0.0067}$	± 0.0045
c_2	-0.0010	± 0.0029	$^{+0.0014}_{-0.0016}$
a_0	0.046	± 0.031	$^{+0.014}_{-0.013}$
a_1	-0.016	± 0.025	$^{+0.010}_{-0.011}$

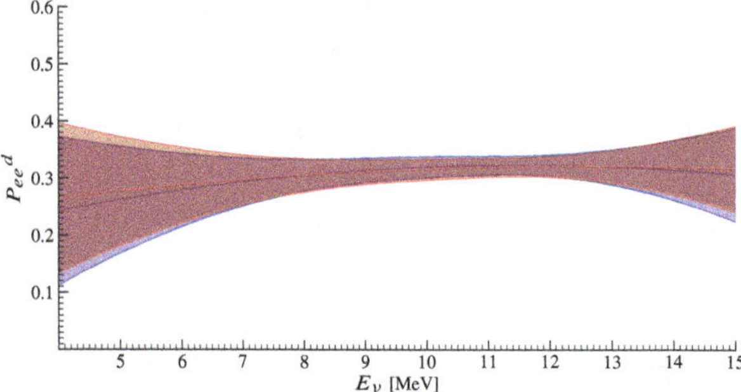

Fig. 1. RMS spread in $P^{\mathrm{d}}_{ee}(E_\nu)$, taking into account the parameter uncertainties and correlations. The red/blue (light/dark, or upper/lower) bands represent, respectively, the results from the maximum likelihood and the Bayesian fit.

Figure 1 shows the RMS spread in $P^{\mathrm{d}}_{ee}(E_\nu)$, taking into account the parameter uncertainties and correlations. This also shows that the maximum likelihood analysis was consistent with the alternative Bayesian analysis.

5. Neutrino oscillations

In previous analyses we used tabulated numerical calculations of neutrino survival probability as a function of the neutrino oscillation parameters. This analysis used an adiabatic approximation when calculating survival probabilities in the LMA region.[14] With the new calculation we could scan values of both Δm^2_{21} and E_ν independently, whereas the previous calculation was carried out at discrete values of $\Delta m^2_{21}/E_\nu$, which resulted in small but observable discontinuities.

The predicted spectrum of E_ν detectable by SNO was scaled by the expected oscillation distortions, and fitted to the same spectrum distorted by Equations 4 and 5. We then calculated the χ^2 between the results from this fit and the results of Table 1, as a function of the neutrino oscillation pa-

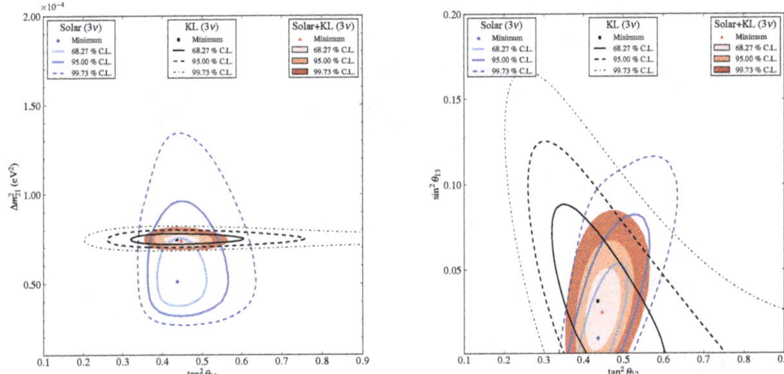

Fig. 2. Three-flavor neutrino oscillation analysis contour using both solar neutrino and KamLAND (KL) results.

rameters, including the results from other solar neutrino experiments (see[2] for details) and minimizing with respect to $\Phi_{\rm B}$. The KamLAND experiment observed neutrino oscillations in $\bar{\nu}_e$s from nuclear reactors and the published lookup table of χ^2 [15] was added directly to the χ^2 from the solar neutrino analysis.

5.1. *Three-flavor neutrino oscillation analysis*

Figure 2 shows the allowed regions of the $(\tan^2\theta_{12}, \Delta m^2_{21})$ and $(\tan^2\theta_{12}, \sin^2\theta_{13})$ parameter spaces obtained from the results of all solar neutrino experiments. It also shows the result of these experiments combined with the results of the KamLAND experiment.

Table 2 summarizes the results from these three-flavor neutrino oscillation analyses. Tests with the inverted hierarchy, i.e. negative values of Δm^2_{31}, gave essentially identical results.[14]

Recent results from long-baseline (LBL) experiments indicate a non-zero θ_{13} with a significance of approximately 2.7σ. A combined analysis of all LBL and atmospheric (ATM) results, and the results from the CHOOZ experiment was performed by Fogli *et al.*[16] (see[2] for full reference list). Table 2 shows the results of combining those results with the ones from solar and KamLAND experiments. LBL+ATM+CHOOZ experiments currently have better sensitivity to θ_{13} than the combined solar and KamLAND experiments, but the combination of all experiments gives a slightly improved

Table 2. Best-fit neutrino oscillation parameters from a three-flavor neutrino oscillation analysis. Uncertainties listed are $\pm 1\sigma$ after the χ^2 was minimized with respect to all other parameters. The global analysis includes Solar+KL+ATM+LBL+CHOOZ.

Analysis	$\tan^2\theta_{12}$	$\Delta m^2_{21}[\mathrm{eV}^2]$	$\sin^2\theta_{13}(\times 10^{-2})$
Solar	$0.436^{+0.048}_{-0.036}$	$5.13^{+1.49}_{-0.98} \times 10^{-5}$	< 5.8 (95% C.L.)
Solar+KL	$0.446^{+0.030}_{-0.029}$	$7.41^{+0.21}_{-0.19} \times 10^{-5}$	$2.5^{+1.8}_{-1.5}$
			< 5.3 (95% C.L.)
Global			$2.02^{+0.88}_{-0.55}$

determination of θ_{13}, hinting at a non-zero value.

6. Acknowledgments

For Portugal, this research was supported by FEDER funds through the COMPETE Program and by Portuguese National funds through FCT - Fundação para a Ciência e a Tecnologia - within the projects CERN/FP/83548/2008 and PTDC/FIS/115281/2009[a].

References

1. H. H. Chen, Phys. Rev. Lett. **55**, 1534 (1985).
2. B. Aharmim *et al.* (SNO Collaboration), nucl-ex/1109.0763, submitted to PRC(2011) .
3. J. Boger *et al.* (SNO Collaboration), Nucl. Inst. & Meth. **A449**, 172 (2000).
4. J. Maneira, Nucl. Phys. B Proc. Suppl. **217**, 1, 50 (2011).
5. J. F. Amsbaugh *et al.*, Nucl. Inst. & Meth. **A579**, 1054 (2007).
6. B. Aharmim *et al.* (SNO Collaboration), Phys. Rev. D **81**, 055504 (2010).
7. B. Aharmim *et al.* (SNO Collaboration), Phys. Rev. Lett. **101**, 111301 (2008).
8. B. Aharmim *et al.*, (2011), arXiv:1107.2901 [nucl-ex] .
9. N. S. Oblath, Ph.D. thesis, University of Washington (2009).
10. K. Boudjemline *et al.*, Nucl. Inst. & Meth. **A620**, 171 (2010).
11. P.-L. Drouin, Ph.D. thesis, Carleton University (2011), to be published.
12. S. Habib, Ph.D. thesis, University of Alberta (2011).
13. A. M. Serenelli, S. Basu, J. W. Ferguson, and M. Asplund, Astrophys. J. Lett. **705**, L123 (2009).
14. N. Barros, Ph.D. thesis, University of Lisbon (2011).
15. A. Gando *et al.* (KamLAND Collaboration), Phys. Rev. D **83**, 052002 (2011).
16. G. L. Fogli, E. Lisi, A. Marrone, A. Palazzo, and A. M. Rotunno, (2011), arXiv:1106.6028v1 [hep-ph] .

[a]Further acknowledgements in.[2]

SEARCH FOR THE θ_{13} MIXING ANGLE WITH THE DOUBLE CHOOZ EXPERIMENT

A. MEREGAGLIA ON BEHALF OF DOUBLE CHOOZ COLLABORATION

IPHC - CNRS - IN2P3,
Strasbourg, France
E-mail: anselmo.meregaglia@cern.ch

In this paper the Double Chooz experiment is presented: a reactor neutrino oscillation experiment for the measurement of the θ_{13} mixing angle.
The experimental concept and the detector design are described, as well as the preliminary results on the neutrino selection based on the first few months of data taking.
The expected sensitivities are shown, namely a limit on $\sin^2(2\theta_{13})$ of 0.032 at 90% C.L. in case of no observation of oscillation and a discovery potential on $\sin^2(2\theta_{13})$ of 0.05 at 3 σ C.L. in case of oscillation measurement.

Keywords: Neutrino; Oscillation; Reactor

1. Introduction

Neutrino oscillation is a phenomenon well established both at the solar and atmospheric scale.[1]
In the last two decades huge improvements have been made in the understanding of this phenomenon and on the knowledge of the mixing parameters, however some open questions remain to be answered. A crucial point is to determine if the mixing angle θ_{13} is vanishing or not, since the discovery of a non zero value would open the way for a search of CP violation in the leptonic sector.
Very recent results by T2K[2] and MINOS[3] indicates that the θ_{13} mixing angle is different from zero and within the discovery potential of the Double Chooz[4] experiment, which aims indeed at the measurement of such a parameter.

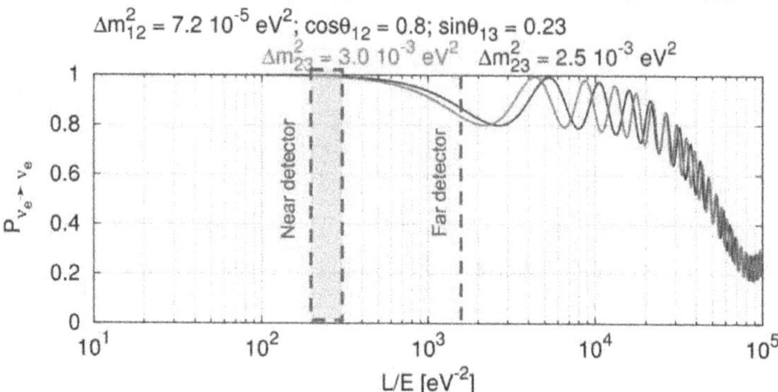

Fig. 1. Oscillation probability $P(\bar{\nu}_e \to \bar{\nu}_e)$ as a function of L/E. The regions covered by the near and the far detector are shown by the red dashed lines.

2. Experimental concept and detector design

Double Chooz is a reactor neutrino oscillation experiment that aims at the observation of the $\bar{\nu}_e \to \bar{\nu}_e$ transition. The probability for such an oscillation can be calculated using the approximated formula given in equation 1 where L is the baseline, E the neutrino energy, θ_{13} the mixing angle that we want to measure, and Δm^2_{23} the mass splitting between the mass eigenstate 2 and 3 (the best fit value from MINOS experiment[5] is 2.43×10^{-3} eV^2).

$$P(\bar{\nu}_e \to \bar{\nu}_e) \cong 1 - \sin^2(2\theta_{13}) \sin^2\left(\frac{\Delta m^2_{23} L}{4E}\right) \tag{1}$$

The advantage of a reactor experiment with respect to long baseline oscillation experiments is that the measurement of the mixing angle θ_{13} is independent of the value of the complex phase δ_{CP}, being a disappearance experiment, and that given the short baseline of about 1 km it is insensitive to matter effects.

The idea of the experiment is to measure the large flux of neutrinos coming from the two cores reactor of Chooz in France (the isotropic generation amounts to about $10^{21}\nu_e$ per second) with two identical detectors. The first detector is located at about 400 m from the reactor cores (where the oscillation probability is very small) whereas the second one is hosted in the former Chooz experiment laboratory at about 1 km from the reactors, at about the first maximum of oscillation as it can be seen in figure 1.
The ratio of the spectra measured at the far and near site gives a direct

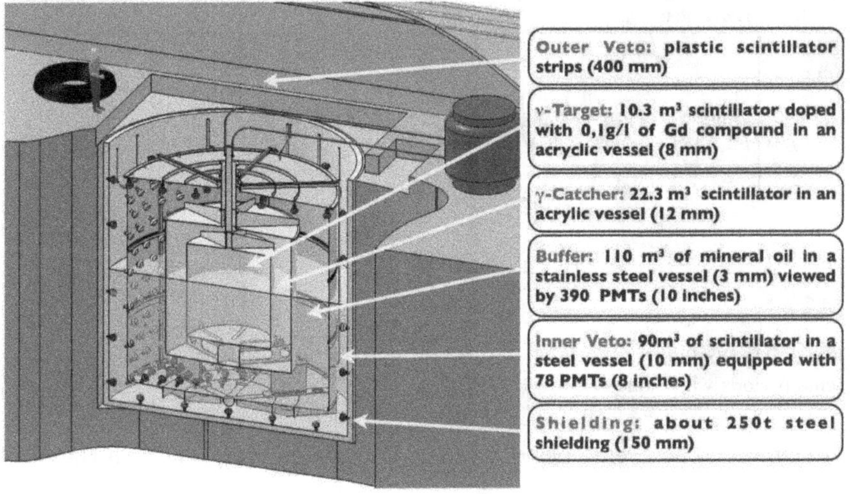

Fig. 2. Cartoon showing the detector design.

measurement of the mixing angle θ_{13}. This evaluation using two identical detectors allows for a cancellation of many systematic errors related mostly to the flux normalization and detector efficiency evaluation.

The detectors are made up of several sub-detector layers and detailed descriptions of all the components can be found in reference 4. A cartoon showing the detector layers with some additional information can be found in figure 2.

3. Signal and background

The neutrino spectrum is a convolution of the flux times the cross section (threshold at 1.8 MeV) and it results in a spectrum between 2 MeV and 8 MeV with a peak at about 4 MeV.
The process we observe is the Inverse Beta Decay (IBD), namely $\bar{\nu}_e + p \rightarrow e^+ + n$, and the signal signature is a prompt signal given by the positron ionization and annihilation (1-8 MeV) and a delayed signal coming from the neutrons absorbed on Gd in the target ($\sim$ 8 MeV). The time correlation between the two signals ($\Delta t \sim 30\mu$s) is also used in the neutrino events selection.

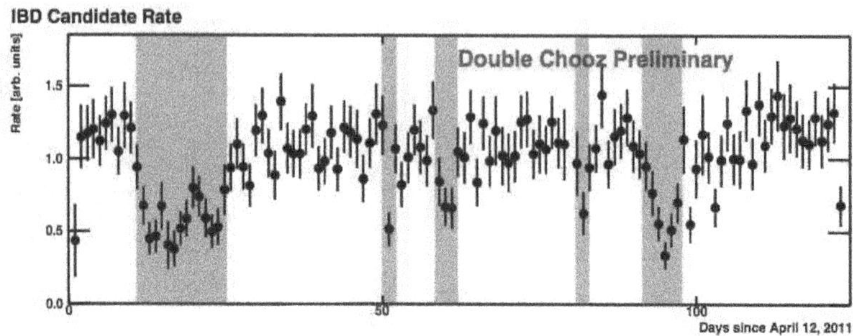

Fig. 3. Preliminary neutrino rate day by day in arbitrary units. The blue bands represents periods where one reactor was off.

We expect $\sim$ 65 events per day in the far detector and $\sim$ 450 in the near one.

We have two types of background: accidental (prompt and delayed signals coming from different sources) and correlated (prompt and delayed signals coming from the same source).
In the former case we have typically prompt signals coming from radioactivity of the rock or PMTs and the delayed one is due to fast neutrons produced by cosmic muons that are thermalized and absorbed in the detector.
In the latter case we can have either fast neutrons that give both prompt and delayed signals or long-lived isotopes such as ^{9}Li. They decay in a $\beta+n$ cascade and that can not be vetoed tagging the muon that produced them since their lifetime is $\sim$ 250 ms and we could not afford such a dead time. The overall number of background events expected per day in the far detector is 3.6 whereas in the near one it is 16.2. The difference is due to the different detector overburden (300 m.w.e. hill topology for the far detector and 115 m.w.e. flat topology for the near one).

4. Status of the experiment

The far detector is taking physics data since April 2011, with a mean efficiency larger than 75 %.
Although the calibration of the detector using radioactive sources is still ongoing, a first energy scale can be computed using the absorption peaks of muon induced neutrons on Hydrogen ($\sim$ 2.2 MeV) and Gadolinium ($\sim$ 8 MeV).

Preliminary results on the neutrino candidates selection have been obtained (see figure 3): more than 4000 candidates have been recorded in about 4 months.

As far as the near detector is concerned, the laboratory should be delivered in April 2012 and we expect the near detector to be ready to take data at the beginning of 2013.

5. Sensitivity

The sensitivity of the experiment at 90 % C.L. on $\sin^2(2\theta_{13})$ is ~ 0.03 in case of no oscillation measured, namely a factor of 5 better than the Chooz limit of 0.15.[6]

The discovery potential at 3 σ C.L., is instead 0.05. This value is well below the best fit value found recently by T2K[2] of 0.11 (N.H.) which could actually be addressed by the Double Chooz experiment using the 2011 statistics only.

In three years of running with two detectors, the Double Chooz experiment could address the whole current range of T2K (0.03 N.H. lower bound).

6. Conclusion

The Double Chooz experiment, aiming at the observation of the $\bar{\nu}_e \rightarrow \bar{\nu}_e$ transition, has been presented.

The far detector is taking data smoothly since April 2011 and we expect the near detector to be fully operational at the beginning of 2013.

Preliminary neutrino candidates have been selected, the detector calibration is being finalised, and the analysis of the first months of data is ongoing. Results are expected very soon.

References

1. C. Amsler et al., Physics Letters B667, 1 (2008) http://pdg.lbl.gov/2009/reviews/rpp2009-rev-neutrino-mixing.pdf
2. K. Abe *et al.* [T2K Collaboration], Phys. Rev. Lett. **107**, 041801 (2011).
3. P. Adamson *et al.* [MINOS Collaboration], [arXiv:1108.0015 [hep-ex]].
4. F. Ardellier *et al.*, arXiv:hep-ex/0405032.
5. A. Habig, Mod. Phys. Lett. A **25** (2010) 1219
6. M. Apollonio *et al.* [CHOOZ Collaboration], Eur. Phys. J. C **27** (2003) 331

A search for neutrinoless double beta decay: recent results from the NEMO-3 experiment and plans for SuperNEMO

F. NOVA

Department of Physics, University of Texas at Austin,
Austin, Texas 78712, USA
**E-mail: nova@physics.utexas.edu*

The observation of neutrino oscillations has proved that neutrinos have mass. This discovery has renewed and strengthened the interest in neutrinoless double beta decay experiments which provide the only practical way to determine whether neutrinos are Majorana or Dirac particles. The recently completed NEMO-3 experiment, located in the Modane Underground Laboratory in the Fréjus Tunnel under the French–Italian Alps, was an experiment searching for neutrinoless double beta decays using a powerful technique for detecting a two-electron final state by employing an apparatus combining tracking, calorimetry and the time-of-flight measurements. We will present latest results from NEMO-3 and will discuss the status of SuperNEMO, the next generation experiment that will exploit the same experimental technique to extend the sensitivity of the current search.

Keywords: NEMO-3, SuperNEMO, double beta decay, Majorana neutrino

1. Neutrinoless double beta decay

Neutrinoless double beta decay ($0\nu\beta\beta$) is a process beyond the Standard Model which, if observed, will imply that neutrinos are Majorana particles, that their mass is non-vanishing and that lepton number is not conserved. The competing two-neutrino double beta decay ($2\nu\beta\beta$) is allowed in the Standard Model: the two decays are distinguished by the distribution of electrons energy-sum. Measuring accurately the $2\nu\beta\beta$ decay is important since it constitutes the ultimate background for the $0\nu\beta\beta$ decay, it is the testing ground for nuclear models and it provides input for the calculations of the Nuclear Matrix Elements (NME).[1]

2. The NEMO-3 experiment

NEMO-3 is a $\beta\beta$ decay experiment, running in the Fréjus Underground Laboratory (4800 m w.e.) since 2003 and finished taking data in 2011.

The detector has the form of a cylinder. A thin source foil ($\approx 50\ \frac{mg}{cm^2}$) contains ~10 kg of multiple isotopes (see table 1): while ^{100}Mo, ^{82}Se, ^{130}Te, ^{116}Cd, ^{150}Nd, ^{96}Zr and ^{48}Ca are used to search for $0\nu\beta\beta$ decay and measure the $2\nu\beta\beta$, sectors with pure Cu and natural Te are used to study the external background. The source is situated in a tracking chamber of 6180 drift cells operating in Geiger mode, providing vertex resolution of 1 cm. The chamber is enclosed by calorimeter walls of 1940 plastic scintillator blocks, with an energy resolution at FWHM of $\frac{14.1-17.6\%}{\sqrt{E}}$ and a time resolution (250 ps) which allows suppression of crossing electron background. The whole detector is covered with iron and boron water shielding.[2]

NEMO-3 is capable to detect e^-, e^+, γ and α particles; measuring the angular distribution and the individual energies of the electrons allows one to study the decay mechanism and helps to reduce and control the background.

Background in NEMO-3 can be classified into three types:[3] *external* (incoming γ's), *radon in the tracker* (reduced by a factor of 6 after installation of a radon trapping facility in 2004), and *internal* (radioactive pollution of the source). MC events are generated using a GEANT-based simulation[4] of the detector with initial kinematics given by the event generator DECAY0.[5]

Table 1. Main results on $2\nu\beta\beta$ and $0\nu\beta\beta$ decays. S/B is a signal to background ratio in the $2\nu\beta\beta$ analysis.

Nuclei	mass (kg)	$Q_{\beta\beta}$ (MeV)	S / B	$T_{1/2}^{2\nu\beta\beta}$ (10^{19} years)	$T_{1/2}^{0\nu\beta\beta}$
^{100}Mo	6.914	3.0348	76	0.717 ± 0.001 (stat) ± 0.054 (syst)	$> 1.0 \times 10^{24}$ y
^{82}Se	0.932	2.9952	4.0	9.6 ± 0.1 (stat) ± 1.0 (syst)	$> 3.2 \times 10^{23}$ y
^{116}Cd	0.405	2.8047	10.3	2.88 ± 0.04 (stat) ± 0.16 (syst)	
^{150}Nd	0.037	3.3671	2.8	0.920 ± 0.025(*stat*) ± 0.063 (syst)	$> 1.8 \times 10^{22}$ y
^{96}Zr	0.0094	3.3500	1.0	2.35 ± 0.14 (stat) ± 0.16 (syst)	$> 9.2 \times 10^{21}$ y
^{48}Ca	0.007	4.274	6.8	$4.4\ ^{+0.5}_{-0.4}$ (stat) ± 0.4 (syst)	
^{130}Te	0.454	2.5289	0.25	$70\ ^{+10}_{-8}$ (stat) $^{+10}_{-9}$ (syst)	$> 1.3 \times 10^{23}$ y

A measurement of $2\nu\beta\beta$ decay was performed for seven isotopes, with unprecedented precision (table 1). The measurement of the decay of ^{100}Mo (Fig. 1) is the most precise and has the biggest statistics collected in the world.[6] The $2\nu\beta\beta$ spectra for other isotopes are shown in Fig. 2.[7,8] In the

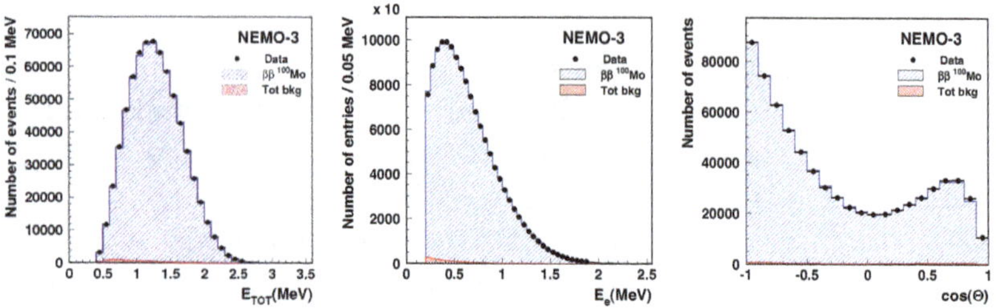

Fig. 1. Total energy, individual energy and angular distributions of the ^{100}Mo $2\nu\beta\beta$ events in the NEMO-3 experiment for the low radon data phase (3.49 years).

case of ^{130}Te this is the first 7.7σ direct observation and provides a reference point to a dispute between geochemical experiments.[9–11]

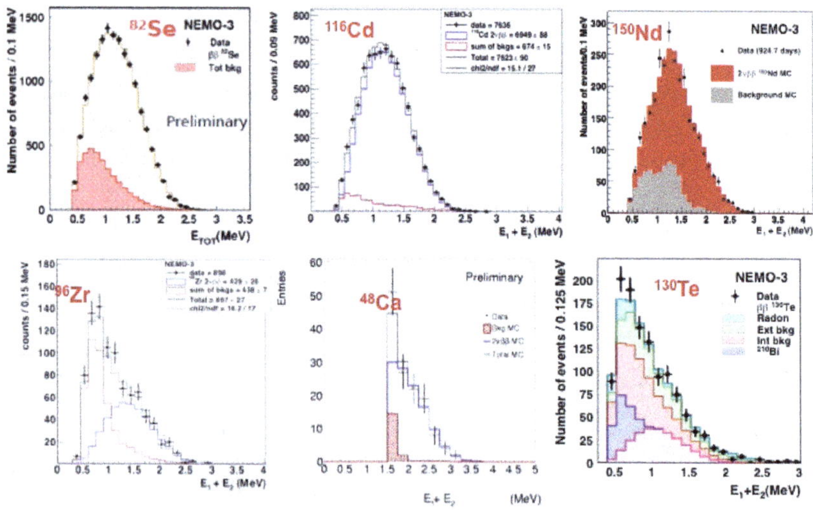

Fig. 2. Total energy of the $2\nu\beta\beta$ events in the NEMO-3 experiment for ^{82}Se, ^{116}Cd, ^{150}Nd, ^{96}Zr, ^{48}Ca and ^{130}Te.

As for the $0\nu\beta\beta$ decay search, Fig. 3 shows the energy sum spectra. No signal was found, therefore a 90% CL lower limit was set on the half-life using a binned log-likelihood test.[12] The results, listed in table 1, correspond, according to different theoretical NME calculations,[1,13–15] to the effective neutrino mass $m_\nu < (0.31-0.96)$ eV for ^{100}Mo and $m_\nu < (0.94-2.6)$ eV for

^{82}Se. Other $0\nu\beta\beta$ mechanisms have also been investigated and limits have been set for decays to excited states, right-handed currents and Majoron emission.

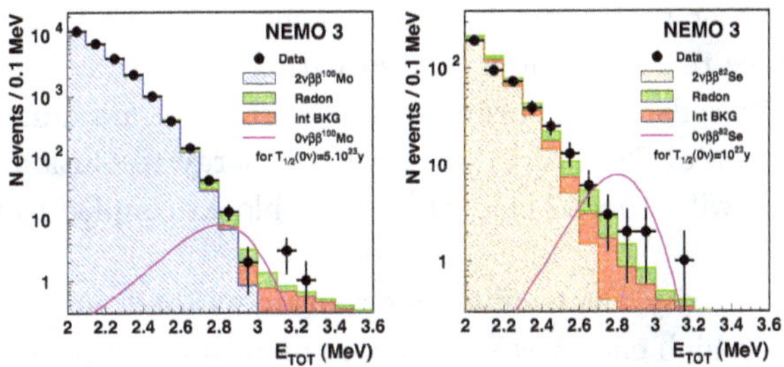

Fig. 3. Total energy spectra of 2 electrons events observed in NEMO-3 after 4.5 years for (a) ^{100}Mo and (b) ^{82}Se. For illustration, the magenta line represents what a $0\nu\beta\beta$ signal would look like with a given half-life.

3. The SuperNEMO experiment

SuperNEMO is a next-generation $0\nu\beta\beta$ experiment based on the technique of tracking and calorimetry of the NEMO-3 detector.[16,17] It will consist of 20 identical modules, each housing 5 kg of source isotope (^{82}Se is the baseline choice), and will be hosted in a new extension of the LSM laboratories.

The SuperNEMO project increases by two orders of magnitude the NEMO-3 $0\nu\beta\beta$ half-life sensitivity (see Table 2), improving the radiopurity of detector components, the energy resolution and the selection efficiency.

Table 2. Characteristics of NEMO-3 and SuperNEMO

Experiment	NEMO-3	SuperNEMO
choice of isotope	^{100}Mo	^{82}Se (or ^{150}Nd or ^{48}Ca)
isotope mass	7 kg	100 kg
radioactive contamination	$A(^{208}\mathrm{Tl}) \approx 100\ \frac{\mu\mathrm{Bq}}{\mathrm{kg}}$	$A(^{208}\mathrm{Tl}) < 2\ \frac{\mu\mathrm{Bq}}{\mathrm{kg}}$
	$A(^{214}\mathrm{Bi}) < 300\ \frac{\mu\mathrm{Bq}}{\mathrm{kg}}$	$A(^{208}\mathrm{Tl}) < 10\ \frac{\mu\mathrm{Bq}}{\mathrm{kg}}$
	$A(^{222}\mathrm{Rn}) \approx 5\ \frac{\mathrm{mBq}}{\mathrm{m}^3}$	$A(^{222}\mathrm{Rn}) < 0.15\ \frac{\mathrm{mBq}}{\mathrm{m}^3}$
energy resolution (FWHM)	$8-10\%$ at 3 MeV	4% at 3 MeV
efficiency	18%	30%
sensitivity	$T_{1/2}^{0\nu\beta\beta} > 2 \times 10^{24}$ y	$T_{1/2}^{0\nu\beta\beta} > 10^{26}$ y
	$m_\nu < (0.31 - 0.96)$ eV	$m_\nu <$ 53–145 meV

3.1. *R & D*

A demonstrator module with all the components of the final design will be ready for competitive physics measurement in 2012 and will be followed by 19 more similar modules.

Several tracker prototypes have been constructed and tested, reaching the target space resolution; a dedicated wiring robot is used for mass production of drift cells. For calorimetry the R&D program aims to reach an energy resolution of 7% FWHM at 1 MeV. Energy resolution of 7.7% has been measured with 10×22 cm^2 Eljen PVT blocks coupled to Hamamatsu PMTs.[18]

As for radiopurity, the troublesome contamination from ^{208}Tl and ^{214}Bi that decay with high energy release can be monitored by the so-called BiPo process (emission of an e followed by a delayed α). The dedicated BiPo-1 detector[19] is running since 2008 to measure the surface radiopurity of the plastic scintillators. A larger detector will qualify the radiopurity of ^{82}Se foil with the required sensitivity in 6 months.

References

1. F. Simkovic, A. Faessler, V. Rodin, P. Vogel, J. Engel, in *Phys. Rev. C*, **77**, 045503 (2008).
2. R. Arnold *et al.*, in *Nucl. Instrum. Meth. A*, **536**, 79 (2005).
3. J. Argyriades *et al.*, in *Nucl. Instrum. Meth. A*, **606**, 449-465 (2009).
4. R. Brun *et al.*, CERN Program Library W 5013, 1984.
5. O. A. Ponkratenko, V. I. Tretyak, Y. .G. Zdesenko, in *Phys. Atom. Nucl.* , **63**, 1282-1287 (2000).
6. R. Arnold *et al.*, in *Nucl. Phys. A*, **781**, 209-226 (2007).
7. J. Argyriades *et al.*, in *Nucl. Phys. A*, **847**, 168-179 (2010).
8. J. Argyriades *et al.*, in *Phys. Rev. C*, **80**, 032501 (2009).
9. A. S. Barabash, in *Eur. Phys. J. A*, **8**, 137 (2000).
10. A. P. Meshik, C. M. Hohenberg, O. V. Pravdivtseva, T. J. Bernatowicz, Y. S. Kapusta, in *Nucl. Phys. A*, **809**, 275-289 (2008).
11. R. Arnold *et al.*, in *Phys. Rev. Lett.* , **107**, 062504 (2011).
12. T. Junk, in *Nucl. Instrum. Meth. A*, **434**, 435 (1999).
13. M. Kortelainen, J. Suhonen, in *Phys. Rev. C*, **76**, 024315 (2007).
14. E. Caurier, J. Menendez, F. Nowacki, A. Poves, in *Phys. Rev. Lett.* **100**, 052503 (2008).
15. J. Barea, F. Iachello, in *Phys. Rev. C*, **79**, 044301 (2009).
16. F. Piquemal, in *Phys. Atom. Nucl.* , **69**, 2096-2100 (2006).
17. R. Arnold *et al.*, in *Eur. Phys. J. C*, **70**, 927 (2010).
18. J. Argyriades *et al.*, in *Nucl. Instrum. Meth. A*, **625**, 20-28 (2011).
19. J. Argyriades *et al.*, in *Nucl. Instrum. Meth. A*, **622**, 120 (2010).

THE NEXT EXPERIMENT AT THE LSC

C. A. B. OLIVEIRA*, A. L. FERREIRA and J. F. C. A. VELOSO

*i3N, Physics Department, University of Aveiro, Aveiro, Portugal, *E-mail: carlos.oliveira@ua.pt*

J. MARTÍN-ALBO, M. SOREL and J. J. GÓMEZ-CADENAS

Instituto de Física Corpuscular (IFIC), CSIC and Universidad de Valencia, Valencia, Spain

on behalf of the NEXT collaboration

The Neutrino Experiment with a Xenon TPC (NEXT) will search for the neutrinoless double beta decay in ^{136}Xe using a 100 Kg, high-pressure xenon, electroluminescent time projection chamber. Such a detector, thanks to its excellent energy resolution and its powerful background rejection, provided by the discrimination of the unique double beta decay topological signature, may become one of the leading experiments of the field. The final detector is approved for operation in the Canfranc Underground Laboratory (LSC), Spain, in 2013. Present status and future developments will be presented.

Keywords: Neutrinoless double beta decay; Electroluminescence; Time Projection Chamber.

1. Introduction

During the last decade, different experiments have shown that neutrinos have mass and mix.[1] This opens the possibility for the Majorana nature of these particles. If the neutrino is a Majorana particle then it is its own antiparticle and the neutrino-less double beta decay ($\beta\beta^{0\nu}$) is possible in some istopes.[2] Since the $\beta\beta^{0\nu}$ decay would violate the lepton number conservation, the detection of such decay would imply physics beyond the Standard Model. Beyond giving an experimental proof of the Majorana nature of neutrinos, the measurement of the half-time of the $\beta\beta^{0\nu}$ decay would allow the determination of the absolute scale of neutrino masses.

In nature, there are different isotopes candidates to be emitters of this decay. Xenon is the only noble gas that has a decaying isotope, ^{136}Xe. Its Q-value ($Q_{\beta\beta} = 2.458$ MeV) is high enough to be used in a $\beta\beta^{0\nu}$ experiment.

The natural abundance of this isotope is 9 %, but it can be enriched by centrifugation at a reasonable cost. In addition, xenon does not have any other long-lived radioactive isotopes and can be easily purified.

2. The NEXT detector concept

NEXT will search for the $\beta\beta^{0\nu}$ decay in ^{136}Xe using a 100 Kg, high-pressure xenon (HPXe), electroluminescent time projection chamber. Such a detector, thanks to its excellent energy resolution and its powerful background rejection, provided by the discrimination of the unique double beta decay topological signature, may become one of the leading experiments of the field. The final detector is approved for operation in the Canfranc Underground Laboratory (LSC), Spain.[3]

2.1. *Energy resolution*

Outstanding energy resolution is essential in $\beta\beta^{0\nu}$ decay searches since the Q value of such process is close to the energy of backgrounds from natural radioactivity that can easily overwhelm the signal peak. The aim of NEXT is to achieve FWHM energy resolutions below 1%. For this, the detector will use gaseous xenon. The fluctuations associated with the production of primary charges in a gas by incident particles define the intrinsic energy resolution that can be achieved and are described by the *Fano factor*, F. The lower the value of F the better the achievable energy resolution. For gaseous pure xenon, various measurements show that $F = 0.15 \pm 0.02$[4] whereas for liquid xenon $F \sim 20$. Thus, better energy resolution can be achieved by using gaseous xenon instead of a liquid phase.[5]

NEXT will use Electroluminescence (EL) for the primary charge signal amplification. The process consists of drifting the primary electrons towards a region where a suitable electric field, below the ionization threshold, is applied. Electrons are then accelerated and can excite atoms of the gas that decay emitting VUV light centered at 173 nm. This type of signal amplification has very low associated fluctuations, when compared to those of traditional avalanche multiplication as demonstrated by the FWHM energy resolutions of 8 % and 4 %, achieved by Gaseous Scintillation Proportional Counters working with xenon, for 5.9 keV and 22 keV X-rays, respectively.[6] By using a recently developed and validated simulation toolkit[7] we estimated the energy resolution achievable with the final NEXT detector. We found that, even considering conservative assumptions where only 0.5 % of the VUV photons produced per $\beta\beta^{0\nu}$ decay would be detected, a FWHM energy resolution bellow 0.4 % can in principle be achieved.[8]

2.2. *Topological signature recognition*

Double beta decay events have a distinctive topological signature in HPGXe: a ionization track, of about 20 cm length at 15 bar, tortuous because of multiple scattering, and with larger depositions in both ends.[9] NEXT will integrate a tracking function in order to reconstruct and recognize the ionization track and use this to further suppress the background.

2.3. *Operation principle*

Ionizing particles interacting in the HPXe will transfer their energy to the medium through ionizations and excitations of the gas atoms. The excited atoms promptly emit VUV scintillation light. The positive ions and free electrons left behind by the particle are prevented from recombination by a suitable electric field. Negative charge carriers drift then toward the TPC anode, entering the EL region, with a more intense electric field. There, further VUV photons are generated isotropically by the EL process. Therefore, both scintillation and ionization produce an optical signal, S_1 and S_2 respectively, to be detected with an array of photosensors (PMTs) located behind the cathode. The detection of the primary scintillation light constitutes the start-of-event (t_0), whereas the detection of EL light provides an energy measurement. The EL light is used also for tracking by detecting it with a second array of photosensors (MPPCs) located behind the anode.[3]

3. NEXT-1 prototypes

Two prototypes were constructed aiming to demonstrate that energy resolutions compatible with the NEXT goal for $Q_{\beta\beta}$ can be achieved at reasonably high energies. They also aim to demonstrate that it is possible to determine the position of interaction of the ionizing particles, essential for the topological signature recognition. One of the prototype is being operated at the *Lawrence Berkeley National Laboratory* (LBNL), California, USA, and the other at the *Instituto de Física Corpuscular* (IFIC), Valencia, Spain.

Currently, both prototypes operate with $\sim$1 kg of xenon and are instrumented with the energy plane consisting of an array of 19 Hamamatsu R7378A PMTs (capable of resisting pressures of up to 20 bar). In both cases, the walls of the light tube are made of uncoated PTFE that provides reflectivity not higher than 50 %. The electric fields are created by applying suitable potentials to parallel meshes.

3.1. *NEXT-1 at LBNL*

The LBNL prototype has an active drift volume of 8 cm long in the drift direction with 14 cm transversal span. The EL gap is 3 mm wide and the 19 PMTs are 13 cm away from it. The waveform resulting from the sum of the 19 PMTs signals is shown in Fig. 1, for a ^{60}Co calibration source and it is clearly visible, with small noise, the S_1 signal due to the primary scintillation and the two S_2 signals: one due to a Compton absorption and the other due to the full photo-absortion of the escape γ-ray.

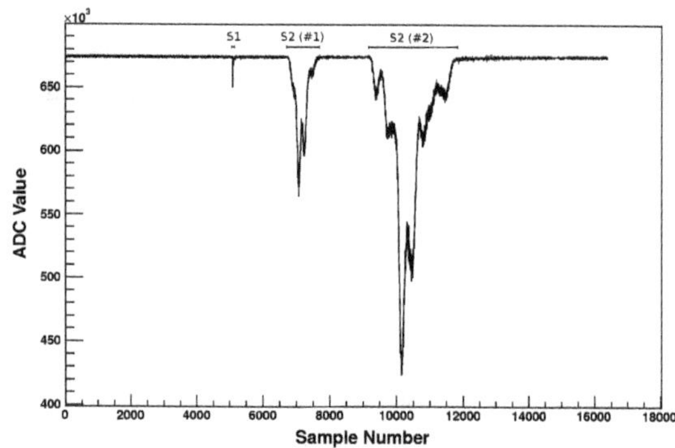

Fig. 1. Typical waveform obtained with a ^{60}Co source, corresponding to the sum of the 19 PMT signals.

The prototype was also operated using a ^{137}Cs 662 keV γ-ray source highly collimated and on the TPC axis, entering the detector through a 2 mm thick stainless steel window. We used a xenon pressure of 10 bar and a EL electric field of $\sim$ 27 kV cm^{-1}. The integrated S_2 signal is shown in Fig. 2a) as a function of the drift time for valid events (an S_1 pulse and one or more S_2 pulses). The highest narrow energy band corresponds to the full photo-absorption of the 662 keV γ-rays. The drift time span of the events corresponds to the maximum drift time, corresponding to the maximum 8 cm drift length. The decrease in the energy bands is due to electron attachment in the drift region. For each valid event, the (x, y) position was determined by weighting the PMT positions by the signal amplitude observed in each. In Fig. 2b) we show the integrated S_2 signal of the full photo-absorption events as a function of the reconstructed radial distance to the TPC axis, $r = \sqrt{x^2 + y^2}$. The systematic deviation effect observed mainly due to the solid angle variation is small in the region with radius smaller than 1.5 cm increasing for higher radial distances. The dependence of the S_2 signal on both the longitudinal coordinate and the radial position can easily be corrected through fits to the data.

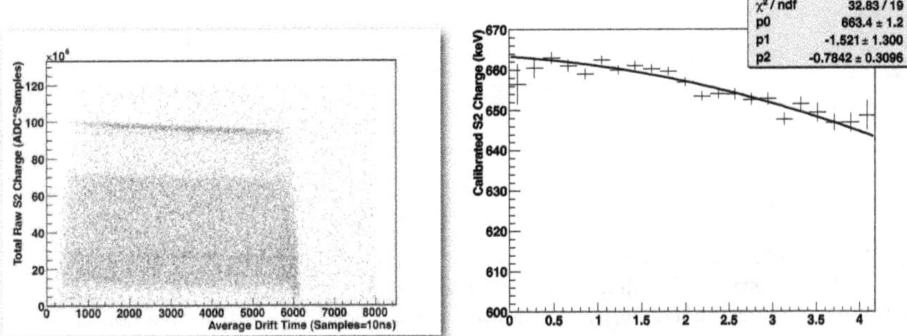

Fig. 2. a) **Left:** Integrated S_2 signal from ^{137}Cs 662 keV γ-rays as a function of the drift time for valid events (an S_1 pulse and one or more S_2 pulses). b) **Right:** Integrated S_2 signal of full photo-absorption events as a function of the reconstructed radial distance to the TPC axis. Both results refer to the LBNL prototype.

After selecting central events (radius smaller than 1.5 cm) and applying the mentioned corrections, we obtained the spectrum presented in Fig. 3a). It shows a FWHM energy resolution of 1.4 % at 662 keV and a well formed gaussian response shape. This result is representative of our good energy resolution and we are currently trying to further improve the operational parameters and also the data correction and analysis in order to approach the 0.9 % limit expected for this energy from the Fano factor.

3.2. *NEXT-1 at IFIC*

The IFIC prototype has been commissioned a few months after that of LBNL, thus the results are still preliminary. The prototype has an active drift volume of 30 cm long in the drift direction with 16 cm transversal span. The EL gap is 5 mm wide and the 19 PMTs are 10 cm away from it. In addition to the energy PMT plane, the detector is instrumented with a tracking plane consisting of an identical array of PMTs. In a second phase (4th quarter 2011) this tracking plane will be substituted by one array of 248 1-mm^2 MPPCs. The detector has been operated at pressures between 4 and 11 bar and with a reduced electric field of ~ 3.5 kVcm^{-1}bar^{-1}.

We present, in Fig. 3b) the (x, y) position of α particles from radioactive contaminants decaying within the gas, reconstructed in a similar way

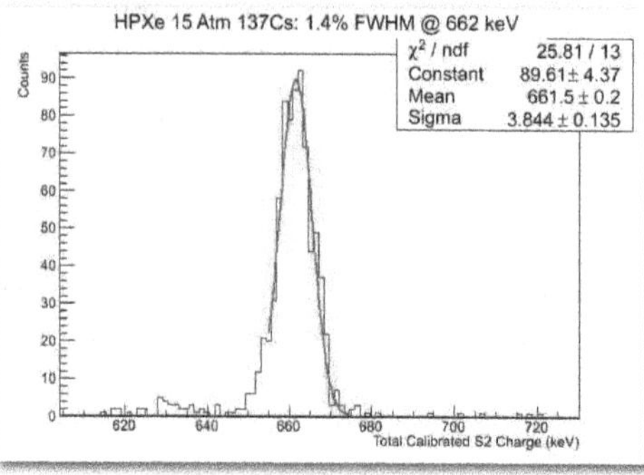

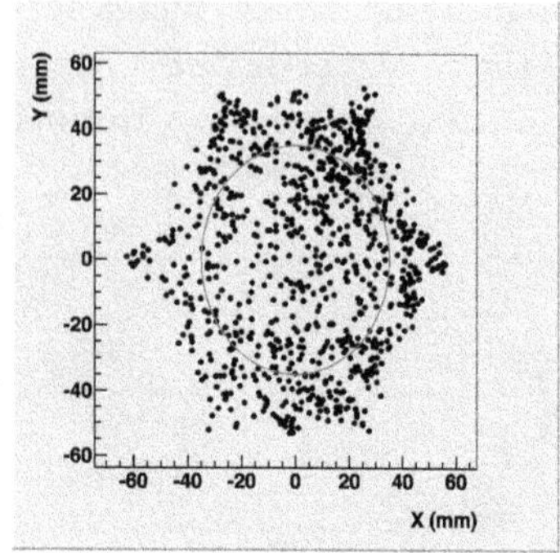

Fig. 3. a) **Left:** Calibrated and attachment corrected spectrum for ^{137}Cs 662 keV γ-rays with reconstructed radius less than 1.5 cm, obtained with the LBNL prototype. b) **Right:** Reconstructed (x, y) position of α particles from radioactive contaminants decaying within the gas of the IFIC prototype.

as described for the LBNL prototype. While the (x, y) reconstruction algorithm will certainly be improved in the future, its current implementation is sufficient to clearly reconstruct the hexagonal shape of the light tube.

4. Conclusions

NEXT is a new-generation of $\beta\beta^{0\nu}$ experiments to be installed in the Canfranc Underground Laboratory, Spain. It will provide both very good energy resolution and topological recognition for high background rejection. Ongoing R&D in low scale prototypes had and/or is demonstrating that the proposed performance can be achieved in the final 100 kg detector. The technical design of the main features of the final detector has finished and the diferent parts are beginning to be constructed. We expect the commissioning and the start of data acquisition by early 2013.

References

1. M. C. Gonzalez-Garcia and M. Maltoni, *Phys. Rep.* **460**, 1 (2008).
2. S. R. Elliott and J. Engel, *J. Phys. G: Nucl. Part. Phys* **30**, p. R183 (2004).
3. The NEXT collaboration, *arXiv:1106:3630v1 [hep-ex]* (2011).
4. D. Nygren, *Nucl. Instrum. Methods Phys. Res., Sect. A* **603**, 337 (2009).
5. A. Bolotnikov and B. Ramsey, *Nucl. Instrum. Methods Phys. Res., Sect. A* **396**, 360 (1997).
6. J. M. F. dos Santos et al, *X-Ray Spectrom.* **30**, 373 (2001).
7. C. A. B. Oliveira et al, *Phys. Lett. B* **703**, 217 (2011).
8. C. A. B. Oliveira et al, *Journal of Instrumentation* **6**, p. P05007 (2011).
9. The NEXT Collaboration, *arXiv:0907.4054v1 [hep-ex]* (2009).

The KATRIN Experiment

Matthias Prall for the KATRIN Collaboration

Institut für Kernphysik, University of Münster,
Wilhelm-Klemm-Straße 9, 48149 Münster, Germany
E-mail: matthias.prall@uni-muenster.de
http://www.uni-muenster.de/Physik.KP/AGWeinheimer/

The **KA**rslruhe **TRI**tium **N**eutrino experiment, KATRIN will determine the neutrino mass scale with a sensitivity of 0.2 eV/c^2 (90% CL) via a measurement of the T_2 β-spectrum near its endpoint at 18.57 keV. The experiment consists of a windowless gaseous Tritium source, a differential- and cryopumping section, the pre- and main-spectrometer, both of the MAC-E filter type and a pixelated silicon detector. A background of less than 10 mHz and an energy resolution of 0.93 eV are necessary to achieve the desired sensitivity within 1000 days of data-taking. The experiment is currently reaching its final commissioning phase. In these proceedings, we focus on the main-spectrometer and its inner wire electrode.

Keywords: neutrino mass, KATRIN, MAC-E filter, wire electrode

1. Overview over the KATRIN Experiment

The **KA**rslruhe **TRI**tium **N**eutrino experiment, KATRIN, will determine $m_{\nu_e}^{(eff)} = \sqrt{\sum |U_{ei}|^2 m_{\nu_i}^2}$ with a sensitivity of 0.2 eV/c^2 (90% CL) via a measurement of the shape of T_2 β-spectrum near its endpoint at $E_0 =$ 18.57 keV.[1,2] Figure 1 depicts a schematic view of the 70 m long KATRIN setup:[1] Gaseous T_2 with an activity of about 10^{11} Bq is injected into the so-called windowless gaseous Tritium source[3,4] (WGTS) at a pressure of $p \approx 10^{-3}$ mbar ①. At $B = 3.6$ T the β-decay electrons follow the B-field towards both ends of the WGTS. Those leaving the WGTS towards the spectrometers are guided through the experiment and are eventually imaged on the detector ⑥. A magnetic flux of $\phi = 191$ T $\cdot$ cm^2 is transported through the experiment. Whereas electrons are guided towards the main-spectrometer ⑤ by the magnetic field, the T_2 must not enter the spectrometers. It is removed in the differential pumping section (DPS) ② and in the cryo pumping section (CPS) ③. The gas collected by the turbomolecular pumps in the DPS is purified and fed back into the T_2-cycle. In the CPS, T_2 is trapped on Argon frost at a temperature in the range of 3-2.5 K.[5,6] The DPS and CPS reduce the T_2-flux into the spectrometers to 10^{-14} mbar · l/s. This is 14 orders of magnitude smaller than the inlet rate of the WGTS.

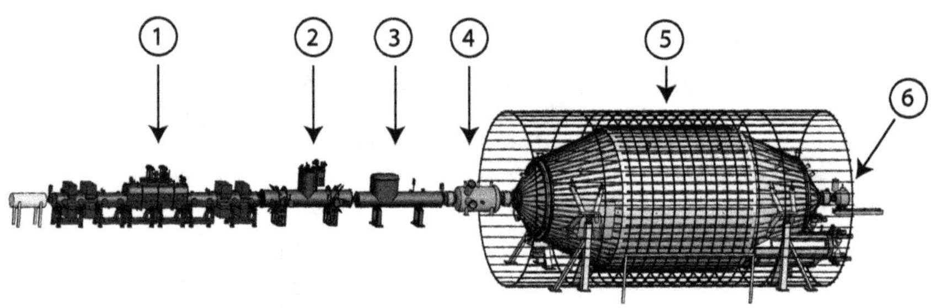

Fig. 1. Overview over the KATRIN experiment. ① Windowless gaseous Tritium source (WGTS), ② Differential pumping system (DPS), ③ Cryo pumping system (CPS), ④ Pre-spectrometer, ⑤ Main-spectrometer with earth field compensation system and 6) detector.

Both the pre-spectrometer (PS) and main-spectrometer (MS) are of MAC-E filter type.[7,8] The PS with an energy resolution of $\Delta E_{\text{PS}} \approx 100$ eV is the first electric potential barrier for the β-decay electrons. The PS will be operated at a fixed potential[9] below the endpoint $E_0 = 18.57$ keV of the T_2 β-spectrum. This reduces the flux of e^- into the MS and therefore also the probability for background production in inelastic collisions with residual gas atoms or molecules in the MS. The MS scans the last 30 eV of the T_2 β-spectrum, which contain the information on the neutrino mass, with a resolution of $\Delta E_{MS} = 0.93$ eV. The partial pressure of T_2 inside the MS (5) will be $p_{\text{tritium}} = 10^{-20}$ mbar equivalent to 10^{-3} decays/s. Finally, the transmitted electrons are counted by a 148 pixel PIN diode with an energy resolution of $\Delta E_{\text{det}} \approx 1$ keV. The segmentation of this detector allows one to correct for inhomogeneities of the electric and magnetic fields inside the spectrometers. Active and passive shields will minimize detector background. Additional post-acceleration by $E_{\text{accel}} = 30$ keV of the transmitted electrons towards the detector to $E \approx E_0 + E_{\text{accel}} \approx 48.6$ keV will reduce the background rate in the region of interest.

2. The MAC-E Filter Principle

The high sensitivity of 0.2 eV for $m_{\nu_e}^{(eff)}$ in KATRIN relies on the MS, being of MAC-E filter type (**M**agnetic **A**diabatic **C**ollimation combined with an **E**lectrostatic Filter). This type of spectrometer was already used for the Mainz[7,10,11] and Troitsk[12,13] neutrino mass experiments. The principle of a MAC-E filter is illustrated in Fig. 2: Two solenoids provide a magnetic guiding field. The β-decay electrons enter the MAC-E filter with a certain starting angle θ_{start} between the magnetic field line and the electron momentum in the left solenoid and are guided by the magnetic field lines along helix-like trajectories against a retarding potential qU_{ret} reflecting electrons with energy E below qU_{ret}. In order to understand the resolution of the MAC-E filter, we split up the kinetic energy of the electron into a component parallel and perpendicular to the guiding magnetic field line: $E_{\text{kin}} = E_{\perp} + E_{||} = E_{\text{kin}} \cdot \sin^2\theta + E_{\text{kin}} \cdot \cos^2\theta$. θ is the angle between the electron momentum and the guiding B-field line (cf. Fig. 2). The electrons are born in the source in β-decays in a field of $B_{\text{source}} = 3.6$ T $< B_{\text{max}}$.

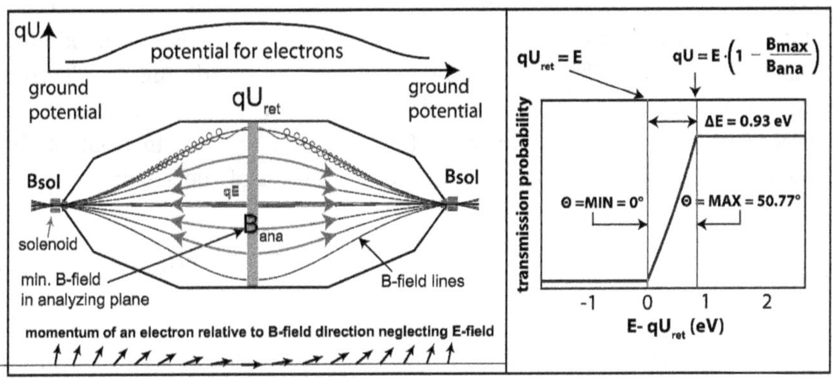

Fig. 2. Working principle (left) and transmission function (right) of the main-spectrometer.

On their way to the detector, they have to pass the field $B_{\text{max}} = 6$ T of the so-called pinch magnet placed between the MS and the detector. Here, electrons are reflected by the magnetic mirror effect if their starting angle θ in the WGTS was above a critical reflection angle $\theta_{\text{mirror}} = \arcsin\sqrt{B_{\text{source}}/B_{\text{max}}} \approx 50.77°$. Like this, electrons which were born with very large starting angles $\Theta > \Theta_{\text{mirror}}$ and thus had a very long trajectory and scattering probability in the WGTS cannot reach the detector. Only the component $E_{\|}$ can be analyzed by the decelerating retarding potential of the MS. The energy resolution of the MS is thus determined by $E_{\perp}$ in the analyzing plane where the retarding potential is maximum. This $E_{\perp}$ is minimized as follows: The quantity $\gamma\mu = E_{\perp}/B$ is conserved if the electron motion is adiabatic, i.e. if the following adiabaticity requirement is fulfilled:[14]

$$\frac{|\Delta\vec{E}|}{|\vec{E}|} << 1 \text{ and } \frac{|\Delta\vec{B}|}{|\vec{B}|} << 1 \text{ within } l_{\text{cyc}} : l_{\text{cyc}} = 2\pi \frac{v_{\|} m_e}{|q_e| B} \tag{1}$$

This requirement was taken into account in the design of the main spectrometer. In KATRIN, where electron energies are at maximum $E_0 =$ 18.57 keV, one has $\gamma \leq 1.04$ and μ - the orbital magnetic momentum of the electrons orbiting the magnetic field lines is a good approximation for a conserved quantity, especially when electrons are slowed down by the electric field in the spectrometers. As μ is conserved, $E_{\perp} = \mu B$ drops while B decreases from the entrance ($B_{\text{sol}} = 4.5$ T) to the analyzing plane ($B_{\text{ana}} =$ $3 \cdot 10^{-4}$ T, cf. Fig. 2). This transfer is indicated by the momentum vectors in fig. 2. The resolution ΔE of a MAC-E filter is given by the maximum transverse energy $E_{\perp,\text{ana}}^{\text{max}}$ in its analyzing plane. Considering an electron from the T_2 endpoint, which has all its kinetic energy E_0 in the transverse component $E_{\perp}$ in the pinch magnet ($B_{\text{max}} = 6$ T) placed in front of the detector, one arrives at a formula for the resolution of the set-up:

$$const. = \mu = \frac{E_0}{B_{\text{max}}} = \frac{E_{\perp,\text{ana}}^{\text{max}}}{B_{\text{ana}}} = \frac{\Delta E}{B_{\text{ana}}} \quad \Rightarrow \quad \Delta E = E_0 \cdot \frac{B_{\text{ana}}}{B_{\text{max}}} = 0.93 \text{ eV} \tag{2}$$

The transmission probability $T(E, qU_{ret})$ of a MAC-E filter is derived by integrating over all electrons, which fulfill $0 < E_{\|} = (E - qU_{ret}) - E_{\perp}$ in the center plane of the MAC-E filter. Electrons with start energy E and θ from 0° to 90° in the entry-side magnet have to be considered. The conservation of μ has to be used to transform between $E_{\perp}$ in the entry-side magnet and center plane.

$$T(E, qU) = 1 - \sqrt{1 - \frac{E - qU}{E} \cdot \frac{B_{\text{max}}}{B_{\text{ana}}}} \tag{3}$$

Thus, the transmission probability $T(E, qU_{\text{ret}})$ depends only on the magnetic field strengths, the energy E of the incoming electron and qU_{ret},

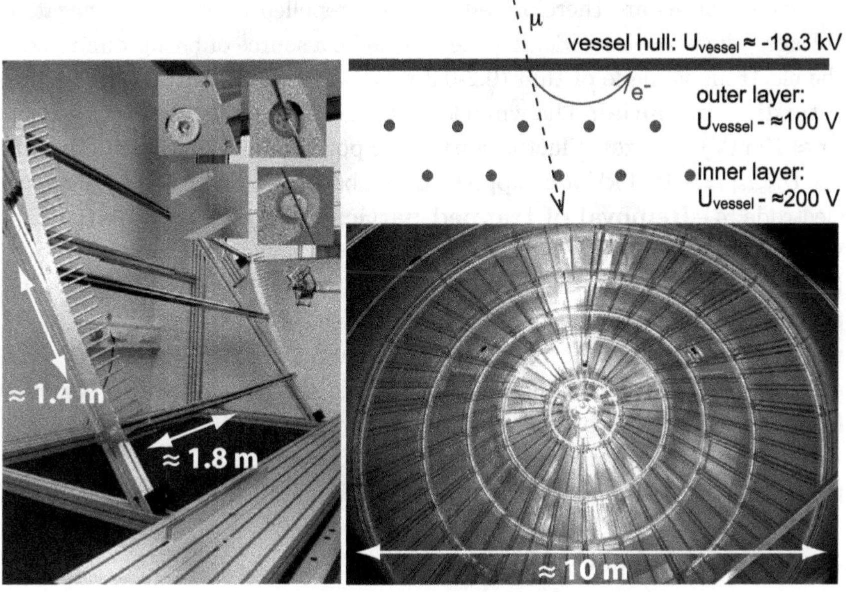

Fig. 3. The wire electrode of the main spectrometer.

the spectrometer potential. Below the interval specified in Fig. 2, the transmission probability is zero. Above this interval it is one.

The actual retardation potential $qU_{\text{ret}} \approx 18.57$ kV inside the main spectrometer is created by the vessel, being on $U_{\text{vessel}} = -18.3$ kV and a system of wire electrodes[15,16] covering the inner surface of the spectrometer vessel (Fig. 3). The wire layers carry negative offset voltages of ≈ -100 V (outer wire layer) to ≈ -200 V (inner wire layer) with respect to the vessel hull. The purpose of this electrode system is four-fold: **1) Shaping of the electric potential:** During a measurement, the 248 wire electrode modules are supplied with 23 independent voltages. This allows one to fine-tune the retarding potential and to avoid Penning traps leading to increased background.[17] **2) Background reduction:** Cosmic muons and residual radioactivity will create secondary electrons with energies mostly below 100 eV and a rate of about 10^5 events/s in the walls (about 650 m^2) of the MS. The rate of secondary electrons accelerated towards the detector by the retarding potential has to be below 10 mHz[1] near the T_2 β-endpoint E_0. The B-field inside the MS already suppresses the secondary electron background by a factor of about 10^5 to the Hz level. An additional background reduction by a factor 100 is necessary. The low energy ($\lesssim$ 100 eV) secondary electrons are therefore additionally repelled by a grid on negative voltage - the wire electrode. In order not to be a source of background itself, the electrode is made of thin (0.2-0.3 mm) wires.**3) Stabilization of the retarding potential:** The wire electrode can be supplied with very clean ($\sigma \approx 10$ mV) voltages. Fluctuations of the potentially less stable tank voltage $U_{\text{vessel}} \approx -18.3$ kV are suppressed by about a factor of ten by the wire electrode. **4) Removal of trapped particles via the ExB-drift:** The wire electrode can be split into a left and right dipole half. An offset voltage ($\Delta U \approx 1$ kV) between the dipole halves can be applied. Electrons passing though the main spectrometer will experience a drift $\vec{v} = (\vec{E} \times \vec{B})/B^2$ perpendicular to the electric dipole field E and the magnetic field B. This effect can be used during measurement breaks to remove trapped particles from the flux tube. Forthcoming publications, currently prepared by the KATRIN collaboration will present the electromagnetic design[17] and the technical realization[18] of the wire electrode in detail.

3. Acknowledgment

The work of the author was paid by the German Ministry for Education and Research (BMBF) under contract number 05A08PM1 and 05A11PM2.

References

1. Angrik J *et al* KATRIN Design Report 2004, http://www-ik.fzk.de/katrin
2. E. Otten and Ch. Weinheimer Rep. Prog. Phys. **71** 086201 (2008)
3. S. Grohmann *et al.* Cryogenics 49, Issue 8 p. 413-420 (2009)
4. S. Grohmann *et al.* Cryogenics 51, Issue 8 p. 438-445 (2011)
5. S. Grohmann *et al.* ICEC 23-ICMC2010 conf. proc. (2010)
6. O. Kazachenko *et al*, NIM A 587 p. 136-144 (2008)
7. A. Picard *et al.* Z. Phys. A 342 p. 71-78 (1992)
8. V.M. Lobashev, Nucl. Inst. and Meth. A 240 p. 305-310 (1985)
9. M. Prall *et al.* publication in preparation
10. Ch. Kraus *et al* Eur. Phys. J. C 40, Issue 4 p. 447-468 (2005)
11. A. Picard *et al.* Nucl. Inst. Meth. B 63, Issue 3 p. 345-358 (1992)
12. V.M. Lobashev, Prog. Part. Nucl. Phys. 48 p. 123-131 (2003)
13. V.M. Lobashev *et. al.*, Nucl. Instr. Meth. A 238 p. 496-499 (1985)
14. B. Lehnert, Dynamics of charged particles, North Holland Publishing (1964)
15. M. Prall *et al.* J. Phys. Conf. Ser. **136** 042090 (2008)
16. K. Valerius *et al.* Prog. Nucl. Part. Phys. 64 Issue 2 p. 291-293 (2010)
17. K. Valerius *et al.* publication in preparation
18. M. Prall *et al.* publication in preparation

DM-Ice: A Direct Dark Matter Search at the South Pole

S. H. Seo*

Oskar Klein Centre and Department of Physics, Stockholm University,
Stockholm, SE 10691, Sweden
for the DM-Ice Collaboration
**E-mail: seo@fysik.su.se*
dm-ice.physics.wisc.edu

DM-Ice is a proposed NaI(Tl) scintillator-based experiment to be located at a depth of 2.5 km in the South Pole ice. It will be designed to look for the annual modulation observed by the DAMA experiment. This experiment complements dark matter search efforts in the northern hemisphere and will probe the observed annual modulation by going to a location where the phase of the many suspected backgrounds are reversed whereas the dark matter signal should remain the same. The unique location will allow for the study of background effects correlated with seasonal variations and the surrounding shielding and environment. In this paper it will be described that DM-Ice-17, a 17 kg detector deployed in December 2010, and the full-size DM-Ice detector capable of checking the DAMA signal.

Keywords: Dark matter, WIMP, Annual modulation, NaI(Tl) crystal, South Pole.

1. Current Landscape for Direct Detection of Dark Matter

DAMA collaboration has claimed a discovery (8.9 σ) of dark matter by observing 13 cycles of annual modulation with 1.17 ton-year data taken with NaI(Tl) crystals in the underground lab of Gran Sasso (3800 m.w.e.).[1] Recent results from CoGeNT collaboration on their low energy excess events[2] and annual modulation (2.7 σ)[3] analyses showed a hint of light WIMP dark matter. Following the CoGeNT, CRESST-II collaboration announced their excess events above known background from 730 kg-day data (4.2 σ at m_χ = 11.6 GeV, 4.7 σ at m_χ = 25.3 GeV).[4] The results of these experiments are consistent with light WIMP although it is difficult to reconcile them with each other with the currently available dark matter and astrophysical models.[5,6] In addition, the CDMS-II[7] and XENON100[8] see no evidence for such dark matter. Figure 1 shows the results from these experiments under the standard MSSM and WIMP halo model.

There are still open issues such as the need for more thorough background studies in DAMA, CoGeNT, and CRESST-II. For example, there are many environmental effects that have annual modulation, though none has been conclusively shown to produce the modulation observed by DAMA. There are also aspects of the DAMA results that seem inconsistent with studies carried out by others.[9]

Although no environmental effects have been shown to be able to fully describe the annual modulation observed by DAMA, many hypotheses have been brought forward (see e.g. Refs.[10,11]). By conducting an experiment with the same detector medium as DAMA in the southern hemisphere, and in particular in the Antarctic ice at the South Pole, many of the possible environmental backgrounds would have seasonal variation that are either reverse in phase or they are absent altogether.

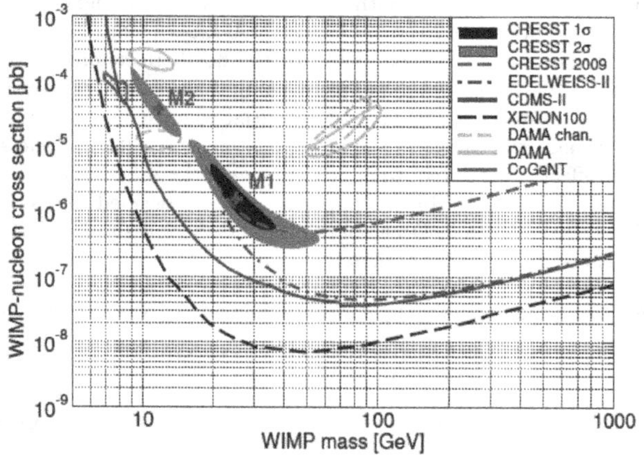

Fig. 1. The spin-independent WIMP-nucleon cross-section as a function of WIMP mass. This plot was taken from Ref.[4]

2. DM-Ice-17 Detector

DM-Ice collaboration deployed 2 modules of encapsulated NaI(Tl) crystals, 8.5 kg each, in December 2010 along with the IceCube[12] string deployment. Each module is housed in a stainless steel pressure vessel containing a NaI(Tl) crystal attached to two light guides and PMTs. The pulses from the two PMTs on a single module are triggered in coincidence and digitized using IceCube mainboards.[13] The NaI(Tl) crystal and attached PMTs in DM-Ice-17 detector are from the NAIAD experiment[14] which had 5 - 10 times higher background than DAMA depending on the crystal. Figure 2 shows a schematic of a single module of DM-Ice-17. The two modules were deployed on two separate IceCube strings roughly 500 m apart, and positioned at −2457 m depth (2200 m.w.e.).

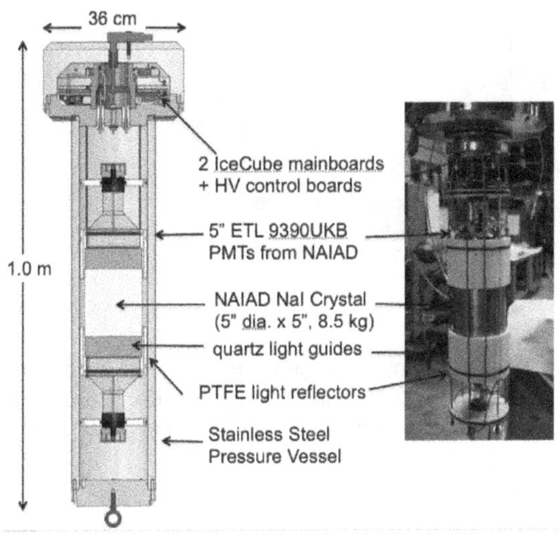

Fig. 2. The DM-Ice-17 detector (only one of two modules is shown). Left: schematic view, Right: photo of the detector without the pressure vessel.

DM-Ice-17 detectors have been taking data continuously since March 2011. The data is being transmitted over the satellite and is currently being analyzed.

3. DM-Ice Sensitivity and Discovery Potential

If DM-Ice does not observe any annual modulation in the low energy region of interest, then an upper limit on the WIMP-nucleon scattering cross-section can be set. Figure 3 shows the sensitivities (90% CL) of DM-Ice for 500 kg-year worth of data with different numbers of background events in 2-6 keV_{ee} energy range. The shaded areas represent DAMA's allowed regions of WIMP signal: dark (light) gray represents 1σ (2σ) allowed region. As shown in the figure, DM-Ice can exclude DAMA with 5 cpd/keV_{ee}/kg (dru), about 5 times the background rate reported by DAMA.

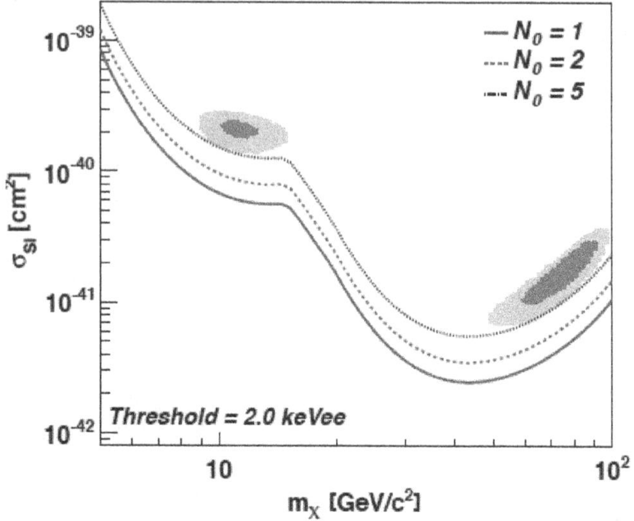

Fig. 3. DM-Ice sensitivities (90% CL) for 2-6 keV$_{ee}$ interval for 500 kg-year worth of data. Each line represents sensitivity with different number of background event scenarios. Background rates, N_0, are in units of dru (= cdp/keV$_{ee}$/kg). Figure from Ref.[15]

On the other hand, if we assume DM-Ice observes a DAMA-like annual modulation signal with the same level of background as DAMA (1 dru), then we can make a discovery with greater than 5σ significance using 2 years of data taken with 250 kg NaI(Tl) crystal.

4. DM-Ice Full Detector Concept and Outlook

The concept of DM-Ice full detector will be similar to the DM-Ice-17 detectors. However, the dimension of a single crystal might be different and several crystals might be housed together inside a single but bigger pressure vessel. At each end of a single crystal two light guides followed by two PMTs will be attached. The two PMTs in a single crystal will be triggered in coincidence condition so that we can reduce background noise events as well as multi-crystal events. The readout electronics will be housed inside stainless steel pressure vessel which might have an optional copper shield inside to prevent photon penetration. We consider to make our own readout electronics rather than using the IceCube mainboards unlike DM-Ice-17. We plan to put the full DM-Ice detector at the bottom (−2457 m depth) ideally near the center of IceCube/DeepCore for muon veto in case it is needed. Depending on the total mass of the crystals housed in a single pressure vessel module, we might need more than one of such a pressure vessel module. Our tentative plan is to have 250 kg of NaI(Tl) crystals or more in total.

The final design of DM-Ice full detector will mainly depend on the commercial production lines of the pure NaI(Tl) crystal. Regardless of the final design, however, it will be located at −2457 m depth to have enough overburden. It is possible that we might have local muon veto consisting of two IceCube-like Digital Optical Modules (DOM)[16] surrounding each pressure vessel module at the top and the bottom.

In view of saving drilling cost at the South Pole, DM-Ice full detector deployment date might depend on the IceCube's South Pole season plan to extend their low energy physics with DeepCore infill strings, which is expected to happen near future. In this circumstance our schedule might need to be nicely coordinated with IceCube's plan.

5. Conclusion

There is a serious experimental tension among recent results from different direct dark matter search experiments: DAMA, CoGeNT, and CRESST-II, CDMS-II and XENON100. By operating a DAMA-like detector at the South Pole we will remove several environmental effects which might cause DAMA modulation. DM-Ice-17 detector has demonstrated that we can do direct dark matter search with NaI(Tl) crystal at the South Pole. This gives us a great opportunity to go ahead with DM-Ice full detector. The R&D of the full detector is on going and we expect that it would be deployed in the near future possibly synchronized with IceCube season plans. Under the successful deployment and operation of DM-Ice full detector, it is highly expected that we could draw an important statement on DAMA claim on dark matter discovery which is not yet confirmed/refuted by any other experiments using the same detector technology.

References

1. R. Bernabei et al., Eur. Phys. J. **C67**, 39 (2010).
2. C. E. Aalseth et al., Phys. Rev. Lett. **106**, 131301 (2011).
3. C. E. Aalseth et al., Phys. Rev. Lett. **107**, 141301 (2011).
4. G. Angloher et al., arXiv:1109.0702 (2011).
5. P. Fox et al., arXiv:1107.0717v2 (2011).
6. M. Farina et atl., arXiv:1107.0715v2 (2011).
7. Z. Ahmed et al., Science **327**, 1619 (2010).

8. E. Aprile et al., arXiv:1104.2549 (2011).
9. Topics in Astroparticle and Underground Physics (TAUP 2009) IOP Publishing Journal of Physics: Conference Series 203, 012039 (2010).
10. J. Ralston, arXiv:1006.5255 (2010).
11. D. Nygren, arXiv:1102.0815 (2011).
12. F. Halzen and S.R. Klein, Rev. Sci. Instrum. **81**, 081101 (2010).
13. R. Abbasi et al., Nucl. Inst. Meth **A601**, 294 (2009).
14. G. J. Alner et al., Phys. Lett., **B616**, 17 (2005).
15. J. Cherwinka et al., arXiv:1106.1156 (2011).
16. A. Achterberg et al., Astropart. Phys. **26**, 155 (2006).

RECENT RESULTS FROM SUPER-KAMIOKANDE

Y. TAKEUCHI* for the Super-Kamiokande Collaboration

Department of Physics, Kobe University,
Kobe, Hyogo 657-8501, Japan
**E-mail: takeuchi@phys.sci.kobe-u.ac.jp*

Super-Kamiokande is a large water Cherenkov detector located at 1000 m underground in Japan. Currently, the fourth phase of the experiment has been running with an upgraded front-end electronics system since September 2008. In this paper, the recent results on atmospheric, solar, and supernova relic neutrinos are reported. A possible future improvement is also reported.

Keywords: neutrino; supernova; water cherenkov.

1. Super-Kamiokande

Super-Kamiokande (Super Kamioka Nucleon Decay Experiment, SK) is a cylindrical water Cherenkov detector with 50-kton of purified water and 11129 of 20-inch PMTs.[1] It is located at 1000 m underground in Kamioka Observatory in Japan.

The main physics targets of the experiment are solar neutrinos, atmospheric neutrinos, supernova neutrinos, and nucleon decays. The fiducial volume of the SK detector is 22.5 kton. SK is the current largest detector in the world for these physics targets. Figure 1 shows a schematic view of the SK detector and the inside of the detector.

The observation with Super-Kamiokande was started in April 1996. Table 1 shows a summary of the experimental phases of SK. After a few reconstruction works and upgraded works, Super-Kamiokande-IV is currently running since September 2008 with the lowest energy threshold in SK at the electron kinetic energy, $E_{kinetic} = 4.0$ MeV. In SK-IV, we have replaced the front-end electronics system in order to collect all the hit information of PMT's.[2] Then, a sophisticated trigger logics are applied by software. For example, appropriate time window of an event is selected as follows; 1.5 μs for the low energy (= high rate) events, 40 μs for the normal events, 540 μs for the high energy ($>\sim$ 10 MeV ($>\sim$ 8 MeV since Sep. 2011)) without

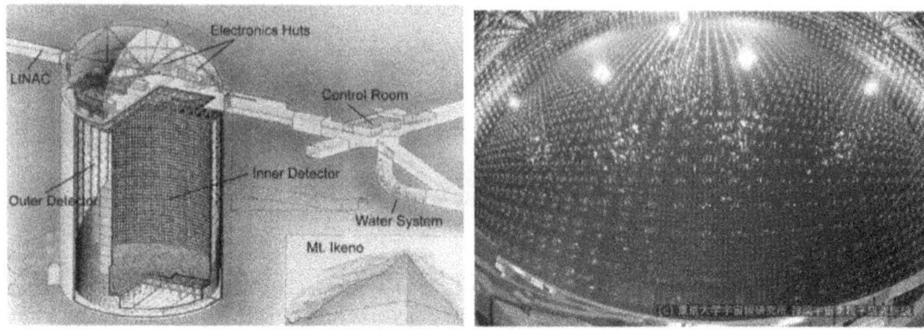

Fig. 1. Right: A schematic view of the Super-Kamiokande detector.[1] Left: Inside of the Super-Kamiokande detector during water filling in June 2006.

Table 1. Summary of the experimental phases of Super-Kamiokande.

Phase	Run period	Number of ID PMT	Photo coverage(%)	Analysis energy threshold Total / Kinetic Energy
SK-I	Apr. 1996 – Jul. 2001	11146	40	5.0 MeV / 4.5 MeV
SK-II	Dec. 2002 – Oct. 2005	5182	19	7.0 MeV / 6.5 MeV
SK-III	Jul. 2006 – Aug. 2008	11129	40	5.0 MeV / 4.5 MeV
SK-IV	Sep. 2008 – today	11129	40	4.5 MeV / 4.0 MeV

OD activity events, and $\pm 512\ \mu$s around the beam spill timing of the T2K experiment. In future, we would like to lower the energy threshold down to $E_{kinetic} = 3.5$ MeV or less.

2. Atmospheric neutrino and related results

Cosmic-ray interactions in Earth atmosphere produce neutrinos. They are called atmospheric neutrinos. The observed energy range in SK detector is 0.1GeV $\sim$ 10TeV. The atmospheric neutrino events are categorized into fully contained (FC) events, partially contained (PC) events, and upward going muons. A clear neutrino oscillation effect is observed especially in the upward muon neutrino events.

We have released the results of the following studies; Full 3-flavor analysis with SK-I,II,III data,[3] CPT violation study with SK-I,II,III data,[4] Non Standard Interaction study with SK-I,II data,[5] and Tau appearance search with SK-I,II,III data.[6]

The results of the nucleon decay searches with SK-I,II,III,IV data are also released. Table 2 shows a list of the results of the typical decay mode.

An Indirect Search for WIMPs in the Sun with SK-I,II,III data is also released.[7]

Table 2. Summary of typical nucleon decay searches in SK.

Mode	Exposure (Kton · year)	90 % C.L. Limit ($\times 10^{33}$ year)
$p \to e^+\pi^0$	219.7	12.9
$p \to \mu^+\pi^0$	219.7	10.8
$p \to \bar{\nu}K^+$	219.7	4.0
$n \to e^-K^+$	140.9	1.0

3. Solar neutrino results

SK detects solar neutrinos through neutrino-electron elastic scattering (ES), $\nu + e \to \nu + e$, where the energy, direction and time of the recoil electron are measured. Due to its large fiducial mass of 22.5 kton, SK gives a precise measurement of the solar neutrino flux, energy spectrum, and time variation via the ES reaction.

The results of the solar neutrino observation in SK-III are reported at the Neutrino 2010 conference.[3,8] After that, we have performed the updated 3-flavor oscillation analysis with new KamLAND data[9] and new Borexino[10] data. Figure 2 shows the allowed region for Δm^2_{12}, $\sin^2\theta_{13}$, and $\sin^2\theta_{12}$ extracted from this analysis. The updated solar global + KamLAND analysis

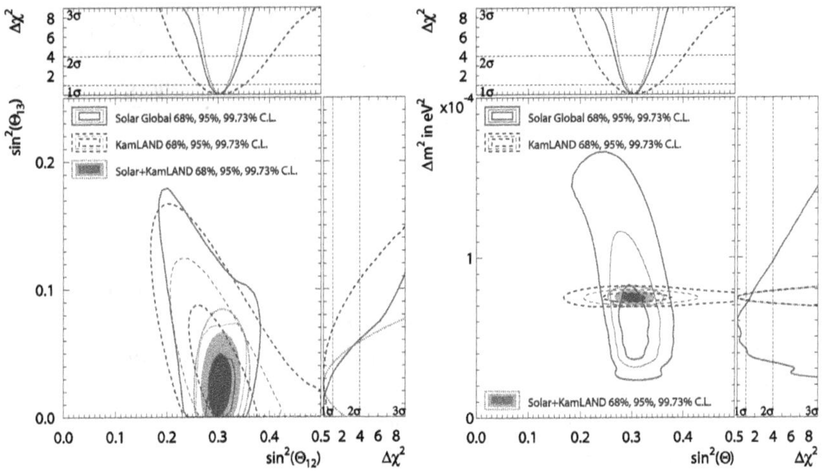

Fig. 2. Allowed region for Δm^2_{12}, $\sin^2\theta_{13}$, and $\sin^2\theta_{12}$. The details of the analysis is described in here.[8] KamLAND and Borexino data are updated from the previous analysis.

finds that the preliminary best fit value of $\sin^2\theta_{13}$ is at $0.025^{+0.017}_{-0.016}$ (stat.).

Figure 3 shows the angular distribution of the current SK-IV final data sample in lowest energy regions. The fiducial volume is central 12.3 kton

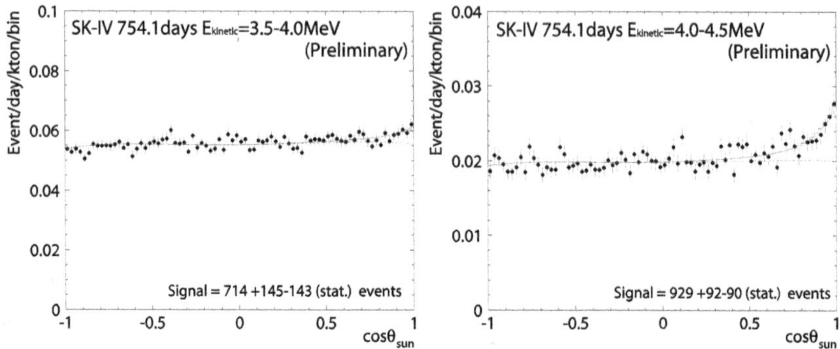

Fig. 3. Solar neutrino data sample in SK-IV in $E_{kinetic} = 3.5 - 4.0$ MeV (Left) and in $E_{kinetic} = 4.0 - 4.5$ MeV (Right). The fiducial volume is central 12.3 kton for both plots. The solid and dashed lines correspond to best-fit signal shape and expected background shape, respectively.

to reduce background events from the detector wall. A clear signal is seen in $E_{kinetic}$ = 4.0–4.5 MeV energy region. Some enhancement is seen in the $E_{kinetic}$ = 3.5–4.0 MeV energy region, though the trigger efficiency is not 100%. We are developing a new software trigger logic with a high performance computer system in order to lower the analysis energy threshold down to $E_{kinetic}$ = 3.5 MeV or less.[11]

4. Supernova relic neutrino results

Supernova relic neutrinos (SRNs) are produced from all core-collapse supernovae in the history of the universe. This signal has never been detected. A study with SK data in 2003 set the world limit on the flux.[12] We have updated the previous analysis as follows; cross section is updated to Strumia-Vissani,[13] improve event selection, energy threshold is lowered from 18 MeV to 16 MeV, increase observation data, and introduce alternative representations.[14]

Figure 4 shows the updated results from 2853 days of SK-I,II,III data. The left plot in Fig. 4 shows the preliminary updated SRN flux limits. In the previous analysis, the results for various relic models were similar. Therefore one unified result was reported. This is not true with 16 MeV threshold and increased data. As a result, we tested five (or six) example cases.[15]

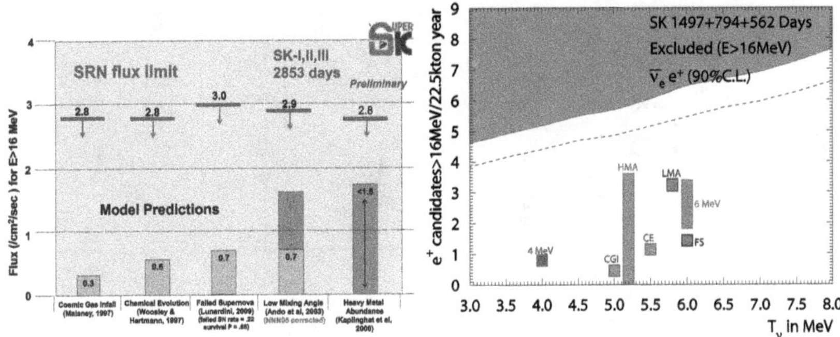

Fig. 4. Left: Updated SRN flux limits (90% C.L.). Right: Excluded area of supernova flux as a function of neutrino temperature. The dashed line shows the 90 % C.L. upper limit obtained for a fixed neutrino temperature. (preliminary) Theoretical models are obtained from here.[15]

The right plot in Fig. 4 is an alternative representation based on two-parameter description of the neutrino emission spectrum.[16] In this calculation, the spectrum is assumed to be Fermi–Dirac distribution and the parameters are the total energy (luminosity) of SN neutrinos and the neutrino average energy (temperature of the Fermi–Dirac distribution). The details of the analysis are reported in here.[14]

4.1. *Possible future improvements on SRN measurement*

In SK detector, SRNs will be observed by inverse beta reaction: $\bar{\nu}_e + p \rightarrow n + e^+$. After the inverse beta reaction, neutron will be captured by proton in water, then generate a 2.2 MeV photon. SK-IV has a special logic to record all the PMT hits within 540 μsec after SRN candidates[a]. So, 2-fold coincidence could be applied to observe SRNs if we could detect the secondary neutrons.

So far, we have developed a tool to select neutron event(=2.2 MeV gamma) candidates in SK-IV. The preliminary signal selection efficiency is 19.8%, with 1.0% background probability in the 540 μsec time window. We have searched for neutron candidates in SK-IV data with this tool. Figure 5 shows the preliminary results from this neutron search. From these plot, we think we do see 2.2 MeV photons in SK-IV data. SRN search with neutron tagging in SK-IV is on going. Further reduction of nucleon decay background with neutron tagging will be studied.

[a]Total energy $>\sim$ 8 MeV ($>\sim$ 10 MeV, since Sep. 2011) and No OD activity

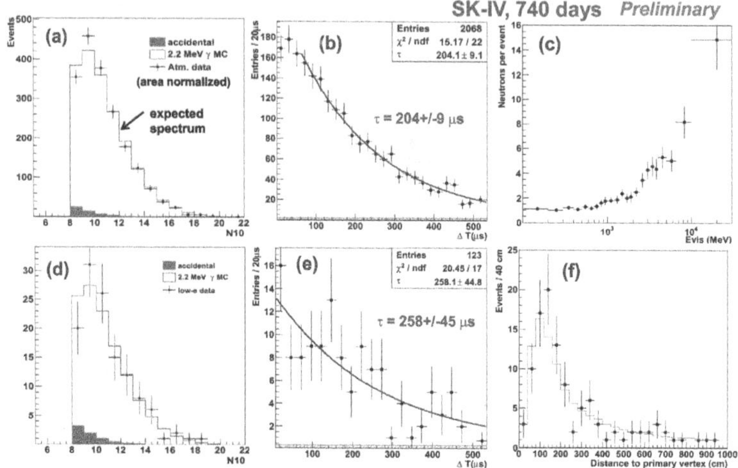

Fig. 5. Preliminary results from the neutron search in SK-IV. Atmospheric FC neutrino events is used for (a),(b),(c), and 16–18 MeV events with no OD activity are used for (d),(e),(f). (a),(d): Number of hit PMT within 10 nsec after time-of-flight(TOF) correction. The solid line is the expected spectrum by MC simulation. (b),(e): Time between the 1st event and the 2nd event (=neutron candidates). (c): Yield of neutron candidates in an atmospheric neutrino event as a function of the neutrino energy. (f): Distance distribution between the 1st event and the 2nd event. The solid line is the expected distribution by MC simulation.

The neutron capture efficiency in SK-IV is not so high. One of possible modifications, to increase the capture efficiency, is adding 0.2% Gd into SK detector (GADZOOKS![17]). To prove feasibility of GADZOOKS!, An R&D, EGADS (Evaluating Gadolinium's Action on Detector Systems), was started from 2009 in Kamioka Observatory, near SK site. Currently, EGADS water systems are in operation and Gd dissolving test is on going. The goal of EGADS is to determine if Gd loading in SK will be safe and effective by mid-2012.

5. Conclusion

Super-Kamiokande-IV is running with the lowest energy threshold in SK. Currently, 100 % efficiency is at $E_{kinetic} = 4.0$ MeV. We are trying lower the analysis energy threshold down to $E_{kinetic} = 3.5$ MeV or less.

Preliminary results and papers on atmospheric neutrinos, nucleon decay searches, astrophysics, solar neutrinos, and relic neutrinos are continuously released.

EGADS, an R&D to prove feasibility of Gd loading in SK detector, is on going in Kamioka Observatory. Results from EGADS are expected in 2012.

References

1. The Super-Kamiokande Collaboration, Nucl. Instr. and Meth. **A501** (2003) 418.
2. S. Yamada *et al.*, IEEE Trans. Nucl. Sci. **57** (2010) 428.
3. Y. Takeuchi, proceedings of the Neutrino 2010 conference, Nucl. Phys. B. (to be appeared)
4. The Super-Kamiokande Collaboration, arXiv:1109.1621, Phys. Rev. D (submitted).
5. The Super-Kamiokande Collaboration, arXiv:1109.1889, Phys. Rev. D (submitted).
6. R. Wendell, talk at the NNN10 conference, Toyama, Japan, December 13-16, 2010.
7. The Super-Kamiokande Collaboration, arXiv:1108.3384, Astrophysical Journal (accepted).
8. The Super-Kamiokande Collaboration, Phys. Rev. D **83** (2011) 052010.
9. The KamLAND Collaboration, Phys. Rev. D **83** (2011) 052992.
10. The Borexino Collaboration, Phys. Rev. Lett. **107** (2011) 141302.
11. G. Carminati, talk at the ICRC11 conference, Beijing, China, August 11-18, 2011.
12. The Super-Kmaiokande Collaboration, Phys. Rev. Lett. **90** (2003) 061101.
13. A. Strumia and F. Vissani, Phys. Lett. B **564** (2003) 42.
14. M. Smy, proceedings of the DPF-2011 Conference, Providence, RI, August 8-13, 2011, arXiv:1110.0012.
15. CE: D.H. Hartmann, S. E. Woosley, Astroparticle Physics 7, 137 (1997); CGI: R. A. Malaney, Astroparticle Physics 7, 125 (1997); FS: C. Lunardini, Phys. Rev. Lett. 102, 231101 (2009); KSW: M. Kaplinghat, G. Steigman, T.P. Walker, Phys. Rev. D62, 043001 (2000); LMA: S. Ando, K. Sato, T. Totani, Astroparticle Physics 18, 307 (2003); 4/6MeV: Horiuchi, Beacom, Dwek, Phys. Rev. D **79** (2009) 083013.
16. Horiuchi, Beacom, Dwek, Phys. Rev. D **79** (2009) 083013.
17. J. Beacom and M. Vagins, Phys. Rev. Lett. **93** (2004) 171101.

THE ENRICHED XENON OBSERVATORY (EXO) FOR DOUBLE BETA DECAY

M. WEBER* on behalf of the EXO collaboration

Einstein Center for Fundamental Physics, LHEP, University of Bern, Bern, 3007, Switzerland
**E-mail: manuel.weber@lhep.unibe.ch*
www.lhep.unibe.ch

The Enriched Xenon Observatory (EXO) is an experimental program designed to search for the neutrinoless double beta decay ($0\nu\beta\beta$) ^{136}Xe. Observation of $0\nu\beta\beta$ would determine an absolute mass scale for neutrinos, prove that neutrinos are massive Majorana particles, and constitute physics beyond the Standard Model. The current phase of the experiment, EXO-200, uses 200 kg of liquid xenon with 80% enrichment in ^{136}Xe. The double beta decay of xenon is detected in an ultra-low background time projection chamber (TPC) by collecting both the scintillation light and the ionization charge. The detector was first fully commissioned with natural xenon and is now taking data in its definitive configuration with enriched xenon. We report on the first observation of $2\nu\beta\beta$ decay in ^{136}Xe after 31 days of data taking using a fiducial mass of 63 kg.

Keywords: Double beta decay; Neutrino mass; Majorana particle; EXO.

1. The EXO-200 Detector

The EXO-200 detector uses 200 kg of xenon enriched to 80% in ^{136}Xe. The xenon is liquified in a time projection chamber[1] (TPC) and used for both the source of nuclear decays and detection medium. The TPC vessel is made out of ultra-low background copper and has the geometry of a cylinder of 40 cm diameter and 44 cm length. It is divided into two identical regions with a common high voltage plane in the middle of the detector. Each end of the cylinder consists of two wire grids and one array of 259 large-area avalanche photo diodes[2] (LAAPDs). The induction and collection wires cross at an angle of 60°, providing 2-dimensional position determination and energy readout of each charge deposit. The time interval between the scintillation signal in the LAAPDs and the charge collection time provides the third coordinate for a full 3-dimensional event reconstruction. For the

data presented here the cathode was biased to -8.0 keV. Field shaping rings in each detector half grade the field and limit the drift length to 19.2 cm yielding a field strength of 376 V/cm. The TPC is installed in a low background cryostat filled with high-purity HFE7000 fluid[3] serving the purpose of heat transfer and shielding from radioactive background. Both the liquid xenon and the HFE7000 are kept at a pressure of 147 kPa and a temperature of 167 K. An active pressure compensating system maintains a very small pressure differential (< 85 torr) across the vessel which is only 1.37 mm thin to keep backgrounds low.
The detector is housed in a clean room module which is surrounded by an array of plastic scintillator panels on four sides, vetoing muons which traverse the lead shield at an efficiency of 95.9%. The muon flux at the installation site, a salt deposit at the Waste Isolation Pilot Plant (WIPP) near Carlsbad, NM, has been measured[4] to be $3.1\times10^{-7}s^{-1}cm^{-2}str^{-1}$.

2. Radon measurements

The radon contamination in the xenon is measured using different techniques. The ^{238}U and ^{232}Th decay chains can be probed after the radon for α/β decays in coincidence with its daughter decay. From the specific activities of isotopes which are in secular equilibrium the radon content is deduced. The preferred method to measure the ^{222}Rn content is to look for ^{214}Bi which β decays into ^{214}Po followed by α decay 164 μs later. This

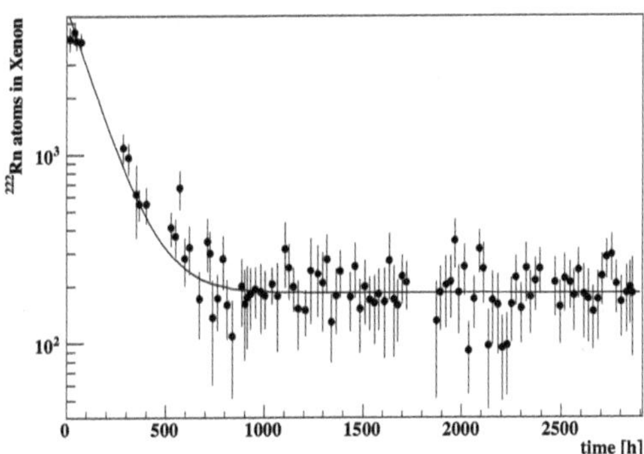

Fig. 1. Radon content in the xenon since the beginning of data taking with enriched xenon. The initial fill fed a large amount of ^{222}Rn from the bottles into the TPC which decayed with a half-life of 3.8 d.

sequence produces a characteristic event topology allowing a detection with high efficiency. β and α particles are discriminated by their ratio of ionization and scintillation yield. α-α coincidences provide a cross check of this method and also allow measuring the ^{220}Rn contamination. A third approach is to perform α spectroscopy and determine the contribution from the various α decays. All measured values agree with each other. Figure 1 shows the measured radon content in the xenon as a function of time since the beginning of data taking on April 30, 2011. The initial fill fed a large amount of radon from the bottles into the xenon which decayed away with a half live of 3.8 day and leveled off at a constant rate of 180 radon atoms.

3. Source calibration

A tubing around the TPC allows one to bring different sources to specific positions outside the vessel. The calibration data obtained from ^{60}Co and ^{228}Th sources is used for various purposes: It allows measuring the purity of the xenon which varies over time highly correlated with the flow rate of the recirculation pump. The electron lifetime is therefore parametrized

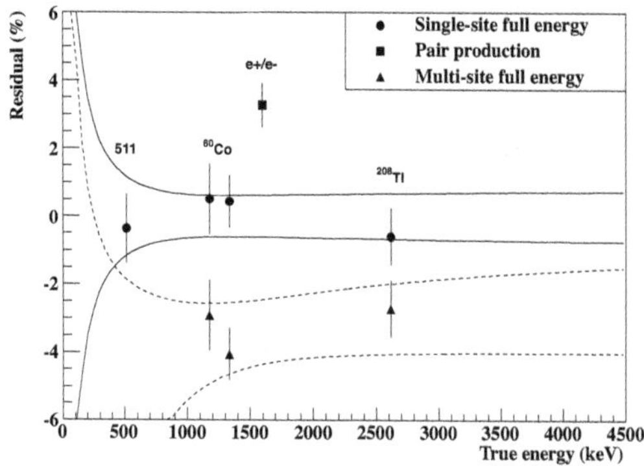

Fig. 2. Fractional residuals between the energy calibration points and the linear model. The single- (solid line) and mulit-cluster (dotted line) uncertainty bands are systematic, resulting from the finite accuracy of the position reconstruction and the electron lifetime correction.

which allows one to apply the appropriate corrections to charge clusters at any given time. The reconstructed energy of four full absorption peaks from ^{60}Co and ^{228}Th are related to their known energies: 1173 keV, 1332 keV, 2625 keV and 511 keV (annihilation radiation). The energy scale fits well to a linear function for both the single-cluster and multi-cluster event samples. The two scales, however, are found to be slightly (~4%) different, as shown in Fig. 2, because of the non-zero charge collection threshold on individual wire triplets. An additional calibration peak at 1592 keV is provided by selecting pair production events from the 2625 keV γs. This type of interaction is similar to that of $\beta\beta$ decays and is found to be slightly different from single-cluster depositions from γs. This shift is well reproduced by the simulation, once the induction between neighboring wire triplets and other electronic effects are taken into account.

A GEANT4[5] based simulation of the detector was developed including the drifting of the charge, propagation of the scintillation light, charge collection and signal digitization. The various sources were simulated and compared to the data which serves as a test of the correctness of the simulated geometry and the fitting program used for the low background data. The worst disagreement between simulated and measured source activity is 8%.

4. Low background spectra

Events are classified as single-cluster or multi-cluster. The energy spectrum obtained after 752.66 hrs of data taking is show in Fig. 3. The spectra are simultaneously fit to PDF's for the $2\nu\beta\beta$ decay signal and various backgrounds using an un-binned maximum likelihood method. For various detector components background models are developed incorporating the material screening performed during detector construction and the estimated cosmogenic activation. The multi-cluster spectrum is dominated by γ backgrounds which constrain their corresponding contribution in the single-cluster spectrum. The single-cluster spectrum is dominated by a large structure with a shape consistent with the $2\nu\beta\beta$ decay of ^{136}Xe. Its contribution to the total number of events is 3886 events whereas the dominant contamination from ^{40}K in the TPC vessel contributes 385 events. The remaining contributions account for a total of less than 650 background events which is consistent with the material screening measurements.[6] Only three candidates were found to be capable of mimicking the $2\nu\beta\beta$ signal requiring a β emitter with no γs, $T_{1/2} > 2$ days and the appropriate shape of the β spectrum: ^{234m}Pa, ^{90}Y and ^{188}Re. ^{234m}Pa is excluded by α spectroscopy while ^{90}Y and ^{188}Re result in a bad χ^2 when incorporating them separately in the fit and it is unlikely that the xenon is contaminated with such isotopes.

The measured half-life of the $2\nu\beta\beta$ decay in ^{136}Xe obtained by the likelihood fit is $T_{1/2} = 2.11 \pm 0.04(\text{stat}) \pm 0.21(\text{sys}) \times 10^{21}$ yr, where the system-

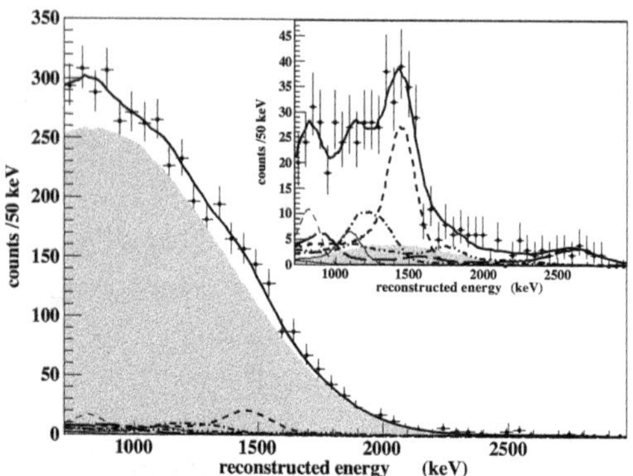

Fig. 3. Energy spectra obtained from 752.66 hrs of EXO-200 data for single-cluster (main panel) and mulit-cluster (inset) events. The fit to a model including the $2\nu\beta\beta$ decay and several background components is shown (solid line). The shaded region is the $2\nu\beta\beta$ contribution and the most prominent background contributions are: ^{232}Th, long dash; ^{40}K, dash and ^{60}Co, dash-dot.

atic uncertainty includes contributions from the energy calibration (1.8%), multiplicity assignment (3.0%), fiducial volume (9.3%) and γ background models (0.6%), added in quadratures.

5. Conclusion

EXO-200 is taking data with enriched xenon and the first data taking period has lead to the observation of the $2\nu\beta\beta$ decay in ^{136}Xe. The measured $T_{1/2}$ is significantly lower than the lower limits published by other experiments[7] and translates to a nuclear matrix element of 0.019 MeV^{-1}, the smallest measured among the $2\nu\beta\beta$ emitters. The detector is being upgraded and prepared for the next phase of data taking and the first $0\nu\beta\beta$ results are expected soon.

References

1. H. Drumm et al. Nucl. Instr. Meth. A 176 (1980) 333.
2. R. Neilson et al. Nucl. Instr. Meth. A 608 (2009) 6875.
3. 3M, see http://products3.3m.com.
4. E.I. Esch et al. Nucl. Instr. Meth. A 538 (2005) 516.
5. S. Agostinelli et al. Nucl. Inst. Meth. A 506 (2003) 250.
6. D. Leonard et al. Nucl. Instr. Meth. A 591 (2008) 490.
7. R. Bernabei et al. Phys. Lett. B 546 (2002) 23,
   Yu. M. Gavriljuk et al. , Phys. Atom Nucl. 69 (2006) 2129.

STATUS OF THE PICASSO EXPERIMENT*

UBI WICHOSKI
FOR THE PICASSO COLLABORATION
Department of Physics, Laurentian University, 935 Ramsey Lake Road
Sudbury, Ontario P3E2C6, Canada

The PICASSO experiment searches for cold dark matter through the direct detection of weakly interacting massive particles (WIMPs) via their spin-dependent interactions with fluorine at SNOLAB, Sudbury - ON, Canada. The detection principle is based on the superheated droplet technique; the detectors consist of a gel matrix with millions of liquid droplets of superheated fluorocarbon (C4F10) dispersed in it. The experiment has been taking data using 4.5-litre detector modules with approximately 80g of active mass per module. In this talk we will give an overview of the experiment, discuss the progress on the understanding of the superheated droplet technique and report on recent developments and future plans.

1. Introduction

1.1. *Dark Matter in the Universe*

Dark matter is believed to account for approximately 85% of the matter content of the Universe. The first observational evidence of its presence dates from the 1930's and has been growing since [1].

However, the nature of dark matter is still unknown. We only know that dark matter exists due to the gravitational effects that it produces, including gravitational lensing of distant galaxies, faster than predicted speed of member galaxies in a cluster of galaxies and the flat profile of the rotation curves of spiral galaxies. Dark matter is also believed to have had a key role in the formation of structures in the Universe [2]. The evidence is consistent with dark matter being made of a yet-to-be-discovered elementary particle called WIMP. WIMPs are hypothesized to be present in the hot soup of elementary particles in the beginning of the Universe. As the Universe expanded and cooled down, the WIMPs decoupled from that hot soup and, due to the freezeout mechanism, their present (relic) abundance in the Universe is predicted to be consistent with that

* This work is supported NSERC and CFI in Canada

of dark matter. Note that the fact the WIMP is expected to have the "right" abundance depends on it having weak-scale interactions [2,3].

Furthermore, in order to be the dark matter particle the WIMP is expected to have the following properties: *i-)* does not interact electromagnetically (does not possess electric charge); *ii-)* Stable; *iii-)* Interacts gravitationally; *iv-)* heavier than the proton and *v-)* is expected to have interactions of order of the weak interaction with ordinary matter.

The last property is crucial for the direct detection of the WIMPs. None of the particles in the Standard Model of Particle Physics has all these properties. Consequently, new physics has to be involved. Theoretically, the best motivated WIMP candidate is the neutralino. The neutralino is a particle that appears in the context of supersymmetric [3] theories – plausible extensions of the Standard Model of Particle Physics.

1.2. *Direct Detection of Dark Matter*

Although dark matter is believed to be present everywhere in the Universe, the best hope for detecting it directly is in our Galaxy.

The profile of the rotational curves of the spiral galaxies indicates that the luminous matter in these galaxies is immersed in a dark matter halo. The dark matter should account for approximately 90% of the total mass of the spiral galaxy.

The direct dark matter detection is based on the idea that WIMPs are constantly traversing the Earth and that they may interact with ordinary matter. In a dark matter detector, the nuclear recoil caused by the elastic interaction between the WIMP and a target nucleus can be detected using various techniques [4]. However, due to the very small cross section expected for the WIMP-ordinary matter interaction, extremely low background detectors have to be employed and be installed in underground laboratories in order to mitigate the cosmic ray induced background.

As the Earth orbits around the Sun and rotates on its axis as it travels through the Galaxy along with the solar system, two types of modulation in the dark matter signal are predicted: the annual and the diurnal modulations. These effects can be used to help to distinguish signal due to dark matter interaction from background [3,5].

2. The PICASSO Experiment at SNOLAB

The PICASSO (Project In CAnada to Search for Supersymmetric Objects) experiment searches for dark matter using the superheated droplet technique.

This technique was originally developed for neutron dosimetry but it has been customized and improved in order to be used in the dark matter search. The main advantages include: *i-)* the possibility of operating the detectors near room temperature as opposed to cryogenic temperatures; *ii-)* achievable insensitivity to minimum ionizing particle (e.g. electrons and gamma rays) at the level of 10^8 to 10^{10} [6]; *iii-)* very low detection threshold (~1 keV) and *iv-)* possible discrimination between events induced by different particle species [9]. The superheated droplet detectors are threshold detectors and their sensitivity is determined by the degree of superheat of the droplets. The degree of superheat is proportional to the operating temperature and pressure. The sensitivity and the threshold energies as a function of the superheat have been determined by the Collaboration using mono-energetic neutron beams at the Tandem accelerator at the Université de Montréal. The PICASSO Special Bubble Detectors (SBD) consist of an acrylic cylinder of 14 cm in diameter and 40 cm in height closed at the bottom by an acrylic lid and at the top by a stainless steel lid. The stainless steel top lid is connected with stainless steel rods to a stainless plate at the bottom for structural reinforcement (Figure 1). The container is filled with a gel matrix holding millions of droplets of the active liquid dispersed in it. The droplet size distribution peaks around 200 μm as also seen in Figure 1.

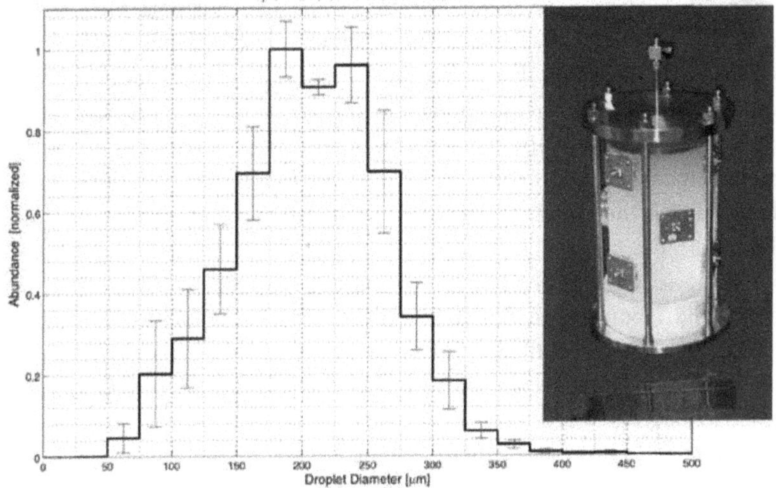

Figure 1. A 4.5-litre SBD and its droplet size distribution.

The active liquid of choice for PICASSO is the perfluorobutane (C_4F_{10}). This liquid, on the one hand, allows the Collaboration to take full advantage of the technique due to its thermodynamical properties. On the other hand, the high fluorine (^{19}F) content makes it a very favorable liquid to explore the spin dependent sector of the interaction with WIMPS (WIMP-proton). The nuclear form factor of ^{19}F enhances the signal by nearly an order of magnitude compared to other frequently used nuclei (Na, Si, Al, Cl). In addition, fluorine is a monoisotopic and mononuclidic element, which translates into no need for isotopic enrichment or depletion.

The Collaboration has designed the detectors and also fabricates them. Very important progress has been made in the fabrication technique and the purification of the ingredients that make up the detector. The main background that the experiment has been fighting, due to α-emitting contaminants, is now down to ~20 cts kg^{-1} d^{-1} in the best detectors.

The detector principle is based on the detection of an acoustic signal created during the explosive phase transition from liquid droplet to gas bubble. The phase transition is triggered by the deposition of a minimum amount of energy in a certain minimum distance according to the Seitz theory [7]. The energy deposition can be done directly by the incident particle, as in the case of background alpha particles being emitted by radioactive contaminants inside the detector or indirectly as in the case of background neutrons. If a neutron hits the target nucleus, the recoiling nucleus deposits the energy. The same process is expected to happen when a WIMP hits the nucleus. Gamma rays' triggering mechanism is different: it is caused by ionization clusters due to δ-rays or Auger electrons created along the tracks of Compton scattered electrons but, as mentioned before, this effect is highly suppressed at operating temperatures and pressures. The response of the detector to different radiations, including the expected signal due to 50 GeV WIMPs (at 1pb), is summarized in Figure 2. In order to detect the acoustic signal, piezo electric sensors are installed on the walls of the detector's container. The sensors are connected to preamplifiers. The PICASSO experiment has had its setup 2 km below the surface, in the underground laboratory at SNOLAB in the Vale Creighton mine in Sudbury, Ontario, Canada, since 2002. The experiment has recently moved to a new location, the Ladder Lab area at SNOLAB, which provides better access and room to expand.

The setup, in the present configuration, can accommodate up to 32 4.5-litres detector modules plus space for test prototypes. The average active mass is 85 g per module. The modules are organized in groups of four. Each group is installed inside a temperature and pressure control system (TPCS) module. The setup contains eight independent TPCS modules. Each TPCS can have its temperature and pressure set independently; this allows the Collaboration to have options and flexibility on how to run the detectors. The Collaboration has

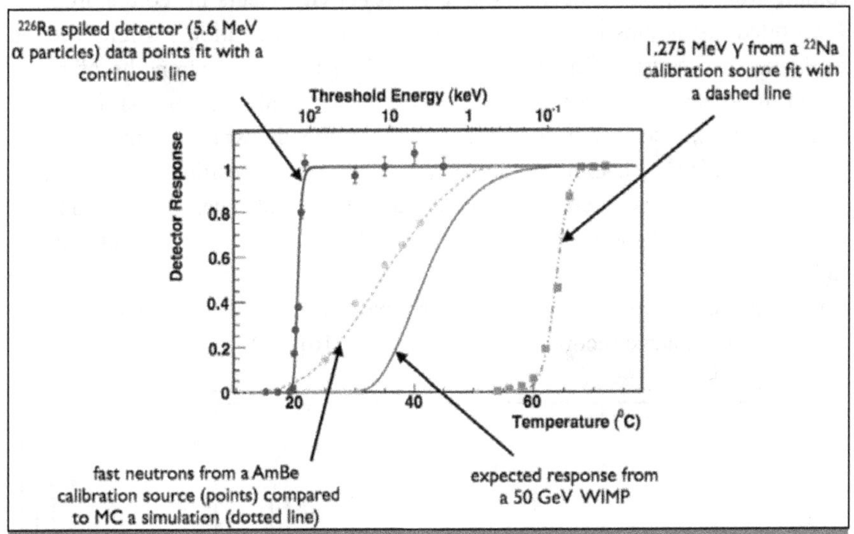

Figure 2. Response of the SBDs to different particles

been operating the detectors keeping the pressure constant and equal to the ambient pressure in the mine during data taking and ramping up the pressure to ~6 bar during recompression runs. The recompression runs are necessary to bring the detectors back to their initial state, i.e., any droplet that might have transitioned to a bubble gas is brought back to the liquid state by the overpressure. Besides that, the increase in pressure "turns off" the detectors by reducing the superheat state (and consequently the sensitivity) to practically zero. Once the detectors undergo recompression for ~ 15 hours, they are ready for a ~ 40-hour physics run.

Aiming to improve the experiment's performance [8], in 2008 the PICASSO Collaboration published a paper reporting for the first time the possibility of discriminating between alpha particle induced phase transitions and neutron (WIMP) induced phase transitions in superheated liquids [9]. The discrimination was based on the statistical difference between the amplitudes of neutron and alpha-particle-induced events. Since then, much progress has been made by the Collaboration in understanding the mechanism of the formation and growth of the bubbles. In 2011, the Collaboration published a second paper on the subject where new tools and parameters were developed to quantify the difference between alpha-induced events and neutron-induced events. Figure 3 shows distributions of events in terms of the acoustic parameter calculated for each event as a function of the temperature. The solid histograms indicate events coming from α-spiked detectors where the α activity occurs inside the droplets. The dotted histograms indicate events coming from calibration sessions with an AcBe neutron source. On the left panel (25°C) the signal strengths of recoil nuclei in α-decays coincide with those from neutron calibrations. On the centre panel (27.5°C), a second peak appears on the high side, which is caused by the joint effect of recoil α-emitting nuclei and the energy deposition by the α-track. There are still events where only α-emitting nuclei recoils nucleate (neutron data were taken at 28°C and are therefore slightly shifted to the right). Finally, on the rightmost panel (45°C), α-particles and α-emitting recoil nuclei contribute simultaneously to the signal that is clearly more intense than the signal coming from neutron-induced recoils (dotted histogram) [10].

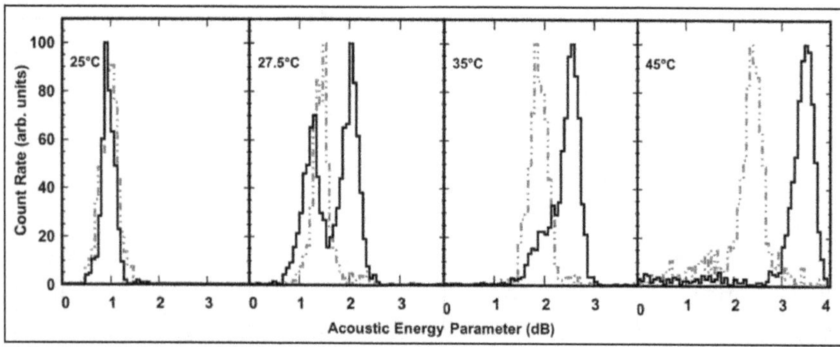

Figure 3. Distribution of the acoustic energy parameter as a function of the temperature

3. Progress in Calibration and Discrimination

Figure 4a shows a calibration curve where recoil energy thresholds for Fluorine obtained with the neutron beam as a function of temperature is plotted – black dots – alongside with alpha data – open (red) circles. The alpha data comes from detectors spiked with ^{226}Ra. ^{222}Rn is produced in the decay chain of ^{226}Ra and diffuses into the droplets. The most energetic recoil in the decay chain of ^{222}Rn is given by the recoil of ^{210}Pb (=146 keV) inside the droplets that defines the threshold. The associate temperature for this recoil energy fits very well with the dashed curve defined by the neutron calibration points. The second red circle at higher temperature also fits very well in the curve and is obtained by spiking the gel matrix with ^{241}Am. The threshold in this case is given by the alphas produced in the decay of the ^{241}Am in the bulk that enter the droplets and deposit energy at the Bragg peak there.

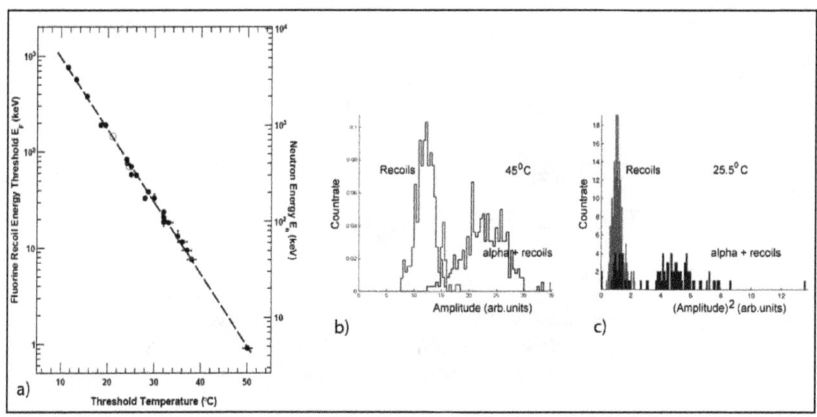

Figure 4. a) calibration curve. b) and c) show the improvements in the alpha-neutron discrimination.

Figures 4b and 4c show the improvement in alpha-neutron discrimination due to optimized hardware (higher sampling rate). The ongoing data analysis reached sensitivity at the level of a few cts/kg$_F$/d in the fall 2011. Due to this sensitivity and the low threshold PICASSO is particularly sensitive to light WIMPS below 15 GeV/c. Taking advantage of the discrimination power, which can reduce the alpha background by a factor of 100 and assuming the present exposure we expect an improvement of sensitivity to better than $\approx 3 \times 10^{-3}$ pb in the SD sector and $\approx 7 \times 10^{-6}$ pb in the SI sector.

Acknowledgments

We would like to thank the organizing committee for the invitation. UW would like to acknowledge the financial support from the National Sciences and Engineering Research Council of Canada (NSERC).

References

1. L. Bergström, New J. Phys. **11,** 105006 (2009).
2. E. W. Kolb and M. S. Turner, *The Early Universe*, Westview Press, 1994.
3. G. Jungman, M. Kamionkowski, and K. Griest, Phys. Rep. **267**, 195 (1996).
4. V. Zacek, arXiv:0707.0472.
5. J. I. Collar, F.T. Avignone III, Phys. Lett. **B275**, 181 (1992).
6. Barnabé-Heider M *et al.,* Nucl. Instrum. Methods A **555**, 184 (2005).
7. F. Seitz: Phys. Fluids **1**, 1 (1958).
8. S. Archambault *et al.*, Phys. Lett. **B682**, 185 (2009).
9. F. Aubin *et al.*, New J. Phys. **10**, 103017 (2008).
10. S. Archambault *et al.*, New J. Phys. **13**, 043006 (2011).

The DarkSide Program at LNGS

A. Wright
Department of Physics, Princeton University
Princeton, NJ 08540, USA
E-mail: ajw@princeton.edu

for the DarkSide Collaboration.

DarkSide is a direct detection dark matter program at LNGS that is based on two phase depleted argon time projection chambers. A combination of low background construction techniques and active background suppression will give the DarkSide detectors very low and extremely well understood rates of background events. A 10 kg prototype detector is currently being operated at LNGS, while DarkSide-50, the first physics detector in the DarkSide program, is currently being constructed.

Keywords: DarkSide, direct dark matter detections, depleted argon

1. Introduction

Overwhelming astrophysical and cosmological evidence now supports the idea that dark matter, an as yet unidentified non-luminous form of matter, makes up a much larger fraction of the mass density of the universe than does "ordinary" baryonic matter. One promising dark matter candidate is the weakly interacting massive particle (WIMP). Evidence for the existence of WIMP-type dark matter is being sought by a global experimental program.[1]

The DarkSide program aims to directly detect WIMP-type dark matter using a series of two-phase depleted argon time projection chambers (TPCs) which will be deployed at Laboratori Nazionali del Gran Sasso (LNGS), Italy. If WIMPs exist, very occasionally one might scatter off of an argon nucleus in the active volume, producing a low energy ($\lesssim$200 keV) nuclear recoil. In order to be sensitive to the very low interaction rates of interest (a few per tonne-yr), it is critical that the DarkSide detectors have both very low and very well understood rates of background events.

In order to accomplish this, the DarkSide experiments will combine ultra-low background technologies with active background suppression. The former will include the use of novel ultra-low background "Quartz Photon Intensifying Device" photo-detectors[2] and argon naturally depleted in ^{39}Ar, which is described in Section 2. The use of active background suppression will both further reduce the important background rates and give DarkSide the capability to measure these rates *in situ.* The DarkSide active background suppression strategy is described in Section 3.

The DarkSide collaboration is currently operating a 10 kg prototype, DarkSide-10, at LNGS. The prototype is described in Section 4, while the future DarkSide program, including the first physics detector, DarkSide-50, is described in Section 5

2. Depleted Argon

One reason that argon is an attractive active medium for dark matter searches is the very strong ability to distinguish between electron- and nuclear-recoil events intrinsic to the argon scintillation pulse shape. Argon scintillation light is produced from two dimer excited states: a singlet state with a ~7 ns lifetime, and a triplet state with a lifetime of ~1.6 μs. The relative amount of light emitted by the two states depends strongly on the ionization density of the excitation, and it has been demonstrated that the resulting difference in pulse shape can be used to suppress electron recoil backgrounds to levels in excess of 1 in 10^8.[3–5] When operated as a two-phase time projection chamber, which records both the energy deposited by and the free charge produced in each event, an additional electron recoil suppression of 10^2-10^3 is obtained from the charge-to-energy ratio.[5,6]

The very strong electron recoil rejection ability of argon is partially offset, in atmospheric argon, by the presence of β-emitting ^{39}Ar at the level of ~1 Bq/kg.[7,8] This limits the ultimate scale achievable in argon detectors with atmospheric argon targets, and increases the analysis threshold of smaller experiments[a].

The ^{39}Ar half life is only 269 years, so the ^{39}Ar activity in atmospheric argon is maintained by cosmogenic production. Argon from underground sources, which is shielded from cosmogenic activation, has been demonstrated[9] to have lower ^{39}Ar activities. One source of underground argon is the Kinder Morgan Doe Canyon Complex in Cortez, Colorado, which ex-

[a]The pulse shape discrimination ("PSD") efficiency is strongly dependent on photon statistics, so higher ^{39}Ar rates may mean that higher analysis thresholds are necessary to ensure sufficient PSD power to suppress the ^{39}Ar background.

tracts CO_2 from natural underground reservoirs. The extracted CO_2 contains about 300 ppm argon. Since February 2010, DarkSide has extracted argon from the Kinder Morgan CO_2 using a vacuum-swing adsorption system:[10] to date (Sept. 2011), 62 kg of argon has been collected.

The ^{39}Ar level in the Kinder Morgan gas has been investigated using a dedicated low background detector containing approximately 0.5 kg of liquid argon as the active target. Two inches of copper shielding and eight inches of lead shielding surround the target, and a plastic scintillator-base muon veto is used to reduce the cosmogenic backgrounds. In depleted argon measurements made on surface at Princeton University, a counting rate of $\sim$0.1 mHz/keV was achieved in the upper ^{39}Ar energy region - if this whole rate is attributed to ^{39}Ar, a limit of 100 mBq/kg can be placed on the ^{39}Ar activity. Fitting the recorded spectrum for the ^{39}Ar rate using a smoothly decaying function to represent the background yields a more stringent 20 mBq/kg limit on the ^{39}Ar rate. The majority of the background in these measurements was assumed to be unvetoed cosmogenic backgrounds, so in the spring of 2011 the detector was moved to the KURF Underground Laboratory in Kimballton, Viginia, USA (at a depth of 1400 meters of water equivalent). The counting rate in depleted argon measurements at KURF was reduced to $\sim$0.02 mHz/keV in the ^{39}Ar energy region, allowing a 20 mBq/kg limit to be placed on the ^{39}Ar rate by direct counting alone. Spectral fits, which are expected to yield even more stringent limits, are currently underway.

3. Background Suppression

Building on the strong electron-recoil suppression intrinsic to argon, the DarkSide experiments will employ strong active suppression techniques against other important background. Radiogenic and cosmogenic neutrons which enter the detector can scatter to produce nuclear recoil which are event-by-event indistinguishable from WIMP-induced events.

To suppress these neutron backgrounds, the DarkSide detectors will be deployed within a 4 m diameter volume of boron-loaded scintillator which is in turn contained in a 11 m diameter by 10 m high water tank, as shown in the left pannel of Fig. 1. The boron-loaded scintillator has a very high efficiency for detecting low energy neutrons either before or after they interact in the argon. Thus, the co-incidence between an event in the neutron veto and a nuclear recoil in the argon detector can be used to reject the neutron-induced background. Simulations suggest that, provided care is taken to avoid the use of materials with high neutron absorption cross section for the inactive components of the argon detectors, veto efficiencies in excess of 99.5% can be achieved for radiogenic neutrons.[11]

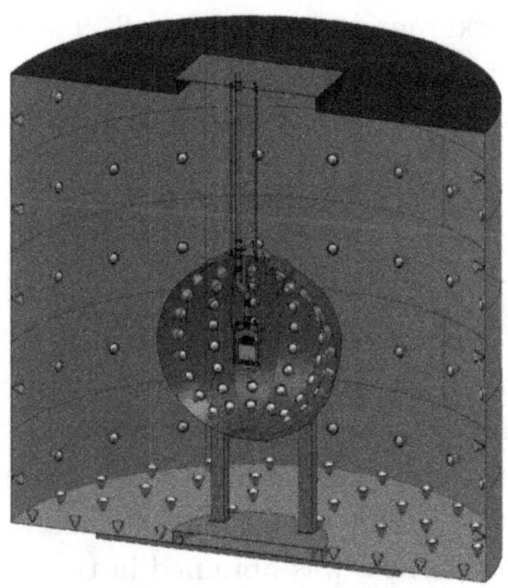

Fig. 1. Left: The DarkSide-50 detector in its active shielding. Right: The DarkSide-10 prototype being assembled for deployment at LNGS.

The higher energy neutrons produced by cosmic ray spallation are more penetrating, and thus less likely to be detected by the neutron veto. To suppress these events, the neutron veto will be constructed within the water tank that was built for the Borexino Counting Test Facility (CTF).[12] The 10 m by 11 m water tank will be instrumented to detect the Cerenkov light produced by muons and other cosmogenic particles, and will thus provide a veto for cosmogenically-induced events in the argon detector. Very conservative simulations suggest that the neutron and cosmogenic vetoes together will suppress the cosmogenic background rate in DarkSide by more than a factor of 1,000.

Perhaps even more important than the direct background reduction provided by the use of active suppression is the degree of confidence that the additional information obtained about the background rates imparts in estimates of the rate of residual backgrounds in the experiment. Once the efficiencies of the active suppression techniques are known[b], the number

[b]The efficiencies of the neutron veto and the PSD based electron recoil rejection can be calibrated directly using radioactive sources, while the efficiency of the cosmogenics veto can be understood by comparing the relative rates observed in the water, liquid scintillator, and argon detectors with predictions from simulation.

of events rejected using each technique can be used to measure the rate of each class of background event. If the number of rejected events is low, and the rejection efficiency is high, it is possible to convincingly argue that the expected number of unvetoed background events is small. Therefore, the extensive use of active background suppression will give the DarkSide collaboration a very robust understanding of the background levels in our detectors.

4. DarkSide-10

The DarkSide collaboration has built and is operating DarkSide-10, a 10 kg prototype detector, which is shown in Fig. 1. DarkSide-10 was operated at Princeton University for a total of about seven months during late 2010 and early 2011. These runs were used to test key features of the DarkSide detector design and to gain experience in the collection and analysis of two-phase data. A light yield of ~4.5 p.e./keV was obtained in these runs.

In order to carefully study the rejection of electron recoil events and backgrounds from radioactive decays on the inner surfaces of the detector, DarkSide-10 was upgraded and moved to LNGS. The detector is currently installed, within a water-based passive shield, in Hall C and is being re-commissioned.

5. The DarkSide Physics Program

The first physics detector in the DarkSide program will be DarkSide-50. This 50 kg detector will be constructed using radio-pure materials and deployed within the active shielding described in Section 3. Detailed Monte Carlo simulations suggest that the total expected number of unvetoed background events in DarkSide-50 is <0.1 events in a 0.1 tonne-year exposure. DarkSide-50 is scheduled to be deployed in late 2012, after which it should achieve a WIMP sensitivity of at least 10^{-45}cm^2 (see Fig. 2) in about three years of operation.

The neutron veto for DarkSide-50 will be 2 m in radius, which is large enough to house a future 5 tonne detector. Other elements of the DarkSide-50 infrastructure, for example the gas handling system, will also be built with enough spare capacity to facilitate this upgrade. As shown in Fig. 2, a 5 T detector could achieve a sensitivity of about 10^{-47}cm^2.

Finally, the technology developed by DarkSide could become part of the MAX ("Multi-tonne Argon and Xenon") program, which is envisioned to consist of twin multi-tonne argon and xenon detectors. This large experiment would probe dark matter interactions to the "ultimate" 10^{-48}cm^2 level, below which the irreducible background from coherent neutrino-nucleus scattering dominates.

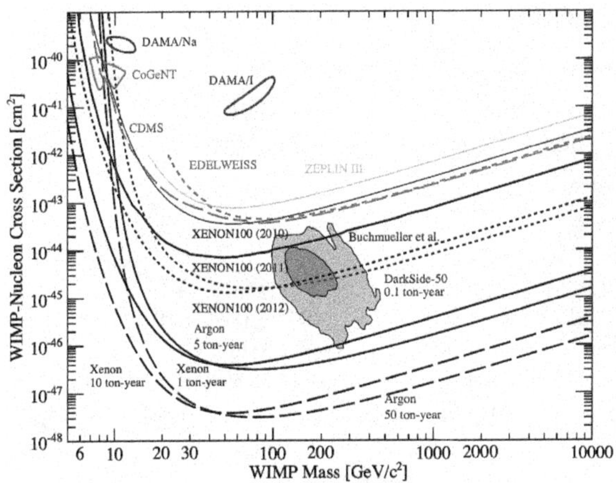

Fig. 2. Projected sensitivities of different detectors in the DarkSide/MAX program. An argon light yield of 6 p.e./keV$_{ee}$, and an ^{39}Ar activity of 40 mBq/kg were assumed in making this Figure. In DarkSide-50, this imposes an analysis threshold of 30 keV$_{\rm recoil}$ to ensure that the PSD (the efficiency of which depends strongly on the total number of detected photoelectrons) is sufficient to suppress this electron recoil background - in the larger detectors the analysis threshold is higher. A 50% PSD nuclear recoil acceptance is assumed. Depending on the true ^{39}Ar activity in the depleted argon and the actual light output achieved, lower energy thresholds may be possible, which would result in improved sensitivity.

6. Summary

DarkSide is a vigorous research program aimed at directly detecting WIMP-type dark matter with two-phase argon time projection chambers. The collaboration is currently operating a 10 kg prototype detector in Hall C of LNGS, and is in the process of constructing DarkSide-50, a 50 kg physics detector. DarkSide-50 should be deployed in late 2012, with future upgrades to tonne-scale detectors possible. Innovative low background techniques, including naturally depleted argon and ultra-low background photo-detectors will be used to ensure a low rate of background events. The extensive use of active background suppression will provide *in situ* measurements of the important background rates, while at the same time suppressing those rates even further. We believe that this careful approach to the understanding of our backgrounds would significantly enhance the credibility of any potential claim of dark matter detection.

Acknowledgments

The DarkSide program is supported in the USA by the NSF and the DoE. We gratefully acknowledge the hospitality of Laboratori Nazionali del Gran Sasso. The author acknowledges the support of the Princeton University PFEP program.

References

1. B. Sadoulet, Rev. Mod. Phys. **71**, S197-S204 (1999).
2. A. Teymourian et al., arXiv:1103.3689 (2011).
3. M.G. Boulay and A. Hime, Astropart. Phys. **25**:179 (2006).
4. M.G. Boulay et. al., arXiv:0904.2930 (2009); C. Jillings, *The DEAP-1 Detector at SNOLAB*, Talk at SNOLAB2010, slides available at http://snolab2010.snolab.ca/program.html.
5. P. Benetti et al. (WARP Collaboration), Astopart. Phys. **28**, 495 (2008).
6. L. Grandi, *WARP: An argon double phase technique for dark matter search.* Ph.D. Thesis, University of Pavia (2005).
7. P. Benetti et al. (WARP Collaboration), Nucl. Instr. Meth. A **574**, 83 (2007).
8. H.H. Loosli, Earth Plan. Sci. Lett. **63**, 51 (1983).
9. D. Acosta-Kane et al., Nucl. Instr. Meth. A, **587**, 46 (2008).
10. H.O. Back et al. AIP Conf. Proc. **1338**, 217-220 (2011).
11. A. Wright et al., Nucl. Inst. Meth. A **644**, 18-26 (2011).
12. G. Alimonti et al. (The Borexino Collaboration), Nucl. Instr. Meth. A **406**, 426 (1998).

High Energy Physics Experiments

S.U. Ahn	Quarkonia Physics with the ALICE Muon Spectrometer
B. Alessandro	Analysis of Cosmic Events with the ALICE Experiment
E. Barberis	Recent Results from the TEVATRON
F. Barile	Identified-particle production at high-momentum with the HMPID detector in the ALICE experiment at the LHC in proton-proton collisions at $\sqrt{s} = 2.76$ TeV
C. Battilana	CMS Muon Detectors and Trigger Performance
J. Berdermann	The South Pole Acoustic Test Setup as a trailblazer towards an acoustic detection of neutrinos
C. Carrillo	The CMS Resistive Plate Chambers system-detector performance during 2011
A. Henriques	ATLAS Detector Overview (Operation Experience, Performance)
P. Jež	Performance and Evolution of the ATLAS Tau Trigger During 2011 Data Taking Period
J. Mašík	Performance of the ATLAS Inner Detector Trigger algorithms in pp collisions at 7 TeV
O. Novgorodova	Performance Results of Assembled Sensor Plane Prototypes for Special Forward Calorimeters at Future E^+E^- Colliders
G. Papotti	LHC Status and Performance in 2011
D. Perevalov	Status of the NOvA Experiment
R. Piandani	A detector for the measurement of the ultrarare decay $K^+ \to \pi^+ \nu \bar{\nu}$: NA62 at the CERN SPS
J.L. Pinfold	The MoEDAL Experiment at the LHC

A. Salamon	The NA62 Liquid Krypton Electromagnetic Calorimeter Level 0 Trigger
S. Spataro	Reconstruction of $\bar{p}p$ events in PANDA
J. Thompson	CMS status and operations

QUARKONIA PHYSICS WITH THE ALICE MUON SPECTROMETER

S. U. AHN* for the ALICE Collaboration
Department of Physics, Konkuk University, Seoul, Korea
Laboratoire de Physique Corpusculaire (LPC), Clermont Université,
Université Blaise Pascal, CNRS-IN2P3, Clermont-Ferrand, France
**E-mail: Sang.Un.Ahn@cern.ch*

ALICE is the only LHC experiment dedicated to the study of the Quark-Gluon Plasma (QGP) which is expected to be created in heavy-ion collisions. Among the most promising observables, heavy quarkonia provide an essential probe for the characterization of the QGP thanks to their sensitivity to the earliest and hottest stages of the collision. Quarkonia measurement can be performed in the dimuon decay channel with the ALICE muon spectrometer at forward rapidity. It has been commissioned by means of cosmic rays and the first LHC beam injections in 2008 and has shown good performances for muon detection. This proceeding presents first quarkonia results obtained in pp and Pb–Pb collisions.

Keywords: QGP; Quarkonia; LHC; ALICE muon spectrometer.

1. Introduction

Suppression of quarkonium production in A-A collisions due to Debye color screening was proposed as a probe of the strongly interacting QCD matter.[1] Quarkonium states expected to be formed at early stage of the A-A collisions can be dissociated while they traverse the hot medium. Different binding energy of the quarkonium states leads to a sequential dissociation of those states depends on the temperature. However, this sequential suppression could be challenged by the statistical regeneration[2] due to the large heavy flavour cross sections at LHC energy. In order to understand the feature of the medium, the Cold Nuclear Matter (CNM) effects, such as modification of parton distribution in the nucleus (shadowing) and dissociation of the quarkonium state due to the nuclear absorption, should be measured in p-A collisions.[3]

Quarkonium production gives rise to a large interest also in pp collisions since the quarkonium production mechanism is still not fully understood. There is no satisfactory theoretical model explaining simultaneously the current experimental results of the production rate and polarization. LO and NLO CSM fail to reproduce the J/ψ production rate measured by CDF at Tevatron.[4] The effective field theory, NRQCD, is successful for reproducing the CDF data, but not the J/ψ polarization.[5] For bottomonium, the NNLO* CSM reproduces the production rate.[6]

2. Quarkonia detection with ALICE muon spectrometer

ALICE[7] is the only detector designed for heavy-ion physics at LHC. It is equipped with a spectrometer dedicated to muon detection in the forward region, $2.5 < y < 4$. The ALICE muon spectrometer consists of 10 tracking chambers, 4 trigger chambers, a warm dipole magnet, and a set of absorbers: frontal absorber, beam shielding and muon filter before the trigger system. It has a large geometrical acceptance and p_t acceptance down to zero for quarkonia measurement. The absorbers help to reduce the background produced from K, π decays and the secondaries from the beam pipe. In order to separate the various states of the Υ resonances, the mass resolution should be better than ~ 100 MeV/c^2. This requirement determined the dipole magnet strength (~ 3.0 Tm) and the tracking chambers spatial resolution (< 100 μm). Also the frontal absorber is important to reach the designed mass resolution. To cope with high multiplicities produced in heavy-ion collisions, high granularity read-out ($> 10^6$ channels) was implemented, and a selective (di-)muon trigger system is installed with different p_t thresholds ($0.5 - 4.2$ GeV/c).

3. Measurement of J/ψ inclusive cross sections in pp collisions at 2.76 TeV and 7 TeV

We present results on the inclusive J/ψ production based on the 2010 pp collision data at $\sqrt{s} = 7$ TeV and preliminary results from the pp data at $\sqrt{s} = 2.76$ TeV as well. The data sample corresponds to events collected with the minimum bias trigger, defined as the logical OR between the requirement of at least one hit in the Inner Tracking System (ITS) pixel layers (SPD), and a signal in one of the two VZERO scintillator hodoscopes (for details on the apparatus, see Ref. 7). For the muon analysis, the minimum bias trigger is required to be in coincidence with a signal in the muon trigger chambers.

The number of J/ψ collected with the ALICE muon spectrometer in pp collisions at 7 TeV in its muon pair decay in 2010 is more than 15000 and the number of J/ψ analyzed in the present analysis is 1924 ± 77. The resolution at the J/ψ mass is ~ 80 MeV/c^2 and well reproduced by detector simulations. For the $\Upsilon \to \mu^+\mu^-$ analysis, pp collision data gave rise to a signal and a clear signal has been revealed in the preliminary analysis of 2011 pp collision data.

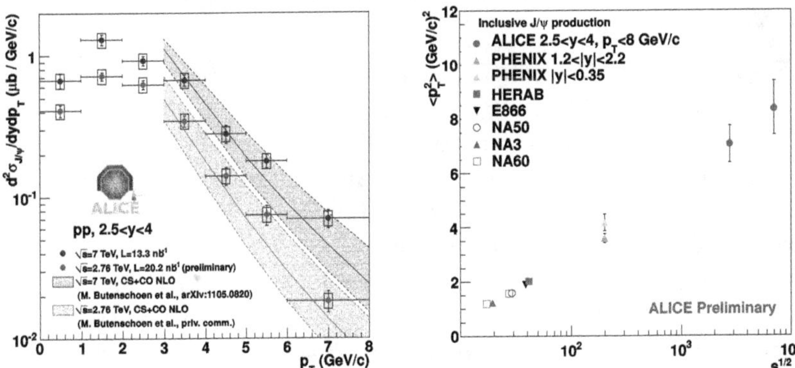

Fig. 1. Left: $d^2\sigma_{J/\psi}/dydp_t$ obtained at $\sqrt{s} = 7$ TeV and 2.76 TeV, compared with NLO NRQCD calculations. Right: J/ψ $\langle p_t^2 \rangle$ versus $\sqrt{s}$.

The inclusive J/ψ production cross sections in pp collisions at $\sqrt{s} = 7$ TeV and 2.76 TeV were measured at forward rapidity with the corresponding integrated luminosities $L_{int} = 15.6$ nb^{-1} and 20.2 nb^{-1} which were normalized with respect to the minimum bias cross section measured using a Van der Meer scan.[8] The inclusive J/ψ production cross section at $\sqrt{s} = 7$ TeV and 2.76 TeV are respectively: $\sigma_{J/\psi}(2.5 < y < 4) = 6.31 \pm 0.25(stat.) \pm 0.80(syst.)^{+0.95}_{-1.96}(pol.)$ μb[9] and $\sigma_{J/\psi}(2.5 < y < 4) = 3.46 \pm 0.13(stat.) \pm 0.42(syst.)^{+0.55}_{-1.11}(pol.)$ μb. Systematic uncertainties include the errors on the signal extraction, on the trigger and reconstruction efficiencies, on the p_t and y input distributions used for the acceptance correction evaluation, and on the luminosity determination. These results are estimated with the assumption of no J/ψ polarization, and the effect of full transverse or longitudinal polarization is quoted as "*pol.*" systematic uncertainty. The result obtained at 2.76 TeV energy is used for the evaluation of the nuclear modification factor R_{AA} to quantify the hot QCD matter effects in Pb–Pb collisions on the J/ψ yield. The large J/ψ statistics allows the evaluation of the differential p_t and y cross sections. Figure 1 (left) shows the p_t distributions obtained in the muon spectrometer acceptance at 7 TeV and 2.76 TeV. At both energies, NRQCD NLO calculations, available for $p_t > 3$ GeV/c, provide a good description of the p_t distributions. Fitting the p_t differential distributions with the function $A \times p_t/(1 + (p_t/p_0)^2)^n$, the J/ψ $\langle p_t \rangle$ and $\langle p_t^2 \rangle$ were extracted. Approximate logarithmic increase of $\langle p_t^2 \rangle$ with $\sqrt{s}$ holds up to the LHC energy, as shown in Fig. 1 (right), and a similar trend is observed also for $\langle p_t \rangle$.

4. Inclusive J/ψ R_{AA} in Pb–Pb collisions at $\sqrt{s_{NN}} = 2.76$ TeV

The Pb–Pb collision data at $\sqrt{s_{NN}} = 2.76$ TeV were collected with a minimum bias trigger, defined as the logical AND between signals from the SPD and a signal in both of the two VZERO detectors. The determination of the collision centrality was based on a Glauber-model fit of the VZERO amplitude.[10] The following collision centrality bins were defined for the J/ψ analysis: 0-10%, 10-20%, 20-40% and 40-80%. The total data sample used in the analysis corresponds to the integrated luminosity $L_{int} = 2.7\ \mu b^{-1}$.

The signal of J/ψ was estimated using a sef of different methods and here we describe one of them. After subtraction of the combinatorial background using the event-mixing technique, the J/ψ signal in each centrality bin is extracted with a Crystal Ball function and the remaining background is fitted by means of a straight line. In order to extract the yield of J/ψ, the number of J/ψ was normalized to the number of minimum bias event in the corresponding centrality bins and corrected for branching ratio of muon decay and the acceptance correction of the detector. To estimate the nuclear modification factor R_{AA} for J/ψ, the yield of J/ψ was normalized to the inclusive J/ψ cross section measured in pp collisions at $\sqrt{s} = 2.76$ TeV and scaled by the nuclear overlap function T^i_{AA} calculated by using the Glauber model. The inclusive J/ψ $R_{AA}^{0-80\%}$ was estimated to $0.49 \pm 0.03(stat.) \pm 0.11(syst.)$ for $p_t > 0$ GeV. Note that our measurement contains a contribution from B feed-down. This contribution was measured to be 10% in pp collisions[11] in our kinematic domain so the effect on R_{AA} is expected to be small. If one assumes a binary scaling for bottom production, one gets a prompt J/ψ R_{AA} lower by 10% with respect to the measured inclusive J/ψ R_{AA}.

The inclusive J/ψ R_{AA} is shown in Fig. 2 as a function of the average number of nucleons participating to the collision ($\langle N_{part} \rangle$ for ALICE has been weighted by N_{coll}, the number of binary nucleon-nucleon collisions). The ALICE measurements in the rapidity range, $2.5 < y < 4$, show less suppression than PHENIX[12,13] ($\sqrt{s_{NN}} = 200$ GeV) in $1.2 < |y| < 2.2$, while they are closer to those measured at midrapidity except in the most central collisions.

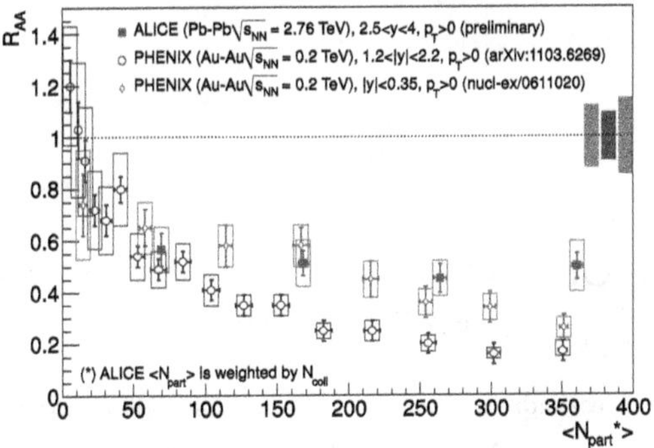

Fig. 2. J/ψ R_{AA} as a function of $\langle N_{\mathrm{part}}\rangle$ compared with PHENIX results in Au-Au collisions at $\sqrt{s_{\mathrm{NN}}} = 200$ GeV.

5. Conclusions

The ALICE experiment measured the inclusive J/ψ production cross section in pp collisions at $\sqrt{s} = 7$ TeV and 2.76 TeV with the forward muon spectrometer in the rapidity range, $2.5 < y < 4$. The inclusive J/ψ R_{AA} was measured in Pb–Pb collisions at $\sqrt{s_{\mathrm{NN}}} = 2.76$ TeV. The comparison with the PHENIX results at $\sqrt{s_{\mathrm{NN}}} = 200$ GeV shows less suppression in the most central collisions that could be a consequence of the regeneration. In order to have a better understanding of this comparison result, p-A collisions are required to measure the CNM effects at the LHC.

References

1. T. Matsui and H. Satz, *Phys. Lett. B* 178, 416 (1986).
2. B. Svetitsky, *Phys. Rev. D* 37, 2484 (1987).
3. R. Vogt, *Phys. Rev. C* 71, 054902 (2005).
4. N. Brambilla et al., arXiv:hep-ph/0412158v2.
5. A. Abulencia et al. (CDF Collaboration), *Phys. Rev. Lett.* 99, 132001 (2007).
6. J. P. Lansberg, *Eur. Phys. J. C* 60, 693 (2008).
7. K. Aamodt et al. (ALICE Collaboration), *JINST* 3, S08002 (2008).
8. K. Oyama for the ALICE Collaboration, arXiv:1107.0692v1 [physics.ins-det].
9. K. Aamodt et al. (ALICE Collaboration), arXiv:1105.038v2 [hep-ex].
10. A. Toia for the ALICE Collaboration, arXiv:1107.1973v1 [nucl-ex].
11. R. Aaij et al. (LHCb Collaboration), *Eur. Phys. J. C* 71, 1645 (2011).
12. A. Adare et al. (PHENIX Collaboration), *Phys. Rev. Lett.* 98, 232301 (2007).
13. A. Adare et al. (PHENIX Collaboration), arXiv:1103.6269 [nucl-ex].

ANALYSIS OF COSMIC EVENTS WITH THE ALICE EXPERIMENT

B. Alessandro on behalf of the ALICE Collaboration

I.N.F.N. sez. di Torino, Italy
E-mail: alessandro@to.infn.it

A large number of cosmic events were recorded in 2009 for the calibration, alignment and commissioning of most of the ALICE (A Large Ion Collider Experiment) detectors. In this paper we present the analysis of these data with a preliminary measurement of the μ^+/μ^- ratio for near-vertical and near-horizontal muons. The muon multiplicity distribution is discussed for the data taken in 2010 and 2011, with particular emphasis on some special events of very high muon density.

Keywords: Cosmic rays; Atmospheric muons; High energy interactions.

1. Introduction

ALICE[1] is one of the four big experiments of the LHC (Large Hadron Collider) at CERN. It is mainly dedicated to study a new phase of matter, called QGP (Quark Gluon Plasma), created in heavy-ion collisions at very high energies. Although its main purpose is the analysis of the collisions at LHC, it can also operate to detect atmospheric muons produced by cosmic ray interactions with the atmosphere.

Since the experiment is located 40 m underground, with 30 m of overburden rock, only muons with an energy larger than 15 GeV can reach the apparatus. The muons with zenith angle in the range $0^o - 60^o$ are detected in the central barrel and tracked with the TPC (Time Projection Chamber), while the Forward Muon Spectrometer (FMS) is employed for the detection of the near-horizontal muons (zenith angle in the range $70^o - 85^o$).

2. Atmospheric muons in the central barrel

Specific triggers have been implemented to detect atmospheric muons crossing the central barrel (detectors inside the magnet) of the ALICE apparatus

(Fig. 1). For this purpose, three detectors have been employed: ACORDE, TOF (Time of Flight) and SPD (Silicon Pixel Detector).

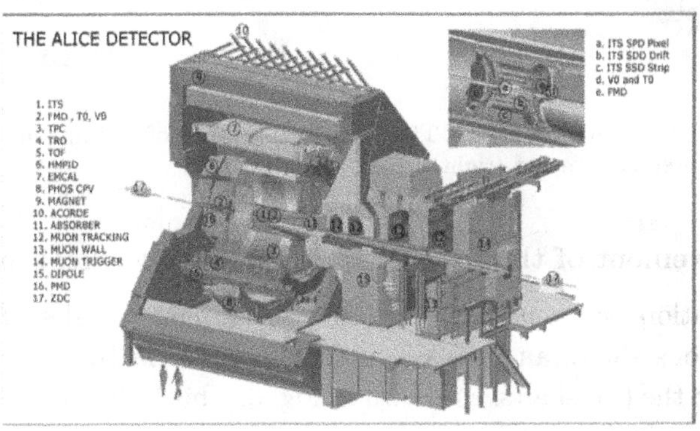

Fig. 1. Layout of the ALICE detector.

ACORDE has 60 scintillator modules located on the three upper faces of the magnet yoke, covering 10% of its area. The trigger is given by the coincidence of the signals in two different modules (two-fold coincidence).

TOF is a cylindrical MRPC (Multi-gap Resistive-Plate Chamber) array with a very large area surrounding completely the TPC. The trigger requires one pad fired in the upper part of the TOF and one pad in the opposite lower part.

SPD is composed of two layers of silicon pixel modules located very close to the interaction point. The trigger is given by the coincidence of two signals in the top and bottom halves of the external layer.

An atmospheric muon crossing the apparatus is reconstructed by the TPC as two tracks: one in the upper semi-cylinder (track up), the other in the lower semi-cylinder (track down), as shown in Fig. 2 (left). A specific algorithm to match the two tracks has been developed for cosmic events in order to count the exact number of muons, and to measure the momentum with a better precision using the whole track length of the particle. Most of the events are single muon or multimuons (Fig. 2 (center)), with a small percentage of interaction events, which occur when a very energetic muon interacts with the iron of the magnet yoke producing a shower of particles crossing the TPC as shown in Fig. 2 (right). Only single and multimuon events are analysed in this paper.

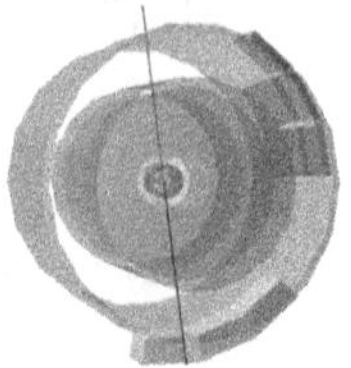

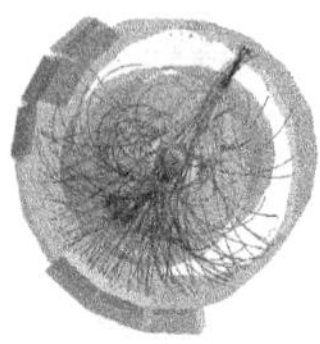

Fig. 2. A single muon crossing the TPC (left), an example of multimuon event (center) and a muon interaction event (right).

3. Measurement of the μ^+/μ^- ratio for near-vertical muons

The correlation between the zenith and the azimuth angles of the atmospheric muons shows an increase in their number in the direction corresponding to the two shafts, as shown in Fig. 3. This is due to a lower energy threshold required to reach the apparatus. Muons with zenith angles in the range $0^o - 20^o$ (near-vertical muons) are not affected by the shaft structures, regardless their azimuth angle, and are choosen to study the μ^+/μ^- ratio. Only single muon events are analysed for this measurement.

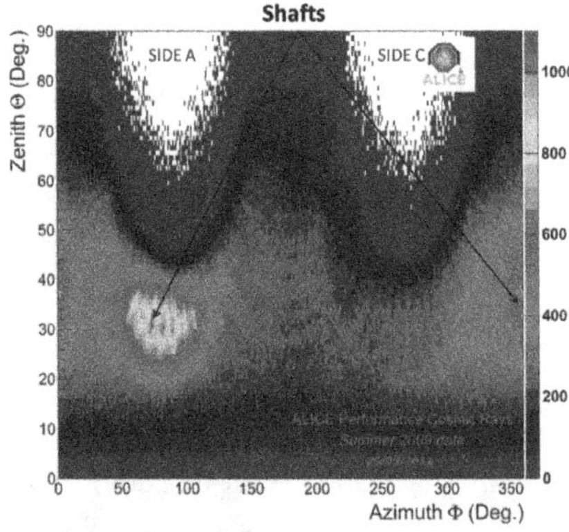

Fig. 3. Zenith vs azimuth angle of the muons in ALICE.

The uncorrected muon momentum distribution of these selected events up to p = 200 GeV/c is shown in Fig. 4 for μ^+ and μ^- at ALICE level.

The measurement of the ratio $R_\mu = \frac{N_{\mu+}}{N_{\mu-}}$ is restricted in the momentum range $10 < p < 100$ GeV/c to remove the threshold effects of the rock above ALICE at low momentum, and to have a reasonable resolution at

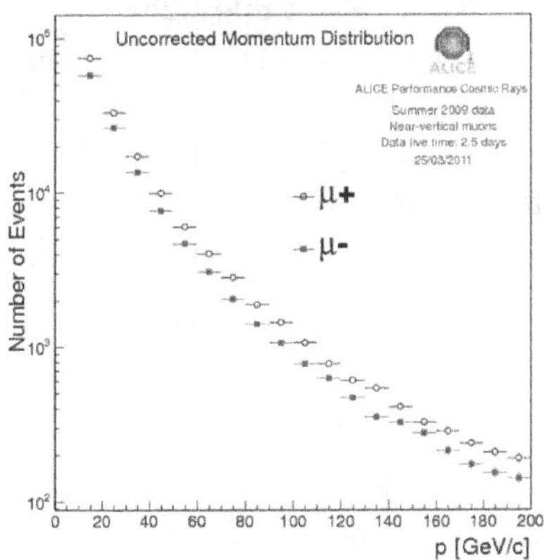

Fig. 4. Uncorrected muon momentum distributions for μ^+ and μ^-.

high momentum ($\sim 30\%$ at p = 100 GeV/c). In Fig. 5 (left) are shown the uncorrected and corrected values of the ratio R_μ as a function of momentum. The corrected value is obtained by detailed Monte Carlo simulations estimating the effect of the migration of entries in different bins due to momentum resolution, the $\mu^\pm$ efficiencies and the charge mis-assignment. The corrected ratio for the whole momentum range $10 < \text{p} < 100$ GeV/c is $R_\mu = 1.275 \pm 0.006(\text{stat.}) \pm 0.01(\text{syst.})$.

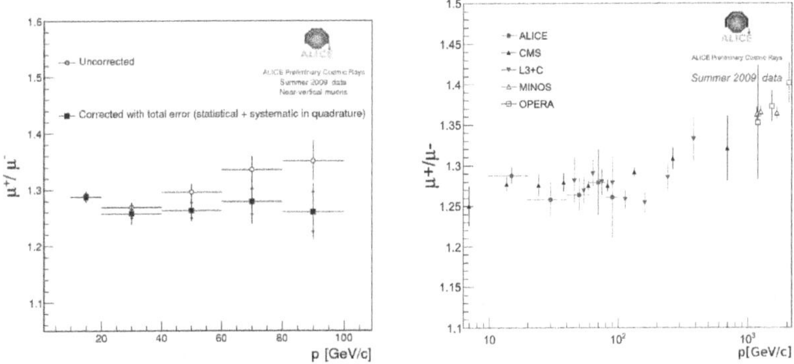

Fig. 5. Corrected μ^+/μ^- ratio as a function of the momentum with statistical and systematic uncertainties (left). The uncorrected ratio is also shown. Comparison of the ratio measured by ALICE with other experiments (right).

The most significant results were the one obtained by L3+C collaboration which published the charge ratio value 1.285±0.003(stat.)±0.019(syst.) for $p < 500$ GeV/c,[2] and a more recent and accurate measurement reported by CMS experiment, with the value of 1.2766±0.0032(stat.)±0.0032(syst.) for $p < 100$ GeV/c.[3] In Fig. 5 (right) our measurement is compared with L3+C and CMS results. We can see that our values are in good agreement with the two quoted experiments although we suffer from larger uncertainties due to a lower statistics.

4. Muon multiplicity distribution

The muon multiplicity distribution for the 2010 and 2011 data is shown in Fig. 6. In 2010 only 2.2 days of data were taken with triggers given by ACORDE and SPD, while until August 2011 we have collected around 10 days of data with triggers given by ACORDE and TOF. We have compared the 2011 distribution with simulations, done with Corsika code[4] and QGSJET-II[5] as hadron interaction model, supposing a pure iron (Fe) composition to get the maximum number of expected muons. We have two events in 2010 and one event in 2011 (see inset in Fig. 6 (right)) with an unexpected high number of muons. The densities of these three events are $\sim 6\mu/m^2$, $\sim 12\mu/m^2$ for 2010 data and $\sim 17\mu/m^2$ for 2011 data. For these densities we expect 1 event in 50 days, 1 year and 3 years of data taking respectively. Further researches and analyses are under way to understand the nature of these events.

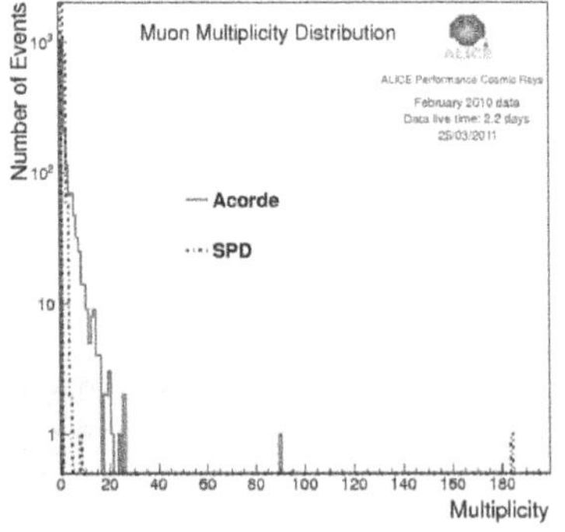

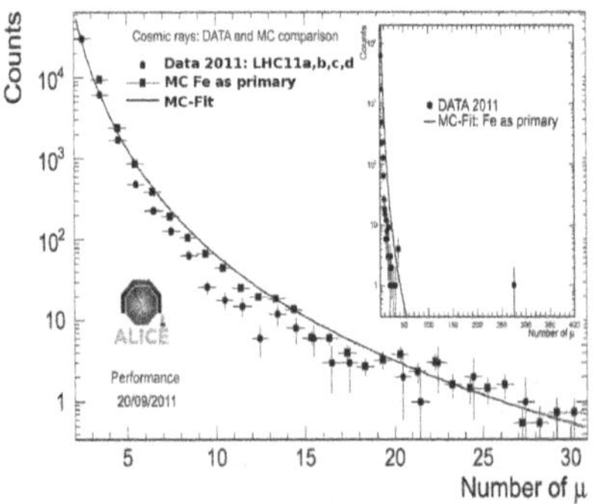

Fig. 6. Muon multiplicity distribution for data taken in 2010 (left) and in 2011 (right).

5. Measurement of the μ^+/μ^- ratio for near-horizontal muons

The FMS consists of a tracking system of 10 detection planes, a large dipole magnet and a muon filter wall followed by 4 planes of trigger chambers (see Fig. 1). Although it was designed for detecting muons produced in p-p or Pb-Pb collisions in the interaction point (IP), we have employed it to study atmospheric muons with large zenith angles. Around 9 days of data have been taken during Summer 2009, collecting a sample of more than 8000 events, reduced to around 4700 after the cuts in the muon directions (only those coming from the IP side were retained) and in the zenith angle ($70^o - 85^o$).

The uncorrected momentum distribution at surface level given in Fig. 7 for $\mu^{\pm}$, is obtained by adding to the momentum measured at ALICE level, the mean momentum loss suffered by the muon crossing the rock above the apparatus.

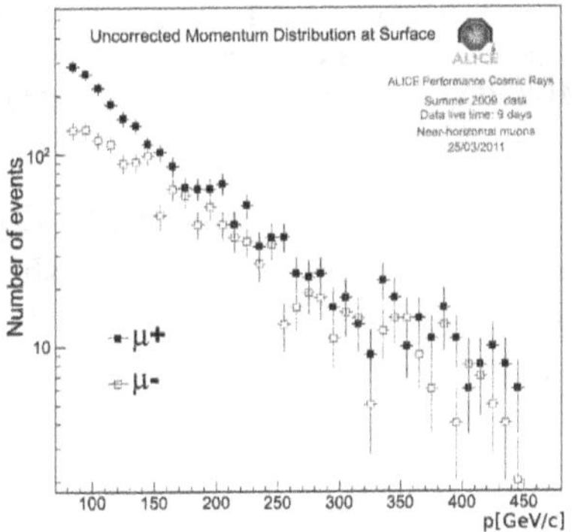

Fig. 7. Uncorrected muon momentum distribution for horizontal μ^+ and μ^- at surface level.

The ratio R_μ is measured in the momentum range $80 < \text{p} < 320$ GeV/c at surface level within momentum bins of size $\Delta\text{p} = 40$ GeV/c. The lower limit is chosen to avoid large differences of efficiency between the two charges, the upper limit to have a reasonable momentum resolution ($\sim 22\%$ at 200 GeV/c measured at ALICE level) and to reduce the effects due to possible mis-alignment of the tracking chambers.

The corrected ratio R_μ has been obtained with the same procedure adopted for near-vertical muons and it is shown in Fig. 8 (left). The value R_μ in the whole momentum range is : $R_\mu = 1.27 \pm 0.04$(stat.) ± 0.1(syst.). The large systematic error is mainly due to the uncertainties in the estimation of the muon chamber mis-alignments.

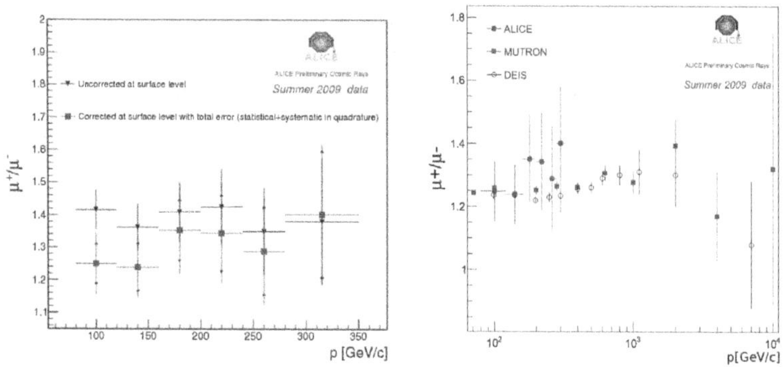

Fig. 8. Corrected μ^+/μ^- ratio as a function of the momentum with statistic and systematic errors compared with uncorrected one (left). Comparison of the ratio measurement with the results of MUTRON and DEIS experiment (right).

The results of MUTRON and DEIS experiments, dedicated to measure the muon charge ratio in the horizontal direction are shown in Fig. 8 (right) and compared with our results. The value reported by MUTRON in the momentum region 100 $<$p$<$ 600 GeV/c is $R_\mu = 1.251 \pm 0.005$(stat.)[6]

References

1. K. Aamodt *et al.*, (ALICE Collaboration), J. Instrum. 3, S08002 (2008).
2. P. Achard *et. al.*, (L3 Collaboration), Phys. Lett. B 598, 15-32, (2004).
3. CMS Collaboration, Phys. Lett. B 692, 83-104, (2010).
4. D. Heck *et al.*, Report FZKA 6019 (1998), Forschungszentrum Karlsruhe.
5. S.S. Ostapchenko, Nucl. Phys. B (Proc. Suppl.) 151, 143, (2006); Phys. Rev. D 74, 014026, (2006).
6. S. Matsuno *et. al.*, Phys. Rev. D 29, 1, (1984).

RECENT RESULTS FROM THE TEVATRON

Emanuela Barberis for the CDF and D0 Collaborations

Department of Physics, Northeastern University,
Boston, MA 02115, USA
**E-mail: e.barberis@neu.edu*
www.physics.neu.edu

This article summarizes recent highlights from the Tevatron physics program.

Keywords: Style file; LaTeX; Proceedings; World Scientific Publishing.

1. The Tevatron and the CDF and D0 detectors

The Tevatron collider at the Fermilab collides protons and anti-protons in Run II at $\sqrt{s}$ =1.96 TeV, with 36 bunch crossings, 396 ns bunch spacing, and a peak luminosity of 4×10^{32} cm^{-2}s^{-1}. Run II successfully ended on September 30th after ten years of operation, delivering more than 10 fb^{-1} of data to its detectors, CDF[1] and D0.[2] CDF and D0 are multipurpose detectors, designed to be highly hermetic. They consist of central tracking volumes surrounded by a solenoidal magnetic field (1.4T and 2.0T for CDF and D0, respectively), electromagnetic and hadronic calorimeters, and outer muon detectors. The detectors performance is well understood; the trigger, reconstruction, and analysis software use optimal and robust algorithms. This article covers only a few selected highlights from the rich menu of physics results that the Tevatron collaborations have produced during Run II. Although the final luminosity recorded by both detectors is in excess of 10 fb^{-1}, the results shown here use up to 9 fb^{-1}.

2. Electroweak Physics

Understanding the production of W and Z bosons and their properties with high precision is an important test of SM predictions. Any deviation from expectations could possibly implicate the presence of new physics. Global electroweak fits currently show good consistency of the measured electroweak (EW) parameters with SM predictions. All variations are below 3σ, with only a couple of measurements (related to the measurement of forward-backward asymmetries) standing out at the $\sim 2\sigma$ level. EW processes also account for the major background sources to the production of a SM Higgs boson and many other processes predicted by theories beyond the SM. Moreover, measuring all EW production cross-sections constitutes a roadmap in testing the ability to measure rare processes, such as the production and decay of a SM Higgs boson decaying into two vector bosons. A few of the most recent EW measurements from the CDF and D0 collaborations are highlighted here.

2.1. Z/γ^* *forward-background asymmetry*

Due to the different couplings of the Z boson to leptons of different handedness, a forward-background charge asymmetry (A_{FB}) can be measured in the angular distribution of the outgoing fermions from the Z decay. Close to the Z pole, A_{FB} can be related to an effective weak mixing angle, while at large invariant masses A_{FB} is sensitive to the presence of new physics. Both collaborations measure a A_{FB}, unfolded to correct for detector effects, which is in good agreement with SM predictions, showing no evidence for new physics at high invariant masses. The latest extraction of the effective weak mixing angle from the A_{FB} measurement by D0[3] is also in agreement with the SM and with measurements of the weak mixing angles made by other experiments. By comparing the measured A_{FB} with templates generated with different couplings of the Z to light quarks, D0 also makes the first measurement of the vector and vector-axial couplings of the Z boson to u and d quarks. The measured couplings are shown in Fig. 1 and are found to be consistent with SM predictions.

2.2. *Di-boson production*

The smallest measured EW cross-section at the Tevatron corresponds to the production of two Z bosons. Its measurement is crucial for assessing the feasibility of a SM $H \to VV$ search and a robust determination of the backgrounds in such channel. D0 recently performed a search in the $ZZ \to 4$ leptons decay channel[4] leading, in combination with $ZZ \to 2l2\nu$, to the current most precise measurement of σ_{ZZ}, i.e. $\sigma_{ZZ} = 1.40^{+0.43}_{-0.37}(stat.) \pm 0.14(syst.)$ pb. Similarly, CDF measures $\sigma_{ZZ} = 1.45^{+0.45}_{-0.42}(stat.)^{+0.14}_{-0.30}(syst.)$ pb and, observing an excess of events in the $4l$ decay channel (not confirmed by the $ZZ \to 2l2\nu$ and $ZZ \to 2l2j$ channels), looks for high mass reso-

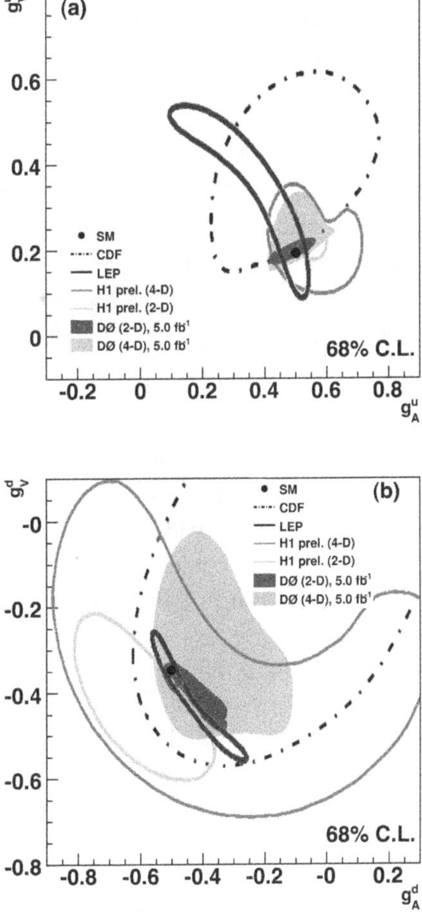

Fig. 1. The 68% C.L. contours for the axial and vector-axial couplings of the Z boson to (a) u, and (b) d quarks as measured by D0, in comparison with other experiments.

nances decaying into ZZ and places a 95% C.L. limit for $\sigma(X \to ZZ)$ at 0.26 pb and 0.28 pb for two signal models. Using a lepton plus jets and transverse missing energy final state signature, both experiments have measured the $WW/WZ \to l\nu jj$ production cross-section. D0 found evidence for $WW/WZ \to l\nu jj$ at 4.2σ and measured $\sigma(WV) = 20.2 \pm 4.5$ pb,[5] and CDF placed the measurement in excess of 5σ at $\sigma(WV) = 18.1 \pm 3.3(stat.) \pm 2.5(syst.)$ pb.[6] Using the same data-set used in,[6] CDF observes a data excess in the 120-160 GeV region of the di-jet invariant mass spectrum,[7] which is further confirmed, at 4.1σ, in a larger data-set (7.3 fb^{-1}). D0 searches for the same structure and finds no evidence of such an excess in their data.[8] Requiring the two jets in the above measurement to be identified as originating from $b-$quarks, leads to a measurement of VZ production, where $Z \to b\bar{b}$. This is a crucial measurement for the understanding of the backgrounds to low mass SM Higgs searches (such as $ZH \to \nu\bar{\nu}b\bar{b}$). The final measurement of VZ requires the inputs from all relevant low mass SM Higgs analyses, but both CDF and D0 have partial measurements (e.g. a CDF measurement of $WW/WZ \to l\nu b\bar{b}$, and CDF and D0 measurements of $WZ/ZZ \to \nu\bar{\nu}b\bar{b}$) which yield results compatible with SM expectations.

3. Top Quark Physics

Since its observation in 1995, the Tevatron has made a vast impact on the knowledge of top quark physics, from the observation of top quark production processes, measurement of mass, width, properties of the Wtb vertex, forward-backward asymmetry, spin correlations, etc. At the Tevatron top quarks are predominantly produced in pairs via quark-quark annihilation. EW production of single top quarks takes place through the s and t channels and was observed in 2009 by both Tevatron collaborations. In the SM, the top quark decays into a W boson and a $b-$quark and its decay modes are therefore classified according to the decay modes of the W.

3.1. *Top Quark Mass*

In the SM, the mass of the top quark is a fundamental parameter that can be related with m_W to the Higgs boson mass. Measurements from all pair production decay modes (some performed with different techniques) from the CDF and D0 experiments are combined to give the Tevatron measurement of the mass of the top quark:[9] $m_{top} = 173.2 \pm 0.9$ GeV. Both CDF and D0 also measure the difference between top and anti-top masses[10,11] which is found to be consistent with SM expectations. A measurement of the top mass in two different theoretical schemes (pole and $\overline{\text{MS}}$) is also extracted by D0 from a measurement of $\sigma_{t\bar{t}}$ as a function of m_t, where $\sigma_{t\bar{t}}$ measured assuming a specific Monte Carlo scheme (either pole of $\overline{\text{MS}}$) is compared with QCD calculations using the pole m_t to extract a measurement of the pole m_t and $m_t(\overline{\text{MS}})$.[12] The measurement of the pole m_t is found to be consistent with the direct m_t measurement.

3.2. *Electroweak Production of Single Top*

In 2009, the CDF and D0 collaborations observed the electroweak production of single top and measured its cross-section (t channel and s channel combined,[13] $\sigma(tb + tqb) = 2.76^{+0.58}_{-0.47}$ pb). Recently, D0 made a model-independent measurement of the t channel single top production using 5.4 fb^{-1} of data[14] (illustrated in Fig. 2). The measured cross-section of $\sigma(tqb+X) = 2.90\pm0.59$ pb corresponds to the the first observation ($> 5.5\sigma$) of t channel single top production.

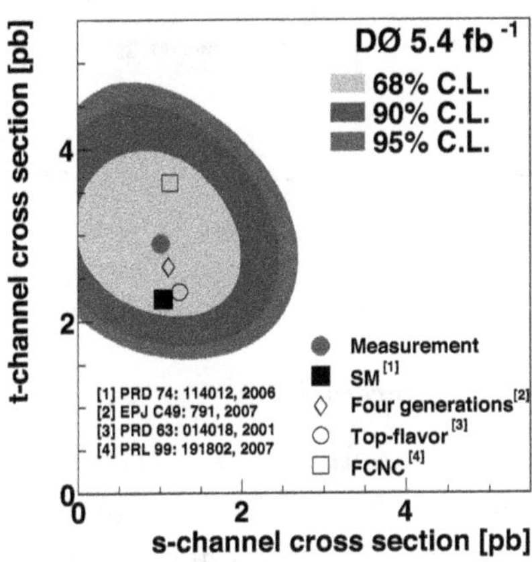

Fig. 2. Posterior probability density for tqb vs tb single top quark production. The measured cross-section and various theoretical predictions are shown.

3.3. *Top anti-top forward-backward asymmetry*

New physics can result in deviations from SM prediction of the forward-background asymmetry (A_{FB}) in top pair production (which is predicted in the SM to be small, i.e. $\sim 5-6\%$ for a qq initial state). CDF recently observed,[15] with 5.4 fb^{-1}, that A_{FB} in the dilepton final state deviates significantly from SM predictions, $A_{FB} = 0.42 \pm 0.15(stat.) \pm 0.05(syst.)$. A second CDF measurement which uses the lepton+jets final state shows deviations which increase for large values of $m_{t\bar{t}}$. D0 recently measured[16] A_{FB} in the dilepton channel, finding a value consistent with the CDF average value (see Fig. 3) but without any clear $m_{t\bar{t}}$ dependence.

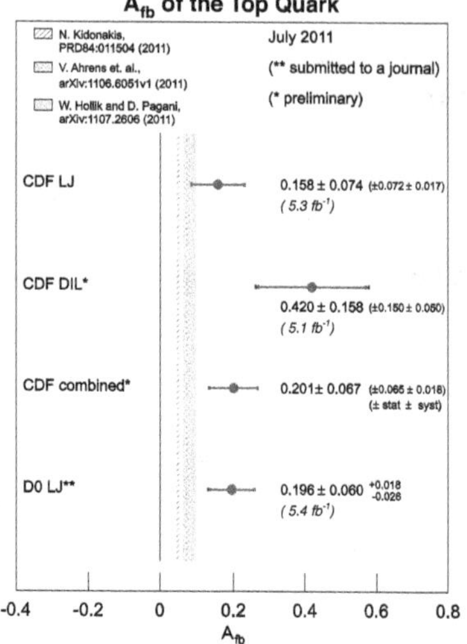

Fig. 3. Top anti-top A_{FB} measurements from CDF and D0.

3.4. *Top anti-top spin correlations*

Top quarks decay before hadronization, and their spin information is preserved in the angular distributions of the decay products. Both CDF[17] and D0 measure the degree at which the top and anti-top spins are correlated and find it to be consistent with SM predictions. The latest D0 measurement[18] results in the first evidence (3.1σ) of spin correlations.

4. Heavy Flavor Physics

The heavy flavor physics program at the Tevatron has been very rich. The focus here is on some of the measurements of CP violation in decays and mixing, where hints for new physics might appear. Due to the mixing between the flavor eigenstates of the B_s^0 system, interference of decays with or without mixing can cause CP violation. Through the challenging analysis of $B_s^0 \to J/\psi\phi$ decays, CDF and D0 measure the CP violation phase and difference in lifetime between mass eigenstates: $\phi_s = -2\beta_s$, and Γ_s. New physics beyond the SM that enters in the mixing box diagram can appear as an additional term to ϕ_s. Previous measurements from D0[19] showed some deviation from SM predictions in ϕ_s, but new results[20] from D0 on 8 fb^{-1} and CDF measurements[21] show a much less significant departure from the SM.

4.1. *Like-sign Dimuon charge asymmetry*

When $B^0_{q=d,s}$ mesons mix with their antiparticles the process is not CP symmetric, but in SM the asymmetry is very small. Any substantial asymmetry in $B^0 - \bar{B}^0$ mixing offers hints of physics beyond the SM. With 9 fb^{-1} and using the semileptonic decays of $B^0_{q=d,s}$, D0 measures[22] the asymmetry in the number of like-sign dimuon events: $A^b_{sl} = \frac{N_b(\mu^+\mu^+)-N_b(\mu^-\mu^-)}{N_b(\mu^+\mu^+)+N_b(\mu^-\mu^-)}$. The measurement, $A^b_{sl} = (-0.787 \pm 0.172 \pm 0.093)\%$, is a 3.9σ deviation from SM predictions. Since A^b_{sl} can be written as a linear combination of the semileptonic charge asymmetries of B^0_d and B^0_s, a^d_{sl} and a^s_{sl}, the measurement can be translated to a constraint in the (a^d_{sl}, a^s_{sl}) plane, which is shown in Fig.4. The results are consistent with previous measurements of flavor-specific asymmetries.

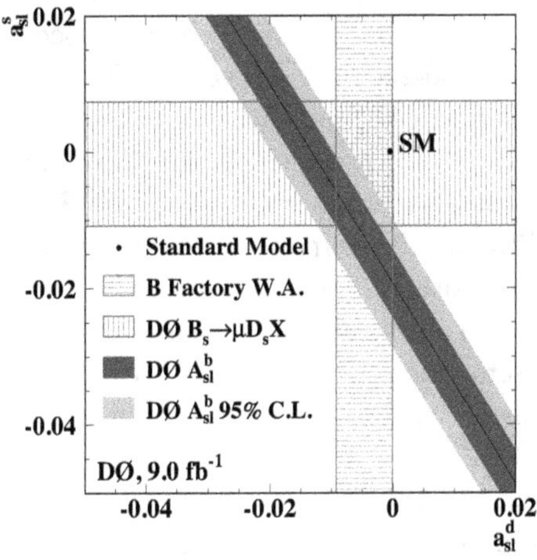

Fig. 4. Measurement of A^b_{sl} in the (a^d_{sl}, a^s_{sl}) plane by D0 (9 fb^{-1}).

5. Searches for Standard Model Higgs Production

In the SM, the Higgs is the EW symmetry breaking (EWSB), mass generating scalar field, whose couplings are proportional to particle masses. At low masses the Higgs boson predominantly decays as $H \to b\bar{b}$, while at high masses it predominantly decays into two vector bosons. The Higgs mass region allowed by the electroweak precision fits is within reach of the

Tevatron experiments, where the search strategy consists of maximizing sensitivity by combining many decay channels, and using all production modes (mainly associate production of W(Z)H, and gluon-gluon fusion). The searches for a low mass Higgs start with events consistent with W/Z decays with jets (Higgs is produced in association with a W or Z, and it decays to $b\bar{b}$). Leptonic triggers or missing transverse energy triggers are used. The main channels are: $W \to l\nu$ ($l = e$ or μ and E_T^{miss}), $Z \to ll$ (ee or $\mu\mu$ consistent with a Z resonance), and $Z \to \nu\nu$ (no charged leptons; two acoplanar jets and E_T^{miss}). For a low mass Higgs search, heavy flavor identification of jets (b-tagging) is essential to improve the significance of a signal. Other important ingredients in optimizing many of the Higgs analyses are: improvements in lepton identification efficiency, improvements in resolution of the di-jet invariant mass, splitting channels according to different S/B content, modeling of backgrounds using data control samples and measurement of SM background cross-sections, and usage of sophisticated multivariate techniques. The main channel used in high mass Higgs searches (in the m_H ~130-200 GeV range) is $WW \to l\nu l\nu$, which benefits from a clean dilepton and E_T^{miss} signature. In the absence of a clear excess of data over SM background predictions, the CDF and D0 collaborations proceed to set cross-section limits with respect to SM predictions. Limits are derived using CL_S and Bayesian methods, where the systematic uncertainties (which affect both the normalization and the shapes of the final discriminant templates) are treated as nuisance parameters taking into account correlated errors among channels and experiments. The most recent CDF and D0 combined[23] exclusion regions on the production of a SM Higgs boson are 100-109 GeV and 156-177 GeV (the expected exclusion regions are 100-108 GeV and 148-181 GeV) as shown in Fig. 5. The limits combine dedicated searches in different channels using up to 8.6 fb^{-1} of data.

6. Searches for Higgs Bosons Beyond the Standard Model

A few scenarios are presented here: the search for a Higgs boson in theories which assume different production and couplings than the SM ones, and the search of Higgs boson within the Minimal Supersymmetric Model (MSSM). 4th generation models see the gluon fusion production mode enhanced by the presence of additional fermions in the fermion loop of the production diagram. The $gg \to H$ production mode can therefore be up by a factor 9 in the 100 GeV $< m_H <$ 300 GeV range. This leads to a combined Tevatron observed exclusion[24] of 124 GeV $< m_H <$ 286 GeV. In fermiophobic Higgs models, a Higgs with different couplings than SM is explored: there is no

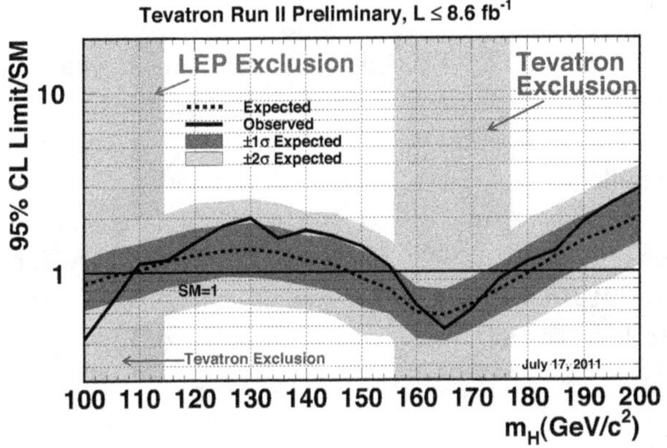

Fig. 5. Observed and expected 95% C.L. upper limits on the ratios to the SM cross-section, as functions of the Higgs boson mass for the combined CDF and D0 SM Higgs analyses. The bands indicate the 68% and 95% probability regions.

production of Higgs via $gg \to H$; only WH, ZH and vector-boson fusion. $\gamma\gamma$, WW, Zγ, and ZZ account for nearly the entire decay width. The $H \to \gamma\gamma$ and ZH/WH analyses with vector boson decays from CDF and D0 are re-optimized, together with the $H \to WW$ analysis from CDF to give a combined Tevatron limit[25] of $m_H < 119$ GeV on the mass of a fermiophobic Higgs boson.

6.1. *Higgs bosons in MSSM*

The MSSM has an extended Higgs sector (two Higgs doublets), which after electroweak symmetry breaking results in four massive scalars (h^0,H^0,$H^{\pm}$) and one pseudo-scalar (A^0). Since the neutral Higgs bosons ($h/H/A$) coupling to b, τ is enhanced at high tanβ, three channels are used at the Tevatron with comparable sensitivity: $H \to \tau\tau$, $bH \to b\tau\tau$, and $bH \to bb\bar{b}$. The Tevatron searches in the $(b)\tau\tau$ decay channels[26,27] show good agreement with SM predictions. The $bH \to bb\bar{b}$ channel is a challenging final state because of the background modeling, and it is not really accessible to the LHC. Both CDF[28] and D0[29] see a broad 2σ excess in this channel around m_H ~120-150 GeV. A combination of results and analysis of larger datasets for this channel is in progress.

7. Conclusions

The Tevatron successfully concluded Run II, delivering >10fb^{-1} to CDF and D0. With such data the experiments have probed and extended our knowledge of the SM and EWSB. Discoveries and measurements in EW, top quark, and heavy flavor physics, Higgs and beyond the SM searches stand as the legacy of the Tevatron program. Data analysis will extend beyond 2012, with significant portion of the data to explore and improvements to be made. Several discrepancies with SM predictions are present, begging for further look. Searches for the Higgs boson will possibly get us close to a discovery by 2012, as the search window narrows, and the Tevatron plays a significant role in the analysis of a low mass Higgs.

References

1. D. Acosta *et al.* (CDF Collaboration), *Phys. Rev. D* **71**, 032001 (2005).
2. V. M. Abazov *et al.* (D0 Collaboration), *Nucl. Instrum. Methods Phys. Res. Sect. A* **565**, 463 (2006).
3. V. M. Abazov *et al.* (D0 Collaboration), *Phys. Rev. D* **84**, 012007 (2011).
4. V. M. Abazov *et al.* (D0 Collaboration), *Phys. Rev. D* **84**, 011103(R) (2011).
5. V. M. Abazov *et al.* (D0 Collaboration), *Phys. Rev. Lett.* **102**, 161801 (2009).
6. T. Aaltonen *et al.* (CDF Collaboration), *Phys. Rev. Lett.* **104**, 101801 (2010).
7. T. Aaltonen *et al.* (CDF Collaboration), *Phys. Rev. Lett* **106**, 171801 (2011).
8. V. M. Abazov *et al.* (D0 Collaboration), *Phys. Rev. Lett.* **107**, 011804 (2011).
9. TEWG (CDF and D0 Collaborations), *FERMILAB-TM-2504-E* (2011).
10. T. Aaltonen *et al.* (CDF Collaboration), *Phys. Rev. Lett* **106**, 152001 (2011).
11. V. M. Abazov *et al.* (D0 Collaboration), *Phys. Rev. D* **84**, 052005 (2011).
12. V. M. Abazov *et al.* (D0 Collaboration), *arXiv:1104.2887v1* (2011).
13. TEWG (CDF and D0 Collaborations), *FERMILAB-TM-2440-E* (2011).
14. V. M. Abazov *et al.* (D0 Collaboration), *arXiv:1105.2788v1* (2011).
15. T. Aaltonen *et al.* (CDF Collaboration), *CDF Note 10584* (2011).
16. V. M. Abazov *et al.* (D0 Collaboration), *arXiv:1107.4995v1* (2011).
17. T. Aaltonen *et al.* (CDF Collaboration), *Phys. Rev. D* **83**, 031104 (2011).
18. V. M. Abazov *et al.* (D0 Collaboration), *Phys. Rev. Lett.* **107**, 032001 (2011).
19. V. M. Abazov *et al.* (D0 Collaboration), *Phys. Rev. Lett.* **101**, 241801 (2008).
20. V. M. Abazov *et al.* (D0 Collaboration), *arXiv:1109.3166v1* (2011).
21. T. Aaltonen *et al.* (CDF Collaboration), *CDF Note 10206* (2010).
22. V. M. Abazov *et al.* (D0 Collaboration), *Phys. Rev. D* **84**, 052007 (2011).
23. TEWG (CDF and D0 Collaborations), *arXiv:1107.5518v2* (2011).
24. TEWG (CDF and D0 Collaborations), *arXiv:1108.3331v2* (2011).
25. TEWG (CDF and D0 Collaborations), *arXiv:1109.0576* (2011).
26. V. M. Abazov *et al.* (D0 Collaboration), *Phys. Rev. Lett.* **107**, 121801 (2011).
27. V. M. Abazov *et al.* (D0 Collaboration), *arXiv:1106.4555* (2011).
28. T. Aaltonen *et al.* (CDF Collaboration), *arXiv: 1106.4782* (2011).
29. V. M. Abazov *et al.* (D0 Collaboration), *Phys. Lett. B* **698**, 97 (2011).

Identified-particle production at high-momentum with the HMPID detector in the ALICE experiment at the LHC in proton-proton collisions at $\sqrt{s} = 2.76$ TeV

F. BARILE* for the ALICE Collaboration
University of Bari and INFN Sezione Bari,
Bari, 70125, Italy
**E-mail: francesco.barile@ba.infn.it*

The ALICE experiment is devoted to the study of heavy-ion collisions at LHC energies. ALICE is also studying proton-proton collisions both as a comparison with lead-lead collisions and in physics areas where ALICE is competitive with other LHC experiments. The ALICE-HMPID detector performs charged particle track-by-track identification (π and K in the momentum range of $1 < p < 3$ GeV/c and p in $1.5 < p < 5$ GeV/c) by means of measurement of Cherenkov angle and momentum information provided by the tracking devices. The current results on hadron transverse momentum spectra measured by the HMPID detector in pp collisions at $\sqrt{s} = 2.76$ TeV, are shown.

Keywords: HMPID; RICH; ALICE; PID.

1. The ALICE detector

In the ALICE (A Large Ion Collider Experiment)[1,2] experiment great care has been devoted to design a high-quality particle identification system in the central region exploiting the combination of several sub-detectors to identify particles with momenta from 0.1 GeV/c up to $\sim$ 100 GeV/c. The hadron identification at high transverse momenta ($1 < p < 3$ GeV/c for π and K, $1 < p < 5$ GeV/c for p) is achieved by the HMPID (RICH) detector.

2. The HMPID detector

The ALICE-HMPID[3,4] detector enhances the PID capability of the ALICE experiment beyond the momentum region available through energy loss (in ITS and TPC) and time-of-flight measurements (in TOF). The detector was optimized to extend the useful range for π/K and K/p discrimination,

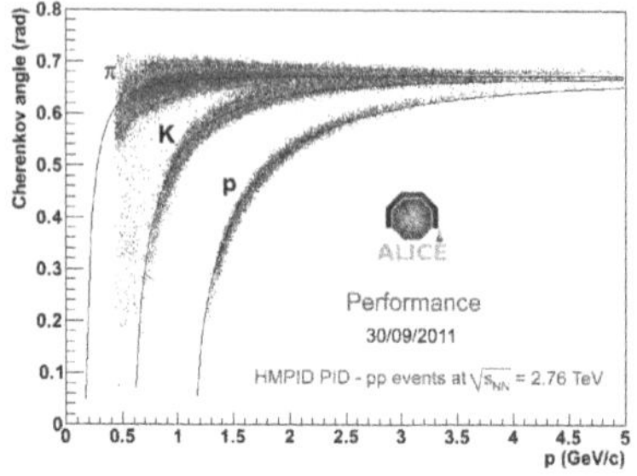

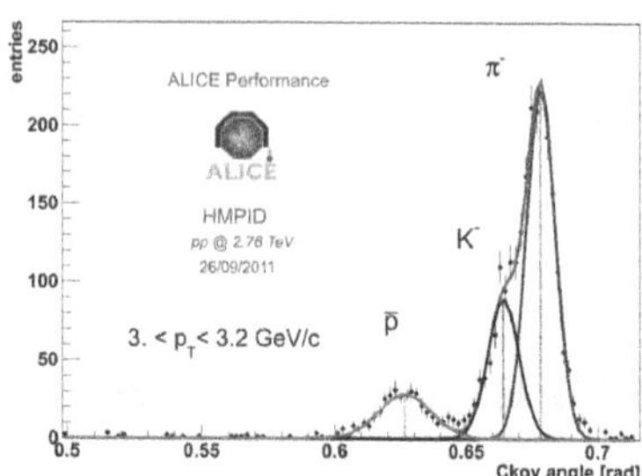

Fig. 1. Left panel: Cherenkov angle distribution as a function of the track momentum in pp collisions at $\sqrt{s}$ = 2.76 TeV; right panel: an example of the Cherenkov angle distribution in $3 < p_t < 3.2$ GeV/c (pp collisions at $\sqrt{s}$ = 2.76 TeV).

on a track-by-track basis, up to 3 and 5 GeV/c respectively. The HMPID is a proximity-focusing Ring Imaging Cherenkov (RICH) counter and consists of seven modules of about 1.4×1.3 m^2 each. The radiator, that defines the momentum range covered by the HMPID, is a 15 mm thick layer of C_6F_{14} (perfluorohexane) liquid with index of refraction $n = 1.2989$ at 175 nm corresponding to $\beta_{min} = 0.77$.[5] The photon and MIP detection is provided by a multiwire proportional chamber with pad-segmented CsI photocatode.

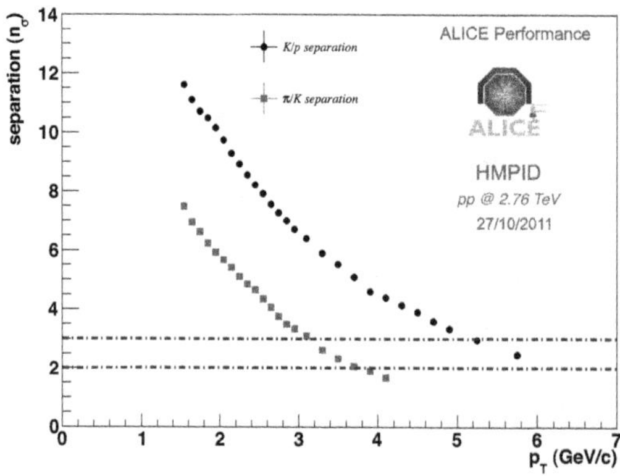

Fig. 2. The n_σ separation for π/K and K/p as a function of the p_t in the HMPID.

The HMPID detector contributes to different physics topics:

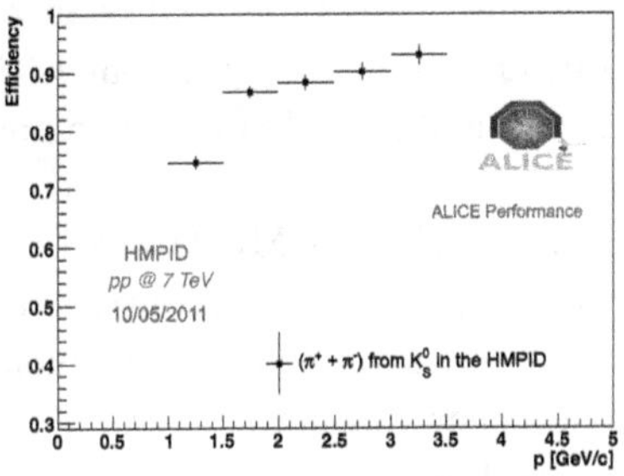

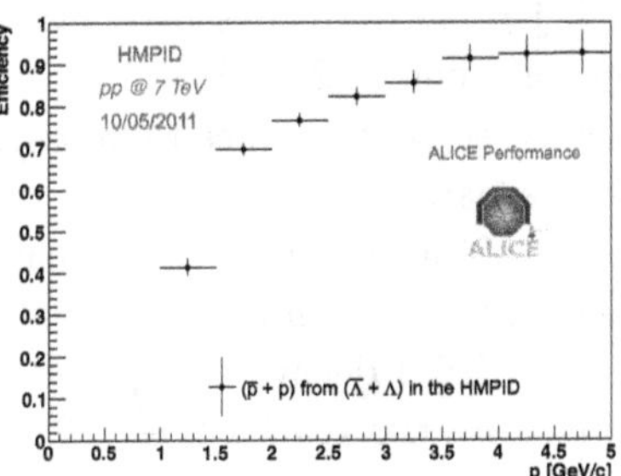

Fig. 3. The PID HMPID efficiency for pions (left panel) and protons (right panel).

- π, K, p spectra to high momentum;
- particle ratios vs p_T ($\bar{p}$/p, p/π, K/π);
- identification of light nuclei (d, t, ^{3}He, α);
- measurement of resonance production such as $\Phi(1020) \longrightarrow K^+K^-$;
- identified particle correlations (HBT) to study source size (dynamics);
- jet physics:
  - study of identified jet fragmentation;
  - study of the flavor of the leading particle;

3. Particle identification in the HMPID

The mass of charged particles is identified from the simultaneous measurement of the track momentum (from the central barrel) and the Cherenkov angle measurement from HMPID.

Figure 1 (left panel) shows the Cherenkov angle as a function of the track momentum in pp collisions at 2.76 TeV. The Cherenkov bands (pions, kaons and protons) are clearly visible and follow the theoretical expectations (black solid lines) for nominal index of refraction.

The Cherenkov angle distribution in a narrow momentum range shows Gaussian behaviour for each particle species, as shown in Figure 1 (right panel). Lines represent a three Gaussian fit to extract the raw yields.

The n-σ separation for π/K and K/p as a function of p_t is shown in Figure 2. The separation, defined as:

$$separation = \frac{\theta_{Ckov}(i) - \theta_{Ckov}(j)}{(\sigma(i) + \sigma(j))/2} \tag{1}$$

with i, j = π, K, p, has been evaluated for the above three Gaussian fit in pp collisions at 2.76 TeV. The n_σ separation is in good agreement with the expected performance represented by the blue dashed lines for two and three σ separation.
A study of the particles identification efficiency of the HMPID has been carried out with samples of protons and pions coming from reconstructed V0 ($\Lambda/\bar{\Lambda}$, K^0_S) decays in p-p collisions at 7 TeV: a pure sample of pions coming from the K^0_S ($K^0_S \to \pi^+ \pi^-$) and protons coming from Λ ($\Lambda \to p\,\pi^-$) have been selected from the Armenteros-Podolanski distribution for V0 candidates in the detector acceptance and the HMPID response has been evaluated. In Figure 3 the PID efficiencies for pions and protons, are shown.
Using the same identified protons coming from the Λ has been possible to measure the C_6F_{14} refractive index of the HMPID radiator(Figure 4). The average value of the $n_{C_6F_{14}}$ has been deducted using the Cherenkov relation :cos $\Theta_C = 1 / \beta\, n_{C_6F_{14}}$. The $n_{C_6F_{14}}$ value is in excellent agreement with the average value quoted in the HMPID Technical Design Report[3] (obtained by optical measurement).

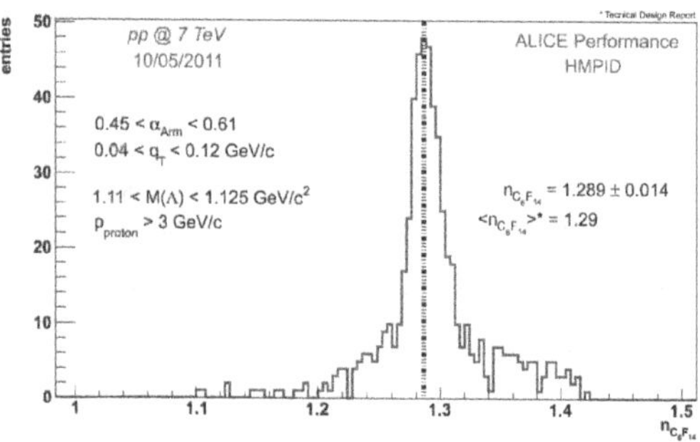

Fig. 4. C_6F_{14} refractive index measurement of the HMPID radiator from identified protons.

Analysis is currently ongoing to finalize the transverse momentum spectra in pp collisions at $\sqrt{s}$ = 7 TeV[6] and 2.76 TeV. HMPID is also involved in analysis in Pb-Pb collisions at $\sqrt{s}$ = 2.76 TeV.
In Figure 5 the identified uncorrected (no efficiency corrected) spectra for negative hadrons are presented: only primary tracks are selected and specific cuts on the detector analysis are used.

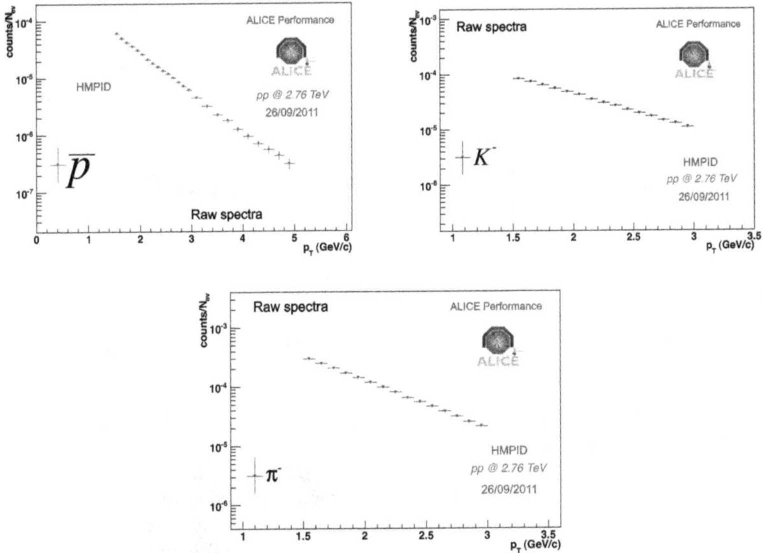

Fig. 5. Uncorrected spectra for negative hadrons measured by the HMPID (pp collisions at $\sqrt{s} = 2.76$ TeV).

4. Conclusions

HMPID enhances the unique PID capability of the ALICE experiment by the identification of π, K and p up to 5 GeV/c in agreement with designed performance demonstrated by the separation. The PID efficiency, the geometrical acceptance and the systematic uncertainties are evaluated to finalize and to combine the identified particle spectra from HMPID with the extensive PID results of the ALICE experiment.

References

1. ALICE Collaboration, Physical Performance Report, Volume I, J.Phys.G: Nucl. Part. Phys. 30 (2004) 1517-1763.
2. ALICE Collaboration, Physical Performance Report, Volume II, J.Phys.G: Nucl. Part. Phys. 32 (2006) 1295-2040.
3. CERN / LHCC 9819, ALICE TDR 1, 14 August 1998.
4. P.Martinengo et al., Nucl.Instrum.Meth. A639 (2011) 7-10.
5. A. Di Mauro et al. ALICE Internal Note 2007-003.
6. F.Barile [ALICE Collaboration], AIP Conf.Proc. 1317 (2011) 71-76.

CMS MUON DETECTORS AND TRIGGER PERFORMANCE

Carlo Battilana on behalf of the CMS collaboration

Centro de Invesigaciones Energéticas, Medioambientales y Tecnológicas (C.I.E.M.A.T), Madrid, Spain, E-mail:carlo.battilana@cern.ch*

The Muon Spectrometer of the CMS experiment is designed to perform precise muon identification, and to measure transverse momentum of crossing particles with high resolution. It provides trigger capabilities and is used, together with the tracking system, in the offline reconstruction, improving overall p_T measurement performance for events with high p_T muons. It consists of different kind of gaseous detectors: Cathode Strip Chambers (CSC) equip the endcaps of the CMS detector, whereas Drift Tubes (DT) cover the barrel region. In addition, a set of Resistive Plate Chambers (RPC) complement the former two, both in barrel and endcaps, providing redundancy to the muon trigger. The Muon System has operated remarkably well during the data-taking periods with LHC colliding beams, both in 2010 and 2011. The overall trigger and detector performance of the Muon System during collision data-taking will be presented, focusing on how this has met design requirements and expectations.

Keywords: LHC; CMS; Muon.

1. Introduction

The Compact Muon Solenoid (CMS)[1] is a general-purpose detector presently operating at the Large Hadron Collider (LHC)[2] at CERN. A large fraction of the CMS physics programme is based on the study of final states containing muons of high transverse momentum. Muons provide a clear signature, relatively easy to identify, of potentially interesting events in the high rate of p–p collisions generated at LHC. A triggering muon system, also able to perform muon identification and precise reconstruction is thus a basic tool for the experiment.

2. The CMS Muon System, Trigger and Performance

The layout of the CMS Muon System[1 3] is shown in Fig. 1. The spectrometer is hosted within the return yoke of the CMS magnet and it consists in a central barrel, divided into five parallel wheels, and two closing endcaps, composed of four disks each. It is built using three different types of gaseous detectors whose design takes into account the radiation environment and the magnetic field strenght at different values of η. Drift Tubes Chambers (DT) are used in the barrel up to $|\eta| < 1.2$, whereas CMS endcaps ($0.9 < |\eta| < 2.4$) are equipped with Cathode Strip Chambers (CSC). Resistive Plate Chambers (RPC) complement DT and CSC, in barrel and endcaps, covering up to $|\eta| < 1.6$, ensuring redundancy to the trigger system.

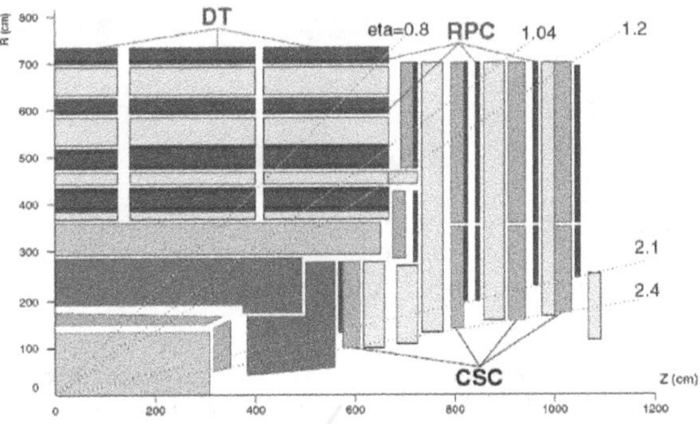

Fig. 1. Longitudinal view of one quarter of the CMS muon spectrometer.

2.1. *Drift Tube Chambers (DT)*

Drift Tube Chambers are used as tracking and triggering devices in the barrel, where magnetic field and track occupancy are low. A total of 250 chambers is equally distributed among the 5 barrel wheels. Each wheel hosts four concentric rings of stations segmented in 12 contiguous sectors. The basic DT detector element is a rectangular drift cell with a transversal size of 4.2cm×1.3cm, filled with a 85/15% Ar/CO_2 gas mixture. Cell anode wires are operated at 3.6 kV, cathode planes are set at -1.2 kV and electrode strips, put at 1.8 kV, are used to shape the electric field lines. This configuration results in an almost constant drift velocity of about 54 μm/ns in a large fraction of the drift volume. Cells are arranged parallely to form detection layers, stacked in group of four to form super-layers (SL). Layers within a SL are staggered to allow left/right ambiguity resolution. Each DT chamber is equipped with two SL measuring the coordinate in the CMS bending plane (r), whereas a single SL measures the coordinate along the beam line (z) in the 3 innermost station rings.

Single DT hit residual distributions were measured with respect to local reconstructed segments, using 2010 and 2011 proton collision data. Hit resolutions were estimated from the σ of Gaussian fits performed on the core of the distributions. Results from the measurement are reported in Fig 2. Resolutions on hits vary between 200 and 350 μm for the radial view resulting in a maximum resolution for local segments from ϕ SLs around 100 μm .

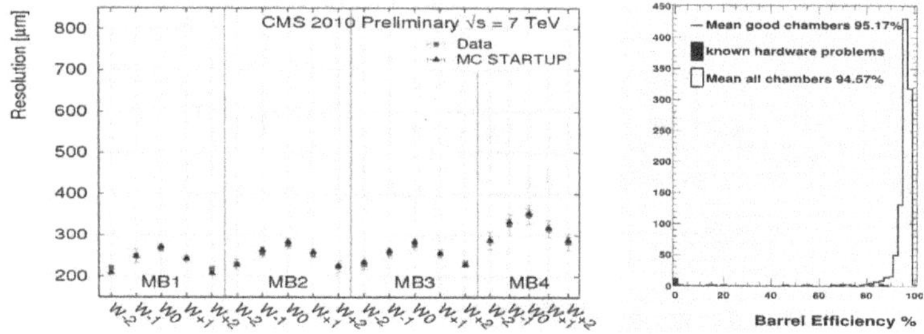

Fig. 2. Map of DT hit resolution in ϕ view for different stations/wheels (left). Distribution of average hit detection efficiency for RPC barrel chambers (right).

2.2. *Cathode Strip Chambers (CSC)*

The endcap region is characterized by high particle rate and magnetic field. Cathode Strip Chambers are used there as tracking and trigger detector due to their short drift length. CSC are organized in trapezoidal chambers arranged, within the four disks of each endcap, in 1, 2 or 3 concentric rings, according to the disk position. There are in total 468 chambers, and rings are composed of 18 or 36 of them. Each chamber is made of 6 layer arrays of 7-9.5 mm thick anode wires enclosed between cathode planes with an anode-cathode spacing of 3.5-4.8 mm. One of the cathodes is segmented with strips whose pitch varies between 4.1 and 16 mm to allow radial (ϕ) coordinate measurement by computing the centroid of the energy deposit in 3 adjacent strips. Anode wires, placed perpendicular to the strips, are used to measure the r coordinate and their fast response is exploited to perform parent bunch crossing identification when building trigger segments. Chambers are filled with a 40/50/10% $Ar/CO_2/CF_4$ gas mixture.

Hit resolution with respect to local reconstruction was computed on collision data in a way similar to the DT, proving that final segment resolution varies between 60 and 150 μm.

2.3. *Resistive Plate Chambers (RPC)*

Resistive Plate Chambers equip both barrel and endcap regions. Due to their good timing resolution are mainly used in the trigger, they also participate to tracking, although with limited resolution. A total of 480 RPC chambers is organized in the five barrel wheels, every wheel consisting of 12 sectors, equipped with 6 RPC layers each. 362 RPC chambers are placed in rings of 36, two of them for each of the three innermost disks on the endcaps. The innermost ring and the outermost endcap disk are not presently instrumented, therefore the RPC subdetector coverage is limited to $|\eta| < 1.6$. CMS uses double gap RPC chambers working in avalanche mode and filled with a 96.2/3.5/0.3% $C_2H_2F_4/Iso - C_4H_{10}/SF_6$ gas mixture. Readout is performed using copper strips, placed between the two gaps. Strips allow the measurement of the bending coordinate (ϕ) with a precision around 1 cm, while the timing resolution is about 2 ns.

Efficiency for hit detection was computed on collision data by propagating DT/CSC segments to RPC layers and searching for geometrically matched hits. Figure 2 (right) shows the average chamber efficiency distribution for barrel RPCs. The mean value of the distribution was found to be around 94%; similar studies were performed on endcap chambers giving a value of about 93%.

2.4. *The CMS Muon Trigger*

The CMS Trigger[4 5] is designed to identify and select for storage about 100 Hz of "physically interesting" events on the basis of their final state, performing an overall rate reduction of about 10^6 with respect to the total rate of p–p collisions at LHC. It operates at different levels called Level 1 Trigger (L1) and High Level Trigger (HLT). The L1 consists of custom programmable electronics and, in the case of muons, performs bunch crossing identification and p_T measurement using information from all the muon subdetectors. DT and CSC trigger segments are matched in full muon tracks by Track Finders that compute the muon p_T. In the case of RPC muon track candidates are built up from single hits matching predefined patterns made, at least, of 3 detector layers. Information from the sub-systems is then processed by the Global Muon Trigger to optimize efficiency and reduce extra-rate induced by poorly reconstructed muons. The HLT is a software system running on a farm of commercial processors and has access to full readout information. In the case of muons, HLT reconstructs them using both spectrometers and tracker data, in order to optimize p_T

assignment, hence improving rate reduction.

Combined L1+HLT efficiencies were computed on 2010 and 2011 collision data for different set of p_T cuts on the basis of a "Tag and Probe" method using Z→ $\mu\mu$ events described in[6] . Figure 3 shows p_T turn-on curves for a cut of 30 GeV, for barrel and endcaps. In 2011 plateau efficiencies were measured to be around 95.0% and 89.9% respectively. Data–Monte Carlo agreement has been found to be better than 98% in the former case and around 99% in the latter.

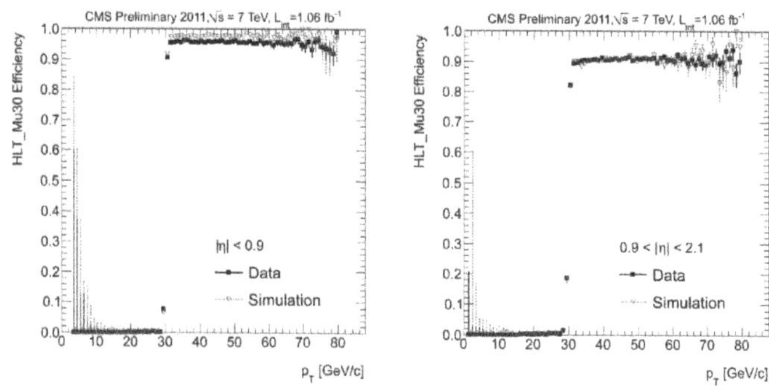

Fig. 3. Combined barrel (left) and endcap (right) muon trigger efficiency turn on curves for a 30 GeV p_T cut.

3. Conclusions

The performance of the CMS Muon Spectrometer and Trigger system in p–p collisions was thoroughly investigated during both 2010 and 2011. The detector performance has been found to be excellent, in line with design expectations and test beam data results[1 3] . The trigger system also behaves as expected[4 5] and results for data are in good agreement with simulation.

References

1. S. Chatrchyan *et al.* [CMS Collaboration], JINST **3** (2008) S08004.
2. L. Evans, (ed.), P. Bryant, (ed.), JINST **3** (2008) S08001.
3. CMS Collaboration, CERN/LHCC **97-32** (1997).
4. S. Dasu *et al.* [CMS Collaboration], CERN/LHCC **2000-038** (2000).
5. P. Sphicas, (ed.) [CMS Collaboration], CERN/LHCC **2002-026** (2002).
6. The CMS Collaboration, CMS PAS **MUO-10-002** (2010).

The South Pole Acoustic Test Setup as a trailblazer towards an acoustic detection of neutrinos

J. Berdermann* for the IceCube Collaboration†
DESY
Zeuthen, D-15735, Germany
**E-mail: jens.berdermann@desy.de*
†*http://icecube.wisc.edu/*

The ability to detect high energetic neutrinos by acoustic means at the South Pole is strongly dependent on local ice properties and the underlying noise floor. The South Pole Acoustic Test Setup (SPATS) has been designed to measure these unknown parameters and to verify the efficiency of a multi-km^3-detector at that location. Since August 2008 SPATS is taking data in a detector mode, which allows identification of transient and static acoustic background in the surrounding volume. Shown are results of finished and ongoing SPATS investigations as well as future perspectives of acoustic neutrino detection.

Keywords: Acoustic neutrino detection; SPATS; Neutrino astronomy; Hybrid detector simulation.

1. Introduction

The detection of neutrinos with ultra high energies provides valuable informations on astrophysics (cosmic ray sources), particle physics (neutrino-nucleon cross section) and cosmology (relic particles).[1] An expected source of such neutrinos are protons with highest energies interacting with the microwave background radiation (GZK-effect[2,3]). A corresponding steep decrease of the charged cosmic ray spectrum above $10^{19.5}$ eV has been observed by the HiRes and Auger experiments.[4,5] To measure the small neutrino flux at highest energies a detector volume of at least 100 km^3 is required, favoring water or ice as medium and acoustic, radio or both as the preferred detection method.

The South Pole Acoustic Test Setup (SPATS)[6] was successfully deployed in the Antarctic ice and records data since four years. SPATS consists of four vertical cables called strings, each instrumented with seven acoustic stations with a transmitter and a sensor module. Each sensor module is made up of three sensors spaced 120 degrees apart on a horizontal steel ring to ensure azimuthal coverage (see Fig. 1). These sensors record acoustic signals from fixed and mobile transmitters and allow conclusions about the speed and the attenuation length of acoustic waves till 500 m depth.[7,8]

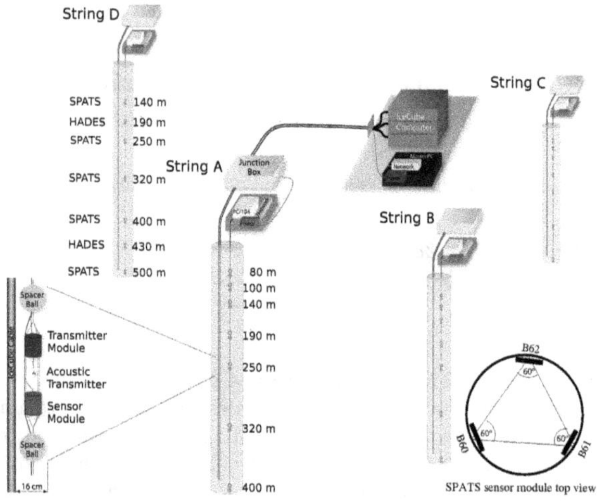

Fig. 1. Shown are a schematic picture of the SPATS detector.

The purpose of SPATS is the test of basic predictions for the acoustic properties of ice at the South Pole by appropriate measurements. Most of the SPATS science goals have been achieved and the physics results have both confirmed and challenged the theoretical predictions. We took the first experimental data on the acoustic attenuation length[7] in the South Pole ice, which is important to determine the spacing of acoustic sensors for an efficient detection of neutrino interactions. The measured pressure and shear wave sound speeds[8] in the bulk ice are important for the expected neutrino signal strength, event localization and reconstruction. For the past three years, SPATS has been running as a detector, allowing us to identify transient and static acoustic background sources in the surrounding volume.[9,10] The absence of any transient event observed from locations other than known sources allow us to set a limit on the flux of ultra high energy ($E_\nu > 10^{20}$ eV) neutrinos.[10] The absolute noise level at the South Pole is still under investigation. It determines the energy threshold of a future neutrino detector as well as the number of transient acoustic signals which could mimic neutrino interactions and may therefore be a serious background source. In the following recent results of ongoing investigation to understand signal and background as well as angular coverage of SPATS sensors are shown.

2. Azimuthal sensor module efficiency

The effect on the azimuthal sensor module sensitivity of low temperatures, high pressure, coupling to ice and the exact freeze-in position of the sensor modules inside the IceCube hole are unknown. Despite the different azimuthal orientation, we expect the angular acceptance of the three sensors in a sensor module to be similar because of previous laboratory tests.[11] Between 28 August 2008 and 20 February 2009 SPATS took data with twelve active sensors, two of them inside the same sensor module. The azimuthal acceptance is analyzed by comparing the responses of these two sensors to signals from reconstructed acoustic transient events coming from different directions (see Fig. 2(a)). In the following convention B6X is used for sensor X on string B at position 6 (X=0,1,2).

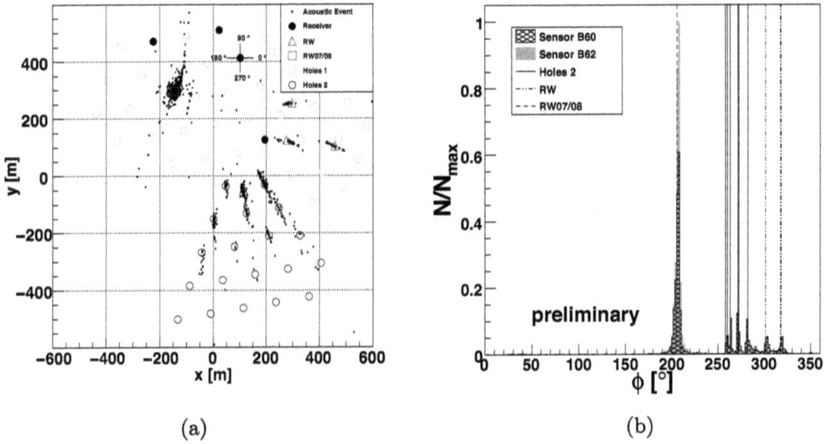

(a) (b)

Fig. 2. Shown are (a) the actual vertex position of all transient events recorded between August 2008 and February 2009. The sources of transient noise are the Rodriguez-wells (RW,RW07/08) and the refreezing IceCube holes. Dark gray circles (Holes 2) indicate positions of IceCube holes drilled in this period of transient data taking and light gray circles (Holes 1) show previously drilled IceCube holes. In (b) the distribution of localized events with a hit on sensor B60 (N_{tot}^{B60} = 5360 events) and/or B62 (N_{tot}^{B62} = 7371 events) in ϕ around their position (x=101.04 m, y=412.79 m) is shown.

The acoustic pulses are produced during the refreezing process at IceCube boreholes or at Rodriguez-Well locations. Rodriguez-wells are large caverns of around 20 m in diameter used for the production and cycling of water for the IceCube hot water drill system. The full position information of localized transient events can be used to calculate their angle in respect to the position of sensor module B6. The two active sensors B60 and B62 behave as expected over a wide azimuthal range which can be seen from Fig. 2(b) by comparing the hit rates for all holes and Rodriguez-wells (RW). The reduction of signals from RW07/08 at sensor B60 compared to B62 in this φ range might come from a shadowing effect as can be caused by the IceCube cable. Both sensors get the same rate of hits even for RW07/08 above a certain signal strength. The azimuthal sensor efficiency is still under investigation and more detailed information might come from the analysis of data taken with a mobile transmitter.

3. Frequency spectrum of transients

The frequency spectrum of all sensors shows a stable behavior during the four years of monitoring noise data. The difference between the single sensors might result from individual refreezing and coupling to the surrounding ice. For all SPATS sensors a dominant peak around 11 kHz (String A,B,C) has been found. SPATS sensors on string D have a different connection to the steel housing and show an additional peak at 51 kHz.

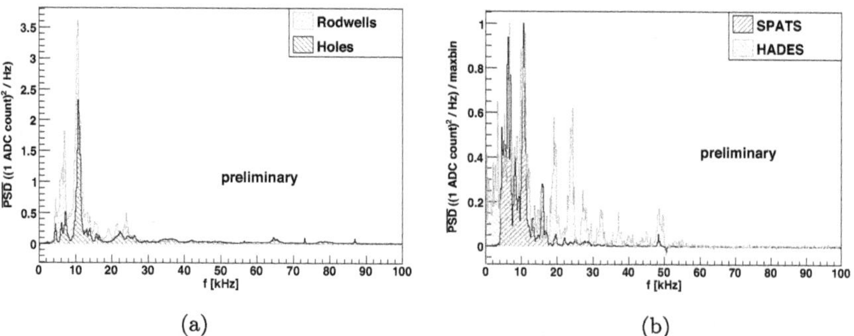

(a) (b)

Fig. 3. Shown are (a) the difference between the mean power spectral density of transient hits from Rodriguez-wells and Holes and (b) a comparison between the normalized mean power spectral density of the SPATS and HADES sensor type.

Comparison of the frequency content between the two transient sources (Fig. 3(a)) shows that acoustic signals from Rodriguez-well caverns have more low frequency components compared to events from cracks near the ice water boundary of refreezing IceCube holes. The relaxation volume in the large Rodriguez-well cavern reduces pressure/stress and can therefore damp an outgoing acoustic signal. Signals from IceCube holes have small peaks at higher frequencies which seems to be a feature from the refreezing process. Two of the sensors at string D, called HADES, had an entirely different mechanical construction[12] , but show peaks at the same frequencies (Fig. 3(b)). Therefore these peaks are rather an indication for a real frequency content from transient events than detector artefacts.

4. Conclusions

The results on sound speed and transient events confirm previous expectations and satisfy the requirements of acoustic neutrino detection at the South Pole. Recent experiments searching for weak particle fluxes of neutrinos with an energy above 10^{18} eV see no or a few events and give only flux limits. Most complications occur due to unknown systematic effects and difficulties in the background separation. This may be overcome by using a large hybrid detector of 100 km^3 size and larger, where radio technology could be complemented by additional acoustic sensors to verify detected events. There are still open questions concerning the absolute noise rate in the ice, but ongoing studies as presented in this work start to separate self noise from the real noise level at the south pole and the ongoing transient analysis increases the confidence in separation between noise and a possible neutrino signal. The detection and investigation of ultra-high energy neutrinos remains a substantial scientific challenge, which will give many important answers to particle and astroparticle questions.

References

1. R. Nahnhauer, *Nucl. Instr. and Meth. A* (2010), arXiv:1010.3082 [astro-ph.IM], doi:10.1016/j.nima.2010.11.010.
2. K. Greisen, *Phys. Rev. Lett.* **16** (1966) 748.
3. G. T. Zatsepin, *JETP Lett.* **4** (1966) 78.
4. R. Abbasi *et al.*, *Phys. Rev. Lett.* **100** (2008) 101101.
5. J. Abraham *et al.*, *Phys. Lett.* **B685** (2010) 239.
6. Y. Abdou *et al.*, arXiv:1105.4339 [astro-ph.IM].
7. R. Abbasi *et al.*, *Astropart. Phys.* **34** (2011) 382.
8. R. Abbasi *et al.*, *Astropart. Phys.* **33** (2010) 277.
9. J. Berdermann, *Nucl. Instr. and Meth. A* (2010), arXiv:1010.2841 [astro-ph.IM], doi:10.1016/j.nima.2010.11.015.
10. R. Abbasi *et al.*, accepted in *Astropart. Phys.*, arXiv:1103.1216 [astro-ph.IM].
11. S. Boeser, Ph. D. thesis (2007) Humboldt University, Berlin.
12. B. Semburg, *Nucl. Instr. and Meth.* **A604** (2009) 215-218.

The CMS Resistive Plate Chambers system-detector performance during 2011

Camilo Carrillo* for the CMS Collaboration

INFN Naples - Italy,
E-mail: camilo.carrillo@cern.ch
www.cern.ch/carrillo

Resistive Plate Chambers are used in the CMS experiment to provide a dedicated muon trigger both in barrel and endcap. About 3000 m^2 of double gap RPCs have been produced and have been installed in the experiment. The RPC system has been studied with millions of muons coming from LHC collisions during 2011. Making use of the redundant muon system composed by Drift Tubes (DT) in the barrel and Cathode Strip Chambers (CSC) in the endcaps that provide independent tracking and trigger informations, the performance of the RPCs has been studied in terms of efficiency, cluster size, spatial resolution and noise rate. Moreover during this long period of detector operations the stability of the system has been monitored to study the relevant detector parameters as a function of time.

Keywords: Resistive Plate Chambers, Compact Muon Solenoid, Large Hadron Collider, Efficiency, Cluster Size, Stability.

1. Introduction

1.1. *The CMS Experiment*

The Large Hadron Collider (LHC),[1] the biggest and most energetic particle accelerator ever built, is a double ring structure that collides beams of protons at a center-of-mass energy of 7 TeV. Located in one of the interaction points is the CMS experiment,[2] where three types of gaseous detectors are used to identify and characterize muons. The choice of the detector technologies has been driven by the very large surface to be covered and by the different radiation environment conditions. In the barrel region ($|\eta| < 1.2$), where the neutron induced background is smaller, the muon rate is lower, and the residual magnetic field is lower compared to the one in the endcaps, drift tubes (DT) chambers are used. In the 2 endcaps, CSC are deployed and cover the region up to $|\eta| < 2.4$. In addition to this, resistive plate chambers (RPCs)[3] are used in both barrel and end-cap regions. These RPCs are operated in avalanche mode to ensure the expected time resolution ($\approx$ 1ns) at rates of the order of 10 kHz/cm^2.

To measure and optimize the performance of the RPCs within CMS, data of both regular runs and a series of dedicated runs during spring 2011 have been studied.

1.2. *The CMS Resistive Plate Chambers*

The CMS-RPC system is composed by double-gap Resistive Plate Chambers; each 2 mm gas gap formed by two parallel bakelite electrodes (bulk resistivity $\rho \approx 10^{10}$ Ω cm). In between the gas-gaps common copper read-out strips are placed as shown in Figure 1. They are operated in avalanche mode.

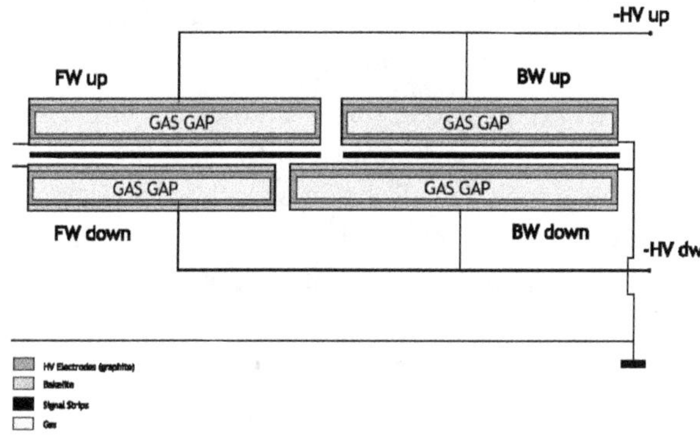

Fig. 1. Resistive Plate Chambers layout.

In the barrel the muon system is made out of four coaxial stations, interleaved with iron yokes. The two outermost stations consist in one layer of RPCs and one layer of DTs. The inner two stations contain a layer of DTs in between two RPC layers in order to trigger and reconstruct low p_T muons.[4] The endcap region consists of three iron disks interlayed with 3 RPC planes and 4 CSC planes interlayed as in the barrel as shown in Figure 2.

The geometry of the RPC strips is mainly driven by the need to have the trigger adjustable on different transverse momentum muons. In the barrel the strip shapes are rectangular while in the endcaps they are trapezoidal.

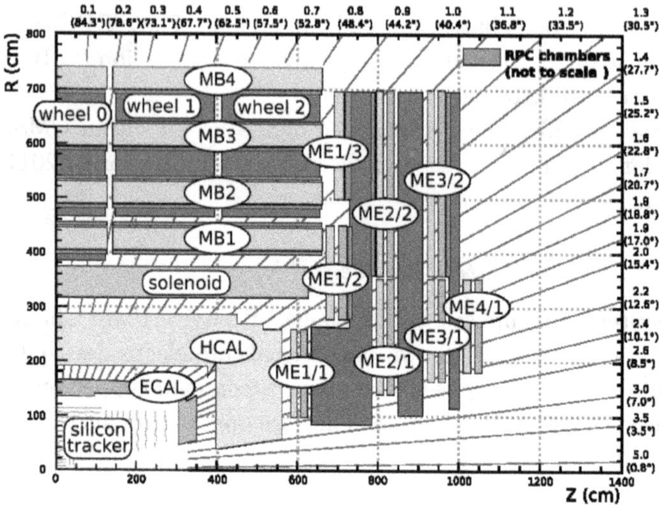

Fig. 2. The CMS Muon System, DT, CSCs and RPCs are shown.

The long side of the RPC strips runs along the beam axis in the barrel and radially in the end-caps. Along the z-direction, each chamber is divided into two to three η-partitions (rolls), resulting in strip lengths from 57 cm to 125 cm in the barrel and 47 cm to 79 cm in the endcap. The total number of RPC strips in CMS is 109608, covering a total area of 2953 m^2.

2. The High Voltage Scan

A high voltage scan (HV scan) was performed during early 2011: collision data was recorded at 11 different High Voltage settings during a series of dedicated runs to define the optimal operating voltage for each chamber. The presence of either a DT or a CSC near each RPC makes it possible to predict hits in the RPCs using only track segments reconstructed within the CSC or DT chamber. A linear extrapolation of every track segment in DTs and CSCs was performed toward the associated RPC strip plane, and then matched to the cluster (a strip or a set of contiguous strips) closest to the extrapolated impact point (Figure 3). This method provides both a measure for the efficiency ϵ and, through the residuals, for the spatial resolution.

One of the most important parameters is the applied High Voltage (HV),[5].[6] The dependence of the avalanche production on the environmental pressure p, the temperature T and the applied HV can be summarized

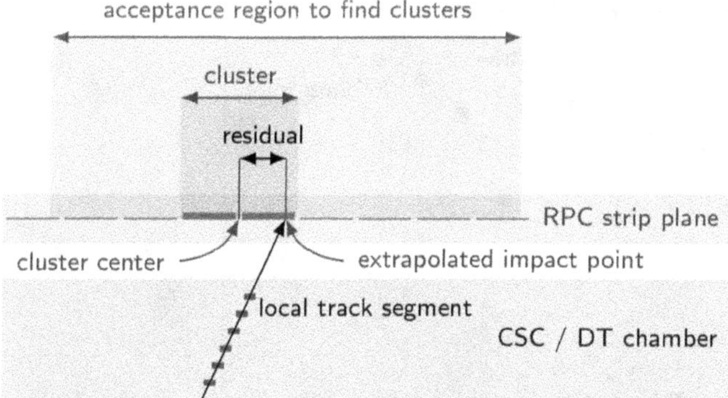

Fig. 3. Sketch of the extrapolation technique used by various RPC detector performance studies

in an effective high voltage following Eq. 1.

$$HV_{eff}(p,T) = HV\frac{p_0}{p}\frac{T}{T_0} \tag{1}$$

where HV_{eff} is the effective high voltage, HV is the applied high voltage, and the reference temperature and pressure are $T_0 = 293\text{K}$ and $p_0 = 965$ mbar respectively. The dependency of the efficiency ϵ with respect to the effective HV follows a sigmoidal shape described by (Eq. 2)[7]

$$\epsilon = \frac{\epsilon_{max}}{1 + e^{s(HV_{eff} - HV_{\epsilon=\frac{\epsilon_{max}}{2}})}} \tag{2}$$

where ϵ_{max} is the asymptothic efficiency when $HV \to \infty$, s is the slope of the sigmoid at the flex point and $HV_{\epsilon=\frac{\epsilon_{max}}{2}}$ is the necessary HV to reach 50% of ϵ_{max}. An example curve for a typical chamber is shown in Figure 4. It is important to operate the detector at a HV in the sigmoid plateau to get good and stable efficiency, all this keeping the cluster size and noise as low as possible.

3. Results from the HV Scan 2011

In the following sections a short description of the results obtained with the HV scan 2011 data, about the efficiency, the cluster size and the noise are presented.

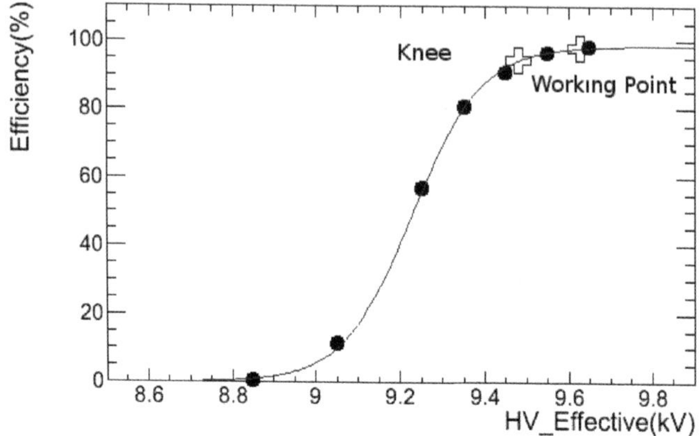

Fig. 4. Efficiency Plateau example, the knee and the working point are shown.

3.1. *Efficiency*

Using the data from the HV Scan, sigmoids (Eq. 2) were successfully fitted for each of the considered roll. A working point HV_{WP} has been defined as:

$$HV_{WP} = HV_{knee} + \begin{cases} 100\text{V} & \text{Barrel} \\ 150\text{V} & \text{Endcaps} \end{cases}$$

where HV_{knee} is HV_{eff} for $\epsilon = 0.95\ \epsilon_{max}$. An averaging procedure was applied for rolls with common HV supply. The agreement between efficiencies measured in subsequent runs and efficiencies predicted using the fitting procedure confirmed the effectiveness of the technique (Figure 5). Monitoring the efficiency during the 2-3 months following the HV scan allowed one to assess the fact that the efficiency remained stable within a plus minus 1% range, due to residual fluctuations due to pressure changes (see Figure 6). To suppress also these variations, an automatic online pressure correction for the HV was introduced in the summer of 2011, and is now under evaluation.

3.2. *Cluster Size*

Cluster size (CLS) is defined as the number of contiguous strips fired when an avalanche is produced in the RPC. Its dependency with the HV was studied during the 2011 HV scan. An example can be found in Figure 7, and

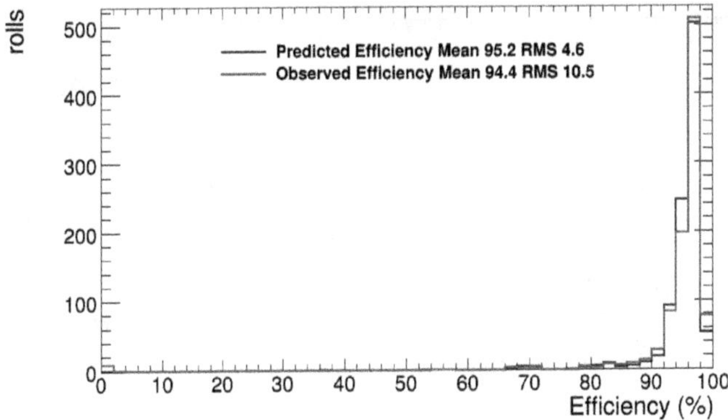

Fig. 5. Measured and predicted distributions of the efficiency at the calculated working point HV_{WP} for the rolls in the barrel. Some chambers were switched off during this measurement, contributing to the difference between the observed and expected average efficiency.

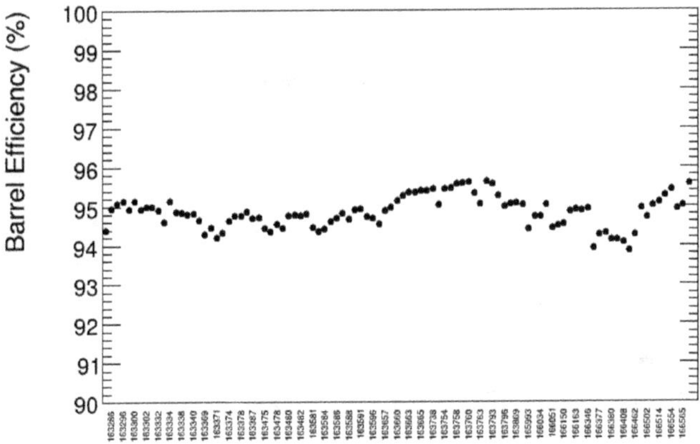

Fig. 6. Evolution of the efficiency after the HV scan for the barrel region. The average efficiency is found to be $\epsilon = 94.9\%$ in the barrel, and $\epsilon = 93.8$ % in the endcaps.

an overall distribution for all the system is presented in Figure 8. Trigger and reconstruction algorithms require $CLS < 2$ for maximum effectivness.

3.3. *Noise*

By measuring the hit rate during no-collision periods, a noise rate well below 0.5 Hz/cm^2 was found for most chambers. This rates is negligible for what concerns accidental muon triggering, since the RPC muon trigger

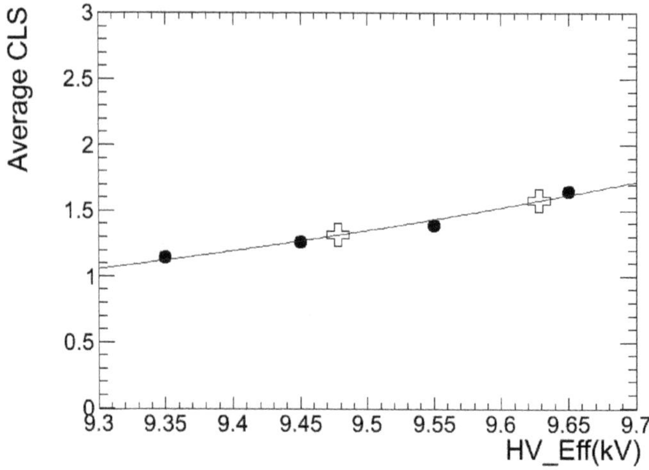

Fig. 7. Average cluster size of a typical chamber for various HV_{eff} in the HV scan.

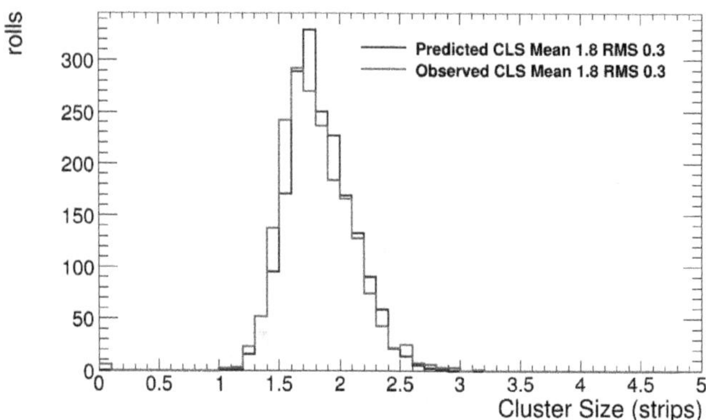

Fig. 8. Measured and predicted cluster size distributions at HV_{WP}.

system requieres a coincidence of fired strips in three or more planes within a 25 ns window and occurring in a limited spatial region.

3.4. *Spacial Resolution*

The residuals (the distance in the RPC plane between the extrapolated impact point and the center of the RPC cluster) computed during the procedure used to measure chamber efficiency, can be used as a measure for the spatial resolution of the RPCs. The standard deviations of Gaussian fits to the distributions of these residuals are then considered as a measure of the spatial resolutions. The measured values stay below $CLS \times pitch/\sqrt{12}$ for all strip pitches.

4. Conclusions

During 2011 different detector parameters, like efficiency, cluster size and noise have been studied. RPCs performance was well understood and tuned using dedicated collision runs. The results show that the RPCs form a stable and reliable sub-detector, contributing to the trigger and reconstruction capabilities necessary for the CMS physics programme.

Acknowledgments

We wish to congratulate our colleagues in the CERN accelerator departments for the excellent performance of the LHC machine. We thank the technical and administrative staff at CERN and other CMS institutes, and acknowledge support from: FMSR (Austria); FNRS and FWO (Belgium); CNPq, CAPES, FAPERJ, and FAPESP (Brazil); MES (Bulgaria); CERN; CAS, MoST, and NSFC(China); COLCIENCIAS (Colombia); MSES (Croatia); RPF (Cyprus); Academy of Sciences and NICPB (Estonia); Academy of Finland, MEC, and HIP (Finland); CEA and CNRS/IN2P3 (France); BMBF, DFG, and HGF (Germany); GSRT (Greece); OTKA and NKTH (Hungary); DAE and DST (India); IPM (Iran); SFI (Ireland); INFN (Italy); NRF and WCU (Korea); LAS (Lithuania); CIN-VESTAV, CONACYT, SEP, and UASLP-FAI (Mexico); MSI (New Zealand); PAEC (Pakistan);SCSR (Poland); FCT (Portugal); JINR (Armenia, Belarus, Georgia, Ukraine, Uzbekistan); MST,MAE and RFBR (Russia); MSTD (Serbia); MICINN and CPAN (Spain); Swiss Funding Agencies(Switzerland); NSC (Taipei); TUBITAK and TAEK (Turkey); STFC (United Kingdom); DOE and NSF (USA).

References

1. L. Evans and P. Bryant. Lhc machine. *JINST*, 3:S08001, 2008.
2. S. Chatrchyan, G. Hmayakyan, V. Khachatryan, AM Sirunyan, W. Adam, T. Bauer, T. Bergauer, H. Bergauer, M. Dragicevic, J. Erö, et al. The cms experiment at the cern lhc. *Journal of Instrumentation*, 3:S08004, 2008.
3. R. Santonico and R. Cardarelli. Development of resistive plate counters. *Nuclear Instruments and Methods in physics research*, 187(2-3):377–380, 1981.
4. CMS Collaboration. The cms muon project:technical design report. cern, geneva. *J. Instrum*, 1, 1997.

5. M. Abbrescia, R. Cardarelli, G. Iaselli, S. Natali, S. Nuzzo, A. Ranieri, F. Romano, and R. Santonico. Resistive plate chambers performances at cosmic rays fluxes. *Nuclear Instruments and Methods in Physics Research Section A: Accelerators, Spectrometers, Detectors and Associated Equipment*, 359(3):603–609, 1995.
6. P. Camarri, R. Cardarelli, A. Di Ciaccio, L. Di Stante, R. Santonico, and M. Wang. Latest results on rpcs for the atlas lvl1 muon trigger. *Nuclear Instruments and Methods in Physics Research Section A: Accelerators, Spectrometers, Detectors and Associated Equipment*, 409(1):646–648, 1998.
7. M. Abbrescia, E. Cavallo, A. Colaleo, G. Iaselli, F. Loddo, M. Maggi, B. Marangelli, S. Natali, S. Nuzzo, G. Pugliese, et al. Cosmic ray tests of double-gap resistive plate chambers for the cms experiment. *Nuclear Instruments and Methods in Physics Research Section A: Accelerators, Spectrometers, Detectors and Associated Equipment*, 550(1):116–126, 2005.

ATLAS DETECTOR OVERVIEW (OPERATION EXPERIENCE, PERFORMANCE)

ANA HENRIQUES

On behalf of the ATLAS collaboration

CERN, Switzerland (Ana.Henriques@cern.ch)

ATLAS is a general-purpose detector located at one of the 4 interaction points of the LHC at the CERN laboratory near Geneva, Switzerland. In 2010 and since March 2011 LHC has been colliding proton beams at the unprecedented centre of mass energy of 7 TeV. During the last month of 2010 operation was dedicated to Pb-ion collisions at a centre of mass energy of 2.76 TeV per nucleon pair. A challenging task of the previous months was to cope with the increasing event rates due to the increasing luminosity delivered by the LHC. A survey of the main ATLAS sub-detector systems, their operating conditions and the performance with colliding beams will be presented in this talk. The operation and results obtained from the data collected so far demonstrate that the detector is robust and functioning very well.

1. Introduction

The ATLAS detector [1] is one of the general purpose experiments installed at the Large Hadron Collider (LHC) at CERN. The ATLAS detector has been built and is maintained by a large collaboration of around 3200 scientists, (including ~1000 PhD students), from 174 institutions and 38 countries. It is designed to study processes at the TeV scale, search for the Higgs boson and for physics beyond the Standard Model as well as to make precision measurements of Standard Model processes. The ATLAS detector covers almost the whole solid angle around the collision point, see Figure 1 [1]. It is ~46 m long and has a diameter of 24 m. Its total weight is 7000 tons and is read out through a total of 88 million channels, which are connected by more than 3000 km of cables.

Pattern recognition, momentum and vertex measurements, and electron identification are achieved with a high-resolution semiconductor pixel and strip detectors (SCT) in the inner part of the tracking volume, and straw-tube tracking detectors (TRT) with the capability to generate and detect transition radiation in its outer part. This whole system was designed to achieve precise tracking and vertexing with transverse momentum $\sigma/ p_T = 0.038\% \, p_T \,(\mathrm{GeV}) \oplus 1.5\%$ ($< 2\% \, p_T < 35\mathrm{GeV}$).

The superconducting solenoid is surrounded by an hermetic calorimeter that covers the pseudorapidity range $|\eta| < 4.9$. A high granularity liquid-argon (LAr) electromagnetic sampling calorimeter, with excellent performance in terms of energy and position resolution ($\sigma/E \sim 10\%/\sqrt{E}(\mathrm{GeV}) \oplus 0.7\%$) covering $|\eta| < 3.2$. The hadronic calorimetry in the range $|\eta| < 1.7$ is provided by a scintillator-tile calorimeter. In the end-caps ($|\eta|>1.5$), LAr technology is also used for the hadronic calorimeters. The LAr forward calorimeter provide both electromagnetic and hadronic energy measurements, and extends the pseudorapidity coverage to $|\eta| = 4.9$. The hadronic energy resolution is ~ $50\%/\sqrt{E}(\mathrm{GeV})\oplus 3\%$ in the barrel region.

Outside the calorimeter, air-core toroids provide a magnetic field for the muon spectrometer up to 7.5Tm (~ 1T field at the center of each coil). Three sets of precision drift tubes (MDT) and cathode strip chambers (CSC) provide an accurate measurement of the muon track curvature in the region $|\eta|<2.7$. Resistive-plate (RPC) and thin-gap chambers (TGC) provide muon triggering capability up to $|\eta| < 2.4$.

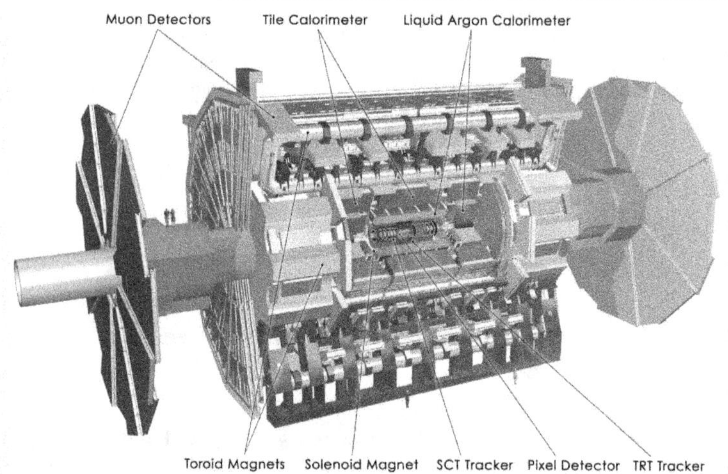

Figure 1. A schematic representation of the different components of the ATLAS detector.

The muon spectrometer provides efficient triggering, identification and momentum measurements. It was designed to measure muon momenta with a resolution of 4% for muons with 3GeV< p_T < 100 GeV and increasing to 10% at 1 TeV. The precision detectors provide ~30-50 μm tracking resolution up to $|\eta|< 2.7$. The muon spectrometer defines the overall dimensions of the ATLAS detector.

Several detectors cover the forward regions on both sides of the ATLAS detector near the beam line, with the aim of measuring and monitoring the LHC luminosity, as well as providing physics measurements in the very forward regions. At 17m from the interaction point there are luminosity monitor detectors based on Cherenkov tubes (LUCID), at 140 m there is a zero degree calorimeter for detecting photons and neutrons as can be produced in heavy Ion collisions, and at 240 m precision tracking detectors in Roman Pots will measure elastic scattering at very small angles for a total cross-section determination.

The trigger system has three levels, the first of which (L1) is fully hardware-based, and relies on information from the calorimeters and the Muon Spectrometer. The other two levels, the level two (L2) and the Event Filter (EF) are software based. The L2 trigger accepts data from defined Regions Of Interests (ROI) of L1 and the EF provides a full event reconstruction on computer farms. The 3-stage trigger system reduces the 40 MHz bunch crossing frequency to a LV1 trigger frequency of 75 kHz and finally to a recording maximum frequency of 200-400 Hz.

---

2. Data Taking Operation, Luminosity and pile-up conditions

From March to end September 2011 the LHC delivered pp collisions at a center-of-mass energy of $\sqrt{s}$=7 TeV. The cumulative luminosity vs. time delivered by LHC, and that recorded by ATLAS are shown in Figure 2 left. The total recorded integrated luminosity is equal to 3.78 fb^{-1} and the overall ATLAS data taking efficiency is 94%. All the sub-detectors are operating with a very high efficiency (>96%). The instantaneous peak luminosity continues to raise, and a value of $3.3x10^{33}$ $cm^{-2}s^{-1}$ was achieved in September. The average pile-up during the 2011 running is ~ 12 interactions per beam crossing, and is growing continuously, as is seen in Figure 2 right. This is one of the biggest challenges in the detector performance in 2011, but well under control. Currently, the systematic uncertainty of the luminosity measurement is 3.7% [2], dominated by the uncertainty in the beam current. Van der Meyer scans are being taken to reduce further this error.

The ATLAS computing infrastructure consists of ~70 sites, organized in a hierarchical structure ("tiers"). It includes the CERN computing center (Tier-0), 10 big regional centers (Tier-1), and various smaller sites (Tier-2). During data-taking, the raw data is processed within 36h to do first calibrations, data quality checks and mask bad cells. Data are usable for physics analysis ~1 week after data taking. ATLAS distribute > 800000 processing jobs/day with a peak rate of 10GB/s. Since the start of LHC ATLAS distributed more than 66 Petabytes (PB) of data all over the globe, to allow the worldwide collaboration to do analysis efficiently.

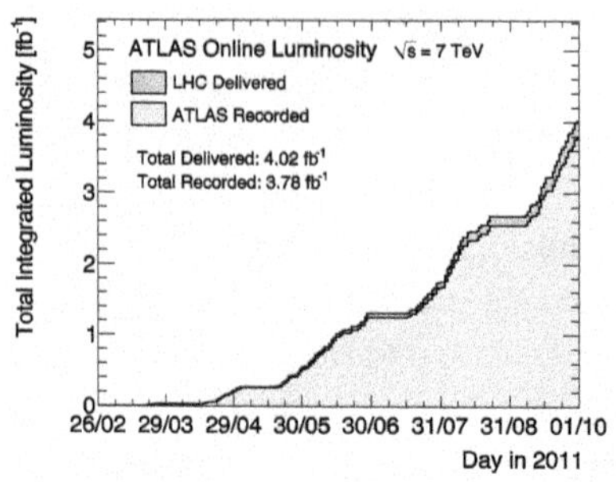

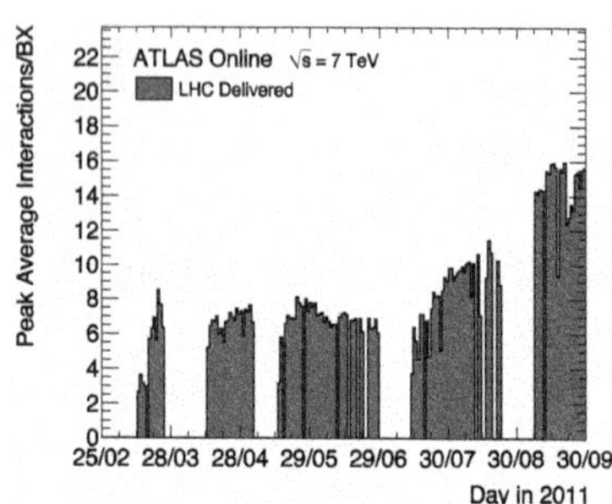

Figure 2. Left: Cumulative Luminosity (delivered by LHC and recorded by ATLAS) versus time during pp collisions at $\sqrt{s}$=7 TeV. Right: The peak average "events per beam crossing" versus time in 2011.

3. Inner tracker performance

The Pixels, SCT and TRT are operating very efficiently and providing high quality data for the physics analysis. A significant effort is put in the precise positioning of the hits recorded by the sensors, to understand the material inside the tracking volume, pattern recognition and track fitting, detector alignment and particle identification. The residuals in the most precise direction of the pixel SCT and TRT sub-detectors are respectively 9μm [rϕ], 25μm [rϕ] and 118μm/straw, with the autumn 2010 alignment. This is in close agreement with the MC residuals obtained for a perfect alignment, reaching values very close to design. The achieved primary vertex x resolution is ≈ 20 μm. Figure 3 shows the reconstructed invariant mass of the $Z \rightarrow \mu^+\mu^-$ comparing spring and summer 2011 alignment and comparing with Monte Carlo (MC) simulations, which were made assuming a perfectly aligned detector. The summer 2011 alignment includes several improvements, in particular a better alignment of ID with respect to the solenoid B field. The improvement is clearly visible and in good agreement with MC. This is also visible in Figure 3 right, in which the Z mass ϕ modulation has disappeared with new alignment corrections. Measurements of the properties of well studied particles, such as the K^0s, φ, D, Ω, Ξ, Λ, and Z, have allowed the momentum scale to be determined to around one part for thousand for lower momentum (< 20 GeV) and to one part per hundred at higher momentum (up to ~100 GeV) [3]. The resolution was found as expected to be dominated by multiple scattering in the low p_T region (<2% at p_T< 35GeV).

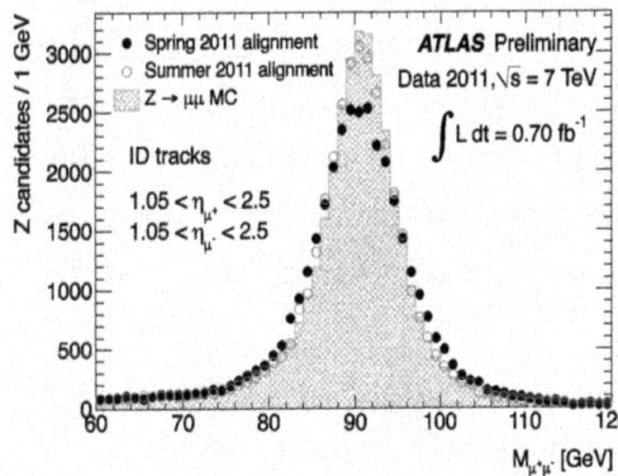

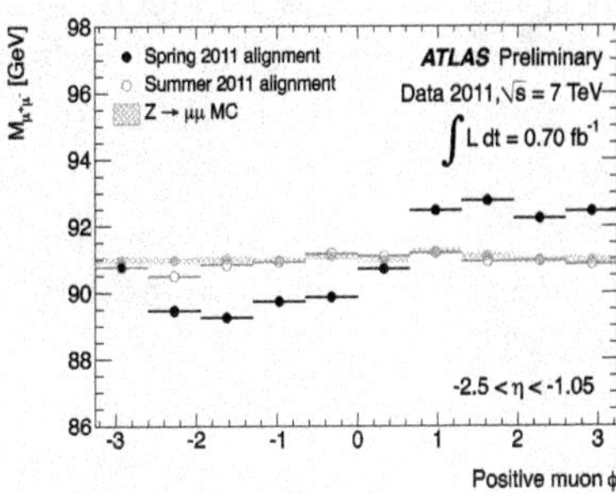

Figure 3. Left: Reconstructed invariant mass distribution of $Z \rightarrow \mu+\mu-$ candidates. Right: Mean Z invariant mass versus φ. The data were processed using alignment information known in spring and summer 2011. MC simulations, which assume a perfect alignment, are also shown. The improvement of the summer 2011 alignment is clearly seen in both figures.

4. Calorimeter, Jet energy scale and E_T miss performance

The validation of the em LAr calorimeter performance in terms of calibration, resolution and linearity was done using well-studied particle decays such as those of π^0, J/Ψ, Z, see Figure 4. The calibration was tuned to describe the $Z \rightarrow e^+e^-$ resonance. The data are compared with the Monte Carlo showing a good agreement, which constitutes a proof of the excellent calibration and response linearity. The linearity is ~ 0.3-1.6% up to 1 TeV for $|\eta|<2.47$. The constant term of the em energy resolution is ~ 1.2% in the barrel, 1.8% in the end-cap and 2.5% in the forward calorimeter [4]. More calibrations will bring these soon to the design target of 0.7%. In 2011 significant improvements were implemented in the relative alignment of the em calorimeter with respect to the inner detector.

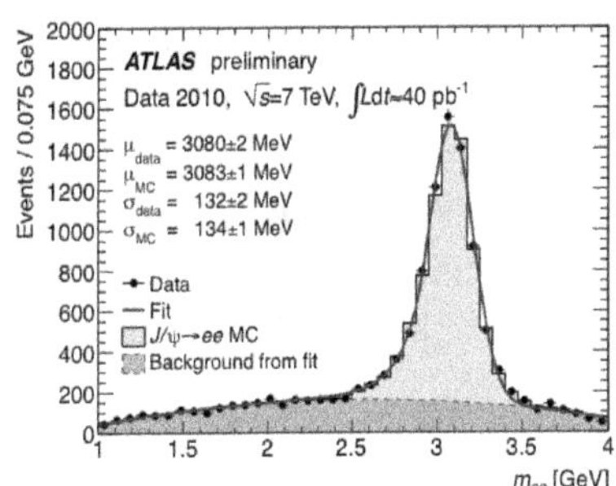

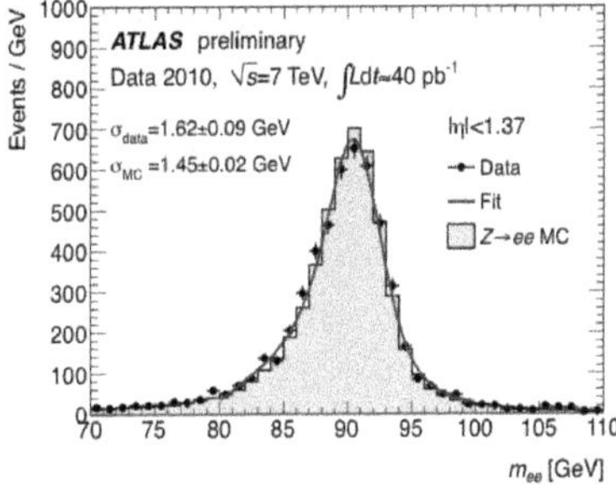

Figure 4. Invariant mass spectrum for electron pairs for J/Ψ and Z decays: left and right plots respectively. A good agreement with the Monte Carlo is observed.

This is illustrated in Figure 5 left, showing the cluster-track matching variables used in the electron and photon reconstruction and identification before and after the alignment (black and red points respectively). After alignment there is a good agreement with MC using W/Z->eυ/e$^+$e$^-$ events. Figure 5 right shows the reconstruction efficiency measured using W→eυ events and predicted by MC as a function of η. The reconstruction ε precision is within ±1% (p_T>35 GeV).

The performance of the hadron calorimeter is important for many physics analyses at the LHC involving jets and missing E_T, such as Higgs, SUSY, etc. The validation of the calibration of the Tile calorimeter at the electromagnetic scale (extracted in the testbeam for 11% of the modules) was obtained with cosmic rays, muons from collisions and using in-situ calibration method (e/p).

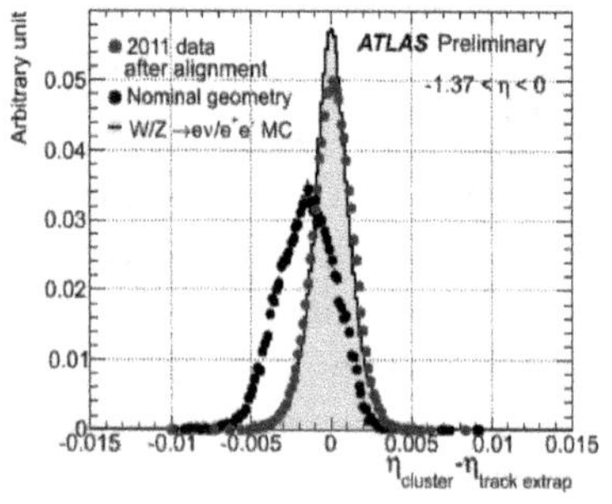

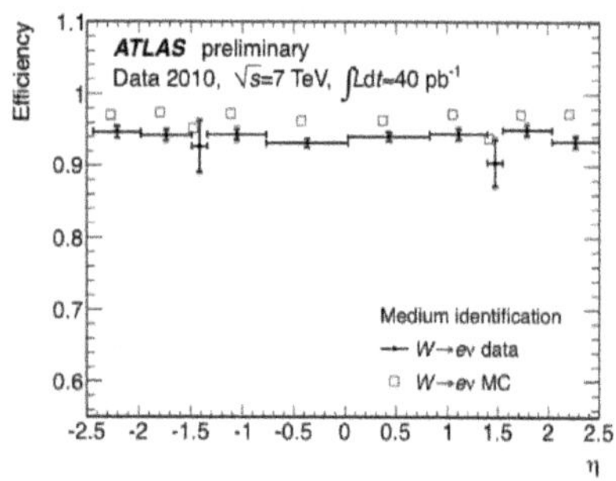

Figure 5 Left: Improvements brought by the relative alignment of the em calorimeter and the ID in 2011 to the cluster-track matching variables used in the e$^-$ and γ reconstruction and identification. Right: Efficiencies measured using W→eυ events and predicted by MC as a function of η.

Figure 6 left shows the mean value of the ratio between the energy deposited in the Tile calorimeter and the track momentum measured by the ID as a function of η, showing a very good agreement with MC. Figure 6 right shows the jet resolution in the barrel region with different calibration methods using 2010 data. The resolution is approaching design and the agreement between data and MC is within 10%. The constant term extrapolated to 2TeV is estimated to ~3% with GCW/LCW/GS calibrations. In 2011 the presence of pile-up worsens the low p_T resolution by ~20%. Improvements are expected after careful pile-up corrections for in-time/out-time bunches, noise threshold tuning in LAr calorimeter and using tracking information.

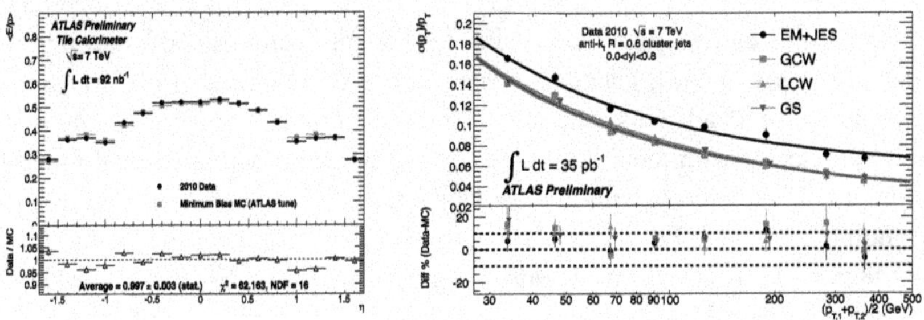

Figure 6. Mean value of the ratio of the e/p ratio in the Tile calorimeter vs h. Right: Jet resolution as a function of the momentum for different calibration schemes.

The jet energy scale error (ΔJES) is the main uncertainty in many physics results (jet/dijet cross section, top, etc). It was evaluated on data for jets with 20 GeV< p_T <3.5 TeV and |η| <4.5 [5]. It was set using the "EM+JES" calibration method (em scale+MC corrections for non compensation + dead material). In 2010 the ΔJES was ~2.5% for 60< p_T <800GeV |η|<2.1 up to 7% max in the forward region as seen in Figure 7 left. The validation of the jet energy scale was done with in situ techniques (γ+jet, multi-jet, track-jet) up to jet p_T ~ 1TeV showing agreement between data and MC within the errors, see Figure 7 right. In 2011 the pile-up worsened ΔJES, it's contribution (~ 5% at p_T=20GeV , ~0% >100GeV) has to be added in quadrature with the 2010 ΔJES. As for the resolution improvements are expected after careful pile-up corrections for in-time/out-of-time bunches and noise threshold tuning in the LAr calorimeter. An increase of statistics in 2011 will allow the use of in situ methods (γ+jet, multi-jet, track-jet) for recalibration and further improvement of the jet scale precision. The goal is to achieve 1%.

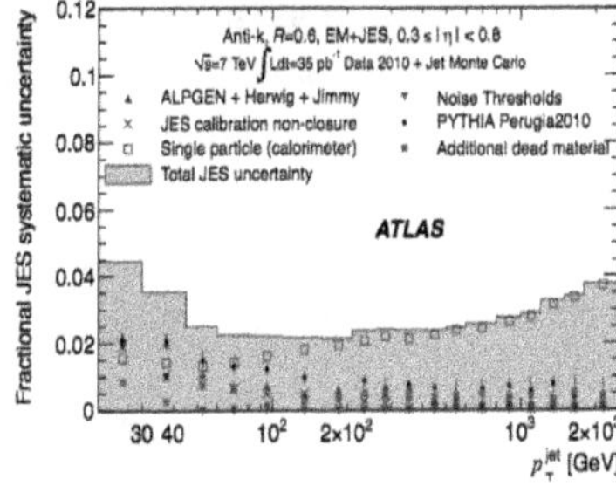

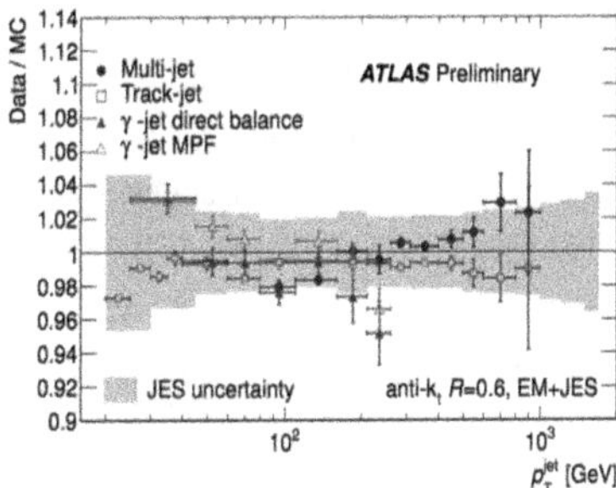

Figure 7 Left: Fractional jet energy scale systematic uncertainty as a function of the jet p_T for jets in the barrel calorimeter (0. < |η | <0.8). The total uncertainty is shown as the solid gray area. Right: Data to Monte Carlo simulation ratios for several in-situ techniques that test the jet energy scale.

The calorimeters also contribute to the measurement of the missing transverse energy ($E_{T\ \mathrm{miss}}$), a quantity sensitive to the calorimeter performance in terms of noise dead cells and mis-calibration. It has been measured in the ATLAS detector with the 7 TeV collision p-p and Pb-Pb data [6] and found to be in good agreement with the simulation, see Figure 8 left. The $E_{T\ \mathrm{miss}}$ resolution in p and Pb collisions is ~ $48\%\sqrt{\sum E_T}$. Figure 8 right shows the distribution of the sum E_T as measured in a data sample of di-jet events selecting two jets with $p_T > 25$ GeV. The expectation from MC simulation, normalized to the number of events in data, is superimposed showing once more a good agreement.

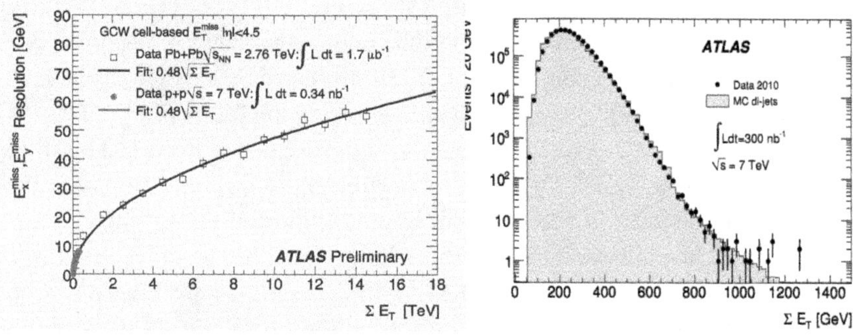

Figure 8 Left $E_{T\ miss}$ resolution versus the total transverse energy for minimum bias data in the pp and Pb-Pb collisions data. Right: Distribution of $\sum E_T$ as measured in a data sample of di-jet events selecting two jets with p_T >25 GeV. The expectation from MC simulation is superimposed.

5. Muon spectrometer performance

Figure 9 left shows the muon spectrometer alignment resolution (in sector 7) as a function of η. The barrel achieves a resolution of 50 μm, close to the design goal, whereas the end-cap gives a resolution of around 110 μm indicating that further improvements are necessary. Further calibration runs with collision data and with magnetic field off will be needed to achieve the desired performance. Figure 9 right shows the muon reconstruction efficiency as a function of p_T; 97% is achieved for p_T >20 GeV in good agreement with MC.

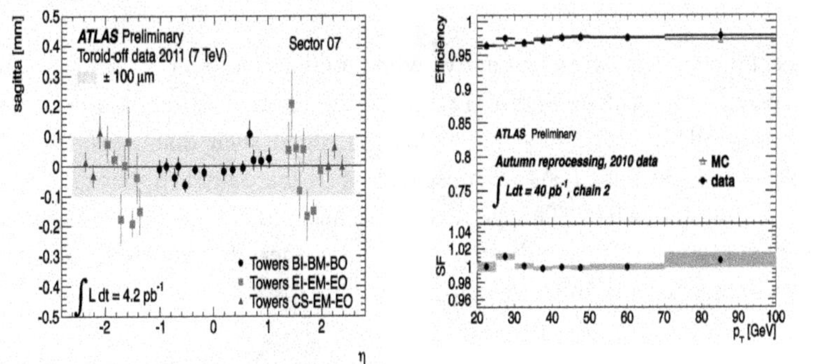

Figure 9 Left: Muon spectrometer alignment resolution as a function of η in sector 7 (black points for the barrel and the red and blue corresponding to the end-caps). Right: Muon reconstruction efficiency as a function of p_T.

The muon spectrometer performance was checked with well know resonance particles decaying into a pair of muons as illustrated in Figure 10 for J/ Ψ and Z decays. The achieved mass resolution is 2.2 % at 91 GeV,1.9% at ~3 GeV, close to design expectations. The response is linear and the absolute momentum scale is known to ~ 0.2% [7,8].

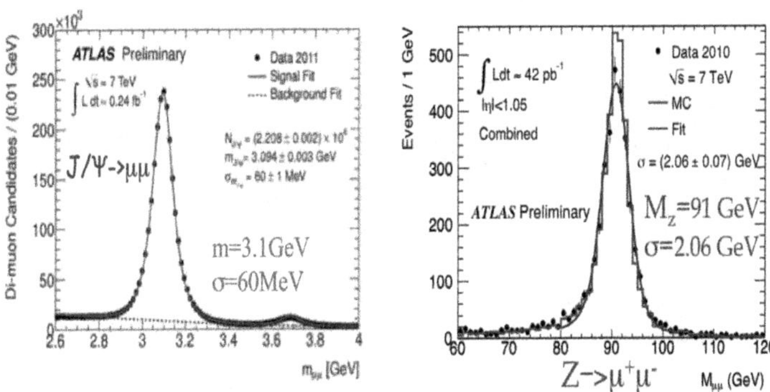

Figure 10. Di-muon invariant mass in data and MC from J/psi and Z decays.

6. Trigger performance

The trigger selection is defined by a trigger menu which consists of more than 300 individual trigger signatures, such as electrons, muons, particle jets, etc. The composition of the trigger menu depends on the instantaneous LHC luminosity, the experiment's goals for the recorded data, and the limits imposed by the available computing power, network bandwidth and storage space. The trigger rates for various levels and data streams agree very well with expectations and scale linearly with luminosity over a wide range of luminosity as shown in Figure 11, giving us reliable extrapolation for higher luminosity/pile-up. Triggers thresholds are tuned such that we keep a major trigger in a physics analysis un-prescaled while the output trigger rate is kept at max of 300-400 MHz. The good performance of the trigger system can be shown in Figure 12 in steeply rising turn-on trigger efficiency curves saturating at high values for jets, muon, electron and taus.

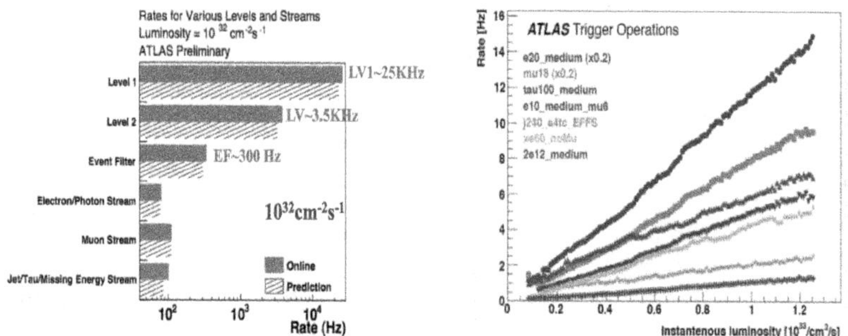

Figure 11 Left: measured and expected rates for LV1,2,EF and physics streams (e/γ, μ, jet/τ/ET_{miss}). Right: Rates of various physics streams as a function of the instantaneous luminosity.

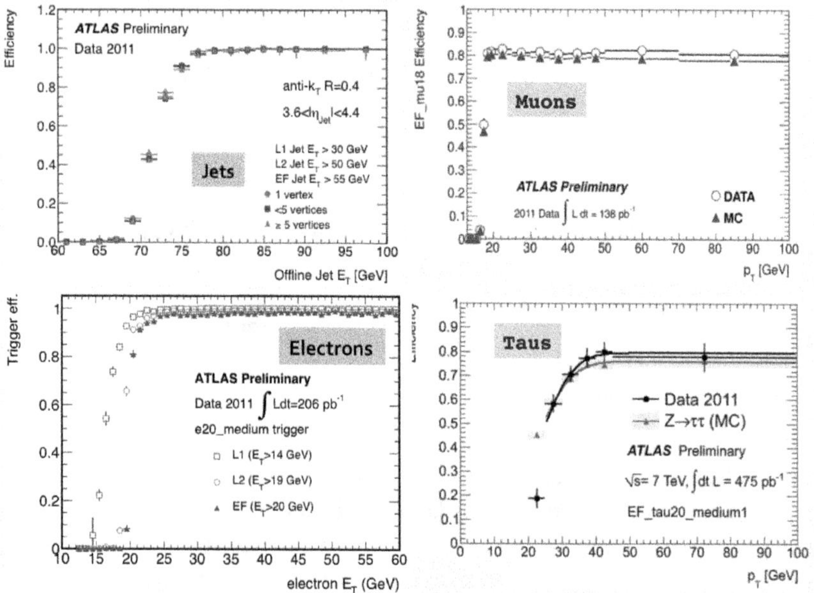

Figure 12. Examples of trigger efficiencies for jets, muon, electron and taus.

7. Conclusions

ATLAS successfully recorded 3.78fb^{-1} with an efficiency of ~94% from data acquisition/processing to use in data analysis. The detector is performing very well, data is well described by the simulation and approaching design performance. The continuously increasing contribution of pile-up in 2011 is a big challenge but is under control. ATLAS is now exploiting the full physics potential of LHC. 73 papers on detector performance and physics results using collision data have been published.

Acknowledgments

These achievements were only possible thanks to the outstanding performance of the LHC machine team and to the many years of dedicated work of the ATLAS collaboration in test beam activities, MC tuning and intense commissioning of the ATLAS detector well before the data taking started, with cosmics rays and now with collision data, in parallel with intense physics programme.

References

1. The ATLAS Collaboration [G. Aad et al.], The ATLAS Experiment at the CERN Large Hadron Collider, JINST 3:S08003, 2008.
2. ATLAS Collaboration, G. Aad et al., Updated Luminosity Determination in pp Collisions at √s = 7 TeV using the ATLAS detector, ATLAS-CONF-2011-011.
3. ATLAS Collaboration, G. Aad et al., Kinematic Distributions of K0s and Λ0 decays in collision data at √s=7 TeV, ATLAS-CONF-2010-033.
4. ATLAS collaboration, Expected electron performance in the ATLAS experiment, ATLAS-PHYS-PUB-2011-006
5. ATLAS Collaboration, G. Aad et al., Jet energy scale and its systematic uncertainty in proton-proton collision at √s=7 TeV with ATLAS 2010 data, ATLAS-CONF-2011-032.
6. ATLAS Collaboration, G. Aad et al., Performance of the Missing transverse Energy Reconstruciton and Calibration in Proton-Proton Collisions at a Centre-of-Mass Energy of √s=7 TeV with the ATLAS Detector, ATLAS-CONF-2010-057
7. ATLAS Collaboration, G. Aad et al., A measurement of the ATLAS muon reconstruction and trigger efficiency using J/ψ decays, ATLAS-CONF-2011-021.
8. ATLASCollaboration,G.Aadetal.,Determination of the muon reconstruction efficiency in ATLAS at the Z resonance in p-p collisions at √s=7 TeV, ATLAS-CONF-2011-008.

PERFORMANCE AND EVOLUTION OF THE ATLAS TAU TRIGGER DURING 2011 DATA TAKING PERIOD

P. JEŽ on behalf of the ATLAS COLLABORATION

Niels Bohr Institute, University of Copenhagen,
2100 Copenhagen, Denmark
E-mail: pavel.jez@cern.ch
www.nbi.ku.dk

The use of the τ leptons in the Standard Model processes and in channels probing for Physics beyond the Standard Model is very important at the LHC. These processes include Higgs production, heavy mass resonances and decays of supersymmetric particles. In the selection of such rare events of interest, the hadronic τ trigger plays a fundamental role. It allows efficient collection of the desired signal events, while keeping the rate of background events within the allowed bandwidth. This contribution summarises the status and performance of the ATLAS tau trigger system during 2011 data taking period and shows the trigger efficiency curves obtained from data.

Keywords: Tau lepton; Trigger; ATLAS; LHC; TDAQ

1. Introduction

The current experimental program of the ATLAS experiment[1] contains many analyses which look for a τ lepton in the final state.[2] The decay of Higgs boson into τ leptons is one of the few decay channels which can be used to observe or exclude a Standard Model Higgs at low mass values ($m_H < 130$ GeV). In the search for more than one neutral Higgs boson, predicted by the Supersymmetric scenarios, Higgs decays into τ lepton pairs become highly relevant over a large range of masses. Additionally, the observation of a charged Higgs boson, which for masses below 200 GeV preferably decays into a τ lepton and a neutrino, would represent a unique clue to both the origin of mass and the deeper symmetries in Nature. Finally many Beyond Standard Model scenarios predict abundant production of τ leptons.

Being the heaviest lepton, τ can decay either to μ or electron ("leptonic τ") or to lighter hadrons ("hadronic τ"). Most of the τ leptons decay hadronically (65 %). In hadronic decays, there is an odd number of charged hadrons possibly accompanied by neutral hadrons (due to charge conservation), forming together so-called τ jet. Finally, there is always at least one neutrino (two for leptonic modes) among the τ decay products.

The ATLAS tau trigger is designed to select hadronic τ's, characterized by the presence of 1 or 3 charged tracks ("1-prong" and "3-prong" decays), possibly accompanied by a few neutral hadrons. This distinguishes τ jets from quark or gluon jets, which typically have much larger particle multiplicities. Because the mass differences involved in a τ jet are smaller compared to τ momentum than in a classical QCD jet, the tau decay products tend to be more collimated with respect to the direction of a mother particle, thus forming very narrow jet.

The goal of the tau trigger is to cope with high luminosity conditions and high input rate of events to the trigger system, demanding a strict online selection of events to be saved for offline analysis. The tau trigger should save the interesting physics as efficiently as possible, while reducing the rate of output events down to an acceptable level.

During 2011 more than 4 fb^{-1} of *pp* collision data has been collected[a] and tau trigger has been actively used in many searches outlined in the previous paragraphs. It has also been used to determine cross section of $W \to \tau\nu$ processes,[3] and in the searches for light charged Higgs boson[4] and neutral MSSM Higgs boson.[5]

2. ATLAS Trigger System

The ATLAS trigger system[1] is designed to cope with an input event rate of 40 MHz and provide an output rate of 300 Hz. The current design includes hardware-based first level trigger (L1), simple and fast firmware logic, using only coarse information, and software-based High Level Trigger (HLT) split into two levels: Levels 2 (L2) and Event Filter (EF). The HLT is implemented in a large farm of processors and applies sophisticated software algorithms to select events using the full granularity of data from the detector. The processing time per event is constrained to the order of microseconds at L1, tens of milliseconds at L2 and seconds at EF.

Level 1 trigger uses information from muon spectrometer and calorimeter to establish so-called Regions of Interest (RoI) within the detector volume. The L2 trigger then inspects only the input from RoI (2-6% of total detector volume). Finally, EF has access to the full detector information and

[a]This is status in October 2011 when data taking is still ongoing

it is possible to execute almost the same algorithms that are used for object reconstruction, identification and selection during the offline analysis.

3. Tau Trigger Implementation

At L1, the tau trigger uses the electromagnetic and hadronic calorimeters to find transverse energy deposits inside the area of 2×2 trigger towers ($\Delta\phi \times \Delta\eta = 0.2{\times}0.2$[b]) that pass given programmable threshold.

Once the L1 tau trigger candidate is found, it is passed on to the HLT. At L2, full detector granularity inside the RoI of size $\Delta\phi{\times}\Delta\eta = 0.6{\times}0.6$ is used and tau candidate is reconstructed from calorimeter cells and inner detector tracks. Since 2011, electronic and pile-up noise suppression is applied when constructing clusters from calorimeter cells. Several selection criteria are then applied that take advantage of calorimeter cluster confinement and low track multiplicity to discriminate taus from the QCD background.

At EF, full detector information is used to reconstruct τ candidate. In particular, a topological clustering algorithm is run on cells in a region of $0.8{\times}0.8$ around the L1 RoI position, and subsequently the same tau reconstruction code as run offline is used, making the EF selection a real online analysis. During the reconstruction noise and pile-up suppression as well as hadronic and tau specific energy calibration are applied.

4. Tau Trigger Performance in 2011

The tau trigger has been active during the whole 2011 run. The instantaneous luminosity has been rising steadily to eventually reach $3.3{\times}10^{33}\text{cm}^{-2}\text{s}^{-1}$, more than 25 times the value at the start of the year. The development of the tau trigger output rates as a function of inst. luminosity is shown on Fig. 1. The rejection ($\frac{1}{\varepsilon_b}$, where ε_b is the background efficiency) for several tau trigger items with different thresholds and different shape cuts is shown on Fig. 2.

The physics analyses using tau trigger to select events needs to know its response to the τ leptons in data. Because Monte Carlo models of signals like W or Z bosons production might not well reproduce data on the new energy frontier of LHC and detector simulation might be incomplete, it is necessary to determine trigger efficiency from real data. The large amount of data collected in 2011, containing also sizable amount of events with τ leptons coming from the decays of Z bosons, makes it possible.

[b]Pseudorapidity η is defined as $\eta = -\ln\tan\frac{\theta}{2}$ where θ is angle measured from the beam axis.

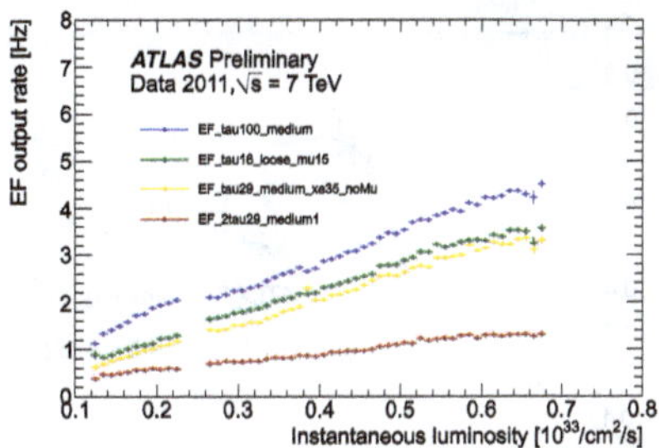

Fig. 1. The EF output rates versus the instantaneous luminosity for 4 selected tau triggers. The numbers in item names correspond to E_T requirements and "medium" identifies the tighthess of identification criteria. One of the triggers is combined with muon trigger and other is combined with missing E_T ("xe") trigger. The vertical ordering in the legend corresponds to the vertical ordering of the plots.

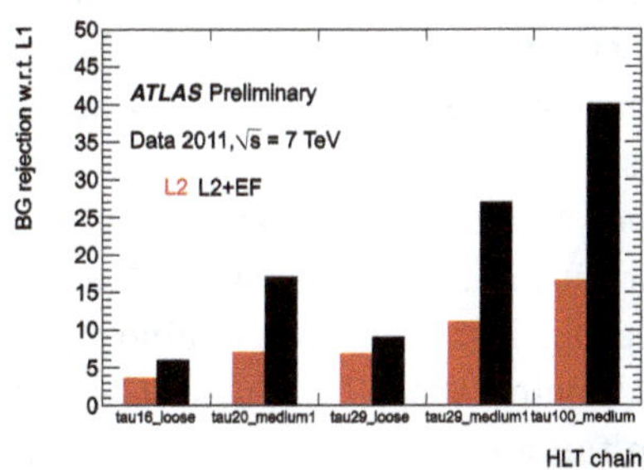

Fig. 2. Rejection of QCD jets for various tau triggers. The numbers are given w.r.t. L1 output of associated tau trigger item. The red (left) and black (right) bars show rejection after L2 and EF, respectively.

The classic method for efficiency measurement is called "tag and probe" and relies on a trigger different from the tau trigger, to tag or save the event to disk, allowing one to probe the performance of the tau trigger on the rest of the event where a τ lepton is expected and its presence is unbiased by tau trigger itself. Example is the production of a Z boson decaying into τ leptons, where one τ lepton tags the event via decay to electron or muon. The result of tau trigger efficiency measurement in such events is shown on Fig. 3. The results in general showed good agreement between data and Monte Carlo simulation and high tau trigger efficiency w.r.t. offline tau candidates passing medium identification criteria.

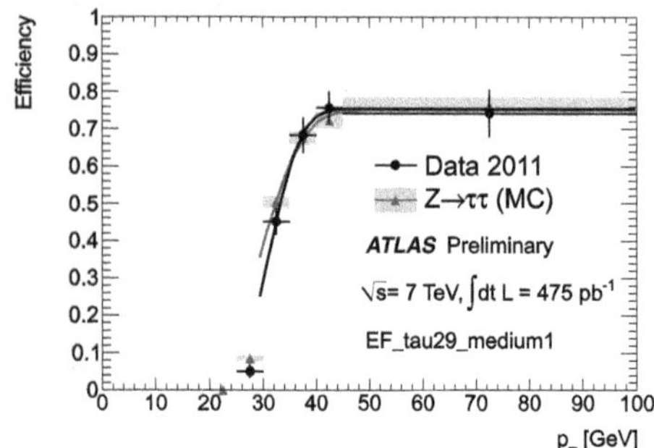

Fig. 3. Efficiency of tau trigger with E_T threshold of 29 GeV w.r.t. reconstructed offline tau candidates passing medium identification criteria as a function of offline p_T.

5. Summary and Outlook

The year 2011 has been another successful year for LHC and ATLAS and the tau trigger has been running smoothly over the whole period. The tau trigger is essential in various Standard Model measurements as well as in the searches for the Higgs boson and the new physics. The performance of the trigger is well described by the Monte Carlo simulation and efficiency has been measured in data using $Z \to \tau\tau$ events. The tau trigger successfully coped with the increasing instantaneous luminosity by rising momentum thresholds and/or tightening the selection on the candidate's shape. The most advanced offline tau identifications algorithms, like Boosted Decision Trees based ID deployed in 2011, are being considered to possibly replace a simple EF cut-based identification in the future. Overall the tau trigger at ATLAS is in very good shape and will continue to select events that might help us to understand more the nature of spontaneous electroweak symmetry breaking or bring some hints about the physics beyond Standard Model.

References

1. ATLAS collaboration, *JINST* **3**:S08003, (2008)
2. ATLAS collaboration, CERN-OPEN-2008-020, (2008)
3. ATLAS collaboration, CERN-PH-EP-2011-122 (submitted to Phys. Lett. B), (2011)
4. ATLAS collaboration, ATLAS-CONF-2011-138, (2011)
5. ATLAS collaboration, ATLAS-CONF-2011-132, (2011)

Performance of the ATLAS Inner Detector Trigger algorithms in pp collisions at 7TeV

J. MAŠÍK*,
on behalf of the ATLAS collaboration.
School of Physics and Astronomy, The University of Manchester,
Oxford Road, Manchester, M13 9PL United Kingdom
**E-mail: Jiri.Masik@hep.manchester.ac.uk*
www.hep.manchester.ac.uk

The ATLAS trigger performs online event selection in three stages. The Inner Detector information is used in the second (Level 2) and third (Event Filter) stages. Track reconstruction in the silicon detectors and transition radiation tracker contributes significantly to the rejection of uninteresting events while retaining a high signal efficiency. To achieve an overall trigger execution time of 40 ms per event, Level 2 tracking uses fast custom algorithms. The Event Filter tracking uses modified offline algorithms, with an overall execution time of 4s per event. Performance of the trigger tracking algorithms with data collected by ATLAS in 2011 is shown. The high efficiency and track quality of the trigger tracking algorithms for identification of physics signatures is presented. We also discuss the robustness of the reconstruction software with respect to the presence of multiple interactions per bunch crossing, an increasingly important feature for optimal performance moving towards the design luminosities of the LHC.

Keywords: ATLAS, Trigger, Inner Detector, Tracking.

1. Introduction

The Large Hadron Collider has been operating successfuly throughout 2011 and has delivered more than 3 fb^{-1} as of September 2011. The peak instantaneous luminosity has been increasing and reached values $3.31\times10^{33}cm^{-2}s^{-1}$ approaching the design value of $10^{34}cm^{-2}s^{-1}$.

The large event rate at the design luminosity requires an efficient trigger system to select events of physics interest and reduce the interaction rate of 40 MHz to about 200 Hz of events which can be stored permanently. The trigger system of the ATLAS[1] experiment is implemented in 3 levels. The first level L1 is a hardware level and its principal inputs are signals from the muon chambers and from the calorimeter. The second (L2) and third (EF) trigger levels which are commonly referred to as High Level Trigger (HLT) are software triggers running on farms of commodity hardware.

2. The Trigger Tracking Algorithms in the Inner Detector

The Inner Detector of ATLAS is described in detail in[2] and its operation discussed in other articles of these proceedings.[3–5] It combines three subdetectors - Pixel, silicon strip (SCT) and the gaseous drift detector with transition radiation detection capabilities (TRT) and provides measurements for high precision tracking and vertexing in the central volume of ATLAS. It is designed as a barrel with 2 endcaps and is located between radii 5 cm and 1 m in a solenoid magnetic field of 2 T within an acceptance $|\eta| < 2.5$.

The track reconstruction in the Inner Detector is an essential component for the event selection in the second and third level of the ATLAS trigger system and is a prerequisite for triggering on electrons, muons, B-physics, taus,[6] and b-jets.

The track reconstruction software in the L2 has to stay within the timing budget of 40 ms allocated to the overall event processing and uses fast custom algorithms for this reason. It has implementations of both histogramming and combinatorial approaches to the pattern recognition. The fit of the track candidates is done with a fast Kalman filter track fit.

In the EF the average event processing time has to stay below 4 s. This allows running a version of the offline reconstruction[7] adapted to the trigger requirements, and the EF reconstruction benefits from sharing the majority of the offline algorithms.

3. Performance

The performance of the trigger tracking was compared with the offline reconstruction. The tracking efficiency is studied using specific monitoring triggers, which are set up for major trigger thresholds and signatures in a similar manner to the selection triggers but they accept events regardless of the outcome of the Inner Detector trigger reconstruction and thus provide an unbiased efficiency measurement.

The tracking efficiencies of finding a track matching an offline muon in triggers with a transverse momentum threshold of 20 GeV and electron candidates in electron monitoring triggers with a transverse energy threshold of 22 GeV are shown in Fig. 1(a) and 1(b) respectively. The efficiencies are very high in both cases, 100% in EF and 99% and 98% in L2 in the plateau region for muon and electron efficiency respectively.

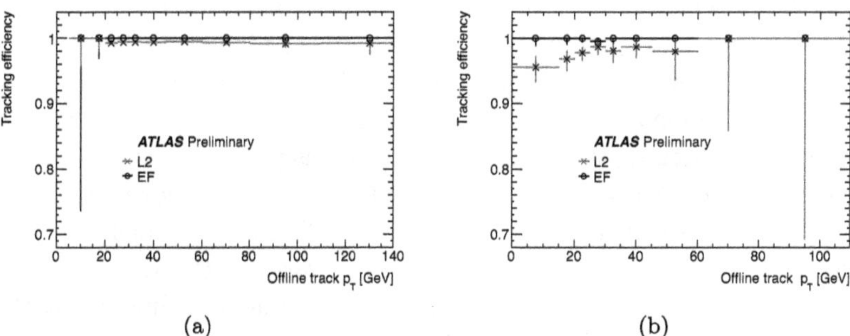

Fig. 1. The tracking efficiencies with respect to the offline muon 1(a) and electron 1(b) candidates found in the monitoring trigger with a transverse momentum threshold of 20 GeV and 22 GeV respectively.

An important aspect of the reconstruction is the robustness with respect to the presence of multiple interactions during the same bunch crossing. The peak value of mean number of interactions has increased from about 3 in the beginning of 2011 up to 15 in the middle of September 2011. This poses a challenge for the pattern recognition which may rely on identification of the primary interaction to reduce the number of considered track candidates and achieve a faster track reconstruction this way.

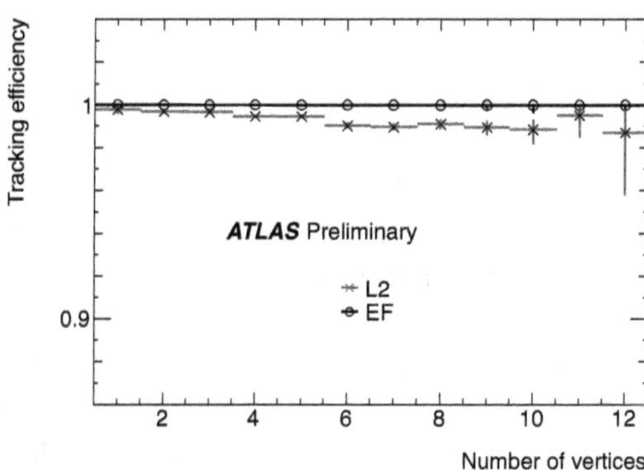

Fig. 2. The tracking efficiency with respect to the offline muon candidates as a function of number of vertices found in the event.

As demonstrated by Fig. 2 the trigger tracking is robust against increased pile-up levels, the efficiency in the EF stays 100% over the range of pileup events up to 12 available in our data sample. There is a small loss of 1% in the L2 which is related to the inefficiency to identify the hard-scatter and it is being addressed in the tuning for higher pile-up levels.

Figure 3 shows how the estimation of the track parameters compares between the online and offline reconstruction. The RMS of $1/p_T$ residuals between the matching trigger and offline track is presented as a function of track η for tracks with transverse momentum spectrum above 6 GeV. The absolute difference in the $1/p_T$ values is small. The agreement in track-parameter values is expected to be slightly worse in the L2 than in the EF due to a simplified material model and also due to a different pattern recognition technique. Another source of differences between online and offline tracks comes a partial access to the detector calibration in the online.

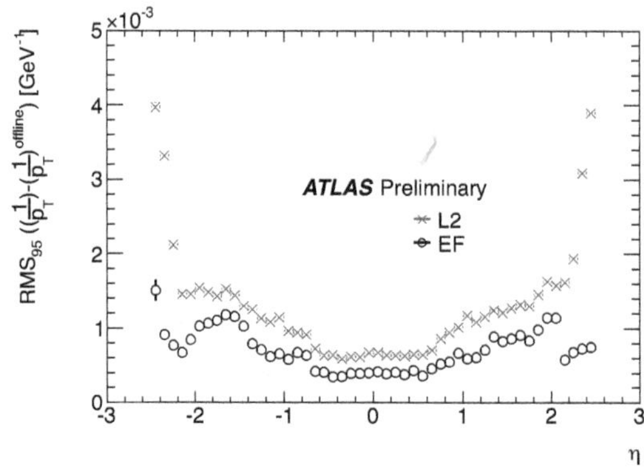

Fig. 3. The RMS of the residuals of $\frac{1}{p_T}$ between the trigger and the matching offline track as a function of track η.

As LHC operation appraches the design luminosity the trigger selection needs to deploy more strict criteria to identify trigger objects with high purity. One of the requirements used for a tight electron selection is a test whether a track has an associated hit in the innermost layer of the pixel detector. In Fig. 4 the efficiency to identify a hit on the EF track in the innermost layer of the pixel detector in case when the offline reconstruction expects and finds such hit is shown as a function of track parameter η. The efficiency is basically 100% with a small inefficiency at the end of η acceptance.

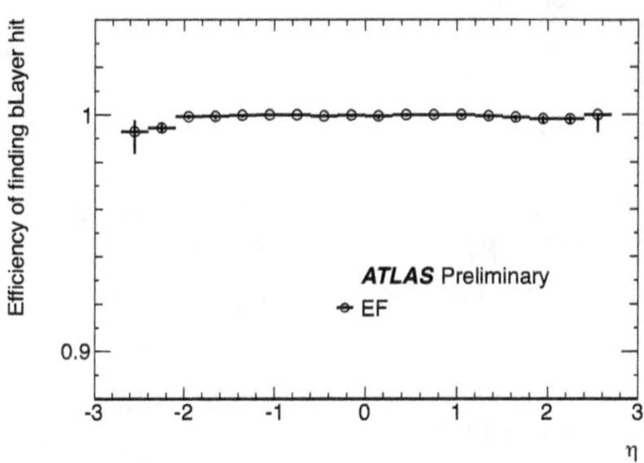

Fig. 4. The efficiency of finding a hit on a trigger track in the innermost layer of the pixel detector when such hit is expected by the offline tracking.

4. Conclusions

The Inner Detector trigger tracking algorithms have been performing very well and contributed significantly to the succesful data taking in 2011. The track reconstruction has dealt remarkably well with the increasing instantaneous luminosity delivered by the LHC operation in 2011 and in particular it has been robust in finding the signal trigger track in the presence of multiple pileup events. A very high efficiency of the track reconstruction in the trigger was presented using as examples muon and electron signatures. The online track reconstruction provides tracks with parameters close to the offline algorithm. This precise estimation of track parameters and properties becomes increasingly important for commissioning tighter object selection as we approach the design luminosity of the LHC.

References

1. The ATLAS Collaboration, *JINST* **3**, p. S08003 (2008).
2. The ATLAS Collaboration, *Eur.Phys.J.* **C70**, p. 787 (2010).
3. T. Ince, for the ATLAS Collaboration, *these proceedings* (2011).
4. V. A. Mitsou, for the ATLAS Collaboration, *these proceedings* (2011).
5. A. Vogel, for the ATLAS Collaboration, *these proceedings* (2011).
6. P. Jež, for the ATLAS Collaboration, *these proceedings* (2011).
7. T. Cornelissen *et al.*, *Concepts, design and implementation of the ATLAS New Tracking (NEWT)*, Tech. Rep. ATL-SOFT-PUB-2007-007, CERN (Geneva, 2007).

PERFORMANCE RESULTS OF ASSEMBLED SENSOR PLANE PROTOTYPES FOR SPECIAL FORWARD CALORIMETERS AT FUTURE E^+E^- COLLIDERS*

O. NOVGORODOVA[1,5**], J. A. AGUILAR[2,3], S. KULIS[2], L. ZAWIEJSKI[3], M. CHRZASZCZ[3], H. HENSCHEL[1], W. LOHMANN[1], S. SCHUWALOW[1], K. AFANACIEV[1,4], A. IGNATENKO[1], S. KOLLOWA[1,5], I. LEVY[6], M. IDZIK[2]

[1]DESY, Plantanenallee 6, Zeuthen, 15738, Germany

[2]AGH University of Science and Technology, ul. Reymonta 19, Krakow 30-059, Poland

[3]Institute of Nuclear Physics PAN, ul. Radzikowskiego 152, Krakow 31-342, Poland

[4]Tel Aviv University, Tel Aviv 69978, Israel

[5]Brandenburg University of Technology, D-03013 Cottbus, Postfach 101344

[6]NCPHEP, Minsk, Bogdanova ulica 153

The FCAL Collaboration prepared two sensor plane prototypes for the Luminosity Calorimeter (LumiCal) and Beam Calorimeter (BeamCal) for a future linear collider detector. For both several challenges appeared. The luminosity measurement has to be done with a precision of 10^{-3}, requiring LumiCal to be a precision device. BeamCal has to operate in a harsh radiation environment and needs radiation hard sensors. Two sensor technologies are considered - Si sensors for LumiCal and GaAs:Cr for BeamCal. A full chain comprising a sensor, fan-out and front-end ASIC was successfully studied in the lab and in a 4.5 GeV electron beam at DESY. Performance parameters like Charge Collection Efficiency (CCE), the Signal to Noise ratio (S/N) were measured. In a second beam test the readout is completed by a multi-channel ADC chip and data concentrator.

1. Introduction

Special Calorimeters are foreseen in the very forward regions of detectors at future linear colliders. In all up-to-date detector models-ILD, SiD or CLIC [1,2,3] detector-the very forward regions are very similar.

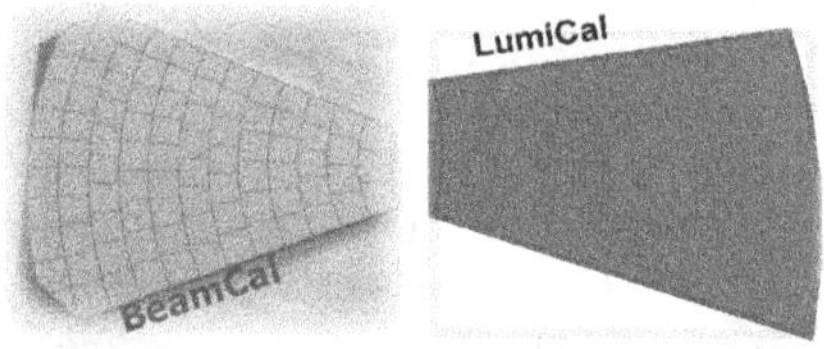

Figure 1. BeamCal GaAs:Cr sensor picture (left), LumiCal Si sensor picture(right).

---

* This work is supported by the 7th Framework Program "Marie Curie ITN", grant agreement number 214560, EUDET and AIDA TA.

** Olga.Novgorodova@desy.de

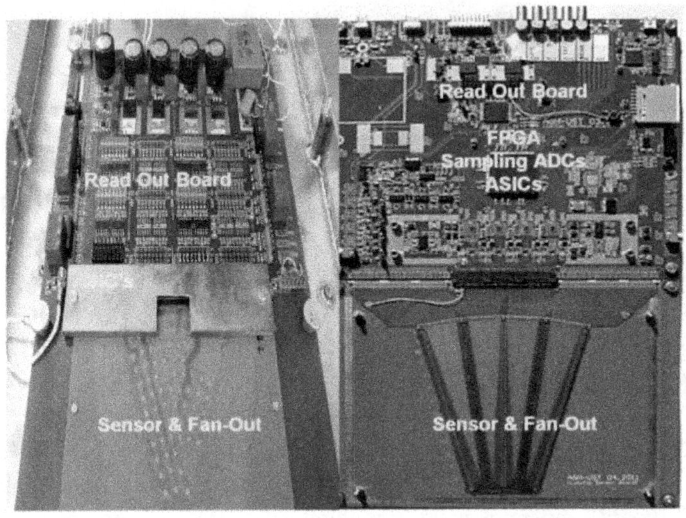

Figure 2. First BeamCal prototype with read-out electronics (left), Second LumiCal prototype (right).

At larger polar angles a LumiCal to measure precisely the luminosity, and at smaller polar angles, just adjacent to the beam-pipe, a BeamCal to improve hermeticity and assist beam tuning will be needed. A pair monitor, just in front of BeamCal, will improve beam diagnostics. For both devices a design as finely segmented sampling calorimeters is elaborated. Sensor prototypes and ASICS are developed. In this article performance measurements in an electron beam are presented for sensors assembled with front-end ASICs.

2. Prototypes Description

2.1. *Sensor material*

For the BeamCal prototype due to expected harsh radiation environment GaAs:Cr was chosen as a sensitive material, tested previously to be radiation hard for electrons up to 1.5 MGy [4]. On the Fig. 1 (left) shown 500 μm thick GaAs:Cr sensor. Sensor is produced by Tomsk University. It has Al metallization on both sides and on one side metallization is segmented in 87 pads square pads. For LumiCal standard p+ on n silicon sensor produced by Hamamatsu is used and shown on Fig. 1 (right). It has 320 μm thick and divided in 4 sectors and each sector has 64 pads with different size.

2.2. *Prototype structure*

The assembled sensor planes have been tested in a 4.5 GeV electron beam at DESY. For each test beam prototypes had different structure of read-out boards shown in Fig. 2. Each prototype has common readout-board on which the sensors are back-side glued to provide HV. On top of the sensor fan-out is mounted and through little holes traces connecting pads to the ASIC's are bonded. Fan-out for LumiCal is made of Kapton foil. For BeamCal a very thin PCB was used. In the 2010 test-beam several regions of 8 channels were read out simultaneously by using stand-alone CAEN ADC v1721 and in 2011 test beam already 32 channels due to implemented ADC developed within FCAL Collaboration [5,6]. In Fig. 2 the sensor plane of BeamCal (left) and the sensor plane of LumiCal (right) are shown in a shielding box prepared for the beam test. The LumiCal sensor plane contains ADC ASICS and an FPGA facilitating the data readout. LumiCal prototype has modular structure (sensor board and read-out board) and implemented multichannel ADC for on board analog to digital conversions. ASICS has two types of readout channels architectures to test difference.

3. Beam test

3.1. *Beam test setup*

Two beam tests took place at DESY II accelerator [7]. To investigate sensors performance as a function of the beam-particle impact positions the sensors were placed within the ZEUS MVD Telescope [8] with two sensor planes upstream and one sensor plane downstream of the sensor plane under test. Each telescope plane consists of two planes of silicon micro strip detectors with 50 mum readout pitch. After careful alignment the position resolution was measured to be 10 μm. Hence, also the response of inter-pad regions of 200 μm size could be investigated.

The telescope and the sensor planes have separate DAQ systems triggered simultaneously by coincidence signal coming from 3 scintillators located before and after beam test setup. More details about the test-beam measurements can be found in Ref. [9].

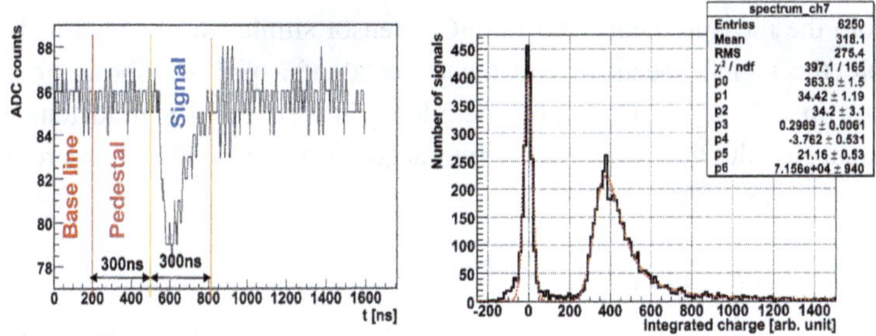

Figure 3. BeamCal signal wave form with baseline, pedestal ans signal windows (left), signal size spectrum example (right).

3.2. *Wave-form analysis*

For each trigger wave form of signal was sampled with 500 MHz flash ADC. Each trigger was analyzed using several time windows of the recorded wave-form. This is illustrated in Fig. 3 (left). From the first 100 ADC values the baseline is determined. In the following time windows the charge is integrated firstly the measure the pedestal and secondly the signal. The distribution of the signal wave-form integrated over time is shown in Fig. 3 (right). The spectrum is nicely described by a Gaussian for the pedestal and Landau convoluted with a Gaussian for the signal. The most probable value (MPV) of signal peak is used to calculate signal to noise ratio (S/N) and charge collection efficiency (CCE).

3.3. *Signal to noise ratio*

S/N is defined as the ratio between MPV of signal and the rms of the pedestal. The analysis of all spectra measured from 16 pads of BeamCal S/N is of about

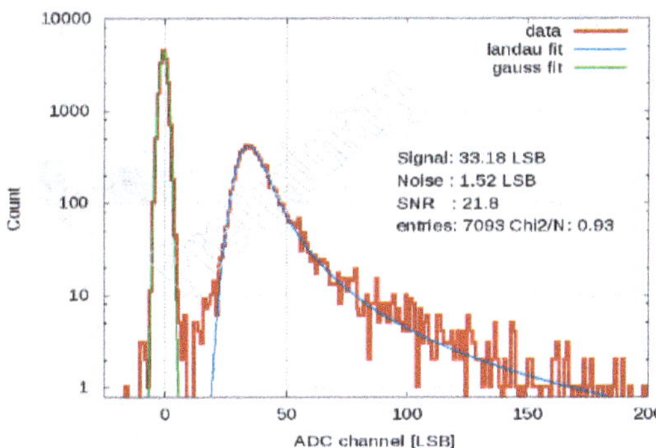

Figure 4. Signal Size spectrum, beam test 2011, LumiCal prototype.

20. From the measurements of a LumiCal sensor similar results obtained. As an example the signal spectrum and the pedestal of a pad are shown in Fig. 4, resulting in a S/N of 21. A more detailed study revealed a mild temperature dependence of the baseline. Taking into account this dependence improves S/N for BeamCal up to 31.

3.4. *Charge collection efficiency*

CCE is defined by ratio of collected charge in the detector to the induced charge by an ionizing particle. The charge collected in the sensor is measured by calibration readout chain by injection of a known charge and integration the signal from the beam. The induced charge was estimated by GEANT3 simulation. For all investigated pads of BeamCal when the sensor was biased with 60 V CCE is ~33%. In addition the CCE was measured in the laboratory using a 90Sr source and triggering for high energy electrons. Measuring the CCE as a function of the applied voltage we found that the CCE can be slightly improved to 42% at 100 V [10].

3.5 . *Investigation of the gaps between pads*

In Fig. 5 the impact points predicted by the telescope are distributed over the sensor area. The color, characterizing the pad, is assigned to each point in case the signal in the pad hit is above a certain threshold. The pad structure both for the BeamCal sensor (left) and the LumiCal sensor (right) is nicely visible. Since the prediction of the impact point is very precise the signal size was studied in several regions of the pad and in the 200 μm region between pads of the BeamCal sensor. The signal size in different areas of the pad was found to be equal. Scanning over the gap between two pads, sharing of the signal between the adjacent pads is observed, and the total signal size drops by about 10%, as can be seen in Fig. 6 (left).

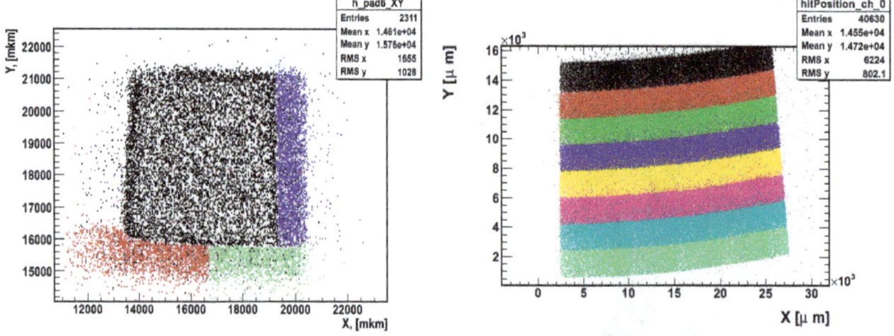

Figure 5. BeamCal (left) and LumiCal (right) prototype sensor pads structure.

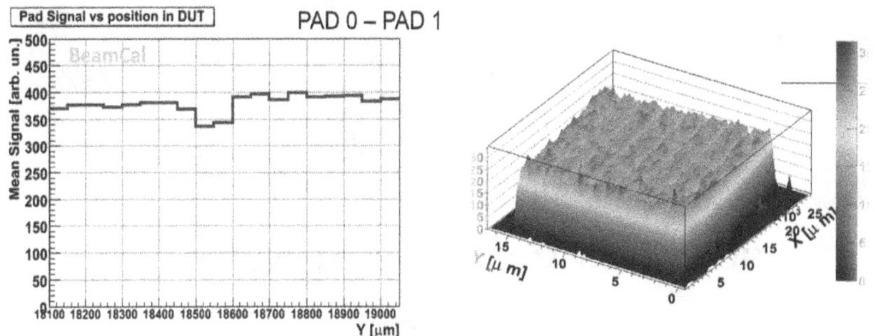

Figure 6. BeamCal edge loss (left) and LumiCal edge effects (right).

In Fig. 6 (right) shown LumiCal sensor signals in MIPs uniformed by beam profile. Areas of smaller signals are seen between pads and losses between pads are estimated around 10%.

4. Conclusions

In 2010 and 2011 fully assembled sensor planes for BeamCal and LumiCal have been investigated in a 4.5 electron beam at DESY. The S/N of the response for single particles has been measured to be about 20. A drop of the signal of about 10% is observed when the particle crosses the gap between two pads.

On the basis of these very good results on the performance of the full chain the design and construction of a full prototype calorimeter is planned.

References

1. International Linear collider Reference Report, http://www.linearcollider.org/about/Publications/Reference-Design-Report.
2. http://newsline.linearcollider.org/2011/06/23/concrete-plans-for-a-platform/ild-and-sid/
3. http://clic-study.web.cern.ch/clic-study/
4. H. Abramowicz et al., Forward instrumentation for ILC detectors, JINST 5 (2010) P12002.
5. M. Idzik, Sz. Kullis, D. Przyborowski, "Development of front-end electronics for the luminosity detector at ILC", Nuclear Instruments and Methods in Physics Research A 608 (2009) 169–174.
6. M. Idzik, K. Swientek, T. Fiutowski, S. Kulis and P. Ambalathankandy, A power scalable 10-bit pipeline ADC for Luminosity Detector at ILC, JINST 6 (2011) P01004.
7. http://adweb.desy.de/home/testbeam/WWW/Description.html
8. http://www.desy.de/~gregor/MVD_Telescope/short_intro.html
9. J.A. Aguilar et al., Luminometer for the future International Linear Collider - simulation and beam test results, Physics Procedia 00 (2011) 1–8.
10. O. Novgorodova et al., Test of sensor-plane prototypes in an electron beam, http://www.ifin.ro/fcal_2011/docs/Proceedings_FCAL_RO_2011.pdf

LHC STATUS AND PERFORMANCE IN 2011

G. PAPOTTI

Beams Department, CERN,
CH-1211 Genève 23, Switzerland
E-mail: giulia.papotti@cern.ch
www.cern.ch

After an initial pilot run in December 2009, the LHC was commissioned in 2010 and saw in 2011 the first year of luminosity production. An excellent performance of all machine components allowed one to achieve a luminosity of over 3E33 $cm^{-2}s^{-1}$ by September 2011 and deliver around 4 fb^{-1} so far at ATLAS and CMS. In this paper, the current performance achievements are discussed, including an outlook for 2012 and possible limiting factors.

Keywords: LHC status 2011.

1. Introduction

The Large Hadron Collider[1] (LHC) is a two-ring accelerator and collider installed in the 26.7 km long LEP tunnel at the European Organization for Nuclear Research (CERN, Geneva). It is designed to study rare events with centre of mass collision energies of up to 14 TeV.

After the first beam threading and capture on 10 September 2008, beam commissioning was abruptly stopped by the incident on 19 September. It required over a year of consolidation and repair to get beam back in the machine on 20 November 2009. Beam commissioning progressed extremely well in the three and a half weeks that followed, including the first ramp to 1.18 TeV. Commissioning stopped over the Christmas period and beam operation started again on 19 February 2010. Collisions at the record beam energy of 3.5 TeV were first established on 30 March 2010.

The choice of a beam energy of 3.5 TeV (compared to the 7 TeV design energy) is dictated by the minimization of the risk of burn-out of the superconducting splices. The design problem that caused the 19 September 2008 incident will be fixed in a long shutdown starting in 2013. The goals for the 2010 and 2011 operation were set in view of the 2013 shutdown. For

2011 the target for integrated luminosity was set to at least 1 fb^{-1} per high luminosity experiment. In order to achieve this, in 2010 a peak luminosity of at least 10^{32} $cm^{-2}s^{-1}$ had to be reached in the high luminosity experiments. Both targets were achieved and exceeded, making commissioning and operation in both years very successful. In this paper, first the LHC layout is quickly recalled, then the 2011 operation is summarized, including the description of some issues which might limit the performance and an overview of options for 2012.

2. LHC layout

The LHC tunnel is located at a depth of 70 to 140 m, across the Swiss and French border in the surroundings of Geneva. It is shaped as 8 arcs interleaved by 8 straight sections. The major experiments are housed in 4 of the straight sections: ATLAS in IR1 (Interaction Region), Alice in IR2, CMS in IR5 and LHCb in IR8.

Key elements of the machine are placed in the other 4 straight sections: the collimation system in IR3 (momentum cleaning) and IR7 (betatron cleaning), the beam dump system in IR6, the Radio Frequency system (RF) and part of the beam instrumentation in IR4. The two beams are injected through IR2 (ring 1, clockwise rotation) and IR8 (ring 2, anti-clockwise rotation). They meet each other only at the experiments while for most of the tunnel length they run through different vacuum chambers in the same magnet (two-in-one magnet design).

3. 2011 timeline

The success of 2011, that is the first year of operation for physics production, is based on the achievements of 2010, which was mainly a commissioning year. In the first part of 2010 the ramp with nominal bunches was set-up, and the first beam instabilities were observed, requiring a good chromaticity control, some octupole current to increase the tune spread and longitudinal emittance blow up counteracting an otherwise inevitable loss of Landau damping. The "squeeze" that reduces the beam size at the interaction point, thus gaining proportionally in luminosity, was first set-up with low intensity beam. During the summer about a month was spent in stable conditions to consolidate and verify machine protection and operational procedures. This was carried out at around 1 MJ of stored beam energy (25 bunches/ring), energy at which other machines have been running routinely in the past (e.g. Hera). After the summer, the injection of bunch trains became mandatory to increase the number of bunches per ring further (150 ns spaced bunches). This implied also the need to introduce crossing angles in the orbit to avoid multiple interactions between bunches while the two rings share a common vacuum pipe around the experimental IRs. The last month of 2010 proton operation was dedicated to the intensity ramp-up, which led to up to 368 bunches circulating per ring and a maximum peak luminosity of $2\times10^{32}\mathrm{cm}^{-2}\mathrm{s}^{-1}$.

The 2011 run started with a small intensity ramp up based on 75 ns spaced beams. This was followed by a "scrubbing" run that allowed one to prepare the machine for 50 ns spaced beams which are known to suffer more from electron cloud effects (see details later). The number of bunches was then increased further with 50 ns spaced beams until the maximum reachable (1380 bunches per ring). Since the beginning of July, the performance was increased by adiabatically changing the beam parameters (emittance and bunch population) to improve the luminosity rates. While the beta function at the interaction point (β^*) was set to 1.5 m at the high luminosity experiments ATLAS and CMS in the beginning of the year, aperture measurements proved that more space than thought was available, and in the beginning of September 2011 the β^* at the two IPs was commissioned further down to 1 m. This granted another 50% in instantaneous luminosity.

It is worth recalling that despite the fact most of the run is dedicated to proton physics production (≈140 days/year), the LHC schedule is divided into many other activities: four weeks are dedicated to lead ion physics, a few days to special runs (e.g., in 2011: run at 1.38 TeV, run with 90 m β^*, luminosity calibration runs, runs for Roman Pots of the TOTEM and ATLAS/ALFA experiments). Some time is allocated for machine commissioning itself in the beginning of the year (3 weeks in 2011). The run is then interleaved with 5-day technical stops in which non-critical repairs and maintenance are scheduled. About 20 days per year are reserved for machine developments and they are usually scheduled before technical stops so to minimize the interruptions to physics production. Highlights from 2011 machine developments are, among others:

- evaluation of the maximum tolerable head-on beam-beam tune shift[2] ;
- exercises of new optics designs that allowed achieving a β^* =30 cm (ATS scheme[3]) ;
- studies to verify the quench limits[4] .

4. Luminosity production

The instantaneous luminosity can be calculated from the machine parameters according to the well known formula:

$$L = \frac{N_1 N_2 n_b f_{rev} R}{2\pi\sqrt{\sigma_{1x}^2 + \sigma_{2x}^2}\sqrt{\sigma_{1y}^2 + \sigma_{2y}^2}}$$

The revolution frequency f_{rev} and the relativistic γ factor that goes into the beam size σ ($\sigma_{x,y}^2 = \beta_{x,y}^* \epsilon_{x,y}/\gamma$, ϵ the normalized emittance) are fixed by the beam energy and the ring circumference. The value of the β function at the interaction point (β^*) was 1.5 m in IR1 and IR5 in the first part of the 2011 run and was lowered to 1 m in September 2011. The number of bunches n_b was 1380 since July 2011. R indicates a reduction factor due to crossing angle, bunch length and others. The bunch intensity (N_1 and N_2) could be increased so far up to 1.35×10^{11}ppb in emittances as low as 2 μm. In particular, the increase in beam brightness was performed during the months of July and August 2011, taking advantage of the potential offered by the LHC injector chain for 50 ns beams.

A comparison between the machine parameters at the time of this conference and the Design Report[1] values can be done: higher bunch intensities in smaller emittance, half the energy and twice β^*. The combination of these factors in the luminosity formula above explains why in 2011 it was possible to exceed one third of the design luminosity ($3.6 \times 10^{33}\text{cm}^{-2}\text{s}^{-1}$ versus the design value of $10^{34}\text{cm}^{-2}\text{s}^{-1}$). Plots of the luminosity evolution, both peak instantaneous and integrated, can be found at the LHC Physics Coordination Pages[5] . At the time of writing of this paper, ATLAS and CMS have integrated around 4 fb^{-1}.

4.1. *Luminosity leveling*

Out of the four major experiments that take data at the LHC, ATLAS and CMS are designed for high luminosity and pile-up $\mu \approx 20$ (number of inelastic interactions per crossing), while Alice and LHCb are designed for lower pile-up values ($\mu < 0.05$ and $\mu < 0.5$ respectively). Note that in 2010 LHCb managed to cope with up to $\mu_{max} < 2.5$ and $\approx 3 \times 10^{32}$ Hz/cm^{-2}.

Given the nevertheless high target in integrated luminosity for LHCb in 2011 (1 fb^{-1}), it was preferable to run whenever possible at a constant instantaneous luminosity, as close as possible to the maximum tolerable one, levelled down from the maximum deliverable from the machine. This motivated the need for a luminosity levelling technique.

In the early part of the run, an experiment was performed to demonstrate the feasibility of luminosity levelling by transversely displacing the beams by a small offset[6] . It was demonstrated that in the absence of strong long range interactions levelling by separation can be performed. Levelling by separation has since become an operational procedure and algorithms and applications have been developed so to help the automation of the process. At the time of writing of this paper, LHCb has reached 1 fb^{-1}, target of the year. It is worth noting that similar conditions are required by the Alice experiment and similar procedures are successfully applied.

5. Machine availability

Given a luminosity lifetime of 16-25 hours, the fill length that optimizes the integrated luminosity taking into account the turnaround time (time from beam dump to declaration of "stable beams") is around 12-15 hours. Up to $\approx$ 120 pb^{-1} per day could be produced ideally, or $\approx$ 800 pb^{-1} per week. So far in 2011 at most $\approx$ 520 pb^{-1} was achieved in a week. This is due to the fact that very few fills in a year get dumped by the operator's choice, and most of them are rather terminated by hardware issues.

The LHC statistics can be found at the link[7] , updated within tens of minutes at the end of each fill. So far in 2011 the machine had beam in for about half of the time, while 23% of the time was spent in "stable beams". It has to be noted that the high intensity beams that circulate in the machine routinely now lower the machine availability through a number of effects, a few of which are explained in the next section.

6. Known issues

Among others three effects are presented in the following as they might present possible limitations in the years to come: radiation induced failures of tunnel electronics (single event effects), losses due to (supposed) dust particles (UFOs) and the electron cloud phenomena.

6.1. *Single Event Effects*

As the peak luminosity increased, the rate of radiation induced failures in the electronics that is installed in the tunnel and some neighbouring underground areas has increased. In fact, Single Event Effects (SEE) failures are now the dominant cause of beam dumps[8] . They for example create errors in the power converters that power the magnets or in the programmable logic computers (PLCs) that control the cryogenic equipment. These errors are detected and initiate a preventive beam dump before the beam can see the effects of the error (e.g. changes of orbit). The number of dumps was quantified in terms of integrated luminosity as approximately 1 dump per 60 pb^{-1}.

Some mitigation measures were put in place during the technical stops, and some will wait for longer stops (2011/12 Christmas stop or 2013 shutdown). When possible the concerned equipment is relocated away from the tunnel (e.g. cryogenic PLCs), otherwise more radiation-tolerant firmware is put in place (e.g. for the quench protection system). In other cases special reset procedures were put in place (e.g. for cryogenics), signal filtering was applied to avoid triggers on spurious signals (e.g. for the RF system). All these actions aim to avoid resets of the system that involve tunnel access. In some locations additional shielding will be put in place (e.g around IR1).

6.2. *UFOs in the LHC*

Since July 2010, 35 fast loss events led to protective beam dumps (18 in 2010, 17 in 2011): 13 were around the injection kickers, 6 triggered the dump through the experiments, only one of them happened at the injection energy. They are rather uniformly distributed around the ring, apart from the injection kickers where many events happen after pulsing, giving a privileged location for studying them[9] . Typically these events have a duration of the order of tens of turns and a rise time of the order 1 ms (too fast to be protected by the collimators multi-stage cleaning). A potential explanation that was found is that of (dust) particles that fall into the beam and create scatter losses and showers that propagate downstream. From this, the name of (Unidentified) Falling Objects (UFOs).

So far none of them led to superconducting magnet quenches, thus the solution adopted so far was to increase the Beam Loss Monitor (BLM) thresholds. The events seemed to have diminished over the course of 2011, but the worry remains for higher energy operation (e.g. 7 TeV) when the generated losses will be higher and the energy sufficient to initiate a quench will be lower (implying a need for lower BLM thresholds). The predictions point to about 100 dumps/year for the future.

6.3. *Electron Cloud*

Electron clouds are generated in accelerators for positively charged particles as an avalanche effect: stray electrons already present in the vacuum chamber get accelerated by the positively charged beam and can hit the wall of the chamber; if the secondary electron yield of the wall is sufficiently higher than one, more electrons will be generated creating the electron "cloud". The build up of the electron cloud depends strongly on a resonance effect that in turn depends on the bunch and batch spacing in the machine.

In the LHC, electron cloud phenomena were observed in both 2010 and 2011[10] . The observables are: pressure rise measured at the vacuum gauges in the warm regions, increase of the beam screen temperature in the cold regions due to an additional heat load, beam instabilities and emittance growth (which could be cured by higher chromaticity or larger transverse emittances). Mitigation measures adopted at the LHC are the installation of solenoids and dedicated scrubbing runs. In 2011, a 10 day period was dedicated to "scrubbing" the machine from electron cloud: 50 ns beams were accumulated up to 1080 bunches and the pressure improved by an order of magnitude all over the machine after 17 effective hours of beam time. It is worth noting that 25 ns operation is expected to be substantially more difficult than 50 ns from the point of view of scrubbing.

7. Conclusions and outlook for 2012

The first two years of LHC operation, 2010 and 2011, saw a very successful commissioning and good transition from commissioning to production. The machine cycle is solid and reproducible and the machine protection works very well. The performance exceeded all most optimistic expectations, as the LHC runs now routinely at $\approx 3 \times 10^{33}\text{cm}^{-2}\text{s}^{-1}$ and 4 fb^{-1} have already been accumulated by ATLAS and CMS, 1 fb^{-1} by LHCb. Possible improvements go in the direction of achieving long fills which are now often limited by SEE on different equipment (mitigation measures are being put in place).

The decisions on the 2012 operational parameters will mostly be held at the Chamonix workshop in the beginning of February 2012. On the table, a possible energy increase from 3.5 TeV to 4 TeV, a further 30% gain in β^* by using tight collimator settings and the decision on the bunch spacing (25 ns or 50 ns). In particular, 50 ns offer a higher potential for brightness from the injectors and will probably be straightforward from the point of view of electron cloud. Concerning 25 ns beams, they would offer cleaner collisions for the experiments (lower pile-up), but they would be more difficult from the machine point of view: twice as many long-range beam-beam encounters, larger stored energy, a scrubbing run would be required in order to be able to run with good beam quality. With both beam spacing options, it seems that 10 fb^{-1} is within reach for 2012.

References

1. O. Bruning, P. Collier *et al.*, "LHC Design Report", CERN-2004-003-V-1, CERN, Geneva (Switzerland), 2004.
2. W. Herr *et al.*. "Head-on beam-beam tune shifts with high brightness beams in the LHC", CERN-ATS-Note-2011-029 MD.
3. S. Fartoukh *et al.*, "An Achromatic Telescopic Squeezing (ATS) Scheme for the LHC Upgrade ", IPAC11, San Sebastian (Spain), 2011.
4. R.W.Assmann *et al.*, "Collimator losses in the DS of IR7 and quench test at 3.5 TeV", CERN-ATS-Note-2011-042 MD (LHC).
5. *http://lpc.web.cern.ch/lpc.*
6. G. Papotti *et al.*, "Experience with Offset Collisions in the LHC", IPAC11, San Sebastian (Spain), 2011.
7. *http://lhc-statistics.web.cern.ch/LHC-Statistics/.*
8. M. Brugger, "Radiation Damage to Electronics at the LHC A First Analysis", TWEPP11, Vienna (Austria), 2011.
9. T. Baer *et al.*, "UFOs in the LHC", IPAC11, San Sebastian (Spain), 2011.
10. G. Rumolo *et al.*, "Electron cloud observations in the LHC", IPAC11, San Sebastian (Spain), 2011.

Status of the NOvA Experiment

Denis Perevalov for the NOvA Collaboration
Fermi National Accelerator Laboratory,
Batavia IL, 60510, USA
**E-mail: denis@fnal.gov*

Denis Perevalov

NOvA is an off-axis long baseline neutrino experiment searching for $\nu_\mu \rightarrow \nu_e$ oscillations using an upgraded NuMI neutrino beam from Fermilab, Batavia, IL. The main physics goal is a measurement or strong limit on the neutrino mixing angle θ_{13}. For sufficiently large values of θ_{13}, NOvA will also be sensitive to measuring CP violation and establishing the neutrino masses hierarchy. A large 14 kton Far detector, comprised of liquid scintillator contained in extruded PVC cells, will also provide an opportunity for other non-accelerator physics searches. While civil construction at the far detector is underway, a smaller prototype near detector has been assembled at Fermilab and is being studied.

Keywords: NOvA, θ_{13}

1. Introduction

NOvA stands for the NuMI Off-Axis ν_e Appearance experiment will study neutrino oscillations at a baseline of 810 km (exploring the region of L/E of ~400 km/GeV). Provided that θ_{13} is as large as early indications from the T2K experiment,[1] NOvA has a possibility of determining the ordering of the neutrino masses and constrain the Dirac CP violating phase. Additionally, NOvA will make a precision measurement of θ_{23} by observing the muon neutrino disappearance. To achieve these goals a 14 kton detector will be constructed in Ash River, Minnesota. This kind of precision measurement requires a capability of suppressing both charge current (CC) and neutral current (NC) backgrounds at the 99% level. Additionally, good detector efficiencies and energy resolution less than 8% of the expected signal width for CC event are required.[2]

2. Experimental Design

2.1. *Neutrino Beam*

To achieve the goal of the NOvA project, two detectors will be placed 14 mrad off-axis to the primary direction of the NuMI (Neutrinos from the Main Injector) source. The off-axis beam provides a relatively narrow neutrino flux at around 2 GeV, which is near the first atmospheric neutrino oscillation maximum and also reduces high energy NC background events.[2] To accommodate the needs of the experiment, the NuMI beam power will be upgraded to 700 kW during the accelerator shutdown from March to December 2012.

2.2. *Detectors*

The NOvA detector system consists of a complementary pair of detectors constructed 14 mrad off-axis to the NuMI source. Both detectors will be highly segmented tracking calorimeters built from PVC cells. The cells are filled with mineral oil based liquid scintillator for the total of 65% of the active volume.[2] The far detector will be a surface-based 14 kTon volume located 810 km from NuMI in Ash River, Minnesota. A smaller 222 Ton detector will be built about 1 km from the target at the Fermilab site in a 105 meters deep underground cavern.

3. Near and Far Detector Status

Beneficial occupancy of the far detector facility was obtained in April 2011. Construction of the far detector at Ash River is scheduled to begin in the first quarter of 2012 with a goal to have a detector segment in place before the Fermilab accelerator shutdown. The full detector is on track for completion in the first half of 2014.

Excavation of the underground cavern for the near detector will also begin following the beam shutdown. In the meantime, a prototype near detector on surface (NDOS) has been completed. A picture of NDOS is shown in Figure 1. This prototype has been taking data since October 2010. The initial study of NDOS data is described in this article below.

3.0.1. *PVC Cells*

The NOvA detector is built up from extruded PVC cells loaded with TiO_2.[2] Each cell is 3.8 cm by 5.9 cm in cross section with 90% reflectivity for light at 430 nm. Extrusions are joined to together to produce a sealed module

Fig. 1. Photograph of the NOvA Near Detector On the Surface.

of 32 cells. In NDOS, the modules are either 4.2 m (vertical orientation) or 2.9 m (horizontal) long while far detector modules are 15.6 m long. These modules are glued together into alternating planes of horizontal or vertical orientation to create a self-supporting 32 layer blocks. ~360,000 cells makeup the 14 kTon far detector.

Six blocks were constructed along with a 1.7 m muon ranger (Figure 1) for NDOS. Building the NDOS fully exercised the quality assurance/quality control (QA/QC) techniques in preparation for full production running for the far detector. This process revealed cracks in ~20% of the manifold covers. These covers have been repaired and a new more robust design will be used for future production. 1200 far detector sized extrusions have been produced to date.

3.0.2. *Liquid Scintillator*

PVC blocks are filled in place with mineral oil containing 5% pseudocumene and wavelength shifters to produce 400-450 nm light. The liquid scintillator makes up 65% of the total detector mass.[2] NDOS required ~30,000 gallons of scintillator while the 14 kTon far detector will use over 3 million gallons.

Oil work at NDOS gained us experience in the filing process. Some internal module obstructions were observed during filling; the causes of these has been resolved. NOvA has currently taken possession of around 100,000 gallons of the oil that will be used for the far site.

3.0.3. *Wavelength Shifting Fiber*

Internal to each cell is a 0.7 mm diameter looped fiber. The fiber shifts the light collected in the scintillator to 490-550 nm.[2] Its ends are routed through the manifold covered to an optical connector where they are available for single sided readout. ~113 km of fiber is used in the near detector design with 13,000 km needed for the far detector.

NDOS fiber handling allowed us to overcome tangling problems related to spooling techniques. We have also learned to measure the fiber performance in realtime as modules are strung. About 50% of the required fiber is already received.

3.0.4. *Avalanche Photodiodes*

The light from the fiber ends is incident on Hamamatsu avalanche photodiodes (APD) which have 85% quantum efficiency for 520-550 nm light.[2] The devices are operated at -15 °C with a gain of 100. For NOvA a 20 photoelectron (pe) signal from a minimum ionizing particle at the far end of a far detector sized module is required with a 10-15 pe threshold applied. Based on initial system verification, we expect 38 pe for such a signal, well above the requirement. 496 APD arrays are required for the near detector and about 12,000 are used in the far detector design.

Surface cleanliness and sealing issues have led to many of the NDOS APDs becoming unusably noisy. 274 installed unit have been removed from the detector for cleaning and study. New surface coatings and installation techniques are under study.

3.0.5. *Data Acquisition System*

The signals from the APDs are processed by front-end electronics (FEBs) which operate in continuous baseline subtraction digitization mode while sampling each channel every 500 ns.[2] 64 FEBs are fed to a Data Concentrator Module which packages and passes the data in 50 μs blocks to a processing buffer nodes. The data is then buffered at the buffer nodes for 20 seconds at which point a software trigger may be issued to record available data in a specified window.

For the near detector a 500 μs trigger window was used with three separate trigger sources; the 0.4 Hz NuMI spill signal, the 1.2 Hz Booster beam signal, and a 10 Hz cosmic pulser. Real throughput capabilities of the DAQ have doubled since initial running. During stress tests of the system, stable running was achieved with a 96% duty factor (80 ms trigger windows at 12 Hz).

4. Results from the Prototype Near Detector on the Surface

Analysis of the data from NDOS is in progress. A full suite of available Monte Carlo (masked to behave like our prototype) together with tracking on real data has allowed us to begin to calibrate and reconstruct.

NDOS has collected 5.6×10^{19} protons on target (POT) worth of data in reverse horn current beam and 8.4×10^{18} POT in forward horn mode from NuMI. Analysis of this sample has yielded 1254 candidate neutrino events with 108 expected cosmic background events. Figure 2(a) shows a peak in the track timing distribution right where the NuMI beam is expected. In Figure 2(b) one can see the excess of tracks pointing back to the NuMI source over the out-of-time cosmic background.[3] Similar distributions have been seen in a Booster neutrino sample of 222 event (with 92 expected background events) from 3×10^{19} POT.[4]

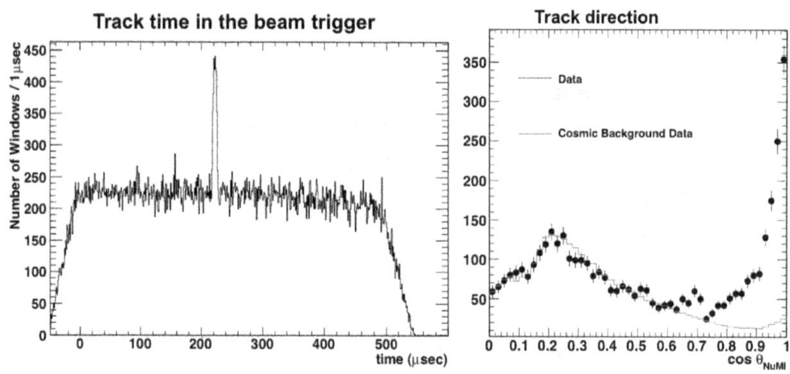

(a) Track time distribution in the Beam events. (b) Track direction distribution for both Beam and Beam-off events.

Fig. 2. Observation of NuMI neutrino events in NDOS.

Additional studies have been performed to understand the energy deposited in the detector and its cell by cell calibrations. Figure 3(a) shows the mean ADC value as a function of the distance from the center of the cell from a cosmic muon sample.[5] A sample fit which could be used to calibrate the detector response is shown. Figure 3(b) shows a sample Michel electron distribution which can be compared against expectation from simulation to provide an electromagnetic energy calibration.[6]

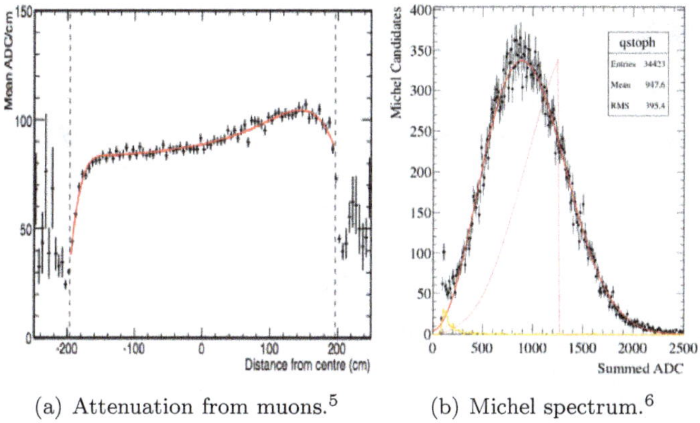

(a) Attenuation from muons.[5]

(b) Michel spectrum.[6]

Fig. 3. Preliminary analysis of events from the NuMI source and cosmic data in NDOS.

5. Conclusion

The NOvA NDOS is taking and analyzing data now. This surface prototype has proved invaluable to all aspects of the experimental program, providing critical feedback for design enhancements and operational experience. Recent results from T2K which hint at a large value for θ_{13}, are very encouraging for the long term physics reach of NOvA and open the opportunity to make real contributions in understanding the neutrino. The far detector is on track to begin data taking in 2012, with the detector hall construction nearly complete and expected beam upgrades running on time. The support for NOvA continues to grow with the collaboration now consisting of 140 physicists from 26 institutions in 4 different countries.

References

1. Abe K, et.al. *Phys.Rev.Lett. 107 (2011) 041801*
2. Feldman G, et. al. 2007 *NOvA Technical Design Report*
3. Betancourt M 2011 *Internal Nova Document NOVA-doc-5874*
4. Johnson C 2011 *Internal Nova Document NOVA-doc-5873*
5. Backhouse C 2011 *Internal Nova Document NOVA-doc-5931*
6. Messier M 2011 *Internal Nova Document NOVA-doc-6023*

A detector for the measurement of the ultrarare decay $K^+ \to \pi^+\nu\bar{\nu}$: NA62 at the CERN SPS

R. Piandani*

University of Perugia and INFN,
Perugia, Italy
** E-mail: roberto.piandani@cern.ch*

The NA62 experiment, which aims to measure the branching ratio of the very rare kaon decay $K^+ \to \pi^+\nu\bar{\nu}$ at the CERN SPS, will be described. The proposed experiment aims to collect $\sim$ 100 $K^+ \to \pi^+\nu\bar{\nu}$ events with a 10% of background. The experimental technique, the detectors and the perspectives for the experiment will be discussed.

1. Introduction

The Standard Model (SM) branching ratio can be computed to an exceptionally high degree of precision: the $O(G_F^2)$ electroweak amplitudes exhibit a power-like GIM mechanism; the top-quark loops largely dominate the matrix element; the sub-leading charm-quark contributions have been computed at NNLO [1]; the hadronic matrix element can be extracted from the branching ratio of the $K^+ \to \pi^0 e^+\nu$ decay, well known experimentally [2]. The SM prediction for the $K^+ \to \pi^+\nu\bar{\nu}$ channel is $(7.81 \pm 0.80) \times 10^{-11}$ [3]. The error comes mainly form the uncertainty on the CKM parameters, the irreducible theoretical uncertainty amounts to $\sim$ 2%. The extreme theoretical cleanness of these decays remains also in new physics scenarios like Minimal Flavour Violation (MFV) [4] and even not large deviation from the SM value (for example $\sim$20%) can be considered as an evidence of new physics.

The decay $K^+ \to \pi^+\nu\bar{\nu}$ has been observed by the stopping kaon experiments E787 and E949 at the Brookhaven National Laboratory and measured branching ratio is $1.73^{+1.15}_{-1.05} \times 10^{-10}$ [5]. However only a measurement of the branching ratio with at least 10% accuracy can be a significant test of new physics. This is the main goal of the NA62 experiment at CERN-SPS [6,7] which aims to collect $O(100)$ $K^+ \to \pi^+\nu\bar{\nu}$ events in about 2 years of data taking, keeping a background contamination around 10%.

2. The NA62 experiment

The requirement of 100 events needs to 10% of signal acceptance and at least $\sim 5 \times 10^{12}$ K^+ decays. To reach the required signal over background ratio demands a background suppression of at least 10^{13}. A high energy kaon beam and a decay in-flight technique are the principles of the experiment. A high acceptance beam line will deliver a 50 times more intense secondary hadron beam with respect the old NA48 beam line. The beam particles will have a positive charge and a momentum of 75 GeV/c ($\pm 1\%$). The average rate is ~800 MHz integrated over an area of 14 cm^2. The beam is positron free and is composed by ~6% of K^+. The average integrated rate on the detectors downstream is ~10 MHz, mainly due to the kaon decay and accidental muons.

The key points of NA62 are: a very powerful kinematical rejection; a system of efficient vetoes to reject events with γ an μ; a precise timing to associate the π^+ to the parent K^+; a particles identification system to identify the kaons among the beam particles and to distinguish π^+ from μ^+ and e^+ in the final state.

3. Kinematical rejection

The main variable in use is the squared missing mass, m^2_{miss}, defined as the difference between the kaon and the charged track 4-momenta assuming the pion mass. This variable separates the signal form more than 90% of the background coming from the main kaon decays (figure 1). The $K^+ \to \pi^+\pi^0$ peak divides 2 regions of m^2_{miss} containing a minimal amount of background.

Against this kind of background low mass and high precision detectors placed in vacuum are mandatory for tracking. A beam tracker along the beam line and a spectrometer downstream to the decay region accomplish this task. The designed beam spectrometer (Gigatracker) consists of 3 Si pixel station matching the beam size. A Si sensor 200 μm thick and a read-out chip 100 μm thick bump bonded on the sensor form one pixel, in total the 3 station have ¡3.5%X_0 of material badget. Test beam on prototypes performed at CERN in 2010 showed a time resolution ¡200 ps per station. Four chambers made by straw tubes (2.1 m long and 9.6 mm mylar tubes) and placed in the same vacuum of the dacy region form the downstream pion spectrometer. They provide the measurement of the coordinates of the

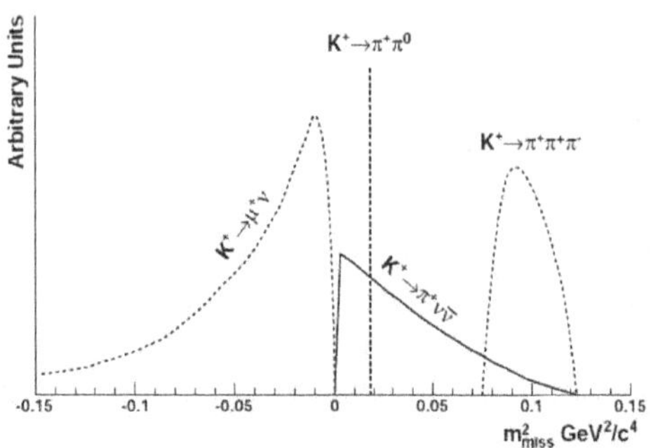

Fig. 1. Squared missing mass for the kaon decays with kinematic thresholds.

impact point of the incoming track. The same NA48 dipole magnet with a $P_{tkick} = 256 MeV/c$ along the y direction, placed after the second chamber, allows the momentum analysis. Full-length plane prototypes operating in vacuum have been tested at CERN using hadron beams in 2007, 2009 and 2010. The tests showed that the single coordinate can be reconstructed with a resolution better than 100 μm.

4. Photon and charged particles vetoes

The photon veto system should guarantee a good level a good level of $K^+ \to \pi^+\pi^0$ suppression already online, in order to reduce the rate for data acquisition. The detectors designed for this goal are: a system of calorimeters (LAV) covering the angle between 8.5 and 50 $mrad$; an electromagnetic calorimeter between 1 and 8.5 $mrad$ and small angle calorimeters covering the region below 1 $mrad$.

Twelve rings surrounding the NA62 decay and detectors regions and placed in vacuum form the LAV system. The lead glass counters of the old LEP experiment OPAL [8] are the building blocks of the rings. They guarantee a level of inefficiency around 10^{-4} down to 0.5 GeV photons, as measured in test beam performed at the BTF in Frascati. The first 2 rings have been mounted and successfully tested at CERN on the hadron beam line in 2009 and 2010. The data showed a time resolution of 700 ps. The electromagnetic liquid Kripton calorimeter of NA48 (LKr) will be reused to veto γ's in the 1–8.5 $mrad$ region. Measurements using $K^+ \to \pi^+\pi^0$ selected on NA48 data have demonstrated the capability of the LKr to reach the required veto performances, a detection inefficiency below 10^{-5} for γ's above 10 GeV and, anyhow, within 10^{-3} down to 1 GeV. The very good online time resolution, 100 ps, makes this detector essential in the trigger. In front of the LKr a small angle calorimeter (IRC) will be positioned to collect the γ's in the region below 1 $mrad$, outside the acceptance of the LKr. After all detectors in order to collect the photons inside the beam pipe another smal angle calorimeter (SAC) has been located. In order to veto the products of the interaction of the kaons in the material of the third station of the GTK, a veto detector for charged particle has been located after the last station of the GTK before the entrance of the decay region.

5. Particle identification

Since 94% of beam particles are protons and π^+, a Cherenkov Threshold Counter (CEDAR) placed on the beam line, which positively recognizes the kaons, allows the rejection of the most part of the beam accidental. The CEDAR is an existing detector built at CERN in 70's [9] and a program of refurbishing both the radiating material and the detection part (readout electronics and PMTs) already started within the NA62 collaboration. A RICH detector has been designed to separate π^+ from μ^+ with inefficiency below 1%. It also provide the timing of the event with a resolution below 100 ps and it should be used in the trigger. A vessel 17 m long placed after the track spectrometer and filled with Ne at atmospheric pressure form the detector. Tests on CERN-SPS secondary beam, performed in 2007 and 2009 [10,11], have shown an average time resolution of 70 ps and an integrated π^+/μ^+ separation of $\sim 5 \times 10^{-3}$ in a momentum range 15–35 GeV/c.The muon detection system will make use of an upgraded vertion of the old NA48 hadron calorimeter (MUV1,2), for the offline rejection of events with muons in the final state and of a plane of fast 22×22 cm^2 pad-scintillators (MUV3) placed at the end of the apparatus after an iron wall, for the trigger.

6. Trigger and data acquisition

The trigger should reduce the 10 MHz detector rate to 10 kHz in order to make the data acquisition feasible. A three-level trigger system should accomplish to this task. The Level 0 (L0) trigger is purely hardware and for his primitives consider only some detectors, the RICH, the LKr and the third station of the MUV. The Level 1 and 2 trigger are based on PC's. The detector PC's process the 1 MHz data rate which passes the L0 trigger and define more complete event information in order to build the L1 trigger decision. The filtered data are sent to a gigabit Ethernet switch and then to a PC farm which provides global event information for the ultimate trigger word (L2). The raw data are finally assembled by dedicated event-building PC's and transferred for storing on tape with a maximum speed of about 100 MB/s.

7. NA62 sensitivity

The sensitivity of the NA62 has been studied with the MC. Assuming a number of kaon decays $\sim 5 \times 10^{12}$, a signal acceptance of $\sim$ 10% and 100% of trigger efficiency, the expected number of $K^+ \to \pi^+\nu\bar{\nu}$ events is 45 events/year with a background contamination ¡13.5%.

8. Conclusions

The ultrarare $K^+ \to \pi^+\nu\bar{\nu}$ is a clear physics case with high sensitivity to new physics beyond the SM. The goal of the NA62 experiment is to collect O(100) $K^+ \to \pi^+\nu\bar{\nu}$ events in 2 years of data taking. The period 2006–2009 has been dedicated to the design and the R&D of the various NA62 detectors. The period 2010–2012 will has been allocated to the construction. At the end of 2012 a Technical run has been planned and the Physics run will be done in 2014/2015.

References

1. A. J. Buras, M. Gorbahn, U. Haisch, U. Nierste, JHEP **0611** (2006) 002. [hep-ph/0603079].
2. C. Amsler *et al.* [Particle Data Group Collaboration], Phys. Lett. **B667** (2008) 1.
3. J. Brod, M. Gorbahn, E. Stamou, Phys. Rev. **D83** (2011) 034030.
4. G. Isidori, F. Mescia, P. Paradisi, C. Smith, S. Trine, JHEP **0608** (2006) 064. [hep-ph/0604074].
5. A. V. Artamonov *et al.* [BNL-E949 Collaboration], Phys. Rev. **D79** (2009) 092004. [arXiv:0903.0030 [hep-ex]].
6. G. Anelli *et al.*, CERN-SPSC-2005-013, SPSC-P-326.
7. NA62 Collaboration, CERN-SPSC-2007-035, SPSC-M-760.
8. K. Ahmet *et al.* [OPAL Collaboration], Nucl. Instrum. Meth. **A305** (1991) 275-319.
9. C. Bovet *et al.*, The CEDAR Counters for Particles Identification in the SPS SEcondary Beams, CERN Report: CERN 82-13 (1982).
10. G. Anzivino *et al.*, Nucl. Instrum. Meth. A **593** (2008) 314.
11. B. Angelucci *et al.*, Nucl. Instrum. Meth. A **621** (2010) 205.

The MoEDAL Experiment at the LHC

J. L. Pinfold*

Physics Department, University of Alberta,
Edmonton, Alberta T6G 2E1, Canada * *E-mail: jpinfold@ualberta.ca*

The MoEDAL experiment is the seventh LHC experiment to be approved by the LHC. It is designed to detect highly ionizing particles such as magnetic monopoles, dyons and singly and multiply electrically charged stable massive particles predicted in a number of theoretical scenarios. MoEDAL consists of an array of ~400 stacks of passive Nuclear Track Detectors deployed around the LHCb intersection region, within the VELO cavern. Spallation product backgrounds will be monitored with an array of MediPix pixel detectors. The design of the detector and its physics reach, which is complementary to that of the large general purpose LHC experiments ATLAS and CMS, will be discussed.

Keywords: LHC, MoEDAL, Dyon, magnetic monopole, stable massive particle, pseudo-stable massive particle, nuclear track detector, timepix chip.

1. Introduction

The MoEDAL Experiment, approved by the LHCC in the Fall of 2009 and the CERN Research Board in March 2010, is the seventh LHC experiment. The MoEDAL collaboration is comprised of ~30 physicists from eight institutes in seven countries from around the world: U. of Alberta (CDN); U. of Bologna/INFN (IT); U. of Cincinatti (US); CERN (CH); Czech Technical U. (CZ); King's College London (UK); Northeastern U. (Boston, US); and, the Inst. of Space Sciences (RO). MoEDAL is a compact low cost experiment by LHC standards but with a huge physics potential that complements the existing LHC experimental program and builds on the existing Canadian LHC involvement.

MoEDAL's prime motivation is to search for the direct production of the magnetic monopoles and dyons at the LHC. Another physics aim is the search for exotic, highly ionizing, stable electrically charged massive particles (SMPs), such as: black hole remnants, double charged Higgs particles, sleptons, R-hadrons, Q-balls, Quirks, etc. A MoEDAL discovery would have revolutionary implications for physics in one or more of the following areas:

magnetic charge, new symmetries such as supersymmetry or technicolor, extra dimensions, the nature of dark matter, and the early universe, etc.

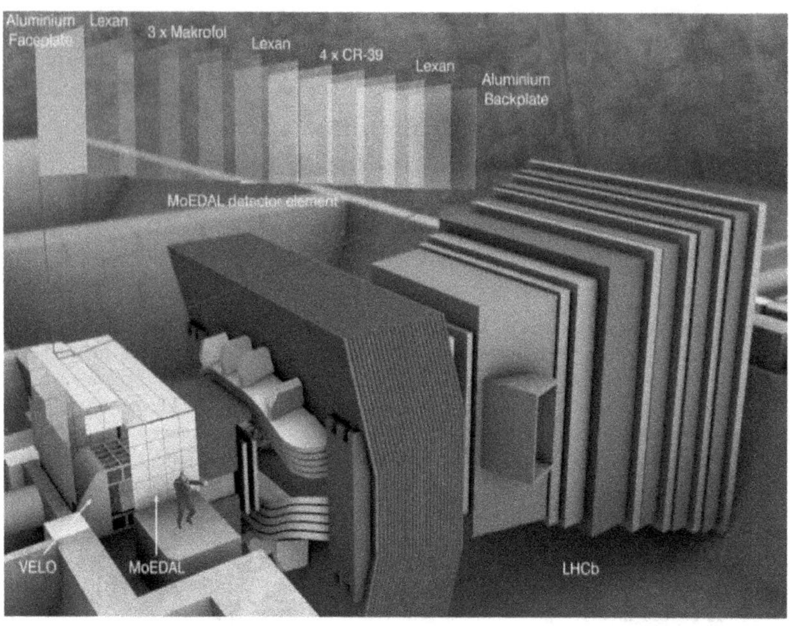

Fig. 1. A visualization of the MoEDAL detector deployed on the walls of the VELO cavern (not shown) surrounding LHCb vertex detector. A detail showing the structure of one of the MoEDAL NTD stacks is shown above the detectors.

The MoEDAL Experiment, approved by the LHCC in the Fall of 2009 and the CERN Research Board in March 2010, is the seventh LHC experiment. The Canadian led MoEDAL collaboration is comprised of 30 physicists from eight institutes in seven countries from around the world: U. of Alberta (CDN); U. of Bologna/INFN (IT); U. of Cincinatti (US); CERN (CH); Czech Technical U. (CZ); King's College London (UK); Northeastern U. (Boston, US); and, the Inst. of Space Sciences (RO). MoEDAL is a compact low cost experiment by LHC standards but with a huge physics potential that complements the existing LHC experimental program and builds on the existing Canadian LHC involvement.

MoEDAL's prime motivation is to search for the direct production of the magnetic monopoles and dyons at the LHC. Another physics aim is the search for exotic, highly ionizing, stable electrically charged massive particles (SMPs), such as: black hole remnants, double charged Higgs particles, sleptons, R-hadrons, Q-balls, Quirks, etc. A MoEDAL discovery would have revolutionary implications for physics in one or more of the following areas: magnetic charge, new symmetries such as supersymmetry or technicolor, extra dimensions, the nature of dark matter, and the early universe, etc.

The MoEDAL detector, deployed at Point 8 around the LHCb intersection region, is the largest array of (400 x 25 cm x 25 cm) plastic Nuclear Track Detector (NTD) stacks ever deployed at an accelerator. A visualization of the MoEDAL detector is provided in Figure 1. Possible radiation induced spallation product backgrounds will be monitored online by a small array of eight TimePix1 silicon pixel imaging chips. A small NTD test array was deployed in Nov. 2009. Plastic from this array will be calibrated at the NASA space radiation facility at Brookhaven National Lab. in the Spring of 2012. A larger MoEDAL NTD test array - one third the size of the full MoEDAL detector - was installed in Jan. 2011. This array will be removed at the end of 2012 and sent for analysis. The first full deployment of the MoEDAL detector will be made in 2013 - during the LHC long shutdown - ready for LHC running at 14 TeV

2. The MoEDAL Physics Program

The search for highly ionizing SMPs at the LHC can be divided into three main categories. The first category is that of massive magnetically charged particles such as the magnetic monopole[1 2] or the dyon.[3] The physics reach for in the case of direct production of a magnetic monopole pair via the Drell-Yan mechanism is shown in Figure 2.

Another open question in modern physics is the existence of SMPs[4] with single electrical charge providing a 2nd category of particle that is heavily ionizing by virtue of its small β. The most obvious possibility for an SMP is that one or more new states exist which carry a new conserved, or almost conserved, global quantum number. SUSY with R-parity, extra dimensions with KK-parity, and several other models fall into this category. The lightest of the new states will be stable, due to the conservation of this new parity, and depending on quantum numbers, mass spectra, and interaction strengths, one or more higher-lying states may also be stable or meta-stable. The third class of hypothetical particle has multiple electric charge such as the black hole remnant,[5] or long-lived doubly charged Higgs boson.[6]

SMPs with magnetic charge, single or multiple electric charge and with Z/β as low as five can in principle be detected by the CR39 nuclear track detectors, putting them within the physics reach of MoEDAL.

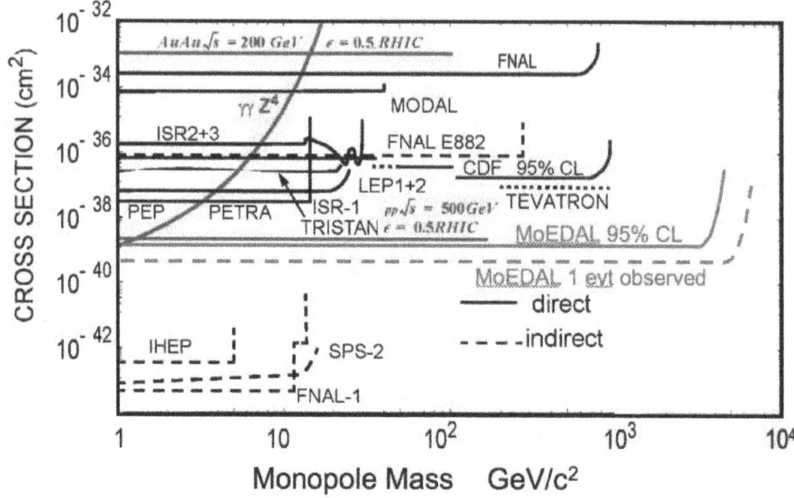

Fig. 2. The expected reach of the search for the direct detection of monopole anti-monopole pair production produced via the Drell-Yan process at the LHC (E_{cm} =14 TeV).

3. The Detector

The MoEDAL detector is comprised of an array of ~ 400 plastic Nuclear Track Detectors (NTDs) stacks deployed around the (Point-8) intersection region of the LHCb detector, in the VELO (VErtex LOcator) cavern as shown in Figure 1. The array consists of NTD stacks, ten layers deep, in Aluminium housings attached to the walls and ceiling of the MoEDAL-VELO cavern. The maximum possible surface area available for detectors is around 25 m^2, although the final deployed area could be somewhat less due to the developing requirements of the infrastructure of the LHCb detector. An array of approximately eight TimePix pixel chips will be used to monitor the spallation product backgrounds in the MoEDAL-VELO cavern. A photograph of TimePix unit similar to the units that we will deploy at Point-8 is given in Figure 3.

3.0.1. *Track-Etch Detectors*

When a charged particle crosses a plastic nuclear track detector it produces damages at the level of polymeric bounds in a small cylindrical region around its trajectory forming the so-called latent track as shown in

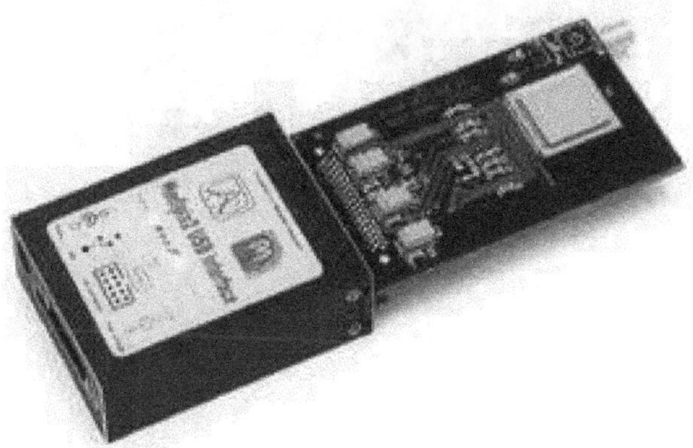

Fig. 3. An example of the TimePix module that will be deployed at Point -8 to monitor the spallation product background and radiation field in the moEDAL-VELO cavern

Figure 4. The damage produced is dependent on the energy released inside the cylindrical region i.e. the Restricted Energy Loss (REL) which is a function of the charge Z and $\beta = v/c$ (c the velocity of light in vacuum) of the incident highly ionizing particle (ion). When the velocity of the incident ion is $< 1\ 10^{-2}$ c the restricted energy loss is equal to the total energy loss of the particle in the medium; otherwise, only a fraction of the electronic energy loss leading to the formation of δ-rays with energies lower than a cut-off energy T_{cut} cut is efficient for the track formation. The REL can be computed from the Bethe-Block formula restricted to energy transfers $T < T_{cut}$ with T_{cut} ($< T_{max}$) a constant characteristic of the medium and taking into account density and shell (charge screening) corrections. Figure 5 shows a sketch of the evolution of the etch-pit cones versus the etching time for a normally incident relativistic particle.

The subsequent etching of the solid nuclear detectors leads to the formation of etch-pit cones, as can be seen in Figure 5. These conical pits are usually of micrometer dimensions and can be observed with an optical microscope. Their size and shape yield information about charge, energy and direction of motion of the incident ion.

The latent track of a highly ionizing particle, such as a that of a magnetic monopole, is manifested by etching, where v_B is the bulk rate and v_T is the faster etch rate along the track. The damage zone is revealed as a cone shaped etch-pit, when the surface of the plastic detector is etched in a controlled manner using an etchant such as hot sodium hydroxide (NaOH) solution. In general the cone base has an elliptical form (circular if the

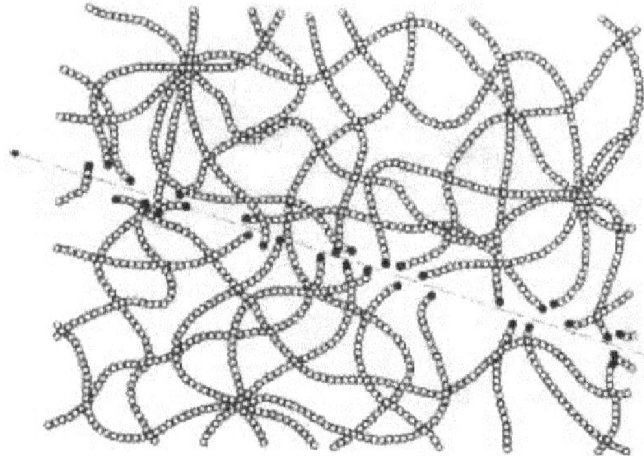

Fig. 4. The breaking of the polymeric bonds by a crossing charged particle.

impinging ion has a normal incidence angle). The response of the detector is given by the etching rate ratio $p = v_T/v_B$. From the measurements of the minor and major axes of the base of etch-pit cones it is possible to determine both p and the angle of incidence θ with respect to the detector surface:

$$p = \sqrt{1 + \frac{4A^2}{(1 - B^2)^2}} \tag{1}$$

$$\theta = sin^{-1}\left(\frac{1}{p}\frac{1 + B^2)}{1 - B^2}\right) \tag{2}$$

where: $A = a/2v_Bt_{etch}$ and $B = b/2v_Bt_{etch}$, a/b are the major/ minor axes of the tracks and t_{etch} is the etching time. The bulk etching velocity can be determined from the measurement of the detector thickness at different times.

By exposing the detector to relativistic heavy ions of known energy and electric charge it is possible to obtain the calibration data expressed as the reduced etching rate $(p-1)$ versus REL. So, the charge of an incoming particle can be determined by the measured p of the corresponding tracks. Only particles releasing a REL above a threshold and incident within a definite angle (which depends on the particle energy loss) will be detected. Specific processing and etching conditions affect the detector threshold REL_{min}, or (Z/β)min that is the minimum charge and speed a particle must have to produce an etchable latent track.

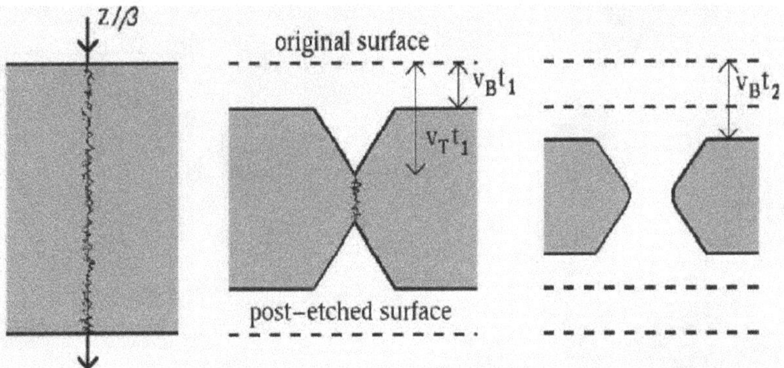

Fig. 5. (a) the latent track for a nucleus with charge Z and velocity $\beta = v/c$; (b) two etch pit cones are formed on both sides of the foil; (c) a prolonged etching can make the two cones connect and form a hole.

Materials commonly used as solid state track etch detectors are polymers and polycarbonates such as polyallyl diglycol carbonate (PADC) also known as CR 39 with the chemical formula ($C_{12}H_{18}O_7n$) or Makrofol (Lexan) with the formula ($C_{16}H_{14}O_3n$).

The basic detector unit of the MoEDAL experiment is a stack of ten sheets of plastic NTDs, consisting of 4 sheets of CR39 (each $\sim$ 0.5 mm thick), 3 sheets of MAKROFOL (each $\sim$0.5mm thick) with Lexan (each $\sim$0.2mm thick) forming the first, middle and ends sheets of the stack. A depiction of a basic MoEDAL NTD stack is given in Figure 6. The damage zone is revealed as a cone shaped etch-pit, when the plastic detector is etched in a controlled manner using an etchant such as hot sodium hydroxide (NaOH) solution. The base area and depth of the etch-pit cones are an increasing function of the Z/β of the particle, where Z is a particle charge and β its velocity.

4. The Current Status of MoEDAL

A small test array of MoEDAL plastic detectors was deployed at the MoEDAL /LHCb intersection point (Point 8) prior to the start of LHC running at 900 GeV centre-of-mass energy (Ecm) in November 2009. After an exposure of over a year the detectors were removed for analysis at INFN/Bologna (etching and scanning) during the shutdown in the winter of 2010/2011. This test array is slated for exposure and subsequent calibration using 1 GeV/nucleon Fe-ion beam at the NASA Space Radiation Labora-

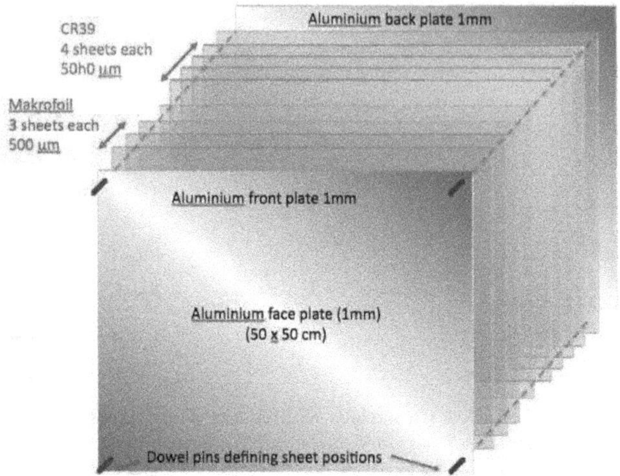

Fig. 6. The basic MoEDAL detector stack consisting of 4 sheets of CR39 and three sheets of MAKROFOL with Lexan sheets at the beginning and end of the stack and separating the CR39 and MAKROFOL stacks. The size of each sheet is 25 cm × 25 cm. The stack is enclosed in a 1mm thick aluminium housing. The total thickness of the stack is 3.6 mm.

tory (NSRL) at the Brookhaven National Laboratory (BNL) in spring.

In January 2011, during this same shutdown, a large test array with a total of 80 m^2 of plastic Nuclear Track Detectors (NTDs) roughly one third the area of plastic in the full detector - was deployed in order to take data at 7 TeV Ecm. A photograph of part of this deployment is shown in Figure 7. This plastic will be removed for analysis at the beginning of the long shutdown of the LHC at the end of 2012. The first full official MoEDAL deployment will take place towards the end of the long LHC shutdown in 2013 when it is expected that the LHC will be running at the full design Ecm of 14 TeV.

As envisaged we recently obtained the agreement of the LHCb in October 2011 to deploy the first of a total of ~ 8 (full deployment) TimePix pixel chips deployed at various chosen points around the MoEDAL NTD array. The purpose of this array is to monitor the radiation field and spallation production background, online.The resulting data record will be continually monitored offline in order for us to prepare for future deployments of plastic prior to the shutdowns currently envisaged at the end of each year of running. This monitoring can be handled remotely via the web. Analysis of the small test stack removed in January 2011 indicates that it would be op-

Fig. 7. A photograph of part of the MoEDAL large test detector array deployment on the MoEDAL-VELO cavern walls at Point-8 on the LHC ring.

timal to remove and replace the complete MoEDAL NTD array at the end of each year of running. In this way the spallation product background will be minimized allowing efficient and timely scanning of the exposed NTDs.

The plastic NTDs exposed during the first full run that will be removed in the winter of 2014/2015 and at that time to also redeploy fresh plastic NTDs for the subsequent year at that time. This procedure of removal, analysis and redeployment will be repeated until we achieve an integrated exposure corresponding to 10 fb-1 of data. The maximum luminosity at Point 8 is a few time 10^{32} cm^{-2} s^{-1} and so this process should take several years.

References

1. P.A.M. Dirac, Proc. R. Soc. London, Set. A, 133, 60 (1931); P.A.M. Dirac, Phys. Tev., 24, 817 (1948).
2. G. Giacomelli and M. Sioli (Astroparticle Physics), Lectures at the 2002 Int. School of Physics, Constantine, Algeria, hep-ex/0211035; G. Giacomelli and L. Patrizii, hep-ex/011209; G.R. Kalbfleisch, Phys. Rev. Lett. 85, 5292 (2000); hep-ex/0005005 ; K. A. Milton et al., hep-ex/0009003; B. Abbott et al., hep-ex/9803023, Phys. Rev. Lett. 81, 524 (1998); G. Giacomelli, Riv. Nuovo Cimento 7 N.12, 1 (1984); M. Acciarri et al., Phys. Lett. B345, 609 (1995); L. Gamberg *et al.*, hep-ph/9906526.
3. J. Schwinger, Science, Volume 165, Issue 3895, 757 (1969).
4. M. Fairbairn et al., Phys. Rep. 438:1.63, (2007).
5. B. Koch, Marcus Bleicher, Sabine Hossenfelder, JHEP 0510, 053, (2005); H. Stoecker, Int. J. Mod. Phys. D16, 185 (2007); S. Hossenfelder, B. Koch, M. Bleicher, e-Print: hep-ph/0507140, (2005).
6. G. B. Gelmini and M. Roncadelli, Phys. Let B99, 411 (1981); R. N. Mohaptra and J. D. Vergados, Phys. Rev. Lett. 47, 1713 (1981); V. barger, H. Baer, W. Y. Keung and R. J. N. Philips, Phys. Rev. D26, 218 (1982); H. F. Gunion, H. E. Haber, G. L. Kane and S. Dawson, "The Higgs Hunters Guide", (Addison Wesley 1990); J. A. Grifols, A. Mendez and G. A. Schuler, Mod. Phys. Lett. A4, 1485 (1989).

The NA62 Liquid Krypton Electromagnetic Calorimeter Level 0 Trigger

A. Fucci, G. Paoluzzi, A. Salamon*, G. Salina

INFN Sezione di Roma Tor Vergata
Via della Ricerca Scientifica, 1 - 00133 Roma Italia
**E-mail: andrea.salamon@roma2.infn.it*

E. Santovetti, F. M. Scarfì

Università degli Studi di Roma Tor Vergata - Dip. di Fisica
Via della Ricerca Scientifica, 1 - 00133 Roma Italia

V. Bonaiuto, F. Sargeni

Università degli Studi di Roma Tor Vergata - Dip. di Ingegneria Elettronica
Via del Politecnico, 1 - 00133 Roma Italia

The NA62 experiment at CERN SPS aims to measure the Branching Ratio of the very rare kaon decay $K^+ \to \pi^+ \nu\bar{\nu}$ collecting $O(100)$ events with a 10% background to make a stringent test of the Standard Model. One of the main backgrounds to the proposed measurement is represented by the $K^+ \to \pi^+\pi^0$ decay. To suppress this background an efficient photo veto system is foreseen. In the 1-10 mrad angular region the NA48 high performance liquid krypton electromagnetic calorimeter is used. The design, implementation and current status of the Liquid Krypton Electromagnetic Calorimeter Level 0 Trigger are presented.

1. The NA62 experiment at CERN SPS

The NA62 experiment[1,2] aims at measuring the very rare kaon decay $K^+ \to \pi^+ \nu\bar{\nu}$ collecting $O(100)$ events with a 10% background in two years of data taking.

$K^+ \to \pi^+ \nu\bar{\nu}$ is an exceptionally clean decay from the theoretical point of view with a Standard Model branching ratio prediction of BR($K^+ \to \pi^+ \nu\bar{\nu}$) = $(8.5 \pm 0.7) \times 10^{-11}$. A precise measurement of BR($K^+ \to \pi^+ \nu\bar{\nu}$) will offer the opportunities of testing the Standard Model and deepening the knowledge of the CKM matrix.

The NA62 detector[3], see Figure 1, currently being installed at the SPS North Area High Intensity Facility is composed of: a differential Cerenkov counter[4] (CEDAR), a beam tracker (GTK) and charged particle detector (CHANTI), a straw chambers magnetic spectrometer, a photon veto system composed of different detectors in the various angular decay regions, a RICH[5], a charged particle hodoscope (CHOD) and a muon detector (MUV).

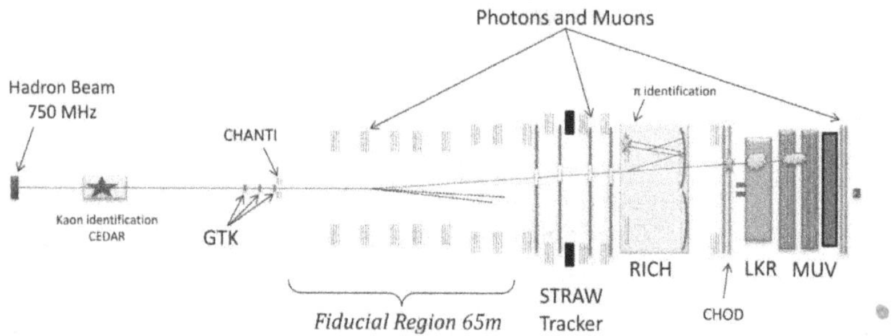

Figure 1. Schematic drawing of the NA62 detector at CERN SPS.

2. Trigger and Data Acquisition system

In order to extract few interesting decays from a very intense flux a complex and performing three level trigger and data acquisition system was designed[6].

The Level 0 trigger algorithm is based on few sub-detectors and is performed by dedicated custom hardware modules, with a maximum output rate of 1 MHz and a maximum latency of 1 ms.

Level 1 and Level 2 software triggers are executed on dedicated PCs. The maximum Level 2 output rate is of the order of 15 kHz.

3. The Liquid Krypton electromagnetic calorimeter

In order to suppress the background from $K^+ \to \pi^+\pi^0$ decay an efficient photon veto system is foreseen. In the 1-10 mrad angular region the NA48 electromagnetic calorimeter is used[7].

This calorimeter is a quasi-homogenous ionization device using liquid krypton as active medium and characterized by excellent time and energy resolution.

The Liquid Krypton calorimeter will be readout by the new Calorimeter REAdout Modules[8] (CREAMs) which will provide 40 MHz 14 bit sampling for all 13248 calorimeter readout channels, data buffering, optional zero suppression and programmable trigger sums for the Level 0 electromagnetic calorimeter trigger processor.

4. The Liquid Krypton Level 0 trigger

The Level 0 Liquid Krypton electromagnetic calorimeter trigger, see Figure 2, identifies electromagnetic clusters in the calorimeter and prepares a time-ordered list of reconstructed clusters together with the arrival time, position, and energy measurements of each cluster. Information on reconstructed clusters is used by the Level 0 Trigger Processor to veto decays with more than one cluster in the Liquid Krypton calorimeter.

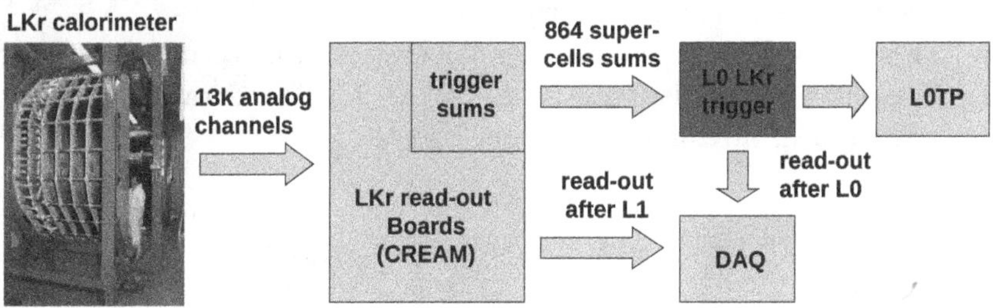

Figure 2. Block diagram of the Liquid Krypton Level 0 trigger inside the Trigger and Data Acquisition system.

The trigger processor also provides a coarse-grained readout of the Liquid Krypton calorimeter that can be used in software triggers and off-line as a cross-check for the CREAM high-granularity readout.

4.1. *Trigger algorithm*

Trigger algorithm is based on energy deposits in tiles of 16 calorimeter cells which are available from the main readout boards.

Electromagnetic cluster search is executed in two steps with two one-dimensional (1D) algorithms, see Figure 3.

From the trigger point of view the calorimeter is divided in slices parallel to the horizontal axis. In the first step peaks in space and time are searched independently in each slice with a 1D algorithm. In the second step different peaks which are close in time and space are merged and assigned to the same electromagnetic cluster.

4.2. *Trigger processor implementation*

The main parameters driving the design of the processor are the high expected instantaneous hit rate (30 MHz), the required single cluster time

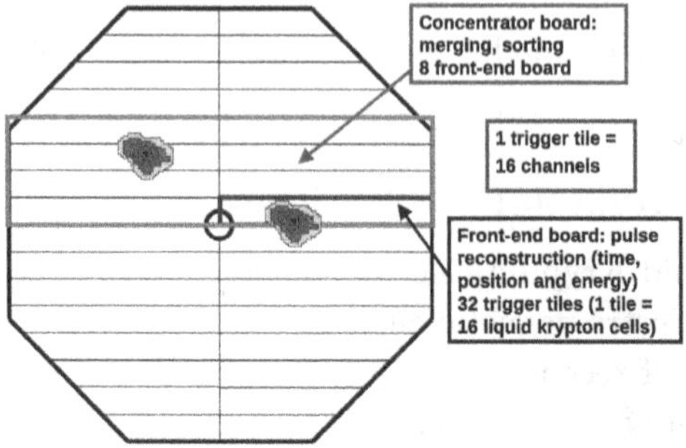

Figure 3. Liquid Krypton electromagnetic calorimeter trigger segmentation.

resolution (1.5 ns) and a maximum allowed latency of 100 μs.

The processor is a three-layer parallel system, composed of Front-End and Concentrator boards, both based on the 9U TEL62 cards[9] equipped with custom dedicated mezzanines, see Figure 4.

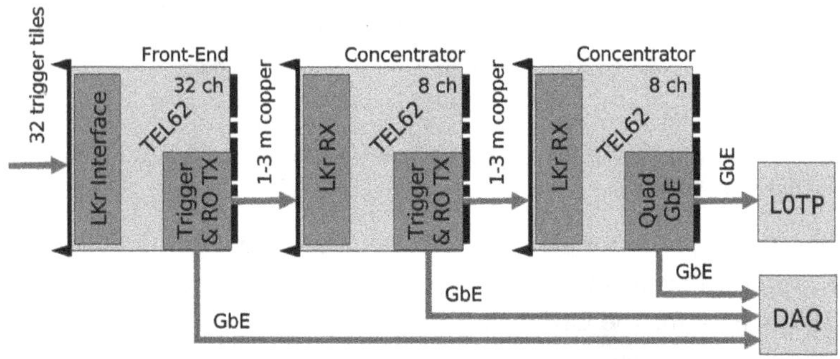

Figure 4. Liquid Krypton trigger processor block diagram. 28 Front-End boards and 8 Concentrator boards are foreseen in the system.

The Liquid Krypton Level 0 trigger continuously receives from the Liquid Krypton readout modules 864 trigger sums[a] each one corresponding to a tile of 16 calorimeter cells.

Each **Front-End board** receives 32 trigger sums and performs peak search in space and computes time, position and energy for each detected peak. In order to extract timing information at the ns level a parabolic interpolation in time around sample maximum and a digital constant fraction discrimination are performed after the peak search algorithms. Information on reconstructed peaks is transferred from the Front-End boards to the Concentrator boards on low-latency high-bandwidth dedicated trigger

[a]Trigger sums are transmitted over shielded copper twisted pairs.

links. Raw data received by the readout modules are also stored in latency memories, to be readout upon request. 28 Front-End boards equipped with 84 custom mezzanines are foreseen in the whole system.

The **Concentrator board** receives trigger data from up to 8 FE boards and combines peaks detected by different front-end boards into a single cluster. Overlap between neighboring Concentrators is foreseen to guarantee that each cluster will be fully contained in at least one Concentrator board with proper logic to avoid double counting. Reconstructed clusters are also stored in latency memories, to be readout upon request. 8 Concentrator boards equipped with 24 custom mezzanines are foreseen in the whole system.

Conclusions

A fast parallel processor for cluster reconstruction and counting in the Liquid Krypton electromagnetic calorimeter of the NA62 experiment has been designed.

In total, the system will be composed of 36 TEL62 boards, 108 mezzanine cards and 215 high-performance FPGAs. The whole system will fit in three 9U crates.

The peak reconstruction algorithm was implemented on FPGA and was proven to fulfill the NA62 timing and rate requirements. Most of the boards and mezzanines have been designed and are currently being tested and assembled together for data taking.

References

1. NA62 Collaboration, Proposal to Measure $K^+ \to \pi^+ \nu\bar{\nu}$ rare decay at the CERN SPS , CERN-SPSC-2005-013, 2005.
2. R. Piandani, These Proceedings
3. NA62 Collaboration, NA62 Technical Design, NA62-10-07, 2010.
4. A. Romano, These Proceedings
5. F. Bucci, These Proceedings
6. M. Sozzi, A concept for the NA62 Trigger and Data Acquisition, NA62-07-03, 2007.
7. NA48 Collaboration, The Beam and Detector for the NA48 neutral kaon CP violation experiment at CERN, Nucl. Instrum. Methods A, A 574, 2007. 433-471
8. V. Ryjov et al, The NA62 Liquid Krypton Calorimeter Readout Module, Proceedings of TWEPP 2011
9. E. Pedreschi, F. Spinella et al, TEL62: an integrated trigger and data acquisition board, Proceedings of TWEPP 2011

Reconstruction of $\bar{p}p$ events in PANDA

S. SPATARO for the PANDA collaboration
Dipartimento di Fisica Generale, Universitá di Torino and INFN, I-10125 Torino, Italy
**E-mail: stefano.spataro@to.infn.it*

The PANDA experiment will study anti-proton proton and anti-proton nucleus collisions in the HESR complex of the facility FAIR, in a beam momentum range from 2 GeV/c up to 15 GeV/c. In preparation for the experiment, a software framework based on ROOT (PandaRoot) is being developed for the simulation, reconstruction and analysis of physics events, running also on a GRID infrastructure. Detailed geometry descriptions and different realistic reconstruction algorithms are implemented, currently used for the realization of the Technical Design Reports. The contribution will report about the reconstruction capabilities of the Panda spectrometer, focusing mainly on the performances of the tracking system and the results for the analysis of physics benchmark channels.

Keywords: Style file; LaTeX; Proceedings; World Scientific Publishing.

1. Introduction

The PANDA[1] experiment will be built at the FAIR facility (Darmstadt, Germany), and it will study anti-proton proton and anti-proton nucleus collisions in a momentum range up to 15 GeV/c. An intensive study of charmonium and open charm states is foreseen, searching also for new exotic states, and studying electromagnetic form factors, Drell-Yan processes, Generalized Parton Distributions and performing single/double hypernuclei spectroscopy. In this contribution the offline software (*PandaRoot*) and the reconstruction capabilities of the Panda experiment will be discussed.

2. The software framework

The detector simulation and reconstruction code of the PANDA experiment is developed inside the *PandaRoot* framework. Basic features, such as the interfaces with simulation, parameter database and the I/O, are handled by the *FairRoot*[2] framework (currently developed by the GSI-IT department), based on the ROOT[3] package with a dynamic data structure (trees and branches); the Virtual MonteCarlo[4] is used for detector simulations, allowing one to use as transport models Geant3 or Geant4 with the same geometry definition and detector code. Inside *PandaRoot* the detector specifics and reconstruction code are developed. At a first stage event generators (EvtGen, DPM, UrQMD, Pythia, Fluka) are used to create signal and background events, sent to the transport model by Virtual MonteCarlo, and finally the detector responses are simulated. The tracking detectors hits are correlated into tracks, which are matched to PID detectors to form charged candidates; calorimeter clusters not matched with tracks are selected as neutral candidates. Particle identification algorythms assign to each candidate the probability of being a well determined particle, and finally all this information is transferred to the analysis stage. The *PandaRoot* software runs on an ALIEN2 based GRID[5] and is currently maintened under different C++ compilers, several Linux distribution and Mac OS, so that it can be installed in a laptop or home institute farm without any restrictions.

3. Detector Geometry

Panda is a fixed target spectrometer, divided into two regions. Around the target region there is the so-called *Target Spectrometer*, with inside a 2T solenoid. A Micro Vertex Detector (MVD), several GEM detector planes, a DIRC Cherenov Detector and an Electromagnetic Calorimeter are placed inside. At present two different detector options are under evaluation for the central tracker, i.e. a Straw-Tube Tracker (STT) or a Time Projection Chamber (TPC); moreover, a Time-Of-Flight detector is under study. Outside the magnet coils, within the return iron yoke several planes of muon tracker (MDT) are placed. The region below 10° in polar angle is covered by the *Forward Spectrometer*, where the tracking is performed by six planes of the Forward Tracking System (FTS), based on straw tubes, placed before, inside and after a dipole magnetic field; moreover TOF detectors, a Shashlik EM calorimeter and muon chambers are foreseen. As event display the ROOT Event Visualization Environment (EVE) is used.

4. Tracking

The PANDA global tracking foreseen the combined used of several detectors of different kind to calculate the momentum of each particle. In the Target Spectrometer, hits from the Micro Vertex Detector (MVD), the central tracker and the GEMs are matched together by means of pattern recognition algorythms. Two different detector concepts were evaluated for the central tracker, a Time Projection Chamber (TPC) and a Straw Tube Tracker (STT). It was recently decided by the PANDA collaboration board that the STT will be the PANDA central tracker. For the patter recognition, as a first step hits from the STT detector are matched together by means of conformal mapping,[6] which transforms circles in the XY plane into straight lines. Hits are added by a *road* technique starting from the external ones and fitted by a helix. In the meantime MVD hits are correlated by means of a Riemann track finder. As a second step, STT and MVD *tracklets* are merged together to improve the original track parameters, to eliminate some spurious hits and/or to include some new hits not considered before. Finally, in the angular region where the particles cross both STT and GEM planes, the tracks are extrapolated from the last point of the central tracker on each plane of the GEM detector. After all the hits are added and a helix fitting is performed, it is necessary to consider the magnetic field unhomogeneities (in particular in the GEM region), energy loss and the different data structure and error calculation of different detectors to improve the momentum reconstruction performance. This is achieved by the Kalman Filter GENFIT,[7] developed within the PANDA collaboration, and using GEANE[8] as track follower. The momentum resolution for muons is shown in Figure 1.

In the Forward Spectrometer particles are tracked by six planes of Straw Tube chambers placed before, in the middle and after a dipole field region. Currently, tracking is performed using MonteCarlo information and realistic pattern recognition algorythms are under development.

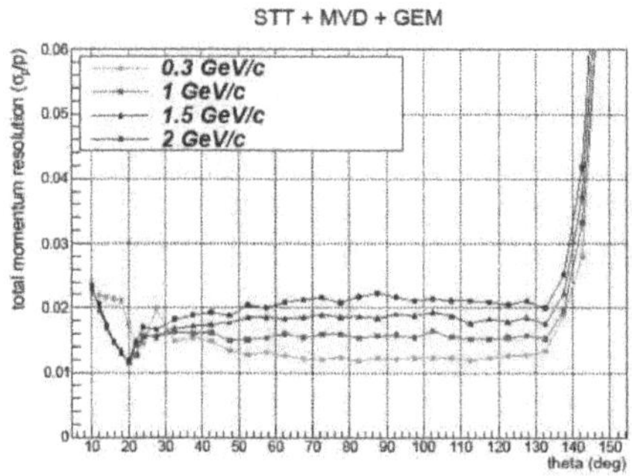

Fig. 1. Momentum resolution for muons with STT+MVD+GEM tracking in the Barrel Spectrometer.

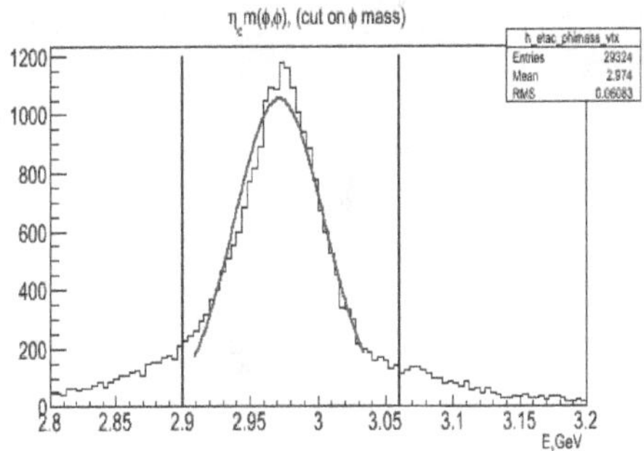

Fig. 2. Reconstruction of the channel $\bar{p}p \to \eta_C \to \phi\phi \to K^+K^-K^+K^-$: invariant mass distribution of the η_C candidates ($\phi\phi$) after the kinematic fitting.

5. Analysis of physics benchmark channels

For the high-level analysis the Rho[9] package has been included inside the PandaRoot framework; the user has the possibility to select *candidates* according to his pid selections and kinematic constraints, and combine them for the physics analysis. At present different kinematic and vertex fitters have been implemented and are under test. For the realization of the Technical Design Report for the central tracker, and to compare the response of the two detector concepts, a list of benchmark channels was selected in order to evaluate the tracking performance, in terms of momentum resolution, reconstruction efficiency and vertex resolution.

The channel $\bar{p}p \to \eta_C \to \phi\phi \to K^+K^-K^+K^-$ allows the reconstruction of the charmonium ground state (η_C) properties in formation mode, whose mass and width earliern measurements are not consistent with the recent ones done by B-factories with higher resolution. Figure 2 shows the $\phi\phi$ invariant mass distribution after a 4C kinematic fitting, with a resolution of 33 MeV/c^2.

Figure 3 shows an example of D meson reconstruction in the channel $\bar{p}p \to \psi(3770) \to D^+D^- \to K^-\pi^+\pi^+K^+\pi^-\pi^-$, important for open charm physics. This channel is more challenging for the tracking, having six charged tracks in the final state and displaced vertices ($c\tau_D = 311.8\mu m$). With the current reconstruction we obtain as D-meson mass resolution 13 MeV/c^2, and as D vertex resolution $\approx 50\mu m$ in XY and $\approx 100\mu m$ in Z coodinates.

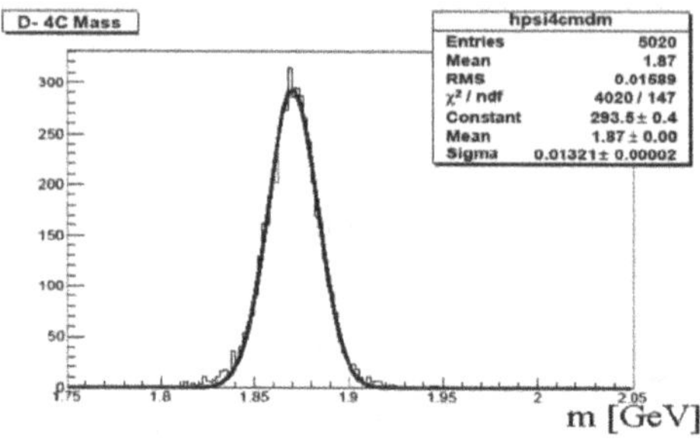

Fig. 3. Reconstruction of the channel $\bar{p}p \to \psi(3770) \to D^+D^- \to K^-\pi^+\pi^+K^+\pi^-\pi^-$: invariant mass distribution of the D$^+$ candidates ($K^-\pi^+\pi^+$) after the kinematic fitting.

6. Conclusions and outlook

PandaRoot is the official framework for the PANDA full simulation, reconstruction and analysis. Tracking in the Target Spectrometer is almost complete and the central tracker Technical Design Report is going to be finalized. Still there are ongoing development activities to improve the global tracking (in particular in the forward part) and the particle identification. One of the most challenging aspects of the PANDA computing is the triggerless Data Acquisiton under the high event rate ($\leq 2 \times 10^7$ evt/s). In this case the *event* is not defined by the DAQ (event builder), but all the signals are stored with time stamps requiring a deconvolution by the software. At present the software structure is going to be redesigned from an event basis to a time ordered simulation, and the reconstruction algorithms are needed to handle the additional time information and do the event building. The basic infrastructre is already present in the framework and the detectors implementation will be finalized in the next months.

References

1. The PANDA Collaboration, *Physics Performance Report for PANDA: Strong Interaction Studies with Antiprotons* (arXiv:0903.3905v1, 2009)
2. http://cbmroot.gsi.de
3. R. Brun and F. Rademakers, *Nucl. Intr. Meth. A* **389**, 81-86 (1997)
4. I. Hrinacova *et al.*, *Proc. of Computing in High Energy and Nuclear Physics (La Jolla)* (THJT006, 2003)
5. Protopopescu D *et al.*, *GSI Scientific Report 2008* 241 (2008)
6. P. Yepes, *Nucl. Inst. Meth. A* **380** 582-585 (1996)
7. C. Höppner *et al.*, *Nucl. Intr. Meth. A* **620** 518-525 (2010)
8. A. Fontana *et al.*, *J. Phys.: Conf. Ser.* **119** 032018 (2008)
9. http://savannah.fzk.de/websites/hep/rho/

CMS status and operations

J. Thompson* for the CMS Collaboration

Laboratory for Elementary-Particle Physics, Cornell University,
Ithaca, NY 14853-5001, USA
**E-mail: joshua.thompson@cornell.edu*

The Compact Muon Solenoid (CMS) experiment at the Large Hadron Collider (LHC) is efficiently collecting data at luminosities above 3×10^{33} cm^{-2} s^{-1}. Recent running has seen increases in the average number of interactions per bunch crossing, testing the capabilities of the tracking and trigger systems. CMS physics results on approximately 1 fb^{-1} of data collected at $\sqrt{s} = 7$ TeV are presented, including results of searches for the Higgs boson.

Keywords: CMS; LHC; detector; particle physics; HEP; higgs; SUSY; top

1. Introduction

The Large Hadron Collider (LHC) at CERN began operations at the world-record energy of 3.5 TeV per beam in the spring of 2010. Operations in 2011 have seen enormous increases in luminosity, such that the entire 2010 dataset is now delivered in less than a day.

The Compact Muon Solenoid (CMS) is one of two general-purpose particle physics detectors on the LHC, designed to support a wide range of physics studies including precision measurements of electroweak and top quark production, and searches for the Higgs boson and beyond-standard model (BSM) processes.

The CMS Collaboration consists of over 3000 scientists, engineers, and graduate students. These individuals are from 173 institutes spanning 40 countries.

2. The CMS Detector

CMS is an approximately cylindrical volume with a diameter of 15 m, a length of 21 m, and a weight of 14000 tons.[1] An all-silicon tracking detector provides precision tracking of charged particles and allows for precise reconstruction of vertices. Closest to the beampipe is a silicon pixel detector, surrounded by a silicon microstrip tracking detector. Photons and electrons are measured in the lead-tungstate crystal electromagnetic calorimeter (ECAL). Hadronic calorimetry is provided by a sandwich of brass and scintillator.

Surrounding these is the centerpiece of the CMS detector, a 3.8 T solenoid magnet. The muon systems and flux return form the outer layer of the detector. Drift tubes (DT) and resistive plate chambers (RPC) provide redundant muon tracking in the barrel region, while in the endcaps the RPCs are supplemented with cathode strip chambers (CSC).

3. Recent CMS operations

The CMS control room in Cessy, France is staffed by a shift crew of five who are responsible for monitoring detector safety as well as data acquisition and quality. As of the end of September 2011, CMS has recorded 3.70 fb^{-1} out of 4.11 fb^{-1} delivered by the LHC, for an efficiency of 90%. Roughly 90% of the recorded data has been certified as "golden" for all physics analysis. An average of 98% of the subdetector channels are operational and in the readout.

One of the key challenges for the LHC experiments is reducing the $\mathcal{O}(10\ \mathrm{MHz})$ collision rate to a rate that can be written to disk ($\mathcal{O}(100\ \mathrm{Hz})$) without compromising physics results. At CMS, the first-level (L1) trigger is implemented in hardware, and combines information from the calorimeters and muon systems to accept < 100 kHz of events. The entire detector is read-out at this rate, providing input to the High Level Trigger (HLT) system, implemented in software and run on a computing farm. The HLT then selects roughly 300 Hz of events to store for offline physics analysis.

The number of bunches in the LHC was slowly increased over the 2011 run, culminating in 1380 bunches (the ultimate number possible at a bunch spacing of 50 ns), with 1318 colliding in CMS. Since July 2011, the luminosity has been increased by decreasing the emittance, increasing bunch intensity, and lowering the β^*. These increases in luminosity at a constant number of bunches increase the number of proton-proton interactions per bunch crossing ("pileup"). At luminosities of $L \sim 3 \times 10^{33}\ \mathrm{cm}^{-2}\ \mathrm{s}^{-1}$, there are about 15 interactions per bunch crossing.

CMS maintains high vertexing efficiency at high pileup. We find that the mean number of reconstructed vertices stays linear as a function of luminosity per bunch. This efficient reconstruction of vertices separated by as little as 1 mm allows for offline objects such as jets and isolated leptons to be corrected event-by-event for pileup effects by removing contamination from vertices not associated with the physics process of interest.

To ensure a high level of physics performance, it is important to maintain low trigger thresholds even in the face of increasing pileup. A primary means to do so is to form triggers from combinations of multiple objects, for example multiple leptons instead of a single lepton. Also, improvements developed and tested on offline data are being migrated to the online environment. These include isolation requirements on leptons, particle flow techniques for jet reconstruction, and rejection of noise in the calorimeters.

4. Recent physics results

4.1. *Electroweak cross sections*

Measurements of production cross sections of W and Z bosons were key early measurements in the LHC data, providing some of the first validation of standard model (SM) predictions at $\sqrt{s} = 7$ TeV. With the 1.1 fb^{-1} dataset, these tests of the SM have been extended to rarer diboson production. W^+W^- candidates are required to have two isolated leptons (e or μ). The primary background is from $W +$ jets events where a jet is misidentified as a lepton. This lepton fake rate is estimated from data. Signal efficiency is estimated from simulation, where the lepton efficiencies are corrected using tag-and-probe measurements in the data. Similar techniques are used for the $WZ \to \ell^{\pm}\nu\ell'^{+}\ell'^{-}$ $(\ell, \ell' = e, \mu)$ and $ZZ \to \ell^{+}\ell^{-}\ell'^{+}\ell'^{-}$ $(\ell, \ell' = e, \mu, \tau)$ modes. Invariant mass distributions are shown in Fig. 1.[2]

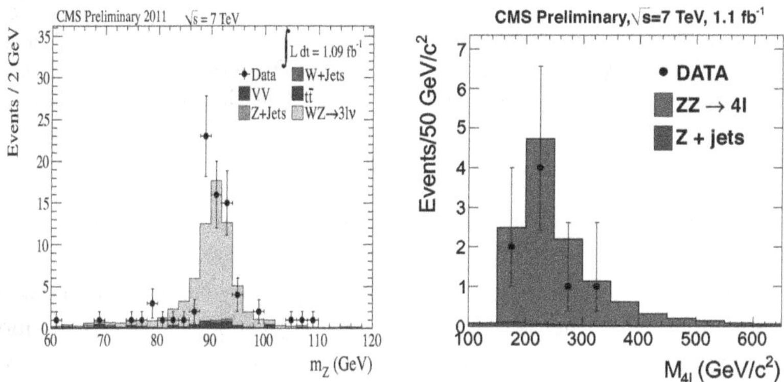

Fig. 1. Reconstructed invariant mass distributions from the measurements of the (left) WZ and (right) ZZ cross sections.

4.2. $t\bar{t}$ *results*

Measurements of the $t\bar{t}$ cross sections in various channels test the NLO predictions of the SM, and are sensitive to various BSM processes that feature anomalous top production. Results on the 1.1 fb^{-1} dataset include the dilepton channel,[3] the all-hadronic channel,[4] and a channel including a hadronically-decaying τ lepton.[5] The channel including a τ is particularly interesting, because BSM scenarios with a light, charged Higgs boson can lead to enhanced τ production. All of these measurements are consistent with the NLO SM prediction of 165 pb. Figure 2 shows various distributions from these measurements.

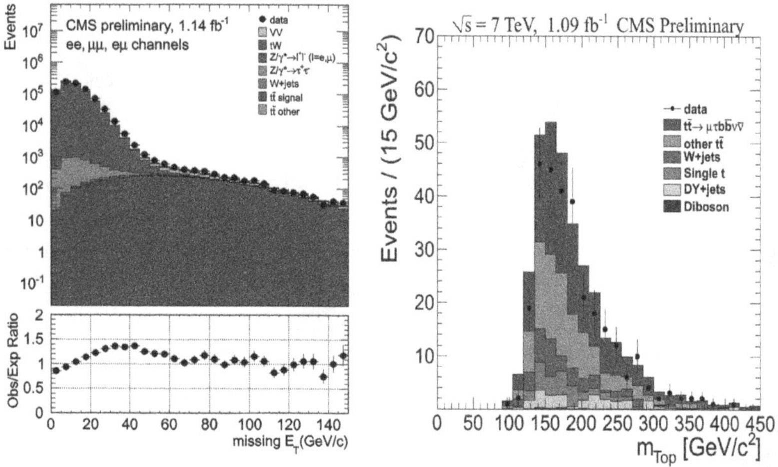

Fig. 2. (Left) Distribution of E_T^{miss} in the dileptonic $t\bar{t}$ cross section measurement;[3] the analysis selection requires $E_T^{miss} > 30\,\text{GeV}$, reducing Drell-Yan backgrounds. (Right) Reconstructed top mass distribution in the $t\bar{t} \to \mu\tau b\bar{b}\nu\bar{\nu}$ channel,[5] using the KINb algorithm.[6]

Other measurements of the $t\bar{t}$ system also provide important tests of the SM. CPT conservation requires that the top and its antiparticle have the same mass. This is tested by reconstructing $t\bar{t}$ events with one top decaying to a muon, while the other top decays hadronically. The charge of the muon is used for top/anti-top identification, while the top mass is reconstructed from a kinematic fit to the jets from the hadronic top decay. The measured mass difference, $\Delta m = -1.20 \pm 1.21\,(\text{stat.}) \pm 0.47\,(\text{syst.})$, is consistent with zero.[7]

4.3. *Searches for supersymmetry*

One of the most frequently mentioned scenarios for BSM physics is supersymmetry (SUSY), which introduces a partner particle for each SM particle with the same quantum number except differing by half a unit of spin.[8] If R-parity is conserved, supersymmetric particles are produced in pairs and will decay to the lightest supersymmetric particle, which will escape the detector undetected. This gives rise to a characteristic signature of large missing transverse energy. The dominant production channels at the LHC are squark-squark, squark-gluino, and gluino-gluino pair production.

Searches for SUSY in CMS emphasize complementarity between analyses, with a variety of signatures and the use of several kinematic variables. Results on the 1.1 fb^{-1} dataset include hadronic searches using the α_T, missing H_T, and M_{T2} variables.[9–11] Searches are also performed with signatures including one or more leptons,[12–14] a Z,[15,16] a b-tagged jet,[11] or photons.[17] These varied signatures provide both model-independence and robustness (via varied SM backgrounds and background estimation methods) to the search program.

Data-driven background estimation is emphasized, to avoid reliance on the tails of SM MC distributions. A variety of methods are used, although they generally involve using a control sample for each background to extract an estimate for the search region.

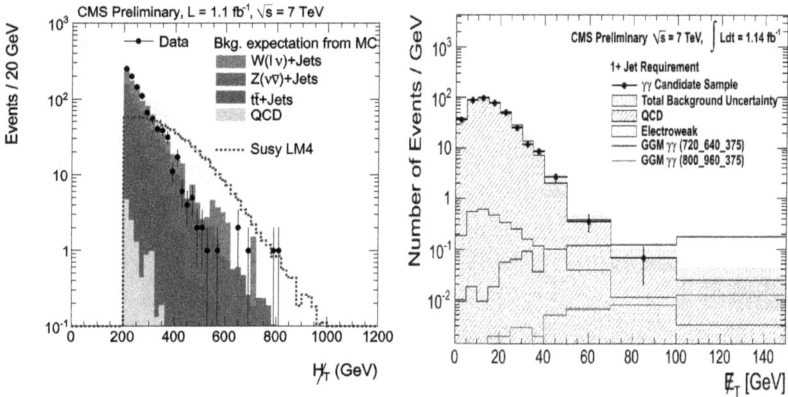

Fig. 3. Distributions of missing transverse energy in (left) the inclusive hadronic SUSY search[10] and (right) the search requiring two photons in the final state.[17] Data and SM expectations agree well in each case, while a SUSY signal would show up as an excess in the high tail.

Representative distributions from SUSY searches are shown in Fig. 3. All SUSY searches find data in agreement with the SM background expectations. Therefore, we interpret our results in the context of various SUSY models. The most popular is the Constrained Minimally Supersymmetric Standard Model (CMSSM), which is characterized by five parameters: m_0, $m_{1/2}$, A_0, $\tan\beta$, and the sign of μ. Because the CMSSM is a very specific model, we also provide results within the framework of so-called Simplified Models, which are very simple models consisting of only a single production and decay mode.[18] A summary of results in the CMSSM is shown in Fig. 4.

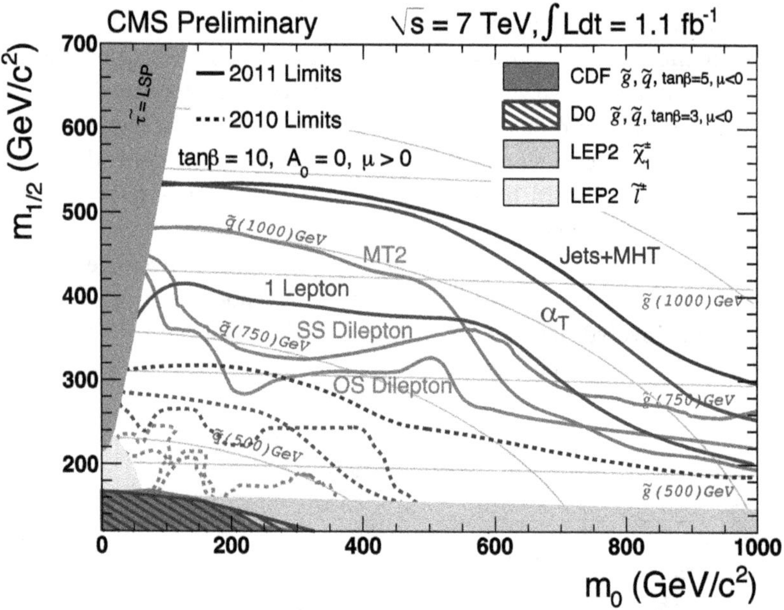

Fig. 4. Summary of regions excluded at 95% CL in the m_0, $m_{1/2}$ plane with $tan\beta = 10$, $A_0 = 0$, and $\mu > 0$.[18]

4.4. *Search for* $B_{(s)} \to \mu\mu$

The decays of the B_s and B mesons to $\mu^+\mu^-$ proceed only via loop diagrams, and are thus highly suppressed in the SM with an expected branching fraction of $(3.2 \pm 0.2) \times 10^{-9}$ in the case of the B_s. The CDF collaboration has recently announced a potentially interesting excess of events in this search, with a central value higher than the SM expectation.[19] With 1.1 fb^{-1} of data, CMS finds the observed data in both the B and B_s signal regions consistent with background. Therefore, we set upper limits on the branching fractions of $\mathcal{B}(B_s \to \mu\mu) < 1.9 \times 10^{-8}$ and $\mathcal{B}(B^0 \to \mu\mu) < 4.6 \times 10^{-9}$ at 95% confidence level.[20] These results have been combined with recent results from the LHCb experiment, which also finds no evidence for a signal.[21]

4.5. *Other measurements and searches*

CMS has an extensive program of searches for a variety of non-SM phenomena, including extra dimensions, fourth generation quarks, W' and Z', and microscopic black holes. At least 19 results were released on the ~ 1 fb^{-1} dataset.[22] All searches found the observed data consistent with the SM background.

4.6. *Searches for the Higgs boson*

One of the flagship goals of the LHC experiments is the discovery or exclusion of the SM Higgs boson. The large dataset collected thus far in 2011 has allowed the LHC experiments to become competitive with the Tevatron experiments in this area.

For SM Higgs masses between the LEP limit of about 115 GeV and about 130 GeV, the most sensitive analysis is in the mode $H \to \gamma\gamma$. The most sensitive analysis then becomes $H \to WW$ until $m_H \sim 190$ GeV, when $H \to ZZ$ takes over. In case of the discovery of the Higgs, all analyses will be relevant, as one will want to confirm that the discovered particle conforms to the SM predictions. Also, some non-SM scenarios favor modes such as $H \to \tau\tau$ that are less sensitive in the SM. The Higgs analyses have been updated with integrated luminosity ranging from $1.1 - 1.7$ fb^{-1}.[23]

At low Higgs masses, the massive QCD background overwhelms the dominant Higgs decay channel $H \to \text{b}\overline{\text{b}}$, hence the importance of $H \to \gamma\gamma$ where the QCD contribution is suppressed. The remaining QCD background is estimated, for a given mass hypothesis, from the $m_{\gamma\gamma}$ sidebands in the data. The data are consistent with background (Fig. 5), yielding an upper limit on the cross-section that is roughly three times the SM cross-section.[24] The $\text{b}\overline{\text{b}}$ channel is viable in the case of Higgs production in association with a vector boson, in which case the resulting W or Z provides separation from QCD background. The sensitivity of this analysis is enhanced with the use of a boosted decision tree discriminant.[25] As noted above, the $H \to \tau\tau$ channel is sensitive not only to a low-mass SM Higgs but also to SUSY scenarios with two Higgs doublets. The data in this search are consistent with the predicted background, leading to limits in a SUSY parameter space.[26]

For Higgs masses between $130 < m_H < 190$ GeV, the decay $H \rightarrow WW \rightarrow 2\ell 2\nu$ provides the best sensitivity. Although the substantial missing energy from the neutrinos precludes the reconstruction of a clean mass peak, the two leptons allow for triggering and background rejection. In addition, the spin-0 nature of the SM Higgs allows for discrimination against the dominant WW background, as shown in Fig. 5. Data agrees with the background prediction, allowing the SM Higgs to be excluded at 95% CL for masses of $147 < m_H < 194$ GeV, compared to an expected exclusion range of $135 < m_H < 200$ GeV.[27]

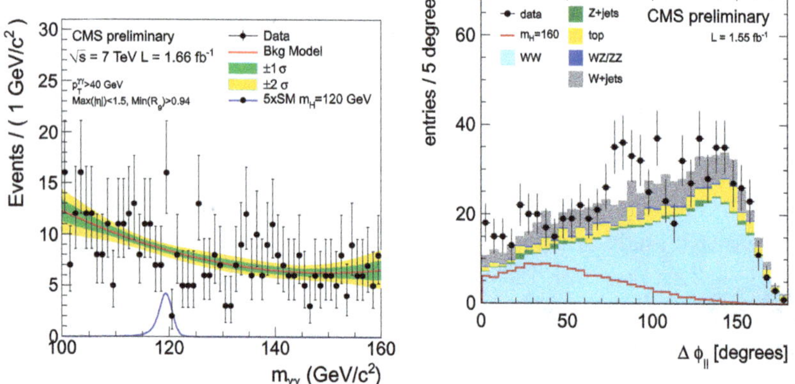

Fig. 5. (Left) Distribution of the diphoton invariant mass $m_{\gamma\gamma}$, showing the potential signal of a 120 GeV SM Higgs boson; note that the signal is normalized to five times the SM cross section. (Right) Distribution of $\Delta\phi_{ll}$ for the mode $H \rightarrow WW$, showing how the scalar Higgs signal peaks at low values, providing good separation from the WW background.

The combined SM Higgs search results are shown in Fig. 6. Masses from 145-216, 226-288, and 310-400 GeV are excluded at 95% confidence level, while the expected exclusion is 130-440 GeV. In the excluded region, the sensitivity now exceeds that of the Tevatron experiments (they remain competitive for low masses).

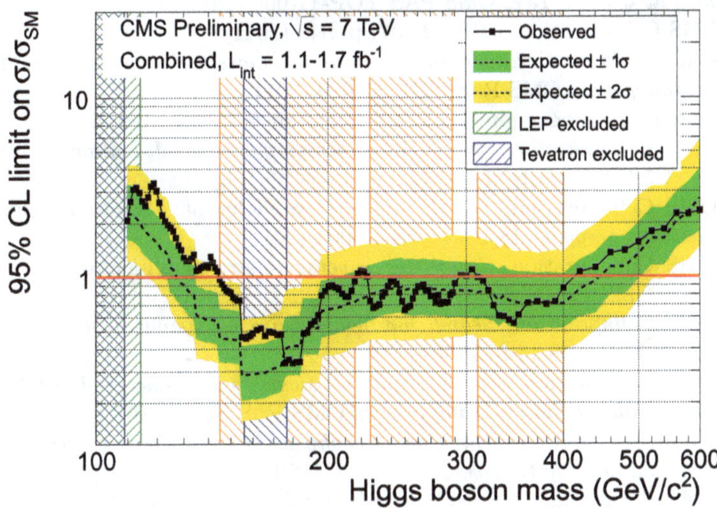

Fig. 6. The combined 95% CL upper limits on σ/σ_{SM} as a function of the SM Higgs boson mass. The observed limits, as well as the expected limits with errors bands, are shown. Regions excluded by CMS are shown in orange (light gray); regions excluded by the LEP and Tevatron experiments are shown for comparison.

5. Conclusion

The CMS detector is efficiently collecting data at instantaneous luminosities above 10^{33} cm^{-2} s^{-1}, with pileup in excess of 10 interactions per bunch crossing. With $> 4\,\text{fb}^{-1}$ of integrated luminosity delivered, CMS reports results of analyses on up to 1.7 fb^{-1}. These results include measurements of cross sections and other properties of SM processes and particles, and searches for BSM physics. The search for the SM Higgs boson now surpasses the sensitivity of the Tevatron experiments at high Higgs masses.

References

1. The CMS Collaboration, *JINST* **3**, p. S08004 (2008).
2. The CMS Collaboration, *Measurement of the WW, WZ and ZZ cross sections at CMS*, CMS PAS EWK-11-010 (2011).
3. The CMS Collaboration, *Top pair cross section in dileptons*, CMS PAS TOP-11-005 (2011).
4. The CMS Collaboration, *Measurement of the ttbar production cross section in the fully hadronic decay channel in pp collisions at 7 TeV*, CMS PAS TOP-11-007 (2011).
5. The CMS Collaboration, *First measurement of the top quark pair production cross section in the dilepton channel with tau leptons in the final state in pp*

collisions at $\sqrt{s} = 7$ *TeV*, CMS PAS TOP-11-006 (2011).

6. The CMS Collaboration, *JHEP* **1107**, p. 049 (2011).
7. The CMS Collaboration, *Measurement of the mass difference between top and antitop quarks*, CMS PAS TOP-11-019 (2011).
8. S. P. Martin, A Supersymmetry primer, hep-ph/9709356, (1997).
9. The CMS Collaboration, *Search for supersymmetry in all-hadronic events with* α_T, CMS PAS SUS-11-003 (2011).
10. The CMS Collaboration, *Search for supersymmetry in all-hadronic events with missing energy*, CMS PAS SUS-11-004 (2011).
11. The CMS Collaboration, *Search for supersymmetry in all-hadronic events with MT2*, CMS PAS SUS-11-005 (2011).
12. The CMS Collaboration, *Search for new physics with single-leptons at the LHC*, CMS PAS SUS-11-015 (2011).
13. The CMS Collaboration, *Search for new physics with same-sign isolated dilepton events with jets and missing energy*, CMS PAS SUS-11-010 (2011).
14. The CMS Collaboration, *Search for new physics in events with opposite-sign dileptons and missing transverse energy*, CMS PAS SUS-11-011 (2011).
15. The CMS Collaboration, *Search for Physics Beyond the Standard Model in Z + MET + Jets events at the LHC*, CMS PAS SUS-11-012 (2011).
16. The CMS Collaboration, *Search for new physics in events with a Z boson and missing energy*, CMS PAS SUS-11-017 (2011).
17. The CMS Collaboration, *Search for Supersymmetry in Events with Photons, Jets and Missing Energy*, CMS PAS SUS-11-009 (2011).
18. The CMS Collaboration, CMS supersymmetry physics results (2011), `https://twiki.cern.ch/twiki/bin/view/CMSPublic/PhysicsResultsSUS`.
19. The CDF Collaboration, Search for $B_s \to \mu^+\mu^-$ and $B_d \to \mu^+\mu^-$ Decays with CDF II, 1107.2304, (2011).
20. The CMS Collaboration, Search for B(s) and B to dimuon decays in pp collisions at 7 TeV, 1107.5834, (2011).
21. The CMS and LHCb collaborations, *Search for the rare decay* $B_s^0 \to \mu^+\mu^-$ *at the LHC with the CMS and LHCb experiments*, CMS PAS BPH-11-019 (2011).
22. The CMS Collaboration, CMS exotica physics results (2011), `https://twiki.cern.ch/twiki/bin/view/CMSPublic/PhysicsResultsEXO`.
23. The CMS Collaboration, *Combination of Higgs Searches*, CMS PAS HIG-11-022 (2011).
24. The CMS Collaboration, *Search for a Higgs boson decaying into two photons in the CMS detector*, CMS PAS HIG-11-021 (2011).
25. The CMS Collaboration, *Search for the Standard Model Higgs Boson Decaying to Bottom Quarks and Produced in Association with a W or a Z Boson*, CMS PAS HIG-11-012 (2011).
26. The CMS Collaboration, *Search for Neutral Higgs Bosons Decaying to Tau Pairs in pp Collisions at* $\sqrt{s} = 7$ *TeV*, CMS PAS HIG-11-020 (2011).
27. The CMS Collaboration, *Search for the Higgs Boson in the Fully Leptonic* W^+W^- *Final State*, CMS PAS HIG-11-014 (2011).

Tracker and Position Sensitive Detectors

A. Bellerive	Timing Resolution Studies of Micro Pattern Gas Detectors Using the Charge Dispersion Signal
S. Colafranceschi	Construction and Performance of full scale GEM prototypes for future upgrades of the CMS forward Muon system
J. Dammer	Micro-Radiography of Living Biological Organisms with MEDIPIX2 Detector and Application of Various Contrast Agents
P. Gianotti	Tracking with Straw Tubes in the $\bar{\text{P}}$ANDA Experiment
J. Grosse-Knetter	Overview of the ATLAS Insertable B-Layer (IBL) Project
S. Heindl	The Belle-II DEPFET Pixel Detector at the SUPERKEKB Flavour Factory
A. Jaeger	Tracking performance of the LHCb spectrometer
G. Kaußen	Performance of the CMS Silicon Tracker at LHC
A. Leflat	Performance of the LHCb Vertex Locator
C. Leroy	Measurement of the Radiation Field in ATLAS with the ATLAS-MPX Detectors
R. Märki	Alignment of the LHCb tracking system
M. Martišíková	Perspectives of the Pixel Detector TIMEPIX for needs of Ion Beam Therapy
A. Messineo	CMS Pixel Detector for the Phase I Upgrade at HL-LHC
M. Minãno	Silicon Strip Detectors for the ATLAS HL-LHC Upgrade
V.A. Mitsou	ATLAS Silicon Microstrip Tracker: Operation and Performance

M. Pesaresi	Upgrade of the CMS Tracker for THE LHC at high Luminosity
N. Pozzobon	A Level 1 Tracking Trigger for the CMS Experiment
G. Bertuccio	Silicon Carbide Microstrip Detectors for high Resolution X-Ray Spectroscopy
B. Storaci	The Performance of the Outer Tracker Detector at LHCb
M. Tacconi	Electrical Characterization of SiPM as a function of Test Frequency and Temperature
S. Taroni	The Alignment of the CMS Silicon Tracker
C. Teyssier	Exploitation of the charge sharing effect in Timepix device to achieve sub-pixel resolution in imaging applications with alpha particles
M. Tobin	The LHCb Silicon Tracker Performance in pp Collisions at the LHC
W. Treberspurg	Measuring Doping Profiles with Spreading Resistance Profiling
A. Tricomi	Performance of the CMS Pixel Detector for the Phase 1 Upgrade at HL—LHC
A. Uras	Performance Study for a Muon Forward Tracker in the ALICE Experiment
A. Vogel	Performance of the ATLAS Transition Radiation Tracker with First High-Energy pp and Pb-Pb Collisions
F. Weyda	Combination of Modern Visualization Techniques for Imaging of Biological Samples
P. Wintz	The Tracking System in the PANDA Apparatus
M. Zientek	A Dedicated Beam Tests of the Full-Scale Prototype of GEMs for CMS in a Strong Magnetic Field

TIMING RESOLUTION STUDIES OF MICRO PATTERN GAS DETECTORS USING THE CHARGE DISPERSION SIGNAL

A. BELLERIVE

Ottawa-Carleton Institute for Physics, Department of Physics,
Carleton University, 1125 Colonel By Drive, Ottawa, K1S 5B6, Canada
E-mail: alain_bellerive@carleton.ca

The International Linear Collider (ILC) will require a large volume Time Projection Chamber (TPC) with transverse space-point resolution of 100 μm for all tracks over the full 2 m drift region. It has been shown that a conventional readout GEM TPC can achieve this resolution using 1 mm or narrower readout pads, at the expense of detector cost and complexity. A new readout technique using the principle of charge dispersion has demonstrated that the transverse resolution goal can be achieved using 2-3 mm wide pads in both small (COSMo) and large (LCTPC) prototype detectors. However, the effect of this new technique on the time resolution was not a part of these studies. Here we present re-analyses of a 4 GeV π^+ beam test at KEK and a high magnetic field cosmic ray test at DESY carried out with the COSMo TPC with charge dispersion. We find the time resolution comparable to conventional MPGD and wire/pad readout TPCs, and consistent with the ILC z-resolution requirements of 500 and 1400 μm at zero and 2 m drift distances, respectively.

Keywords: Gaseous Detectors; Position-Sensitive Detectors; Micro Pattern Gas Detectors; Gas Electron Multiplier; Micromegas; ILC.

1. Introduction

The International Large Detector (ILD) is one of two proposed all-purpose detectors for the future International Linear Collider (ILC). To meet the stringent resolution requirements, the ILD proposes a Time Projection Chamber (TPC)[1] for central tracking, with a goal of measuring 200 track points with single hit resolutions of $\sim$100 μm in the transverse direction for all tracks, and $\sim$1400 μm in the longitudinal direction, after 2 m of drift in a 3.5 T magnetic field.[2]

The transverse resolution goal represents an order of magnitude improvement over the conventional proportional wire/cathode pad TPC performance, which are limited by the intrinsic $\mathbf{E} \times \mathbf{B}$ effect,[3] and approaches the fundamental limit imposed by diffusion. ILD will forego the conventional wire/pad readout for recently developed Micro-pattern Gas Detectors (MPGD); either the GEM[4] or Micromegas,[5] both of which offer many advantages. MPGDs have a smaller material budget, which is important in a high background environment such as the ILC, and naturally reduce space charge build up in the drift volume by suppressing positive ion feedback from the amplification region. Of greatest importance however, is that the $\mathbf{E} \times \mathbf{B}$ effect is negligible for an MPGD; the holes have $< 100\ \mu$m spacing and rotationally symmetric distribution, thus there is no preferred track angle.

MPGDs have performed well in small drift chambers, however the number of readout channels that would be required in a large TPC makes them impractical. If not for the E×B effect, conventional wire/pad TPCs could have achieved excellent resolution of the avalanche position using relatively wide pads from the centroid of induced signals on cathode pads. For comparable accuracy, an MPGD-TPC would require readout pads that are much narrower to allow for charge sharing, significantly increasing the required number of readout channels and complexity.

Recently, a new MPGD readout technology called *charge dispersion* has been developed and shown to allow the use of wide pads, while still achieving excellent transverse resolution.[6] Charge dispersion and its application to MPGD-TPC readouts is well understood, and has been described previously in.[7,8] The conventional MPGD anode is replaced by a highly resistive thin film bonded to the readout pad plane with an intermediate insulating spacer, forming a distributed 2-dimensional RC circuit. Thus, when an electron avalanche arrives at the anode, it disperses according to the system RC time constant, and effectively covers a larger area, allowing for the use of wider pads without loss of accuracy.

The charge dispersion MPGD-TPC readout concept was tested using a small 15 cm drift TPC with $2{\times}6\text{mm}^2$ readout pads in high energy hadron beam tests carried out at KEK[9] and in 5 T magnetic field cosmic-ray tests at DESY.[10] This work was part of a collaboration between research groups at Carleton, Orsay, Saclay, and Montreal (COSMo). More recently, a 1 m large prototype was constructed by the LCTPC collaboration with $3{\times}6\text{mm}^2$ readout pads, and tests with a charge dispersion readout are ongoing at DESY, showing good results.[11] Both GEM and Micromegas MPGDs have been tested with the resistive anode, and excellent transverse resolution has been obtained. However, the longitudinal (time) resolution has not been looked at until now.

In this paper, we present an initial study of the single-hit time resolution using raw charge pulse data from the COSMo TPC tests carried out at KEK and DESY, and compare our findings with the requirements of the proposed ILD-TPC.

2. Experimental Setup

2.1. *KEK*

In October of 2005, the COSMo prototype Micromegas-TPC was tested with a 4 GeV π^+ beam in the PS experimental hall of KEK in Tsukuba, Japan. The gas mixture was Ar:C_4H_{10} (95:5), chosen for the low transverse diffusion of 126 $\mu m/\sqrt{cm}$ in the 1 T magnetic field at 70 V/cm. The drift velocity was ~26 μm/ns, and the longitudinal diffusion, D_L, was $\sim$ 479 $\mu m/\sqrt{cm}$, calculated by Magboltz.[12] The transverse resolution, ϵ_T, was measured in the original analysis: at zero drift distance it was found that $\epsilon_{T_0} = 50 \pm 2$ μm.[9]

2.2. *DESY*

In November of 2006, the COSMo prototype TPC was again tested at DESY in Hamburg, Germany, where a magnetic solenoid provided fields up to 5 T, comparable to the ILD-TPC proposal. Cosmic ray data was collected over a period of four weeks. The gas mixture was the so-called T2K gas, Ar:CF_4:C_4H_{10} (95:3:2), which is considered a possible candidate for the ILC TPC. It has an electron drift velocity of 74 μm/ns at a moderate 200 V/cm electric field, and a large $\omega\tau \sim$ 20 at 5 T, which reduces transverse diffusion to $D_T \simeq 19$ $\mu m/\sqrt{cm}$. The resulting longitudinal diffusion was 248 $\mu m/\sqrt{cm}$. Transverse resolution was measured to be independent of z: $\epsilon_T(z)_{[B=5\ T]} \simeq \epsilon_{T_0} \simeq 50$ μm.[10]

3. Time Analysis

The z-coordinate of a track point in a TPC is reconstructed from the drift velocity and the drift time of ionization electrons to a row of anode pads, thus one requires a consistent method of measuring the arrival time of charge on a pad. Since the spatial distribution of drifting electrons is primarily determined by diffusion, the arrival time distribution of electrons must be about Gaussian, which is mirrored by the current induced in the readout pads - the peak of the current pulse corresponding to the *mean* arrival time of electrons. The charge pre-amplifiers integrate the current, therefore we fit the leading edge of the raw charge pulse, $Q(t)$, with an Error function of the form:

$$Q(t) = \int I(t)\,dt \approx \frac{1}{\sigma_{\mathrm{L}}\sqrt{2\pi}} \int_0^t exp\left[-\frac{1}{2}\left(\frac{t-t_0}{\sigma_{\mathrm{L}}}\right)^2\right] dt\,, \tag{1}$$

where the mean arrival time is obtained directly from the fit parameter t_0. The parameter σ_{L} is the characteristic rise-time of the pulse, and therefore should be proportional to the spatial width of the charge distribution in the z direction.

The pulse fitting method is equivalent to obtaining the Gaussian current pulse by numerical differentiation of the charge pulse, however we found the process relatively slow and more sensitive to baseline noise in the signal, so it was abandoned. In the past it has been commonplace to take the arrival time as the time that the pulse initially begins to rise - the start time, rather than the mean time. We believe, however, that the mean time to be a better choice for the arrival time because it is determined by more electrons.

The fitting procedure is repeated for each primary pulse (the pulse with the greatest amplitude) from each of the 7 pad rows of the readout, after which a linear least-squares fit is performed over all rows which yields the dip angle, θ, and the z-position of the track through the TPC for a given drift velocity.

To measure the longitudinal resolution, so-called *inclusive* residuals from the initial track fit over all rows are added to a histogram. Without knowledge of the true track positions, the fit is repeated, each time with a single row removed, resulting in *exclusive* residuals which are added to a separate histogram. An unbiased estimate of the resolution is then:[13]

$$\epsilon = \sqrt{\epsilon_{inc}\,\epsilon_{exc}}\,, \tag{2}$$

from the geometric mean of the standard deviations of the inclusive, ϵ_{inc}, and exclusive, ϵ_{exc}, residual distributions.

4. Results of Time Resolution

The expected z-dependence of the longitudinal resolution is given by:

$$\epsilon_{\mathrm{L}}(z) = \sqrt{\epsilon_{\mathrm{L}_0}^2 + \xi\, z}\,, \tag{3}$$

where ϵ_0 is the resolution at zero drift distance, and ξ is a fit parameter equal to $D_{\mathrm{L}}^2/\left(v_{\mathrm{drift}}^2\, N_{\mathrm{eff}}\right)$. N_{eff} is the effective number of electrons contributing to

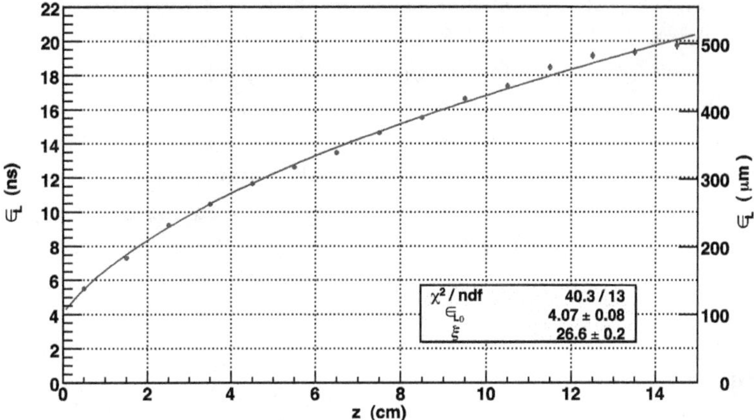

Fig. 1. Single-hit time resolution plotted against drift distance, z, in 1 cm wide bins for KEK. The fit with the expected behaviour due to longitudinal diffusion and electron statistics is given by Equation (3).

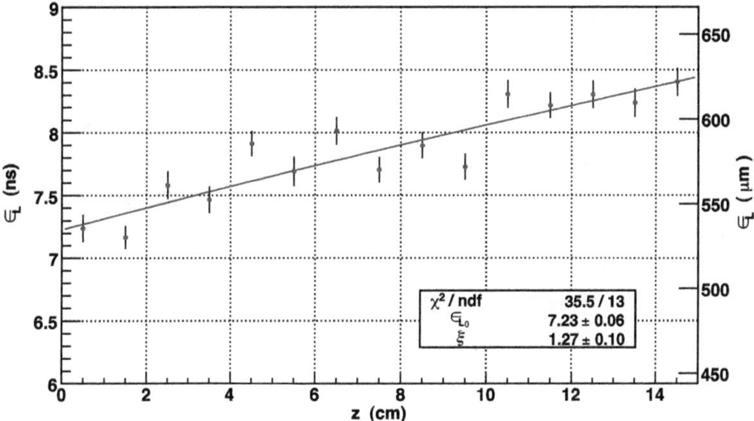

Fig. 2. Single-hit time resolution plotted against drift distance, z, in 1 cm wide bins for DESY. The fit with the expected behaviour due to longitudinal diffusion and electron statistics is given by Equation (3).

each point measurement over the length of a pad and it is obtained from the resolution fit using the longitudinal diffusion constant $D_{\rm L}$ as calculated by Magboltz.[12]

Figures 1 and 2 show plots of the resolution against z for both KEK and DESY data samples. Both plots are fit with Equation (3), which gives $\epsilon_{\rm L_0} = 4.1 \pm 0.1$ ns (100 μm) for KEK, and 7.2 ± 0.1 ns (550 μm) for

DESY. The superior resolution in the KEK test is expected because of the 3× slower drift velocity 26.8 μm/ns. N_{eff} is approximately 12 electrons for KEK and 9 electrons for DESY; this difference is probably due to the higher energy loss of a 4 Gev π^+ compared to a cosmic μ^- (a minimum ionizing particle), passing through argon gas. The curve fit allows for an extrapolation of the resolution to larger drift distances. For the maximum drift length of the proposed ILD-TPC, using T2K gas we could expect to achieve $\epsilon_{\text{L}}(2\,\text{m}) \simeq 17.5$ ns (1300 μm).

5. Summary and outlook

We have made the first measurements of the time resolution of a prototype MPGD-TPC with charge dispersion readout, which is a possible candidate for the future ILD-TPC for the International Linear Collider. A new algorithm was developed for measuring the average arrival time of charge on a pad, by fitting the leading edge of the unshaped Flash-ADC pulse with an error function.

The longitudinal (time) resolution was measured using the geometric mean of inclusive and exclusive residual distributions from track fits. For the KEK 4 GeV π^+ beam test, with an argon-isobutane gas mixture (95:5) in a 1 T magnetic field, we found the zero drift time resolution to be $\epsilon_{\text{L}} = 4.1 \pm 0.1$ ns (100 μm). For the DESY 5 T cosmic ray test using T2K gas, a possible candidate for the ILD-TPC, we measured $\epsilon_{\text{L}_0} = 7.2 \pm 0.1$ ns (540 μm); an extrapolation to 2 m gives $\epsilon_{\text{L}}(2\,\text{m}) \simeq 17.5$ ns (1300 μm). Both are consistent with the requirements of the proposed ILD-TPC for the future International Linear Collider.

6. Acknowledgments

This article builds upon the careful and detailed work of M.Sc. student R. Woods, as well as co-workers D. Attié, K. Boudjemline, P. Colas, M. Dixit, I. Giomataris, J.-P. Martin, E. Rollin, K. Sachs, Y. Shin, and S. Turnbull. The author has been financially supported in Canada by the Canada Research Chair (CRC) Program, and the Canadian Foundation for Innovation (CFI). The research was supported by a project grant from the Natural Science and Engineering Research Council of Canada (NSERC). Partial support by DESY is also gratefully acknowledged.

References

1. D.R. Nygren, A time projection chamber, Presented at 1975 PEP Summer Study, PEP 198, 1975 and included in Proceedings.
2. ILD Concept Group, The International Large Detector: Letter of Intent (2010). URL http://arxiv.org/abs/1006.3396v1.
3. C.K. Hargrove, et al., The Spacial Resolution of the Time Projection Chamber at TRIUMF, Nucl. Instr. and Meth. in Phys. A 219 (1984) 461-471.
4. F. Sauli, Nucl. Instr. and Meth. in Phys. A 386 (1997) 531-534.
5. Y. Giomataris, et al., Nucl. Instr. and Meth. in Phys. A 376 (1996) 29-35.
6. M. Dixit et al., Nucl. Instr. and Meth. in Phys. A 518 (2004) 721-727.
7. M. Dixit and A. Rankin, Nucl. Instr. and Meth. in Phys. A 566 (2006) 281-285.
8. M. Dixit, JINST 5 P03008 (2010).
9. K. Boudjemline et al., presented at IEEE NSS/MIC Conf, San Diego, California, Oct. 2006.
10. M. Dixit et al., Nucl. Instr. and Meth. in Phys. A 581 (2007) 254-257.
11. M. Dixit, Review talk presented at the International Workshop on Future Linear Colliders (LCWS11), Grenada, Spain, September 26-20, 2011.
12. S. Biagi, Magboltz 2, version 8.6 (2009) CERN library.
13. R. Carnegie et al., Nucl. Instr. Meth. in Phys. A 538 (2005) 372-383.

Construction and Performance of full scale GEM prototypes for future upgrades of the CMS forward Muon system

M. Abbrescia, A. Colaleo, G. de Robertis, F. Loddo, M. Maggi, S. Nuzzo, S. A. Tupputi
Politecnico di Bari, Università di Bari and INFN Sezione di Bari - Bari, Italy

Y. Ban, J. Cai, H. Teng
Peking University - Beijing, China

A. Mohapatra, T. Moulik
NISER - Bhubaneswar, India

A. Gutierrez, P. E. Karchin
Wayne State University - Detroit, USA

L. Benussi, S. Bianco, S. Colafranceschi, D. Piccolo, G. Raffone, G. Saviano
Labortori Nazionali di Frascati INFN - Frascati, Italy

D. Abbaneo, P. Aspell, S. Bally, J. Bos, J. P. Chatelain, J. Christiansen, A. Conde Garcia, E. David, R. De Oliveira, S. Duarte Pinto, S. Ferry, F. Formenti, A. Marchioro, H. Postema, A. Rodrigues, L. Ropelewski, A. Sharma, N. Smilkjovic, M. Zientek
Physics Department, CERN - Geneva, Switzerland

A. Marinov, M. Tytgat, N. Zaganidis
Department of Physics and Astronomy Universiteit Gent - Gent, Belgium

K. Gnanvo, M. Hohlmann, M. J. Staib
Florida Institute of Technology - Melbourne, USA

G. Magazzu, E. Olivieri, N. Turini
INFN Sezione di Pisa - Pisa, Italy

K. Bunkowski, T. Fruboes
Warsaw University - Warsaw, Poland

In the prospect of an upgrade of the CMS Experiment, an international collaboration is performing feasibility studies on employing large-area triple-GEM detectors for the high-η region (1.6-2.4) of the CMS Endcap, which is currently not instrumented. Given their good spatial resolution, high rate capability, and radiation hardness, these micro-pattern gas detectors are an appealing option for simultaneously enhancing muon tracking and triggering capabilities in this region. A detailed review of the development and characterization of small and full-size (1m x 0.5m) prototypes will be presented. These full-size GEM foils are produced using a novel single-mask etching technique developed at CERN. In addition, we discuss the performance of a full-size trapezoidal triple-GEM detector in a strong magnetic field during a dedicated beam test campaign to address CMS requirements on the detectors.

Keywords: Tracker gems muon CMS endcap upgrade.

1. Introduction

The GEMs for CMS collaboration is performing feasibility studies about the possible instrumentation of the high-η region of the CMS Endcap. While in the CMS[1] barrel the muon system relies on Drift Tubes (DT) and Resistive Plate Chambers[2] (RPC), the Endcaps are instrumented with Cathode Strip Chambers (CSC) and RPCs. RPCs provide excellent time resolution and contribute to the muon trigger making the system very robust and redundant. In the Endcaps, the RPCs instrument the region extending to $\eta = 1.6$. RPC detectors were initially foreseen for the high-η region, but there have been several concerns about the performance of RPC detectors in such a hostile environment. In this scenario Gas Electron Multipliers (GEMs)[3] are an appealing technology and the GEMs for CMS collaboration is carrying out detailed studies to equip the vacant high-η region with GEMs as in Figure 1. In this feasibility study, as foreseen in the RE11 project, each chamber will cover 10°; also the collaboration is working on the possibility to install two chambers face-to-face to dramatically improve the tracking resolution.

In the high-η region ($|\eta| > 1.6$) particle rate (kHz/cm^2), the low energy photons, the thermal neutrons and the magnetic field constitute strong requirements that must be fulfilled. The GEMs for CMS collaboration is proposing GEMs for the CMS upgrade since this technology can stand the hostile described environment, providing precision tracking and fast trigger information simultaneously. This means that GEMs could improve the muon trigger efficiency and additionally reconstruct tracks as well. At the moment CMS is actually lacking robustness at high-η because of missing redundancy, GEMs could meet this weakness and improve the overall muon system.

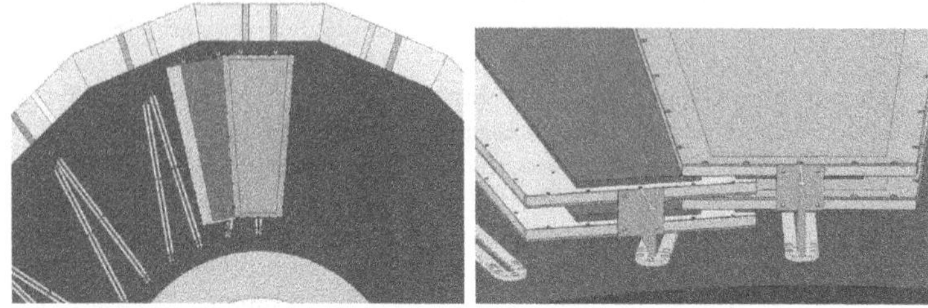

Fig. 1. The GEMs for CMS collaboration is studying the detector integration in the CMS Endcap high-η region. Each chamber will cover an area of 10° for a total amount of 36 detectors. A possible idea under investigation is to double the number of detectors installing a second layer of chambers face-to-face to improve the tracking resolution.

2. Work summary

Since 2009 this collaboration has been working on small[4] and full-size[5] prototypes performing feasibility studies, based on GEM detectors, for a possible future upgrade of the CMS high-η region. In 2009 the collaboration started designing and testing small 10×10cm GEM chambers while already in 2010 the first full-size detector was ready and tested in the SPS-H4 muon (150GeV) beam showing excellent performances. In 2011 still a lot of work has been done on small prototypes trying new stretching technique and new readout electronics; the final goal was to build a real candidate for the CMS high-η region. The state-of-the-art of full-size chambers is represented by the so-called GE11_II (March 2011) which is an enhanced version with respect of GE11_I (April 2010). Both GE11_I and GE11_II share the same geometry (trapezoid with dimensions 990mm $\times$ $(220 - 455)$mm) since they are thought to fit into the same CMS Endcap slot. The GEM foil production relies, as usual, on the photolithographic processes developed at CERN using 50μm thick kapton sheet with 5μm copper clad on both sides; the single mask technique is used to overcome the alignment problems which become critical once a GEM foil dimension exceeds 40cm. The collaboration is investigating the possibility to produce foils outside of CERN, specifically for the future mass production. Preliminary results on small prototypes produced outside of CERN are positive and promising.

The GE11_II drift electrode, identical to the GE11_I, is part of the aluminum chamber envelope itself and it is produced by gluing a 300μm kapton layer with 5μm copper cladding to a 3mm aluminum plate. The readout electronics adopted is the VFAT[6] chip for both GE11_I and GE11_II, while the amount of channels increased from 1024 to 3184 using a reduced strip pitch (varying from 0.6mm to 1.2mm).

The collaboration has been working heavily on the electronic readout system in parallel with the detector construction; several constituted groups are coordinated to develop the electronic future readout system CMS compliant. Several architecture are under study as the design of a new chip with enhanced features with the respect to the VFAT2. Also in August beam test the collaboration decided to take data also with the APV chip and the Scalable Readout System (SRS) developed and supported by the RD51 collaboration. This readout system will be used next year for testing new and old prototypes in the RD51 - CMS beam tests replacing the TURBO front-end electronics developed by INFN (Siena and Pisa) used so far.

The other very important challenging parameter that has been tuned, in the GE11_II, is the gap configuration (drift, transfer 1, transfer 2, induction gap size): 3/1/2/1 mm. This had an impact on the HV divider that had to be modified, also because due to severe timing requirements we optimized fields and GEM voltages according to the optimal values obtained from the 2010 test beam. The divider that has been built is still differing from the designed one due to the lack of HV SMD resistors on the market. To overcome such problems, the final solution is to produce a ceramic HV divider with desired resistor values; moreover this permits to have a very compact design which will allow one to optimize also the cooling system for the divider itself. The collaboration is going to build soon a ceramic divider for the GE11_II. As for the GE11_I, also for the GE11_II each foil was thermally stretched using a special oven with temperature of 37°C for 24 hours; in particular, for the latter the stretching was challenging since transfer gap and induction gap were just 1mm. Once the stretching process is completed the foil is ready to be glued together with the frame, while the last step is the glue curing, again performed using the same oven. In every step of the process GEM foils are tested with careful optical inspections and sector-by-sector HV test in the dedicated clean room.

3. Preliminary results

The GE11_II was tested in the June-July, August and September beam test at the H8 and H4 muon beam line at the SPS CERN. In the first campaign also a strong magnetic field, provided by the CMS magnet M1, was used in order to validate the detector in such environment similar to the CMS Endcap high-η. Figure 2 shows a preliminary HV and threshold scan with $Ar : CO_2$ 70:30. In 2011 test beams, as usual, we have used several gas mixtures and different configurations, data analysis is still ongoing.

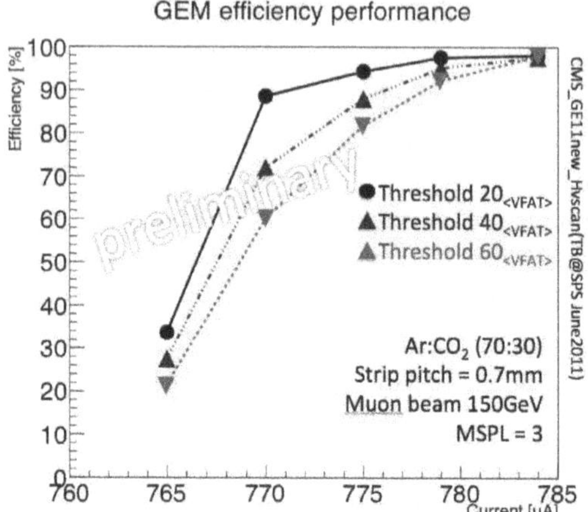

Fig. 2. The new GE11_II prototype performance during the June test beam at the RD51 SPS-H4 beam line.

4. Summary and outlook

Concerning the construction techniques the GEMs for CMS collaboration made a significant improvement building the world's largest GEM based full-size detector (GE11_II) with a transfer 1, induction gap size of 1mm and optimized HV divider. Moreover the collaboration acquired deeper knowledge about the detector performance thanks to several focused beam periods during 2011. Data-analysis is still ongoing due to the large amount of data taken.

References

1. CMS Collaboration, "The CMS experiment at the CERN LHC", JINST **3** (2008) S08004.
2. R. Santonico, "RPC: Status and perspectives", *In *Pavia 1993, Proceedings, The resistive plate chambers in particle physics and astrophysics* 1-11*
3. F. Sauli, "GEM: A new concept for electron amplification in gas detectors", Nucl. Instrum. Meth. A **386**, 531 (1997).
4. D. Abbaneo *et al.*, "Characterization of GEM Detectors for Application in the CMS Muon Detection System", 2010 IEEE proceeding, 10.1109/NSS-MIC.2010.5874006
5. D. Abbaneo *et al.*, "Construction of the first full-size GEM-based prototype for the CMS high-η muon system,", 2010 IEEE, 10.1109/NSSMIC.2010.5874107
6. P. Aspell *et al.*, "The VFAT production test platform for the TOTEM experiment".

MICRO-RADIOGRAPHY OF LIVING BIOLOGICAL ORGANISMS WITH MEDIPIX2 DETECTOR AND APPLICATION OF VARIOUS CONTRAST AGENTS

JIRI DAMMER, VIT SOPKO AND JAN JAKUBEK

Institute of Experimental and Applied Physics, Czech Technical University in Prague
Horska 3a/22, CZ-12800 Prague 2, Czech Republic
e-mail:jiri.dammer@utef.cvut.cz

FRANTISEK WEYDA

Biological center of the Academy of Sciences of the Czech Republic,
Institute of Entomology, Branisovska 31, CZ-37005 Ceske Budejovice, Czech Republic

JIRI BENES

Charles University in Prague, First Faculty of Medicine,
Salmovska 1,CZ-120 00 Prague 2, Czech Republic

JULIAN ZAHOROVSKY

University Hospital Na Bulovce, Department of Radiological Physics,
Budinova 2, CZ-180 81 Prague 8, Czech Republic

We describe a newly developed radiographic system equipped with Medipix2 semiconductor pixel detector and a micro-focus FeinFocus X-ray tube tabletop. The detector is used as an imager that counts individual photons of ionizing radiation, emitted by the X-ray tube. The digital pixel detectors of the Medipix family represent a highly efficient type of imaging devices with high spatial resolution better than 1μm, and unlimited dynamic range allowing single particle of radiation and to determine their energies. The setup is particularly suitable for radiographic imaging of small biological samples, including in vivo observations with various contrast agents (iodine and lanthanum nitrate). Along with the description of the apparatus we provide examples of application of iodine and lanthanum nitrate contrast agents as tracers in various insects as model organisms. The iodine contrast agent increases the absorption of X-rays and this leads to better resolution of internal structures of biological organisms, and especially the various cavities, pores, etc. Micro-radiographic imaging helps to detect organisms living in a not visible environment, visualize internal biological processes and also to resolve the details of their body (morphology). Tiny live insects are an ideal object for our studies.

1. Introduction

Micro-radiography is an imaging technique using X-rays in the studies of internal structures of objects. This fast and sensitive imaging tool is based on differential X-ray attenuation by various tissues and structures within the biological sample. The non-absorbed radiation is detected with a suitable detector and creates a radiographic image. In order to detect the differential properties of X-rays passing through structures of various compositions in the sample, a high resolving power imaging detector is needed. At the same time, a high spatial resolution of the resulting image is required.

The digital pixel detectors of the Medipix[a] family represent a highly efficient imaging device with a high scanning speed, high density of detection units (pixels) resulting in good spatial resolution and unlimited dynamic range allowing to detect single particles of ionizing radiation and to evaluate their energy.

1.1. *Detector Medipix2*

In the semiconductor pixel detector, such as Medipix2 detector, the ionization charge is compared using the noise suppression effect with a certain discrimination level (threshold). If the charge is larger than this threshold, a digital pulse is created and recorded by a digital counter. The advantage of these detectors is that the digital information contained in the counter does not change in time; therefore, the image does not suffer with the problem of dark current. For the setting of the discrimination level, only particles of certain energies are counted and the others can be ignored. Exposure time of these detectors is unlimited, allowing us to reach a high signal to noise ratio, and thereby also images of a high contrast and quality[1].

The Medipix2 detector consists of two chips: the sensor chip and CMOS chip counting electronics. The sensor is a standard semiconductor detector (Si, GaAs, CdTe, HgI). The rear contact sensor is divided into a matrix of 256 × 256 cells (pixels) with the edge of 55 μm. The size of the active detector area is 14.11 × 14.11 mm (Fig. 1). Counting chip electronics is connected to the sensor using a bump-bonding technology. This chip is equipped with a complete data acquisition route consisting of preamplifiers, one channel analyzer and 13-bit counter for each pixel sensor[2]. Data can be transferred from the Medipix device via a 32-bit serial or parallel interface. Due to the high-speed

---

[a] The Medipix2 detector was developed at CERN by the Medipix Collaboration. The Institute of Experimental Physics of the Technical University in Prague participates on the R&D and applications.

communication (up to 100 MHz for the last type), the acquisition time of one image can be as fast as 10-32 miliseconds[1].

1.2. *The X-ray μ-imaging system*

Our X-ray micro-radiographic setup is using a geometric magnification of the object, which is obtained due to the use of a point X-ray source. In the case of Feinfocus tungsten Tube, with copper or molybden anode[b], the spot size is of less than 1 μm, operating voltage 10-140 kV, and current of 10-1000 mA. The Medipix2 detector is used as an imager whith detection threshold is adjustable from above 5 keV[3]. The whole setup for X-ray micro-radiography is shown in Fig. 1. Exposure time of one frame is from 100 ms to 10s, depending on the type of the object, working distance, beam parameters and detector settings. The quality of radiographic images strongly depends on the beam hardening. The correction of the acquired data consists of the calibration in each detector pixel at different levels of hardening of the spectrum with a beam hardening of attenuating aluminium[4].

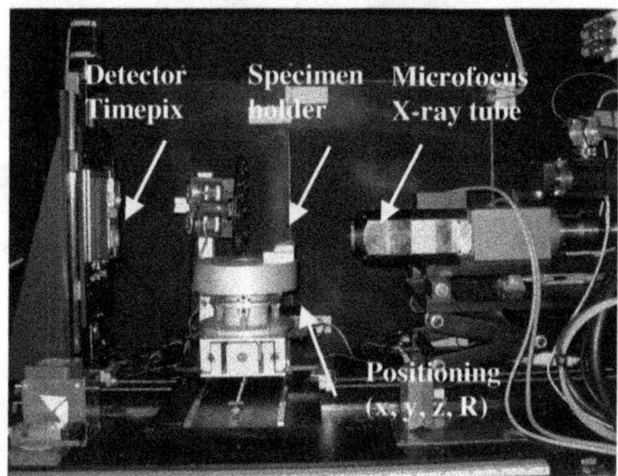

Fig.1. Experimental setup for X-ray microradiography and microtomography: microfocus FeinFocus X-ray tube, calibration carrousel, automatic sample holder and Medipix2

2. Results

Our micro-radiographic system can be used for the examination of internal structures and processes in biological objects where iodine or lanthanum nitrate is used as tracer, including in vivo imaging. The opportunity of the in vivo observation allows a real-time observation of physiological events or a long-term observation of dynamic processes (Fig. 2).

---

[b]Anode can be changed as needed. The choice of the anode depends on the type of the imaged material.

Application of contrast agents increases the absorption of X-rays and this leads to better resolution of internal structures of biological organisms, and especially the various cavities, cores, etc. These structures appear with higher contrast and are therefore more visible (Fig. 3). In our case we used a commercially available contrast agent (Optiray). More information on use of iodine in biology can be found in ref. 5. Lanthanum nitrate solution as tracer has been tested for myelinated nerve fibers in a vivo study – e.g., see in ref. 6. Later, several authors use it as electron-dense tracer in transmission electron microscope studies (TEM) of tissue permeability. In micro-radiographic studies of fire bug we found mild commutation of lanthanum in some parts of body and malpighian tubules (excretory organs) as shown in (Fig. 4). But we need to continue in these studies for more data.

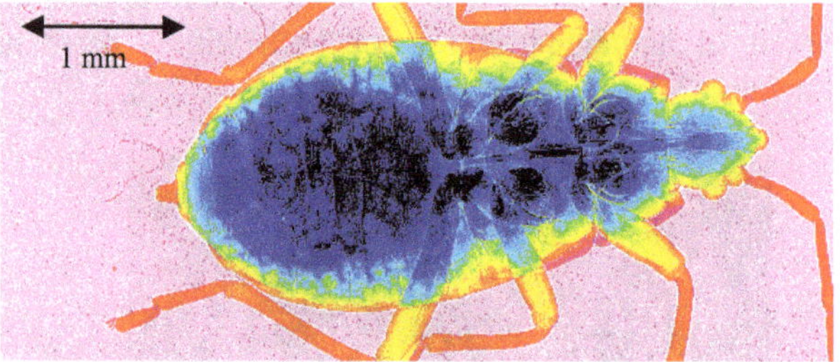

Fig.2. Micro-radiography of imago of fire bug, Pyrrhocoris apterus. Control specimen.

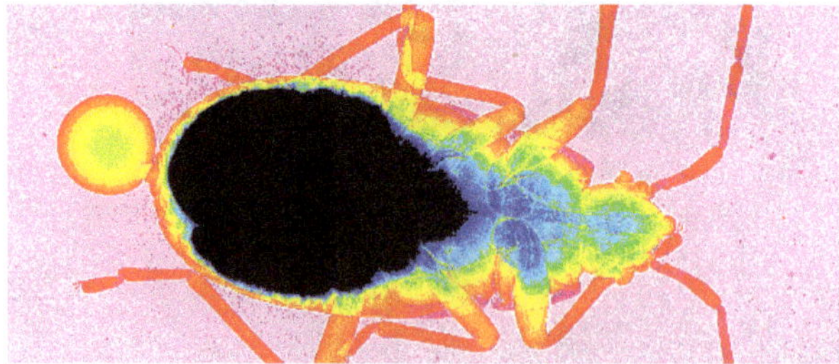

Fig.3. The fire bug has been for 20 hrs in very low concentration of contrast solution Optiray based on iodine in fresh water. Whole abdomen reveals strong contrast.

3. Conclusions

The described experimental setup, based on a semiconductor pixel X-ray detector Medipix2, stands as an appropriate imaging tool for biological studies of internal anatomy of insects. The major advantages of such an approach are as follows: high contrast due to the full dynamic range, high spatial resolution which is limited only by the size of the X-ray source spot. At the same time,

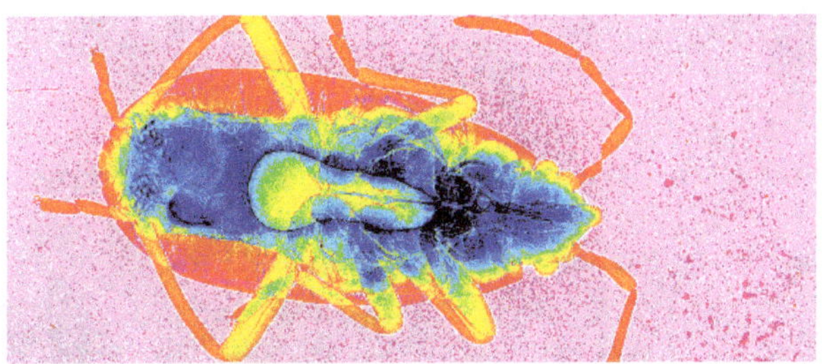

Fig.4. The fire bug has been placing for 20 hrs in very low concentration of contrast agent lanthanum in fresh water. Some parts of abdomen as well as Malpighian tubules reveal low contrast.

the technique is fast, inexpensive and non-destructive, allowing real-time *in vivo* observations of various physiological and/or developmental processes[7]. The application of contrast agents opens new possibilities to perform non-invasive observations of dynamical processes inside living organisms.

Acknowledgments

This work was realized out in frame of the CERN Medipix Collaboration and was supported in part by the Research Grant Collaboration of the Czech Republic with CERN No. 1P04LA211, by the Fundamental Research Center Project LC06041 and the Research Programs 6840770029 and 6840770040 of the Ministry of Education, Youth and Sports of the Czech Republic.

References

1. X. Llopart et al, *IEEE Trans. Nucl. Sci.* **49**, 2279–2283 (2002).
2. P. Frallicciardi et al., Conf. Proc. IEEE NSS/MIC 2008, M10-112 (2008).
3. J. Jakubek, *Nucl. Instrum. Methods Phys. Res* **A 576**, 223-234 (2007).
4. J. Jakubek et al, *Nucl. Instrum. Methods Phys. Res.* **A. 563**, 278-271(2006).
5. Iodine in biology. Wikipedia. http://en.wikipedia.org/wiki/Iodine_in_biology#cite_note-34
6. M.L.Mackenzie et al, J.Anat. **138**, 1-14 (1984).
7. J. Dammer, F. Weyda, V. Sopko, J. Jakubek, JINST **6**, 1-6 (2011).

Tracking with Straw Tubes in the $\overline{\text{P}}$ANDA Experiment

P.Gianotti*, V.Lucherini and E.Pace

INFN, Laboratori Nazionali di Frascati
Via E. Fermi 40, 00044, Frascati, Italy
**E-mail: paola.gianotti@lnf.infn.it*

K.Kozlov, H.Ohm, S.Orfanitski, M.Mertens, J.Ritman, M.Roeder, V.Serdyuk, P.Wintz

IKP1 Forschungszentrum Jülich GmbH
Wilhelm-Johnen-Strasse, 52428 Jülich, Germany

M.Idzik and D.Przyborowski

AGH University of Science and Technology
30 Mickiewicza Av., 30-059 Krakow, Poland

S.Jowzaee, M.Kajetanowicz, G.Korcyl, P.Salabura, J.Smyrski

Jagiellonian University
ul. Golebia 24, 31-007 Krakow, Poland

P.Kulessa, K.Pysz

Institute of Nuclear Physics PAN
ul. Radzikowskiego 152, 31-342 Krakow, Poland

G.Boca, S.Costanza, L.Lavezzi, P.Montagna, A.Rotondi

Università di Pavia and INFN,
Via A. Bassi 6, 27100 Pavia, Italy

M.Savriè

Università di Ferrara and INFN
Via Saragat 1, 44122 Ferrara, Italy

O.Levitskaya and A.Kashchuk

PNPI RAS Gatchina
Leningrad district 188300, Russia

Tracking charged particles is one of the essential tasks of the $\overline{\text{P}}$ANDA experiment, providing information about primary and secondary decay vertices, momenta and types of charged particles emitted after antiproton-proton annihilation. Different tracking devices are under construction for the $\overline{\text{P}}$ANDA spectrometer and among them the two straw tube trackers. A new technique, based on the use of straw tubes operated at over-pressure has been adopted allowing the construction of self-supporting modules avoiding heavy mechanical frames.

Keywords: Gaseous detector; straw tubes; hadronic physics.

1. $\overline{\text{P}}$ANDA straw tubes

$\overline{\text{P}}$ANDA is a new experiment that will be installed at HESR, the new antiproton storage ring under construction as part of the new FAIR facility at Darmstadt, Germany.[1] This experiment will investigate QCD in the charmonium mass range, and other aspects of particle and nuclear physics. It will be a fixed target detector with a Target Spectrometer (TS), surrounding the interaction point, and a Forward Spectrometer (FS) for detecting particles emitted at forward angles.

Tracking charged particles is one of the essential tasks of $\overline{\text{P}}$ANDA and for this reason different detectors will be used. Among them $\overline{\text{P}}$ANDA will have 2 straw tube trackers: the first, located in the TS, is a cylindrical device; the second, in the FS, consists of a set of planar detectors. Independently from their layout, the 2 systems will use self-supporting straw tube modules made of similar straws of 1 cm diameter. The cathode material is a thin aluminized mylar film (thickness 27 μm) with a gold-plated tungsten-rhenium wire, of 20 μm diameter, as anode. The anode wire is stretched by a weight of 50 g and crimped in copper, gold-plated pins. Cylindrical precision end-plugs, made from ABS with a wall thickness of 0.5 mm, close the tubes at both ends and provide the pin and the gas pipe housing. They are glued to the mylar film leaving a small 1.5 mm film overlap on both ends. There, a gold-plated copper-beryllium spring wire is inserted to provide either the cathode grounding and to allow a 2 mm tube elongation induced by the over-pressure of the gas mixture. The total weight of a fully assembled straw is only 2.5 g. The straw tubes are operated with an over-pressure of about 1 bar to allow the construction of self-supporting modules, a technique that has been developed and successfully implemented for the first time for the COSY-TOF straw tube tracker.[2] In the following sections the details of the designs of the $\overline{\text{P}}$ANDA Central and Forward straw tube trackers are presented together with experimental results of the R&D phase.

2. Self-supporting straw tube modules

Straw tubes are single channel drift detectors consisting of an anode wire surrounded by a thin cylindrical cathode. With a wire tension[a] of about 50 g inside a 1.5 m long straw tube, for the 4636 straws of the $\overline{\text{P}}$ANDA central tracker this adds up to a wire tension equivalent to about 230 kg which must be maintained. Usually, this is done by fixing the straw tubes inside a strong and massive support frame or by adding reinforcement structures. These methods inevitably increase the detector thickness given in radiation length and are not acceptable for the $\overline{\text{P}}$ANDA experiment. Therefore, it has been decided to adopt a technique based on self-supporting straw layers, with intrinsic wire tension, developed for the COSY-TOF straw tracker.[2] The straw tubes are assembled and the wire is stretched by 50 g with an over-pressure of 1 bar. Then the tubes are close-packed and glued together to planar multi-layers on a reference table which defines a precise tube to tube distance of 10.1 mm. At the gas over-pressure of 1 bar the double-layer not only maintains the nominal wire tension but it also become self-supporting.

Figure 1 shows a pressurized straw multi-layer. The system is supporting the weight of a lead brick of 3 kg.

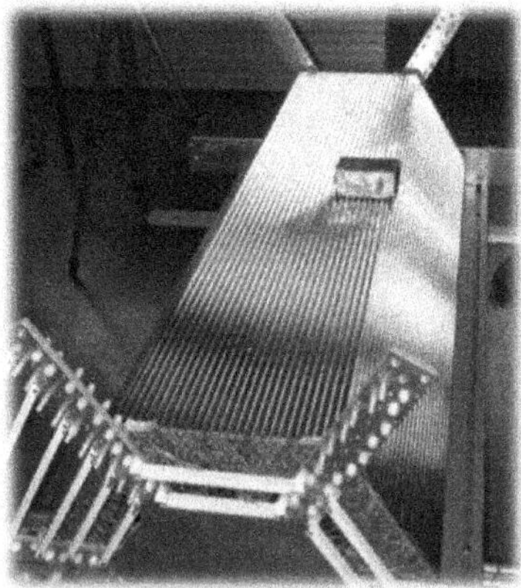

Fig. 1. Pressurized, close-packed straw layers show strong rigidity as demonstrated here by a 3 kg Pb-brick.

The gas mixture used is Argon based with 10% CO_2 as quencher. This gas mixture is known as being one of the best for high-rate hadronic envi-

[a]Usually given as the mass weight used to stretch the wire.

ronments due to the absence of polymeric reactions of the components once there is a clean gas environment including all materials and parts of the detector and gas supply system. The HV is set to have a gas gain not greater than 10^5 in order to warrant long term operation. With these parameters, a spatial resolution, in the $r - \phi$ plane, of less than 100 μm is expected. This value has been extrapolated from the measured value obtained with straw tube prototypes of 1 m length operated at an over-pressure of 250 mbar (see Fig. 2).

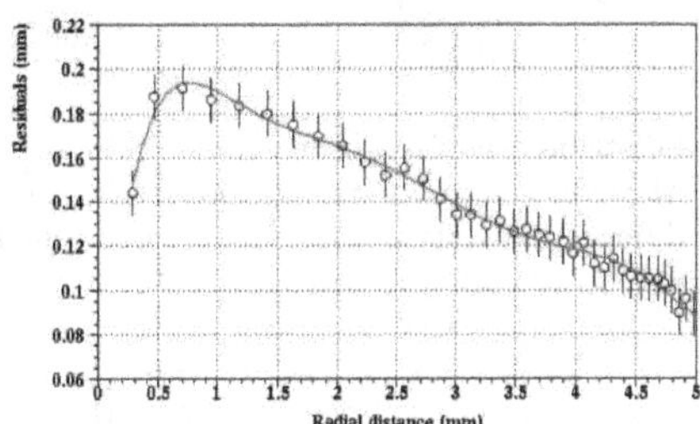

Fig. 2. Spatial resolution measured with self-supporting straw tubes of length 1 m, operated at 1.250 bar.

3. Energy loss measurements

A special requirement for the $\overline{\text{P}}$ANDA straw tubes is that of helping in the process of particle identification in the TS for particle momenta below 1 GeV/c. This is something not routinely done in other straw tube detectors, therefore this possibility has been deeply studied with Montecarlo simulations and experimental tests. Figure 3 (left) shows the distribution of the simulated specific energy losses for different particles, plotted versus the momentum. The radial path has been reconstructed by the measured drift radius and by the dip angle resulting from the fit. The dE/dx has been calculated with the truncated mean at 30% in order to cut out the higher dE/dx tails. Figure 3 (right) reports the separation power of the central straw tube tracker for the different particle species; particle's identification is feasible below 0.8 GeV/c momenta.

To test experimentally these results a dense array of 128 straw tubes, arranged in four double-layers of 32 straws each, has been exposed to the proton beam of the COSY synchrotron of the Jülich Research Center.[3] Straw signals have been fed into a 16-ch 240 MHz flashADC. For each event

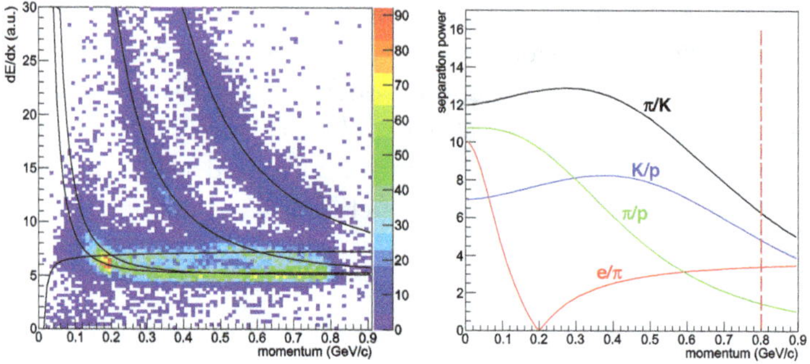

Fig. 3. (left) Simulated distribution of dE/dx vs momentum for different particles. The superimposed lines are the mean value of the bands. (right) Separation power in the straw tube detector for different particle species. The vertical line at 0.8 GeV/c indicate the threshold below which identification is feasible.

the signals from the fired straws have been summed up to build the energy loss distribution and, as for the simulations, the truncated mean technique has been applied to cut out high energy tails. The energies of the truncated distributions have then to be divided by the appropriate reconstructed track lengths. Truncated energy loss distributions for different proton momenta are shown in Fig. 4 for 2.95, 1.0 and 0.64 GeV/c momenta. The results for

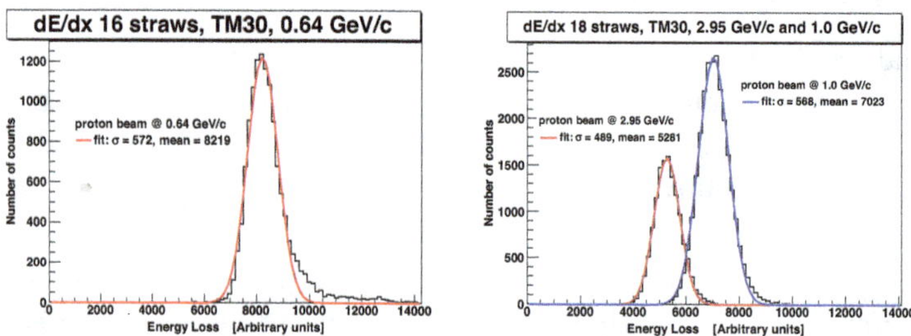

Fig. 4. (left) dE/dx distribution for protons of 0.640 MeV/c with Gaussian fit. Only events with 16 straw tubes per reconstructed track are included. (right) dE/dx distributions for protons of 2.95 GeV/c and 1.0 GeV/c with Gaussian fits. Only events with 18 straw tubes per reconstructed track are included. Truncated mean of 30 % is applied to all histograms.

the 0.64 GeV/c protons cannot be presented on the same energy scale of the results of 1.0 and 2.95 GeV/c due to the lower number of hit straws (16 instead of 18), and because a higher threshold was used during the analysis of this data set. The results of these measurements give an energy resolution of 8% for 1.0 GeV/c protons. This improves at lower momenta with the increase of the particle energy deposit. At 0.64 GeV/c, with only 16 straws per track, the resolution is equal to 7%. For tracks inclined by 45° a systematical deterioration of the resolution of about 1% is observed. Nevertheless, it must be pointed out that in the $\overline{\text{P}}$ANDA central tracker the mean value of hit tubes will be around 23 improving energy resolution measurements.

4. The central tracker

The $\overline{\text{P}}$ANDA TS straw tube tracker (CT) occupy a cylindrical volume with an internal diameter of 150 mm and an external one of 418 mm. Due to the presence of the target pipe, in the x, y plane this volume is divided in two halves, with a gap of 42 mm in between. Along z, the allowed space is 1500 mm, plus 150 mm in the upstream region for electronics, gas supplies, and other services. To fill up this volume, it has been decided to use planar layers mounted in a hexagonal shape as shown in Fig. 5. The proposed arrangement consists of 4 double-layers parallel to the detector axis, 4 skewed double-layers, with an angle, with respect to the beam axis, of about $\pm 3^o$, and further 2 straight double-layers. Finally, to approach the cylindrical shape, 7 layers with a decreasing number of straws are placed in the outer region. In total the detector consists of 4636 straws divided in two identical semi-chambers held by two light mechanical frames (8.2 kg in case of Aluminum). The overall detector will result in a material budget of 1.2% of a radiation length.

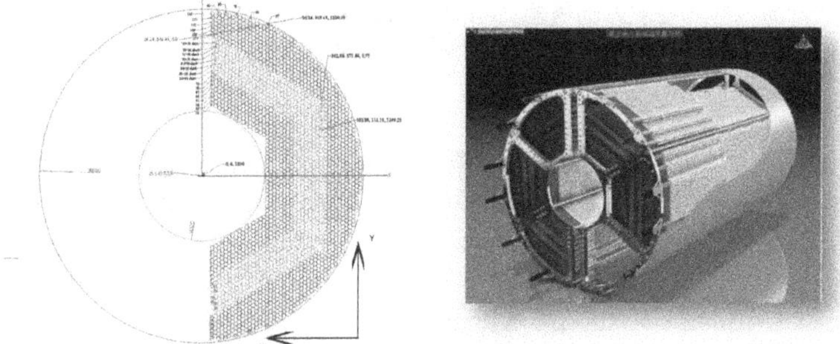

Fig. 5. (left) Arrangement of the straw tubes within the $\overline{\text{P}}$ANDA central tracker. (right) CAD drawing of the whole detector

5. The forward tracker

The particle trajectories in the FS will be measured by the Forward Tracker (FT) consisting by three pairs of chambers made of straw tubes similar to those proposed for the CT. The first pair will be placed in front, the second within and the third behind a dipole magnet. The detection planes are built of separate modules, each containing 32 straws arranged in two layers as shown in Fig. 6. Each of the six stations of the FT contains four double-

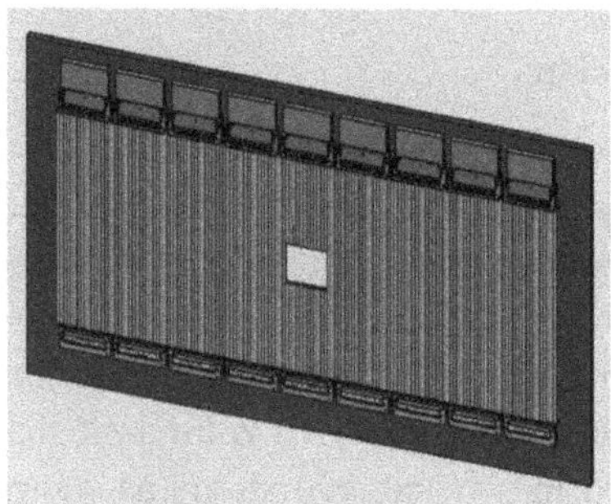

Fig. 6. CAD drawing of the first chamber of the $\overline{\text{P}}$ANDA FT. Nine modules, made of 32-straw-tubes, are mounted on a rectangular support frame. The opening in the middle of the detector is foreseen for the passage of the beam pipe.

layers: the first and fourth layer contain vertical straws (0°) and the two intermediate layers - the second and the third one - contain straws inclined respectively at $+5°$ and $-5°$. This arrangement allows a three dimensional reconstruction of events.

The straw plane of the FS will have a small gap in the center to allow the passage of the beam pipe.

References

1. $\overline{\text{P}}$ANDA (antiProton ANnihilation at DArmstadt) `http://www-panda.gsi.de/`
2. P. Wintz, "A Large Tracking Detector In Vacuum Consisting Of Self-Supporting Straw Tubes", AIP Conf. Proc **698**, (2004) 789-792.
3. COSY is a cooler synchrotron and storage ring for protons `http://www2.fz-juelich.de/ikp/cosy/en/`

Overview of the ATLAS Insertable B-Layer (IBL) Project

J. GROSSE-KNETTER*
on behalf of the ATLAS IBL Collaboration
II. Physikalisches Institut, Universität Göttingen,
Göttingen, Germany
**E-mail: jgrosse1@uni-goettingen.de*

The upgrades for the ATLAS Pixel Detector will be staged in preparation for high luminosity LHC. The first upgrade for the Pixel Detector will be the construction of a new pixel layer which will be installed during the first shutdown of the LHC machine, foreseen in 2013-14. The new detector, called the Insertable B-layer (IBL), will be installed between the existing Pixel Detector and a new, smaller radius beam-pipe. The IBL will require the development of several new technologies to cope with increased radiation and pixel occupancy and also to improve the physics performance through reduction of the pixel size and a more stringent material budget. Two different and promising silicon sensor technologies, planar n-in-n and 3D, are currently under investigation for the IBL. An overview of the IBL project, of the module design and the qualification for these sensor technologies with particular emphasis on irradiation and beam tests is given.

Keywords: ATLAS upgrade; Insertable B-layer; pixel detectors; silicon detectors; radiation hard detectors.

1. Introduction

The Pixel Detector[1] is the innermost layer of the tracking system of the ATLAS Experiment[2] and contributes significantly to the ATLAS track and vertex reconstruction. The detector consists of identical sensor-chip-hybrid modules, arranged in three barrels in the centre and three disks on either side for the forward region. The expected particle fluence of 50 Mrad ionising dose and $10^{15} n_{\mathrm{eq}}/\mathrm{cm}^2$ non-ionising dose in the Pixel Detector during LHC operation will result in a degradation of sensors and electronics performance of these modules. In particular, the innermost layer ("B-layer") is expected to gradually reduce tracking efficiency with consequences for the track and vertex reconstruction of the entire detector.

In order to maintain and improve physics performance until the high-luminosity upgrades of the LHC and the ATLAS detector around 2020, the 3-layer barrel system of the Pixel Detector will be upgraded to a 4-layer system by inserting a new layer between the current B-layer and a new beam pipe. Installation of this "Insertable B-layer" (IBL)[3] is planned for the LHC shutdown 2013-2014 ("phase-1 upgrade"). Due to the smaller radius of the IBL module position around the beam line and the expected enhanced performance of LHC after the phase-1 upgrade, the IBL has to cope with a higher occupancy and an increased particle fluence of 250 Mrad ionising dose and $5 \cdot 10^{15} n_{\mathrm{eq}}/\mathrm{cm}^2$ non-ionising dose. Both aspects require new approaches to sensor and front-end (FE) readout technology, which is described in more detail in Sec. 2.

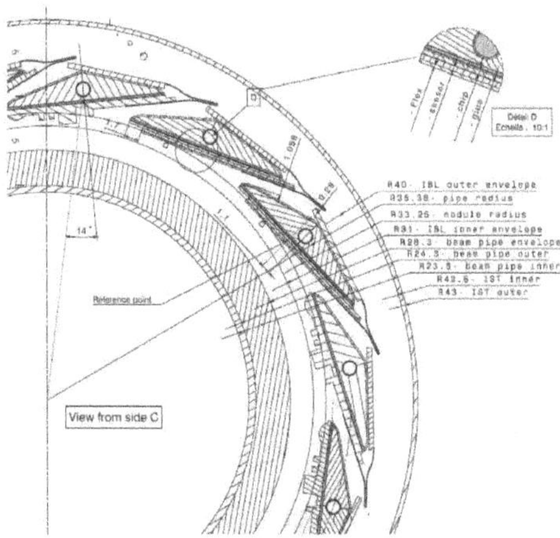

Fig. 1. Cross-sectional view of a section of the IBL with new beam pipe. Radii of the envelopes are given in mm

The general layout of the IBL is shown in Fig. 1. The modules are mounted on 14 staves directly attached to the beam pipe. The radius of the beam pipe will be reduced to 2.4 mm. Due to the inclination of the staves to allow for azimuthal overlap, the sensors are located at a radius between 3.1 and 3.8 mm. Each stave is loaded with sensor-FE hybrids such that it holds 32 FE chips. This corresponds to a total of 448 FE chips on all 14 staves.

2. IBL module layout

IBL considers two sensor technologies: 3D sensors and n-in-n planar sensors. While modules with a 3D sensor will have one FE chip per module, those with a planar sensor will have 2 FE chips per tile. Some of the IBL design requirements for modules are as follows:

- Max. power dissipation $200\,\mathrm{mW/cm^2}$ at $-15°\mathrm{C}$.
- High tracking efficiency of better than 97% after full dose ($5 \cdot 10^{15} n_{\mathrm{eq}}/\mathrm{cm}^2$).
- Minimal active edge, typically around $200\,\mu\mathrm{m}$.

In order to reduce the occupancy per pixel, the pixel size for both module types is reduced to $50 \times 250\,\mu\mathrm{m}^2$, i.e. the long side was shortened compared to the pixel size of $50 \times 400\,\mu\mathrm{m}^2$ as used in the current detector.

2.1. *Readout chip*

The readout chip of the current detector, the FE-I3, was found to be too limited in terms of radiation hardness and readout speed to be operated in the IBL environment. This lead to the design of a new readout chip, the FE-I4.[4,5] The re-design was realised in a smaller feature size, leading to an increased radiation tolerance. Each pixel cell contains a two-stage amplification-shaping circuit followed by a discriminator, similar to the general concept in FE-I3. However, the digital readout inside the chip was largely modified and based on a "4-Pixel Digital Region", in which a 2×2 pixel region in a double-column share a common digital processing stage. This involves hit buffering at the pixel level which largely increased the data transfer speed which in the previous design was limited by data transfer to the end of column for buffering. It also reduces power consumption due to lower digital activity, helping to meet the design criterion on module power dissipation. In addition to the design changes, the FE-I4 is organised in 80 columns by 336 rows, thus covering a larger area of $2.02 \times 1.88\,\mathrm{cm}^2$ and increasing the fractional active area of the chip.

Several 10 wafers of the prototype of FE-I4 were produced and tests demonstrated the correct analogue and digital functionality of the chip. Typically, a yield of 40 out of 60 chips on a wafer is observed.

2.2. *Sensor technologies*

Two sensor technologies are under consideration for IBL, planar and 3D. One important aspect is the stringent requirement on radiation tolerance of the material. In addition, the size of inactive regions at the edge must be minimised in order to keep inefficiencies low because modules are mounted without overlap in the direction of the beam line.

The planar n-in-n sensors are produced on diffusion oxygenated float-zone silicon. The design is similar to that of the current detector,[1] but the extension of the inactive area could be largely reduced to 215 μm by shifting the guard rings partially under the active pixel area.

3D sensors[6] use p^+ and n^+ electrodes that are edged into the p^+ bulk. The sensor is biased between the two electrode types so that the distance for charge collection is reduced. This is beneficial after irradiation to avoid signal loss due to trapping. The IBL 3D design features an inactive guard ring area of approx. 200 μm width.

2.3. *Bump bonding*

In order to reduce the amount of inactive material in the detector, FE chips are thinned to 90 μm. The bump bonding process that connects the sensor pixels with each readout cell of the FE chip involves a heating step in which remaining mechanical stress in the thinned FE results in a bending that is larger than what can be tolerated by the bump bonding process. In a process employed at IZM[a], the thinned FE is attached to a glass support that provides mechanical stability throughout the heating step. After the sensor-FE-assembly is completed, the glass support is removed with a laser.

3. Module test measurements

Prototype modules composed of the FE-I4 prototype and either of the two sensor types were mounted on boards and extensively tested in calibration measurements, source tests, and in testbeams. Several modules were irradiated to the IBL design dose specified above. Irradiations were carried out using thermal neutrons in the TRIGA reactor at Jozef Stefan Institute, Ljubljana, and 26 MeV protons at Karlsruhe Institute for Technology. Irradiation with low-energy protons resulted in a total ionising dose of the devices being about a factor of three larger than the IBL design dose. As a consequence, a small but non-negligible number of pixels in the FE-I4 fail and were excluded from further analysis.

Beam tests are performed[7] at the DESYII synchrotron at DESY, Hamburg, using 4 GeV positrons, and at the SPS at CERN using 180 GeV/c pions. The setup consists of a beam telescope[8] for the position measurement, and several modules under test placed between the telescope modules for best track extrapolation.

[a]Institut für Zuverlässigkeit und Mikrointegration, Berlin, Germany

The hit efficiency is calculated as the fraction of extrapolated tracks that have a close pixel hit cluster. The measured hit efficiencies for unirradiated planar and 3D devices are 99.9% and 99.6%, respectively, at normal incidence. The slightly lower efficiency of the 3D samples originates from tracks passing through the electrodes and increases to 99.9% for inclined tracks.[9] Fully depleted irradiated devices show a slightly reduced hit efficiency in the range of 95.3% to 99.0%.

The hit efficiency measured against the distance from the sensor edge yields the width of the inactive region at the edge of the sensors. An effective inactive width of 215 μm was measured for planar devices. 3D devices show an inactive width of 170 μm. Both technologies thus clearly pass the IBL criterion.

4. Summary

The IBL project is designed to maintain and enhance performance of the ATLAS Pixel Detector by inserting a new layer with a new beam pipe inside the current pixel system. A new FE readout chip was prototyped successfully. Module prototypes with both sensor technologies under consideration, planar n-in-n devices with slim edge and 3D devices, show good performance. In particular testbeam measurements before and after irradiation to the IBL design dose show a high hit efficiency. Other technical challenges like the bump bonding of thin chips were successfully tackled. The production of components has started and the construction is expected to be completed for installation in the LHC shut down in 2013.

References

1. G. Aad et al., *ATLAS Pixel Detector electronics and sensors*, JINST 3 P07007 (2008).
2. The ATLAS collaboration, *The ATLAS Experiment at the CERN Large Hadron Collider*, JINST 3 S08003 (2008).
3. The ATLAS collaboration, *ATLAS Insertable B-Layer Technical Design Report*, CERN-LHCC-2010-013 / ATLAS-TDR-019 (2010).
4. M. Barbero et al., *Nucl. Instr. Meth. A* **650**, 111 (2011).
5. M. Barbero et al., *The FE-I4 Pixel Readout Chip and the IBL Module*, talk at Vertex 2011, Rust 2011, to be published in *Proceedings of Science.*
6. S. Parker, C. J. Kenney, J. Segal, *Nucl. Instr. Meth. A* **395**, 328 (1997).
7. J. Weingarten, *ATLAS IBL Sensor Qualification*, talk at PSD9, Aberystwyth 2011, to be published in *The Journal of Instrumentation.*
8. A. Bulgheroni, *Nucl. Instr. Meth. A* **623**, 399 (2010).
9. P. Grenier et al., *Nucl. Instr. Meth. A* **638**, 33 (2011).

THE BELLE-II DEPFET PIXEL DETECTOR AT THE SUPERKEKB FLAVOUR FACTORY

STEFAN HEINDL
on behalf of the DEPFET Collaboration
Institut für Experimentelle Kernphysik, Karlsruhe Institute of Technology (KIT)
76131 Karlsruhe, Germany
stefan.heindl@kit.edu
www.kit.edu

The ongoing upgrade of the asymmetric electron positron collider KEKB also requires extensive detector upgrades to cope with the new design luminosity of $8 \cdot 10^{35}\, cm^{-2} \cdot s^{-1}$. Of critical importance is the new silicon pixel vertex tracker, which will significantly improve the decay vertex resolution, crucial for time dependent CP violation measurements. This new detector will consist of two layers of DEPFET pixel sensors very close to the interaction point. These sensors combine both particle detection and amplification of the signal by embedding a field effect transistor into a $75\,\mu m$ thick fully depleted silicon substrate, providing very high signal to noise ratios and excellent spatial resolution. Using this technology satisfies the given requirements of extremely low material and high radiation tolerance at the new Belle II experiment. The power dissipation due to continuous readout at high rate and spatial constraints also give strict requirements for the mechanical support and cooling of the new detector. We will discuss the overall concept of the pixel vertex tracker, its expected performance and the challenging mechanical integration.

Keywords: Belle II; DEPFET; Pixel Detector; SuperKEKB.

1. Introduction

During the last decade, the Belle detector has been operated successfully at the KEKB collider, a B-meson factory at the High Energy Accelerator Research Organization (KEK) in Tsukuba, Japan. The machine is an asymmetric e^+e^- collider, which is running at the $Y(4S)$ resonance at $10.58\, GeV$ and holding the world record in terms of instantaneous luminosity produced ($L = 2.1 \cdot 10^{34}\, cm^{-2} \cdot s^{-1}$). During its lifetime, KEKB has delivered more than 700 million $B\overline{B}$-meson pairs to the Belle detector, which helped to explore the CKM matrix and to understand the CP violation mechanism in the Standard Model of Physics. The importance of such studies was pointed out by the Nobel Prize awarded to Kobayashi and Maskawa in 2008.

Despite of the great success after collecting more than $1\,ab^{-1}$ of data in the Belle detector, the CKM picture and the sources of CP violation are still not completely revealed and higher statistics, together with better precision measurements, are required. Therefore, a next-generation Super Flavour Factory will be realized by upgrading the existing KEKB accelerator. The new machine called *SuperKEKB* will achieve an increase in luminosity by a factor of 40 by an extreme reduction of the beam size at the interaction point and a moderate increase of the beam currents (the so-called *nano beam* option). Its completion is currently scheduled for the end of 2014.

The new luminosity will result in a higher number of interactions, which also leads to higher occupancy and radiation dose. In order to cope with these conditions, the existing Belle detector will be upgraded[1] to Belle II. Therefore, some of the subdetector systems are replaced and new ones are added, like the *pixel vertex detector* (PXD). Excellent vertexing is a key component for time dependent measurements of CP violation, which require precise reconstruction of the decay vertices of the B-mesons. The new PXD will provide both improved spatial resolution compared to the existing detector and the ability to handle the increased luminosity.

2. The pixel vertex detector (PXD)

The Belle II PXD group has decided to use the DEPFET (***DE**pleted **P**-channel **F**ield **E**ffect **T**ransistor*) active pixel[2] as baseline technology for the new detector. Each DEPFET pixel is a field effect transistor (FET) integrated on a fully depleted bulk [Fig. 1, left]. A deep n-implant under the channel of the FET acts as an *internal gate* by creating a potential minimum for the electrons produced by the impinging particles. Due to the electric field configuration, they drift towards the surface of the pixel cell and get accumulated in the internal gate, where they modulate the transistor's drain current. Therefore, a first stage of signal amplification with a gain of $\approx 400\,pA/e^-$ is achieved. After the readout of the current, all accumulated charge is removed via a n^+-contact called *clear*.

The DEPFET technology has shown its excellent performance in the results of multiple test beam campaigns[3] during the past years. $450\,\mu m$ thick prototype sensors have achieved signal-to-noise ratios >100 and single point resolutions of $\approx 1\,\mu m$. We expect a reasonably high SNR with the smaller signal charge in sensors thinned down to $75\,\mu m$, which will be used for the PXD in order to minimize multiple scattering.

Other features of the DEPFET pixel are its low power consumption and material budget, because only pixels whose currents are read need to be activated while all others remain off without loosing charge. The material budget can be as low as $\approx 0.18\%$ of X_0 per layer due to the production process, which allows thinning down to only 50 μm.

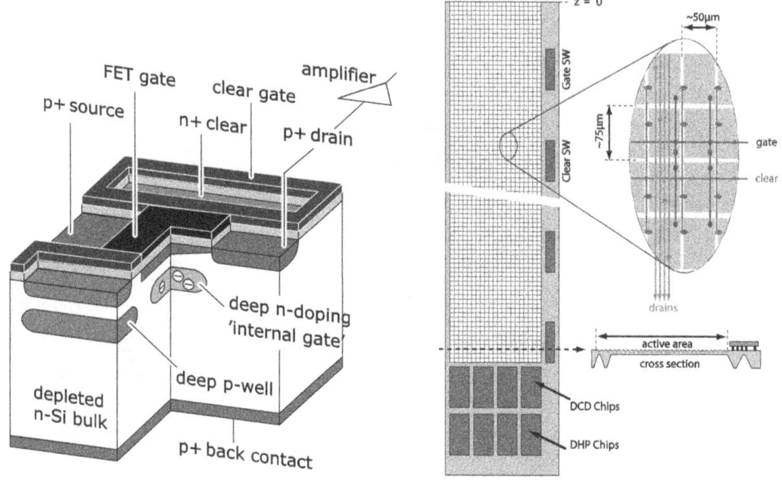

Fig. 1. Schematics of the PXD detector. *Left:* DEPFET pixel. *Right:* All-silicon ladder including ASICs.

2.1. *PXD ladder and ASICs*

The PXD ladder [Fig. 1, right] is an *all-silicon ladder* where the central part is the active area consisting of a matrix of 1536×250 DEPFET pixels, thinned down to 75 μm. The full ladder is $170 \times 15.4\, mm^2$ in size and due to the limited wafer size produced in two parts, which are then glued together in the middle. Its mechnical stiffness is provided by an unthinned frame (450 μm, with etched grooves to further reduce material) surrounding the active area and providing the necessary space for the readout chips, which are bump-bonded directly to the silicon. Three different ASICs are needed for the readout of the matrix:

(1) **SwitcherB18:** The Switcher is the steering chip controlling the row-wise readout of the matrix (*rolling shutter mode*) by sending the GateOn and Clear signals. Twelve of them are placed along the longitudinal side of the matrix on the so-called *balcony*.

(2) **DCDB:** The **D**rain **C**urrent **D**igitizer (**B**elle) is the analog frontend with integrated ADCs for sampling of the DEPFET currents. Four of them are placed at each end of the matrix.
(3) **DHP:** The **D**ata **H**andling **P**rocessor is the processor placed behind each DCD for storing the raw data and performing the common mode correction, pedestal subtraction and zero suppression.

A kapton cable is connected at both ends of the ladder to send out the data from the DHP to the DAQ, whose main components are compute nodes[4] with FPGAs housed in an ATCA shelf.

2.2. *PXD mechanics*

The detector consists of two cylindrical layers of PXD ladders, which will be mounted very close to the interaction point. The main design parameters were fixed after a full simulation of the detector performance using the ILC framework[5] and are given in Table 1:

Table 1. PXD ladder design parameters. Longitudinal pixel sizes are larger at the ladders' ends due to the differing angles of the particle tracks.

	Inner layer	Outer layer
Number of ladders	8	12
Radius[a] [mm]	14	22
Number of pixels	1536×250	1536×250
Pixel sizes [μm^2]	$(512 \cdot 55 + 2 \cdot 512 \cdot 60) \times 50$	$(512 \cdot 70 + 2 \cdot 512 \cdot 85) \times 50$
Thickness [μm]	75	75

Note: [a] For comparison: outer beam pipe radius is 12.5 mm.

Since the particles created by the interactions have a very low momentum in their final state ($< 1\,GeV$) and also because the other tracking detectors outside must not be disturbed, the material budget of the PXD has to be kept very low. The ladders themselves will be thinned down to 75 μm and only contribute about 0.18% of X_0 per layer. The 450 μm thick frame makes them self-supporting and they will be mounted to specially designed endflanges placed outside the acceptance angle at both ends. Each support endflange consists of two identical stainless steel halfshells produced by a rapid prototyping technique (direct LASER metal sintering). That manufacturing process was chosen due to the small size and high detail of the objects needed. The material budget of the full detector including support and services is calculated to only 0.7% of X_0 in the active region.

2.3. *PXD cooling*

Despite the low power consumption of the DEPFET pixels, the total power dissipated by the PXD detector will be $\approx 400\,W$. The use of cooling pipes within the acceptance region was not allowed because of the material budget, but thermal simulations using FEA software allowed finding a suitable way of cooling. The end regions of the ladders, where the ASICs produce most of the heat, will be cooled by the endflanges, while the center region will have to rely on forced convection using cold dry air. The special production process of the endflanges makes it possible to directly include the cooling pipes into the material without any additional tubing. The foreseen cooling system[6] is a two-phase system based on CO_2, which uses the high latent heat to cool the detector.

The feasibility of this cooling concept has already been demonstrated[7] with a thermal mockup using a prototype endflange.

3. Conclusion

A new B-meson factory called *SuperKEKB*, designed to deliver a luminosity never achieved before, is currently under construction at KEK (Japan). In order to fully exploit the wide physics field, a new detector (*Belle II*) will also be built. For the innermost subdetector, a two-layer pixel detector based on DEPFET technology is foreseen. This technology offers high spatial resolution and a first stage of internal amplification while keeping power consumption and material budget low.

The DEPFET Collaboration has now entered the construction phase and the new subdetector will be ready for first data taking at the end of 2015. Two major project milestones have been reached in the last months: the production and full-speed readout of thinned Belle II-type sensors and the proof of feasibility of the CO_2-based detector cooling scheme.

References

1. T. Abe *et al.*, *Belle II Technical Design Report*, tech. rep., KEK (2010), arXiv:1011.0352v1 [physics.ins-det].
2. J. Kemmer and G. Lutz, *Nucl. Instr. and Meth.* **A 253**, 365 (1987).
3. L. Andricek *et al.*, *Nucl. Instr. and Meth.* **A 638**, 24 (2011).
4. Hao Xu *et al.*, *16th IEEE-NPSS Real Time Conference Record* , 571 (2009).
5. URL: `ilcsoft.desy.de/portal`.
6. A. van Lysebetten *et al.*, *PoS (Vertex 2007)* , 9 (2007).
7. S. Heindl, CO2 Cooling of PXD Endflange: Results of CERN Cooling Test, Talk presented at 7th International Workshop on DEPFET Detectors and Applications, Ringberg Castle, (2011).

Tracking performance of the LHCb spectrometer

A. Jaeger*†, P. Seyfert*, M. De Cian*,
J. van Tilburg*, Stephanie Hansmann-Menzemer*
**on behalf of the LHCb collaboration,[1] † Speaker, E-mail: andreas.jaeger@cern.ch*

LHCb is an experiment running at the Large Hadron Collider (LHC). It is designed to search for evidence of new physics effects through precise measurements of B- and D-meson decays. During the 2010 run, a total integrated luminosity of 38 pb^{-1} was collected, on which many results have been published. Until the summer of 2011, a data-set of 340 pb^{-1} has been collected. A crucial element in many analyses is the track reconstruction. The tracking system provides excellent spatial and mass resolutions which are essential to reduce the enormous background from the LHC collisions. For all analyses which measure production cross sections or branching fractions a good knowledge of the tracking efficiency is important. Tracking efficiencies are measured with a dedicated tag-and-probe technique using J/ψ mesons. A total uncertainty on the tracking efficiency below 1% is obtained with this method.

1. The LHCb spectrometer

The LHCb experiment is a single arm forward spectrometer, which covers the pseudo-rapidity region of $1.9 < \eta < 4.9$. Due to the different momentum fractions of the interacting partons, $b\bar{b}$ pairs are mainly boosted in forward direction at LHC energies of $\sqrt{s} = 7\,\mathrm{TeV}$, which makes the forward region interesting for B physics. One important ingredient for these measurements is an efficient and high-resolution tracking system.

1.1. *Tracking system*

The most important part for the track reconstruction is the vertex locator (VELO). It is directly located at the interaction point and provides information on the position and the flight direction of the particles. This silicon strip device can measure vertices very precisely and provides a good separation of primary (PV) and secondary vertices. For a typical value of 25 reconstructed tracks used in the vertex reconstruction, the PV resolution transverse to the beam is $\sigma_{x,y} = 13\,\mu\mathrm{m}$ and in direction of the beam

$\sigma_z = 69\,\mu\text{m}$. In front of the magnet with an integrated B field of 4 Tm the Tracker Turicensis (TT) is located. It consists of silicon strips and is important for the reconstruction of long-lived particles. Downstream the magnet, there are three tracking stations (T stations), consisting of silicon strips and straw drift chambers. They provide a momentum resolution of 0.4%, resulting in a mass resolution for J/ψ mesons of $12.3\,\text{MeV}/c^2$ [1] .

2. Track finding and fitting

The track finding algorithms in LHCb start with a search in the VELO detector for straight-line tracks. There are two algorithms[2–5] to promote these VELO tracks to so-called long tracks, which are the most valuable tracks for physics analysis. Both algorithms combine VELO tracks with information from the T stations. They are described in detail elsewhere[2,4,5] .Finally the TT measurements are added to the track to improve the momentum resolution. After the track is found it is fitted using a Kalman filter that takes multiple scattering and energy loss into account. Details about track finding and fitting at LHCb are described in the reference[6] .

The efficiency of the track reconstruction is the main monitoring performance indicator. Average efficiencies for long tracks are found to be above 96% and depend on analyses and their phase space selection. Another important indicator is the ghost rate. Ghosts are fake tracks that can occur due to random matching of measurements or due to wrong combination of subdetector tracks. The latter is the dominating source of ghosts in LHCb.

3. Strategy for efficiencies

Measuring the tracking efficiency is crucial for many physics analyses, especially those that aim to measure production cross sections or branching ratios. Often the tracking efficiency is taken from Monte Carlo (MC) simulation. Nevertheless, the simulation might not always describe the data within the required accuracy. This introduces a potential bias in the final measurement. To measure the efficiency in data a tag-and-probe approach is used. The efficiency is determined as a function of several parameters such as the momentum of the particle, the pseudo-rapidity and the multiplicity of the event. The final result is given as a ratio of the tracking efficiency between data and MC as a function of the phase space. This ratio table is than used to weight the MC according to the phase space covered in each analysis. Using $J/\psi \to \mu^+\mu^-$ decays, the momentum spectrum is close to the one from B daughters.

3.1. *Tag-and-probe method*

The tag-and-probe method uses two-body decays (e.g. J/ψ, Z^0, K^0_S), where one of the daughter particles, the "tag" leg, is fully reconstructed, while the other particle, the "probe" leg, is only partially reconstructed. The tracking efficiency is then obtained by trying to match the partially reconstructed probe leg to a fully reconstructed long track. If a match is found the probe leg is defined as efficient. The three methods described below all use $J/\psi \to \mu^+ \mu^-$ decays, but a different combination of tracking detectors for the partial reconstruction of the probe leg.

In the VELO method, which measures the efficiency that a VELO track is found, the probe tracks consist of measurements from the TT detector and the T stations. These tracks are called downstream tracks and are found by the downstream tracking algorithm.[7] In the T station method, which measures the efficiency that a T station track is found, the probe tracks consist of measurements from the VELO and the muon stations. These VELO-muon tracks are found by a dedicated reconstruction of tracks in the muon stations[8] which are subsequently matched to VELO tracks. In the long method, which measures the efficiency that a long track is found, the probe tracks consist of measurements from the TT and the muon stations. These TT-muon tracks are found by a dedicated reconstruction of tracks in the muon stations which are subsequently matched to TT measurements.

All methods have different mass widths and background fractions. Therefore, the uncertainty on the efficiency is different. The tracking efficiency is calculated as the fraction of events for which the probe track can be matched to a reconstructed long track:

$$\varepsilon = \frac{\#\text{probes matched to long track from } J/\psi \text{ decay}}{\#\text{probes from } J/\psi \text{ decay}} \tag{1}$$

To estimate the number of J/ψ decays, the background contribution under the J/ψ mass peak is removed by subtracting the number of events in the sidebands.

The efficiencies from the VELO and T station method are multiplied to obtain the combined long-track finding efficiency. Having two independent methods provides an important cross-check for systematic studies.

4. Results

The tracking efficiency obtained from the 2011 data and according Monte Carlo samples for the long method as a function of the momentum (p) and as a function of the pseudo-rapidity (η) is shown in Fig. 1. A reasonable

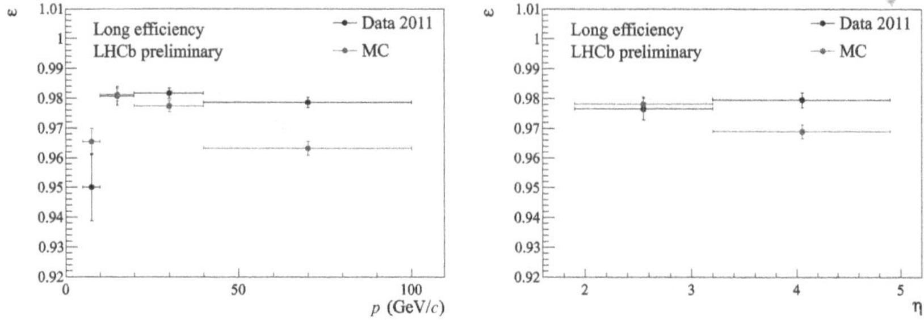

Fig. 1: Tracking efficiency from the 2011 data for the long method as a function of the momentum, p and as a function of the pseudo-rapidity, η.

agreement between data and simulation is observed. The tracking efficiency measured in data divided by the tracking efficiency measured in MC is determined in several bins of phase space. From Fig. 2 (left) can be concluded that the efficiency dependence versus number of primary vertices and also number of tracks is well described in the MC. Therefore, it is decided to determine the tracking efficiency only as a function of p and η. Nevertheless, this means that the MC sample will need to be reweighted to describe the data in terms of number of tracks. Figure 2 (right) shows the efficiency ratio versus p and η for 2011 data. The overall efficiency ratio and its uncertainty

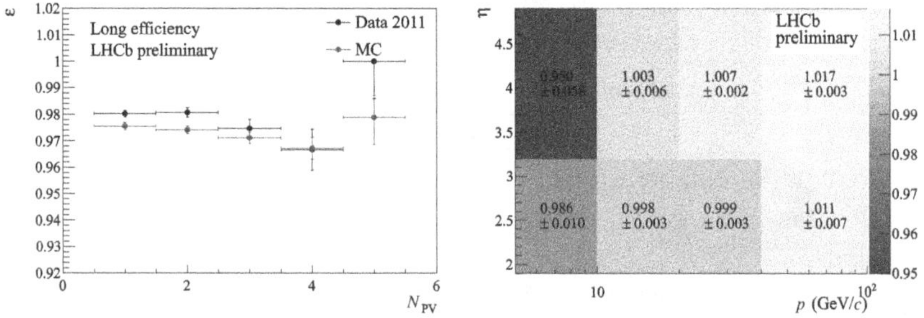

Fig. 2: Efficiency vs. number of primary vertices (left) and tracking efficiency ratio between data and MC 2011 (right)

depends on the particle distribution of the data in terms of p and η. Using the distribution of the muons from the selected $J/\psi \rightarrow \mu^+\mu^-$ decays used in this study, an efficiency ratio of 1.000 ± 0.001 is found for 2011 data.

4.1. *Systematic uncertainty*

In order to evaluate the different contributions to the systematic uncertainty several tests are performed. By reweighting the MC simulation in different parameters to match the data small differences in the ratio are seen. The contribution from the treatment of the background using a sideband subtraction method is evaluated by performing several mass fits to the data with different signal and background functions. Finally, the difference between combined and long method is evaluated. The overall systematic uncertainty of the method presented here is 0.7%.

The tracking efficiency only measures the efficiency for particles that traversed the full tracking system. In particular, it does not account for particles that had an interaction in the detector material. For muons in our momentum region this effect is negligible, but not for hadrons. Assuming that the total material budget of the detector is known up to ±10%, the additional systematic uncertainty for hadrons due to material interactions is 1.5%.

5. Conclusion

Tracking efficiencies have a direct impact on physics analyses. Therefore, agreement between data and MC is very important. We provide a ratio table, which can be used in any physics analysis in LHCb to estimate the uncertainties due to differences in data and MC. Applying the results on $J/\psi \to \mu^+\mu^-$ shows that the simulation describes the data very well with a statistical uncertainty of 0.1% and a systematical uncertainty of 0.7%. For hadrons an additional systematical uncertainty of 1.5% has to be taken into account.

References

1. LHCb, R. Aaij et al., *Measurement of J/ψ production in pp collisions at* $\sqrt{s} = 7\,\text{TeV}$, *Eur. Phys. J.* **C71** (2011) 1645, [arXiv:1103.0423].
2. O. Callot and S. Hansmann-Menzemer, "The Forward Tracking: Algorithm and Performance Studies." CERN-LHCb-2007-015.
3. M. Needham, "Performance of the track matching." CERN-LHCb-2007-129.
4. O. Callot, "FastVelo, a fast and efficient pattern recognition package for the Velo." CERN-LHCb-PUB-2011-001.
5. D. Hutchcroft, "VELO Pattern Recognition." CERN-LHCb-2007-013.
6. O. Callot and M. Schiller, "PatSeeding: A Standalone Track Reconstruction Algorithm." CERN-LHCb-2008-042.
7. O. Callot, "Downstream Pattern Recognition." CERN-LHCb-2007-026.
8. A. Satta, "Muon identification in the LHCb HLT." CERN-LHCb-2005-071.

Performance of the CMS Silicon Tracker at LHC

Gordon Kaußen, on behalf of the CMS Collaboration
Institut für Experimentalphysik, Universität Hamburg,
22459 Hamburg, Germany
E-mail: gordon.kaussen@cern.ch

After nearly two years of operation at the Large Hadron Collider with proton-proton collisions at a center of mass energy of 7 TeV, the performance of the CMS silicon tracker is reviewed. Both the status and basic properties of the pixel and the strip detector, as well as the performance of the tracking such as vertex reconstruction and b-tagging are discussed.

Keywords: LHC; CMS; Detector; Silicon Tracker; Performance.

1. Introduction

The Compact Muon Solenoid (CMS) experiment is one of the four large detectors operated at the Large Hadron Collider (LHC) at the CERN laboratory in Switzerland. It is a multipurpose high energy physics experiment composed of (from inside-out) a silicon pixel tracker, a silicon strip tracker, an electromagnetic calorimeter, a hadronic calorimeter, a superconducting magnet and a muon system embedded in the iron return yoke of the magnet.

The tracking detector, which is subject to this article, consists of two major parts. The first one is the pixel detector comprising about 66 million pixels arranged in three barrel layers and two endcap disks on either side. The rectangular pixels with a cell size of $100\,\mu\text{m} \times 150\,\mu\text{m}$ cover an area of about $1\,\text{m}^2$ ranging from a radius of $r = 4.4\,\text{cm}$ to $r = 10.2\,\text{cm}$ with respect to the beam pipe. The second one is the strip tracker containing about 10 million strips subdivided into about 15,000 silicon modules. The strip tracker surrounds the pixel detector, adding 10 layers in the barrel and 12 disks on either side of the endcaps to the tracking system. It ranges from $r = 25.5\,\text{cm}$ to $r = 110\,\text{cm}$ and covers a surface of about $198\,\text{m}^2$. Both rectangular (in the barrel) and trapezoidal modules (in the endcaps) are used with a typical strip cell size of the order $100\,\mu\text{m} \times 10\,\text{cm}$.

The data used in this article were collected in proton-proton collisions at

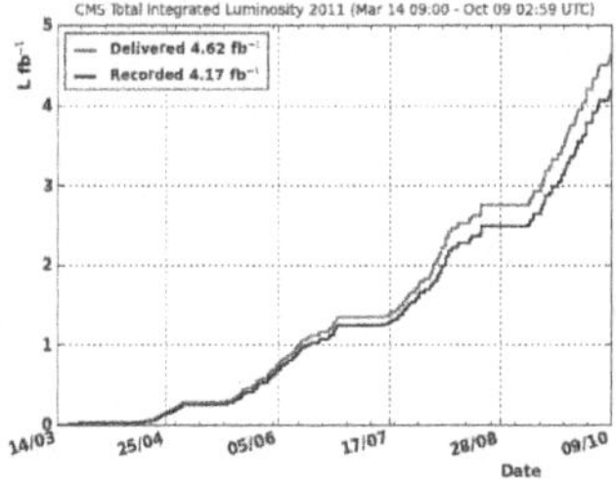

Fig. 1. Integrated luminosity vs. time delivered to (red) and recorded by (blue) CMS during stable beams at 7 TeV center of mass energy[1] (status as of October 9th, 2011).

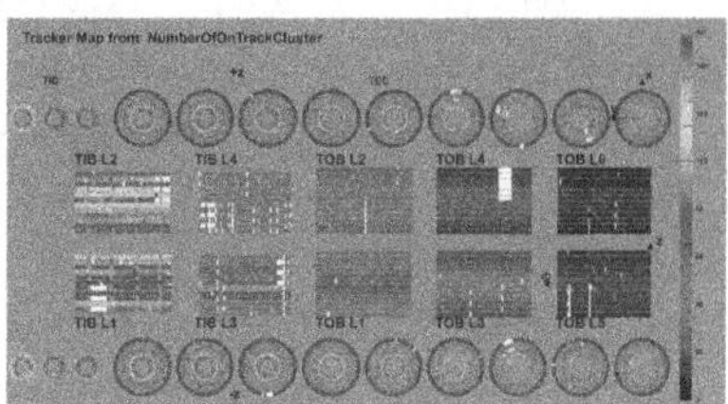

Fig. 2. Occupancy map of reconstructed clusters attached to a particle track in the strip tracker. The white areas are modules that are not in data acquisition.

the LHC at a center of mass energy of 7 TeV. Although the total integrated luminosity recorded by CMS is about $4.17\,\mathrm{fb}^{-1}$ as of October 9th 2011, see Fig. 1, the presented results are based only on a subset of the whole data sample, as indicated in the dedicated plots. Nevertheless, the analyses are representative also for the current detector operation since the performance of the tracking system has not changed with time.

2. Performance of the Pixel and Strip Tracker

In the following sections, some basic properties of both the pixel and strip tracker are highlighted. If the analysis/calibration techniques are the same or comparable for the two detector parts, they will be explained only for one exemplarily.

2.1. *Operational Status*

The first proton-proton collisions at $\sqrt{s} = 7\,\mathrm{TeV}$ occurred in spring 2010, which means that the detector has been running for nearly two years now. During this time, few issues resulted in a permanent removal of some tracker parts from the data acquisition. The main reasons are the loss of control rings disabling the configuration of the modules, faulty front-end drivers disabling the data acquisition, closed cooling loops disabling the active cooling of the sensors and the readout electronics and thus a proper operation of the modules, and faulty power groups disabling the low or high voltage. A detailed description of the tracker layout can be found in Ref. 2.

Figure 2 shows the occupancy of reconstructed clusters which could be matched to a trajectory in a nominal collision run for the different layers of the strip tracker. The barrel region is unrolled in the $z - \phi$ plane, while the endcap disks are displayed in $x - y$ (In CMS, x is pointing to the LHC center, y is pointing to the surface and z along the beam pipe completing a right-handed frame. ϕ is the azimuthal and θ the polar angle). The white areas correspond to silicon modules not in the DAQ. Since the tracker has several redundant layers and the track reconstruction algorithms are robust enough to treat missing layers properly, these holes are not serious for the event reconstruction. In the strip tracker, a total of 97.8 % of all readout channels is operational. A similar result holds for the pixel detector, where a total of 97.1 % operational readout channels is achieved.

The signal-to-noise ratio (S/N) measured in the strips after correcting for the track incidence angle with respect to the module surface results in a Landau distribution as expected for thin absorbers. The 320 μm modules have a most probable S/N of about 18, while the 500 μm modules have $S/N \approx 22$. The numbers quoted belong to the so-called deconvolution readout mode,[2] which is the nominal operation and provides a weighted sum of three consecutive signal samples. In peak mode, these values are about 1.7 times larger for the sake of a worse time resolution.

2.2. *Hit Efficiency and Resolution*

To measure the hit reconstruction efficiency, a particle trajectory is extrapolated to the layer under investigation and a reconstructed hit compatible with the expected position is searched for. It is required that the trajectory has a valid hit on either side of this layer and that the track extrapolation does not cross the border of a silicon module but that it is well within the sensitive area of the silicon. The layer efficiency is then calculated by the number of found hits divided by the number of expected hits. In this study, known bad components such as disconnected modules are excluded in order to measure the efficiency only for the operational part of the tracker. Figure 3 shows the efficiency of all pixel layers in the barrel and endcap regions. It is well above 99 %. The same measurement was done for the strip tracker, also resulting in a hit efficiency of > 99 % for all layers.

The spatial hit resolution is obtained using overlapping modules in the same layer. This reduces the effects of multiple scattering and track extrapolation uncertainties, since the overlapping modules are close to each other with only little material in between. In addition, the impact of misalignment is negligible for those module pairs. Since the hit resolution depends on the

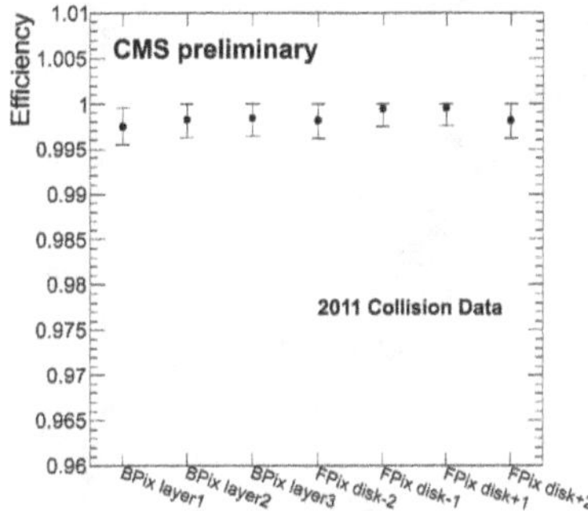

Fig. 3. Hit reconstruction efficiency[3] as measured for all layers of the pixel tracker.

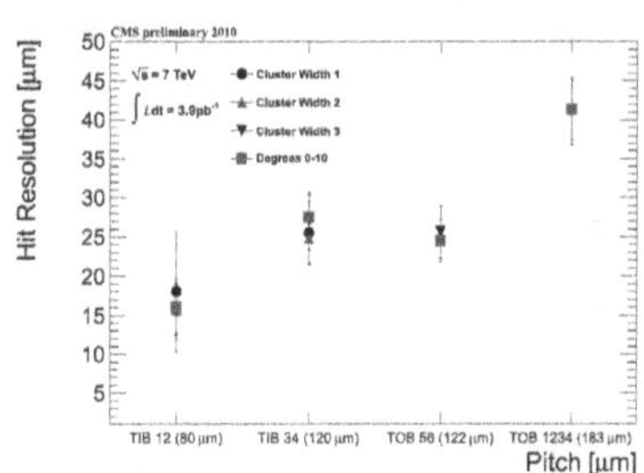

Fig. 4. Spatial hit resolution[3] as measured for the different pitches in the strip tracker barrel. The results are shown for different cluster widths and track incidence angles.

strip or pixel cell size, the cluster width (charge sharing) and the track incidence angle, the results can be determined in different ways, binning the data for example in strip pitch and cluster width or track angle, as shown in Fig. 4. As expected, a linear dependance on the strip pitch is observed with a resolution in the barrel between $\sim 15\,\mu$m and $45\,\mu$m. In the pixel detector, the resolution is about $10\,\mu$m in the transverse and $20\,\mu$m in the longitudinal direction.

2.3. *Timing and Gain Calibration*

The correct timing between the tracker readout and the bunch crossing is vital to achieve the best detector performance. In dedicated commissioning runs the timing of each single strip module was randomly varied around the nominal operation point so that the reconstructed signal could be studied at several delays putting the information of all modules together. In Fig. 5 such a time profile is shown for the different strip tracker parts. The signal amplitude is displayed as a function of the readout delay with respect to the nominal working point. From these measurements, a signal width of 12 ns, as expected from the deconvolution pulse shape, is obtained. The maximum should peak at zero if the nominal working point is the best. Otherwise, each deviation can be used to optimize the readout timing. Both for the strip and the pixel detector the best sampling time could be adjusted with a precision of about 1 ns.

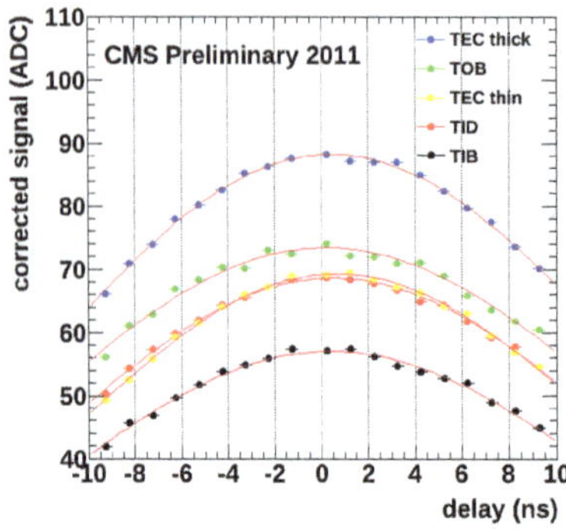

Fig. 5. Time profile of the signal[3] in different parts of the strip tracker.

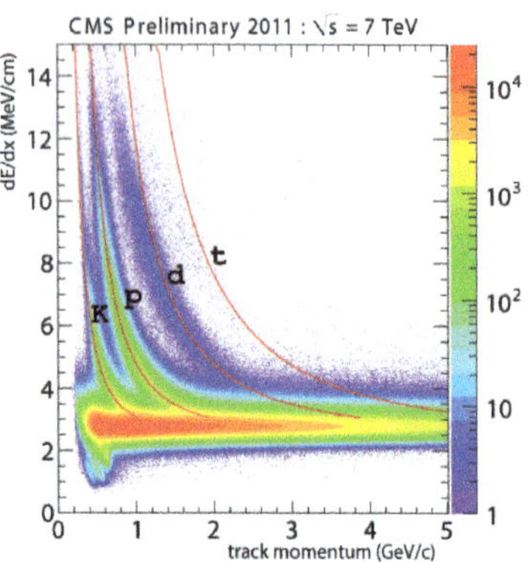

Fig. 6. Energy loss dE/dx measured in the strip tracker versus track momentum.[3] Kaons, Protons, Deuterons and Tritium are visible.

Another important ingredient after the adjustment of the timing is the gain calibration of the readout chains. First, a synchronization pulse of the readout chips, called tickmark, is used as measure for the electronics gain and the signals of all strip modules are normalized to a default tickmark height. Then, minimum ionizing particles are used to equalize the response of all silicon sensors by applying a particle gain calibration factor to the measured signals. Afterwards, all sensors and readout chains have equal behaviour and the measured charge can be used for particle identification exploiting the energy loss (dE/dx). Figure 6 shows the measured energy loss for kaons, protons, deuterons and tritium as a function of track momentum. The red lines correspond to Bethe-Bloch distributions, where the theoretical prediction was fitted to the proton data and then extrapolated to the other particles. The deviations at large dE/dx are due to saturation effects in the readout.

3. Performance of the Tracking

3.1. *Nuclear Interactions*

A way to investigate the material inside the tracker volume is the use of nuclear interactions. Either neutral or charged hadrons like pions or kaons emerging from the primary vertex can interact with the tracker material leading to secondary vertices. The abundance of such vertices is proportional to the amount of material in the respective place. Thus, their re-

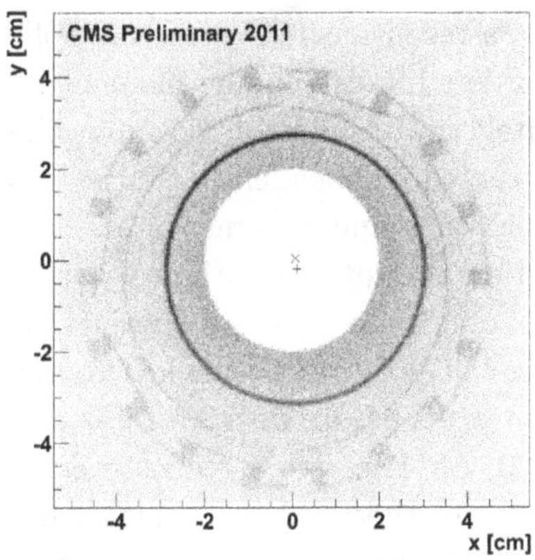

Fig. 7. $x-y$ view of reconstructed nuclear interaction vertices[3] for $-20\,\text{cm} < z < 20\,\text{cm}$.

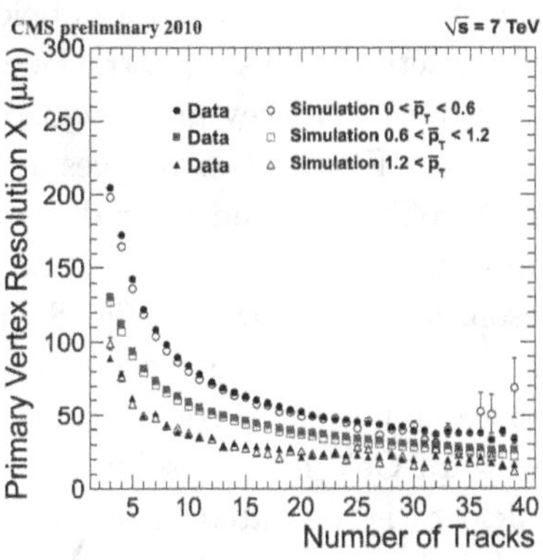

Fig. 8. Primary vertex resolution in x as a function of the number of tracks in the vertex[4] for different average track momenta $\bar{p}_T$.

construction with a slightly modified tracking procedure relaxing the track quality criteria (the tracks have usually low momentum and are not pointing to the primary vertex) can provide a picture of the material distribution similar to radiography. Such a result is shown in Fig. 7, where the position of nuclear interaction vertices is displayed in the $x - y$ plane for the innermost part of the pixel detector (the white region is an artifact of the event selection). Both the first pixel layer and the beam pipe (dark black circle) are clearly visible. A fit of these data shows on the one hand that the beam pipe is displaced with respect to the pixel layer, and on the other hand that the average beam spot position (the "x" in the picture) and the fitted beam pipe center (the "+" in the picture) are close to each other.

Another study of the beam pipe position along z for two different magnetic fields has shown that the variation of its position is very small, less than 0.5 mm in both x and y direction. This is important for a future upgrade of the pixel tracker, where the innermost layer will be even closer to the beam pipe compared to the current layout. Further details about nuclear interaction reconstruction in CMS can be found in Ref. 5.

3.2. *Vertex Reconstruction*

A crucial ingredient for a proper event reconstruction and interpretation is the vertex reconstruction. Its performance can be measured in terms of primary vertex resolution and efficiency. The resolution is obtained using a split method where all tracks belonging to a reconstructed vertex are split in two sets. To do so, the tracks are first ordered in descending momentum and afterwards they are assigned alternately and randomly to one or the other set. Finally, two vertices are reconstructed independently for the two track collections and their distance in both x, y and z is measured. The gaussian width of the respective distribution is quoted as primary vertex resolution, as shown in Fig. 8 for the x coordinate. It can be seen that the resolution strongly depends on the number of tracks attached to the vertex and on the average track momentum $\bar{p}_T$. A vertex with 30 tracks and $\bar{p}_T > 1.2\,\mathrm{GeV}$ has a resolution of about $20\,\mu\mathrm{m}$. The vertex reconstruction efficiency is computed with a similar splitting method and performing a tag and probe analysis afterwards. As for the resolution, the efficiency depends on the number of tracks and is close to $100\,\%$ if more than two tracks with $p_T > 0.5\,\mathrm{GeV}$ are contained in the vertex.

In addition to the vertex itself, the track impact parameter (IP), which is the point of closest approach of a track with respect to the vertex, is an important measure used for example in b-tagging algorithms. The impact parameter resolution can be obtained by removing the track under investigation from the vertex fit and calculating afterwards its distance from the vertex in the three space dimensions. This measurement is a convolution of the IP resolution, the vertex resolution and the contamination with gen-

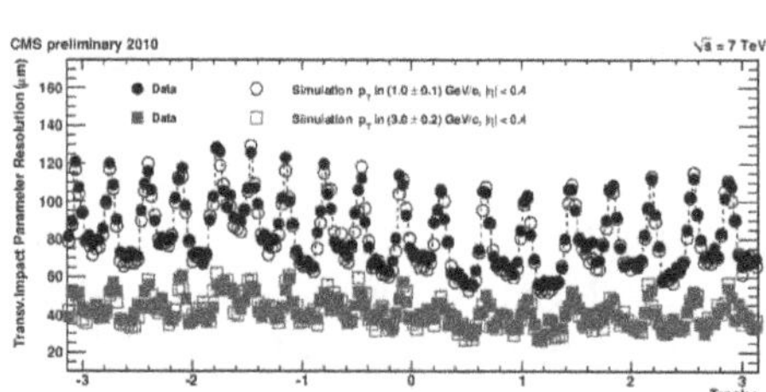

Fig. 9. Transverse impact parameter resolution[4] as a function of the track ϕ for different track momenta p_T.

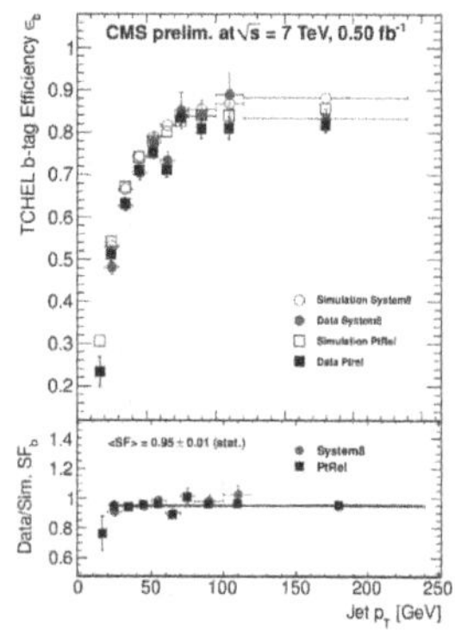

Fig. 10. b tagging efficiency[6] for a specific b identification algorithm as a function of the jet p_T. The lower panel shows the Data/MC scale factors.

uinely displaced particle tracks. The plain IP resolution, which depends on the track momentum and direction (different material budget in different $\eta - \phi$ regions) was measured to be about 20 μm in the transverse and 40 μm in the longitudinal direction for tracks with $p_T > 10\,\text{GeV}$. In Fig. 9 the transverse IP resolution is shown for different momenta as a function of the track azimuthal angle ϕ. For low track momenta, 18 peaks corresponding to 18 cooling pipes in the first pixel layer are clearly visible. In addition, a slight $\sin\phi$ modulation can be observed, which is due to the displacement of the beams with respect to the innermost tracking layer (compare Fig. 7). A detailed description of vertex and IP reconstruction can be found in Ref. 4.

3.3. *b-Tagging*

Different algorithms and working points are in place to identify jets originating from b quarks in CMS (see Ref. 6 for details). Their performance strongly depends on the ability to reconstruct secondary vertices and track impact parameters. The track counting algorithm for example counts the number of tracks in a jet that have an IP significance above a certain threshold. This provides a discriminator to tag b jets with either high efficiency or high purity. In Fig. 10 the b tag efficiency for a specific algorithm (Track Counting High Efficiency Loose) and two different measurement methods ("PtRel" and "System8") is shown as a function of the jet p_T. An efficiency of roughly 85 % is achieved for jets with $p_T > 80\,\text{GeV}$. The mistag rate for light flavour jets is for the same tagger measured to be about 15 % for $p_T \approx 80\,\text{GeV}$ jets.

4. Conclusions

After nearly two years of operation, the CMS silicon tracker has an operational fraction of about 97 % of the readout channels. The recorded luminosity of $\sim 4.17\,\text{fb}^{-1}$ is high quality data with the tracker fulfilling both the performance and the physics requirements. Local hit and global track reconstruction provide excellent precision needed for vertex and b jet identification.

References

1. CMS Luminosity Collision Data, `https://twiki.cern.ch/twiki/bin/view/CMSPublic/LumiPublicResults`.
2. The CMS Collaboration, *The CMS Experiment at the CERN LHC*, JINST 3 S08004 (2008).

3. CMS Tracker Detector Performance Results, `https://twiki.cern.ch/twiki/bin/view/CMSPublic/DPGResultsTRK`.
4. The CMS Collaboration, *Tracking and Primary Vertex Results in First 7 TeV Collisions*, CMS Physics Analysis Summary, CMS PAS TRK-10-005 (2010).
5. The CMS Collaboration, *Studies of Tracker Material*, CMS Physics Analysis Summary, CMS PAS TRK-10-003 (2010).
6. The CMS Collaboration, *Performance of b-jet identification in CMS*, CMS Physics Analysis Summary, CMS PAS BTV-11-001 (2011).

PERFORMANCE OF THE LHCb VERTEX LOCATOR

ALEXANDER LEFLAT
on behalf of the LHCb Vertex Locator Group
*Skobeltsyn Institute of Nuclear Physics, Moscow State University
Moscow 119991, 1(2), Leninskie gory, GSP-1, Russian Federation
E-mail: Alexander.Leflat@cern.ch*

LHCb is a dedicated experiment to study new physics in the decays of beauty and charm hadrons at the Large Hadron Collider (LHC) at CERN. The VELO is the silicon detector surrounding the LHCb interaction point, and is located only 7 mm from the LHC beam during normal operation. The VELO is moved into position for each fill of the LHC, once stable beams are obtained. The VELO consists of two retractable detector halves with 21 silicon micro-strip tracking modules each. A module is composed of two n^+-on-n 300 micron thick half disc sensors with R-measuring and Phi-measuring micro-strip geometry, mounted on a carbon fiber support. The VELO has been successfully operated for the first LHC physics run. Operational results show a signal to noise ratio of around 20:1 and a cluster finding efficiency relative to the design of 99.5%.

1. Introduction

The VErtex LOcator (VELO) provides precise measurements of track coordinates close to the interaction region, which are used to identify the displaced secondary vertices which are a distinctive feature of b and c-hadron decays [1]. The VELO consists of a series of silicon modules, each providing a measure of the R and Φ coordinates, arranged along the beam direction (Figure 1). Two planes perpendicular to the beam line and located upstream of the VELO sensors are called the pile-up system. The VELO sensors are placed at a radial distance from the beam which is smaller than the aperture required by the LHC during injection and the system must therefore be retractable. The detectors are mounted in a vessel that maintains vacuum around the sensors and is separated from the machine vacuum by a thin walled corrugated aluminum sheet. This is done to minimize the material traversed by a charged particle before it crosses the sensors and the geometry is such that it allows the two halves of the VELO to overlap when in the closed position.

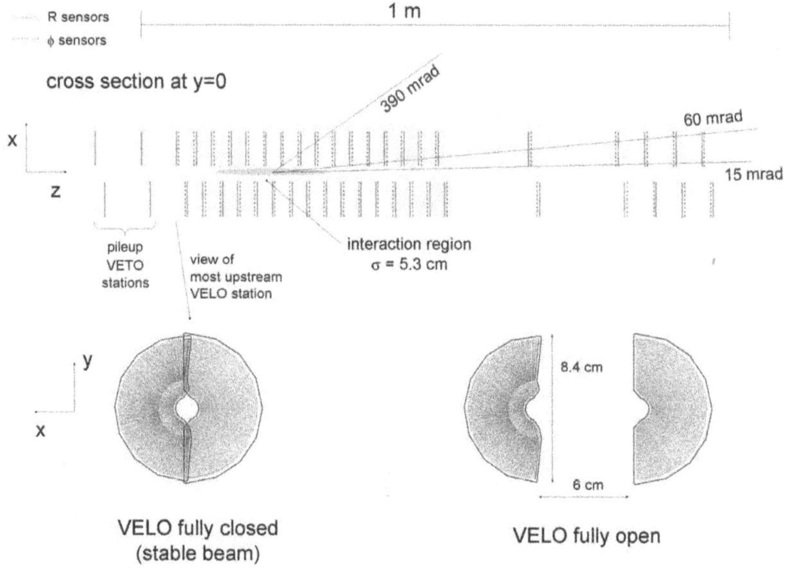

Figure 1. Cross section in the (x,z) plane of the VELO silicon sensors, at y = 0, with the detector in the fully closed position. The front face of the first modules is also illustrated in both the closed and open positions. The two pile-up veto stations are located upstream of the VELO sensors.

2. VELO Description

2.1. *Sensors*

The severe radiation environment at 8 mm from the LHC beam axis required the adoption of a radiation tolerant technology. The choice was n-implants in n-bulk technology with strip isolation achieved through the use of a p-spray. The minimum pitch achievable using this technology was approximately 35 μm, depending on the precise structure of the readout strips. For both the R and Φ-sensors the minimum pitch is designed to be at the inner radius to optimize the vertex resolution. The layout of strips on the sensors is illustrated in Figure 2.

The technology utilized in both the R- and Φ-sensors is identical. Both sets of sensors are 300 μm thick. Readout of both the R- and Φ-sensors is at the outer radius and requires the use of a second layer of metal (a routing layer or double metal) isolated from the AC-coupled diode strips by approximately 3 μm of chemically vapour deposited (CVD) SiO2. The strips are biased using polysilicon 1MΩ resistors and both detectors are protected by an implanted guard ring structure. Both R and Φ sensors have 2048 strips. R-sensors are segmented into four sectors, the inner strip pitch is 40 μm, outer strip pitch is 92 μm. Φ-sensors have an inner strip pitch of 37 μm and outer strip pitch of 98 μm.

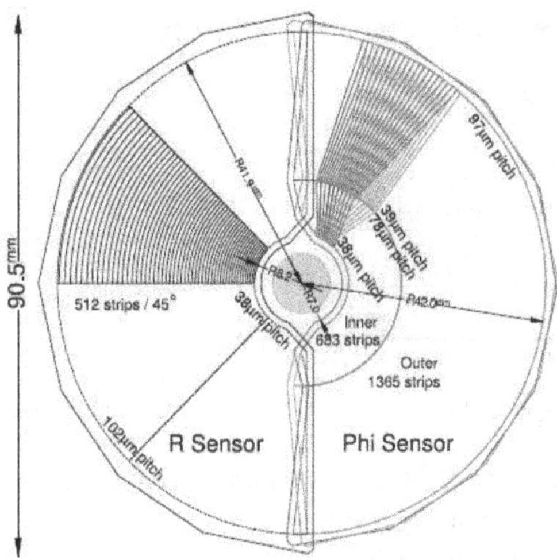

Figure 2. R-Φ geometry of the VELO sensors.

2.2. *Modules*

The module has three basic functions. Firstly it must hold the sensors in a fixed position relative to the module support. Secondly it provides and connects the electrical readout to the sensors. The module serves as the base for the electronics containing 16 FE Beetle chips [2] mounted on a kapton substrate. Finally it enables the thermal management of the modules, which are operating in vacuum. The module allows the heat from the front-end chips and sensors to be conducted away via the module substrate and a connection to a cooling bracket.

2.3. *Readout*

The electronics layout is summarized in Figure 3, which shows the readout chain for one side of one module. The individual components are discussed in this section.

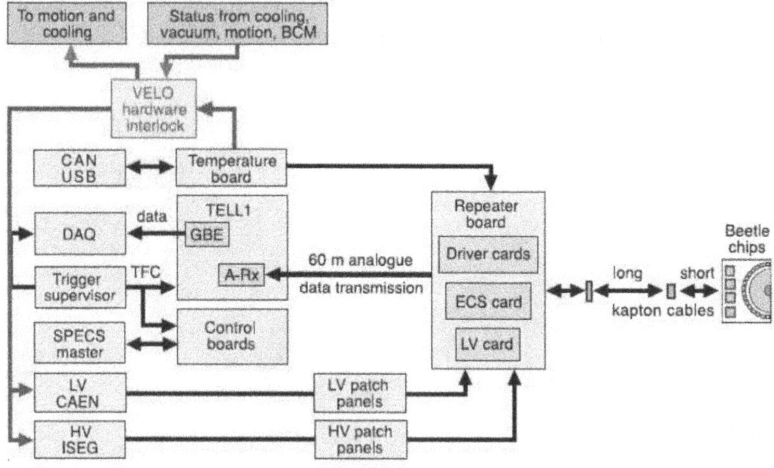

Figure 3. VELO readout chain for one side of one module.

The VELO uses the Beetle, a custom designed radiation hard ASIC based on 0.12 mm CMOS technology with an analog front-end, as the front-end chip [2]. The chip was designed with a peaking time and sampling frequency to match the LHC bunch crossing rate of 40 MHz, and to be able to readout at a speed which can match the first level trigger (L0) accept rate of 1 MHz. Each of the 128 channels consists of a charge sensitive preamplifier/shaper and an analog pipeline of 160 stages designed to match the L0 4 ms latency. The data are brought off the chip at a clock frequency of 40 MHz, with 32 channels multiplexed on each of 4 output lines. This allows the readout of one event to be achieved in 900 ns.

The repeater board (RPT) is located directly outside of the VELO tank inside the repeater crates. The RPT function is mainly a repeater for the differential signals arriving at the board, including data signals, Time and Fast Control and front-end chips configuration signals.

Specific to the VELO is the digitization of the data on the TELL1 and the complex pre-processing of the data. Each TELL1 board deals with the data from one sensor, i.e. 2048 channels or 64 analog links, and features 4 A-Rx cards. After digitization on A-Rx cards, the TELL1 performs data processing in FPGAs before sending the zero-suppressed data to the trigger farm [3]. The steps in this data processing include pedestal subtraction, channel re-ordering, common mode suppression, and clustering. Pedestal subtraction is implemented with a running average pedestal following algorithm available, if required, to calculate the value for each channel.

2.4. *Cooling*

Since the detectors and read-out electronics are operated inside the vacuum system, active cooling is required. Furthermore, in order to limit the effects of radiation damage of the silicon sensors, the irradiated sensors are operated at temperatures -8°C and kept cold throughout shutdowns. The refrigerant in the system is two-phase CO2 cooled by a conventional freon cooler. The liquid CO2 is transported via a 55 m long transfer line to the VELO, where it is distributed over 27 capillaries per detector half. Each capillary is thermally connected to five cooling blocks that are attached to each detector module.

2.5. *Motion*

Before the LHC ring is filled, the detectors have to be moved away from the interaction region by 30 mm in order to allow for beam excursions during injection and ramping [4]. After stable beam conditions have been obtained, the detectors are placed into an optimized position centered in x and y around the interaction region. This position is not exactly known beforehand; it may vary in both x and y, even from fill to fill. Therefore, a procedure has been developed to determine the beam position with the detectors in the retracted position using the data. The detectors are then moved in stages, remeasuring the vertex position at each stage, to reach the optimal position. The motion is performed with a mechanism that can bring the detectors to their position with an accuracy of better than 10 μm by means of a stepping motor with resolver readout.

Hence, the VELO is closed and opened every fill. The closing procedure takes approximately 4 minutes. Safety monitoring is performed during and after the closing. The list of parameters checked includes: Beam Position Monitors from the accelerator, the vertex position measurement from the VELO data, Beam Condition Monitors (BCM) that measure machine backgrounds, the RF foil temperature, HV current fluctuations. In total 32 parameters are monitored. If one of them does not satisfy the safety conditions then a 2-minutes "Grace period" is started, and the VELO is retracted.

2.6. *VELO safety*

Apart from the inbuilt sub-systems safety functions, there are two main tools to keep the VELO safe. The first is the alert system, this is a software (PVSS) monitoring and interlock system. The alert system issues a severity level for all problems: Warning, Error and Fatal. Over 1370 parameters are monitored by the alert system. The second system is the hardware interlock. This is an FPGA based protection system that depends on input conditions that include those from the cooling, vacuum, and BCM systems, and temperatures. This interlock box is a reliable tool that has so far been most commonly activated in the cases of partial power cuts.

3. Physics performance, Hit, PV and IP resolution

The average Signal to Noise ratio on the VELO clusters is approximately 20:1, while the cluster finding efficiency is 99.5% relative to the full designed number of strips. The detector hit resolution depends on the strip pitch, and on the projected angle of the tracks. A best hit resolution of 4 μm has been obtained for a 40 μm strip pitch and 7-10 degree angle. The primary vertex and impact parameter resolution are two critical measures of the detector performance for physics. The primary vertex resolution depends on number of tracks in the vertex. The average number of tracks in LHCb is around 25. The X, Y and Z PV resolution for 25 tracks are 13, 13 and 69 μm respectively. The impact parameter is defined as the closest distance of approach of each track to the primary vertex, assuming the tracks originate from the primary interaction point. The impact parameter resolution is only 13 μm in the transverse plane for high transverse momentum tracks. This is illustrated on Figure 4.

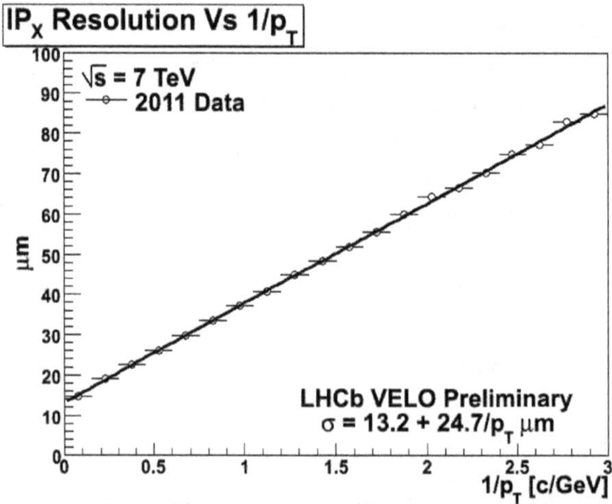

Figure 4. Impact Parameter in X. Measured in 2011 data from a minimum bias sample of tracks with respect to the z of the fitted primary vertex. Events with one reconstructed PV only are used.

4. Conclusions

The Vertex Locator has operated smoothly throughout the 2010-2011 physics runs. 40 pb^{-1} was delivered in 2010, and 1 fb^{-1} has been delivered in 2011. The nominal instantaneous LHCb luminosity was achieved in May 2011. In the VELO the average Signal to Noise ratio is the order of 20, while the cluster finding efficiency is 99.5%. The VELO has the best hit, primary vertex and impact parameter resolution among the LHC tracking detectors.

References

1. LHCb collaboration, P. R. Barbosa-Marinho et al., Vertex locator technical design report, CERN-LHCC-2001-011, http://cdsweb.cern.ch/record/504321.
2. S. Löchner and M. Schmelling, The Beetle Reference Manual, Note LHCb-2005-105, http://cdsweb.cern.ch/record/1000429.
3. G. Haefeli, Contribution to the development of the acquisition electronics for the LHCb experiment, LPHE Master thesis, Lausanne Switzerland, 2004.
4. LHCb collaboration, The LHCb detector at LHC. 2008 JINST 3 S08005, http://iopscience.iop.org/1748-0221/3/08/S08005.

MEASUREMENT OF THE RADIATION FIELD IN ATLAS WITH THE ATLAS-MPX DETECTORS

MICHAEL CAMPBELL[a], ERIK HEIJNE[a, c], CLAUDE LEROY[b,*], JEAN-PIERRE MARTIN[b], GIUSEPPE MORNACCHI[a], MARZIO NESSI[a], STANISLAV POSPISIL[c], JAROSLAV SOLC[c], PAUL SOUEID[b], MICHAL SUK[c], DANIEL TURECEK[c] AND ZDENEK VYKYDAL[c]

[a] *CERN, CH-1211 Geneva 23, Switzerland*

[b] *Laboratoire R.-J.A. Lévesque, Université de Montréal, 2905 Chemin des Services Montréal, QC H3T 1J4, Canada*

[c] *Institute of Experimental and Applied Physics, Czech Technical University in Prague, Horská 3a/22, 12800 Praha 2, Czech Republic*

*E-mail: * corresponding author, leroy@lps.umontreal.ca*

A network of 16 ATLAS-MPX (silicon pixelated) detectors has been installed by the ATLAS-MPX Collaboration at various positions within the ATLAS detector and its environment. The ATLAS-MPX detectors allow real-time measurements of spectral characteristics and composition of the radiation field inside and around the ATLAS detector during its operation. Results obtained with the ATLAS-MPX detectors are reported in this article. They include luminosity measurement obtained with van der Meer luminosity scans and measurement of induced radioactivity in between/after collision.

1. Introduction

Sixteen pixelated silicon ATLAS-MPX detectors are operated at various positions within the ATLAS detector and cavern [1,2]. These detectors allow precise measurement of the spectral characteristics and composition of the radiation field within the ATLAS detector and its environment. These detectors have the capability of providing quantitative real-time information on fluxes and flux distributions of the main radiation species in the experiment, including slow and fast neutrons. The ATLAS-MPX detector (MPX) network has been operated continuously starting from early 2008 (natural radiation background, including cosmic muons measurement), during 2010 and 2011 (collision data) up to nowadays. Distinctive images (frames) are acquired continuously by each detector station, with an exposure time (0.1 ms -- 600 s) adaptable to local particle flux. The ATLAS - MPX network provides independent measurements of LHC luminosity. The MPX detectors close to the beam directly visualize van der Meer luminosity scans. That allows the measurement of the effective overlapping beam sizes and maximum collision rate with MPX detectors. The flux of low energy transfer, high energy transfer particles and MIPs are measured with the MPX detectors. In particular, the MPX detectors provide measurement of the thermal and fast neutrons component of the mixed radiation field as each detector is covered with a mask of converter materials dividing its area into regions sensitive to the type of neutron (slow and fast). Maps of fluences and doses of neutrons and gamma for a given luminosity are obtained from the MPX measurement of the radiation field. The MPX detectors measure the background generated by LHC collisions and the induced radioactivity in between/after collision. The induced activation as measured with an MPX detector is analyzed in terms of short and long decay components. The real-time measurement of the LHC-generated background radiation, allows the validation of background radiation simulation studies.

2. The ATLAS-MPX detectors network in ATLAS

The Medipix2 hybrid silicon pixel device [3] is a position-sensitive device that is composed of a 300μm thick silicon detecting layer with a charge collecting electrode plane divided in 256x256 pixels, each of 55μm x 55μm area. Each pixel is connected to its respective readout chain (preamplifier, double discriminator and digital counter) integrated on the Medipix2 readout chip [3]. The detector is controlled by the Universal Serial Bus (USB) [4] through the "Pixelman" software package [5] using a PC (or a laptop). Settings of the pulse height discriminators provide noise suppression and at the same time determine the input energy of radiation to be detected. Using the adjustable energy threshold, the devices can be used for position and energy sensitive spectroscopic detection of radiation, from energy of 5 keV up to tens of MeV, deposited in the silicon sensitive volume. The ATLAS-MPX detectors can be operated in two modes: the cluster-tracking mode (shortly tracking mode) and the pixel-hits counting mode (shortly counting mode). Both modes are based on simultaneous exposure of all the 256 x 256 pixels to the incoming radiation quanta, during a well-defined, adjustable exposure time. The tracking mode is based on the observation and analysis of shapes of individual clusters generated by individual interacting particle. It is applicable at low detection rates with shorter exposure time to avoid track overlap and enable cluster shape recognition. The clusters are classified according to several criteria: their area (number of adjacent pixels); roundness (comparing cluster area to length of its border); linearity (consistency of activated pixels in the cluster with straight-line track); width of the straight track [6]. Heavy charged particles (such as protons and alpha-particles) deposit energy in several pixels. The shape of their tracks depends on their energy and incidence angle. Typically, heavy charged particles appear as heavy blobs when striking the pixelated plane perpendicularly and as heavy tracks when striking almost parallel to the pixelated plane. For the heavy tracks, the Bragg peak energy deposition can be observed, giving the entry position of the particle. Light particles, such as electrons and positrons with a kinetic energy in the keV range are diffused in the material and have random trajectories. They will leave tracks of different shapes depending on their energy. Minimum ionizing particles (MIPs) produce continuous long straight tracks [6]. The counting mode is used in a high intensity radiation environment (at higher events count rates, above $5\text{x}10^3$counts.$\text{cm}^{-2}\text{s}^{-1}$). In this mode, the number of interactions/hits in individual pixels is counted at selectable threshold settings.

The 16 detectors are positioned as follows [2]: Two devices are between ID and JM plug in front of LAr calorimeters (one on side A and one on side C). Three devices are between TILECAL barrel and TILECAL EB (two on side A and one on side C). One device is on the top of TILECAL barrel and one is on the top of TILECAL EB, both on side A. Three devices are on the muon small wheel chambers (side A) and one is attached to EIL4 part of the muon spectrometer (side A). One device is placed within JF shielding at the back of LUCID detector close to the LHC vacuum chamber. Four devices placed outside of the experiment volume are measuring the radiation environment during LHC operation. One detector sits on top of the JF shielding, two on cavern walls and the last one is in the USA15 cavern.

3. Measurements performed with ATLAS-MPX detectors network

3.1 *ATLAS radiation field measurement*

Each ATLAS-MPX detector can measure in real-time at its position the radiation field that results from any proton-proton collision occurring at the ATLAS interaction point (IP1). Since their operation start up in 2008 until now, it has been demonstrated that the ATLAS-MPX detectors have the capability to measure composition and characteristics of the radiation fields of intensities changing significantly from natural background up to maximal achieved luminosity in "cold" (i.e., far from IP1 and beam line) and "hot radiation" positions (i.e., close to IP1 and beam line), as exemplified in Figure 1 with data from all MPX detectors recorded from 22 to 24 May 2011, showing for each LHC fill the decrease in local flux. The y-axis in Figure 1 shows the number of registered clusters of all types, normalized per cm^2 but scaled for different exposure times of N seconds, for all detectors with their position given by the distance R (in m) from the beam line and the distance D (in m) from IP1 along the Z-axis (N, R, and D are explicitly specified for 6 detectors in Fig. 1). The black contours in the back plane of the figure indicate the number of clusters per cm^2 counted with an ATLAS-MPX detector. This number follows a decreasing exponential behavior from the time of the start until the end of the run indicating the evolution of LHC luminosity.

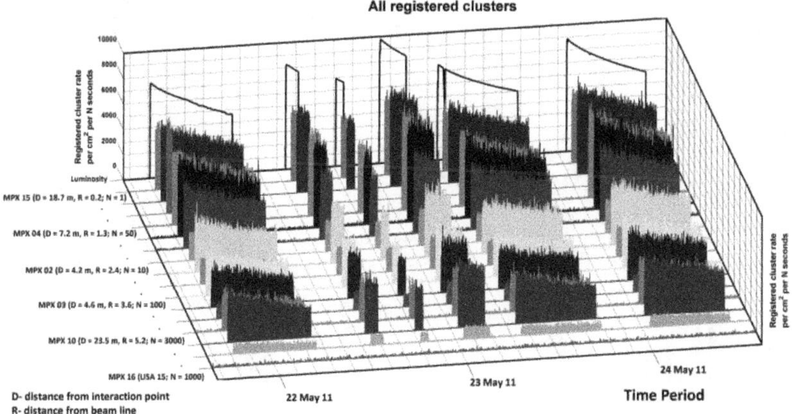

Figure 1. Registered clusters of all types (number of registered events per cm^2 per N seconds of measurement) are shown for all detectors with their positions given by their distance from beam line (R, in m) and distances from IP1 along the Z-axis (D, in m). For clarity, N, D, and R are explicitly given to the left of each plane only for MPX15, MPX04, MPX02, MPX03, MPX10 and MPX16 ordered by increasing value of R. The black contours in the back plane of the figures indicate the LHC luminosity decreasing exponential behavior with time during the selected period from 22 to 24 May 2011.

3.2 *Luminosity measurement (van der Meer luminosity scan method)*

In the case of beam-beam collisions, the ATLAS-MPX detector response is expected to be proportional to the product of the two beam intensities resulting from paired bunches collided in IP1. Then, the ATLAS-MPX detector response (number of clusters.cm^{-2}) is a linear function of the machine integrated luminosity (nb^{-1}) as shown in Figure 2 for the MPX03 detector, located between TILE and EB [2]. ATLAS-MPX detectors positioned close to the beam axis, namely MPX01 (between ID and JM plug), MPX14 (between ID and JM plug on side C) and MPX15 (on the back of LUCID) [2] can directly visualize van der Meer luminosity scans [7,8] enabling an evaluation of the effective overlapping beam sizes and maximum collision rate. Figure 3 shows, as a function of time, the cluster rates of events recorded with the MPX15 detector continuously during the Van der Meer (VdM) transverse scans. The cluster rates measured during these scans as function of time appears as bell-shaped curves with their maximum achieved at a time corresponding to zero beam separation. Then, for all ATLAS-MPX detectors considered, a Gaussian with a constant background is used to fit the cluster rates of events recorded during the horizontal and vertical luminosity scans. Application of the fit procedure gives the value of the effective horizontal (Σx) and vertical (Σy) profiles of the overlapping beams. The results obtained for Σx and Σy from fitting a Gaussian with a constant background to MPX15, MPX01 and MPX14 Van der Meer scans data recorded during fills 1059, 1089 and 1783 are summarized in Figure 4.

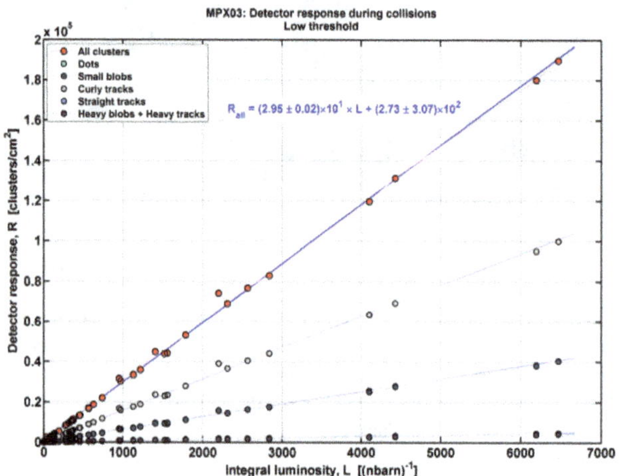

Figure 2. The MPX03 detector response (cluster.cm^{-2}) recorded at low threshold during collisions is shown as function of the integrated luminosity (nb^{-1}). The response recorded for the various types of clusters is a linear function of the integrated luminosity over a large range of values [2].

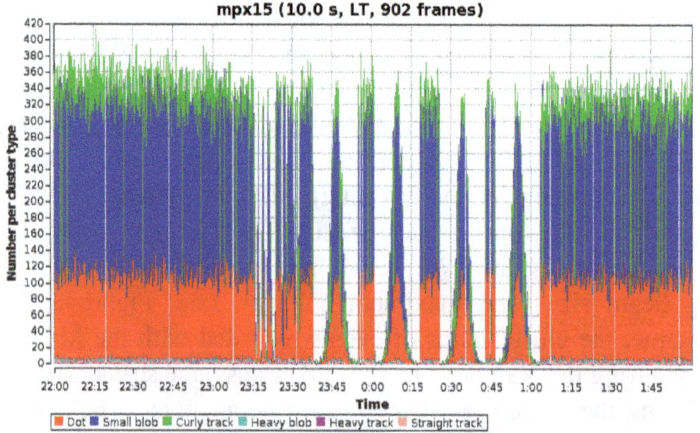

Figure 3. Visualization with MPX15 of two horizontal van der Meer scans followed by two vertical scans during fill 1089 on 9th May 2010 (GMT) with a frame length time of 10.0 s.

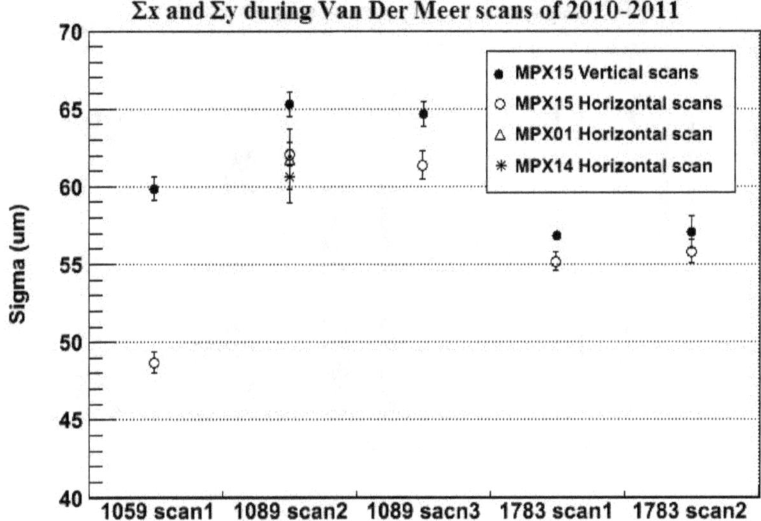

Figure 4. Σx and Σy obtained from fitting a Gaussian with a constant background to MPX15, MPX01 and MPX14 van der Meer scans data recorded during fills 1059, 1089 and 1783.

3.3Measurement of activation in ATLAS environment

The residual material activation in the ATLAS environment due to the high flux of radiation during collisions is measured through the detection of gamma decay of radioisotopes produced by the interaction of collision particles with these materials. The signal recorded with MPX13, located between ID and JM shielding plug is shown as an example in Figure 5 for several fills from 08-09-2011 and 16-09-2011. The detector is operated in tracking mode for an exposure time of 0.1 s. Colors represent all tracks (red) and straight tracks (blue). The latter are only present during collision periods (excepted for a small number of muon tracks from cosmic background) and indicate the successive fills. The signal due to activation is clearly visible at the end of runs and the activation signal increases with time as a result of adding up long decay components. *Signal during no runs is caused almost only by photons and electrons (in red - Figure 5).*

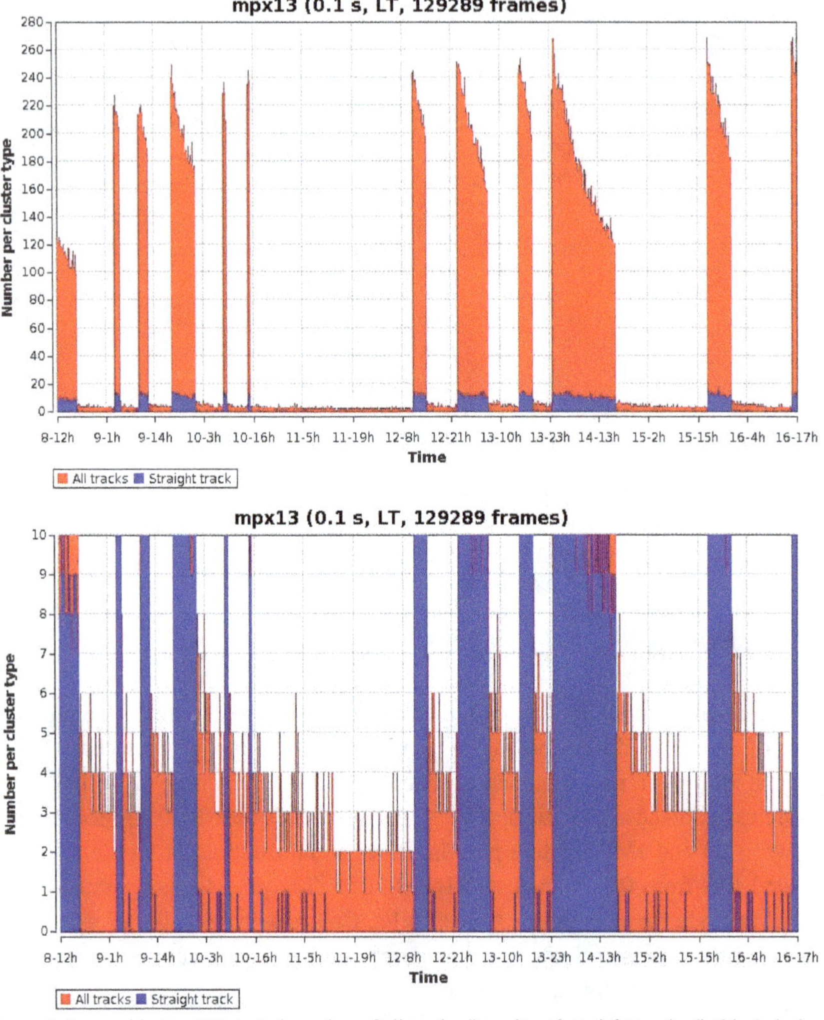

Figure 5. (top and bottom) Detected number of all tracks (in red) and straight tracks (in blue) during beam time versus background with the detector MPX13 during the period from 08-09-2011 to 16-09-2011. The straight tracks are only present during collision periods and indicate the successive fills (see text). Signal during no runs is caused almost only by photons and electrons (in red). The difference between the two figures is only the y-axis scale

Comparison with dose rate calibration measurements with known gamma sources (^{60}Co and ^{137}Cs) permits an estimate of a mean photon ambient dose equivalent rate (in μSv.h^{-1}). The induced activation measured with an MPX detector is analyzed in terms of short and long decay components with the count rate measured after the end of collisions. It usually shows up to four decay components with the following half-lives: a few minutes, a few hours, several tens of hours and longer than 50 days.

4. Conclusions and outcome

The network of 16 ATLAS-MPX (MPX) detectors is providing real-time measurement of composition and characteristics of the radiation environment inside and around the ATLAS detector during its operation. These measurements include, among others, measurement of the composition and characteristics of the radiation fields of intensities from natural background up to maximal achieved luminosity, measurement of luminosity obtained with van der Meer luminosity scans, measurement of induced radioactivity in between/after collision. The measurements of the thermal neutron fluence per luminosity unit during collisions measured for each ATLAS MPX detector are found in good agreement with Monte Carlo simulations [9]. Reconciliation of simulated proton and muon currents with HETP and MIPs measurement, respectively, requires further studies with detailed Monte Carlo simulations and removal of ambiguity between energetic protons and neutron signals, better determination of the direction of entering MIPs on a sensor area and discrimination of orthogonally incident MIPs from low energy electrons and photons. This can be possibly achieved by adding TMX detectors [10] to the network of ATLAS-MPX detectors. The TMX detector with its time over threshold (TOT) capability is analogous to a Wilkinson type ADC and allows direct energy measurement in each pixel. The ATLAS-MPX network will possibly provide absolute fluxes of fast neutrons and distinguish directionality (backward from forward) of neutron motion at their position. Then, maps of fluences could be provided possibly enabling the calculation of safety factors to be used in the future for assessment of the survival of the ATLAS electronics exposed to increasingly high radiation levels, as foreseen in future ATLAS and LHC upgrades The upgraded network will possibly further contribute to estimates of SEE effects affecting the ATLAS and LHC machine electronic devices [11]. The TMX detectors positioned close to Tile and Liquid Argon calorimeters could provide measurement of the missing transverse energy and study of hadron leakage into the muon detector system. In addition an increase of the data readout speed by factor of 50 is foreseen [12], thus reducing the dead time of the measurements significantly. This improved readout system will use a single-wire communication, simplifying the present cable structure.

References

1. M. Campbell, C. Leroy, S. Pospisil, M. Suk, Measurement of Spectral Characteristics and Composition of Radiation in ATLAS by MEDIPIX2 USB Devices, Project proposal at https://edms.cern.ch/document/815615.
2. M. Campbell et al. (ATLAS-MPX Collaboration Report), " ATLAS MPX Report on Data Analysis of the Radiation Field in ATLAS" (2011), in progress.
3. Medipix collaboration http://medipix.web.cern.ch/MEDIPIX
4. Z. Vykydal, J. Jakubek, S. Pospisil, Nucl. Instr. and Meth. A563 (2006) 112.
5. D. Turecek et al., "Pixelman: a multi-platform data acquisition and processing software package for Medipix2, Timepix and Medipix3 detectors, 12th International Workshop on Radiation Imaging Detectors Juy 11th-15th 2010, Robinson College, Cambridge U.K. 2011 JINST 6 C01046; T. Holy et al., Nucl. Instr. and Meth. A 563 (2006) 254.
6. J. Bouchami et al., Nucl. Instr. Methl A 633 (2011), 187 ; T. Holy et al., Nucl. Instr. Meth A 591 (2008), 287.
7. S. van der Meer, Calibration of the effective beam height in the ISR, CERN-ISR-PO-68-31; 44 ISR-PO-68-31 (1968).
8. ATLAS Collaboration, Luminosity Determination in pp Collisions at using the ATLAS Detector in 2011, ATLAS-CONF-2011-130.
9. Mike Shupe, private communication, 2010.
10. X. Llopart et al., Nucl. Instr. and Meth. A 581, (2007), 485.
11. G. Papotti, "LHC status and performance in 2011", presentation at the 13th ICATPP Conference, Villa Olmo, Como, October 2011, these proceedings.
12. V. Kraus et al., " FITPix - fast interface for Timepix pixel detectors" 12th International Workshop on Radiation Imaging Detectors July 11th-15th 2010, Robinson College, Cambridge U.K. 2011 JINST 6 C01079

Alignment of the LHCb tracking system

R. Märki on behalf of the LHCb alignment group
Ecole Polytechnique Fédérale de Lausanne (EPFL),
Lausanne, Switzerland,
E-mail: raphael.marki@epfl.ch
www.epfl.ch

The LHCb is an experiment at LHC dedicated to precision measurements of CP violation and rare decays in the b and c sectors. The experiment features a tracking system consisting of silicon strip detectors and straw tube drift chambers up- and downstream of the magnet to precisely measure the vertex position and the momentum resolution of the particles travelling through the detector. An important ingredient to the track parameter resolution is the spatial alignment of the tracking system on which we report here.

Keywords: LHCb; alignment; tracking system

1. The LHCb tracking system

The LHCb experiment is a single-arm forward spectrometer designed to measure CP violation and rare decays in the b and c quark sector. The tracking system of LHCb is composed of four subdetectors, namely: the VErtex LOcator (VELO),[1] a silicon strip detector which is the subdetector closest to the collision point, the Tracker Turicensis (TT),[2] another silicon strip detector situated between the VELO and the magnet, the Inner Tracker (IT),[2] the third silicon strip detector covering the innermost part of three tracking stations down-stream magnet, the Outer Tracker (OT),[3] straw tube drift chambers covering the outer part of the stations downstream the magnet.

2. Track alignment at LHCb

All tracking subdetectors have been surveyed during assembly and after installation. To measure their position with an even higher precision, real tracks are used in a procedure which is called track alignment.

At LHCb, track alignment is using a sample of selected good quality tracks to calculate a χ^2 based on the track fit residuals[4] (ie. the distance between the hit in the detector and the reconstructed track). The χ^2 also gets a contribution proportional to the difference between the new alignment constants and the survey constants in order to take into account the initially measured position. In a second step, an algorithm minimizes this χ^2 as a function of alignment parameters which are called degrees of freedom.

The degrees of freedom of the alignment are chosen to be either translations (Tx, Ty, Tz) or rotations (Rx, Ry, Rz) in a cartesian coordinate system. At each level of the detector (whole detector, layer, sensor, etc.) the degrees of freedom which are aligned are chosen differently depending on the sensitivity one has with the selected tracks.

Degrees of freedom for which the alignment algorithm has very low sensitivity are called weak modes. Movements of detector elements along such degrees of freedom leave the χ^2 invariant. A remarkable example of a weak mode is the translation along the beam axis since many tracks are parallel to this direction. Weak modes can also be scaling and shearing patterns. Most weak modes are successfully suppressed by the survey contribution to the χ^2 since this contribution never stays invariant with respect to alignment parameter changes.

Different kind of data sets are used at LHCb for alignment purposes. Each of them giving different constraints to the system and allows some different degrees of freedom to be aligned. The first category are tracks from TED runs[a] or from collision of the beam with the residual beam in the beam pipe. They usually have small angles almost parallel to the beam axis. The second kind of tracks are those from standard proton-proton collisions happening within the VELO. These samples contain various angle tracks.

Another very important constraints are the overlapping regions between subdetectors and between sensors within a subdetector. They help to link different detector parts to each other during the alignment procedure.

3. Internal VELO alignment

At LHCb, the first step of the track alignment is the internal VELO alignment where only the elements of this subdetector are considered and aligned. In the VELO, two sensors are fixed to set the global position of LHCb and the scale.

The first alignment[5] was performed starting from the survey constants

[a]particles induced by the interaction of LHC beam with a beam absorber located in the injection line at about 300 m upstream the detector

on the TED data using a method based on Millepede.[6] Later the alignment was evaluated using proton-proton collision and proton collision with the residual gas in the VELO region by an alignment method based on Kalman filter.[4]

Since the VELO is frequently moved in and out, it is essential to monitor the stability of the misalignment of one half of the detector with respect to the other (called 2 half alignment). The method used is the so called PV left-right method. The principle is to compute the primary vertices with each half of the detector separately and then to calculate the average distance between left and right PVs. This distance evaluates the 2 half misalignment. During the full 2010 data taking the 2 half alignment, in the direction of the main movement along the x axis, is measured to be stable within $\pm$ 5 micron.

In the track alignment, one also uses explicitely primary vertices to align the VELO halves. The tracks comming from primary vertices are constraint to have a common origin and the χ^2 contribution is recalculated with these new tracks. In addition, many overlap tracks are added to "link" the two halves.

One of the main weak modes that could affect the VELO is a twist at the module level around the beam axis (z axis). This kind of misalignment biases the impact parameter (IP) of the tracks with a clear dependence on the sensor position, the other overall performance being only slightly affected. Selecting tracks which cross many or all modules helps to constrain this weak mode. Such tracks are very common in the beam-gas selection which includes standart p-p collision, satellite collisions happening at $\pm$ 700 mm around the nominal collision point as well as beam-gas events. The results on the IP show a clear reduction of this weak mode when one uses the alignment obtained with this data sample.

3.1. *VELO alignment performance*

In the VELO, the sensor alignment evaluated by the bias of the residual is better than 4 μm which leads to a PV resolution (x,y,z) with 25 tracks of (13.0, 12.5, 68.5) μm in data and (10.7, 10.9, 58.1) μm in Monte Carlo studies, as shown in Fig. 1. Thus there is still room for improvement.

4. Global alignment

The second step of the LHCb alignment procedure[7,8] is the alignment of TT, IT and OT by χ^2 minimization with the same method based on a

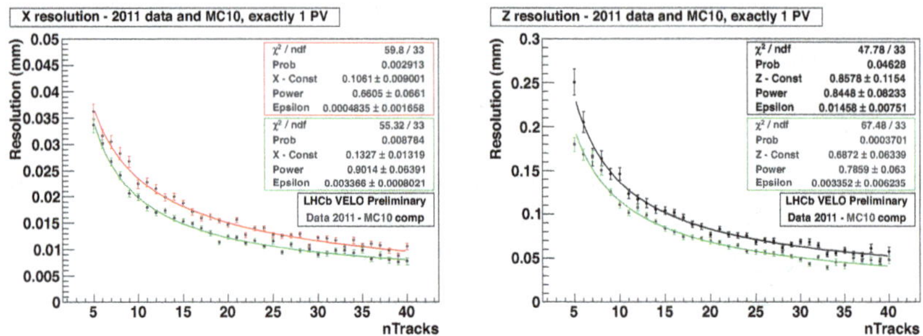

Fig. 1. PV resolution on x (left) and z (right) with data and Monte Carlo simulation.

Kalman filter used in the VELO. At this stage, the alignment uses a samples of tracks crossings the whole detector requiring some additional kinematic cuts (eg. high momentum, high angle, good quality, etc.).

Like in the VELO, some elements are fixed along z to survey constants since the alignment has a low sensitivity to this degree of freedom for most detector parts. More precisely one fixes the z position of the whole TT subdetector and also of two layers in OT which constrain the global position of IT and OT.

The momentum scale at LHCb depends considerably on the description of the magnetic field and also on the x (bending plane) position of IT and OT. Therefore even small misalignments and shearings along x lead to significative momentum and q/p biases. To constrain the momentum scale and get rid of the q/p bias, mass constraints are implemented in the alignment algorithm. The used reasonances are the ones of $J/\psi \to \mu\mu$ and $D^0 \to K\pi$. To constrain the masses, the daughter particle tracks are modified to have a common origin vertex and to have the known invariant mass of the mother particle (the PDG value is used). With these new tracks, the residuals are computed again and are then used in the χ^2 minimization.

The track alignment of IT, TT and OT has also a very low sensitivity along the y direction (vertical axis) since the stereo angle between the four planes of each station is only 5 degrees. However, it is very important to know their vertical position since this determines the detector acceptance, crucial in track finding and simulation. To evaluate the alignment along this direction, one can extrapolate VELO tracks in events without magnetic field to TT, IT or OT and search for corresponding hits in the detector. By plotting the y distribution of these hits, edges and gaps are determined which correspond to the edges and gaps of the sensors and give their exact position.

4.1. *Global alignment performance*

The alignment precision can be evaluated by the residual bias and is 15.6 μm for IT and 16.7 μm for TT as shown in Fig. 2. Also for the TT and the tracker station alignment, there is room for further improvements.

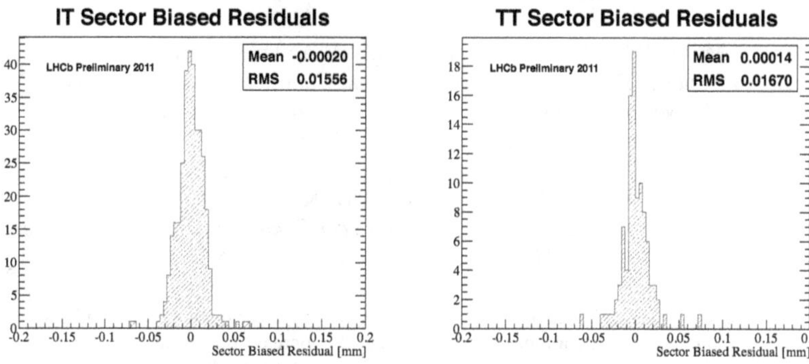

Fig. 2. Mean bias of residuals distributions for IT (left) and TT (right).

5. Summary

A performant alignment method has been developped which helps to determine accurately the position of the detector parts at LHCb. Due to geometrical reasons it is not sensitive to a few degrees of freedom (eg. the z position) which are fixed to the surveyed constants. Where it would be possible an alternative method is used for the degree of freedom to which the track alignment is not sensitive. Finally, the position of the silicon strip detectors are known with a precision of 4 μm for the VELO, 15.6 μm for IT and 16.7 μm for TT.

References

1. A. Leflat, *these proceedings* (2011).
2. M. Tobin, *these proceedings* (2011).
3. B. Storaci, *these proceedings* (2011).
4. W. Hulsbergen, *Nucl. Instr. and Meth. A* **600**, p. 471 (2009).
5. S. B. et al., *Nucl. Instr. and Meth. A* **618**, p. 108 (2010).
6. M. G. S. Viret, C. Parkes, *Nucl. Instr. and Meth. A* **596**, p. 157 (2008).
7. L. N. et al., *CERN-LHCB-2008-066* (2008).
8. A. H. et al., *CERN-LHCB-2008-065* (2009).

PERSPECTIVES OF THE PIXEL DETECTOR TIMEPIX FOR NEEDS OF ION BEAM THERAPY

M. MARTIŠÍKOVÁ*, B. HARTMANN AND O. JÄKEL
Medical Physics in Radiation Oncology, German Cancer Research Center DKFZ, Im Neuenheirmer Feld 280, D-69120 Heidelberg, Germany
**m.martisikova@dkfz.de*

C. GRANJA, J. JAKUBEK
Institute of Applied and Experimental Physics, Czech Technical University in Prague 12800 Prague, Czech Republic

Radiation therapy with ion beams is a highly precise kind of cancer treatment. In ion beam therapy the finite range of the ion beams in tissue and the increase of ionization density at the end of their path, the Bragg-peak, are exploited. Ions heavier than protons offer in addition increased biological effectiveness and decreased scattering. In this contribution we discuss the potential of a quantum counting and position sensitive semiconductor detector Timepix for its applications in ion beam therapy meaurements. It provides high sensitivity and high spatial resolution (pixel pitch 55 μm). The detector, developed by the Medipix Collaboration, consists of a silicon sensor bump bonded to a pixelated readout chip (256 × 256 pixels with 55 μm pitch). An integrated USB-based readout interface together with the Pixelman software enable registering single particles online with 2D-track visualization. The experiments were performed at the Heidelberg Ion Beam Therapy Center (HIT), which is a modern ion beam therapy facility. Patient treatments are performed with proton and carbon ions, which are accelerated by a synchrotron. For dose delivery to the patient an active technique is used: narrow pencil-like beams are scanned over the target volume.

The possibility to use the detector for two different applications was investigated: ion spectroscopy and beam delivery monitoring by measurement of secondary charged particles around the patient. During carbon ion therapy, a variety of ion species is created by nuclear fragmentation processes of the primary beam. Since they differ in their biological effectiveness, it is of large interest to measure the ion spectra created under different conditions and to visualize their spatial distribution. The possibility of measurements of ion energy loss in silicon makes Timepix a promising detector for ion-spectroscopic studies in patient-like phantoms.

Unpredictable changes in the patient can alter the range of the ion beam in the body. Therefore it is desired to verify the actual ion range during the treatment, preferably in a non-invasive way. In order to overcome the limitations of the currently used PET technique, in this study we investigate the possibility to measure secondary charged particles emerging from the patient during irradiation. It was demonstrated that the Timepix detector is able to resolve and visualize this emerging radiation. The investigated dependence of the signal on the beam energy between 89 and 430 MeV/u shows that for all the investigated energies some signal was registered. Its pattern corresponds to ions. Differences in the total amount of signal for different beam energies were observed. The time-structure of the signal was moreover correlated with that of the incoming beam. This shows that we register products of prompt processes, which are less likely to be influenced by biological washout processes than the signal registered by the PET techniques coming from decays of beam-induced radioactive nuclei.

The studies discussed in this contribution demonstrate that the Timepix detector provides measurements attractive for needs of ion beam therapy. To fully exploit its capabilities further research is needed.

1. Ion Beam Therapy

Radiotherapy with ion beams is motivated by the finite range of ion beams and the presence of the Bragg-peak in the dose distribution at the end of their range. Ion beams are used when critical structures close to the irradiated target volume have to be spared from radiation and at the same time high dose is required to be delivered to the target. Ions heavier than protons offer in addition lower scattering and increased biological effectiveness due to the higher local ionization densities achieved. On the other hand, they undergo nuclear interactions with the nuclei of the tissue which can lead to creation of nuclear fragments. Those fragments exhibit different biological effectiveness than the primary ions.

The medical application of ion beams started at the University of Berkeley, California, USA in 1954 by proton radiotherapy. Until the end of 2010 more than 84,000 patients have been treated world-wide with hadron beams, most of them with protons [1]. The Heidelberg Ion Beam Therapy Center (HIT) located in Heidelberg, Germany is an example of a modern ion beam therapy facility. It started clinical operation in November 2009. Currently proton and carbon ion treatments are provided, however it is designed for ion species from protons to oxygen. For ion acceleration a synchrotron is used, providing energies of ion beams with a range in water between 2 and 30 cm. For dose delivery to the patient a dynamic beam delivery technique is used [2]. Narrow pencil-like beams are scanned over the target volume and different depths in tissue are accessed by changing the beam energy.

1.1. *Investigated Applications of Pixel Detectors with High Spatial Resolution*

Although the first patient treatment with ion beams took place almost half a century ago, it is still a vital research field which aims to further improve this kind of therapy. We identified applications which could benefit from the capabilities of the Timepix detector. They include ion spectroscopy and beam delivery monitoring by measurement of secondary charged particles around the patient.

During carbon ion therapy, a variety of ion species is created by nuclear fragmentation processes of the primary beam. Since they differ in their biological effectiveness, it is important to measure the ion spectra created under different conditions. Especially, high precision spectroscopy inside patient-like phantoms is of large interest. The advantage of the Timepix detector is its small size in comparison to the large apparati previously used for ion identification [3].

Due to unpredictable internal biological changes during the therapy (swelling, shrinkage of the tumor), organ motion or incorrect patient positioning, the delivered dose distribution can differ from the planned one [4]. Therefore it is desired to verify the actual ion range and to monitor the spatial dose distribution in the patient during the treatment, preferably in a non-invasive way. The current approach uses a PET camera to register photons coming from decays of $ß^+$-radioactive nuclei created in nuclear interactions of the beam with tissue. The PET-method, however, suffers mainly from low induced activities in comparison to PET diagnostics [5] and the spatial resolution is limited by biological washout processes like blood flow which can move the activated nuclei [6,7]. Charged particles emerging from the patient during the therapy are an alternative to photons [8]. The goal of this approach is to increase the signal and to improve the spatial resolution, since ions are predicted to be created in direct processes. In this initial study we investigate the possibility to measure the emerging secondary charged particles with the Timepix-detector.

2. Timepix Detector

The Timepix detector [9] was developed by the Medipix Collaboration as a position sensitive quantum counting radiation detector with high sensitivity and enhanced signal-to-noise ratio. We used a 300 μm thick silicon sensor bump-bonded to the pixelated readout chip with 256 × 256 pixels with 55 μm pitch. The detector provides a sensitive area of 1.4 x 1.4 cm^2. It was read out using an USB based interface [10] and the Pixelman software [11], which provide also online 2D-visualization of the measured signal. The pixels of the detector can operate in three different modes: counting mode, time-over-threshold mode for energy measurements and time mode measuring the time of particle interaction with the detector. The response of all the detector pixels in time-over-threshold mode is calibrated in terms of energy [12]. Analysis of the obtained data was performed with self-written software based on Matlab and the C programming language. This includes also a custom made pattern recognition algorithm.

The high spatial resolution allows operation of the detector as an active nuclear emulsion [13], registering single particles online. We investigated the response of the Timepix detector to therapeutic ion beams and the possibility to differentiate between protons and carbon ions of different energies. Furthermore we studied the possibility to register secondary charged radiation leaving the patient during the therapy and properties of this radiation.

3. Results

3.1 Towards Ion Spectroscopy

Detectors from the Medipix-family were previously shown to be able to differentiate between different types of radiation like photons, electrons and heavy charged particles [14]. Within this study [15] differences in properties of pixel-clusters produced by protons and carbon ions were analyzed. When the beam impinges on the detector perpendicularly, round clusters arise, as shown in figure 1 a) and b). Here proton and carbon ion beams with the same range in water were used. The cluster size shows strong differences between the two particle types. Moreover, the measured distribution of the cluster energy in the time-over-threshold more is also different (see Figure 1 c), what is given by the difference in stopping power. Thus the cluster size and cluster energy are promising parameters to distinguish different ion types [15].

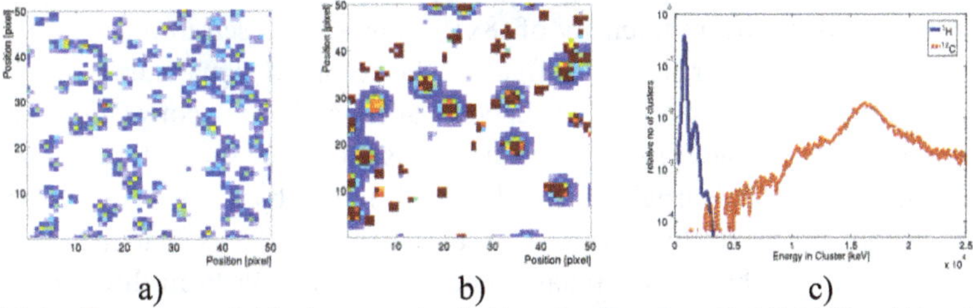

FIGURE 1. Clusters recorded in time-over-threshold mode: a) protons, E=143 MeV and b) carbon ions, E=271 MeV/u with the same range in material. c) Quantitative distribution of the cluster energy. Both images from [15].

3.2 Measurement of Secondary Charged Particles Emerging from a Phantom During Carbon Ion Beam Therapy

In this study [16] we investigate the possibility to measure the emerging secondary charged particles with the Timepix-detector with the aim to monitor the dose delivery to the patient in a non-invasive way. To simulate a patient, a head phantom from Alderson Phantom (Radiology Support Devices Inc., Long Beach, CA, USA) was used. It contains human bones and rubber to simulate soft tissues. One point in the brain region was chosen and positioned in the isocenter. During irradiations a static pencil beam was directed to the chosen point of the phantom. The beam has an approximately Gaussian fluence profile with a FWHM of about 10 mm. The water-equivalent thickness of the phantom in the beam direction was about 16 cm. The emerging secondary radiation was measured with the Timepix detector positioned on the beam axis, directly behind the phantom. The detector was operated in time-over-threshold mode. The signal obtained behind the phantom (in the axis of the primary beam) was compared for energies of 89, 250, 265 and 430 MeV/u beam. Furthermore, the time-structure of the signal was investigated, looking for a correlation of the signal with the time-structure of the initial beam.

In figure 2 the images of signals registered by Timepix are shown [16]. For all ion beam energies some signal is registered. Moreover, the signal differs for different energies. In figures a) – c) the range of the primary carbon ions is shorter than the water equivalent thickness of the phantom at the irradiated position (approx. 16 cm). Therefore we observe here secondary radiation. According to [14] round clusters correspond to ions, in this case products of nuclear fragmentation of the primary beam. At the highest energy level (Fig. 2 d) the range of the primary beam of 30 cm in water is longer than 16 cm. Therefore in this case we register also primary carbon ions.

The time structure of the signal was investigated [16] at the lowest energy of 89 MeV/u, with a corresponding water-equivalent range of 2 cm. The highest currently available beam intensity of $8x10^9$ ions/s was used in order to increase the measured signal per frame. In figure 3 the recorded signal summed over the whole detector area is shown as a function of time. A clear correlation with the time structure of the primary beam is visible: the spill structure of the synchrotron is periodical with 5 s of beam, followed by a few seconds without beam before the next spill starts. This measurement shows that the registered signal is prompt. This is beneficial, since signals coming from direct processes are less likely to be influenced by movement of the source and biological washout processes.

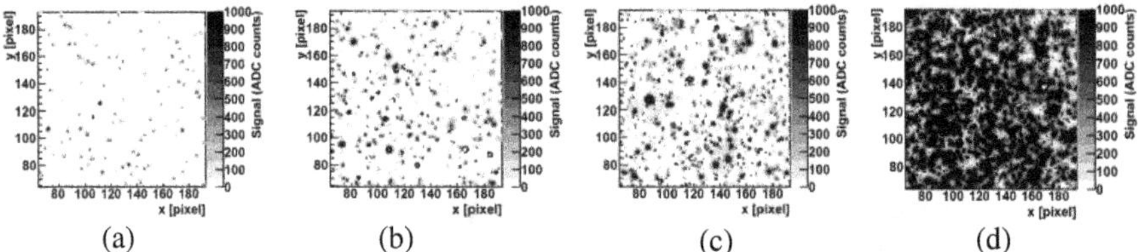

Figure 2. Charged particle radiation leaving the head phantom recorded by the Timepix in time-over-threshold mode at different energies over the whole energy range of carbon ions available at HIT: a) E=88.8 MeV/u (r_{water}=2cm), b) E=250.1 MeV/u (r_{water}=12.5cm), c) E=265.0 MeV/u (r_{water}=13.8cm) and d) E=430.1 MeV/u (r_{water}=30cm). The images [16] show the central quarter of the detector and correspond to 7 x 7 mm^2. The beam intensity was set to the currently lowest nominal value of 5×10^6 ions/s. The only exception was the measurement at the lowest energy, where the intensity was set to 8×10^7 ions/s in order to increase the number of registered events.

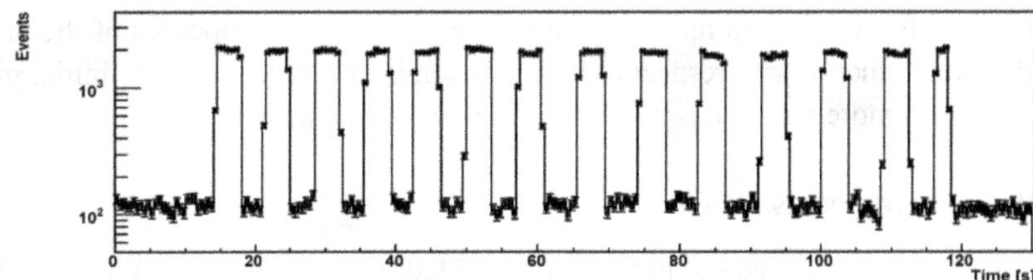

Figure 3. Time structure of the signal recorded behind the phantom for the lowest beam energy of 89 MeV/u with water-equivalent range of 2 cm. Image from [16].

4. Summary and Conclusions

Ion beam radiation therapy using scanning ion beams is a high precision treatment technique. Two dimensional silicon-based detectors provide high spatial resolution and online readout and are therefore attractive for improvement of quality measurements which are needed to safely perform this kind of therapy. We have shown that the Timepix detector based on crystalline silicon, provides valuable measurements with a potential of a significant benefit for ion beam therapy.

The Timepix-detector enables to visualize single ion tracks and nuclear interactions in the sensor. The size of the measured clusters and the energy of the cluster were found to differ for protons and carbon ions of the same range in the material. Therefore we conclude that Timepix is a promising device for ion spectroscopic studies, which can provide useful information for calculation of biologically effective dose.

Due to the high precision of the dose application by ion beams in ion beam therapy, it is desired to monitor the ion range in the patient and to determine the spatial dose distribution in-vivo. For the purpose of beam range monitoring, radiation leaving a patient-like phantom during irradiation by carbon ion beams in a therapy-like setting was investigated. The Timepix detector was used for measurements of energy deposition in silicon. Radiation leaving the head, with a water equivalent thickness in the beam direction of about 16 cm, was measured for four beam energies with ranges in water between 2 and 30 cm. For all energies secondary radiation could be registered. Furthermore, it was possible to correlate the measured radiation to the time-structure of the beam delivery, even at the lowest energy with the corresponding water-equivalent range of 2 cm. This shows that the detector response is caused by prompt radiation. Shape and size of the measured clusters correspond to ions. What information on the beam range can be gained by measurements of those ions will be investigated in further work.

Finally, we have to mention that the investigation of properties of the pixel detector Timepix with respect to its use in ion beam therapy is in its initial phase and much more research is needed to fully exploit its potential.

Acknowledgments

This work was carried out in frame of the Medipix Collaboration. We thank HIT for providing the beam time and support. M. Martišíková was funded by "Deutsche Forschungsgemeinschaft" (German Research Foundation), contract number MA4437/1-2.

References

1. Home page of the Particle Therapy Co-Operative Group http://ptcog.web.psi.ch
2. T. Haberer et al., *Nucl. Instr. Meth.* A **330,** 296-305 (1993).
3. E. Haettner, H. Iwase and D. Schardt., *Rad. Prot. Dosim.* **122**, 485-487 (2006).
4. K. Parodi et al., Nucl. Instrum. Meth. A **591** 282-286 (2008)
5. W. Enghardt, et al., Radiotherapy and Oncology **73** S96-98 (2004).
6. F. Pönisch, et al., Phys. Med. Biol. **49** 5217-5232 (2004).
7. A. Knopf, et al., Phys. Med. Biol. **54** 4477-4495 (2009).
8. P. Henriquet, PhD Thesis, Insitut de physique nucleaire de Lyon, University of Lyon (2010).
9. X. Llopart et al., *Nucl. Instrum. Meth.* A **581,** 485-494 (2007).
10. Z. Vykydal et al., *Nucl. Instrum. Meth.* A **563,** 112-115 (2006).
11. D. Tureček et al., *JINST* **6,** C01046 (2011).
12. J. Jakubek et al., *Nucl. Instrum. Meth.* A **633,** S262-S266 (2011).
13. Z. Vykydal at al., Proceedings of the 9th ICATPP Conference, World Scientific Publishing, Singapore, 779-784 (2006).
14. J. Bouchami, et al., Nucl. Instr. Methods A **633**, S187-189 (2011).
15. B. Hartmann et al., *Presentation on the Int. Workshop on Rad. Imaging Det., Abstract book,* p 189 (2011).
16. M. Martišíková et al., *Presentation on the Int. Workshop on Rad. Imaging Det.*, Abstract book p 200 (2011).

CMS PIXEL DETECTOR FOR THE PHASE I UPGRADE AT HL-LHC

ALBERTO MESSINEO*, *On behalf of the CMS Collaboration*

Physics Department "E.Fermi", University of Pisa,
L.go B.Pontecorvo n.3 Pisa, 56127, Italy
**E-mail: alberto.messineo@df.unipi.it*

The CMS pixel detector is part of the complex tracking system of the CMS experiment at the LHC collider (CERN, Geneva CH). It has been designed for a stable operation and optimal performance at the LHC instantaneous luminosity $L_{max} = 10^{34}$ $cm^{-2}s^{-1}$. The future plans of the LHC collider envisage an increase of the luminosity up to $2.2 \times L_{max}$ with 7 TeV per proton beam, namely the high luminosity upgrade (HL-LHC) phase I * . In order to maintain the high level of accuracy and efficiency of the tracking in this new challenging condition, a CMS pixel detector upgrade phase I program has been set up. The main goals of the upgrade activity are the material budget reduction in the tracking volume and the increase of the number of hits associated to a charged track. This paper gives an overview of the upgrade project, describes the planned R&D activities and focuses on the expected improvements of the new CMS pixel detector system.

Keywords: LHC; CMS; Tracking; Pixel Detector; Upgrade.

1. Introduction

The CMS pixel detector is the tracking device closest to the interaction point, crucial for the efficient track finding and the accurate reconstruction of secondary vertices[1,2] . It has been installed and commissioned in the CMS experiment in 2008, after about 15 years needed for design, development and construction[3] . The pixel detector system covers hermetically the tracking volume, up to pseudorapidity $|\eta| \leq 2.5$, and the current layout consists of three barrel layers (BPIX) and two disks symmetrically placed on each side of the interaction point (FPIX). The performances of this detector during the CMS data taking periods are discussed in reference[4] . The

*S.Myers, contribution to "Europhysics Conference on High-Energy Physics 2011".

pixel detector has been designed to operate at an instantaneous luminosity $L_{max}=10^{34}$ $cm^{-2}s^{-1}$ and to be radiation tolerant up to an integrated fast hadrons fluence corresponding to a two years operation at L_{max} for the innermost barrel layer.
The LHC machine will upgrade the injectors chain in order to reach at least twice L_{max} after the shut-down foreseen in 2017. The degradation of detector performance by the accumulated radiation damage and by the increase of instantaneous luminosity implies a full replacement of the pixel detector, with an improved design instead of a simple replacement of damaged parts with identical components. The CMS pixel upgrade phase I project has been set up[5]. Given the allocated resources and the short time available for design, construction and commissioning, this upgrade will address technical solutions based on robust and already available technologies.
The organization involves the collaboration of national production centers that share the planned activity[a]. The present schedule foresees the completion of the R&D phase in the fall of 2013, the starting of modules construction in mid-2014 and a long commissioning of the full assembled pixel detector in 2016 in order to minimize the impact on the CMS experiment during following data taking periods. The installation in the tracking system is foreseen in 2017.

2. The Upgrade Project

The replacement of the current pixel detector with an improved design is justified by technical and physics driven arguments. The main motivations are:

(a) The track impact parameters resolution is affected by the multiple scattering, especially at low momentum. A global reduction of the pixel material budget in the tracking volume is mandatory to improve the pixel performance.
(b) Parts of the current detector need to be replaced due to radiation damage of sensors and electronics.
(c) The performance of the current Read Out Chip (ROC) is excellent and the radiation tolerance is enough for HL-LHC Phase I[6]. Nevertheless detailed ROC simulation indicates that pixel data losses become un-

[a]The centers involved in the project are: Central European Consortium (DE), CERN (CH), Desy Laboratory (DE), INFN Consortium (IT), National Taiwan University (TW) and Paul Scherrer Institute (CH) for the BPIX upgrade; FermiLab (US) and Purdue University (US) for the FPIX upgrade.

acceptable at luminosity 2 × L_{max}, reaching a level of about 15% and even worse in case of LHC operation at 50 ns. This affects the track reconstruction with inefficiency that appears in the central η region of the tracking volume.

(d) In collision events with high track density the track reconstruction is time consuming. The fake track rate increases rapidly due to the long extrapolation distance between current pixel and the rest of the tracking system. A new intermediate pixel layer between the two systems is desirable.

Moreover the upgrade must observe the following constraints:

(a) The detector performance should not be degraded and must be improved despite the new challenging operational conditions.
(b) The large fraction of CMS tracking services (i.e. readout optical lines, power electrical lines, main cooling piping) will remain unchanged. The present equipment should fulfill all the technical needs for the new pixel detector.
(c) The new pixel system should fit inside the present tracker envelope since no other upgrade is considered.
(d) The upgraded detector should have a minimal impact on CMS and the operation of the CMS tracking system must not change significantly after the upgrade.

3. Upgraded Pixel Detector

The main targets of the upgrade project are: layout, material budget and readout electronics. Details are discussed in the following.

3.1. *Layout*

The upgraded pixel detector layout is sketched in fig.1. The sensitive volume includes four barrel layers and three disks on each side of the collision point. The proposed layout allows, by design, four hits associated to a charged track over the whole tracking volume within the range $|\eta| \leq 2.5$. The new BPIX layout has an optimized radial position of each layer. The design of a new beam pipe allows the position of the innermost layer, the first layer, at distance of 39 mm from the beam axis. The option for a smaller beam pipe is under investigation with a further shrink of the first layer up to a radius of 29.5 mm. Both options improve the tracking and the extrapolation of the tracks towards the interaction point. The other layers are placed at distances of 68 mm, 109 mm and 160 mm from the beam axis. The position of the fourth layer is located nearby the first layer of the next tracking system, the silicon micro-strip detector. This will allow a precise measurement of track position improving both the track interpolation between the two tracking systems and the track reconstruction efficiency. The upgraded FPIX disks cover the radial area between 45 mm and 161 mm from the beam axis, and the three disks are placed at longitudinal distances of 291 mm, 356 mm, and 516 mm from the interaction point. Each disk is made of two independent rings with modules arranged in a turbine like geometry, in order to maximize the point resolution, and with extra inclination as can be seen in fig.1, in order to assure a hermetic acceptance up to η=2.5. The layers in BPIX and the disks in FPIX are designed to

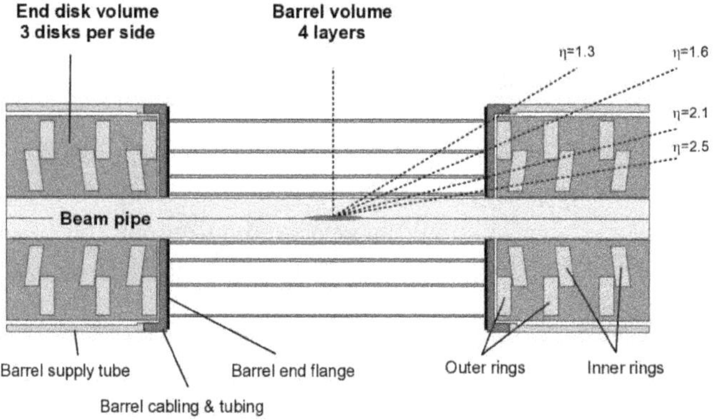

Fig. 1. Layout of the upgraded pixel detector. The tracking volume includes four barrel layers and three disks on each side of the collision point.

allow easy replacement of each single part and are equipped with one sensor flavor only. The pixel sensor geometry is identical to the current one with sizes 100 μm $\times$ 150 μm; with such a choice the new pixel sensor has no impact on the design of the upgraded electronics readout.
The upgraded pixel layout will be equipped with a larger number of pixel sensors, a factor 1.9 compared to the current system, for a total of about 126M channels.

3.2. *Material Budget*

Each layer of the current pixel detector represents about 2% of the radiation length χ_0, 1/3 approximately is given by the silicon material and the readout, and the remaining part by supporting structure, cooling system and cables. The reduction of the material budget in the tracking volume is achieved with two choices: the optimization of the location of service material and the development of a new cooling system.

The passive material (i.e. end-flanges with electrical connectors, power boards, electrical-digital converters, cables), today located in the service tube at $1.2 \leq |\eta| \leq 2$, will be moved to $|\eta| \geq 2$, almost 1 meter away from the interaction point. This solution is made possible with a redesign of the supply tube and after the development of low mass micro twisted pair electrical cables designed to transmit detector data.

The current pixel system cooling, based on C_6F_{14} fluid, contributes with passive material (i.e. coolant fluid, pipes, manifolds) distributed all over the sensitive volume, and this affects the pixel detector resolution as shown in fig.2 (left). The peaks in the transverse impact parameter resolution, well reproduced by Monte Carlo simulation, are localized in the positions of the 18 cooling tubes present in the current BPIX. As expected the degradation is larger for low momentum particles.

The pixel upgrade foresees a new cooling system based on a bi-phase evap-

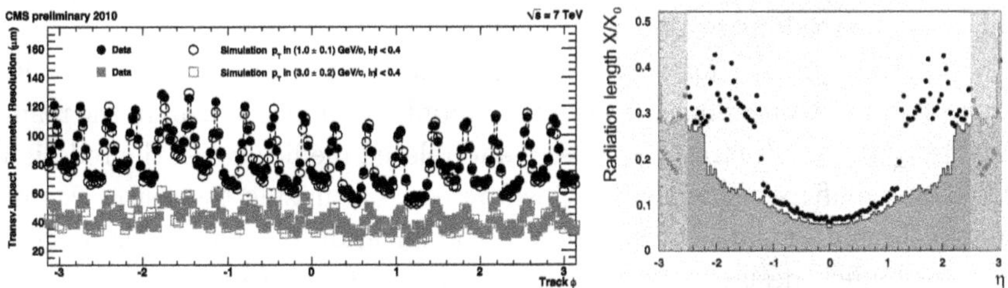

Fig. 2. Left: Transverse impact parameter resolution measured in BPIX as a function of the azimuthal angle ϕ. Comparison between data (filled symbols) and Montecarlo simulation (open symbols) for two selected particle transverse momenta: 1 GeV/c (circles) and 3 GeV/c (squares). Right: Fraction of material radiation length for the BPIX, current pixel (dots) and upgraded pixel (solid histogram). The shaded areas indicate regions outside the fiducial tracking volume.

orative CO_2 with a cooling power of about 10 KW and detector operating temperature at -20 ^oC. Clear benefits come from the use of the CO_2 cooling: the fluid low mass and low density decrease the passive material budget and the higher heat transfer coefficient allows good efficiency also with smaller thermal contacts. The high latent heat allows the design of longer cooling paths and the serial cooling of multiple ladders avoiding the use of both

stiff and robust end-flanges and manifolds for the coolant distribution. The high pressure operation needed for CO_2 cooling allows the use of smaller diameter pipes (1.6/1.8mm diameter) with thinner walls (50 μm) compared to current ones, providing a further reduction of material budget. The upgraded cooling system made possible the use of thin (200 μm) carbon fiber to design ladders needed to support pixel modules and Airex foam material reinforced by thin carbon fiber for the end-flanges.
An example of the expected reduction in material budget for BPIX is shown in fig. 2 (right) where a sizable improvement of the radiation length is visible, a similar trend is observed for the interaction length. Analogous reduction of the material budget is expected for the FPIX. Such a reduction will have a large impact on charged particle tracking efficiency as well as on electron and photon identification.

3.3. *Readout Electronics*

Operation at high luminosity needs the upgrade of the pixel ROC. The detailed Montecarlo simulation has shown that the limiting factors are not inherent to the readout architecture nor to the core of the ROC design (i.e. amplifier, shaper, threshold and trimming circuits), and notably this includes the readout pixel cell[7].
The upgrade phase I foresees the redesign of peripheral parts of the ROC circuits, leaving unchanged the double column drain mechanism in the readout. In order to minimize data losses and related inefficiencies, the redesign of readout buffers inside the ROC is foreseen, optimizing their size according to the higher data rates. The first modification is made in the trigger latency buffers designed to store inside the ROC data during the level 1 trigger decision time (about 4 μs). The addition of a second buffer stage, in order to remove hits faster from the double column, will decrease the readout dead time. The hits from the double columns are written into this readout buffer instead to be transmitted directly, and the double column is quicker resumed for next data taking. Simulation shows that after modifications the residual data loss at a luminosity of $2 \times L_{max}$ has a peak value of 4.7% (with the average over a fill of 2.1%) for the innermost layer[8].
The upgraded pixel layout needs 67% more readout optical lines, today not available in the pixel services: the increase of the bandwidth usage of the existing optical lines is the possible solution. In order to have a more robust design with current available technologies digital optical transmission of data is planned, discarding the present analog links that will be replaced by 160/320 MHz digital links. This implies the design in the periphery of

the each ROC of one 8-bit ADC, in order to digitize the pixel pulse height, a data serializer and line drivers, in order to manage data packing, and a PLL to generate clock frequencies higher than the basic 40 MHz. The conversion to digital transmission requires also modifications of the module controller chip (TBM) that will have to read the two half of a module in parallel, at 160 MHz, and after multiplexing should send data at 320 MHz[8].
Serialized pixel data are sent at such high frequency over a path of 1 m using a low power electrical link, micro twisted pair cables, already developed and prototyped[9]. After an electrical to optical conversion data are sent through existing optical links. Minor modifications are needed to adapt the full acquisition chain to the upgraded pixel readout[10].
The larger number of pixel modules increases the power demand: the solution implies the use of DC-DC conversion nearby the pixel detector in order to limit losses in the main distribution power cables[11].

4. Conclusion

The project for a CMS pixel upgrade phase I has been set up. The modification addresses the limiting factors of the current system at HL-LHC phase I. The upgrade foresees the increase of the number of hits per charged track, the decrease of the material budget in the tracking volume and the redesign of peripheral parts of the pixel readout electronics. A large community in CMS is involved in the upgrade and according to the present schedule the installation in the CMS experiment of the upgraded pixel system is foreseen in 2017.

References

1. CMS Physics TDR, Volume I & II : CERN-LHCC-2006.
2. W. Erdmann, IJMPA Volume: 25, Issue: 7(2010) pp. 1315-1337
3. The CMS tracker system project: TDR CERN-LHCC-98-006
4. The CMS Collaboration, Eur.Phys.J.C70:1165-1192,2010
5. Technical Proposal for the Upgrade of the CMS detector through 2020, CERN-LHCC-2011-006
6. H.C. Kaestli et al, Nucl. Instr. and Meth. A 565 (2006) p.188
7. H.C. Kaestli, PoS arXiv:1101.5977v1
8. H.C. Kaestli, PoS arXiv:1001.3933v1
9. B. Meier, Proceeding of TWEPP 2008, CERN report CERN-2008-008.
10. M. Friedl et al, JINST 5 C12054, 2010
11. L. Feld et al.,CERN-CMS-CR-2011-201 PoS(Vertex 2011)041

SILICON STRIP DETECTORS FOR THE ATLAS HL-LHC UPGRADE

M. MIÑANO*
on behalf of the ATLAS Collaboration
Instituto de Física Corpuscular, CSIC- University of Valencia, Valencia, Spain
**E-mail: mercedes.minano@ific.uv.es*
http://ific.uv.es

The Large Hadron Collider (LHC) at CERN is planning an upgraded machine called High Luminosity LHC (HL-LHC). The upgrade is foreseen to increase the LHC design luminosity up to 5×10^{34} $cm^{-2}s^{-1}$. For the ATLAS experiment, this implies the complete replacement of its internal tracker to cope with the increase in pile-up backgrounds and the higher radiation doses.
In this paper an overview of the ATLAS tracker upgrade project is given. The development of n-on-p silicon sensors with sufficient radiation hardness is the subject of an international R&D programme for the strip region of the future ATLAS tracker. Irradiated sensors were tested to study the radiation-induced degradation and determine their performance after irradiation of up to a few 10^{15} 1 MeV n_{eq}/cm^2. Results from a wide range of irradiated silicon detectors and layout concepts are presented.

Keywords: ATLAS; Upgrade; strip detectors; radiation.

1. Introduction

The Large Hadron Collider (LHC) at CERN started its operation in 2009. Since then it has provided proton-proton collisions with a continued increase in luminosity. ATLAS[1] is one of two general purpose detectors recording the collision products to exploit the full physics potential of the LHC up to its nominal luminosity of 10^{34} $cm^{-2}s^{-1}$.
An upgrade of the LHC towards higher luminosities[2] is foreseen in order to extend the physics program and allow precision measurements of any potential discoveries made at the LHC. An increase of the instantaneous peak luminosity up to 5×10^{34} $cm^{-2}s^{-1}$ is foreseen in the so-called High-Luminosity LHC (HL-LHC) upgrade planned to take place in around 2023. This upgrade is expected to deliver $\sim$ 3000 fb^{-1} of total recorded inte-

grated luminosity per experiment. This scenario is planned to be achieved in two phases. The Phase-I upgrade around 2017 to go to a luminosity of $2 \times 10^{34}\ cm^{-2}s^{-1}$ accumulating 300 fb^{-1} of integrated luminosity and the Phase-II upgrade to prepare the collider and experiments for HL-LHC operation.

The Inner Detector (ID) is the ATLAS internal tracking system. It combines high-resolution silicon detectors (pixel and microstrip technologies) in the innermost and intermediate layers and a continuous gaseous straw drift-tube detector in the outermost radii of the tracker. The HL-LHC conditions will lead to a huge increase of the event rate and higher radiation damage mainly in the ID compared to the conditions at the LHC. The number of pile-up events per bunch crossing will increase from ~ 23 at the nominal LHC luminosity to ~ 200 at the HL-LHC. The radiation dose accumulated by the silicon detectors will increase by one order of magnitude. As an example, the innermost strip layers ($R = 38\ cm$) will receive approximately $1.2 \times 10^{15}\ n_{eq}/cm^2$ whilst the current SCT layers at $R = 30\ cm$ are designed to withstand a maximum total fluence of $2 \times 10^{14}\ n_{eq}/cm^2$. The current ID not designed to provide the required tracking performance for the HL-LHC. Therefore, a new inner tracking detector must be designed and built for HL-LHC operation. While ATLAS plans for the Phase-I upgrade are limited to the pixel detector, the entire inner detector is foreseen to be replaced for the Phase-II upgrade which is the focus of this paper.

2. New ATLAS Inner Detector

The 2 main requirements for the new ID are a finer granularity to keep the channel occupancy acceptably low and enhanced radiation hardness. The new ID is foreseen to be an all silicon-based system. Figure 1 shows the current *strawman* layout. The inner barrel region will consist of four pixel layers at radii [3.7-20.9] cm. The central barrel region will be extended to five strip layers at radii [38-100] cm. The inner three layers are designed to have 2.4 cm-long strips (*short strip layers*). The outer two layers called *long strip layers* have 4.8 cm-long strips. Three outer strip layers will replace the barrel TRT. The forward region is covered by six pixel and five strip discs in each endcap. This layout ensures 9-hit coverage in the pseudorapidity range $\mid \eta \mid \leq 2.5$ and is expected to keep the occupancy below 1.6% at the innermost radius, which is considered to be adequate.

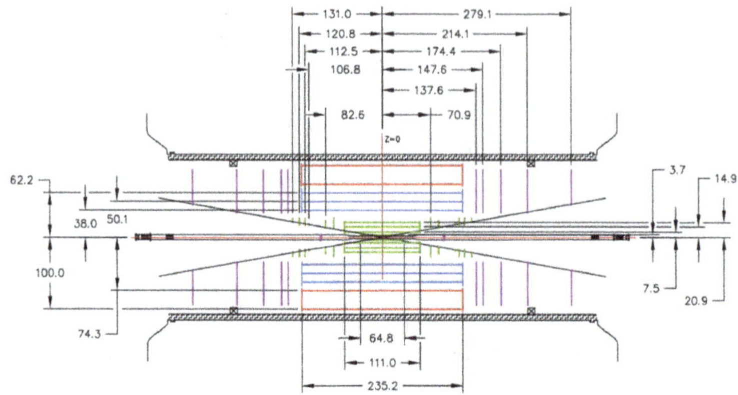

Fig. 1. *Strawman* layout (v14) for the upgraded ATLAS tracker. It combines 4 barrel pixel layers + 2 × 6 discs (green), 3 short strip (2.4 cm) layers (blue) and 2 long strip (4.8 cm) layers (red) with 2 × 5 discs (purple).[3]

3. Barrel & Endcap Integration Concepts

Several R&D projects are well advanced in terms of the design and assembly of strip modules, integrating sensors, hybrids with readout electronics, cooling and services. Figure 2 shows two alternative integration concepts for the short-strip barrel region. The baseline design is the *stave*; It consists of a central core composed of a spacing material (carbon-foam or honeycomb) which integrates the cooling circuit and carbon fiber facings and a bus-tape glued on both sides. Single-side silicon strips sensors are glued on the bus-tape and hybrids carrying the front-end electronics are then glued on the top of the sensitive side of the sensors.

The alternative option is the *supermodule*; Double-sided silicon strip modules are assembled on a lightweight with integrated services which are thus decoupled from the modules. In both options, modules prototypes have been constructed and tested, showing excellent electrical behaviour. Multi-module prototypes (4) have been also produced and tested showing no degradation in electrical performance with respect to a single module.

The developed concept for the endcap strip integration is the *petal stave* which is based on adapting the barrel *stave* option to a petal shape. The geometry of the petal requires six differents sensors (Fig. 3). In addition, the pitch variations (67 to 106 μm) lead to high bonding angles, requiring the use of pitch adapters.

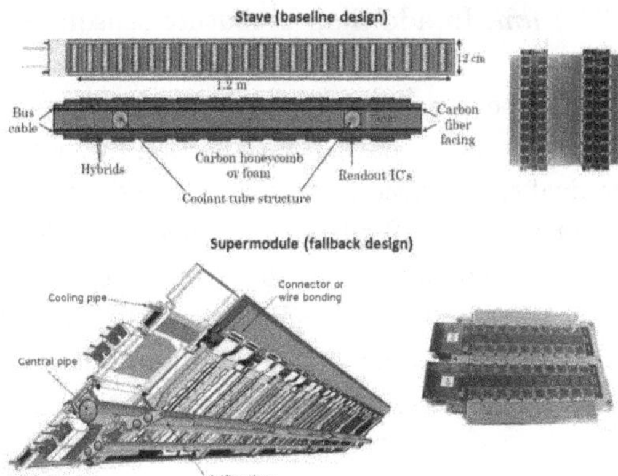

Fig. 2. Stave (top left) and Supermodule (bottom left) barrel strip integration concepts. To the right, a module which is the basic component of the structure is shown. The *stave* module is single-sided and the *supermodule* module is double-sided.

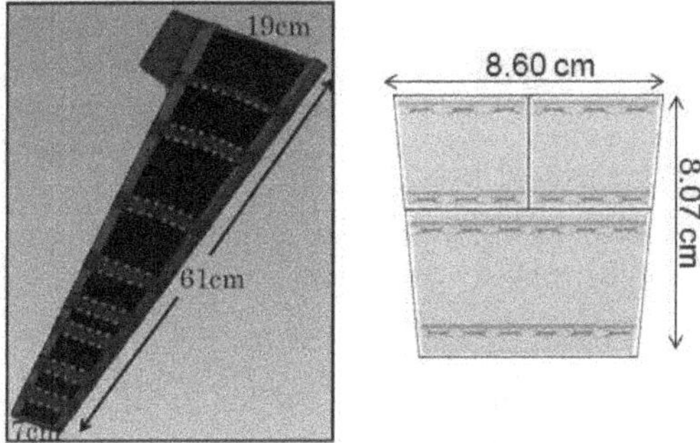

Fig. 3. Petal layout (left) for the endcap region and one of the six designed sensors (right).

4. Silicon Strip Sensors

Silicon sensors for the new tracker have been designed. The choice of n^+ strips in a p-type substrate was found to be more suitable and cost-effective.[4] n^+-on-p silicon strip sensors have been developed by the ATLAS collaboration and produced by Hamamatsu Photonics in a 6-inch wafers.[5] The wafers contain a full-size $97.54 \times 97.54\ mm^2$ sensor of 320 μm thick. The bulk is p-type FZ silicon and the isolation structure is p-stop. The sensor has four 2.4 cm strip rows, two with axial strips parallel to the sensor edges and two with stereo strips (40 mrad). There are 1280 strips per row

with a pitch of 74.5 μm. In addition, miniature sensors with 104 strips of 8 mm length are included on the edges of the wafer.
Full size sensors have been extensively tested before irradiation. Both the bulk and strip characteristics have been thoroughly evaluated to achieve the required technical specifications both before and after irradiation.[6] The characterization gives results all within specifications. Modules for the two design concepts have been built with these full size sensors. One module was irradiated in the PS facility at CERN with a beam of 24 GeV protons up to a dose of 1.9×10^{15} n_{eq}/cm^2. After irradiation the module was shown to be fully functional and the noise increase was consistent with shot-noise expectations.

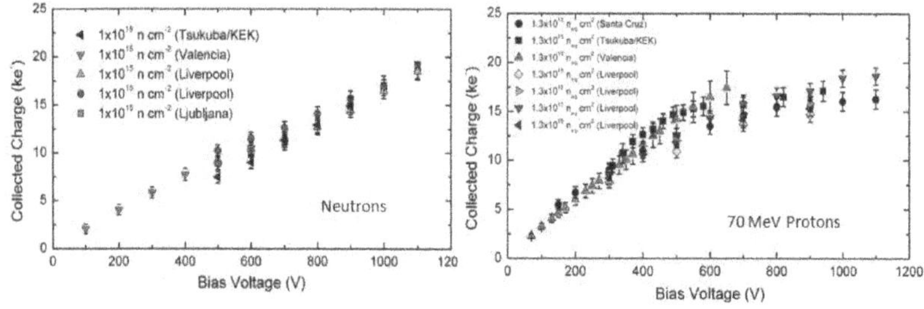

Fig. 4. Collected charge of neutrons (left) and protons (right) in irradiated miniature sensors as a function of the applied bias voltage.[7]

Miniature sensors have been mainly used for irradiation studies. Sensors were irradiated with protons, pions and neutrons at different irradiation facilities. The doses were all up to or beyond 10^{15} n_{eq}/cm^2. As shown in Fig. 4, a very good agreement was found in the collected charge measurements among the different samples and the different testing laboratories. More detailed results can be found in reference.[7]

5. Conclusions

The ATLAS collaboration is already designing a new all-silicon tracker for HL-LHC operation. Silicon strip sensors for this tracker have been produced in n-in-p technology which meet the required technical specifications before irradiation and show sufficient functionality in terms of charge collection after irradiation. Module integration concepts have been defined and an extenside R&D program of components is being carried out.

References

1. The ATLAS Collaboration, *JINST3* **S08003** (2008).
2. F. Gianotti et al., *Eur. Phys. J.* **C39**, 293 (2005).
3. P.P. Allport et al., *Nucl. Instr. and Meth. A* **636**, 90 (2011).
4. G. Casse et al., *Nucl. Instr. and Meth. A* **487**, S465 (2002).
5. Y. Unno et al., *Nucl. Instr. and Meth. A* **636**, S24 (2010).
6. J. Bohm et al., *Nucl. Instr. and Meth. A* **636**, S104 (2011).
7. K. Hara et al., *Nucl. Instr. and Meth. A* **636**, S83 (2011).

ATLAS SILICON MICROSTRIP TRACKER: OPERATION AND PERFORMANCE

V. A. MITSOU, for the ATLAS SCT Collaboration

Instituto de Física Corpuscular (IFIC), CSIC – Universitat de València, P.O. Box 22085, E-46071, Valencia, Spain
E-mail: vasiliki.mitsou@ific.uv.es

The Semiconductor Tracker (SCT) is a silicon strip detector and one of the key precision tracking devices in the Inner Detector of the ATLAS experiment at CERN LHC. The completed SCT has been installed inside the ATLAS experimental cavern since 2007 and has been operational since then. Calibration data has been taken regularly and analyzed to determine the performance of the system. In this paper the current status of the SCT is reviewed, including results from data-taking periods in 2010 and 2011. We report on the operation of the detector including overviews on services, connectivity and observed problems. The main emphasis is given to the performance of the SCT with the LHC in collision mode and to the performance of individual electronic components.

Keywords: Silicon detectors; LHC; Microstrip sensors.

1. Introduction

The ATLAS detector,[1] one of the two general-purpose experiments at the Large Hadron Collider (LHC) at CERN, has been taking proton-proton collision data at a centre-of-mass energy of 7 TeV since March 2010. The ATLAS Inner Detector (ID)[2] combines silicon detector technology (pixels[3] and microstrips[4]) in the innermost part with a straw drift detector[5] with transition radiation detection capabilities (Transition Radiation Tracker, TRT) on the outside, operating in a 2-T superconducting solenoid.

The microstrip detector (Semiconductor Tracker, SCT), as shown in Fig. 1, forms the middle layer of the ID between the Pixel detector and the TRT. The SCT system comprises a barrel[6] made of four nested cylinders and two end-caps[7] of nine disks each. The barrel layers carry 2112 detector units (modules)[8] altogether, while a total of 1976 end-cap modules[9] are mounted on the disks. The whole SCT occupies a cylinder of 5.6 m in length and 56 cm in radius with the innermost layer at a radius of 27 cm.

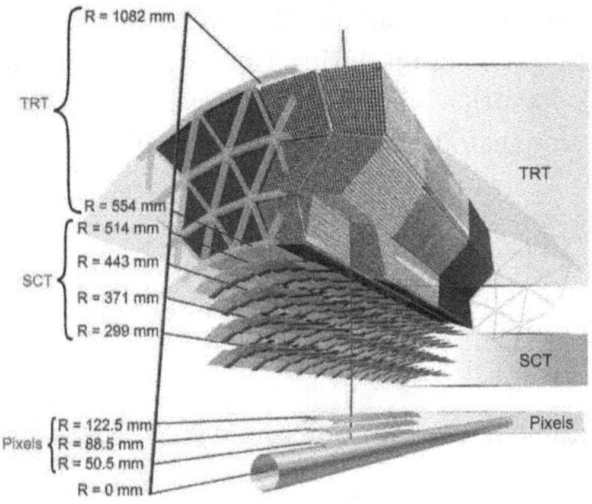

Fig. 1. Layout of the ATLAS Inner Detector: it comprises the Transition Radiation Detector, the Semiconductor Tracker and the Pixel system.

The silicon modules[8,9] consist of one or two pairs of single-sided p-*in*-n microstrip sensors glued back-to-back at a 40-mrad stereo angle to provide two-dimensional track reconstruction. The 285-μm thick sensors[4] have 768 AC-coupled strips with an 80 μm pitch for the barrel and a $57 - 94$ μm pitch for the end-cap modules. Barrel modules follow one common design, while for the forward ones four different types exist according to their position in the detector. The readout[10] of each module is based on 12 ABCD3TA ASICs[11] manufactured in the radiation-hard DMILL process mounted on a copper/kapton hybrid. Each module is designed, constructed and tested to operate as a stand-alone unit, mechanically, electrically, optically and thermally.

2. Operation stability

The performance of the SCT modules and substructures have been repeatedly tested in the past in dedicated beam tests (before 2004), during the Combined Test Beam with other ID components in summer 2004 and with cosmic rays[12] both on the surface and after installation in the ATLAS cavern. Since autumn 2009 various aspects of the SCT operation have been studied continuously while recording physics data in *pp* collisions at 900 GeV, 2.36 TeV and, since March 2010, at 7 TeV. The SCT performs very well with increasing collision rate, as demonstrated in Fig. 2, where the fraction of SCT module sides giving errors as a function of time is shown. It is stressed that the time interval spans an increasing instantaneous luminosity of five orders of magnitude. The SCT data-quality inefficiency is mainly due to HV ramping up during LHC stable-beam declaration, with an overall 99.9% operation efficiency. The total fraction of data with SCT errors remained less than 0.25% throughout 2010.

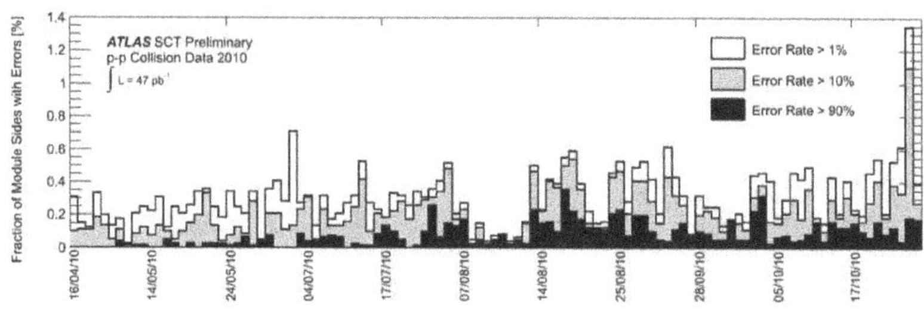

Fig. 2. Fraction of SCT module sides that are reporting errors as a function of time (run number) for the data taken in 2010.

The only issues encountered affecting the — otherwise stable — SCT configuration are: (a) a faulty connection to a cooling loop[13] discovered during commissioning, that cannot be repaired since it is located behind the endplate; and (b) some unexpected failures[14] of off-detector optical transmitters (TX-plugins)[15] that are being replaced by humidity-resistant plugins. In total, more than 99% of all SCT modules are fully operational.

3. SCT performance

Noise measurements are performed by two methods:[16] injection of calibration pulses and noise estimation from response-curve fit; and measuring noise occupancy as a function of threshold to extract the input noise. The measured values are well correlated and, as shown in Fig. 3 (left), the noise occupancy is well below the required level of 5×10^{-4}.

Good agreement has been observed between Monte Carlo and 900 GeV data in the number of hits per module side, as shown in Fig. 3 (right). The discrepancy at low N is due to the lower noise assumed in simulation compared to data by approximately a factor of three.

As far as the SCT timing[10] is concerned, per-module adjustments are made to account for the optical fibre length and the time-of-flight from the interaction point though timing scans. The related measurements show that the SCT is well synchronized with the Level-1 trigger.

The Lorentz angle as determined by finding the angle corresponding to the minimum cluster width is shown in Fig. 4 (left) for the barrel layers

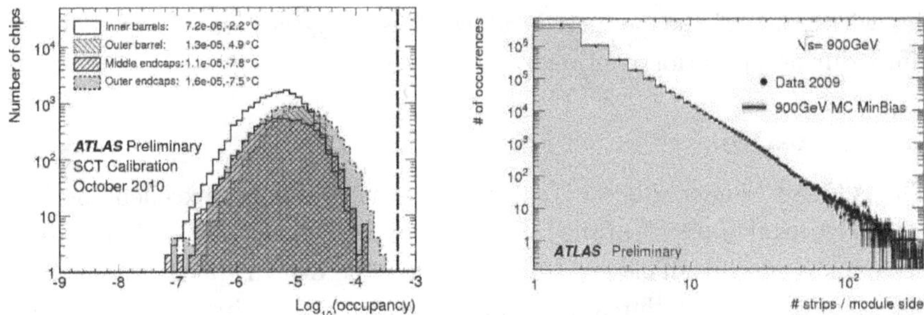

Fig. 3. Left: Distributions of noise occupancy for each chip split up according to the module type. Right: Number of hit strips per module side for 900 GeV data compared with a minimum bias Monte Carlo sample.

and for various data/conditions. The measurements agree well with the predictions, proving the good signal digitization in the full simulation.

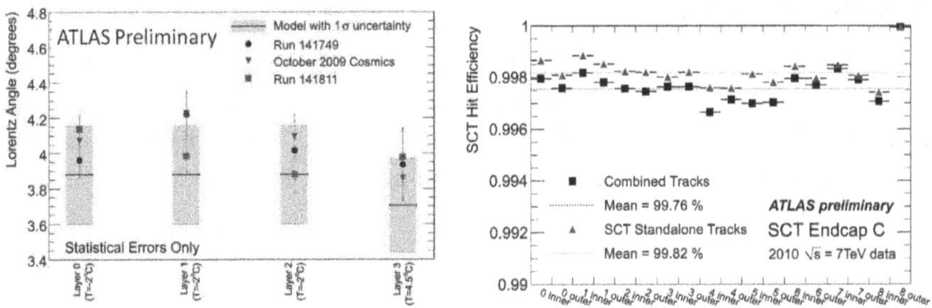

Fig. 4. Left: Lorentz angle extracted from the cluster-size vs. angle plot for collision and cosmic ray data compared to model predictions. Right: Intrinsic module efficiency for tracks measured in the SCT end-cap C.

The intrinsic strip efficiency is computed by counting missed hits in well-reconstructed tracks with $p_T > 1$ GeV, excluding dead modules and chips. The SCT hit efficiency for all three SCT parts (shown in Fig. 4, right, for the end-cap C) is well above the 99% specification.

4. Radiation damage

The radiation damage is being monitored through the leakage current. The comparison between measurements and FLUKA[17] predictions of the fluence shows good agreement in the barrel region, whilst a discrepancy in the low-R forward region is under study. The leakage current evolution with time lies within 1σ of the predicted values. The long-term monitoring of the effect during operation continues.

5. Conclusions

The ATLAS Semiconductor Tracker performance shows excellent performance. A total of 99.1% of the SCT modules are used for data taking. Noise occupancy and hit efficiency are well within the design specifications. The optical transmitter failures are understood and do not impair data taking efficiency. The Monte Carlo simulation reproduces accurately the detector geometry and material budget. The radiation damage is being monitored and is in good agreement with expectations. The SCT is a key precision tracking device in ATLAS and we are taking more and more good physics data every day. The SCT commissioning and running experience is used to extract valuable lessons for future silicon strip detector projects.

Acknowledgments

The author acknowledges support by the Spanish Ministry of Science and Innovation (MICINN) under the project FPA2009-13234-C04-01, by the Ramón y Cajal contract RYC-2007-00631 of MICINN and CSIC, and by the Spanish Agency of International Cooperation for Development under the PCI project A/030322/10.

References

1. ATLAS Collaboration, *JINST* **3**, p. S08003 (2008).
2. ATLAS Collaboration, *Eur.Phys.J.* **C70**, 787 (2010).
3. G. Aad *et al.*, *JINST* **3**, p. P07007 (2008).
4. A. Ahmad *et al.*, *Nucl.Instrum.Meth.* **A578**, 98 (2007).
5. E. Abat *et al.*, *JINST* **3**, p. P02013 (2008).
6. A. Abdesselam *et al.*, *JINST* **3**, p. P10006 (2008).
7. A. Abdesselam *et al.*, *JINST* **3**, p. P05002 (2008).
8. A. Abdesselam *et al.*, *Nucl.Instrum.Meth.* **A568**, 642 (2006).
9. A. Abdesselam *et al.*, *Nucl.Instrum.Meth.* **A575**, 353 (2007).
10. A. Abdesselam *et al.*, *JINST* **3**, p. P01003 (2008).
11. F. Campabadal *et al.*, *Nucl.Instrum.Meth.* **A552**, 292 (2005).
12. E. Abat *et al.*, *JINST* **3**, p. P08003 (2008).
13. D. Attree *et al.*, *JINST* **3**, p. P07003 (2008).
14. M. S. Cooke, Preprint arXiv:1109.6679 [physics.ins-det] (2011).
15. A. Abdesselam *et al.*, *JINST* **2**, p. P09003 (2007).
16. V. A. Mitsou, *IEEE Trans.Nucl.Sci.* **53**, 729 (2006).
17. G. Battistoni *et al.*, *AIP Conf.Proc.* **896**, 31 (2007).

UPGRADE OF THE CMS TRACKER FOR THE LHC AT HIGH LUMINOSITY

MARK PESARESI* (on behalf of the CMS collaboration)

Blackett Laboratory,
Imperial College, London
SW7 2AZ, United Kingdom
**E-mail: mark.pesaresi@imperial.ac.uk*

The LHC is expected to increase its luminosity above the original nominal value of $10^{34}\,\mathrm{cm}^{-2}\mathrm{s}^{-1}$, eventually achieving an order of magnitude increase after major upgrades will be performed after 2020. This configuration of the machine is known as the High Luminosity LHC (HL–LHC). The CMS experiment will require a completely new tracking system in order to maintain adequate performance in the HL–LHC environment as well as to provide tracking information for the Level–1 trigger decision. Innovative solutions are being studied to improve tracking resolution, reduce the material budget, increase the sensor granularity and provide useful information for an upgraded trigger system. The most relevant requirements and constraints are summarised here, along with highlights from R&D activities.

Keywords: ICATPP 2011; LHC; CMS; Tracker; Upgrade; Track Trigger

1. Introduction

The proposed luminosity upgrade[1] for the Large Hadron Collider (LHC) is expected to take place in a series of machine shutdowns over a 10–15 year period after LHC start-up. The first shutdown, starting in 2013, will be primarily for ugrades to allow proton collisions at a center-of-mass energy of 14 TeV. A collimator upgrade during this period should also see that the machine reaches its design luminosity of $10^{34}\,\mathrm{cm}^{-2}\mathrm{s}^{-1}$ with 25 ns bunch spacing (~20 interactions/bunch crossing) before the second long year-long shutdown in 2017. With an new injector chain and further collimator upgrades, it is expected that the LHC luminosity during the following run (also known as Phase I) will reach $2\times10^{34}\,\mathrm{cm}^{-2}\mathrm{s}^{-1}$. After accumulating an estimated $300\,\mathrm{fb}^{-1}$ of integrated luminosity, a final shutdown around 2021–2022 should allow the machine to reach its ultimate luminosity of

$>5\times10^{34}\,cm^{-2}s^{-1}$ (Phase II). The Phase II luminosity could be achieved with a bunch spacing of 25 or 50 ns and as such may lead to pileup conditions of up to 250 interactions/bunch crossing under a worst case scenario.

With this increase in luminosity, the LHC experiments will also require various upgrades in order to cope with the increased particle fluxes and data rates. The CMS experiment[2] proposes to replace its entire tracking system in preparation for Phase II. The current silicon strip tracker has been designed to tolerate particle fluences equivalent to an integrated luminosity of $\sim$500 fb^{-1} at the LHC and hence tracking performance will have degraded after years of radiation damage. The inner pixel detector will also have to be replaced, after having been replaced once before for Phase I.[3]

2. Phase II Tracker Requirements

The main requirements and constraints for the upgraded tracker are summarised here.

(a) cabling & power — tracking services (optical links for readout and control, cooling pipes, power lines) are constrained and will not increase. With greater powering demands, the upgraded tracker will have to implement DC–DC[4] conversion on-detector in order to limit losses in the main power lines. Minimising the on-detector power consumption wherever possible will also help to reduce the cooling requirements in the tracker.

(b) granularity — a new tracker will have to cope with the congested HL–LHC environment. This requires increased granularity, since channel occupancies at intermediate radii (r$\sim$25 cm) will be at least 15–30% and maintaining occupancies at a few % would be desirable for efficient tracking. As such, the new tracker should implement shorter strips of 2–5 cm — a reduction factor of $\sim$5 in length.

(c) bandwidth — an increased channel count will require higher bandwidth digital readout links as opposed to the all-analogue readout system employed by the current tracker. The links must satisfy the constraints on power and cabling requirements.

(d) radiation tolerance — the upgraded tracker will need to be radiation tolerant to survive the higher fluences at HL–LHC. In the intermediate and outer tracking regions (r>20 cm) the sensor technology used in the current pixels would be sufficiently radiation hard for this environment. An R&D programme is investigating materials which could withstand the extreme fluences in the inner region. However, it may be the case that the inner pixel layers will still have to be replaced at regular intervals.

(e) material — tracker mass should be reduced at every opportunity in order to improve the performance of the detector by minimising the effects of multiple scattering, electron bremsstrahlung and secondary interactions. This could be achieved by optimising the tracker layout and module design, better placement of services and by implementing a new cooling system based on liquid CO_2.

(f) trigger information — the Level–1 (L1) hardware trigger will not be able to keep its rate below the maximum permitted 100 kHz in the high pileup environment. Increased pileup will adversely affect the ability to perform effective isolation in the calorimeter. Higher thresholds are undesirable and would not be necessarily be able to reduce the trigger rate enough. A future tracker could provide tracking information to the L1 trigger in order to reduce the rate in much the same way as it is used in the higher level software triggers.

3. Stacked Tracking for a L1 Trigger

Collisions at the LHC produce a large number of low momentum particles that make up a significant fraction of hit data generated by the tracker (Figure 1). Charged particles with transverse momentum $p_T < 0.7\,\mathrm{GeV}/c$ are considered uninteresting for the purposes of triggering since they fail to reach the outer sub-detectors due to the bending power of the 4 T magnetic field.

By correlating hits between closely spaced ("stacked") sensors, this low p_T background can be rejected by only selecting hits that lie within a few pixels of each other in the bending plane (r–ϕ). In a 4 T magnetic field, studies show that for a layer of stacked sensors placed at 25 cm and a radial separation between sensors of ~1 mm, a pixel pitch of order 100 μm in r–ϕ can be used to select tracks with transverse momentum greater than a few GeV/c.[5,6] In this way, the on-detector data rate can be reduced by at least an order of magnitude before tracking information is forwarded to the L1 trigger for matching to other trigger objects.

While the concept is simple, the challenges are to implement a design within the constraints of the system as described in Sec. 2. The most significant issues are the interconnect scheme (between the top and bottom sensors) which should be low mass and low power, the readout electronics and the bandwidth constraints for transmission of trigger data. Simulation studies show[6] that the correlation algorithm is effective for occupancies

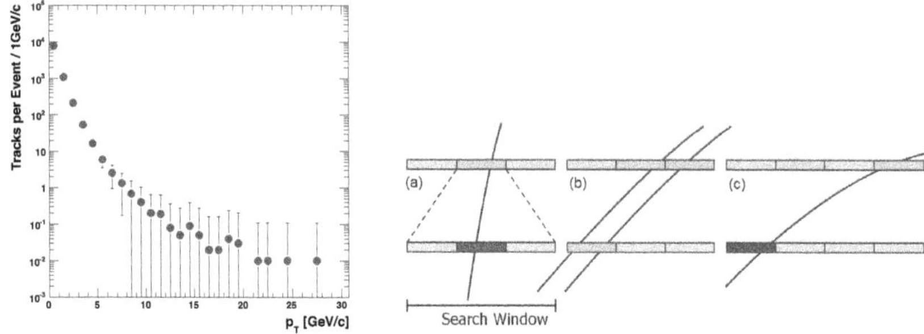

Fig. 1. Left: The p_T spectrum (averaged per event) for all minimum bias particles that leave hits in a sensitive layer placed at a radius of 25 cm and at an average pileup of 400 p–p interactions per event. Events were simulated within a magnetic field of 4 T and a coverage of $|\eta| < 2.5$. Right: Conceptual example of a stacked pixel layer where high p_T tracks pass the correlation of ± 1 pixels (a) and low p_T tracks fail (c). The region where tracks may or may not pass the correlation is dependent on the track p_T, sensor pitch and sensor separation (b).

of a few %, hence stacked strip sensors may be used at outer radii while intermediate radii would benefit from stacks of pixel sensors.

4. Current R&D Activities

With an increased channel count and move towards digital optical links,[7,8] the upgraded tracker will no longer be able to utilise an all-analogue readout system. In addition, in order to minimise the required bandwidth and power consumption on-detector, a future readout system will not retain pulse height information from the strips. A prototype binary front-end ASIC has been fabricated in 130 nm CMOS technology for possible use in the CMS HL–LHC outer tracker (Fig. 2). The CMS Binary Chip (CBC) prototype is a 128–channel readout chip for short (2–5 cm strips) featuring a binary unsparsified readout scheme.[9] This maintains the simplicity of the entirely synchronous front-end currently employed in the CMS tracker with a fixed data throughput for every trigger.

The prototype has been tested successfully under a range of conditions. Most recently, the chip has been used to readout a 5 cm strip sensor placed in 400 GeV/c proton beam at the CERN H8 beam line. The analogue behaviour of the CBC is within specifications — the gain is ~50 mV/fC and the noise has been measured to be ~800 electrons (for a 5 pF strip sensor). The chip could operate with sensor leakage currents of up to 1 μA for either polarity of signal. The power consumption has been measured to be

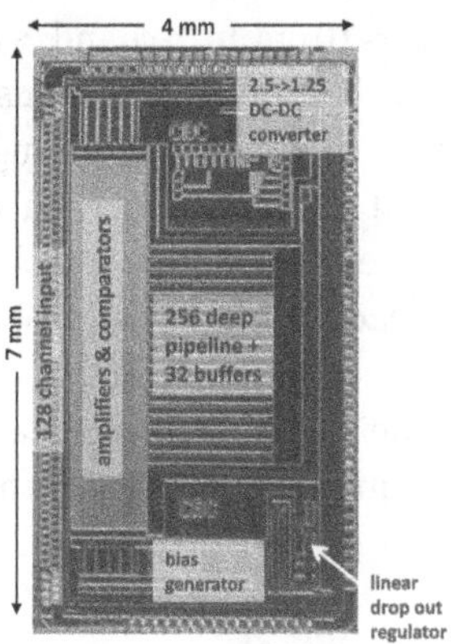

Fig. 2. Schematic of the CBC prototype ASIC.

<300 μW per channel (for a 5 pF strip sensor) — a factor of 5 reduction compared to the APV25[10] ASIC used in the current CMS strip tracker.

One of the most significant challenges for the design of the new tracker is in the implementation of stacked triggering modules. As a result, much effort has been placed into viable module concepts with a view to demonstrating working prototypes within a few years.

The "Stacked Strip" (2S) module (Fig. 3) focusses on currently available technologies using layers of closely separated 5 cm long strip sensors. As such, modules of this type could only be used in layers at radii 50–120 cm, and possibly in the endcaps. From both sensors, the strips of pitch 90 μm could be wire-bonded at the edges to a high-density substrate which would handle the interconnection between top and bottom sensors. This novel module hybrid also eliminates the need for pitch adaptors between the sensors and the front-end ASIC which would, in this case, be bump-bonded directly to the hybrid. The front-end chip would not only buffer the event before L1 readout, but could also correlate the information from both sensors in order to identify high p_T tracks before sending this data off-detector to the L1 trigger. A bump-bondable version of the CBC with triggering logic is being designed for use in a stacked strip module prototype.

A variant of the "Stacked Strip" module with edge readout is the "Stacked Pixel Strip" (PS) module (Fig. 3) which replaces the lower strip sensor in the 2S module with a pixel sensor of equivalent pitch in r–ϕ but ~1.3 mm in z.[11] This means that z information can be provided for both tracking offline, removing the need for stereo modules, as well as to the L1 trigger. The module size is ∼10 cm in r-ϕ and ∼5 cm in z (half the length of the 2S module). In this scheme, the strip sensor would be wire-bonded to the high density substrate and read out through a bump-bonded binary front-end ASIC. The digital output from each channel would be routed through the hybrid and wire-bonded to a pixel readout chip situated on the underside. The pixel readout chip would perform both the hit correlation for the trigger and the buffering and readout of the pixel hits on receipt of a L1 trigger. While there is no significant mass penalty with such a design, careful study is required to mimimise the power consumption (especially the analogue power in the pixels) and to satisfy the bandwidth constraints.

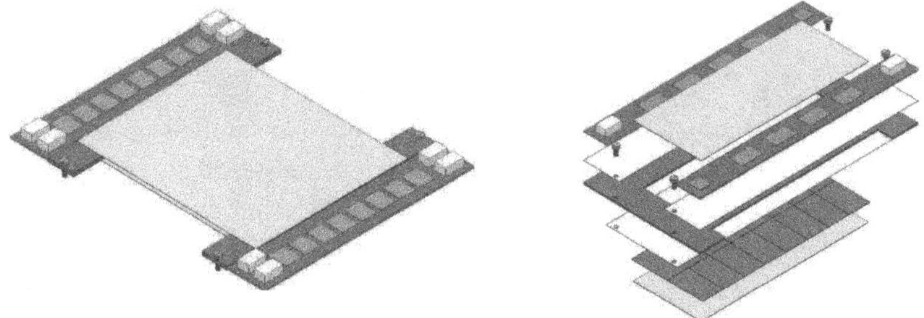

Fig. 3. Left: Visualisation of a Stacked Strip (2S) module, Right: Visualisation of Stacked Pixel-Strip (PS) module.

The "Vertically Integrated Hybrid Module"[12] is based on 3D-interconnection technology with data transfer between stacked sensors via a ∼1 mm low mass PCB interposer layer. The pixel cell sizes are approximately 100 μm×1 mm. Analogue data is transferred from the upper sensor through the interposer where a single 3D correlator and readout ASIC resides, on top of the lower sensor. While the technology is new, this design would allow for the implementation of a flexible local trigger logic with low power consumption per pixel, providing large modules can be constructed with adequate yields. A 3D demonstrator chip (VICTR) has been fabricated and is currently under test.

5. Conclusions

The CMS experiment plans to upgrade its tracking system in expectation of the LHC luminosity upgrade. The detector design will be driven by the requirements of the unique operating conditions at HL–LHC, a need to reduce material for improvement in detector performance and to provide tracking data to the L1 trigger. The stacked tracking concept has demonstrated viability for use at HL–LHC but significant challenges still remain in the realisation of such a system due to constraints in power and bandwidth. A variety of module concepts are under consideration and prototypes are expected in the next few years. R&D is progressing quickly: the next generation of ASICs have already been prototyped and studied while further designs will be submitted next year; aggressive packaging technologies such as high density substrates and C4 bump-bonding are under investigation; an extensive sensor testing campaign is underway; and developments such as DC–DC conversion and CO_2 cooling to be implemented for the Phase I pixel upgrade are being studied for use in a new tracker.

References

1. W. Scandale and F. Zimmermann, *Nuclear Physics B - Proceedings Supplements* **177-178**, p207 (2008).
2. The CMS Collaboration, *Journal of Instrumentation* **3**, p. S08004 (2008).
3. W. Erdmann *et al.*, *Nuclear Instruments and Methods in Physics Research Section A: Accelerators, Spectrometers, Detectors and Associated Equipment* **617**, 534 (2010).
4. M. Weber, *Nuclear Instruments and Methods in Physics Research Section A: Accelerators, Spectrometers, Detectors and Associated Equipment* **592**, p44 (2008).
5. J. A. Jones, Development of Trigger and Control Systems for CMS, PhD thesis, London Univ.2006.
6. M. Pesaresi, Development of a new silicon tracker at CMS for Super-LHC, PhD thesis, London Univ.2009.
7. P. Moreira and A. Marchioro, The GBT: A Proposed Architecure for Multi-Gb/s Data Transmission in High Energy Physics, in *Topical Workshop on Electronics for Particle Physics*, 2007.
8. F. Vasey, Versatile Link Status, `http://indico.cern.ch/conferenceOtherViews.py?view=standard&confId=47853`.
9. M. Raymond, The CMS Binary Chip for microstrip tracker readout at the SLHC, in *Topical Workshop on Electronics for Particle Physics*, 2011.
10. M. Raymond *et al.*, **CERN-LHCC/2000-041** (2000).
11. A. Marchioro, A hybrid architecture for a prompt momentum discriminating tracker for SLHC, in *20th International Workshop on Vertex Detectors*, 2011.
12. R. Lipton, 3D Detector and Electronics Integration Technologies: Applications to ILC, SLHC, and beyond, in *7th International Hiroshima Symposium on Development and Applications of Semiconductor Tracking Devices*, 2009.

A Level 1 Tracking Trigger for the CMS Experiment

Nicola Pozzobon (on behalf of the CMS Collaboration)
Dipartimento di Fisica "G. Galilei", Università degli Studi di Padova, and Istituto Nazionale di Fisica Nucleare, Sezione di Padova
Padova, 35131, ITALY
**E-mail: nicola.pozzobon@pd.infn.it*

The LHC machine is planned to be upgraded in the next decade in order to deliver a luminosity about 5 to 10 times lager than the design one of 10^{34} cm^{-2}s^{-1}. In this scenario, a novel tracking system for the CMS experiment is required to be conceived and built. The main requirements on the CMS tracker are presented. Particular emphasis will be given to the challenging capability of the tracker to provide useful information for the Level 1 hardware trigger, complementary to the muon system and calorimeter ones. Different approaches based on pattern hit correlation within closely placed sensors are currently under evaluation, making use of either strips or macro-pixels. A proposal to optimize the data flow at the front-end ASIC and develop a tracking algorithm to provide tracks at Level 1 will be presented.

Keywords: ICATPP 2011; Proceedings; LHC; CMS; Tracker; Upgrade; Level 1 Trigger.

1. Introduction

A luminosity upgrade of the CERN Large Hadron Collider[1] is expected to take place in two phases.[2,3] The typical collider luminosity will then range from $\mathcal{L} \simeq 5 \times 10^{34}$ cm^{-2}s^{-1} to 10^{35} cm^{-2}s^{-1}, instead of the design one of 10^{34} cm^{-2}s^{-1}. The current CMS tracker[4] was not designed to operate in such an environment: radiation damage and data losses will need the replacement of the Pixel Detector in about 5 years, while novel trigger strategies will be needed when the event rate starts exceeding the scale of GHz. As an example, the maximum allowed Level 1 (L1) muon trigger bandwidth, for nominal LHC conditions, is 12.5 kHz, a rate that can be kept under control at $\mathcal{L} \simeq 5 \times 10^{34}$ cm^{-2}s^{-1} using full Muon System information but which will become much larger at $\mathcal{L} \simeq 10^{35}$ cm^{-2}s^{-1}, requiring also the use of information from the tracker to overcome this limitation. Without upgrading the L1 trigger, the 12.5 kHz rate will be obtained imposing muon

p_T thresholds at values larger than 60 GeV/c, which will eventually reject all the interesting events.

The main goals of L1 tracking trigger are the reduction of the overall data rate and the completion of muon and calorimeter L1 triggers to identify relevant trigger objects such as leptons. The first one can be achieved building L1 trigger primitives, independent on the tracker layout, rejecting hits from low p_T tracks, while the accomplishment of the second one will need dedicated online tracking and vertexing algorithms.

2. On-Detector Data Rate Reduction

One promising strategy to reject hits from low p_T tracks relies on pattern hit correlation in closely placed sensors.[5] Simulations of pp collisions at $\sqrt{s} = 14$ TeV show that the 95% of tracks leaving hits in silicon sensors at radius $R \simeq 30 - 50$ cm from the beam have $p_T < 2$ GeV/c. The lateral displacement of hits from a $p_T = 2$ GeV/c track in sensors at $R \simeq 50$ cm separated by $\Delta R \simeq 1$ mm is about 100-150 μm, corresponding to the typical pitch of current pixels and strips of the CMS tracking system. Silicon tracker modules capable of performing the necessary hit correlation to identify low p_T tracks based on this concept, called "stacked modules" or "p_T-modules", are currently being designed.

Two experimental proofs-of-principle took place in 2010. Spare strip modules of current CMS tracker were bonded together 2 mm apart from each other and used to record the passage of cosmic rays. The angle of incidence of tracks could then be clearly correlated to the separation between measured clusters up to 20°, an angle analogous to the one of $p_T = 2$ GeV/c tracks at $R \simeq 108$ cm in a 4 T magnetic field. Another interesting proof of this concept came from pp collision data at $\sqrt{s} = 7$ TeV. An offline analysis of hits in stereo double-sided strip modules at $R \simeq 70$ cm showed that a cut on lateral displacement between clusters could be translated in a p_T threshold of few GeV/c with a clear turn-on curve, as shown in Fig. 1.[6]

The L1 tracking trigger primitives built by p_T-modules are called *track stubs*. This functionality is being studied both in terms of design of the chip logic (FNAL proposal) and simulations within the framework of CMS analysis software (CMSSW). One of the major proposals within CMS makes use of well-established designs already used in advanced computing electronics. The main requirements for the pipelines used for data transfer are the capability to sustain input data every 25 ns and to be asynchronous in order to avoid the use of additional clocks. A submission of a test chip through MOSIS is expected for spring 2012.[7]

Simple clustering algorithms are needed to reduce the number of combinatory hit pairs to be matched in order to build stubs. To have them implemented in fast boolean logic, all the strips or pixels over threshold count as a logic "1". The simplest of these algorithms compares the signals from adjacent strips or pixels in the plane transverse to the beam line, requiring that no more than three consecutive positive signals are found between two cells with null content. Other algorithms under preliminary study may be eligible for ~ 1 mm long pixels and try to include in the cluster also signals from adjacent rows in the sensor.
The production of track stubs, according to the FNAL proposal, is based on the comparison between position of clusters in sensors with different length: the outer sensor has longer pixels or strips which are used to give the actual p_T discrimination while the shorter ones in the inner sensor are used to locate the stubs in in the direction of the beam. The size of search windows are defined by look-up tables (LUT's) depending mainly on the distance from the beam line. LUT-based algorithms, even if easy to implement, are not currently being used within CMSSW in favor of procedures based on p_T threshold and backprojection to the luminous region, until the final tracker layout is chosen and LUT's can be optimized.[8]

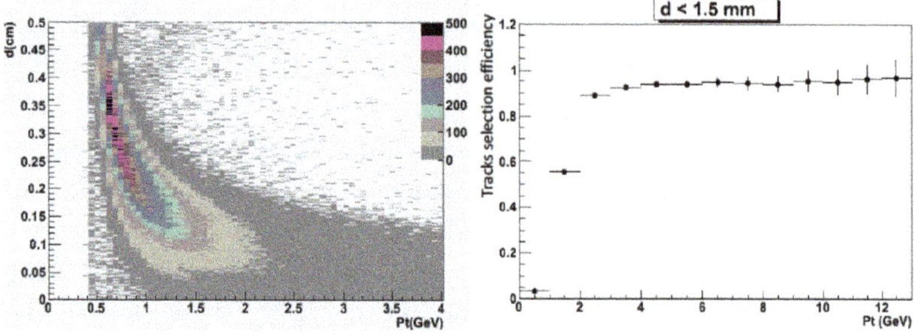

Fig. 1. Proof-of-principle of low p_T track rejection with stacked sensors. The correlation between lateral separation of Si strip clusters in the two sides of second layer of current CMS Tracker Outer Barrel is shown on the left. The efficiency in selecting tracks as a function of track p_T is shown on the right for lateral separations smaller than 1.5 mm.[6]

3. Off-Detector Online Tracking and Vertexing

The track stubs are the actual L1 tracking trigger primitives wich can be combined in a projective geometry to produce objects that can be used in lepton triggers to improve muon system or calorimeter resolutions or to

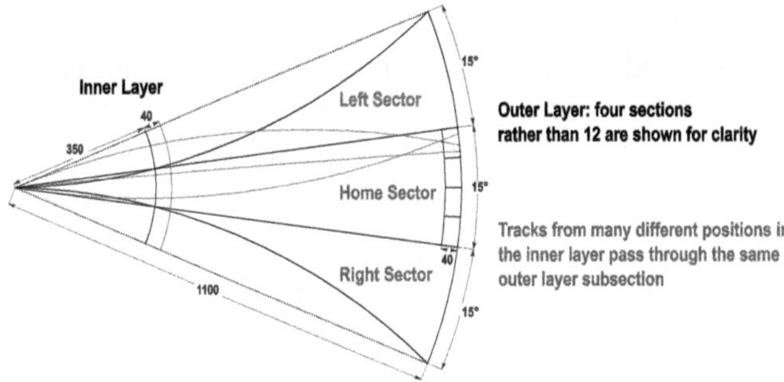

Fig. 2. A 15 degree home sector and the two neighboring sectors used to define the search algorithm to match stubs to existing seed tracklets in the innermost double stack of a tracker layout in the long barrel style. By using both p_T and position in the innermost layer, one can sort the tracklets according to their position in the outermost one. Distances are expressed in mm.

isolate trigger objects from the same detectors. The concept tracker layout used to study this aspect is the so called "long barrel", which consists of the Phase 1 Pixel Detector and of an Outer Tracker composed of 6 barrel-like layers of p_T modules. Barrels of stacks are mounted on ladders featuring a lever arm between stacks of ~ 4 cm and arranged in a hermetic fashion so that data flow to combine stubs together into consecutive stacks is mantained within each ladder, allowing also for redundancy in overlap regions. If the stub production corresponds to the current local trigger, the matching of pairs of stubs together within a ladder into *tracklets* is analogous to the regional trigger while the full tracking at L1 corresponds to the global trigger idea. There are also other layouts under evaluation and a variety of tools are being developed to help in the decision of the final design.
Tracking algorithms at L1 for the long barrel tracker are currently being studied both in the context of the FNAL proposal and CMSSW. In both cases, stubs are used as inputs for track finders and tracklets seed the search. To guarantee a parallel approach, the FNAL proposal requires that the tracker is divided into 15° sectors so that the bending radius of a track spanning also an adjacent sector (maximizing the search window) corresponds to $p_T \simeq 2.4$ GeV/c, which then becomes a sort of "intrinsic" threshold within the tracking algorithm, as shown in Fig. 2. Straight line approximation of tracklet projection is used to program a cluster of about 20 FPGA's per sector which actually will compose the hardware of the online tracking. On the CMSSW simulations side, the collaboration is currently facing some of the relevant problems in the developement of L1

tracking trigger tools, such as optimizing the tracklet builder, defining the seed propagation and the matching windows, handling and removing duplicate tracks, defining the required vertex resolution, assigning momentum information to a track minimizing the number of encoding bits.[9] Moreover, once these simulation tools are realistic in their implementation, an accurate evaluation of fake rates will be needed in order to rely on the candidate physical objects which the L1 trigger will output.

4. Summary

The effort in developing a L1 tracking trigger for the CMS experiment at the LHC Phase 2 luminosities is dealing with many challenging aspects of the final task to be accomplished. The rejection of low p_T tracks by means of pattern hit correlasion in closely placed sensors has already undergone proof-of-principle also with collision data. Different options for the trigger modules and the front-end electronics are being designed for test and evaluation. Also the design of an online tracking algorithm to be implemented in FPGA's and coping with a 40 MHz bunch crossing is in progress and based on sound ideas which can be realized with commercial electronics. Different ongoing studies aim at the definition of physical objects to trigger on, and eventually this will be of major interest in L1 tracking trigger because of its impact on CMS scientific production.

References

1. L. Evans, P. Bryant *et al.*, *JINST* **3**, p. S08001 (2008).
2. F. Zimmermann, *PoS* **EPS-HEP2009**, p. 140 (2009).
3. W. Scandale and F. Zimmermann, *Nucl. Phys. B (Proc. Suppl.)* **177-178**, p. 207 (2008).
4. S. Chatrchyan *et al.*, *JINST* **3**, p. S08004 (2008).
5. *Technical Proposal for the Upgrade of the CMS Detector Through 2020* (CERN, CMS-DOC-2717, v.15 2011).
6. G. Broccolo, *Physics Procedia* **(TIPP 2011 proceedings, in press)** (2011).
7. R. Lipton and M. Johnson, *Private communications*
8. A. Ryd, L. Fields and E. Salvati, *Private communications*
9. N. Pozzobon, *A Level 1 Tracking Trigger for the CMS Experiment at the LHC Phase 2 Luminosity Upgrade* (Ph.D. Thesis, Università degli Studi di Padova, CERN Document Server Records: CERN-THESIS-2011-029, CMS-TS-2011-008 2011).

SILICON CARBIDE MICROSTRIP DETECTORS FOR HIGH RESOLUTION X-RAY SPECTROSCOPY

GIUSEPPE BERTUCCIO[†,*], DONATELLA PUGLISI

Department of Electronic Engineering and Information Science, Politecnico di Milano Como Campus, Via Anzani 42, 22100 Como, Italy

National Institute of Nuclear Physics (INFN), Milan, Italy

[*]E-mail: Giuseppe.Bertuccio@polimi.it

CLAUDIO LANZIERI

SELEX Sistemi Integrati S.p.A., Via Tiburtina km 12,400 Rome, Italy

Silicon Carbide (SiC) is a wide bandgap semiconductor with outstanding physical properties for realizing ionizing radiation detectors. We present the manufacturing, electrical and spectroscopic characterization of a prototype SiC microstrip detector constituted by 32 strips, 2 mm long, 25 μm wide with 55 μm pitch. The detectors have been fabricated on 115 μm thick undoped epitaxial 4H-SiC using Ni-SiC Schottky junctions. The measured leakage currents are below 5 fA at +25°C and 0.6 pA at +107°C with internal electric fields up to 30 kV/cm. X-ray spectra from ^{55}Fe and ^{241}Am with energy resolution of 224 eV FWHM and 249 eV FWHM (12-13.5 electrons r.m.s.) have been acquired at +20°C and +80°C, respectively.

1. Introduction

In the last decades, wide bandgap semiconductors like GaAs, CdTe, CdZnTe, HgI_2 and SiC have obtained increasing interest in the research oriented to overcome the intrinsic limitations of the traditional Silicon and Germanium on ionizing radiation detectors. In particular, epitaxial Silicon Carbide (SiC) – polytype 4H, due to a high purity of its crystal structure and of its considerable thickness (>100 μm) has become from promising candidate to real competing alternative to silicon in radiation detection, imaging and spectroscopy. Some of the peculiar advantages of 4H-SiC with respect to silicon are its wide bandgap (3.2 eV), high critical breakdown field (2 MV/cm vs. 0.3 MV/cm for Si), high carrier saturation velocity (200 μm/ns vs. 100 μm/ns for Si) and its high thermal conductivity (5 W/Kcm vs. 1.5 W/Kcm for Si). The higher average energy requested for generating electron-hole pairs in SiC (7.8 eV [1]) with respect to

[†] Work partially supported by Italian Space Agency (ASI) and Italian Institute of Nuclear Physics (INFN).

Si (3.7 eV [2]) is well compensated by the ultra low leakage currents of the junctions on SiC which permit the detector to operate without any cooling system and maintaining an excellent signal-to-noise ratio in a wide range of temperature. This leads to notable advantages in terms of lower costs, small sizes and high performance of the detection systems.

This paper presents a prototype of 4H-SiC microstrip detectors suitable for high resolution X-ray spectroscopy with imaging capability in a wide range of operating temperature from -20°C to +107°C.

2. Detector Layout and Manufacturing

SiC microstrip detectors have been fabricated on the top of a 2 inches epitaxial 4H-SiC wafer. Each detector consists of 32 Ni-SiC Schottky junction strips with length of 2 mm, width of 25 μm and pitch of 55 μm. Photograph and cross-section of the microstrip device used in this work are shown in Figure 1.

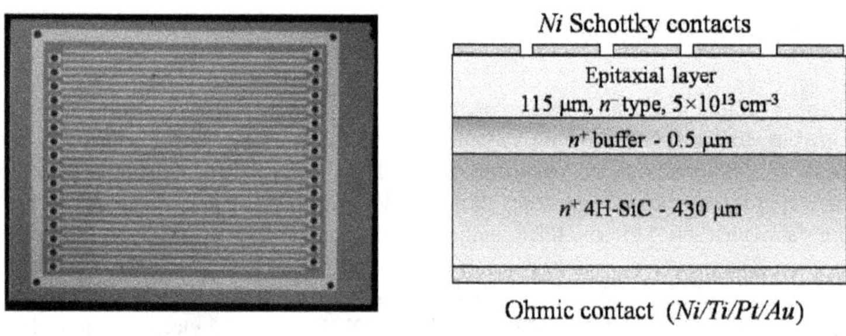

Figure 1. Photograph and cross-section of the SiC microstrip detector used in this work.

The active region of the detector is an undoped high purity epitaxial SiC layer 115 μm thick grown on a highly doped 430 μm thick substrate. An n^+-type buffer layer of 0.5 μm is grown between the substrate and the epitaxial layer. Nickel contacts were realized on the silicon front side of the SiC epitaxial layer for the formation of rectifying Schottky contacts. A Ni/Ti/Pt/Au ohmic contact covers the back surface of the wafer. The SiC epitaxial layer has been produced by LPE-ETC [3-4] and the wafer was processed at Selex Sistemi Integrati [5].

3. Electrical Characterization

3.1. *Statistical leakage current distribution*

The leakage current of each strip has been measured at +100 V and +200 V applied at the ohmic back contact keeping the strips at 0V. Such bias voltages generate inner electric fields of about 15 kV/cm and 30 kV/cm, respectively. Figure 2 shows the current and the current density for each of the 32 strips as measured at +25°C. Values of few fA (few pA/cm^2) have been measured on all the strips. Such ultra-low current values make these detectors practically noiseless (ENC < 1 electron r.m.s. up to 20 μs peaking times).

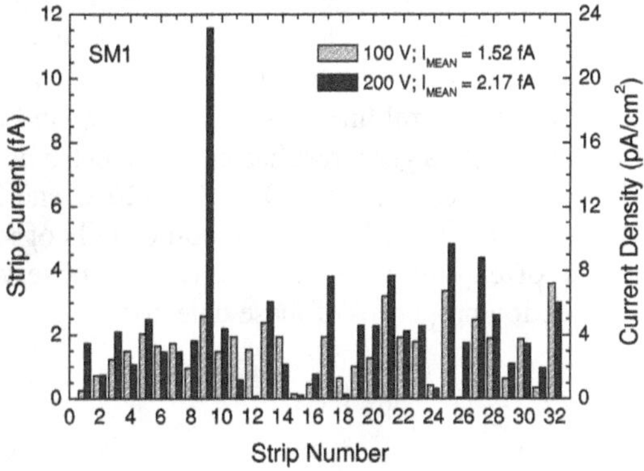

Figure 2. Current and current density (right y-axis) at 100 V and 200 V measured on all 32 strips of the microstrip SM1 at room temperature.

3.2. *Temperature dependence*

The leakage current of the strip no. 23 has been measured at +100 V and +200 V as a function of temperature from +27°C to +107°C. As can be observed in Figure 3, the current remains below 1 pA up to +107°C and at a reverse voltage bias of +200 V. It is worthwhile to note that the current density measured at +100°C is about 1 nA/cm^2, that is the typical value that Silicon detectors show at room temperature.

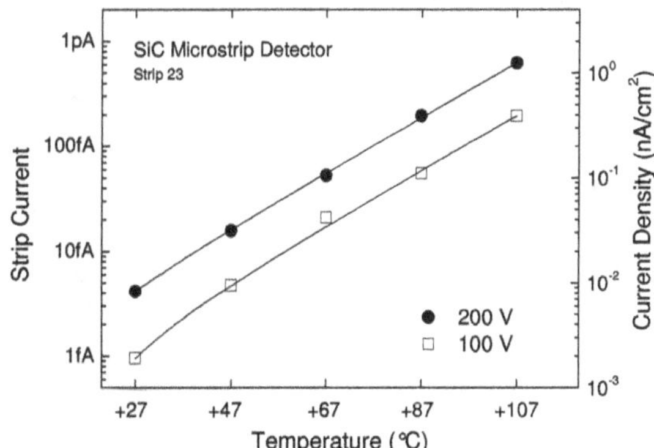

Figure 3. Strip current and current density as a function of temperature.

4. X-ray Spectroscopy

X-ray spectra from ^{55}Fe and ^{241}Am sources have been acquired at different bias voltages and temperatures coupling the detector to a custom ultra low noise CMOS charge preamplifier [6]. As shown in Figure 4, energy resolution of 228 eV FWHM corresponding to 12 electrons r.m.s. has been measured at room temperature. X-ray spectra from ^{241}Am were acquired for a few hours also at -20°C and +80°C. Several spectral lines from Mn, Cu, Np and Ag can be clearly distinguished in Figure 5 with a good resolution. The electronic noise in terms of eV FWHM has been measured between 205 eV at -20°C and 249 eV at +80°C corresponding to 11-13.5 e^- r.m.s.. No degradation or tails of spectral lines with temperature and time of exposure to X-rays have been observed, revealing the excellent charge collection properties of these detectors.

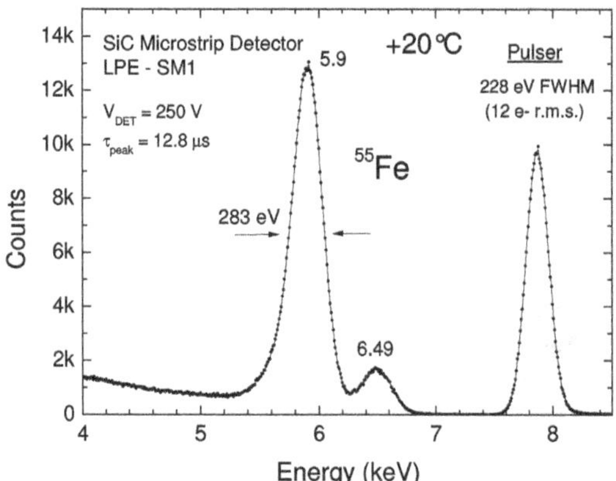

Figure 4. X-ray spectrum from ^{55}Fe source acquired at +20°C with a SiC microstrip detector.

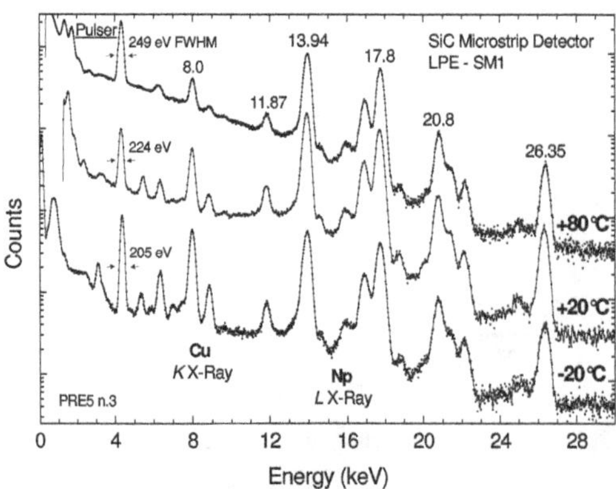

Figure 5. ^{241}Am spectra acquired at -20°C, +20°C and +80°C.

5. Conclusion

A high purity epitaxial SiC microstrip detector has been characterized at room temperature and in the range -20°C to +107°C revealing outstanding performance in terms of ultra low leakage currents, high energy resolution and stability as a function of detector bias, time of exposure to X-ray radiation and operating temperature.

Acknowledgments

Authors would like to thank S. Caccia for his collaboration in the design of the CMOS preamplifier, S. Masci for device bonding and P. Ferrari for his helpful contribution to the electrical characterization.

References

1. G. Bertuccio and R. Casiraghi, *IEEE Trans. Nucl. Sci.* **50**(1), 175 (2003).
2. F. Scholze, H. Rabus and G. Ulm, *Journ. Appl. Phys.* **84**(5), 2926 (1998).
3. LPE S.p.A., Via Falzarego, 8, 20021 Baranzate (MI), Italy.
4. E.T.C. Epitaxial Technology Center S.r.l., 16a Strada - Pantano d'Arci, 95030 Catania, Italy.
5. Selex Integrated Systems S.p.A., Via Tiburtina km 12, Rome, Italy.
6. G. Bertuccio and S. Caccia, *Nucl. Instr. and Meth.* **A579**, 243 (2007).

The Performance of the Outer Tracker Detector at LHCb

B. Storaci*

on behalf of the LHCb Outer Tracker Collaboration

NIKHEF,
Science Park 105, 1098 XG Amsterdam, The Netherlands
**E-mail: Barbara.Storaci@cern.ch*

The LHCb experiment is a single arm spectrometer, designed to study CP violation in B–decays at the Large Hadron Collider (LHC). It is crucial to accurately and efficiently detect the charged decay particles, in the high–density particle environment of the LHC. For this, the Outer Tracker was constructed, consisting of ~55,000 straw tubes, covering in total an area of 360 m^2 of double layers. A precise drift-time measurement results in a single hit resolution of 220 μm, at an average occupancy up to 10% and at 1 MHz trigger rate. At the time of the conference, the detector has been commissioned with almost two years of LHC beam collision data. After dedicated studies to establish timing and spatial alignment, the first results on the detector performance (efficiency, resolutions, etc.) have been obtained.

Keywords: Outer Tracker, Drift-chambers, Straw–tubes, ageing, performance, commissioning, LHCb

1. Introduction

The LHCb detector is a single arm spectrometer.[1] Its tracking system is divided in a silicon detector close to the interaction region, a dipole magnet, and a tracking system behind the magnet. By measuring the deflection of the charged particles by the magnetic field, the momentum of the particles is determined. The tracking system behind the magnet is divided in two parts: a small silicon detector at high rapidity in the highest particle flux region, and a gaseous straw–tube detector, the Outer Tracker (OT), covering most of the LHCb acceptance. The OT has a modular design: 168 long F–modules (500×34 cm^2), and 96 short S–modules above and below the beam–pipe. One F-module consists of two staggered layers with a total of 256 straws. The channels are electrically floating at the centre, and seperately read out at the two ends, resulting in a total of 256 channels per F-module. The anode is 25μm thick gold–plated tungsten wire; the cathode is made of a carbon–doped (XC) kapton straw with a diameter of 4.9 mm on the inside, and aluminium at the outside for electrical grounding and shielding. The OT detector operates with a gas mixture of $Ar/CO_2/O_2$ 70/28.5/1.5 at a nominal high voltage of 1550 V, corresponding to a gain of approximately 5×10^4.[2]

2. Detector Production and Installation

The straw–tube modules (185 F–modules and 110 S–modules) were produced from 2004 to 2006. The quality of modules was checked after mass–production allowing a detailed qualification of the modules. The dark current after HV training was typically 1 nA per wire, and the gas–tightness of the entire module was of the order of 10^{-4} l/s at an overpressure of 7 mbar. An automatised test–setup was built to irradiate the full width of each module with a ^{90}Sr line source (a β source, emitting electrons of energies up to 2.3 MeV) in steps of 1 cm along the entire module length. The corresponding current response of each wire was recorded and used to detect bad wire–locator positioning, straw deformations, abnormally high wire currents, etc. Typically the uniformity of the module responses were measured to be better than $\pm 10\%$, while less than 1% of the total channels were not functional (mostly wires deliberately disconnected due to shortcuts, and mostly located in the first few modules produced).

3. Ageing

The OT–modules were designed to withstand a large irradiation dose during the planned ten years of operation.[1] However, the modules have shown a rapid gain loss under mild irradiation (few nA/cm) in the laboratory. Under the influence of irradiation, a small insulating layer of hydrocarbon–containing substance is deposited on the anode wire, thereby reducing the signal response of the detector. Moreover, the ageing pattern showed several unusual characteristics rarely reported in literature by other drift chamber detector studies: most of the damage is placed upstream the source, producing a not symmetric pattern (half–moon shape) and the maximum gain loss is not the largest directly below the source, where the intensity is highest. The ageing is caused by contamination of the counting gas due to outgassing of the plastifier di–isopropyl–naphthalene in araldite AY103–1 used at construction.

The ageing process was extensively studied in the laboratory.[3] Irradiation tests were carried out on a small selection of final modules using a 2 mCi ^{90}Sr source. The gain loss is quantified by comparing the current profile before and after the irradiation.

A number of preventive actions to reduce the deterioration of the detector response were devised, like heating the modules for 2 weeks at 40°C to increase the outgassing rate, adding a few percent of oxygen to the counting gas to decrease the ageing rate through the enhancement of ozone formation in the avalanche, and lowering the gas flow rate in order to limit efficient removal of the ozone formed in the avalanche. In addition to these preventive measures, a treatment has been devised to remove the insulating deposits on the anode wire, consisting in the application of large high voltage values that take the OT drift–tubes in the discharges regime and produce high dark currents up to 20 μA/cm per wire.[4] This techniques restores the gain and even prevents from further irradiation damage up to a certain dose, see Figure 1. Inspection under a scanning electron–microscope (SEM) and analyses with energy dispersion X–ray spectroscopy (EDX) were performed on HV–trained wires. No mechanical damage to the gold plating of the anode wire were seen and no extra depositions on the wire surface were present.[4]

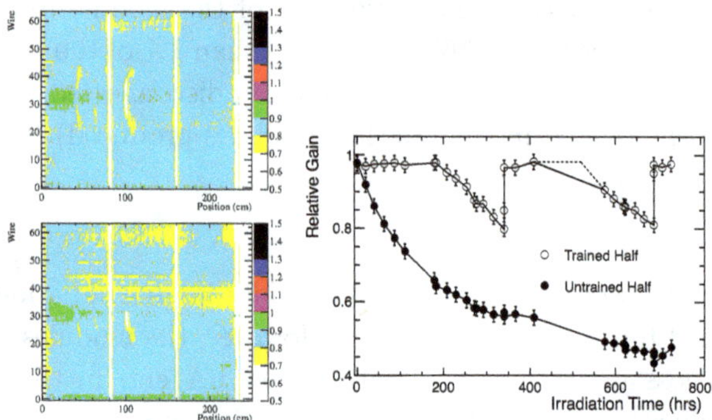

Fig. 1. Single scan of a module in situ before (**Left-Top**) and after (**Left-Bottom**) the HV training process. The channels between 1–32 were used as a reference and no HV training was applied. **Right**:The relative gain is deteriorating as a function of irradiation time. However, the HV trained half of the module recovers the gain. Furthermore, the three HV training procedures (at 0, 300, and 700 hours, respectively) all show a prevention of further gain loss.

A system to scan in situ the modules was commissioned. To obtain comparable performance with the laboratory system two ^{90}Sr sources with an activity of 74 MBq each are used to scan the module. Scans up to eighteen half–modules are performed every technical stop and up to now no sign of gain loss is observed.

A complementary method to check the gain deterioration in situ is through periodic threshold scans. The hit–efficiency versus threshold is described by an error function and the variation in the point at 50% of the efficiency, the half–efficiency point (HEP), is a sign of gain loss. The stability of this parameter for the full detector is monitored every 200 pb^{-1}, and after an integrated delivered luminosity of $\sim$ 800 pb^{-1} on September 7, 2011, no sign of gain loss is observed.

4. Readout Monitoring

During commissioning and operation of the LHCb detector, the stability and the quality of the OT FE–electronics performances was monitored. For the calibration tasks, a test–pulse facility (injecting pulses with adjustable heights and time phases into the FE–electronics preamplifiers) has been provided. A number of monitoring and data–quality analysis tasks have been developed, e.g. to determine dead and noisy channels, or even to detect defects in the timing of the OT channels, as non linear response. The analyses described above were continuously used during 2009–2011 period to detect and replace noisy or broken components. This resulted in a total of 99.7% functional channels in June 2011.

5. Detector Performance

The study of the beam–beam collision data, started in 2009, allowed one to constantly check the OT performance with LHC beam. The key quantity is the drift–time of the straw–tubes, extracted from the TDC time after correction for time–of–flight and propagation time along the wire. The comparison between the hit predictions from the tracking and the actual drift–times demonstrated the validity of the $T(r)$ relation extracted from the test–beam data[5] (see Fig. 2 left). The tracking prediction for the distance of a hit from the anode wire was also used to study the hit efficiency along the cell profile. A flat efficiency distribution with a plateau between 99.2–99.5% for all modules was found, as expected from the test–beam data.[5] Figure 2 shows an example of the residuals obtained from track minimisation in data events: a double–gaussian model was fitted to the data, from which an intrinsic spatial resolution of about 220 μm was estimated. A momentum resolution $\Delta p/p$ between 0.3 and 0.5% is obtained and it is reflected in an excellent mass resolution of 24 MeV for 2–body B–decays.

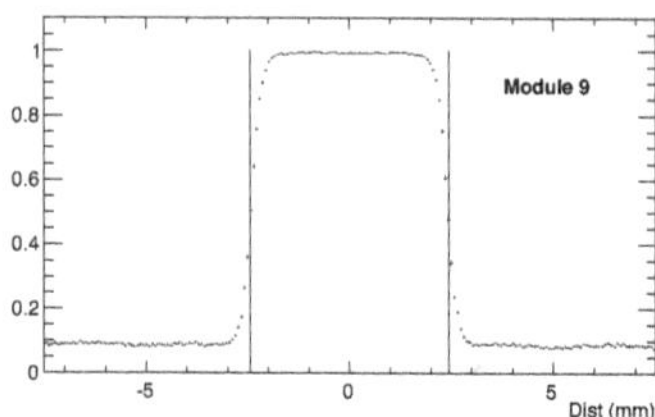

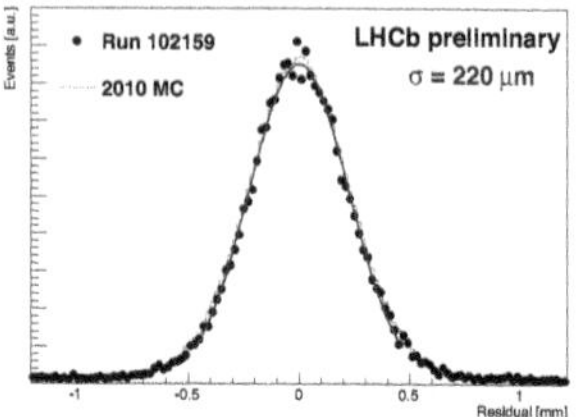

Fig. 2. **Left**: Cell efficiency profile for one OT–module. Fit performed in the plateau region obtaining an average efficiency inside the cell of 99.5%. **Right**: Residuals from track minimisation (beam–data events). The curve denotes the double–gaussian model fitted to the data.

6. Conclusions

The LHCb Outer Tracker detector was constructed to accurately and efficiently detect the charged decay particles and measure their momenta. It consists of ~55,000 straw tubes, covering in total an area of 360 m^2 of double layers. At the time of the conference the detector has operated successfully for almost two years. To date no signs of ageing have been observed. Dedicated studies to monitor the FE–electronics as well as the performance (like hit–efficiency, hit–resolution, etc.) have been presented. These studies confirmed that the Outer Tracker detector is fully operational and performs as expected providing an excellent momentum and mass resolution.

References

1. LHCb Collaboration, The LHCb Detector at the LHC, 2008, JINST, 3 S08005;
   LHCb collaboration, LHCb reoptimized detector design and performance: Technical Design Report, [CERN-LHCC-2003-030], CERN Geneva, September 2003.
2. G. van Apeldoorn et al., Outer Tracker Module Production at NIKHEF - Quality, [CERN-LHCb-2004-078], CERN Geneva, October 2004.
3. S.Bachmann et al., Ageing in the Outer Tracker: Phenomenon, Culprit and Effect of Oxygen, NIM A617, 202 (2010)
4. N.Tuning et al., Ageing in the Outer Tracker: Aromatic Hydrocarbons and Wire Cleaning, NIM A656, 45 (2011)
5. G. van Apeldoorn et al., Beam Tests of Final Modules and Electronics of the LHCb Outer Tracker in 2005, [CERN-LHCb-2005-076], CERN Geneva, October 2005.

ELECTRICAL CHARACTERIZATION OF SiPM AS A FUNCTION OF TEST FREQUENCY AND TEMPERATURE

M.J. Boschini[1,3], C. Consolandi[1], P.G. Fallica[4], M. Gervasi[1,2], D. Grandi[1], M. Mazzillo[4], S. Pensotti[1,2], P.G. Rancoita[1], D. Sanfilippo[4], M. Tacconi[1]* and G. Valvo[4]

[1]*Istituto Nazionale di Fisica Nucleare, INFN Milano-Bicocca, Milano (Italy)*
[2]*Department of Physics, University of Milano Bicocca, Milano (Italy)*
[3]*CILEA, Segrate (MI) (Italy)*
[4]*STMicroelectronics, Catania (Italy)*

E-mail: mauro.tacconi@mib.infn.it

Silicon Photomultipliers (SiPM) represent a promising alternative to classical photomultipliers for the detection of photons in high energy physics and medical physics, for instance. In the present work, electrical characterizations of test devices - manufactured by STMicroelectronics - are presented. SiPMs with an area of $3.5 \times 3.5\,\mathrm{mm}^2$ and a cell pitch of $54\,\mu$m were manufactured as arrays of 64×64 cells and exhibiting a fill factor of 31%. The capacitance of SiPMs was measured as a function of reverse bias voltage at frequencies ranging from about 20 Hz up to 1 MHz and temperatures from 310 K down to 100 K. Leakage currents were measured at temperatures from 410 K down to 100 K. Thus, the threshold voltage - i.e., the voltage above a SiPM begins to operate in Geiger mode - could be determined as a function of temperature. Finally, an electrical model capable of reproducing the frequency dependence of the device admittance is presented.

1. Introduction

In recent years Silicon Photomultipliers (SiPM) has been developed for usage in photon detection. The high gain, insensitivity to magnetic field and low reverse bias voltage of operation make SiPM a promising candidates as replacement of classical photomultipliers[1] in several of their applications. The SiPM is an array of parallel-connected single photon avalanche diodes (SPAD)[2] . Every SPAD operates in Geiger mode with a quenching resistor in series to prevent an avalanche multiplication process from taking place. The overall output of a SiPM depends on how many SPADs are simultaneously ignited. [3,4]

The current SiPM device was manufactured by STMicroelectronics (details of the technology can be found in Refs.[2,5]). It consisted of an array of 64×64 SPAD cells with $3.5 \times 3.5\,\text{mm}^2$ effective area, a cell pitch of $54\,\mu\text{m}$ and a fill factor of $\approx 31\%$.

In the present article, the electrical characteristics of these devices are shown as function of test frequency from 1 MHz down to about 20 Hz and temperature from 410 K down to 100 K (Sects. 2–2.2). In addition, in Sects. 3 and 3.1 an electrical model is discussed and compared with data obtained from dependencies of capacitance and resistance on the test frequency for both a photodiode and a SiPM devices.

2. Electrical Characteristics of SiPM Devices

The electrical characteristics of SiPM devices were investigated as function of applied reverse bias voltage (V_r) and temperature. The capacitance response was also studied as a function of test frequency of the capacimeter employed for such a measurement (Sect. 2.1).

Furthermore (Sect. 2.2), the measurements of the SiPM leakage current allowed one to determine (as a function of temperature) the value (and its temperature dependence) of the so-called *threshold voltage* (V_{th}), i.e., the reverse bias voltage above which a SiPM begins to operate in Geiger mode.

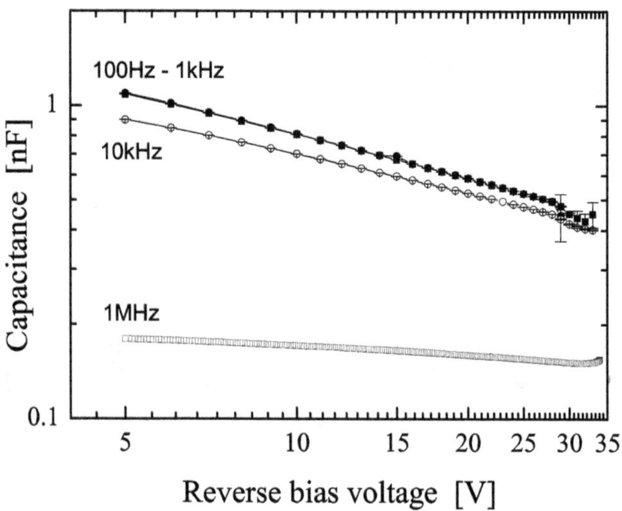

Fig. 1. Capacitance (in nF) as a function of reversed bias voltage (in V) using the LCZ meter with test frequencies of 100 Hz (•), 1 kHz (■), 10 kHz (○) and BC with 1 MHz (□).

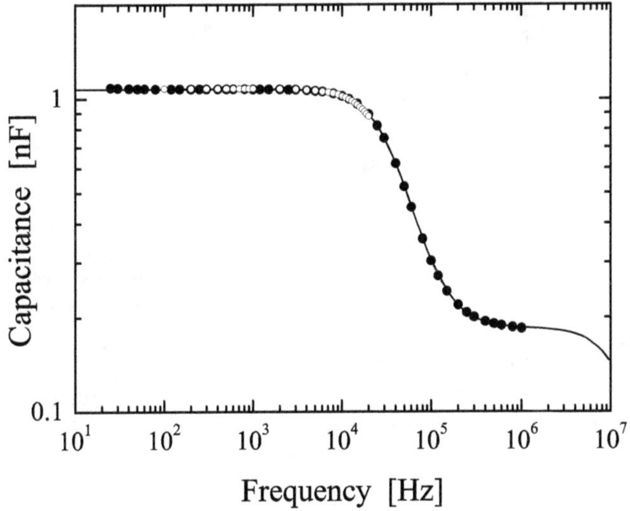

Fig. 2. Capacitance (in nF) as a function of the frequency (in Hz) for a sample operated with a reverse bias of 5 V: data points indicated with • (○) were obtained using the LCR (LCZ) meter, the continuous line is obtained from Eq. (9) (see discussion in Sect. 3.1).

2.1. *Capacitance Response*

The capacitance response of SiPM devices was systematically measured as a function of reverse bias voltage applied and test frequency at 300 K (i.e., at room temperature). Moreover, using a liquid nitrogen cryostat, at a few fixed frequencies the capacitance was also measured as a function of reverse bias voltage and temperature from 310 K down to 100 K. For these purposes three instruments were employed, i.e., a Boonton capacimeter (BC) (using its internally provided power source and a test frequency of 1 MHz), an LCZ (Agilent Technologies 4276A) and LCR (Agilent Technologies 4284A) meters with the reverse bias supplied to the device from an external power supply. The LCZ meter employed test frequencies from 100 Hz up to 20 KHz; the LCR meter from $\approx$ 20 Hz up to 1 MHz. Furthermore, the measurements were performed selecting the parallel equivalent-circuit mode of the LCR and LCZ meters for the device under test (DUT). Thus, the device response is assumed to be that one from a parallel capacitor-resistor circuit (see discussion in Sect. 3). Finally, the currently reported experimental errors are the standard deviations obtained using 50 subsequent measurements of

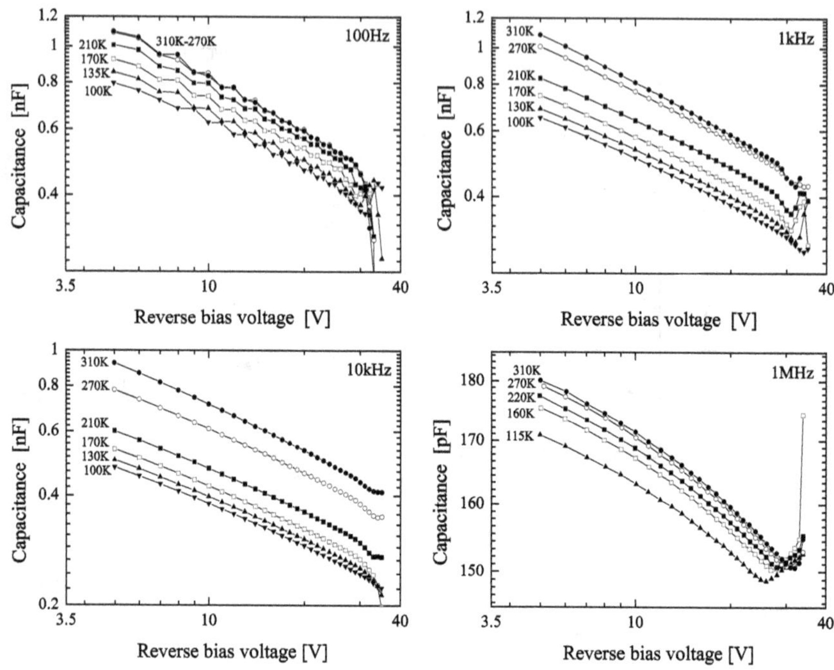

Fig. 3. Capacitance response (in nF or pF) obtained using the LCZ meter as a function of reverse bias applied (in V) and temperature from 310 K down to 100 K: 100 Hz (upper left), 1 kHz (upper right), 10 kHz (bottom left) and - using the BC - 1 MHz (bottom right).

the device capacitance under the same experimental conditions.

In Fig. 1, the capacitance response of a SiPM using frequencies of 100 Hz,

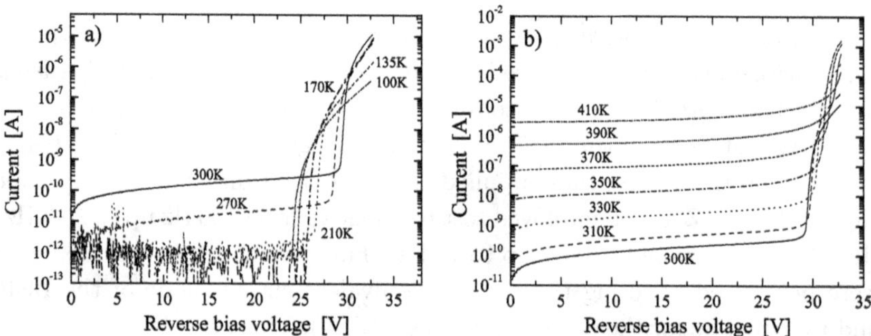

Fig. 4. (a) Leakage current (in A) as a function of reverse bias voltage (in V) from 310 down to 100 K; (b) leakage current (in A) as a function of reverse bias voltage (in V) from 300 up to 410 K.

1 kHz, 10 kHz and 1 MHz is shown as a function of V_r at 300 K. Above 1 kHz, the measured capacitance decreases with the increase of the test frequency. In Fig. 2, the capacitance response - determined using both LCZ and LCR meters - is shown for a device operated with a reverse bias voltage of 5 V at 300 K. It can be observed that above $\approx$ 10 kHz the measured capacitance largely decreases with the increase of the test frequency of the instrument up to achieving an almost constant response above a few hundreds of kHz.

In Fig. 3, the capacitance response using 100 Hz, 1 kHz, 10 kHz and 1 MHz is shown as a function of V_r (in V) and temperature from 310 K down to 100 K. The measured capacitance is observed to decrease with decreasing temperature. This behavior is also observed for frequencies lower than $\approx$ 10 kHz; while a similar capacitance response was exhibited at 300 K (see Fig. 1).

2.2. *Current-Voltage Characteristics*

The dependence of leakage current on reverse bias was studied as a function of temperature from 410 K down to 100 K as shown in Fig. 4. With increasing V_r the leakage current does not vary by more than an order of magnitude until the multiplication regime is ignited, i.e., so far the *threshold voltage*, V_{th}, is reached. V_{th} is the voltage corresponding to the value of leakage current at the intercept between the two I versus V curves obtained, the first when the multiplication regime is sharply starting and the second one at lower voltages, i.e., before the multiplication occurs: an example regarding the determination of the value of V_{th} at 270 K is reported in Fig. 5(a). As shown in Fig. 5(b), the threshold voltage lowers - thus, SPADs turn into

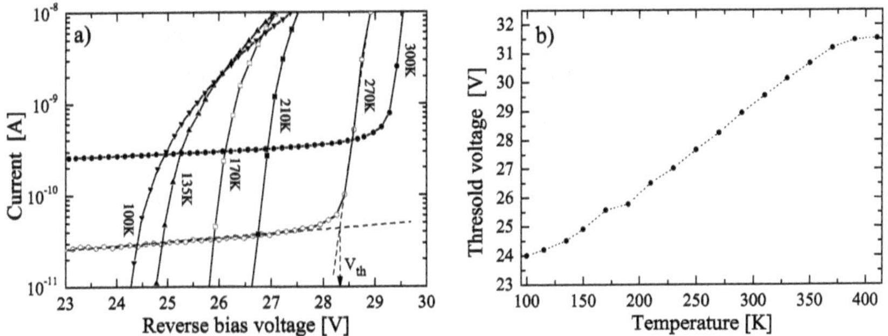

Fig. 5. (a) Leakage current (in A) as a function of reverse bias voltage applied (in V) and an example of threshold voltage, V_{th} in V, determination at 270 K; (b) threshold voltage (in V) as a function of temperature (in K).

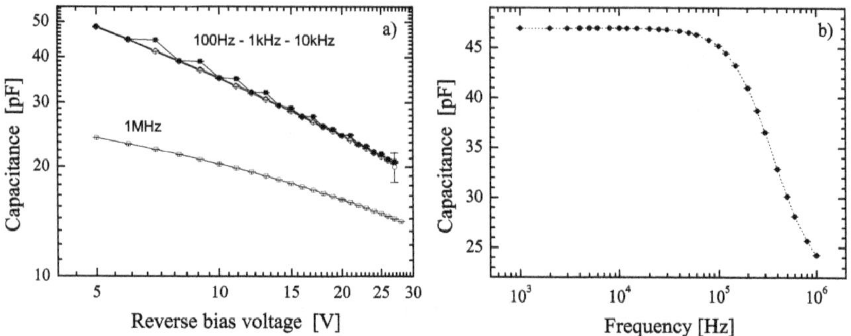

Fig. 6. (a) Capacitance (in pF) of a photodiode as a function of reverse bias voltage (in V) at 100 Hz, 1 kHz, 10 kHz (using the LCZ meter) and 1 MHz (using the BC); (b) capacitance (in pF) of a photodiode as a function of test frequency in Hz (using the LCR meter) with an applied reverse biased voltage of 6 V.

a Geiger-mode regime at progressively lower reverse bias voltages - with lowering temperature at the rate of $\approx -29\,\mathrm{mV/K}$ from ≈ 360 down to $\approx 130\,\mathrm{K}$.

3. Electrical Model

At room temperature, the electrical frequency response of SiPMs was further investigated using a test device manufactured by STMicroelectronics. The latter device was a photodiode (PD) with a structure similar to that of a SPAD cell, an active area of $\approx 0.2\,\mathrm{mm}^2$, but no quenching resistor in series. In Fig. 6(a), the capacitance of the PD is shown as a function of reverse bias voltage at 100 Hz, 1 kHz, 10 kHz and 1 MHz. Furthermore, one can see that the device exhibits almost no dependence on the test frequency below $\approx 100\,\mathrm{kHz}$ [Fig. 6(b)].

For silicon photodiodes and radiation detectors, the electrical response down to cryogenics temperatures is usually modeled using the so-called *small-signal ac impedance of junction diode* (SIJD) operated under reverse

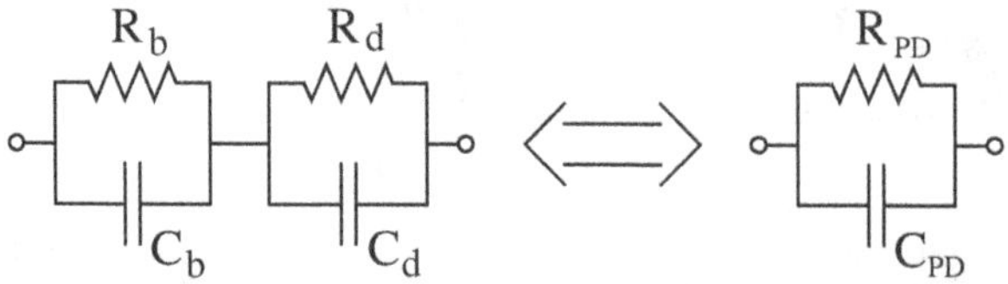

R_b R_d C_b C_d $\Longleftrightarrow$ R_{PD} C_{PD}

Fig. 7. SIJD model for silicon photodiodes and radiation detectors: C_{d} (C_{b}) and R_{d} (R_{b}) are the capacitance and resistance of the depleted (field free) region, respectively; C_{PD} and R_{PD} are the overall capacitance and resistance of the photodiode, respectively.

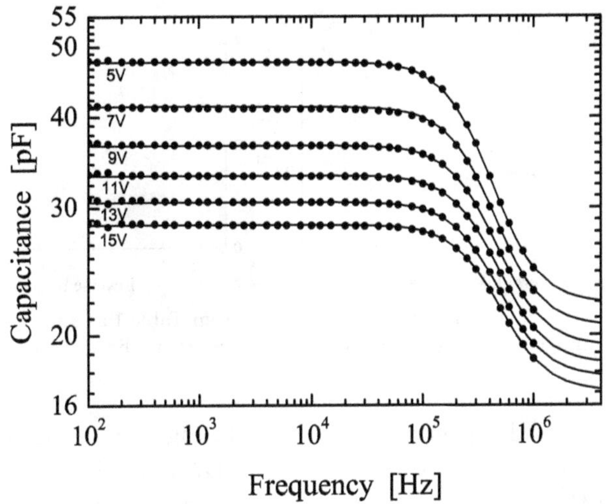

Fig. 8. Measured capacitance (•), C_{PD}, (in pF) as a function of test frequency (in Hz) with superimposed continuous lines obtained from the SIJD model [Eq. (3)].

bias (e.g., see Refs.[6,7], Section 4.3.4 of Ref.[8] and references therein). In the framework of the SIJD model, the device consists of the depleted and field free regions connected in series. Each one of these regions, in turn, consists of

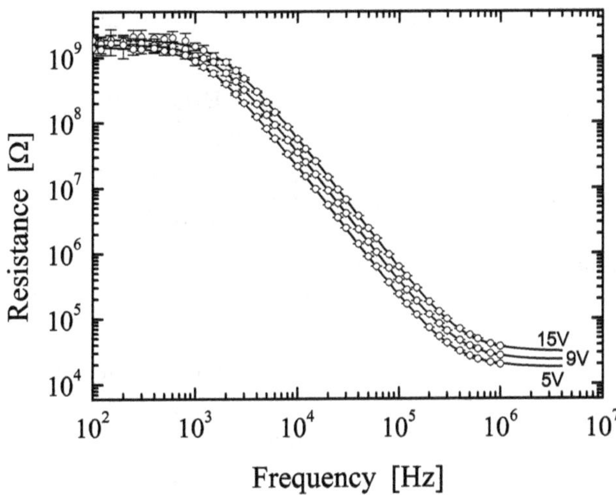

Fig. 9. Measured resistance (○), R_{PD}, (in Ω) as a function of test frequency (in Hz) at 5, 9 and 15 V with superimposed continuous lines obtained from the SIJD model [Eq. (2)].

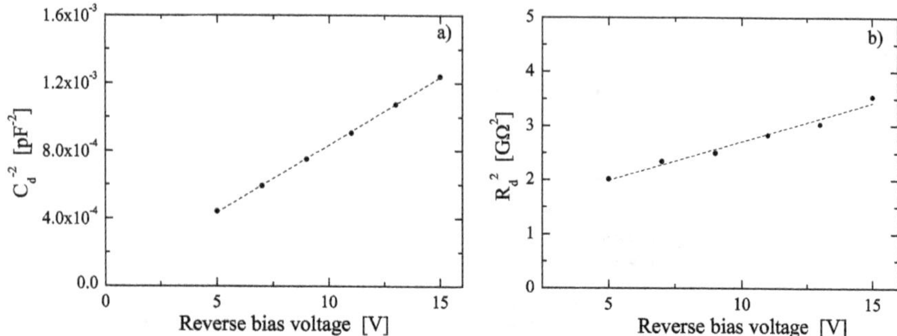

Fig. 10. (a) $1/C_{\mathrm{d}}^2$ (in pF^{-2}) and (b) R_{d}^2 (in $\mathrm{G}\Omega^2$) from Table 1 as a function of applied reversed bias voltage in V. The dashed lines are those from a linear fit to the data.

a parallel-connected capacitor and resistor. In Fig. 7, the equivalent circuit of a photodiode is shown: C_{d} (C_{b}) and R_{d} (R_{b}) are the capacitance and resistance of the depleted (field free) region, respectively.

Labeling (Fig. 7) the overall[§] capacitance and resistance of a photodiode with C_{PD} and R_{PD}, respectively, one can express the admittance of the device as:

$$\begin{aligned} Y_{\mathrm{PD}}(\omega) &= G_{\mathrm{PD}}(\omega) + j\,B_{\mathrm{PD}}(\omega) \\ &= \frac{1}{R_{\mathrm{PD}}} + j\,\omega C_{\mathrm{PD}} \end{aligned} \tag{1}$$

with $\omega = 2\pi f$, where f is the test frequency. The conductance (i.e., the real part of the admittance) and susceptance (the imaginary part of the admittance) are respectively given by:

$$G_{\mathrm{PD}}(\omega) = \frac{1}{R_{\mathrm{PD}}} = \frac{R_{\mathrm{d}} + R_{\mathrm{b}} + \omega^2 R_{\mathrm{d}} R_{\mathrm{b}} (R_{\mathrm{d}} C_{\mathrm{d}}^2 + R_{\mathrm{b}} C_{\mathrm{b}}^2)}{(R_{\mathrm{d}} + R_{\mathrm{b}})^2 + \omega^2 R_{\mathrm{d}}^2 R_{\mathrm{b}}^2 (C_{\mathrm{b}} + C_{\mathrm{d}})^2} \tag{2}$$

$$B_{\mathrm{PD}}(\omega) = \omega C_{\mathrm{PD}} = \omega \left[\frac{\omega^2 R_{\mathrm{d}}^2 R_{\mathrm{b}}^2 C_{\mathrm{d}} C_{\mathrm{b}} (C_{\mathrm{b}} + C_{\mathrm{d}}) + R_{\mathrm{d}}^2 C_{\mathrm{d}} + R_{\mathrm{b}}^2 C_{\mathrm{b}}}{(R_{\mathrm{d}} + R_{\mathrm{b}})^2 + \omega^2 R_{\mathrm{d}}^2 R_{\mathrm{b}}^2 (C_{\mathrm{b}} + C_{\mathrm{d}})^2} \right] \tag{3}$$

[e.g., see Equations (4.184, 4.185) at page 455 of Ref.[8]].

It has to be remarked that the values of both C_{PD} and R_{PD} are those which can be determined, for instance, selecting the parallel equivalent-circuit mode for the DUT using the LCR meter. In the SIJD model, the values of C_{d}, R_{d}, C_{b} and R_{b} do not depend on the test frequency. In addition, C_{d}, R_{d} are expected to depend on the applied reverse bias voltage, because the depleted layer width increases with increasing V_{r}.

[§] C_{PD} and R_{PD} are the quantities directly measured selecting the parallel equivalent-circuit mode of the LCR and LCZ meters.

Table 1. C_d, R_d, C_b and R_b obtained from a fit of the SIJD model for a photodiode to experimental data as a function of reverse bias voltage (V_r).

V_r [V]	C_d [pF]	R_d [GΩ]	C_b [pF]	R_b[kΩ]
5	47.6	1.42	41.1	5.2
7	41.1	1.53	40.8	5.2
9	36.5	1.58	40.8	5.2
11	33.2	1.68	40.4	5.2
13	30.5	1.74	40.9	5.3
15	28.4	1.88	40.5	5.3

C_d, R_d, C_b and R_b were determined (Table 1) as a function V_r, by a fit to the measured quantities C_{PD} and R_{PD} (obtained using the LCR meter) using the corresponding expressions [e.g., see Eqs. (2, 3)] obtained from the SIJD model for a photodiode. For instance, in Fig. 8 (Fig. 9) the experimental data and fitted curves are shown for the capacitance (resistance) measurements as a function of test frequency and applied reverse bias voltage. As expected[6] (Table 1), the field free region is almost independent of V_r, while both the capacitance and resistance of the depleted region exhibit a dependence on the reverse bias. In Fig. 10, $1/C_d^2$ [Fig. 10(a)] and R_d^2 [Fig. 10(b)] values from Table 1 are shown as a function of applied reverse voltage. Although the junction photodiode cannot be considered as a one-sided step junction[¶], the the dashed curves (Fig. 10) obtained from fits to the reported data - assuming a linear dependence of $1/C_d^2$ and R_d^2 on V_r - are well suited for reproducing the dependence on the reverse bias voltage.

3.1. *Electrical Model for SiPMs*

A SiPM device consists of a set of SPAD devices (4096 in the current SiPM under test) connected in parallel. The equivalent electrical circuit of the elemental cell (i.e., a SPAD cell) is shown in Fig. 11 and differs from that of a photodiode (discussed in Sect. 3) by a resistance (R_s) added in series. In Fig. 11, C_d (C_b) and R_d (R_b) are the capacitance and resistance of the depleted (field free) region of the photodiode, respectively. In the present technology of STMicroelectronics, R_s is typically about (0.2–1.0) MΩ.

¶The depletion layer characteristics as a function of reverse bias voltage can be found treated, for instance, in Chapter 6-2 of Ref.[10] (see also Ref.[9]).

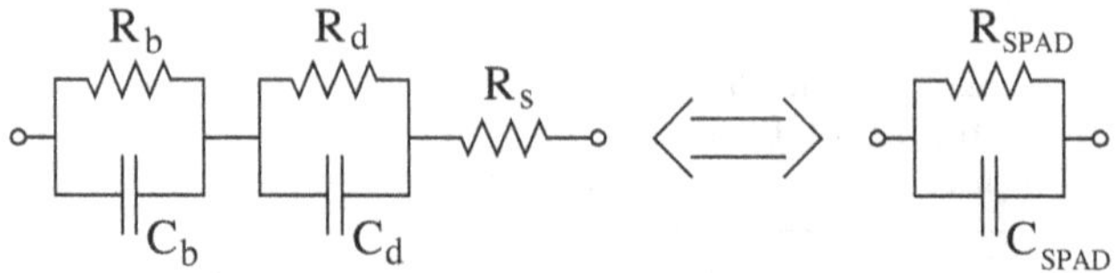

Fig. 11. SIJD model for a SPAD cell: $C_{\rm d}$ ($C_{\rm b}$) and $R_{\rm d}$ ($R_{\rm b}$) are, respectively, the capacitance and resistance of the depleted (field free) region with a series resistance $R_{\rm s}$; $C_{\rm spad}$ and $R_{\rm spad}$ are the overall capacitance and resistance of a SPAD cell, respectively.

The admittance ($Y_{\rm spad}$) of a SPAD cell is given by

$$\begin{aligned} Y_{\rm spad}(\omega) &= G_{\rm spad}(\omega) + j\,B_{\rm spad}(\omega) \\ &= \frac{1}{R_{\rm spad}} + j\,\omega C_{\rm spad} \end{aligned} \tag{4}$$

with $C_{\rm spad}$ and $R_{\rm spad}$ respectively expressed in terms of $C_{\rm d}$, $C_{\rm b}$, $R_{\rm d}$, $R_{\rm b}$ and $R_{\rm s}$ as

$$\begin{aligned} C_{\rm spad} &= \frac{B_{\rm spad}(\omega)}{\omega} \\ &= \frac{\omega^2 R_{\rm d}^2 R_{\rm b}^2 C_{\rm d} C_{\rm b}(C_{\rm b} + C_{\rm d}) + R_{\rm d}^2 C_{\rm d} + R_{\rm b}^2 C_{\rm b}}{D_1 + D_2}, \end{aligned} \tag{5}$$

$$\begin{aligned} R_{\rm spad} &= \frac{1}{G_{\rm spad}(\omega)} \\ &= \frac{D_1 + D_2}{R_{\rm b} + (1 + \omega^2 C_{\rm b}^2 R_{\rm b}^2)(R_{\rm s} + R_{\rm d}) + \omega^2 R_{\rm d}^2 C_{\rm d}^2(\omega^2 C_{\rm b}^2 R_{\rm s} R_{\rm b}^2 + R_{\rm s} + R_{\rm b})}, \end{aligned} \tag{6}$$

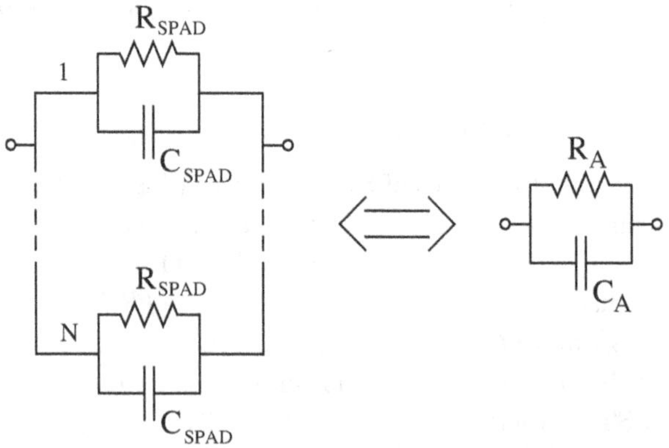

Fig. 12. SIJD model for a SiPM device consisting of N SPAD devices connected in parallel: $C_{\rm spad}$ and $R_{\rm spad}$ are, respectively, the capacitance and resistance of the equivalent circuit of a SPAD cell (Fig 11), $C_{\rm A}$ and $R_{\rm A}$ those of the SiPM device.

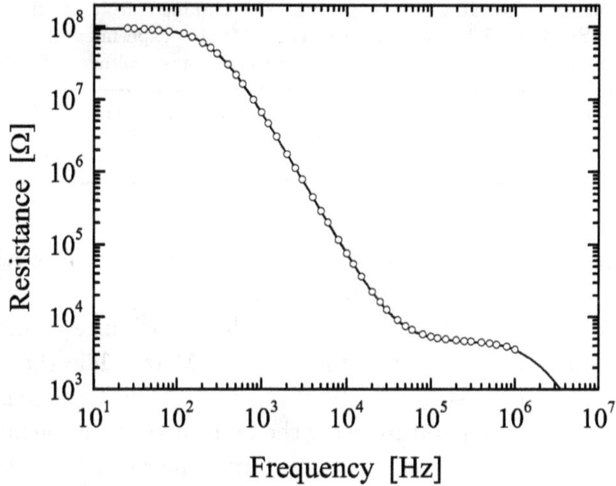

Fig. 13. Measured resistance (○), R_A, (in Ω) as a function of frequency (in Hz) with superimposed the continuous line obtained from Eq. (8).

where D_1 and D_2 are

$$D_1 = \omega^2 R_b^2 C_b^2 (R_s + R_d)^2 + \omega^2 R_d^2 C_d^2 (R_s + R_b)^2$$
$$D_2 = (R_d + R_b + R_s)^2 + \omega^2 R_d^2 R_b^2 C_d C_b (2 + \omega^2 R_s^2 C_d C_b).$$

The equivalent electrical circuit* of a SiPM is shown in Fig. 12. Assuming that each SPAD cell has the same admittance, one finds that the overall admittance of an array is determined by:

$$\begin{aligned} Y_A(\omega) &= N\, Y_{\text{spad}}(\omega) \\ &= G_A(\omega) + j\, B_A(\omega) \\ &= \frac{1}{R_A} + j\, \omega C_A \end{aligned} \tag{7}$$

with

$$G_A(\omega) = \frac{N}{R_{\text{spad}}} \tag{8}$$
$$B_A(\omega) = \omega N\, C_{\text{spad}}, \tag{9}$$

*It has to be remarked that one can also add a series resistance to R_A. This was the case for another SiPM device with 3600 SPAD cells manufactured by STMicroelectronics; for such a device a small additional series resistance of about 1 kΩ was needed.

Table 2. C_d, R_d, C_b, R_b and R_s obtained from a fit of the SIJD model (adapted to a SiPM) to experimental data with the device operated at a reverse bias voltage of 5 V.

V_r [V]	C_d [pF]	R_d [GΩ]	C_b [pF]	R_b[MΩ]	R_s[kΩ]
5	0.262	385	0.056	12.5	195

where C_{spad} and R_{spad} are obtained from Eqs. (5, 6), respectively; finally, N = 64 × 64 = 4096 for the present device.

In Fig. 2 (13), the measured values of the capacitance (resistance) are shown as a function of test frequency up to 1 MHz, while the continuous line is that obtained using Eq. (9) [Eq. (8)]. These measurements were carried out at room temperature using the LCR meter with the SiPM device operated at a reverse bias voltage of 5 V. Furthermore, in Table 2 the values of C_d, R_d, C_b, R_b and R_s obtained from such a fit of the SIJD model (adapted to a SiPM) to experimental data are reported. It can be remarked that the value obtained for R_s is in agreement with one of those typically used in the current technology.

Finally, one can point out that the frequency dependence of the present SIJD model[||] for SiPM and photodiode devices is well in agreement with measurements.

4. Conclusions

The electrical characteristics of SiPM devices were investigated as a function of applied reverse bias voltage (V_r) and temperature from 410 down to 100 K. The capacitance response was also studied as a function of test frequency. One finds that the measured capacitance decreases i) with increasing the test frequency above 10 kHz at 300 K and ii) with the decreasing of the temperature. Furthermore, the measurement of the leakage current allowed one to determine the value (and its temperature dependence) of the threshold voltage (V_{th}) above which a SiPM begins to operate in Geiger mode: V_{th} decreases with lowering temperature at the rate of ≈ -29 mV/K from 300 down to 100 K.

It was developed an electrical model to treat the frequency dependence of a SiPM device operated under reverse bias voltage. This model is based on the so-called small-signal ac impedance of junction diode (SIJD) used

[||]The reader can found modelizations of SiPM devices adapted for a time based readout[11] or physical parameters measured at fixed frequencies (e.g., at 100 kHz in Ref.[5] and 1 MHz in Ref.[12]).

for radiation detectors and photodiodes. The model was found to be well suited to account for the SiPM dependencies of capacitance and resistance on frequency.

References

1. V.D. Kovaltchouk et al., *Nucl. Instr. and Meth. in Phys. Res. A* 538 (2005), 408–415.
2. M. Mazzillo et al., *Nucl. Instr. and Meth. in Phys. Res. A* 591 (2008), 367–373.
3. F. Zappa et al., *Sens. Actuat. A* 140 (2007), 103–112.
4. M. Mazzillo et al., *Sens. Actuat. A* 138 (2007), 306–312.
5. G. Condorelli et al., *Nucl. Instr. and Meth. in Phys. Res. A* 654 (2011), 127–134.
6. C.H. Champness, *J. of App. Phys.* 62 (1987), 917.
7. C. Leroy and P.G. Rancoita, Particle Interaction and Displacement Damage in Silicon Devices operated in Radiation Environments, *Rep. Prog. in Phys.* 70 (2007), 403–625, doi: 10.1088/0034-4885/70/4/R01.
8. C. Leroy and P.G. Rancoita, Principles of Radiation Interaction in Matter and Detection - 3rd Edition - (2011), World Scientific, Singapore, ISBN-978-981-4360-51-7.
9. R.B. Fair, *J. Electrochem. Soc.: Solid State Sci.*, 118 (1971), 971.
10. Wolf, H.F. (1971). Semiconductors Wiley-Interscience, New York.
11. P. Jarron et al., *IEEE Nuclear Science Symposium Conference Record (NSS/MIC)* (2009), 1212–1219, doi: 10.1109/NSSMIC.2009.5402391.
12. F. Corsi et al., *Nucl. Instr. and Meth. in Phys. Res. A* 572 (2007), 416.

THE ALIGNMENT OF THE CMS SILICON TRACKER

S. TARONI for the CMS Collaboration

Dipartimento di Fisica, University of Perugia,
Perugia, Italy
**E-mail: Silvia.Taroni@cern.ch*

The CMS all-silicon tracker consists of 16588 modules. In 2010 it has been successfully aligned using tracks from cosmic rays and pp-collisions, following the time dependent movements of its innermost pixel layers. Ultimate local precision is now achieved by the determination of sensor curvatures, challenging the algorithms to determine about 200000 parameters. Remaining alignment uncertainties are dominated by systematic effects that can bias track parameters by an amount relevant for physics analyses. These effects are controlled by adding further information, e.g. the mass of decaying resonances. The orientation of the tracker with respect to the magnetic field of CMS is determined with a stand-alone χ^2 minimization procedure.

Keywords: CMS; Silicon; Tracker; Alignment.

1. Introduction

The CMS all silicon tracker consists of 16588 modules[1]. In order to measure the charged particle trajectories with excellent momentum, angle and position resolution, the position of these modules needs to be known with a precision of a few micronss. To pursue this, the alignment is done using large amount of track data with different topologies. The track-based alignment can be formulated as a linear least squares problem where the following expression needs to be minimized:

$$\chi^2(\mathbf{p}, \mathbf{q}) = \Sigma_j^{tracks} \Sigma_i^{hits} \mathbf{r}_{ij}^T(\mathbf{p}, \mathbf{q}_j) \mathbf{V}_{ij}^{-1} \mathbf{r}_{ij}(\mathbf{p}, \mathbf{q}_j) \qquad (1)$$

where $\mathbf{r}_{ij}$ is the residual vector containing all residuals from the tracks used and their hits, defined as $\mathbf{r}_{ij}$ = track-model prediction − measured hit. The residuals are a function of $\mathbf{p}$, the vector containing all alignment parameters describing the actual geometry and $\mathbf{q}_j$, the track parameters of the jth track. $\mathbf{V}_{ij}^{-1}$ is the inverse covariance matrix containing all information on the measurement precision and the correlations. [a]

2. Algorithm and input data

The alignment has been performed with the *Millepede-II*[2,3] algorithm, exploiting an implementation of the *Broken lines*[4] approach as internal tracking model. The simultaneous estimation of $\sim 200k$ free alignment parameters has been carried out. Position and orientation of the detector module contribute with 5 (6) degrees of freedom, while the remaining 3 parameters describe the bowing of the sensors[5]. For the current alignment $\sim$1 fb^{-1} of data, taken in 2011, and a sample of cosmic tracks have been used to determine the positions of the sensors, using the geometry from the long cosmic run in 2008 (CRAFT08)[6] as starting point. The high statistics of the low momentum tracks (3M) and of the loosely isolated muons (15M) samples allows one to reach a statistical accuracy of 5 (10) μm in the tracker barrel (endcap). The mass and the vertex in the Z sample (375k muon pairs) and the one-leg tracks connecting the opposite sides of the detector of the cosmic sample (3.6M) are exploited for the detection and the constraint of systematic distortions. The current alignment accounts also for the movements of the large structures of the pixel detector. These movements are monitored using unbiased vertex - track residuals: for each track, the residual with respect the primary vertex, reconstructed with the other tracks in the event, is computed. These movements are recovered by performing the alignment procedure with time dependent parameters.

3. Studies of the sensor deformations

The study of the refitted tracks from cosmic rays showed a strong dependence of the $\langle\chi^2\rangle$ on the distance of closest approach to the beamline d_0 (figure 1, flat module points). The trend of these points is due to the combination of a non flat detector with a large incident angle which characterizes the cosmic rays having $d_0 \gg 0$. Green circles in figure 2 clearly show that the flat sensor assumption is not a realistic description. Using a curved

[a] A local right-handed coordinate system is defined for each module with the origin at the geometric center of the sensor: the u-axis is defined along the more precisely measured coordinate, the v-axis orthogonal to the u-axis and in the module plane, pointing away from the readout electronics, and the w-axis normal to the module plane. In addition, CMS uses a right-handed coordinate system, with the origin at the nominal collision point, the x-axis pointing to the center of the LHC, the y-axis pointing up (perpendicular to the LHC plane), and the z-axis along the anticlockwise-beam direction.

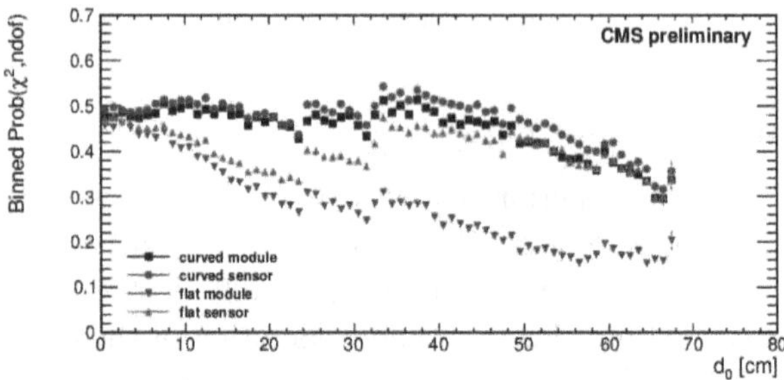

Fig. 1. Distribution of the probability of the χ^2 vs d_0. The results are shown for different levels of module and sensor description.

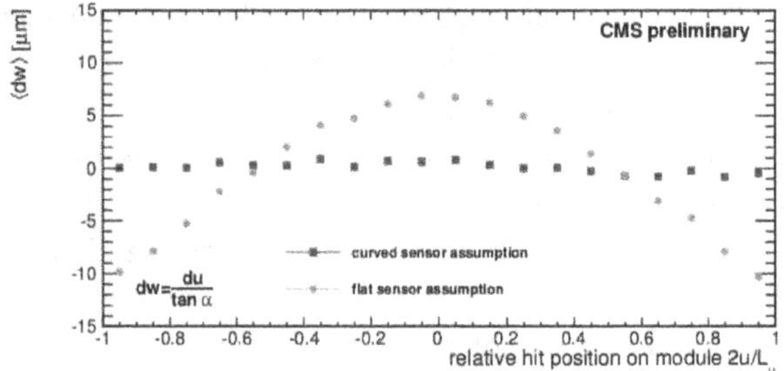

Fig. 2. Residuals perpendicular to the surface in the two innermost layers of the strip barrel (TIB), expressed as $dw = du/tan\alpha$, where α is the track incident angle with respect to the normal to the sensor, as a function of the relative hit position module (L is the module width). Green circles: flat sensor assumption; blue squares: curved sensors.

surface parametrization, the residuals as a function of the hit position recovered to the expected flat distribution. In addition, at larger radii of both barrel and endcap, there are composite modules: two sensors are mounted in one module frame and daisy-chained to one readout electronic block. A residual tilt of the second sensor with respect to the plane of the first can be present: the typical kink is small ($\alpha \sim 0.8$ mrad) but produces an effect larger than the sensor bow. With this more detailed surface parameterization, the χ^2 dependence from d_0 in the cosmic sample is completely recovered (purple circles in figure 1), showing only substructures produced by the effect of the material of the tracker and the services.

4. Tracker - B field orientation

A residual tilt angle between the nominal position of the tracker and the magnetic field direction can be measured in CMS. To estimate its size, a scan of the horizontal and vertical tilt angles has been performed. The aligned value is obtained through a χ^2 minimization procedure. The estimated tilts, shown in figure 3, are $\sim$ 300 μrad around the x_{CMS} (θ_x) direction and $\sim$ 0 μrad for the y (θ_y). The possible systematics have been in-

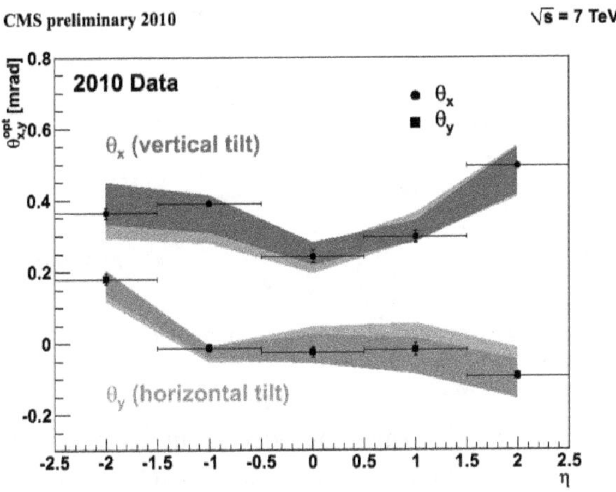

Fig. 3. Tracker - B field tilt angle, around x (θ_x) and y (θ_y) directions, obtained from a χ^2 minimization procedure using 2010 p-p collision data.

vestigated varying the p_T threshold of the tracks used in this study (0.5 and 1.2 GeV), the estimator of the track quality ($\langle \text{prob}(\chi_2/\text{ndof}) \rangle$, $\langle \chi_2/\text{ndof} \rangle$) and the B-field map; the analysis of simulated data with no tilt served as an additional check of the procedure: the obtained optimal θ_x and θ_y in Monte Carlo data are compatible with null angles.

5. Momentum changing "weak" mode

The biggest challenge for the alignment is the detection and constraint of the "weak" modes, distortions which do not, or only weakly, influence the χ^2 but still affect the track parameters. The topology of the cosmic tracks can be exploited to keep under control many weak modes; as well illustrated in[6], an artificial rotation of the layers can be fully recovered using cosmic rays. Despite the power of this kind of tracks in solving these misalignments, some deformations need other events topologies. An example is the twist deformation that affected the CMS tracker when only cosmics and low momentum tracks from p-p collisions were used for the alignment. Figure 4 shows the Z mass dependence on the muon pseudorapidity if no additional

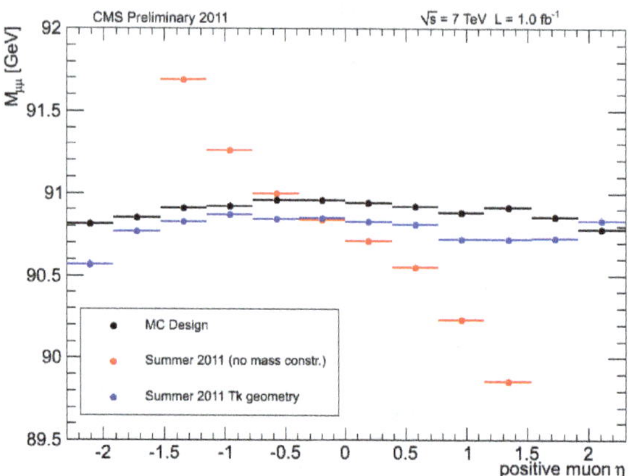

Fig. 4. Distribution of the invariant mass of the two muons in the Z sample as a function of the positive muon η.

constraint is used. Exploiting the knowledge of the Z mass in the alignment, the twist can be controlled: no dependence is shown by the blue points which behave as the design condition ones (black) obtained from data simulated with a perfectly aligned detector.

6. Conclusions

The alignment of the CMS tracker is a big challenge: $\sim 200k$ parameters need to be determined simultaneously with high precision.

Using *Millepede-II* with a *Broken lines* approach, a statistical accuracy of less than 5 (10) μm for the barrel (endcap) in the aligned position of the modules has been reached. Thanks to high statistics and topologically different samples, it was achieved an accurate description of the deformation of the surface of the sensors and of the tracker - B field tilt.

References

1. CMS Collaboration, *JINST* **3**, p. S08004. 361 p (2008).
2. V. Blobel, *Nucl. Instrum. and Meth.* **A566**, 5 (2006).
3. `https://wiki.terascale.de/index.php/Millepede_II`.
4. V. Blobel, *Nucl. Instrum. and Meth.* **A566**, 14 (2006).
5. C. Kleinwort and F. Meier, *Nucl. Instrum. and Meth.* **A650**, 240 (2011).
6. CMS Collaboration, *JINST* **5**, p. T03009. 41 p (2009).

Exploitation of the charge sharing effect in Timepix device to achieve sub-pixel resolution in imaging applications with alpha particles.

C. Teyssier[1,4,*], P. Allard Guérin[1], G. Bergeron[1], F. Dallaire[1], C. Leroy[1], S. Pospisil[2], Y.B. Trudeau[3]

1. Laboratoire R.-J.A. Lévesque, Université de Montreal, Montreal, QC H3T 1J4, Canada

2. Institute of Experimental and Applied Physics, Czech Technical University in Prague, Horská 3a/22, 12800 Praha 2, Czech Republic

3. ANIQ R&D Inc., Laboratoire R.J.A. Lévesque, Montreal, QC H3T 1J4, Canada

4. Université de Lyon, F-69003, Lyon, France, Université Lyon 1, Villeurbanne, France, CNRS/IN2P3, UMR5822, Institut de Physique Nucléaire de Lyon, F-69622 Villeurbanne, France

* *E-mail: teyssier@lps.umontreal.ca*

The Timepix device is a pixelated silicon detector. Because of its structure, an incoming particle can deposit its energy in several adjacent pixels as a result of the charge sharing effect. The distribution of energy in the pixels activated by a heavy charged particle can be exploited to determine the entering point of the particle with a precision better than the pixel dimensions. This is experimentally illustrated by images of different samples obtained with alpha particles. This work was carried out within the CERN Medipix Collaboration.

Keywords: charge sharing; sub-pixel resolution; imaging.

Introduction

The Timepix device [1] consists of a 300 μm thick silicon sensor matrix of 256×256 cells (64k p-n diodes) bump-bonded to a pixelated read-out chip. Each matrix element (55×55 μm^2) is connected to its respective read-out chain integrated on the read-out chip. For each pixel, a counter measures the time during which the signal is above a preset threshold. With this time over Threshold (TOT) capability, the Timepix detector allows energy measurement in each pixel. When hitting the silicon layer, the particles deposit energy by creating electron-hole pairs. Those charge carriers diffuse before being collected [2]. The lateral spread of the charge carriers can cause a sharing of the charge among adjacent pixels. Thus, a particle can activate one or several adjacent pixels, forming a cluster of pixels. That is the case for heavy ionizing particles such as alpha particles. Because of their short range in silicon, clusters from alpha particles have a form of a 2D symmetric blob.

The idea here is to exploit the distribution of TOT in a blob caused by an alpha particle, to find its energy and entering point with a precision better than the dimension of the pixels. Because the energy loss of a particle depends on the nature and thickness of the crossed sample, an image of a thin sample can be produced by measuring the energy of each interacting particle after crossing of the sample together with its impact point in the sensor. In such a way, 3D information about the sample is obtained (its thickness on a x-y position). The data are recorded with the software Pixelman [3] as frames that contains the TOT of all the pixels (65536) after a given exposure time and an event-by-event analysis is possible (blob-by-blob). At first, the relation between the TOT and the energy deposited in a pixel has to be found for each pixel as we will see in section 1. Then, in section 2, the experimental set-up will be described and the method of analysis explained. Finally in section 3, we will show images obtained in our experiments and conclude on the achieved resolution.

1. Calibration

A particle induces a charge in the silicon that can be shared between adjacent pixels, creating a blob. The collected charge in each pixel from a blob is measured in TOT units by the Timepix. The sum of these fractional charges is called volume of the blob.

If no calibration is performed, the resulting spectrum of cluster volumes presents some distorsion, as illustrated in Fig. 1-left for the exposure of the detector to an ^{241}Am source (gamma source, alpha particles shielded): the two peaks of ^{241}Am at 26 keV (2.3%) and 59.5 keV (35.9%) cannot be identified. In Fig. 1-right, the spectrum is shown for different sizes of blobs. The measured volume appears to be dependent on the number of activated pixels. Thus, the pixels needs to be individually calibrated.

To do so, only the blobs of one pixel will be used for the calibration. In the

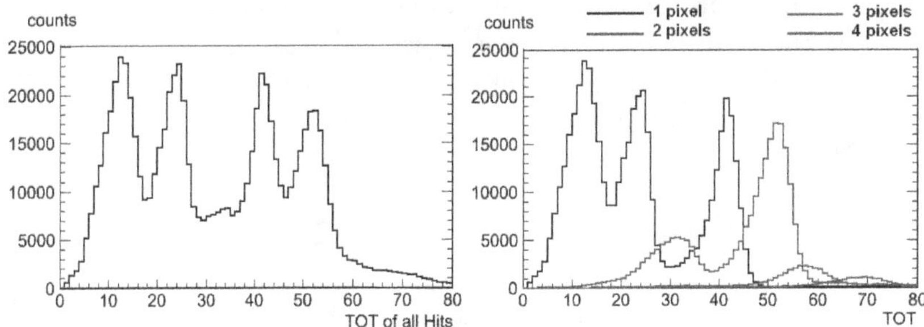

Fig. 1. Distribution of cluster volumes in TOT units measured by a Timepix device exposed to a ^{241}Am source (gamma source, alpha particles shielded). Left: for all clusters. Right: For different sizes of clusters

process of its calibration, the detector has been exposed to a ^{241}Am source (26 and 59.5 KeV gamma emission) and a ^{133}Ba source (30.6keV electrons). The spectrum of single hits is generated for each pixel and three peaks are found. With these 3 points, a calibration can be done using a linear model with a threshold function (Fig. 2). In Fig. 2, the resulting calibration of a random pixel (pixel number 32891) is shown. A more precise calibration could be done but requires specific equipment [4].

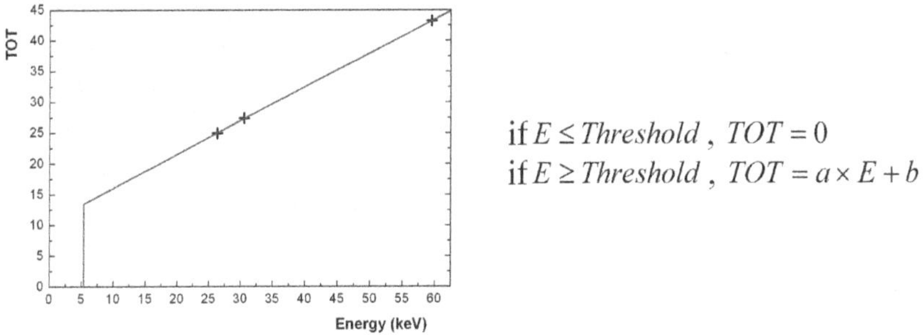

Fig. 2. Calibration of pixel 32891 for a threshold set at 5.36keV: $a = 0.549 \pm 0.039$ keV^{-1} and $b = 10.54 \pm 1.63$.

The set of parameters of all pixels was obtained for a threshold set at 5.36 keV (lowest above the noise for our detector), a frequency of 10 MHz for the Timepix counter and a applied voltage of 100 V. Figure 3 shows the new spectrum of the volume of blobs in keV for the different sizes of blobs. One can see that the peaks are well aligned and found at 24.7 ± 2.2 keV and

58.0 ± 2.7 keV. This calibration was established with particles of relatively low energy ($\leq$ 60KeV). It has been shown [1,4] that the pixels response is linear with the deposited charge up to an equivalent of 0.9 MeV. The present calibration will be used up to this energy.

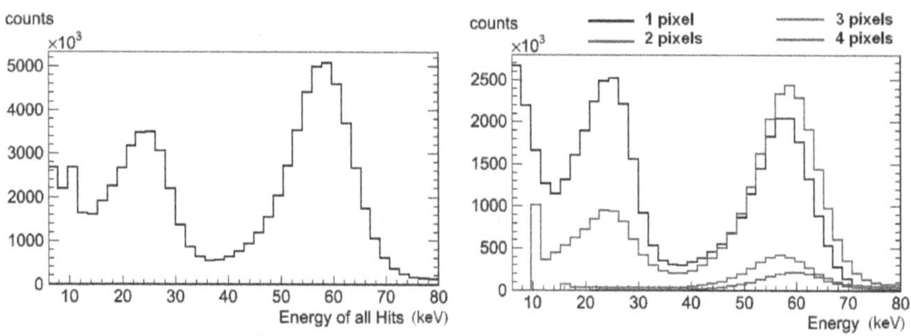

Fig. 3. Distribution of cluster volumes in keV measured by a Timepix device exposed to a source of 241Am. Left: for all blobs. Right: for different sizes of blobs.

2. Experimental set-up

The experiments were done in a vacuum chamber. Imaging has been done using the following setup: the material to be imaged was held above the Timepix device set in horizontal position. The samples to be imaged were exposed to alpha particles produced by an ^{241}Am source. The source was positioned above the detector on a support allowing the adjustment of the distance between the detector and source (scheme on Fig. 4 left).

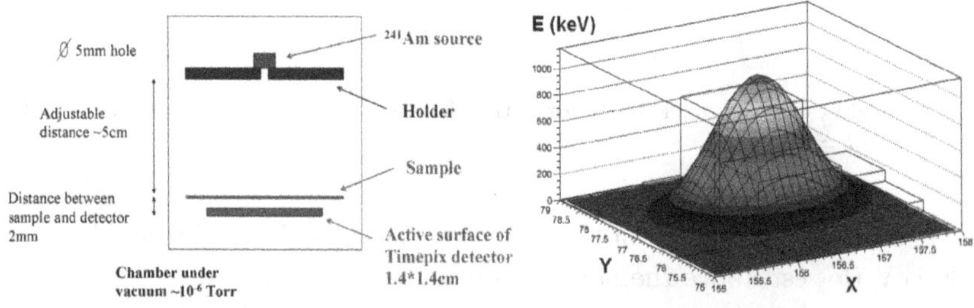

Fig. 4. Left: Experimental setup. Right: 2D Gaussian fit on a blob from an alpha particle of 1.9 MeV.

A pattern recognition algorithm developed in the context of the Medipix Analysis Framework MAFalda [5] uses the blob shape to recognize the different types which can be associated to a specific type of particle [6]. These associations are really efficient for heavy particles recognition in this range of energy. Then, on each blob associated to a heavy particle, the energy is calculated for each pixel and a 2D Gaussian fit is performed. The Gaussian fit gives good results for the modelisation of the charge sharing [2,7] as illustrated in Fig. 4 right. The data are stored in a ROOT file that contains the values of fit parameters and the total energy of blobs. To build the image, blobs are sorted by their center (given by the Gaussian fit) and their mean energy is calculated for each subpixel (number of subpixels set by the user).

3. Results

A first measurement was performed on a sample made of Mylar foils (24.3μm thick) and hair (scheme Fig. 5-left) with alpha particles from a ^{241}Am source. An image is shown in Fig. 5-middle for a choice of 3 subpixels per pixels (subpixels of 18.3 $\times$ 18.3μm). The three regions of different thickness of Mylar (0, 1 or 2 for the number of Mylar foils) are clearly visible (the vertical line on the right is a line of dead pixels of our detector).

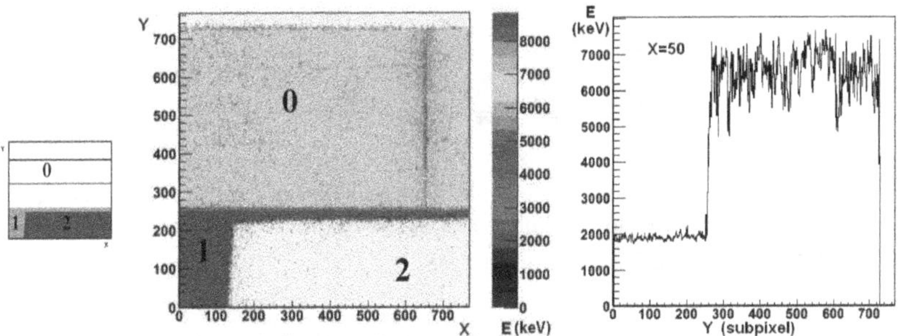

Fig. 5. Imaging with alpha particle from ^{241}Am source. Left: scheme of the sample. Middle: image generated with a choice of 3 subpixels (590k). Right: profile of energy along Y-axis for X=50.

Region 0 in the picture is less homogenous than region 1. This is due to the high energy of the particles coming directly from the source into the detector (mean energy: 5.480MeV). The deposited energy in the central pixels is higher than our limit for the linear behaviour of the pixels. The energy of particles in region 1 can be determined and found equal to 1905$\pm$

45keV. The thickness of the Mylar foil evaluated with this measurement is 24.6 ± 0.4 μm [8] and is in good agreement with the measured thickness of the Mylar foil with a micrometer: 24.3 ± 0.1 μm. The profile of the mean energy allows one to find the border between region 0 and 1. For example at $X = 50$ (Fig. 5 right) the border is found at $Y = 257 \pm 2$. The resolution depends on the statistic (here ≈ 10 particles per subpixel).

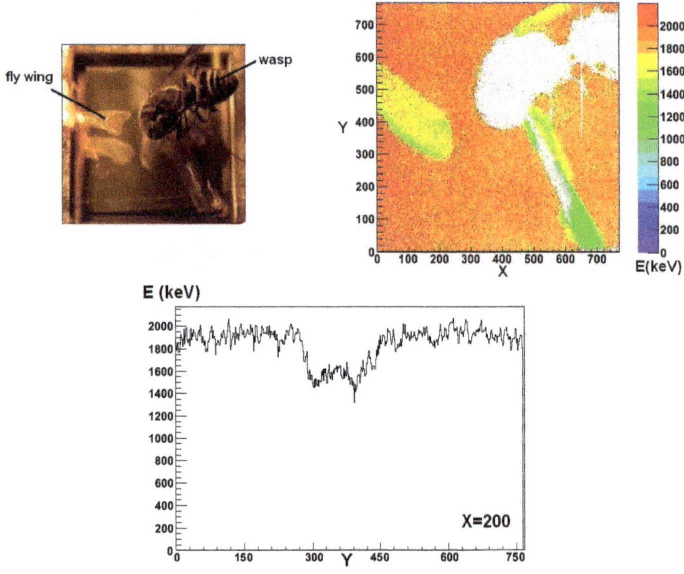

Fig. 6. Imaging with alpha particle from ^{241}Am source. Up left: picture of the sample. Up right: image generated for a choice of 3 subpixels (590k). Down:Profile of energy along Y-axis for X=200.

Then, our imaging method was tested with another sample consisting of a wasp and a fly wing (photo on Fig. 6-up left). The image with a choice of 3 subpixels (image of 768*768 pixels) is presented in Fig. 6-up right. The outline of the objects is well defined. The profil of energy for X = 200, where the wing is larger, is shown on Fig. 6-down. The energy of particles crossing the fly wing at its thicker part is evaluated at 1575±43keV. Then the energy loss in the fly wing can be determined (by comparison with the energy of particles crossing only the Mylar foil): 185 ± 25keV [8]. The wings of insects are principally composed of chitine, an estimate of the thickness of the wing [8] is possible assuming that it is entirely made of chitine: $10.0 \pm 1.4\mu$m.

Conclusion

Imaging with alpha particle allows one to explore very thin objects ($\leq$ 50μm). Using the Timepix detector with the exploitation of the charge sharing effect, the resolution strongly depends on statistics. In our present data, the achieved resolution was 1.4μm for thickness resolution (z) and 36μm for spatial resolution (x,y).

Acknowledgments

This work has been done in the framework of the CERN Medipix Collaboration. It has been supported by the Natural Sciences and Engineering Research Council of Canada (NSERC), the Canada Foundation for Innovation (CFI) and the Ministry of Education, Youth and Sports of the Czech Republic under Research Projects MSM 6840770029 and LA08015. C. Teyssier acknowledges the French Ministry of research for doctoral fellowship and the region *Rhône-Alpes* for her mobility fellowship.

References

1. X. Llopart, R. Ballabriga, M. Campbell, L. Tlustos, W. Wong, *Timepix, a 65k programmable pixel readout chip for arrival time, energy and/or photon counting measurements*, Nucl. Instr. Meth. A 581 (2007), 485.
2. J. Bouchami et al., *Study of the charge sharing in silicon pixel detector by means of heavy ionizing particles interacting with a Medipix2 device* Nucl. Instr. Meth. A 633 (2011), 117.
3. D. Turecek et al., *Pixelman: a multi-platform data acquisition and processing software package for Medipix2, Timepix and Medipix3 detectors* 12th International Workshop on Radiation Imaging Detectors Juy 11th-15th 2010, Robinson College, Cambridge U.K. 2011 JINST 6 C01046.
4. J. Jakubek, *Precise Energy Calibration of Pixel Detector Working in Time-Over-Threshold Mode*, Nucl. Instr. Meth A 633 (2011), 262.
5. J. Bouchami et al., *User-extensible implementation of a pattern recognition algorithm for imprints produced by ionizing radiation in a device from the Medipix family*, submitted for publication in Nucl. Instr. Meth. A (2010).
6. J. Bouchami et al., *Measurement of pattern recognition efficiency of tracks generated by ionizing radiation in a Medipix2 device*, Nucl. Instr. Meth. A 633 (2011), 187.
7. J. Jakubek et al., *Pixel detectors for imaging with heavy charged particles*, Nucl. Instr. and Meth. A591 (2008), 155.
8. J. F. Ziegler *SRIM* http://www.srim.org/

The LHCb Silicon Tracker Performance in pp Collisions at the LHC

Mark Tobin
E-mail: Mark.Tobin@cern.ch
On behalf of the LHCb Silicon Tracker Group*
Physik Institut der Universität Zürich,
Winterthurerstrasse 190, CH-8057 Zürich, Switzerland

The LHCb experiment performs high-precision measurements of CP violation and searches for New Physics using the enormous flux of beauty and charmed hadrons produced at the LHC. The LHCb detector is a single-arm spectrometer with excellent tracking and particle identification capabilities. The Silicon Tracker is part of the tracking system and measures very precisely the particle trajectories coming from the interaction point in the region of high occupancies around the beam axis. It covers a total sensitive area of about 12 m^2 using silicon micro-strip technology. This paper reports on the operation and performance of the Silicon Tracker during the Physics data taking at the LHC. First measurements of radiation damage are also shown.

Keywords: LHCb; Silicon Tracker;

1. Introduction

The LHCb experiment[2] is a single arm spectrometer designed to study heavy flavour physics in decays of B-mesons. It will constrain the parameters of the CKM matrix[3] by measuring CP-violation and probe for physics beyond the Standard Model by studying rare decays. The detector covers polar angles in the range 15 to 300 mrad and exploits the fact that $b\bar{b}$ pairs are produced in the same forward cone at the LHC.

The LHCb Silicon Tracker is a silicon micro-strip detector with a sensitive area of approximately 12 m^2 and a total of 272k readout channels. It consists of two detectors: the Tracker Turicensis (TT), a 150 cm wide and 130 cm high tracking station covering the full LHCb acceptance upstream of the LHCb magnet; and the Inner Tracker (IT) which, despite covering

*The full author list is given in Ref. 1.

only 1.2% of the total acceptance in a 120 cm by 40 cm cross-shaped region in the centre of three planar tracking stations downstream of the magnet, reconstructs 20% of all tracks passing through LHCb.

The Tracker Turicensis has four detection layers orientated at (0°, +5°, -5°, 0°) with respect to the vertical axis. The detector is constructed from 500 μm thick p-on-n sensors with a pitch of 183 μm. Sensors are bonded together to provide readout sectors with 1, 2, 3 or 4 sensors depending on their position relative to areas of higher particle flux. This leads to four different length readout strips (up to 37 cm). There are 280 readout sectors in the TT and a total of 143600 readout channels.

The three Inner Tracker stations each consist of four independent boxes arranged around the beam pipe. Each box contains four detection layers with the same orientation as those in the TT. Detector modules in the boxes either side of the beam pipe use 410 μm thick sensors and have a length of 22 cm while those in the boxes above and below the beam pipe are 11 cm long and 320 μm thick. There are 336 readout sectors in IT and a total of 129k readout channels.

The number of working channels in the TT and IT is 99.77% and 98.22% respectively. The main source of inefficiency is caused by the failure of VCSEL diodes used in the optical readout. The broken VCSELs can easily be replaced during short shutdowns in the TT but access and repairs to the IT electronics is much harder due to its location close to the beam pipe.

2. Detector Performance

The first proton-proton collision events were used to time align the detector with respect to the LHC collisions. It is important to ensure that the trigger and control signals are synchronised across the whole of LHCb. The internal time alignment was made to account for different length cables and different time of flight for particles passing each station. The optimal delay settings were determined from a scan using runs taken where the delay between the sampling time of the front-end electronics and the trigger time with respect to the LHC clock was varied in 6.5 ns steps. For each delay setting, the most probable value (MPV) was determined by fitting a Landau convolved with a Gaussian to the distribution of the charge for groups of up to 12 detector modules. The peak of the pulse shape as a function of time was determined by fitting the expected front-end signal shape to the distribution of the MPV versus the delay. An example is given in figure 1. After this procedure, the detector was time aligned to better than 1 ns precision.

The signal to noise ratio (S/N) was determined after the time alignment

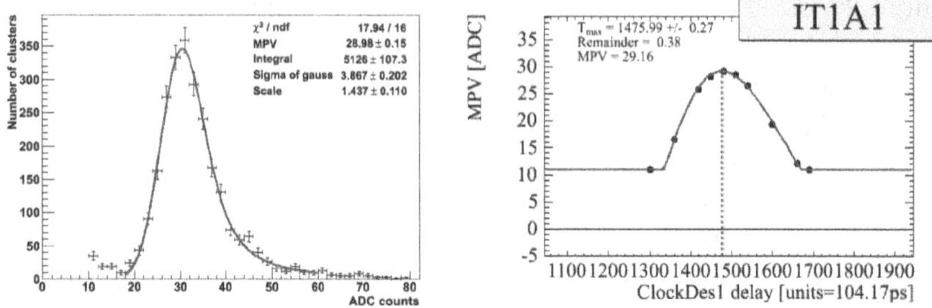

Fig. 1. Charge distribution measured for one ladder in IT (left). The Most Probable Values obtained from fits to the ADC distributions for different trigger delays for a set of short ladders in IT (right).

procedure using tracks reconstructed with momentum greater than 5 GeV. The S/N for the TT was found to be in the range 12 to 15 and is shown in figure 2 for the different strip capacitances. The long and short ladders in IT have a S/N of 16.5 and 17.5 respectively. The distribution of the S/N for all ladders in the IT is shown in figure 2. The values obtained are within 10-20% of those expected from prototype measurements.[4]

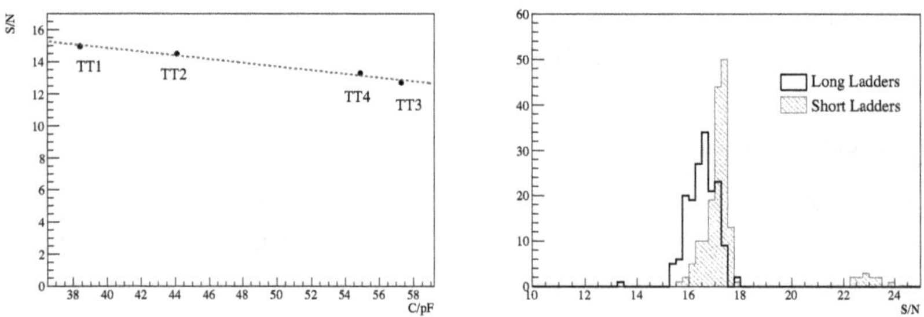

Fig. 2. Measured signal to noise ratio as a function of strip capacitance in TT (left) and for different strip lengths in IT (right). There is a peak with S/N around 23 for the short ladders in IT as some 410 μm sensors were used during the module production.

The data was used to make a spatial alignment of the detector. The alignment procedure is based on a closed-form alignment method using tracks fitted with a Kalman filter and more details can be found in Ref. 5. The TT and IT were aligned using (long) tracks which pass through the full LHCb tracking system. The unbiased residuals for hits on tracks are shown in figure 3 and the hit resolution was found to be 62 and 58 μm in TT and IT respectively. The resolution is expected to be around 50 μm

from the simulation.

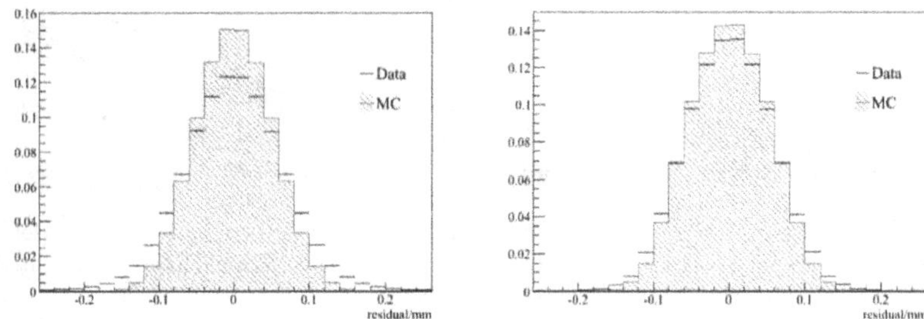

Fig. 3. Unbiased residual distributions for TT (left) and IT (right) shown for data and simulation.

The intrinsic detector efficiency was measured using isolated high momentum tracks. A search was made for clusters in a window around the track and the hit efficiency is defined to be the ratio of the number of hits found to the number of hits expected. It was measured to be 99.3% for IT and 99.7% for TT. The noise cluster rate was below 10^{-5}.

The data was also used to tune the detector simulation. Many different effects were measured and then included in the simulation of the detector. The details are described in Ref. 6.

3. Radiation Damage

The performance of the Silicon Tracker is expected to degrade with increasing radiation dose. The leakage current increases linearly with the fluence. The expected change in the leakage current, ΔI_{leak}, can be calculated using the relation $\Delta I_{leak} = \alpha V \Phi_{eq}$ where α is the damage constant for 1-MeV neutrons,[7] Φ_{eq} is the average 1-MeV neutron equivalent fluence, and V is the volume of the sensor. A value of $\alpha = 4 \times 10^{-17}$ A/particle/cm was assumed.[8] The evolution of the leakage current as a function of the delivered luminosity is shown in figure 4 for all sectors in the stereo layers of TT and for all sectors with long readout strips in IT . The expected fluence was calculated using Fluka simulations[8] and the expected evolution of the leakage current is also shown in figure 4 for the sensors closest to the beam.

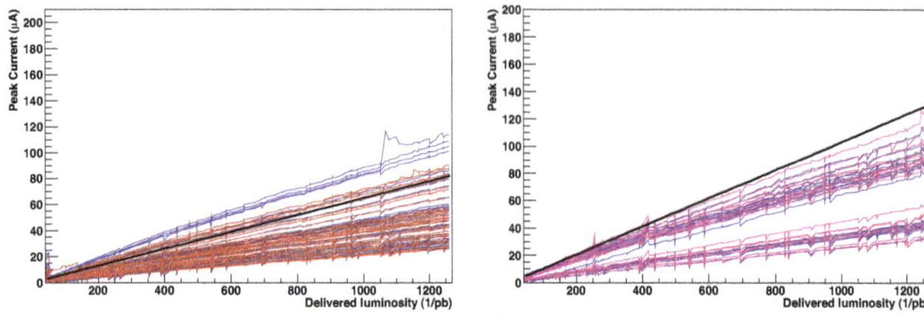

Fig. 4. Evolution of the leakage current for all sectors in the stereo layers of TT (left) and for all sectors with long readout strips in IT (right). The expected change is indicated by a straight line for the modules closest to the beam pipe.

4. Conclusions

The detector was fully installed by summer 2008. A programme of commissioning without beam followed to test the robustness of the readout chain. The fraction of the detector which is fully working is 99.77% and 98.22% for TT and IT respectively.

The first proton-proton collision data was used to time align the detector with an accuracy of 1 ns. The signal to noise ratio was found to be in the range 12 to 15 for different strip lengths in the TT. The long and short IT ladders have signal to noise ratios of 16.5 and 17.5 respectively.

Spatial alignment of the detector was performed. The hit resolution was found to be 62 and 58 μm for TT and IT respectively. It is slightly worse in data compared to simulation. The hit efficiency was measured using high momentum tracks to be 99.3% in TT and 99.7% in IT.

The LHCb Silicon Tracker is performing extremely well. The effect of radiation damage has been measured and is in line with expectations.

References

1. M. Tobin *et al.*, The LHCb Silicon Tracker Perfomance in pp Collisions at the LHC (2011), LHCb-PROC-2011-051.
2. A. A. Alves *et al.*, *JINST* **3** (2008), S08005.
3. C. Amsler *et al.*, *Physics Letters* **B667**, p. 1 (2008).
4. M. Agari *et al.*, (2002), LHCb-2002-058.
5. W. Hulsbergen, *Nucl. Inst. Meth.* **A600**, 471 (2009).
6. J. van Tilburg, Studies of the Silicon Tracker resolution using data (2010), LHCb-PUB-2010-016.
7. M. Moll *et al.*, *Nucl. Inst. Meth.* **A426**, p. 87 (1999).
8. M. Siegler *et al.*, Expected particle fluences and performance of the LHCb Trigger Tracker (2004), LHCb-2004-070.

MEASURING DOPING PROFILES WITH SPREADING RESISTANCE PROFILING

W. TREBERSPURG[†], T. BERGAUER, M. DRAGICEVIC, M. KRAMMER

Institute of High Energy Physics, Austrian Academy of Sciences, Vienna, Austria

A Spreading Resistance Profiling measurement station has been developed to measure doping profiles with limited effort, alternatively to more sophisticated methods usually found in semiconductor industry. By modifying a simple setup, which is commonly used at HEP institutes to characterize silicon detectors, data of good quality could be obtained. Doping profiles, with penetration depths between 1 and 200 microns have been analyzed and compared to Scanning Electron Microscopy images and Capacitance Voltage measurements.

1. Spreading Resistance Profiling

The Spreading Resistance Profiling (SRP) technique has been developed in the 1960s and is nowadays mainly used by commercial companies with sophisticated and specific equipment. Although simple characterizations of doping profiles are often needed there are no publications describing measurements using cheap but reliable setups. The spreading resistance profiling technique is capable of characterizing very shallow junctions into the nm regime and has a very high dynamic range of concentrations (10^{12}–10^{21} cm^{-3}). As it is a comparative technique a calibration is necessary to obtain absolute doping concentrations.

The concept of SRP is illustrated in figure 1. During the measurement two carefully aligned probes are stepped along the beveled semiconductor surface and measure the resistance at each position. The different components of the total measured resistance (R_T) are the resistance of the probes (R_P), the contact resistance (R_C), the spreading resistance (R_{SP}) and the resistance of the material (R_M).

$$R_T = 2R_P + 2R_C + 2R_{SP} + R_M \tag{1}$$

Unlike the material resistance the spreading resistance is not generated by the linear distance the current has to cover but by spreading effects (Fig.1).

[†] Corresponding Author

Address: Nikolsdorfer Gasse 18, 1050 Vienna, AUSTRIA

Email: Wolfgang.Treberspurg@oeaw.ac.at

Basically the current is concentrated at the small probe tips of radius a and spreads out radially into the material of resistivity ρ [1].

$$R_{SP} = \rho/2a \quad (2)$$

In the case of tip distances of about five times the contact radius, approximately 80% of the potential drop occurs due to current spreading. For further details see [2].
For two reasons this technique is especially convenient to realize measurement with limited effort:

- According to eq. 2 R_{SP} is strongly correlated to the probe tip radius. By decreasing the radius very high resistance values can be measured and the influence of R_C and R_M is reduced.
- As the measured resistance is just slightly influenced by fluctuations of the probe spacing the alignment of the probes is not critical.

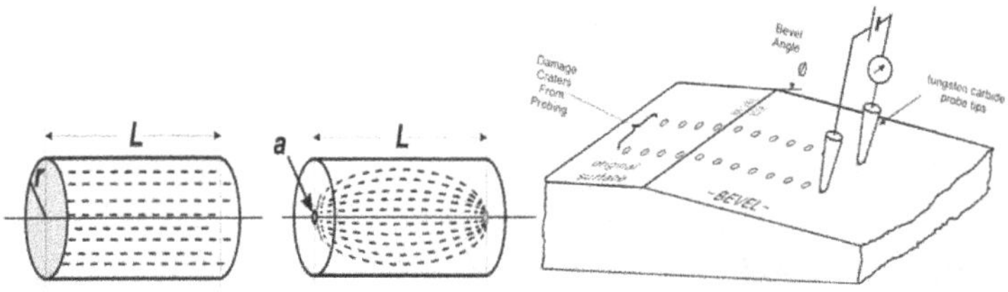

Figure 1. **Left:** Illustration of the current spreading effect [3]. **Right:** At SRP measurements the probes are stepped along the beveled semiconductor surface [2].

2. The Measurements

2.1. *Preparation*

The preparation of samples is a very important part preceding any SRP measurement. The samples have been grinded and polished with commercial wheels containing diamonds of different grain size. The last applied grinding wheels need to have a peak to peak roughness of 10 nm in order to prevent scratches. The samples are mounted with melted wax onto a well calibrated bevel block and are heated up for several minutes to 100°C.
As the profile should be measured by 100 to 150 data points an adequate bevel angle has to be chosen. The influence of surface defects and temperature fluctuations can be reduced by using small areas and short measuring times.

Mainly angles between 0.5° and 1° have been used. To verify the achieved bevel angle a commercial coordinate measurement machine*.

2.2. *Measurement Setup*

The measurement station consists of a movable xyz table, which is able to move in minimal steps of 0.5 μm and which carries two positioners holding the probe tips (Fig. 2). Successful measurements require tungsten carbide or osmium probes of high hardness in order to penetrate the natural silicon oxide and reduce the contact resistance. For the following measurements tungsten carbide probes with a radius of 7 μm have been used. The sample is mounted onto a digital scale with adhesive tape. The scale is used to ensure a constant surface pressure of the probe tips to the sample surface. By considering the small contact area a weight of a few grams is sufficient to cause a localized phase transformation of the silicon bulk and generates probe marks. A picture of the setup is shown in figure 2.

Figure 2. The facility of the measurement station includes a weighing scale on the left hand side of the moveable table. The sample is mounted on the weighing scale with double-sided adhesive tape.

The non-grinded flat region of the sample is used to align the probes before the actual measurement starts on the beveled surface. Probe spacings of 30 μm to 50 μm have been used.

For each measurement point the table lowers down in steps of 1 μm each 500 ms. After each movement the weight is measured, which is loaded on the sample by the probe tips. As soon as the desired weight of 5.5 g is reached, the resistance between the probe tips is measured. Since the resistance depends on the loaded weight (Fig. 3) the desired weight has to be as exact as possible, the

* Mitutoyo, Euro-Apex Euro-C776

step size of the table is changed to 0.5 μm after a threshold weight (4 g) is reached. Finally the probes are lifted, and the table moves in x and y to the next point and starts lowering again. The measurement is performed inside a dark, temperature and humidity controlled box. The results turned out to be highly reproducible between two individual measurements of the same sample (Fig.3).

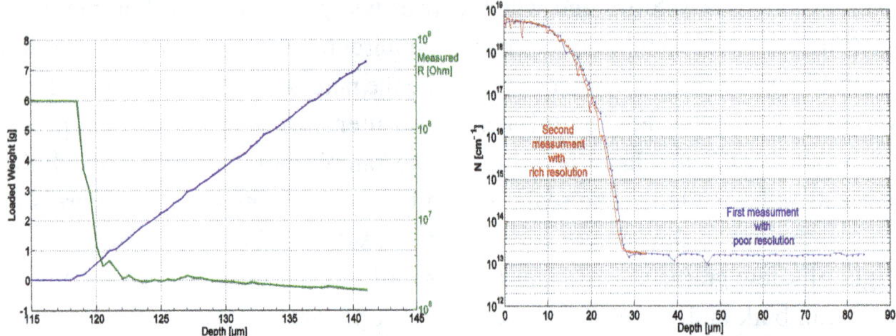

Figure 3. **Left:** The loaded weight (blue) and the resistance (green) has been measured at each position of the table. **Right:** Two independent measurements with different resolution have been made on the same sample.

3. First Results

SRP measurements have been performed on typical strip implants of silicon strip sensors, with implant depths of approximately 1.5 μm. By using enhanced chemical preparation the doping profile could be made visible by changes of the local topographical contrast (Fig. 4) in electron microscopy images [a] [3]. With the SRP technique the pn junction can be identified by a resistance peak, caused by the depletion of free charge carrier.

Other samples from different vendors have been measured as well. Special focus was made to large-area diodes. Their deep backside diffusion implants, with penetration depths of up to 150 μm, are used to narrow down the active zone of the semiconductor detectors. Three samples with different penetration depths have been characterized. The comparison between CV profiling and SRP data revealed that the profiles both show similar results in respect to the penetration depth (Fig. 5).

---

[a] The sample preparation and the Scanning Electron Microscopy Images have been made in cooperation with the "Universitäre Service-Einrichtung für Transmissions-Elektronenmikroskopie" (USTEM) at Vienna University of Technology.

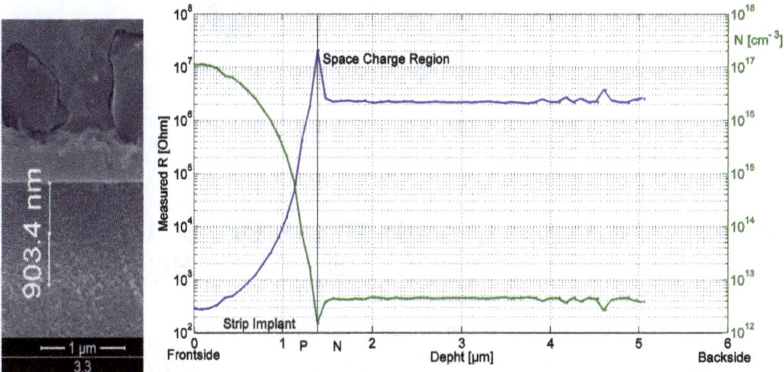

Figure 4. **Left:** Scanning Microscope Image of the sample. **Right:** Doping profile of the strip implant. The measured resistance is plotted blue and the calculated concentration green.

The smooth changeover between the bulk concentration and the deep backside implant of sample FZ 120p (blue) is confirmed by the CV profiles (Fig. 5).

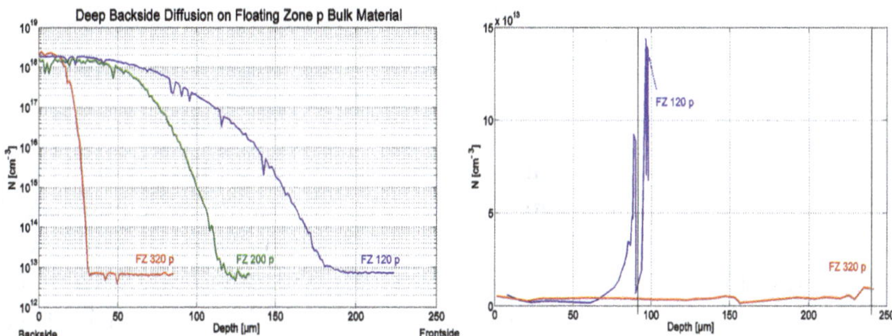

Figure 5. **Left:** Spreading Resistance results of deep backside diffusion profiles (the back side is located on the left hand side). **Right:** Doping profiles of Capacitance Voltage measurements of the same samples (the back side is located on the right hand side).

References

1. R. Brennan, D. Dickey, "Determination of Diffusion Characteristic Using Two- and Four-Point Probe Measurements", Technical note Solecon Labs.
2. D. Schroder, "Semiconductor Material and Device Characterization", Wiley, 2006.
3. W. Treberspurg, "Manufacturing Process of Silicon Strip Sensors and Analysis of Detector Structures", Master Thesis, TU Wien, 2011.

PERFORMANCE OF THE CMS PIXEL DETECTOR FOR THE PHASE 1 UPGRADE AT HL–LHC

A. Tricomi* on behalf of the CMS Collaboration

Physics and Anstronomy Department, University of Catania and INFN Catania
Catania, I-95123, Italy
**E-mail: alessia.tricomi@ct.infn.it*

Operation of the CMS detector at Phase 1 HL-LHC (2×10^{34} cm^{-2}s^{-1}), will require the upgrade of the pixel detector. Simulation studies aimed to address the performance of the new pixel detector are discussed.

Keywords: Pixel detector; LHC; High-Luminosity.

1. Introduction

The CMS detector[1] was originally designed to run at an instantaneous luminosity of 10^{34} cm^{-2}s^{-1}. Operation at the highest Phase 1 luminosity (2×10^{34} cm^{-2}s^{-1}), foreseen to start in 2018, after a two year shutdown, will require upgrades to detector elements that will deteriorate because of radiation damage and/or will be affected by the higher occupancies that complicate pattern recognition or lead to increased dead time. A replacement of the current pixel detector is foreseen for Phase 1 run. After a brief description of the new layout and the motivation for upgrade, the results of simulation studies aimed to understand the achievable performance are extensively discussed. A description of the mechanics and electronics foreseen for the new pixel detector can be found elsewhere in this Proceedings.[2]

2. The Pixel Detector

The present CMS pixel detector,[3] conceived over 10 years ago, is a first generation hybrid pixel detector. It consists of three cylindrical barrel layers (BPIX) in the central region supplemented by two forward disks (FPIX). The main motivation for an upgraded pixel[4] detector is justified by the sizable reduction of the data loss which the current system will suffer. At the design luminosity the buffer size and readout speed of the current pixel

Read Out Chip (ROC) will lead to a dynamic inefficiency in the pixel detector of 4% (16%) with 25 ns (50 ns) bunch spacing. The inefficiency increases exponentially with luminosity. At $2 \times 10^{34}\text{cm}^{-2}\text{s}^{-1}$ with 25 ns bunch spacing the ROCs in the inner region will suffer an inefficiency of 15% leading to an overall degradation of tracking performance, as can be seen in Figure 1.

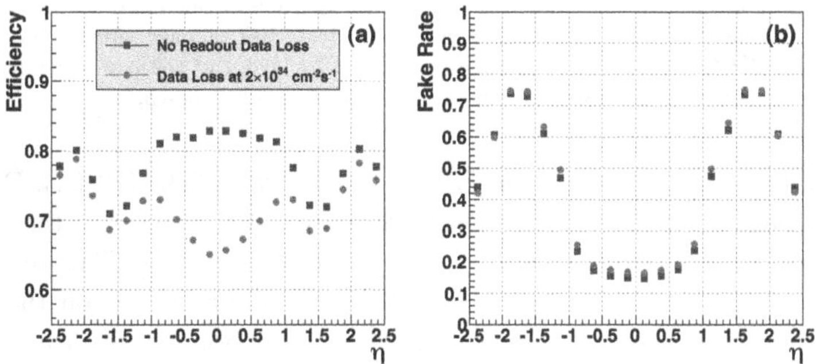

Fig. 1. Performance of the current pixel detector with $t\bar{t}$ events at 2×10^{34} $\text{cm}^{-2}\text{s}^{-1}$ with no data loss (blue) and with the estimated 15% data loss (red): (a) tracking efficiency vs pseudorapidity; (b) fake rate vs pseudorapidity.

Furthermore the radiation hardness of the detector is not sufficient for operation to the end of Phase 1, when LHC luminosity will exceed the design value. The current detector contains also significant passive material that degrades measurements due to multiple scattering, photon conversions and nuclear interactions. Last but not least the three-hit coverage of the detector is not completely hermetic, leading to $10 \div 15\%$ inefficiencies at $|\eta| < 1.5$ and larger inefficiencies in the region $1.5 < |\eta| < 2.5$.
The goal of the upgraded Phase 1 pixel detector is to be fully efficient at a luminosity of 2×10^{34} $\text{cm}^{-2}\text{s}^{-1}$, with less material and with four hit coverage up to $|\eta| < 2.5$. For this reasons a new pixel detector with a fourth barrel layer, an extra disk on each side and a new ROC has been proposed. The amount of passive material has been significantly reduced by moving readout electronics and connectors further out. Bi-phase CO_2 cooling will also replace the C_6F_{14} single phase cooling currently used, allowing smaller heat exchanger pads and pipes. The layout of the current pixel detector compared to the Phase 1 one is shown in Fig. 2.

The reduction in material inside the pixel tracking volume are shown

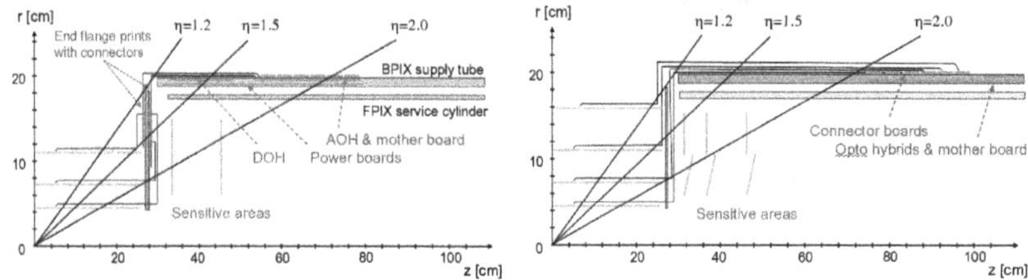

Fig. 2. Schematic view of the current (left) and upgrade layout (right) which consists of an additional barrel layer and an additional endcap disk on each side. The disks are placed in order to maximize the 4-hit η coverage. The layout of the new services, which are moved further downstream in z in the new design, is also shown.

in Fig. 3 for the barrel system (BPIX) and forward disks (FPIX). Despite the increase in the number of pixels, a factor 2.4 reduction in the weight of the BPIX is foreseen and a 40% reduction for the FPIX. This reduction in the amount of passive material will have a large impact on the track reconstruction efficiency as well as electron and photon identification and resolution, thus playing an important role on the reconstruction of final state signatures involving electrons and photons.

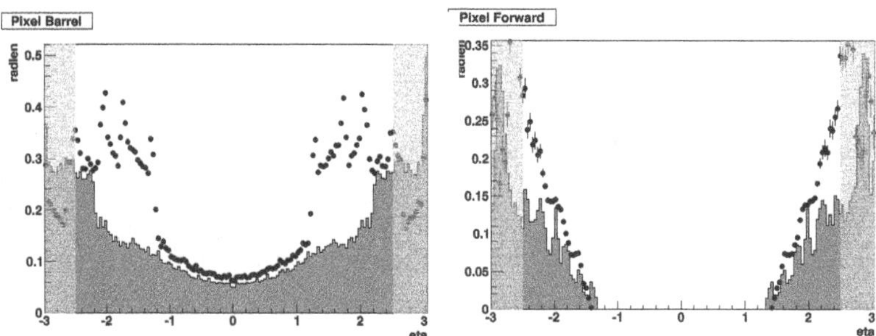

Fig. 3. Radiation length of barrel (left) and forward (right) pixel detector for the current system (dots) and the proposed upgrade (histogram). The shaded region shows the material distribution outside the fiducial tracking volume.

3. Performance studies

The reduced mass and increased three-hit coverage as well as the enhanced pixel lever arm and the first layer moved closer to the beam pipe, all have a net effect on pattern recognition, track parameter resolution, vertexing,

and b-tagging performance of the upgraded detector. To estimate the foreseen improvements in the performance wrt the current pixel system, full GEANT4 based simulations have been carried on. An iterative tracking algorithm has been used for the track reconstruction which uses triplet seeds and quadruplet and triplet seeds for the current geometry and for the upgraded geometry, respectively.
Figure 4 shows the tracking efficiency and rate of fake tracks as function of pseudo-rapidity and transverse momentum for the current and the upgraded detector for $t\bar{t}$ events at a luminosity of 2×10^{34} cm^{-2}s^{-1} with 25 ns bunch spacing (50 pile up events). A clear improvement in the tracking

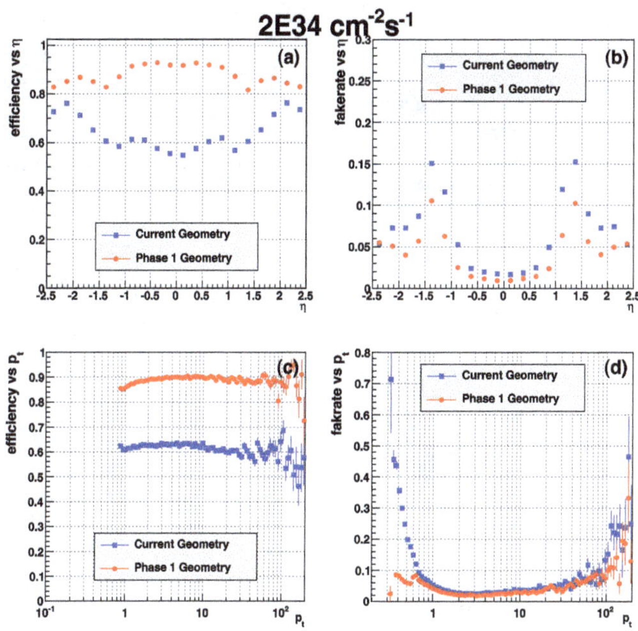

Fig. 4. Comparison of tracking efficiency and fake rate for the current (blue) and upgraded (red) detectors in $t\bar{t}$ events at 2×10^{34} cm^{-2}s^{-1} with 25 ns bunch spacing.[4]

efficiency is seen in the whole η and momentum region. Also significant reduction in the rate of fake tracks, due to the reduced multiple scattering, especially for low momentum tracks, is found. Fake tracks are caused by the incorrect association of hits and are much more likely in regions with more passive material.

The same study has been done also for single muon events. Indeed a number of important physics channels involve final states with high energy isolated muons, like the CMS golden channel $H \to 4\mu$. At 50 pile up events, the current detector suffers a 15% drop in efficiency wrt the upgraded detector over the whole momentum range.

The new pixel detectors shows a significant improvement, of the order of $30 \div 40\%$, in the transverse and longitudinal impact parameter resolutions (Fig. 5 left) as well as in primary (Fig. 5 right) and secondary vertex resolutions (about 20% improvement).

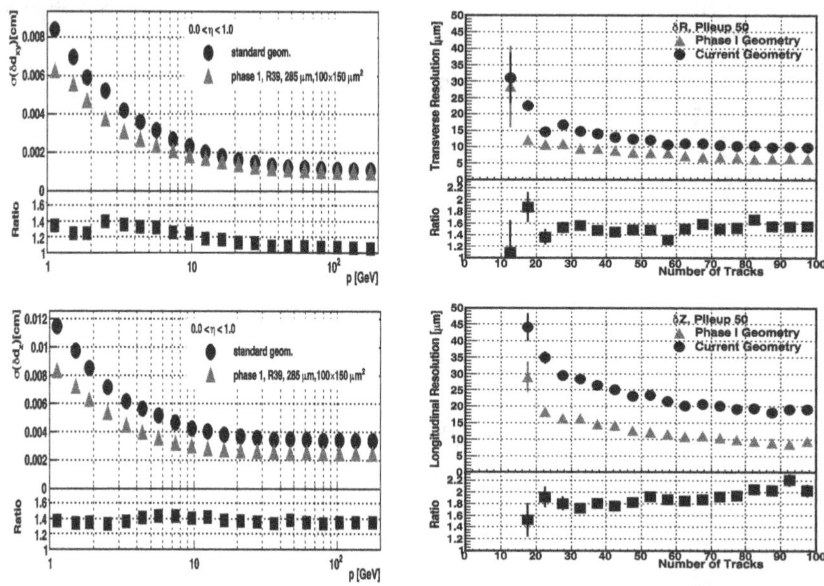

Fig. 5. The transverse (top two) and longitudinal (bottom two) impact parameter (left) and primary vertex (right) resolutions at 50 pileup events.[4] The ratio of the two resolutions is also plotted to illustrate the relative improvement.

The improvement in tracking efficiency, fake rate, parameter resolution, and vertexing all contribute to significantly increase the b-tagging performance of the new detector. The tagging of b jets is a key ingredient of many physics analyses and, in particular, of searches for Higgs boson and New Physics, those improving the b-tagging capability is of crucial importance. Figure 6 shows the b tagging performance with no pileup and 50 pileup events. The b-tagging performance of the present detector is seriously degraded by the large number of overlapping interactions in each bunch crossing. The upgraded detector would reduce the light quark background of the Combined Secondary Vertex Tag by more than a factor of 6 for a b-efficiency of 60%, or conversely it would achieve a relative 40% improvement in b-tagging efficiency for a fixed mistag rate of 1%. At high pileup the upgraded detector is able to almost regain the performance of the current detector at 0 pileup.

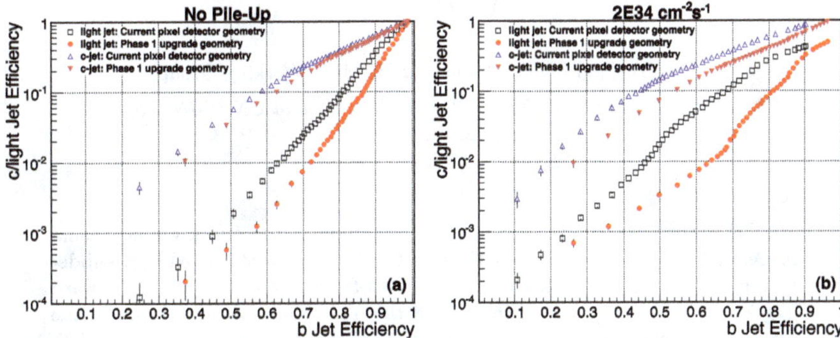

Fig. 6. The b quark efficiency of the Combined Secondary Vertex Tag is plotted versus the light quark (and gluon) efficiency for the current tracker (black) and the Phase 1 upgrade (red) for $t\bar{t}$ events in two different luminosity scenarios: (a) the instantaneous luminosity is assumed to be low enough that there are no multiple collisions; (b) the instantaneous luminosity is assumed to be 2×10^{34} cm^{-2}s^{-1} with 25 ns bunch spacing.

References

1. The CMS Collaboration *JINST* **3**, S08004 (2008).
2. A. Messineo, The CMS Collaboration, *This Proceedings.*
3. The CMS collaboration, *The CMS tracker system project: technical design report*, CERN-LHCC-98-006, CMS-TDR-005 (1998). The CMS collaboration, *The CMS tracker: addendum to the technical design report*, CERN-LHCC-2000-016, CMS-TDR-005-add-1 (2000).
4. The CMS Collaboration, *Technical Proposal for the Upgrade of the CMS detector through 2020*, CERN-LHCC-2011-006. LHCC-P-004 (2011).

Performance Study for a Muon Forward Tracker in the ALICE Experiment

A. Uras* on behalf of the ALICE MFT Working Group
IPNL, Université Claude Bernard Lyon-I and CNRS-IN2P3, Villeurbanne, France
**E-mail: antonio.uras@cern.ch*

ALICE is the experiment dedicated to the study of the quark gluon plasma in heavy-ion collisions at the CERN LHC. Improvements of ALICE subdetectors are envisaged for the upgrade plans of year 2017. The Muon Forward Tracker (MFT) is a proposal in view of this upgrade, motivated both by the possibility to increase the physics potential of the muon spectrometer and to allow new measurements of general interest for the whole ALICE physics. In order to evaluate the feasibility of this upgrade, a detailed simulation of the MFT setup is being performed within the AliRoot framework, with emphasis on the tracking capabilities as a function of the number, position and size of the pixel planes, and the corresponding physics performances. In this report, we present preliminary results on the MFT performances in a low-multiplicity environment.

Keywords: ALICE; Muon Forward Tracker; MFT; muons; dimuons.

1. Introduction: Current Muon Arm Setup and Physics

Lepton-pair measurements always played a key-role in high energy nuclear physics: leptons arrive to the detectors almost unaffected, being not sensible to the strong color field dominating inside the QCD matter, thus providing an ideal tool to probe the whole evolution of nuclear collisions. The ALICE experiment at the CERN LHC, representing the most recent effort in this field, measures single lepton and lepton pair production both at central rapidity, in the electron channel, and at forward rapidity, in the muon channel.

Identification and measurement of muons in ALICE are performed in the Muon Arm,[1] covering the pseudo-rapidity region $-4 < \eta < -2.5$[a]. Starting

[a] Being the collision symmetrical, we indicate the muon spectrometer acceptance using

from the nominal interaction point (IP) the Muon Arm is composed of the following elements. (i) A hadron absorber made of carbon, concrete and steel, between $z = 0.9$ and $z = 5.03$ m; its material budget corresponding to ten hadronic interaction lengths, it provides a reliable muon identification. (ii) A dipole magnet 5 m long providing a magnetic field of up to 0.7 T in the horizontal direction, corresponding to a field integral of 3 Tm. (iii) A set of five tracking stations, each one composed of two cathode pad chambers with a space resolution of about $\sim 100\ \mu$m in the bending direction: the stations are located between $z = 5.2$ and $z = 14.4$ m, the first two ones upstream of the dipole magnet, the third one in its gap and the last two ones downstream. (iv) A 1.2 m thick iron wall, corresponding to 7.2 hadronic interaction lengths, placed between the tracking and trigger systems, which absorbs the residual secondary hadrons emerging from the front absorber. (v) The muon trigger system, consisting of two detector stations, placed at $z = 16.1$ and $= 17.1$ m, respectively, each one composed of two planes of resistive plate chambers, with a time resolution of about 2 ns.

The ALICE experiment has already an intense physics program based on muon measurements, since the very start of its data taking. Within this program, currently active both in p–p and Pb–Pb collisions, three main directions can be identified: study of quarkonia production,[2] of open Heavy Flavors (HF) production,[3] of low mass dimuons.[4] The study of quarkonia production in p–p collisions allows one to investigate perturbative and non-perturbative aspects of QCD, by means of the analysis of the kinematic distributions of the resonances;[5] quarkonia production in nuclear collisions, on the other hand, is of primary importance to test quarkonia suppression/recombination mechanisms possibly being effective in deconfined QCD matter and already intensively investigated at the SPS and RHIC heavy-ion facilities. Open charm and beauty production, besides giving information on the initial stage of the nuclear collisions thanks to the short formation time of the heavy quarks, also provides sensitivity to the energy density of the deconfined matter through the mechanism of in-medium energy loss of heavy quarks. Low mass dimuons, finally, provide insight to soft QCD processes in the LHC energy regime, and allow one to characterize the properties of the deconfined medium through the analysis of the properties and the production cross sections of the light vector mesons.

---

positive values in the following, both for the (pseudo-)rapidity and the z coordinate.

2. The Muon Forward Tracker Proposal

The current ALICE muon physics program suffers from several limitations, basically because of the multiple scattering induced on the muon tracks by the hadron absorber. The details of the vertex region are then completely smeared out: in particular, this prevents us to disentangle prompt and displaced J/ψ production (the production of J/ψ from b accounting for $\sim 20\,\%$ of the prompt cross section) as well as to disentangle open charm and open beauty without making assumptions relying on physics models (thus introducing systematic uncertainties on the measurement). In addition, we have only very limited possibilities to reject muons coming from semimuonic decays of pions and kaons, representing an important background both in single muon and dimuons analyses, in particular at low masses and low p_{T}. Finally, the degradation of the kinematics, imposed by the presence of the hadron absorber, plays a crucial role in determining the mass resolution for the resonances, especially at low masses.

To overcome these limitations, better exploiting the unique kinematic range accessible by the ALICE Muon Arm, the Muon Forward Tracker (MFT) was proposed in the context of the ALICE upgrade plans, to take place in the years 2017/2018 during the LHC shutdown. The MFT is a silicon pixel detector added in the Muon Spectrometer acceptance ($2.5 < \eta < 4$) upstream of the hadron absorber. The basic idea, motivating the integration of the MFT in the ALICE setup, is the possibility to match the extrapolated muon tracks, coming from the tracking chambers *after* the absorber, with the clusters measured in the MFT planes *before* the absorber; the match between the muon tracks and the MFT clusters being correct, muon tracks should gain enough pointing accuracy to permit a reliable measurement of their offset with respect to the primary vertex of the interaction. The measurement of the muons' offset should then allow one to: (i) disentangle prompt (quarkonia, thermal photons) from displaced (open HF) dimuons; (ii) distinguish open charm and open beauty dimuons on the basis of the analysis of the pairs' offset; (iii) study HF via single muons down to $p_{\mathrm{T}} \approx 1$ GeV/c with limited model dependence; (iv) study beauty production down to zero p_{T} via J/ψ from b, a unique feature at the LHC in A-A collisions. In addition, applying quality cuts on the matching between the extrapolated tracks and the MFT clusters, it should be possible to reject a large fraction of background coming from semimuonic decays of primary and secondary pions and kaons, as well as punch-through hadrons arriving at the tracking chambers without being stopped in the absorber.

3. MFT Design and Preliminary Performance Studies

The MFT setup, as described in the simulation studies considered in the present report, is composed of five tracking planes placed at $z = 50$, 58, 66, 74, 82 cm from the IP, before the hadron absorber and the scintillator interaction counter (VZERO) placed in front of it. Each plane is composed of a $0.2\,\%$ x/X_0 disk-shaped support element, and a 50 μm-thick assembly of silicon sensors and readout elements arranged in the front and back part of the support. Each plane thus contribute with $0.3\,\%$ x/X_0, leading to a total of $1.5\,\%$ x/X_0 for the whole MFT. The material budget traversed by the muons before arriving the MFT depends on the geometry of the beryllium beam pipe. In the simulations presented here we considered a cylindrical beam pipe having an internal radius of 2 cm and a thickness of 500 μm, a realistic scenario for the 2017 upgrade. It should be noted here that a cylindrical beam pipe induces at high rapidity multiple scattering effects not at all negligible; for this reason, present studies are on-going on possible setups with a modified conical beam-pipe. Conclusions on such investigations will be available in the next future, and within this report no comparison will be established between concurrent geometry setups of the beam pipe.

For the active elements covering the MFT planes we assumed a 20×20 $\mu\mathrm{m}^2$ pixel segmentation, already available for a CMOS technology. Possibly occurring charge-dispersion effects may cause the activation of side pixels, in addition to the one actually traversed by the tracked particle: this is taken into account in the clusterization algorithm, which defines the center of the cluster averaging over the centers of the pixels composing it. In this way it is possible to recover a spatial resolution of the order of 20 $\mu\mathrm{m}/\sqrt{12}$.

The tracking strategy starts from the muon tracks reconstructed after the hadron absorber. These are extrapolated back to the vertex region, taking into account both the energy loss and the multiple scattering induced by the hadron absorber. Each extrapolated track is then evaluated at the last plane of the MFT (the one closest to the absorber) and, for each cluster in this plane, its compatibility with the extrapolated track is evaluated in terms of the quantity:

$$\chi^2_{\mathrm{clust}} = \frac{\Delta x^2 \sigma_y^2 + \Delta y^2 \sigma_x^2 - 2 \cdot \Delta x \Delta y \, \mathrm{cov}(x,y)}{\sigma_x^2 \sigma_y^2 - \mathrm{cov}^2(x,y)} \,, \tag{1}$$

where Δx and Δy represent the distance between the track position and the cluster along x and y, while σ_x^2, σ_y^2 and $\mathrm{cov}(x, y)$ are the covariance matrix elements accounting for the combined uncertainty on the cluster and the track position. For each compatible cluster a new candidate track is created, whose parameters and their uncertainties are updated with the information given by the added cluster by means of a Kalman filter algorithm. Each candidate track is then extrapolated back to the next MFT plane, where a search for compatible clusters is performed in the same way as before. As the extrapolation proceeds towards the vertex region, the uncertainties on the parameters of the extrapolated tracks become smaller, the number of compatible clusters decreases and the number of candidate tracks converges. If more than one final candidate is found at the last extrapolation, the best one is chosen according to the global fit quality.

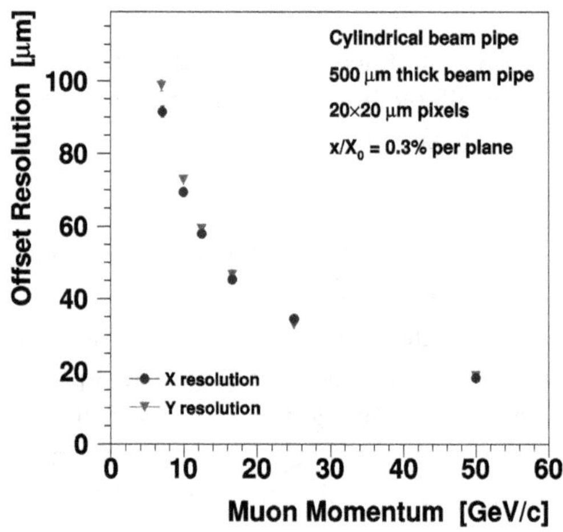

Fig. 1. Offset resolution for single muons as a function of the momentum, along the x and y directions. Error bars reflect the data sample available for each point.

With a spatial resolution of the order of $\sim 20\ \mu\mathrm{m}/\sqrt{12}$ in the MFT planes, we have a reliable measurement of the muons' offset, i.e. the transverse distance between the primary vertex and the muon track. To investigate the MFT performances in terms of the offset resolution for the single muons, single muons were generated at a fixed position in (0, 0, 0). The offset of the reconstructed muon tracks was then evaluated assuming the production vertex to be measured by the ALICE Internal Tracking System

(ITS) with a precision of 100 μm along the z direction. The observed offset resolution has been studied as a function of the muon momentum, see Fig. 1. The resolution is as good as ~ 100 μm even for muon momenta down to ~ 7 GeV/c. Slight differences between x and y directions probably reflect a remnant of the $x - y$ asymmetry introduced by the dipole magnetic field after the absorber. Studies are ongoing to characterize the offset resolution as a function of the muon rapidity, too, from which significant differences between a cylindrical and a conical beam pipe geometry could emerge.
The improved pointing accuracy gained by means of the matching between the extrapolated muon tracks and the MFT clusters, allows one to have a better evaluation of the opening angle for prompt muon pairs. This results in better mass resolutions for all the resonance measured in the dimuon channel, in particular for the lightest ones (η, ω and ϕ) whose soft muons suffer more from the degradation of the kinematics induced by the hadron absorber. A comparison between the currently available resolution and the one resulting from the preliminary MFT simulations is shown in Fig. 2, for the ω and ϕ mesons. With a simple Gaussian fit on the peaks obtained with the MFT simulations, one finds $\sigma_\omega \approx 16$ MeV/c^2 and $\sigma_\phi \approx 20$ MeV/c^2, an improvement up to a factor ~ 3 with respect to the resolutions available with the current Muon Arm setup.

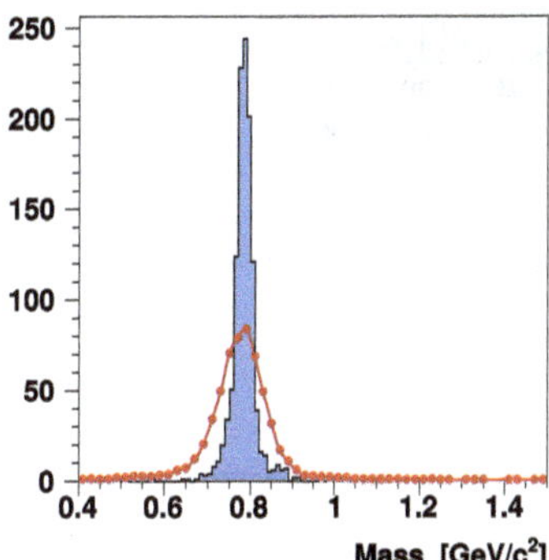

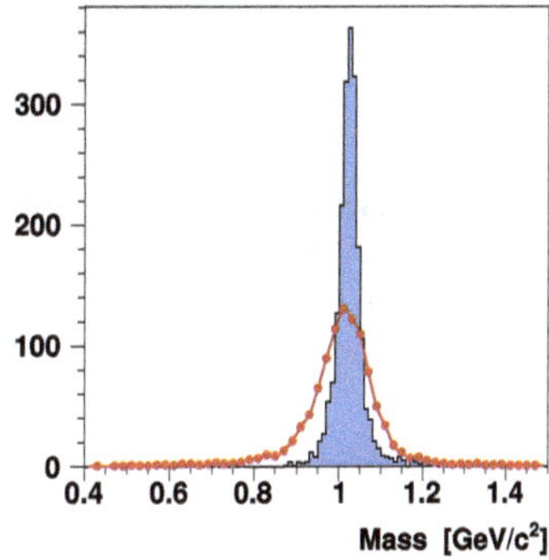

Fig. 2. Comparison between the mass resolution available with the current Muon Arm setup (red points and line) and the one achievable by means of the MFT (blue profile) for the ω (left panel) and ϕ (right panel) resonances.

The last point addressed in these preliminary investigations on the MFT performances, concerns the possibility to distinguish prompt muon pairs from open charm and open beauty contributions on the basis of the analysis of the weighted offset of the dimuons, defined as $\Delta_{\mu\mu} = \left[0.5 \cdot (\Delta_{\mu 1}^2 + \Delta_{\mu 2}^2)\right]^{0.5}$, with $\Delta_\mu = \left[0.5 \cdot (\delta x^2 V_{xx}^{-1} + \delta y^2 V_{yy}^{-1} + 2\delta x \delta y V_{xy}^{-1})\right]^{0.5}$, where δx and δy are the x and y offset of the muon track, and V^{-1} is the inverse of the covariance matrix accounting for the combined uncertainty on the track and the vertex position. From the definition above, we expect to find wider $\Delta_{\mu\mu}$ distributions as we go from prompt to displaced dimuons: this is exactly what we observe in Fig. 3, where the prompt dimuons (in this case the ϕ meson) have a narrower distribution than the open charm dimuons ($c\tau \sim 150\ \mu$m), which in turn have a narrower distribution than the open beauty dimuons ($c\tau \sim 500\ \mu$m). This should permit to establish a reliable separation between the signal components, on a statistical basis, without any model dependence.

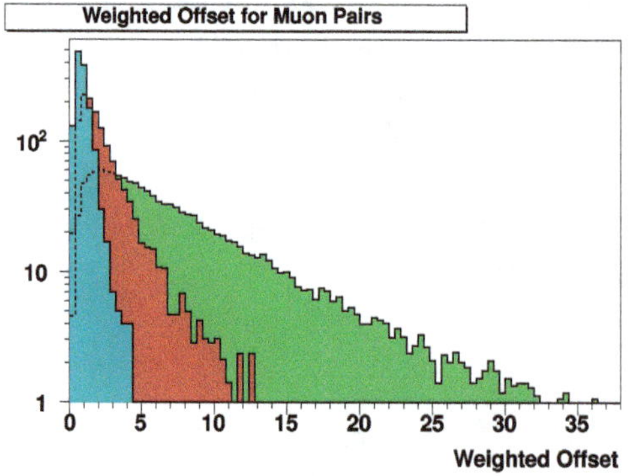

Fig. 3. Comparison between the weighted dimuon distributions for prompt (cyan profile), open charm (red profile) and open beauty (green profile) dimuons.

4. Conclusions

The muon physics program of the ALICE experiment is conditioned by the intrinsic limitations of the current setup of the Muon Arm. The addition of a silicon Muon Forward Tracker in the acceptance of the Muon Spectrometer should overcome these intrinsic limitations increasing the physics potential of the muon spectrometer, giving at the same time the possibility to perform new measurements of general interest for the whole ALICE physics. In this report we discussed the preliminary results of the simulations including the MFT, obtained in a low-multiplicity environment (matching efficiency $\sim 100\,\%$) with a focus on the offset resolution for single muons, the improvement of the mass resolution for the light mesons ω and ϕ and the weighted offset distributions for prompt and displaced muon pairs. These preliminary results are encouraging, and motivate an additional effort aiming to investigate the matching performances in high multiplicity, both as a function of the muons' kinematics and the MFT setup.

References

1. K. Aamodt *et al.*, *The ALICE experiment at the CERN LHC* (2008), JINST **3**, S08002.
2. G. Martinez Garcia, for the ALICE Collaboration (2011), `arXiv:1106.5889`.
3. A. Dainese, for the ALICE Collaboration (2011), `arXiv:1106.4042`.
4. A. De Falco, for the ALICE Collaboration (2011), `arXiv:1106.4140`.
5. K. Aamodt *et al.*, *Rapidity and transverse momentum dependence of inclusive* J/ψ *production in pp collisions at* $\sqrt{s} = 7$ *TeV* (2011), `arXiv:1105.0380`.

PERFORMANCE OF THE ATLAS TRANSITION RADIATION TRACKER WITH FIRST HIGH-ENERGY pp AND Pb-Pb COLLISIONS

A. VOGEL on behalf of the ATLAS Collaboration

Department of Physics, University of Bonn
Nussallee 12, 53115 Bonn, Germany
E-mail: adrian.vogel@cern.ch

The ATLAS Transition Radiation Tracker (TRT) is the outermost of the three sub-systems of the ATLAS Inner Detector at the Large Hadron Collider at CERN. It consists of close to 300000 thin-wall drift tubes (straws) providing on average 30 two-dimensional space points with 0.12–0.15 mm resolution for charged particle tracks with $|\eta| < 2$ and $p_T > 0.5\,\mathrm{GeV}$. Along with continuous tracking, it provides particle identification capability through the detection of transition radiation X-ray photons generated by high-velocity particles in the many-polymer fibres or films that fill the spaces between the straws. Custom-built analog and digital electronics is optimised to operate as luminosity increases to the LHC design. In this article, a review of the commissioning and first operational experience of the TRT detector will be presented. Emphasis will be given to performance studies based on the reconstruction and analysis of LHC collisions. The first studies of the TRT detector response to the extremely high track density conditions during the November 2010 heavy-ion LHC running period will be presented. These studies give interesting insight to the expected performance of the TRT in future high-luminosity LHC proton-proton runs.

Keywords: ATLAS; Transition Radiation Tracker; TRT; Performance.

1. Introduction

The ATLAS experiment[1] is one of the four large particle detectors at the Large Hadron Collider (LHC) at CERN. It is a multi-purpose detector and consists of three main components: Inner Detector, calorimeters, and Muon Spectrometer. The Inner Detector in turn comprises three subsystems: the silicon Pixel detector, the Semiconductor Tracker (SCT), and the Transition Radiation Tracker (TRT). Immersed in a solenoidal magnetic field of 2 Tesla, the Inner Detector measures trajectories and momenta of charged particles with $p_T > 0.5\,\mathrm{GeV}$ up to a pseudorapidity of $|\eta| = 2.5$.

2. Design of the TRT

The TRT is a straw-tube tracker. It consists of drift tubes with a diameter of 4 mm that are made from wound Kapton and reinforced with thin carbon fibres. In the centre of each tube there is a gold-plated tungsten wire of 31 μm diameter. With the wall kept at a voltage of -1.5 kV and the wire at ground potential, each tube acts as a small proportional counter. The tubes are filled with a gas mixture of 70 % Xe, 27 % CO_2, and 3 % O_2.

The TRT barrel region contains 52 544 straw tubes of 1.5 m length, parallel to the beam axis. They cover a radius from 0.5 m to 1.1 m and a pseudorapidity range of $|\eta| < 1$. The central wires are electrically split and read out at both ends of the straw. The endcaps contain radial 0.4 m long straws that are arranged perpendicular to the beam axis. Each side consists of 122 880 straws, covering the geometrical range $0.8\,\mathrm{m} < |z| < 2.7\,\mathrm{m}$ and $1 < |\eta| < 2$. The endcap straws are read out at their outer end.

When a charged particle traverses the TRT, it ionises the gas inside the straws. The resulting free electrons drift towards the wire where they are amplified and read out. The front-end electronics sample the incoming signal in 24 time bins of 3.12 ns and compare it against a threshold corresponding to 300 eV, resulting in a 24-bit pattern that gets buffered in a digital pipeline and then passed on to the central ATLAS data acquisition.

The spaces between the straws are filled with polymer fibres (barrel) and foils (endcaps) to create transition radiation, which may be emitted by highly relativistic charged particles as they traverse a material boundary. This effect depends on the relativistic factor $\gamma = E/m$ and is strongest for electrons. Typical photon energies are 5–30 keV. These soft X-rays can be absorbed by Xe atoms, depositing additional energy in the gas and leading to significantly higher readout signals. Such signals are detected by comparing them against an additional high threshold of 6 keV that is sampled in three 25-ns time bins alongside the pattern described before.

This design makes the TRT complementary to the silicon-based tracking devices: the single-point resolution of 120 μm is larger than that of the silicon trackers, but this is compensated by the large number of hits per track (typically more than 30) and the long lever arm.

3. Tracking Performance

The TRT readout data merely contains time information, which needs to be calibrated to be useful for tracking.[2] The first step is the T_0 calibration, defining the offset between the start of the readout and the arrival of par-

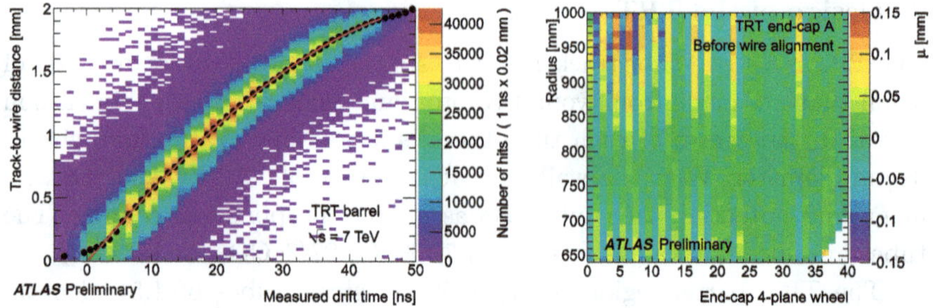

Fig. 1. Left: The $R(t)$ relation for the TRT barrel. The line shows the relation that is used to determine the drift distance based on the measured drift time. Right: Mean of a Gaussian fit to the TRT track residuals vs. radius and wheel before the wire-by-wire alignment.

ticles. It accounts for the time of flight, the signal propagation, and clock offsets. Its results are subject to small daily variations on the level of 100 ps, which are mainly caused by a drift of the central ATLAS clock.

The $R(t)$ calibration relates the measured drift time with a particle's distance of closest approach to the readout wire. It depends on the properties of the active gas (mixture, pressure, temperature), the voltage that is applied to the tube, and the magnetic field. The $R(t)$ relation is modelled by a third-order polynomial, as shown in Fig. 1, left. The resulting coefficients turn out to be very stable on the time scale of months. This is due to the TRT's "Gas Gain Stabilization System", which automatically adjusts the applied voltages to compensate for small variations of the other gas parameters, and also a precise monitoring of the composition of the gas mixture.

A key ingredient for maximum tracking performance is the alignment of all detector elements.[3] Figure 1, right, shows the spatial distribution of the Gaussian mean of the track residuals—the visible structures are apparently caused by a misalignment of some detector modules and a structure deformation. Such effects are cured by a track-based wire-by-wire alignment after which the distribution in Fig. 1, right, becomes nearly uniform.

This improvement is also reflected in Fig. 2: using the latest alignment data of autumn 2010, residual widths of 118 μm and 132 μm can be achieved for the barrel and endcap regions respectively, applying a cut of $p_T >$ 15 GeV. Providing an average of 30 such position measurements, the TRT contributes significantly to the tracking performance of the Inner Detector as a whole, particularly at high p_T.

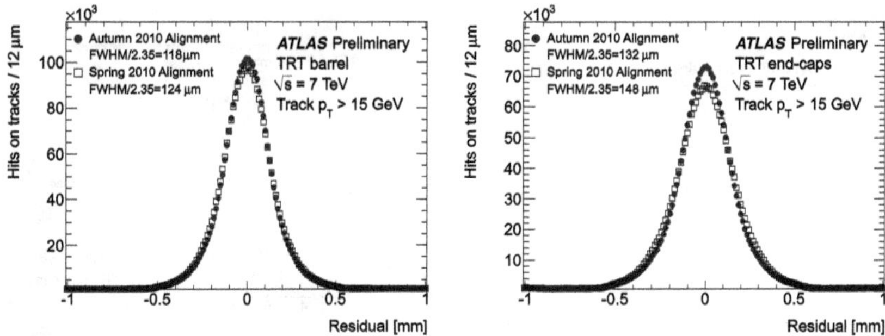

Fig. 2. Unbiased tracking residual distributions for the TRT barrel (left) and endcaps (right), comparing the alignment of spring 2010 (open squares) with that of autumn 2010 (solid circles).

4. Particle Identification Performance

The fact that the emission of transition radiation is much more likely for an electron than for a pion of the same momentum can be used to discriminate these particle types.[4] Figure 3, left, shows the high-threshold turn-on curve, i. e. the probability of getting a high-threshold hit as a function of a particle's relativistic γ factor. This probability is low for pions over a large momentum range (and almost entirely due to Landau fluctuations), but it rises quickly for electrons with momenta of only few GeV. This allows electron–hadron discrimination up to energies of 150 GeV.

Another source of information is the time over threshold (ToT), i. e. the number of time bins for which a readout signal exceeds the (low) threshold. This quantity depends on the particle's specific energy loss dE/dx, which in turn depends on the relativistic velocity β according to the Bethe–Bloch law. The combination of high-threshold and ToT information is shown in Fig. 3, right, which displays the pion misidentification probability as a function of momentum. One can see that the ToT is particularly helpful for low momenta, where the high-threshold turn-on curve is still rising.

5. Heavy-Ion Collisions

Heavy ions are a demanding challenge for the TRT: they yield thousands of tracks and average occupancies around 50 %, peaking at values up to 90 % for the most central collisions. The TRT readout electronics needed to employ lossless data compression to handle the enormous event data sizes, and the Inner Detector tracking and reconstruction software had to be adapted to cope with high-occupancy events.

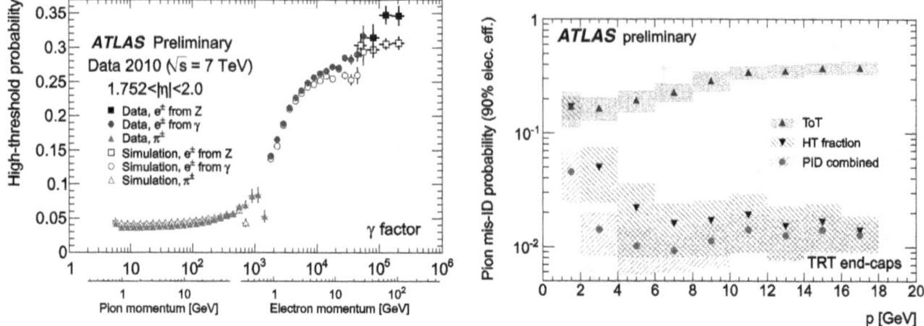

Fig. 3. Left: The high-threshold turn-on curve, shown for the outer endcap wheels. Right: The pion misidentification probability for selection criteria that give 90 % electron efficiency.

It turns out that the TRT is able to improve the track finding and the momentum resolution of the overall tracking system also for heavy-ion collisions: tracking studies show that the TRT still contributes an average of more than 30 hits per track over almost the entire covered η range. This is also a promising prospect for the conditions that will be encountered as the LHC delivers higher and higher luminosities in the future.

6. Summary

The ATLAS TRT provides tracking and identification of charged particles. With the latest alignment and calibration constants, spatial residuals of 118 μm (barrel) and 132 μm (endcaps) can be achieved. The TRT improves the overall momentum resolution of the Inner Detector, and by using its unique feature of transition radiation it contributes significantly to electron identification and background rejection. The experience gained from heavy-ion collisions shows that the TRT is ready for the high occupancies that have to be expected as the LHC's luminosity increases further.

References

1. The ATLAS Collaboration, *The ATLAS Experiment at the CERN Large Hadron Collider*, JINST 3 (2008) S08003.
2. The ATLAS Collaboration, *Calibration of the ATLAS Transition Radiation Tracker*, ATLAS-CONF-2011-006.
3. The ATLAS Collaboration, *Alignment of the ATLAS Inner Detector Tracking System*, ATLAS-CONF-2011-012.
4. The ATLAS Collaboration, *Particle Identification Performance of the ATLAS Transition Radiation Tracker*, ATLAS-CONF-2011-128.

COMBINATION OF MODERN VISUALIZATION TECHNIQUES FOR IMAGING OF BIOLOGICAL SAMPLES

FRANTISEK WEYDA

Faculty of Science, University of South Bohemia, Branisovska 31, CZ-37005 Ceske Budejovice, Czech Republic, e-mail: weydafk@seznam.cz

JIRI DAMMER

Institute of Experimental and Applied Physics, Czech Technical University in Prague Horska 3a/22, CZ-12800 Prague 2, Czech Republic

We have used several visualization techniques to characterize biological objects. A micro-radiography with the hybrid single photon counting silicon pixel detector Medipix2 (matrix 256 x 256 sq. pixels of 55 μm pitch) is an imaging technique using X-rays in the studies of internal structures of objects. The detector Medipix2 is used as an imager of an ionizing radiation, emitted by X-ray tubes (micro or nano-focus FeinFocus). An unlimited dynamic range of the Medipix2 detector and a high spatial resolution below 1μm is particularly suitable for a non-destructive and non-invasive radiographic imaging of small biological samples in a living state, including in vivo observations and a micro-tomography. Contrast agents (based on iodine or lanthanum) could be used for dynamic studies inside of organisms. Infrared digital photography has ability to shot still photographs or movies in complete dark. Is it also possible to use it for studies of internal organs and structures inside of living biological objects. Field emission scanning electron microscopy (FESEM) in low temperature mode is sophisticated recent technique successfully used in biological laboratories. The main advantage is ability to study details of tissues and cells close to living state at very high magnification. Special cryo-transfer system connected to FESEM allows deeply frozen samples to be prepared in way like freeze-fracturing followed by freeze-etching for observation directly inside of electron microscope. Combination of information from all above mentioned techniques could give us very powerful visualization tool for complex studies of biological specimen.

1. Introduction

Recently, in biology we have at disposal a big amount of analytical data from various techniques realized often separately. Comparison of data obtained with use of various visualization techniques applied to the same biological objects and their synthesis is very desirable. Arthropods (mostly insects) living totally or partially in hidden form or internal processes inside various biological objects are primary interest for us last years. Generally, it is not easy to study such problems using classical scientific visualization techniques. In addition, we prefer non-invasive and non-destructive visualization techniques. Non-invasive techniques enable us to visualize life of biological systems without influencing them, while non-destructive techniques enable us to visualize life of biological systems without destructing them.

Mostly, we try to obtain complex information on biological specimen based on combination of various scientific techniques like optical microscopy, infrared digital photography, micro-radiography and electron microscopy.

2. Methods

2.1. *Micro-radiography*

Micro-radiographic apparatus for imaging as well as for the observation of real-time in-vivo processes in living organisms equipped with detector Medipix2[a] was developed in the Institute of Experimental and Applied Physics, Czech Technical University (Czech Republic). Micro-radiography provides the fast and easy tool for imaging of hidden form of life as well as inner structure of biological samples. This type of imaging is based on attenuation of X-rays that pass through the object to a detector. One part of radiation is absorbed by the scanned object and the rest of (non-absorbed) radiation is detected by suitable detector as Medipix2[1].

Hybrid semiconductor pixel detector Medipix2 (MEDical Imaging PIXel detector 2nd generation) is used as an imager, which counts individual particles of ionizing radiation. It consists of 256x256 square pixels with pitch of 55 μm giving a total sensitive area of 14.08 mm x 14.08 mm. Each pixel has in its cell complete electronic chain of single channel analyzer. Detector Medipix2 has high sensitivity to low energy X-ray photons; position sensitive and noiseless single photon detection with preselected photon energies; photon counting in each pixel performed by digital counter; digital integration; high speed digital communication and data transfer[2].

The response of each individual detector pixel was calibrated for different absorber thicknesses to suppress the beam hardening effect in the given object and to compensate pixel in homogeneity across the whole pixellated sensor[3]. For the calibration we have used aluminum filters.

---

[a] See also www.cern.ch/medipix and www.utef.cvut.cz/medipix

2.2. *Optical microscopy*

Optical microscopy is basic technique for biologists working with small organisms. We use it routinely for control and manipulation with biological objects. Modern version is digital optical microscopy having chip as imager.

2.3. *Infrared digital photography*

Infrared digital photography has ability to shot still photographs or movies in complete dark. Due to its ability to penetrate dense tissues it is also possible to use it for studies of internal organs and structures inside of some living biological objects[4,5].

2.4. *Electron microscopy*

Conventional scanning (SEM) and transmission electron microscopy (TEM) enable for us to describe fine details of tissues and cells at very high magnification. Some disadvantage is necessity to prepare specimen in complicated manner (fixation, dehydration or desiccation, embedding into resin or coating with metal) causing various artifacts. Field emission scanning electron microscopy (FESEM) in low temperature mode is sophisticated recent technique successfully used in biological laboratories. The main advantage over conventional techniques is ability to study tissues and cells close to living state at very high magnification. Special cryo-transfer system connected to FESEM allows deeply frozen samples to be prepared in way like freeze-fracturing followed by freeze-etching for observation directly inside of electron microscope.

3. Examples of application of visualization techniques to selected biological objects

Let us to demonstrate several examples of complex studies using above mentioned visualization techniques.

3.1. *The horse-chestnut leafminer, Cameraria ohridella- modern invasive insect pest*

The horse-chestnut leafminer, *Cameraria ohridella* is a small species of leaf-mining moth from the family *Gracillariidae* (Insecta, Lepidoptera). First recorded in 1984 and spread very quickly all over the Balcan peninsula and occupied Middle and later whole Europe. It is insect pest living mostly in hidden state (larvae and pupae inside of leaf mines). Infrared digital photography[4] and micro-radiography are suitable to study hidden life of that pest species (Figure 1).

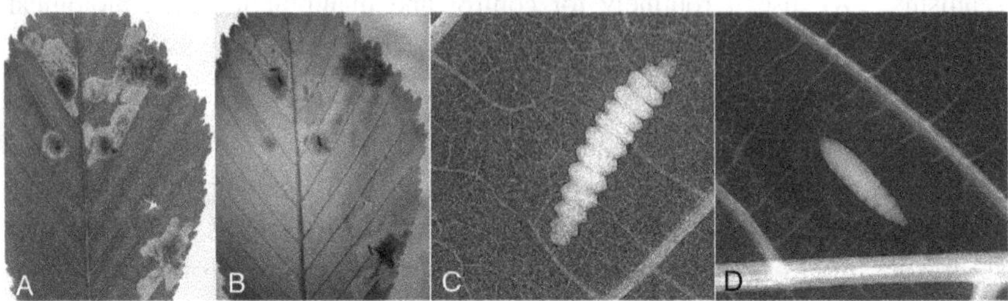

Fig.1. Leaf of horse-chestnut with mines of horse-chestnut leafminer. A- photography at visible light; B- near-infrared light at 750 nm; C, D- micro-radiography of larva and pupa of horse-chestnut leafminer

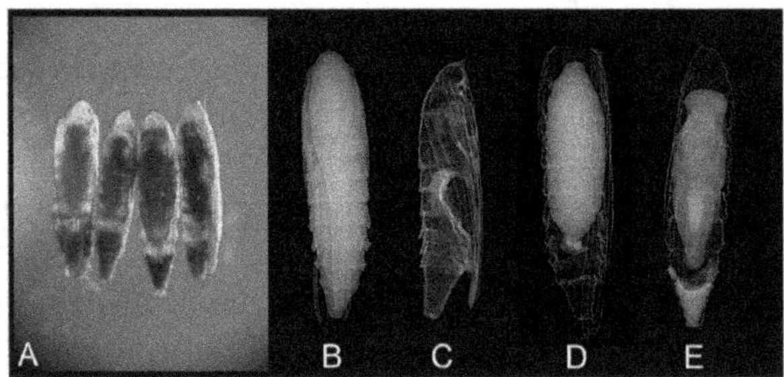

Fig.2. Pupae of horse-chestnut leafminer. A- near-infrared light at 750 nm; B-C micro-radiography; B- living pupa; C- death pupa; D- pupa infected with larva of parasitoid; E- pupa of horse-chestnut leafminer containing pupa of parasitoid.

Micro-radiography is superior technique, because it is able to distinguish larva as well as pupa and imago of wasp parasitoid living inside of horse-chestnut leafminer, while infrared digital photography is able to distinguish pupa and imago of wasp parasitoid only (Figure 2). Field emission scanning electron microscopy (FESEM) visualize all postembryonal stages of wasp parasitoid when open with scalpel (Figure 3).

The horse-chestnut leafminer overwinter as diapausing pupae. They are present in two forms: Living diapausing and death diapausing pupae. Death of those pupae is caused by natural mortality, or some are parasitized, some are infected by fungal infection. For serious study of horse-chestnut leafminer it is necessary to distinguish both forms of pupae, living and death. Death pupae contain cavities filled with gas from destruction of tissue. Micro-radiography distinguish it very well (Figure 2C).

For most insect pest exist predators and parasites. Larvae and pupae of horse-chestnut leafminer, *Cameraria ohridella* are parasitized with approx. 30 parasitoids (small hymenopteran wasps). They live inside of larvae and pupae destroying their tissue. Infrared digital photography and micro-radiography (superior) allow us to study those hidden parasitoids (see above).

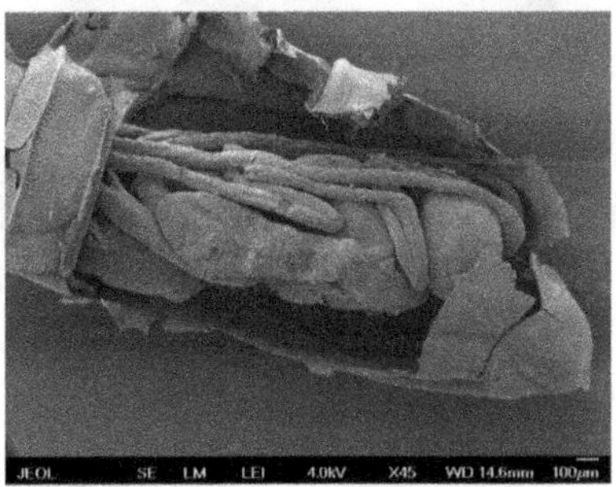

Fig.3. Field emission scanning electron microscopy (FESEM) of imago of parasitoid wasp inside of mechanically open pupa of horse-chestnut leafminer.

Studies based on modern techniques concerning horse-chestnut leafminer and its parasitoids are very important, because they bring new information about such invasive insect pest.

3.2. *The warehouse moth (Ephestia elutella) of the family Pyralidae infesting almonds and chocolate*

The warehouse moth (*Ephestia elutella*) is a small pest moth causing damage of agricultural crops and food like almonds, chocolate etc. Inside of almonds it is not visible in optical microscope. Infrared digital photography is insufficient to visualize damages inside of food. On the opposite, micro-radiography visualize

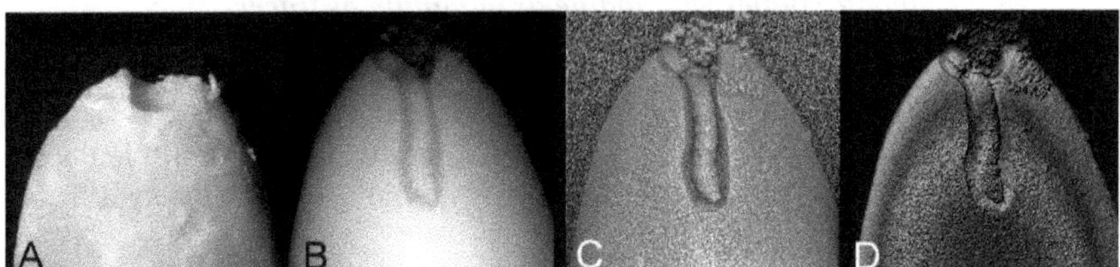

Fig.4. The warehouse moth larva in almond. A- macrophotography; B- micro-radiography; C,D- image analysis of figure B made with ACC Image Structure and Object Analyser.

warehouse moth larvae easily (Figures 4,5). Similar situation is when chocolate is infested (Figure 5).

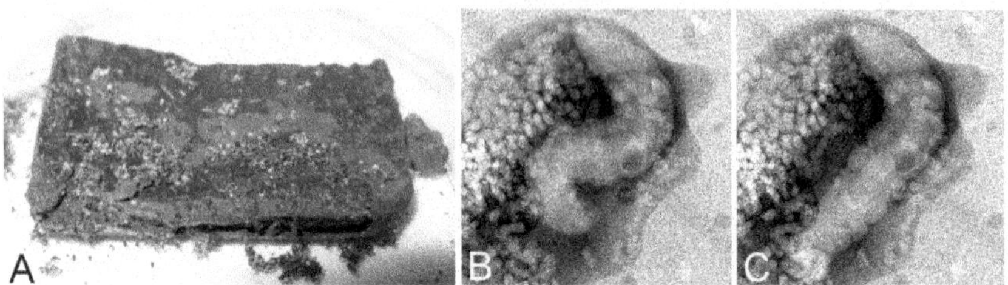

Fig.5. The warehouse moth larvae in chocolate. A- macrophotography; B,C- micro-radiography.

3.3. *Fire bug imago* **Pyrrhocoris apterus** *(Insecta, Heteroptera)*

Fire bug, *Pyrrhocoris apterus*, is well known laboratory model species for physiological studies. We have prepared micro-tomographic reconstruction of body of fire bug. 3D image is computed from 180 projections. Various anatomical details like digestive and excretory systems, fat body etc. is possible to study on virtual sections.

3.4. *Fossil arthropods in amber*

Amber, fossil resin from trees (usually several tens of million years old), contains various inclusions including fossil arthropods, mostly insects. Every visualization technique (optical and electron microscopy, infrared photography and micro-radiography), brings partial results. Synthesis of results from those techniques help us to understand morphology of those fossil insects and process of fossilization.

3.5. *Dynamic studies with iodine or lathanum as tracers*

Chemical tracers revealing high contrast in micro-radiographic apparatus enable us to realize dynamic studies concerning digestion, transport of substances throughout body and their storage, excretion, etc. We use iodine and lanthanum as contrast tracers and larvae of mosquito *Culex pipiens* as model species[6].

Micro-radiography successfully completes results from optical (infrared) and electron microscopy.

4. Conclusions

Synthesis of results from optical and electron microscopy, infrared digital photography and micro-radiography represents could give us very powerful visualization tool for complex studies of biological specimen. Especially the micro-radiographic imaging with detectors Medipix2 represents modern non-invasive and non-destructive method of investigation suitable for inspection of hidden biological objects (for example pests living inside of plants) as well as for study of invisible internal structures of tiny biological samples. Living organisms survive repeated scanning in micro-radiographic apparatus. So, these types of detectors enable 2D and 3D imaging and bring new possibilities for researchers to perform high resolution studies of hidden biological objects and structures[7,8].

Acknowledgments

This work was realized out in frame of the CERN Medipix Collaboration and was supported in part by the Research Grant Collaboration of the Czech Republic with CERN No. 1P04LA211, by the Fundamental Research Center Project LC06041 and the Research Programs 6840770029 and 6840770040 and Grant No. 2B06005 of the Ministry of Education, Youth and Sports of the Czech Republic.

References

1. J. Jakubek, D. Vavrik, T. Holy, M. Jakubek, Z. Vykydal, *Nucl. Instrum. Methods Phys. Res.* **A. 563**, 278-271(2006).
2. X. Llopart, M. Campbell, R. Dinapoli, D. San Segundo, & E. Pernigotti, *IEEE Trans. Nucl. Sci.* **49**, 2279–2283 (2002).
3. J. Jakubek, *Nucl. Instrum. Methods Phys. Res* **A 576**, 223-234 (2007).
4. F. Weyda, I. Hrdy, *Antenna*, **26**, 249-252 (2002).
5. F. Weyda, *Proceedings „Applied Optics and Microscopy" (In Czech),* 61-68 (2007)
6. J. Dammer, F. Weyda, V. Sopko, J. Jakubek, JINST **6**, 1-6 (2011)
7. F. Weyda, *Proceedings of the XVII. Czech and Slovak Plant Protection Conference),* 541-544 (2006).
8. P. Frallicciardi et al., *Conf. Proc. IEEE NSS/MIC 2008,* M10-112 (2008).

THE TRACKING SYSTEM IN THE PANDA APPARATUS

P. WINTZ for the PANDA collaboration
Institut für Kernphysik, Forschungszentrum Jülich GmbH,
Wilhelm-Johnen-Strasse, 52428 Jülich, Germany
**E-mail: p.wintz@fz-juelich.de*

The $\bar{P}$ANDA experiment at the new FAIR facility at Darmstadt (Germany) will investigate antiproton collisions on proton and nuclear targets in the charm quark mass regime. The wide-range physics program requires a universal detector concept, combining state-of-the-art and novel techniques in particle measurements and data readout. This paper gives an overview of the detector setup and summarizes the status, in particular of the charged particle tracking detectors in the $\bar{P}$ANDA spectrometer.

Keywords: $\bar{P}$ANDA experiment; antiproton beam; hadron physics; tracking detector.

1. The $\bar{P}$ANDA experiment

The new Facility for Anti-proton and Ion Reseach (FAIR)[1] at Darmstadt (Germany) will provide antiproton and ion beams with unprecedented quality and intensity. The PANDA experiment (Anti-**P**roton **AN**nihilation at **DA**rmstadt) will be installed in the High Energy Storage Ring (HESR), a slow ramping synchrotron and storage ring, which provides up to 10^{11} circulating antiprotons with a momentum varied between 1.5 GeV/c and 15 GeV/c. Two different operation modes of the HESR are foreseen: (i) the high intensity mode, with the highest luminosity of $2\times10^{32}\,\mathrm{cm}^{-2}\mathrm{s}^{-1}$ at a slightly higher beam momentum spread of $\Delta p/p = 1\times10^{-4}$, and (ii) the high resolution mode with $\Delta p/p = 4\times10^{-5}$ at slightly reduced luminosity of $2\times10^{31}\,\mathrm{cm}^{-2}\mathrm{s}^{-1}$. The excellent beam energy definition is achieved by stochastic and electron cooling elements in the ring and will allow the measurement of the excitation function of mass and width of a single produced hadronic state with an accuracy down to $E_{CM} \simeq 50$ by changing the beam momentum in small steps in the region of interest.

The PANDA experiment[2] investigates antiproton collisions on proton, deuteron and nuclear targets in the charm quark mass regime with total center-of-mass energies in the $\bar{p}$-nucleon system between about 2.2 GeV and 5.5 GeV. Since in $\bar{p}p$ annihilation states with all possible quantum numbers allowed for fermion-antifermion pairs can be directly formed the complete spectrum is accessible, in contrast to e^+e^- annihilation for instance.

Main topics of the PANDA program are charmonium and D-meson spectroscopy with a precise determination of masses, widths, quantum numbers and branching ratios. Still, the existing data for some of the known states (e.g. $\eta_c' \, (2^1S_0)$ or $h_c \, (1^1P_1)$) are inconsistent or uncertain and need clarification. New states, recently discovered[3] by experiments at B-factories like BaBar, Belle and CLEO and usually quoted as XYZ states, still have undetermined and not understood properties and need further investigation. The search for new exotic states like glueballs, hybrids or multi-quarks is of particular interest and should be favored by the gluon rich $\bar{p}p$ annihilation. The spectroscopy of (multi-)strange and charmed baryons at PANDA has discovery potential, especially for the Ξ spectrum where large cross-sections of the order of μb are expected.[3]

Antiproton-nucleus collisions allow the study of the properties of hadrons in nuclear matter with respect to their vacuum properties. In medium modifications of masses and widths of the hadrons will be investigated, for instance by measuring a subthreshold production of the D meson by its lowered mass in medium. Also a possible absorption of J/ψ mesons and other charmonia in nuclear matter will be studied. Further topics are investigations of the nucleon structure and the production of single and double hypernuclei (e.g. ${}^{A}_{\Xi^-}Z$, ${}^{A}_{\Lambda\Lambda}Z$), which give access to study for the first time the Ξ-nucleon and hyperon-hyperon interaction. A complete description of the whole physics program of the PANDA experiment can be found in Ref. 4.

2. The $\bar{P}$ANDA detector

The broad physics program demands a universal detector concept, combining state-of-the-art and novel detection techniques and data readout concepts. A close to 4π solid angle coverage for charged and neutral particles is needed for a partial-wave-analysis of the reaction final state. In addition, Dalitz plot analyses of three-body final states require a smooth acceptance function across the full phase-space with minimized discontinuities in the transition between adjacent sub-detectors, by gaps or mechanical structures. The tagging of charmed hadrons with decay lengths in the order of 100 μm requires a vertex reconstruction around the target with highest resolution of about 50 μm. A relative momentum resolution (σ_p/p) of about 1-2 % for charged particles is needed to reach a resolution of about a few 10 MeV for reconstructed masses and widths. Furthermore, it is essential to identify $e^{\pm}$, $\mu^{\pm}$, $\pi^{\pm}$, $K^{\pm}$ and $p/\bar{p}$ in a broad momentum range from about a few 100 MeV/c to 8 GeV/c. Electromagnetic calorimetry with high resolution in a wide photon energy range is required. In general, the material budget of the tracking detectors should be as low as possible to reduce multiple scattering, electron bremsstrahlung and photon conversion. All detectors must be able to cope with high rates in an environment of about $6\times10^7\,s^{-1}$ charged particle tracks at full luminosity with an average hadronic reaction rate of $2\times10^7\,s^{-1}$. The instantaneous rates can be even about a factor of three higher.

Figure 1 shows the layout of the PANDA detector which will be installed as a fixed target spectrometer system in the HESR. It consists of a central target spectrometer inside a 4 m long, 2 Tesla superconducting solenoid for particles emitted at larger polar angles and a forward spectrometer with a 2 Tm dipole magnet for particles with small polar angles ($\theta \leq 5°$ vertically, $\theta \leq 10°$ horizontically). Details about the magnets can be found in Ref. 5.

Various target systems are planned. For $\bar{p}p$ and $\bar{p}n$ collisions a cluster gas jet or pellet target system will be used and for $\bar{p}A$ collisions fiber or foil target will be installed. The high luminosity of $2\times10^{32}\,cm^{-2}s^{-1}$ is reached by a frozen pellet beam target (diameter $\sim$25 μm, rate 10 kHz) with a high density of about 4×10^{15} atoms/cm^2.

The charged particle tracking includes a Micro-Vertex-Detector (MVD) around the target interaction region as inner tracker for precise vertex reconstructions, a large barrel-shape Straw-Tube-Tracker (STT) as Central Tracker for momentum reconstruction and particle identification, a stack of three Gas-Electron-Multiplier (GEM) disks at the forward end of the target spectrometer (TS), and a set of six tracking stations consisting of planar straw layers in the forward spectrometer (FS) around the dipole magnet.

Charged particle identification is done by imaging Cerenkov detectors (Barrel DIRC, Endcap Disc DIRC, Forward RICH), a time-of-flight system (Barrel TOF, Forward TOF) and a muon detection system embedded in the iron yoke of the target spectrometer solenoid and a muon range system in the forward spectrometer. The DIRC detectors are able to separate charged pions, kaons and protons with momenta above 1 GeV/c. For lower momenta the charged particle identification will be based on the measurement of the energy loss inside the STT using the $\frac{dE}{dx}$ information, in particular to separate K/p below $\sim$0.8 GeV/c and π/K below $\sim$0.5 GeV/c. An electromagnetic calorimeter (EMC) made of PWO crystals measures

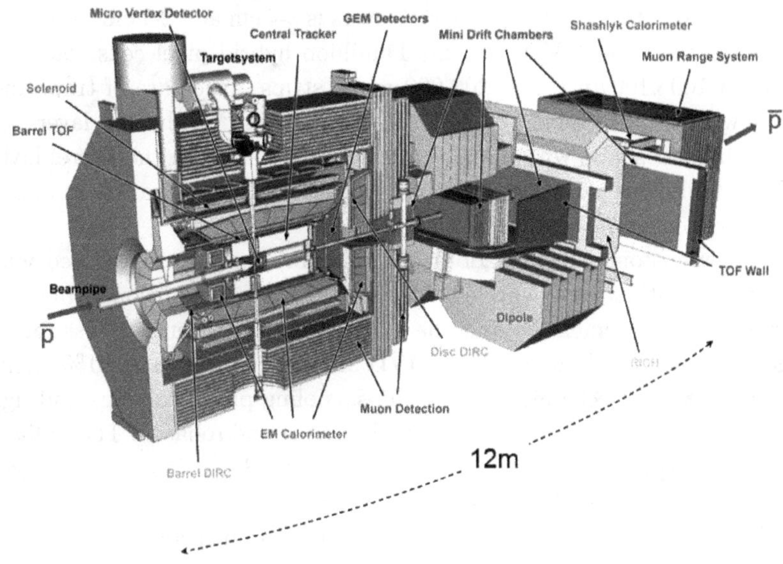

Fig. 1. The layout of the $\bar{P}$ANDA spectrometer with the pellet target system. The antiproton beam is entering from the left. A detailed description of all detector components is included in the text.

photons in a wide energy range from 1 MeV to 10 GeV and consists of a barrel part and a backward and forward endcap with an energy resolution better than 2 % ($\frac{\sigma_E}{\sqrt{E}}$). Different techniques including the operation at a low temperature of -25° C have been developed to increase the light output of the crystals by several factors compared to their employment in other experiments like CMS for instance. More details about the EMC detector can be found in Ref. 6. The forward spectrometer includes an electromagnetic calorimeter of shashlyk type.

3. Charged particle tracking

The charged particle tracking includes the MVD as inner tracker, the STT as Central Tracker, the GEM stations and tracking stations in the forward spectrometer (FS) around the dipole magnet. The MVD consists of silicon hybrid pixel detectors and double-sided micro strip detectors which are arranged in four barrel layers around the beam/target interaction point and six disc layers in the forward direction. The first barrel has a radius of 25 mm and the first disc layer is placed in a 20 mm z-distance to the target. The spacing between adjacent barrel layers is 3-4 cm and 2-7 cm for the disc layers. In total the MVD contains 11 million hybrid pixel cells, each with a size of $100\times100\,\mu m^2$, and 200,000 micro strips, which are of trapezoidal shape with a pitch and stereo angle of $70\,\mu m$ and 15° in the disc layers, and of rectangular shape with $130\,\mu m$ pitch and 90° angle in the barrel layers. The MVD layout has been optimized to get for most of the tracks four hit points for vertex reconstructions.

A new custom sensor readout chip (ToPix) has been developed which features a continuous, triggerless data readout and an additional time-over-threshold measurement to add the information of the energy loss for pid. The total material budget of the MVD amounts to less than 10 %, including the sensors, electronic readout chips, cooling pipes, cablings and lightweight support structures based on carbon fiber and rohacell. The radiation damage is expected from simulation to be below $3\times10^{13}\,n_{eq}/cm^2$ per year. Details about the MVD can be found in Ref. 7.

The MVD is surrounded by the STT with an inner and outer radius of 150 mm and 420 mm, and a length of 1650 mm. The STT consists of 4636 straw tubes, made of aluminised mylar film with a thickness of $27\,\mu m$, diameter of 10 mm and length of 1500 mm. End plugs at both ends made of thermoplastics (ABS) close the tube, fix the crimp pin for the wire (W/Re, $20\,\mu m$ diameter) and provide the gas in- and outlet by small diameter gas tubes. The assembled tubes are close-packed to planar layers, pressurized and glued to double- and 4-layer modules which are self-supporting, i.e. sustaining the wire and tube tension equivalent to 50 gr and 0.8 kg per straw. By this novel technique, being first developed for the COSY-TOF Straw Tracker,[8] dedicated mechanical frame or reinforcement structures for stretching the tubes can be avoided and a very low material budget for the detector can be reached. The tubes are operated with an $Ar/CO_2(10\%)$ gas mixture at 2 bar pressure. The planar straw modules are arranged in six hexagonal sectors to achieve the cylindrical STT volume. In radial direction 15 to 19 axial layers oriented in beam direction and 8 stereo layers, skewed by $\pm3^\circ$ measure the helical trajectories in 3d-space with a single hit resolution of $\sigma_{r\phi} \leq 150\,\mu m$ and $\sigma_z \leq 2.8\,mm$. The readout of the straw signal time and amplitude will provide both, the drift time for spatial isochrone reconstruction and charge information to measure in addition the energy loss ($\frac{dE}{dx}$) for particle identification. In particular the separation of p/K and π/K below 800 MeV/c can be only done by the STT. The low material budget of about 1.2 % (X/X_0) and dense straw filling with a high number of up to 27 hits allow the measurement of the energy loss with a resolution better than 8 % ($\frac{\sigma_E}{E}$). All quoted resolutions are conservative and have been confirmed by prototype test measurements. In a continuous, triggerless data acquisition the event time (t_0) can be derived from the reconstruction of single tracks without exact isochrone information in the beginning. The method was checked by simulation to yield a resolution better than 2 ns (σ_{t0}). More details about the STT can be found in Ref. 9.

At the forward end of the STT a tracking station consisting of three disc-shape drift detectors gives additional track hits in the polar angle range from 2° to 21°. Each disc consists in the center of a double-sided readout plane containing pad electrodes, followed on each side by a stack of GEM (Gas Electron Multiplier) foils for gas amplification, then a drift electrode and a window foil. The material budget for each GEM station is less than 0.5 %. The spatial resolution is better than 100 μm, with a double-track resolution of ~10 mm. The total number of readout channels is ~100.000.

The forward spectrometer contains six charged particle tracking stations, which are placed before, inside and after the dipole magnet to reconstruct the particle track and momentum in a polar angle range below 5° in vertical, and below 10° in horizontal direction. The distance to the target of the stations vary from 2.9 m to 7.5 m. The momentum resolution at the nominal B-field of 2 Tm is about 0.5% ($\frac{\sigma_p}{p}$) in a high accessible momentum range down to about 0.3 GeV/c. Each tracking station consists of two straw double-layers in vertical direction and two stereo double-layers, skewed by ±5°. The total number of straws is 10752, each with a diameter of 10 mm. The different size of the tracking stations requires different straw lengths ranging from 0.64 m to 1.48 m. The used gas mixture is Ar/CO_2 with a pressure of 2 bar similar to the central STT.

4. Results

The performance of the charged particle tracking system (MVD, STT, GEM) in the target spectrometer has been checked by a simulation of single muon tracks and a full analysis of certain benchmark channel reactions, including helix and momentum reconstruction, vertex fitting, particle identification and mass reconstructions. For single tracks a momentum resolution of 1–2 % ($\frac{\sigma_p}{p}$) in a large polar angle range from 10° to 140° has been obtained, down to a low particle momentum of 0.3 GeV/c. This will allow the measurement of masses or widths of states with a precision down to a few 10 MeV.

The benchmark channel analyses included the reactions $\bar{p}p \to \pi^+\pi^-\pi^+\pi^-$, $\bar{p}p \to \eta_C \to \phi\phi \to K^+K^-K^+K^-$, and $\bar{p}p \to \Psi(3770) \to D^+D^- \to K^-\pi^+\pi^+K^+\pi^-\pi^-$. As an example, the latter benchmark could be reconstructed with a high efficiency of 5 %. A vertex resolution of the D-meson decays of about 50 μm ($\sigma_{x,y}$) and 100 μm (σ_z) was obtained and the D mass could be reconstructed with a precision of 13 MeV (σ).

5. Summary

The proposed detector system for the wide-range physics program of the PANDA experiment combines state-of-the-art detection techniques with novel features. The charged particle tracking system combines a very low material budget with efficient and high resolution measurements of the particle trajectory, inner decay vertices, particle momentum and energy loss for pid. The continuous, triggerless data readout in the high rate and high particle multiplicity environment requires efficient and fast algorithms for online event reconstruction and selection.

Various test measurements with detector prototype setups and a full simulation and analysis of the most challenging physics benchmark reaction channels confirm that the proposed detector layout fulfills the required performance for the charged particle tracking. For most of the detectors the technical design reports are foreseen to be finished in the next months. Then a few years construction phase will start, followed by a commissioning of the detectors with beam and a PANDA pre-installation starting in 2015 in Jülich (Germany). The final installation of the PANDA detector in the HESR at the FAIR facility will be in 2016/2017.

References

1. http://www.fair-center.org. W.F. Henning, *FAIR and its experimental program J. Phys. G* **34**, 551 (2007).
2. http://www-panda.gsi.de.
3. Particle Data Group, *Review of Particle Physics*, *J. Phys. G* **37** 075021 (2010).
4. The PANDA Collaboration, *Physics Performance Report for PANDA*, arXiv:0903.3905.
5. The PANDA Collaboration, *Technical Design Report for the PANDA Solenoid and Dipole Spectrometer Magnets*, arXiv:0907.0169.
6. The PANDA Collaboration, *Technical Design Report for PANDA Electromagnetic Calorimeter (EMC)*, arXiv:0810.1216.
7. D. Calvo (PANDA MVD Collaboration), *Nucl. Phys. B* **215** 192 (2011).
8. P. Wintz, "A Large Tracking Detector in Vacuum Consisting of Self-Supporting Straw Tubes", in *AIP Conf. Proc.* **698** 789 (2004).

9. P. Gianotti, *et al.*, "Tracking with Straw Tubes in the PANDA Experiment", *these proceedings*.

A DEDICATED BEAM TESTS OF THE FULL-SCALE PROTOTYPE OF GEMS FOR CMS IN A STRONG MAGNETIC FIELD

D. ABBANEO, S. BALLY, H. POSTEMA, A. CONDE GARCIA, J. P. CHATALAIN, G. FABER, L. ROPELEWSKI, S. DUARTE PINTO, G. CROCI, M. ALFONSI, M. VAN STENIS, A. SHARMA, M. VILLA, M. ZIENTEK*

Physics Department, CERN
Geneva, Switzerland
**E-mail: Michal.Zientek@cern.ch*

L. BENUSSI, S. BIANCO, S. COLAFRANCESCHI, F. FABBRI, L. PASSAMONTI, D. PICCOLO, D. PIERLUIGI, G. RAFFONE, A. RUSSO, G. SAVIANO

Laboratori Nazionali di Frascati dell'INFN
Frascati, Italy

A.MARINOV, M. TYTGAT, N. ZAGANIDIS

Department of Physics and Astronomy, Universiteit Gent
Gent, Belgium

M. HOHLMANN, K. GNANVO

Dept. of Physics and Space Sciences, Florida Institute of Technology,
Melbourne, FL, United States of America

M.G. BAGLIESI, R. CECCHI, N. TURINI, E. OLIVERI, G. MAGAZZU

INFN Pisa, Universita Degli Studi di Siena,
Siena, Italy

In the high-eta (1.6 - 2.4) region of the CMS endcap, Gas Electron Multipliers (GEM) present an interesting option for a future upgrade of the forward region of the CMS muon system. Large GEM detectors are challenging due to technological issues; in the view of the CMS upgrade we have designed and built the largest full-size triple GEM-based muon detector to-date. This prototype meets the stringent requirements of the hostile forward environment of CMS at high-luminosity LHC. Dedicated test beam measurements have been performed at the SPS in June 2011 to study efficiency, space resolution, and timing performance with different inter-electrode gap configurations and gas mixtures and in a strong magnetic field of 3T (as at CMS). Preliminary results of these experimental tests will be presented.

1. Introduction

Muon tracking and triggering system of the Compact Muon Solenoid (CMS) detector at CERN Large Hadron Collider (LHC) is based on three technologies Drift Tubes, Catode Strip Chambers and Resistive Plate Chambers (RPCs) [1]. While Drift Tubes and Catode Strip Chambers are responsible for muon tracking in the barrel and endcap region, RPCs provide level 1 muon trigger in both the barrel and endcap detector parts. Present RPC endcap system is not fully completed and to maintain the best performance during the future runs at full luminosity and beam energy, an extension of the RPC endcap system is forseen. Although the future improvement will not apply to the very forward region $\eta > 1.6$ and all the endcap disks will remain vacant. This creates an opportunity to equip them with technology, that would be able to sustain hostile environment and be suitable for operation at the LHC and it is future upgrades. For the planned LHC luminosity upgrade ($10^{34\text{-}35}$ cm^2/s), the expected particle flux in that particular detector region amounts to several tens of kHz with a total integrated charge over 10 years of several C/cm^2. To be able to perform in such a conditions, new approaches must be found.

2. Case for GEMs

Since 2009 when the dedicated R&D program was launched Gas Electron Multipliers (GEMs) [2] aspire to be the chosen technology. The main advantages that GEMs can offer are excellent spatial resolution < 100 μm, high efficiency > 98%, time resolution below 5 ns and the rate capability of order 10^6 Hz/mm^2. Usage of non-flammable gas mixtures and the potential to build large-area constructions may be considered as additional benefits. All mentioned above make GEMs an appropriate candidate to handle the expected particle flux at the LHC.

3. Small and full-scale prototypes

Up to know, several different triple-GEM prototypes have been successfully developed and studied. These constructions differ mainly in size, gap configuration and read-out layouts. Among them small 10 x 10 cm^2 and full-scale 990 x 200 - 450 mm^2 constructions can be distinguished. Small prototypes were produced by using either so called standard-double or single-mask technique. Full-scale prototypes as distinct from small prototypes can only be produced by single-mask technique, since it is the only one that allows one to avoid upper and lower mask pattern alignment problem and in consequence obtain large active areas. Since it was proved that small prototypes made by the single-mask technique can achieve similar performance results as the standard ones [3], the single-mask technique was used to produce first (GE1/1 I) [4] and second (GE1/1 II) full-scale prototype. Both constructions share the same trapezoidal geometry but differ in the gap size configuration, pitch size and the number of channels. GE1/1 possess 3/2/2/2 mm gap configuration and it is divided in 4 η regions containing 256 radial readout strips each oriented along the long side of the detector. Every region is read out by two digital VFAT chips [5], the strip pitch varies from 0.8 at narrow end to 1.6 mm at the wide end of the detector. The GEM foils are sectorized into 35 high voltage sectors transverse to the strip direction to decrease the discharge probability. GE1/1 II is an improved version of GE1/1 I and possess reduced gap size configuration of 3/1/2/1 mm and pitch size ranging from 0.6 to 1.2 mm. Moreover, the detector is divided into 8 η regions containing 384 strips giving a total number of channels equal to 3072.

4. GE1/1 II Beam-tests overview and preliminary results

At the end of June 2011 one standard, double-mask GEM and two full-scale prototypes were taken to H2 beam line of SPS [6] for the full set of tests in strong magnetic field. The goal of this campaign was to evaluate the performance of above mentioned detectors under conditions most similar to those in high-η region in CMS. Standard GEM and two prototypes were positioned in Y-Z plane inside M1 magnet solenoid (Fig. 1). The RD51 GEM tracking telescope was located 5 m before the magnet along X axis. 150 GeV pion and muon beams were extracted along X axis. The detectors under tests were filled with $ArCO_2CF_4$ gas mixture while the tracking GEMs with standard $ArCO_2$ mixture.

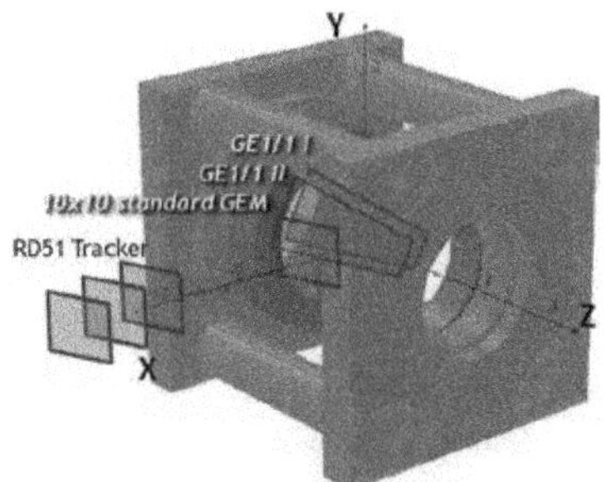

Figure 1. Experimental set up at H2-SPS

Influence of external magnetic field on the electrons velocity trajectory was examined and results as a characteristic of electron displacement in the function of magnetic field strength (Fig. 2). The tests were carried out in the configurations where vectors of magnetic field and detectors electric field were perpendicular to each other, as well as the angle formed 60 °.Thus, in conditions far worse than those prevailing in the region of high-η. A similar study was carried out for the average number of fired clusters. The results (Fig. 3) showed small divergence in the number of fired clusters from the trend at 1T. However, the obtained values do not differ significantly, therefore magnetic field effect is not clear.

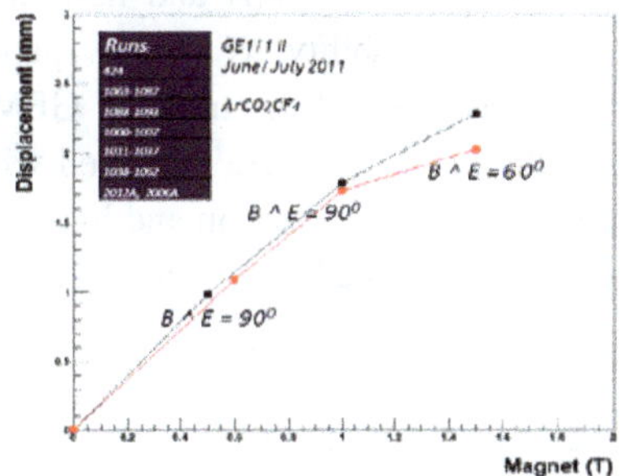

Figure 2. Displacement vs magnetic field.

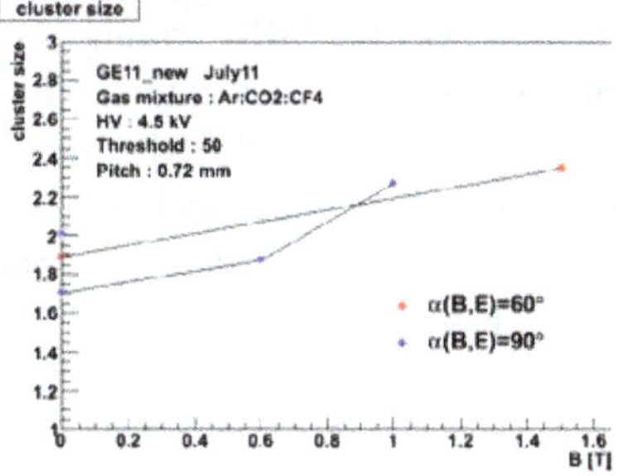

Figure 3. GE1/1 II cluster size vs magnetic field.

For the timing studies, one scintillator was placed in the front and two in the back of the RD51 tracking telescope. These scintillators were used to generate the trigger to signal the beam passing through the detector. Spread in arrival time of the particle signaled by the GEM VFAT chip with respect to the scintillator trigger was measured by the TDC module.

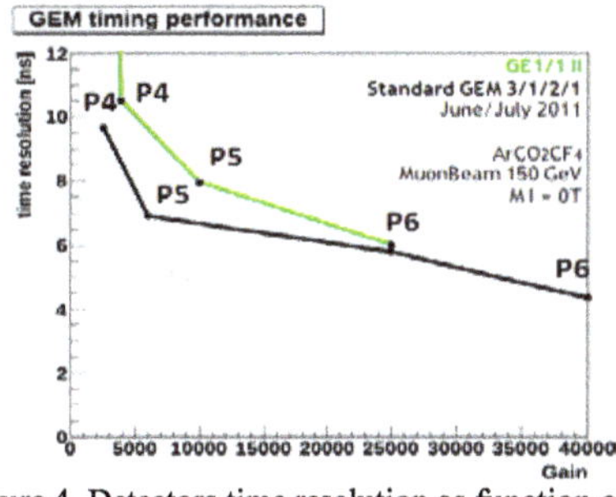

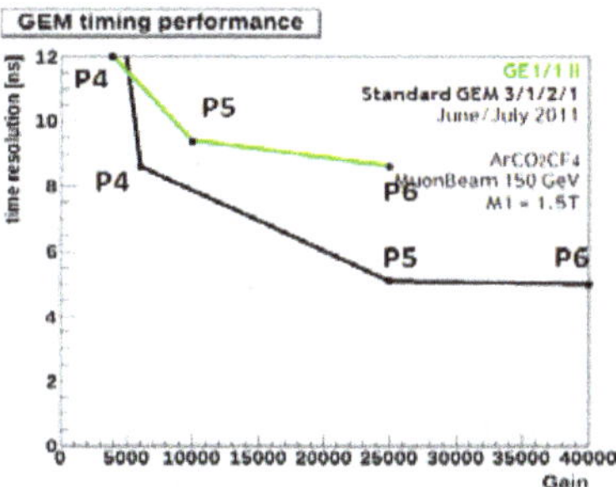

Figure 4. Detectors time resolution as function of gain without (left) and with (right) magnetic field.

The results (Fig. 4) showed better performance of standard GEM. It is caused by custom-made, software high voltage divider, which allows one to modify the fields across each gap of the detector independently. Unfortunately full-scale prototype is not adapted to be used with such a divider, thus it is equipped with standard one made by discrete components. The main disadvantage of this solution is being dependent on existing values of SMD high voltage resistors. What prevents from obtaining the desired field and voltage drops configuration and in consequence reaching lower time resolution.

In September 2011 GE1/1 II was tested again in H4 SPS beam line [7]. This time the goal was to test the detector with analog APV chips and new data acquisition system SRS. During those tests the capability to obtain spatial resolution by puls height information was showed. Two of the tracking GEMs running with $ArCO_2$ and GE1/1 II running with $ArCO_2CF_4$ were equipped with analog APV chips. Since both trackers share the same construction and have the same spatial resolution, the upper bound of GE1/1 II spatial resolution can be calculated as follows:

$$\sigma^2_{\Delta y} = \sigma^2_{y5} + \sigma^2_{y1} = 2\sigma^2_y \text{ , or } \sigma_y = \sigma_{\Delta y} / \sqrt{2} = 53 \tag{1}$$

The strip pitch of the tracker is the same in x and y, therefore $\sigma_y = \sigma_x$

$$\sigma_{GE1/1\ II} \leq \sqrt{(\sigma^2_{\Delta x} - \sigma^2_{x5})}, \text{ i.e. } \sigma_{xGE1/1\ II} \leq 103\ \mu m \tag{2}$$

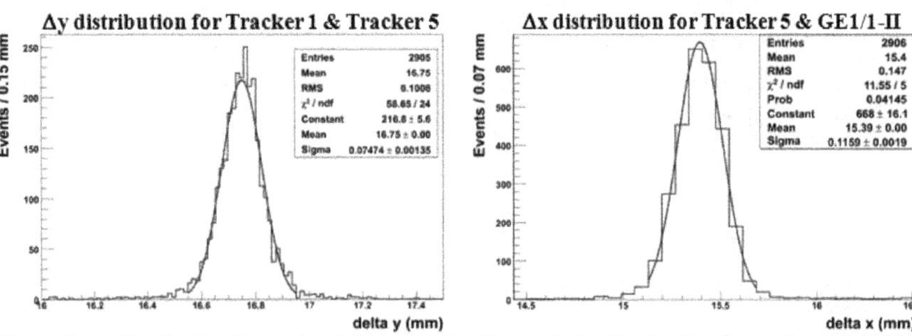

Figure 5. Δy distribution for tracker 1& tracker 5. Figure 6. Δx distribution for tracker 5 & GE1/1 II.

5. Summary

Fully operational GEM detectors 990 x 445 – 220 mm^2 have been designed and produced after long intense work on small size prototypes. By the test-beams at RD51 and CMS setup with small and full-scale prototypes we demonstrated that the candidate prototype is addressing all the requested requirements in terms of high efficiency and gain, stable safe and reliable operation at CMS-LHC environment.

6. References

1. CMS Collaboration, “The CMS experiment at the CERN LHC", JINST 3 (2008) S08004.
2. F. Sauli, “GEM: A new concept for electron amplification in gas detectors", Nucl. Instrum. Meth. A 386, 531 (1997).
3. D. Abbaneo et al. “Characterization of GEM detectors for application in the CMS Muon Detection System”, Nuclear Science Symposium Conference Record (NSS/MIC), 2010 IEEE proceeding, 10.1109/NSSMIC.2010.5874006
4. D. Abbaneo et al., “Construction of the first full-size GEM-based prototype for the CMS high-η muon system”, Nuclear Science Symposium Conference Record (NSS/MIC), 2010 IEEE, 10.1109/NSSMIC.2010.5874107
5. P. Aspell et al., “The VFAT production test platform for the TOTEM experiment"
6. http://nahandbook.web.cern.ch/nahandbook/default/h2/1%20General.htm
7. http://nahandbook.web.cern.ch/nahandbook/default/h4/1%20General.htm

Calorimetry

The ALICE Electromagnetic Calorimeters

Terry C. Awes, for the ALICE Collaboration
Oak Ridge National Laboratory,
Oak Ridge, TN 37831 USA
**E-mail: awestc@ornl.gov*
www.ornl.gov

ALICE is the general purpose experiment at the LHC dedicated to the study of heavy ion collisions. ALICE includes two different electromagnetic calorimeters: a high resolution, modest acceptance PHoton Spectrometer (PHOS) and a large acceptance, moderate resolution electromagnetic calorimeter (EMCal). The electromagnetic calorimeters are designed to trigger on high energy gamma-rays and jets, and to enhance the capabilities of ALICE for these measurements. The PHOS is a $PbWO_4$ crystal calorimeter while the EMCal is a Pb/Scintillator sampling shish-kebab type calorimeter. The PHOS and EMCal construction, readout, and performance are described.

Keywords: ALICE experiment, Heavy ion collisions, Electromagnetic calorimeter, Shower trigger, Jet trigger

1. Introduction

The ALICE[1] experiment at the LHC has been designed to carry out comprehensive measurements of nucleus-nucleus collisions with the goal to study matter under extreme conditions. In particular, a major goal is to explore the phase transition between confined matter and QCD matter[2,3] and characterize the properties of deconfined matter at high density and temperature. For this purpose, ALICE was constructed with a wide variety of detector subsystems for measurements of hadrons, leptons, and photons in the high multiplicity environment of heavy-ion collisions.

The Photon Spectrometer (PHOS)[4] is a high resolution electromagnetic calorimeter designed to provide excellent measurements of photons and neutral mesons up to moderate transverse momenta with primary objective to characterize the initial temperature of the produced matter via measurement of thermal gamma radiation. The EMCal[5] is a large acceptance electromagnetic calorimeter that was a late addition to ALICE and which complements the high resolution, but limited acceptance, of the PHOS electromagnetic calorimeter. The primary design goal of the EMCal was to improve the capabilities of the ALICE experiment for jet measurements. The main EMCal design criterion was to provide as much electromagnetic calorimetry coverage as possible within the constraints of the existing ALICE detector systems, with moderate energy resolution. The space available to the EMCal was limited to about 110° of azimuthal coverage.

2. The ALICE Electromagnetic Calorimeters

The physical parameters of the PHOS and EMCal are summarized in Table 1. The PHOS is constructed of individual $PbWO_4$ crystals assembled in strip units of 16 crystals (2×8) further assembled into supermodules of 56×64 crystals.[4,6] To increase the scintillation light output of the crystals the volume of the supermodule containing the crystals is maintained at an operating temperature of -25°±0.1°.Three PHOS supermodules have been installed with each supermodule subtending 20° in ϕ and ±0.13 in η. The small Moliere radius of $PbWO_4$ together with the installation of PHOS at a distance of 4.6m from the interaction point insures low hit occupancies even in the highest multiplicities of central Pb+Pb collisions.

Table 1. The physical parameters of the PHOS and EMCal electromagnetic calorimeters.

Quantity	PHOS Value	EMCal Value
Tower Size (cm^3)	2.2 × 2.2 × 18	6.0 × 6.0 × 24.6
Tower Size ($\Delta\phi \times \Delta\eta$)	0.0048 × 0.0048	0.0143 × 0.0143
Sampling Ratio	-	1.44 mm Pb / 1.76 mm Scint.
Number of Layers	-	77
Sampling Fraction	-	10.5
Radiation Length X_o (cm)	0.89	1.23
Moliere Radius R_M (cm)	2.0	3.20
Effective Density (g/cm^3)	8.28	5.68
Number of Radiation Lengths	20.2	20.1
Number of Towers	10752	11,520
Number of Modules	-	2880
Number of Supermodules	3	10
Total Coverage ($\Delta\phi \times \|\eta\|$)	60° × 0.13	100° × 0.7

The EMCal is a Pb/Scintillator sampling calorimeter of shish-kebab design, with wavelength-shifting (WLS) fiber readout.[5,7] The EMCal is constructed of 2880 identical individual modules. Each module consists of 77 layers of Pb and scintillator with the scintillators of each layer comprised of four individual optically isolated tiles, to provide four independent readout towers per module. Twelve modules are attached to a strongback to form a strip module, with 24 strip modules inserted into a crate to form a supermodule. A full supermodule subtends 20° in ϕ and 0.7 in η. Two supermodules installed end-to-end provide the full $-0.7 < \eta < 0.7$ acceptance of the ALICE TPC. The modules have a 1.5° taper in the η direction to provide an approximately projective geometry in η.

3. Calorimeter Readout

The PHOS and EMCal share the same readout system with minor differences. Both subsystems use 5×5 mm^2 Avalanche Photodiodes (APD) (Hamamatsu S8664-55 or Perkin Elmer C30739ECERH-1) as active photosensors with the same low noise JFET charge sensitive preamplifier (CSP). In the case of PHOS, the APD with CSP is attached to the crystal. For the EMCal, the APD and CSP are attached to the WLS fiber bundle at the back of the tower, within the strongback of the strip module. The APD signals are routed via cable to the Front End Electronics (FEE) boards located at the backside of the PHOS supermodules, or at the large η end of the EMCal supermodules. The readout and trigger electronics[8–13] make use of elements of the ALICE TPC readout electronics.[14] The readout is shown schematically in Figure 1. A single FEE card[8–10] provides readout of 32 towers. The signals are shaped with 100ns shaping time for EMCal (1 μs for PHOS) in dual shaper channels differing by a factor of 16 in gain, and sampled at 10MHz with the 10-bit ALICE TPC Readout (ALTRO) chip[14] for 14-bits effective dynamic range. The APDs are operated at a nominal gain of M=30 for EMCal (M=50 for PHOS) which gives a full scale range of 250 GeV (100 GeV), with a least significant bit of 16 MeV (6 MeV).

The FEE also provides individual bias control between 210 and 415 volts for each of the 32 APDs read out by the FEE. The setup, control, and readout of the FEE cards occurs over a custom GTL bus, developed for the ALICE TPC. The Readout and Control Unit (RCU) provides control of the GTL bus and carries a Detector Control System (DCS) daughter card which provides control via a linux operating system via an ethernet connection, and provides the interface to the LHC trigger and timing control system. The RCU carries a standard ALICE Detector Data Link (DDL) daughter card for transmission of data via optical fiber to the ALICE DAQ and High Level Trigger.

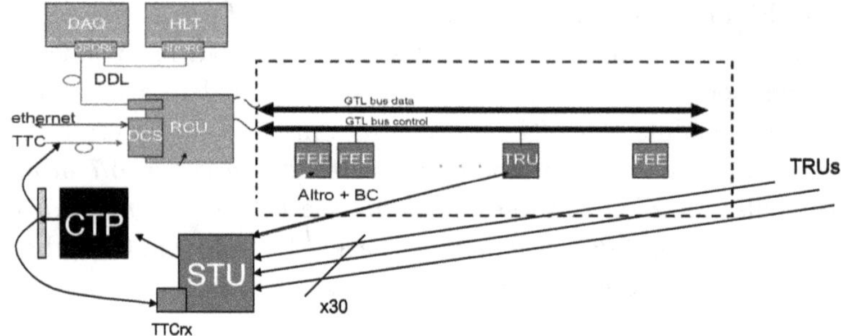

Fig. 1. Overview of the calorimeter readout and trigger electronics.

4. Calorimeter Triggers

The ALICE calorimeters provide input to the Level 0 (L0) and Level 1 (L1) trigger decisions in ALICE.[11,13] The signals of 2×2 groups of adjacent towers are analog summed in the FEE boards and transmitted to a local Trigger Region Unit (TRU)[12] board where the 2×2 tower sums of up to 14 FEE cards (112 2×2 sums) are digitized at the LHC clock frequency of 40 MHz. The digitized 2×2 tower sums are summed over time samples with pre-sample pedestal subtraction to provide an integral energy measurement. Finally, overlapping 4×4 tower digital sums are formed within each TRU. Each 4×4 sum is then compared against a threshold to provide an L0 trigger output that indicates the presence of a high energy shower. The L0 trigger decision from each TRU is passed to a Summary Trigger Unit (STU)[5] (or a Trigger Or module for PHOS[11]) which performs the logical OR of the L0 outputs from all TRUs to provide a single L0 input to the ALICE Central Trigger Processor (CTP) within 800 ns after the interaction.

Upon receipt of an accepted L0 trigger from the CTP, the digitized time-summed 2×2 tower sums from each TRU are passed to the STU over two LVDS serial data lines operating at 400Mbits/s. In the STU the 4×4 overlapping tower sums are formed again, but across TRU boundaries over the full acceptance to provide an improved L1 high energy shower trigger. At the same time, sums over large $N \times N$ overlapping tower regions are also formed to provide an L1 jet trigger. The STU simultaneously receives event multiplicity information from the ALICE V0 detector which can be used to provide multiplicity dependent (impact parameter dependent) thresholds for the L1 single-shower and jet energy sum triggers.

The L0 shower triggers were commissioned during the 2010 ALICE run and have been in operation throughout the 2011 run period. The L1 shower and L1 jet triggers of the EMCal are now used as main components of the mix of ALICE rare triggers.

5. Test Beam Performance

The PHOS and EMCal have been thoroughly studied in a series of beam tests at the CERN PS and SPS using beams of identified electrons and pions. The data have been used to extract information on the shower shape, electron/hadron rejection, non-linearity, and position and energy resolution for comparison with Monte Carlo simulations. The energy resolution of the PHOS[4,15] and EMCal[5,16] can be parameterized as:

$$\frac{\sigma(E)}{E}[\%] = \begin{cases} \frac{2.8}{\sqrt{E(\mathrm{GeV})}} \oplus 1.0 & \text{for PHOS} \\ \frac{10.5}{\sqrt{E(\mathrm{GeV})}} \oplus 1.9 & \text{for EMCal} \end{cases} \tag{1}$$

The measured dependence of the position resolution on deposited electromagnetic energy can be parameterized as:

$$\sigma(x)[\mathrm{mm}] = \begin{cases} \frac{2.6}{\sqrt{E(\mathrm{GeV})}} \oplus 0.3 & \text{for PHOS} \\ \frac{5.3}{\sqrt{E(\mathrm{GeV})}} \oplus 1.5 & \text{for EMCal} \end{cases} \tag{2}$$

These results highlight the superior performance of the high resolution PHOS detector.

6. Initial Calibration and Performance

Due the known temperature dependence of the APD gains (-1.7%/° C at M=30), it is essential to monitor and corrected for gain variations. As noted above, the PHOS detector is operated at -25°±0.1° C. In addition, an LED system with one LED per crystal is used to monitor the PHOS. The EMCal is operated at ambient temperature inside the ALICE magnet. The EMCal temperature is monitored across the supermodule, and in addition, the time dependence of the APD gain is monitored using a custom built LED system. The light from a single ultra-bright blue LED is transmitted to an EMCal strip module via a 3 mm optical fiber. At the strip module it is split into twelve 500 μm fibers that bring the light to a hole between the four towers at the back of each module. A small diffuser reflects the LED light back up into the WLS of each tower. The LED is triggered by an avalanche pulser system to provide an intense pulse of several ns duration.[8] The 24 LEDs are themselves monitored by 24 unit-gain photodiodes that are readout with an extra FEE card. The LED system for each supermodule is located at the end of the supermodule.

Since it was not practical to calibrate all towers with electrons in a test beam, the individual gain curves for each APD were determined and used to set the initial biases for each readout channel and equalize the gain for all towers (M=50 for PHOS, M=30 for EMCal). This allowed the equalization of the initial tower calibrations to better than ~20%. In the case of the EMCal, each supermodule was calibrated prior to installation using the peak of the energy deposit spectrum of cosmic muons traversing the calorimeter. This allowed one to obtain an initial tower relative energy calibration with *rms* variation of less than 10%. PHOS used first data from minimum bias p+p collisions to match the calibrations of individual towers by matching the shape of the single tower energy spectra, which provided a relative calibration variation of less than 7%.

The final relative tower-by-tower energy calibrations are obtained from measurements of the π^0 mass peak in the two-gamma invariant mass spectrum, where one of the showers is centered on the tower to be calibrated. This data has been obtained during the ALICE p+p datataking, and will continue to be taken and analyzed to improve the relative energy calibrations with the goal to obtain final relative calibrations of better than 1%.

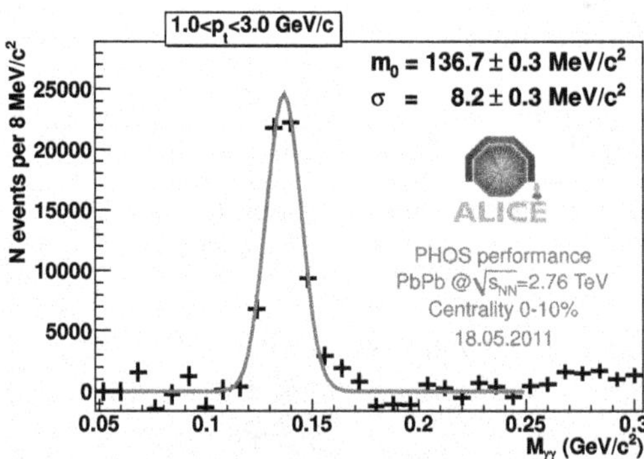

Fig. 2. Photon pair invariant mass peaks for 2.76 A TeV Pb+Pb collisions measured with PHOS (after mixed event background subtraction).

Figure 2 shows the photon pair invariant mass spectrum measured in PHOS for central 2.76 A TeV Pb+Pb collisions after mixed-event background subtraction demonstrating the feasibility of the π° measurement even at low transverse momenta in the highest occupancy environment.

7. Summary

ALICE includes two complementary electromagnetic calorimeters: a high resolution $PbWO_4$ calorimeter of limited acceptance optimized for photon, π^0, and electron measurements at low transverse momenta, and a large acceptance, moderate resolution sampling electromagnetic calorimeter, EMCal. By virtue of its significantly larger acceptance, the EMCal will extend the capabilities of ALICE for photon, π^0, and electron measurements to higher p_T, complementing the PHOS. The EMCal will also improve the capabilities of ALICE for jet measurements by measuring the neutral energy component of jets, and allowing one to trigger on jets. The calorimeters provide trigger input at L0 to trigger on single showers (photons, π^0's, and electrons) of moderate energy. The EMCal additionally provides an L1 trigger on high energy showers and on jets.

References

1. ALICE-Collaboration, *A Large Ion Collider Experiment*, tech. rep., CERN/LHCC-1995-71.
2. ALICE-Collaboration, *J. Phys. G.; Nucl. Part. Phys.* **30**, 1517 (2004).
3. ALICE-Collaboration, *J. Phys. G.; Nucl. Part. Phys.* **32**, 1295 (2006).
4. ALICE-Collaboration, *Technical Design Report of the Photon Spectrometer (PHOS)*, tech. rep., CERN/LHCC-1999-004.
5. ALICE-Collaboration, *Technical Design Report of the ALICE Electromagnetic Calorimeter*, tech. rep., CERN/LHCC-2008-014.
6. D. Aleksandrov *et al.*, *Nucl.Instrum.Meth.* **A550**, 169 (2005).
7. T. C. Awes, *Nucl.Instrum.Meth.* **A617**, 5 (2010).
8. H. Muller *et al.*, *Nucl.Instrum.Meth.* **A565**, 768 (2006).
9. Z.-B. Yin, H. Muller, R. Pimenta, D. Rohrich, Y. Sibiriak *et al.*, *Nucl.Instrum.Meth.* **A623**, 472 (2010).
10. Y.-P. Wang *et al.*, *Nucl.Instrum.Meth.* **A617**, 369 (2010).
11. H. Muller, R. Pimenta, L. Musa, Z. Yin, D. Rohrich *et al.*, *Nucl.Instrum.Meth.* **A518**, 525 (2004).
12. H. Muller, *Trigger Region Unit for the ALICE PHOS calorimeter*, tech. rep., CERN/LHCC-2005-038.
13. H. Muller, T. C. Awes, N. Novitzky, J. Kral, J. Rak *et al.*, *Nucl.Instrum.Meth.* **A617**, 344 (2010).
14. R. Esteve Bosch, A. Jimenez de Parga, B. Mota and L. Musa, *IEEE Trans.Nucl.Sci.* **50**, 2460 (2003).
15. G. Conesa, H. Delagrange, J. Diaz, M. Ippolitov, Y. Kharlov *et al.*, *Nucl.Instrum.Meth.* **A537**, 363 (2005).
16. J. Allen *et al.*, *Nucl.Instrum.Meth.* **A615**, 6 (2010).

STATUS OF THE ATLAS LIQUID ARGON CALORIMETER AND ITS PERFORMANCE

T. Barillari
Max-Planck-Institut für Physik, Föhringer Ring 6, 80805 München, Germany
E-mail: barilla@mppmu.mpg.de

On behalf of the ATLAS Collaboration

The liquid argon (LAr) calorimeter is used in ATLAS for all electromagnetic and for hadron calorimetry. The LAr calorimeter consists of an electromagnetic barrel calorimeter and two end-caps with electromagnetic, hadronic and forward calorimeters. The overall status of the LAr detectors operating over a long time period will be summarized in this note. Selected topics showing the performance of the LAr calorimeter using data from 2010 will be also shown.

1. Introduction

The ATLAS detector,[1] is a toroidal apparatus built to study the high energetic proton-proton (p-p) collisions at the CERN Large Hadron Collider (LHC),[2] and to provide excellent discovery potentials. The ATLAS detector covers almost the entire solid angle around the nominal interaction point and it consists of an inner tracking system that operates inside an axial magnetic field of 2 Tesla and covers the region $|\eta| < 2.5$ [a], an hybrid calorimeter system that expands up to $|\eta| < 4.9$, a large muon spectrometer which provides coverage out to $|\eta| < 2.7$, and specialized detectors in the forward region. In the region of $|\eta| < 1.7$ the hadronic calorimetry is provided by a steel-scintillator tile calorimeter. The remainder of the calorimetry uses liquid argon (LAr calorimeter) as the active medium. Installation of the LAr calorimeter in the ATLAS experimental hall was completed in early

[a]The coordinate system in the ATLAS detector uses as origin the nominal interaction point. The positive x axis is defined as pointing from the interaction point to the center of the LHC ring, the positive y axis is pointing upwards and the beam direction defines the z axis of a right handed coordinate system. The polar (θ) and azimuthal (ϕ) angles are measured with respect to this reference system. The pseudorapidity is $\eta = -\ln\tan(\theta/2)$, and the rapidity is $y = \frac{1}{2}\ln\frac{(E+p_z)}{(E-p_z)}$, where E denotes the energy and p_z is the component of the momentum along the beam direction

2008. The 20 months separating the completion of the installation from the first LHC collisions have been used to commission the LAr calorimeter. ATLAS started recording data in December 2009 at the center-of-mass energies $\sqrt{s} = 0.9\,\text{TeV}$ and $2.36\,\text{TeV}$. Throughout 2010 and 2011, the LHC delivered p-p collisions at $\sqrt{s} = 7\,\text{TeV}$. For a rapid understanding of the experimental signatures of the LHC collisions the LAr calorimeter plays a central role in many physics measurements. It is required to have a good performance in identifing electrons, photons and taus, to reconstruct jets, and to perform excellent missing transverse energy measurements. This note will summarize aspects of operating the LAr detectors over a time period of a few years and will focus on the performance of the LAr calorimeter using ATLAS collisions data from 2010.

2. The LAr Calorimeter and its Readout System

The LAr calorimeter is composed of sampling detectors housed in one barrel and two end-cap cryostats kept at about 87 K. It consists of a highly granular electromagnetic (EM) sampling calorimeter with accordion-shaped electrodes and lead absorbers, and contains a barrel part (EMB) and an end-cap part (EMEC). The EMB is segmented in depth into three samplings, the first being very finely segmented in η in order to provide very good position resolution. The EMEC consists of two concentric wheels, an outer and an inner wheel. A presampler (PS), consisting of an active LAr layer and installed directly in front of each EM calorimeter up to $|\eta| < 1.8$, provides a measurement of the energy lost upstream. Behind the EMEC is located a coarser hadronic end-cap calorimeter (HEC) which has a more conventional parallel-plate structure, with copper absorbers. Each HEC is composed of two wheels consisting each of 32 wedge-shaped modules. In the bore of the HEC is situated the forward calorimeter (FCal) made of copper/tungsten. It consists of three disk-shaped modules, one electromagnetic module and two hadronic modules. Some of the details of the LAr calorimeter are summarized in Table 1. All the LAr detectors correspond to a total of 182,468 readout cells, i.e. 97.2% of the full ATLAS calorimeter readout, of which 99.79% are up to today operational.

The signals coming from the calorimeters are processed by front-end boards (FEBs) that are located on the front-end crates (FECs) mounted on the cryostats directly on the cryogenic feedthroughs. These signals are then amplified (except for the HEC where the amplification is done inside the cryostats), split into three gain scales, and shaped with a bipolar shaping function in the FEBs. The shaped signals are sampled every 25 ns, with the

Table 1. Details of the LAr calorimeter sub-systems.

LAr system	Coverage	Absorber	Channels	Design Resolution (E in GeV)
EMB	$0 < \|\eta\| < 1.475$	Pb	109568	$\sigma(E)/E = 10\%\sqrt{E} \oplus 0.7\%$
EMEC	$1.375 < \|\eta\| < 3.2$	Pb	63744	$\sigma(E)/E = 10\%\sqrt{E} \oplus 0.7\%$
HEC	$1.5 < \|\eta\| < 3.2$	Cu	5632	$\sigma(E)/E = 50\%\sqrt{E} \oplus 3\%$
FCal	$3.1 < \|\eta\| < 4.9$	Cu/W	3524	$\sigma(E)/E = 100\%\sqrt{E} \oplus 7\%$

samples being initially stored in an analog pipeline until receipt of a Level-1 trigger signal, then are read out and transmitted via optical fibres to the back-end electronics located outside the detector cavern. Readout Drivers (RODs) processes the received samples using a digital filtering algorithm.[3]

3. LAr Calorimeter Operation

The first performance studies with the complete LAr calorimeter coverage have been done using cosmic muon data and LHC beam splash events from September 2008. These data have provided a check of the first level trigger energy computation, and the timing of the electronics. Temperature and purity stability checks have been done using these and also more recent data. In addition to ensure the stability and the good perfomance of each channel pedestal, gain and noise calibration runs are regularly taken at every LHC fill. The calibration constants needed for the digital filtering algorithm are also monitored over long periods. All the analyses performed so far have shown a robust operation of the LAr calorimeter: The measured values are all consistent with design values,[4] see Table 1. No extra contribution has been found to the global resolution constant term.

Despite the good LAr calorimeter performance, hardware problems have been encountered during the last years of operation. The most serious of these has been related to the low-voltage power supplies (LVPS) that provide the various low voltages required by the front-end electronics, and by the Optical Transmitter (OTx). The uncertainty on the long term performance of these LVPS has brought to the decision to build new supplies that will be installed in ATLAS during the long LHC shutdown, foreseen for 2013. The OTx devices also began failing in early 2008 and through 2009. The cause of this problem has been investigated but no certain explanation has been found. During the December 2010 shutdown all the problematic OTx devises have been replaced by spare OTx. There have been no more failures since then. In addition noise bursts have been also observed, mostly correlated with the presence of p-p collisions. The cause of these noise bursts is not really understood. In general the number of bursts stays low with constant luminosity. Noise bursts can affect a large number of channels, which are then masked and corrected for in data analysis.

4. LAr Calorimeter Performance Studies with Data

The LAr calorimeter performance has been assessed using the 2010 collisions data. In ATLAS the electromagnetic energy scale for the LAr calorimeters and the hadronic calibration schemes are based on extensive standalone and combined beam-tests. Corrections for the electromagnetic energy scale and resolution have been determined using the 2010 data.[5] It has been found that the electron identification efficiency integrated over the electron momentum (p_T) between 20 and 50 GeV and $|\eta| < 2.47$ varies between 94.7 and 80.7% depending from the electron selection criteria (medium/tight selection). In general the good understanding of an experimental measurement, i.e. $Z \to$ ee, is assessed by comparing the measured with the simulated distributions. The measured dielectron invariant mass distribution, m_{ee}, is shown in Fig.1 (left plot). The resolution obtained using a set of 2010 data is here only slightly worse than that predicted by the Monte Carlo (MC). The performance of the missing transverse

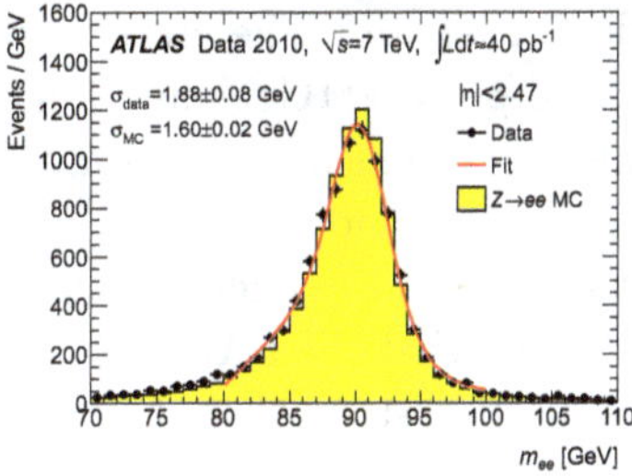

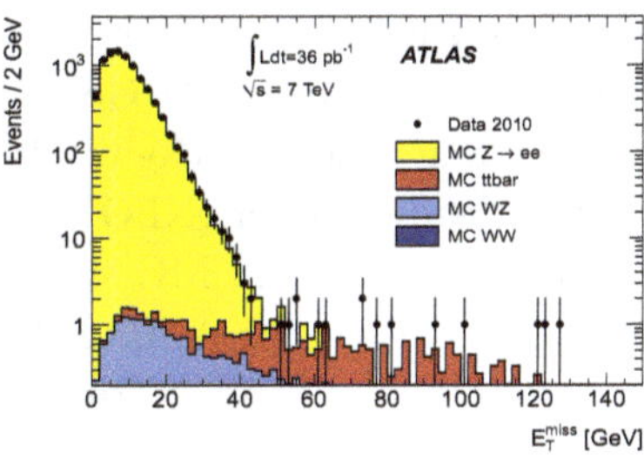

Fig. 1. Dielectron invariant mass m_{ee} distribution[5] for the central $Z \to$ ee analysis (left plot). Distribution of E_T^{miss} measured in a data sample of $Z \to$ ee events.[6] The expectation from MC simulation is superimposed and normalized to data (right plot).

energy (E_T^{miss}) in ATLAS for data and MC simulation is shown in Fig 1 (right plot) for selected $Z \to$ ee events.[6] The MC simulation expectations, from $Z \to ll$ events and from the dominant Standard Model (SM) backgrounds, are in this figure superimposed. Reasonable agreement between data and MC simulation is observed. The ATLAS calorimeter performance in measuring high energetic jets[7] is shown in Fig.2 where the inclusive jet multiplicity for jets with the momentum of the jet $p_{\mathrm{T}} > 30$ GeV and rapidity $|y| < 2.8$ is plotted on the left. The number of events decrease with

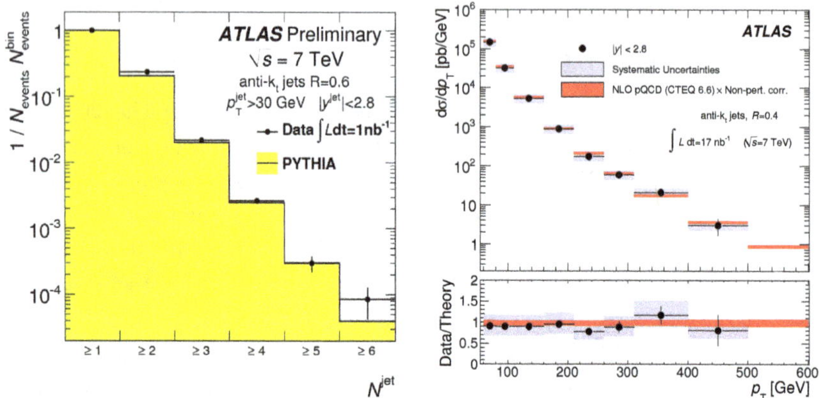

Fig. 2. Observed inclusive jet multiplicity (N^{jet}) distribution[7] (black dots) for jets compared to PYTHIA MC prediction (yellow histogram), where $N^{\text{bin}}_{\text{events}}$ denotes the number of events in a given jet multiplicity bin (left plot). Inclusive jet differential cross section[8] as a function of jet p_T integrated over the full region $|y| < 2.8$ (right plot).

increasing jet multiplicity and events with six and more jets in the final state are observed. The observed distribution is described well by the MC simulation. The differential inclusive jet cross section in 7 TeV p-p collisions is shown in Fig.2 on the right, as a function of jet p_T.[8] The cross section nicely extends from $p_T = 60\,\text{GeV}$ up to around $p_T = 500\,\text{GeV}$, and falls by more than four orders of magnitude over this range. The data are compared to next-to-leading order (NLO) pQCD calculations, and reasonable agreement between data and MC simulation is observed. In conclusion all the studies performed using ATLAS data have shown that the LAr calorimeter is performing well. The performances are well understood and close to the design expectation.

References

1. ATLAS Collaboration, *JINST* **3**, (2008).
2. L. Evans and P. Bryant, *LHC Machine JINST* **3**, (2008).
3. W.E. Cleland, E.G. Stern, *Nucl. Instrum. Methods* **A 338**, p. 467 (1994).
4. ATLAS Collaboration, *Eur. Phys. J.* **C 70**, p. 723 (2010).
5. ATLAS Collaboration, arXiv:1110.3174 [hep-ex].
6. ATLAS Collaboration, arXiv:1108.5602 [hep-ex], submitted to EPJC.
7. ATLAS Collaboration, *ATLAS-CONF-2010-043*, (2010), http://cdsweb.cern.ch/record/1277683.
8. ATLAS Collaboration, *Eur. Phys. J.* **C71**, p. 1512 (2011).

The TileCal/ATLAS calorimeter calibration systems

J. Carvalho
LIP-Coimbra, University of Coimbra, 3004-516 Coimbra, Portugal
E-mail: jcarlos@coimbra.lip.pt
(on behalf of the ATLAS Collaboration)

The calibration and monitoring systems of the ATLAS hadronic tile calorimeter (TileCal), are presented. Special attention is given to the experience gained so far with the results obtained on the analysis of the LHC collision data.

Keywords: hadron calorimetry; calibration; monitoring.

1. Introduction

The ATLAS central hadronic calorimeter[1] is a sampling calorimeter built from steel absorber plates and scintillator tiles. It consists of three cylinders along its axis, each divided in 64 wedges in ϕ, and it covers a pseudorapidity up to $|\eta| = 1.7$. Each tile is read out by two wavelength shifting fibers, coupled to photomultiplier tubes (PMT). The fibers are grouped in bundles forming projective towers pointing to the interaction region. The calorimeter is readout by 10k channels for 5k cells, of size $\Delta\eta \times \Delta\phi = 0.1 \times 0.1$ with twice coarser granularity in η for the outer of the three longitudinal layers.

Different systems are used to calibrate and monitor the calorimeter. The charge injection system is used to calibrate the relative response of the PMT signal readout electronics and to track any variation with time. The laser system is designed to calibrate and monitor the response of the PMTs, in particular the stability of their gains, their global linearity and for timing studies. A Cesium radioactive source system is designed to measure the quality of the optical response of each calorimeter cell, to equalize the signal response and to monitor it with time. Finally the current produced by minimum bias interactions, integrated over thousands of bunch crossings, is used to continuously monitor the calorimeter response during collision runs. This system is called the Minimum Bias (MB) calibration system. About 12% of the modules were previously calibrated in a testbeam with electrons, muons and pions.[2]

2. Calibration and monitoring of the TileCal

Each TileCal cell can be divided, for monitoring and calibration purposes, into: the optical part (the scintillating tiles and the wavelength shifting fibers); the PMT that converts and amplifies the optical signal; and the readout electronics that shape, amplify and digitize the signal. For each one of these parts a dedicated monitoring and calibration system was designed and built as described in the next sections. Figure 1 shows a diagram of the sequence of the TileCal optical and electronic readout, and the inputs of the calibration and monitoring systems.

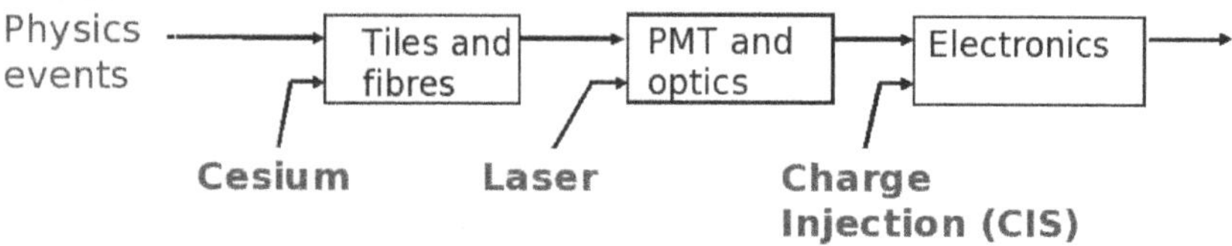

Fig. 1. Diagram of the optical and electronic readout, and of the calibration and monitoring systems.

2.1. *Charge Injection System*

The charge injection system (CIS) is designed to calibrate the relative response of the pulse readout electronics for all calorimeter PMTs and to monitor their variations with time. Each PMT channel has two analogue paths (bi-gain), the high and low gain (82 counts/pC and 1.3 counts/pC, respectively), digitized by a 10-bit ADC, covering a range of 800 pC (corresponding to a energy deposition of about 700 GeV).

Each channel is equipped with calibration capacitors, charged from a high precision voltage source and discharged into the input electronics. The measurement of the charge injected at the input stage of the electronic channels allows to obtain the CIS constants used to convert the signal from ADC counts to pC.[3]

In ATLAS there are periodic CIS runs scanning the full dynamic range during beam-off periods (between LHC fills and maintenance periods).

The variation of the calibration constants for individual ADC channels over the first half of 2011 data taking period has a RMS of about 0.1% and channels with a variation greater than 1% are tagged for recalibration.

2.2. *Laser system*

The laser system was designed to calibrate and monitor the PMT response. It is also used to check the PMT linearity, for studies on pulse saturation recovery and in calorimeter timing.

A laser is used to produce pulses with a wavelength of 532 nm and a 10 ns width, synchronized with the bunch-crossing clock. The light is split, with part of it being sent to a set of precise photodiodes for relative intensity monitoring, and the remaining being delivered to each calorimeter PMT.

The short and long term monitoring of the PMT gain stability is done in special calibration runs from the ratio of the charge measured by each PMT to the photodiodes response.[4]

The linearity of the PMT can be studied varying, in a well defined range, the intensity of the light delivered.

Figure 2 shows the result of a PMT high voltage (HV) scan to check the system sensitivity to the gain change, since the PMT response in this case scales linearly to the applied HV for small HV changes. The agreement between the measured and the expected gain variation is very good, showing that the gain monitoring sensitivity is adequate for the response follow-up in time.

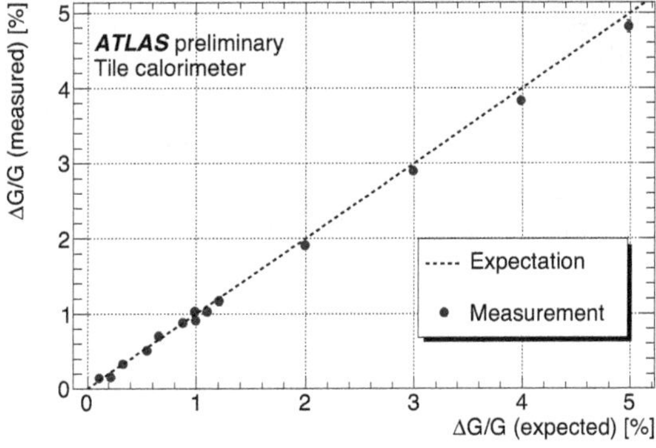

Fig. 2. Measured relative gain variation as a function of the expected relative gain variation, from a PMT high voltage (HV) scan.

2.3. *Cesium source system*

Cs calibration is designed to determine the quality of the optical response of each calorimeter cell, to adjust the PMT high voltage in order to equalize the response from all cells and to monitor it with time.[5] The system uses a ^{137}Cs γ source which moves perpendicularly to the tiles' surface through a hole in the scintillating tiles. This source is transported by a hydraulic system in a series of straight paths along the calorimeter modules, so it passes through holes in every single scintillating tile and absorber plate. The current from each PMT is measured with an integrator during the time of the passage of the source capsule through each cell.[6,7]

In ATLAS the Cesium system is used to monitor the long term stability of the calorimeter. It allows not only to monitor the PMT stability but also to detect bad tile–fiber couplings, scintillator aging and optical problems in general.

Figure 3 shows the TileCal response to the Cesium source as a function of time, in all four partitions (the activity of the three sources used are different). The response follows, in a reasonable way, the expected Cesium activity decay curve in the two years shown. The small deviation is due to the expected up-drift of the PMT gain during the initial phase of light exposure.

2.4. *Minimum bias events*

The minimum bias (MB) events in ATLAS are inelastic *pp* collisions with low momentum transfer, whose rate is proportional to the LHC luminosity.[7] These events produce a non-negligible occupancy of the TileCal cells, with rates uniform in the azimuthal angle ϕ and moderately dependent on the pseudo-rapidity η. Since the MB current, averaged over milliseconds, is almost constant and proportional to the interaction rate, it is used to monitor the calorimeter response and the relative luminosity during physics runs.

Figure 4 shows the average anode current for a TileCal cell as a function of the instantaneous luminosity, in 2010. The correlation is excellent, being linear within 0.5% for higher luminosity. The modulation on the deviation from linearity is due to systematic errors not yet taken into account on the measurement.

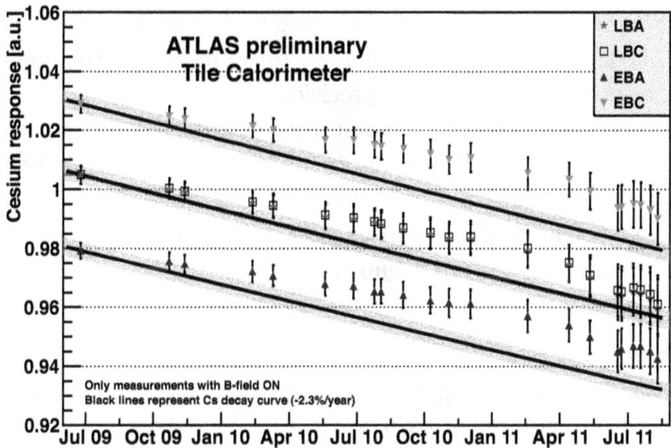

Fig. 3. TileCal response to the Cesium source as a function of time, in all the four partitions. The black lines represent the expected values considering the Cs decay lifetime (decrease of 2.3% per year).

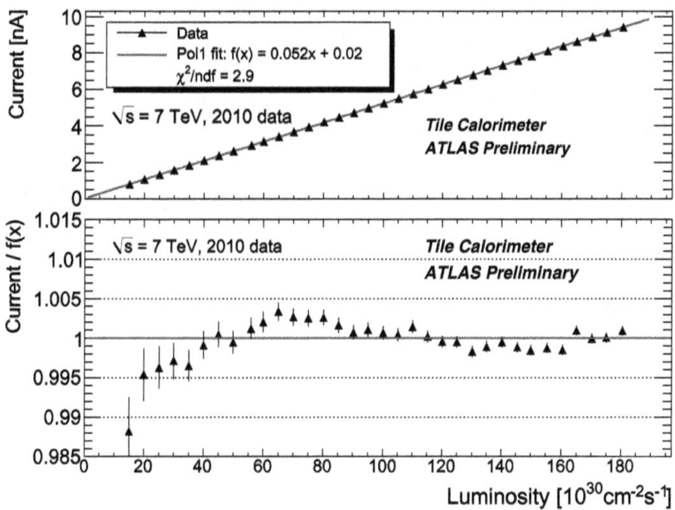

Fig. 4. Average anode current for a TileCal cell as a function of the instant luminosity, in 2011. The bottom plot shows the deviation from the linear behaviour.

3. Conclusions

The ATLAS hadronic calorimeter monitoring and calibration systems were presented. The CIS, laser and Cesium allow to calibrate and to monitor the calorimeter response with a 0.5–1% precision.

The different calibration and monitoring systems have been commissioned, and perform according to the design requirements while being exploited during data-taking and regular calibration runs. These systems are fundamental to reach the design parameters on the calorimetry energy resolution and energy scale linearity and stability.

References

1. ATLAS/Tile Calorimeter Collaboration "Tile Calorimeter - Technical Design Report" CERN/LHCC 96-42, 1996.
   The ATLAS Collaboration, G. Aad et al., "Readiness of the ATLAS Tile Calorimeter for LHC collisions" Eur. Phys. J. C71 (2011) 1193-1236.
2. P. Adragna et al. "Testbeam studies of production modules of the ATLAS Tile Calorimeter", Nucl. Instr. Meth. A606 (2009) 362-394.
3. K.J. Anderson et al. "Design of the Front-end Analog Electronics for the ATLAS Tile Calorimeter" Nucl. Instr. Meth. A551 (2005) 469-476.
4. V. Giangiobbe "The TileCal Laser calibration system" ATL-TILECAL-PROC-2011-007, CERN, 2011.
5. E. Starchenko et al. "Cesium monitoring system for ATLAS Tile Hadron Calorimeter" Nucl. Instr. Meth. A494 (2002) 381.
6. N. Shalanda et al. "Radioactive source control and electronics for the ATLAS Tile Calorimeter cesium calibration system", Nucl. Instr. an Meth. A508 (2003) 276.
7. G. Gonzalez Parra "Integrator based readout in Tile Calorimeter of the ATLAS experiment" ATL-TILECAL-PROC-2011-010, CERN, 2011.

CMS HADRON CALORIMETER OPERATIONS IN THE 2011 LHC RUN AND THE UPGRADE PLANS

PAWEL DE BARBARO

On behalf of the CMS Collaboration

Department of Physics and Astronomy, University of Rochester, Rochester, New York, 14627, USA

Email: Pawel.de.Barbaro@cern.ch

We have witnessed amazing performance of LHC accelerator in 2010 and its even more amazing performance in 2011. By early October 2011, LHC has delivered more than 4 fb^{-1} of integrated luminosity, with peak instantaneous luminosities reaching above $3*10^{33}$ cm^{-2}/s. In this note, we review CMS Hadron Calorimeter operations during the 2011 LHC run. In particular, we describe HCAL calibration methods and discuss main sources of calorimeter noise and development of noise filters. Finally, we present plans for HCAL upgrade..

Keywords: Calorimetry, LHC, CMS, Hybrid Photo Detectors

1. Introduction

CMS Hadron Calorimeter (HCAL) consists of four distinctive parts. In the central region (|eta| < 3), Hadron Barrel (HB) and Hadron Endcap (HE) calorimeters are sampling scintillator/brass absorber calorimeters and operate inside 3.8T CMS magnet. As a read-out they use Hybrid Photo-detectors (HPDs). Outside of the magnet, a Hadron Outer calorimeter is placed as a tail-catcher and it uses one or two scintillator layers read out by HPDs. In the forward direction (3 < |eta| < 5), Hadron Forward (HF) calorimeter uses steel absorber and Cherenkov light from scintillating quartz fibers read out with conventional photo-multipliers (PMTs) to detect signals[1].

Hybrid Photo-detectors[2] (HPDs) used as readout devices in HB, HE and HO calorimeters were designed to operate in high magnetic fields. They are proximity focusing devices with 3.5mm gap between photo-cathode and silicon diode, subdivided into 18 pixels. They operate at HV of approximately 8kV, with electric field direction aligned up with magnetic field lines of CMS magnet.

With gain of ~ 2000, they provide linear response over wide range of energies from minimum ionizing particles (muons) up to 3 TeV hadron showers. CMS HCAL calorimeters operated stably during 2010 and 2011 LHC runs with over 99% of channels alive.

2. HCAL Calibration methods

Initial calibration of HCAL was based on several complementary methods. Charge injection calibration (fC/ADC) conversion factors were measured during incoming quality control of front-end electronics Charge Integrate and Encode (QIE) boards. Co^{60} radioactive sources were used to provide initial values for relative energy calibration constants. Source measurements were performed after assembly of HCAL detector in the CMS surface assembly hall (SX5) in 2006 and 2007. Absolute energy scale of HCAL was determined on selected wedges using pion and electron test beams (H2 area at CERN). The energy scale was transferred over to all assembled wedges using radioactive sources.

Prior to LHC commissioning, CMS HCAL calibrations were further improved using cosmic ray muons. At the start-up of LHC operations in 2008 and 2009, additional energy calibration was performed using so-called 'splash' events. Such events were recorded during one of the stages of LHC commissioning, when proton bunches hitting closed collimators upstream of CMS would result in large flux of muons traversing detectors horizontally. In addition, analysis of 'splash' events provided HCAL timing calibration (synchronization at the hardware level of calorimeter trigger and signals with respect to other CMS triggers and readout clock).

In order to monitor long term stability of energy response of HPDs (in HB, HE and HO) and PMTs (in HF), HCAL uses LED calibration system. Data is collected during LHC inter-fill periods. In addition, stability of pedestals and signal synchronization is monitored using random and Laser triggers. This data is recorded during abort gap period within LHC orbit.

Starting in 2010, LHC proton-proton collisions data was used to further improve and validate energy and timing calibration constants. Two methods were employed to obtain relative response of HCAL channels at fixed detector eta ring and estimate response corrections to produce a uniform response in phi. One of the methods (variance method) was based on minimum bias events recorded in 'no Zero Suppression' mode, the other method (iterative method) was based on analysis of energy spectra in each HCAL channel using photon-triggered events. The relative response corrections did not change absolute energy scale of HCAL.

In 2011, LED gain monitoring was employed to correct for long term drifts in response of HPD pixels. Most of the HCAL Barrel and Endcap channels were stable at the level of 2-3% over the period of one year. However, approximately 4% of HB and HE channels exhibited gain drifts at the level of ~ 10% during running period from October 2010 to September 2011. Phi-symmetry calibration for 2011 data was extracted after applying time-dependent LED corrections. As shown on Figure 1, LED corrections compensate the gain drift and reduce the spread of 2011 phi-symmetry calibration close to the phi-symmetrized 2010 data. Note that conservative estimate on error of phi-symmetry calibration is 1.5%.

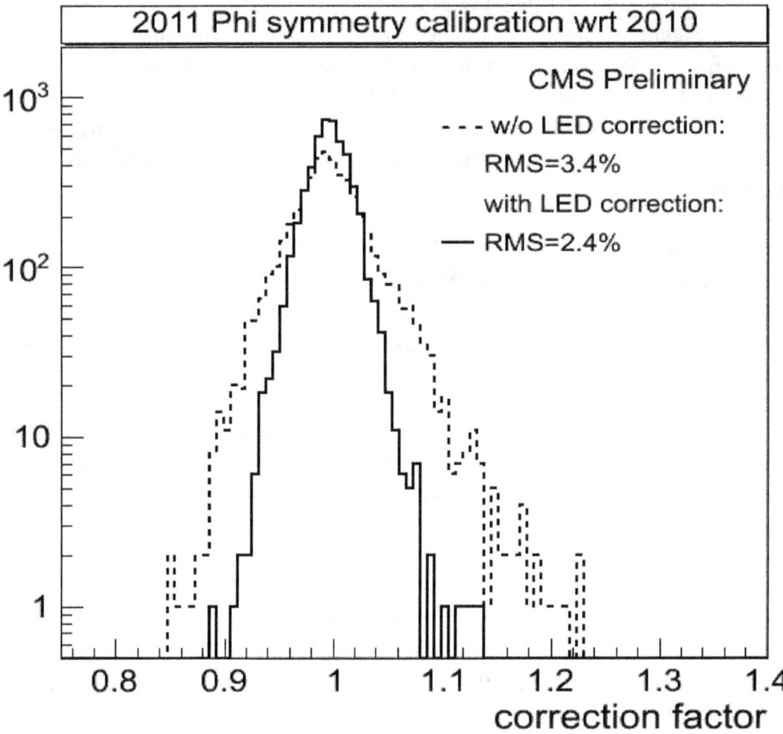

Figure 1. Gain corrections are based on monitoring of response of HPDs to LED signals. Final phi-symmetry corrections are calculated with LED gain corrections. Conservative estimate of error on phi-symmetry calibration is 1.5%. LED corrections compensate the gain drift and reduce the spread of 2011 phi-symmetry calibration close to the phi-symmetrized 2010 data.

In addition to phi-symmetry methods, isolated hadron calibration was used to extract eta-dependent absolute energy scale corrections. Dedicated trigger was used to select events with isolated tracks with p_T >38 GeV. These corrections could only be calculated for |eta| < 2.4. Further methods are being developed to obtain corrections for section of HCAL Endcap without tracker coverage. For HF, energy balancing in Z->ee events, with one electron depositing energy in Barrel or Endcap and the other one in HF was employed.

3. Calorimeter noise and noise filtering

The noise backgrounds in HB and HE are almost entirely dominated by photo-detector or electronics effects causing trigger to randomly overlap with collision physics triggers. Rates of noise generated triggers are independent of collision rates and overlap between noise and physics triggers is low (~ 1 per 10^5 triggers). However, at high ET or MET thresholds, trigger rates and noise contamination becomes an issue. Therefore, development of efficient noise filtering both at High Level Trigger (HLT) and at the reconstruction level is very important.

There are three main sources of noise in central HCAL sub-detectors (HB, HE, HO) which use HPDs as readout. In case of HPD ion feedback noise, it affects typically one or two HPD pixels at a time. Photoelectrons induce liberation of ions from silicon diode which accelerate across the HV gap and interact with the photocathode freeing additional photoelectrons. HPD discharge noise is caused by dielectric flashover from the wall (operating at 8kV). Such discharges can produce large signals in multiple pixels within same HPD (up to 18). Third source of noise can affect up to 72 channels of four HPDs located in a single Readout Box (RBX). It is possibly due to external noise coupling to the common HV line.

HPD and RBX noise produce distinctive patterns in HCAL. Filters have been developed making use of hit patterns, timing, pulse shape and electromagnetic fraction of calorimetric energy. In 2011 data taking, noise filtering is applied at two stages. Conservative filters are applied at HLT level to remove events identified as having HPD/RBX noise, but accept triggers for which noise overlaps with potential physics signal. Filter requires pulse shape consistent with signal and rejects pattern of energy deposits consistent with coming from HPD or RBX noise.

Additional noise filter algorithms were developed for offline reconstruction. HCAL reads out signal in ten 25 ns integration windows (Time Slices). The noise filter code looks at signal shape as a function of time and forms discriminator by comparing fit to a noise hypothesis with fit to the ideal pulse shape. Filter is used to flag (or remove) hits and remove event if many hits are flagged. Estimate of over-cleaning by offline filters is done by comparing MET before with respect to MET after cleaning was applied. Cleaned MET is almost always lower that RAW MET, with over-cleaning rate of 0.01%.

It is important to note that efficiency (and over-efficiency) of HCAL offline noise filters is sensitive to potential synchronization problems of front-end clocks. In addition, as LHC continues to increase its instantaneous luminosity, elevated pile-up levels in HCAL lead to gradual reduction of performance of HCAL noise filters, requiring further development. Finally, potential operation of LHC with 25ns bunch crossing in 2012 would also introduce additional challenges to noise filtering in HCAL.

4. Upgrade plans

HCAL upgrade activities up to end of LHC 2nd Long Shutdown (LS2), planned to 2018/19 will address present known HCAL issues: HPD discharges in HB, HE and HO, as well as beam-induced backgrounds in HF photomultipliers (PMTs). Replacement of HO HPDs with Silicon Photomultipiers (SiPMs) is scheduled for 2013/14 LHC shutdown (LS1). At the same time, present, thick window HF PMTs (Hamamatsu R7725) will be replaced with thin window, multi-anode PMTs, (Hamamatsu R7600U-200-M4).

Replacement of HB and HE HPDs with SiPM readout is scheduled for LS2. Front-end (FE) and back-end (BE) upgrades, including FE/BE communication are scheduled for 2018/19 as well. Thus, the planned upgrades, in addition to reducing HCAL noise, it will improve HCAL performance by allowing for increased longitudinal and transverse segmentations and introduction of hardware timing.

5. Conclusion

CMS HCAL worked stably during first two years of LHC operation. With more than 99% of channels life, HCAL contributed to excellent performance of entire CMS detector. Coming years will be challenging for HCAL, as LHC will reach its nominal instantaneous luminosity. HCAL upgrades planned till the end of this decade will allow HCAL to improve its performance.

References

1. HF is the topic of another contribution to this conference and will not be discussed in this paper.
2. P. Cushman et al., NIM A 504 (2003) 502

OVERVIEW OF THE CMS CALORIMETRY

A. GHEZZI* on behalf of the CMS collaboration

Physics Department, University of Milano-Bicocca,
Milano, Italy
**E-mail: alessio.ghezzi@mib.infn.it*

The calorimeter system of the CMS detector is made of a high precision fine-grained electromagnetic calorimeter (ECAL) composed of lead tungstate crystals and a sampling hadronic calorimeter (HCAL) consisting of plates of brass absorbers and scintillator tiles with wavelength shifting fibers. A pre-shower detector composed of lead layers interleaved with silicon strips is installed in front of the ECAL End-cap crystals, and a forward calorimeter, made up of quartz fibers embedded within steel absorber, extend the coverage in the pseudorapidity region up to $|\eta| < 5$. The commissioning of the detector and the performance of the calorimeters during runs at the LHC with pp collisions will be reviewed. The resulting performance in the reconstruction of jets, electrons and photons will be briefly presented.

1. Introduction

The Compact Muon Solenoid (CMS) detector[1] is a general purpose detector installed at the CERN Large Hadron Collider (LHC).

The calorimeter system of CMS is designed to achieve an excellent energy resolution the invariant mass of di-photons, essential for the detection of the Higgs boson through its electromagnetic decay ($H \to \gamma\gamma$), and to provide good measurement of central jets. Hermeticity and coverage up to large pseudorapidity are required for a good measurement of the missing transverse energy (missing E_T).

After construction and commissioning, CMS started to collect data from pp collisions at $\sqrt{s} = 14$ TeV at the end of March 2010. In the following, after a description of the calorimeter system of CMS, the results of the commissioning and the performances obtained with the data collected in 2010 will be reported.

2. Description of the calorimeter system

The calorimeter system of CMS must be able to survive the hostile radiation environment of LHC and must be able to operate in the 3.8 T magnetic field of the CMS solenoid. These considerations were crucial in the design of the detector.

The electromagnetic calorimeter (ECAL) is an homogeneous calorimeter built of lead tungstate ($PbWO_4$) scintillating crystals, which are chosen for their radiation hardness, for the short radiation length and Moliere radius and for their fast response. ECAL is divided into a barrel and two endcaps. The barrel covers the pseudorapidity region $|\eta| < 1.479$ and consists of 61200 channels with a granularity $\Delta\eta \times \Delta\phi = 0.0187 \times 0.0187$. The scintillation light is read out by avalanche photo diodes (APD). The two endcaps cover the region $1.479 < |\eta| < 3$ and consist of 14648 crystals read out by vacuum photo-triodes (VPT), with a granularity that varies with η from 0.021×0.021 to 0.050×0.050. A pre-shower detector, made of two lead absorber and two silicon strip layers, is installed in front of the endcaps. ECAL is designed to achieve an excellent energy resolution with a constant term below 0.5% for the measurement of the energy of photons that do not interact in the tracker material. Variations in the crystal or APD temperature, or changes in the crystal transparency due to irradiation, directly contribute to the constant term. The specifications require a thermal stability better than 0.05 C (0.1 C) in the barrel (endcaps), which is obtained by an extensive feedback system employing about 7000 thermistors. The change in the crystal transparency is measured by injecting laser or LED light into each crystals.[2]

The hadron calorimeter (HCAL) is made of layers of brass absorber and plastic scintillator tiles (BC408 produced by BiCron). The light is collected by wavelength shifter fibers and read out by hybrid photo-diodes (HPD). HCAL is divided into a barrel covering the pseudorapidity region $|\eta| < 1.4$ with a granularity $\Delta\eta \times \Delta\phi = 0.087 \times 0.087$, two endcaps, covering the region $1.3 < |\eta| < 3.0$, and a forward calorimeter extending the coverage up to $|\eta| < 5$. In the barrel layers of scintillators are installed outside the solenoid to catch the tails of the shower of high energy hadrons. Due to the high radiation level in the forward region a different design has been chosen for the forward calorimeter: it is made of a steel absorber with embedded quartz fibers where Cerenkov light is produced by the particles in the shower. Fibers of two different length are used in order to distinguish electromagnetic and hadronic showers from the difference in the longitudinal profile of their showers.

3. Commissioning before LHC collisions

In the years before the collisions the performance of the detector were extensively studied by exposing modules of the detector to hadrons and electron beams. ECAL was thus able to study in detail the response of the detector and to demonstrate that the energy resolution and the performance of the laser monitoring system are within the design requirements.[3] Different techniques were used to obtain the calibration of the relative channel-to-channel response before the LHC collisions:[4] one quarter of the barrel was calibrated with an electron beam with high precision (0.3%), the light yield and the photodetector gain for each channel were measured in the laboratory before the construction, all the barrel modules were exposed to cosmic ray muon in a dedicated setup before final assembly in CMS, events from the beam dump in collimators 150 m upstream the detector were used to locally uniform the response in the endcaps. Combining the different methods the precision of the calibration varies with η from 1% to 2% for the barrel channels that were not calibrated with the electron beam, and it is about 5% for the endcaps.
The combined ECAL and HCAL is a non compensating calorimeter. The non-linearity and energy resolution of the system has been measured by studying the response of the systems to pion beams with momenta ranging from 2 to 350 GeV/c.[5] The relative calibration of the HCAL channels is obtained by measuring the response of the scintillator tiles to γ rays from a ^{60}Co source.

After the final installation of the detector, long data-taking exercise with cosmic ray muons were conducted, with the goal of commissioning the experiment for extended operation.[6] They represented a valuable opportunity to test and debug the full trigger and data acquisition chain and to test the integration and stability of the system. The response to cosmic muon in the HCAL was also employed to improve the relative calibration of the detector reaching a precision at the start-up of about 5% in the barrel, better than 10% in the endcap and about 12% in the forward calorimeter.

4. Results and performance with pp collisions

In the data taking with the first collisions the stability of the detector in its final configuration could be measured and measurements of the basic detector performances were doable already with few nb^{-1}s of integrated luminosity. A good agreement is observed in the basic detector observables between the measurement in data and the prediction of a detailed Monte-Carlo simulation of the apparatus,[7],[8] indicating a good understanding of the calorimeters already at the start-up. As examples the distributions of the energy spectrum in the ECAL barrel and endcaps are reported in figure 1 and the ratio of calorimeter energy to the hadron track momentum is reported in figure 2. In both cases a good agreement between data and MonteCarlo can be seen.

The data from the pp collisions can be employed to calibrate the detector. Symmetry of the energy deposition along the azimuthal angle (ϕ-symmetry) can be exploited to calibrate the relative response of channels placed at the same pseudorapidity. The mass of the π^0 reconstructed from photon pairs can be employed to calibrate the ECAL detector and the peak position of the Z boson decaying into two electrons can be used to set the energy scale. The response to isolated tracks can be used to calibrate HCAL along η and to set the energy scale. The details on the calibration procedure and are reported in separate contributions in these proceedings[9].[10]

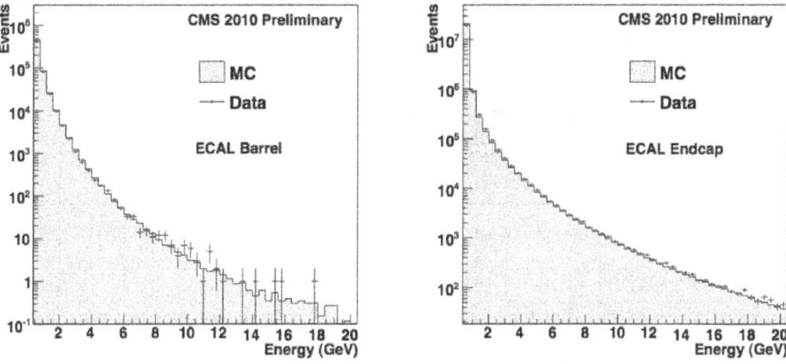

Fig. 1. Energy distribution of the most energetic channel in the barrel (left) and in the endcaps (right).

4.1. *Reconstruction of electrons and photons*

The reconstruction of electrons and photons starts from the reconstruction of their energy deposit in ECAL. The emission of brehmstrallung photons by the electrons and the photon conversion in the tracker material produce

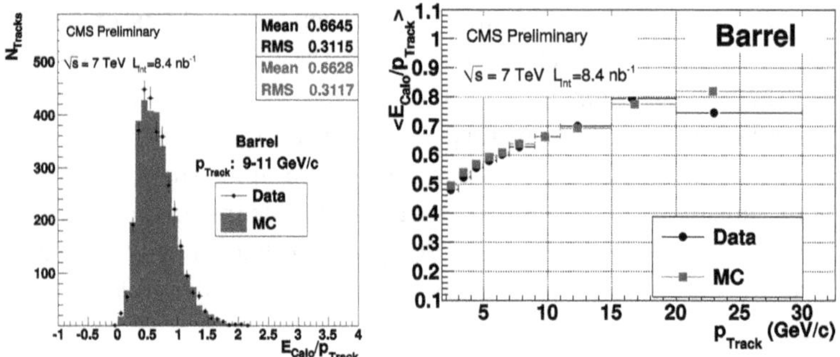

Fig. 2. Left: Distribution of the ratio of the calorimeter energy over the track momentum (E_{calo}/P) for tracks with momentum. Right: Average E_{calo}/P as a function of the track momentum.

a spread of the energy reaching the calorimeter. A clustering procedure have been developed in order to collect it. The sample of reconstructed clusters is dominated by QCD-induced background and the purity of the photon sample can be increased by applying selections based on the shape of the cluster and on its isolation, both from other energy deposits in the calorimeters and from tracks reconstructed in the tracker. The identification of electron exploits also variables based on the matching between the electron track and its cluster in the ECAL.

The good agreement between data and MonteCarlo on the key variables at the different steps validates the algorithms for the reconstruction and identification of electrons[11] and photons.[12]

4.2. *Reconstruction of jets and missing transverse energy*

Different approaches can be adopted in the jet reconstruction depending on the way the individual contributions from subdetectors are combined: calorimeter jets are reconstructed using only the information from energy deposits in the calorimeter, particle-flow jets are based on the particles reconstructed and identified in the event by a particle-flow algorithm[13] that combines the information from all relevant CMS sub-detectors. The reconstructed jet energy is corrected for the $\eta - p_T$ dependence of the detector response as predicted in the MonteCarlo. Residual corrections and the energy scale are determined from data, using balance in di-jets and Z/γ+jet events.[14] The jet energy resolution will depend on the η and p_T of the jet and can be measured by studying the distribution of the asymmetry in di-jet events.[14] The resolution as a function of the jet p_T for jets in the central region is reported in figure 3, both for the calorimeter and particle-flow jets, showing a significant improvement in the resolution from the particle-flow approach.

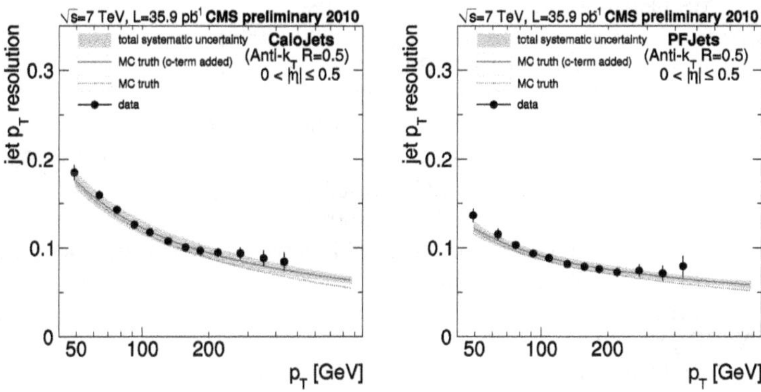

Fig. 3. Jet energy resolution as a function of the jet p_T. Left:calorimeter jets, right: particle-flow jets.

Similarly to the jet reconstruction, also the missing transverse energy can be reconstructed either using only the calorimeter information or with a particle-flow approach. The performance in the missing E_T reconstruction can be evaluated with collision data: di-jet events can be used to measure the resolution on the x-y component of the missing E_T, and events with a reconstructed photon or Z boson can be used to study the component of the recoil orthogonal and parallel to the boson transverse momentum.[15] In both case the particle-flow approach significantly improves the performances of the missing E_T reconstruction.[15]

5. Conclusions

The calorimeter system of CMS is fully operative and well understood since the beginning of data taking at the LHC thanks to the careful design, the detailed test beam studies and the long commissioning phase. The reconstruction of high level objects involving the calorimeters (electrons, photons, jets, missing E_T) is well understood and validated with respect to the expectations from the MonteCarlo simulation of the detector at each step of the reconstruction. The reconstruction of jets and missing E_T based on a particle-flow approach brings significant improvements in the resolution compared to reconstructions based only on calorimetric information.

References

1. The CMS Collaboration, 2008 JINST 3 S08004
2. M. Anfreville et al., NIM A, 594, pp 292-320 (2008)
3. P. Adzic et al., JINST 2 (2007) P04004
4. P. Adzic et al., JINST 3 (2008) P10007
5. CMS HCAL/ECAL Collaboration, Eur. Phys. J. C (2009) 60: 359373
6. The CMS Collaboration, JINST 5 (2010) T03010, JINST 5 (2010) T03012,
7. The CMS Collaboration, PAS EGM-10-002
8. The CMS Collaboration, PAS JME-10-008
9. R. Paramatti, "Performance of the CMS Electromagnetic Calorimeter at the LHC", 13th ICATPP proceedings
10. P. DeBarbaro, "Performance and Upgrade plans of CMS HCAL", 13th ICATPP proceedings
11. The CMS Collaboration, PAS EGM-10-004 (2010)
12. The CMS Collaboration, PAS EGM-10-005 (2010)
13. The CMS Collaboration, PAS PFT-09-001(2009), PAS PFT-10-002 (2010)
14. The CMS Collaboration, CMS-JME-10-011 (2011)
15. The CMS Collaboration, CMS-JME-10-009 (2011)

Performance of the CMS-Forward (HF) Calorimeter in the 2010 LHC Run and the Upgrade Plans

E. Gülmez
On behalf of the CMS Collaboration
Physics Department, Bogazici University,
Bebek, Istanbul, 34342 TURKEY
E-mail: gulmez@boun.edu.tr

Performance of the CMS-Forward (HF) calorimeter during the 2010 data taking period at the LHC will be briefly discussed. Some details about the PMT window hit events, mostly by muons, will be given. Results of the tests done on the new 4-anode PMTs that are planned to be installed during the 2013 shutdown period will be shown. The overall plans for the upgrade will be mentioned.

Keywords: Calorimetry, LHC, CMS, Forward Calorimetry, PMT, Multi-anode PMT, Cherenkov radiation

1. Introduction

The CMS-HF Calorimeter was designed[1] to detect the forward particles over a pseudorapidity (η) range of 3 to 5. Such a detector will optimize the detection of those processes that produce forward jets, especially processes involving heavy Higgs and SUSY particles. An added benefit is to improve the determination of the missing transverse energy. The two cylindrical HF units are placed at each end of the detector at the beam height. The HF is a sampling calorimeter with plastic clad quartz fibers embedded into the iron absorber. It has an active radius of 1.4 m. Each unit is 1.65 m long and composed of 18 slices of 20-degree sections. Long (1.65 m) and short fibers (1.43 m) in the calorimeter sample the energy in the showers. Light produced in each fiber through Cerenkov radiation goes into its own PMT. Phototubes used in the HF have been selected and tested at the University of Iowa PMT test station,[2,3] where the new 4-anode PMTS are also being tested.

At the beginning of 2010, the HF Calorimeter was ready for data taking. It was placed in the garage position during the shutdown after the 2009 initial startup of the LHC. Data taken during the beam circulation tests in 2009 were used for adjusting the timing offsets so that the signals would fall in the second time slice of 25 ns. Calibration constants that were determined with the help of the single photo electron peak and LED measurements were also checked using the collision data.

During 2010, the anomalous events that had been seen in the HF calorimeter during test beam runs continued to be a problem. These events were mostly due to the particles hitting the PMT windows and producing Cherenkov radiation,[4] thereby creating large false energy signals (Figure 1-left). One of the suggestions to eliminate these false signals is to discard the data that do not show any correlation in the long and short fibers.[5] In principle, long and short fibers should measure correlated energies. However, using the correlations in the long and short fibers eliminates some of the anomalous events after the data acquisition, but it does not prevent the signals to be produced in the first place. Because of these spurious signals, jet energy estimates done in the trigger calculations turn out to be erroneous and produce extra trigger signals.

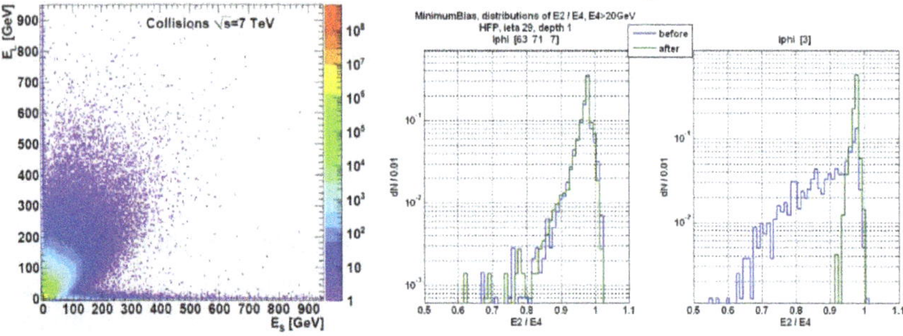

Fig. 1. Energy measurement in long fibers versus short fibers (left).[6] Events along the axes are due to the particles hitting the PMT windows directly, i.e. anomalous signals. E2/E4 ratio for the reference (middle and right)[7] and tyvek (right) sleeve before (blue) and after (green) the sleeve replacement. E2(E4) is the integrated signal over two (four) time slices.

In addition to these so called anomalous signals, it was observed that there were longer signals in some channels due to the scintillation occurring in the light guide reflective material and the sleeves (Figure 1- middle and right). These were also affecting the jet energy measurements. Measurements obtained from the two new types of sleeve material, tyvek and teflon, showed that the tyvek material had the lowest scintillation effect

(Figure 1-right). During the technical shutdown in January 2011, all the sleeves were replaced with the new tyvek sleeves.

To improve the signal to noise ratio, two time slices are now used in the new versions of the CMSSW software. This means a better alignment in time for the HF channels is needed so that the signals from the two time slices could be used to determine the energy. More phase scans for the HF were done at the beginning of the 2011 run period for the time alignment and to understand the noise contribution to the L1 jet triggers.

Even though the correlations between the long and short fibers seem to give us a good way of eliminating most of the anomalous signals, but not all of them. Hence, it does not completely eliminate the erroneous trigger signals. Also, with the increased luminosity, even the correlation method is useless since many towers are occupied with real signal. The solution has to reduce these signals at the source such as replacing all the PMTs with thinner window 4-anode PMTs. Thinner window will produce less Cherenkov scintillation in the window and the multianode readout will enable the use of online algorithms to eliminate such window signals. The real Cherenkov light coming from the calorimeter towers will produce similar signals in all four anodes. This will not be the case for the Cherenkov light produced in the window.

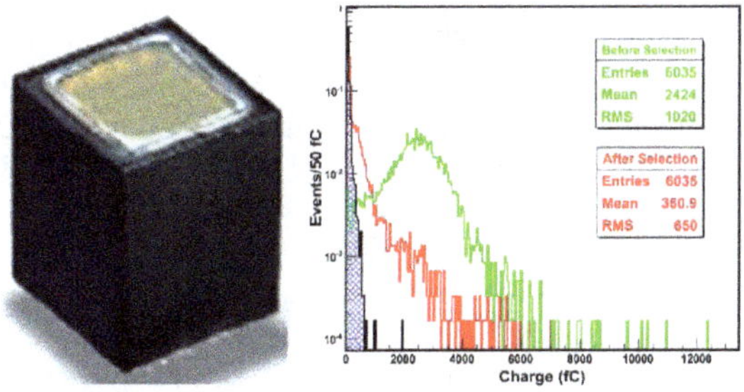

Fig. 2. A picture of the 4-anode square PMT, Hamamatsu 7600 (left). Reduction of the unwanted window events after selecting those signals with high correlation between the channels (right-red distribution). Window events will produce signals in some of the channels but not all of them (uncorrelated) as seen in the event distribution before the selection (green).

Hamamatsu 4-anode PMTs (R7600) are chosen to be the replacement PMTs for the HF (Figure 2-left). Initial tests done with these PMTs in the test beam at Fermilab and CERN show that the Cherenkov light produced by particles hitting the PMT windows could be distinguished easily and clearly from the real signals (Figure 2- right).[8] During the 2013 shutdown, new thin-window high-efficiency 4-anode PMTs will be installed instead of the current PMTs to reduce the anomalous signals to negligible levels. These new PMTs are being tested at the Iowa PMT test station prior to the installation, as they are delivered by Hamamatsu.

To conclude, the HF detectors performed very well during the 2010 run period. In 2013, all the existing PMTs will be replaced with the 4-anode PMTs.

Acknowledgements

Valuable comments and helpful suggestions of B. Bilki, Y.Onel, A. Penzo, and T. Yetkin are greatly appreciated. This work was partially supported by Bogazici University Scientific and Technological Research Fund (#5883).

References

1. "Design, Performance and Calibration of the CMS Forward Calorimeter Wedges," CMS HCAL Collaboration, Eur. Phys. J. C **53 (2008) 139-166.**
2. "Selection and Testing of 2000 Photomultiplier Tubes for the CMS-HF Forward Calorimeter," E. Gülmez, U. Akgun, A. S. Ayan, P. Bruecken, F. Duru, A. Mestvirishvili, M. Miller, J. Olson, Y. Onel, I. Schmidt, E. W. Anderson, and D. Winn, *Proceedings of IMTC 2004 Instrumentation and Measurement Technology Conference, Como, Italy, May 18-20, 2004, (2004)1870.*
3. "Complete tests of 2000 Hamamatsu R7525HA phototubes for the CMS-HF Forward Calorimeter, U. Akgun, A. S. Ayan, P. Bruecken, F. Duru, E. Gülmez, A. Mestvirishvili, M. Miller, J. Olson, Y. Onel, I. Schmidt, Nucl. Instr. And Meth. **A550, 145(2005).**
4. "A Study of Anomalous Events in CMS-HF PMTs," A. Halu, E. Gülmez, M. Deliomeroglu, CMS CR-2008/107 and the Proceedings of the International Conference on High Energy Physics In Memoriam Engin Arik and Her Colleagues, Oct. 27-31, 2008, Istanbul, Turkey, Balkan Physics Lett., **17,** 17021, 138-141(2009).
5. "Beam Test Results for the Anomalous Large Energy Events Removal in Hadronic Forward Calorimeter," U. Akgun, E. A. Albayrak, B. Bilki, K. Cankocak, W. Clarida, P. Debbins, A. Mestvirishvili, A. Moeller, Y. Onel, I. Schmidt, J. Wetzel, T. Yetkin M. K. Carleton, A. R. Clough, P. D. Lawson N. Sonmez D. R. Winn J. Freeman A. Penzo, F. Ozok, CMS DN-2009/005.
6. T. Yetkin, private communication.
7. V. Epshteyn, P. de Barbaro and T. Yetkin, private communication.
8. B. Bilki, private communication.

LIQUID ARGON CALORIMETER PERFORMANCE AT HIGH RATES

VICTOR KUKHTIN

Joint Institute for Nuclear Research
Joliot-Curie 6, 141980 Dubna Russian Federation
E-mail: kukhtin@sunse.jinr.ru

on behalf of the HiLum ATLAS Liquid Argon Endcap collaboration

The performance of the ATLAS liquid argon endcap and forward calorimeters has been projected at the planned high luminosity LHC option HL-LHC by exposing small calorimeter modules of the electromagnetic, hadronic, and forward calorimeters to high intensity proton beams at IHEP/Protvino accelerator. The results of HV current and of pulse shape analysis, and also the dependence of signal amplitude on beam intensity are presented.

Keywords: LHC, LH-LHC, LAr calorimetry, space charge effect

1. Introduction

The increase of instantaneous and integral luminosity by a factor of up to 5 with respect to the nominal LHC value has serious consequences for the signal reconstruction, operation and radiation hardness requirements of the electromagnetic endcap (EMEC) [1], hadronic endcap (HEC) [2], and forward (FCal) [3] calorimeters. Small modules of each calorimeter type have been build , placed in separate cryostats , and exposed to high intensity proton beams at IHEP. By placing absorber elements in-between the cryostats the ratios of particle and energy fluxes have been adapted to the ATLAS situation. In liquid argon calorimeters at high enough ionization rates space-charge effects, instigated by the build-up of the slowly drifting positive argon ions, can degrade the signal. The impact of the space-charge effects on the calorimeter signal and on the high voltage (HV) current are discussed.

2. Set-up, beam and beam instrumentation

The general set-up of the three calorimeter modules in the beam (see Fig.1) has been optimized based on the results of the detailed Monte Carlo studies.

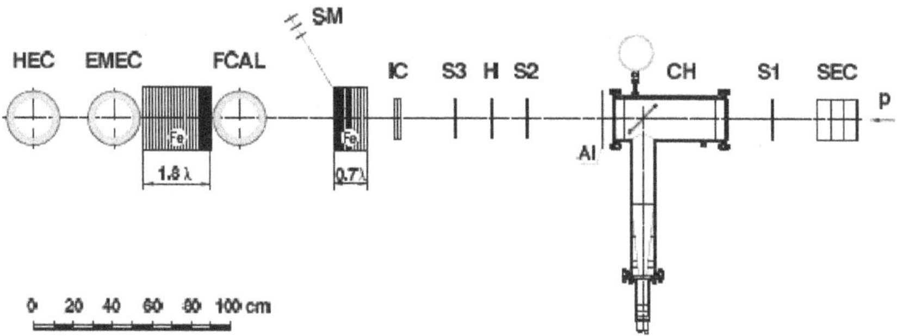

Figure 1. Schematic top view of the overall test beam set-up. Shown are the three cryostats with the FCal, EMEC and HEC module and the related Fe absorbers as well as the beam instrumentation: the scintillation counters S1,S2,S3 and the scintillation counter hodoscope H. The beam intensity is monitored with the secondary emission chamber SEC, The Cerenkov counter CH, the ionization chamber IC, the scintillation counter monitor SM and the aluminium foil Al.

The proton beam of 50 GeV is extracted from U-70 synchrotron via channeling with bent crystal technique [4]. The accelerator has been operated with only 5 out of 30 bunches filled. The spacing between filled bunches almost 1 μs. It allows the study of the high flux response of the calorimeters not affected by pile-up from previous bunches. Using the maximum deposited energy density from MC minimum bias events in ATLAS for endcap calorimeters and the corresponding numbers for IHEP beam setup for one incident proton the correspondence between LHC luminosity and beam intensity was found for each modules. The uncertainty in the beam intensity numbers is at least at the level of a factor two.

3. Calorimeter modules

A replica of a small section of electromagnetic module of the ATLAS FCal (see Fig.2) contains two groups of 16 electrodes each. One group has 119 μm liquid argon gaps as proposed for the upgraded ATLAS FCal at the HL-LHC and another one – 269 μm gaps as it is now in ATLAS.

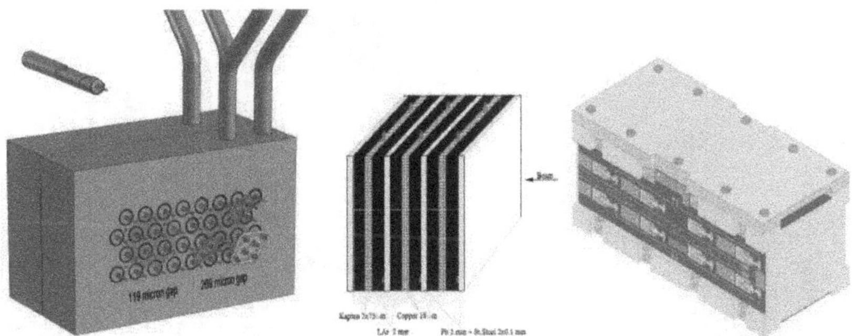

Figure 2.Schematic view of calorimeter modules installed each in separate cryostat (from the left to the right) – FCal with two different liquid argon gap sizes, EMEC and HEC .

Either half of the FCal module can be centered on the beam to study each option individually.

The EMEC and HEC modules are both of planar structure (rather than the accordion geometry for the EMEC) with the argon gap width 2 mm. The positive HV up to 2 kV was applied to both modules.

The read-out electronics is rather close to that used in the early ATLAS HEC test beam tests at CERN. But instead of operating preamplifiers in cold as for HEC in ATLAS, warm preamplifiers were mounted on the front-end board located outside the cryostat. The total number of read-out channel for all three modules was equal to 16.

4. Results of HV current analysis

Charged particles, passing through the LAr gaps of the calorimeter and producing drifting ions and electrons, induce also a DC current in the HV system. This current is correlated with the average beam intensity, resp. the luminosity at LHC or HL-LHC. An additional to the standard HV supply system external measurement device [5] has been constructed. The logging rate of this device is at 10 Hz per channel and the precision of the time stamp is 10 ms. It is connected either to the FCal or EMEC HV supply system. Electronic noise limits the effective resolution to about 25 nA. Shown at Fig.3 FCal HV current behaviour as function of proton beam intensity was fitted by the straight line. The non-linear contribution was estimated to be less that 0.36% at the beam intensity corresponding to the nominal LHC luminosity.

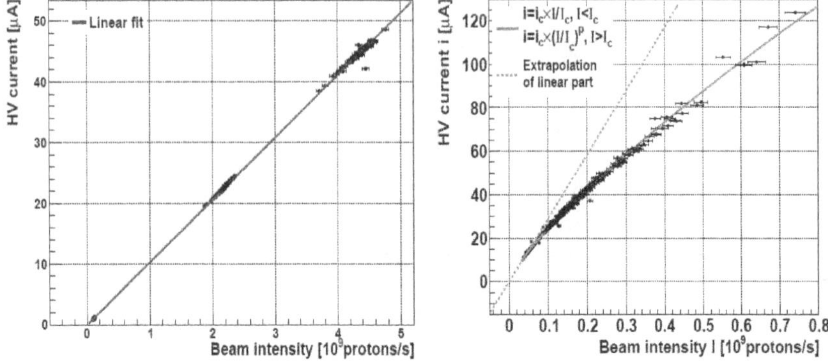

Figure 3. Total FCal (on the left) and EMEC (on the right) HV current as function of the beam intensity. The beam measurements were done by ionization chamber for FCal and Cherenkov counter for EMEC.

With the larger liquid argon gap it is expected that the linear relation between the beam intensity and HV current changes to a power dependence as it is seen for the EMEC on Fig.3 .

5. Dependence of the pulse shape and the signal amplitude on beam intensity

To compare the signals, data were taken at each beam intensity varying the HV setting for calorimeter modules. The difference in signal length due to the variation of the electron drift velocity with the electric field is found at low intensities for the modules with larger LAr gaps. The further reduction of the signal length for intensities above the critical value follows closely the expectation from the space-charge simulation.

The dependence of the signal amplitude on the beam intensity has been studied for all modules. For EMEC and FCal-100 modules it is shown at Fig. 4.

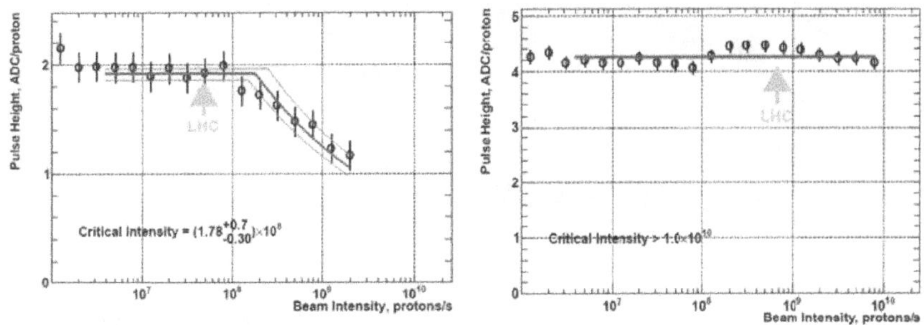

Figure 4 Dependence of pulse height of HEC (on the left) and FCal (119) (on the right) modules on beam intensity

A model for small bulk recombination rate constant assumes constant response below the critical intensity and a falling power dependence above. The data for the 119 μm LAr gap FCal module do not show any dependence on beam intensity

6. Conclusions

The observed pulse shapes follow closely the prediction of simulation. The space-charge effects are visible in the signal amplitude for the calorimeter modules with larger LAr gap size but they are absent for 119 μm gap width. The dependence of the HV current on the beam intensity is in agreement with the expectations.

References

1. M. Aleksa et al., Construction, assembly and tests of the ATLAS electromagnetic end-cap calorimeters, JINST **3** (2008) P06002..
2. D.M. Gingrich et al., Construction, Assembly and Testing of the ATLAS Hadronic End-Cap Calorimeter, JINST **2** (2007) P05005.
3. A.Artamonov et al., The ATLAS Forward Calorimeter, JINST 3 (2008) P02010.
4. V.I.Kotov et al., Application of bent crystals at IHEP 70 GeV accelerator to enhance the efficiency of its usage, Proc. of EPAC 2000, Vienna, Austria, pp.364-366.
5. A.Afonin et al., Relative luminosity measurement of the LHC with the ATLAS forward calorimeter, JINST 5 (2010) P05005.

THE ATLAS FORWARD CALORIMETER

ROBERT S. ORR
Department of Physics, University of Toronto, Toronto, Ontario M5S 1A7, Canada
Email: Orr@physics.utoronto.ca

On Behalf of the ATLAS LIQUID ARGON Collaboration

Complete calorimetric hermeticity is important for many physics studies at the LHC. In order to extend the pseudorapidity coverage up to 4.9 units, ATLAS uses a liquid argon forward calorimeter integrated into the endcap cryostat. In this region only a modest stochastic term is required in the energy resolution. The main challenge is survivability and reliability in this hostile environment. We discuss the development of the forward calorimeter from construction to its utility in physics studies.

1. Introduction

1.1. *Motivation for FCal Design*

Full solid angle detector coverage is important for general purpose collider detectors such as ATLAS. Much discovery physics depends on the ability to accurately measure missing transverse energy (E_T^{miss}), and, in practical terms, this can only achieved by 4π calorimeter coverage. Among the new physics areas where E_T^{miss} is important are the search for non-interacting particles resulting from the decay of SUSY particles. 4π calorimeter coverage is also important in forward jet tagging, which is central to, for example, vector boson fusion production of the Higgs.

The ATLAS forward calorimeter (FCal) covers the pseudorapidity range $3.1 < |\eta| < 4.9$. Since the FCal is situated so close to the circulating beams, the jets seen in this detector tend to be of high energy. This results in a rather modest requirement on the stochastic term in the energy resolution, the effective energy resolution being dominated by the constant term. The energy resolution measured[1], for single test beam particles, was parameterized as $\sigma_E/E = a/\sqrt{E} \oplus b$; $a = (28.5 \pm 1.0)\% \cdot \sqrt{GeV}$, $b = (3.5 \pm 0.1)\%$ was measured for electrons and $a = (94.2 \pm 1.6)\% \cdot \sqrt{GeV}$, $b = (7.5 \pm 0.4)\%$ for pions, between 20 GeV and 200 GeV.

1.2. *FCal Technology*

The main challenge is designing the FCal[2] was to ensure that it would function reliably in the extremely hostile environment close to the LHC beams. In order to have some measurement of longitudinal shower development, the FCal is divided into three sections. In all three sections the construction is such that only radiation hard materials are used, and there are no adhesives used in joints. All three modules use a novel electrode structure. This consists of copper tubes parallel to the beam axis, which contain electrode rods. In the FCal1 the electrode rods and the calorimeter matrix are copper. This serves to ensure a good thermal conductivity close the shower maximum, and avoids local heating, and perhaps boiling of the liquid argon. In order to have an extremely dense detector, the FCal2 and FCal3 modules have tungsten electrode rods, and the matrix consists of small sintered tungsten slugs. The liquid argon gap between the electrode rods, and the copper electrodes tubes is maintained by a spiral of radiation hard PEEK plastic in all three modules. The general arrangement of the FCal , and the electrode details are given in Fig.1. The disposition of the FCal in the ATLAS detector may be found in [3].

Layer	Absorber	LAr gap	$N_{\text{electrodes}}$	N_{channels}
FCal1	Cu	269 μm	24,520	2,016
FCal2	W	376 μm	20,400	1,000
FCal3	W	508 μm	16,448	508

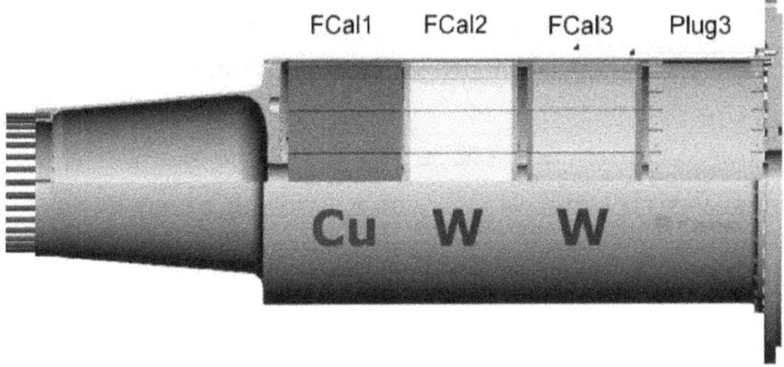

Figure 1. The general arrangement of the FCal, and details of the electrode spacing.

2. FCal Performance

Various approaches have been employed in setting the absolute energy scale of the ATLAS calorimeter system. The electromagnetic energy scale was initially set using the test beam data, and this was then optimized using $Z \to e^+e^-$.

2.1. *In-situ Pseudorapity Intercalibration of Jet Energy Scale.*

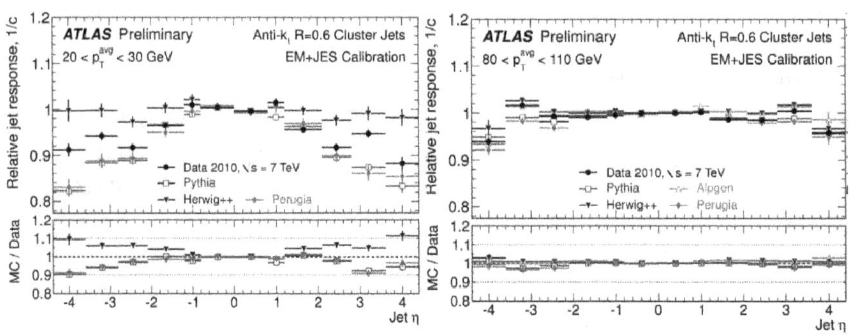

Figure 2. Relative jet response as a function of η in two typical p_T bins.

Due to the different technologies employed in the different regions of the calorimeter system, we have used an *in situ* intercalibration method [4] to ensure a uniform jet energy scale. The method adopted is a tag-and-probe approach, based on requiring a p_T balance between a reference jet in the reference region $|\eta| < 0.8$, and a probe jet. The p_T balance is characterized by an asymmetry $A = \left(p_T^{probe} - p_T^{ref}\right)/p_T^{avg}$, with $p_T^{avg} = \left(p_T^{probe} + p_T^{ref}\right)/2$. The asymmetry is then used to measure an η dependent inter-calibration factor, via the relation

$$\frac{p_T^{probe}}{p_T^{ref}} = \frac{2+A}{2-A} = \frac{1}{c}.$$

In fact, the analysis is in bins of jet η and p_T, yielding a binned correction

$$c_{ik} = \frac{2-\langle A_{ik}\rangle}{2+\langle A_{ik}\rangle}.$$

The relative jet response after correction is shown in Fig. 2 as a function of η for a low and a high p_T bin. The Monte Carlos indicate that the response of the FCal is well understood for high p_T. For lower p_T there is some disagreement in the Monte Carlos on the treatment of soft effects. This results in more uncertainty in our knowledge of the response of the FCal.

2.2. *Jets in the Forward calorimeter.*

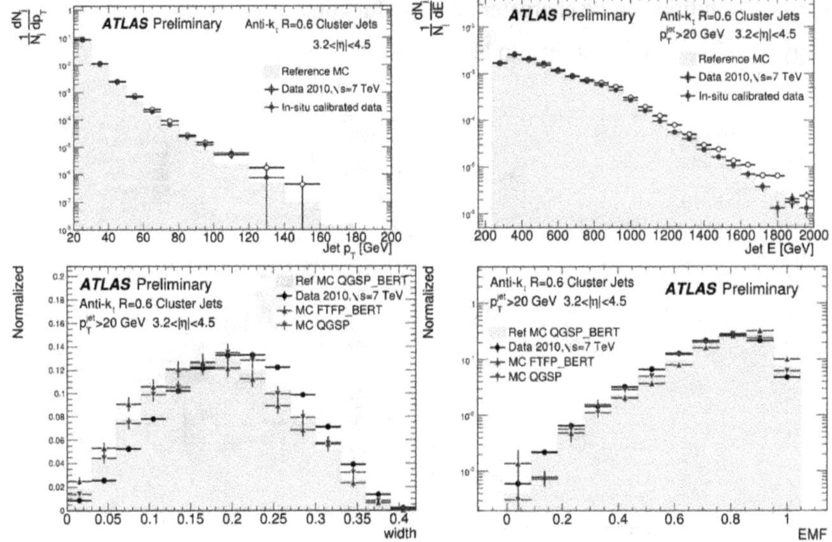

Figure 3. Measured characteristics of jets in the pseudo-rapidity region of the forward calorimeter.

The tag and probe method also allowed us to investigate the characteristics of jets in the region of the forward calorimeter. These are shown in Fig. 3. EMF is the fraction of the jet energy deposited in FCal1, while the jet width is defined by the transverse energy weighted average distance in $\eta \times \phi$ space of the constituent clusters from the jet axis.

2.3. *Physics Object Based Study of* E_T^{miss} .

E_T^{miss} is the magnitude of the vector quantity $\overline{E_T}^{miss}$, the missing transverse momentum. Due to the importance of this quantity in discovery physics, we have performed [5] a detailed analysis of the performance of the calorimeter system using full event reconstruction and an energy calibration specific to the reconstructed physics objets. In this approach the components of E_T^{miss}, are given by,

$$E_{x(y)}^{miss,calo} = E_{x(y)}^{miss,e} + E_{x(y)}^{miss,\gamma} + E_{x(y)}^{miss,\tau} + E_{x(y)}^{miss,jets} + E_{x(y)}^{miss,softjets} + \left(E_{x(y)}^{miss,calo,\mu} \right) + E_{x(y)}^{miss,CellOut}$$

, in a self-explanatory notation. For isolated muons, the term $E_{x(y)}^{miss,calo,\mu}$ is not added in order to avoid double counting. In studying the calorimeter objects we have used both minimum bias events, and QCD di-jets. The performance of the calorimeter system is shown in left panel of Fig. 4. This figure includes studies performed using $Z \to l^+l^-$, as described in [5]. One expects that the E_T^{miss} resolution can be characterized as a function of the total transverse energy, $\sum E_T$, and this is seen to be the case.

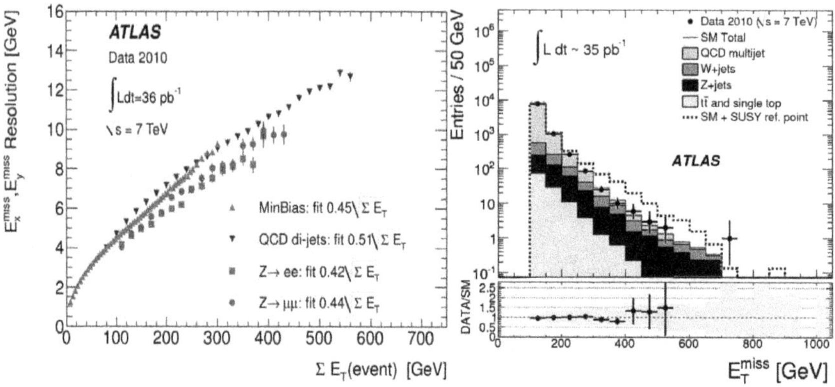

Figure 4. An object based study of the resolution in E_T^{miss} (left). A typical SUSY signal as it would show up in the E_T^{miss} distribution (right)

3. FCal In Physics Analysis.

3.1. *SUSY Analysis using E_T^{miss} .*

As already mentioned, resolution in E_T^{miss} is an important requirement for any physics analysis which depends on the detection of non-interacting particles. As an illustrative example we show one figure from a recent ATLAS SUSY search [6] based on the signature of jets and E_T^{miss} in Fig. 4

3.2. *FCal Used to Measure Centrality in Pb-Pb Collisions .*

While the FCal was designed to ensure full pseudo-rapidity coverage in E_T^{miss} , it has also proven of utility in characterizing the centrality of heavy ion collisions[7]. A peripheral heavy ion collision is one in which the ions just have a grazing collision. These collisions involve only one, or a small number of, nucleons in each ion. Of more interest are central collisions, where the ions collide " head-on" and many nucleons are involved. Here one expects optimum conditions for the formation of a hot, dense, medium, as depicted in the left panel in Fig. 5. Several decades ago, Bjorken [8] suggested that jets might lose energy when traversing this medium. This phenomenon is known as "jet quenching". The observation in ATLAS of events such as that depicted in the

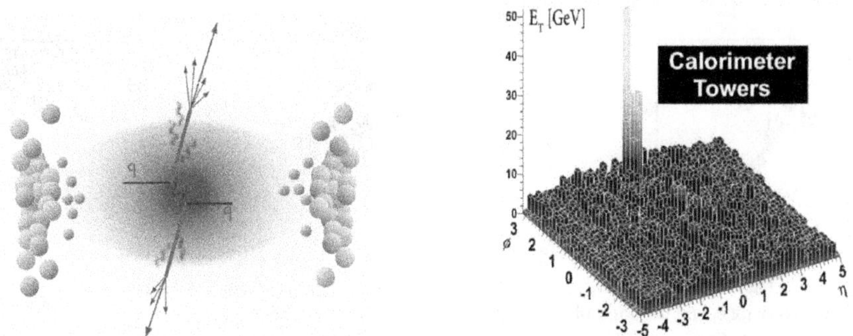

Figure 5. The panel on the left depicts the formation of a hot, dense medium in heavy ion collisions. The panel on the right is a typical candidate for a central collision in Pb-Pb collisions at ATLAS.

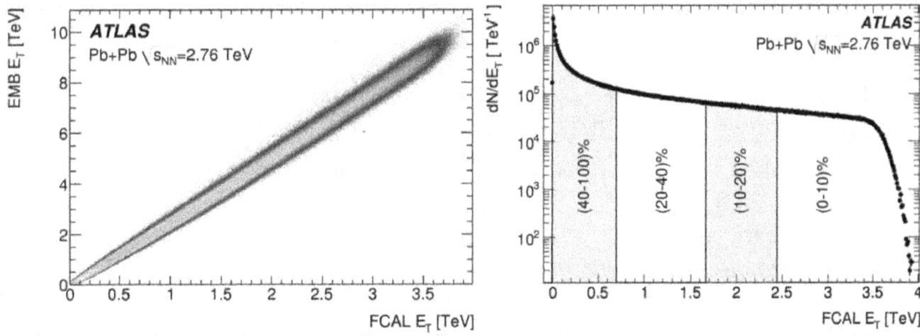

Figure 6. The panel at left shows how the transverse energy measured in the central region is strongly correlated with the total transverse energy measured in the FCal. The panel on the right indicates how the FCal transverse energy is used to characterize the fraction of the total cross section, and hence the centrality of events.

right panel of Fig. 5, was strongly suggestive of this phenomenon. To search for this phenomenon, we defined the jet transverse energy asymmetry as $A_J = (E_{T1} - E_{T2})/(E_{T1} + E_{T2})$. The issue was to find a way of characterizing the centrality of events in a way that did not employ the jets being used in the asymmetry measurement. In Fig. 6 the two panels show, on the left, the correlation between the transverse energy of events measured in the central region and the total transverse energy in the FCal. On the right is shown how the total FCal transverse energy is used to characterize the cross section fraction, and hence the centrality, of the events. In Fig.7 the asymmetry in di-jet transverse energy is plotted for four bins of FCal total transverse energy. The black points are the data, open circles are proton-proton collisions, and the solid histogram is a Monte Carlo with no jet quenching. It is apparent that the di-jet

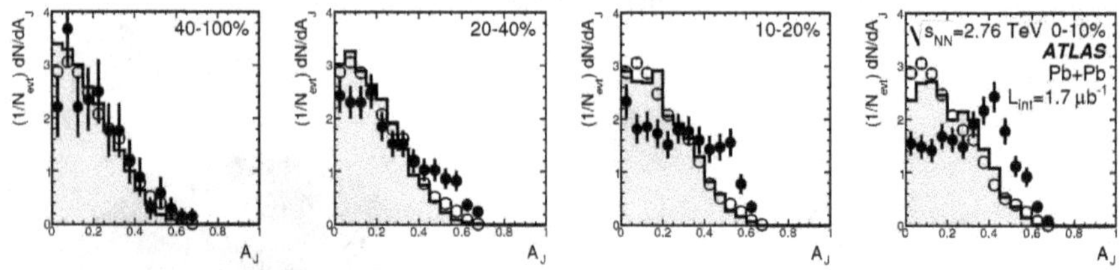

Figure 7. The measured di-jet asymmetry The black points are the data, open circles are proton-proton collisions, and the solid histogram is a Monte Carlo with no jet quenching.

asymmetry in the data increases with increasing centrality, as expected for jets with a greater path length through the hot, dense medium.

References

1. J.P Archambault *et al., Energy Calibration of the ATLAS Liquid Argon Forward Calorimeter,* JINST 3 P02002, 2008.
2. A. Artamonov *et al., The ATLAS Forward Calorimeters,* JINST 3 P02010, 2008.
3. T. Barillari, *Status of the ATLAS Liquid Argon Calorimeter and its Performance after One Year of Operation,* these proceedings.
4. G. Aad *et al., In-situ pseudorapidity intercalibrationfor the evaluation of jet energy scale uncertainty using dijet events in proton-proton collisions at* $\sqrt{s} = 7\ TeV$, ATLAS-CONF-2011-014
5. G. Aad *et al., Performance of Missing Transverse Momentum Reconstruction in Proton-Proton Collisions at* $\sqrt{s} = 7\ TeV$ *with ATLAS,* CERN-PH-EP-2011-114, submitted to Eur. Phys. J. C. August 2011
6. G. Aad *et al., Search for squarks and gluinos using final states with jets and missing transverse momentum with the ATLAS detector in* $\sqrt{s} = 7\ TeV$ *proton-proton collisions.* Phys. Lett. B 701 (2011) 186-203
7. G. Aad *et al., Observation of a Centrality-Dependent Dijet Asymmetry in Lead-Lead Collisions at* $\sqrt{s_{NN}} = 2.76\ TeV$ *with the ATLAS Detector at the LHC.* Phys. Rev. Lett. 105, 252303. (2010)
8. J.D. Bjorken, FERMILAB-PUB-82-059-THY (1982)

Upgrade Plans for ATLAS Forward Calorimetry for the HL–LHC

Koloina Randrianarivony*, on behalf of the ATLAS Liquid Argon Calorimeter Group

Carleton University,
Ottawa, ON K1S5B6, Canada
**E-mail: krandria@physics.carleton.ca*

An overview of the ATLAS Forward Calorimeter system is provided. Upgrade plans for this system for the proposed High–Luminosity upgrade of the LHC are discussed.

Keywords: ATLAS, Calorimetry, Forward, FCal, LAr, HL–LHC.

1. The ATLAS Forward Calorimeters

The ATLAS detector[1] is a multi-purpose detector located at the Large Hadron Collider (LHC[2]). The ATLAS forward calorimeters (FCals) are situated inside the endcap cryostats together with the electromagnetic endcap and the hadronic endcap calorimeters. The FCals cover the forward regions of $3.1< |\eta| <4.9$, where the energies and density of particles are very high. Each FCal consists of three disk-shaped modules located around the beam pipe. The active part of the detectors are arrays of thin annular gaps filled with Liquid Argon (LAr) that allow operation in the high-flux region. The absorber material is copper for the FCal1 electromagnetic calorimeter module and tungsten for the two hadronic modules, FCal2 and FCal3.

2. Forward Calorimeter High–Luminosity Upgrade Plans

The LHC is designed to run with an instantaneous luminosity of 10^{34} $cm^{-2}s^{-1}$. A High Luminosity (HL–LHC) upgrade is planned for the machine with the goal of providing a peak luminosity of 5×10^{34} $cm^{-2}s^{-1}$ and an integrated luminosity of 250 fb^{-1} per year.

HL–LHC luminosities will provide challenges for the operation of the FCal. Charge build up in the gaps can distort the electric field particularly in the region of highest particle flux at small radii close to the beam line.

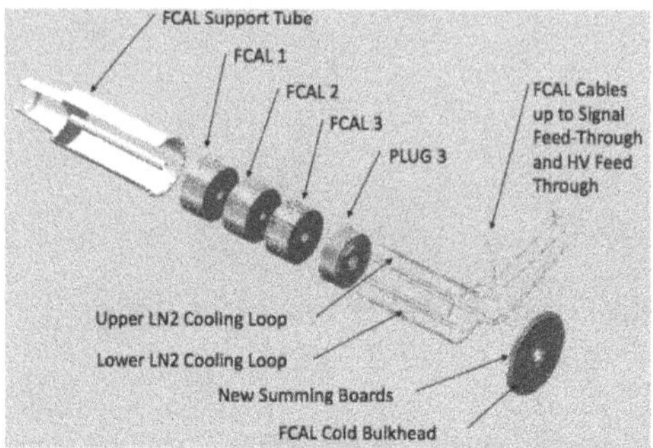

Fig. 1: Exploded view of super FCal design. Note the additional cooling loops and proposed location of new summing boards. The interaction point is towards the top left.

This effect is expected to be compounded by a voltage drop across high voltage protection resistors located on "summing boards" in an inaccessible region in the cryostat. There is also the potential for the Liquid Argon to boil in the FCal due to the energy deposited by the high particle flux hitting the detectors.

There are two main solutions proposed to maintain calorimeter performance. The first one is the replacement of each FCal with a super FCal (sFCal) built with smaller LAr gaps, cooling loops to ameliorate any beam heating problems, and a new HV protection system. It has been shown in a high luminosity test[3] that 100 micron gaps for a new FCal1 will be more than adequate for LHC luminosities.

An engineering sketch of the sFCal concept is shown in Figure 1, with the new set of summing boards positioned in front of the FCal cold bulkhead. It has also been proposed that a portion of the space occupied by Plug 3 be used to house the new summing boards. The installation of the sFCal will require the removal of the current FCal and opening of part of the end-cap cryostat. A second solution for the operation of the FCals at HL–LHC is the installation of a new calorimeter in front of the existing FCals, known as the Mini–FCal.

3. Mini–FCal

The Mini–FCal would absorb a large portion of the flux that would otherwise reach the innermost part of the FCal, reducing it to a level at which the existing FCal can operate normally.

Fig. 2: The left hand illustration details the position of the Mini–FCal with respect to the other End-Cap Calorimeters. The right hand diagram shows the Mini–FCal with the first absorber plate removed, revealing the first layer of diamond sensors.

3.1. *Warm Calorimeter based on Diamond Detectors*

The Mini–FCal has a transverse size that is constrained by the dimensions of the existing cryostat support structure into which it must be installed. The current proposal is a cylindrically shaped sampling calorimeter 35 cm in diameter and 30 cm in depth, with a parallel-plate structure using copper absorbers and layers of diamond sensors as the active layer, as shown in Figure 2. It is located directly in front of the FCal and is supported by the warm wall of the end-cap cryostat. The remaining volume of the region in front of the Mini–FCal will be lined with a neutron absorbing material such as borated polyethylene to reduce neutron backsplash into other detectors. The installation of the Mini–FCal is relatively simple as the cryostat will not need to be opened, however we still need to verify that the warm tube can support the Mini–FCal's weight.

A crucial aspect of the Mini–FCal design is an understanding of the radiation tolerance of the diamond sensors proposed for the active layers. Ten years of running at the HL–LHC is assumed to correspond to about $2{\times}10^{17}$ neutrons/cm^2. We performed irradiation tests on polycrystalline Chemical Vapour Deposition diamond detectors with 500 MeV protons at TRIUMF[4] to a fluence of $2.25{\times}10^{17}$ protons/cm^2. The detectors remained operational, but provided 5% of their original signal amplitude. We also performed spatial resolution tests and the results will be used in further simulations of the Mini–FCal response. We are also investigating two other options for the Mini–FCal concept, one using high pressure xenon gas and the other using liquid argon. These options are described in the next two sections.

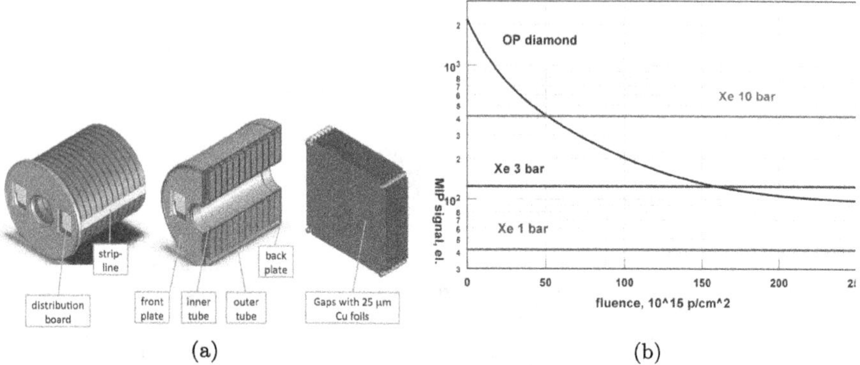

(a) (b)

Fig. 3: (a) Sketches of the Warm High–Pressure xenon Mini–FCal and (b) Comparison of the expected signal from diamond sensor from optical grade with 300 μm thickness. The curve assumes an initial charge collection efficiency (CCE) of 16% and the degradation curve from the proton test at TRIUMF.[4] The xenon option response assumes a CCE of 5% from electons only and no degradation with intgrated luminosity.

3.2. *Warm High–Pressure Xe Mini–FCal option*

This option uses a similar design to the diamond detector Mini–FCal except that the gaps between the copper absorbers are filled with high pressure (up to 10 bar) Xenon gas. Readout of the ionization signal is accomplished with a stack of foils 25 μm apart that are alternately at ground and high voltage. The foils are separated by ceramic washers. Each stack is 1 cm square and is the smallest volume that can be read out. This arrangment of foils is referred to as a "micro-gap" and is illustrated in the right hand diagram of Figure 3a. A comparison of the expected signal between diamond sensors and the xenon option is shown in Figure 3b. Note that at high integrated fluences the signal from this option is expected to exceed that from diamond detectors.

3.3. *Liquid Argon Cryogenic Mini–FCal Option*

The third technology considered for the Mini–FCal is the use of a liquid argon calorimeter with the same basic annular electrods of the existing FCal1, except that 100 μm gaps are used rather than 250 μm gaps. The overall dimensions of this option are similar to the diamond Mini–FCal. This option would be installed in a cryogenic vessel that would be located in the vacuum space formed between the existing cold vessel of the EndCap cryostat and a new inner warm tube, illustrated in Figure 4. The new warm

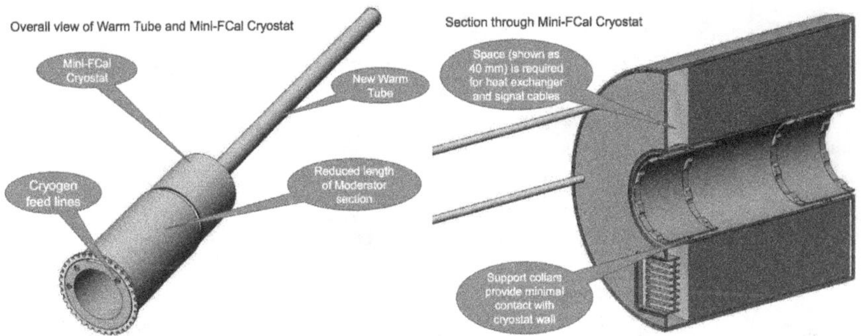

Fig. 4: Conceptual drawings of the Mini–FCal Cryogenic option.

tube will be redesigned to provide space in the vacuum region for this option and will provide for the cryogenic feedthroughs required. Note the use of special support collars at the inner edge to minimize heat transfer. This option requires considerable amount of upstream material in the form of feedthroughs and cryogenics and we will need to evaluate the effect of these on physics performance.

4. Summary

Two main solutions have been explored to prepare for the effect of HL–LHC luminosities on ATLAS Forward Calorimeters: a complete replacement of the current FCals and the insertion of a small calorimeter in front of the FCals. Three options for the small calorimeter have been considered, one using diamond detectors, a warm high pressure gas xenon option and a liquid argon calorimeter. These options continue to be evaluated.

References

1. The ATLAS Collaboration, *The ATLAS experiment at the CERN Large Hadron Collider*, JINST **3** (2008) S08003.
2. L. Evans, P. Bryant, *LHC Machine*, JINST **3** (2008) S08001.
3. A. Glatte, et al., *Liquid Argon Calorimeter performance at high rates*, submitted to Nucl. Instr. and Meth. A, September 2011.
4. D. Axen, et al., *Diamond detector irradiation tests at TRIUMF*, JINST **6** (2011) P05011.

The Crystal Calorimeter with Timing for the KLOE-2 experiment

M. Cordelli, S. Giovannella, F. Happacher, A. Luca', S. Miscetti,
A. Saputi, I. Sarra*, G. Pileggi and B. Ponzio

Laboratori Nazionali di Frascati, INFN,
Frascati, 00044, Italy
**E-mail: ivano.sarra@lnf.infn.it*

To improve the reconstruction of rare η and K meson decays in the KLOE-2 experiment we have designed a low angle calorimeter to improve the acceptance by covering the region in front of the first inner quadrupole. The proposed solution consists on an homogeneous calorimeter, CCALT, based on a new generation of crystals, LYSO. A matrix prototype granting a total transverse coverage of 2.8 R_M has been built. We tested this matrix with cosmic rays and to electron and photon beams. We report the measurements done with a tagged photon beam at the MAMI facility in the energy range between 40 to 380 MeV. An energy resolution of 5.5% at 100 MeV still dominated by leakage has been achieved. A position resolution between 3.5 mm is also observed at 100 MeV. A detailed comparison between data and MC performances has been carried out by simulating the matrix prototype with Geant4.

Keywords: Calorimetry; Crystals; Avalanche Photodiode.

1. Introduction

In the last years, a new machine scheme based on the Crab-waist and a large Piwinsky angle has been proposed and tested [1] to improve the reachable luminosity at the Frascati ϕ-factory, DAΦNE, an e^+e^- collider running at the center of mass energy of the ϕ resonance. The success of this test motivated the startup of a new experiment, named KLOE-2 [2], which aims to complete the KLOE [3] physics program and perform a new set of interesting measurements. In the new machine layout of DAΦNE the position of the inner quadrupole, QD0, at 30 cm from IP, reduces to 21° the minimum polar angle of the photons accepted by EMC. This opens the possibility to insert new calorimeters in this volume that will allow one to lower the minimum polar angle for photon detection down to 8°. This will enhance the multiphoton detection capability of the detector for the search of rare decays of kaons, η and η' mesons.[4]

2. CCALT: a Crystal Calorimeter with Time

The discussion of the previous section indicates that this calorimeter has to be very dense, with a small value of radiation length, X_0, and Moliere radius, R_M, not hygroscopic and with a large light output to improve photon detection efficiency at low energy (from 20 to 500 MeV). Moreover, the calorimeter has to be extremely fast in order to allow prompt photon reconstruction and to reject accidental/machine Touschek background (100 kHz per channel). Simulation studies indicate the need of a time resolution of $300 \div 400$ ps for 20 MeV photons. A suitable solution is offered by a crystal calorimeter with good timing performances, named CCALT. The detector consists of two calorimeters, one per side with respect to IP, each one composed by 48 projective crystals, Fig. 1. The best crystal choice matching the

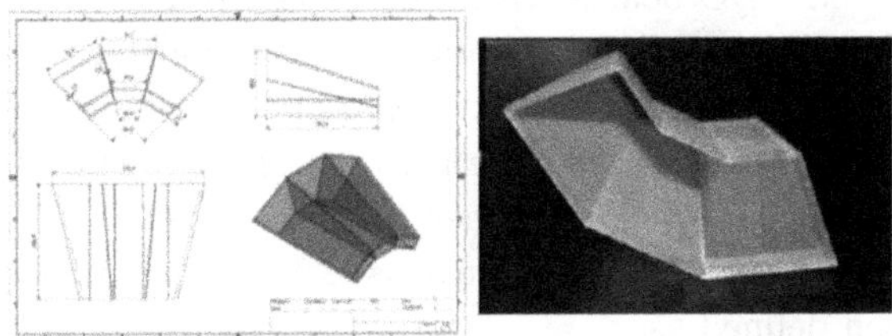

Fig. 1. CCALT: each calorimeter is composed by 4 wedges. Carrying 3 sectors each for a total subdivision of 12 modules in ϕ. Each module has a granularity of 4 crystals, with final transverse dimension in the readout plane of ~1.4x1.8 mm^2.

requirements is provided by the new generation of Cerium doped Lutetium Yttrium Orthosilicate, LYSO, which shows X_0 (R_M) values of 1.1 (2 cm), with a light yield of about 80% relative to NaI. LYSO has also a scintillation emission time of 40 ns that, for the high light yield, should grant a time resolution of $300 \div 400$ ps for 20 MeV photons. The presence of an axial magnetic field of 0.52 kGauss forces the usage of silicon based photo detectors. The readout with APD/Sipm is a valid solution.

In February 2011, we have built a crystals matrix prototype with transverse of ~2.8 R_M, longitudinal dimensions being constrained by budget to be 13-15 cm (corresponding to 11 12 X_0 of longitudinal containment). The prototype consists of an inner matrix of 9 LYSO crystals by SICCAS (20x20x150 mm^3) readout by APD S8664 (10x10 mm^2 by Hamamatsu) and an outer matrix, for leakage recovery, composed by 8 $PbWO_4$ crystals readout by standard Hamamatsu Bialcali photo multipliers of 1 inch of diameter.

3. Test results with photon beams

We have taken data at the MAMI (Mainz Microtron) electron beam facility, for four days in March 2011. In the facility hall A2 the electron beam is converted to an intense beam of real photons through bremsstrahlung in a thin metal foil radiator. The scattered electrons in this process are momentum analyzed by plastic scintillator spectrometer which provides a determination of the energy of the associated bremsstrahlung photon with a resolution of few per mil. We used the photon beam at 10 kHz in the energy range from 40 to 380 MeV. We have triggered by using a coincidence between the OR signal from the inner matrix and the signal from the spectrometer. We acquired data with a Camac system, reading out LeCroy ADC and TDC boards with a sensitivity of 250 fC/count and 100 ps/count respectively. We have calibrated the calorimeter response of each channel with minimum ionizing particles, m.i.p., crossing the calorimeter orthogonally to the crystal axis. We get $\sigma_{\rm ped}$ of 3 (2) counts and a m.i.p. peak around 120 counts for the inner (external) crystals. The statistical precision on the peak determination is $\sim$ 2%. The total response of the detector is then defined as:

$$Q_{TOT} = \sum_i (Q_i - P_i) \cdot 1/M_i \tag{1}$$

where Q_i and P_i are the collected charge and the pedestal of the i-th channel and M_i is the minimum ionization peak. We have fit the distributions with a logarithmic gaussian function [5], to quote the energy response and resolution. To understand the different terms of the energy dependence of the energy resolution, we have carried out a full simulation of the prototype based on Geant-4. We added a 4% gaussian smearing for each channel in MC to reproduce data (to take into account crystal mis-calibration, longitudinal not-uniformity/not-linearity). We obtained an excellent data-MC agreement for Inner Matrix, Fig. 2 (left). While there is still a marginal data-MC agreement on the overall response. In Fig. 2 (right), we show the energy dependence of the energy resolution which has been fit with the following equation:

$$\sigma_E/E = a/\sqrt[4]{E[GeV]} \oplus b/E[GeV] \oplus c \tag{2}$$

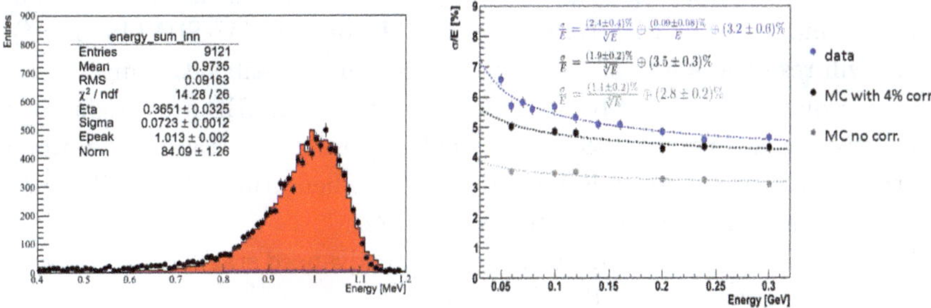

Fig. 2. (Left) Data MC Comparison, sum of the Inner Matrix crystals. Data in black; MC in red. (Right) Dependence of the energy resolution on beam momentum. Blue points data, black (gray) points MC with (without) the 4% correction.

We found the stochastic term, a, to be 2.4±0.4% and a constant term, c, 3.2±0.6% for data in good agreement with MC expectation. The noise term b is almost negligible.

We have then determined the position resolution with centroid method defined as $X_{pos} = \sum Q_i \cdot X_i / Q_{tot}$. We observe a position resolution of $\sim$ 3.4 mm RMS. We performed a position scan along X; data-MC comparison is excellent, Fig. 3.

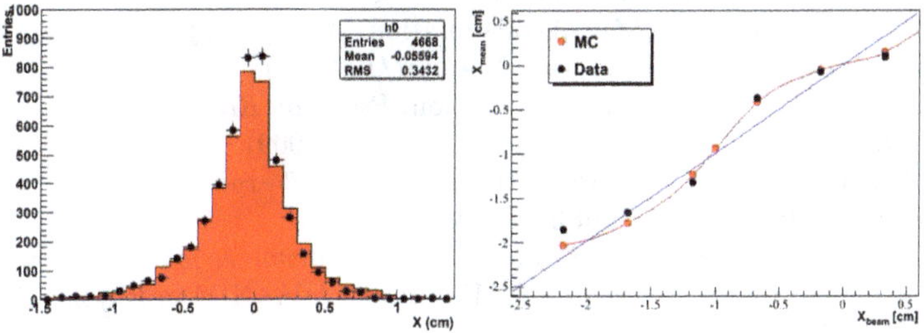

Fig. 3. Energy weighted centroid summing all cells, left panel. Position scan along X, right panel; on X-axis, the value 0 is the centrum of the central inner matrix crystal.

We could not determine the timing resolution at Mami, due to the large jitter of the trigger (1 ns). We determined it at BTF of Frascati, with a test performed on march 2009 with a smaller matrix prototype. We obtained σ_t = 250 (49) ps at 500 MeV, 291 (120) ps at 100 MeV without (with)

correction for trigger jitter [6].
Due to the constraints on thermal and position requirements, we will readout the final detector with Large Area SIPM from ADVANSID (4x4 mm^2). This will result in a loss of $\sim$ 2-4 in time resolution, still satisfying the detector requirements. First tests on single crystals with LED are satisfactory. We obtained a time resolution with single crystal 20x20x150 mm^3 readout with a 4x4 mm^2 SIPM of 400 ps at 20 MeV, when firing with a UV LED. Energy vs led light was calibrated with MIPs.
Construction of the mechanical support is in advanced stage. It will be done at the mechanical shop of INFN Napoli.

4. Acknowledgments

The authors are indebted to many people for the successful realization of the matrix. In particular, we thank all the LNF mechanical shop for the realization of the support and APD boxes, expecially G. Bisogni, U. Martini and A. De Paolis. We also thank the MAMI staff for providing the beam time and A. Thomas for great support during the test beam data taking. The realization of the preamplifiers was done in collaboration with E.Reali from Roma-2 university.

References

1. P.Raimondi, in *Crab Waist Collisions in DAΦNE AND SUPER-B DESIGN*, (Proceedings of EPAC08, Genoa, Italy 2008).
2. F. Bossi et al, for the KLOE-2 collaboration, *A proposal for the Roll-In of the KLOE-2 detector*, (LNF-Internal Note, 07/19 2007).
3. F. Bossi et al, for the KLOE collaboration, *Precision Kaon and Hadron Physics with KLOE*, (Riv. Nuovo Cimento 031:531-623 2009).
4. G. Amelino-Camelia et al, *Phisycs with the Kloe-2 experiment at the upgraded DAΦNE*, (EPJ C, vol. 68, n. 3-4, 619-681, 2010).
5. H. Ikeda et al, *A detailed test of the CsI(Tl) calorimeter for BELLE with photon beams of energy between 20 MeV and 5.4 GeV*, (NIM in Physics Research A 441 (2000) 401-426).
6. M. Cordelli et al, *Test of a LYSO matrix with an electron beam between 100 and 500 MeV for KLOE-2*, (NIM A671: 109-112, 2010).

THE ATLAS TILE CALORIMETER PERFORMANCE AT THE LHC IN PP COLLISIONS AT 7 TEV

ALEXANDER SOLODKOV
ON BEHALF OF THE ATLAS COLLABORATION
Institute for High Energy Physics, 1, Pobeda street, Protvino, 142281, Russia
E-mail: Sanya.Solodkov@cern.ch

The Tile Calorimeter (TileCal), the central section of the hadronic calorimeter of the ATLAS experiment, is a key detector component to detect hadrons, jets and taus and to measure the missing transverse energy. Due to the very good muon signal to noise ratio it assists the spectrometer in the identification and reconstruction of muons. TileCal is built of steel and scintillating tiles coupled to optical fibers and read out by photomultipliers. The calorimeter is equipped with systems that allow one to monitor and to calibrate each stage of the read-out system exploiting different signal sources: laser light, charge injection and a radioactive source. The performance of the calorimeter has been measured and monitored using calibration data, random triggered data, cosmic muons, splash events and more importantly LHC collision events. The results presented assess the absolute energy scale calibration precision, the energy and timing uniformity and the synchronization precision. The results demonstrate a very good understanding of the performance of the Tile Calorimeter that is well within the design expectations.

1. Introduction

The ATLAS experiment at CERN [1] is successfully taking data at the LHC at 7 TeV center-of-mass energy since 2009. The Tile Calorimeter (TileCal) [2] is the central hadronic section of the ATLAS Calorimeter. TileCal is a sampling calorimeter made of scintillating tiles as active medium and steel plates as absorbers. It is divided into four partitions, two barrels (LB) and two extended barrels (EB), covering in total a pseudorapidity range of $|\eta| < 1.7$ and is segmented into 64 modules along the azimuth ϕ. Wavelength shifting fibers collect the light generated in the tiles and carry it to photomultipliers (PMT) (see Fig. 1). Two fibers, attached to every tile from different sides in ϕ, go to different PMTs, providing redundant double readout of a signal. Each PMT receives signal from multiple tiles which are grouped into cells of different size depending on their pseudorapidity and depth. Three longitudinal layers A, BC, D are defined inside the modules and the dimensions of the cells are optimized to obtain a structure of projective towers with granularity of $\Delta\eta \times \Delta\phi = 0.1 \times 0.1$

in the first two layers and $\Delta\eta \times \Delta\phi = 0.2 \times 0.1$ in the last layer (see Fig. 2). Cells of an additional special layer E, attached to extended barrel modules, are read out by a single PMT each. In total, TileCal has 5182 cells and 9852 channels.

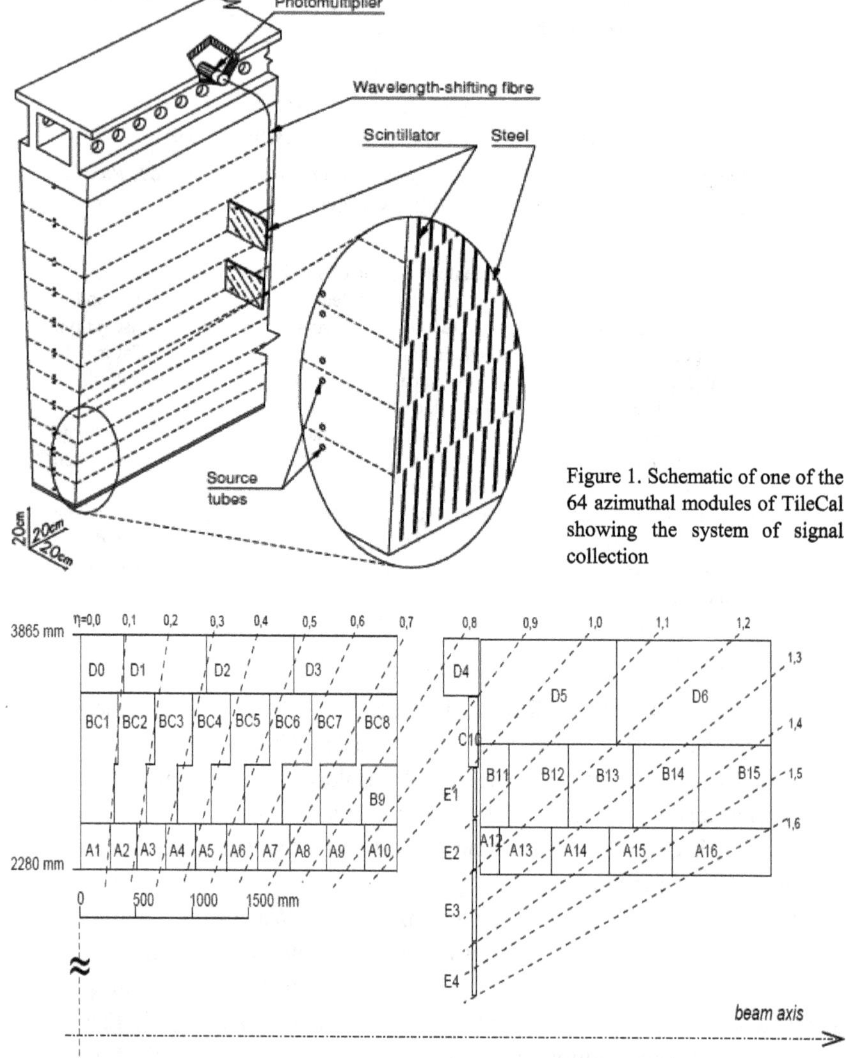

Figure 1. Schematic of one of the 64 azimuthal modules of TileCal showing the system of signal collection

Figure 2. Drawing of a half of the calorimeter divided into a barrel and an extended barrel part with the cell division scheme depicted.

Thanks to the double readout of most of the cells, single channel failure does not cause significant problem. More serious failures in on-detector electronics, which cannot be fixed without intervention, might require to disable the readout from a whole module – 22 cells in barrel and 18 cells in extended barrel are masked in this case. Thanks to the good calorimeter spatial granularity, energy in masked cells can be recovered offline using interpolation between working neighbors. Before the maintenance period of winter 2010 3.8% of the cells were unusable for physics and all of them were repaired. Similar amount of cells will be repaired again at the end of 2011 (see Fig. 3).

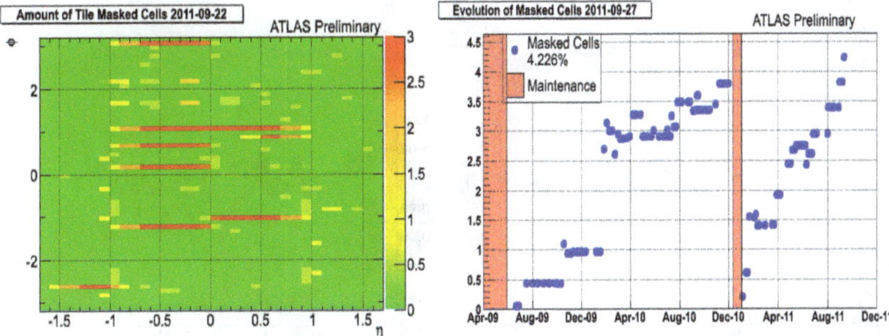

Figure 3. Left: two dimensional (η;ϕ) map showing the number of cells masked per tower, each tower being composed by three cells in the three layers A, BC, D. Right: evolution in time of the percentage of masked cells; red bands represent maintenance periods.

2. Inter-calibration and electromagnetic scale

A fraction of 11% of the TileCal modules were calibrated in the beam tests in 2001-2003 [3]. Electron beams with energies between 20 and 180 GeV were used to establish the electromagnetic scale for the first calorimeter layer, while two other layers were inter-calibrated with respect to the first one using muons.

After installation of the whole Tile Calorimeter in the ATLAS experimental hall cell inter-calibration was done with the help of the Cesium calibration system. The Cesium source was moved through every calorimeter cell and high voltage of every PMT was adjusted to have the cell response equal to the response measured during beam tests. The comparison between cosmic data and Monte Carlo (MC) prediction and between beam test muons and cosmic muons has confirmed that propagation of the electromagnetic scale from the beam tests to ATLAS was successful. Non-uniformity within one layer as seen by muons turned out to be at the level of 2-3% (see Fig.4), the maximal difference between layers is 4% and EM scale is stable over a 3-year period (see Table 1).

Table 1. dE/dx for cosmic muons data divided by dE/dx from simulation for 3 barrel and 3 extended barrel layers

Layer	2008	2009	2010
LB-A	0.966 ± 0.012	0.972 ± 0.015	0.971 ± 0.011
LB-BC	0.976 ± 0.015	0.981 ± 0.019	0.981 ± 0.015
LB-D	1.005 ± 0.014	1.013 ± 0.014	1.010 ± 0.013
EB-A	0.964 ± 0.042	0.965 ± 0.032	0.996 ± 0.037
EB-B	0.977 ± 0.018	0.966 ± 0.016	0.988 ± 0.014
EB-D	0.986 ± 0.012	0.975 ± 0.012	0.982 ± 0.014

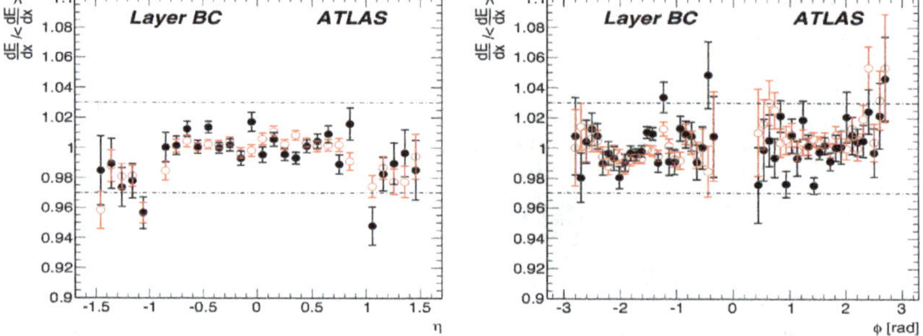

Figure 4. Normalized truncated mean (dE/dx)/<dE/dx > as a function of η (left) and φ (right), for data (black, full dots) and MC (red or light grey in b/w, open circles), showing the uniformity of the response to cosmic muons in 2009. Dotted lines delimit a ±3% variation from unity. Bigger errors or no data close to φ=0 and φ=±π are due to limited statistics for horizontal cosmic muons.

3. Timing

In order to reconstruct the signal amplitude correctly [4], the peak pulse time with respect to the electronic sampling clock has to be known with good precision. In 2008 the cell times were synchronized to single reference channel in every partition using the laser calibration system [5]. Inter-calibration between partitions was also performed in 2008 using single-beam splash events. In such events some protons from the beam collide with collimators placed at about 140 m from the nominal interaction point and produce a very large number of minimum ionizing particles reaching the detector parallel to the beam axis and depositing a large amount of energy in the whole TileCal. Later on the timing was validated with cosmic events, with 2010 splash events and in collisions. As it is demonstrated in Fig. 5, all the TileCal cells are synchronized with the precision of better than 1 ns.

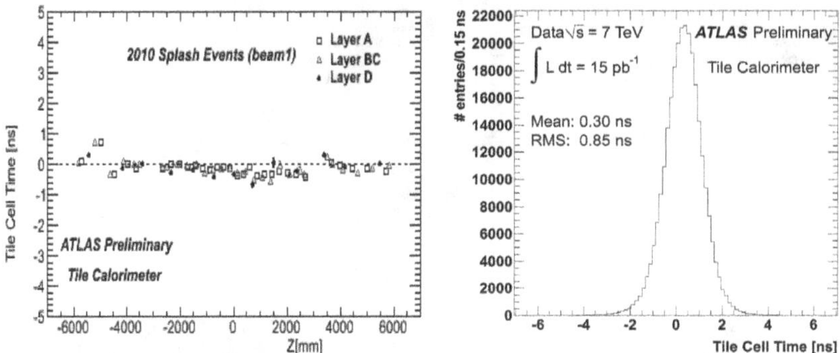

Figure 5. Left: time of cells, averaged over ϕ in single beam splash events. Different layers are shown by different markers. Right: time of the TileCal cells in jets coming from pp collisions; only cells with energies above 20 GeV are shown.

4. Performance in collisions

The LHC began to run in 2009 providing collisions at $\sqrt{s}$ = 900 GeV center-of-mass energy, later switching to 2.36 TeV and finally operating at 7 TeV. Figure 6 (left) shows the distribution of energy deposition in TileCal from Minimum Bias events for the three different $\sqrt{s}$ points. The response of TileCal to single pions was also studied. Isolated tracks of momentum **p** were required to deposit little energy, consistent with minimum ionizing particles in the electromagnetic calorimeter in front of TileCal, to be sure their whole energy **E** is deposited in the hadronic calorimeter. The mean value of the consequently defined **E/p** ratio is shown in Fig. 6 (right) and in Fig.7. In all the cases good agreement between data and MC was found.

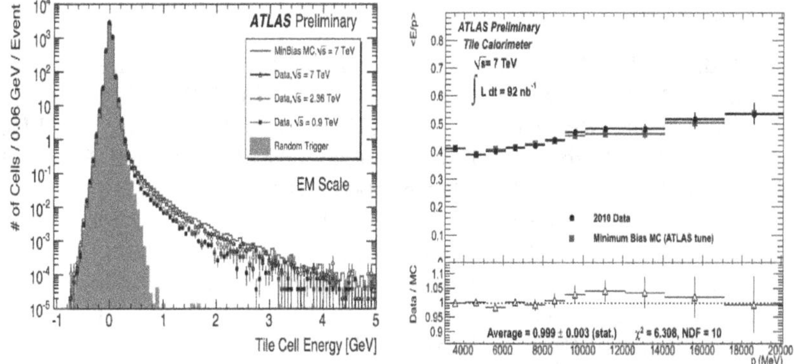

Figure 6. Left: energy deposition in TileCal cells for different $\sqrt{s}$ values, Minimum Bias simulation ($\sqrt{s}$ = 7 TeV) and randomly triggered events. Right: mean value of the ratio between energy deposited in TileCal and the track momentum **p** (measured by the Inner Detector) as a function of **p**.

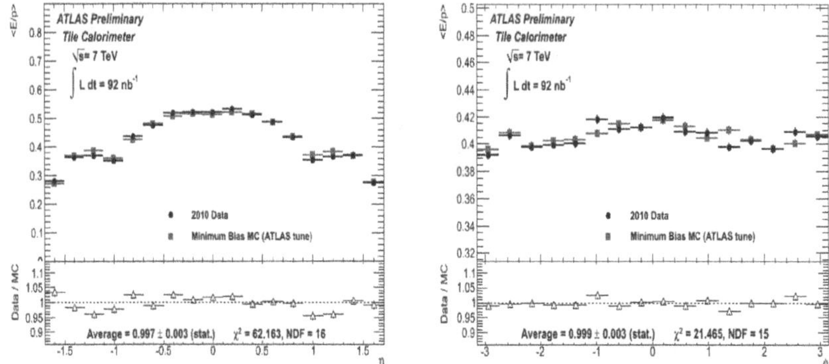

Figure 7. Mean value of the ratio between energy deposited in TileCal and track momentum (measured by the Inner Detector) as a function of η (left) and ϕ (right), for isolated pions showering in TileCal.

5. Conclusions

The first years of data taking at the LHC demonstrate good performance of the ATLAS Tile Calorimeter, which is well within the design specifications. The TileCal calibration and monitoring systems guarantee stability in time of the calorimeter response, uniformity within 2-3% in η and ϕ. The energy scale uncertainty, which was successfully extrapolated from the beam tests to ATLAS, is conservatively considered to be 4%. The time synchronization between cells is well below 1 ns and has been verified with single beam, with cosmic muons and in collisions.

References

1. G. Aad, et al., The ATLAS Experiment at the CERN Large Hadron Collider, JINST 3 (2008) S08003. doi:10.1088/1748-0221/3/08/S08003
2. G. Aad, et al., Readiness of the ATLAS Tile Calorimeter for LHC collisions, The European Physical Journal C - Particles and Fields 70 (2010) 1193–1236, doi:10.1140/epjc/s10052-010-1508-y.
3. P. Adragna et al. Testbeam studies of production modules of the ATLAS Tile Calorimeter, Nucl. Insturm. Methods A 606, 362-394 (2009)
4. G. Usai et al., Signal Reconstruction of the ATLAS Hadronic Tile Calorimeter: implementation and performance, 2011 J. Phys.: Conf. Ser. 293 012056 doi:10.1088/1742-6596/293/1/012056
5. J. Carvalho et al., The TileCal/ATLAS calorimeter calibration systems, these proceedings

VALIDATION OF THE LOCAL HADRONIC CALIBRATION SCHEME OF ATLAS WITH COMBINED BEAM TEST DATA IN THE ENDCAP AND FORWARD REGIONS OF ATLAS

A.E. KIRYUNIN

Max-Planck-Institut für Physik, Werner-Heisenberg-Institut
Föhringer Ring 6, 80805 Munich, Germany
E-mail: kiryunin@mppmu.mpg.de

P. STRIZENEC

Institute of Experimental Physics of the Slovak Academy of Sciences
Watsonova 47, 04001 Kosice, Slovakia
E-mail: pavol@mail.cern.ch

on behalf of the ATLAS Liquid Argon Endcap Collaboration

The local hadronic calibration developed for the energy reconstruction and the calibration of jets and missing transverse energy in ATLAS, has been validated using data obtained during combined beam tests of the ATLAS endcap and forward calorimeters. The analysis has been performed using special sets of calibration weights and corrections obtained with the GEANT4 simulation of a detailed beam test setup.

Keywords: ATLAS; Hadronic calibration; Beam tests; Simulations.

1. Introduction

For many physics processes studied with the ATLAS detector[1] at the Large Hadron Collider (LHC) at CERN, it is very important to have accurate measurements of the energy of jets and of the missing transverse energy. A lot of efforts were done by the ATLAS collaboration to develop methods of the hadronic calibration for the calorimeter system, allowing the reconstruction of true energies of hadrons. Such techniques have to correct for the invisible energy due to the non-compensating nature of ATLAS calorimeters, for the energy losses in regions not instrumented with read-out and for energy losses due to unavoidable (at LHC conditions) threshold cuts in the signal reconstruction. The local hadronic calibration (LC) is one of the calibration schemes used in ATLAS. The subject of this talk is the validation of this scheme with data obtained during combined beam tests of modules of the ATLAS endcap and forward calorimeters.

2. Local Hadronic Calibration

The local hadronic calibration scheme is described in detail in Ref. 2. The input for this calibration are three-dimensional topological clusters[3] reconstructed at electromagnetic (EM) scale[a]. The calibration starts by classifying clusters as mostly electromagnetic or mostly hadronic depending on cluster shape variables and on the cluster energy. The following LC steps are applied: 1) weighting of the cluster energy to account for the non-compensating nature of the calorimeters, 2) out-of-cluster corrections to correct for lost energy deposits inside the calorimeter due to the noise thresholds, 3) dead material corrections to compensate for energy deposits in materials outside of active calorimeter volumes. LC-weights and corrections are determined from detailed Monte Carlo (MC) simulations.

3. Beam Test of ATLAS Endcap and Forward Calorimeters

A combined setup of modules of the ATLAS electromagnetic and hadronic endcap calorimeters and the forward calorimeter was exposed to beams of charged pions, electrons and muons in the H6 beamline at the CERN SPS in 2004.[4] Beams covered a zone corresponding to the ATLAS pseudorapidity region of $2.5 < |\eta| < 4.0$. To validate the local hadronic calibration, data from energy scans with charged pions in two impact points were used. These two points represented the endcap region ($\eta \simeq 2.75$ in ATLAS) and the forward region ($\eta \simeq 3.6$ in ATLAS). Scans covered the range of the beam energies (E_{BEAM}) from 10 to 200 GeV.

4. Simulations

The GEANT4 toolkit[5] (version 9.2) was used to simulate the response of the beam test setup to beam particles. Two sets of MC samples were produced. The first set contained simulations of energy scans with charged pions in studied impact points. This set was used to compare MC predictions with available experimental data. The second set of MC samples was used to derive weights and correction coefficients for the local hadronic

---

[a]This scale correctly measures the energy deposited in calorimeters by EM showers.

calibration. It contained simulations of charged and neutral pions with energies distributed in the interval from 1 to 400 GeV logarithmically flat and with uniform coverage of the front face of the calorimeter modules. From the standard physics lists available for the hadronic shower simulation in GEANT4,[6] two physics lists QGSP-BERT and FTFP-BERT were used.

5. Results

Results of the validation of the local hadronic calibration, presented in this contribution, are based on the analysis of the energy response and resolution for charged pions in the endcap and forward regions. The reconstructed energy is defined as the sum of energies of all clusters in an event. Gaussian curves are fitted to the reconstructed energy distributions in the interval $\pm 3\sigma$ around the peak value E_0. The parameters E_0 and σ from this fit are used to determine the response (E_0/E_{BEAM}) and the resolution (σ/E_0). Results are presented for the EM-scale and for the LC-scale after application of QGSP-BERT-based corrections only.

The energy dependencies of the response in the endcap region are shown at the top of Fig. 1 for both experimental data and simulations. The bottom of this figure presents the ratio of the simulated response and the experimental response: $R = (E_0/E_{\rm BEAM})_{\rm MC}/(E_0/E_{\rm BEAM})_{\rm Data} = E_0^{\rm MC}/E_0^{\rm Data}$, as a function of the beam energy. The non-linear response at EM-scale indicates the non-compensating nature of the endcap calorimeters. For $E_{\rm BEAM} \geq 40\,{\rm GeV}$, QGSP-BERT overestimates the experimental response by 2%, while FTFP-BERT overestimates data by 4%. The response at LC-scale is rather stable for $E_{\rm BEAM} \geq 60\,{\rm GeV}$, where it is close to 1.0 for data and QGSP-BERT-based simulations. The agreement between data and MC predictions is better at LC-scale than at EM-scale.

To assess the systematic uncertainty of the local hadronic calibration, the EM-scale difference between data and simulations should be factored out. This can be achieved by analysing the double ratio of the energy response. It is determined as the ratio of the parameters R, obtained at LC- and at EM-scale $(R_{\rm LC}/R_{\rm EM})$. This double ratio equals to 0.987 and 0.990 for QGSP-BERT and FTFP-BERT physics lists, respectively. Its deviation from unity by $1.0 - 1.3\%$ can be considered as the uncertainty of the local hadronic calibration for charged pions in the endcap region.

The energy resolution at EM-scale for pions in the endcap region decreases with the beam energy from $(44.6 \pm 0.2)\%$ at $10\,{\rm GeV}$ down to $(11.29 \pm 0.04)\%$ at $200\,{\rm GeV}$, for the data. The two physics lists give close results and predict a too good resolution compared to experimental values

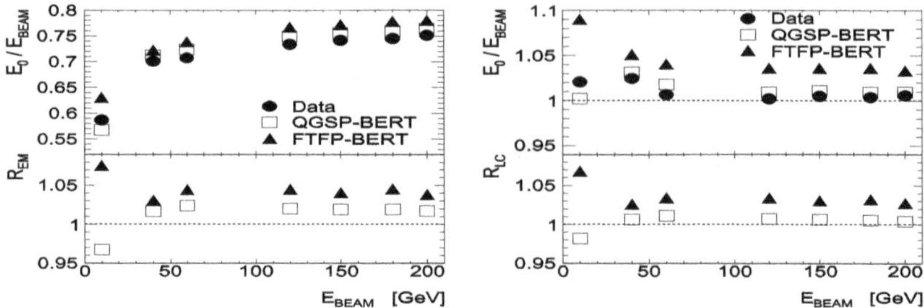

Fig. 1. Top: Response for pions in the endcap region at EM-scale (left) and at LC-scale (right) as a function of the beam energy. Bottom: Ratio between simulations and experimental data for the pion response.

by 10 – 15%. The improvement of the energy resolution after application of the local hadronic calibration is about 5 – 15% for data and simulations. The weighting step improves the energy resolution at higher beam energies, while out-of-cluster corrections improve it at lower energies. The improvement of the energy resolution at LC-scale with respect to EM-scale is slightly better for simulations than for experimental data.

The energy dependencies of the response in the forward region are presented in Fig. 2 for experimental data and for simulations. The non-linear response at EM-scale reflects the non-compensation of the forward calorimeter. FTFP-BERT describes the experimental response at EM-scale rather well. QGSP-BERT underestimates the response by 3 – 8%. The application of QGSP-BERT-based LC-weights and corrections yields an overestimate of the experimental response by 3 – 4%. A good agreement between experimental response and the response predicted by FTFP-BERT at EM-scale becomes worse at LC-scale.

As for the endcap region, the double ratio of the response in the forward region is also analysed. Results obtained for the two physics lists are very similar. The double ratio is smaller than 1.0, and the deviation from unity is up to 3%. This is taken as the uncertainty of the local hadronic calibration for charged pions in the forward region.

The pion energy resolution at EM-scale varies with the beam energy between (44.1 ± 0.3)% at 10 GeV and (12.23 ± 0.03)% at 200 GeV, for the data in the forward region. The two physics lists give similar results and predict worse energy resolution at EM-scale compared to experimental values by 5 – 15% (for $E_{\mathrm{BEAM}} \geq 40\,\mathrm{GeV}$). After applying LC-weights and corrections the resolution in general improves by 10 – 25% for both data and simula-

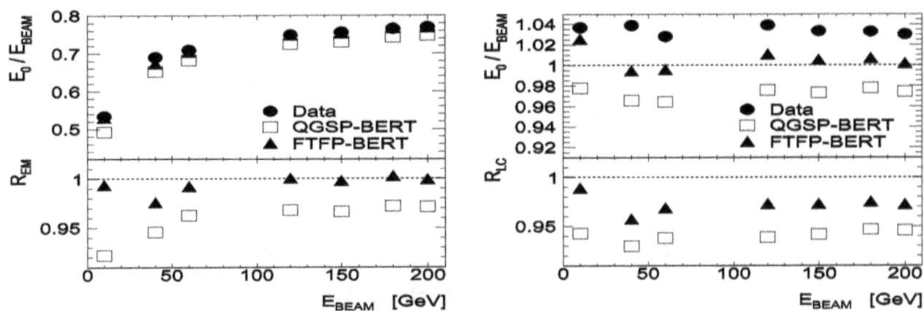

Fig. 2. Top: Response for pions in the forward region at EM-scale (left) and at LC-scale (right) as a function of the beam energy. Bottom: Ratio between simulations and experimental data for the pion response.

tions. The local hadronic calibration significantly improves the agreement between experimental and predicted values of the resolution, especially for the FTFP-BERT physics list.

6. Conclusions

The local hadronic calibration scheme, an advanced method for reconstructing hadronic signals in the ATLAS calorimeter system, is validated with data obtained during beam tests of modules of the ATLAS endcap and forward calorimeters. The scheme allows the reconstruction of the initial energy of charged pions in the endcap region with uncertainties below 1.5%. In the forward region the uncertainty of the response is about 3%. One of the important features of the local hadronic calibration is a significant improvement of the energy resolution in all studied calorimeter regions.

References

1. The ATLAS Collaboration, *Expected Performance of the ATLAS Experiment: Detector, Trigger and Physics*, Tech. Rep. CERN-OPEN-2008-020, CERN (2009), arXiv:0901.0512.
2. T. Barillari *et al.*, *Local Hadronic Calibration*, Tech. Rep. ATL-LARG-PUB-2009-001-2, CERN (2009), `http://cdsweb.cern.ch/record/1112035`.
3. W. Lampl *et al.*, *Calorimeter Clustering Algorithms: Description and Performance*, Tech. Rep. ATL-LARG-PUB-2008-002, CERN (2008), `http://cdsweb.cern.ch/record/1099735`.
4. J. Pinfold *et al.*, *Nucl. Instrum. Meth.* **A593**, 324 (2008).
5. S. Agostinelli *et al.*, *Nucl. Instrum. Meth.* **A506**, 250 (2003).
6. A. Ribon *et al.*, *Status of Geant4 hadronic physics for the simulation of LHC experiments at the start of LHC physics program*, Tech. Rep. CERN-LCGAPP-2010-02, CERN (2010).

On the Challenge of Keeping ATLAS Tile Calorimeter Raw Data

V. K. TSISKARIDZE (on behalf of the ATLAS Collaboration)

Physikalisches Institut, Albert-Ludwigs-Universität Freiburg,
3, Hermann-Herder-Str., Freiburg, 79104, Germany
Physics Faculty, Iv. Javakhishvili Tbilisi State University,
1 Chavchavadze ave., Tbilisi, 0128, Georgia
E-mail: vakhtang.tsiskaridze@cern.ch

The Tile Calorimeter (TileCal) for the ATLAS experiment at the CERN Large Hadron Collider (LHC) is currently taking data with proton-proton collisions. The TileCal read-out system was initially designed to reconstruct the data in real-time and to store for each channel the signal amplitude, time and quality factor at the required high rate. This approach implied discarding 80% of the raw data that correspond to noise or small signals. Practical experience operating in this scheme with increasing rate have led to several modifications and understanding that some kind of data compression is helpful during data processing and storing.

An alternate approach is to use online reconstruction for Level 2 triggering only and to implement a data flow lossless compression scheme for further offline analysis. A new version of the lossless compression algorithm is proposed which allows to both save the complete raw data and to feed the trigger with the reconstructed signal amplitude and time. It does not increase the data flow as compared to the existing approach and the size of the data fragments transmitted is more stable. We will describe the lossless compression algorithm as a possible upgrade of the Tile data acquisition and highlight some details of the implementation. We will report on its testing and validation and on the overall performance measured on high rate tests, calibration and $\sqrt{s} = 7$ TeV proton-proton collisions runs.

Keywords: ATLAS; TileCal; LAr; Lossless Compression; Frag5; DSP.

1. Introduction

Data processing in ATLAS Tile Calorimeter[1,2] consists of online and offline phases. Online processing is effectuated in the fixed-point arithmetic Digital Signal Processors (DSP). Operation environment limits output bandwidth to 400 (32 bits) words and the processing time to 10 μs assuming the ATLAS Level 1 trigger rate of 100 kHz. The initial approach comprised providing reconstructed Amplitude and Time (using Reco fragment[3]) for High Level Trigger (HLT) as well as storing the complete raw data for up to 8 (out of 45) selected channels for further offline analysis. An appropriate threshold is set to select the channels for which the complete raw data is stored (indicated as Frag1 fragment[3]). Practical experience operating with this scheme has shown that the constant increase of the collision rate requires a frequent threshold tuning. This indicated that some kind of data compression would be highly desirable during data processing and storing.

An alternate approach is to use online reconstruction for Level 2 triggering only and to implement a lossless compression scheme to record all the raw data. Saving all the raw data has several advantages. It makes possible full offline reprocessing as well as debugging and validation; may help to increase performance for physical quantities used in the analysis (jets, missing transverse energy) and enforce efforts on small signals like those produced by muons traversing the calorimeter. Saving the raw data increases the possibility to cope with Minimum Bias pile-up or unforeseen problems. Furthermore, low signals may appear helpful for exotic searches. Keeping all the raw data allows analyzing background noise and inter-channel dependencies to increase precision of the measurement. It simplifies data collection as there is no need of complicated estimates and threshold adjustments. It has already inspired various optimizations in the current working scheme. Shortly, with all the raw data the offline processing is always open to further improvements.

2. Lossless Compression

The initial idea for lossless compression was to use the fact that pulses in different channels are soundly correlated. Thus the real amount of information to store would be much less than it looks at the first sight and compression may help to improve the performance. The first version of lossless data compression comprised the pedestal compression only and soon was replaced by a more complicated and powerful compression scheme presented on CHEP09 (Praga, March 2009).[4] According to this scheme the channels are processed in an appropriate order and the differences between samples in consecutive channels are recorded. The method substantially used the geometry of the Calorimeter as it was sensitive to the channels ordering used during the compression. It needed *no information* about the signal pulse shape and proved to be able to compress piled-up and other non-standard signals from *all* TileCal channels fitting within the bandwidth constraint.

While the proposed algorithm was able to pack all the raw data and fit within the bandwidth limitations, it could not fit within the tight time constraints imposed by the Level 1 trigger rate of 100 kHz. To be competitive with the existing processing scheme meeting the following requirements was considered mandatory for new compression tool called later Frag5:

- compression formats should be simple enough to enable software fitting within the DSP Level 1 trigger time constraints;
- reconstructed magnitudes should be easily accessible for HLT (no unpacking should be needed);
- the energy should be reconstructed with the same precision as in the currently used fragment;
- a reliable theoretical model should exist to evaluate the algorithm performance under various circumstances;
- in case of "good" signal in *all* or almost all channels the compression should be able to fit within the bandwidth limit;
- it should be possible to compress effectively both piled-up and "unexpected" signals.

To improve DSP performance all possible precalculations were moved outside the loops, Optimal Filtering constants[5] were rearranged to tune them for DSP commands format, specific DSP commands were used to increase the performance, such as built-in support for rounding. DSP uses a robust Software-Pipelined Loop mechanism that can significantly speed-up loops without branching. This appeared an important resource: eliminating 'if' statements from the loops and replacing them by arithmetical operations allowed to twice speed-up the code. Besides the other benefits, these tricks allowed to include energy sums Sum(Et, Ez, E) calculation into the DSP.

To simplify calculations in lossless compression tool it was decided to use some *standard*, previously measured pulse shape as a reference and store differences between samples and reference shape rather than differences between consecutive samples. Frag5, the third version of the lossless compression algorithm, fully benefits of these improvements. The closer recorded signal to the standard pulse shape assumed by the algorithm, the higher the compression of the data. Worst case analysis shows that:

- any kind of data in all channels can be recorded at 72 kHz rate at least;
- "reasonable" signals (including pile-up) with energy deposits in all channels, even significantly out of time with respect to the reference signal ($\pm$25 ns), can be stored at 80 kHz rate;

- the standard pulse shape in all channels, even if somewhat spread in time, can be recorded at 95 kHz rate.

3. Implementation: Pros and Cons

By the end of February 2011 a fully functional version of Frag5 was successfully tested demonstrating the feasibility of the digits lossless compression approach. It was discussed within the TileCal community and later briefly reported at TIPP11 conference[3] as TileCal R&D project.

Since it was clear that the increasing luminosity and energy of the LHC would pose ever more challenging conditions to the signal reconstruction, it was therefore considered very important to be ready with realistic upgrades of the Tile data acquisition that can support lossless compression schemes.

Even lacking urgent motivations (like clear physics cases or operational bottleneck in the current model) to abandon the current scheme, continuation of studies and validation tests of the lossless compression scheme was encouraged aiming at a realistic goal to have it implemented in the system.

Recent physics runs demonstrated that the current scheme requires considerable modification. No single fixed threshold could be set to select required limited amount of channels for recording under conditions of ever increasing collision rate. It was decided to change current (Frag1) format and to implement some level of data compression similar to that of Frag5. Nevertheless, new Frag1 format is still dropping a signal below the predefined threshold while it is compressing all sufficiently high pulses using two formats for small and for large signals. Quality Factor (QF) calculations are also ceased for dropped signals as they are time consuming. This means discarding potentially valuable information about the signals below the threshold. In parallel with these changes, additional layer was added to Frag5 to ensure particular handling of weak signals.

Here is a brief comparative study of Frag5 and Frag1 approaches:
Reconstruction: Frag5 uses exactly the same standard code for online amplitude reconstruction as the current scheme, which is already validated.[5]
Size: Frag5 reduces fragment size by about 10% compared to Frag1 (for Threshold = 6 ADC counts) and shows more stability in fragment size. The same remains valid comparing the new versions of Frag5 and Frag1.
Payload: Frag5 reduces data network payload and improves its stability.
Scaling: Frag1 performance is scaling with the increasing luminosity, while Frag5 is much less affected by this factor.

Example 1: Laser calibration run (Fig. 1, signal ~70 ADC counts, time jittering (i.e. phase variation) ~25 ns): Frag1 will be forced to raise the threshold up to 70 ADC counts or drop the rate down to 55 kHz (for both versions), while Frag5 has recorded all raw data at 93 kHz rate. For the TileCal Long and Extended Barrels[2] two peaks appeared because some Tile modules[2] were temporarily switched off.

Example 2: Maximal average bandwidth load (Fig. 2) for TileCal Long Barrel (the average number of inelastic interactions per bunch crossing $\mu = 15.5$, bandwidth threshold for Frag1 is 6 ADC counts, Frag5 stores all raw data). For each of 32 Tile Read-Out Driver (ROD)[5] fragments produced by 16 Long Barrel RODs, the moving average of fragment sizes over 16 consecutive events is calculated. Among these 32 numbers corresponding to the ROD fragments the maximal is selected and the histogram created.

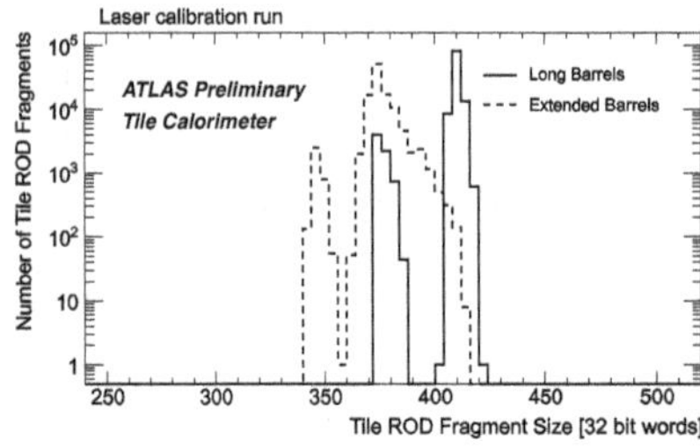

Fig. 1. Laser calibration run

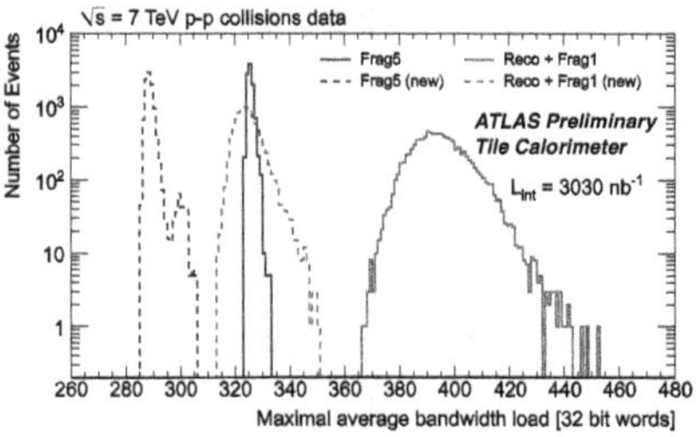

Fig. 2. Bandwidth load

The bandwidth limit for 100 kHz Level 1 trigger rate is 398 words.

To help ensure smooth incorporation of lossless compression scheme into present and operational system and to avoid possible (if any) impact on data taking during the transition, it was proposed to make Frag5 interchangeable on-the-fly with the current scheme (Reco+Frag1). This approach called Twin Mount Framework (TMF) was successfully implemented into the system. It increases the safety, stability and overall performance of the system:

- has no impact on current data taking;
- provides the opportunity of unobstructed development and validation of lossless compression, as well as Frag1, with a realistic goal to have lossless compression safely implemented in the system.

High Rate tests have shown, that installation of Twin Mount Framework does not affect the performance, i.e. Reco+Frag1 works at the same rate both with and without TMF.

The TileCal operation experience and the outcome, as well as the evolution of currently used strategy indicate that using lossless compression is very likely in the nearest future. Whether it happens by directly installing Frag5, which seems preferable, or gradually importing its solutions into Frag1, the mission of the first lossless compression tool will be fulfilled. Moreover, the similar approach may appear useful in the ATLAS Liquid Argon Calorimeter (LAr) which has very similar data acquisition scheme and uses the same Optimal Filtering approach for data reconstruction.[6] The LAr data exceed significantly those of the TileCal and take up almost a half of the ATLAS event size, thus compression here may appear particularly attractive. Lossless compression approach may be applied to the LAr to study the possibility of storing *all* the raw data without increasing the currently used data-flow and with a minor (if any) increase of the currently used capacity. Should it happen we will have managed to record all the raw data for ATLAS Calorimetry (LAr and TileCal). This will be done without installing additional hardware to upgrade the subdetectors.

4. Conclusions

A lossless compression tool (Frag5) is a fully functional software able to store all the TileCal raw data at 100 KHz rate fitting both within bandwidth and time limitations of the DSP. Some details of implementation have been highlighted and results of testing presented. Comparative study with respect to the existing approach is performed considering Frag5 as possible upgrade of current data reconstruction and storing strategy. Evolution and current state of both systems have been traced indicating the importance of compression schemes in the ATLAS Calorimetry data processing.

References

1. The ATLAS Collaboration, *The ATLAS Experiment at the CERN Large Hadron Collider*, JINST 3 (2008) S08003
2. *ATLAS Tile Calorimeter: Technical Design Report*, CERN-LHCC-98-015
3. A. Valero and the ATLAS Tile Calorimeter, *Implementation and performance of the signal reconstruction in the ATLAS Hadronic Tile Calorimeter*, http://cdsweb.cern.ch/record/1357024
4. V. Tsiskaridze and the ATLAS Tile Calorimeter, *Lossless compression of ATLAS Tile Calorimeter raw data*, 2010 J. Phys.: Conf. Ser. 219 022046
5. G. Usai and the ATLAS Tile Calorimeter, *Signal Reconstruction of the ATLAS Hadronic Tile Calorimeter: implementation and performance*, 2011 J. Phys.: Conf. Ser. 293 012056
6. The Liquid Argon Back End Electronics collaboration, *ATLAS liquid argon calorimeter back end electronics*, 2007 JINST 2 P06002

Advanced Detectors, Particles Identification, Devices and Materials in Radiation

M. Antonelli	Bunch by Bunch X-Ray Beam Position and Intensity Monitoring Using a Single Crystal Diamond Detector
C. Aramo	Recent Results in Silicon-CNT Photodetectors
R.L. Bates	3D silicon detectors
F. Bucci	The Ring Imaging Cherenkov Detector of the NA62 Experiment
A. Cemmi	Gamma Irradiation Effects on the Lead-Free Bi_2O_3-B_2O_3-SiO_2 Glasses
E. Coniavitis	Tau Reconstruction And Identification Performance in ATLAS
R. Eber	Silicon Sensor Developments for the CMS Tracker Upgrade
R. Giordano	Testing Radiation Tolerance of Electronics for the SuperB Experiment
L. Mazzaferro	The Resistive Plate Chambers of the ATLAS experiment: performance studies on Calibration Stream
L. Molnar	Cherenkov light in ALICE: performance of the HMPID detector
A. Morici	Radiation damage in soft ferromagnetic materials for enhanced inductor cores operating in extreme environments
M. Nagel	Neutron and Proton Tests of Different Technologies for the Upgrade of Cold Readout Electronics of the ATLAS Hadronic Endcap Calorimeter
M. Riva	Power Distribution Architecture for High Energy Physic Hostile Environment
A. Romano	Kaon identification in the NA62 experiment at CERN

A. Sparkes	Performance of the Rich Detectors at LHCb with 2011 Data
G. Ventura	Low Temperature Thermal Conductivity of Polymers and Composites: Recommended Values
M. Zeller	Overview of the R&D program on Liquid Argon TPCs under development at the University of Bern
A. Zibell	High Rate Proton Irradiation of 15mm Muon Drifttubes

BUNCH BY BUNCH X-RAY BEAM POSITION AND INTENSITY MONITORING USING A SINGLE CRYSTAL DIAMOND DETECTOR

M. ANTONELLI[§,@], A. TALLAIRE[#], J. ACHARD[#], S. CARRATO[§], G. CAUTERO[&], A. DE SIO[‡], M. DI FRAIA[‡], D. GIURESSI[&], R. H. MENK[&], E. PACE[‡]

@ Presenting Author, e-mail: matias.antonelli@phd.units.it

§ DI[3], University of Trieste, Italy

 # LSPM-CNRS, Université Paris 13, France

& Detectors and Instrumentation Laboratory, Sincrotrone Trieste, Italy

‡ Department of Physics and Astronomy, University of Florence, Italy

Diamond is an outstanding material for the production of semitransparent *in situ* photon beam monitors which can withstand the high dose rates occurring in new generation synchrotron radiation storage rings and in free electron lasers. Here we report on the development of a 500 μm thick freestanding, single-crystal chemical vapor deposited diamond detector with segmented electrodes; it exhibits a high resistivity of some 10^{15} Ω cm which allows charge integration operations. Using the latter at a frame rate of 8.33 kHz in combination with a needle synchrotron radiation beam and mesh scans, the inhomogeneity of the sensor was found to be of the order of 2%. With a measured electronics noise of 2 pA / Hz$^{1/2}$ a 0.05% relative precision in the intensity measurements (at 1 μA) and a 0.1 μm resolution in the position encoding have been estimated. Moreover, the high electron–hole mobility of diamond compared with other active materials enables very fast charge collection. This allowed us to utilize single pulse integration to simultaneously detect the intensity and the position of each synchrotron radiation photon bunch generated by a bending magnet.

1. Introduction

In the 3rd generation synchrotron radiation (SR) sources, the high brightness beamlines using undulator radiation are the most sensitive to electron beam oscillations. Therefore, local bump orbit feedback systems are developed to improve the stability of the delivered radiation.

The insertion of photon Beam Position Monitors (pBPM) is useful in order to provide those systems with important information, such as photon beam position and absolute intensity [1]; furthermore, several beamline experiments require these data to be known in order to give quantitative results [2].

Electron beam position stability has been intensively addressed in the past years by the use of Fast Orbit Feedback (FOFB) based on electron Beam Position Monitors (eBPM) [3]; conversely, beamlines have not been provided so far with a fast local control system based on the information collected by the pBPMs. Those measures are rarely used to compensate for long term thermal drifts (of the order of magnitude of minutes) or to adjust experimental data; recently, integration of pBPM information into the FOFB has been proposed [4]. Hence, for both diagnostics and calibration issues, several SR applications require an *in situ* detector showing high transparency, high radiation hardness, fast response and homogeneity.

Diamond is an outstanding material for the production of semitransparent *in situ* pBPMs [5]: because of its high bond energy it can withstand the high dose rates occurring in new generation SR storage rings; its low atomic number makes it semitransparent under certain conditions (involving both thickness and photon energy); besides, due to its high energy gap, intrinsic diamond is an insulator with low thermal noise at room temperature, while its high electron–hole mobility allows charge to be collected faster than in other active materials.

With these characteristics, diamond might be the only suited material to be utilized as pBPM for X-ray free electron lasers (FEL), which provide very short intense photon pulses with a peak power of some tens of MW.

This paper reports on the development of a pBPM based on a single-crystal chemical vapor deposited (CVD) diamond detector; experimental results are presented to witness the capabilities of this device for both low frequency and *bunch by bunch* beam monitoring.

2. Diamond Detector

A single-crystal CVD diamond layer has been grown homoepitaxially by microwave plasma-assisted CVD. After the synthesis, the sample was removed from the substrate by laser cutting and polished on both surfaces, resulting in a 500 μm thick freestanding layer with a 5×5 mm^2 area. It has been cleaned in class 1000 environment and provided with Cr–Au contacts deposited using standard thermal evaporation techniques.

As shown in Fig. 1, the two front electrodes are semicircles with a radius of 1 mm spaced by 100 μm; a guard ring surrounds these electrodes, but it has never been used in the presented experiments; on the rear side, a 4×4 mm^2 back plane electrode has also been deposited. The diamond sensor has then been connected to a gold-plated copper PCB using silver glue and wire bonds respectively for rear and front electrodes.

This detector is meant to be transversally inserted in the monitored beam; for radiations above 8 keV, a minor part of the incident photons is absorbed and electron–hole pairs are generated. When a bias voltage is applied between rear and front electrodes, free charge can be collected; consequently, the measured currents can provide information about beam position and intensity.

3. Experimental Setup

The described detector has been tested at the Microfluorescence bending magnet-beamline of the Elettra synchrotron in both *multi-bunch* mode and *single-bunch* mode; in the former, 432 electron bunches are distributed along the ring, spaced by 2 ns; in the latter, a single bunch is filled and accelerated with a revolution period of 864 ns.

After being generated by the accelerator, broad band radiation (infrared ÷ hard X-rays) propagates along evacuated pipes passing through several absorption stages: a 300 μm Be window, a 1.6 μm Al endcap, a 10 cm air gap and the 250 μm kapton endcap of the detector's housing. Since no monochromator has been inserted in the light path, the presented experiments have been performed with a radiation showing a maximum flux at energy of 20 keV and a spread of about 15 keV. At the end of the evacuated pipes a double-slit collimator has been used to obtain a needle-beam with adjustable cross section dimensions ranging from 75 μm to 2 mm.

The diamond detector has been mounted on an XY movable stage which is housed in an evacuated chamber provided with stepper motors, bias and signal cables; thus mesh scans of the sensor can be performed.

The diamond has been biased with voltages ranging from 15 V to 500 V and the generated photocurrents have been read by means of two different multi-channel acquisition systems, which work respectively at low and high sampling frequencies. In the first case charge integration and analog to digital conversion (ADC) at 8.33 kHz have been performed, while in the second one the signals are sent to 10 GHz ADCs directly or through radio-frequency pre-amplifiers.

The motor drivers and the acquisition electronics are controlled by a PC, which allows real-time data storage as well.

4. Results

At first the low frequency arrangement has been used to obtain sensitivity maps of the active area of the detector. Utilizing a needle-beam with a section of $70 \times 350\ \mu m^2$ in combination with the stepper motors, several mesh scans have been carried out; with such measures the inhomogeneity, defined as the standard deviation from the mean response, in the region of interest (i.e. the central part) has been found to be of the order of 2%. Moreover, with an electronic noise of $2\ pA / Hz^{1/2}$ measured at 1 μA, a 0.05% intrinsic relative precision in the intensity monitoring has been estimated. Figure 2 shows a map of the detector for a bias voltage of 300 V; the gray-scale intensity, representing the total measured current, is reasonably uniform in the central region.

Figure 1. Diamond detector.

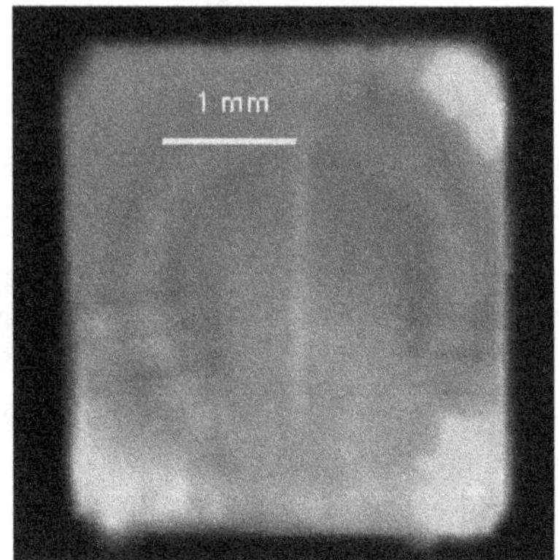

Figure 2. Mesh scan.

Separate current acquisition from each front electrode allows the position to be monitored as well. *Difference over sum* of left and right signals has been used to calculate the beam centroid position during horizontal scans; if those results are compared to the known ground-truth displacements, the detector proves to be capable of monitoring position in the central linear region with a precision of 150 nm (estimated using error propagation of standard deviation of the signals).

The high frequency setup has been used to exploit the fast response of the diamond; the monitoring system has been able to detect the signals generated by each photon bunch hitting the sensor in both multi-bunch and single-bunch operating modes.

In the first case, the presented detector has produced waveforms in which the bunches can be clearly distinguished from each other, though they are spaced by 2 ns only; Figure 3 shows a trace acquired during multi-bunch operations with a bias voltage of 300 V and a beam section of 2×2 mm^2.

In the second case, single pulses were spaced by 864 ns (i.e. the revolution period); since the machine bunch had a FWHM of about 150 ps, the acquired pulses can also be considered as impulse responses of the whole pBPM. An example of such a response is reported in Fig. 4, denoting rise times below 1 ns. Horizontal linear scans have also been performed in single-bunch mode; using the areas of the pulses generated by each channel, the beam position has been estimated with a precision of less than 6 μm.

The presented results can be compared with those produced by similar experiments on mono-crystalline [5] and poly-crystalline [6] CVD diamond detectors with regard to low frequency resolution and high frequency response respectively, while single-shot SR characterization has not been reported yet.

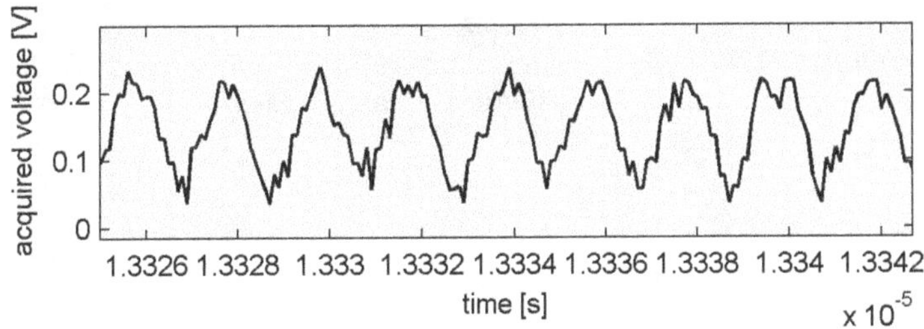

Figure 3. Acquired waveform in multi-bunch mode.

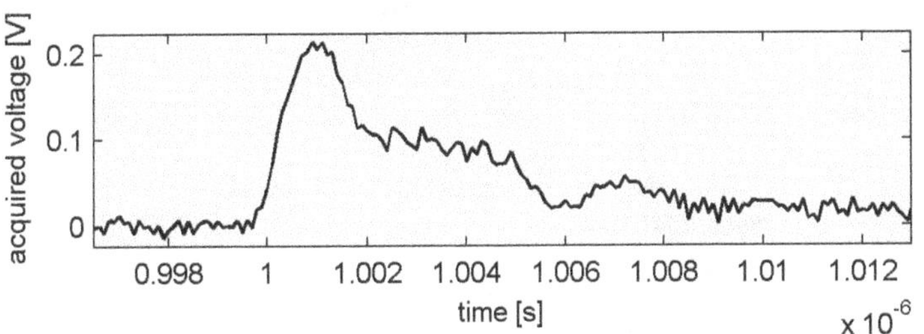

Figure 4. Single-bunch pulse response of the diamond detector.

5. Final Remarks

A single-crystal CVD diamond detector has been developed and characterized by means of multi-bunch and single-bunch SRs. The reported performances show that diamond detectors are well suited to be used as pBPMs in synchrotron beamlines; in particular, rise times below 1 ns and resolutions of less than 6 μm allow bunch by bunch position monitoring. Moreover, diamond has proved to be one of the most promising materials to be used as pBPM in FEL applications.

References

1. A. Galimberti *et al*, *Nucl. Instrum. Meth. A*, 477, 2002.
2. R. H. Menk *et al*, *AIP Conf. Proc*, 879, 2007.
3. M. Lonza *et al*, *10th ICALEPCS*, Geneva, Switzerland, October 2005.
4. T. Schilcher *et al*, *DIPAC '05*, Lyon, France, June 2005.
5. J. Morse *et al*, *Journal of Synchrotron Radiation*, 17, 2010.
6. P. Bergonzo *et al*, *Journal of Synchrotron Radiation*, 13, 2006.

RECENT RESULTS IN SILICON-CNT PHOTODETECTORS

C. ARAMO*
I.N.F.N. Sezione di Napoli
Dipartimento di Scienze Fisiche, Università degli Studi di Napoli Federico II,
Complesso Universitario di Monte Sant'Angelo
Napoli, 80126, Italy
**E-mail: carla.aramo@na.infn.it*

A. Ambrosio
CNR-SPIN U.O.S. di Napoli

M. Ambrosio, M. Cilmo
INFN Sezione di Napoli

F. Guarino, P. Maddalena
Dipartimento di Fisica, Università "Federico II" di Napoli and INFN

V. Grossi, M. Passacantando, S. Santucci
Dipartimento di Fisica, Università dell'Aquila and INFN

E. Nappi, A. Tinti, A. Valentini
Dipartimento di Fisica, Università di Bari and INFN

E. Fiandrini, G. Pignatel
INFN Sezione di Perugia e Dipartimento di Fisica, Università degli Studi di Perugia

P. Castrucci, M. De Crescenzi, M. Scarselli,
Dipartimento di Fisica, Università degli Studi di Roma Tor Vergata

It has been demonstrated from various authors that a Si-Carbon Nanotube heterojunction can be obtained by growing MultiWall Carbon nanotubes (MWCNT) over a silicon substrate. The electron transport characteristics of hybrid Si-CNT structures have been also largely investigated. Among the wide spectrum of nanotube characteristics, an important rule is determined by their capability to absorb light quanta creating a couple electron-hole that can be separated applying an external electric field. A few mm^2 nanotube layers contains an extremely large number of active elements that can convert incident light into electrons and generate an electrical signal both in case of pulsed light and of continuous radiation. This opens the way to the use of MWCNT for realizing a new kind of radiation detector to be used both for high energy and spatial physics and for sensor applications. In this paper we report on a new detector device realized using MWCNT growth over a silicon substrate. This device presents peculiar characteristics, low noise, good conversion efficiency of photons into electrical current and good signal linearity in a wide range of radiation wavelength from UV to IR at room temperature. The spectral behaviour reflects the silicon spectral range with a maximum at about 880 nm.

1. Introduction

Among all the new materials employed in nanotechnology applications, in the last twenty years, Carbon Nanotubes has been largely studied due to their innovative characteristics. One of the key features of such nanostructures for electronic and optoelectronic applications is that their metallic or semi conductive character depends on the chirality [1, 2, 3]: nanotubes can cover all the range from conductive to semi conductive to insulator simply changing his chiral index. In the case of more complex structures (Multi Wall Carbon Nanotubes – MWCNT) individual elements are composed by tens of SWCNT (Single Wall Carbon Nanotubes). In this case it is possible to suppose that a layer of MWCNT covers all the range of conductivity from metal to insulator and an electrical signal generated inside the layer can be transmitted and collected simply applying a draining voltage.

In this paper we report on a new photodetector characterized by a conduction threshold of a few volts. It is based on carbon nanostructures (CNTs), multi walled carbon nanotubes (MWCNTs) or carbon nanofibers, grown on silicon substrates by means of chemical vapour deposition (CVD) technique. The characteristics of this device seem to be very appealing in a variety of technological fields, from high energy physics to space applications.

2.. Carbon Nanotubes Detectors

2.1 Layout

A sketch of the device is reported in Figure 1a [4]. As silicon substrate we use 500 mm thick n-doped silicon substrates 5x7.8 mm^2 sized with an average carrier concentration of 3.45 10^{15} cm^{-3} and 3.2 Ωcm resistivity, covered on both its faces by a 120 nm silicon nitride (Si_3N_4) layer. MWCNTs are grown in the black area around electrodes, covering an area of 5x4 mm^2. On one side two AuPt electrodes, 1x1 mm^2 sized, are sputtered. On the other side a large area AuPt electrode covers the silicon surface and allows to apply an electrical field between the two faces. Then, carbon nanostructures were grown directly on the front surface by CVD. A 3 nm nickel (Ni) film was thermally evaporated on the substrate. Ni-deposited substrates were introduced into a CVD reactor (base pressure: 10^{-6} Torr) and heated at either 500 or 700°C for 20 min in NH_3 gas at a flow rate of 100sccm. During this period, Si_3N_4 layer prevented nickel diffusion into the silicon wafer so avoiding nickel silicides formation. CNTs were obtained by adding C_2H_2 at a flow rate of 20sccm for 10min at the same temperature of the NH_3 thermal treatment (500 or 700°C). During both the annealing and the CNTs growth, the pressure inside the CVD reactor was kept fixed at 5.5 Torr. Figure 1b and Figure 1c show scanning electron microscopy (SEM) images of the samples synthesized at 500°C and 700°C, respectively. Figure 1b exhibits CNTs very short, bended and characterized by different diameter along the structure.

The average diameter of these nanostructures was of 48±9nm. In Figure 1c, entangled carbon nanotubes (18±7nm in mean diameter), grown as a vertical carpet (Figure 1c, inset), are easily recognizable. High resolution TEM images, not reported here, were also recorded confirming for the former sample the presence of carbon nanofibers and for the latter the formation of MWCNTs.

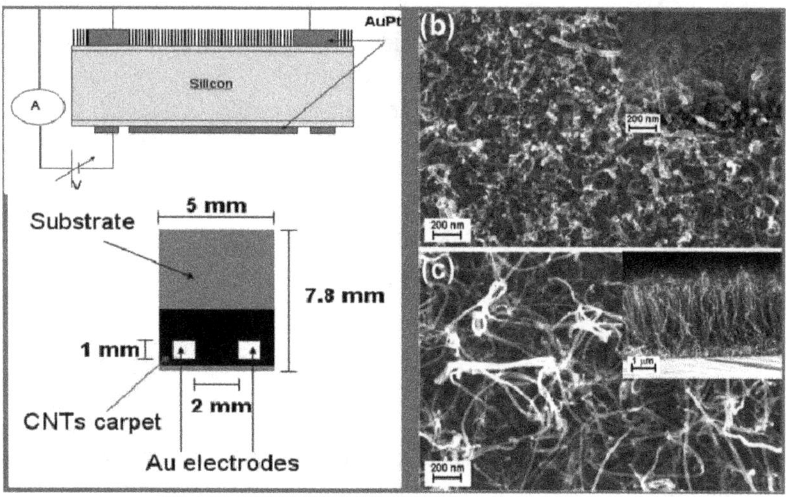

Figure 1. (a) Section and top view of the multilayer structure constituting the detector. Measurement layout is also depicted. Top and side (inset) SEM view of a CNS sample grown at either 500 (b) or 700°C (c).

2.2 Measurements

Current-voltage (I-V) measurements were performed at room temperature for the sample namely C2 at 500 °C. The drain voltage was applied between the two electrodes in the top and the electrode on the back (Fig. 1a). The detector shows a very low dark current, as reported in Fig. 2, and it is evident the presence of a Schottky junction.

To investigate the device properties as radiation detector we used diode laser emitting at different wavelengths. The laser light intensity was controlled by a low voltage power supply and measured with a power meter. Measurements were performed steeping the laser intensity from 0.1 mW to 1 mW with a step 0.1 mW, the draining voltage from -30 to 50 V step 1 V and the light wavelength at fixed values of 405, 532, 650, 685, 785, 880, 980 nm. No control on room temperature has been applied. Current were measured with a Keitley 2635 nanoamperometer that provides also the draining voltage applied to the Si-CNT detector. The measurement procedure was controlled automatically by a LabView program installed in a PC.

Figure 2 shows also the photocurrent measured in the described condition when the sample C2 is illuminated with a 650 nm continuous laser beam at various light intensities. The charge is drained along the silicon substrate by means of the applied electric field. The plot shows clearly the rectifying behavior of Si-CNT Nanotube Schottky junction. The sample operates as a Schottky diode with a very low reverse current and a direct current proportional to the laser light intensity. As the applied voltage increases, the current remains null up to a few volt threshold, then increases linearly until a saturation level (Fig.3).

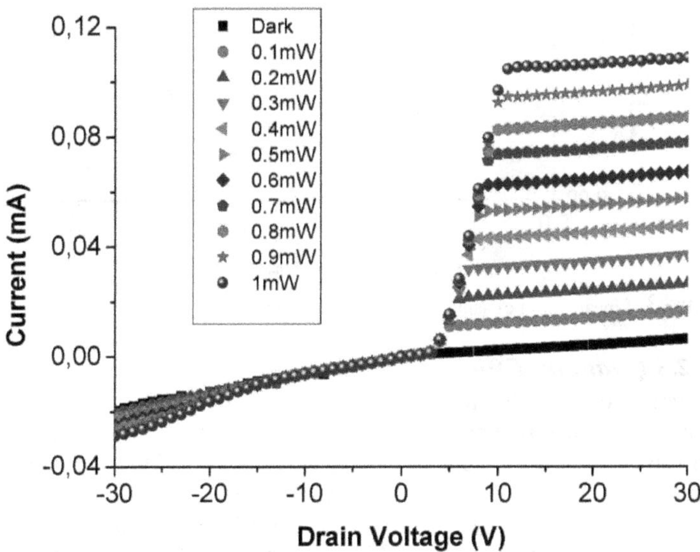

Figure 2. Photocurrent of sample C2 for various laser intensity and wavelength of 650 nm.

It can be seen as the generated signal is typical of a detector: the drained current increases with the drain voltage, up to a voltage value from which the current becomes constant.

This nearly constant current means that all photo-generated carriers have been collected at the electrodes. In the saturation region our device works as an ideal photodetector, in which the output signal depends only on the radiation intensity. Moreover, the proportionality between photocurrent in the plateau and light power is linear for all used wavelength, as reported in Fig. 3 for drain voltage of 25 V and wavelength of 785 nm.

The junction threshold has been measured to be 3.55 V for all intensities of illumination, while the threshold for current plateau obviously depends from the current to be drained, as shown in Fig. 4.

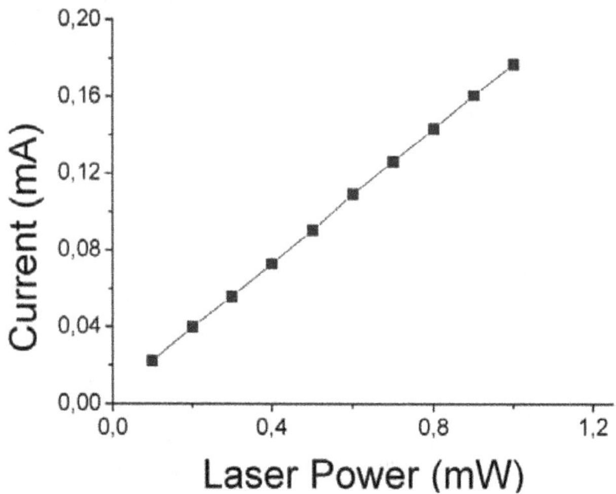

Figure 3. Linearity plot of sample C2 @V = 25 V and λ = 785 nm. The line is the linear fit curve.

2.3 Quantum Efficiency

The ratio between the number of electrons collected at saturation and the number of incident photons (QE - quantum efficiency) is reported in Fig. 5. For each wavelength, the efficiency obtained averaging values of 20 measurements is reported with experimental errors. As we expect due to the high linearity of detector, efficiency at a fixed wavelength doesn't depend on laser light intensity.

This is clearly shown in Fig. 5 where the QE measured at ten light intensities are perfectly superimposed. To take into account the fraction of light reflected by surfaces we measured reflected light at 650 nm. This is 0.7% of incident light, so that the bias due to surface light reflection is negligible. Note that we never use amplification devices. The QE of 31.1% measured at λ=880 nm and the QE of 25% measured at 980 nm are very interesting results for the use of this detector in the infrared wavelength region.

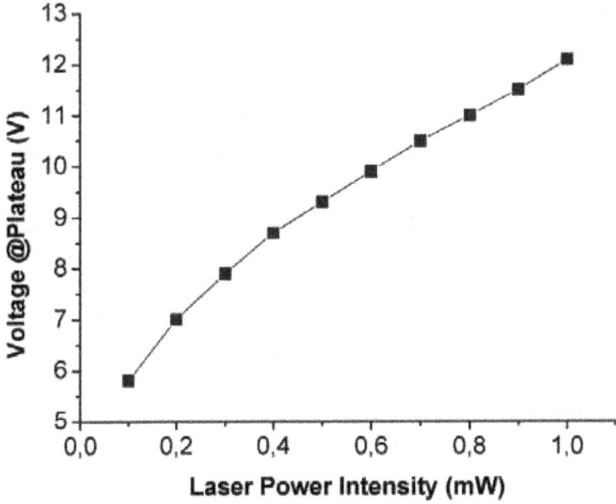

Figure 4. The voltage at the plateau depends from light intensity of laser.

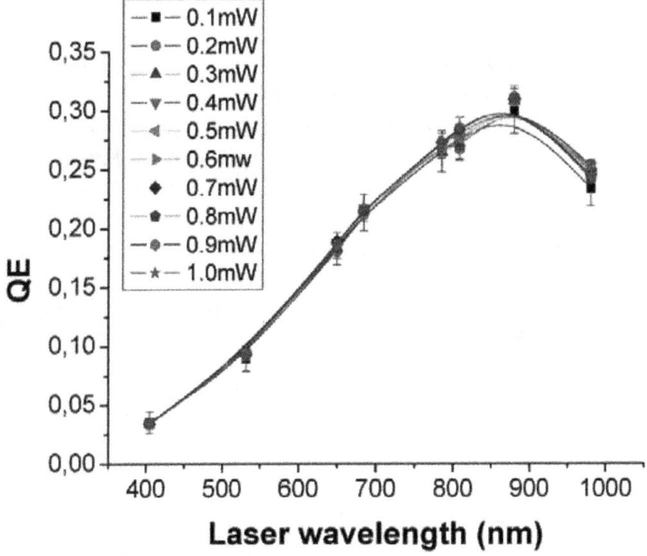

Figure 5. QE as a function of laser light wavelength for ten laser light intensities.

3. Conclusion

A novel photon detector made of Silicon and CNT has been realized. The detector is based on the effect of an heterojunction [5,6] created during the MWCNT deposition on a silicon substrate with CVD process. The main characteristics of this detector are: low threshold, low dark current, large plateau region, high linearity, stable at room temperature, quantum efficiency depends from light wavelength and CNTs growth temperature.

The detector surface is uniformly sensitive due to the uniform layer of nanotubes and no signal is obtained except in the area covered with CNTs. The described detector can be immediately produced industrially and used at room temperature as light sensor.

Acknowledgments

The authors would like to thank P. Di Meo, A. Pandalone e A. Vanzanella for their precious assistance for data acquisition and experimental instrumentation and B. Alfano for data acquisition programs.

References

1.Saito, R.; Dresselhaus, G. & Dresselhaus M.S. *Physical Properties of Carbon Nanotubes,* Imperial College Press, ISBN 978-1-86094-093-4, London (1998).

2.Dresselhaus, M.S.; Dresselhaus G & Avouris, P. *Carbon Nanotubes,* Springer-Verlag, ISBN 978-3-540-41086-7, Berlin (2001).

3.Jorio, A., Dresselhaus, G. & Dresselhaus M.S.. *Carbon Nanotubes, Advanced Topics in the Synthesis, Structure, Properties and Applications,* Springer-Verlag, ISBN 978-3-540-72864-1, Berlin (2008).

4.Ambrosio, M. *et al.* (2010) A novel photon detector made of silicon and carbon nanotubes. *Nuclear Instruments and Methods in Physics Research,* **Volume 617, Issues 1-3**, (May 2010), pp. 378-380, , ISSN 0168-9002.

5. Ambrosio A. and Aramo C. (2011). Carbon Nanotubes-Based Radiation Detectors, Carbon Nanotubes Applications on Electron Devices, Jose Mauricio Marulanda (Ed.), ISBN: 978-953-307-496-2, InTech, Available from http://www.intechopen.com/articles/show/title/carbon-nanotubes-based-radiation-detectors

6. Tinti A. et al. Electrical analysis of carbon nanostructures/silicon heterojunctions designed for radiation detection. Nuclear Instruments and Methods in Physics Research A 629, 377–381 (2011).

3D silicon detectors

R. L. Bates*
SUPA School of Physics and Astronomy, The University of Glasgow
Glasgow, G12 8QQ, UK
**E-mail: richard.bates@glasgow.ac.uk*
http://ppewww.physics.gla.ac.uk/~batesr/

Significant process in 3D detectors has taken place since Sherwood parker proposed the 3D silicon detector in 1997. The 3D detector was conceived as a method to overcome the radiation induced reduction in carrier lifetime in heavily irradiated silicon detectors via the use of advanced MEMS device fabrication techniques. This paper reviews the state of the art in 3D detectors. Work performed within the major fabrication institutes will be discussed, including modifications to the original design to reduce complexity and increase device yield. Characterization of 3D detectors up to the maximum radiation fluence expected at the high luminosity LHC operation will be presented. Results from both strip and pixel devices will be shown using characterization methods that include 90-Sr betas, focused laser and high-energy pions.

Keywords: silicon detector, 3D detector, radiation hard, pixel detector

1. Introduction

The LHC saw a change in the paradigm in the use of silicon detectors, in particle physics, from small scale vertex detectors to their use as large area trackers using both silicon strip and pixel detector assemblies. The predicted radiation fluence of the LHC experiments drove the development of silicon sensors that can be operable at fluence levels of 10^{15} cm^{-2} 1MeV n_{eq}, which was unprecedented at the time of planing the LHC experiments.

Over the next ten years the LHC based experiments will undergo upgrades to extend their physics reach. For example, the ATLAS experiment will place an additional pixel layer inside the existing pixel subdetector at a minimum radius of 31 mm from the interaction point, known as the Inner B–layer[1] (IBL). This will expose the silicon detectors to a fluence of 3 x 10^{15} cm^{-2} 1MeV n_{eq}. Including necessary safety factors, the design requires the silicon detectors to be operable at a fluence of

5×10^{15} cm^{-2} 1MeV n_{eq}. In addition the LHC will be upgraded, to the high luminosity LHC (HL-LHC), to deliver a factor of ten increase in the integrated luminosity. This will, in turn, increase the radiation field inside the experiments by the same ratio. As a consequence the silicon pixel detectors will have to survice a fluence up to 10^{16} cm^{-2} 1MeV n_{eq}. Which is a new radiation tolerance frontier. To this end new operational modes and device structures are being investigated for silicon detectors. The 3D detector is such a new device structure.

2. 3D silicon detector design

The 3D detector was first suggested by S. Parker[2] as a method to overcome the radiation induced reduction in the charge carrier mean free lifetime[3,4] which becomes a problem after a radiation fluence of order 10^{15} cm^{-2} 1MeV n_{eq}. At such a high radiation level full charge collection in planar silicon detector is not expected.

The 3D detector has an array of n- and p-type electrode columns passing through the silicon substrate rather than being implanted on its surface, as shown in Fig. 1. These electrodes are realised by a combination of deep reactive ion etching to realise the holes and low pressure vapour deposition to fill them with a dopant.[5,6] Standard detector technologies are used for surface structures. The design allows the combination of a standard substrate thickness with a lateral electrode spacing of a few tens of micrometers. Therefore the depletion and charge collection distances are reduced, without reducing the sensitive thickness of the detector. This implies that the device is extremely fast, has high charge collection and a low operating voltage, and therefore low power consumption, even after a high irradiation dose. The enclosed structure of the unit cell of the 3D detector will also reduce the amount of charge sharing, which could be advantageous in increasing the signal in a given pixel after heavy irradiation. These features should make 3D detectors substantially more radiation hard than standard planar devices.[7]

2.1. *Full 3D sensors*

The original 3D detector, known as the full 3D detector, is fabricated using one sided processing. It requires the sensor wafer to be wafer bonded to a support wafer and for the electrode holes to be completely filled with doped polysilicon. Both electrode types are connected together on the top side of the device as the fabrication is a single sided process. The back side

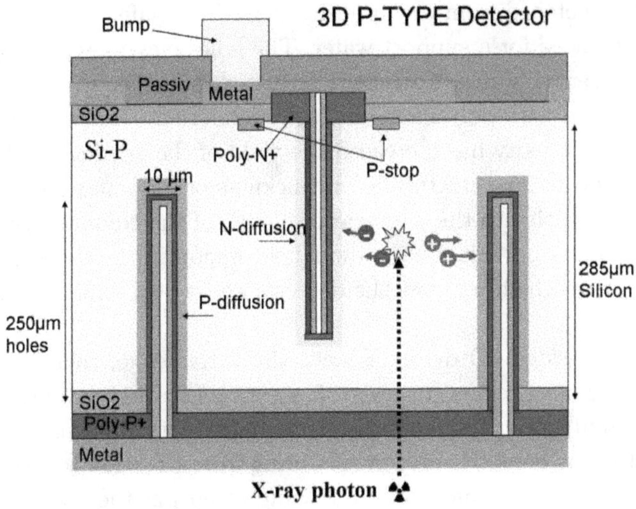

Fig. 1. Schematic of a 3D detector.

handeling wafer maybe removed to reduce material. The full 3D detector allows the addition of an electrode around the full detector matrix that contains the electric field and allows the detector to be active all the way to the edge of the device, known as an active edge.[8] Full 3D detectors have been fabricated at both Sintef[9,10] and Stanford.[5]

Sensors that are compatible with the ATLAS FE-I3[11] front-end pixel readout chip have been fabricated at Sintef, assembled and tested at CERN.[12] They show good electrical characteristics and full charge collection. The carrier lifetime in the filled column should be sufficient to collect significant charge from radiation incident upon the column. However, the charge collected in this region has been observed, with a fined focused x-ray beam, to be reduced from that expected. It is believed that the reason is that during the fabrication process an oxide is formed on the edge of the column which forms a barrier to carrier collection from the column.[13] A change in the dopant chemistry removing oxygen should reduce the trapping and increase charge collection from inside the column.

2.2. *Double-sided 3D sensors*

An alternative fabrication process, know as the double-sided 3D detector, was introduced[6,14] to reduce 3D detector fabrication complexity. This utilizes double-sided mask alignment technology which enables the columns of one type to be etched from the opposite side of the wafer to the other type, removing the need for a support wafer. The holes may pass all or part-way through the sensor wafer. To reduce stress in the wafer the holes are not completely filled with polysilicon and are therefore not active. The reduced fabrication complexity has increased the yield of the 3D device. However if the columns do not penetrate the full thickness of the wafer the sensor performance is sensitive to the penetration depth of the columns which needs to be well controlled. There is a lower field region in the detector directly above a column which requires the detector to be over depleted to be fully active.[15]

The double-sided 3D detector lacks the active edge, therefore to minimise the edge dead region the guard fence structure was developed.[16] This consists of multiple ohmic columns around the active silicon. The ohmic columns stop the depletion region spreading from the pixel to the cut edge. With a guard fence ohmic column spacing of 50 μm the cut edge can be as close as 100 μm from the active silicon without adversely affecting the device current-voltage characteristics.

3. 3D for CMS

Full 3D pixel sensors compatible with the CMS pixel readout chip have been fabricated by Sintef.[17] The pixel unit cell is 150 μm x 100 μm in size. Two different device configurations have been investigated: four junction columns (4E) and two junction columns (2E) per pixel. The distances between the centers of neighbouring junction and bias columns of the 2E and 4E sensors are 62.5 μm and 45 μm, respectively, while the diameter of the columns are 14 μm. The matrix is surrounded by an active edge. The sensors have been integrated into assemblies and tested.[18]

The device breakdown takes place at 100 V, well above the full depletion voltage of less than 40 V. The four electrode device has, as expected, higher noise than the 2E device due to the higher capacitance load. The noise is observed to fall at full depletion to between 250 and 300 electrons for the 2E configuration, compared to 100 electrons for a planar device.

From the preliminary beam test studies with 120 GeV protons, a signal to noise ratio (S/N) of 80 and a signal to threshold ratio (S/T) of 3 have been obtained for the non-irradiated 285 μm thick 2E detector. This S/T value is too low for the successful operation in the CMS detector and more work is being undertaken to lower the threshold value.

4. ATLAS IBL campaign

The double-sided 3D sensor has been recommended to be used for either 25% or 50% of the sensors in the ATLAS IBL. Both FBK and CNM are processing the sensors using a mask set that is as common as possible. Pixel sensors compatible with the ATLAS FE-I4[19] front-end pixel amplifier chip have been fabricated and extensively tested both before and after irradiation.[20,21] The sensor show a good uniformity over the matrix for threshold, noise and detection response. A noise performance of 150 electrons at a threshold of 3200 electrons have been routinely obtained before irradiation. After an irradiation of 5 x 10^{15} cm^{-2} 1 MeV n_{eq} from reactor neutrons the module threshold and noise performance is only slightly altered, with a noise increase of only 5 electrons to 155 electrons, when measured at -15°C. The current of the device increased as expected.

Test beam results show good detection efficiency and charge collection efficiency over the matrix before irradiation with and without a magnet field.[20] After irradiation to 6 x 10^{15} cm^{-2} 1 MeV n_{eq} 3D and planar sensors coupled to the FE-I4 pixel chip show similar hit detection efficiencies of approximately 97%. For tracks normal to the device low efficiency regions are observed around the electrodes. These appear as smeared regions of reduced efficiency for non-normal track angles.[22]

5. 3D detector test beam analysis

Many test beams have been performed on 3D detector modules. This paper concentrates on a test beam performed in a 120 GeV pion beam with CNM double-sided 3D pixel detectors coupled to the TimePix[23] readout chip, with pixel size of 55 μm x 55 μm. The columns have a diameter of 10 μm, with a depth of 250 μm in a 285 μm thick substrate. The devices collect holes with an n-type substrate and p-doped columns connected to the readout electronics. Using electrical characterisation lateral depletion was measured at 2 V and full depletion at 10 V. The devices was operated over depleted at 20 V, where it had a leakage current of 3.8 mA at room temperature.

The 3D detector assembly was placed on a precision x-y-theta stage at the centre of a beam telescope made from 6 TimePix planar silicon detector assemblies. The telescope has a pointing resolution for reconstructed tracks at the device under test of 2.3 ± 0.1 μm.[24]

The energy collected in the pixel unit cell was reconstructed as a function of incident position of the pion beam, as shown in Fig. 2. Full charge collection was observed for regions distant from the columns. Areas at the edge of the pixel cell exhibited charge sharing, manifested as a low energy peak in the pulse height spectra, see Fig. 2d. When the energy deposited in the neighbouring pixel is combined into a cluster, the cluster shows full charge collection all the way to the edge of the pixel cell, see Fig. 2e. The centre of the pixel cell shows full charge collection from the silicon above the column, that is 35 μm of silicon. The charge collected at the corner of the pixel is low as the charge collected in the 35 μm of silicon above the ohmic columns are shared between four pixels and falls below the pixel threshold. The average detection efficiency of the full pixel was found to

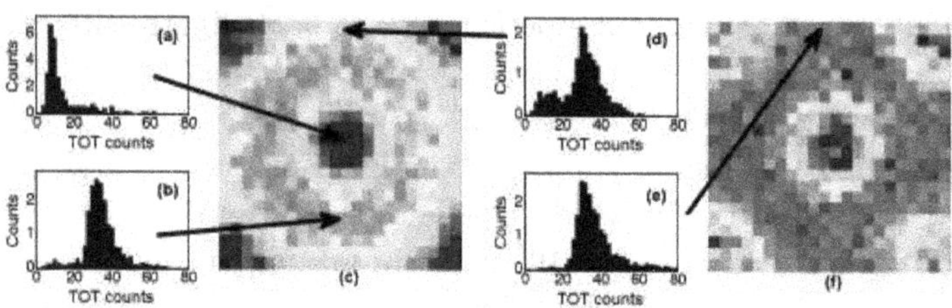

Fig. 2. The pulse height spectra as a function of incident position, obtained from a double-sided 3D detector fabricated by CNM connected to a TimePix pixel chip and illuminated by a high energy pion beam. The histograms are of the ToT counts in the central electrode region (a), away from the central electrode and pixel edges (b). Pixel maps showing the mean energy deposition across the pixel matrix, for a single pixel (c) and the energy in clusters (f). (d) and (e) show the histograms of the energy deposited at the pixel edges for the single pixel and the clusters.

be 93% for normal incidence due to the low efficiency in the corners of the pixel unit cell. As the angle of incidence was increase to 10 degrees the detection efficiency increased to 99.8 $\pm$ 0.5% across the pixel matrix. At an angle of 10 degrees the track traverses a full pixel within the thickness of the sensors. This is also the angle at which the best spatial resolution of 9.18 $\pm$ 0.1μm[24] is obtained. At normal incidence the spatial resolution is 15.8 $\pm$ 0.1μm which is in agreement with the binary resolution of the pixel of pitch 55 μm. This is due to the low charge sharing present in the 3D detector structure. More details of the test beam analysis can be found in Ref. 25.

6. Charge collection in 3D detectors

Short 285 μm thick 3D double-sided strip sensors have been fabricated to study charge collection after heavy irradiation. The short strip device has an 80 μm pitch between columns of the same type, strips of pitch 80 μm and 4 mm strip length. They were irradiated up to a fluence of 2 x 10^{16} cm^{-2} 1 MeV n_{eq}. They were tested with a 90-Sr source and readout with the LHC speed analogue front-end Beetle chip[26] integrated into the Alibava data acquisition system.[27] The detectors were cooled to -13.4 °C. For a full description see Ref. 28.

The results of the charge collections as a function of fluence for the device biased to 150 V in shown in Fig. 3a. The 3D detectors give full charge collection up to a fluence of 10^{15} cm^{-2} 1 MeV n_{eq}. The collected charge falls to 47% of the expected charge deposition after a fluence of 10^{16} cm^{-2} 1 MeV n_{eq}. The noise remains constant as a function of fluence and therefore the signal-to-noise ratio (S/N) as a function of fluence follows the collected charge curve. The collected charge in the 3D detector is well modelled by TCAD simulation without any high field effects being required. The collected charge is more than that observed in a 300 μm thick planar device operated at a bias voltage of 1000 V, which collected 30% of the expected deposited charge after a fluence of 10^{16} cm^{-2} 1 MeV n_{eq}. The larger collected charge in the 3D detector is due to the higher electric fields inside the 3D sensor and the shorter collection distances compared to the planar devices.

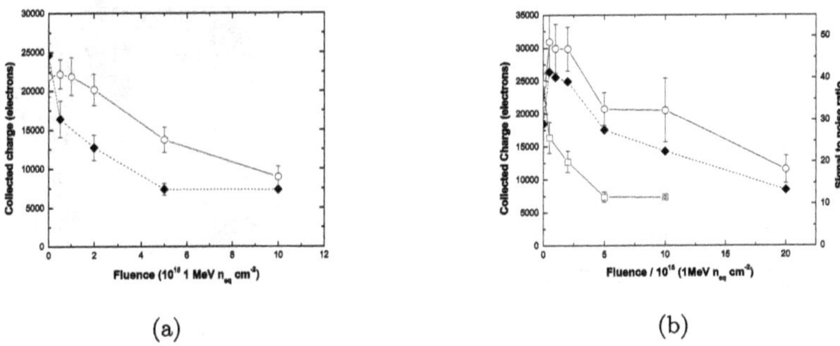

(a) (b)

Fig. 3. The collected charge as a function of fluence for a 285 μm thick 3D sensor and a 300 μm thick planar sensor operated at 1000 V (a) 3D operated at 150 V (open circles), planar as closed diamonds. (b) 3D operated at 250–350 V (open circles), planar as open squares. The 3D sensor's S/N is shown (closed diamonds).

The collected charge in the 3D detector increases greatly with a increase in bias voltage.[28] This is shown in Fig. 3b for bias voltages up to 350 V. The increase in charge is due to charge multiplication, which is evident for devices irradiated to less than 2 x 10^{15} cm^{-2} 1 MeV n_{eq}. The multiplication is believed to be due to impact ionisation which take place in the high field regions around the electrodes which extend through the device thickness. At these higher bias voltages a charge of 50% of that expected is collected after a fluence of 2 x 10^{16} cm^{-2} 1 MeV n_{eq}. The charge multiplication also leads to increased noise at the highest bias voltages and therefore lower S/N than expected, see Fig. 3b.

The Sr-90 measurements described above probe the average response of the detector, allowing the absolute charge collection efficiency and signal to noise ratios to be extracted. To probe the relative charge collection efficiency as a function of position in a unit cell of the device a focused laser system is employed. A 4 μm diameter focused infrared laser spot is raster scanned in 2 μm steps across the front surface of the sensor.[29] As the absorption depth of the 974 nm wavelength light in silicon is about 100 μm the charge collection process in the upper portion of the detector is investigated preferentially.

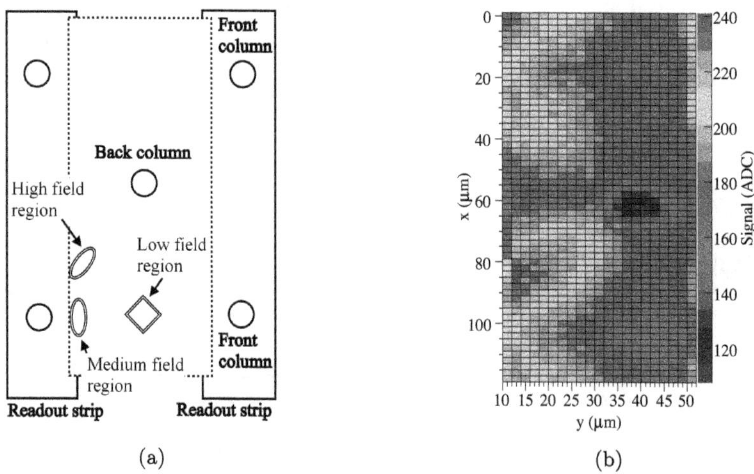

Fig. 4. (a) Diagram of the 3D strip detector under test with the scanned area showed as a dashed square. Three areas of differing electric field magnitude are labelled. (b) The response of the 3D detector biased to 260 V to collimated laser light after a fluence of 2 x 10^{15} cm^{-2} 1 MeV n_{eq}.

Laser scans on un-irradiated sensors show a uniform collection of charge outside of the electrode regions. In contrast to this large areas of non-uniformity can be seen in the unit pixel of irradiated sensors. Figure 4 shows the charge collected when the laser is scanned across a part of the unit cell illustrated. The sensor was irradiated to 2 x 10^{15} cm^{-2} 1 MeV n_{eq}) and biased at 260 V. Less charge is collected in regions of low field between columns of the same doping type and an enhanced signal in the region of high field between columns of opposite doping is observed. A higher field region will result in faster charge collection and less time for charge trapping resulting in greater charge collection. The relative signal from these regions was investigated for increasing bias voltage. The shape of the signal as a function of bias voltage was the same for all regions with a difference between the high and low field regions of about 30%. Below 150V an increasing voltage leads to an increase in the volume of depletion region throughout the device and therefore an increase in the charge collected in both regions. However, the increase in charge collection is modest. When the voltage is increased from 150 V to 260 V the charge collection increases dramatically to be over twice that collected at 150 V. This is due to charge multiplication effects as mentioned above.

7. Conclusions

3D silicon detectors are now well developed. The full 3D and the double-sided 3D detector can both be fabricated in industry. The double-sided device to date has a better yield due to the less demanding processing technology. The double-sided 3D detector has been chosen as a sensor solution for the first upgrade to the ATLAS silicon system; namely the IBL.

Precision scans of the pixel have been preformed in high energy test beams. The charge deposition in a unit cell has been mapped and full charge collection is observed in the majority of the cell outwith the electrode columns. High detection efficiency across the 3D pixel matrix at zero degree beam incident angle has been observed. This increases to 100% at an angle of incidence of 10 degrees where charge is never deposited only in the electrodes. The charge sharing has been shown to be less than for a planar device resulting in lower spatial resolution (consistent with binary readout) however, this might result in more charge being collected in the hit pixel after irradiation.

90-Sr source test shows higher charge collected in the 3D detector compered to the planar device. After an irradiation dose of 10^{16} cm^{-2} 1 MeV n_{eq} a 3D detector operated at a modest bias voltage of 150 V collected charge of 47% of the deposited charge compared to 30% in the planar device operated at 1000 V. At higher bias voltages, (typically 300 V), charge multiplication was observed in the 3D irradiated device. Spatially resolved laser scanning showed a uniform response across the unit cell outside the column region for an un-irradiated sensor when operated above full depletion. After heavy irradiation the response was non-uniform with an area of enhanced signal in the region of high field between columns of opposite doping.

References

1. M. Capeans, G. Darbo, K. Einsweiller, M. Elsing, T. Flick, M. Garcia-Sciveres, C. Gemme, H. Pernegger, O. Rohne and R. Vuillermet, *ATLAS Insertable B-Layer Technical Design Report*, CERN technical report CERN-LHCC-2010-013. ATLAS-TDR-019, CERN (Geneva, Switzerland, 2010).
2. S. I. Parker, C. J. Kenney and J. Segal, *Nucl. Instr. and Meth A* **395**, 328 (1997).
3. G. Kramberger, V. Cindro, I. Mandic, M. Mikuz and M. Zavrtanik, *Nucl. Instr. and Meth A* **476**, 645 (2002).
4. A. G. Bates and M. Moll, *Nucl. Instr. and Meth A* **555**, 113 (2005).
5. S. Parker, C. Kenney, J. Segal and C. Storment, *IEEE Trans. Nucl. Sci.* **46**, 1224 (1999).
6. G. Pellegrini, M. Lozano, M. Ulln, R. Bates, C. Fleta and D. Pennicard, *Nucl. Instr. and Meth A* **592**, 38 (2008).
7. C. DaVia and S. J. Watts, *Nucl. Instr. and Meth A* **603**, 319 (2009).
8. C. J. Kenney, S. Parker and E. Walckiers, *IEEE Trans. Nucl. Sci.* **48**, 2405 (2001).
9. T.-E. Hansen, A. Kok, T. A. Hansen, N. Lietaer, M. Mielnik, P. Storas, C. DaVia, J. Hasi, C. Kenney and S. Parker, *JINST* **4**, p. P03010 (2009).
10. A. Kok, T. E. Hansen, T. A. Hansen, N. Lietaer, S. Anand, C. Kenney, J. Hasi, C. DaVia and S. I. Parker, Fabrication of 3d silicon sensors, in *Proceedings of Science, the 19th International Workshop on Vertex Detectors*, (Loch Lomand, UK, 2010).
11. I. Perica, L. Blanquart, G. Comes, P. Denes, K. Einsweiler, P. Fischer, E. Mandelli and G. Meddeler, *Nucl. Instr. and Meth A* **565**, 178 (2006).
12. A. Kok, Fabrication of full 3d active edge sensors(March 2011), 6th "Trento" workshop on advanced silicon radiation detectors. `https://indico.cern.ch/contributionListDisplay.py?confId=114255`.
13. J. Hasi, Status of 3d sensors processing at stanford(February 2010), 5th "Trento" workshop on advanced silicon radiation detectors. `http://agenda.hep.manchester.ac.uk/conferenceDisplay.py?confId=1181`.
14. A. Zoboli, M. Boscardin, L. Bosisio, G.-F. Betta, S. Piemonte, C. Ronchin and N. Zorzi, *IEEE Trans. Nucl. Sci.* **55**, 2775 (2008).
15. D. Pennicard, G. Pellegrini, M. Lozano, R. Bates, C. Parkes, V. O'Shea and V. Wright, *IEEE Trans. Nucl. Sci.* **54**, 1435 (2007).
16. G. D. Betta, A. Bagolini, M. Boscardin, L. Bosisio, P. Gabos, G. Giacomini, C. Piemonte, M. Povoli, E. Vianello and N. Zorzi, Development of modified 3d detectors at fbk, in *IEEE Nuclear Science Symposium Conference Record*, (Knoxville, USA, 2010).
17. O. Koybasi, D. Bortoletto, T. Hansen, A. Kok, T. Hansen, N. Lietaer, G. U.

Jensen, A. Summanwar, G. Bolla and S. W. L. Kwan, *IEEE Trans. Nucl. Sci.* **57**, 2897 (2010).

18. O. Koybasi, E. Alagoz, A. Krzywda, K. Arndt, G. Bolla, D. Bortoletto, T.-E. Hansen, T. A. Hansen, G. U. Jensen, A. Kok, S. Kwan, N. Lietaer, R. Rivera, I. Shipsey, L. Uplegger and C. DaVia, *IEEE Trans. Nucl. Sci.* **58**, 1315 (2011).
19. M. Garcia-Sciveres, D. Arutinov, M. Barbero, R. Beccherle, S. Dube, D. Elledge, J. Fleury, D. Fougeron, F. Gensolen, D. Gnani, V. Gromov, T. Hemperek, M. Karagounis, R. Kluit, A. Kruth, A. Mekkaoui, M. Menouni and J.-D. Schipper, *Nucl. Instr. and Meth A* **636**, 155 (2011).
20. P. Grenier, G. Alimonti, M. Barbero, R. Bates, E. Bolle, M. Borri, M. Boscardin, C. Buttar, M. Capua, M. Cavalli-Sforza, M. Cobal, A. Cristofoli, G.-F. D. Betta, G. Darbo, C. D. Vi, E. Devetak, B. DeWilde, B. D. Girolamo, D. Dobos, K. Einsweiler, D. Esseni, S. Fazio, C. Fleta, J. Freestone, C. Gallrapp, M. Garcia-Sciveres, G. Gariano, C. Gemme, M.-P. Giordani, H. Gjersdal, S. Grinstein, T. Hansen, T.-E. Hansen, P. Hansson, J. Hasi, K. Helle, M. Hoeferkamp, F. Hgging, P. Jackson, K. Jakobs, J. Kalliopuska, M. Karagounis, C. Kenney, M. Khler, M. Kocian, A. Kok, S. Kolya, I. Korokolov, V. Kostyukhin, H. Krger, A. L. Rosa, C. Lai, N. Lietaer, M. Lozano, A. Mastroberardino, A. Micelli, C. Nellist, A. Oja, V. Oshea, C. Padilla, P. Palestri, S. Parker, U. Parzefall, J. Pater, G. Pellegrini, H. Pernegger, C. Piemonte, S. Pospisil, M. Povoli, S. Roe, O. Rohne, S. Ronchin, A. Rovani, E. Ruscino, H. Sandaker, S. Seidel, L. Selmi, D. Silverstein, K. Sjbk, T. Slavicek, S. Stapnes, B. Stugu, J. Stupak, D. Su, G. Susinno, R. Thompson, J.-W. Tsung, D. Tsybychev, S. Watts, N. Wermes, C. Young and N. Zorzi, *Nucl. Instr. and Meth A* **638**, 33 (2011).
21. A. Micelli, K. Helle, H. Sandaker, B. Stugu, M. Barbero, F. Hgging, M. Karagounis, V. Kostyukhin, H. Krger, J.-W. Tsung, N. Wermes, M. Capua, S. Fazio, A. Mastroberardino, G. Susinno, C. Gallrapp, B. D. Girolamo, D. Dobos, A. L. Rosa, H. Pernegger, S. Roe, T. Slavicek, S. Pospisil, K. Jakobs, M. Khler, U. Parzefall, G. Darbo, G. Gariano, C. Gemme, A. Rovani, E. Ruscino, C. Butter, R. Bates, V. Oshea, S. Parker, M. Cavalli-Sforza, S. Grinstein, I. Korokolov, C. Pradilla, K. Einsweiler, M. Garcia-Sciveres, M. Borri, C. D. Vi, J. Freestone, S. Kolya, C. Lai, C. Nellist, J. Pater, R. Thompson, S. Watts, M. Hoeferkamp, S. Seidel, E. Bolle, H. Gjersdal, K.-N. Sjoebaek, S. Stapnes, O. Rohne, D. Su, C. Young, P. Hansson, P. Grenier, J. Hasi, C. Kenney, M. Kocian, P. Jackson, D. Silverstein, H. Davetak, B. DeWilde, D. Tsybychev, G.-F. D. Betta, P. Gabos, M. Povoli, M. Cobal, M.-P. Giordani, L. Selmi, A. Cristofoli, D. Esseni, P. Palestri, C. Fleta, M. Lozano, G. Pellegrini, M. Boscardin, A. Bagolini, C. Piemonte, S. Ronchin, N. Zorzi, T.-E. Hansen, T. Hansen, A. Kok, N. Lietaer, J. Kalliopuska and A. Oja, *Nucl. Instr. and Meth A* **650**, 150 (2011).
22. J. Weingarten, Irradiation and beam tests qualification for atlas ibl pixel modules(September 2011), The 9th international conference on position sensitive detectors. `http://indico.cern.ch/conferenceDisplay.py?confId=48618`.
23. X. Llopart, R. Ballabriga, M. Campbell, L. Tlustos and W. Wong, *Nucl.*

Instr. and Meth A **581**, 485 (2007).

24. K. Akiba, M. Artuso, R. Badman, A. Borgia, R. Bates, F. Bayer, M. van Beuzekom, J. Buytaert, E. Cabruja, M. Campbell, P. Collins, M. Crossley, R. Dumps, L. Eklund, D. Esperante, C. Fleta, A. Gallas, M. Gandelman, J. Garofoli, M. Gersabeck, V. V. Gligorov, H. Gordon, E. H. Heijne, V. Heijne, D. Hynds, M. John, A. Leflat, L. F. Llin, X. Llopart, M. Lozano, D. Maneuski, T. Michel, M. Nicol, M. Needham, C. Parkes, G. Pellegrini, R. Plackett, T. Poikela, E. Rodrigues, G. Stewart, J. Wang and Z. Xing, *Nucl. Instr. and Meth A* **PP** (2011).
25. A. M. Raighne, K. Akiba, L. Alianelli, R. Bates, M. van Beuzekom, J. Buytaert, M. Campbell, P. Collins, M. Crossley, R. Dumps, L. Eklund, C. Fleta, A. Gallas, M. Gersabeck, E. N. Gimenez, V. V. Gligorov, M. John, X. Llopart, M. Lozano, D. Maneuski, J. Marchal, M. Nicol, R. Plackett, C. Parkes, G. Pellegrini, D. Pennicard, E. Rodrigues, G. Stewart, K. J. S. Sawhney, N. Tartoni and L. Tlustos, *JINST* **6**, p. P05002 (2011).
26. S. Lochner and M. Schmelling, *The Beetle Reference Manual - chip version 1.3, 1.4 and 1.5*, CERN technical report LHCb-2005-105. CERN-LHCb-2005-105, CERN (Geneva, Switzerland, 2006).
27. R. Marco-Hernandez, *IEEE Trans. Nucl. Sci.* **56**, 1642 (2009).
28. R. L. Bates, C. Parkes, B. Rakotomiaramanana, C. Fleta, G. Pellegrini, M. Lozano, J. P. Balbuena, U. Parzefall, M. Koehler, M. Breindl and X. Blot, *Nuclear Science, IEEE Transactions on* **PP**(December 2011).
29. A. Zoboli, G.-F. D. Betta, M. Boscardin, L. Bosisio, S. Eckert, S. Khn, U. Parzefall, C. Piemonte, S. Ronchin and N. Zorzi, *Nucl. Instr. and Meth A* **604**, 238 (2009), ¡ce:title¿PSD8¡/ce:title¿ ¡ce:subtitle¿Proceedings of the 8th International Conference on Position Sensitive Detectors¡/ce:subtitle¿.

The Ring Imaging Cherenkov Detector of the NA62 Experiment

F. BUCCI*

INFN - Sezione di Firenze
Via G. Sansone 1, 50019 Sesto Fiorentino
Firenze, Italy
**E-mail: francesca.bucci@fi.infn.it*

The NA62 experiment is designed to measure the branching ratio of the decay $K^+ \to \pi^+ \nu\overline{\nu}$ with a 10% accuracy at the CERN SPS. To suppress the main background coming from the $K^+ \to \mu^+\nu$ decay, a Ring Imaging Cherenkov detector (RICH), able to separate π and μ in the momentum range between 15 and 35 GeV/c with a muon contamination in a pion sample $< 10^{-2}$ is needed. The RICH must also have an unprecedented time resolution (100 ps) to disentangle accidental time associations of beam particles with pions. The last updates of the detector layout are presented along with the results of the beam tests of the RICH prototype: the muon misidentification probability was found to be 0.7% and the time resolution $<$ 100 ps in all the momentum range.

Keywords: PID; RICH; timing; kaon rare decays

1. The NA62 experiment

NA62 is a fixed target experiment which aims at measuring the branching ratio of the $K^+ \to \pi^+ \nu\overline{\nu}$ decay with a precision of 10% by collecting $\mathcal{O}(100)$ signal events with a signal to background ratio $S/B \approx 10$ in two years of data taking.[1] The contributions to this process due to the Standard Model are extremely suppressed and can be calculated with excellent precision ($BR_{SM} = (7.81 \pm 0.75 \pm 0.29) \times 10^{-11}$) .[2] This makes the decay a favoured place to look for possible new physics. The measurement is experimentally challenging and its present value, $(1.73^{+1.15}_{-1.05}) \times 10^{-10}$, has an uncertainty of more than 50%.[3]

A 75 GeV/c momentum beam derived from a flux of 400 GeV/c protons hitting a Be target is employed. 6% of particles are kaons, the rest are pions and protons. The NA62 detector (Fig. 1) is described elsewhere.[4] An incoming kaon and an outgoing pion in time with the incoming kaon are the signal signature. To match the downstream track to the correct

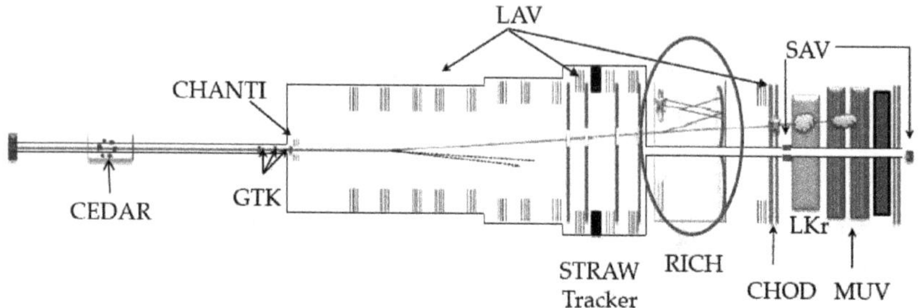

Fig. 1. The NA62 layout

kaon the time of both the upstream and downstream tracks need to be measured at 100 ps level. The main background, due to the $K^+ \to \mu^+\nu$ decay where the muon is misidentified as a pion, requires a 10^{12} rejection factor. According to MC simulation, with a kinematical cut and a muon veto system a total 10^{10} suppression factor can be achieved. Thereby, a Ring Imaging Cherenkov detector (RICH) must provide a further rejection of at least 10^2.

2. RICH

The RICH (Fig. 2) is composed of a 18 m long cylindrical vessel filled with neon at atmospheric pressure. Two semispherical mirrors are placed at the downstream end of the vessel. The Cherenkov photons are reflected on two focal regions equipped with $\sim$ 1000 photomultiplier tubes (PMTs) each. Since the area to be covered by each mirror is very large, a mosaic of smaller segments is used.

2.1. *Vessel and radiator*

The vessel is a vacuum proof segmented tube with a diameter varying from 4 to 3.4 m. No gas recirculation and purification system is foreseen. Since the momentum range over which π and μ must be identified goes from 15 to 35 GeV/c, neon gas at roughly atmospheric pressure is chosen as radiator ($(n-1) = 63.7 \times 10^{-6}$ at $\lambda = 300$ nm and 20° C; $p_{th} = 12.3$ GeV/c for π).

2.2. *Mirrors*

A mosaic of 20 mirrors with a focal length of 17 m is used, 18 hexagonal and 2 semi-hexagonal. The mirrors are made of a 2.5 cm thick glass substrate

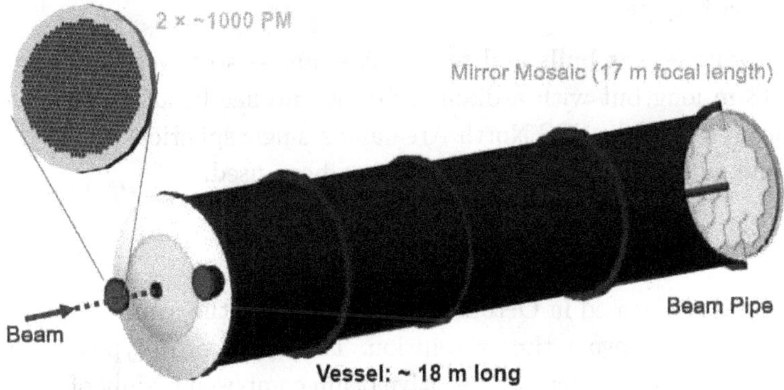

Fig. 2. The RICH detector

coated with aluminium. Each mirror must be supported and adjustable for alignment. A dowel with a spherical head is inserted in the hole drilled on the not-reflecting surface and used to sustain the mirror. Two actuating ribbons give the horizontal and vertical alignment. Piezo motors, located out of the acceptance, are used to pull the ribbons. An aluminium honeycomb structure, 5 mm thick, has been chosen as mirror support structure to minimize the material in front of the electromagnetic calorimeter.

2.3. *Photon Detection*

The best compromise between fast response, small dimensions and cost seems to be the PMT Hamamatsu R7400U-03. It has a good response up to the near ultraviolet (185 nm) with a peak quantum efficiency of $\sim 20\%$ at 420 nm. The PMT will be operating at 900 V with a gain of $\sim 1.5 \times 10^6$. The transit time jitter is 280 ps (FWHM). To convey the Cherenkov light to the active area of the PMT a Winston cone[5] covered with an aluminized mylar foil is used. The PMTs are separated from the neon by 1 mm thick quartz windows and are mounted on an external aluminium window. A cooling system is required to avoid a local heat source on the radiator due to the power dissipation of the PMT HV divider.

To exploit the fast PMT response, the NINO[6] ASIC was chosen as discriminator. To match the optimal NINO performance region, the PMT output is sent to a current amplifier with a differential output. The readout system relies on HPTDCs embedded in a TEL62 board, which is a development of the TELL1 board designed for the LHCb experiment.[7]

3. RICH prototype

A RICH prototype was built and tested. A stainless steel vessel, vacuum resistant, 18 m long but with a diameter of 60 cm, has been placed along the K12 beam line in the SPS North Area and a single spherical mirror, 2.5 cm thick and with a focal length of 17 m, has been used.

3.1. *2007 test beam*

A first test was performed in October 2007 to measure the average number of hit PMTs and the event time resolution. The RICH prototype was exposed to a 200 GeV/c momentum negative beam composed mainly of pions. The detector was equipped with a limited number of PMTs (96) placed in the region where the Cherenkov ring of a 200 GeV/c pion was expected. The average number of hit PMTs per event was found to be 17 and the RMS of the average event time was measured to be $\sim$ 70 ps.[8]

3.2. *2009 test beam*

The RICH prototype was tested again in May-June 2009.[9] The improvements with respect to the previous test were: the larger number of PMTs (414), a new readout electronics based on the TELL1 board and a water cooling system. Positive hadron beams with different momenta in the 10 to 75 GeV/c range were used.

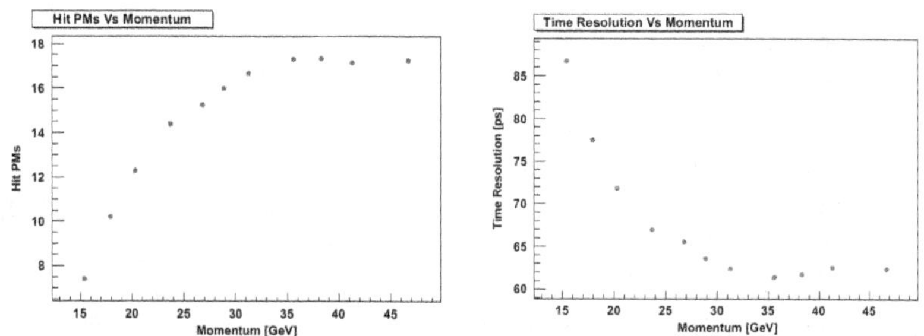

(a) Average number of hit PMTs for a single ring as a function of the momentum. (b) Time resolution as a function of the momentum.

Fig. 3.

The number of hit PMTs per ring as a function of the momentum is presented in Fig. 3(a). Figure 3(b) shows the event time resolution: a value

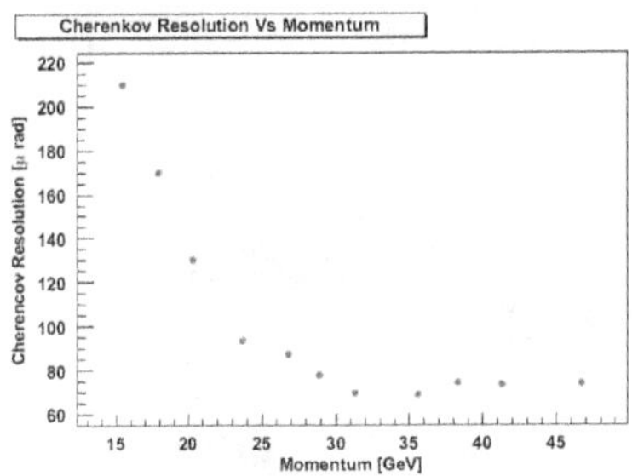

(a) Cherenkov angle resolution as a function of the momentum

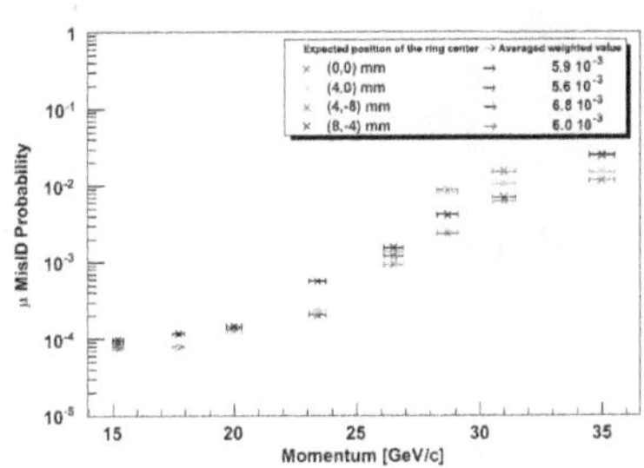

(b) Muon misidentification probability measured for 4 alignment positions of the mirror as a function of the momentum.

Fig. 4.

below 100 ps has been measured over the momentum range of interest. The Cherenkov angle resolution (Fig. 4(a)) decreases to a constant value of about 70 μrad for $\beta = 1$ particles. The π/μ separation in the momentum range 15-35 GeV/c was also measured. Figure 4(b) shows the muon misidentification probability as a function of the particle momentum, measured for four different positions of the mirror with respect to the beam line. The overall integral of the measurements gives a muon misidentification probability of $< 0.7\%$.

4. Conclusions

The design parameters of the NA62 RICH have been validated by the results of the beam tests performed in 2007 and 2009. The project matches the requirements expected for the NA62 experiment. The final detector is under construction and the first physical run is foreseen in Spring 2014.

References

1. G. Anelli et al., CERN-SPSC-2005-013, SPSC-P326;
2. J. Brod, M. Gorbahn and E. Stamou, *Phys. Rev D* **83**, 034030 (2011).
3. A. Artamonov et al. E949 Collaboration, *Phys. Rev. Lett.*, **101**, 191802 (2008).
4. F. Hahn (ed.) et al. NA62-10-07, December 2010.
5. R. Winston, *J. Opt. Soc. Am.* **60**, 245-247 (1970).
6. F. Anghinolfi et al., *Nucl. Instr. and Meth. A* **533**, 183-187 (2004).
7. C. Avanzini et al., *Nucl. Instr. and Meth. A* **623**, 543-545 (2010).
8. G. Anzivino et al., *Nucl. Instr. and Meth. A* **593**, 314-318 (2008).
9. B. Angelucci et al., *Nucl. Instr. and Meth. A* **621**, 205-211 (2010).

GAMMA IRRADIATION EFFECTS ON THE LEAD-FREE Bi_2O_3-B_2O_3-SiO_2 GLASSES

S. BACCARO, A. CEMMI
ENEA-UTTMAT-IRR, Via Anguillarese 301, 00123 Rome, Italy

<alessia.cemmi@enea.it>

T. WANG, G. CHEN
Key Laboratory for Ultrafine Materials of Ministry of Education, School of Materials Science and Engineering, East China University of Science and Technology, Shanghai 200237, China

In the present work, gamma irradiation effects on lead-free Bi_2O_3-B_2O_3-SiO_2 heavy metal oxide glasses are investigated. By doping Ti^{4+} ions, irradiation resistance of glasses is improved, especially in the case of the higher radiation dose. However, traditionally used stabilizing agent Ce^{4+} ions exert a negative effect on glasses by increasing the value of irradiation induced absorption coefficient, μ. This is maybe due to the different role of Ce^{4+} ions in the present Bi_2O_3-B_2O_3-SiO_2 glass structure..

1. Introduction

Heavy metal oxide glasses based on PbO have been extensively studied due to their excellent infrared transmission, increased third-order nonlinear optical effects, and high radiation resistance [1,2]. As the development of corresponding environmental regulations, the wide use of PbO containing HMO glasses has been restricted due to their harmful effects on the health and environment. Due to the similarity of properties between Bi^{3+} and Pb^{2+}, Bi_2O_3 is considered as a suitable substitute material which can replace PbO in the preparation of Pb-free HMO glasses [3-5]. S. Baccaro et al. compared gamma irradiation effects on Bi_2O_3-B_2O_3-SiO_2 and PbO-B_2O_3-SiO_2 glasses which showed that the former had a much lower irradiation resistant ability [6]. This limits to a great extent the application of bismuth oxide glasses for some fields like radiation shielding.

It is known that Ce^{3+}-doping plays a positive role in improving the radiation hardness of the glasses [7-10]. While combining use of CeO_2 and TiO_2 has been already applied to improve irradiation resistance of solar covering glasses [8-9] and ZnO-based scintillating glasses [10]. The aim of the present work is to investigate the irradiation resistance of Bi_2O_3 based borosilicate glasses via the introduction of CeO_2 and/or TiO_2.

2. Experimental

Commercial grade chemicals of Bi_2O_3, B_2O_3, SiO_2, TiO_2, CeO_2 were used as starting materials for preparation of the glasses samples by conventional melt-quenching technique. As bismuth oxide glasses usually present orange, brown or even dark-brown colors with increasing Bi_2O_3 concentration, $Ba(NO_3)_2$ and Sb_2O_3 were added (0.5 wt % each) as oxidant to prevent the process of darkening [11].

These starting materials were mixed in an appropriate proportion according to glass compositions as shown in Table 1, and melted in alumina crucibles at 1200℃-1250℃ for 40min in air. After melting, molten glass liquid was poured directly into a preheated stainless-steel mold for quenching and annealed at 400℃ in an oven for 4h to release thermal stress. Then the samples were cooled down to room temperature and then cut into small pieces with dimensions about 10 mm× 10 mm and a thickness of about 1.6 mm after polishing.

Table1 Compositions of glasses

Sample Code	Oxide (mol %)				
	Bi_2O_3	B_2O_3	SiO_2	TiO_2	CeO_2
G21_Base	45	10	45	-	-
G21-1	45	10	45	-	0.2
G21-2	45	10	45	0.2	-
G21-3	45	10	45	0.1	0.1

Glasses were irradiated with γ-radiation at the CALLIOPE irradiation plant (ENEA Casaccia Research Centre, Rome, Italy) at room temperature. Absorbed doses of 0.5kGy, 1.0kGy, 5.0kGy, 15.5kGy were applied at a dose rate of 1.0kGy/h. Transmission spectra in the UV-Vis region were measured by a Lambda 950 UV-Visible Spectrophotometer (Perkin Elmer) in order to compare the optical transmittance of all glass samples before and immediately after irradiation. The thicknesses of the samples were measured by Micro 2000 (Sheffield England).

3. Results and discussion

Transmission spectra of glass samples before and after irradiation at different doses are shown in Figures 1 and 2, respectively. By comparison between base glass and Ce^{4+} doped glasses before irradiation, an obvious reduction in transmission at the near UV region is observed which becomes m remarkable with the Ce^{4+} ion concentration increasing from 0.1 mol% (G: to 0.2 mol% (G21-1). This arises from the charge transfer transition of ions, resulting in formation of broad absorption band at UV region extendi the longer visible wavelength [12]. On the other hand, Ti^{4+} ions doping exer effect on the near UV cut-off position of glass.

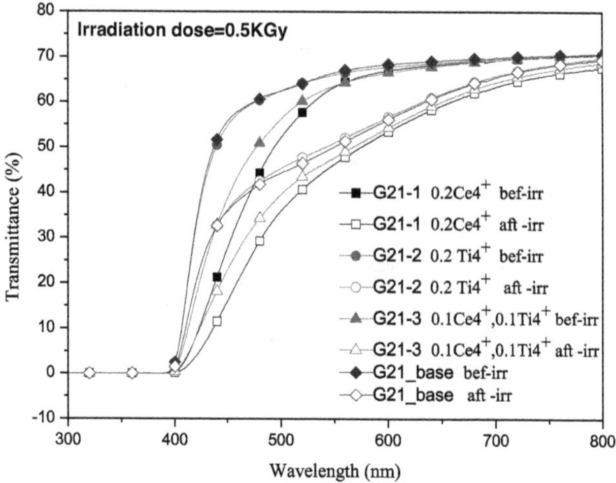

Figure 1 Transmission spectra of samples before and after irradiation at 0.5KGy

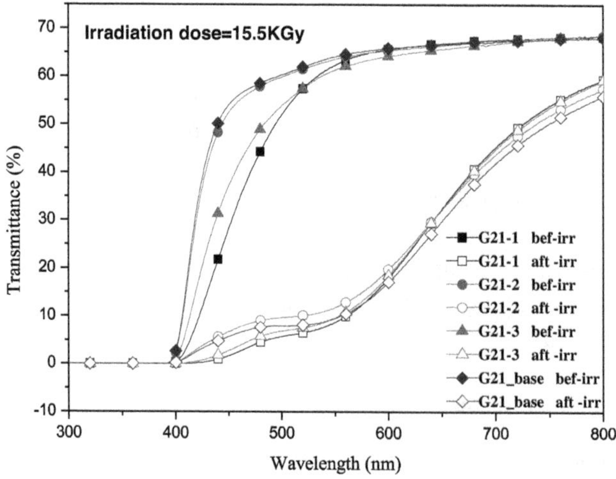

Figure 2 Transmission spectra of samples before and after irradiation at 15.5kGy.

As can be seen from these figures, a general decrease in transmittance occurs after irradiation in the whole visible range which becomes more remarkable as the irradiation dose increases. The properties of glasses are usually subject to a variety of changes under the influence of γ ray irradiation, due to partial rupture of chemical bonds and/or destruction of the network as well as introduction of defects [6]. In particular, most of the radiation damage in glass results from the formation of color centers due to the presence of the induced defects which are trapped in the glass network [7]. It is reported that defects in glasses are formed in pairs of negative electron centers (EC) and positive hole centers (HC) [6-7]. As we observed in Figure 1, the reduction of transmittance of glass samples is just related to the formation and accumulation of irradiation induced defects which acted as color centers. It is known that HC and EC were generated by the interaction of defects in the glass structure due to the irradiation, where the absorption of EC occurs in the ultraviolet region, while the absorption of HC occurs in the visible part [7]. Therefore, the transmission decrease of the glasses in the visible region can be attributed to HC. It can also be seen from the figures that the decrease of transmittance becomes more remarkable as the increase of irradiation doses, which can be interpreted that high irradiation doses generated more HC defects in glasses.

In order to calculate the irradiation damage on glass samples, the so-called radiation-induced absorption coefficient (μ) is introduced which is defined as $\mu=(1/d)\ln(T_0/T)=\alpha_0-\alpha$, where d is the length of light path through the measured sample, T_0, T are transmittance before and after irradiation, and α_0, α are the corresponding absorption coefficients before and after irradiation, respectively. From above definition it is known that the μ value is proportional to the density of color centers (EC and HC) induced by irradiation [7].

As an example, the calculated μ values of all samples after irradiation at 15.5 kGy are plotted against wavelength in Figure 3. At all radiation doses, Ti^{4+} ions show positive effect on radiation resistance of glasses, especially at the higher doses. Ce^{4+} doping (G21-1) or Ce^{4+}/Ti^{4+} co-doping (G21-3) increases the μ values of glasses by comparison with that of the matrix. In particular, Ce^{4+}/Ti^{4+} co-doping shows the lower μ value than Ce^{4+} doping, most likely due to the Ti^{4+} involvement and/or lower Ce^{4+} concentration. Above phenomena indicate that Ce^{3+} ions exert a negative effect on the radiation resistance of the present glasses which is absolutely different from the previously reported work on glasses in different systems [6-7, 10]. This is maybe attributed to the different role of Ce^{4+} ions in the present Bi_2O_3-B_2O_3-SiO_2 glass while further work on structure of samples before and after irradiation, for example, by FTIR spectra, have to be done to offer the supporting evidence.

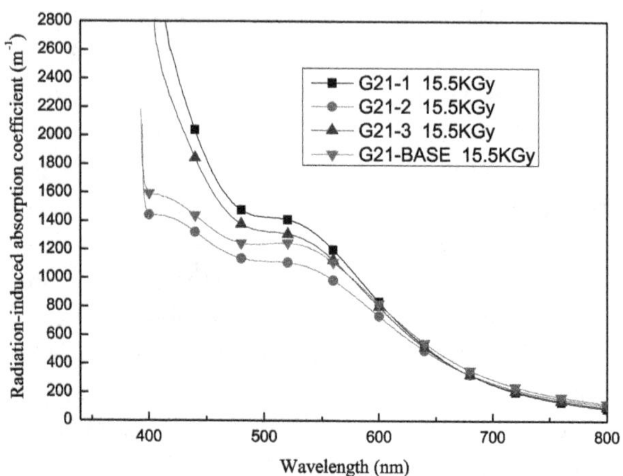

Figure 3 Radiation induced absorption coefficient of samples after irradiation at 15.5 kGy

Acknowledgments

This study is supported by Shanghai Leading Academic Discipline Project, Project B502, Shanghai Key Laboratory Project (08DZ2230500) and National Natural Science Foundation of China, No. NSFC 51072052.

References

1. A.F. Zatsepin, A.I. Kukharenko, D.A. Zatsepin et al., *Opt. Mater.*, **33(4)**, 601 (2011).
2. A. Pan and A. Ghosh, *J. Non-Cryst. Solids.*, **271** , 157(2000).
3. C. Stone, A. Wright, R. Sinclair et al., *Phys. Chem. Glasses* **41**, 409(2000).
4. Y. Zhang, Y. Yang, J. Zheng et al., *J. Am. Ceram. Soc.*, **92 (8)**, 1881(2009).
5. S. Bale, S.Rahman, A. Awasthi et al., *J. Alloy. Compd.*,**460(1-2)**, 699(2008).
6. Y. Ou, S. Baccaro, Y. Zhang et al., *J. Am. Ceram. Soc.*, **93(2)**, 338(2010)
7. S. Wang, G. Chen, S. Baccaro et al., *Nucl. Instr. and Meth.Phys. Res. B* **201/3** , 475(2003).
8. B. Yale, K.M. Fyles, US Pat. No. 5017521(1991).
9. P.S. Danielson, US Pat. No. 4746634 (1987).
10. G. Qian, S. Baccaro, A. Guerra et al., *Nucl. Instr. Meth. Phys. Res. B*, **262(2)**,276(2007).
11. Y. Zhang, Y. Yang, J. Zheng et al., *J. Am. Ceram. Soc.*, **91(10)**,3410(2008).
12. H. Ebendorff-Heidepriem and D. Ehrt, *Opt. Mater.*, **15(1)**,7(2000).

Tau Reconstruction And Identification Performance in ATLAS

Elias Coniavitis
on behalf of the ATLAS Collaboration
Department of Physics, University of Oxford, UK
E-mail: elias.coniavitis@cern.ch

Tau leptons play an important role in the physics program at the LHC. They are used in searches for new phenomena like the Higgs boson or Supersymmetry and in electroweak measurements. They can also be used for detector-related studies like the determination of the missing transverse energy scale. Identifying hadronically decaying tau leptons requires good understanding of the detector performance, combining the calorimeter and tracking detectors. We present the current status of the tau reconstruction and identification at the LHC with the ATLAS detector. The identification efficiencies are measured in $W \rightarrow \tau\nu$ events, and compared with the predictions from Monte Carlo simulations. The misidentification probability from electrons and jets is also estimated from dedicated control samples in data.

Keywords: tau reconstruction, tau identification, ATLAS, LHC

1. Introduction

Tau leptons play an important role in the physics program of the ATLAS experiment[1] at the LHC, as they are relevant both for Standard Model measurements as well as for searches for new physics. A tau lepton decays to leptons 35.2% of the time and to hadrons the remaining 64.8%.[2] As the leptonic decays are extremely hard to distinguish from prompt leptons, the focus here will be on the reconstruction and identification of the hadronic decays. These are dominated by the production of charged pions along with a neutrino and possibly one or more neutral pions. There are also decay modes involving kaons, but they are substantially rarer. Experimentally, the hadronic decays can be separated by the number of *"prongs"* (charged decay products), with *1 prong* decays being the most common (49.5%), followed by *3 prong* decays (15.2%). Besides the number of prongs, the experimental signature consists of collimated energy deposits in the calorimeter, since the hadronic tau decay produces a relatively narrow *tau-jet*, often with a relatively large electromagnetic component due to the neutral pions decaying to photons. Additionally, the hadronic tau decay can be characterized by a large leading track momentum fraction and, possibly, a secondary vertex.

Two components of the ATLAS detector are crucial to tau reconstruction[a]: the tracker and the calorimeter. The tracker is immersed in a 2 T magnetic field generated by the central solenoid, and consists of three subsystems: the Pixel detector, the Semi-Conductor Tracker (SCT), and the Transition Radiation Tracker (TRT). The first two subsystems cover a region of $|\eta| < 2.5$ in pseudorapidity, while the TRT reaches up to $|\eta| = 2.0$. The electromagnetic (EM) and hadronic calorimeters cover the range $|\eta| < 4.9$, with the η region matched to the inner detector having a finer granularity in the EM section. The EM calorimeter uses liquid argon (LAr) as the active material and lead as absorber, the hadronic calorimeter uses steel and scintillating tiles in the barrel region, while the end-caps use LAr as the active material and copper as the absorber.

2. Reconstruction and Identification

The reconstruction of hadronic tau decays in ATLAS is seeded by calorimeter jets. These jets are reconstructed using the anti-k_t algorithm,[3,4] with a distance parameter $R = 0.4$, using three-dimensional topological calorimeter energy clusters[5] as inputs. Only jets with $E_T > 10$ GeV and within $|\eta| < 2.5$ are retained. Reconstructed tracks are then associated to the seeds, to form tau candidates. In order to be considered a track has to be within the *core cone*, defined as the region within $\Delta R < 0.2$ around the axis of the seed jet, have $p_T > 1$ GeV, and pass the following quality criteria[b]:

- Number of pixel hits ≥ 2,
- Number of pixel hits + number of SCT hits ≥ 7,
- $|d_0| < 1.0$ mm and $|z_0 \sin\theta| < 1.5$ mm.

Since hadronic tau decays consist of a specific mix of charged and neutral pions, the energy scale of hadronic tau candidates is calibrated inde-

[a]ATLAS uses a right-handed coordinate system with its origin at the nominal interaction point (IP) in the center of the detector and the z-axis along the beam pipe. The x-axis points from the IP to the center of the LHC ring, and the y axis points upward. Cylindrical coordinates (r, ϕ) are used in the transverse plane, ϕ being the azimuthal angle around the beam pipe. The pseudorapidity is defined in terms of the polar angle θ as $\eta = -\ln\tan(\theta/2)$. The distance ΔR in the $\eta - \phi$ space is defined as $\Delta R = \sqrt{(\Delta\eta)^2 + (\Delta\phi)^2}$.

[b]d_0 is the distance of closest approach of the track to the reconstructed primary vertex in the transverse plane. z_0 is the longitudinal distance of closest approach.

pendently of the jet energy scale. As a first step, the clusters associated with the seed jet are calibrated using the local hadron calibration (LC).[6] The cells of the clusters are weighted to correct the measured energies for the non-compensation of the calorimeters and for energy deposits outside reconstructed clusters and in un-instrumented (dead) material, based on the local properties of the clusters. As a second step, an additional correction specific for hadronic tau decays is applied. This correction is derived from simulation of various physics processes involving tau leptons, is binned in p_T, η and track multiplicity of the tau candidate, and is applied on top of the LC calibration. The uncertainties on the tau energy scale are evaluated by comparing the responses in different Monte Carlo (MC) samples (varying a number of conditions, such as the detector geometry, the hadronic shower model etc.) and an indicative example of the uncertainties as a function of the true tau lepton's p_T can be seen in Fig. 1(a).

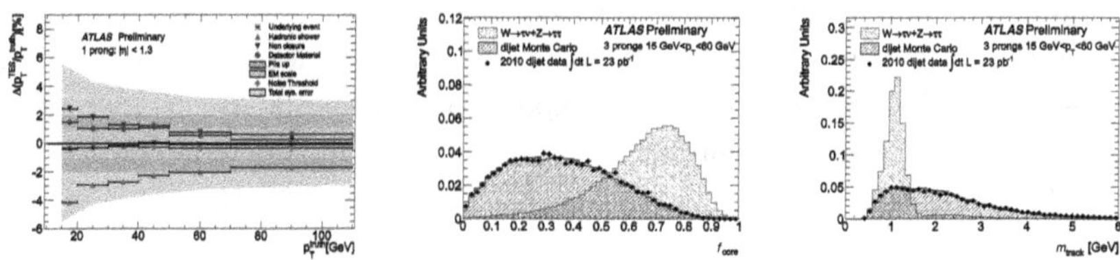

Fig. 1. Tau energy scale systematics as a function of the true tau p_T (left). Examples of two discriminating variables used for the tau identification: the fraction of energy in the *core cone* (centre) and the invariant mass of the track system (right). The filled histograms are MC, the points are data with a dijet selection.

The reconstruction process described above has a high efficiency, but also suffers from a very low purity, the main background coming from QCD jets, with a secondary background from electrons. A second step is therefore necessary to reduce these backgrounds, referred to as the *Tau Identification.* Using information both from the calorimeter and the tracker, several discriminating variables are constructed and then used by three different algorithms, developed to discriminate between signal and backgrounds: a cut-based selection, a projective likelihood discriminant, and a boosted decision tree. Figure 1 shows two examples of such variables – the full listing and definitions of all variables can be found in.[7] The performance of the three methods is compared in Fig. 2(a), for *3 prong* tau candidates. The performance of one of the methods is also demonstrated on a $Z \to \tau\tau$ selection, shown before and after tau identification has been applied.[8]

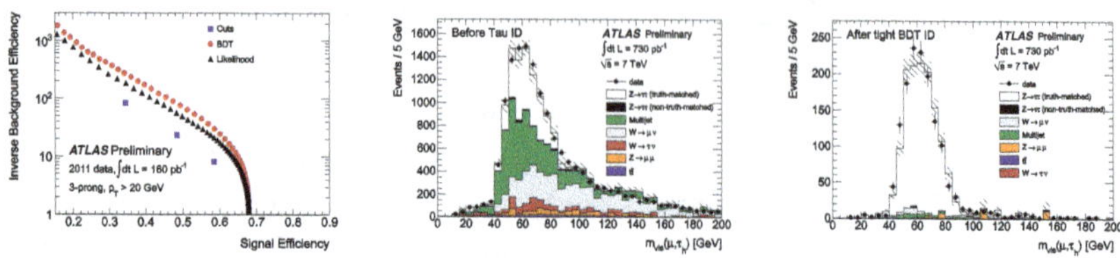

Fig. 2. Left: Comparison of the tau identification methods. MC is used for the signal, data with a dijet selection for the background. Centre and right: the visible decay products' invariant mass in a $Z \to \tau\tau$ selection, before and after tau identification.

3. Efficiency and Mis-identification Measurements

The efficiency of the tau identification is measured in $W \to \tau\nu$ events[9] using two different methods. In the tag & probe method, the $E_{\mathrm{T}}^{\mathrm{miss}}$ significance, $S_{E_{\mathrm{T}}^{\mathrm{miss}}}$,[c] is used to tag events. The track multiplicity spectrum of jets and tau candidates is fitted, to extract the $W \to \tau\nu$ spectrum, with the template for the QCD multijet background extracted from a low $S_{E_{\mathrm{T}}^{\mathrm{miss}}}$ control region. This method is cross-checked with a second method, where the measured $W \to \tau\nu$ yield is compared to the expectation from the W cross-section measured in the e and μ channels. In both measurements, the result is in agreement with the efficiency expected from MC (Fig. 3(a)).

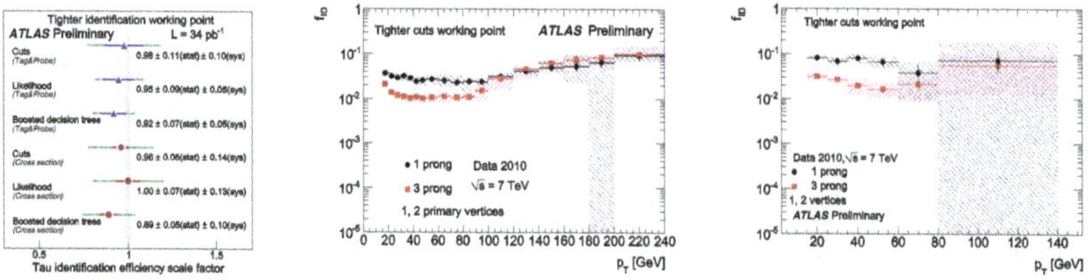

Fig. 3. Left: Ratio of the τ efficiency measured in data over that in MC. Centre and right: mis-identification probabilities as a function of p_{T} measured in a QCD dijet and in a γ+jet sample. In all three plots, the values are for a selection yielding tau signal efficiencies of $55 - 60\%$.

The mis-identification probability has also been measured in data, using a number of different processes.[10] Depending on whether a jet is quark- or gluon-initiated the probability of it faking a hadronic tau decay is expected

---

[c] $S_{E_{\mathrm{T}}^{\mathrm{miss}}} = E_{\mathrm{T}}^{\mathrm{miss}}/(0.5\ \mathrm{GeV}^{1/2}\sqrt{\Sigma E_{\mathrm{T}}})$, where $E_{\mathrm{T}}^{\mathrm{miss}}$ is the transverse component of the vector sum of all energy deposits in the calorimeter, calibrated to their respective energy scale, and ΣE_{T} the scalar sum of the E_{T} of all topological clusters.

to be different. Measurements have therefore been done using both gluon-dominated (QCD dijet and trijet events) and quark-dominated (γ+jet, Z+jet) processes, and the results are indeed different (Fig. 3). The electron mis-identification probability has also been measured, using a tag & probe method on $Z \to ee$ events – data and MC expectations were found to be in agreement, except in the calorimeter transition region, for which correction factors for the MC were calculated based on this measurement.

4. Summary and Outlook

The tau reconstruction and identification in ATLAS is well-performing and reliable. Efficiency and mis-identification probabilities have been measured in data, using a number of different methods and processes. A lot of interesting physics results involving taus have already been released by ATLAS: Standard Model measurements[11–13] that complement measurements in other channels and at the same time demonstrate the good performance of the tau reconstruction, as well as searches for new physics[14–16] that are already excluding substantial parts of the relevant parameter space. Further studies of the ID efficiency, energy scale and other tau properties, such as resolving the tau-jet substructure, are underway with the large amount of data being collected during 2011, and many more interesting physics results with taus are to be expected.

References

1. The ATLAS Collaboration, *JINST* **3**, p. S08003 (2008).
2. K. Nakamura *et al.*, *J. Phys.* **G37**, p. 075021 (2010).
3. M. Cacciari and G. P. Salam, *Phys. Lett.* **B641**, 57 (2006).
4. M. Cacciari, G. P. Salam and G. Soyez, FastJet http://fastjet.fr.
5. W. Lampl *et al.*, ATL-LARG-PUB-2008-002, (2008).
6. Barillari, T et al.(Jun 2008), ATL-LARG-PUB-2009-001-2.
7. The ATLAS Collaboration, ATLAS-CONF-2011-077, (2011).
8. https://twiki.cern.ch/twiki/bin/view/AtlasPublic/TauPublicCollisionResults
9. The ATLAS Collaboration, ATLAS-CONF-2011-093, (2011).
10. The ATLAS Collaboration, ATLAS-CONF-2011-113, (2011).
11. The ATLAS Collaboration, arXiv:1108.2016 [hep-ex], (2011).
12. The ATLAS Collaboration, arXiv:1108.4101 [hep-ex], (2011).
13. The ATLAS Collaboration, ATLAS-CONF-2011-119, (2011).
14. The ATLAS Collaboration, arXiv:1107.5003 [hep-ex], (2011).
15. The ATLAS Collaboration, ATLAS-CONF-2011-132, (2011).
16. The ATLAS Collaboration, ATLAS-CONF-2011-138, (2011).

SILICON SENSOR DEVELOPMENTS FOR THE CMS TRACKER UPGRADE

Robert Eber on behalf of the CMS Tracker Collaboration

Institut für Experimentelle Kernphysik, Karlsruhe Institute of Technology
76131 Karlsruhe, Germany
E-mail: eber@kit.edu
www.kit.edu

As the high luminosity phase of the LHC is approaching, the CMS collaboration started research for the future silicon sensor baseline for the CMS Tracker phase II upgrade. Wafers of various materials (Float-zone, Magnetic Czochralski and Epitaxial), thicknesses from 320μm down to 50μm and n-bulk or p-bulk doping have been ordered at one manufacturer, HPK, for good comparison. Alongside, the feasibility of processing sensors with double metal routing on 6" wafers is explored. Different structures answer different questions covering aspects from radiation hardness to layout issues in this evaluation. A mixed irradiation program with protons and neutrons probes radiation hardness representing a mixture of charged and neutral hadrons expected in the CMS tracker after an integrated luminosity of 3000fb^{-1} at several radii. This contribution gives an overview of the first proton irradiated diodes. Furthermore, a sensor with integrated pitch adapter on a second metal layer is characterised and presented for the first time.

Keywords: CMS; Tracker Upgrade; Silicon Sensors; Radiation hardness

1. Campaign overview

The LHC at CERN will be upgraded to the high luminosity LHC after 2020. The CMS Tracker therefore has to deal with a harsher radiation environment. More radiation damage is introduced in the silicon sensors which leads to higher leakage current, higher depletion voltage and lower signal to noise. In addition, more proton interactions produce more tracks and lead to a higher occupancy. The CMS Collaboration started a campaign to identify the future silicon sensor technology baseline for the phase II upgrade of the CMS tracker. To find irradiation hard material, a large irradiation program has been set up for the materials of interest. The mixed irradiations with protons and neutrons represent the radiation environment of charged and neutral hadrons with different ratios at an integrated luminosity of 3000fb^{-1} at five radii inside the CMS Tracker.[1] To reduce the material budget in the tracker, sensors with new readout designs have been developed.

1.1. *Choice of materials*

The silicon base material is one of the important parameters for radiation hardness of silicion sensors. In this campaign three different processes for growing sensor grade silicon are compared: Float-zone (FZ), Magnetic Czochralski (MCz) and Epitaxial (Epi) grown material. All materials are available as n-bulk and p-bulk versions. The n-type strip implants of p-bulk sensors require additional isolation which is done with two different techniques, also under investigation in this campaign. The p-spray method is a homogeneous p-doping of lower concentration all over the strip side of the sensor. P-stop on the other hand is a p-type strip implant with a higher doping concentration between the n-type readout strips.[2] In addition, wafers with a second metal layer were ordered. Six wafers of each material and bulk doping together with the double metal wafers sum up to 158 wafers in total.

This contribution covers deep diffused FZ with a thickness of 320μm and 200μm as well as MCz at $200\mu m$. Thinner material is not foreseen for outer radii in the tracker. The double metal sensor under investigation is FZ 200μm thick.

1.2. *Deep diffusion material*

An advantage of thinner sensors is the lower depletion voltage compared to thicker sensors. After irradiation also trapping of charge carriers is reduced due to the reduced thickness and a higher electrical field. A reduction of the active thickness can be achieved with deep diffusion: a high concentration of dopants is diffused deep into the wafer. This leads to a smooth change of doping concentration compared to wafer bonding, but the process is cheaper. In the left side of figure 1 a fitted doping profile used for the CV simulation of 320μm and 120μm thick FZ diodes can be seen, which leads to a good agreement between simulation and measurement (fig. 1 right).

2. Irradiated structures: Diodes

Diodes provide a good insight into the material itself. To identify radiation hard material for the future CMS Tracker, current (IV) and capacitance

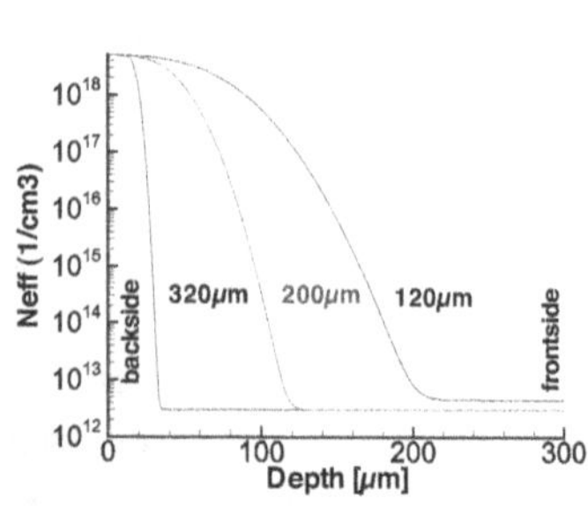

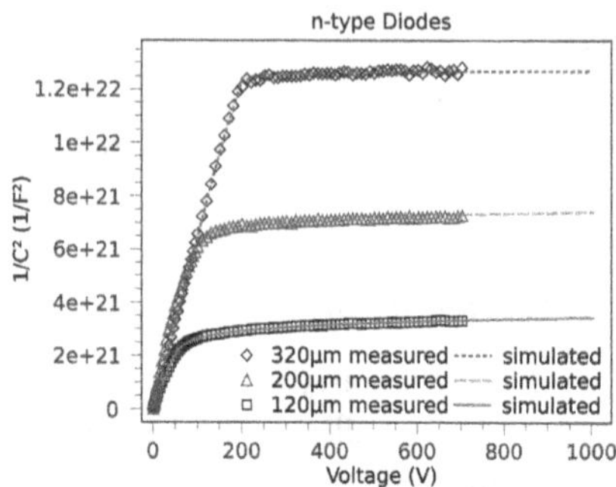

Fig. 1. (*Left*): Deep diffusion doping profiles for 320μm, 200μm and 120μm n-type diodes. (*Right*): Simulation of diodes' capacitance-voltage curve with shown doping profiles.

(CV) as function of bias voltage as well as charge collection and TCT[a] measurements are performed. Irradiations of the diodes presented here are all performed with 24MeV protons from the Cyclotron in Karlsruhe (ZAG).

IV is used to determine the current related damage parameter $\alpha = \frac{\Delta \mathrm{I}}{\mathrm{F} \cdot \mathrm{V}}$ with $\Delta \mathrm{I} = \mathrm{I}_{\mathrm{irrad}} - \mathrm{I}_0$ where $\mathrm{I}_{\mathrm{irrad}}$ is the current after irradiation and I_0 is the current before irradiation. F is the fluence in 1MeV neutron equivalent and V is the active volume of the diode. For the leakage current the value at 20% over depletion was taken. For diodes irradiated to $1.1 \cdot 10^{14} \mathrm{n_{eq}cm^{-2}}$ α is below the expected[3] value of $5 - 5.5 \cdot 10^{-17}$A/cm after an annealing of ten minutes at 60°C. Diodes irradiated to $3.0 \cdot 10^{14} \mathrm{n_{eq}cm^{-2}}$ lie above the expected value with $(6.0 \pm 0.3) \cdot 10^{-17}$A/cm (fig.2). The error bars include a temperature uncertainty of 0.5°C. However, the current of diodes with larger fluence was measured at 0°C and scaled to 20°C while the current of the other diodes was measured directly at 20°C. For the current scaling, an activation energy of $E_a = 1.21eV$ was used.[4] Errors on the irradiated fluence and the activation energy, which was not measured for these samples so far, are not included here.

The depletion voltage (fig. 2 *right*) of the irradiated diodes was taken from the $1/C^2$ vs. voltage plot. All n-type diodes show type inversion after irradiation which is seen in TCT measurements. p-type diodes show increasing depletion voltage after irradation, FZ increasing slightly more than MCz. No difference between diodes with p-spray(Y) and p-stop(P) is observed.

[a]Transient Current Technique

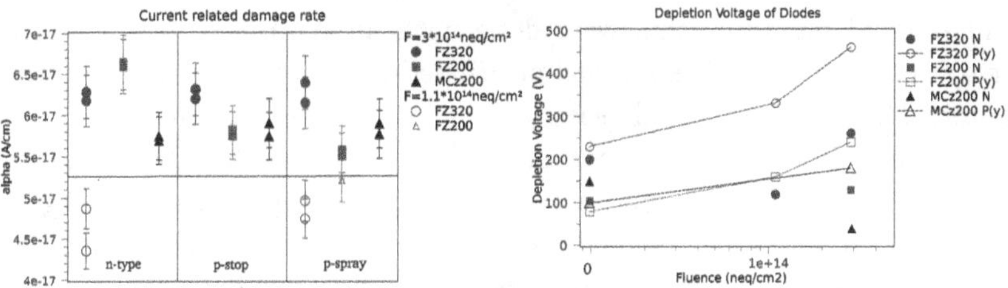

Fig. 2. (*Left*): Current related damage parameter for two diodes per material. (*Right*): Depletion voltage of diodes after irradiation with 24MeV protons.

3. New designs: Double metal sensor

An idea to save material in the tracker is to implement the pitch adapter, used to connect to the readout with wirebonds, directly on the sensor. The corresponding test structure, which is called **B**aby **P**itch **A**dapter (BPA)[5] , adapts the pitch of the readout on the first metal layer. The **D**ouble **M**etal **P**itch **A**dapter sensor (DMPA) with a second metal layer advances this idea by implementing the routing of aluminum lines on a second metal layer separated from the first metal layer by a 1.3μm thick oxide. It was shown that processing a second metal layer on 6" wafers is working. The electrical characterization of the FZ 200μm DMPA showed comparable results with respect to a baby standard sensor looking at interstrip resistance, bias resistance, strip leakage current, pinhole measurement and coupling capacitance. The interstrip capacitance however increases towards the middle of the pitch adapter, where the pads on the second metal layer lay directly over neighbouring strips of the first metal layer peaking at an increase of 50%. An increased capacitance leads to higher noise on affected strips.

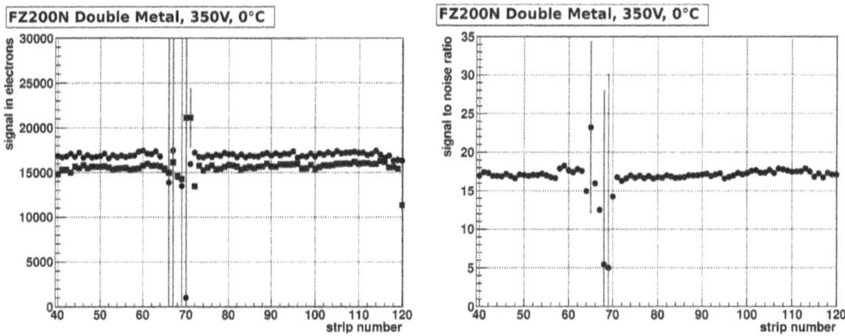

Fig. 3. Cluster (circles) and seed (squares) signal (*left*) and signal to noise (*right*) for the FZ 200μm double metal sensor.

The signal measurements were done with an ALiBaVa[6] setup and a ^{90}Sr source. The source's spot was positioned over the region of the pitch adapter. The collected signal (see fig. 3 *left*) is at its maximum of 17000 electrons for a 200μm thick sensor and is flat over the sensor. The signal to noise ratio (fig. 3 *right*) is good with a value of 17 over the sensor. Only 7 strips show an increased cluster noise of about 25% in the inner region of the pitch adapter from strip 65 upwards, which leads to a drop in the signal to noise ratio and arises from a higher interstrip capacitance. Yet, in other regions (especially till strip 64) the pitch adapter on the second metal layer has no effect on the signal to noise ratio and is working flawlessly.

Compared to a FZ 320μm BPA[5] , which shows a drop in the signal to noise ratio in the pad region (strip 27 upwards), the signal to noise for the FZ 200μm DMPA is 13% to 70% (pad region) higher and almost flat over the sensor.

4. Conclusions and outlook

Two first irradiations with protons have been performed and a first evaluation of depletion voltage and leakage current scaling with fluence was done. The double metal sensor with integrated pitch adapter was analyzed and performed better than a sensor with pitch adapter on the first metal layer.

In the coming months, more irradiations on the structures will be performed. The operability of an irradiated double metal sensor with integrated pitch adapter will be investigated as well.

Acknowledgements

Many thanks go to the whole collaboration and all people who did measurements presented here: J. Erfle, K.-H. Hoffmann, A. Kornmayer and Th. Pöhlsen.

References

1. K.-H. Hoffmann, *Nucl. Instr. Meth. A* (2011), doi:10.1016/j.nima.2011.05.028.
2. A. Dierlamm, Silicon Sensor Developments for the CMS Tracker Upgrade (2011), in proceedings of PSD9, Aberystwyth.
3. M. Moll, Radiation Damage in Silicon Particle Sensors, PhD thesis, Universitaet Hamburg, (Hamburg, Germany, 1999).
4. RD50, rd50.web.cern.ch.
5. J. Erfle, Silicon Sensor Developments for the CMS Tracker Upgrade (2011), in proceedings of RD11, Florence.
6. M.-H. Ricardo, *Nucl. Instr. Meth. A* **623**, 207 (2010).

Testing Radiation Tolerance of Electronics for the SuperB Experiment

A. ALOISIO and R. GIORDANO
Università degli Studi di Napoli "Federico II" and INFN Sezione di Napoli
Napoli, Via Cinthia, I-80126, Italy
E-mail: aloisio@na.infn.it, rgiordano@na.infn.it

V. BOCCI
INFN Sezione di Roma
Roma, Piazzale Aldo Moro 2, I-00185, Italy
E-mail: vbocci@na.infn.it

SuperB is a novel, high luminosity ($10^{36}cm^{-2}s^{-1}$), asymmetric e+e- collider to be built at the University of Rome Tor Vergata, Italy. A detector and its associated electronics (ETD) will be installed in this facility. High-speed serial links will be used for trigger and clock distribution and for data read-out. Given the high luminosity of the accelerator, the on-detector ends of the links will have to cope with the expected radiation levels.

In this work, we present the results of irradiation tests on some candidate components for the electrical part of the links. We performed tests with a 62-MeV proton beam at INFN Laboratori Nazionali del Sud (Catania, Italy) and with a 60Co γ-ray source at ENEA, La Casaccia (Rome, Italy).

Keywords: SuperB, SerDes, links, serial, optical, radiation, TID, SEU, SEE.

1. Introduction

SuperB[1] is a very high luminosity ($10^{36}cm^{-2}s^{-1}$), asymmetric e+e- flavour factory, which will permit to observe CP-violating asymmetries in very rare b and c quark decays and to measure branching fractions of heavy quark and lepton elusive decays.[2] Also, the studies at such a factory will complement the research program foreseen at hadronic colliders in the coming decade, making it possible to investigate signals of new physics, should they be revealed. The SuperB factory has been approved and funded in 2010 by the Italian Government as one of the flagships projects in the National Research Program. The facility will be built in the campus of University of Rome Tor Vergata.

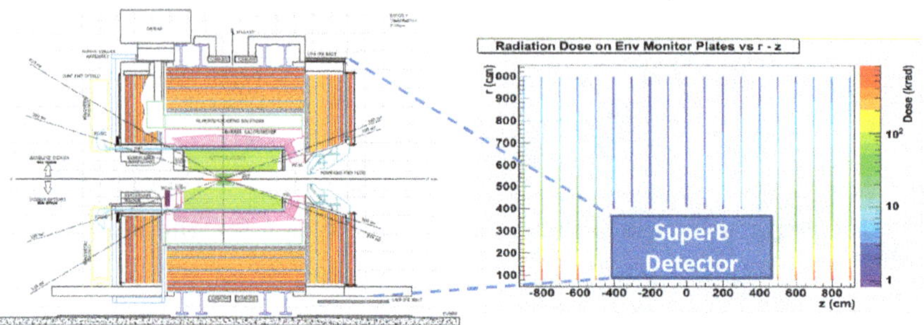

Fig. 1. Left: SuperB Detector. Right: radiation map estimated by means of simulations.

An international Collaboration including, among the others, research groups from USA, Canada, France and Italy is presently designing an experiment[3] to be installed in this facility. Such a design includes five sub-detectors (a silicon vertex detector, a drift chamber, a particle identification detector, an electromagnetic calorimeter and an instrumented flux return) plus the overall electronics trigger and DAQ (ETD) system.[4] Given the high luminosity of the machine, all the sub-detectors and the ETD system will have to withstand a high level of radiation. The expected total ionizing dose (TID) for the ETD, per effective year (10^7 s), estimated by preliminary simulations of the luminosity scaling background component,[5,6] might range from ~ 5 kGy(Si) to ~ 10 Gy(Si) moving from 1 m to 10 m from the beam pipe (Fig. 1).

The trigger and front-end side of the ETD is implemented as a synchronous pipeline. Both the distribution of synchronization and trigger signals to the whole system and the transfer of detector data to DAQ are performed by means of high-speed serial links (from 1 Gbps to 3 Gbps). Therefore, the links are a key element of the ETD and their electrical and optical components must cope with the expected radiation levels. Within the SuperB collaboration, there is an R&D activity aimed at identifying off-the-shelf devices, meeting the requirements of the links subsystem, and at qualifying their tolerance to radiation.

In this work, we present the results of irradiation tests on two SerDeses, the Texas Instruments TLK2711A[7] and the National Semiconductors DS92LV18.[8] We focus on the electrical part of the links, i.e. the SerDes device, the optical components will be identified and qualified at a later stage. We present a versatile test bench we have designed, which, during irradiation, measures the Bit Error Ratio (BER) and logs the dynamic and static power consumption of the device under test (DUT). We performed tests with a 62-MeV proton beam at INFN Laboratori Nazionali del Sud (Catania, Italy) and with a ^{60}Co γ-ray source[9] at ENEA, La Casaccia (Rome, Italy). We discuss the results of the tests in the view of the deployment of the tested components in the SuperB experimental apparatus.

2. Test bench

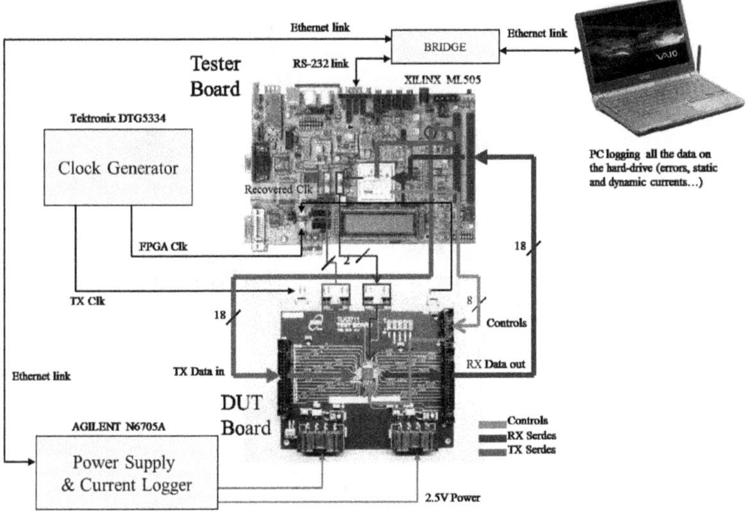

Fig. 2. Block diagram of the test bench.

The set-up (Fig. 2) includes an off-the-shelf board, the "tester" board, hosting a Xilinx V5LX50T FPGA,[10] which sends/receives data from the DUT and controls/monitors its status bits. A multi-channel power supply provides the DUT with the required voltages at the different power inputs. Each output channel of the supply drives a single input of the DUT with a four wire scheme for sensing the actual voltage at the load. The supply is controlled by a remote personal computer which logs the voltage and the current drawn on each channel. A clock generator provides both the tester board and the DUT with the required clock signals. For the test of the DS92LV18, an additional generator provides the de-serializer with a reference clock, according to the specifications of the device.

The test is divided in time units (TU) of the duration of 10 s each, repeating one after the other. Each TU is divided in two parts: during the first part the tester FPGA executes a BER test of the DUT ($\sim$ 9.5s), while during the second one ($\sim$ 0.5s) it holds the DUT in powerdown. Soft errors happening in one TU do not affect subsequent TUs, as they are removed by the powerdown. Moreover, with this strategy, the current log contains both dynamic ($I_d(t)$) and static current ($I_s(t)$) measurements in sequence as function of the time and therefore there is no need for separate dedicated measurements of the two currents. We developed an on-line script which extracts $I_d(t)$ and $I_s(t)$ from the raw log of current and plots them.

Given our architecture, in our setup we record all the parameters (BER and the currents at each power domain) as a function of time and therefore of the absorbed dose.

3. DS92LV18: Test Results

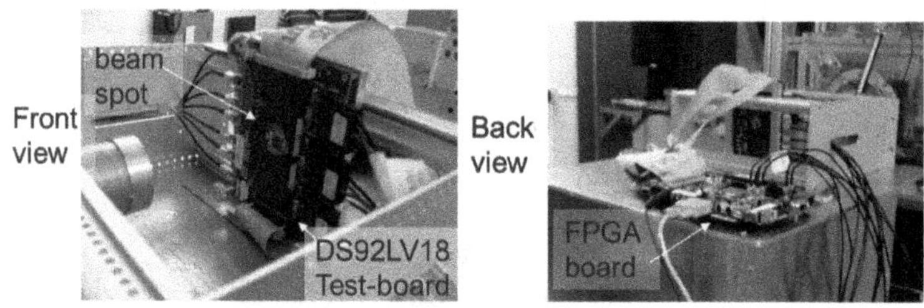

Fig. 3. Test setup for the proton irradiation of the DS92LV18.

The test setup for proton irradiation tests of the DS92LV18 is shown in Fig. 3. During the test, we set the beam current at the maximum possible intensity ($\sim$ 500 pA). One sample has been exposed to the beam for 6 ks, corresponding to a total ionizing dose of $\sim$ 4 kGy (Si) and to a total number of protons impacting on the die of $\sim 5{\cdot}10^{13}$. The clock frequency of the SerDes was 62.5 MHz (corresponding to a line rate of 1.25 Gbps, 2.5 Gbps full-duplex). We set the power supply at the maximum value allowed by the specifications (3.45 V), in such a way to achieve the highest sensitivity to single event effects (SEEs). During the test we observed both single and multiple bit errors (for a total of 25) in the transmitter part of the device. We observed just one single bit error in the receiver and 6 losses of locks (i.e. conditions when the received data is corrupted and the recovered clock stops toggling). After a loss of lock, the tester board sends a special training sequence in order to recover the link within the minimum time permitted by the device (800ns). We did not experience any single event functional interrupt.

Taking into account the fluence of protons absorbed by the sample and the measured errors, the SEE cross sections for the serializer, the deserializer and the loss-of-lock were respectively, $\sigma_{tx} = 2.5 \cdot 10^{-12} cm^2$, $\sigma_{rx} = 10^{-13} cm^2$, $\sigma_{link} = 6 \cdot 10^{-13} cm^2$. The order of magnitude of the cross sections agrees with previously published measurements[11,12] on similar chips (DS65LV1023/1224), which were implemented with the same process. After irradiation, we left the system annealing at room temperature and under measurement for 8 hours. We did not observe any error.

We have tested one sample of the DS92LV18 for TID tolerance with gamma rays up to a total dose of 5 kGy (Si). We irradiated the sample with a dose rate of 0.36 kGy (Si)/h ± 1% (certified by ENEA) for 14 hours. We did not observe neither transmission errors nor losses of lock. The currents absorbed by the device on all the power inputs were steady within 1 mA during the whole test. We report a comparison of the current drawn by the PLL of the device during proton and gamma irradiation. We show the current trends for two samples (Fig. 4), the first has been irradiated with gammas, the second with protons. Even if the currents at the beginning of the two tests are different (very likely due to different temperature and process variations between the considered samples), the trends are very similar and they both show an increase of the order of 0.6 mA.

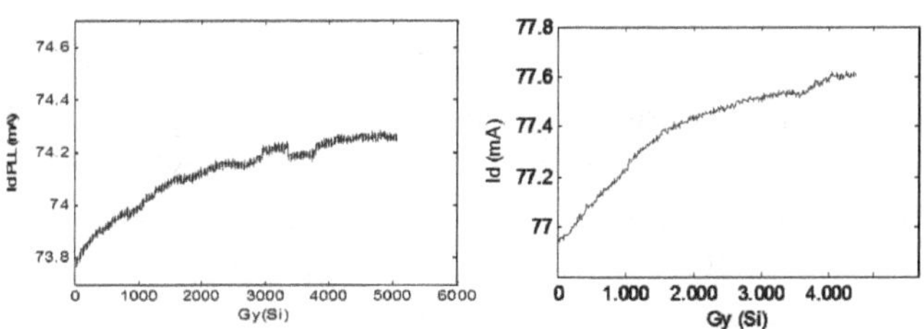

Fig. 4. DS92LV18: dynamic PLL current versus dose for two different samples. Left: sample irradiated with gammas. Right: sample irradiated with protons.

4. TLK2711A: Test Results

We performed proton irradiation of three samples of TLK2711A at different dose rates, in order to measure if dose-rate-dependent effects were present. During the tests, the clock frequency has been set to 100 MHz, in order

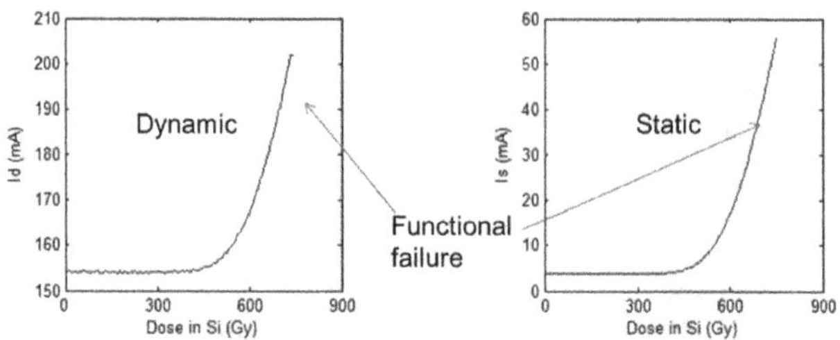

Fig. 5. TLK2711-A: current for the digital section versus dose. Left: Dynamic. Right: Static.

Table 1. Proton irradiation test conditions for TLK2711A samples.

Sample number	#1	#2	#3
I_{beam} (pA)	300	85	40
Dose rate Gy (Si)/min	36	10	4
Time on beam (min)	20	103	176
Total dose (Gy (Si))	760	1060	762
Protons on die	$1.2{\cdot}10^{10}$	$1.6{\cdot}10^{10}$	$1.2{\cdot}10^{10}$

to achieve a serial line rate of 2 Gbps (4 Gbps full-duplex). Table 1 summarizes the test conditions for all the samples. This device showed a much higher sensitivity to radiation than the DS92LV18. The dynamic and static currents exhibited well recognizable trends, showing an increment with the absorbed dose. We measured both single and burst (i.e. consecutive) word errors. The longer bursts were of the order of $\sim$ 10 words. If we define "functional failure" the condition of having more than 1000 transmission errors in the same TU, then all the samples we tested failed functionally at a dose of $\sim$ 700 Gy (Si). The signature of all the failures was

(1) at the serializer, an incorrect data pattern from the very beginning of each TU;
(2) at the deserializer, a data pattern correct for $\sim$ 1-2 s, then continuous errors.

For all the samples, the dynamic current of the "digital" power domain increased rapidly when the failure occurred (Fig. 5). Correspondingly, we also observed a severe power down current increment. We measured a few single (~10) errors in the serializer part before functional failure and the first one occurred at a dose of ~ 400 Gy (Si).

5. Conclusions

We presented TID and SEEs radiation tests on off-the-shelf SerDes candidate to be used in serial links for SuperB. The National DS92LV18 has shown a good tolerance both to SEEs and TID. The currents were steady even after 5 kGy (Si) of proton irradiation and after 5 kGy (Si) of gamma irradiation. The currents absorbed by the Texas TLK2711A during proton irradiation were very dependent upon the absorbed dose. For this device we tested 3 samples tested until breakdown (which happened at 0.7 kGy (Si)). Being the radiation tolerance very low, compared to the requirements of SuperB, there was no reason to proceed to TID testing for this component.

We are going to test other National SerDes devices implemented with a process potentially more resistant than the one of the DS92LV18. We also intend to perform some radiation tests on Xilinx Virtex-5 and Virtex-6 FPGAs with embedded SerDes.

References

1. SuperB Collaboration (M.E. Biagini et al.), "SuperB Progress Reports: The Collider.," Sep. 2010. 162pp. e-Print: arXiv:1009.6178 [physics.acc-ph]
2. SuperB Collaboration, "SuperB Progress Reports – Physics". e-Print: arXiv:1008.1541v1 [hep-ex]
3. SuperB Collaboration: E. Grauges et al., Francesco Forti, Blair N. Ratcliff, David Aston, "SuperB Progress Reports – Detector" . e-Print: arXiv:1007.4241v1 [physics.ins-det]
4. D. Breton, 'The ETD status', XII SuperB General Meeting - LAPP – Annecy (F), March 2010
5. R. Cenci, "SVT backgrounds and ETD rates," SuperB Kick-off Meeting, Isola d'Elba 2011
6. E. Paoloni, "Report from the detector side," SuperB Kick-off Meeting, Isola d'Elba 2011
7. Texas Instruments, "1.6 TO 2.7 GBPS TRANSCEIVER", July 2008
8. National Semiconductor, "DS92LV18 18-Bit Bus LVDS Serializer/Deserializer - 15-66 MHz," June 2006
9. S. Baccaro, A. Cecilia, A. Pasquali, "Gamma irradiation facility at Enea- Casaccia Centre (Rome)," 2005 [On-line] Available:

http://www.iaea.org/inis/collection/NCLCollectionStore/_Public/37/054/37054956.pdf
10. "ML505/ML506/ML507 Evaluation Platform User Guide," UG347, v3.0.1, Xilinx, 2008
11. Hamilton, B.J., Turflinger, T.L., "Total Dose Testing of 10-Bit Low Voltage Differential Signal (LVDS) Serializer and Deserializer", Proceedings of 2001 IEEE Workshop on Radiation Effects Data, pp. 177 – 181
12. Ichimiya, R. et al., "Radiation Qualification of Commercial-Off-The-Shelf LVDS and G-link Serializers and Deserializers for the ATLAS Endcap Muon Level-1 Trigger System", Proceeding of 10th Workshop on Electronics for LHC and Future Experiments, Boston, MA, USA, 13 - 17 Sep 2004, pp.389-393

The Resistive Plate Chambers of the ATLAS experiment: performance studies on Calibration Stream

Luca Mazzaferro*
on the behalf of the ATLAS collaboration
Department of Physics, University of Rome Tor Vergata and INFN Roma 2.
Rome, 00133, Italy
**E-mail: luca.mazzaferro@roma2.infn.it*
http://www.fisica.uniroma2.it/

ATLAS (A Toroidal LHC ApparatuS) is one of the four experiments installed on the hadron-hadron collider LHC at CERN. It is a general purpose experiment, with a physics program which spans from the search for the Higgs Boson to the search of physics Beyond the Standard Model (BSM). An integrated luminosity of about 5 fb^{-1} is expected to be reached by the end of 2011. The Resistive Plate Chambers, installed in the barrel region, are used to provide the first muon level trigger, and cover an area of 16000 m^2, readout by about 350000 electronic channels.

To ensure optimal trigger performance, the RPC operational parameters like cluster size, efficiency and spatial resolution are constantly monitored.

In order to achieve the desired precision, the data used for the analysis are extracted directly from the second level of the trigger, hence assuring very high statistics. This dedicated event stream, called Calibration Stream, is sent automatically to the RPC Calibration Center in Naples.

Here the analysis is performed using an automatic tool tightly integrated in the `ATLAS GRID` environment, the Local Calibration Data Splitter (LCDS), which configures and manages all the operations required by the analysis (e.g. software environment initialization, grid jobs configuration and submission, data saving and retrieval, etc).

The monitored RPC operational parameters, the performance analysis and the LCDS will be presented.

Keywords: RPC; performance studies; lcds; grid; computing

1. The Large Hadron Collider

The Large Hadron Collider (LHC)[1] is a two-ring superconducting hadron-hadron collider installed in the same tunnel of the LEP electron-positron collider. With a circumference of about 27 km it was designed to collide two proton beams with a center of mass energy up to 14 TeV, a luminosity of 10^{34} $cm^{-2}s^{-1}$ and a time interval between two bunches of about 25 ns.

Four experiments are installed on the interaction points: ATLAS, CMS, ALICE and LHCb. At the time of writing this document the LHC operates like proton-proton collider with 7 TeV center of mass energy and a luminosity greater than 10^{33} $cm^{-2}s^{-1}$.

2. A Toroidal LHC Apparatus

ATLAS[2] is a general purpose experiment focused on the search of the Higgs boson and the physics beyond the Standard Model. It is 44 m long, 25 m high and it weights 7000 *ton*.

Describing the structure from the interaction point to the outwards we find: the *Inner Detector* (ID) designed, mainly, to identify the primary vertexes and for precision tracking of the short life particles. It consists of a Pixel Detector, a Semiconductor Tracker (SCT) and a Transition Radiation Tracker (TRT) inside a 2 T solenoidal magnetic field. Around the ID the *Electromagnetic and Hadronic Calorimeters* are installed; they are used to measure electromagnetic/hadronic shower reconstruction and the missing transverse energy, E_t^{miss}, measurement. The active material is LAr while the passive ones are steel and copper. In the outer region, the *Muon Spectrometer* is used to identify and reconstruct the muons tracks, to measure their transverse momentum and to associate the events to the correct bunch crossing, it is immersed into a toroidal magnetic field with a bending power of about 3 Tm in the *barrel region*, and of about 6 Tm in the *endcaps*.

Four detector technology composes the muon spectrometer covering different regions: the Muon Drift Tube (MDT) chambers, installed both in the barrel and endcaps region, and Cathode Strips Chambers (CSC), in the endcaps near the beam-pipe, are are used for precise tracking in the bending direction; the Resistive Plate Chambers (RPC), in the barrel region, and Thin Gap Chambers (TGC), in the endcaps, provide the first muon level trigger and the measure of the non-bending coordinate.

3. RPC chambers and performance studies

The RPC,[3] figure 1, is a gaseous detector based on the detection of the gas ionization produced by a charged particle traversing the chamber. The electrodes, 2 mm thick, are made of resistive material (Bakelite) with a volume resistivity of about $510 * 10^{10}$ Ωcm coated, on the external surfaces, with a thin layer of graphite painting (surface resistivity of about 100 $K\Omega$) allowing a uniform distribution of the high voltage along the plates. The electrodes are spaced by a 2 mm gas gap filled with a gas mixture of $C_2H_2F_4 - C_4H_{10} - SF_6$ in proportion of 94.7/5/0.3. This allows the detector to be used in avalanche mode with a voltage between electrodes of about 9.8 kV. The signals, generated by ionizing particles, are readout by two orthogonal series of strips in capacitive coupling, with a typical width of about 30 mm.

The ATLAS RPCs[4] have a time resolution of about 1.5 ns and a space resolution of about 1 cm satisfying the ATLAS first muon level trigger requirements.

Performance studies[5,6] have been developed to monitor the behavior of the detector and to provide information for RPC chambers fine tuning of the working parameters.

The main operational parameters being monitor are:

- Efficiency: measured using precision tracks and requiring the presence of an RPC hit within 70 mm. It is a key parameter for the overall trigger efficiency.
- Cluster Size distribution: distribution of the numbers of adjacent strips fired within a time interval of 15 ns is crucial in assessing the optimal working point of the detector from the point of view of both high and low voltage settings.
- Space Resolution: estimated using the residual distribution[a] fitted with a Gaussian distribution.

In figures 2, 3 and 4 are shown some example plots deriving from the analysis.

In order to achieve the desired precision the data for this analysis are extracted from the second muon level trigger and organized in a dedicated stream, called *Calibration Stream*, assuring an acquisition rate at least of 1 kHz.

After the extraction, the data stream is sent from the Tier0 to the *Naples Tier2 Calibrations Center*[7] where the Local Calibration Data Splitter (LCDS) is installed and runs.

4. The Local Calibration Data Splitter

The LCDS[9] is a software framework designed to manage tasks within the `ATHENA` framework.[7] Already used for the DQA and calibration analysis of

[a]The residual is defined as the position of the RPC cluster with respect to the position of the MDT track extrapolation on the RPC plane.

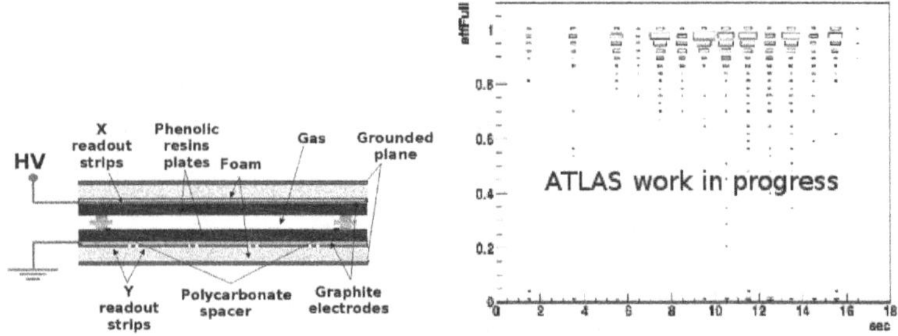

Fig. 1. Transverse view of an RPC chamber

Fig. 2. Panel efficiency per sector in ϕ view

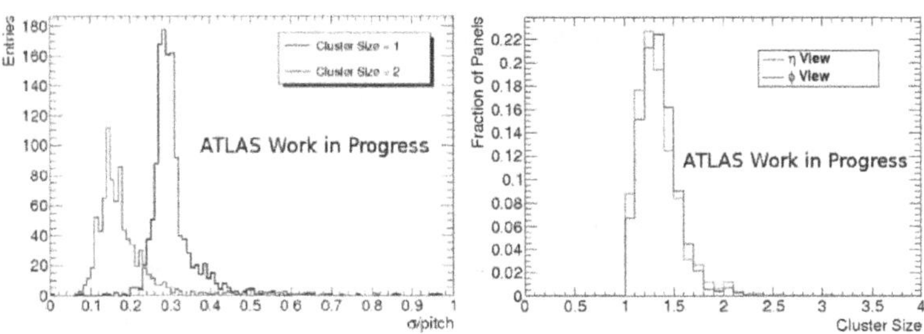

Fig. 3. Distribution of spatial resolution divided by the strip pitch.

Fig. 4. Cluster size distribution for η and ϕ panels.

the ATLAS MDT and TGC chambers, it is now modified to operate with the RPC performance analysis and runs into the *Naples Calibration Center.*

The LCDS is tightly integrated into the **ATLAS GRID** environment using the **pAthena** and **pRun** tools.[8] A web-interface, with **HTTPS** secure protocol and an Authorization System which check the validity of the personal digital certificate, allows the user[b] to start and to manage the analysis operations and to follow all the steps:

(1) As soon as a new dataset is saved into the *Tier2*, the LCDS allows the *shifter* to start the analysis clicking on a web button. This first step, called *ntuple creation*, produces ntuple files with all the relevant informations. At the same time a web page shows the operation's status and enables the user to access to the log file. To perform this analysis the LCDS uses the **pAthena** tool, splitting the job in sub-jobs running concurrently;

[b]possibly a *shifter*

(2) from the ntuples the summary files are created. The RPCs informations are organized into txt files in a useful way to be merged together. This step uses the `pRun` tool to run the jobs in the `GRID` environment;
(3) the last operation is performed in the local area[c] and produces the performance plots. The results are both saved into a `ROOT` file and able to be downloaded as images files directly from the web interface.

The web-interface allows the user also to kill and restart the *ntuple creation* and the *summary creation* jobs.

The main advantages of using the LCDS to perform the RPC analysis are:

- the operation latency reduction and resources optimization deriving both from the multi-threading approach which permits to perform the three operations steps at the same time for different datasets and to use the web interface together, and from the `pAthena` and `pRun` tools for jobs splitting and parallelizations;
- the automation of the all operations which reduces the latency and enhances the analysis reproducibility, allowing, in the future, the organizations of RPC performance shifts;
- the system expandability to other analysis.

References

1. CERN,*LHC Design Report*, http://ab-div.web.cern.ch/ab-div/Publications/LHC-DesignReport.html
2. ATLAS Collaboration, *Atlas Detector and Physics Performance: Technical Design Report*, Volume 1-2, ATLAS TDR 14, CERN/LHCC 99-14
3. R.Carderelli, R. Santonico, *Development of Resistive Plate Counters*, Nuclear Instruments and Methods 187 (1981) 377-380
4. ATLAS Muon Collaboration, *ATLAS Muon Spectometer: Technical Design Report*, CERN/LHCC 97-22
5. G.Cattani, in *The Resistive Plate Chambers of the ATLAS experiment: performance studies*, J. Phys.: Conf. Ser. 280 012001, 2011)
6. A. Di Simone, *The calibration of the resistive plate chambers of ATLAS*, 2010 J. Phys.: Conf. Ser. 219 032036
7. ATLAS Computing Group, *Computing Technical Design Report.*, s.l. : CERN-LHCC-2005-022, 2005.
8. P. Nilsson et al, *The PanDA System in the ATLAS Experiment*, s.l. : PoS(ACAT08)027 , 2008
9. G. Carlino et al, *ATLAS Muon Calibration Framework*, ATLASL-SOFT-PROC-2011-030, 2011

[c]The area where the LCDS runs into the Naples Calibration Center

Cherenkov light in ALICE: performance of the HMPID detector

L. MOLNAR* for the ALICE Collaboration
CERN,
CH-1211, Geneve 23, Switzerland
**E-mail: levente.molnar@cern.ch*
www.cern.ch

The ALICE High Momentum Particle Identification detector (HMPID), a proximity focusing Ring Imaging Cherenkov counter, is the largest of its kind with the 11 m^2 CsI-based photon detector total area. The HMPID is designed and optimized for identification of charged π, K, p and $\overline{p}$ on a track-by-track basis in heavy-ion collisions at LHC energies. To achieve track-by-track identification, 3 σ separation is required for π/K and K/p in the momentum range of 1-3 GeV/c and 1.5-5 GeV/c respectively. The HMPID is part of the ALICE data taking since the first proton-proton collisions and was also successfully operated in the first heavy-ion run in 2010. In this talk, we present the detector performance at the present time and its evolution over the past two years of operation in the LHC environment.

Keywords: ALICE, HMPID, RICH, CsI photocathode

1. Introduction

The High Momentum Particle Identification Detector[1,2] (HMPID) in ALICE is dedicated to the inclusive measurement of charged particles, both in heavy-ion and proton-proton collisions. Since its installation in ALICE in 2006, HMPID is successfully operated and collecting data: first in the cosmics campaign during commissioning, then in the LHC splash events during LHC injection tests and finally in beam operation since the end of 2009. In this paper we focus on the characteristic HMPID performance from proton-proton and heavy-ion collisions from 2010 and 2011.

2. The High Momentum Particle Identification Detector

The High Momentum Particle Identification Detector is a proximity focusing type Ring Imaging Cherenkov Counter (RICH). It consists of seven identical modules (1.4 m × 1.3 m each), installed on an independent support cradle and mounted at the two o'clock position of the ALICE space frame in a cupola-like structure, thus pointing to the nominal interaction point at 4.7 m distance. The HMPID acceptance covers the region $-0.6 \leq \eta \leq 0.6$ in pseudo-rapidity and 57.6^o in azimuth.[2] Each chamber is equipped with three NEOCERAM radiator vessels with fused silica optical windows, filled and emptied by gravity flow at a constant rate by the liquid circulation system, containing C_6F_{14} liquid as radiator (with index of refraction of n=1.2989 at λ =175 nm). Water and oxygen contamination is monitored and kept below 5 ppm.[3,4] Each chamber is equipped with six 60 cm × 40 cm CsI photocathodes (quantum efficiency ~25% at λ =175 nm).[5] The pad cathodes are segmented to 8 mm × 8.4 mm pads for a total of 161280 channels in the 7 chambers. A positive voltage of 2050 V is applied on the anode wires providing $\sim 4 \times 10^4$ gas gain.

The front-end electronics is based on two dedicated ASIC chips, GASSIPLEX and DILOGIC.[6] The GASSIPLEX chip is a low-noise signal processor, with measured noise of 1000e r.m.s. and dispersion less than 50e. The readout time after L2 trigger arrival is in the order of 300 μs for the estimated large 12% module occupancy. Each hardware element of the HMPID is controlled and monitored by the Detector Control System (DCS),[7] any change in the hardware parameter is archived and fed to the offline reconstruction.

The 2011 hardware configuration of the HMPID is close to the nominal acceptance. There is no loss on the electronics channels: only ~300 pads are masked out of 161280. As a conservative estimate ~81 % of the detector is at nominal settings (filled radiator with two high voltage sectors at 2050 V).

3. Data selection and HMPID Performance

The photon production and detection in HMPID depend on following factors: the applied chamber gain, the radiator properties and the CsI photocathode Quantum Efficiency (QE) and electronics noise. In contrast to test beam measurements, where detector parameters can be changed individually, determination of detector performance in beam operation comprises simultaneous changes in the detector configuration over time.

The HMPID performance study was carried out on minimum bias events selected by online trigger[8] and offline physics event selection from $\sqrt{s} = 2.76$ and 7 TeV p-p collisions and $\sqrt{s_{NN}} = 2.76$ TeV Pb-Pb collisions in the LHC periods: 10b - 11a. The z component of the primary vertex is required to fall within ± 10 cm. Good tracks from ITS and TPC are selected based on the common primary track definition of ALICE[9] and propagated up to the HMPID. The propagated tracks are required to point within 3 cm of the measured charged particle position in HMPID, that is known with sub-mm precision along the anode wire direction and with less than 4 mm precision perpendicular to the anode wires.[2]

The cluster charge distribution of the charged particles (MIPs) is extracted from the matched tracks for each high-voltage sector of the HMPID. Figure 1 left panel shows the most probable value and sigma of the MIP charge distribution extracted from Landau fit. Figure 1 right panel shows the electronics threshold corrected mean single-photon pulse height for each photocathodes extracted from exponential fit to the photon cluster charge distribution of the rings. In the measured LHC periods, from LHC10b (2010 March) to LHC11a (2011 March) the different high voltage sectors / photocathodes show gain variation, but the individual high voltage sectors / photocathodes exhibit stable gain over time.

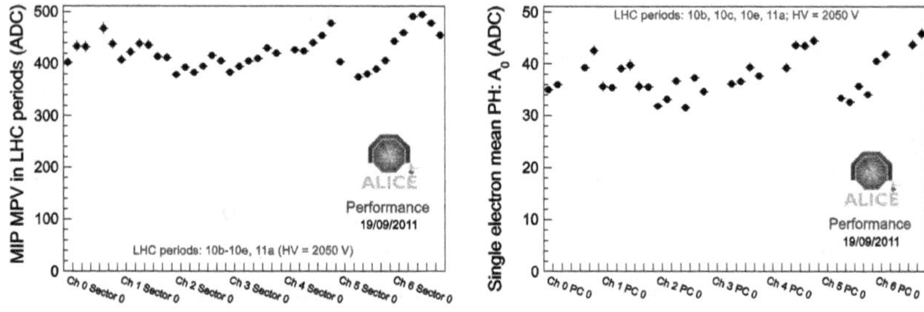

Fig. 1. Left panel: mean and RMS of the most probable value extracted from the track matched charge particle cluster for each HV sector in the indicated LHC periods. Right panel: mean and RMS of the electronics threshold corrected single electron mean pulse height for each photocathode in the indicated LHC periods.

Figure 2 left panel shows the number of photon clusters from data and simulation as a function of $\sin^2$ of the Cherenkov angle for rings fully contained in the photocathodes. In the simulation, the nominal gain (35 ADC) nominal QE and nominal C_6F_{14} radiator transparency are used.[2] The trend and the absolute number of the photon cluster number show good agreement between data and simulation, where the number of photon cluster depends on the radiator thickness L, detector factor of merit N_0 and the

$\sin^2$ of the Cherenkov angle,[10] as shown in Eq. 1.

$$N_{photons} = 370 \cdot L \cdot N_0 \cdot sin^2\Theta_C \tag{1}$$

Figure 2 shows the average number of photon clusters extracted for rings at saturation for each photocathode in different LHC periods (2010 March - 2011 March). While the different photocathodes show variation among each other, the average number of photon clusters per ring at saturation is stable for the individual photocathodes. The variation in the average number of photon clusters per ring on the photocathodes does not show monotonically decreasing trend, rather fluctuations, that might come from the variation in the radiator liquid transparency over the studied one year period. The photon production and detection is working efficiently, the QE of the photocathodes did not decrease during the measured period, even 6 years after their production.

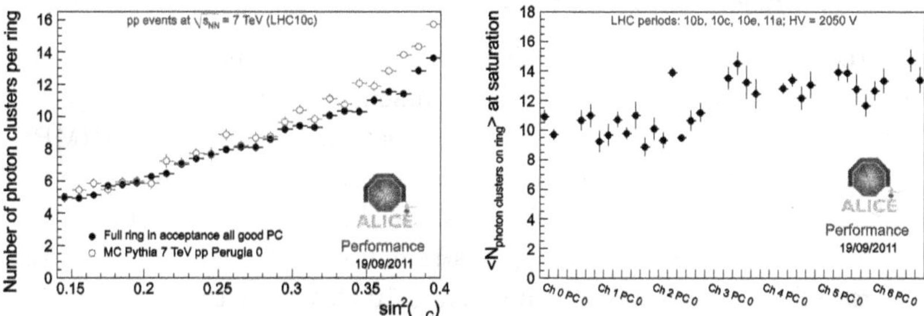

Fig. 2. Left panel: number of photon clusters in data and in simulation as a function of $\sin^2$ of the Cherenkov angle for rings fully contained in the photocathodes. Right panel: average number of photon clusters extracted for rings at saturation for each photocathode.

While the above parameters are strictly determined by the HMPID performance, the resolution of single-photon Cherenkov angle and particle identification are influenced by the properties of the propagated track to the HMPID (located 4.7 m from the interaction point with significant material budget in front with respect to other RICH detectors, see HMPID Physics in this proceedings[11]). Figure 3 left panel shows the single-photon Cherenkov angle resolution obtained for $\pi^{\pm}$ at p > 1.5 GeV/c with $d_{MIP-track} < 3$ cm. The measured resolution: 14.4 mrad is close to the nominal value of 12 mrad obtained with ideal tracking in beam test.[2] Figure 3 right panel shows the Cherenkov angle resolution as a function of the number of photons on the

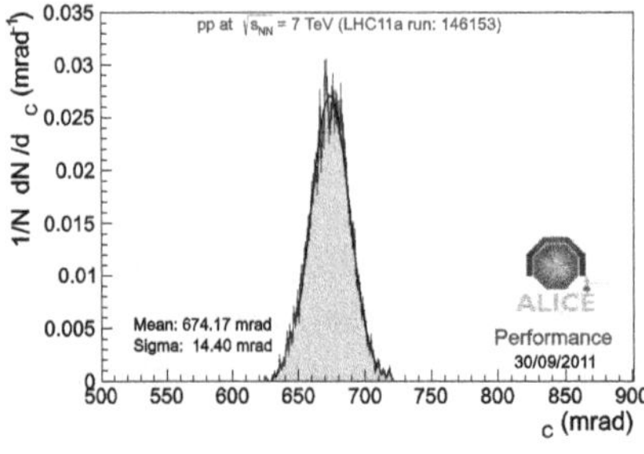

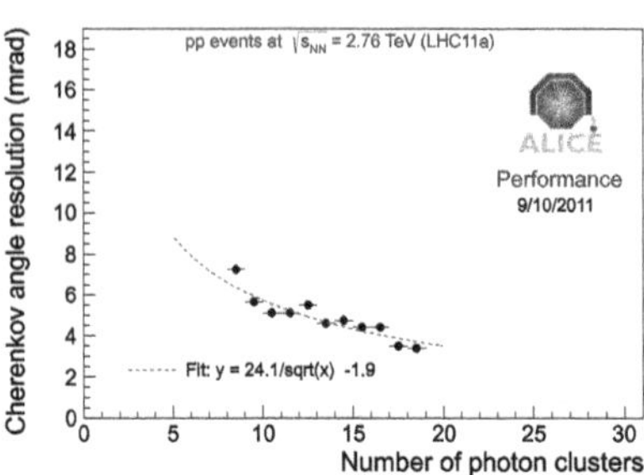

Fig. 3. Left panel: single-photon Cherenkov angle resolution of $\pi^{\pm}$ at p $>$ 1.5 GeV/c with $d_{MIP-track} < 3$ cm. Right panel: Cherenkov angle resolution as a function of the number of photon clusters.

fully contained ring. Fit shows the projection to single-photon Cherenkov angle resolution (note: data set is different from the left panel).

4. Future direction: VHMPID

Motivated by the flavor and jet physics of the intermediate momentum region and the suppression of the high momentum particles a new particle identification detector: the Very High Momentum PID detector (VHMPID) is proposed in ALICE to provide PID in the momentum region 5-30 GeV/c .[12] The VHMPID is proposed as an ALICE upgrade in a focusing RICH setup operating with C_4F_{10} gas radiator, that could be pressurized to improve PID in the momentum region 5-10 GeV/c, with length ~100 cm due to the limited space in ALICE. In total, 4 super-modules and one central module around the existing PHOS detector would cover ~10% of the ALICE central barrel acceptance. To improve tracking and Cherenkov angle reconstruction, dedicated charged particle detection layers would enclose the Cherenkov sub-module. Due to the limited acceptance, high momentum triggers are necessary from the existing TRD detector or from dedicated VHMPID trigger layers. The photon detector would consist of segmented CsI photocathodes covering ~3 m^2 depending on the mirror segmentation with the HMPID like MWPC setup. Beam-test measurement and R&D are ongoing to study alternative radiator gas eg.: C_4F_8O, alternative photon-detector: CsI-coated TGEM.

5. Summary

Since its installation in 2006 HMPID is successfully taking data. Evaluation of the detector performance in beam operation shows that photon detection is at the expected level, the CsI photocathodes are efficient after 6 years of production. The resolution of particle identification and the single-photon Cherenkov angle resolution is at the design values. While the HMPID is efficiently taking data, R&D is ongoing on the proposed VHMPID to extend the ALICE particle identification capabilities to new physics regions.

References

1. F. Carminati *et al.* [ALICE Collaboration], J. Phys. G **30**, 1517 (2004).
2. ALICE Collaboration, ALICE HMPID Technical Design Report, CERN/LHCC 98/19.
3. C. Pastore, I. Sgura, G. De Cataldo, A. Franco and U. Fratino, Nucl. Instrum. Meth. A **639** (2011) 231.
4. A. Di Mauro *et al.* CFP11IWI-USB ISBN: 978-1-4577-0622-6.
5. A. Di Mauro *et al.* ALICE Internal Note 2007-003.
6. J. C. Santiard, K. Marent, CERN-ALICE-PUB-2001-49; J. C. Santiard, Nucl. Instrum. Meth. A **518**, 498 (2004);
   Witters H., Santiard J. C., Martinengo P., Dilogic-2 : CERN-ALICE-PUB-2000-010.
7. G. De Cataldo, A. Franco, C. Pastore, I. Sgura and G. Volpe, Nucl. Instrum. Meth. A **639** (2011) 211.
8. K. Aamodt *et al.* [ALICE Collaboration], Phys. Rev. Lett. **106** (2011) 032301. [arXiv:1012.1657 [nucl-ex]].
9. A. K. Aamodt *et al.* [ALICE Collaboration], Phys. Rev. Lett. **105** (2010) 072002 [arXiv:1006.5432 [hep-ex]].
10. J. Seguinot, Ecole J. Curie de Physique Nucleaire, Mabuisson, France, September 1988.
11. F. Barile, in these proceedings: Identifiedparticle production and spectra at highmomentum with the HMPID detector in the ALICE experiment at the LHC in protonproton collisions at 2.76 TeV.
12. A. Di Mauro *et al.*, Nucl. Instrum. Meth. A **639** (2011) 274.

Radiation damage in soft ferromagnetic materials for enhanced inductor cores operating in extreme environments

Stefania Baccaro†, Enrica Ghisolfi*, Loredana Mannarino* and Andrea Morici‡*

†ENEA-UTTMAT Casaccia Res. Center, via Anguillarese 301,
00060 S.Maria di Galeria, Rome, Italy
*FN S.p.A. Nuove Tecnologie e Servizi Avanzati, Strada per Crescentino 41,
13040 Saluggia, Vercelli, Italy
‡ENEA Saluggia Res. Center, Strada per Crescentino 41,
13040 Saluggia, Vercelli, Italy
{stefania.baccaro,andrea.morici}@enea.it,
{enricaghisolfi,loredanamannarino}@fnspa.com

FeSi compounds are suitable for the construction of inductor cores in electronic DC-DC switching applications. This study provides directions for the development of the core material intended to be used in extreme condition environments, like particle accelerators or nuclear facilities, where the reliability of devices must be high. A *FeSi* based soft ferromagnetic alloy has been implemented and analyzed under the effect of γ-rays irradiation tests at ENEA Casaccia Res. Center, Rome, Italy. Prototypes have been exposed for a total 5 kGy dose. In order to obtain high magnetization field and low coercive field, special techniques of fabrication have been adopted. Pre- and post-damage SEM and EDS micro-analyses are provided.

Keywords: Ferromagnetic Materials; Radiation Damage; Radiation Hardness; Inductor Cores.

1. Introduction

The aim of this research is to develop high performance inductor cores for DC-DC switching power supplies, targeting applications in which radiations are present, e. g. particle colliders or nuclear facilities. In these environments, the hard conditions risk to alter the functionality of electronic devices, in particular the active ones, like power MOSFETs and ICs.[1,2]

The optimization of the inductor core material consists in reaching a high internal magnetic field under workload in presence of ionizing radiations and external magnetic fields, by increasing the magnetic permeability of the core material. This means to achieve a high B_r field, while keeping H_c correspondingly low. Besides, this permits to lower the physical inductor dimensions. The smaller exposed area increases the inductor immunity to ionizing radiations — the dose rate acquired is less, moreover the space occupation within the experiments is optimized.

FeSi alloy was found to have suitable features for this purpose.[3,4] The properties of the *FeSi* alloy and the optimized manufacturing process give high-quality magnetic features and good immunity to radiation damage, thus keeping, over the lifespan of the device in the hostile environment, the same features without significant degradation of them.

2. State-of-the-art

The *Si* presence in the mixture introduces, from the point of view of electrical properties, a better high-frequency inductor core performance, due to the reduction of induced current loops in the material under workload, and from the point of view of mechanical ones, the avoidance of magnetostriction phenomena.[5] Moreover, the *Si* presence increases anisotropy thus enhancing soft magnetic properties.[6] Besides outstanding soft magnetic properties are obtained when the *Si* concentration is around 6.5%, Ruiz *et al.*[4] confirms that the cold ductility of such material makes its industrial manufacture by conventional methods hard — the cold rolling is very difficult and the industrial mass production is not feasible.

The damage by γ-rays can lower the saturation B_r value due to the induction of microcrystalline defects. Ionizing radiations are found to be an effective way to change surface and interface microstructure,[7] however this has the major effect on thin films of material. On bulky ones, like the prototype that is going to be presented, γ radiations are expected to have a minor impact on the microstructure.

3. Specimen preparation

By means of F.N. laboratories know-how, an optimal recipe for the prototype fabrication has been found. It was decided to develop the magnetic cores by Metal Injection Moulding (MIM) with a properly designed mould using an injection moulding press (mod. Negri e Bossi NB 100). In comparison to conventional production routes, MIM offers advantages for component of complex shapes and on large scale production. The process of fabrication was started from 20 - 30 μm atomized powder (Figure 1) of *Fe* enriched with a nominal 6.8 % of *Si*.[3] The energy dispersed analysis (EDS) of the commercial powder showed a pure composition (Figure 1(a)). The powder has been prepared for extrusion by adding organic additives to be evacuated by debinding in the latter part of the process.

A homogeneous compound (that is a pelletized mixture of metal powder and thermoplastic binder made by extrusion) was produced. The extruded compound shows a very high percentage of metallic powder (90 wt%). To get the best density value in the injected part, called "green" due to the presence of polymer, the proper process conditions were studied and imple-

(a) EDS analysis (b) 20 μm grain detail (c) Atomized powder

Fig. 1. *Fe-6.8%Si* powder at SEM analysis.

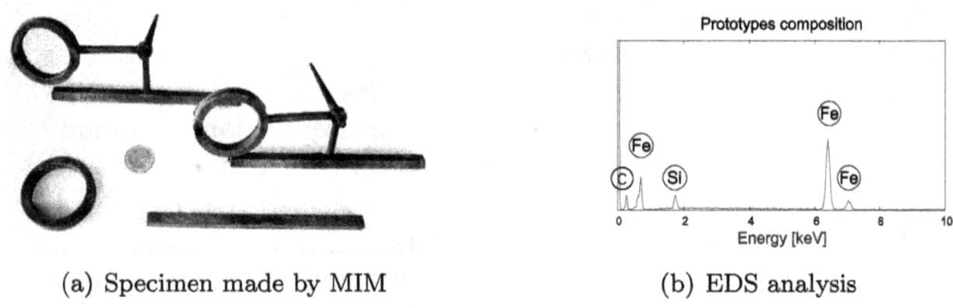

(a) Specimen made by MIM (b) EDS analysis

Fig. 2. Prototype ring characterization.

mented. The following, proprietary, parameter were determined: temperature of the material, of the cylinder and mould, pressure of injection, hold pressure, injection speed and screw feeding speed, time for the injection, time for hold pressure, time of cooling. The specimen shows a "green density" of about 55% theoretical density of the metallic powder. To allow the magnetic properties analysis, standard dimensions prototypes where made (Figure 2(a)): in particular a parallelepiped of 8 x 8 x 160 mm and a ring of 52 mm external radius, 44 mm inner radius with square section.[8]

In order to flush away the plastic part of the prototype, the debinding process is necessary: in this step the temperature raise to $600\,^oC$, then the sintering process will reach a temperature of $1260\,^oC$. This phase is necessary to bring the grains nearer, and to assure the minimum space between them, in order to get the best density and consequently the best magnetic characteristics. The process needs to be slow enough to avoid defects and gaps. A challenge is the use of a hydrogen atmosphere to avoid Fe oxidation.

The debinding and sintering process optimizations are in progress. It was decided to make a preliminary characterization of the "green" specimen, as shown by the presence of the carbon in Figure 2(b), and to investigate their behavior before and after γ irradiation.

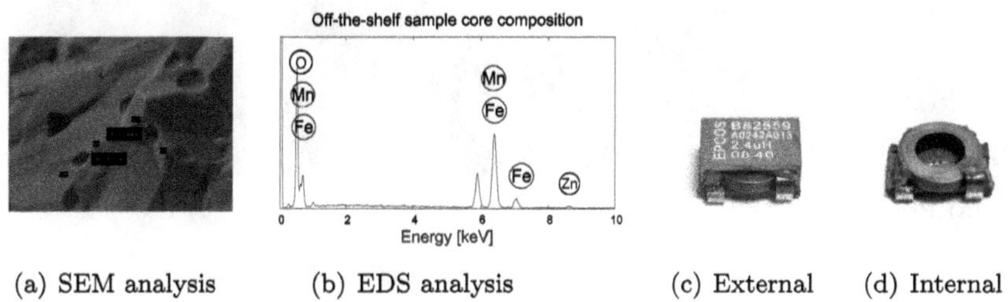

(a) SEM analysis (b) EDS analysis (c) External (d) Internal

Fig. 3. Off-the-shelf sample characterization.

4. Measures

In order to step up in performance, over the already present off-the-shelf inductor cores, an investigation of commercial components has been done (Figure 3). The SEM (ZEISS, mod. EVO 40) photograph (Figure 3(a)), taken in the internal part of the sample core, highlights that the microscopic structure is compact: grains show a dimension varying between 10 to 22 μm, and gaps between them of an average of 5 μm. These gaps are the first point of improvement for the cores. The EDS analysis (Figure 3(b)) shows that the commercial inductor core has an average composition of Fe, O, Mn and Zn, respectively in weight 46 %, 28 %, 19 % and 3 %, typical of a soft ferrite core. Other contaminant species are present in minor part.

The prototype ring was measured by permeameter DC measurements as specified in IEC 60404-04 and ASTM A773, by the use of AMH-20K-S measure system. The magnetic measurement determines both the initial normal magnetization curve and the hysteresis loop. Tests was replicated to obtain a good interpolation between measurements. The targets are to overcome a B_r of more than 0.8 T, with a H_c of 80 A/m. Measured values have still not reached the magnetic properties targets due to the presence of the polymer in the ring.

The radiation hardness tests under γ-rays irradiation were conducted at ENEA Casaccia Res. Center, Rome, Italy, in the ^{60}Co radioisotope "Calliope" source. Tests have been carried on at dose rate conditions of 100 Gy/h, during 2 days of exposure in the 24 hours. The dose of 5 kGy is enough to model, in first approximation, the absorbed dose in a particle collider experiment.[1] The effect of ionizing radiation was probed by SEM and EDS micro analyses, performed before and after irradiation on all samples. As shown in Figure 4, the SEM images of the internal sections of the irradiated ring sample reveal an initial local fusion of the polymeric component and a less crystallinity of the same component with respect to the non-irradiated one.

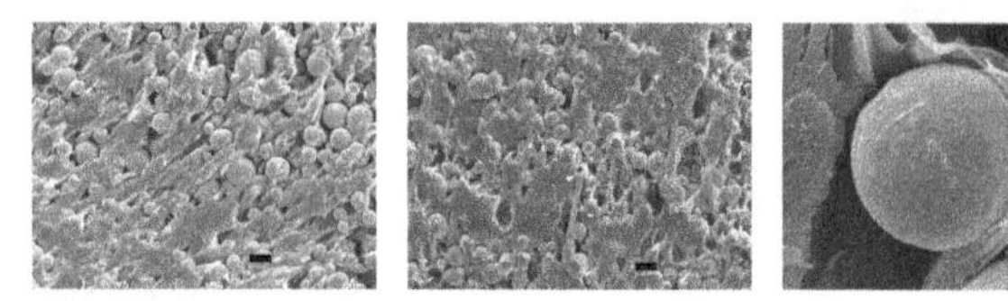

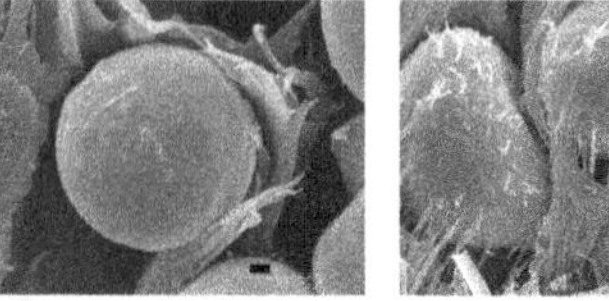

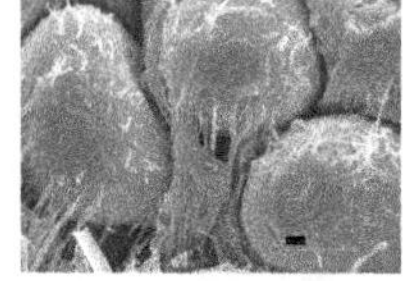

(a) Pre-damage (b) Post-damage (c) Pre-damage (d) Post-damage

Fig. 4. Radiation damage for the internal section of the prototype ring.

5. Conclusions

In this text a preliminary study for enhancing inductor cores has been presented. The specimen preparation and fabrication process has been developed. The final shape of the inductor will be designed when the magnetic targets will be reached. The magnetic properties should be investigated also on temperature sweep.

The γ-rays irradiation does not alter significantly the material. FTIR and X-rays analyses can give deeper information on the polymeric ageing. Neutron based radiation tests have to be performed in future extensions, because they can damage differently the material.

Acknowledgments

Authors want to thank INFN for financing within the APOLLO experiment, in particular Dr. Agostino Lanza. A special thank to all FN S.p.A. laboratory staff.

References

1. P. Tenti, G. Spiazzi, S. Buso, M. Riva, P. Maranesi, F. Belloni, P. Cova, R. Menozzi, N. Delmonte, M. Bernardoni, F. Iannuzzo, G. Busatto, A. Porzio, F. Velardi, A. Lanza, M. Citterio and C. Meroni, *Journal of Instrumentation* **6**, p. P06005(jun 2011).
2. S. Baccaro, A. Cecilia, G. Serra and ENEA, *Radiation effects on electronic components* (ENEA, 2002).
3. H. Gavrila and V. Ionita, *Journal of Optoelectronics and Advanced Materials* **4**, 173(jun 2002).
4. D. Ruiz, T. Ros-Yanez, L. Ortega, L. Sastre, L. Vandenbossche, B. Legendre, L. Dupre, R. Vandenberghe and Y. Houbaert, *Magnetics, IEEE Transactions on* **41**, 3286 (oct 2005).
5. M. Takahashi, N. Kato, T. Sato and T. Wakiyama, *Magnetics, IEEE Transactions on* **23**, 3068 (sep 1987).
6. J. Diaz, N. Hamdan, P. Jalil, Z. Hussain, S. Valvidares and J. Alameda, *Magnetics, IEEE Transactions on* **38**, 2811 (sep 2002).
7. D. Aurongzeb, K. B. Ram and L. Menon, *Applied Physics Letters* **87**, 172509(oct 2005).
8. R. Strnat, M. Hall and M. Masteller, *Magnetics, IEEE Transactions on* **43**, 1884 (may 2007).

NEUTRON AND PROTON TESTS OF DIFFERENT TECHNOLOGIES FOR THE UPGRADE OF COLD READOUT ELECTRONICS OF THE ATLAS HADRONIC ENDCAP CALORIMETER

M. NAGEL

Max-Planck-Institut für Physik,
Föhringer Ring 6, 80805 Munich, Germany
E-mail: nagel@mppmu.mpg.de
www.mppmu.mpg.de

on behalf of the HECPAS Collaboration
(IEP Košice, Univ. of Montréal, MPI Munich, IEAP Prague, NPI Řež)

The expected increase of total integrated luminosity by a factor of ten at the HL-LHC compared to the design goals for LHC essentially eliminates the safety factor for radiation hardness realized at the current cold amplifiers of the ATLAS Hadronic Endcap Calorimeter (HEC). New more radiation hard technologies have been studied: SiGe bipolar, Si CMOS FET and GaAs FET transistors have been irradiated with neutrons up to an integrated fluence of $2.2 \cdot 10^{16}\,\mathrm{n/cm^2}$ and with 200 MeV protons up to an integrated fluence of $2.6 \cdot 10^{14}\,\mathrm{p/cm^2}$. Comparisons of transistor parameters such as the gain for both types of irradiations are presented.

Keywords: ATLAS, Liquid Argon calorimeter, HL-LHC, cold readout electronics, radiation hardness

1. The ATLAS Hadronic Endcap Calorimeter

The hadronic endcap calorimeter (HEC) of the ATLAS experiment[1] at the CERN Large Hadron Collider (LHC) is a copper-liquid argon sampling calorimeter in a flat plate design.[2,3] The calorimeter provides coverage for hadronic showers in the pseudorapidity range $1.5 < |\eta| < 3.2$. The HEC shares each of the two liquid argon endcap cryostats with the electromagnetic endcap (EMEC) and forward (FCAL) calorimeters, and consists of two wheels per endcap.

A HEC wheel is made of 32 modules, each with 40 liquid argon gaps, which are instrumented with active read-out pads. The signals from the read-out pads are sent through short coaxial cables to preamplifier and summing boards (PSB) mounted on the perimeter of the wheels inside the liquid argon cryostat. The PSB boards carry highly-integrated preamplifier and summing amplifier chips in Gallium-Arsenide (GaAs) technology. The signals from a set of preamplifiers are summed to one output signal, which is transmitted to the cryostat feed-through.[4] Figure 1 shows a PSB board mounted and connected on the perimeter of a HEC wheel.

Fig. 1. A detailed view of a PSB board mounted on the perimeter of a HEC wheel.

2. Requirements of the HEC cold electronics for the HL-LHC upgrade

The GaAs technology currently employed in the HEC cold electronics has been selected for its excellent high frequency performance, stable operation at cryogenic temperatures, and radiation hardness. The radiation hardness specifications were defined for ten years of operation at the LHC design luminosity of $10^{34}\,\mathrm{cm}^{-2}\,\mathrm{s}^{-1}$, including a safety factor of ten. For the high-luminosity upgrade of the LHC (HL-LHC), the luminosity is foreseen to increase by a factor of 5–10, effectively eliminating the safety factor. The ATLAS collaboration therefore decided to re-examine the radiation hardness of the current HEC cold electronics and of potential alternative technologies.[5] Detailed studies of the expected radiation levels after ten years of running under HL-LHC conditions yielded the following requirements (including a safety factor of 10) for the HEC cold electronics:

- Neutron fluence of $2 \cdot 10^{15}\,\mathrm{n/cm^2}$
- Proton fluence of $2 \cdot 10^{14}\,\mathrm{p/cm^2}$
- Gamma dose of 20 kGy

3. Tests

The neutron irradiation tests were performed at the Řež Neutron Physics Laboratory near Prague in the Czech Republic, up to an integrated fluence of $2.2 \cdot 10^{16}\,\mathrm{n/cm^2}$. A 37 MeV proton beam incident on a D_2O target created a somewhat divergent beam of neutrons with a flux density up to $10^{11}\,\mathrm{n/cm^2/s}$. The proton irradiation test were performed at the Proton Irradiation Facility at the Paul-Scherrer-Institut in Switzerland with a 200 MeV proton beam up to an integrated fluence of $2.6 \cdot 10^{14}\,\mathrm{p/cm^2}$. The transistors were bonded in ceramic casings and mounted on boards, which were then aligned in the particle beams. The three different transistor technologies being tested were Si CMOS FET, SiGe Bipolar HBT (both SGB25V 250nm from IHP), and Triquint CFH800 250nm GaAs FET pHEMT, respectively. The performance of the transistors was monitored during irradiation with a vector network analyzer recording a full set of S-parameters. Beam current measurements and radiation films placed at various distances along the beam were used to determine the particle flux and the beam profile.

4. Results

The various transistor parameters were calculated from the measured S-parameters using standard small signal circuit models. The transistor parameters were averaged over a certain frequency range to obtain their mean and rms-values for every set of S-parameters. This frequency range extended from 300 kHz to 100 MHz, unless a certain parameter was unstable at low or high frequencies, in which case appropriate cuts were applied.

These averaged transistor values were then used to characterize their behaviour as a function of radiation.

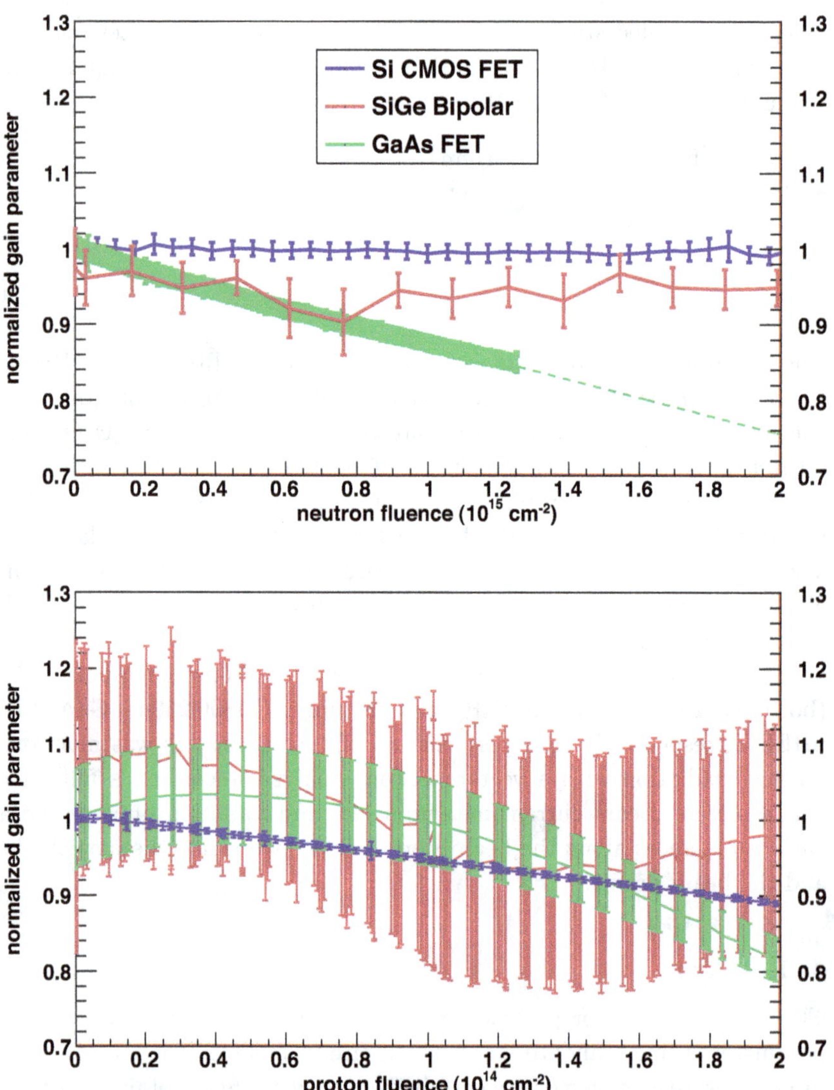

Fig. 2. Relative gain loss as a function of neutron (top) and proton (bottom) fluence for various technologies.

Figure 2 summarizes the results for the neutron (top) and the proton (bottom) tests in terms of the appropriate gain parameters. Displayed are the various gain parameters, i.e. the real part of the transconductance g_m for the FET transistors and the current gain β for the Bipolar transistors, as a function of the corresponding particle fluence up to the required limit, normalized to the corresponding value before irradiation. We apply a linear fit to the observed radiation dependence, and the relative change of the gain parameter resulting from the fit is quoted in Table 1.

Table 1. Loss of gain of various technologies under neutron and proton irradiation.

Technology	Neutron fluence $2 \cdot 10^{15}\ n/cm^2$	Proton fluence $2 \cdot 10^{14}\ p/cm^2$
Si CMOS FET	-1 %	-11 %
SiGe Bipolar	-3 %	-10 %
GaAs FET	-24 %	-21 %

The results show that the alternative technologies under consideration are more radiation hard than the current GaAs technology, in particular for neutron irradiation. However, the loss in gain for the GaAs FETs is only moderate, and could be considered acceptable, especially if occuring uniform across the HEC. Gamma irradiation tests are planned for the near future to complete the radiation hardness tests.

References

1. *ATLAS: technical proposal for a general-purpose pp experiment at the Large Hadron Collider at CERN* LHC Tech. Proposal, LHC Tech. Proposal (CERN, Geneva, 1994). CERN-LHCC-94-43, LHCC-P-2.
2. ATLAS Collaboration, *ATLAS Liquid Argon Calorimeter: Technical Design Report*, tech. rep., CERN (1996), CERN-LHCC-96-41, ATLAS TDR 2.
3. D. M. Gingrich *et al.*, *J.Inst.* **2**, p. P05005 (2007), ATL-LARG-PUB-2007-009, ATL-COM-LARG-2007-006, MPP-2007-237.
4. J. Ban, H. Brettel, W. D. Cwienk, J. Fent, L. Kurchaninov, E. Ladygin, H. Oberlack, P. Schacht, H. Stenzel and P. Strizenec, *Nucl.Instrum.Meth.A* **556**, 158 (2006), MPP-2005-193.
5. P. Schacht, ATLAS Liquid Argon Endcap Calorimeter R & D for sLHC, in *Proceedings of the 11th ICATPP Conference on Astroparticle, Particle, Space Physics, Detectors and Medical Physics Applications*, (Villa Olmo, Como, 2009).

POWER DISTRIBUTION ARCHITECTURE FOR HIGH ENERGY PHYSIC HOSTILE ENVIRONMENT*

M. ALDERIGHI(1,6), M. CITTERIO(1), S. LATORRE(1), M. RIVA(1,8,X), P. COVA(3,10), N. DELMONTE(3,10), A. LANZA(3), M. BERNARDONI(10), R. MENOZZI(10), A. COSTABEBER(2,9), A. PACCAGNELLA (2,9), F. SICHIROLLO(2,9), G. SPIAZZI(2,9), M. STELLINI(2,9), P. TENTI(2,9), S. BACCARO(4,5), F. IANNUZZO(4,7), A. SANSEVERINO(4,7), G. BUSATTO(7), V. DE LUCA(7)

(1) INFN Milano, (2) INFN Padova, (3) INFN Pavia, (4) INFN Roma, (5) ENEA UTTMAT, (6) INAF, (7) University of Cassino, (8) Università degli Studi di Milano, (9) University of Padova, (10) University of Parma,

In the high luminosity phase of the Large Hadron Collider (LHC) the selection of the most suitable architecture able to supply the instrumentation of the experiments represents a critical task today. The power conversion units will have to supply low voltages and high currents to the loads with reduced transmission losses and, moreover, their design will have to face the critical demand of efficiency, robustness and limited size together with the need to operate in hostile environment. The paper discusses the most promising solutions in the power supply distribution networks which could be implemented in the upgraded detectors at the High Luminosity LHC collider. The proposed topologies have been selected by considering their tolerance to high background magnetic field and nuclear radiations as well as their limited electromagnetic noise emission. The analysis focuses on the description of the power supplies for noble liquid calorimeters, such as the Atlas LAr calorimeters, though several outcomes of this research can be applied to other detectors of the future LHC experiments.

1. Energy Distribution in Hostile Environment

1.1. *Introduction*

The planned increase of instrumentation sensitivity in the ATLAS experiment of the Large Hadron Collider (LHC) at CERN, and the increase of the radiation background that will characterize the future LHC luminosity upgrade, about a factor of 10 above the nominal value, are indeed incompatible with the current capability of the distribution systems in use in the experiments. Electronic devices and equipment must face a highly hostile environment, basically for the

---

* This work is supported by I.N.F.N. Italy Apollo collaboration.

X Corresponding author Marco Riva is with Department of Physics, Università degli Studi di Milano, 20133 Milano - Italy (e-mail: riva@unimi.it)

very high background of both charged and neutral particles ([1]-[2]) and for the presence of a non negligible magnetic field up to 100 mT, thus opening a severe tolerance issue for component selection and system design. The last generation of microprocessors, recently introduced and planned to be deployed in the front-end boards, has shown an increase in the operating frequency and, of course, in the supply demands. The increased need for power poses necessary a severe design challenges for the traditional isolated topology and pushes towards converters more suitable to meet the new high efficiency and high power requirements. The presented investigation proposes a new power supply distribution network for the front-end and read-out electronics of the ATLAS Liquid Argon (LAr) calorimeters. The solution presented is based upon a main converter (MC) and many Point of Load converters (POLs) switching at up to 200 kHz and 280 kHz respectively. The topology chosen for the MC, the Switch In Line Converter (SILC), is a DC to DC phase-shifted converter characterized by a disposition in line of the MOSFETs and good soft switching performances. The converter, acting as a charge generator controlled period by period, has a first order dynamics, well suited for active loads, for instance low-voltage, high-current point-of-load converters. The PoL converters are step down topologies that are required to supply very low voltage loads (0.9V-5V) with currents that can range from few amps up to tens of amps. Besides the classical buck converter topology, an Interleaved Buck with Voltage Divider (IBVD) is proposed with the aim of increasing the step down ratio while, at the same time, increasing the current capability through interleaving of switching cells.

1.2. *Distributed Power Supply*

Several configurations of the power supply distribution systems are usually employed in commercial applications. Innovations are determined by the evolution of digital circuits, whose switching frequencies moved to the GHz range and supply voltages reduced down to the level of 1 V with a correspondent increase of the current demand. Some of the most common applied power distribution strategies are here briefly described.

Centralized Power Architecture (CPA) - Introduced in the early 80's, it's based upon an single main DC/DC converter, that generates the whole set of regulated DC voltages (typically ±12V, ±5V, and +3.3V) supplying the circuit subsystems. This architecture does not fully meet the requirements of sub-micron CMOS requiring the generation of voltage levels in the range between 1 V and 2.5 V.

Distributed Power Architecture (DPA) - A single intermediate DC bus (typically +48V or +12V) is generated by a main DC to DC converter. The bus voltage is reduced to lower levels by means of point-of-load DC/DC converters located on the motherboards.

Intermediate Bus Architecture (IBA) - In addition to the generation of a voltage bus, like in the DPA but typically higher (48V-76V), a second set of PCB main voltages is provided (8V-14V). Then lower voltages are generated by the point-of-load converters. The number of converters employed imposes a penalty to the efficiency.

Spot Power Architecture - It is an intermediate solution between CPA and DPA. In addition to the voltage bus typical of DPA, one or more voltage lines supply the circuits directly.

In the range of 800 W to one 1.5 kW, which this paper refers to, the DPA power architecture seems the most suitable to the authors and Figure 1 shows two possible implementations.

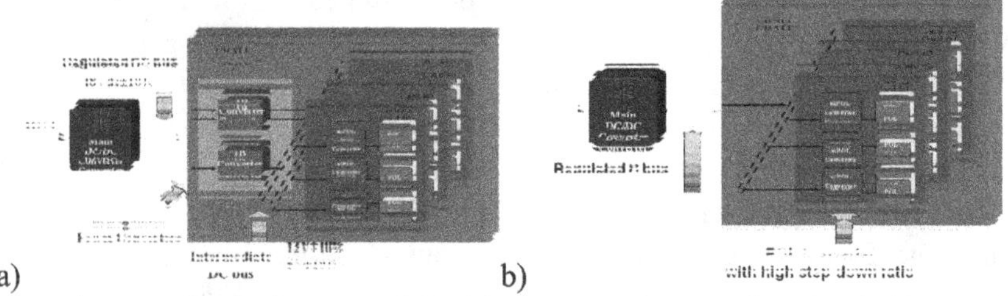

Figure 1. Power Distribution proposal for LAr detectors: a) Intermediate Bus Architecture, b) Distributed Power Architecture.

1.3. *Switch in line converter for the DPA*

The schematic of the main converter is shown in figure 2a. Due to the particular disposition of the active switches it is called "Switch In Line Converter". The circuit consists in the series connection of two half bridge converters (lower HB and upper HB). In figure 2b, the ideal switching sequence and the current waveforms into the windings T1' and into Tl are represented.

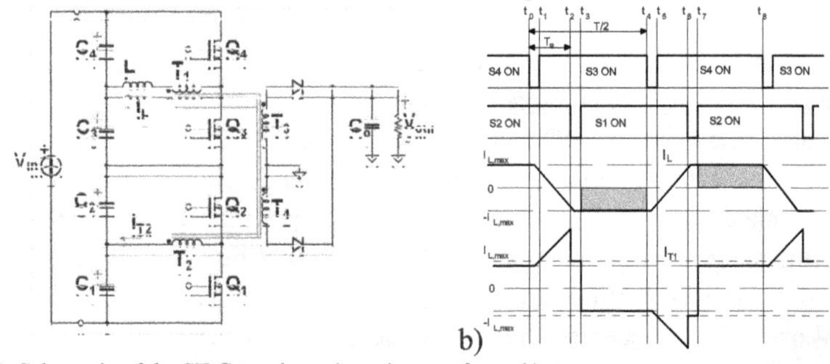

Figure 2. Schematic of the SILC topology a); main waveforms b)

Connections in parallel are allowed because the SILC power cell acts like a controlled current generator. The input voltage partition due the series connection of the half bridges, reduces the voltage stress of the MOSFETs and the voltage gap to span during the resonant transition. This allows stretching the soft switching persistence for a wider power range as compared with other PWM phase shifted topologies.

A first order dynamics emerges from this linear approximated model. A more accurate model, based upon a state space description in the discrete time, confirms the pseudo-first order behaviour of the SILC phase-shift controlled power cell [3].

The characterizations make reference to a 1.5 kW distributed power architecture. Efficiency measurements, neglecting the losses of the point-of-load converters are above 85% at full load and remain high until the load falls below one third of full load.

1.4. *POL Converter*

Among the non-isolated PoL converter topologies that have been investigated for the application, a solution employing a high step-down ratio converter operating at high frequency is selected. The high step-down ratio is motivated by the need of pushing the DC bus voltage to the highest possible value so as to increase the power capability of the distribution system, the current being limited by DC bus cabling. The chosen topology is shown in figure 3a and its operation has been analyzed in [4]-[6].

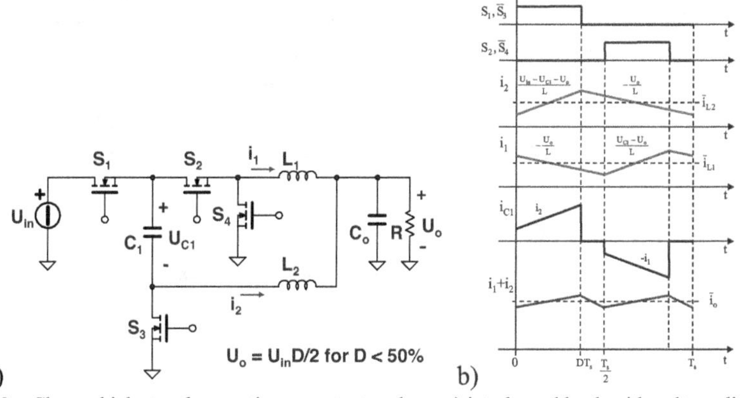

Figure 3 – Chosen high step down ratio converter topology: a) interleaved buck with voltage divider (IBVD): b) main waveforms for D < 0.5

When operating with duty-cycle values lower than 0.5 this topology features: high step-down ratio with reasonable duty-cycle values, reduced switch voltage stress ($U_{in}/2$), interleaved operation with automatic current sharing, and output current ripple cancellation.

A prototype was built and tested with the following specifications: $U_{in} = 12$ V, $U_o = 2$ V, $I_o = 20$ A, $f_s = 280$ kHz. The converter was designed to operate in Continuous Conduction Mode with standard ferrite core inductors (2.2 μH each). A Photo of the prototype together with its measured efficiency curve is shown in figure 4. The tests were performed in the absence of any external magnetic field.

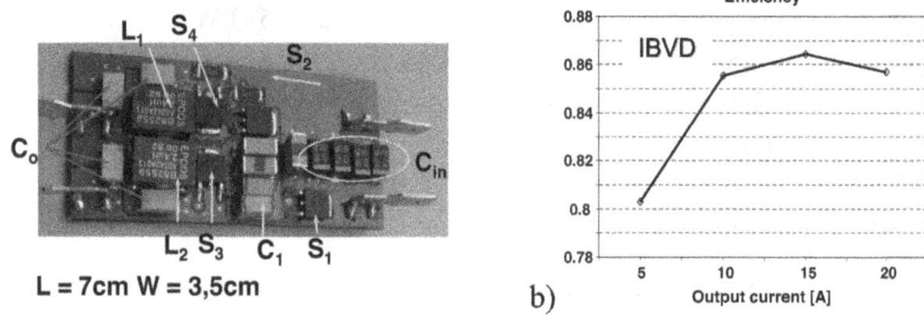

Figure 4 – a) Converter layout; b) efficiency as a function of output current

References

1. Schultes J., Andreazza A., Becks K.-H., Citterio M.; Einsweiler K., Kersten S., Kind P., Latorre S., Mattig P., Meroni C., Sabatini F., *The power supply system for the ATLAS pixel detector*, 2004 IEEE, Nuclear Science Symposium Conference Record, Volume 3, 16-22 pp. 1711-1715 Vol. 3.
2. S. Majewski (on behalf of the ATLAS Liquid Argon Calorimeter Group), *Electronic Readout of the ATLAS Liquid Argon Calorimeter: Calibration and Performance*, June 2010 ATL-LARG-PROC-2010-002.
3. Belloni F., Maranesi P.G., Riva M., *Distributed Power Supply for Payloads Aboard Space Vehicles*, 2006 IEEE Power Electronics Specialists Conference, pp. 1-5.
4. Y. Jang, M. Jovanovic, *Interleaved Boost Converter With Intrinsic Voltage-Doubler Characteristic for Universal-Line PFC Front End*, in IEEE Trans. Power Electronics, vol. 22, no. 4, July 2007, pp. 1394-1401.
5. Y. Jang, M. Jovanovic, United States Patent Application US2006/0087295 A1, April 27, 2006.
6. S. Buso, G. Spiazzi, F. Faccio, S. Michelis, *Comparison of dc-dc converter topologies for future SLHC experiments*, in IEEE Energy Conversion Congress and Expo (ECCE), San Jose California, September 2009, pp. 1775-1782.

Kaon identification in the NA62 experiment at CERN

A. Romano* on behalf of the NA62 collaboration
School of Physics and Astronomy, University of Birmingham,
Birmingham, B15 2TT, United Kingdom
**E-mail: angela.romano@cern.ch*

The main purpose of the NA62 experiment at the CERN SPS is to measure the Branching Ratio of the ultra-rare kaon decay $K^+ \to \pi^+ \nu\bar{\nu}$ with 10% accuracy. This will be achieved by collecting about 100 events with a Signal to Background ratio of 10/1. NA62 will use a 75 GeV/c unseparated charged hadron beam and a kaon decay-in-flight technique. For the kaon identification a hydrogen gas-filled differential Cherenkov counter (CEDAR) is placed in the incoming beam. The CEDAR detector is required to achieve a kaon identification efficiency of at least 95% with a time resolution of about 100 ps.

Keywords: Kaon decays; Cherenkov detectors; Gas filled counters.

1. The NA62 experiment at CERN SPS

The NA62 experiment[1] at the CERN SPS is designed to use rare kaon decays to probe physics Beyond the Standard Model in a complementary way respect to the direct searches for potential new particles at the LHC. The main goal of NA62 is the measurement of the Branching Ratio (BR) of the decay $K^+ \to \pi^+ \nu\bar{\nu}$ with 10% accuracy. The study of this process provides a determination of the CKM matrix element $|V_{td}|$ and the theoretical computations of its rate can reach an exceptionally high degree of precision[2]. The $K^+ \to \pi^+ \nu\bar{\nu}$ decay is a flavor-changing neutral-current channel forbidden at a tree-level. The leading SM contribution to the matrix element is dominated by short-distance processes mediated by quark loops, where the top quark exchange is the dominant component. The required hadronic matrix elements can be extracted from the accurately measured leading semileptonic decay $K^+ \to e^+ \pi^0 \nu$ via isospin rotation[3]. The experimental status is based on 7 events collected by the E787/949 collaborations at BNL[4]: $BR(K^+ \to \pi^+ \nu\bar{\nu}) = 1.73^{+1.15}_{-1.05} \times 10^{-10}$. The measured value is compatible with the SM prediction[5]:$(7.81 \pm 0.80) \times 10^{-11}$. The NA62 strategy is to collect about 100 events of the rare kaon decay $K^+ \to \pi^+ \nu\bar{\nu}$, keeping the background contamination lower than 10%, in two years of data taking starting after LHC shutdown in 2013.

2. The kaon identification in NA62

NA62 is a fixed-target experiment that uses 400 GeV/c protons from the SPS to produce a 75 GeV/c secondary unseparated hadron beam. The main beam components are pions ($\sim$ 60%) and protons ($\sim$ 20%), while kaons correspond to the 6%. NA62 exploits a kaon decay-in-flight technique free from the background due to multiple scattering of kaons impinging on the stopping target. The advantages of using a high energy beam is that the kaon production cross-section increases with the proton energy and that the background rejection improves for high energy decay products. The disadvantage is that kaons cannot be efficiently separated from pions and protons at the beam level: upstream detectors, which tag the kaons and measure their momentum and direction, are then exposed to a particle flux about 17 times larger than the useful (kaon) one. It is crucial to make a positive identification of the minority particles of interest, kaons, in the high rate beam environment before they decay. This is achieved by placing a *ChErenkov Differential counter with Achromatic Ring focus* (CEDAR) in the incoming beam; the CEDAR is insensitive to pions and protons with minimal accidental mis-tagging. The CEDAR detector is required to achieve a kaon identification efficiency of at least 95%. In addition, a time resolution of $\sim$ 100 ps, in conjuction with timing information from other detectors, is necessary to reconstruct the $K^+ \to \pi^+ \nu\bar{\nu}$ decay and ensure the rejection of the background due to the accidental overlap of events in the NA62 detector.

3. The CEDAR counter

The counter was built in the early 80s at CERN for application in the SPS secondary beams[6]. Two different versions exist: a "CEDAR-North" optimized for operation with helium gas and high energies (K/π separation up to 300 GeV/c); a "CEDAR-West" optimized for operation with nitrogen gas and used with low momentum beam (K/π separation up to 150 GeV/c). A simulation programme verified that the West version will function well for NA62 purposes. A test beam on a CEDAR-West counter was performed in 2006 and its results validated the counter ability to distinguish kaons from pions and protons in NA62.

The CEDAR counter (Fig. 1) is a $\sim$ 6 m long vessel filled with gas of controlled pressure. For a given beam momentum, the Cherenkov angle of

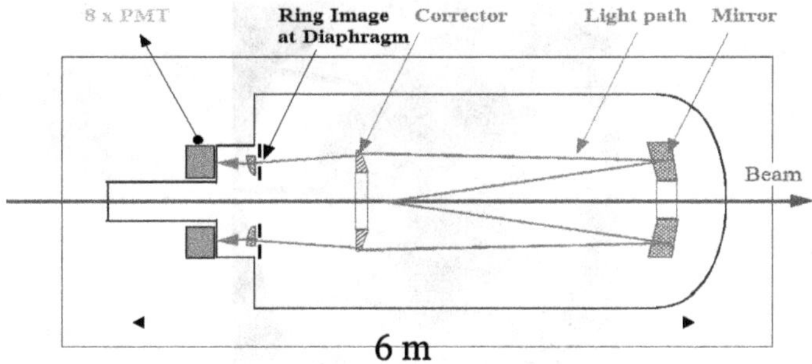

Fig. 1. The CEDAR layout

the light emitted by a charged particle traversing the radiator is a unique function of the mass of the particle and the wavelength of the emitted light. A spherical Mangin mirror, located at the end of the vessel, reflects the Cherenkov light back on a ring-shaped diaphragm with an adjustable aperture width. A chromatic corrector system reduces the severe effect of chromatic dispersion. Under the assumptions of a minimal beam divergence ($< 80\mu$rad) and a precise alignment of the optical and beam axis, the Cherenkov light produced by the particles of interest (kaons) is transmitted by the internal optics through the diaphragm aperture. Behind the diaphragm eight condenser lenses focus the light on eight photomultipliers (PMT). Although kaons are the minority particles ($\sim 6\%$) of the beam, the kaon rate in the high-intensity beam for NA62 will be $\sim$ 50 MHz (average). The Cherenkov light yield expected at the exit windows of the CEDAR vessel is about 250 photons per kaon, which translates in a photon rate of $\sim$ few MHz/mm^2. The existing CEDAR PMT and readout electronics are inadequate to cope with this illumination and need to be replaced.

4. The CEDAR upgrade design

An upgraded CEDAR-West design has been proposed to positively tag incident kaons at the required rate of 50 MHz. It uses hydrogen instead of nitrogen as radiator to reduce the beam scattering in the gas and it is equipped with new photomultipliers (PMT) and readout electronics. The eight light spots produced by the internal optics on the exit windows of the vessel are projected onto new PMT planes by means of external spherical mirrors and a 90-degrees reflection (Fig. 2). The achieved light spots on new PMT planes are enlarged and uniform.

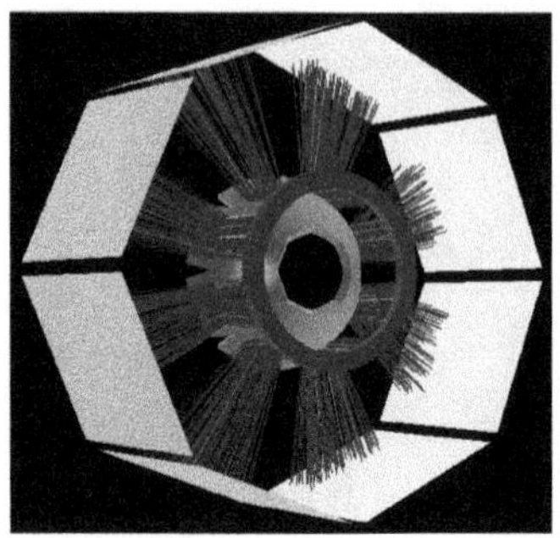

Fig. 2. New design of the CEDAR external optics: photons exit from the vessel windows, reflect on spherical mirrors and hit the PMT planes.

The technology choice for the CEDAR photon detector consists of metal package photomultipliers of the HAMAMATSU R7400[7] series, U-03 (UV glass window) type, which were chosen for their smallness, capability of single photon counting, ability of standing at high rate per unit area, UV/blue light sensitivity with the highest efficiency, excellent time resolution, limited anode current, feasibility under radiation exposure. The light collection system design is an array of cones and PMTs packed in a compact configuration to reduce the effect of dead areas. A safety requirement resulting from the mechani cal modifications and the use of hydrogen gas is a nitrogen-filled environmental casing around the optical-readout electronics and HV, which is mandatory to eliminate any possibility of an explosion in the event of a hydrogen leak from the CEDAR.

5. The CEDAR simulation

In the proposed design the photon flux on single PMT device is ≤ 5 MHz and the PMT anode current is kept within a safe limit (two order of magnitude lower than the sustainable value stated in the datasheet). The photon rate implies several limitations on the performances of the photon detection involving event pile-up, electronics dead time and smearing effects on the readout system. A FLUKA[8] simulation and dedicated studies of the beam halo have been used to evaluate the expected neutron and muon doses on the CEDAR, which are ~ 0.4 Gy/year and ~ 0.3 Gy/year, respectively. A Geant4 MC simulation programme is used to study the optimal configuration for the CEDAR photon detector, which is the one providing a maximum light collection and Kaon ID efficiency as well as a manageable rate for PMTs and readout channels (≤ 5 MHz). The Geant4 simulation allows one to perform several studies with the hydrogen-filled CEDAR-West:

(a) evaluate the kaon-pion separation achievable with the counter;
(b) optimize the diaphragm aperture width for kaons;
(c) maximize the kaon ID efficiency;
(d) minimize the pion suppression;
(e) control the number of detected photons per kaon;

The CEDAR-West internal optics is not optimized for operation with the hydrogen gas, as a consequence the Cherenkov photon spectrum produced by kaons peaks at two different radii on the diaphragm plane and the low wavelength part is mostly unusable being in a region not free from pions. Preliminary simulations have showed that the hydrogen-filled CEDAR-West counter can work optimally at diaphragm apertures > 1.5 mm (standard aperture width previously fixed for the nitrogen-filled version). This modification is necessary to compensate the significant photon loss at low wavelengths due to the pion contamination. By requiring at least a 6-fold coincidence among the eight light spots a kaon ID efficiency $> 95\%$ is achievable with a diaphragm aperture width of ~ 3.5 mm. For a single photon time resolution[9] of ~ 300 ps, the mean value of detected photons per kaon provides a kaon time resolution better than 100 ps.

6. CEDAR test beam at CERN

A test beam of a prototype of the CEDAR-West will be performed in October 2011. The validation of the Cherenkov counter assigned for application in NA62 and the test of new PMTs, front-end and readout electronics will be achieved during the test.

References

1. G. Anelli et al., CERN-SPSC-2005-013, SPSC-P-326 (2005).
2. G. Buchalla and A. J. Buras, *Phys. Lett. B* **333**, 221 (1994); *Phys. Rev. D* **54**, 6782 (1996).
3. F. Mescia and C. Smith, *Phys. Rev. D* **76**, 034017 (2007).
4. A. V. Artamonov et al., *Phys. Rev. D* **79**, 092004 (2009).
5. J. Brod, M. Gorbahn and E. Stamou, *Phys. Rev. D* **83**, 034030 (2011).
6. C. Bovet et al., CERN Report: CERN 82-13 (1982).
7. HAMAMATSU PHOTONICS K. K., (314-5, Shimokanzo, Toyooka-village, Iwatagun, Shizuoka-ken, 438-0193, Japan) `http://www.hamamatsu.com`
8. G. Battistoni et al., Proceedings of the Hadronic Shower Simulation Workshop 2006, Fermilab 6-8 September 2006, M. Albrow, R. Raja eds., AIP Conference Proceeding 896, 31-49, (2007).
9. G. Anzivino et al., *Nucl. Instr. and Meth. A* **593**, 314-318 (2008).

PERFORMANCE OF THE RICH DETECTORS AT LHCB WITH 2011 DATA

A.SPARKES

Physics and Astronomy Department, University of Edinburgh, Edinburgh, EH9 3JZ, United Kingdom
E-mail: a.sparkes@sms.ed.ac.uk

The Large Hadron Collider Beauty Experiment aims to precisely measure CP violation and rare decays, both of which require accurate charged particle identification. The Ring Imaging Cherenkov detectors are vital components of the system. Charged particles travelling through radiators at relativistic speeds produce Cherenkov photons which are detected by arrays of Hybrid Photon Detectors. Known resonances, such as ϕ, K_s, Λ and D^*, provide samples of pions, kaons and protons, which are used to assess the performance of the detectors. Methods for alignment and calibration of the detectors lead to an optimal Cherenkov angle resolution. The particle identification performance on recent data is presented.

Keywords: LHCb, RICH

1. LHCb Detector

The LHCb detector[1] is a forward arm spectrometer which is designed to precisely measure CP violating parameters and make observations of rare decays. It has excellent vertexing, tracking and triggering capabilities as well as precise identification and separation of charged particles using the RICH system.[2] It has now been taking data for over a year and at the time of writing has recorded $\sim 1\ \text{fb}^{-1}$ with a centre-of-mass energy of 7 TeV and an instantaneous luminosity of $3.5 \times 10^{32}\ \text{cm}^{-2}\text{s}^{-1}$. This is greater than the design luminosity of $2 \times 10^{32}\ \text{cm}^{-2}\text{s}^{-1}$, but despite this the detector has been performing well. LHCb has already started performing some of the world's most precise measurements involving B and D meson decays.

2. RICH system

With first data, the resonance $\phi \to K^+K^-$ would not have been observed without particle identification (PID) information from the RICH detectors.

In order to cover the full momentum range $1-100$ GeV/c the RICH system consists of two RICH detectors. RICH 1 is situated before the magnet, covers a horizontal angular acceptance of $\pm 25 - \pm 300$ mrad and a momentum range of $1-60$ GeV/c. It has two radiators, aerogel with a refractive index of 1.03 at $\lambda = 400$ nm and C_4F_{10} gas with refractive index of 1.0014 at $\lambda = 400$ nm. RICH 2 is located after the magnet and covers an angular acceptance of approximately $\pm 15 - \pm 120$ mrad with a momentum coverage of $15-100$ GeV/c. Its radiator is CF_4 gas which has a refractive index of 1.0005 at $\lambda = 400$ nm. Both RICH detectors have flat and spherical mirrors to reflect and focus the Cherenkov light from the radiators onto arrays of photon detectors (HPDs) which are placed outside the LHCb acceptance.

3. Hybrid Photon Detectors

There are 196 Hybrid Photon Detectors[3] (HPDs) in RICH 1 and 288 in RICH 2. Each has a quartz window coated with an S20 multi-alkali photocathode, which converts the incoming Cherenkov photons to photoelectrons. These are accelerated through the vacuum tube by up to 18kV and focused onto the 32×32 silicon pixel chip at the base.[4] They are sensitive to a wavelength range of $200-600$ nm. Data is read out at 1 MHz from a total of 500,000 pixels. The real-time behaviour of the HPDs is monitored from the LHCb control room during collisions.

The HPD vacuum degrades over time, and this is visible through a process called ion feedback. Residual gas molecules in the tubes get ionised by the photoelectrons and are accelerated onto the photocathode where they generate secondary electrons. Ion feedback is regularly monitored using light from a continuous wave laser such that the evolution of individual HPDs can be accurately predicted. Those HPDs approaching the threshold to rapid degradation are exchanged and repaired before being deployed in the RICH detectors again. A fraction of $< 5\%$ of such HPDs are replaced per year. Consequently, the PID performance is not affected.

4. RICH Alignment

After the initial alignment of the mirrors and HPDs within the RICH detectors following installation, the alignment is now regularly improved and updated using collision data. The flat and spherical mirrors are aligned using an iterative method which minimises the distortion of the Cherenkov angle distribution in spherical ring images on the detector plane. Alignment parameters are automatically updated for every fill of the LHC. Cherenkov angle resolutions of 1.59 mrad (C_4F_{10}) and 0.66 mrad (CF_4) have been achieved for the two gas radiators. This meets the expectation from the LHCb detector Monte-Carlo of 1.53 mrad (C_4F_{10}) and 0.66 mrad (CF_4). In the HPDs, photoelectron trajectories are affected by the residual stray magnetic field of the LHCb dipole magnet. Magnetic Distortion Correction Systems have been implemented in both RICH detectors. They project a known pattern of light spots onto the HPDs. From the recorded responses with the two polarities of the bending magnet and without magnetic field, magnetic distortion maps are extracted and automatically fed into the detector database.

5. Identification Efficiency

For each track detected in the tracking system the corresponding expected number and Cherenkov radii of the resulting photons are calculated for each particle hypothesis and for each of the three radiators. In a global maximum likelihood fit the hypotheses are tested against the recorded photon hits and are compared with each other by differences in their logarithmic likelihood (Delta Log Likelihood, DLL). Pure samples of kaons, pions and protons are needed to calculate the identification and mis-identification efficiencies. The decays $K_s^0 \to \pi^+\pi^-$ and $\Lambda \to p\pi^-$ as well as the charm decay $D^{*+} \to D^0(\to K^-\pi^+)\pi^+$ can be selected with a high purity using kinematic cuts only. These channels provide the information for identification efficiency plots shown in Fig. 1. The two curves represent the two different DLL cuts, $DLL > 0$ and $DLL > 5$. The first is used to measure the cut efficiency, the second to select a high purity data sample. A drop in efficiency for kaons at high momentum will improve during the reprocessing of the current data when the updated alignment will have been fully implemented.

6. Efficiency with Multiple Primary Vertices

LHCb was designed to run with a mean number of visible interactions per bunch crossing (μ) of 0.4. It is now running with a typical μ of 1.5 and a maximum of 4. Multiple primary vertices lead to a reduction in PID efficiency. Figure 2 shows that the probability for a kaon to be misidentified as a pion increases when the requirement for the kaon identification efficiency is held constant for varying numbers of primary vertices. The PID performance is found to be quite robust against multiple interactions for up to four Primary Vertices.

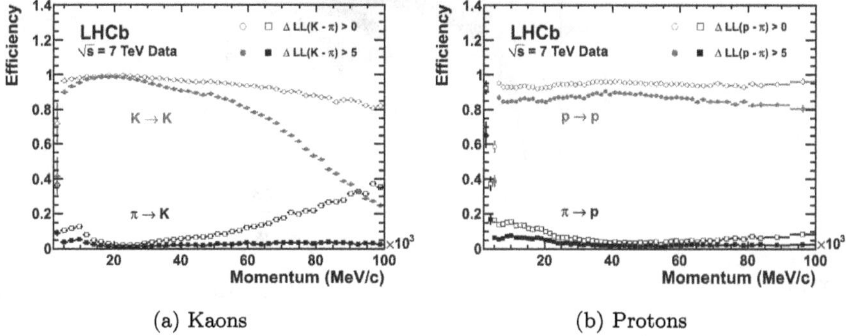

(a) Kaons (b) Protons

Fig. 1: Identification Efficiencies

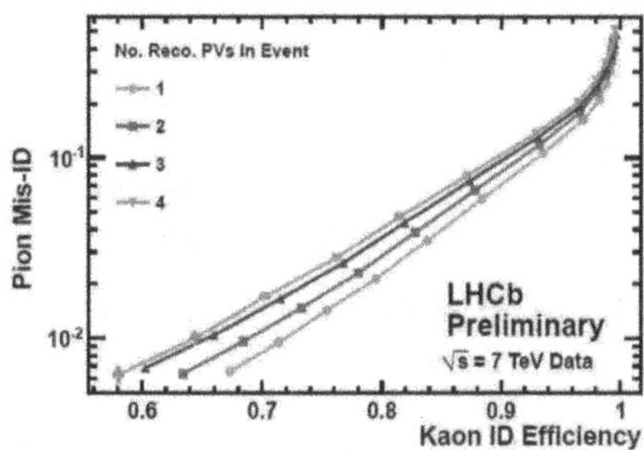

Fig. 2: Identification efficiency for multiple primary vertices. As the number of PVs increases, the ID efficiency drops

7. Performance of the RICH detectors

Plotting the Cherenkov angle against the momentum as in Fig. 3, illustrates the particle separation power of the combined RICH detectors. This is essential for many LHCb physics analyses such as "Charmless charged two-body B decays at LHCb with 2011 data".[5] The analysis takes tracks with a large distance from the Primary Vertex, uses the RICH detectors to select pions and kaons, creates pairs with a good vertex and computes the invariant mass. First the decay channel $B^0 \to K^+\pi^-$ is selected and there is a clear signal peak. Then just by changing the RICH PID requirements, selection of $B^0 \to \pi^+\pi^-$ and $B_s \to K^+K^-$ is possible (see Fig. 4).

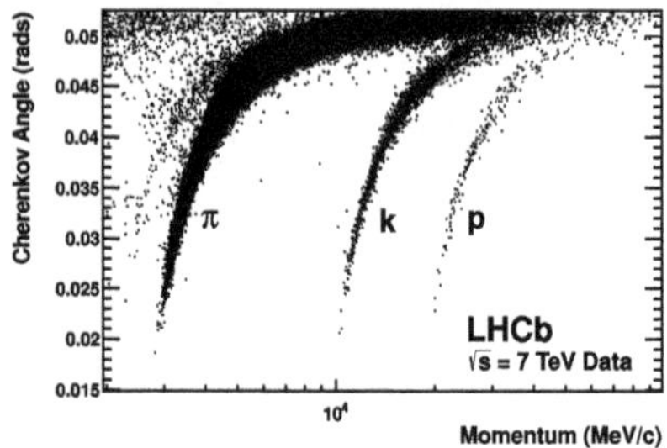

Fig. 3: Cherenkov angle as a function of Momentum

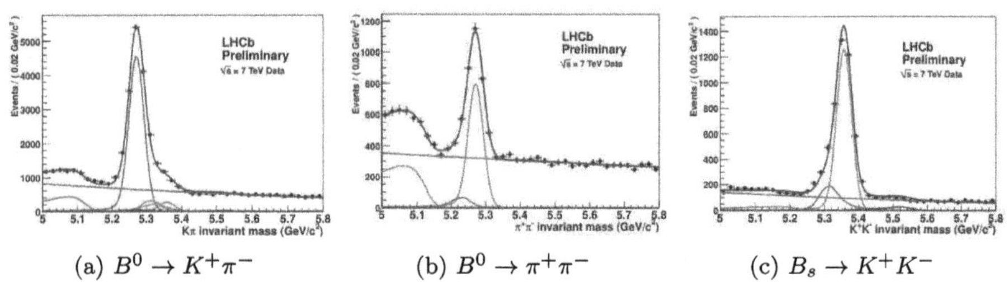

(a) $B^0 \to K^+\pi^-$ (b) $B^0 \to \pi^+\pi^-$ (c) $B_s \to K^+K^-$

Fig. 4: Invariant mass plots of different B decays

8. Summary and Conclusion

The LHCb RICH detectors have been operating well and consistently. The PID performance is excellent and consistent with expectations despite the increased instantaneous luminosity. It is invaluable to many of the key physics analyses. The RICH detectors will be upgraded for the LHCb upgrade in 2018 by replacement of the HPDs and going to a readout of 40 MHz.

References

1. CERN/LHCC 2003-030, *LHCb TDR 9* (9 Sept 2003).
2. The LHCb Collaboration, *JINST 3*, S08005 (2008).
3. M.Moritz et al., IEEE Transactions on Nuclear Science, Vol 51, No.3 (June 2004)
4. K.Wyllie et al., Proc. of the Fifth Workshop on LHC Electronics, Snowmass, CO, USA (20- 24 September 1999) CERN/LHCC/99-33.
5. The LHCb Collaboration, *LHCb-CONF-2011-042*

LOW TEMPERATURE THERMAL CONDUCTIVITY OF POLYMERS AND COMPOSITES: RECOMMENDED VALUES

V. MARTELLI

Lens Institute, University of Florence, Via Sansone 1, 50019 Sesto Fiorentino, Florence, Italy, martelliv@fi.infn.it

G. VENTURA

INFN, Section of Florence, Via Nello Carrara 1, 50019, Sesto Fiorentino, Florence, Italy, ventura@fi.infn.it

D. MICELA

Department of Physics, University of Florence,Via Sansone 1, 50019, Sesto Fiorentino, Florence, Italy, duccio.micela@unifi.it

We present low temperature thermal conductivity data of 14 polymeric materials and 12 composites, along with their integrated conductivity data usually up to room temperature. The latter information is precious in the initial phase of a project for a rapid evaluation of thermal power flowing through a mechanical part (of known geometrical factor) when its ends are at different temperatures.

1. Introduction

Constructing low temperature instruments requires knowledge of the properties of materials involved in their construction, in particular the thermal conductivity. Unfortunately, this is poorly known for many materials. Collections of data in text books are incomplete and in some cases misleading. For most materials, information is scattered through the literature. Searching out these data is time consuming, and often results in conflicting information. A systematic critical analysis of thermal conductivity measurements from the literature has been often called upon [1]. NIST (National Institute of Standard and Technology) has been contributing to such challenge supplying recommended data of a few materials usually in the 4-300 K temperature range, see e.g. [2].

In this paper we present low temperature thermal conductivity data of 14 polymeric materials and 12 composites, along with their integrated conductivity data up to room temperature, when possible. The latter information is precious in the starting phase of a project for a rapid evaluation of thermal power flowing through a mechanical part (of known geometrical factor) when its ends are at different temperatures.

2. Recommended data

Table 1 reports low temperature thermal conductivity data (W/cmK) of 14 polymeric materials at 9 different temperatures for most of them. For each material it is also reported the integrated thermal conductivity (W/cm), evaluated from 0 K to the reported temperatures.

Table 2 reports data of thermal conductivity and integrated conductivities of 12 composites.

For what concerns thermal conductivity (k), our recommended data are taken from published experimental data (references are cited in tables); when more than one set of values was available, our preference went to the one which best overlapped the other measurements in different ranges of temperature. When data did not cover the investigated range at high temperature, we used room temperature data provided by manufacturers, if a reasonable interpolation was possible. If it was not, we just reported the data available from literature.

We estimated that the maximum error introduced by our extraction of values of k from published data is 5%. Since the uncertainty of original published data is usually of the order of 5%, a conservative error estimate is 10%.

For what concerns integrated thermal conductivity, it was obtained from the same data used to get k. Note that for the evaluation of the integrated conductivity, data of k were extrapolated down to $T=0\ K$. Such procedure was performed using the existing experimental data down to the minimum of the data temperature range and then extrapolating data down to $T=0\ K$ by means of a reasonable fit function. The latter calculation is quite simple for polymers, since a $k=T^n$ formula describes well k at temperatures below 1 K, see e.g. [3],[4]. Moreover, the contribution of the integrated k in the sub-kelvin range is so small that it does not appreciably modify the accuracy of integrated k data at and above 4 K.

In summary, the idea of fixing the temperature $T=0\ K$ as starting point for integrated thermal conductivity data allows for a fast comparison among the thermal conductance of samples made by any of the listed materials, without introducing appreciable errors in a preliminary project calculation.

It is worth noting that data of k are given at fixed temperatures. This means that the reconstruction of $k(T)$ from data reported in the tables is in general not possible or very inaccurate, since polymeric materials show peculiar behaviours such as plateaux around 10 K. Supplying detailed $k(T)$ plots was not the goal of this paper. Should be more accurate values of k needed, the reader is referred to the reported literature.

TABLE 1

THERMAL CONDUCTIVITY (W/CMK) OF 14 POLYMERIC MATERIALS AT 9 DIFFERENT TEMPERATURES (UPPER ROW OF EACH MATERIAL) AND INTEGRATED THERMAL CONDUCTIVITY BETWEEN 0 K AND THE CORRESPONDENT TEMPERATURE (W/CM) (LOWER ROW OF EACH MATERIAL)

Material	Ref.	4 K	10 K	20 K	50 K	100 K	150 K	200 K	250 K	300 K
Nylon 6,6	[2],[5]	$1.2\ 10^{-4}$	$3.9\ 10^{-4}$	$9.7\ 10^{-4}$	$3.2\ 10^{-3}$	$7.8\ 10^{-3}$	$1.3\ 10^{-2}$	$1.9\ 10^{-2}$	$2.6\ 10^{-2}$	$3.3\ 10^{-2}$
		$2.8\ 10^{-4}$	$1.7\ 10^{-3}$	$8.3\ 10^{-3}$	$6.9\ 10^{-2}$	0.34	0.87	1.7	2.8	4.3
Kevlar 49	[6],[7]	$5.3\ 10^{-4}$	$2\ 10^{-3}$	$4.3\ 10^{-3}$	$1.1\ 10^{-2}$	$1.9\ 10^{-2}$	$2.6\ 10^{-2}$	$3.1\ 10^{-2}$	$3.5\ 10^{-2}$	$3.9\ 10^{-2}$
		$9.9\ 10^{-4}$	$8.6\ 10^{-3}$	$3.9\ 10^{-2}$	0.27	1	2.1	3.6	5.3	7
PC (polycarbonate)	[8],[9]	$3.1\ 10^{-4}$	$4.8\ 10^{-4}$	$6.9\ 10^{-4}$	$1.2\ 10^{-3}$	$1.6\ 10^{-3}$	$1.8\ 10^{-3}$	1.910^{-3}	$2.0\ 10^{-3}$	$2.05\ 10^{-3}$
		$5.7\ 10^{-4}$	$3.1\ 10^{-3}$	$9\ 10^{-3}$	$3.9\ 10^{-2}$	0.11	0.2	0.29	0.38	0.49
PB (polybutadiene – 18RU)	[10]	$9.2\ 10^{-4}$	$9.2\ 10^{-4}$	$9.9\ 10^{-4}$	-	-	-	-	-	-
		$1.9\ 10^{-3}$	$7.9\ 10^{-3}$	$1.8\ 10^{-2}$	-	-	-	-	-	-
PB (polybutadiene – 86RU)	[10]	$2.9\ 10^{-4}$	$4.3\ 10^{-4}$	$5.9\ 10^{-4}$	-	-	-	-	-	-
		$8.1\ 10^{-4}$	$3\ 10^{-3}$	$8.7\ 10^{-3}$	-	-	-	-	-	-
PP (polypropylene)	[11],[12]	$1.2\ 10^{-4}$	$3.8\ 10^{-4}$	$8.1\ 10^{-4}$	$1.5\ 10^{-3}$	$1.9\ 10^{-3}$	$2.1\ 10^{-3}$	$2.2\ 10^{-3}$	$2.25\ 10^{-3}$	$2.3\ 10^{-3}$
		$2.1\ 10^{-4}$	$1.6\ 10^{-3}$	$7.1\ 10^{-3}$	$4.2\ 10^{-2}$	0.13	0.23	0.34	0.45	0.56
Polyethylene (low density)	[13],[14]	$1.4\ 10^{-4}$	$2.5\ 10^{-4}$	$4\ 10^{-4}$	$7.8\ 10^{-4}$	$1.3\ 10^{-3}$	$1.8\ 10^{-3}$	$2.1\ 10^{-3}$	$2.6\ 10^{-3}$	$2.7\ 10^{-3}$
		$3.4\ 10^{-4}$	$1.7\ 10^{-3}$	$5.4\ 10^{-3}$	$2.3\ 10^{-2}$	$7.7\ 10^{-2}$	0.16	0.26	0.37	0.5
Polyethylene (high density)	[5],[15]	$2.5\ 10^{-4}$	$1.4\ 10^{-3}$	$5\ 10^{-3}$	-	-	-	-	-	-
		$3.3\ 10^{-4}$	$5\ 10^{-3}$	$3.6\ 10^{-2}$	-	-	-	-	-	-
PET 40% crystallinity (polyethylene terephthalate)	[12],[16]	$8.1\ 10^{-5}$	$2.9\ 10^{-4}$	$7.1\ 10^{-4}$	$1.4\ 10^{-3}$	$2\ 10^{-3}$	$2.2\ 10^{-3}$	$2.3\ 10^{-3}$	$2.4\ 10^{-3}$	$2.4\ 10^{-3}$
		$1.7\ 10^{-4}$	$1.3\ 10^{-3}$	$6.5\ 10^{-3}$	$3.9\ 10^{-2}$	0.13	0.23	0.34	0.46	0.58
PET film (Mylar)	[14],[17], [18]	$1.1\ 10^{-4}$	$2\ 10^{-4}$	$3.1\ 10^{-4}$	$4.9\ 10^{-4}$	$7.7\ 10^{-4}$	$9.9\ 10^{-4}$	$1.2\ 10^{-3}$	$1.3\ 10^{-3}$	0.15
		$2.6\ 10^{-4}$	$1.4\ 10^{-3}$	$4.4\ 10^{-3}$	$1.7\ 10^{-2}$	$5\ 10^{-2}$	$9.4\ 10^{-2}$	0.15	0.21	0.28
POM (isotropic sample)	[12]	$2.8\ 10^{-4}$	$9.3\ 10^{-4}$	$2.3\ 10^{-3}$	$4.7\ 10^{-3}$	$4.9\ 10^{-3}$	$4.3\ 10^{-3}$	$3.7\ 10^{-3}$	$3.3\ 10^{-3}$	$3\ 10^{-3}$
		$5.4\ 10^{-4}$	$4.1\ 10^{-3}$	$2.1\ 10^{-2}$	0.13	0.37	0.61	0.8	0.98	1.1
POM 68% crystallinity (parallel to extrusion direction)	[12]	$4.5\ 10^{-4}$	$9.4\ 10^{-4}$	$2.7\ 10^{-3}$	$8.7\ 10^{-3}$	$1.1\ 10^{-2}$	0.01	0.0093	0.0086	0.008
		$5\ 10^{-4}$	$4\ 10^{-3}$	$2.2\ 10^{-2}$	0.2	0.7	1.0	1.2	1.7	2.1
PTFE (Teflon)	[2],[19]	$4.2\ 10^{-4}$	$9.5\ 10^{-4}$	$1.4\ 10^{-3}$	$2.1\ 10^{-3}$	$2.4\ 10^{-3}$	$2.6\ 10^{-3}$	$2.7\ 10^{-3}$	$2.7\ 10^{-3}$	$2.7\ 10^{-3}$
		$4.8\ 10^{-4}$	$4.9\ 10^{-3}$	$1.7\ 10^{-2}$	$7\ 10^{-2}$	0.19	0.3	0.44	0.58	0.7
PVC	[20],[21]	$8\ 10^{-5}$	$1\ 10^{-4}$	$2.7\ 10^{-4}$	$4.2\ 10^{-4}$	$7.3\ 10^{-4}$	$1\ 10^{-3}$	$1.3\ 10^{-3}$	$1.4\ 10^{-3}$	$1.5\ 10^{-3}$
		$2.5\ 10^{-4}$	$7.9\ 10^{-4}$	$2.5\ 10^{-3}$	$1.3\ 10^{-2}$	$4.4\ 10^{-2}$	$8.7\ 10^{-2}$	0.14	0.21	0.3
RTV silicone	[22]	$1.1\ 10^{-4}$	$3.4\ 10^{-4}$	$7.4\ 10^{-4}$	$1.3\ 10^{-3}$	$2.2.\ 10^{-3}$	$2.7\ 10^{-3}$	$3\ 10^{-3}$	$2.7\ 10^{-3}$	$2.7\ 10^{-3}$
		$2.1\ 10^{-4}$	$1.5\ 10^{-3}$	$7\ 10^{-3}$	$3.9\ 10^{-2}$	0.13	0.25	0.39	0.54	0.67
Torlon 4301	[23],[24]	$1.3\ 10^{-4}$	$1.6\ 10^{-4}$	$6.1\ 10^{-4}$	$9\ 10^{-4}$	$1.2\ 10^{-3}$	$1.5\ 10^{-3}$	$1.8\ 10^{-3}$	$2.1\ 10^{-3}$	$2.6\ 10^{-3}$
		$2.8\ 10^{-4}$	$1.1\ 10^{-3}$	$5.1\ 10^{-3}$	$2.9\ 10^{-2}$	$8.1\ 10^{-2}$	0.15	0.23	0.32	0.44

TABLE 2. THERMAL CONDUCTIVITY (W/CMK) OF 12 POLYMERIC COMPOSITES AT 9 DIFFERENT TEMPERATURES (UPPER ROW OF EACH MATERIAL) AND INTEGRATED THERMAL CONDUCTIVITY BETWEEN 0 K AND THE CORRESPONDENT TEMPERATURE (W/CM) (LOWER ROW OF EACH MATERIAL)

Material	Ref.	4 K	10 K	20 K	50 K	100 K	150 K	200 K	250 K	300 K
EPOXY + Ag (filled 25-50%)	[25], [26]	$1.1\ 10^{-3}$	$3.1\ 10^{-3}$	$6\ 10^{-3}$	-	-	-	-	-	-
		$2.2\ 10^{-3}$	$1.5\ 10^{-2}$	$5.8\ 10^{-2}$	-	-	-	-	-	-
EPOXY + Glass (G10 9.1% filled)	[2], [27]	$6.3\ 10^{-4}$	$1.1\ 10^{-3}$	$1.6\ 10^{-3}$	$2.4\ 10^{-3}$	$3.1\ 10^{-3}$	$3.7\ 10^{-3}$	$4.5\ 10^{-3}$	$5.2\ 10^{-3}$	$6.1\ 10^{-3}$
		$1.3\ 10^{-3}$	$6.9\ 10^{-3}$	$2\ 10^{-2}$	$8\ 10^{-2}$	0.22	0.39	0.60	0.83	1.1
EPOXY + Cu (filled 13%)	[28]	$7.5\ 10^{-4}$	$1.1\ 10^{-3}$	$1.5\ 10^{-3}$	-	-	-	-	-	-
		$1.6\ 10^{-3}$	$7.3\ 10^{-3}$	$2\ 10^{-2}$	-	-	-	-	-	-
EPOXY + Cu (filled 53.5%)	[28]	$1.1\ 10^{-3}$	$3\ 10^{-3}$	$5.7\ 10^{-3}$	-	-	-	-	-	-
		$1.9\ 10^{-3}$	$1.5\ 10^{-2}$	$5.7\ 10^{-2}$	-	-	-	-	-	-
EPOXY + EPOXY	[29]	$7.4\ 10^{-4}$	$8.2\ 10^{-4}$	$9.6\ 10^{-4}$	$1.3\ 10^{-3}$	-	-	-	-	-
		$1.5\ 10^{-3}$	$6.4\ 10^{-3}$	$1.5\ 10^{-2}$	$5\ 10^{-2}$	-	-	-	-	-
EPOXY + HTSC (filled 20%)	[30]	$9\ 10^{-4}$	$1.5\ 10^{-3}$	$2.1\ 10^{-3}$	$3.3\ 10^{-3}$	$3.9\ 10^{-3}$	-	-	-	-
		$1.7\ 10^{-3}$	$9.3\ 10^{-3}$	$2.3\ 10^{-2}$	0.11	0.29	-	-	-	-
EPOXY + Pb (filled 52%)	[31]	$3\ 10^{-4}$	$5.5\ 10^{-4}$	$7.9\ 10^{-4}$	$1.3\ 10^{-3}$	$1.8\ 10^{-3}$	$1.9\ 10^{-3}$	$2.1\ 10^{-3}$	$2.2\ 10^{-3}$	$2.2\ 10^{-3}$
		$4.6\ 10^{-4}$	$3.3\ 10^{-3}$	$9.9\ 10^{-3}$	$4.3\ 10^{-2}$	0.12	0.22	0.32	0.42	0.53
EPOXY + HM (high modulus) CFRP parallel to fiber direction	[32]	$5.8\ 10^{-5}$	$1.6\ 10^{-4}$	$6.4\ 10^{-4}$	$2.7\ 10^{-3}$	-	-	-	-	-
		$8.2\ 10^{-5}$	$8\ 10^{-4}$	$5.6\ 10^{-3}$	$5.4\ 10^{-2}$	-	-	-	-	-
EPOXY + HM (high modulus) CFRP perpendicular to fiber direction	[32]	$2.3\ 10^{-4}$	$6.4\ 10^{-4}$	$2.7\ 10^{-3}$	$2.4\ 10^{-2}$	-	-	-	-	-
		$4.6\ 10^{-4}$	$2.9\ 10^{-3}$	$2.0\ 10^{-2}$	0.42	-	-	-	-	-
EPOXY + HT (high modulus) CFRP parallel to fiber direction CFRP	[32]	$1.3\ 10^{-4}$	$5\ 10^{-4}$	$1.9\ 10^{-3}$	-	-	-	-	-	-
		$2.3\ 10^{-4}$	$2.1\ 10^{-3}$	$1.5\ 10^{-2}$	-	-	-	-	-	-
EPOXY + HT (high modulus) CFRP perpendicular to fiber direction CFRP	[32]	$1.8\ 10^{-4}$	$4.4\ 10^{-4}$	$1.2\ 10^{-3}$	$9.5\ 10^{-3}$	-	-	-	-	-
		$4.3\ 10^{-4}$	$2.3\ 10^{-3}$	$1.1\ 10^{-2}$	0.17	-	-	-	-	-

REFERENCES

1. Woodcraft A L and Gray A, AIP Conf. Proc. Vol. 1185 Thirteenth International Workshop on Low Temperature Detectors (LTD13) **1185**, 681-4 (2010).
2. Marquardt E D et al., Proc. 11th Int. Cryocooler Conference, Keystone, Co. June 20-22, (2000).
3. Barucci M et al., *Cryogenics,* **48**, 166-168 (2008).
4. Ventura G, Risegari L: *The art of cryogenics: Low-Temperature Experimental Techniques:* Elsevier Science (2003).
5. Scott T A et al., *J. Appl. Phys.* **4**), *(1973).*
6. Ventura G, Martelli V, *Cryogenics* **49**, 376-377 (2009).
7. Ventura G, Martelli V, *Cryogenics* **49**, 735-737 (2009).
8. Jaeckel M et al., *Cryogenics, 38,* **1**, 105-108, (1998).

9. Zaitlin M P and Anderson A C, *Physical Review B*, **12**, 4475-4486 (1975).
10. Matsumoto D S and Anderson A C, *Journal of non-crystalline solids*, **44**, 171-180 (1981).
11. Barucci M et al., *Cryogenics* **42**, 551-555 (2002).
12. Choy C L and Greig D, *J. Phys. C: Solid State Physics* **10**, 169-179 (1977).
13. Reese W, Tucker J E, *Journal of Chemical Physics*, **43**, 105-114 (1965).
14. Steere R C, Thermal properties of thin-film polymers by transient heating, *Journal of Applied Physics* **37**, *(1966)*.
15. Kolouch R J and Brown R G, *Journal of Applied Physics* **39**, 3999-4003 (1968).
16. Greig D and Sahota M S, *J. Phys. C: solid state physics* **16**, L1051-L1054 (1983).
17. Hays K M et al., Low thermal conductivity superconducting metallized Mylar leads, *Rev. Sci. Instruments* **55**, 1660-1662 (1984).
18. Peterson R E and Anderson A C, *The review of scientific instrumen*, **11**, 834-835 (1972).
19. Scott T, Giles M, *Physical Review Letters* **29**, 642-643 (1972).
20. Risegari et al., *Journal of Low Temperature Physics* **144**, 49-59 (2006).
21. Ventura G, Martelli V, Proc. 11th Conf. ITCCP, Villa Olmo, Como, Italy, 5 - 9 October 145-149 (2009).
22. A. Baudot, J. Mazuer and J. Odin, *Cryogenics* **38**, 227-230 (2009).
23. Barucci M et al., *Cryogenics* **45** 295–299 (2005).
24. Ventura G et al., *Cryogenics* **39** 481–484 (1999).
25. Araujo F F T, and Rosenberg H M, *J. Phys. D* **9**, 665-675 (1976).
26. Reynolds C L Jr. and Anderson A C, *Rev Sci lnstrum* **48**,1715 (1977).
27. Runyan M C, Jones W C,, *Cryogenics* **48** 448-454 (2008).
28. Schmidt C, *Cryogenics* **15**, 17-20 (1975).
29. Scheibner W and Jäckel M, *Phys. Stat. Sol. (a)* **87**, 543 (1985).
30. Jaeckel M, *Cryogenics* ***35(11)*** 713-716 (1995).
31. Ustjushanin E E, *Cryogenics* **31**, 241-243 (1991).
32. Radcliffe D J and Rosenberg H M, *Cryogenics* **22**, 245-249 (1982).

Overview of the R&D program on Liquid Argon TPCs under development at the University of Bern

I. Badhrees, A. Ereditato, S. Janos, I. Kreslo, M. Messina, S.Haug, C. von Rohr, B.Rossi, T. Strauss, M. Weber, M. Zeller*

Laboratory for High Energy Physics, University of Bern,
Bern, 3012 Bern, Switzerland
**marcel.zeller@lhep.unibe.ch*
www.lhep.unibe.ch

The Liquid Argon Time Projection Chamber (TPC) technique is a promising technology for future neutrino detectors. At LHEP of the University of Bern (Switzerland), R&D projects towards large detectors are on-going. The main goal is to prove long drift paths of the order of 10 m. Therefore, a liquid Argon TPC with 5m of drift distance is being constructed. Many other aspects of the liquid Argon TPC technology are also under investigation, such as a new device to generate high voltage in liquid Argon (Greinacher circuit), a recirculation filtering system and the multi photon ionization of liquid Argon with a UV laser. Two detectors are being built: a medium size prototype for specific detector technology studies, and ARGONTUBE, a 5 m long device.

Keywords: Time Projection Chamber (TPC), liquid Argon, Large mass detectors, Neutrinos.

1. Introduction

In the last decades the neutrino oscillation phenomenon has been established in many sectors such as solar [1], atmospheric [2], and reactor neutrinos [3]. Almost all parameters of the mixing matrix of the flavour and mass eigenstates have been determined apart from θ_{13}. Recently the T2K experiment [4] has shown an indication of a non-vanishing value of θ_{13} with a statistical significance of 2.5 sigma. Future measurements envisaged in the field of neutrino oscillations must provide a precise determination of θ_{13} investigating possible CP violating effect in the lepton sector and adress the problem of the hierarchy of the mass eigenvalues. All measurements mentioned so far are expected to show only tiny effects. For this reason large sample of events are needed. Such large samples can only be obtained with a synergic use of high flux neutrino beams and large mass detectors. Our research program is focused on the development of detector technology that allows one to conceive detectors of very large masses and use of Liquid Argon Time Projection Chambers (LArTPC). One possibility to build a large-mass LArTPC has been discussed in [5] where the conceptual idea of the GLACIER detector is presented. In particular, the GLACIER concept is based on a single cylindrical volume of 70 m diameter and 20 m in height. In this volume a LArTPC would be installed, with the drift coordinate parallel to the height of the cylinder. The main challenge of such a detector is the very long drift distance, whose feasibility is not established. So far, the longest drift path has been shown by the ICARUS collaboration with 1.5 m length. Very long drift distances require high purity LAr (impurities of the level 0.1ppb) to prevent large electron yield depletion because of attachment of electrons with electro-negative impurities (i.e. O2, H2O). Furthermore, a very high voltage is needed to make a sizable drift field, 1kV/cm over such a distance.

2. LAr TPC technology

Before proceeding to describe the ARGONTUBE project we will introduce the LArTPC detector concept and shortly summarize its features and peculiarities. A LArTPC consists of a LAr volume, where the Argon acts simultaneously as target and active medium, well suited for neutrino physics given its high density ($1.4g/cm^3$). In the instrumented volume an electric field is applied to prevent the electron-ion pairs from recombination and to drift the electrons towards the read out planes. The latter consist of two wire planes (or more like in the ICARUS detector [6]). A uniform field is obtained thanks to a cathode and field shaper rings installed at fixed distance and biased at a given voltage. The wire planes acting as anode. The voltage is provided by a voltage divider that can be realized with several technique and later we will show one possible implementation. The geometrical reconstruction of particle tracks is realized by recording the electric signal generated by ionization electrons on the wires, whose position corresponds to the transverse coordinates, and the third coordinate is given by means of the measurement of the drift time (reference time $t0$ minus arrival time to the wires t) by assuming that the drift velocity is known at a given electric field. A photomultiplier is used to provide the $t0$ and the trigger signal for the data aquisition system. A TPC offers the possibility to have a homogenous sensitive volume, with a three dimensional reconstruction capability and a charge read out that allows one to measure the energy released by particles [7].

3. ARGONTUBE

The ARGONTUBE is a LAr TPC with 5 m drift to prove the feasibility of large mass detectors with long drift. In Fig. 1 the drawings of the ARGONTUBE detector are shown. In particular, in Fig.1 (a) the dewar is visible together with the inner tube where the TPC will be hosted. In Fig. 1 (b) a part of the field cage (a few rings out of 125 in total) are visible, the size and shape of the rings have been optimized by finite element method calculation in order to get a drift field as uniform as possible. In Fig. 1 (c) details of the field cage already assembled are shown.

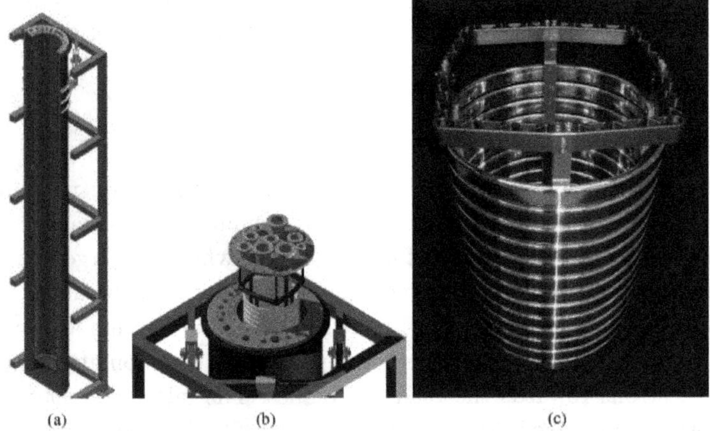

(a) (b) (c)

Fig. 1. (a) Drawing of the ARGONTUBE cryostat with the inner vessel. (b) Drawing of the top flange with all the feed troughs, a part of the field shaping rings and some details of the fastening system to the top flange. (c) Picture of the detector with field shaping rings

The desired purity of the LAr is reached by making a vacuum at the level of 10^{-6} mbar in the inner tube before the LAr filling. The LAr is filled through cryogenic filter, to further increase the LAr purity and to fight against possible out-gassing of detector components in the warm phase. The inner volume is equipped with cryogenic pumps capable of recirculating the LAr through a filter analogous to the one mentioned above, about 100 l/h of pumping speed can be reached.

In such a long drift TPC a critical point is the high voltage needed to

Fig. 2. ARGONTUBE detector with reduced drift length of 76 cm in order to test the design and the implementation of the GCW circuit

generate the drift field. Given the difficulty to feed the required voltage (500kV) into the LAr without provoking discharges, we have chosen to install in the LAr a Greinacher/Cockroft-Walton (GCW) circuit consisting of a chain of rectifying cells. The installation on the detector of a GCW is shown in the Fig. 2, an AC input voltage is required ($2V_0$ peak-to-peak voltage) and each cell of the circuit rectifies the AC signal providing an overall voltage of $n \times 2V_0$, where n is the number of cells (in our case n =125). We have chosen a configuration in which each stage of the GCW will bias one field shaper . With a distance between each ring of 4 cm and an input voltage of 2 kV (4 kV peak-to-peak) we obtain the required field of 1 kV/cm.

In order to test the design of the ARGONTUBE detector one module of 76 cm drift distance has been installed in a smaller cryostat. The field cage, the read-out wire planes, the recirculation circuit and all the cryogenic parts of the ARGONTUBE have been produced and tested. The detector showed very good performance and 76 kV in liquid Argon has been reached. Cosmic ray event have been recorded as shown in Fig 3., the purity was measured with a high intensity UV-laser which gives uniforme ionisation tracks shown in [8], [9]. A charge lifetime of a about 2000 μs was optained which is corresponds to a purity of approximately $\sim$0.1 ppb.

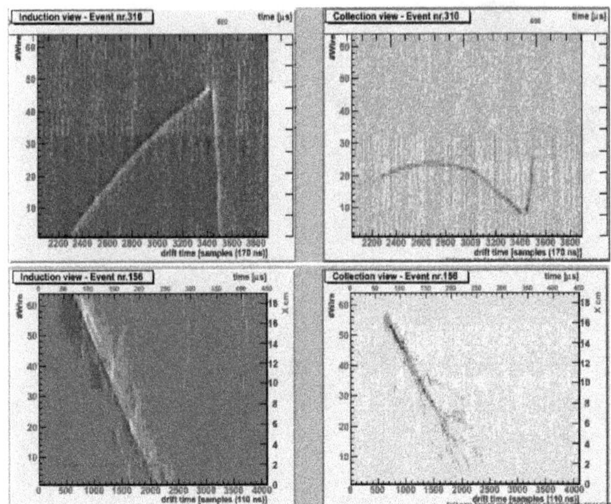

Fig. 3. Cosmic particle tracks of the 76 cm module test

4. Conclusion

In this paper we described the R&D program under development at LHEP, University of Bern, where the main goal is to establish a LArTPC with the longest drift distance ever built. One module with 76 cm drift distance of the ARGONTUBE detector has been successfully tested and showed good performance. We expect results from ARGONTUBE soon.

References

1. B. Aharmim et al., by SNO collaboration, Phys. Rev. C 81, 055504 (2010).
2. R. Wendell et al., by Super-Kamiokande collaboration, Phys. Rev. D 81, 092004 (2010).
3. S. Abe et al., by KamLAND collaboration, Phys. Rev. Lett. 100, 221803 (2008).
4. K. Abe et al., by T2K collaboration, Phys. Rev. Lett 107, 041801 (2011).
5. A. Rubbia, J.Phys.Conf.Ser. 171 12020 (2009); A. Ereditato and A. Rubbia, Proc.Workshop on Physics with a Multi-MW proton source held at CERN, Switzerland (2004).
6. S. Amerio et al., by ICARUS collaboration, Nucl. Instrum. Meth. A 527 329 (2004).
7. M. Zeller et al., JINST 5 10009 (2010).
8. B. Rossi et al., JINST 4 07011 (2009).
9. I. Badhrees et al., New J. Phys. 12 113024 (2010).

HIGH RATE PROTON IRRADIATION OF 15mm MUON DRIFTTUBES

A. ZIBELL, O. BIEBEL, R.HERTENBERGER, A. RUSCHKE, CH. SCHMITT

Fakultät für Physik, Ludwig-Maximilian-Universität,
Am Coulombwall 1, 85748 Garching, Germany
**E-mail: andre.zibell@physik.uni-muenchen.de*
www.etp.physik.uni-muenchen.de

H. KROHA, B. BITTNER, P. SCHWEGLER, J. DUBBERT, S. OTT

Max-Planck-Institut für Physik,
Föhringer Ring 6, 80805 München, Germany

Future LHC luminosity upgrades will significantly increase the amount of background hits from photons, neutrons and protons in the detectors of the ATLAS muon spectrometer. At the proposed LHC peak luminosity of $5{\cdot}10^{34}\frac{1}{cm^2 s}$, background hit rates of more than $10\frac{kHz}{cm^2}$ are expected in the innermost forward region, leading to a loss of performance of the current tracking chambers.

Based on the ATLAS Monitored Drift Tube chambers, a new high rate capable drift tube detecor using tubes with a reduced diameter of 15mm was developed.

To test the response to highly ionizing particles, a prototype chamber of 46 15mm drift tubes was irradiated with a 20 MeV proton beam at the tandem accelerator at the Maier-Leibnitz Laboratory, Munich. Three tubes in a planar layer were irradiated while all other tubes were used for reconstruction of cosmic muon tracks through irradiated and nonirradiated parts of the chamber. To determine the rate capability of the 15mm drifttubes we investigated the effect of the proton hit rate on pulse height, efficiency and spatial resolution of the cosmic muon signals.

Keywords: LHC, ATLAS, MDT

1. Introduction

At the propsed LHC peak luminosity of $5{\cdot}10^{34}\frac{1}{cm^2 s}$, the innermost foreward region of the ATLAS muon spectrometer will be confronted with a background hit density of about 10 $\frac{kHz}{cm^2}$, as the rates scale with the luminosity,[1] leading to a loss of performance of the current tracking chambers caused by high occupancy and space charge effects.

An upgrade candidate are MDT[a]-chambers with a reduced tube diameter of 15 mm.[2] To study the high-rate capability, a prototype chamber with 46 tubes was irradiated by 20 MeV protons at the MLL Tandem-accelerator in Garching.

2. Experimental setup

Figure 1 shows the experimental setup. A fast coincidence of the scintillation counters serves as trigger on cosmic muons. The proton beam was defocussed to a beam-spot of 3·0.5 cm^2 and wobbled with 800 Hz over a horizontal distance of 7 cm. Only the tubes 20 - 22 were irradiated by protons, that are - consistent with SRIM[b] simulations - stopped by the second or third tube wall, depending on small angle scattering in the first two tubes.

The triggering scintillators are segmented, thus one can distinguish between muons crossing the irradiated and non-irradiated sections along the 1 m tubes, see fig. 1. Three different beam intensities were used during the measurements, 200 kHz, 1100 kHz and 1300 kHz, corresponding to an effective hit-density of 19, 105 and 124 $\frac{kHz}{cm^2}$. Reference runs with no beam were taken.

3. Signalheight

The MDT electronics provides information on drifttime (time between trigger and anode wire signal to reconstruct the distance of the track from the wire) and signalheight of the analog pulse from the wire. This signalheight is measured in ADC-counts, and the signal height spectra for the first two tubes in the irradiated detector layer are shown in fig. 2 for the different proton irradiation levels.

In the case of no proton irradiation, all four tubes show a similar spectrum with Landau like energy distribution. At 19 $\frac{kHz}{cm^2}$ kHz proton irradiation, the tubes 22 and 21, being the first ones hit by the beam, show an overlay of the muon signals with artificial proton coincidences. The energy spectrum of these proton hits is very broad at high values, even above the dynamic range of the ADC, where the spectrum is cut off. At the highest irradiation levels, the ADC spectra of tubes 22 and 21 are shifted to the left, where they are cut off by the ADC threshold for small signals. This is

[a]Monitored Drift Tube
[b]The Stopping and Range of Ions in Matter, http://www.srim.org

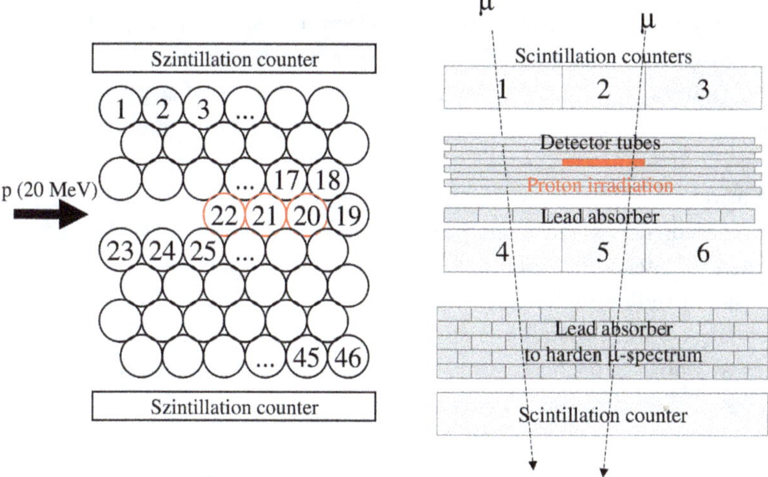

Fig. 1. Schematic side-view of the detector and view from beam direction. The lead absorber and the additional trigger scintillator are used to harden the muon spectrum.

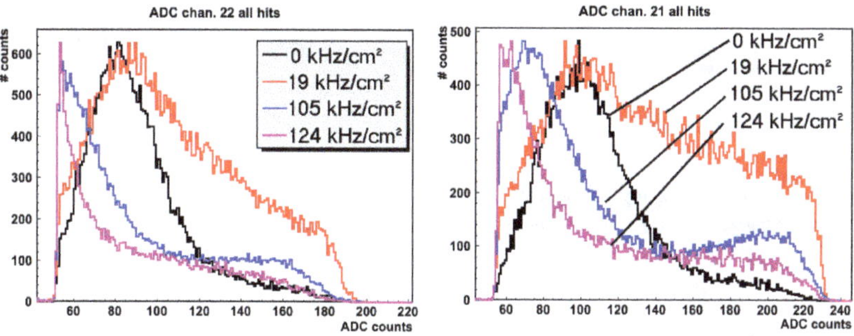

Fig. 2. Pulse height spectrum of the first (left) and second (right) irradiated tube at different irradiation levels.

due to the reduction of the gas amplification close to the anode wire, caused by the positive ions, drifting slowly to the tube walls. The electric field of these space charges shields the area around the wire and reduces the effective electric field for gas amplification. Protons reaching the gas volume of the third tube are stopped there, depositing several MeV of energy. As the impact on the ADC spectrum is different, this tube is excluded from further consideration. Tube 19 shows identical ADC spectra for all irradiation levels, as all protons are stopped before entering this tube.

Figure 3 (left) shows the probability of a detected signal in each tube for a muon trigger. In the case of no irradiation, the acceptance of the detector can be observed, with its periodical 8-layer structure with lower acceptance at the chamber edges. Due to random coincidences of a proton hit with a cosmic muon, at 200 kHz there is a 1:1 weighting of muon and proton hits in the irradiated layer, that reaches a ratio of up to 2.5:1 at 1300 kHz irradiation.

If one considers the most probable ADC value for hits, that could be matched to tracks of cosmic muons through the respective tube, this can be taken as an indicator for the amount of gas amplification. Figure 3 (right) shows these values for all tubes and irradiation levels, normalised to their values at non-irradiated sections of the tubes, except for 1300 kHz, where the lack of statistics did not allow this analysis.

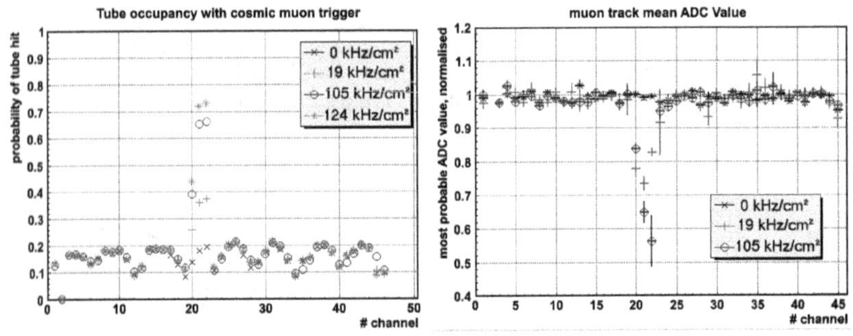

Fig. 3. Tube occupancy including muon and proton hits (left) and mean ADC values for muon tracks through the tubes (right) at different irradiation levels.

4. Efficiency and Tracking resolution

The spatial resolution of a drifttube can be obtained by comparing its radius prediction with the one derived from the track fit through the non-irradiated tube layers (all beside layer 4). Comparing these two different radii introduces the 3-sigma-efficiency, giving the fraction of tracks where this difference lies within 3 times the spatial resolution of the tube.

Radius dependent resolution is shown in fig. 4. The resolution is predicted to deteriorate only by a few ten μm due to space charge effects under irradiation.[2] Efficiency drops at 19 $\frac{kHz}{cm^2}$ due to space charge effects in the irradiated section and due to occupancy effects, dependent on the total hit rate per tube.

The exact analysis depends strongly on the definition of the 3-sigma-efficiency, this is still ongoing work.

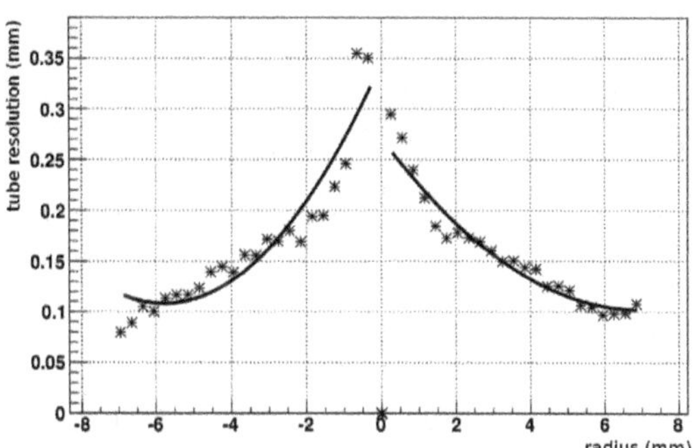

Fig. 4. Spatial resolution for non irradiated tubes.

5. Conclusion

With the Garching Tandem accelerator, the threefold of the predicted values for background irradiation with highly ionizing particles at high-luminosity-LHC could be simulated. Due to dead-time and space charge effects the tracking efficiency and gas amplification is reduced.

These effects are small compared to detectors of the currently used drifttubes with 30mm diameter, where the occupancy is about seven times higher, and the degradation of efficiency and resolution is much worse. The number of tube layers per 15mm MDT chamber will be doubled compared to a 30mm tube chamber. Thus the overall segment tracking efficiency of the system at 19 $\frac{kHz}{cm^2}$, about twice the expected hit-rate at high luminosity LHC, meets the requirements for the forward region of the ATLAS muon spectrometer.

References

1. S. Baranov et al., in *Estimation of Radiation Background, Impact on Detectors, Activation and Shielding Optimization in ATLAS* (ATL-GEN-2005-001).
2. B.Bittner et al., in *Development of fast high-resolution muon drift-tube detectors for high counting rates* Nucl. Instr. and Meth. A 628 (2011) 154, doi:10.1016/j.nima.2010.06.306.

Broader Impacts Activities, Treatments and Software Application

M. Pizzichemi	Image Fusion Software in the ClearPEM-Sonic Project
T. Sloan	Cosmic Rays and the Environment
R. Soluk	A Proposal for the ATLAS Cosmic Muon and Exotics Detector (ACME) at the LHC
N. Turini	The Trigger System of the TOTEM Experiment at LHC

THE LEDERMAN SCIENCE CENTER: PAST, PRESENT, FUTURE*

MARJORIE G. BARDEEN†
mbardeen@fnal.gov

Fermilab Education Office, Fermilab MS 226, Box 500
Batavia, Illinois, 60510 United States

For 30 years, Fermilab has offered K-12 education programs, building bridges between the Lab and the community. The Lederman Science Center is our home. We host field trips and tours, visit schools, offer classes and professional development workshops, host special events, support internships and have a strong web presence. We develop programs based on identified needs, offer programs with peer–leaders and improve programs from participant feedback. For some we create interest; for others we build understanding and develop relationships, engaging participants in scientific exploration. We explain how we created the Center, its programs, and what the future holds.

1. Past

The story of the Lederman Science Center began in 1986 after a group of middle school teachers asked to bring their students to Fermilab for a tour. They wanted the students to learn about the work of scientists and visit a scientific laboratory. The Education Office was given permission to try, but the students could not disturb the work environment.

Although students would not be able to visit the most interesting places—the accelerator tunnel and the collision halls—we knew the students should be prepared for their visit. Education Office staff worked with a few teachers and physicists to develop an instructional unit to be taught before the field trip. We called the unit *Beauty and Charm*. The unit was a success, but the traditional tour was designed for adults, not for 12-year-olds.

We wanted a field trip to offer students something they could only get at the Lab. We also wanted them to learn. As Ben Franklin said, "Tell me and I forget. Teach me and I remember. Involve me and I learn." Ways to involve students

* This work is supported by the U.S. Department of Energy, the U.S. National Science Foundation and Fermilab Friends for Science Education.
† Work supported by the U.S. Department of Energy.

were a Q&A with a scientist and hands-on experiments or exhibits on the tour. But the latter required dedicated space that we did not have.

What the Education Office needed was a building for students and teachers, but we were not sure how to proceed. Cynthia Yao, the wife of a member of the Fermilab Board of Overseers, gave us the answer when she shared her experience starting the Ann Arbor Hands-on Museum. With support from the director and sufficient funds, could we follow her example?

Getting Director Leon Lederman on board and in turn the Board of Overseers was easier than getting the money. We raised $25,000 to host an expert panel and community focus groups for needs assessments. The money came with the stipulation that the new building be named for Leon, hence the Leon M. Lederman Science Education Center. The expert panel members came from several important science museums such as the Exploratorium, the Museum of Science and Industry and the St. Louis Science Center. They thought our idea an excellent one and recommended that we use a national park visitors' center as our model. That would mean content focused on Fermilab's science and technology, and a facility that included a spacious lobby to greet visitors and host events, a small "theater," exhibit space, an information kiosk and store.

Our community focus groups told us about programming. They wanted programs for children, students, teachers—everyone ages 3 to 90 in an aesthetic and functional building! They wanted creative and dynamic interactive exhibits and an educational resource center with an R&D program for education kits and programs.

With this input, Leon was able to allocate Department of Energy (DOE) funds to design the building and make plans to build the Center using special DOE funds for general physics plant limited to $2M per project. Former Director Bob Wilson had a design in his hip pocket for an art museum that featured a large cantilevered roof over glass exterior walls and an open design that allowed visitors to view the art day or evening, even when the building was not open. Working from that concept, Fermlab architects came up with an aesthetic, dynamic design for the new science center. The initial cost was over budget, so the footprint became smaller, the basement and loading dock disappeared, the lobby shrank and the theater and meeting rooms were moved to a second building, to be constructed at a later date. In 2011, we are still in the original building, but we are able to schedule rooms in Wilson Hall for sessions with a scientist and for some teacher workshops.

Once we knew the size of the exhibit space, we formed a working group of physicists and physics teachers to brainstorm exhibit concepts for our field trips. The group developed 12 key ideas that we wove into a story called *Quarks to*

Quasars. Some exhibit ideas lent themselves to photos or posters; others were hands-on. The target audience remained ~120 middle school students on a school field trip guided by trained docents. We felt the experience could be extended down to 10-year-olds who knew the atomic model and up to high school students in physical science classes. High school physics students who tended to come in smaller groups would go "out and about" to see some of the technical areas of the Lab such as the magnet facility, CDF or DZero work areas, or the detectors themselves when available, etc.

We created an exhibit matrix based on concept areas such as accelerators and detectors and exhibit types, interactive or not. However, we met a physicist, a former student of Leon's who worked at the Rubin H. Fleet Science Center in San Diego who took one look at the matrix and said, "Ideas, methods, tools; that's the way to organize your exhibit matrix." And so we did.

We had a modest budget but found an exhibit designer who thought working with us an interesting way to expand his portfolio. He also had middle school boys who could be his own focus group as he developed the exhibit ideas. We later found we had something else in common: our interest to have cartoons as exhibit signage matched his lifelong dream to be a cartoonist. He knew how to make our signs look like professional cartoons. We installed our exhibition that serves thousands of students each year. When we discovered that families also stopped by, we added introductory videos and printed guides for parents and grandparents.

2. Present

The Lederman Science Center has over 35 exhibits in four areas: Ideas—What We Study; Methods—How We Work; Tools—What We Use to Do Our Work; Accelerators and Detectors. The exhibit story and annotated list of exhibits is online at: [http://ed.fnal.gov/lsc_exhibits/qtoq.html]

But the Lederman Science Center is more than exhibits for field trips. The Teacher Resource Center, science lab and technology room allow the Education Office to offer a wide range of programs associated with Fermilab physics and with Fermilab's restored prairie, a DOE National Environmental Research Park.

Today, *Beauty and Charm* for grades 6-8 includes the fourth edition of the original instructional unit, a teacher workshop to prepare teachers to teach the unit, the docent-led field trip that includes a visit to Wilson Hall, the Linac and Main Control Room, the *Quarks to Quasars* exhibits with study guides, an opportunity to meet a scientist or engineer and online materials such as *Fermilabyrinth*. [http://ed.fnal.gov/projects/labyrinth/games/]

We host children in grades K-5 in our science lab after they have studied one of four topics: electricity and magnetism, light, heat, or mechanics. Scientists, engineers and computing specialists also take activities on the road visiting classrooms with interactive presentations on these and other topics. A workshop helps K-5 teachers understand and feel comfortable with these physical science concepts through facilitated exploration of a series of simple, open-ended experiments, and through discussion and reflection with a master elementary school teacher and Fermilab scientists.

Fermilab offers classes from Saturday Morning Physics for high school students in Wilson Hall to prairie rangers and science adventures for younger students at the Lederman Science Center. Typical adventures might include *Lego Engineering, Magnetic Magic* or *Animal Architecture.* Also, scout troops can complete one of 12 badges in a two-hour session. Special events include an open house and outdoor fair for families, the *Wonders of Science* show and a career expo for high school students.

Table 1: Current Fermilab K-12 Education Programs [http://ed.fnal.gov]

Research Experiences	**Teacher Resource Center**	**Events for Kids & Families**
Academic Year High School	Resource Collections	DUSEL Education Program
Interns	Workshops	QuarkNet Masterclass
QuarkNet Summer Research	Chem West	STEM Career Fair
(Student/Teacher Teams)	**Classroom Resources**	Wonders of Science
TARGET	Classroom Presentations	Family Open House
TRAC	Data for Students	Family Outdoor Fair
Field Trips/High School Tours	I2U2 e-Labs & Science	**Professional Development**
Lederman Science Center	Investigations	**for Teachers**
Physics Science Experiences	Online Resources	QuarkNet Center
Beauty and Charm	What is scientific research?	I2U2 Workshops
Phriendly Physics	**Classes for Kids**	Physical Sciences
Prairie Science Experiences	Prairie Rangers	Workshops
Insects at Work in Our World	Saturday Morning Physics	Prairie Workshops
Prairie - Our Heartland	Science Adventures	QuarkNet Boot Camp
Particles and Prairies	Scout Programs	QuarkNet Outreach
Tours		Summer Secondary Science
		Institutes

When focusing on teachers, the Education Office extends its influence farther than if all participants were students. The Teacher Resource Center provides a preview collection of K-12 instructional materials. TRC services include professional development workshops, consultation assistance, bibliographies and reference assistance. Educators have access to curriculum materials, books, multimedia, educational supply catalogs, periodicals and newsletters. The collection also includes reports on science and mathematics education, standards, assessment, equity and other topics. Visitors also have access to selective, password-protected websites. Staff is available to assist educators with curriculum and instruction issues.

Our high school program for teachers includes *Summer Secondary Science Institutes* in biology, chemistry and physics. Also, we are part of the national infrastructure for QuarkNet and Interactions in Understanding the Universe funded in part by the National Science Foundation and DOE.

3. Future

As Fermilab looks beyond the Tevatron to new research at the Intensity, Cosmic and Energy Frontiers, the Education Office looks forward to new opportunities for teachers and students—new exhibits, a new focus on discovery science as a journey, not an event.

While decades of observations and results have gone into the Standard Model, it leaves fundamental questions unanswered. What are the neutrinos telling us? Are there extra dimensions of space? What is dark matter? What is dark energy? We start with unanswered questions that stretch the imagination of physicists and students alike. We look at instruments scientists build to test their ideas and the results that support or refute them, and then follow along as physicists unravel new mysteries of particle physics. The launch of the LHC and new tours of the Tevatron, CDF and DZero allow Fermilab to be a gateway to scientific discovery, bringing students into a new era of particle physics.

Combined with the Lederman Science Center hands-on exhibits, no tour of Fermilab will be complete without a stop at the fabulous instruments that particle physicists built here. Standing next to CDF, climbing inside DZero, walking down the Tevatron tunnel will motivate students as never before.

The Leon M. Lederman Science Education Center began as a dream and has become a facility that offers opportunities for students and families to explore the science and technology of Fermilab through hands-on exhibits and programs that inspire the next generation of scientists and engineers and the citizens who support their work.

Acknowledgments

The author would like to thank Pier Oddone for continuing the tradition of Fermilab directors to give wholehearted support for the Lab's K-12 educational programming at the Lederman Science Center and to thank all those who contribute to the success of the programs—the Education Office staff, hundreds of volunteers from Fermilab's technical staff and our partners in the education community. The Fermilab Education Office is supported by the U.S.

Department of Energy. Additional programmatic support comes from the U.S. National Science Foundation and Fermilab Friends for Science Education.

THE INTERNATIONAL PARTICLE PHYSICS OUTREACH GROUP (IPPOG): AIMS AND ACTIVITIES

DAVID BARNEY[†]

CERN, CH-1211 Geneva 23, Switzerland
E-mail: David.Barney@cern.ch

The International Particle Physics Outreach Group, IPPOG, is a network of particle physics communication and education experts. IPPOG's principle aim is to maximize the impact of education and outreach efforts related to particle physics through information exchange and the sharing of expertise. IPPOG has initiated several major European and Worldwide activities, such as the "International Particle Physics Masterclasses" where each year thousands of high school students in more than 20 countries come to one of about 120 nearby universities or research centres for a day in order to unravel the mysteries of particle physics. IPPOG has also initiated a global database of education and outreach materials, aimed at supporting other particle physicists and education professionals. The aims and activities of IPPOG will be described, as well as plans to include more countries & laboratories in the network.

1. Overview of IPPOG

On 10th September 2008 the startup of the Large Hadron Collider ("LHC") at CERN was witnessed by hundreds of millions of people worldwide, intrigued by the idea that scientists are continuing a quest that mankind has been pursuing for millennia, trying to answer some of the most basic questions about the Universe. Why is the Universe the way it is? What is it made from? How will it evolve? The world's largest science experiment has captured the imagination of the media and public alike, in a way that only the "Space race" has done before. The "Large Hadron Collider", "Higgs Boson" and "Particle Physics" are now everyday terms and, crucially, this subject is inspiring young people to study not only physics, but science in general, at school/university. IPPOG has the explicit aim of spreading the key messages of this incredible research and helping others engaged in similar activities worldwide.

1.1. *The IPPOG members*

IPPOG [1, 2] is a network of scientists, researchers, science educators and explainers engaged in informal science education and outreach for particle

---

[†] On behalf of IPPOG

physics. It began life in the late 1990's as "EPPOG"* and was an initiative from then CERN Director General Chris Llewellyn-Smith, to share ideas and improve standards of particle physics education & outreach (E&O) in the run-up to the LHC. IPPOG includes representatives from member states of CERN, each of the four major experiments at CERN's LHC as well as representatives from CERN and DESY. The group reports to ECFA† and EPS-HEPP‡. Many delegates are closely linked to national science networks and have even, in some instances, instigated these national networks. Figure 1 shows some of the delegates, who they represent and their main activities.

Figure 1. Five of the present IPPOG members, showing the diversity of primary activities – one of the strengths of the group.

1.2. *IPPOG as a discussion forum*

The members of the IPPOG group share ideas and experience in an effort to strengthen understanding and support of particle physics and related sciences. This is achieved through twice-yearly group meetings supported by smaller working groups and regular contact through conventional and Web 2.0 methods. We discuss a wide variety of E&O methods and activities, including exhibitions, multimedia, printed materials, public events. In all cases we try to determine

---

* European Particle Physics Outreach Group

† European Commission for Future Accelerators

‡ European Physical Society – High Energy Particle Physics Group

what things worked well, and why; how they could be improved; how they could be used in other countries/contexts etc.

2. Activities Instigated by IPPOG

The breadth of knowledge and experience of IPPOG has led to several worldwide activities being instigated by the group. A couple of them are described in the following sections.

2.1. *International Particle Physics Masterclasses*

Inspired by similar activities within the UK, in 2005 IPPOG developed the concept of "International Particle Physics Masterclasses" [3], providing high school students around Europe the opportunity to "become physicists for a day." Over the course of about two weeks, groups of students aged between 16-18 went to nearby universities/institutes for a day to have lectures and tours in the morning, followed in the afternoon by a realistic "enquiry-based learning exercise" with real data and analysis tools. The exercise was followed by a video conference with CERN and the other groups participating that day, to discuss their results – just like real physicists! Initially the exercise was based on LEP data (from OPAL and DELPHI) and used an event-display package called "Hands on CERN"[§] that was translated into nearly 20 languages by IPPOG members. Although the exercise was a relatively simple "event identification and counting" procedure, it probed fundamental physics – the branching ratio of the Z boson and allowed a comparison with the real LEP result and a discussion of the scientific method including statistical & systematic errors. During the years that followed more institutes and countries joined the program – including Brazil, Israel, USA, China, Canada and Japan. Figure 2 shows how the number

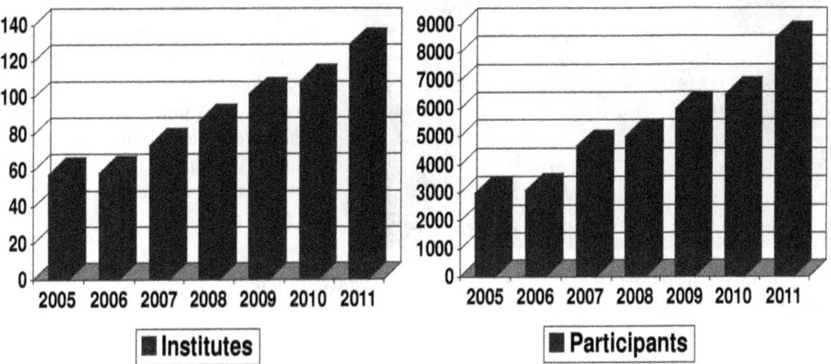

Figure 2. Increase in participation to the International Particle Physics Masterclasses since its inception in 2005.

[§] Created by ex IPPOG chair Prof. Erik Johansson of Stockholm University of institutes and individuals participating in the masterclasses has increased over the past 7 years.

In 2011 a major change took place: the analysis of real data from the LHC (ALICE, ATLAS and CMS). Initially a daunting prospect due to the complicated nature of hadron collisions, software tools were developed to allow display of the events and subsequent simple analyses and, once again, touched upon rather deep physics. This included:

1. Examination of the inner structure of the proton
2. Re-discovery of the Z-boson and measurement of its mass
3. Study of di-muon events to identify J/Ψ candidates
4. Identification of K^0, Ξ, Λ based on event topology and calculations of invariant mass

In some cases a few simulated events were added to the mix, to allow the students to search for new physics – Higgs bosons and Z'. It was remarkable that the young students were quickly able to come to terms with the software and get to the underlying physics, as shown in Figure 3. Refinements to the software and analyses are now underway in order to make 2012 masterclasses an even more memorable experience.

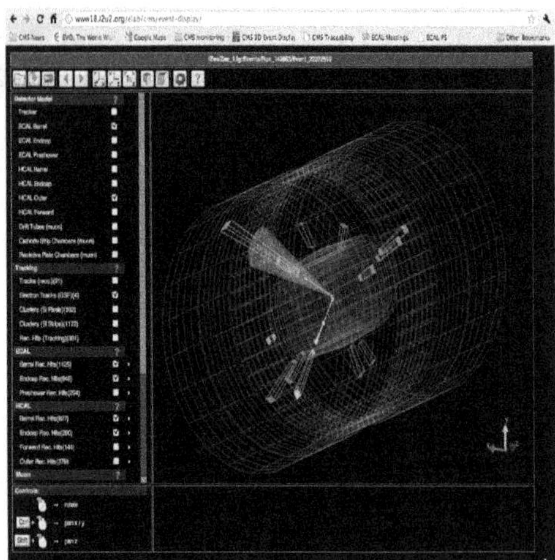

Tom Jordan (QuarkNet):
'I was truly pleased with how quickly the students were able to "get to the physics." There were the 7-8 minutes of necessary questions like:
"How do I load an event?"
"Where is the zoom?"
"Can I rotate the display?"
They quickly started arguing about the physics. It was very cool. Then they asked each other:
"Are those the same charge?"
"That doesn't look global to me. What do you think?"
"Are those muon tracks even leaving the same point?"'

Figure 3. Image of the CMS web-based event display software with quotes on its usage to explore LHC physics.

2.2. *The IPPOG education & outreach database*

In 2010 IPPOG started to develop the first global database of education and outreach materials related to particle physics. The primary audience of the database is other physicists, explainers, educators etc. who wish to develop their own public talks, exhibitions and informal education activities. Although still a fledgling, the database already contains some three hundred "items," each of which may contain multiple files and is accompanied by a comprehensive description of the item, the audience, the objective, languages available etc. The database contains a wide range of item types, including books/journals/articles, coaching resources for scientists, exhibits and exhibitions, ideas for public events, teaching tools and education kits. Figure 4 shows the diversity of some of the existing items.

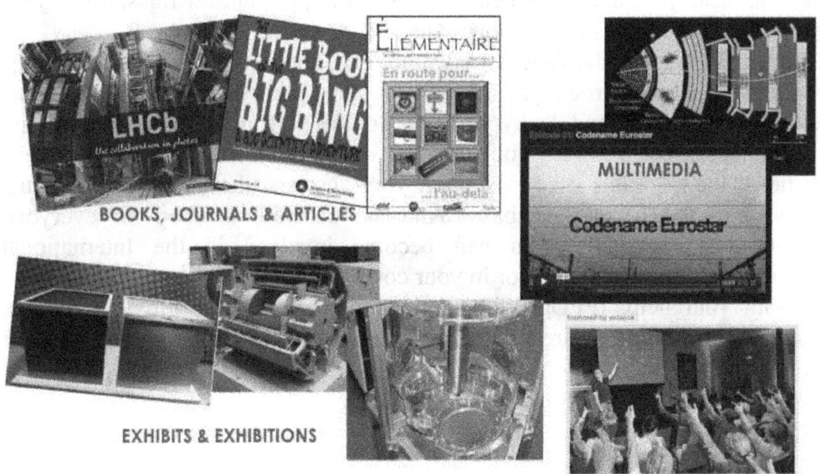

Figure 4. Some of the hundreds of items currently in the IPPOG E&O database

The database is currently undergoing a major transformation, to provide a modern attractive interface and to use the CERN Document Server (CDS) in order to take advantage of automatic document conversion, video streaming etc. We hope to have the new version available by the end of 2011.

3. Expansion of the IPPOG Group

IPPOG evolved from EPPOG at the beginning of 2011, to reflect the evolving need for membership from outside of the CERN member states[**] and LHC experiments. Indeed CERN itself is undergoing expansion too, so new CERN member states (e.g. Israel) would be obvious candidates to join IPPOG. Major laboratories such as FNAL and KEK, as well as other countries/continents and related disciplines (neutrino physics, particle astrophysics etc.) are also

[**] The USA had representation on EPPOG since 2004

candidates for membership. A working group was initiated within IPPOG in order to examine the possibilities and methodologies for expansion and should reach its conclusions soon.

4. IPPOG and You!

IPPOG can share with you recommended materials to help communicate effectively with a wide range of audiences in either informal or formal settings. It can also put you in touch with current outreach and science education programs near you, and even help you explore opportunities for starting your own outreach initiatives through our extensive network.

But IPPOG, and the field as a whole, can also benefit greatly from your involvement. Every particle physicist has at some point answered the questions "what do you do?" and "why do you do this?" from family, friends and acquaintances. The answers you give, the methods you use to answer these questions, the materials used to support your answers: these are all things that can go into the IPPOG database! And the database is open for everyone to contribute. Secondly, you can become involved in the International Masterclasses, either at CERN or in your country/institute.

With your help, IPPOG will continue to contribute to global efforts in strengthening cultural awareness, understanding and support of particle physics and related sciences.

Acknowledgments

We gratefully acknowledge the support from the host states, laboratories and experiments of the IPPOG members. We would particularly like to thank the CERN Directorate, the European Physical Society, the Helmholtz Foundation, the Technical University Dresden, QuarkNet, the National Science Foundation and Department of Energy, without whom the Masterclasses and database would not be possible.

References

1. IPPOG web site (including database): http://ippog.web.cern.ch
2. Facebook page: http://facebook.com/IPPOG
3. Masterclasses: http://www.physicsmasterclasses.org

Reconstruction Software for High Multiplicity Events in GEM Detectors

M. Berretti[1*], V. Avati[3], E. Bossini[1], P. Brogi[1], E. Brücken[2,2a], S. Giani[3], V. Greco[1,3], S. Lami[1a], G. Latino[1], E. Oliveri[1], F. Oljemark[2,2a], K. Österberg[2,2a], A. Scribano[1], N. Turini[1], J. Welti[2,2a].

[1] *Università degli Studi di Siena and Gruppo Collegato INFN di Siena, Italy.* [1a] *INFN Sezione di Pisa, Italy.* [2] *Department of Physics, University of Helsinki, Finland.* [2a] *Helsinki Institute of Physics, Finland.* [3] *CERN, Geneva, Switzerland.*

We present a general description of the offline software developed for the reconstruction of inelastic events by the TOTEM T2 telescope at the LHC. Tracking reconstruction in the CMS forward region, where T2 is installed, is challenged by a large amount of charged particles generated by the interaction with the material placed between the IP and T2. In this contribution we describe the simulation of the T2 GEM chambers as well as the reconstruction procedures employed to track the particles in such severe environment. The strategy for the telescope alignment and the measurement of the charged particle pseudorapidity distribution is finally presented.

1. The TOTEM experiment and the T2 detector

TOTEM is an experiment dedicated to the measurement of total cross section, elastic scattering and diffractive processes at the LHC.[1] The experimental apparatus, shown in fig. 1, is made up of three subdetectors: Roman Pots are utilized for the elastic scattered and diffractive protons measurement, while two inelastic detectors (T1 and T2) are utilized for the forward charged particle flow and inelastic rate measurements. T2, placed at about 14 meters from the interaction point (IP), covers the pseudorapidity region between 5.3 and 6.5. There are two T2 telescopes, one for each side of the IP, consisting of 2 quarters made by 10 planes. The planes are made up of semiround triple-GEM (Gas Electron Multipliers) chambers,[2] each one with an azimuthal acceptance of 192°.[3] The T2 plane Read-Out (RO) con-

*Speaker and corresponding author. Phone: +39-0577-234683. E-mail: mirko.berretti@cern.ch

sists of a matrix of 1560 pads (having an area $\Delta\eta \times \Delta\phi = 0.06 \times 0.018$) and 2 columns of 256 strips (pitch:$400\mu m$, width:$80\mu m$). The signal generated by the front.end electronics is digital i.e. only the list of the activated pads and strips is saved.

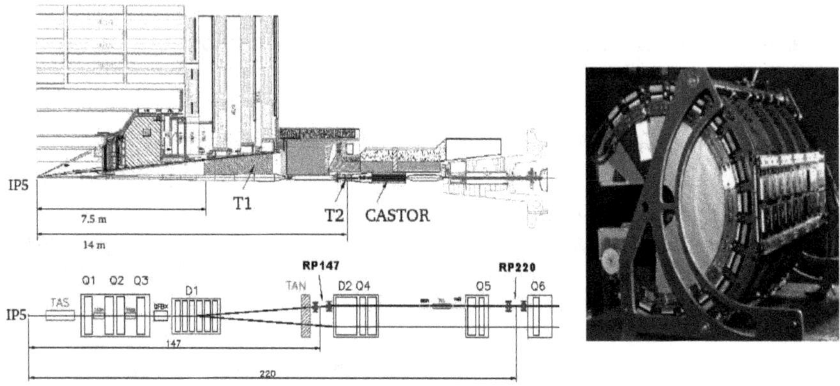

Fig. 1. Left: the TOTEM experimental apparatus. Right: one of the T2 quarters.

2. T2 detector simulation

The main program used to simulate the T2 detector triple GEM is Garfield[a]. A precise definition of the electrostatic configuration inside the detector is obtained interfacing Garfield with Maxwell 2D SV[b], a finite-elements based software. The simulation of the transport and ionization properties in gas mixtures is done using the Garfield interface with Magboltz, Imonte and HEED.[4] These dedicated software tools allow the study of the response of the detector in terms of spatial charge distribution and signal induced on the electrodes. Unfortunately this detailed simulation cannot be used to study the detector response on proton-proton (pp) Monte Carlo (MC) events, since it would take too much CPU-time. Therefore, a parametrization of the detailed simulation allowing one to reproduce detector response both on data and on the detailed simulation is needed.[4] To create a model reproducing the detailed simulation, a geometrical approach has been used:

---

[a] `http://garfield.web.cern.ch/garfield/`
[b] `http://www.ansoft.com/products/em/maxwell/`

the contribution of one ionizing particle (called C) to the charge in the α-th electrode is given by [c]:

$$N^\alpha = \sum_C N_C^{e^-} \cdot G \cdot n^\alpha(X_C, Y_C, \sigma_{ch}(Z_C)) \tag{1}$$

$$\sigma_{ch} = \sqrt{Z_C} \times \sigma_0\,, \quad \sigma_0 = \text{gas diffusion coefficient} \tag{2}$$

$$n^\alpha(X_C, Y_C, \sigma_{ch}) = \frac{1}{100} \times n(d_{x1}^\alpha, d_{x2}^\alpha, \sigma_{ch}) \cdot n(d_{y1}^\alpha, d_{y2}^\alpha, \sigma_{ch}) \tag{3}$$

$d_{i1,i=x,y}^\alpha$ = Transversal distance between the first ionization cluster and the first border of the α-th pad.

$d_{i2,i=x,y}^\alpha$ = Transversal distance between the first ionization cluster and the second border of the α-t pad.

$$n(d_1, d_2, \sigma_{ch}) = 50 \times |1 \mp 1 \pm [(1 - erf(\frac{d_1}{\sqrt{2}\sigma_{ch}})) \mp (1 - erf(\frac{d_2}{\sqrt{2}\sigma_{ch}}))]|$$

The upper (bottom) sign is used when the first ionization cluster is inside (outside) the α-t pad.

Where $N_C^{e^-}$ is the number of primary electrons released in the drift space by the ionizing particle, G is the gain of the triple GEM and $n(d_{y1}^\alpha, d_{y2}^\alpha, \sigma_{ch})$ is an error function (erf) combination performing the integration in the range of the electrode area of the gaussian cloud generated by one of the primary electrons after the amplification. Notice that the integral depends on the X_C, Y_C, Z_C position of the primary electron (Z_C is the electron distance with respect to the RO board). After taking into account some particular electrostatic effects which give a larger effective area to the strips, we found that this simple model reproduces well the detailed simulation. In order to tune the model with the data, two free parameters are needed: the effective strip area and the equivalent chip-threshold. The former is needed in order to tune the cluster size of the strips, the latter is used in order to properly simulate the efficiency of each RO chip. Single particle event simulation has been utilized in order to convert the chip efficiency measured on data in an equivalent threshold of the chip to be applied in the simulation [d]. The results on the comparison between data and tuned simulation for the pad

[c]The formula is valid for the pad case, after the charge of the overlapping strips has been subtracted. Similar formula are applied for the strips simulation.

[d]This technique allows one to maintain the properties for which the higher is the charge in the electrode the higher is the probability to switch it on

efficiency and cluster size are shown in fig. 2. Similar comparison have been obtained for the strip RO.

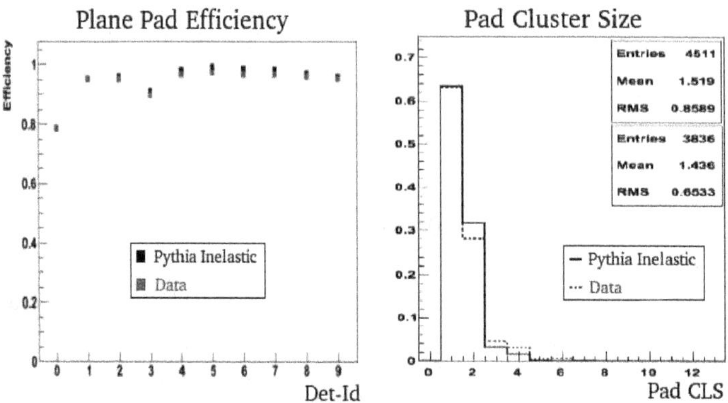

Fig. 2. 7 TeV data and MC (Pythia) inelastic events comparison. Plane pad efficiency (left) and cumulative pad cluster size (right) for one of the T2 quarter

3. Track reconstruction in T2

The amount of particles produced by the interaction of primary particles with the material in front of and around T2 was found to be particularly challenging both for the detector performances and the physics analysis. The modelization of the forward region has been simulated with GEANT4 and properly tuned with the data[5e].A large amount of secondary particles (which roughly constitute 90% of the signal in T2) is produced mainly in the vacuum chamber walls in front of the detector, in the beam pipe (BP) cone at $\eta = 5.53$ and in the lower edges of the CMS Hadron Forward (HF) calorimeter. Secondaries are the main reponsable of high-track multiplicity events in T2, producing a strip occupancy larger than 40% for $\sim$ 10% of the events. Due to the fact that the local magnetic field is weak and almost collinear with the track direction we have no selecting power for the lowest energy particles. The track finding algorithm is composed by three subprocedures:

[e]CMS found an important discrepancy between HF measurement and Pythia simulations for M.B. events[6] (the latter underestimating the energy). Due to the dependence of the secondary multiplicity to the forward energy (not yet measured at the LHC in the T2 acceptance) we still have a discrepancy of $\sim$ 30% between data and the common tunes of Pythia inelastic simulations.

(1) *Road Finding:* using pad clusters (preferred on the strip clusters since they have smaller occupancy, smaller noise, higher efficiency and the possibility to measure X,Y,Z position at the same time) the algorithm looks for 3D collinear cluster tubes (pad-roads) through the 10 planes of the quarter. The algorithm is inspired to a seedless Kalman Filter technique, where noise and multiple scattering sources can be assumed negligible.
(2) *Track Finding:* for each pad road, the overlapping strips are associated to each pad cluster and all the possible combinations of strip and pad cluster hits are generated. The best combinations are chosen by a minimum χ^2 criteria. Simulation studies have shown that a minimum of 4 pad clusters (with at least 3 overimposed strip clusters) can be required as a quality criteria of the road.
(3) *Track Fitting:* the hits in a road are fitted, the geometrical and quality track parameters are computed and outliers reduction is eventually performed.

The primary tracks hitting T2 are reconstructed with an efficiency of $\sim$ 95%, the η resolution is 0.07 in the central acceptance region assuming a vertex constraint.

4. Alignment correction

Particular effort has been devoted to correct for misalignment biases, found to be dominated by global T2 quarter displacements respect to the IP. The most important internal alignment parameters which are possible to resolve within the T2 hit resolution are the shift of the planes in the X and Y direction. Two different methods (iterative and MILLIPEDE approaches) have been developed in order to correct for such displacements. The relative alignment between the two quarters of an arm has been obtained using tracks reconstructed in the overlap regions, while the global alignment has been derived by studying the expected symmetry in the track parameters distributions and the position on each T2 plane of the "beam pipe shadow" (very low track efficiency radial zone due to primary particles absorbed by the $\eta \sim 5.53$ beam pipe cone).

5. Analisis of the $dN_{\mathrm{CH}}/d\eta$

A preliminary measurement of the forward charged particle η distribution has also been performed using the data taken in special 2011 runs at low luminosity with an inclusive T2 trigger. Secondary track rejection has been derived from data analysis, while primary track efficiency and smearing effects correction have been obtained from MC studies. The results are reported in fig. 3, where the black unjoined points show the experimental measurements with the uncertainties related to statistical effects and the solid curves show the MC prediction.

The coloured band represents the overall systematic uncertainty, mainly related to the estimation of the track efficiency, to the effect of detector misalignment and to the secondary track contribution subtraction.

Work is in progress for the determination of the residual systematics related to the MC modeling of the forward particle energy spectrum and to the simulation of magnetic field effects. These last, less relevant uncertainties, are expected to give an additional contribution at the level of few percent.

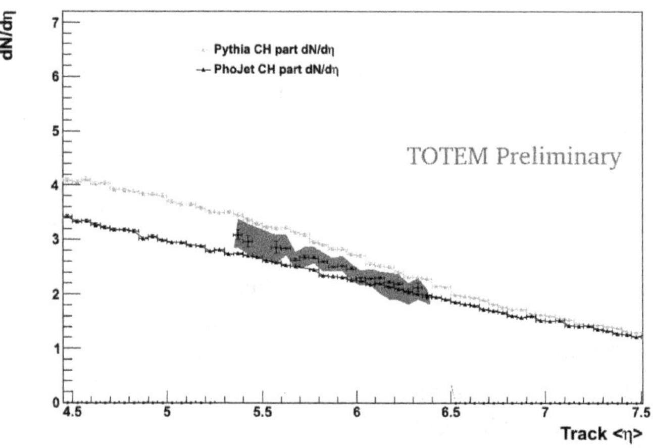

Fig. 3. Preliminary measurement of $\frac{dN}{d\eta}$ (black unjoined points with statistical error), compared to predictions from Pythia6 (gray solid line) and Phojet (black solid line). The coloured band represents the combination of the main systematic uncertainties.

References

1. The TOTEM Collaboration, *CERN-LHCC-2004-002* (2004).
2. F. Sauli, *Nucl. Instrum. Methods A* **386**, p. 531 (1997).
3. G. Anelli et al. (TOTEM Collaboration), *JINST* **3:S08007** (2008).
4. E. Oliveri, The forward inelastic telescope t2 for the totem experiment at lhc, PhD thesis, University of Siena, (Italy, 2009).

5. P. Brogi, The forward inelastic telescope t2 for the totem experiment at lhc, PhD thesis, University of Pisa, http://cdsweb.cern.ch/record/1384794, (Italy, 2011).
6. The CMS Collaboration, *CMS-PAS-FWD-10-01* (2011).

Recent results on neutral particles spectra from the LHCf experiment

M. BONGI* for the LHCf Collaboration

Istituto Nazionale di Fisica Nucleare (INFN), Sezione di Firenze,
Sesto Fiorentino (FI), 50019, Italy
**E-mail: massimo.bongi@fi.infn.it*

LHCf is an experiment designed to study the very-forward production of neutral particles produced in collisions at the Large Hadron Collider (LHC). Its results will be useful to calibrate the hadron interaction models of the Monte Carlo (MC) codes which are used for the interpretation of energy spectrum and composition of high-energy cosmic rays as measured by air-shower ground arrays. The experiment has already completed taking p-p data at $\sqrt{s} = 0.9$ TeV and at $\sqrt{s} = 7$ TeV in 2009–2010. The detectors are now being upgraded and they will be installed again in the LHC tunnel for operation with protons at $\sqrt{s} = 14$ TeV, for p-ion collisions, and possibly for ion-ion collisions too. Comparisons between the single photon spectra and MC simulations, and a preliminary result about π^0 energy measurement are reported.

Keywords: ultra-high-energy cosmic rays; hadron interaction models; LHC

1. The LHCf experiment

Several measurements of the highest part of the cosmic ray spectrum has been performed in the last years by detecting extensive air showers produced in the atmosphere.[1–9] This indirect detection technique relies on the use of Monte Carlo (MC) simulation codes in order to estimate the energy and the composition of primary particles, and the comparison between experimental data and simulations can lead to different interpretations depending from the details of the hadronic interaction model used in the MC. The aim of the LHCf experiment[10,11] is to provide experimental results useful in testing and calibrating hadronic interaction models, by measuring the energy spectra and the transverse momentum of neutral particles in a very high pseudo-rapidity region ("forward" region) at the Large Hadron Collider (LHC). The knowledge of the characteristics of the forward region is crucial for understanding air-shower development and the LHC gives us the

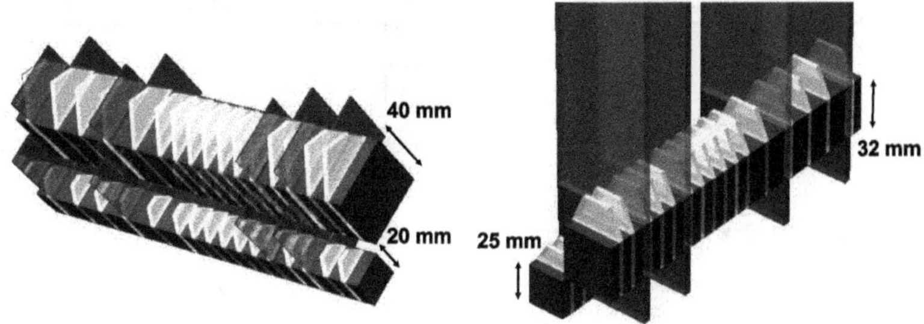

Fig. 1. Schematic view of the LHCf detectors (Arm1 on the left, Arm2 on the right). Plastic scintillators (light blue color) are interleaved with tungsten blocks (dark gray color). Four couples of position sensitive layers (scintillating fibers in Arm1, dark blue color; silicon micro-strip detectors in Arm2, purple color) are present in each calorimeter.

unique opportunity of studying these processes up to equivalent fixed-target energies of 10^{17} eV (at its design center-of-mass energy $\sqrt{s} = 14$ TeV).

Two detectors (Arm1 and Arm2) have been symmetrically installed 140 m away at the sides of the ATLAS detector (IP1) at zero-degree collision angle, in the regions where the single beam pipe coming from the interaction point splits to two separate tubes. Each detector is composed of two towers (see Fig. 1) which are sampling and imaging calorimeters containing tungsten absorbers and plastic scintillator layers (44 radiation lengths, 1.55 hadron interaction lengths). The smaller towers cover the zero-degree collision angle and the accessible pseudo-rapidity region ranges from 8.4 to infinity. Four X-Y layers of position sensitive detectors (scintillating fibers in Arm1, silicon micro-strip detectors in Arm2) provide measurements of the transverse profile of the showers. The experimental set-up allows only neutral particles such as photons and neutrons to reach the calorimeters, as D1 dipole magnets sweep away charged particles: in particular, in case both the photons from the decay of neutral pions reach the towers of one detector, invariant mass and energy of the π^0 can be reconstructed (see Fig. 2).

In the range $E > 100$ GeV, the LHCf detectors have energy and position resolutions for electromagnetic showers better than 5% and 200 μm, respectively.

2. Data analysis and results

The energy of photons is determined from information on the light produced in the scintillators, after applying corrections for the non-uniformity

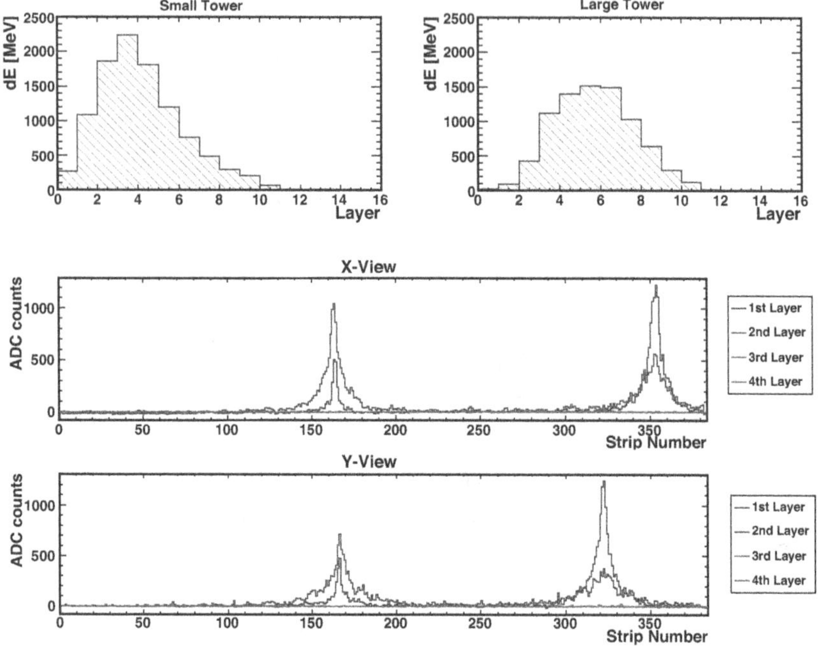

Fig. 2. A typical π^0 event in the Arm2 detector, showing one photon in each of the calorimeter towers. Top: longitudinal profile of the showers in the small tower (left) and the large tower (right). Bottom: superposition of X and Y transverse profiles of the two showers measured by the four layers of silicon micro-strip detectors.

of light collection and for particles leaking out of the edges of the calorimeter towers. The transverse impact position of showers is determined using the information provided by the position sensitive detectors. This information is used also to exclude from the analysis events with more than one shower inside the same tower in a detector. Events produced by neutral hadrons are eliminated by simple identification criteria based on the longitudinal development of the showers, which is different for electromagnetic and hadronic particles. In order to combine the spectra measured by the two calorimeters, which have different geometrical configurations, in this analysis we include only events detected in a common pseudo-rapidity and azimuthal range: $\eta > 10.94$ and $\Delta\phi = 360°$ for the small towers, $8.81 < \eta < 8.99$ and $\Delta\phi = 20°$ for the large towers. Further details about the analysis can be found in Ref. 12.

Figure 3 shows the single-photon energy spectra obtained from a subset of data taken in 2010 for proton-proton collisions at $\sqrt{s} = 7$ TeV (inte-

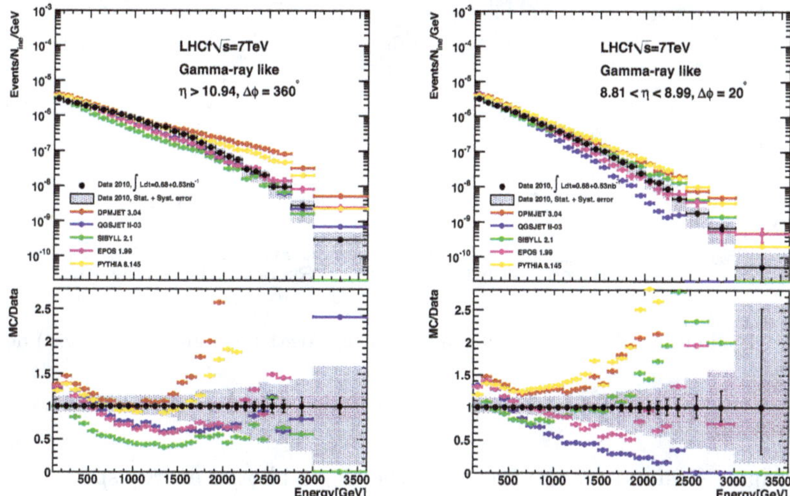

Fig. 3. Comparison between the measured single-photon energy spectra (black dots) and the predictions of the following MC codes: DPMJET 3.04 (red), QGSJET II-03 (blue), SIBYLL 2.1 (green), EPOS 1.99 (magenta) and PYTHIA 8.145 (yellow). Top panels show the spectra and bottom panels show the ratio of MC results to experimental data. Left and right panels refer to different pseudo-rapidity ranges, $\eta > 10.94$ and $8.81 < \eta < 8.99$ respectively. Error bars show the statistical error and gray shaded areas the systematic error for experimental data. Magenta shaded areas indicate the statistical error associated to MC simulations (using EPOS 1.99 as a representative of all the models).

grated luminosity $\int L\ dt = 0.68\ \mathrm{nb}^{-1}$ and $0.53\ \mathrm{nb}^{-1}$ for Arm1 and Arm2, respectively). On the vertical axis, the number of inelastic collisions N_{ine}, is calculated as $N_{ine} = \sigma_{ine} \int L\ dt$ assuming the inelastic cross section $\sigma_{ine} = 71.5$ mb. The spectra are obtained by combining the results of both Arm1 and Arm2 detectors in the two pseudo-rapidity regions and are compared with results predicted by MC simulations using different models: DPMJET 3.04,[13] QGSJET II-03,[14] SIBYLL 2.1,[15] EPOS 1.9[16] and PYTHIA 8.145.[17,18]

Figure 4 shows the preliminary energy spectra of neutral pions as measured by Arm1 and Arm2 detectors. Events plotted are those selected in a ± 10 MeV range around the reconstructed π^0 invariant mass peak.

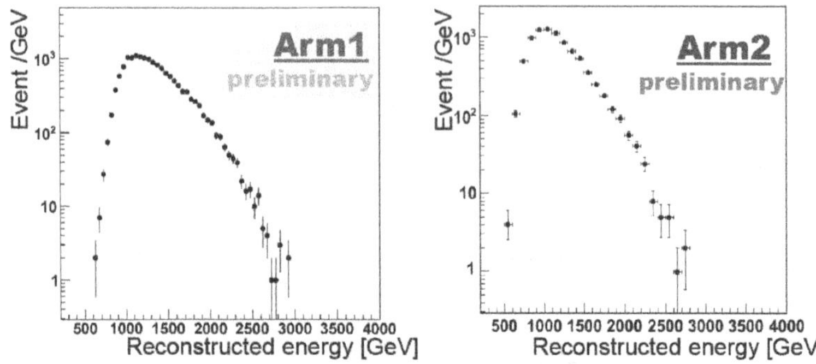

Fig. 4. Preliminary π^0 energy spectra measured by Arm1 (left) and Arm2 (right) detectors.

3. Summary

LHCf has measured for the first time the single-photon energy spectra for $E > 100$ GeV in the very-forward region of proton-proton collisions at LHC. By comparing the result with the prediction of various hadron interaction models we found that none of them shows a perfect agreement within the statistical and systematic errors. Comparison of π^0 energy spectra with MC is in progress, as well as additional studies such as measurement of neutron spectra and neutral particle transverse momentum spectra.

References

1. J. Abraham, *et al.*, *Phys. Rev. Lett.* **101**, 061101 (2008).
2. P. Abreu, *et al.*, *Astropart. Phys.* **34**, 314 (2010).
3. J. Abraham, *et al.*, *Phys. Rev. Lett.* **104**, 091101 (2010).
4. A. Castellina, *et al.*, in the Proceedings of this Conference.
5. R. Abbasi, *et al.*, *Astropart. Phys.* **32**, 53 (2009).
6. R. Abbasi, *et al.*, arXiv:1002.1444v1 [astro-ph.HE], (2010).
7. R. Abbasi, *et al.*, arXiv:0910.4184v2 [astro-ph.HE], (2009).
8. H. Sagawa, *et al.*, *AIP Conf. Proc.* **1367**, 17 (2011).
9. G. Thomson, *et al.*, in the Proceedings of this Conference.
10. O. Adriani, *et al.*, *JINST* **3**, S08006 (2008).
11. O. Adriani, *et al.*, *JINST* **5**, P01012 (2010).
12. O. Adriani, *et al.*, *Phys. Lett. B* **703**, 128–134 (2011).
13. F.W. Bopp, *et al.*, *Phys. Rev. C* **77**, 014904 (2008).
14. S. Ostapchenko, *Phys. Rev. D* **74**, 014026 (2006).
15. E.-J. Ahn, *et al.*, *Phys. Rev. D* **80**, 094003 (2009).
16. K. Werner, *et al.*, *Phys. Rev. C* **74**, 044902 (2006).
17. T. Sjöstand, *et al.*, *JHEP* **0605**, 026 (2006).
18. T. Sjöstand, *et al.*, *Comput. Phys. Comm.* **178**, 852 (2008).

LHC DATA FOR TEACHERS AND STUDENTS

KENNETH CECIRE
kcecire@nd.edu
Department of Physics, University of Notre Dame, 225 Nieuwland Science Hall
Notre Dame, Indiana 46556, USA

The U.S. QuarkNet program began in 1999 to involve high school students and teachers in authentic particle physics investigations using real data. This took various forms from the use of cosmic ray detectors to Z decay exercises with Hands-on-CERN. In 2010, QuarkNet opened a new chapter with the use of real data from the LHC. In collaboration with I2U2, QuarkNet staff and select teachers developed an e-Lab and a masterclass using CMS data. This development continues with the release by the CMS collaboration of over 100,000 events for education. Students and teachers have used the CMS e-Lab and masterclass as well as the ATLAS masterclass, also with real data, with very encouraging results. Working with IPPOG, QuarkNet has made these opportunities available internationally as well as within the U.S. text.

1. Introduction

1.1. *QuarkNet*

From the beginning of the QuarkNet program in 1999, a major goal was to place real data from the Large Hadron Collider (LHC) into the hands of students and teachers in a form that is meaningful for student learning and investigation. Prior to the availability of such data, the QuarkNet staff sought alternative ways to get data that would meet some of the same goals, whether it was from cosmic ray detectors or more current particle physics experiments (e.g., the Tevatron at Fermilab).

1.2. *Organization*

This paper is organized from the point of view of a QuarkNet student looking backwards and drilling down on her particle physics experience. Thus the text starts from ongoing, sustained particle physics investigations in an e-Lab and then dwells on a powerful introduction to particle physics research in the masterclass. It ends at the beginning; the student's teacher formed in particle physics investigation in the Boot Camp.

2.1. *e-Lab Concept*

QuarkNet and I2U2 developed the Cosmic Ray e-Lab through the first decade of the century so that students and teachers could potentially share data from all QuarkNet-supplied CRMDs. In the e-Lab, it is possible to upload data from a detector as a text file, analyze any data that has been uploaded from anywhere, and publish results in the form of online posters. The Cosmic Ray e-Lab provides structure and resources for students to pursue investigations as well as an online logbook into which students can place notes, observations, and questions; their teachers can read and respond to student entries [1].

The CRMD and Cosmic Ray e-Lab enables students and teachers to do experiments and investigations with real data on cosmic ray muons. The LIGO e-Lab was designed to allow students and teachers to analyze seismic data that is background to the search for gravitational waves by the Laser Interferometer Gravitational Wave Observatory (LIGO). In this case there is no data for students to collect and upload; it is all taken by LIGO [2].

2.2. *CMS e-Lab*

In order to enable students and teachers to carry on e-Lab investigations with real data from the LHC, QuarkNet and I2U2 undertook to develop an e-Lab based on the Compact Muon Solenoid (CMS) detector in the LHC.

The CMS e-Lab starts, as do the other e-Labs, with a Project Map to enable students to plan and then navigate through a research project. Between built-in resources and a logbook to which teachers can respond, students have ample scaffolding in the e-Lab to learn how to do an analysis of CMS data and put it in context.

At the heart of the CMS e-Lab experience is the opportunity to build plots from CMS data. Students are able to select which particle candidates they would like to study (e.g., Z boson) and which decay data to examine (e.g. $\mu+\mu-$). They can then create a mass plot and apply cuts to isolate what they might be looking for. They can change parameters on their plots, create different plots to compare, and eventually come to a conclusion regarding their research question. This gives students information for creating an online poster shared with the whole community of CMS e-Lab users.

The last stage is also the first. Students can, at the conclusion of an investigation, create an online poster which explains their research and conclusions. This poster can be viewed by anyone who logs on to the e-Lab. As more posters are added, they become a library of ideas and questions to launch new investigations; thus, as student who is beginning the process can examine posters and start to form questions to be investigated in the e-Lab [3].

3. Masterclasses

3.1. *Masterclass Concept*

The particle physics masterclasses are based conceptually on masterclasses in the arts. For example, a piano student might prepare a piece to bring to such a masterclass so that he may work with a "master" who has deeper expertise than his normal piano teacher, perhaps a well-known concert pianist. The master works with the student on the piece. The student's ability to play that piece improves; more importantly, he learns a great deal more about the instrument and technique. At the end of the masterclass, the student may, along with others, play a public recital of the music studied.

The analogy takes hold on the day of the particle physics masterclass, when students go to a university or laboratory to participate. There, students are given some background and then a data set in the form of event displays to analyze. Particle physicists and tutors (university physics students and/or prepared high school physics teachers) act as the "masters," helping students not only to perform the analysis but also to learn about the process under study, the accelerator, and the detector. The students in the masterclass under the tutelage of the mentor draw conclusions and then share these in a videoconference with other masterclass institutes, with moderators from CERN.

Masterclasses in particle physics first started in the United Kingdom in 1996, using data from the OPAL detector of the Large Electron-Positron Collider (LEP). Hands-on CERN was developed and the masterclass concept evolved. In 1995, European masterclasses began with both OPAL and DELPHI data; Erik Johannsen of the University of Stockholm contributed a JAVA-based online event display for the study of the decays of W and Z bosons from DELPHI that was particularly popular [4]. Masterclasses grew; the first U.S. Masterclass Institute (Brookhaven National Laboratory) joined in 1996. In 1998, the U.S. created its own masterclass program with videoconferences hosted by Fermilab. The U.S. program continued, however, to work closely with the European program, even as it grew itself [5]. By 2010, the last year of the LEP-based masterclasses, there were about 110 masterclass institutes (24 in the U.S. program) with a total of about 6,000 students. In 2011, this grew to about 130 institutes, 8,000 students moving into a new era [6]. Masterclasses 2011 were the first to offer exclusively LHC-based masterclasses in three flavors: ATLAS, ALICE, and CMS.

3.2. *ATLAS Masterclasses*

Two masterclasses were offered by the ATLAS collaboration. One is the ATLAS Z-path exercise, created by Farid Ould-Saada and Maiken Pedersen of the University of Oslo. In this exercise, students use the HYPATIA program (Christine Kourkomelis, University of Athens, et al.) to sort between Z→dilepton candidate events and various sorts of background events. Students in the masterclass institute collect and contribute their invariant masses calculated by HYPATIA to make a mass plot of the Z boson, which is shared in the videoconference. A highlight of the Z-path exercise is the inclusion of Monte Carlo Z events, allowing the students to "discover" an unexpected particle.

The W-path exercise, designed by Konrad Jende and Michael Kobel of the Technical University of Dresden, uses the MINERVA program (Monika Weilers, Rutherford-Appleton Laboratories, et al.) to distinguish between W+ and W- candidate events. Students then derive the W+/W- ratio as a probe of the structure of the proton.

Both ATLAS exercise programs are based on the ATLANTIS event display used by the ATLAS collaboration [7].

3.3. *ALICE Masterclasses*

The ALICE masterclass exercise uses data from 900 GeV proton collisions in the LHC to probe strangeness. Students calculate yields of hadrons with strange quarks (K^0_s, Λ, anti-Λ, Ξ^-). The result shows that the yield is greater than predicted in Monte Carlo calculations, in line with actual results published by the ALICE collaboration [8].

3.4. *CMS Masterclasses*

Masterclasses 2010 featured a J/Ψ exercise for CMS, developed by QuarkNet and I2U2 collaborators (Cecire, Hategan, Nguyen, et al.). The central piece is the online event display, iSpy-online, written in Javascript and based on the iSpy program used in the CMS collaboration. iSpy-online is three-dimensional and can be manipulated in useful ways: snap to different views (x-y, x-z, y-z), rotate, zoom, turn physics objects and detector subsystems off and on, etc. Data obtained from the CMS collaboration by Tom McCauley et al. was in the form of 2,000 dimuon events with calculated invariant masses between 2 and 5 GeV. Most of the events are not good J/Ψ meson candidates but enough are to make for a clear mass plot if the weak candidates can be eliminated. The CMS masterclass does this by having students rate the events 0-3 (3 is best) and then decide which ratings are allowable for the plot. Most institutes were able to generate a clean peak [9].

In 2012, CMS will have a new W/Z exercise. Students will study event displays to sort between W+, W-, and Z. The W+/W- and W/Z ratios will be calculated with a main goal of building a mass plot of the remaining Z candidates. Additional J/Ψ and Y candidate events can be found in the data; these look similar to Z bosons in the event display but will show peaks at much lower energies.

3.5. *Recent Results of Masterclass Experience*

There were some initial problems in the implementation of masterclasses in 2011 due to the newness of the exercises; there was, however, overall satisfaction with the result. Students surveyed in the U.S. showed a higher level of satisfaction than in 2010, although both were high, with a higher percentage scoring complete agreement with a positive statement.

In addition, students reported gains in learning and increased interest in physics as a result of the masterclasses in 2011. Teachers reported that classroom preparation activities offered by QuarkNet for the masterclasses helped students to do well and enjoy the experience. These findings are generally in line with previous years [10].

4. Particle Physics Boot Camp

4.1. *Lead Teacher Institutes and Boot Camps in QuarkNet*

QuarkNet started with 20 lead teachers in 1999; each of ten new QuarkNet centers would have two of these teachers as initial co-leaders along with the mentors. That year, the QuarkNet staff put together an interim program that would give these teachers some introduction to how particle physics works. The following year, in which another 20 lead teachers came from another 10 new centers, QuarkNet coordinator Tom Jordan designed a central exercise around which he and the QuarkNet staff built the Lead Teacher Institute. The heart of this program was to give teachers data in the form of spreadsheets from a hypothetical electron-positron collider. Teachers were given little instruction; their task was to work in teams to make sense of the data and then report [11].

As the Lead Teacher Institutes developed over the years, milestones were added to help the teachers find their direction but built-in resources were withdrawn or made optional. The practice was guided inquiry with maximum freedom. As QuarkNet matured, the need to recruit new centers diminished; the Lead Teacher Institute was made into the Particle Physics Boot Camp and opened for all QuarkNet teachers to apply to participate. (The few new lead teachers were also fed into the Boot Camp.) Boot Camp became an established, highly successful feature of QuarkNet.

4.2. *LHC Boot Camp*

With the start of operation of the LHC in 2009, QuarkNet Principal Investigator Randal Ruchti suggested a large LHC workshop, using real data, for as many QuarkNet teachers as possible in summer 2010. In that year, it turned out that neither QuarkNet nor the LHC were ready. The new workshop was scheduled for summer 2011; Tom Jordan took the lead in designing it and patterned it after the Boot Camp. The LHC Boot Camp, using CMS data, was offered to approximately 40 QuarkNet teachers that July.

Data was given in spreadsheet form to five groups, each specializing in a particular process observed in CMS: $J/\Psi \rightarrow \mu\mu$, $W \rightarrow e\nu$, $W \rightarrow \mu\nu$, $Z \rightarrow ee$, and $Z \rightarrow \mu\mu$. The teachers were given minimal information, a simplified set of milestones, and support from QuarkNet staff and fellows. In previous versions of the Boot Camp, there would be two analysis groups who gave talks; in the LHC Boot Camp, the five groups all presented posters [12].

5. Conclusion

In three programs using real data from the LHC: Particle Physics Boot Camp, Masterclasses, and the CMS e-Lab, there is the beginning of a path in which teachers and students can be invited into the particle physics community in a meaningful way. In Boot Camp, teachers experience a mode of learning that works well for deep understanding of physics by students and begin to gain an understanding of particle physics research. The masterclass introduces students to particle physics and to physics research by making them "physicists for a day." This leads to increased engagement in physics and excitement about particle physics. With the CMS e-Lab, students can enter into sustained particle physics investigations. In this process, students and teachers become an extended part of the LHC community. The next challenge is to create more pathways of investigation for teachers and students with real, cutting-edge data and to deepen the international collaboration that is an important feature of particle physics and, in particular, of the masterclass. The goal is to create an international community that is a continuum: for many students to be physicists for a day, a large number of these to be physicists for a year or more, and then for a smaller number of particularly motivated and focused students to become physicists for a lifetime.

Acknowledgments

This work is funded by the U.S. National Science Foundation, the U.S. Department of Energy, Office of Science, and the University of Notre Dame. The author wishes to acknowledge the following:

- Michael Fetsko, Thomas Gallo, and Laura Akesson of Mills Godwin High School
- Michael Kobel and Uta Bilow of the Technical University of Dresden
- The QuarkNet and I2U2 collaborations
- The International Particle Physics Outreach Group
- M.J. Young and Associates
- The teachers, staff, physicists, and students worldwide who have worked on the projects described in this paper
- LaMargo Gill of Fermilab for editorial assistance

References

1. Cosmic Ray e-Lab, http://www.i2u2.org/elab/cosmic.
2. LIGO e-Lab, http://www.i2u2.org/elab/ligo.
3. CMS e-Lab, http://www.i2u2.org/elab/cms.
4. Hands-on CERN, http://www.physicsmasterclasses.org/exercises/hands-on-cern/hoc_v21en/index.html.
5. K. Cecire, *International Conference on Hands-on Science* (2010).
6. M. Kobel, *ATLAS March Outreach Meeting* (2011).
7. LHC@InternationalMasterclasses, https://kjende.web.cern.ch/kjende/en/index.htm.
8. ALICE Masterclass Exercise Description (PDF), http://www.physicsmasterclasses.org/exercises/ALICE/alice-exercise-en.pdf.
9. CMS International Masterclasses, http://www.physicsmasterclasses.org/exercises/CMS/cms.html.
10. K. Cecire, *Proceedings of the DPF 2011 Conference* (2011).
11. Lead Teacher Institute Projects, http://quarknet.fnal.gov/projects/summer00/index.shtml.
12. CMS Data Week, http://quarknet.fnal.gov/projects/summer11/index.shtml.

RecPack, a general reconstruction toolkit

A. Cervera-Villanueva* and J.J. Gómez-Cadenas
IFIC, CSIC and University of Valencia,
Valencia, Spain
**E-mail: acervera@ific.uv.es*

J. A. Hernando
University of Santiago de Compostela,
Santiago de Compostela, Spain

A general solution for the problem of reconstructing the evolution of a dynamic system from a set of experimental measurements is presented. This solution has been realised in a C++ toolkit that can incorporate different methods for fitting, propagation, pattern recognition and simulation. The RecPack functionality is independent of the experimental setup, what allows one to apply this toolkit to any dynamic system.

Keywords: reconstruction, kalman filter, propagation, fitting

1. Introduction

In high energy physics (HEP), as in other fields, one frequently faces the problem of reconstructing the evolution of a dynamic system from a set of experimental measurements. Most of reconstruction programs use similar methods. However, in general they are reimplemented for each specific experimental setup. Some examples are fitting algorithms (i.e. Kalman Filter[1]), equations for propagation, random noise estimation (i.e. multiple scattering), model corrections (i.e. energy loss, inhomogeneous magnetic field, etc.), etc. Similarly, the data structure (measurements, tracks, vertices, etc.), which can be generalised as well, is not reused in most of the cases.

RecPack tries to avoid that by providing a setup–independent data structure and algorithms, which can be applied to any dynamic system. The package follows an "interface" strategy, that is, all the classes that could have a different implementation have their own interface, in such a way that the rest of the classes do not depend on such a specific implementation. This modular structure allows a great flexibility and generality.

RecPack was born in the HARP experiment at CERN-PS[2] and is currently being developed mainly for T2K.[3] Other experiments using it are MICE, MuScat, MIPP, NEMO and SuperNemo.

2. Structure of the package

RecPack distinguishes between data classes (passive) and tools (active). The tree of data classes is shown in Fig. 1. *EVector* and *EMatrix* are just a typedef of CLHEP's *HepVector* and *HepMatrix* respectively (these are vectors and matrices of *double*'s with variable dimension). This establish the only RecPack external dependence. However, the user can replace the CLHEP classes by its favorite vector and matrix classes.

A structure that appears in several levels of the data model is the pair formed by a vector of parameters (*EVector*) and its covariance matrix (*EMatrix*). Thus, a new class called *HyperVector* has being introduced to hold this repeated structure. For example, experimental measurements (*Measurement*) can be always reduced to a *HyperVector*, and the same is true for the fitting or propagation parameters (*State*). Before the track fit-

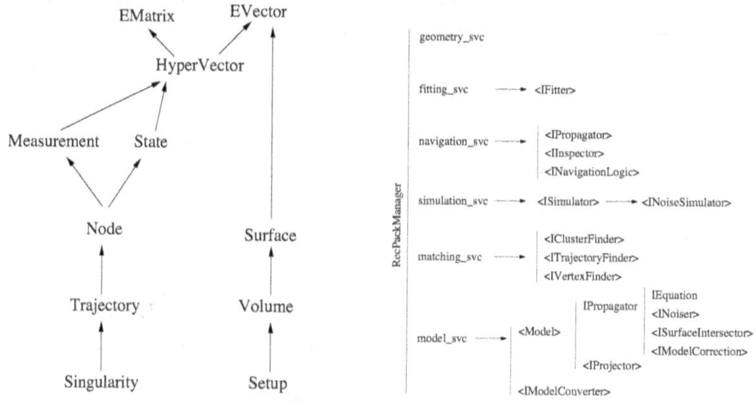

Fig. 1. Architectural design of the data classes (left) amd tools.

ting occurs, a *Trajectory* is essentially defined as a collection of uncorrelated *Measurement*'s. Track fitting results are stored in *State*'s. In the most general case, the fitting parameters are local and therefore each *Measurement* must have a *State* associated to it. An intermediate object, called *Node*, has being introduced to accommodate the *Measurement*, the *State* and the quantities that relate both objects (the residual *HyperVector*, the local χ^2 of the fit,etc). Consequently, a *Trajectory* can be seen as a collection of *Node*'s, plus a set of global quantities as the total χ^2 of the fit, the number of degrees of freedom, etc. The classes *Surface*, *Volume* and *Setup* are treated in Sec. 3.

The tools (see Fig. 1-right) manipulate the data. Most of them are pure interfaces (hence the I), allowing them to have different implementations. Some of the tools contain sub-tools (*ISimulator, Model, IPropagator*, etc.). If a tool has several sub-tools of the same type (i.e. *Model* has several *IProjector*'s), these are stored in associative containers (<...> in the graph), which permits the access by key.

Tools are stored in services (_svc in Fig. 1), which not only actuate as containers for the tools, but also as managers. The user interacts with the *RecpackManager* class via its services. In the following sections each of the services is treated individually.

3. Geometry

The RecPack geometry service provides the methods for the definition of the experimental setup (*Setup*), which is built through the addition of volumes (*Volume*) and surfaces (*Surface*) with well defined position and axes inside the setup. One can distinguish between base surfaces, which have no boundaries, and finite surfaces, which extend the base class by incorporating a well defined size. RecPack provides some predefined volume (Box and Tube –two concentric cylinders–) and surface types (Plane: Rectangle and Ring; and Cylinder), but adding new types is straight forward. As an example, the following code defines a box called "tracker" placed inside the "mother" volume, and a surface called "wall" inside the tracker:

```
setup.add_volume("mother", "tracker", "box", pos, ax1, ax2, s1,s2,s3 );
setup.add_surface("tracker", "wall", "rectangle", pos, ax1, ax2, s1,s2 );
```

where pos, ax1 and ax2 are 3D *EVector*'s, and s1,s2 and s3 are doubles defining the size. In this way, complicated setups as the ones of Fig. 2 can be build.

Geometrical objects may have properties, which are indirectly associated to them in the *Setup* class. Typical volume properties are the ones that influence the evolution of the system (magnetic field, radiation length, energy loss rate, etc.). The following c++ code sets $x0$ (a *double*) as the

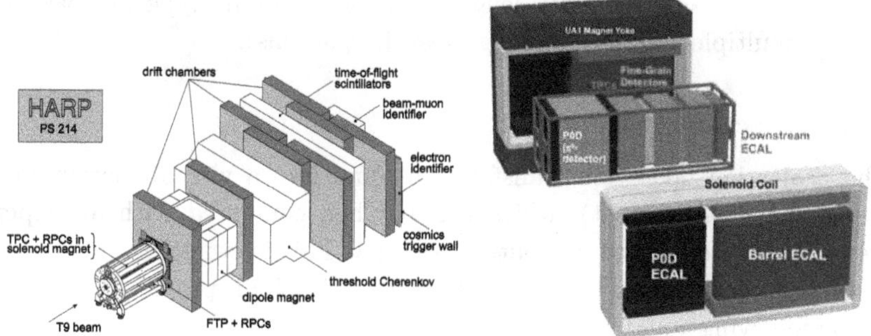

Fig. 2. Example of experimental setups corresponding to the HARP[2] detector at CERN-PS (left) and the T2K-ND280[3] detector at JPARC (right).

radiation length of the "tracker":

setup.set_volume_property("tracker", RP::X0, x0);

where $x0$ must be a data member or global variable since it is saved by reference.

4. Model

The model service is the container and manager for model related equations: intersection with surfaces, propagation and projection of states, random noise computation, etc. It contains an extensible collection of *Model*'s. Each model performs two major operations, propagation and projection:

4.1. *Propagation*

A *State* can be propagated to a given surface or length by the interface class *IPropagator*. Intermediate calculations are delegated to smaller tools:

- <*ISurfaceIntersector*>: it is a collection of *ISurfaceIntersector*'s, each of which calculates the path length to a different base surface type (Plane or Cylinder).
- *IEquation*: it computes the state vector at a given length (either provided externally or by a *ISurfaceIntersector*).
- <*IModelCorrection*>: applies a small correction to the propagation done by the *IEquation* (i.e. energy loss).
- <*INoiser*>: each of them computes the random noise covariance

matrix for the given length and for a specific type of noise (i.e. multiple scattering, energy loss fluctuations).

4.2. *Projection*

The projection operation transforms a *State* into a virtual measurement (predicted–measurement), which can be then compared with an experimental measurement to compute a residual. This is crucial for fitting and matching algorithms. The state *HyperVector* ($\vec{\mathbf{v}}$, $\mathbf{C}_v$) is projected according to the following equations:

$$\vec{\mathbf{m}}^{pred} = \vec{\mathbf{h}}(\vec{\mathbf{v}}), \quad \mathbf{C}_m^{pred} = \mathbf{H}\mathbf{C}_v\mathbf{H}^T \ , \tag{1}$$

where $\vec{\mathbf{h}}$ is the projection function, which depends on the measurement type, $\mathbf{H} = \partial\vec{\mathbf{h}}/\partial\vec{\mathbf{v}}$ is the projection matrix, $\vec{\mathbf{m}}^{pred}$ is the predicted–measurement, and $\mathbf{C}_m^{pred}$ its covariance matrix.

Several measurement types ("xy", "uv", "xyz"," $r\phi$", etc.) may coexist in a single trajectory, which can be fitted to an unique model. To do so, each *Model* must contain an extensible collection of *IProjector*'s, each of which corresponds to a different measurement type.

5. Fitting

The fitting service is in charge of fitting clusters, trajectories and vertices via its fitters (*IFitter*). The user can either use one of the existing fitters or provide his own. Two fitters for trajectories ,least squares and Kalman filter,[1] are available. In the case of a Kalman Filter fit a seed state must be provided.

Fitting algorithms, called fitters, need the two setup–dependent operations described above: the prediction of the trajectory at the next measurement surface (propagaton) and the comparison betweed this prediction and the actual experimental measurement, which requires the "projection" of the predicted state to form a residual. Fitting equations can be kept independent of the model and measurement type(s) if these two operations are external to the fitter. As described above, propagation and projection are performed by each *Model*.

6. Navigation

In *Setup*'s with more than one *Volume* the propagation functionality is provided by special *IPropagator*'s, called *Navigator*'s, which are able to handle the volume hierarchy. A default *Navigator* is provided, but others can be added easily (i.e. Geant4[4]). The default *Navigator* propagates a *State* in several steps. Propagation in each step is performed by the *IPropagator* associated to the *Model* of the *State*. Before and after each step a list of *IInspector*'s (associated to volumes and surfaces) is called. An *IInspector* is a tool that performs a concrete action: set the properties of the entering volume, sum up intermediate path lengths, set the length of the next step (dynamic stepping), etc. User defined *IInspector*'s can be added to any surface or volume.

Two important features of the default navigator are: i) the intersection with surfaces [a] is done analytically whenever is possible (and numerically otherwise) and ii) user defined *INavigationLogic*'s allow one to establish the sequence in which volumes and surfaces must be traversed.

7. Matching

This generic name refers to the methods that are related with pattern recognition (PR) problems. In general, the purpose of PR algorithms is to distribute the existing measurements into clusters, these into trajectories and these into vertices. Two types of PR algorithms are provided by RecPack: matching functions, which serve to estimate the probability of two objects of being related to each other (trajectory–trajectory, trajectory–measurement), and PR logics, which define the sequence in which such a relations are established. The first are always general, while the second may have a strong setup dependence. PR logics for *Trajectory*'s are introduced via the *ITrajectoryFinder* interface class. Currently, the RecPack matching service provides a *CellularAutomatom* trajectory finder.

8. Simulation

Some times reconstruction programs must operate over simulated measurements. However, in general the user must provide the classes and methods that allow the interface between simulation and reconstruction, which is not always an easy task. The RecPack simulation service solves this problem by generating simulated measurements with the data format required by the rest of the services.

The user must declare the active *Volume*'s and *Surface*'s (the ones that produce measurements), and specify the measurement type in each of

[a]The problem of intersecting a volume is always reduced to the intersection with its outer walls.

them. Active surfaces produce a measurement when they are intersected, while active volumes produced measurements through dynamic stepping (see Sec. 6).

Given a simulation seed (*State*), the simulation service uses the navigation service to produce an ideal trajectory inside the setup. Then, a special *IInspector* ("MeasSimulator") creates ideal measurements (according to Eq. 1) in the active volumes or surfaces by calling the *IProjector* corresponding to the measurement type in that volume or surface, and adds the propagation noise (i.e. multiple scattering) and experimental errors.

This simple simulator does not attempt to be a full simulator (i.e. Geant4). Instead, its main purpose is to serve as a debugging tool or as a fast simulator. Existing simulation toolkits, as Geant4, could be easily integrated into RecPack by implementing the *IInspector*'s that generate measurements in the different subdetectors. Such an inspectors should be able to access the Geant4 information and then use it to create specific measurements.

9. Conclusions

In summary, RecPack is a modular and extensible reconstruction toolkit, which provides the basic data structure and most of the common methods needed by any reconstruction program: matching, fitting and navigation. It also has functionality to perform a quick interface with simulation packages.

References

1. R.E. Kalman, J. Basic Eng. 82 (1960) 35
   R.E. Kalman, R.S. Bucy, J. Basic Eng. 83 (1961) 95
   R. Fruhwirth, M. Regler, Nuc. Inst. Meth. A241 (1985) 115.
   R. Fruhwirth. Nucl. Inst. Meth. A262 (1987) 444
2. http://harp.web.cern.ch/harp/
3. arXiv:1106.1238v2. Accepted for publication in Nucl. Inst. Meth. A
4. A. Dell'Acqua *et al.*, GEANT4 Collaboration, Nucl. Inst. Meth. A506 (2003) 250.

NUCLEAR AND NON-IONIZING ENERGY-LOSS OF ELECTRONS WITH LOW AND RELATIVISTIC ENERGIES IN MATERIALS AND SPACE ENVIRONMENT

M.J. Boschini[1,2], C. Consolandi*[,1], M. Gervasi[1,3], S. Giani[4], D. Grandi[1],
V. Ivanchenko[4], P. Nieminem[5], S. Pensotti[3], P.G. Rancoita[1] and M. Tacconi[1]

[1]*INFN-Milano Bicocca, P.zza Scienza,3 Milano, Italy*
[2]*CILEA Via R. Sanzio, 4 Segrate, MI-Italy*
[3]*Milano Bicocca University, P.zza della Scienza, 3 Milano, Italy*
[4]*CERN, Geneva, 23, CH-1211, Switzerland*
[5]*ESA, ESTEC, AG Noordwijk (Netherlands)*
**E-mail: cristina.consolandi@mib.infn.it*

The treatment of the electron–nucleus interaction based on the Mott differential cross section was extended to account for effects due to screened Coulomb potentials, finite sizes and finite rest masses of nuclei for electrons above 200 keV and up to ultra high energies. This treatment allows one to determine both the total and differential cross sections, thus, subsequently to calculate the resulting nuclear and non-ionizing stopping powers. Above a few hundreds of MeV, neglecting the effect due to finite rest masses of recoil nuclei the stopping power and NIEL result to be largely underestimated. While, above a few tens of MeV, the finite size of the nuclear target prevents a further large increase of stopping powers which approach almost constant values.

1. Introduction

Nuclei and electrons populate the heliosphere. Most of the nuclei are galactic cosmic rays (GCR), while electrons can additionally be originated by the Sun and Jupiter's magnetosphere, which is a major source of relativistic electrons in the heliosphere (e.g., see Ref.[1,2] and references therein). Protons and electrons are also major constituents of the Earth's radiation belts. These particles can interact with materials and onboard electronics in spacecrafts, inducing displacements of atomic nuclei, thus inflicting permanent damages. As the particle energy increases, for instance above $\approx 20\,\text{MeV}$ for protons and ≈ 130 MeV/nucleon for α-particles (e.g., see Section 4.2.1.4 and Figure 4.26 at page 418 of Ref.[3]), the dominant mechanism for displacement damage is determined by hadronic interactions; for electrons and low-energy nuclei the elastic Coulomb scattering is the relevant physical process to induce permanent damage.

The non-ionizing energy-loss (NIEL) is the energy lost from particles traversing a unit length of a medium through physical processes resulting in permanent atomic displacements. The displacement damage is mostly responsible for the degradation of semiconductor devices - like those using silicon - where, for instance, depleted layers are required for normal operation conditions (e.g. see Ref.[4]). The nuclear stopping power and NIEL deposition - due to elastic Coulomb scatterings - from protons, light- and heavy-ions traversing an absorber were previously dealt[5,6] with (see also Sections 1.6, 1.6.1, 2.1.4–2.1.4.2, 4.2.1.6 of Ref.[3]). In the present work, the nuclear stopping power and NIEL deposition due to elastic Coulomb scatterings of electrons are treated up to ultra relativistic energies.

The developed model (i.e., see Sects. 2–2.4) for screened Coulomb elastic scattering up to relativistic energies is included into Geant4 distribution[7] and is available with Geant4 version 9.5 (December 2011). In Sects. 3, 4, the nuclear and non-ionizing stopping powers for electrons in materials are treated, while a final discussion is found in Sect. 5.

2. Scattering Cross Section of Electrons on Nuclei

The scattering of electrons by unscreened atomic nuclei was treated by Mott[8] (see also Sections 4–4.5 in Chapter IX of Ref.[9]) extending a method of Wentzel[10] (see also Born[11]) and including effects related to the spin of electrons[8] . Wentzel's method was dealing with incident and scattered waves on point-like nuclei. The differential cross section (DCS) - the so-called *Mott differential cross section* (MDCS) - was expressed by Mott[8] as two conditionally convergent infinite series in terms of Legendre expansions. In Mott–Wentzel treatment, the scattering occurs on a field of force generating a radially dependent Coulomb - unscreened (screened) in Mott[8] (Wentzel[10]) - potential. It has to be remarked that Mott's treatment of collisions of fast electrons with atoms (e.g., see Chapter XVI of Ref.[9]) involves the knowledge of the wave function of the atom, thus, in most cases the computation of cross sections depends on the application of numerical methods (see a further discussion in Sect. 2.2). Furthermore, the MDCS was derived in the laboratory reference system for infinitely heavy nuclei initially at rest with negligible spin effects and must be numerically evaluated for any specific nuclear target. Effects related to the recoil and finite rest mass of the target nucleus (M) were neglected. Thus, in this framework the total energy of electrons has to be smaller or much smaller than Mc^2.

As discussed by Idoeta and Legarda[12] (e.g., see also Refs.[13,14]), Mott provided an "exact" differential cross section because no *Born approximation*[a] of any order is employed in its derivation. Various authors have approximated the MDCS for special situations, usually expressing their results in terms of ratios, $\mathcal{R}$, of the so-obtained approximated differential cross sections with respect to *that one for a Rutherford scattering* (RDCS) - the so-called *Rutherford's formula*, see Section 1.6.1 of Ref.[3] - for an incoming particle with $z = 1$ given by:

$$\frac{d\sigma^{\mathrm{Rut}}}{d\Omega} = \left(\frac{Ze^2}{p\beta c}\right)^2 \frac{1}{(1-\cos\theta)^2} = \left(\frac{Ze^2}{2\,p\beta c}\right)^2 \frac{1}{\sin^4(\theta/2)} = \left(\frac{Ze^2}{2\,mc^2\beta^2\gamma}\right)^2 \frac{1}{\sin^4(\theta/2)}, \tag{1}$$

where m is the electron rest mass, Z is the atomic number of the target nucleus, $\beta = v/c$ with v the electron velocity and c the speed of light; γ is the corresponding *Lorentz factor*; p and θ are the momentum and scattering angle of the electron, respectively; finally, since the interaction is isotropic with respect to the azimuthal angle, it is worth noting that $d\Omega$ can be given as

$$d\Omega = 2\pi \sin\theta \, d\theta. \tag{2}$$

The MDCS is usually expressed as:

$$\frac{d\sigma^{\mathrm{Mott}}(\theta)}{d\Omega} = \frac{d\sigma^{\mathrm{Rut}}}{d\Omega} \mathcal{R}^{\mathrm{Mott}}, \tag{3}$$

where $\mathcal{R}^{\mathrm{Mott}}$ (as above mentioned) is the ratio between the MDCS and RDCS. In particular, Bartlett–Watson[15] determined cross sections for nuclei with atomic number $Z = 80$ and energies from 0.024 up to 1.7 MeV (see also Ref.[16]). McKinley and Feshbach[17] expanded Mott's series in terms of power series in αZ (with α the fine-structure constant) and $(\alpha Z)/\beta$; these expansions, which give results accurate to 1% up to atomic numbers $Z \approx 40$ (e.g., see discussions in Refs.[18,19]), were further simplified to obtain an approximate analytical formula with that accuracy for $\alpha Z \leq 0.2$. Feshbach[20] tabulated values of the differential cross section as a function of scattering angle for nuclei with atomic number up to 80 and electrons with kinetic

[a]In quantum mechanical potential scattering, the scattered wave may be obtained from the so-called Born expansion. The Born approximation is the first term of the Born expansion (see, for instance, references indicated in Section 1.6.1 Ref.[3]).

energies larger than 4 MeV. Curr[18] reported values of the differential cross section as a function of scattering angle accurate at 1% for $(\alpha Z)/\beta \lesssim 0.6$; while Doggett and Spencer[21] tabulated the MDCS for energies from 10 down to 0.05 MeV. Recently, Idoeta and Legarda[12] provided a further series transformations and made a systematic comparison with those from McKinley and Feshbach[17], Curr[18], Doggett and Spencer[21]. For electrons with kinetic energies from several keV up to 900 MeV and target nuclei with $1 \leqslant Z \leqslant 90$, Lijian, Quing and Zhengming[22] provided a practical interpolated expression [Eq. (16)] for $\mathcal{R}^{\mathrm{Mott}}$ with an average error less than 1%; in the present treatment, that expression - discussed in Sect. 2.1 - is the one assumed for $\mathcal{R}^{\mathrm{Mott}}$ in Eq. (3) hereafter.

The analytical expression derived by McKinley and Feshbach[17] - mentioned above - for the ratio with respect to Rutherford's formula [Equation (7) of Ref.[17]] is given by:

$$\mathcal{R}^{\mathrm{McF}} = 1 - \beta^2 \sin^2(\theta/2) + Z\,\alpha\beta\pi \sin(\theta/2)\,[1 - \sin(\theta/2)] \tag{4}$$

with the corresponding differential cross section (McFDCS)

$$\frac{d\sigma^{\mathrm{McF}}}{d\Omega} = \frac{d\sigma^{\mathrm{Rut}}}{d\Omega}\,\mathcal{R}^{\mathrm{McF}}, \tag{5}$$

where $d\sigma^{\mathrm{Rut}}/d\Omega$ is from Eq. (1). It has to be remarked that for positrons, the ratio $\mathcal{R}^{\mathrm{McF}}_{\mathrm{pos}}$ becomes

$$\mathcal{R}^{\mathrm{McF}}_{\mathrm{pos}} = 1 - \beta^2 \sin^2(\theta/2) - Z\,\alpha\beta\pi \sin(\theta/2)\,[1 - \sin(\theta/2)] \tag{6}$$

(e.g., see Equation (6) of Ref.[23]). Furthermore, for Mc^2 much larger than the total energy of incoming electron energies the distinction between laboratory (i.e., the system in which the target particle is initially at rest) and center-of-mass (CoM) systems disappears (e.g., see discussion in Section 1.6.1 of Ref.[3]). Furthermore, in the CoM of the reaction the energy transferred from an electron to a nucleus initially at rest in the laboratory system (i.e., its recoil kinetic energy T) is related to the maximum energy transferable T_{max} as

$$T = T_{\mathrm{max}} \sin^2(\theta'/2) \tag{7}$$

[e.g., see Equations (1.27, 1.95) at page 11 and 31, respectively, of Ref.[3]], where θ' is the scattering angle in the CoM system. From Eqs. (2, 7) one obtains

$$dT = \frac{T_{\mathrm{max}}}{4\pi}\,d\Omega'. \tag{8}$$

Since θ is $\approx \theta'$ for Mc^2 much larger than the electron energy, one finds that Eq. (7) can be approximated as

$$T \simeq T_{\max} \sin^2(\theta/2), \tag{9}$$

$$\Longrightarrow \sin^2(\theta/2) = \frac{T}{T_{\max}} \tag{10}$$

and

$$dT \simeq \frac{T_{\max}}{4\pi}\, d\Omega. \tag{11}$$

Using Eqs. (4, 10, 11), Eqs. (1, 5) can be respectively rewritten as:

$$\begin{aligned} \frac{d\sigma^{\text{Rut}}}{d\Omega} &= \frac{T_{\max}}{4\pi}\frac{d\sigma^{\text{Rut}}}{dT} \\ \Longrightarrow \frac{d\sigma^{\text{Rut}}}{dT} &= \left(\frac{Ze^2}{p\beta c}\right)^2 \frac{\pi T_{\max}}{T^2}, \end{aligned} \tag{12}$$

$$\begin{aligned} \frac{d\sigma^{\text{McF}}}{T} &= \left(\frac{Ze^2}{p\beta c}\right)^2 \frac{\pi T_{\max}}{T^2} \\ &\quad \times \left[1 - \beta^2 \frac{T}{T_{\max}} + Z\,\alpha\beta\pi\sqrt{\frac{T}{T_{\max}}}\left(1 - \sqrt{\frac{T}{T_{\max}}}\right)\right] \\ \Longrightarrow \frac{d\sigma^{\text{McF}}}{T} &= \left(\frac{Ze^2}{p\beta c}\right)^2 \frac{\pi T_{\max}}{T^2}\left[1-\beta\frac{T}{T_{\max}}(\beta+Z\alpha\pi)+Z\alpha\beta\pi\sqrt{\frac{T}{T_{\max}}}\right] \\ &= \left(\frac{Ze^2}{p\beta c}\right)^2 \frac{\pi T_{\max}}{T^2}\, \mathcal{R}^{\text{McF}}(T) \end{aligned} \tag{13}$$

with

$$\mathcal{R}^{\text{McF}}(T) = \left[1-\beta\frac{T}{T_{\max}}(\beta+Z\alpha\pi)+Z\alpha\beta\pi\sqrt{\frac{T}{T_{\max}}}\right] \tag{14}$$

[e.g., see Equation (11.4) of Ref.,[24] see also Ref.[19] and references therein]. Similarly, for positrons one finds

$$\frac{d\sigma^{\text{McF}}_{\text{pos}}}{T} = \left(\frac{Ze^2}{p\beta c}\right)^2 \frac{\pi T_{\max}}{T^2}\left[1-\beta\frac{T}{T_{\max}}(\beta-Z\alpha\pi)-Z\alpha\beta\pi\sqrt{\frac{T}{T_{\max}}}\right]$$

[e.g., see Refs.[19,23] and references therein]. Finally, in a similar way the MDCS [Eq. (3)] is

$$\begin{aligned} \frac{d\sigma^{\text{Mott}}(T)}{dT} &= \frac{d\sigma^{\text{Rut}}}{dT}\, \mathcal{R}^{\text{Mott}}(T) \\ &= \left(\frac{Ze^2}{p\beta c}\right)^2 \frac{\pi T_{\max}}{T^2}\, \mathcal{R}^{\text{Mott}}(T) \end{aligned} \tag{15}$$

with $\mathcal{R}^{\rm Mott}(T)$ from Eq. (18).

2.1. *Interpolated Expression for* $\mathcal{R}^{\rm Mott}$

As mentioned in Sect. 2, Curr[18] derived $\mathcal{R}^{\rm Mott}$ as a function the atomic number Z of the target nucleus and velocity βc of the incoming electron at several scattering angles from $\theta = 30^\circ$ up to 180°. Recently, Lijian, Quing and Zhengming[22] provided a practical interpolated expression [Eq. (16)] which is a function of both θ and β for electron energies from several keV up to 900 MeV, i.e.,

$$\mathcal{R}^{\rm Mott} = \sum_{\rm j=0}^{4} a_{\rm j}(Z,\beta)(1-\cos\theta)^{\rm j/2}, \tag{16}$$

where

$$a_{\rm j}(Z,\beta) = \sum_{\rm k=1}^{6} b_{\rm k,j}(Z)(\beta-\overline{\beta})^{\rm k-1}, \tag{17}$$

and $\overline{\beta}\, c = 0.7181287\, c$ is the mean velocity of electrons within the above mentioned energy range. The coefficients $b_{\rm k,j}(Z)$ are listed in Table 1 of Ref.[22] for $1 \leqslant Z \leqslant 90$.

At 10, 100 and 1000 MeV for Li, Si, Fe and Pb, values of $\mathcal{R}^{\rm Mott}$ were calculated using both Curr[18] and Lijian, Quing and Zhengming[22] methods and found to be in a very good agreement. It has to be remarked that with respect to the values of $\mathcal{R}^{\rm McF}$ obtained from Eq. (4) at 100 MeV one finds an average variation of about 0.2%, 3.2% and 8.8% for Li, Si and Fe nuclei, respectively. However, the stopping power determined using Eq. (52) (i.e., with $\mathcal{R}^{\rm Mott}$) differs by less than 0.5% with that calculated using Eq. (53) (i.e., with $\mathcal{R}^{\rm McF}$). $\mathcal{R}^{\rm Mott}$ obtained from Eq. (16) at 100 MeV is shown in Fig. 1 for Li, Si, Fe and Pb nuclei as a function of the scattering angle. Furthermore, it has to be pointed out that the energy dependence of $\mathcal{R}^{\rm Mott}$ from Eq. (16) was studied and observed to be negligible above ≈ 10 MeV [as expected from Eq. (17)].

Finally, from Eqs.(7, 16) [e.g., see also Equation (1.93) at page 31 of Ref.[3]], one finds that $\mathcal{R}^{\rm Mott}$ can be expressed in terms of the transferred energy T as

$$\mathcal{R}^{\rm Mott}(T) = \sum_{\rm j=0}^{4} a_{\rm j}(Z,\beta)\left(\frac{2T}{T_{\rm max}}\right)^{\rm j/2}. \tag{18}$$

2.2. *Screened Coulomb Potentials*

As already mentioned in Sect. 2, a complete treatment of electron interactions with atoms (e.g., see Chapter XVI of Ref.[9]) involves the knowledge of the wave function of the target atom and, thus - as remarked by Fernandez-Vera, Mayol and Salvat[14] -, a relevant amount of numerical work when the kinetic energies of electrons exceed a few hundreds of keV.

The simple scattering model due to Wentzel[10] - with a single exponential screening function [e.g., see Equation (2.71) at page 95 of Ref.[3], Equation (21) in Ref.[25] and Ref.[10]] - was repeatedly employed in treating single and multiple Coulomb scattering with screened potentials (e.g, see Ref.[25] - and references therein - for a survey of such a topic and also Refs.[5,6,26–28]). Neglecting effects like those related to spin and finite size of nuclei, for proton and nucleus interactions with nuclei it was shown that the resulting elastic differential cross section of a projectile with bare nuclear-charge ez on a target with bare nuclear-charge eZ differs from the Rutherford differential cross section (RDCS) by an additional term - the so-called *screening parameter* - which prevents the divergence of the cross section when the angle θ of scattered particles approaches 0° [e.g., see Refs.[5,6,26–28] (see also references therein) and Section 1.6.1 of Ref.[3]]. It has to be remarked that the RDCS for $z = 1$ particles can also be employed to describe the scattering of non-relativistic electrons with unscreened nuclei (e.g, see Refs.[8,12] and references therein). As derived by Molière[26] for the single Coulomb scattering using a *Thomas–Fermi potential,* for $z = 1$ particles the *screening parameter* $A_{s,M}$ [e.g., see Equation (21) of Bethe[27]] is expressed as

$$A_{s,M} = \left(\frac{\hbar}{2p\,a_{TF}}\right)^2 \left[1.13 + 3.76 \times \left(\frac{\alpha Z}{\beta}\right)^2\right] \tag{19}$$

where α, c and $\hbar$ are the fine-structure constant, speed of light and reduced Planck constant, respectively; p (βc) is the momentum (velocity) of the incoming particle undergoing the scattering onto a target supposed to be initially at rest - i.e., in the laboratory system -; a_{TF} is the screening length suggested by Thomas–Fermi (e.g., see Refs.[29,30])

$$\mathrm{a}_{TF} = \frac{C_{TF}\,\mathrm{a}_0}{Z^{1/3}} \tag{20}$$

with

$$\mathrm{a}_0 = \frac{\hbar^2}{me^2}$$

the Bohr radius, m the electron rest mass and

$$C_{\rm TF} = \frac{1}{2}\left(\frac{3\,\pi}{4}\right)^{2/3} \simeq 0.88534$$

a constant introduced in the Thomas–Fermi model [e.g., see Equations (2.73, 2,82) - at page 95 and 99, respectively - of Ref.[3] and Ref.[6], see also references therein]. The modified Rutherford's formula $[d\sigma^{\rm WM}(\theta)/d\Omega]$, i.e., the *differential cross section* - obtained from the Wentzel–Molière treatment of the single scattering on screened nuclear potentials - is given by [e.g., see Equation (2.84) of Ref.[3], Section 2.3 in Ref.[25] and Ref.[6] (see also references therein)]:

$$\frac{d\sigma^{\rm WM}(\theta)}{d\Omega} = \left(\frac{zZe^2}{2\,p\,\beta c}\right)^2 \frac{1}{\left[A_{\rm s,M} + \sin^2(\theta/2)\right]^2} \tag{21}$$

$$= \frac{d\sigma^{\rm Rut}}{d\Omega}\frac{\sin^4(\theta/2)}{\left[A_{\rm s,M} + \sin^2(\theta/2)\right]^2}$$

$$= \frac{d\sigma^{\rm Rut}}{d\Omega}\,\mathfrak{F}^2(\theta). \tag{22}$$

with

$$\mathfrak{F}(\theta) = \frac{\sin^2(\theta/2)}{A_{\rm s,M} + \sin^2(\theta/2)}. \tag{23}$$

$\mathfrak{F}(\theta)$ - the so-called *screening factor* - depends on the scattering angle θ and screening parameter $A_{\rm s,M}$. As discussed in Sect. 2.4, in the DCS the term $A_{\rm s,M}$ cannot be neglected [Eq. (22)] for scattering angles (θ) within a forward (with respect to the electron direction) angular region narrowing with increasing energy from several degrees (for high-Z material) at 200 keV down to less than or much less than a mrad above 200 MeV.

An approximated description of elastic interactions of electrons with screened Coulomb fields of nuclei can be obtained factorizing the MDCS, i.e., involving Rutherford's formula $[d\sigma^{\rm Rut}/d\Omega]$ for particles with $z=1$, the screening factor $\mathfrak{F}(\theta)$ and the ratio $\mathcal{R}^{\rm Mott}$ between RDCS and MDCS:

$$\frac{d\sigma^{\rm Mott}_{\rm sc}(\theta)}{d\Omega} \simeq \frac{d\sigma^{\rm Rut}}{d\Omega}\,\mathfrak{F}^2(\theta)\,\mathcal{R}^{\rm Mott} \tag{24}$$

[e.g., see Equation (1) of Ref.[12], Equation (A34) at page 208 of Ref.[13], see also Ref.[14] and citations from these references]. Thus, the corresponding

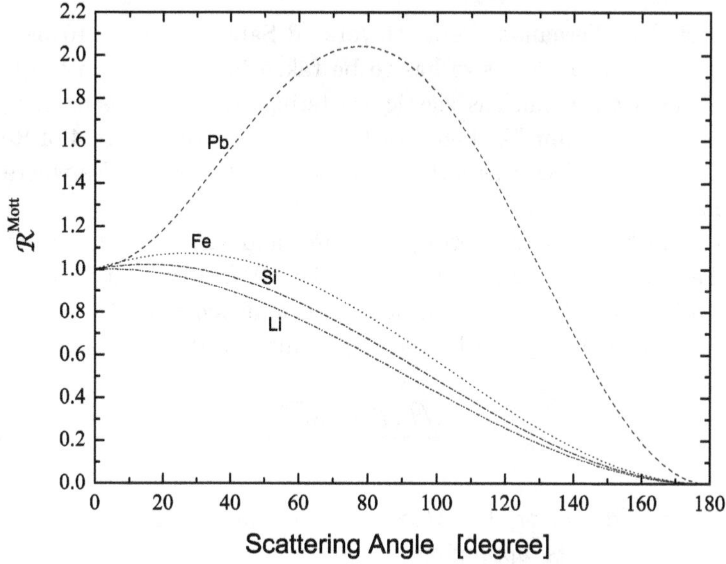

Fig. 1. $\mathcal{R}^{\mathrm{Mott}}$ obtained from Eq. (16) at 100 MeV for Li, Si, Fe and Pb nuclei as a function of scattering angle.

screened differential cross section derived using the analytical expression from McKinley and Feshbach[17] can be approximated with

$$\frac{d\sigma_{\mathrm{sc}}^{\mathrm{McF}}(\theta)}{d\Omega} \simeq \frac{d\sigma^{\mathrm{Rut}}}{d\Omega}\, \mathfrak{F}^2(\theta)\, \mathcal{R}^{\mathrm{McF}}. \tag{25}$$

It has to be remarked - as derived by Zeitler and Olsen[31] - that spin and screening effects can be separately treated for small scattering angles; while at large angles (i.e., at large momentum transfer), the factorization is well suited under the condition that

$$2Z^{4/3}\alpha^2 \frac{1}{\beta^2\gamma} \ll 1$$

(e.g., see Refs.[12,31]). Zeitler and Olsen[31] suggested that for electron energies above 200 keV the overlap of spin and screening effects is small for all elements and for all energies; for lower energies the overlapping of the spin and screening effects may be appreciable for heavy elements and large angles.

2.3. *Finite Nuclear Size*

As suggested by Fernandez-Vera, Mayol and Salvat[14], above 10 MeV the effect of the finite nuclear size has to be taken into account in the treatment of the electron–nucleus elastic scattering. With increasing energies, deviations from a point-like behavior (see, for instance, Figure 4 of Ref.,[14] Ref.[32,33] and references therein) were observed at large angles where the screening factor [Eq. (23)] is ≈ 1.

The ratio between the actual measured and that expected from the point-like differential cross section (e.g., the MDCS) expresses the square of the *nuclear form factor* ($|F|$) which, in turn, depends on the momentum transfer q, i.e., that acquired by the target initially at rest:

$$q = \frac{\sqrt{T(T+2Mc^2)}}{c}, \tag{26}$$

with T from Eq. (7) or, for Mc^2 larger or much larger than the electron energy, from its approximate expression Eq. (9) [e.g., see Equations (31, 57, 58) of Ref.[33], Section 3.1.2 of Ref.[3], Refs.[14,28,32,34]].

The factorized differential cross section for elastic interactions of electrons with screened Coulomb fields of nuclei [Eq. (24)] accounting for the effects due to the finite nuclear size is given by:

$$\begin{aligned} \frac{d\sigma^{\rm Mott}_{{\rm sc},F}(\theta)}{d\Omega} &= \frac{d\sigma^{\rm Mott}_{\rm sc}(\theta)}{d\Omega}\,|F(q)|^2 \\ &\simeq \frac{d\sigma^{\rm Rut}}{d\Omega}\,\mathfrak{F}^2(\theta)\,\mathcal{R}^{\rm Mott}\,|F(q)|^2 \end{aligned} \tag{27}$$

[e.g., see Equation (18) of Ref.[14], Ref.[28] and also references therein]. Thus, using the analytical expression derived by McKinley and Feshbach[17] [Eq. (4)] one obtains the corresponding screened differential cross section [Eq. (25)] accounting for the finite nuclear size effects, i.e.,

$$\begin{aligned} \frac{d\sigma^{\rm McF}_{{\rm sc},F}(\theta)}{d\Omega} &= \frac{d\sigma^{\rm McF}_{\rm sc}(\theta)}{d\Omega}\,|F(q)|^2 \\ &\simeq \frac{d\sigma^{\rm Rut}}{d\Omega}\,\mathfrak{F}^2(\theta)\,\mathcal{R}^{\rm McF}\,|F(q)|^2 \end{aligned} \tag{28}$$

$$\begin{aligned} &= \frac{d\sigma^{\rm Rut}}{d\Omega}\,\mathfrak{F}^2(\theta)\,|F(q)|^2 \\ &\quad \times \left\{1-\beta^2\sin^2(\theta/2) + Z\,\alpha\beta\pi\sin(\theta/2)\,[1-\sin(\theta/2)]\right\}. \end{aligned} \tag{29}$$

In terms of kinetic energy, one can respectively rewrite Eqs. (27, 28) as

$$\frac{d\sigma^{\rm Mott}_{{\rm sc},F}(T)}{dT} = \frac{d\sigma^{\rm Rut}}{dT}\,\mathfrak{F}^2(T)\,\mathcal{R}^{\rm Mott}(T)\,|F(q)|^2 \tag{30}$$

$$\frac{d\sigma^{\rm McF}_{{\rm sc},F}(T)}{dT} \simeq \frac{d\sigma^{\rm Rut}(T)}{dT}\,\mathfrak{F}^2(T)\,\mathcal{R}^{\rm McF}(T)\,|F(q)|^2 \tag{31}$$

with $d\sigma^{\rm Rut}/dT$ from Eq. (12), $\mathcal{R}^{\rm Mott}(T)$ from Eq. (18), $\mathcal{R}^{\rm McF}(T)$ from Eq. (14) and, using Eqs. (7, 9, 23),

$$\mathfrak{F}(T) = \frac{T}{T_{\rm max}A_{\rm s,M} + T}.$$

The nuclear form factor accounts for the spatial distribution of charge density probed in the electron–nucleus scattering [e.g., see Equation (58) of Ref.[33], Section 3.1.2 of Ref.[3], Refs.[14,28,32,34] and references therein]. For instance, among those spherically symmetric treated in literature, one finds that for i) an *exponential charge distribution* ($F_{\rm exp}$) [e.g., see Equation (6) of Ref.[28], Equation (93) at page 252 of Ref.[33] and references therein], ii) a *Gaussian charge distribution* ($F_{\rm gau}$) [e.g., see Equation (6) of Ref.[28] and references therein] and iii) an *uniform–uniform folded charge distribution* over spheres with different radii ($F_{\rm u}$) [e.g., see Equation (22) of Ref.[14], Ref.[32] and references therein]. For instance, the form factor $F_{\rm exp}$ is

$$F_{\rm exp}(q) = \left[1 + \frac{1}{12}\left(\frac{q r_{\rm n}}{\hbar}\right)^2\right]^{-2}, \tag{32}$$

where $r_{\rm n}$ is the nuclear radius [e.g., see Equation (6) of Ref.[28]]. To a first approximation, $r_{\rm n}$ can be parameterized by

$$r_{\rm n} = 1.27A^{0.27}\ {\rm fm} \tag{33}$$

with A the atomic weight [e.g., see Equation (7) of Ref.[28]]. Equation (33) provides values of $r_{\rm n}$ in agreement up to heavy nuclei (like Pb and U) with those available, for instance, in Table 1 of Ref.[34]. The nuclear form factor is 1 for $q = 0$ and rapidly decreases with increasing q [e.g., see Eq. (32), Equation (6) of Ref.[28] and Equation (22) of Ref.[14] for $F_{\rm exp}$, $F_{\rm gau}$ and $F_{\rm u}$, respectively]. Furthermore, from inspection of Eqs. (7, 9, 26) small q are those corresponding to scattering angles within the forward (with respect to the electron direction) angular region which, in turn, narrows with increasing electron energy. For instance, in lithium the square values ($|F(q)|^2$) of these form factors are in agreement within 1% up to $\theta' \lesssim 124.1^\circ$ (2.4°) at 20 MeV (1 GeV); in silicon up to $\theta' \lesssim 138.4^\circ$ (2.4°) at 20 MeV (1 GeV); in iron up to $\theta' \lesssim 108.0^\circ$ (2.1°) at 20 MeV (1 GeV). However, as

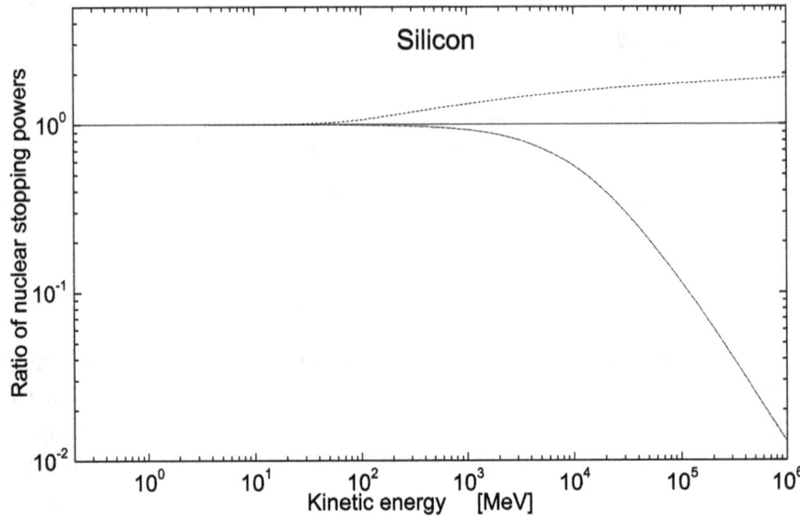

Fig. 2. As a function of the kinetic energy of electrons from 200 keV up to 1 TeV, ratios of nuclear stopping power of electrons in silicon calculated neglecting i) nuclear size effects (i.e., for $|F_{\rm exp}|^2 = 1$) (dashed curve) and ii) effects due to the finite rest mass of the target nucleus (dashed and dotted curve) [i.e., in Eq. (53) replacing $d\sigma^{\rm McF}_{{\rm sc},F,{\rm CoM}}(T)/dT$ with $d\sigma^{\rm McF}_{{\rm sc},F}(T)/dT$ from Eq. (31)] both divided by that one obtained using Eq. (53).

discussed in Sect. 2.4, these upper angles are larger or much larger with respect to those required to obtain 99% of the total cross section. Thus, the usage of any of the above mentioned nuclear form factors - e.g., $F_{\rm exp}$ as in the present treatment - is expected to be appropriate in the treatment of the transport of electrons in matter, when single scattering mechanisms are relevant, for instance in dealing with the nuclear stopping power and non-ionization energy-loss deposition.

2.4. *Finite Rest Mass of Target Nucleus*

The DCS treated in Sects. 2–2.3 is based on the extension of the MDCS to include effects due to interactions on screened Coulomb potentials of nuclei and their finite size. However, in the treatment, the electron energies were assumed to be small (or much smaller) with respect to that (Mc^2) corresponding to the rest mass (M) of target nuclei.

The Rutherford scattering on screened Coulomb fields - i.e., under the action of a central force - by massive charged particles at energies larger or much larger than Mc^2 was treated by Boschini et al.[5,6] in the CoM system (e.g., see also Sections 1.6, 1.6.1, 2.1.4.2 of Ref.[3] and references therein). It was shown that the differential cross section [$d\sigma^{\rm WM}(\theta')/d\Omega'$ with θ' the scattering angle in the CoM system] is that one derived for describing the interaction on a fixed scattering center of a particle with i) momentum $p'_{\rm r}$ equal to the momentum of the incoming particle (i.e., the electron in the present treatment) in the CoM system and ii) rest mass equal to the *relativistic reduced mass* $\mu_{\rm rel}$ [e.g., see Equations (1.80, 1.81) at page 28 of Ref.[3]]. $\mu_{\rm rel}$ is given by

$$\mu_{\rm rel} = \frac{mM}{M_{1,2}} \tag{34}$$

$$= \frac{mMc}{\sqrt{m^2c^2 + M^2c^2 + 2\,M\sqrt{m^2c^4 + p^2c^2}}}, \tag{35}$$

where p is the momentum of the incoming particle (the electron in the present treatment) in the laboratory system: m is the rest mass of the incoming particle (i.e., the electron rest mass); finally, $M_{1,2}$ is the invariant mass - e.g., Section 1.3.2 of Ref.[3] - of the two-particle system. Thus, the velocity of the interacting particle is

$$\beta'_{\rm r}c = c\sqrt{\left[1 + \left(\frac{\mu_{\rm rel}c}{p'_{\rm r}}\right)^2\right]^{-1}} \tag{36}$$

[e.g., see Equation (1.82) at page 29 of Ref.[3]]. For an incoming particle with $z = 1$, $d\sigma^{\rm WM}(\theta')/d\Omega'$ is given by

$$\frac{d\sigma^{\rm WM'}(\theta')}{d\Omega'} = \left(\frac{Ze^2}{2\,p'_{\rm r}\,\beta'_{\rm r}c}\right)^2 \frac{1}{\left[A_{\rm s} + \sin^2(\theta'/2)\right]^2}, \tag{37}$$

with

$$A_{\rm s} = \left(\frac{\hbar}{2\,p'_{\rm r}\,a_{\rm TF}}\right)^2 \left[1.13 + 3.76 \times \left(\frac{\alpha Z}{\beta'_{\rm r}}\right)^2\right] \tag{38}$$

the screening factor [e.g., see Equations (2.87, 2.88) at page 103 of Ref.[3]]. Equation (37) can be rewritten as

$$\frac{d\sigma^{\rm WM'}(\theta')}{d\Omega'} = \frac{d\sigma^{\rm Rut'}(\theta')}{d\Omega'}\,\mathfrak{F}^2_{\rm CoM}(\theta') \tag{39}$$

with

$$\frac{d\sigma^{\rm Rut'}(\theta')}{d\Omega'} = \left(\frac{Ze^2}{2p'_{\rm r}\beta'_{\rm r}c}\right)^2 \frac{1}{\sin^4(\theta'/2)} \tag{40}$$

the corresponding RDCS for the reaction in the CoM system [e.g., see Equation (1.79) at page 28 of Ref.[3]] and

$$\mathfrak{F}_{\mathrm{CoM}}(\theta') = \frac{\sin^2(\theta'/2)}{A_{\mathrm{s}} + \sin^2(\theta'/2)} \tag{41}$$

the screening factor. Using, Eqs. (7, 8), one can respectively rewrite Eqs. (40, 41, 39, 37) as

$$\frac{d\sigma^{\mathrm{Rut}'}}{dT} = \pi \left(\frac{Ze^2}{p'_{\mathrm{r}}\beta'_{\mathrm{r}}c}\right)^2 \frac{T_{\mathrm{max}}}{T^2} \tag{42}$$

$$\mathfrak{F}_{\mathrm{CoM}}(T) = \frac{T}{T_{\mathrm{max}}A_{\mathrm{s}} + T} \tag{43}$$

$$\frac{d\sigma^{\mathrm{WM}'}(T)}{dT} = \frac{d\sigma^{\mathrm{Rut}'}}{dT}\mathfrak{F}_{\mathrm{CoM}}(T) \tag{44}$$

$$\frac{d\sigma^{\mathrm{WM}'}(T)}{dT} = \pi \left(\frac{Ze^2}{p'_{\mathrm{r}}\beta'_{\mathrm{r}}c}\right)^2 \frac{T_{\mathrm{max}}}{(T_{\mathrm{max}}A_{\mathrm{s}} + T)^2} \tag{45}$$

[e.g., see Equation (2.90) at page 103 of Ref.[3] or Equation (13) of Ref.[6]].

As already mentioned (Sect. 2.2), the screening parameter A_{s} prevents the DCS to diverge - see last term in Eq. (37) -, i.e., for θ' of the order of or smaller than

$$\theta'_{\mathrm{sc}} = \arcsin\left(2\sqrt{A_{\mathrm{s}}}\right)$$

effects due to screening of the nuclear Coulomb field have to be accounted for. θ'_{sc} rapidly decreases with increasing the kinetic energies of electrons. For instance, in iron θ'_{sc} is $\approx 1.7°$ (0.03 rad) at 200 keV and $\approx 0.004°$ (7.0×10^{-2} mrad) at 200 MeV; in silicon, it is $\approx 1.3°$ (0.022 rad) at 200 keV and $\approx 0.003°$ (5.5×10^{-2} mrad) at 200 MeV; while, in lithium, it is $\approx 0.75°$ (13 mrad) at 200 keV and $\approx 0.002°$ (3.3×10^{-2} mrad) at 200 MeV. Therefore, in Eq. (39) the term A_{s} (i.e., the screening parameter [Eq. (38)]) cannot be neglected for scattering angles within a forward angular region narrowing with increasing energies from a few degrees (for low-Z material) at about 200 keV down to less than or much less than a mrad above 200 MeV. It is worthwhile to remark that in silicon, for instance, θ' can be approximated with θ up to a few hundred MeV.

To account for the finite rest mass of target nuclei, the factorized MDCS

[Eq. (27)] has to be re-expressed in the CoM system as:

$$\begin{aligned}\frac{d\sigma^{\rm Mott}_{{\rm sc},F,{\rm CoM}}(\theta')}{d\Omega'} &= \frac{d\sigma^{\rm Mott}_{\rm sc}(\theta',\beta'_{\rm r},p'_{\rm r})}{d\Omega'}\,|F(q)|^2 \\ &\simeq \frac{d\sigma^{\rm WM'}(\theta')}{d\Omega'}\,\mathcal{R}^{\rm Mott}_{\rm CoM}(\theta')\,|F(q)|^2 \\ &\simeq \frac{d\sigma^{\rm Rut'}(\theta')}{d\Omega'}\,\mathfrak{F}^2_{\rm CoM}(\theta')\,\mathcal{R}^{\rm Mott}_{\rm CoM}(\theta')\,|F(q)|^2, \end{aligned} \tag{46}$$

where $F(q)$ is the nuclear form factor (Sect. 2.3) with q the momentum transfer to the recoil nucleus [Eq. (26)]; finally, as discussed in Sect. 2.1, $\mathcal{R}^{\rm Mott}$ exhibits almost no dependence on electron energy above $\approx 10\,{\rm MeV}$, thus, since at low energies $\theta \backsimeq \theta'$ and $\beta \backsimeq \beta'_{\rm r}$, $\mathcal{R}^{\rm Mott}_{\rm CoM}(\theta')$ is obtained replacing θ and $\beta'_{\rm r}$ with θ' and $\beta'_{\rm r}$, respectively, in Eq. (16).

Using the analytical expression derived by McKinley and Feshbach[17], one finds that the corresponding screened differential cross section accounting for the finite nuclear size effects [Eqs. (28, 29)] can be re-expressed as

$$\frac{d\sigma^{\rm McF}_{{\rm sc},F,{\rm CoM}}(\theta')}{d\Omega'} \simeq \frac{d\sigma^{\rm Rut'}(\theta')}{d\Omega'}\,\mathfrak{F}^2_{\rm CoM}(\theta')\,\mathcal{R}^{\rm McF}_{\rm CoM}(\theta')\,|F(q)|^2 \tag{47}$$

with

$$\mathcal{R}^{\rm McF}_{\rm CoM}(\theta') = \left\{1-\beta^2_{\rm r}\sin^2(\theta'/2)+Z\,\alpha\beta'_{\rm r}\pi\sin(\theta'/2)\left[1-\sin(\theta'/2)\right]\right\}. \tag{48}$$

It has to be remarked that scattered electrons are mostly found in the forward or very forward direction. For instance, using Eq. (48) one can derive that in lithium $\approx 99\%$ of electrons are scattered with $\theta' \lesssim 0.27°$ ($0.007°$) at 20 MeV (1 GeV); in silicon with $\theta' \lesssim 0.46°$ ($0.009°$) at 20 MeV (1 GeV); in iron with $\theta' \lesssim 0.6°$ ($0.013°$) at 20 MeV (1 GeV).

In terms of kinetic energy T, from Eqs. (7, 8) one can respectively rewrite Eqs. (46, 47) as

$$\frac{d\sigma^{\rm Mott}_{{\rm sc},F,{\rm CoM}}(T)}{dT} = \frac{d\sigma^{\rm Rut'}}{dT}\,\mathfrak{F}^2_{\rm CoM}(T)\,\mathcal{R}^{\rm Mott}_{\rm CoM}(T)\,|F(q)|^2 \tag{49}$$

$$\frac{d\sigma^{\rm McF}_{{\rm sc},F,{\rm CoM}}(T)}{dT} \simeq \frac{d\sigma^{\rm Rut'}(T)}{dT}\,\mathfrak{F}^2_{\rm CoM}(T)\,\mathcal{R}^{\rm McF}_{\rm CoM}(T)\,|F(q)|^2 \tag{50}$$

with $d\sigma^{\rm Rut'}/dT$ from Eq. (42), $\mathfrak{F}_{\rm CoM}(T)$ from Eq. (43) and $\mathcal{R}^{\rm McF}_{\rm CoM}(T)$ replacing β with $\beta'_{\rm r}$ in Eq. (14), i.e.,

$$\mathcal{R}^{\rm McF}_{\rm CoM}(T) = \left[1-\beta'_{\rm r}\frac{T}{T_{\rm max}}(\beta'_{\rm r}+Z\alpha\pi)+Z\alpha\beta'_{\rm r}\pi\sqrt{\frac{T}{T_{\rm max}}}\right]. \tag{51}$$

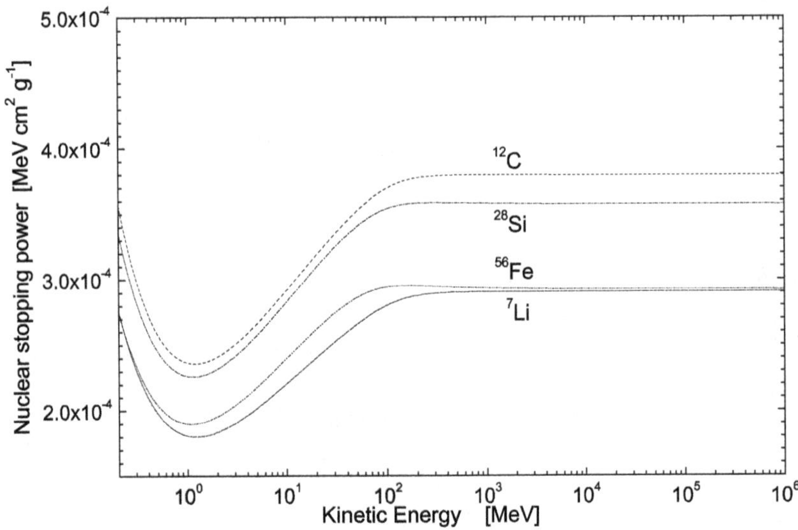

Fig. 3. Nuclear stopping powers (in $\mathrm{MeV\,cm^2/g}$) in ^{7}Li, ^{12}C, ^{28}Si and ^{56}Fe - calculated from Eq. (53) - and divided by the density of the material as a function of the kinetic energy of electrons from 200 keV up to 1 TeV.

Finally, as discussed in Sect. 2.1, $\mathcal{R}^{\mathrm{Mott}}(T)$ exhibits almost no dependence on electron energy above $\approx 10\,\mathrm{MeV}$, thus, since at low energies $\theta \simeq \theta'$ and $\beta \simeq \beta'_{\mathrm{r}}$, $\mathcal{R}^{\mathrm{Mott}}_{\mathrm{CoM}}(T)$ is obtained replacing β with β'_{r} in Eq. (18).

3. Nuclear Stopping Power of Electrons

Using Eq. (49), the nuclear stopping power - in $\mathrm{MeV\,cm^{-1}}$ - of Coulomb electron–nucleus interaction can be obtained as

$$-\left(\frac{dE}{dx}\right)^{\mathrm{Mott}}_{\mathrm{nucl}} = n_A \int_0^{T_{\max}} \frac{d\sigma^{\mathrm{Mott}}_{\mathrm{sc},F,\mathrm{CoM}}(T)}{dT}\, T\, dT \tag{52}$$

with n_A the number of nuclei (atoms) per unit of volume [e.g., see Equation (1.71) of Ref.[3]] and, finally, the negative sign indicates that energy is lost by electrons (thus, achieved by recoil targets). Using the analytical approximation derived by McKinley and Feshbach,[17] i.e., Eq. (50), for the nuclear stopping power one finds

$$-\left(\frac{dE}{dx}\right)^{\mathrm{McF}}_{\mathrm{nucl}} = n_A \int_0^{T_{\max}} \frac{d\sigma^{\mathrm{McF}}_{\mathrm{sc},F,\mathrm{CoM}}(T)}{dT}\, T\, dT. \tag{53}$$

As already mentioned in Sect. 2.4, the large momentum transfers - corresponding to large scattering angles - are disfavored by effects due to

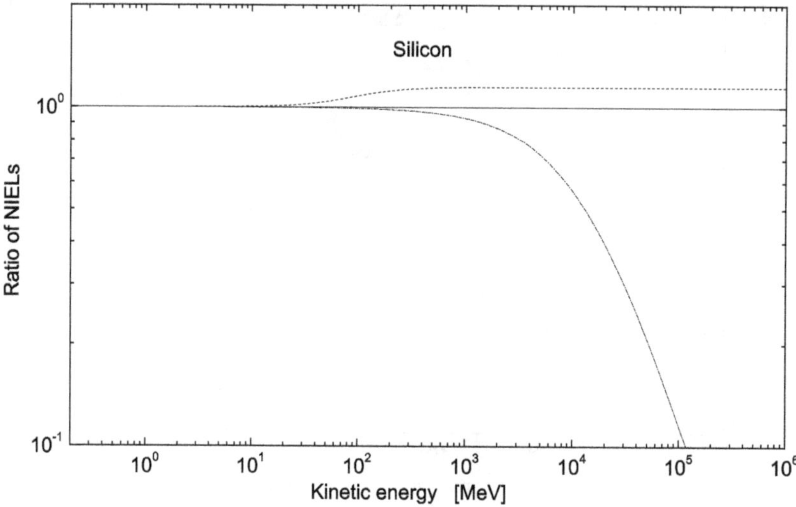

Fig. 4. For $T_d = 21\,\mathrm{eV}$, ratios of NIELs of electrons in silicon calculated as a function of the kinetic energy of electrons from 220 keV up to 1 TeV neglecting i) nuclear size effects (i.e., for $|F_{\mathrm{exp}}|^2 = 1$) (dashed curve) and ii) effects due to the finite rest mass of the target nucleus (dashed and dotted curve) [i.e., in Eq. (55) replacing $d\sigma^{\mathrm{McF}}_{\mathrm{sc},F,\mathrm{CoM}}(T)/dT$ with $d\sigma^{\mathrm{McF}}_{\mathrm{sc},F}(T)/dT$ from Eq. (31)] both divided by that one obtained using Eq. (55).

the finite nuclear size accounted for by means of the nuclear form factor (Sect.2.3). For instance, in Fig. 2 the ratios of nuclear stopping powers of electrons in silicon are shown as a function of the kinetic energies of electrons from 200 keV up to 1 TeV. These ratios are the nuclear stopping powers calculated neglecting i) nuclear size effects (i.e., for $|F_{\mathrm{exp}}|^2 = 1$) and ii) effects due to the finite rest mass of the target nucleus [i.e., in Eq. (53) replacing $d\sigma^{\mathrm{McF}}_{\mathrm{sc},F,\mathrm{CoM}}(T)/dT$ with $d\sigma^{\mathrm{McF}}_{\mathrm{sc},F}(T)/dT$ from Eq. (31)] both divided by that one obtained using Eq. (53). Above a few tens of MeV, a larger stopping power is found assuming $|F_{\mathrm{exp}}|^2 = 1$ and, in addition, above a few hundreds of MeV the stopping power largely decreases when effects due to the finite nuclear rest mass are not accounted for.

In Fig. 3 , the nuclear stopping powers in ^{7}Li, ^{12}C, ^{28}Si and ^{56}Fe are shown as a function of the kinetic energy of electrons from 200 keV up to 1 TeV. These nuclear stopping powers in MeV cm^2/g are calculated from Eq. (53) and divided by the density of the medium. The flattening of the high energy behavior of the curves is mostly due to the nuclear form factor which prevents the stopping power to increase with increasing T_{max}. As expected, the stopping power are slightly (not exceeding a few percent)

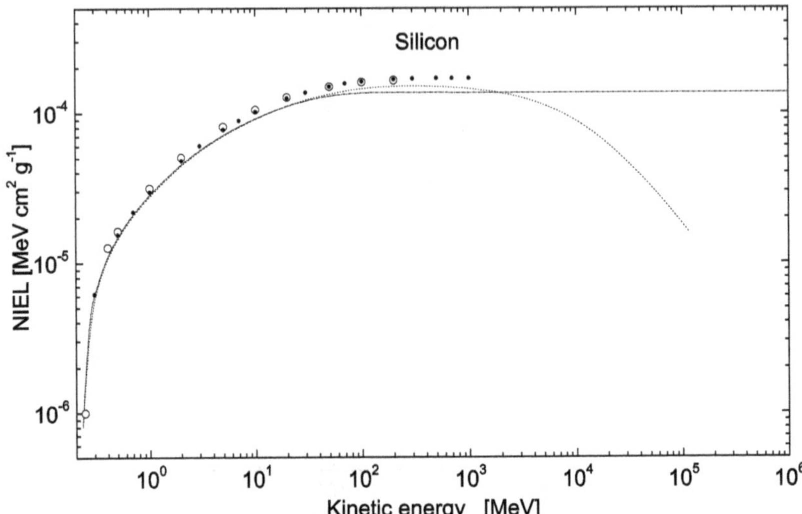

Fig. 5. For $T_d = 21\,\mathrm{eV}$, NIEL (in $\mathrm{MeV\,cm^2\,g^{-1}}$) in silicon calculated using Eq. (55) as a function of the kinetic energy from 220 keV up to 1 TeV (dashed and dotted curve); NIEL values from Messenger et al.[42] (∘) and Jun et al.[43] (•) calculated in the laboratory system without accounting for the effects due to the screened Coulomb potential, finite size and rest mass of recoil silicon; the dotted curve is obtained replacing $d\sigma^{\mathrm{McF}}_{\mathrm{sc},F,\mathrm{CoM}}(T)/dT$ with $d\sigma^{\mathrm{McF}}(T)/dT$ [Eq. (13)] in Eq. (55).

varied at large energies replacing F_{exp} with F_{gau} or F_{u} (Sect. 2.3). However, a further study is needed to determine a most suited parametrization of the nuclear form factor[35–37] particularly for high-Z materials; for instance, in lead the stopping power results to be depressed at energies of about (20-40) MeV, while in medium and light nuclei this occurs at energies of the order or above 100 MeV.

4. Non-Ionizing Energy-Loss of Electrons

A relevant process - which causes permanent damage to the silicon bulk structure - is the so-called *displacement damage* (e.g., see Chapter 4 of Ref.,[3] Refs.[4,6,38] and references therein). Displacement damage may be inflicted when a *primary knocked-on atom* (PKA) is generated. The interstitial atom and relative vacancy are termed Frenkel-pair (FP). In turn, the displaced atom may have sufficient energy to migrate inside the lattice and - by further collisions - can displace other atoms as in a collision cascade. This displacement process modifies the bulk characteristics of the device and causes its degradation. The total number of FPs can be estimated calculating the energy density deposited from displacement processes. In turn, this energy density is related to the *non-ionizing energy loss* (NIEL), i.e., the energy per unit path lost by the incident particle due to displacement processes.

In case of Coulomb scattering of electrons on nuclei, the non-ionizing energy-loss can be calculated using (as discussed in Sect. 2–3) the MDCRS or its approximate expression McFDCS [e.g., Eqs. (49, 50), respectively], once the screened Coulomb fields, finite sizes and rest masses of nuclei are accounted for, i.e., in MeV/cm

$$-\left(\frac{dE}{dx}\right)^{\mathrm{NIEL}}_{\mathrm{n,Mott}} = n_A \int_{T_d}^{T_{\max}} T\,L(T)\,\frac{d\sigma^{\mathrm{Mott}}_{\mathrm{sc},F,\mathrm{CoM}}(T)}{dT}\,dT \tag{54}$$

or

$$-\left(\frac{dE}{dx}\right)^{\mathrm{NIEL}}_{\mathrm{n,McF}} = n_A \int_{T_d}^{T_{\max}} T\,L(T)\,\frac{d\sigma^{\mathrm{McF}}_{\mathrm{sc},F,\mathrm{CoM}}(T)}{dT}\,dT \tag{55}$$

[e.g., see Equation (4.113) at page 402 and, in addition, Sections 4.2.1–4.2.1.2 of Ref.[3]], where T is the kinetic energy transferred to the target nucleus, $L(T)$ is the fraction of T deposited by means of displacement processes. The *Lindhard partition function*, $L(T)$, can be approximated using the so-called *Norgett–Robintson–Torrens expression* [e.g., see Refs.[39,40] and/or Equations (4.121, 4.123) at pages 404 and 405, respectively, of Ref.[3] (see also references therein)]. $T_{\mathrm{de}} = T\,L(T)$ is the so-called *damage energy*, i.e., the energy deposited by a recoil nucleus with kinetic energy T via displacement damages inside the medium. In Eqs. (54, 55) the integral is computed from the minimum energy T_d - the so-called *threshold energy for displacement*, i.e., that energy necessary to displace the atom from its lattice position - up to the maximum energy $T_{\max}$ that can be transferred during a single collision process. For instance, T_d is about 21 eV in silicon (e.g., see Table 1 in Ref.[41] and references therein) requiring electrons with kinetic energies above ≈ 220 keV [e.g., see Equation (4.142) at page 412 of Ref.[3]].

As already discussed with respect to nuclear stopping powers in Sect. 3, the large momentum transfers (corresponding to large scattering angles) are disfavored by effects due to the finite nuclear size accounted for by the nuclear form factor. For instance, in Fig. 4 the ratios of NIELs for electrons in silicon are shown as a function of the kinetic energy of electrons from 220 keV up to 1 TeV. These ratios are the NIELs calculated neglecting i) nuclear size effects (i.e., for $|F_{\mathrm{exp}}|^2 = 1$) and ii) effects due to the finite rest mass of the target nucleus [i.e., in Eq. (55) replacing $d\sigma^{\mathrm{McF}}_{\mathrm{sc},F,\mathrm{CoM}}(T)/dT$ with

$d\sigma^{\mathrm{McF}}_{\mathrm{sc},F}(T)/dT$ from Eq. (31)] both divided by that one (Fig. 5) obtained using Eq. (55). Above $\approx$ 10 MeV, the NIEL is $\approx$ 20% larger assuming $|F_{\mathrm{exp}}|^2 = 1$ and, in addition, above (100–200) MeV the calculated NIEL largely decreases when the effects of nuclear rest mass are not accounted for. Finally, it has to be remarked that similar results can be obtained neglecting the screening factor: already at energies lower that 200 keV, $T_d \approx$ 21 eV is much larger than $T_{\mathrm{max}}A_{\mathrm{s}}$.

In Fig. 5, for $T_d = 21$ eV the non-ionizing energy loss (in MeV cm^2 g^{-1}) calculated using Eq. (55) in silicon is shown (dashed and dotted curve) as a function of the kinetic energy from 220 keV up to 1 TeV and is compared with that one tabulated by Messenger et al.[42] (Jun et al.[43]) from $\approx$ 240 keV up to 200 MeV (1 GeV). For the laboratory system, Messenger et al.[42] used the approximate MDCS found by Doggett-Spencer[21] and Lindhard's partition function numerically obtained by Doran[45] without accounting for the effects due to screened Coulomb potential (i.e., $\mathfrak{F}^2 = 1$), finite size (i.e., $F^2 = 1$) and finite rest mass of the silicon target; while, Jun et al.[43] followed the approach discussed in Ref.[22] (see the treatment in Sect. 2.1) to determine an approximate expression of the MDCS and dealt the Lindhard partition function using the modified Norgett–Robintson–Torrens expression found[b] by Akkerman and Barak (2006). The dotted curve is obtained replacing $d\sigma^{\mathrm{McF}}_{\mathrm{sc},F,\mathrm{CoM}}(T)/dT$ with $d\sigma^{\mathrm{McF}}(T)/dT$ [Eq. (13)] in Eq. (55): at 100 MeV–1 GeV, the agreement between the latter calculation and values from Messenger et al.[42] and Jun et al.[43] is within several percents. It has to be remarked (see also Fig. 4) that i) above (100–200) MeV effects due to screened Coulomb potentials, finite sizes and finite rest masses of nuclei have to be taken into account and ii) for energies between $\approx$ 100 MeV and $\approx$ 1 GeV the effects of neglecting the nuclear form factor and finite rest mass of nuclei almost compensate each other.

5. Conclusions

The treatment of electron–nucleus interactions accounting for effects due to screened Coulomb potentials, finite sizes and finite rest masses of nuclei allows one to determine both the total and differential cross sections, thus, to calculate the resulting nuclear and non-ionizing stopping powers from low (about 200 keV) up to very high energy (1 TeV).

[b] Jun et al. (2009) determined that the usage of the Norgett–Robintson–Torrens expression or, alternatively, the one modified by Akkerman and Barak (2006) yields similar NIEL values.

Above a few hundreds of MeV, neglecting the effects of finite rest masses of recoil nuclei, the stopping power and NIEL result to be largely underestimated. Above a few tens of MeV the finite size of the nuclear target prevents a further increase of both stopping power and NIEL, which approach almost constant values. The flattening of the high energy behavior of the nuclear and non-ionizing energy-losses is mostly due to the nuclear form factor which prevents stopping powers to increase with increasing $T_{\max}$. However, a further study is needed to determine a most suited parametrization of the nuclear form factor able to provide a satisfactory trend in the energy region below about hundred MeV also for high-Z materials.

Finally, at 100 MeV–1 GeV an agreement to within several percents was obtained between the present calculation with respect to the NIEL values from Messenger et al.[42] and Jun et al.[43] .

References

1. P. Meyer and R. Vogt, *Phys. Rev. Lett.* 8 (1962), 387–389.
2. M.J. Owens, T.S. Horbury and C.N. Arge, *Astrophys. J.* 714 (2010), 1617, doi: 10.1088/0004-637X/714/2/1617.
3. C. Leroy and P.G. Rancoita, *Principles of Radiation Interaction in Matter and Detection*, 3rd Edition, World Scientific (Singapore) 2011.
4. C. Leroy and P.G. Rancoita, Particle Interaction and Displacement Damage in Silicon Devices operated in Radiation Environments, *Rep. Prog. in Phys.* 70 (no. 4)(2007), 403–625, doi: 10.1088/0034-4885/70/4/R01.
5. M. Boschini et al., Geant4-based application development for NIEL calculation in the Space Radiation Environment, Proc. of the 11th ICATPP, October 5–9 2009, Villa Olmo, Como, Italy, C. Leroy, P.G. Rancoita, M. Barone, E. Gaddi, L. Price and R. Ruchti Editors, World Scientific, Singapore(2010), 698–708, IBSN: 10-981-4307-51-3.
6. M. Boschini et al., Nuclear and Non-Ionizing Energy-Loss for Coulomb Scattered Particles from Low Energy up to Relativistic Regime in Space Radiation Environment, Proc. of the 12th ICATPP, October 7–8 2010, Villa Olmo, Como, Italy, S. Giani, C. Leroy and P.G. Rancoita, Editors, World Scientific, Singapore (2011), 9–23, IBSN: 978-981-4329-02-6.
7. S. Agostinelli et al., Geant4 a simulation toolkit, *Nucl. Instr. and Meth. in Phys. Res. A* 506 (2003), 250–303.
   See also the web site: *http://geant4.web.cern.ch/geant4/*
8. N.F. Mott, *Proc. Roy. Soc. A* 124 (1929), 425–442; *A* 135 (1932), 429–458.
9. N.F. Mott and H.S.W. Massey, The Theory of Atomic Collisions - 3rd Edition- (1965), Oxford University Press, London.
10. G. Wentzel, *Z. Phys.* 40 (1927), 590–593.
11. M. Born, *Z. Phys.* 38 (1926), 803.
12. R. Idoeta and F. Legarda, *Nucl. Instr. and Meth. in Phys. Res. B* 71 (1992), 116–125.

13. M.J. Berger, Monte Carlo Calculation of the Penetretion and Diffusion of Fast Charged Particles, in Methods in Computational Physics **vol. 1** (1963), B. Alder, S. Fernbach and M. Rotenberg Editors, Acdemic Press, New York, 135–215.
14. J.M. Fernandez-Vera, R. Mayol and F. Salvat, *Nucl. Instr. and Meth. in Phys. Res. B* 82, (1993) 39–45.
15. J.H. Bartlett and R.E. Watson, *Proc. Am. Acad. Arts Sci.* 74 (1940), 53.
16. N. Sherman, *Phys. Rev.* 103 (1956), 1601–1607.
17. A.Jr. McKinley and H. Feshbach, *Phys. Rev.* 74 (1948), 1759–1763.
18. R.M. Curr, *Proc. Phys. Soc.* (London) A68 (1955), 156–164.
19. J.H. Cahn, *J. of Appl. Phys.* 30 (1959), 1310–1316.
20. H. Feshbach, *Phys. Rev.* 88 (1952), 295–297.
21. J.A. Doggett and L.V. Spencer, *Phys. Rev.* 103 (1956), 1597-1601.
22. T. Lijian, H. Quing and L. Zhengming, *Radiat. Phys. Chem.* 45 (1995), 235–245.
23. O.S. Oen, *Nucl. Instr. and Meth. in Phys. Res. B* 33, (1988) 744–747.
24. F. Seitz and J.S. Koehler, *Solid State Physics* **vol. 2**, edited by F. Seitz and D. Turnbull, Academic Press Inc., New York (1956).
25. J.M. Fernandez-Vera et al., *Nucl. Instr. and Meth. in Phys. Res. B* 73 (1993), 447–473.
26. von G. Molière, *Z. Naturforsh.* A2 (1947), 133–145; A3 (1948), 78–97.
27. H.A. Bethe, *Phys. Rev.* 89 (1953), 1256–1266.
28. A.V. Butkevick, *Nucl. Instr. and Meth. in Phys. Res. A* 488 (2002), 282–294.
29. L.H. Thomas, *Proc. Cambridge Phil. Soc.* 23 (1927) , 542.
30. E. Fermi, *Z. Phys.* 48 (1928), 73–79.
31. E. Zeitler and A. Olsen, *Phys. Rev.* 136 (1956), A1546-A1552.
32. R.H. Helm, *Phys. Rev.* 104 (1956), 1466-1475.
33. R. Hofstadter, *Ann. Rev. Nucl. Sci.* 7 (1957), 231.
34. H. De Vries, C.W. De Jager, and C. De Vries, *Atomic Data and Nuclear Data Tables* **36** (1987), 495.
35. M.A. Nagarajan and L. Wang, *Phys. Rev. C* 10 (1974), 2206-2209.
36. G. Duda, A. Kemper and P. Gondolo, *J. Cosm. Astrop. Phys.* 04 (2007), 012, doi:10.1088/1475-7516/2007/04/012
37. U.D. Jentschura and V.G. Serbo, *E. Phys. J. C* 64 (2009), 309–317.
38. C. Consolandi et al., *Nucl. Instr. and Meth. in Phys. Res. B* 252 (2006), 276.
39. I. Jun, *IEEE Trans. on Nucl. Sci.* 48 (2001), 162–175
40. S.R. Messenger et al., *IEEE Trans. on Nucl. Sci.* 50 (2003), 1919–1923.
41. I. Jun, M.A Xapsos, S.R. Messenger, E.A. Burke, R.J. Walters, G.P. Summers and T. Jordan, *IEEE Trans. on Nucl. Sci.* 50 (2003), 1924–1928.
42. S.R. Messenger et al., *IEEE Trans. on Nucl. Sci.* vol. 46, no. 6 (1999), 1595-1601.
43. I. Jun, W. Kim and R. Evans, *IEEE Trans. on Nucl. Sci.* vol. 56 (2009), 3229–3235.
44. A. Akkerman and J. Barak, *IEEE Trans. on Nucl. Sci.* vol. 53 (2006), 3667–3674.
45. D.G. Doran, *Nucl. Sci. Eng.* 49 (1972), 130–144.

3D SIMULATION FOR MAXIMIZING ELECTRON TRANSFER EFFICIENCY IN THICK GEMS

BAISHALI GARAI[1], K RAJANNA[2]

Department of Instrumentation and Applied Physics, Indian Institute of Science
Bangalore-560012, India
[1]*baishali@isu.iisc.ernet.in*, [2]*kraj@isu.iisc.ernet.in*

V. RADHAKRISHNA

Space Astronomy Group, ISRO Satellite centre.
Bangalore-560017, India
rkrish@isac.gov.in

Thick GEM for UV detector applications must provide high detection efficiency for a single photoelectron produced by UV light. Electron Transfer Efficiency (ETE) of GEM detector determines the detection efficiency. We have used GARFIELD simulation for estimation of ETE at various operating parameters, which are to be optimized for high detection efficiency.

1. Introduction

Thick GEM (THGEM) detectors [1, 2] with longer avalanche length offer much higher gain. THGEMs are used for applications such as in UV detectors and gas PMTs, which require high gain without discharge. In these detectors, semitransparent or reflective CsI photocathode coupled to single or double THGEMs are adopted. When UV light falls on the photocathode, photoelectrons are produced, which when passes through the THGEM hole, undergo avalanche multiplication due to the high electric field inside the hole to produce detectable output. Presently, these detectors are used for UV-photon imaging in RICH detectors [3]. GEM based RICH detectors have become routine tools for relativistic-particle identification in particle and heavy-ion experiments. The high counting rate and large multiplicities in some experiments impose harsh requirements on the design and operation of RICH system and on the photon detector in particular. They are required to have fast response and excellent detection efficiency for single photons, and at the same time minimum sensitivity to ionizing background radiations [1]. One of the most important performance parameters of the photon detector in RICH system or gas PMT is the detection efficiency. This depends on the quantum efficiency of the photocathode and the single electron detection efficiency of GEM structures. Single electron detection efficiency depends on the ability of the photoelectron, produced by incident (UV photons) to reach the GEM hole for multiplication. This is referred to as Electron Transfer Efficiency (ETE). ETE is defined as the ratio of the number of primary electrons reaching the amplification region to the total number of primary electrons produced by photon/particle interaction in the gas. The detection efficiency as a whole is given by

Detection efficiency (DE) = Quantum efficiency (QE)* ETE.

THGEM design parameters, such as hole diameter and pitch affect the ETE. In addition, operating parameters such as dipole field, drift field, gas mixture and its pressure also affect the ETE [1, 4]. In this paper, we present simulation work, which is useful in optimizing these parameters for maximizing ETE. THGEMs manufactured from PCB boards by mechanical drilling has a limitation on hole diameter and pitch. This limits the maximum achievable ETE as the optical transparency is limited [5]. So other means of maximizing ETE needs to be considered and optimized.

1.1. *Modeling and simulation*

3D GEM structures have been designed by ANSYS and the actual detector performance has been simulated using GARFIELD simulation code [6]. Monte Carlo technique has been used for ETE calculations. The ETE has been computed generating a matrix of large number of electrons uniformly distributed on the drift area and following their path as they drift down the channel [5]. In our studies, we have taken GEMs with thickness of 250 μm and hole diameter of 200 and 300 μm. GEM structures have hole edge rim sizes of 100 μm and a pitch of 1000 μm is maintained. Figure 1 shows the Garfield plot for the path of the electrons starting from the drift electrode. It is seen that not all the electrons starting from the drift electrode reach the GEM hole for multiplication. Many electrons end at different portions of the GEM structure including the top metal electrode and the sidewalls of the GEM hole. Factors affecting the number of electrons reaching the hole, such as drift and dipole field, gas mixture and gas pressure are discussed in the following section.

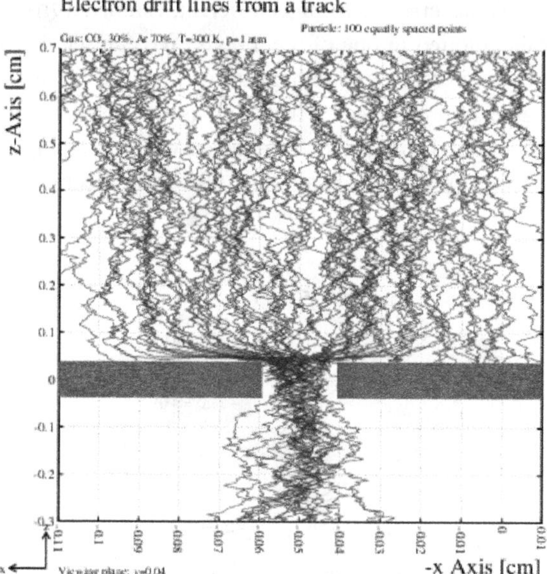

Fig. 1. 2D view of Garfield plot showing the end point of electrons at different portions of the GEM structure.

1.2. *Results and discussion*

The operating parameters that affect ETE are drift field, dipole field present inside the GEM hole, the gas mixture and pressure. Figure 2 shows the variation of ETE with drift field and dipole field. Earlier experimental studies too report on this variation [4]. ETE increases marginally with dipole field. Significant variation in ETE is observed for changing drift field. This is due to the fact that transport property of the electrons in the gas medium depends on the drift field. It is also reported that the quantum efficiency of the photocathode increases with increasing drift field [3]. Thus high quantum efficiency and high ETE need conflicting requirements for drift field.

Choice of gas mixture for THGEM based gas PMT and UV detector is usually governed by the photoelectron backscattering to the photocathode. P10, $Ne:CF_4$ (90:10), $Ar:CO_2$ (70:30), and $Ar:N_2$ (90:10), are often used in gas PMT and UV detectors [7]. Here, ETE is studied for these gas mixtures at different pressures. Drift field value of 0.3 kV/cm is maintained considering higher ETE (Fig. 2) and lower backscattering [3]. The ETE estimated using GARFIELD is plotted in Fig. 3. Transverse diffusion coefficient, estimated from MAGBOLTZ program [8] is also plotted in one of the Y-axis. We can see that ETE increases with increase in gas pressure and is highest for Ar:CO_2 mixture. The diffusion plotted in the same figure indicates that ETE is improved because of reduced diffusion. Even though, diffusion is also reduced with increasing drift field, ETE reduces as electrons are driven towards the GEM top electrode by higher drift field.

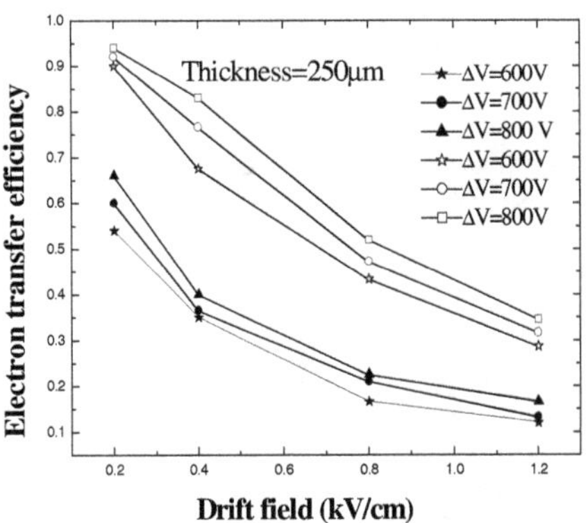

Fig. 2. Effect of drift field and dipole field on ETE (open symbols are for hole diameter of 300 μm and solid symbols for 200 μm.) Gas mixture used is Ar: CO_2 (70:30) at 1 atm. pressure. Pitch of the GEM is 1000 μm.

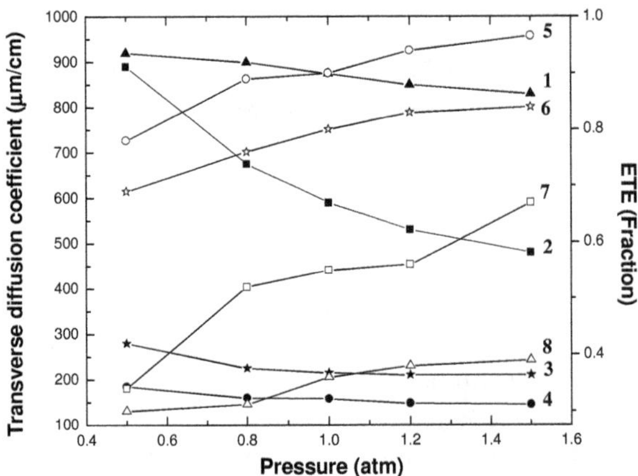

Fig. 3. Effect of gas mixture and pressure on ETE. (Open symbols show ETE while closed symbols show transverse diffusion coefficient variation for different gas mixtures. (1,8) Ar:N_2 (90:10); (2,7) P10; (3,6) Ne:CF_4 (90:10); (4,5) Ar:CO_2 (70:30) with pressure. The GEM model used here has a thickness of 250 μm and hole diameter of 300 μm. The drift field is maintained at 0.3 kV/cm.

2. Conclusion

There are several parameters which affect ETE and they are quite contradictory. GARFIELD simulation on the effect of these parameters on ETE indicates that the drift field, gas mixtures and pressure should be chosen suitably for best detection efficiency by optimizing quantum efficiency and ETE. GARFIELD is a useful tool to estimate ETE at any given operating parameter.

References

1. R. Chechik, A. Breskin, C. Shalem Nucl. Instr. and Meth. A **553** 35 ((2005).
2. R. Chechik, *et al.* Nucl. Instr. and Meth. A **535**, 303 (2004).
3. S. Dalla Torre, Nucl. Instr. and Meth. A **639**, 111 (2011).
4. C. Richter, *et al.* Nucl. Instr. and Meth. A **478**, 538 (2002).
5. A. Sharma, Nucl. Instr. and Meth. A **454**, 267 (2000).
6. GARFIELD-Version-9, by Rob Veenhof, CERN, Geneva, Switzerland.
7. A. F. Buzulutskov, Physics of Particles and nuclei, vol. **39**, No. **3**, 424 (2008).
8. MABOLTZ- by Steve Biagi, University of Liverpool, Great Britain.

DMMW: A tool for multi-wavelength Dark Matter searches

I. Gebauer*

Institut für Experimentelle Kernphysik, Karlsruhe Institute of Technology,
Karlsruhe, 76133, Germany
**E-mail: gebauer@kit.edu*
www.kit.edu

We present DMMW, a publicly available code, which computes the **D**ark **M**atter **M**ulti-**W**avelength emission spectrum for generic Dark Matter models. We briefly discuss a few applications to a variety of astrophysical systems within and beyond the Galaxy. In particular we constrain the averaged diffusion in the Cosmic Ray source regions of the Large Magellanic Cloud. DMMW calculates the secondary emission produced during the propagation of DM-induced leptons from the steady state distribution of these particles, as well as the prompt γ-ray emission produced directly during annihilation or decay. We believe it is extremely timely to introduce DMMW: a natural step needed to unveil the possibly exotic nature of some of Fermi unidentified sources will consist of follow-up multi-frequency observational campaigns. DMMW enables users to easily make theoretical predictions for the radio, UV, X-ray and soft γ-ray emissions associated with the relativistic electrons and positrons produced in Dark Matter annihilation or astrophysical sources. The DMMW code can be interfaced to spectral fitting packages relevant to various wave-lengths, e.g. XSPEC for X-rays, and the Fermi Science Tools. The code has been tested by comparison to numerical solutions obtained with the GALPROP code.

Keywords: DMMW, indirect dark matter searches, cosmic ray transport.

1. Introduction

Secondary emission from relativistic e^+ e^- pairs produced in DM annihilation can form an important constraint for indirect Dark Matter (DM) searches. Previous studies have shown that the amount of secondary emission can be comparable to the prompt γ-ray emission.[1] In fact, for the Coma cluster the strongest limit on a possible DM annihilation signal is currently obtained from the radio data (see the left side of Fig. 1). Specifically, at radio frequencies the DM-induced emission is dominated by the synchrotron radiation of the relativistic electrons and positrons with a flux depending on environment (*e.g.* magnetic field and thermal electron plasma density). Inverse Compton (IC) scattering of the non-thermal $e^{\pm}$ on target CMB, starlight IR and possibly other photon background populations give rise to a spectrum of up-scattered photons stretching from below the extreme ultra-violet up to the soft gamma-ray band, peaking in the X-ray energy band. The last relevant contribution to the photon emission due to DM-induced relativistic electrons and positrons is the process of non-thermal bremsstrahlung, *i.e.* the emission of gamma-ray photons in the deflection of the charged particles by the electrostatic potential of ionized gas. Finally, a hard gamma-ray component arises from prompt emission in WIMP pair annihilations. Knowledge of the steady-state distribution of the DM-induced electron-positron population $n_e(E, \vec{r})$ allows one to compute the WIMP-induced secondary emission, provided the magnetic field structure and strength, as well as the gas and starlight densities are known. The steady-state distribution itself crucially depends on the diffusion coefficient. The right side of Fig. 1 illustrates the dependence of the secondary emission for the Draco dwarf galaxy on the diffusion coefficient and the magnetic field strength. In the following we will focus on the emission expected from extragalactic objects, such as galaxy clusters, dwarf galaxies and external galaxies. For these objects only very general arguments about the transport parameters can be made. In order to keep the number of free parameters small we neglect ill-constrained transport modes like diffusive reacceleration, convection and any spatial dependence of the transport parameters and describe the transport of Cosmic Rays (CRs) by the reduced transport equation:

$$Q_e(E, \vec{r}) + \nabla \left[K(E, \vec{r}) \nabla \frac{dn_e}{dE} \right] + \frac{\partial}{\partial E} \left[b(E, \vec{r}) \frac{dn_e}{dE} \right] = 0 \, , \qquad (1)$$

where dn_e/dE is the electron/positron density per unit of kinetic energy, $b(E) = \dot{p}(E)\beta$ is the momentum loss rate, which encodes the energy losses and $K(E, \vec{r}) = \beta^{\eta} K_0 \left(\frac{E}{E_0} \right)^{\delta}$ is the spatial diffusion coefficient.

2. DMMW: A brief Overview of the Code

DMMW allows the user to fit the DM particle properties for a variety of different objects and CR transport environments. The code solves the steady-state transport equation Eq. 1 for spherically symmetric boundary conditions for $e^{\pm}$ from DM annihilation or astrophysical sources, p from astrophysical sources, as well as secondary $e^{\pm}$, p and $\bar{p}$ from interactions of primary p with the gas. The secondary emission from synchrotron radiation, IC, bremsstrahlung and π^0-decay is calculated from the steady-state

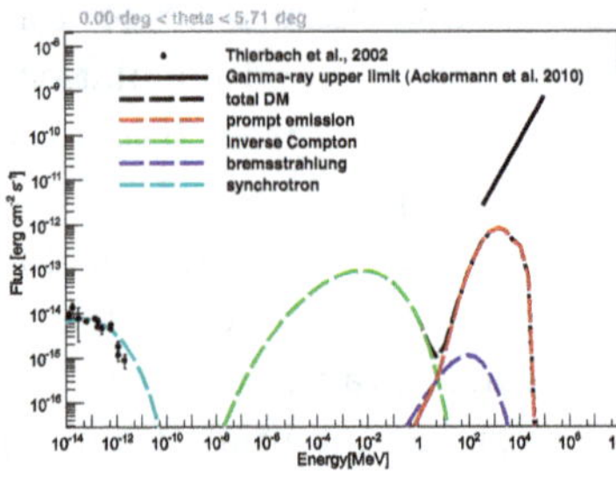

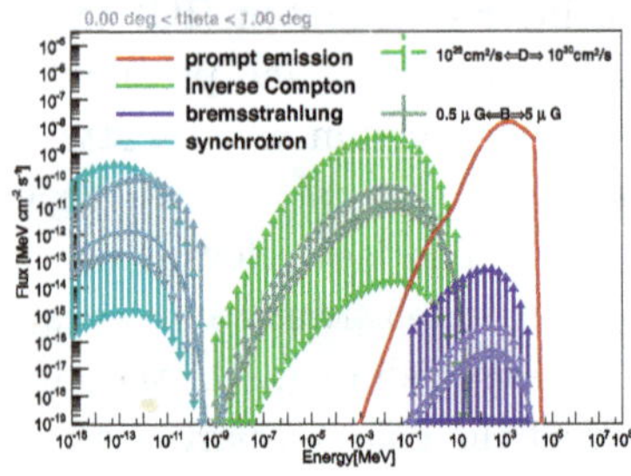

Fig. 1. **Left:** The multi-wavelength emission of the Coma cluster for a 40 GeV particle annihilating into $\mu^+\mu^-$ using the best-fit model from[1] ($K = \beta \cdot 2.1 \cdot 10^{29}(\rho/1\text{ GV})^{0.33}\text{ cm}^2/\text{s}$, $B = 1.2\ \mu\text{G}$, ISRF is CMB only). The strongest limits are given by the radio observations of the extended halo. Data: radio,[2] γ-ray upper limits.[3] **Right:** Variations in the secondary emission of the Draco dwarf galaxy due to variations of the diffusion coefficient ($10^{26} - 10^{30}\text{ cm}^2/\text{s}$) and the magnetic field strength ($0.5 - 5\ \mu\text{G}$). DM particle properties as on the left side of Fig. 1.

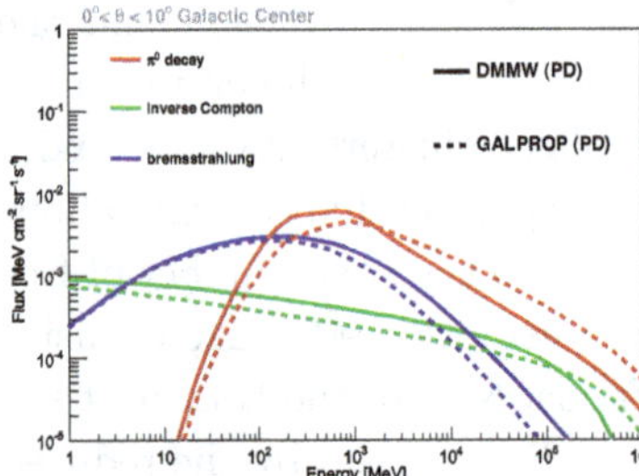

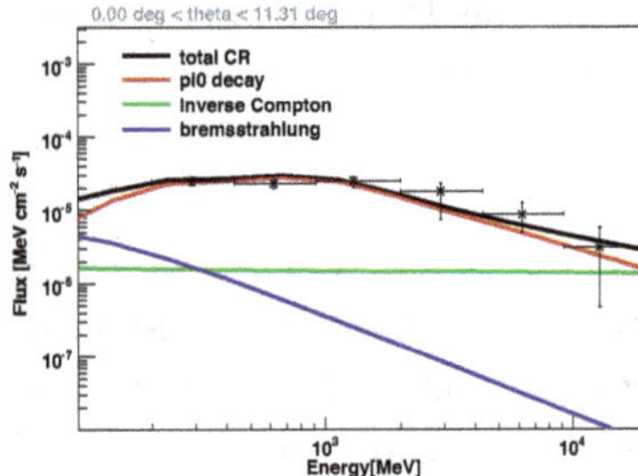

Fig. 2. **Left:** DMMW prediction of the diffuse γ-ray emission from a region of 10° around the Galactic center compared to the GALPROP prediction. For this comparison a constant gas density of $n_{HI} = n_{HII} = n_{H2} = 0.03/\text{ cm}^3$, a constant magnetic field of 3 μG, only the CMB component of the ISRF and the transport parameters of the *plain diffusion* (PD) model[4] have been used. The boundary has been set to $R_h = 10$ kpc in DMMM and $z_{max} = R_{max} = 10$ kpc in GALPROP. **Right:** Diffuse γ-ray emission of the LMC for a diffusion coefficient of the form $K(\rho) = \beta^{-2} \cdot 3 \cdot 10^{25}(\rho/1\text{ GV})\text{cm}^2/\text{s}$ (as a function of particle rigidity ρ) and $\delta = 0.33/0.6$ below/above 10^4 GV compared to the Fermi-LAT data.[5] We model the HI distribution within the LMC as a Gaussian with a width of 3.5 kpc and a total HI mass of $4.8 \times 10^8 \text{M}_\odot$,[6] the H_2 distribution as a Gaussian with a width of 3 kpc and a total mass of $5 \times 10^7 \text{M}_\odot$.[7] For the ionized hydrogen component we assume a total mass of $4 \times 10^6 \text{M}_\odot$ and a Gaussian width of 2 kpc. Since the emission primarily originates from 30 Doradus, these values are not representative for the emission regions. To account for this the HI and $H2$ distributions have been scaled by a factor 0.1. The source distribution is taken as a Gaussian with a width of 3 kpc, the total magnetic field strength on large scales is 4.3μG, where we assumed a regular component of 1.1μG and a random component of 4.1 μG.[8] For the ISRF we use an averaged value of 10^{-2}eVcm^{-3} between 0.2μm and 200μm for radii smaller than 4 kpc on top of the CMB component, which roughly corresponds to scaling the contribution from starlight and dust in the Milky Way by a factor of 0.1.

solution. The solution is obtained for a sphere with radius R_h, beyond R_h free escape is assumed. Details on our boundary condition treatment can be found in an upcoming publication.[9] The basic input for DMMW are the transport parameters, as defined by $K(E)$ and $b(E)$ (via the ISRF, the gas density and the magnetic field) in addition to the source distributions and initial energy spectra $Q(r, E)$. The energy spectrum of the DM annihilation products is obtained from the DMFIT code,[10] which calculates the differential γ-ray and $e^{\pm}_{DM}$ yields for a given DM mass; all parameters relevant to DMFIT (the decay channels, DM mass and interpolation options) are specified via the DMMW input file. The code outputs skymaps calculated from the steady state solution, as well as the CR fluxes. Figure 2 shows a comparison between the diffuse γ-ray emission in DMMW and GALPROP[a] for the Galactic center.

3. The Large Magellanic Cloud

In this section we apply DMMW to CR transport in the Large Magellanic Cloud (LMC). Diffuse γ-ray emission from the LMC has been observed by the EGRET telescope aboard the Compton Gamma-Ray Observatory (*CGRO*, 1991-2000)[11] and most recently with much higher angular resolution by the Fermi-LAT telescope.[5] The integrated flux above 100 MeV amounts to $1.6 \pm 0.1 \times 10^{-10}$ erg cm^{-2}s^{-1}, while the spectrum of the γ-ray emissivity is very similar to the diffuse γ-ray emissivity of the Milky Way. It was demonstrated, that the emission can be described by scaling the GALPROP predictions for the Milky Way to the LMC data.[5]

Here we build a model for CR transport within the LMC, by estimating the diffusion coefficients and source luminosities which are compatible with the Fermi-LAT observations of the LMC, assuming realistic gas densities. The γ-ray emission from the LMC shows only little correlation to the gas and is rather correlated to the massive star forming region 30 Doradus. In particular, Fermi does not observe any significant enhancement in the γ-ray emission from the gas ridge, which runs over $\sim 3^o$ along $\alpha_{J2000} \sim 05^h40^m$. This region is expected to contain about 20% of the total gas mass in the LMC[12] and mainly consists of giant molecular clouds. Since these H_2 clouds are not prominent in γ-rays, the GeV protons from 30 Doradus have to be efficiently confined to the source region and hence their average propagation length is expected to be significantly smaller than in the Milky Way. From here we estimate that an inital 5 GeV proton may not travel further than

[a]http://galprop.stanford.edu

~ 2.5 kpc before it drops below 0.5 GeV. This limits the diffusion coefficient in the GV range to values below $(2.8-4.7)\cdot 10^{25}\text{cm}^2/\text{s}$, which in turn means that the source luminosity has to be $\sim 10^{39}$ erg/s. This value is considerably lower than the upper limits from the SN rate estimate [b], but it is consistent with the assumption that the sources in 30 Doradus dominate the observed γ -ray emission. The right panel of Fig. 2 shows the γ-ray spectrum between 100 MeV and 20 GeV. The required source luminosities are $L_p = 1.5 \cdot 10^{39}$ erg/s and $L_e = 2.3 \cdot 10^{38}$ erg/s.

4. Conclusion

We briefly presented DMMW, a novel tool for the calculation of secondary γ-ray emission from both, CRs and DM annihilation/decay products in extragalctic objects. DMMW can be used for a wide range of different applications in the context of indirect DM searches or generic studies of transport parameters. Here we have demonstrated the exemplatory application to the Milky Way, the Draco dwarf galaxy, the Coma cluster and the LMC. The code uses semi-analytical solutions to the simplified CR transport equation,[1,9] which makes the simultaneous fit of the unknown transport parameters and DM properties feasible. DMMW is publicly available from http://dmmw.ucsc.edu.

References

1. S. Colafrancesco, S. Profumo and P. Ullio, *A&A* **455**, 21(August 2006).
2. M. Thierbach, U. Klein and R. Wielebinski, *A&A* **397**, 53(January 2003).
3. M. Ackermann *et al.*, *APJL* **717**, L71(July 2010).
4. I. V. Moskalenko *et al.*, *APJ* **565**, 280(January 2002).
5. A. A. Abdo *et al.*, *A&A* **512**, A7+(March 2010).
6. L. Staveley-Smith *et al.*, *MNRAS* **339**, 87(February 2003).
7. Y. Fukui *et al.*, *APJS* **178**, 56(September 2008).
8. B. M. Gaensler *et al.*, *Science* **307**, 1610(March 2005).
9. I. Gebauer and S. Profumo, *in prep.* .
10. T. E. Jeltema and S. Profumo, *JCAP* **11**, 3(November 2008).
11. P. Sreekumar *et al.*, *APJL* **400**, L67(December 1992).
12. T. Luks and K. Rohlfs, *A&A* **263**, 41(September 1992).
13. D. Maoz and C. Badenes, *MNRAS* **407**, 1314(September 2010).

[b]The supernova rate in the LMC and SMC is approximately 1 SN/300 yrs, about a factor 6 smaller than the SN rate in the Milky Way.[13] Assuming that the averaged source luminosity obtained from the GALPROP models is representative for the CR sources in the LMC, an upper limit for the expected source luminosity of CR electrons and protons is given by $L_e = 4.2 \cdot 10^{39}$ erg/s and $L_p = 1.8 \cdot 10^{40}$ erg/s.

THE GREATER IMPACT OF THE LHC: WHAT'S IN IT FOR THE REST OF THE WORLD?

STEVEN GOLDFARB[†]
Steven.Goldfarb@cern.ch

Department of Physics, University of Michigan
Ann Arbor, MI 48109, USA

As with any fundamental scientific endeavour, the true and enduring impact of the LHC physics program will not be measured in our lifetime. Rather, generations of future achievements will follow from the combination of scientific knowledge and methods born from the research. The unprecedented size, scope and innovative nature of the LHC experiments, however, are already leaving a broad and significant impact on our society. A few of the more impressive, far-reaching, and amusing examples of these contributions are presented in this document.

What remains is the task of communicating this progress to those who support it; that is, the rest of the world. Coordinated efforts by the outreach and educational programs of CERN and the experiments, coupled with a significant increase in media coverage, have succeeded to put the spotlight on the LHC. In addition, the rapidly growing popularity of social media has provided these programs with new tools to focus this spotlight on the benefits of our work. I present means to use these tools effectively as we enter a new era of public scientific communication based on dialogue and discussion.

1. Introduction

1.1. *Questions*

As a particle physicist participating in outreach, I spend a significant amount of time answering questions about the LHC physics program. These questions come from visitors, students, media representatives, government officials, rock stars, airplane passengers, and other scientists, to name a few.

Of the questions asked, some are simple and straightforward, others are a little more difficult, and some are very difficult – if not impossible – to answer (without misunderstanding). Yet, all are deserving of our best effort. After all, the askers of the questions are the very same people who make the decisions and provide the resources that allow us to carry on our work.

[†] Dr. Goldfarb currently serves as Outreach Coordinator for the ATLAS Experiment at CERN.

1.2. *Easy, Tough, Tougher*

Here are some examples of what I call the "easier" questions: *What is the LHC? How does it work? Have you found anything yet?* Each of these has a fairly clear answer that can be provided to the satisfaction of the audience and myself. Slightly "tougher" questions include: *Will you find the Higgs? Will you create a Black Hole? What is Dark Matter?* These require explanations of the underlying physics and can be a little challenging, depending on the background and capabilities of the audience and on my own ability to synthesize the information. Among the "toughest" questions are: *Why do you want to destroy the earth? Where do you hide the antimatter? Shouldn't we cure cancer instead?* I consider these as "tough", as a poorly worded or incomplete answer risks not only to confuse the audience, but also to leave them convinced that the LHC is either not safe or a waste of effort and resources.

1.3. *My Favorite*

There are, in fact, good answers to all of these questions and I get a lot of pleasure in presenting them to the various audiences. Unfortunately, this note has a limited size and scope, so I will not present the answers here. Rather, I will focus on my favorite question: *What do we get from the LHC?* That is, *What's in it for us?* This document will provide some answers to that question, recognize our inability to answer it completely, discuss current efforts to bring these answers to the public, and finally address the need and potential means for more physicists to contribute to these efforts.

2. Answers

2.1. *It's the Physics, Stupid*

First and foremost, the LHC is for physics exploration. It is already producing collisions at energy levels never attained before, allowing the most complex and precise detectors ever constructed to make measurements that address our most fundamental questions about the universe.

Any attempt to justify the existence of the LHC that does not address this purpose falls short. Human beings exist today because of our ability to study and learn about the world that surrounds us. It is a fundamental part of our nature and we would not survive without that trait. Furthermore, the long-term effects of basic research are evident everywhere we go. From agriculture to industry, transportation to communication, medicine to education, all of our great accomplishments can be traced back to science and the pursuit of knowledge.

So, why don't we just leave it at that? Well, truth be told, we have no idea what the physics of the LHC will lead to. We simply do not know what answers will be found, what new questions will be produced, what knowledge we will bring to the world, how we will affect the way we think, or how we will all be changed in the long run.

2.2. *Timeless Brilliance, Timeless Ignorance*

But this is nothing new. Thales of Miletus did not foresee the construction of 500 kV power grids when he experimented with amber and lodestone. Sir Isaac Newton might have pondered new means to calculate planetary orbits, but did not consider the advantages of geo-synchronous satellites for communication, when he saw the apple fall. And, as brilliant as they were, I doubt that Paul Dirac and Carl Andersen could imagine the precision imaging of PET scans, when they were proposing and discovering the positron. It has taken years - often generations - to see the full impact of their research.

Unfortunately, the world is not so patient when it comes to measuring the success of scientific investment. While we must focus on the necessity and excitement of our work for the future of our field and for the future of mankind, the world of short-term budgets, frequent political campaigns, instant news and short attention spans often demands immediate results. So, what can we say about the contributions of the LHC, today? Well, in fact, quite a bit.

3. What the LHC Has Already Contributed

3.1. *Introduction*

When you assemble a group of highly focused, well educated, and extremely motivated individuals from around the world, bring them together to a laboratory that features top-notch computing infrastructure, world-leading engineering facilities, a rich scientific history, and a view of Mont Blanc, things happen. Repeat that four times (in fact, more than that), and you have the LHC.

And things have happened. From contributions to computing and software methods, to advances in accelerator and detector technology, to new methods for university-level education and research training, to the development of collaborative tools, social science and entertainment, the LHC has already born measurable results.

These advancements came from a mixture of projects essential to LHC operations, connections made by participants or outsiders between these specific developments and other related fields, and whimsical ideas from people completely disconnected from the actual work. Yet, the ideas all share a common thread: they might never have born fruit had it not been for the LHC. In the following sections, I present a few notable examples.

3.2. *Computing*

Grid Computing, as proposed by Ian Foster and Carl Kesselman [1], is designed to connect heterogeneous computers, memory, and storage, and then to present them to the user through a simple interface. The name came by its comparison to the electrical grid, as users only need know how to use the socket to get power.

While the LHC is not attributed to the Grid's creation, it is certainly one of the most prolific users. And the lessons learnt by a group of physicists focused on crunching huge amounts of data (petabytes) at incredible rates (starting with a billion collisions per second), have quickly propagated to the rest of the world.

Worldwide applications for Grid Computing have included genetic mapping, disease response simulation, urban planning, and economic modeling, to name a few. One source for following recent scientific applications is the weekly newsletter called "International Science grid this week (isgtw) [2]. A good example is this article on research into Alzheimer's disease by performing data analysis on a 2,000-cpu grid of computers called LINGA [3].

3.3. *Medicine*

Meanwhile, at CERN, the application of advanced accelerator technology developed for the LHC and its upstream devices was discussed at the European Network for Light Ion Hadron Therapy (ENLIGHT) annual meeting [4]. Ideas include the usage of commercial radionuclides for medical imaging, such as PET scans. Investigations are also underway to look into the usage of the LEIR facility for radiobiology and medical accelerator research and development.

Detector technologies developed for LHC pixel detectors, used to track particles with micron precision, are now used for a variety of medical imaging applications. The Medipix Collaboration [5] and the PIXSCAN Computer Tomography scanner [6] are capable of making precision measurements of soft tissue in animals to a better spatial resolution than ever obtained before.

3.4. *Other Optical Imaging Applications*

New precision optical imaging techniques exploit robotic devices developed for testing the LHC silicon detectors. Accurate, low-noise digital reproductions of historical recordings are made without any physical contact with the original devices. Imagine Robert Johnson without snaps and crackles!

3.5. *Education*

Arguably, the most tangible and important product coming out of the LHC factory are the thousands of students imbedded in the research program. Around 500 Ph.D. theses will come out of the LHC every year. In addition, technical student [7], summer student [8] and study abroad programs will give hundreds of undergraduate students the opportunity to take courses with physicists who are leaders in the field, to interact with LHC research teams from all over the world, and to participate and contribute to cutting-edge research.

If there were a recipe for the creation of peace in the world, it would certainly include the assembly of students from all corners of the globe, with diverse backgrounds, cultures and languages, put together to work on a common goal: understanding nature. These students carry their skills and their enthusiasm with them for the rest of their lives, regardless of their choice of future career, and the LHC is providing them with plenty of opportunity for growth.

3.6. *Collaborative Tools*

The largest of the LHC collaborations, ATLAS and CMS, feature 3000-4000 participants, each. ATLAS, for example, comprises 174 institutions from 38 nations, spanning the globe from Australia to Morocco to Stockholm to Rio de Janeiro. Given this distribution, coordination of the construction, integration and final operation of the detectors themselves would have been impossible without the implementation of flexible and reliable collaborative tools.

At the onset, few tools were available to handle these needs, and the diversity of the computing and networking available at the institutes made it impossible to rely on commercial solutions. An initiative by the California Institute of Technology called VRVS (and eventually EVO) [9] was launched to address these challenges through IP-based video conferencing. This system has succeeded in effectively connecting all of the LHC institutions, not only making it possible for the detectors to be constructed remotely, but allowing remote experts to supervise their installation and running at CERN, and then to continue on to participate actively in the analysis of data.

New methods were also developed at CERN to support event and meeting management [10] and webcasting [11] for the LHC. The University of Michigan ATLAS Collaboratory Project [12] has worked with CERN to develop, test and promote a variety of collaborative technologies, including the archival of lectures [13], a technology adopted by CERN IT and in heavy use, today. Nearly all of these technologies have been adapted by commercial enterprise.

3.7. *Social Science*

The success of these large global collaborations depends on much more than communication infrastructure, of course, and many studies have been performed to try to understand what makes them work. One study, by a team of physicists and social scientists, for example, has led to the publication of a book called "Collisions and Collaborations" [14], in which the seemingly inefficient flat organizational structure of an LHC collaboration is found to have strengths not found in hierarchical business management models. Other relevant studies have been made by psychologists, sociologists and even anthropologists. Most lead to the conclusion that the common self-motivation of scientists seeking to understand the inner workings of the universe is at the heart of our ability to work together effectively. Most physicists are not surprised by this finding.

3.8. *Entertainment*

There have been a number of interesting and amusing interactions by members of the art and entertainment industry with the LHC and its experiments. On 30 March 2010, the LHC turned into a star by webcasting its first high-energy collisions to a worldwide audience of over 2 million viewers. In the time leading up to that event and in the year and a half that has followed, a number of talented artists, musicians, filmmakers and others have enjoyed the possibility of associating themselves with the LHC.

Ron Howard, director of "Angels & Demons", chose ATLAS as the source of antimatter for a devious plot. The director of the opera Les Troyens by Hector Berlioz chose the design of the ATLAS toroid as a model for its main stage prop. Popular bands "Black-Eyed Peas" and "Dr. Feelgood" enjoyed visits to the LHC, and a new "Muppet Movie" (Nov 2011) will feature a visit to ATLAS.

One might argue that these events are only opportunities for stars to mutually benefit from each other's fame. They have the dual effect, however, of influencing the arts while simultaneously exposing the LHC and its scientific program to a very large audience and, in particular, before a public that would not normally seek out information on the frontiers of scientific research. This is a golden opportunity, indeed, but how do we take advantage of it?

4. How We Get The Word Out

4.1. *The Outreach Recipe*

As with any effective project, HEP Outreach and Education rely on human and monetary resources (ingredients), a well-focused strategy (cooking instructions) and tools for implementation (cooking). Given limited resources and the desire to focus the majority of those resources on research, communication teams face the continuous challenge of stretching the dinner budget. Yet, the food is getting better all the time. How do they do it?

4.2. *The Ingredients*

The key ingredients of any Outreach and Education effort are enthusiastic and inventive members of the collaborations: some at CERN, working on centrally organized activities, and others at home institutes, working on locally, nationally or internationally organized projects. While these groups work together as often as possible, resources and coordination come from different areas.

CERN-based efforts typically include small teams of physicists, science communicators and media specialists working part-time and coordinated by physicists. They use limited resources to maintain platforms, such as web sites or visitor centers to support activities, such as VIP and group visits, and communication with media and the public.

Institutional projects range from open houses or exhibits at community events or universities to large-scale, international projects. European programs, such as "Discovering the Cosmos" [15] bring educational tools and methods to students and teachers, receiving support from funding agencies. The U.S.-based project, QuarkNet [16], offers physics training programs for high school science teachers, in order to help them to bring the excitement of LHC physics to students in their classrooms. Large or small, these projects all profit from the core infrastructure provided by the CERN-based teams.

4.3. *Cooking Instructions*

Effective communication of the goals, results and impact of the LHC relies on the implementation of target-oriented communication plans. The CERN DG Communication group and each of the LHC experiments have drafted complementary plans, specifying goals and target audiences. As an example, for the ATLAS experiment, these include:

- Fostering public appreciation of the scientific goals and achievements of ATLAS and of particle physics (Target: Media and General Public);
- Maintaining support for ATLAS, the LHC and particle physics research (Target: decision makers and members of industry);
- Attracting and retaining the next generation of scientists and science educators (Target: teachers and students age 13-18).

These goals and targets form the basis of a strategy that allows the experiment to use its limited Education and Outreach resources effectively.

4.4. *Cooking It Up*

The imagination and productivity of the teams involved in outreach and education have given rise to a significant quantity and an amazing diversity of projects and material. Here is a sample:

- Brochures on LHC physics, computing, technical transfer, and specific topics, such as anti-matter or the Standard Model;
- Playing cards with particles or detector components as background, pop-up books, photo books, cartoon books, and 3d viewing devices;
- Animations of the LHC, the detectors, and live events; videos of special events, or interviews with topical specialists;
- Current reporting on major conferences, key publications, public events at CERN or at the experiments;
- Public installations of varying scale, from small desk-top exhibits to animations on buildings or very large screens;
- Music videos about the LHC or the detectors, music CD's with performances by collaboration members;
- Murals and photographs printed on five-story buildings;
- Multimedia installations travelling through museums;
- Science fairs and major media events;
- Masterclasses teaching high school students to analyze real LHC data;
- Programs for local residents and students, including the "Big Bang Passport" [17] and "Researchers Night" [18], organized by CERN;
- Visits by students and the public to locations around CERN, including the LHC and detector control rooms (30,000 visitors in 2010);
- Remote visits to control rooms, using video conferencing to allow the participation of audiences located around the world.

Each of the individual projects is typically managed by the CERN-based teams, in many cases through the coordination of the LHC Outreach Group (LOG) or through the International Particle Physics Outreach Group (IPPOG) [19].

4.5. *Serving It Up*

For many years, the content described above, was delivered to the public either directly, via web portals or local activities, or by traditional media, including newspapers, journals, TV and radio, through interviews and the coverage of major events. Over the past several years, however, social media platforms, such as Facebook [20] and Twitter [21], have changed things dramatically.

First of all, these new delivery platforms cannot be ignored. Nearly all experiments have set up sites and are actively engaging these newfound audiences. Secondly, one should note the impact these platforms have on content style. The general public in not only the final target audience, but plays an important role as a vector, much like the traditional media. In fact traditional media has become a target of the social media, in addition to being a source of information. It is thus important to model content in a form that is easily picked up and understood by the general public, in order to maximize coverage.

The important feedback loops created by the injection of social media into the communication stream provides an outstanding opportunity for the LHC to get quality content exposed to a new and ever larger audience. A significant effort, however, is required to ensure correct information is delivered accurately to the target audience. Vigilant participation by the communication groups and by members of the collaborations is necessary, as this worldwide game of "telephone" can often generate wrong and misleading statements.

5. Conclusion: Join the Conversation

All physicists would do well to offer at least a small part of their precious time to the communication effort. In addition to fulfilling their social obligation, as responsible scientists, serving the people from who they receive support, the exercise is an excellent and rewarding learning process. Learning to write blogs, give public talks, guide visitors, and to be interviewed by the media, teaches a physicist the ability to synthesize her/his normally complex and specialized work. It gives one a sense of perspective and, in fact, forces one to understand how that work fits into the big picture. That is, in explaining to the world what physics brings to them, one often finds what is in it for her/himself.

Acknowledgments

I would like to acknowledge and thank my colleagues in the world of physics education and outreach for their hard work and dedication toward communicating the excitement and necessity of particle physics to the world. In particular, I extend my appreciation to the ATLAS team, our friends around the ring in the LHC Outreach Group, and those working with similar rings and devices in the International Particle Physics Outreach Group. Finally, I would like to thank Randy Ruchti and the ICATPP organizers for having the foresight to create the Broad Impact session and for extending their invitation, as well as ATLAS management for giving me the support to accept it.

References

1. I. Foster, C. Kesselman, "The Grid: Blueprint for a new computing infrastructure" (2004).
2. International Science Grid This Week (ISGTW): http://www.isgtw.org
3. “Hat trick for Alzheimer’s grand challenge,” ISGTW, Sep. 28, 2011: http://www.isgtw.org/visualization/hat-trick-alzheimer’s-grand-challenge
4. European Network for Light Ion Hadron Therapy (ENLIGHT): http://cern.ch/enlight
5. Medipix Collaboration home page: http://cern.ch/medipix
6. PIXSCAN home page at imXgam: http://imxgam.in2p3.fr/pixscan.php
7. CERN Technical and Doctoral Student Programs: http://cern.ch/hr-recruit/Tech-Doct/default.asp
8. CERN Summer Student Program: https://cern.ch/hr-recruit/summies/default page
9. Enabling Virtual Organizations (EVO): http://evo.caltech.edu
10. Indico Event Management System: https://indico.cern.ch
11. CERN Webcast Service: http://cern.ch/webcast
12. University of Michigan ATLAS Collaboratory Project: http://atlascollab.umich.edu
13. University of Michigan Web Lecture Archives: http://lecb.physics.lsa.umich.edu/CWIS/SPT--Home.php
14. “Collisions and Collaboration” by M. Boisot, S. Tami, B. Nicquevert, and M. Nordberg, 2011. http://collisionsandcollaboration.com
15. “Discover the COSMOS” project: http://www.cosmosportal.eu/cosmos
16. QuarkNet: http://quarknet.fnal.gov
17. CERN “Big Bang Passport” project: http://alice.cern.ch/record/1368903
18. “European Researchers Night”: http://cdsweb.cern.ch/record/1296255
19. International Particle Physics Outreach Group: http://ippog.web.cern.ch/ippog
20. Facebook: http://www.facebook.com
21. Twitter: http://www.twitter.com

R&D on the Geant4 Radioactive Decay Physics

S. Hauf, D.H.H. Hoffmann, P.Lang and S. Neff

Institut für Kernphysik, Technische Universität Darmstadt,
Darmstadt, 64289 , Germany
E-mail: steffen.hauf@astropp.physik.tu-darmstadt.de

M. Kuster

European XFEL GmbH,
Hamburg, 22761, Germany,
Email: markus.kuster@xfel.eu

M. Batič and M.G. Pia

INFN Sezione di Genova,
Genova, 16136, Italy
E-mail: MariaGrazia.Pia@ge.infn.it

Z.W. Bell

Oak Ridge National Laboratory,
Oak Ridge, TN 37831, USA

G. Weidenspointner

Max-Planck-Institut Halbleiterlabor, Munich, 81739, Germany
Max-Planck-Institut fuer extraterrestrische Physik, Garching, 85748, Germany
E-mail: Georg.Weidenspointner@hll.mpg.de

A. Zoglauer

University of California at Berkeley,
Berkeley, CA 94720-7450, USA

We present validation measurements for the Geant4 radioactive decay simulation following a self-consistent approach. The validation is based on gamma spectroscopy measurements with HPGe and NaI detectors. In addition we present a re-designed radioactive decay simulation for Geant4, with extended functionality, such as support for long term activation, and programmed to modern coding standards.

Keywords: Geant4; Radioactive Decay; Monte-Carlo Simulation; Validation

1. Introduction

The usage of Monte Carlo codes for the design, verification and analysis of particle-, photon science, bio- and astrophysical experiments has become increasingly important: (high energy) particle and photon science experiments increasingly rely on ultra-sensitive detectors. The construction of these often requires good estimates of the expected background flux throughout the design phase, often before physical hardware is actually available. This estimate must usually also include the radiation due to the buildup of radioactive material produced by cosmogenic activation. Additionally all detector components need to be able to cope with the expected radiation levels. This requires an accurate estimate of the radiation dose deposited i.e. in the detector's logic circuits. Correct dose estimates are important for other research fields as well. In medical science and human space flight a correct estimate of the radiation doses and their distribution can be useful for the planning of the therapy or the mission profile and minimizes unnecessary exposure of the patient or the crew.

The above short list of applications is far from complete, but it exemplary highlights the versatility required from a simulation toolkit which is capable of handling such scenarios. Geant4[1,2] is such a versatile toolkit. It is based on object oriented C++ design principles and as such is easily extensible to address the requirements of a broad range of applications including the previously mentioned.

One example of highly sensitive detectors placed in a harsh radiation environment are astrophysical space-born observatories such as the planned IXO/ATHENA X-ray observatory.[3] For faint sources the performance of the Wide Field Imager, ATHENA's imaging spectrometer, will depend on the background flux due to cosmic rays and secondary particles resulting from their interaction with the satellite components. In addition to the prompt background, cosmic ray protons will lead to activation of the satellite material. The resulting radioactive isotopes will subsequently decay and emit β and γ radiation resulting in a delayed background component.

Thus for ATHENA and similar space missions but also for many ground based experiments and medical applications the simulation of radioactive decay processes can be of importance.

Comparisons between the decay simulation of Geant4 and measurements exist (see i.e. Hurtado *et al.*,[4] Bissaldi *et al.*[5]) and often show a good agreement, although a tuning of the simulation parameters is sometimes required.

Furthermore, long term activation, of which a correct modeling is of importance to many high sensitivity experiments, can not easily be realized within the existing Geant4 framework. Although extensions which overcome this limitation exist (for Geant3 MGGPOD,[6] for Geant4 Cosima[7]), a redesign of the Geant4 decay code, would allow for easier integration into Geant4 projects.

2. Validation of the Geant4 Radioactive Decay Physics

For our verification measurements we used a simple experimental setup consisting of an isotope source installed in front of a HPGe detector or NaI detector. The experimental setup which includes shielding components, the collimator, the dewar and the lab room, was transfered to a Geant4 geometry model. The measured spectra were compared to spectra simulated with Geant4 version 4.9.3, which were broadened to match the measured energy resolution and normalized to the source activity. As is exemplary shown in Fig. 1 we find a general qualitative agreement but quantitative deviation is observed.

In order to explain these deviations we compared the primary output of the Geant4 decay implementation with evaluated data. This comparison did not show any significant deviation of the photo peak intensities for the measured isotopes. The full details of this comparison go beyond the scope of this paper and will thus be given in a later journal publication. Because the initial decay simulation output is reasonably accurate we assume that

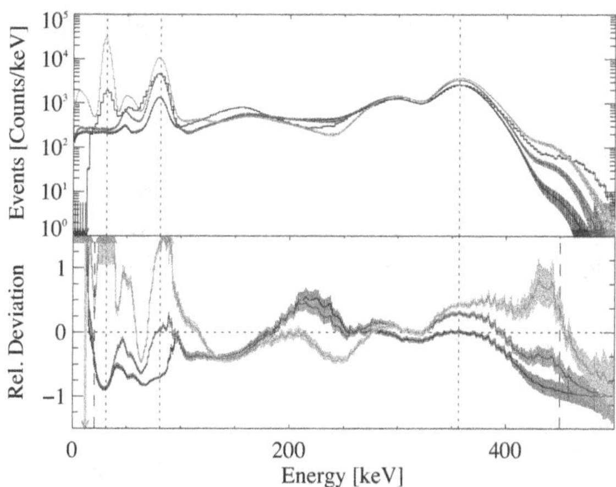

Fig. 1. A comparison of a simulated (grey lines) NaI gamma spectroscopy of ^{133}Ba to measured and background subtracted data (black line). The entry window thickness was varied from measured thickness (dark grey) over half thickness (medium gray) to not present (light gray). The bottom panel shows the relative deviation of the simulated from the experimental data.

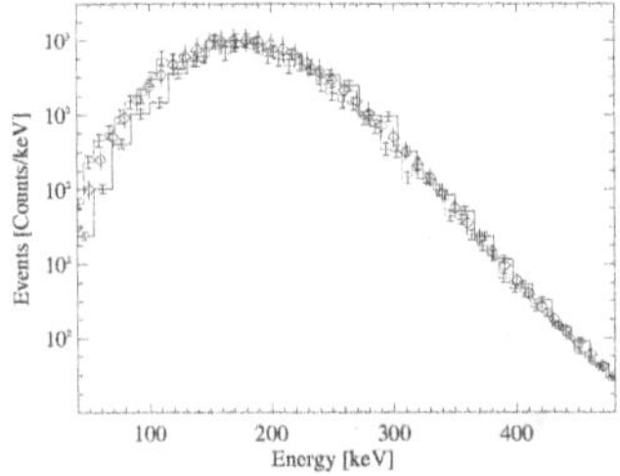

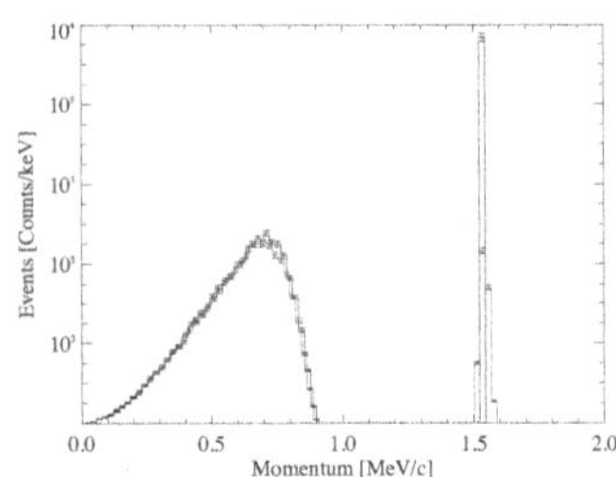

Fig. 2. Left: The energy distribution of the positrons from a simulated ^{22}Na decay with the new (red) and old (blue) decay code implementation compared to experimentally determined data from.[8] Right: The post decay nucleus momentum distribution from a simulated ^{22}Na decay with the new (red) and old (black) decay code implementation. In the new implementation the momentum is broadened due to the recoil from the emission of Auger-electrons which is not the case in the old implementation.

the observed deviations need to be explained by either systematic effects, as e.g. uncertainties in the detector modeling and systematic uncertainties originating in the experimental setup or by effects occurring in the simulated tracking of the initial gamma ray and the resulting secondary particles. Also a combination of both can be the case. Investigations on this matter are still ongoing, thus a final conclusion on the reason for the observed deviations has yet to be reached.

3. Improving the Geant4 Radioactive Decay Simulation Code

As was laid out in the introduction a large versatility of the Monte-Carlo code is beneficial for its use in varying scenarios. Based on this design philosophy we decided to reorganize most of the Geant4 radioactive decay code with a focus on modularizing it in such a way that code parts can easily be exchanged and simple but transparent interfaces to the overall Geant4 code are implemented. This easily allows for the inclusion of new features such as long term activation, user supplied Beta-Fermi functions and radioactive (background) sources attached to volumes in the *DetectorConstruction.cc* file. Typical speed improvements we could achieve during our tests are of the order of a factor of 2 for isotope decays in an empty geometry consisting only of a vacuum filled box when compared to the decay implementation of Geant4 version 4.9.3. During testing it became apparent that especially the process used for simulating the decay of excited nuclei dominates the overall computing speed and should be the focus of future improvements.

Figure 2 shows results obtained with our new code implementation in comparison to Geant4 version 4.9.3. Both implementations yield results which compare well to experimental data of the energy spectrum of positrons as a result of β^+-decay.

4. Conclusions and Outlook

We have shown preliminary results of a validation of the Geant4 radioactive decay simulation. These currently do not quantitatively reproduce the measured data. Additional work is necessary to understand the observed deviations and reach a final conclusion, which will be detailed upon in a future journal publication. We have also presented an early version of an improved decay simulation with added functionality and modularity. This code will be made available to the community after testing and validation.

This work is supported by the Bundesministerium fuer Wirtschaft and Technologie and the Deutsches Zentrum fuer Luft und Raumfahrt - DLR.

References

1. J. Allison *et al.*, *IEEE Trans. Nucl. Sci.* **53**, 270(February 2006).
2. S. Agostinelli *et al.*, *Nucl. Instrum. Methods Phys. Res., Sect. A* **506**, 250(July 2003).
3. ESA, IXO/ATHENA Mission Summary http://sci.esa.int/science-e/www/area/index.cfm?fareaid=103(October, 2011).
4. S. Hurtado, M. Garca-Len and R. Garca-Tenorio, *Appl. Radiat. Isot.* **61**, 139 (2004), Low Level Radionuclide Measurement Techniques - ICRM.
5. E. Bissaldi, A. von Kienlin, G. Lichti, H. Steinle, P. Bhat, M. Briggs, G. Fishman, A. Hoover, R. Kippen, M. Krumrey, M. Gerlach, V. Connaughton, R. Diehl, J. Greiner, A. van der Horst, C. Kouveliotou, S. McBreen, C. Meegan, W. Paciesas, R. Preece and C. Wilson-Hodge, *Exp. Astron.* **24**, 47 (2009), 10.1007/s10686-008-9135-4.
6. G. Weidenspointner, M. J. Harris, C. Ferguson, S. Sturner and B. J. Teegarden, *New Astron. Rev.* **48**, 227 (2004), Astronomy with Radioactivities IV and Filling the Sensitivity Gap in MeV Astronomy.
7. Zoglauer, A. and Weidenspointner, G. and Galloway, M. and Boggs, S. E. and Wunderer, C. B., Cosima - the cosmic simulator of MEGAlib, in *Nuclear Science Symposium Conference Record*, (IEEE, 2009).
8. H. Wenninger, J. Stiewe and H. Leutz, *Nucl. Phys. A* **109**, 561 (1968).

PUBLIC RELEASE AND ANALYSIS OF DATA FROM THE CMS EXPERIMENT AT THE LHC

T. MCCAULEY

Fermi National Accelerator Laboratory,
Batavia, IL 60510, USA
E-mail: thomas.mccauley@cern.ch

The CMS collaboration has made several large event datasets public for educational and outreach purposes, including more than 300k events containing pairs of electrons, muons and jets.

The data are prepared and published using an extensible, text-based (JSON) data format. With these datasets students learn about scientific analysis through searches and studies of J/ψ, Υ, W and Z particles. Students explore the data using an experiment-independent online event display and histogram tool, often as part of international educational programs in Europe, the USA, and the rest of the world such as I2U2, the IPPOG Masterclasses, and QuarkNet.

I describe the current status of these activities, the positive feedback from the students, and the bright future outlook, including the great potential to broaden these activities to a wider range of experiments and audiences.

1. Introduction

Since the beginning of proton-proton collisions at $\sqrt{s} = 900$ GeV in December 2009, from lead-lead collisions at $\sqrt{s_{NN}} = 2.76$ TeV and proton-proton collisions at $\sqrt{s} = 7$ TeV at the Large Hadron Collider (LHC) at CERN, the CMS experiment[1] has collected (as of this writing) more than 4 fb^{-1} of data. From this data it has published just over 100 papers describing results from searches for the Higgs boson, supersymmetry, and other exotic phenomenon, and studies of QCD, electroweak, top, heavy ion, forward, and B physics.[2]

In December 2010 and April 2011 the CMS collaboration agreed to release a fraction of its collected data to the public for use in education and outreach. The datasets are as follows, where those appearing in **bold** are already delivered and/or in use[a]:

- 2000 events each of $J/\psi \rightarrow \mu\mu$, $J/\psi \rightarrow ee$
- 2000 events each of $\Upsilon \rightarrow \mu\mu$, $\Upsilon \rightarrow ee$
- 500 events each of $Z \rightarrow \mu\mu$, $Z \rightarrow ee$
- 1000 events each of $W \rightarrow \mu\nu$, $W \rightarrow e\nu$
- 100,000 events each of **dimuon**, dielectron, and dijet events in the energy range 2-110 GeV

The main users of this data have been high school students around the world who explore the data as part of programs organized by groups such as I2U2[3] (Interactions in Understanding the Universe), IPPOG[4] (The International Particle Physics Outreach Group), and Quarknet.[5]

2. Educational groups and programs

I2U2 is an "educational virtual organization" that has created and maintains educational programs and tools aimed at fostering public awareness and understanding of science. These tools include so-called e-Labs which provide Web-based tools for student and public exploration of real scientific from experiments such as CMS and LIGO as well as cosmic ray detectors installed in schools. More description of the e-Lab for CMS is given in Sec. 4.

IPPOG is a network of particle physicists and educators that seek to raise public awareness and understanding of particle physics. To help achieve this goal IPPOG has been since 2005 organizing so-called masterclasses.[6] Each year thousands of students in countries all over the world travel to nearby universities or research laboratories for one day. The day begins with lectures from particle physicists and students then move on to analysing real data from particle physics experiments. The day concludes with a video conference with other student groups in the world to discuss results.

Quarknet is an organization that provides programs and training for teachers and students in the USA in particle physics: from the fundamentals, to the experiments, and eventually to analysis of real data (often in the context of masterclasses). All three organizations collaborate, share tools and resources, and together provide a rich educational experience for the public and students alike.

---

[a]the data is available via the links found here:
`https://twiki.cern.ch/twiki/bin/view/Main/CMSPublicData`

3. Data preparation and format

The datasets themselves were produced with several requirements in mind:

- The data format must be easy-to-use and not require complicated software in order to read and manipulate it
- The content of the datasets should be readily useable in student exercises but not restrictive in scope, to allow for flexibility and surprises
- The main users are to be students, supervised by teachers, studying the data in the context of masterclasses and e-Labs
- The exercises based on the datasets shouldn't be too difficult - one can't expect students to do a full-blown physics analysis - but shouldn't be trivial either

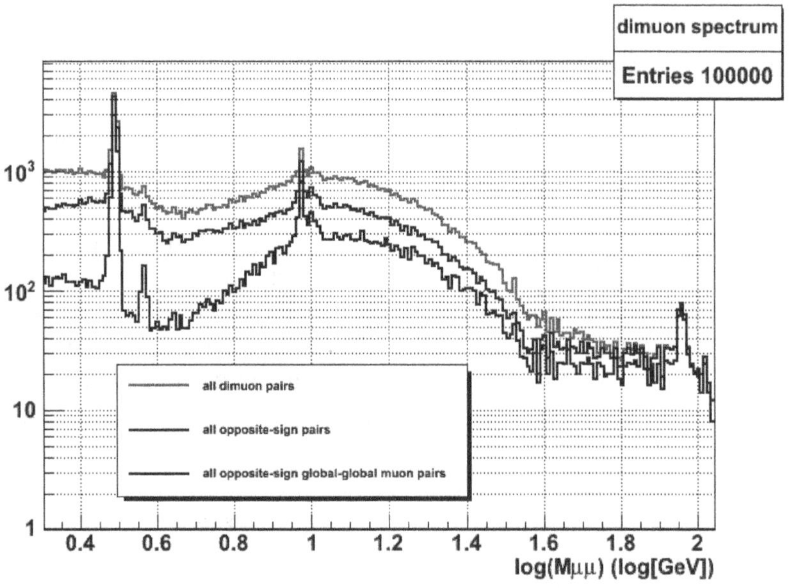

Fig. 1. Invariant mass spectra of the dimuon dataset. The top spectrum is from all muon pairs. The middle spectrum is from selection of opposite-sign muon pairs. The bottom spectrum is from opposite-sign, global (*i.e.* high-quality) muon pairs.

The basis format of the released datasets is the so-called ig format,[7] which was originally developed for the iSpy event display[8] used in the CMS experiment. An ig file is simply a zip archive with a simple directory structure where in each file multiple runs contain multiple events. The event files at the bottom of the directory structure are written in JSON (JavaScript Object Notation)[9] format, which is a human-readable, text-based file format. Physics event information and graphics information (such as positions of objects in global coordinates) are encoded as JSON objects in the files. Internally in the file there is a simple relational database model between these objects.

The ig file format has several beneficial features. For one, the files are self-documenting, containing a schema describing the contents. The JSON format is easily parsed and written using C++, python, and JavaScript. This allows for easy production of summary files (*e.g.* containing lines of four-vectors) in csv (comma-separated-variable) and JSON format. In addition, the zip archives can be trivially reorganized to suit the purposes of an exercise; one can easily for example mix events from the dielectron and dimuon datasets into ig files containing 100 events each. The ig files themselves are created using the software framework of the CMS experiment, converting the CMS format into ig format. Users therefore require no special knowledge of CMS software and are also insulated from possible differences in versions of CMS software and event formats. This feature, along with its flexibility and extensibility, allows the ig format to be in principle experiment-independent.

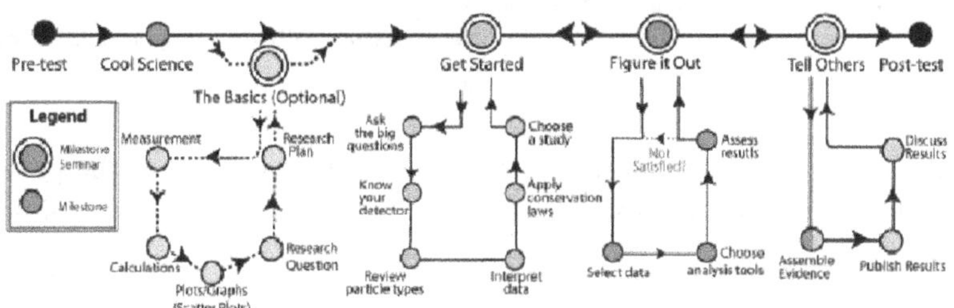

Fig. 2. Screen capture of the I2U2 CMS e-Lab project map. Online the project map is interactive and provides material at each stage.

A closer look at the dimuon dataset illustrates fulfillment of the last requirement listed above. Presented with the dataset, students can select pairs of muons and calculate an invariant mass. The spectrum for all muon pairs can be seen in Fig. 1 (the top spectrum). Simple selections on only opposite-sign muon pairs and then on only high-quality muon pairs bring out the J/ψ, ψ', and Υ peaks at the lower end of the spectrum. Students in this exercise learn about the CMS detector, invariant mass, and how to do simple selections to better extract signal from background.

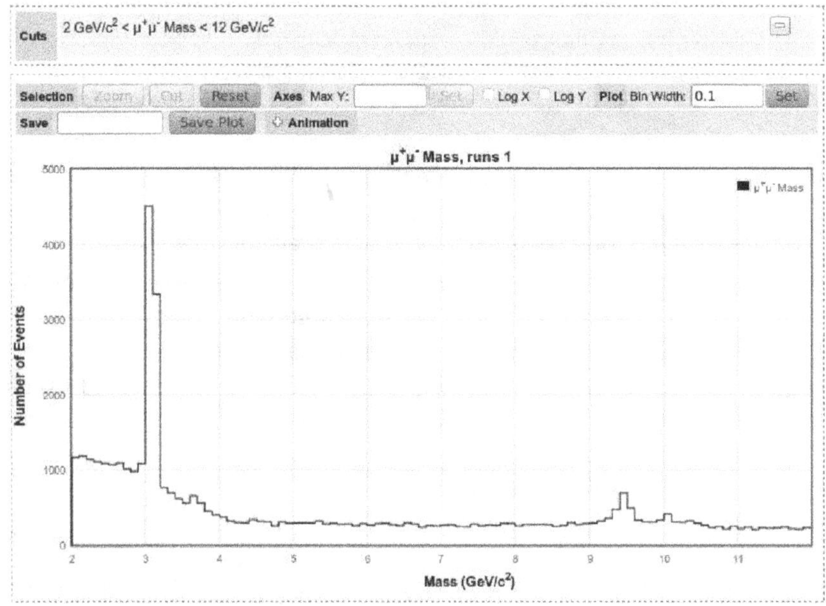

Fig. 3. Screen capture of a dimuon invariant mass spectrum using the I2U2 online histogram tool.

Further, closer examination of this dataset can also reveal the presence of several cosmic ray muons: an example of one of the "surprises" mentioned above.

4. Analysis tools and their usage

All of the ig files released can be viewed in the iSpy event display which provides full 2D (ρz and $r\phi$) and 3D views of the event and detector as well as a tabular view. Distribution and installation is trivial as it is distributed as a fully-bound executable for Mac OSX and Linux (a Windows installer has been developed but not released as of this writing).

I2U2 provides a comprehensive online program for study of CMS data in e-Labs. A schematic of the program can be seen in Fig. 2. After passing through the first two milestones students are invited to explore data, to "Figure It Out". All of the datasets are available and with the aid of an online histogram tool various physical and kinematic properties such as invariant mass, transverse momentum, and missing transverse energy can be studied. One can also study correlations between these properties, for example cut on transverse momentum and see how it affects the invariant mass distribution. An example of an invariant mass plot can be seen in Fig. 3. An indication of the controls available in the tool (*e.g.* cutting and zooming) can be seen in the figure.

I2U2 also provides a browser-based event display.[10] The display renders events stored in ig format. This event display is written in JavaScript and works on any Web browser that supports HTML5 canvas. A screen capture of a $J/\psi \to \mu\mu$ event in the CMS detector can be seen in Fig. 4.

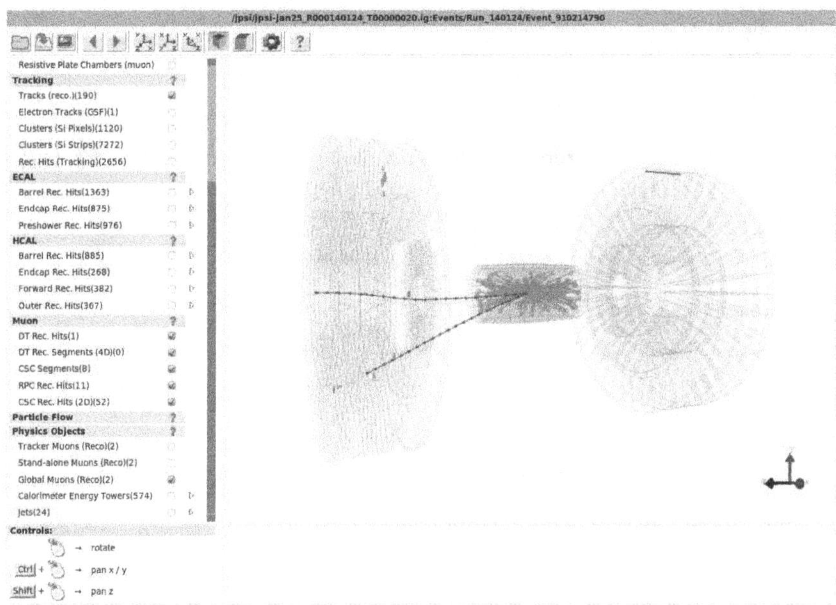

Fig. 4. Screen capture of a $J/\psi \to \mu\mu$ event in the I2U2 browser-based event display.

Masterclasses have made use of the I2U2 event display as well as iSpy in visual selection of dimuon events. Event files in csv format can be easily read and analyzed using standard spreadsheet software. Since its initial release thousands of high school students from over twenty countries have analyzed CMS data in the context of the masterclasses and e-Labs.

5. Conclusions and future plans

The first release of data from the CMS experiment at the LHC has been very successful. The format in which the data has been released and the tools provided in which to study it have allowed for easy distribution and analysis by the public. In the context of masterclasses and e-Labs thousands of students all over the world have learned about particle physics, the LHC, the CMS experiment, and have gained a flavor of how physicists analyze real data.

Future plans include the preparation and release of the rest of the datasets, preparation of more exercises based on the current datasets , and improvement of the analysis tools used to study the data. The extensibility of the data format may even allow for collaboration between other LHC experiments that have released data to the public such ATLAS and ALICE.

6. Acknowledgements

I wish thank the organizers of ICATPP. I also wish to thank my collaborators in I2U2, Quarknet, and CMS, and acknowledge the support of the US DOE and NSF.

References

1. CMS Collaboration, *J. Instrum.* **3**, S08004 (2008).
2. `http://cms.web.cern.ch/org/physics-cms` and J. Thompson, these proceedings.
3. `http://www18.i2u2.org`
4. `http://ippog.web.cern.ch/ippog` and D. Barney, these proceedings.
5. `http://quarknet.fnal.gov`
6. `http://www.physicsmasterclasses.org` and K. Cecire, these proceedings.
7. `http://iguana.cern.ch/ispy/ig-specs.htm`
8. `http://iguana.cern.ch/ispy`
9. `http://json.org`
10. `http://www18.i2u2.org/elab/cms/event-display`

Software package for the characterization of Tracker layouts

S. Mersi[†*], D. Abbaneo[†], N. De Maio[†], G. Hall[‡]
* *stefano.mersi@cern.ch*

On behalf of the CMS collaboration

† *European Organization for Nuclear Research*
CERN CH-1211
Genève 23, CH

‡*Imperial College London*
South Kensington Campus
London SW7 2AZ, UK

The high luminosity operation of the LHC will require an upgrade of the CMS Silicon Strip Tracker, possibly implementing trigger capabilities. In order to evaluate the possible options and geometries, a standalone software package has been developed (tkLayout) to generate detector layouts, evaluate the effect of inactive material and provide an a priori estimate of the tracking performance. The package can be used to compare the performance of different options, and then to optimise the chosen detector concept; tkLayout is not specific to CMS, thus it can be adapted to design studies for other tracking detectors. The technology of tkLayout is presented, along with some results obtained in the context of the CMS Tracker design studies.

Keywords: Tracking; Design; CMS; HL-LHC.

1. Overview

The Silicon Strip Tracker[1,2] currently operating in CMS[3] has just collected data for an integrated luminosity of about $5\,\mathrm{fb}^{-1}$ and the LHC performance is expected to grow in the next year. After delivering about $500\,\mathrm{fb}^{-1}$ the machine is expected to undergo a major upgrade in the early 2020s, after which its instantaneous luminosity should exceed the design goal, eventually reaching $5 \times 10^{34}\,\mathrm{cm}^{-2}\mathrm{s}^{-1}$. This scenario is known as High-Luminosity LHC (HL-LHC).

This scenario represents a challenge for the CMS detector due to the high radiation dose and number of pile-up events (up to 100 to 200), which will adversely affect the L1-Trigger.

Together with the HL-LHC upgrade the Silicon Strip Tracker of CMS should be replaced to cope with the charged particle density and the collaboration is therefore studying the option of instrumenting the upgraded tracker with detector modules capable of measuring a track transverse momentum[4,5] (p_T) locally and sending high-p_T hits to a real-time tracking processor embedded in the Level-1 trigger.

With this design, the upgraded tracker should provide more functionality than the present one, and yet the tracking resolution of the current detector is limited by its amount of material. This poses a key question in the design of the tracker upgrade: what is the optimal trade-off between adding functionality (like high-p_T real-time tracking) and improving tracking performance? What is the impact of different design choices on the final detector performance?

2. A detector layout design software

When designing a new detector for high energy physics it is common practice to rely on detailed (and complex) Monte Carlo simulations. While this cannot be avoided for the qualification of a detector design, this approach needs a lot of effort to understand simulation details and to optimise event reconstruction algorithms.

It is therefore useful to evaluate the potential performance of a design by estimating the track parameter resolution from first principles.

For this reason a software tool was developed capable of creating the full three-dimensional description of a tracker starting from a small set of design parameters as described in Section 2.1. This was complemented with a customisable model of material, as shown in Section 2.2. Finally, the combined information on material amount and sensor properties is used to estimate the detector's potential performance in terms of tracking (Section 2.3).

2.1. *Layout creation*

The first functionality of this software is to place detector modules in three-dimensional space, so hermeticity is assured. A small number of plots are then produced for consistency checks.

The detector geometry is built starting from a few basic parameters. Modules can be arranged either in barrel layers or in end-cap disks. Here a right-handed system is used with the y axis vertical and the z axis aligned with the beam line.

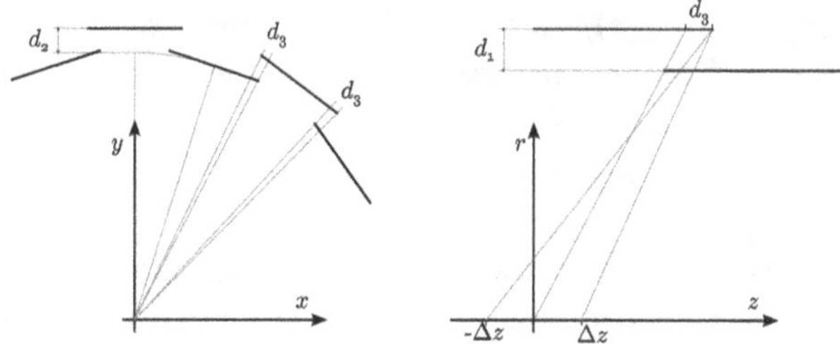

Fig. 1. Placement of modules in the transverse (left) and longitudinal (right) planes. Figure is not to scale.

Barrel modules are represented by rectangles of arbitrary size and are arranged in an arbitrary number of layers (optionally automatically adjusting the layer radii to make an optimal use of the detector surface). The length of the barrel is determined by the number of consecutive modules arranged in the z direction.

Detectors are staggered in the radial direction to avoid volume collisions, by requiring a small radial gap d_1 between modules on the same layer at the same φ and different z and a larger radial gap d_2 between modules on the same layer at the same z and different φ.

Hermetic coverage along z direction is guaranteed by placing adjacent modules so that an overlap of at least d_3 would be seen from the origin $(0, 0, 0)$ and no gap would be seen from $(0, 0, \pm\Delta z)$, where Δz is the expected variance of the z coordinate of the primary interaction points. Hermetic coverage in φ is guaranteed by placing adjacent modules with an overlap of at least d_3 along this direction (see Figure 1).

End-cap modules are represented by rectangles or wedge shapes and are placed in rings staggered in z by a small gap d_1. Rings are staggered in z between each other by a larger gap d_2. All the disks are built so that a small overlap d_3 is seen between all adjacent modules by straight tracks passing through the origin. All disks are built with the same algorithm (to simplify the construction) and they are designed providing hermetic coverage for tracks starting from the origin, regardless if the ring is placed near to or far from the interaction point (within the specified z range).

All modules are then assigned a set of parameters (number of strips, power consumption, etc.). and the expected detector occupancy is estimated with a simple extrapolation from the present CMS Silicon Strip Tracker. A *specific occupancy* parameter was defined ($\pi_{B,E}(r)$) for barrel and end-cap modules separately as the number of hits per unit of polar angle per unit of eta $\pi_{B,E}(r) = n_{B,E}(r)/(\Delta\varphi\,\Delta\eta)$. Using a standard simulation of the current tracker with minimum bias events, a second order polynomial was used to fit $\pi_{B,E}$. This fit (scaled by the expected number of minimum bias events per collision) is then used to predict the expected occupancy in an arbitrary layout.

Hermetic coverage is confirmed by using straight lines originating from vertices placed on the beam line around the origin with a Gaussian distribution in z.

All summary information (number of modules, total power consumption, total number of channels, expected number of measurement points as a function of η, expected occupancy, ...) is stored in form of a mini-web site.

2.2. *Model for the material budget*

The material amount implied by detector modules is assigned to each module volume without any detail of the geometric distribution of material within the module itself. Material due to services running inside detector layers is also assigned to the module volumes, as these are hermetically covering the detection surfaces.

Several objects can be defined for each module type to represent the services reaching the module from the end-flange (A_i, like power cables and optical connection) and the components of the module itself (B_j, like sensors, supporting mechanics, cooling pipes). Each object is defined by its material composition (1 g Copper + 0.5 g PVC + ...).

The actual material assigned to a module depends on its position in the supporting structure: in a barrel layer the amount of material M is

$$M(n) = n\sum_i A_i + \sum_j B_j \tag{1}$$

where n is the module position on a rod ($n = 0$ for modules at $z \simeq 0$).

This parametrisation takes into account the accumulation of running services towards the end of a barrel layer: a module next to the end-flange will contain in its volume as many service lines as the number of modules it has in front. A similar computation is done for end-cap rings, with an additional scaling factor for the accumulation of services due to the fact that service density decreases when these spread outward from one module ring to the next.

Additionally each material A_i and B_j can be defined as “local” if it only adds to the material inside detecting layers (like support mechanics or

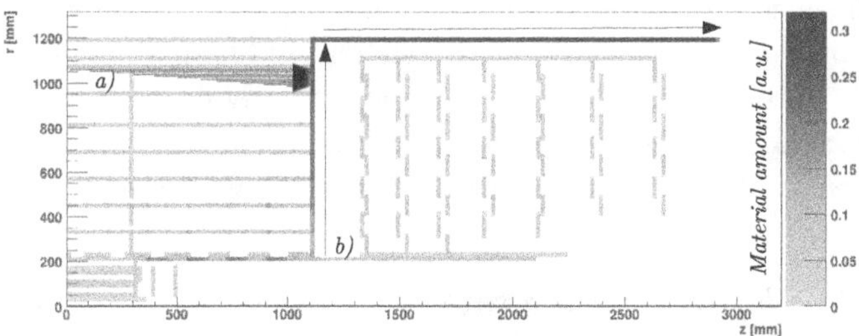

Fig. 2. Distribution of material (interaction length) in an (r, z) section of a tracker model. The accumulation of material along the barrel layers (a) is indicated by the arrows (not evident in this gray scale). The accumulation of routed services (b) is evident.

silicon sensors) or "exiting" if it implies the presence of some other services coming from outside (like cooling pipes or power lines).

Once the material of modules and services in the detecting layers is in place, additional volumes representing support structures and services running out of the detecting surfaces are created.

Materials in the service volumes are automatically created depending on the amount of material previously defined as "exiting" in the facing detecting layers with some configurable conversion rules (for example many small cooling pipes will join through a manifold into fewer larger pipes exiting). Part of these services are automatically routed up to the edge of the tracking volume with the same mechanism described above for services inside the detecting layers (for example the cooling pipes will be propagated, while the material of their manifolds will not). Additionally, some fixed amount of material can be added to the service volumes to represent specific objects. The routing procedure is sketched in Figure 2.

After the material assignment is performed, some summary plots are created: the η distribution of material, photon conversion and nuclear interaction probabilities, etc.

2.3. *Performance estimation*

The accuracy of the track parameters derived from a fitting procedure is described here, taking into account the precision of the measurement points and multiple scattering. This is done by considering two distinct fits: a circle in the (r, φ) plane and a straight line in the (r, z) plane. In reality these are not independent, but this approximation was proven to be valid, *a posteriori*, by a comparison of the results derived with a full simulation of the

CMS Silicon Strip Tracker. These calculations closely followed those published[6,7] by Karimäki, with two main differences: first multiple scattering is taken into account here, while the author explicitly neglects it in the cited article; second we are only interested in the general solution of the problem, as tkLayout performs the computation for each particular case.

2.3.1. *Fit in the (r, φ) plane: a circle*

We consider a system of cylindrical coordinates (r, φ, z), centred on the interaction point. In the hypothesis of uniform magnetic field and no multiple scattering, the projection of the track on the plane with $z = 0$ will be a circle. If the actual measured N points are $P_i = (r_i, \varphi_i)$ the error ε_i of the measurement point i will be given by:

$$\varepsilon_i = \frac{1}{2}\rho r_i^2 - (1 + \rho d) r_i \sin(\varphi_i - \varphi_0) + \frac{1}{2}\rho d^2 + d \qquad (2)$$

where r_i is the sensor radial position, φ_0 is the initial track polar angle in the transverse plane, φ_i is the polar angle of each hit, ρ is the track curvature ($\rho = 1/R$, with R radius of curvature) and d is the distance of closest approach of the track to the z axis (transverse impact parameter).

The track fitting procedure is intended to obtain the estimate $\widehat{\alpha}_i$ of the track parameters $\alpha_i = \{\rho, \varphi_0, d\}$ from the set of measured points P_i. Here we will derive the explicit formulation of the covariance matrix of the estimated parameters $U_{ij} = \mathrm{cov}\,[\widehat{\alpha}_i, \widehat{\alpha}_j]$: we are interested in the error measurement $\sigma(\widehat{\alpha}_i) = \sqrt{U_{ii}}$.

It is assumed here that the best fit of the trajectory to the measured points will be given by minimising $\chi^2 = \sum_{i,j} \varepsilon_i W_{ij} \varepsilon_j$ where $W = U^{-1}$ is the weight matrix. This is given by $W = D^T C^{-1} D$, where $D_{ij} = \partial\varepsilon_i / \partial\alpha_j$ and C is the covariance matrix of the measured points $C_{ij} = \mathrm{cov}\,[\varepsilon_i, \varepsilon_j]$

The matrix D can be derived from (2): $\partial\varepsilon_i / \partial\alpha_j \simeq \{r_i^2/2, -r_i, 1\}$. The approximation is not strictly necessary for the computation, but it is valid for tracks with bending radii much larger than the tracker radius ($p_T \gg 1\,\mathrm{GeV/c}$ for CMS) and it makes the matrix D only dependent on the detector geometry as seen by the track.

If one rotates the reference frame by φ_0, the track is directed along the new x axis at the origin and the coordinate measured by the sensors is y. If the sensor radial positions r_i are taken to be error-less and for high-momentum tracks ($\rho d \ll 1$) we have $\mathrm{d}\varepsilon_i \simeq -\mathrm{d}y_i$ and thus $C_{ij} = \mathrm{cov}\,[y_i, y_j]$.

The particle multiple scattering can be considered a deviation from the ideal track to be measured, and thus it can be treated as a measurement error. Hence, the matrix C can be written as $C = C^M + C^R$, with C^M the covariance matrix due to multiple scattering and $(C^R)_{ij} = \delta_{ij}\sigma(y_i)$ the covariance matrix due to the intrinsic resolution of the measurement point P_i, which is taken as an input by tkLayout.

The matrix elements $(C^M)_{ij}$ can be evaluated by first computing the larger matrix $(\widetilde{C}^M)_{mn} = \mathrm{cov}\left[\widetilde{y}_m, \widetilde{y}_n\right]$ representing the covariance matrix of the impact points of a straight track traversing M perpendicular planes, where it can interact through multiple scattering (not only the measurement planes). The matrix C^M can be obtained from $\widetilde{C}^M$ by dropping the $M-N$ lines and columns corresponding to the interaction of the particle with a non-sensitive element.

Given the radial positions of interactions $r_n = r_1, r_2, \ldots, r_M$ with scattering angles $\vartheta_n = \vartheta_1, \vartheta_2, \ldots, \vartheta_M$, the deviation from the ideal path is $\widetilde{y}_n \simeq \sum_{i=1}^{n} (r_n - r_i)\,\vartheta_i$ and, since the scattering angles ϑ_n are uncorrelated

$$\widetilde{C}_{mn} = \langle \widetilde{y}_m, \widetilde{y}_n \rangle \simeq \sum_{i=1}^{\min(n,m)} (r_m - r_i)\,(r_n - r_i)\,\langle \vartheta_i^2 \rangle \tag{3}$$

With this method a number of sample tracks are generated for each layout and for each of them the list of crossed materials and modules is obtained, so that $C = C^M + C^R$ and D can be computed. Finally the expected resolution on the track parameters can be obtained from the covariance matrix $U = \left[D^T C^{-1} D\right]^{-1}$.

2.3.2. *Fit in the (r, z) plane: a line*

The fit in the (r, z) plane is actually the first to be evaluated for each selected track. In this plane the tracks are taken to be straight lines (again in the $\rho d \ll 1$ approximation), thus the track equation is

$$r_i = \frac{z_i - z_0}{\mathrm{ctg}(\theta)} \tag{4}$$

The resolution on the last two track parameters (the longitudinal impact parameter z_0 and the polar angle θ) can be evaluated with the same method described above on the simpler linear fit. The error on z is taken from the longitudinal resolution Δz for the barrel modules. For the end-cap modules an effective $\Delta z = \Delta r \cdot \tan(\theta)$ is assigned, where Δz is the module's longitudinal resolution.

2.3.3. *Notes on end-cap modules*

In Section 2.3.1 it was assumed that r_i is known for all the hits and y_i is the only source of uncertainty. This is a good approximation only for barrel modules, while for end-cap modules z_i is the known coordinate of the sensor. In the actual implementation we assumed that during the fit procedure the parameter $\mathrm{ctg}(\theta)$ is obtained first (with its error) and then r_i is obtained through (4). To the first order approximation this can be translated into an effective $(\Delta y)_1$ error on y_i through:

$$(\Delta y)_1 = \rho r_i \left[r_i \frac{\Delta \mathrm{ctg}(\theta)}{\mathrm{ctg}(\theta)} \right] \quad (5)$$

Also in case of wedge-shaped end-cap modules, the actual measured coordinate is $y = r\varphi$, as for barrel modules, but in case of square-shaped modules an additional error is present (also depending on $\Delta\mathrm{ctg}(\theta)$) and is taken into account in a similar way to $(\Delta y)_2$. For end-cap modules the effective resolution is taken to be $\sigma^2_{\mathrm{eff}}(y_i) = \sigma^2(y_i) + (\Delta y)_1^2 + (\Delta y)_2^2$.

3. Model validation

The current CMS Silicon Strip Tracker was used as a bench-mark to test the tracking performance predicted by tkLayout. A layout was created with the same number of barrel layers and end-cap disks. The material model was tuned in order to reproduce the Outer Barrel material and it was then applied to the whole tracker layout. A smaller inner tracker was also generated to represent the pixel[8] detector. The correct strip pitch p_i was assigned to all of the strip tracker sensors and the resolution $\sigma_i^2 = p_i^2/12$ was taken to be that of a binary readout system. The resolution of the pixel detector was instead assigned explicitly to match the actual detector. No further tuning was performed.

The modelling done by tkLayout produces a layout similar to that of the actual CMS Outer Barrel, so it fails to correctly model the Inner Barrel (these details are described in Ref. 2). Even more notably the peculiar design of the CMS Tracker, with a smaller end-cap inserted inside the Outer Barrel is not intended to be modelled in tkLayout, which thus fails to reproduce the actual service routing for this sub-detector.

The actual CMS tracker is accurately simulated in the official software of the Collaboration, based on GEANT4, which was validated against the collision data collected. The material amount (measured in radiation lengths) obtained from the full simulation is compared in Figure 3 with the same

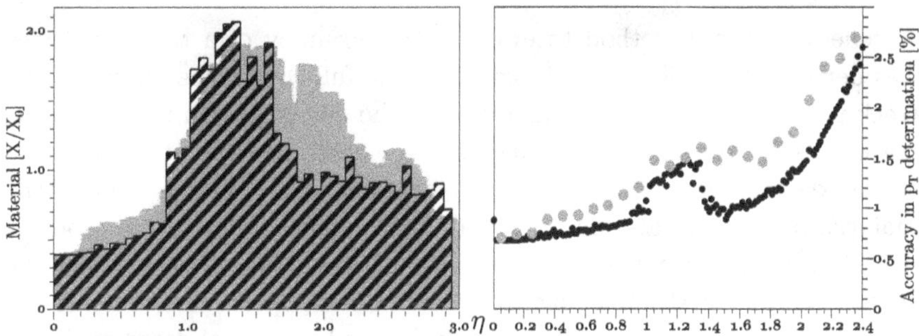

Fig. 3. Material amount (left) and expected resolution on p_T for muons with p_T = 10 GeV/c (right) as a function of pseudorapidity η. Comparison between official CMS software full simulation (gray) and tkLayout estimate (black).

quantity estimated with tkLayout. The material amount is correctly reproduced at low η and at the material peak ($\eta = 1$ to 1.4). The accuracy in measuring p_T of a muon with $p_T = 10\,\text{GeV/c}$ is shown in the same figure. The tkLayout estimation matches closely the full simulation in the η range where the material amount is correctly reproduced.

A complete comparison between the resolution on track parameters estimated by tkLayout and by the full simulation is shown in Table 1. Two bench-mark cases are considered here: muons with $p_T = 10$ and $100\,\text{GeV/c}$. The estimated uncertainty of the parameter fit is averaged for tracks with η in the ranges shown below. The uncertainty is correctly reproduced by tkLayout with an error of around 10% to 20%.

Table 1. Residuals = (tkLayout / FullSim) - 1

	$p_T = 10\,\text{GeV}$			$p_T = 100\,\text{GeV}$		
η	0 to 0.8	0.8 to 1.6	1.6 to 2.4	0 to 0.8	0.8 to 1.6	1.6 to 2.4
p_T	-12 %	-19 %	-20 %	11 %	-0.5 %	-5 %
d_0	7 %	13 %	15 %	17 %	35 %	41 %
φ	12 %	12 %	10 %	13 %	16 %	7 %
ctg(θ)	14 %	10 %	14 %	13 %	17 %	56 %
z_0	13 %	7 %	5 %	17 %	15 %	22 %

4. Conclusions

A generic analytic method to evaluate the accuracy of a tracking device was derived. This allows one to compute the full covariance matrix of the track parameters for any configuration of the detector with respect to the track. A software tool (tkLayout) was developed, capable of describing a tracker detector geometry from few basic parameters. A simple model of the material budget was also implemented, which allows a simple and coherent definition of detector materials and automatically takes into account the service routing in the tracking volume. TkGeometry also implements the mentioned estimation of tracking resolution and it was validated against a full simulation of the CMS tracker. The accuracy of tkLayout was proven to be around 10% to 20%.

This software is currently used within CMS to evaluate possible tracker layout concepts, and it proved to be specially useful in order to make a fair comparison between different design approaches.

References

1. *CMS: The Tracker Project Technical Design Report* 1998. CERN-LHCC-98-06.
2. *Addendum to the CMS Tracker TDR* 2000. CERN-LHCC-2000-016.
3. G. L. Bayatian *et al.* CERN-LHCC-2006-001.
4. M. Pesaresi, *PoS* **VERTEX2010**, p. 047 (2010).
5. M. Pesaresi and G. Hall, *JINST* **5**, p. C08003 (2010).
6. V. Karimäki, *Nuclear Instruments and Methods in Physics Research Section A* **305**, 187 (1991).
7. V. Karimäki, *Nuclear Instruments and Methods in Physics Research Section A* **410**, 284 (1998).
8. C. Amsler, K. Bosiger, V. Chiochia, W. Erdmann, K. Gabathuler *et al.*, *JINST* **4**, p. P05003 (2009).

ASTROFIT:
AN INTERFACE PROGRAM FOR EXPLORING COMPLEMENTARITY IN DARK MATTER RESEARCH

N.NGUYEN* and D.HORNS

Institute for Experimental Physics, University of Hamburg,
Luruper Chaussee 149, 22761 Hamburg, Germany
**E-mail: nelly.nguyen@desy.de*

T.BRINGMANN

II. Institute for Theoretical Physics, University of Hamburg
Luruper Chausee 149, 22761 Hamburg, Germany
E-mail: torsten.bringmann@desy.de

AstroFit is an interface adding astrophysical components to programs for fitting physics beyond the Standard Model (BSM) to experimental data from collider searches. The project aims at combining a wide range of experimental results from indirect, direct and collider serarches for Dark Matter (DM) and confronting it with theoretical expectations in various DM models. Here, we introduce AstroFit and discuss first results.

Keywords: BSM physics, Dark Matter, Complementarity.

1. Introduction

Various different experiments explore the properties of Dark Matter (DM) while a plethora of theories offer explanations to its nature. It is the goal of the AstroFit project to constrain DM models by combining experimental data from both astrophysics and collider physics, and thus find the best fit regions for the parameter space of the underlying theory. AstroFit itself is a Fortran program, serving as an interface between programs used in particle physics for fitting physics beyond the Standard Model, such as Fittino,[1–3] and programs like DarkSUSY[4,5] that can be used to calculate theoretical predictions for direct and indirect detection experiments. An overview of experimental input usable with AstroFit will be given as well as an example of how to use AstroFit in combination with the Fittino program. Here, the first results from a Constrained Minimal Supersymmetric Standard Model

(CMSSM) fit including information from latest collider (i.e. Large Hadron Collider (LHC)), direct detection and indirect detection instruments are presented.

2. Complementarity of Experiments

Indirect searches for Dark Matter concentrate on finding products from DM annihilation, such as photons of different energy ranges (e.g. radio, X-ray, γ-ray), antiprotons, positrons and neutrinos with specially designed experiments, ranging from ground-based Cherenkov telescopes like H.E.S.S.,[6] MAGIC[7] and VERITAS[8] over satellite experiments like Fermi-LAT[9] and PAMELA[10] to balloon experiments as ATIC,[11] among many others. For this analysis, we used photon flux upper limits from dwarf galaxies by H.E.S.S.[12] and Fermi-LAT[13] as observables, which have been calculated by AstroFit as

$$\frac{\mathrm{d}\Phi(\Delta\Omega, E_\gamma)}{\mathrm{d}E_\gamma} = \frac{1}{8\pi}\frac{\langle\sigma v\rangle}{m_\chi^2}\frac{\mathrm{d}N_\gamma}{\mathrm{d}E_\gamma} \times \bar{J}(\Delta\Omega)\Delta\Omega\,, \tag{1}$$

where $\mathrm{d}\Phi(\Delta\Omega, E_\gamma)/\mathrm{d}E_\gamma$ is the differential photon flux, E_γ the photon energy, $\langle\sigma v\rangle$ the velocity weighted annihilation cross-section and m_χ the mass of the DM particle, $\mathrm{d}N_\gamma/\mathrm{d}E_\gamma$ the differential number of gammas produced per annihilation per energy and $\bar{J}(\Delta\Omega) = (1/\Delta\Omega)\int_{\Delta\Omega}\mathrm{d}\Omega\int_{l.o.s.}\mathrm{d}l\ \rho_{\mathrm{DM}}^2(l)$ the integral over the DM density squared along the line of sight. Instead of using the photon flux upper limit as direct observable, the thermally averaged cross-section times the relative velocity can be used if desired, as some experimental limits are preferably presented in terms of $\langle\sigma v\rangle_{\max}$.

Direct detection instruments such as DAMA/LIBRA,[14] CoGeNT[15] and Xenon,[16] located in underground laboratories, measure signals from DM interactions with target elements like Xenon, Germanium or compounds such as NaI. Signals are detected via scintillation, phonons or ionization, depending on the experiment. Assuming a scattering of a weakly interacting massive particle (WIMP) as DM candidate with the target material, the spin-independent scattering cross-section per nucleon, conventionally adopted for comparison between experimental results, is calculated as follows in AstroFit:

$$\sigma_{nucleon}^{SI} = \frac{(Z\sqrt{\sigma_p} \pm (A-Z)\sqrt{\sigma_n})^2}{A^2}\,, \tag{2}$$

with σ_p and σ_n being the spin-independent cross-section for one proton or neutron, respectively, and Z and A being the atomic and mass number of the target element. Latest results from direct detection have recently shown a conflict between measurements from different experiments. While the DAMA, CoGeNT and CRESST[17] collaborations each published detections of signals, the Xenon collaboration has shown upper limits lower than the signal regions of the previous experiments for the WIMP-nucleon cross-section. Possible reasons for this diversity are still being discussed.

As particle physics input from *collider experiments*, information on B- and Z-physics, like masses, edges in mass spectra, widths, asymmetries, etc. have been used in this analysis as well as the anomalous magnetic moment of the muon and latest event rates from the ATLAS experiment at LHC (see 1–3 for details). Also, the relic density of Cold Dark Matter (CDM) provided by WMAP,[18] $\Omega_{\rm DM}h^2 = 0.1123 \pm 0.0035$ which can be directly compared to the theoretical prediction for thermally produced DM from DarkSUSY, is used as an observable in AstroFit.

3. Structure of AstroFit

As depicted in figure 1, AstroFit provides an extensive database of relevant experimental data to add to a fit process, at the moment implemented in Fittino. While the steering of AstroFit is done by a user-friendly text input file, information on the particle spectrum is given via a SUSY Les Houches Accord (SLHA) file directly from Fittino. From this spectrum file, AstroFit subroutines are designed to calculate the model predictions by using various DarkSUSY functions. The theoretically estimated observables are then compared to actual measurements from astrophysical experiments. From the comparison, a $\Delta\chi^2$-contribution is calculated for each individual observable and is handed to Fittino in each step of the minimization process to be used together with the information from particle physics. In a stand-alone subroutine of AstroFit, the $\Delta\chi^2$ is calculated by applying $\Delta\chi^2 = \sum\left((O_{\rm exp} - O_{\rm theo})/\sigma_{\rm exp}\right)^2$ for data points or continuously in parabolic increase by extrapolation using the given confidence level for limits or containment regions for claimed signals for any observable. A minimization strategy in Fittino then determines the lowest global χ^2, and the results comprise the best fit regions for all model parameters.

4. Analysis and Results for CMSSM Model

Exemplarily, this analysis focussed on the CMSSM, using a Markov Chain Monte Carlo algorithm to fit the following parameters (defined at the Grand Unification Theory scale): M_0 – the universal scalar mass, $M_{1/2}$ – the uni-

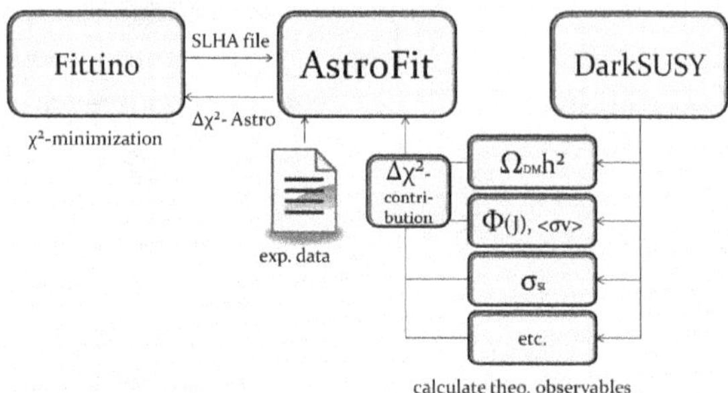

Fig. 1. AstroFit flowchart: Minimization process taking place in Fittino, calculations of theoretical observables using subroutines from DarkSUSY and comparison to astrophysical observables done in AstroFit.

versal gaugino mass, A_0 – the common trilinear coupling and $\tan\beta$ – the ratio of the vacuum expectation values of the two Higgs fields. We assume the neutralino to be the DM particle and set $\mu = +1$.

For our analysis, the data from CoGeNT has been used in one fit and the data from Xenon100 and Xenongoal in two others (compare 19) . Additionally, fits with data for Xenon1T and CRESST[20] are in preparation.

While claimed DM signals from the CoGeNT experiment could not find agreement with a CMSSM fit, upper limits from the Xenon experiment give constraints in this scenario. In figure 2, fit results are shown using only particle physics information (including latest 2fb^{-1} results from LHC) and the relic density of CDM. Adding further constraints from indirect and direct detection, i.e. photon flux upper limits and results from Xenon100 data, leads to the results shown in figure 3. In the fit, the bulk region is already excluded by all other input conditions, while adding Xenon100 limits provides an additional constraint on both the coannihilation and the funnel region.

Photon flux upper limits do not constrain the CMSSM parameter space so far. However, using the latest joint likelihood results for dwarf spheroidal galaxies from Fermi-LAT (see 13) could show first effects.

5. Conclusion and Outlook

Using all available results from DM searches can help confirm, constrain or exclude regions in parameter space of DM models remarkably, making it possible to edge closer to understanding physics beyond the Standard Model in general and the nature of DM in particular. In this study, observables from particle physics have been combined with the relic density of

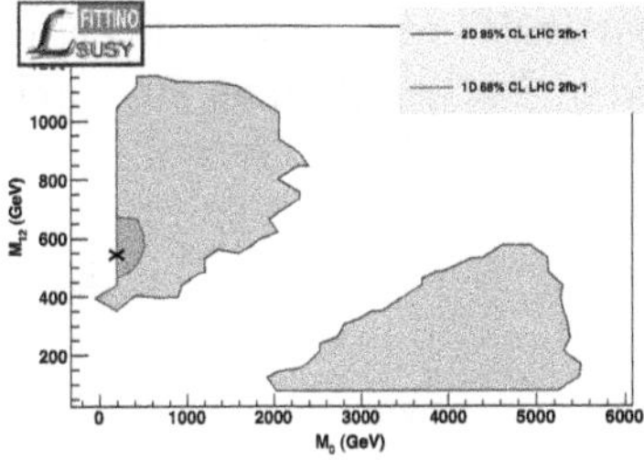

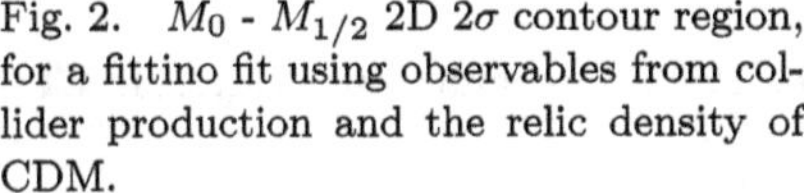
Fig. 2. M_0 - $M_{1/2}$ 2D 2σ contour region, for a fittino fit using observables from collider production and the relic density of CDM.

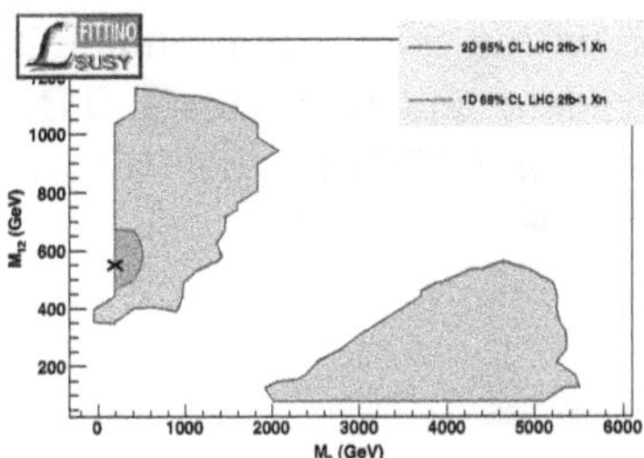

Fig. 3. Same as Fig. 2, including also Xenon100 11d limits. The additionally constrained models lie mostly in the coannihilation and funnel region.

CDM, photon flux upper limits from dwarf spheroidal galaxies as well as direct detection signals and upper limits, showing considerable impact in constraining the CMSSM parameters. While claimed DM signals from the CoGeNT experiment could not be fit within a CMSSM model, Xenon100 upper limits contributions have been shown. With upcoming fits using AstroFit and Fittino, the parameter space of DM models can be constrained even further, using latest results from the Xenon100 (as well as predictions from Xenongoal and Xenon1T) and CRESST direct detection instruments, incoming results from the LHC and indirect detection information. As numerous results in the field of experimental DM physics are expected in the near and mid-term future, it is extremely important to strengthen integrative approaches, which AstroFit facilitates. Within AstroFit, it is planned to provide subroutines for all major observables in DM research, such as photon fluxes in different energy regimes and from different sources, antiproton, positron and neutrino fluxes in addition to the relic density of Cold Dark Matter and data on the scattering cross-section from direct detection, which can be used together with measurements from collider searches. The inclusion of available information in particle physics, astroparticle physics and cosmology in a combined DM search is an important tool to help interprete state-of-the-art physics in these disciplines. As such, it is planned to release a public version of AstroFit for a larger community of researchers to use.

Acknowledgements

NN acknowledges financial support through the DFG funded collaborative research center SFB 676. Also, NN acknowlegdes the distinguished collaboration within the Fittino community with particular regards to P. Bechtle, X. Prudent and B. Sarrazin. TB is supported by the German Research Foundation (DFG) through the Emmy Noether grant BR 3954/1-1.

References

1. P. Bechtle, K. Desch, P. Wienemann, Comput. Phys. Commun. **174** (2006) 47-70. [hep-ph/0412012].
2. P. Bechtle, K. Desch, M. Uhlenbrock, P. Wienemann, Eur. Phys. J. **C66** (2010) 215-259. [arXiv:0907.2589 [hep-ph]]
3. P. Bechtle, B. Sarrazin, K. Desch, H. K. Dreiner, P. Wienemann, M. Kramer, C. Robens, B. O'Leary, Phys. Rev. **D84** (2011) 011701. [arXiv:1102.4693 [hep-ph]]
4. Gondolo, P., Edsjö, J., Ullio, P., et al. 2003, Identification of Dark Matter, 256 [arXiv:astro-ph/0211238]
5. P. Gondolo, J. Edsjö, P. Ullio, L. Bergström, M. Schelke, E.A. Baltz, T. Bringmann and G. Duda, `http://www.darksusy.org`
6. F. Aharonian *et al.* [H.E.S.S. Collaboration], Astron. Astrophys. **457** (2006) 899-915. [astro-ph/0607333].
7. J. Albert *et al.* [MAGIC Collaboration], Astrophys. J. **674** (2008) 1037-1055. [arXiv:0705.3244 [astro-ph]].
8. T. C. Weekes, *et al.*, Astropart. Phys. **17** (2002) 221-243. [astro-ph/0108478].
9. W. B. Atwood *et al.* [LAT Collaboration], Astrophys. J. **697** (2009) 1071-1102. [arXiv:0902.1089 [astro-ph.IM]].
10. M. Boezio, *et al.*, New J. Phys. **11** (2009) 105023.
11. Isbert, J., Adams, J. H., Ahn, H. S., et al. 2009, Bulletin of the American Astronomical Society, 41, #475.25
12. A. Abramowski *et al.* [HESS Collaboration], Astropart. Phys. **34** (2011) 608-616. [arXiv:1012.5602 [astro-ph.HE]].
13. The Fermi-LAT Collaboration: M. Ackermann, Ajello, M., Albert, A., et al. 2011, [arXiv:1108.3546]
14. R. Bernabei *et al.* [DAMA Collaboration], Nucl. Instrum. Meth. **A592** (2008) 297-315. [arXiv:0804.2738 [astro-ph]].
15. C. E. Aalseth, *et al.*, Phys. Rev. Lett. **107** (2011) 141301. [arXiv:1106.0650 [astro-ph.CO]].
16. E. Aprile *et al.* [XENON100 Collaboration], [arXiv:1107.2155 [astro-ph.IM]].
17. Jochum, J., Angloher, G., Bauer, M., et al. 2011, Progress in Particle and Nuclear Physics, 66, 202
18. E. Komatsu *et al.* [WMAP Collaboration], Astrophys. J. Suppl. **192** (2011) 18. [arXiv:1001.4538 [astro-ph.CO]].
19. E. Aprile *et al.* [XENON100 Collaboration], [arXiv:1104.2549 [astro-ph.CO]].
20. Angloher, G., Bauer, M., Bavykina, I., et al. 2011, [arXiv:1109.0702]

QUANTIFYING THE UNKNOWN: ISSUES IN SIMULATION VALIDATION AND THEIR EXPERIMENTAL IMPACT

M. G. PIA *, M. BATIČ, G. HOFF and P. SARACCO

INFN Sezione di Genova,
Genova, 16136, Italy
**E-mail: MariaGrazia.Pia@ge.infn.it*
www.ge.infn.it

M. BEGALLI

State University of Rio de Janeiro,
Rio de Janeiro, RJ 20550-013, Brazil
E-mail: begalli@fnal.gov

M. HAN, C. H KIM and H. SEO

Hanyang University,
Seoul, 133-791, Korea
E-mail: chkim@hanyang.ac.kr

S. HAUF and M. KUSTER

Technical University Darmstadt,
Darmstadt, 64289, Germany,
E-mail: steffen.hauf@astropp.physik.tu-darmstadt.de

L. QUINTIERI

INFN Laboratori Nazionali di Frascati,
Frascati, 00044, Italy
E-mail: Lina.Quintieri@lnf.infn.it

G. WEIDENSPOINTNER

Max-Planck-Institut Halbleiterlabor,
Munich, 81739, Germany
E-mail: Georg.Weidenspointner@hll.mpg.de

A. ZOGLAUER

University of California at Berkeley,
Berkeley, CA 94720-7450, USA
E-mail: zog@ssl.berkeley.edu

The assessment of the reliability of Monte Carlo simulations is discussed, with emphasis on uncertainty quantification and the related impact on experimental results. Methods and techniques to account for epistemic uncertainties, i.e. for intrinsic knowledge gaps in physics modeling, are discussed with the support of applications to concrete experimental scenarios. Ongoing projects regarding the investigation of epistemic uncertainties in the Geant4 simulation toolkit are reported.

Keywords: Monte Carlo; Simulation; Validation; Geant4.

1. Introduction

The investigation and quantification of epistemic uncertainties[1] is well established in the domain of deterministic simulation, but it is a relatively new domain of research in the context of Monte Carlo simulation. It concerns the issue of how epistemic uncertainties, i.e. uncertainties due to lack of knowledge, namely in modeling physics processes, affect the outcome of Monte Carlo simulation. Epistemic uncertainties are present in Monte Carlo codes, when the absence of experimental data, or inconsistencies in available measurements, prevent the achievement of firm conclusions regarding the correct values of physics parameters or the validity of physics models used in the simulation. Epistemic uncertainties can induce systematic effects in the simulation; this issue is especially important, since can negatively affect the accuracy and reliability of simulation results.

Due to their intrinsic nature, related to lack of knowledge, epistemic uncertainties are difficult to quantify. Despite their importance in complex systems, there is no generally accepted method of measuring epistemic uncertainties and they contribute to the reliability of the whole system. A variety of mathematical formalisms[2] has been developed for this purpose; the most common methods adopted in the context of deterministic simulations are interval analysis and applications of Dempster-Shafer theory of evidence.[3] Nevertheless, these techniques may not always be directly applicable in identical form to the treatment of epistemic uncertainties in Monte Carlo simulations.

Sensitivity analysis[4] is a tool for exploring how uncertainties influence the model output. This approach is adopted in two exploratory projects, which intend to evaluate possible methods for uncertainty quantification related to the Geant4[5,6] simulation toolkit. Epistemic uncertainties are usually represented in statistical analyses as a set of discrete possible or plausible choices; in the exploratory analyses described here the possible choices concerned the values of physical parameters or a set of alternative physics models.

2. Proton depth dose simulation

This study assesses the impact of epistemic uncertainties associated with various physics models and parameters relevant to Monte Carlo codes through the simulation of a concrete use case: the depth dose profile in water generated by a proton beam as in a typical therapeutical facility. For this purpose the geometry of a realistic hadrontherapy facility[7] publicly available as a Geant4 example was utilized.

A sensitivity analysis has examined the response of the system to a wide set of modeling approaches; this method plays a conceptually similar role to the interval analysis method applied in deterministic simulation, where parameters subject to epistemic uncertainties are varied within bounds. The environment for this analysis has been realized in the context of a Geant4-based application; the characteristics of Geant4 as a toolkit, encompassing a wide variety of physics models, allow the configuration of the simulation with a large number of different physics options in the same software environment. The outcome associated with the various models subject to investigation has been compared by means of rigorous statistical analysis methods to quantitatively estimate the effect of physics-related systematic uncertainties.

Epistemic uncertainties are associated with parameters used by the simulation models: proton stopping powers and the water mean ionization potential, for whose values a consensus has not yet been achieved in the scientific community. The interval analysis has highlighted a shift in the position of the Bragg peak related to range of variability of these parameters.

Nuclear interactions, both eleastic and inelastic, affect the shape of the depth dose distribution. Epistemic uncertainties in this domain derive from the still incomplete validation of the hadronic models used by the simulation. No statistically significant effects on the depth dose profile have been identified as a result of the interval analysis; nevertheless, significant systematic differences deriving from epistemic uncertainties are observed in other features of the simulation outcome, such as secondary particle production. Multiple scattering modeling also plays an important role in the evaluation of possible sources of systematic effects.

Sensitivity analysis as applied to this simulation topic contributes to identify and quantify possible systematic effects in the simulation; it cannot infer anything about the validity of any of the physics models, for which experimental data would be needed.

The analysis of the proton depth dose simulation shows that the appearance of systematic effects generated by epistemic uncertainties in the physic models depends not only on the intrinsic characteristics of the uncertainties, but also on the characteristics of the simulation environment.

3. Atomic binding energies

General purpose Monte Carlo codes use a variety of compilations of atomic electron binding energies, either deriving from theoretical calculations or from empirical evaluations of direct and indirect experimental data. Despite the fundamental character of these atomic parameters, there is no consensus among the various Monte Carlo systems and physics models about their values: the differences across the binding energies reported in the various compilations range from the electronvolt scale to several hundred electronvonvolts.

The analysis adopted two complementary approaches: on one side direct validation based on binding energies measurements, on the other side the evaluation of how different compilations of these parameters contribute to the accuracy of physics observables calculated by the simulation with respect to experimental data.

Reference experimental values for direct validation of atomic binding energies are relatively limited: the main issue for direct validation consists of discrepancies in experimental values due to calibration effects, for instance when measurements are taken in different laboratories and exploit different experimental techniques. Two sets of reference data concerning core shells, which have been subject to a process of recalibration and evaluation, have been assembled by Powell[8] and NIST (United States National Institute of Standards),[9] which encompass respectively only 65 and 81 binding energy values. In addition, NIST reports reference ionization energies for all elements.[10]

Direct comparison of the binding energies in the various compilations by means of statistical methods has identified the compilation by Williams[11] as the one, among those considered in this study, exhibiting the best compatibility with Powell and NIST reference data. Regarding ionization energies, the compilation by Carlson[12] appears to best reproduce NIST reference values.

Characteristic K and L-shell X-ray transition energies are more accurately calculated by using binding energies compiled by Larkins.[13]

The values of atomic binding energies can significantly affect the accuracy of ionization cross section calculations, both for electron and proton impact ionization. Among the compilations subject to analysis, EADL

(Evaluated Data Library)[14] contributes to deteriorate the accuracy of ionization cross sections with respect to empirical compilations.

No significant effect depending on the choice of binding energies is observed in the photon spectrum in Compton scattering accounting for Doppler broadening.

4. Conclusions

The exploratory analysis of epistemic uncertainties in two Monte Carlo simulation domains has highlighted their contribution to simulation accuracy and their capability of generating systematic effects in simulation results. Further investigations are in progress to identify and quantify epistemic uncertainties in Geant4 physics models.

Due to the limited page allocation in these conference proceedings, the detailed results of the analysis cannot be reported here; they can be found in dedicated publications.[15,16]

References

1. W. L. Oberkampf et al., *Reliab. Eng. Syst. Safety* **75**, 333 (2002).
2. J. C. Helton, *Reliab. Eng. Syst. Safety* **84**, 1 (2004).
3. G. Shafer, *A Mathematical Theory of Evidence* (Princeton Univ. Press, 1976).
4. T. G. Trucano et al., *Reliab. Eng. Syst. Safety* **91**, 1331, (2006).
5. S. Agostinelli et al., *Nucl. Instrum. Meth. A* **506**, 250 (2003).
6. J. Allison et al., *IEEE Trans. Nucl. Sci.* **53**, 270 (2006).
7. G. A. P. Cirrone et al., *IEEE Trans. Nucl. Sci.* **52**, 262 (2005).
8. C. J. Powell, *Appl. Surf. Sci.* **89**, 141 (1995).
9. J. R. Rumble Jr. et al., *Surf. Interface Anal.* **19** 241 (1992).
10. W. C. Martin and W. L. Wiese *Atomic, Molecular and Optical Physics Handbook*, ed. by G. W. F. Drake (AIP, Woodbury, NY, 1996)
11. D. R. Lide ed., *CRC Handbook of Chemistry and Physics* (CRC, Boca Raton, FL, 2009).
12. T. A. Carlson, *Photoelectron and Auger Spectroscopy* (Plenum, NewYork, 1975).
13. F. P. Larkins, *Atom. Data Nucl. Data Tables* 20, 311 (1977).
14. S. T. Perkins et al., *Tables and Graphs of Atomic Subshell and Relaxation Data Derived from the LLNL Evaluated Atomic Data Library (EADL), Z=1-100*, UCRL-50400 Vol. 30 (1991).
15. M. G. Pia et al., *IEEE Trans. Nucl. Sci.* **57**, 2805 (2010).
16. M. G. Pia et al., *IEEE Trans. Nucl. Sci.* **58**, in press (2011).

IMAGE FUSION SOFTWARE IN THE CLEARPEM-SONIC PROJECT

M. PIZZICHEMI[†], N. DI VARA, G. CUCCIATI, A. GHEZZI and M. PAGANONI

Dipartimento Fisica, Università Milano - Bicocca,
Piazza della Scienza, 3 - 20126, Milano, Italy
[†]*E-mail: marco.pizzichemi@mib.infn.it*

F. FARINA*

Dipartimento di Informatica, Sistemistica e Comunicazione
Università Milano - Bicocca, Viale Sarca 336/14 - 20126, Milano, Italy

B. FRISCH

CERN
CH-1211 Geneve 23, Switzerland

R. BUGALHO

Laboratorio de Instrumentação e Fisica Experimental de Particulas
Av. Elias Garcia 14 - 1° 1000-149 Lisboa, Portugal

ClearPEM-Sonic is a mammography scanner that combines Positron Emission Tomography with 3D ultrasound echographic and elastographic imaging. It has been developed to improve early stage detection of breast cancer by combining metabolic and anatomical information. The PET system has been developed by the Crystal Clear Collaboration, while the 3D ultrasound probe has been provided by SuperSonic Imagine. In this framework, the visualization and fusion software is an essential tool for the radiologists in the diagnostic process. This contribution discusses the design choices, the issues faced during the implementation, and the commissioning of the software tools developed for ClearPEM-Sonic.

Keywords: Image Visualization, ClearPEM-Sonic, 3DSlicer

*Now at Consortium GARR, Italian NREN, Via dei Tizii, 6 - 00185 Roma, Italy

1. Introduction

ClearPEM-Sonic is a Positron Emission Tomograph (developed by the Crystal Clear Collaboration) specialized for mammography[1] , that is combined with a 3D ultrasound system that can acquire both B-mode and elastographic images. The ultrasound device is called Aixplorer and is provided by SuperSonic Imagine.[2] The exam modalities provide a set of three images, that are referred to a common reference frame with the use of a magnetic probe provided by TrackStar.[3] As the main goal of this system is to improve early stage detection of breast cancer by combining metabolic and anatomical information, the visualization and fusion software plays an essential role in the whole diagnosis process. The machine is installed in a dedicated area of Hopital Nord in Marseille.

2. Requirements

The software requirements include both device control and image processing functionalities. Superimposition of the three volumes in a common reference frame is mandatory as well as their fine registration in a single volume. The patient and exam database is managed with the acquisition software AcqTool, developed by Laboratorio de Instrumentação e Fisica Experimental de Particulas (LIP). The reconstruction and visualization software needs to deal with it in order to load and save the single exams. An interface has to be provided to allow the medical operators to deal with the reconstruction scripts developed by LIP.[4] In addition, the ability to control the magnetic positioning system and to export images into DICOM format is required. Finally, the software has to retain sufficient user-friendliness in its graphical interfaces to be used in a clinical environment.

3. Solution

Currently, a significant number of frameworks and applications are available to visualize and manage medical images, both commercial and open source. During the design phase of the software for ClearPEM-Sonic, a number of development platforms have been evaluated according to different aspects, like the support of multiple OS platforms, the features they provide in terms of image fusion algorithms and the extendability of new modules through de-facto standard libraries or plug-ins. The result of the survey highlighted that the best candidate is the 3DSlicer framework.[5] 3DSlicer is an open source application and framework for medical image visualization and manipulation that can be expanded with plug-ins coded in C++ and Python. The image manipulation methods have been implemented as C++ plug-ins using the Insight Segmentation and Visualization Toolkit (ITK).[6] Additional functionalities, like image reconstruction control and image conversion to DICOM,[7] have been coded in C++ modules. Python scripting language expressiveness has provided benefit in implementing both the graphical user interfaces of the modules and the codes that orchestrate the single tasks of the interactive diagnosis process. The developed modules have been designed to retain high level of independence, allowing extension of new features and high adaptability to the changes in the diagnostic work flow.

4. Plug-in description

Four plug-ins were built to support the clinical operators in the management of the patients archive, image reconstruction and image visualization. The software architecture is shown in Fig. 1.

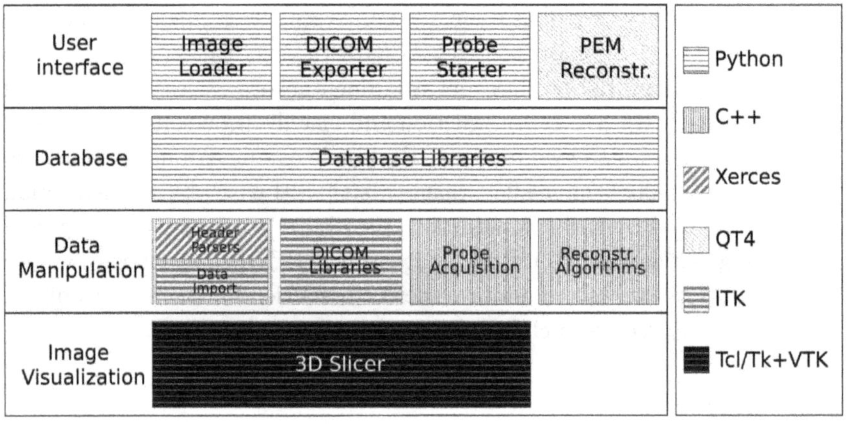

Fig. 1. Architecture of the visualization and fusion software

4.1. *Magnetic probe control*

The plug-in controlling the magnetic positioning system relies on C++ dedicated libraries and drivers provided by the manufacturer. It is implemented via an XML description that is used to generate the C++ command line code and the GUI (Graphical User Interface). The program reads the data coming from the probe, average them, and produces an header file, complying with the Interfile standard, containing the necessary information about the measurement. The plug-in also allows one to choose the output directory that will be used by the image loader to superimpose the PET-US images.

4.2. *PEM reconstruction*

A stand-alone GUI, written in Qt4, has been developed to control the PEM reconstruction process. It is connected with the database and it is able to process the data acquired by the PEM scanner using the reconstruction algorithms developed by LIP. The GUI allows the user to choose the reconstruction parameters (e.g. voxel size, number of iterations, type of filter) and constantly monitor the progress of computation. The output files are clinical images written in RAW format and stored in the appropriate patient folder of the system database.

4.3. *Image loader*

A custom plug-in has been developed to superimpose the images acquired through the different exam modalities in the same frame of reference in 3DSlicer. Relying on C++ libraries ITK and Xerces,[8] the program is capable of automatically loading in the correct relative position the PEM and ultrasound RAW images, exploiting the positional information acquired from the magnetic tracking system.

4.4. *DICOM exporter*

Exporting images to DICOM format is mandatory to integrate in hospitals' PACS. Therefore, each image imported into 3DSlicer can be exported to DICOM format by means of a dedicated plug-in implemented in ITK. The user has also the possibility to modify each DICOM parameter related to the patient before the creation of the image series.

5. Early validation

Software validation has been carried out through extensive tests on pre-existing datasets, as well as on combined PET, echographic and elastographic images on breast phantoms[9] acquired by ClearPEM-Sonic. Furthermore, during the delivery and commissioning phase all the plug-ins were tested, in order to overcome integration and different version issues due to the modular structure of the framework. An example of a combined PEM, echographic and elastographic visualized into 3DSlicer is shown in Fig. 2.

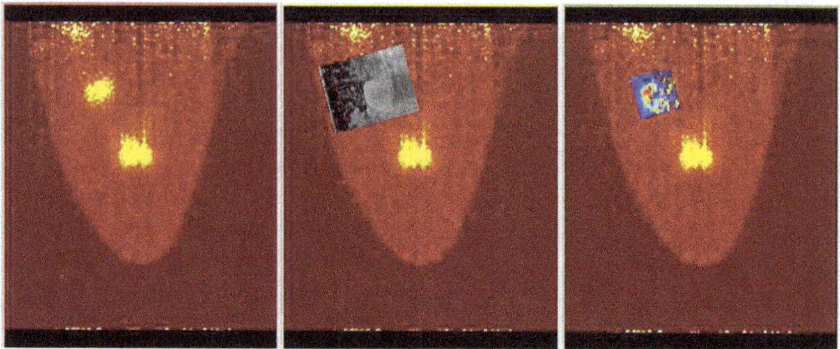

Fig. 2. Sagittal views of the combined dataset on a breast phantom. The images depict the PEM image, the overlapping of PEM and echographic images, and elastographic over PEM images (from left to right)

6. Conclusion

We developed an image fusion software for the ClearPEM-Sonic project capable to visualize the combined dataset produced by the PET-US scanner. The framework provides a user-friendly GUI, the possibility to control all exam parameters and the image reconstruction, and a seamless integration with the hospital PACS. Various tests carried out on gelatin breast phantoms have shown the reliability of the software in accurately superimpose the images acquired in different modalities. The high degree of software automation, combined with the innovative characteristics of this mammography device, provides a useful tool for early detection of breast cancer.

References

1. P. Lecoq and J. Varela, *Nuclear Instruments and Methods in Physics Research Section A: Accelerators, Spectrometers, Detectors and Associated Equipment* **486**, 1 (2002).
2. SuperSonic Imagine website. `url:http://www.supersonicimagine.fr/`.
3. Ascension Technology Corporation : 3D Guidance medSAFE. `url:http://www.ascension-tech.com/`.

4. M. V. Martins and N. Matela, *IEEE Nuclear Science Symposium Conference Record* (2005).
5. 3DSlicer. url:http://www.slicer.org.
6. Insight Segmentation and Registration Toolkit. url:http://www.itk.org.
7. DICOM Digital Imaging and COmmunication in Medicine. url:http://medical.nema.org/.
8. Xerces-C++ XML Parser url:http://xerces.apache.org/xerces-c/.
9. B. Frisch and J. Dang, *IEEE Transactions on Nuclear Science* **58**, 660 (2011).

COSMIC RAYS AND THE ENVIRONMENT

T. SLOAN

Physics Department, Lancaster University,
Lancaster, UK
**E-mail: t.sloan@lancaster.ac.uk*

The changing cosmic ray rate has been proposed as a mechanism which could have contributed significantly to the global warming seen during the last century. The evidence for this is discussed and upper limits for the effect are derived from the evidence. The effects of cosmic rays on lightning and its effect on the global electric circuit will also be discussed.

Keywords: Cosmic rays; global warming; lightning.

1. Introduction

Since industrialisation in the 19th century the carbon dioxide (CO2) concentration in the atmosphere has increased from 280 ppmv to 380 ppmv and at present this is rising exponentially. At the same time the mean global surface temperature has increased by roughly 0.8 degrees C. Climatologists' models of the atmosphere predict that the measured rise in the CO2 concentration will cause the observed rise in temperature. Hence the hypothesis of man-made global warming.

The Intergovernmental Panel on Climate Change (IPCC), representing the view of the majority of climatologists, consider that it is "very likely" that the observed global warming is man-made. The term "very likely" is defined to be 90% probable. However, this is disputed by some, mainly non-climatologists, mostly claiming that the effect could be natural since the Earth has seen larger swings in temperature in the more distant past than the one observed since industrialisation. For this to be case, the climatologists' models have to be wrong. In addition, there has to be an as yet undiscovered phenomenon of the right sign to cause the global warming seen during the last 150 years.

Assuming that the climatologists' models are wrong, one such mechanism proposed for a new undiscovered effect is the influence of cosmic rays on clouds. Ney[1] first proposed such an effect. More recently, the effect has received much publicity following the observation of a correlation between a change in the low level cloud cover and the cosmic ray rate during the solar maximum in solar cycle 22 (1985-1997).[2,3] The latter authors went on to hypothesise that the decreasing cosmic ray rate over the last century caused less ionization and so less low cloud. This would lead to smaller reflection of the solar radiation and hence warming of the Earth.

In the first part of this paper this observation is discussed in some detail and the attempts made to corroborate the hypothesis are described. In the second part of the paper the mechanism by which cosmic rays are thought to cause lightning will be discussed.

2. The influence of cosmic rays on cloud formation.

The condensation of water droplets is a familiar process in cloud chambers used to display charged particle tracks. However, the conditions in such chambers are roughly a 4 times supersaturated vapour pressure in an ultra-clean environment. The condensation process in clouds in the atmosphere is a much different process. In the atmosphere dust particles, salt particles from sea spray and many other impurities are present and inside a cloud the supersaturation level is unlikely to exceed a fraction of 1%.

Water molecules have a large electric dipole moment so that two molecules can readily cohere. However, it is less clear how more molecules come together to form larger aerosol particles. It is thought that such particles containing multiple water molecules are formed by their attraction to microscopic impurity particles made from deliquescent materials such as sulphuric acid or sea salt. It has been shown that ionization could enhance such formation.[4] Indeed, the latest results from the CLOUD experiment at CERN[5] and from Enghoff et al.[6] show that ionization helps the formation of aerosol particles in the presence of sulphuric acid. However, the experiments show that this is only one process which affects cloud formation in the presence of many others which are not yet identified.

Aerosol particles, once formed, diffuse in the atmosphere until they reach a region where the water vapour pressure is large enough so that they can grow either by condensation or by merging with other aerosol particles. In this way the droplets seen in clouds (radius of order 10 μm) are formed. In clouds these continue to grow until they are large enough to fall as raindrops (radius $\sim$ mms).

If ionization from cosmic rays influences this process then a change in the cosmic ray rate would imply a change in the total cloud cover. Indeed the cosmic ray rate has been observed to decrease between the years 1850 to 1950,[7] presumably due to decreases in the solar wind. Solar radiation is reflected from low level clouds. If cosmic rays influence low cloud cover, as hypothesised in Marsh and Svensmark,[3] then the decreasing cosmic ray rate leads to decreasing low level cloud cover which allows more solar energy to fall on the Earth leading to more warming.

Marsh and Svensmark noticed a correlation between low level cloud cover (measured from the ISCCP satellite data) and the cosmic ray neutron monitor count rates in solar cycle 22 which lasted from 1985 to 1995. These authors showed that if they assume that the dip due to the decreased cosmic ray rate at solar maximum (1992) is a global phenomenon the decreasing cosmic ray rate, seen over the last century, could have produced radiative forcing (i.e. warming) at a similar level to that produced by CO2.

Much work has gone into attempting to corroborate this hypothesis.[8] Detailed analysis shows that there are many problems with the hypothesis as follows.

(1) The correlation is not present if the ISCCP Vis-IR cloud data are used rather the IR data i.e. another set of data does not show the correlation. Furthermore the IR data now available through to 2010 clearly shows the decrease peaking in 1992 but there is no discernible dip in the low level cloud cover in the following solar cycle which peaked in 2003. I.e. another solar cycle does not exhibit the dip which should have been there if there had been a significant contribution of cosmic rays to the global cloud cover. This can be seen from the plots on the ISCCP web site, http://isccp.giss.nasa.gov/climanal7.html.
(2) If the minimum in 1992 is caused by the decreasing cosmic ray ionization as the solar activity reaches a maximum then this minimum should vary with the magnetic latitude of the Earth since the influence of the solar activity is much smaller near the magnetic equator than near the poles. The amplitude of the dip in 1992 is observed to be constant with magnetic latitude. A statistical analysis shows that less than 23% of the dip comes from the variation with the cosmic ray rate at 90% confidence level.
(3) Only the low cloud cover shows the dip at 1992. The middle and higher level cloud covers, where the variation in cosmic ray ionization rate is larger, show no discernible correlation.
(4) The was a large burst of solar activity on 29 September 1989. This so called ground level event (GLE) was visible in neutron monitors around the World and it was seen in the Nagoya muon telescope. Hence

it should have produced a significant change in the atmospheric ionization. The was no significant change in the global cloud cover from before to after the GLE event.

(5) There was no significant change in cloud cover in the region of the Chernobyl reactor from before to after the nuclear accident in 1986 when large amounts of radioactivity were released into the atmosphere.

(6) There is a wealth of data on the effects of nuclear bomb tests on the weather. Some effects are felt in the region of the bomb blast. However, at large distances where the ionization level is still high but the effects of the blast are small, there are no recorded effects of ionization on cloud cover or the weather.

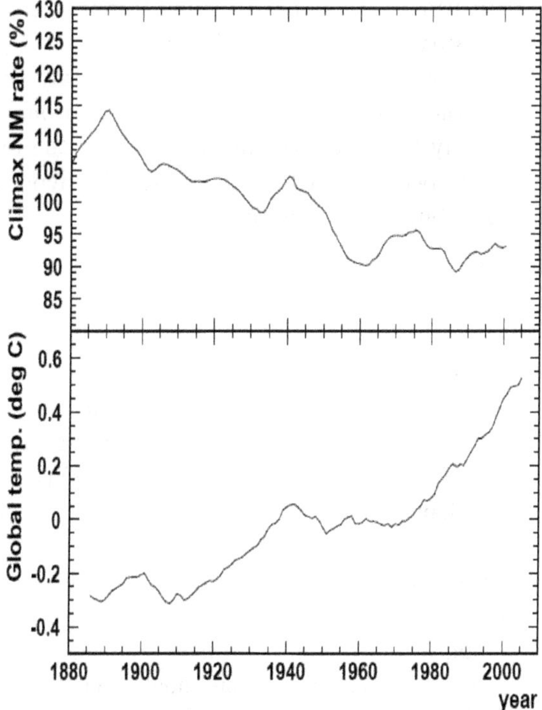

Fig. 1. **Upper** plot shows the equivalent Climax neutron monitor rate as a function of time, after 11 year smoothing has been applied. **Lower** plot shows the mean global surface temperature with the same smoothing.

None of these effects corroborate the hypothesis that a significant amount of the cloud cover is caused by ionization from cosmic rays.

Further attempts were made to correlate the long term variation in the global mean surface temperature with the ionization from cosmic rays and with solar activity via both the observed sun spot numbers and the reconstructed irradiance. A small correlation was observed from which it was deduced that less than 14% of the global warming seen since 1950 comes from the variation in solar activity.

Furthermore, the cosmic ray data were compared with the mean global surface temperature (see figure 1). It can be seen from figure 1 that the changing cosmic ray rate occurred mainly in the first half of the twentieth temperature when there was little change in the mean temperature. However, the cosmic ray rate has been almost constant since 1950 while the temperature has increased rapidly. A statistical analysis of this effect shows that less than 10% of the global warming seen in the twentieth century comes from the variation in the cosmic ray rate.

In summary, much work has been done to corroborate the hypothesis that solar activity, either through cosmic ray rate variations or otherwise, contributes significantly to the global warming seen in the twentieth century. No evidence has been found to corroborate the hypothesis and we deduce that the contribution of such an effect is less than 10% of the observed warming.

3. Lightning

Lightning is an electrical discharge between clouds or between a cloud and the Earth. The observed electric field gradients in the vicinity of charged clouds is of the order of kV/cm i.e. small compared to the breakdown potential of air (30 kV/cm). Hence lightning discharges must be caused by a different mechanism than the one responsible for ordinary spark discharges in the air.

The principle of the mechanism is thought to originate from the fact that the gain in energy between collisions with atoms of air of a particle approaching minimum ionizing is greater than the energy loss in the collision. This is illustrated in figure 2 which shows a curve of the stopping power, dE/dx, of a charged particle as a function of the kinetic energy of the particle. At an energy greater than a threshold, K_{th}, the energy gained per centimetre, which is proportional to the electric field E, is greater than the energy lost due to collisions with atoms i.e. the particle accelerates. Eventually it will gain enough energy to knock out an electron from an atom which has energy greater than the threshold K_{th}. In this way a particle in

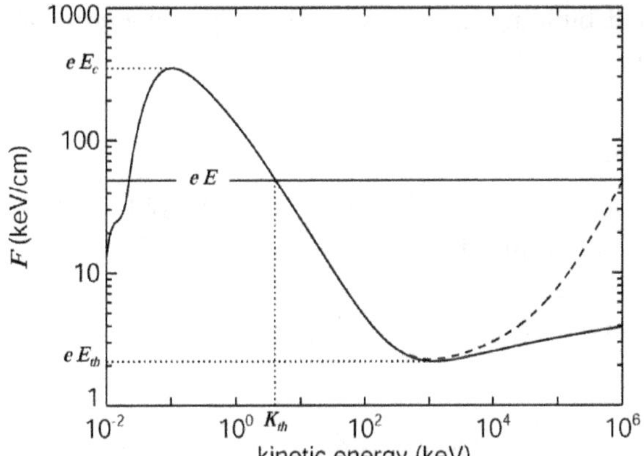

Fig. 2. The value of the stopping power $F = dE/dx$ as a function of electron kinetic energy. For kinetic energies above the threshold K_{th} the energy gained from the electric field in the thunder cloud is greater than that lost by collision losses. Hence the particle accelerates.

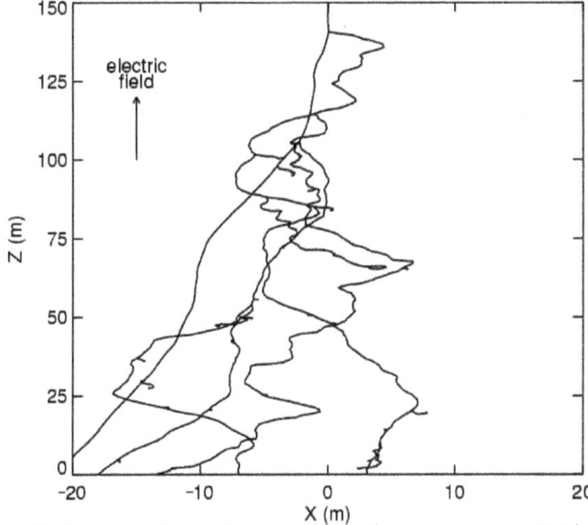

Fig. 3. Monte Carlo simulation of an electron (uppermost particle) of energy large enough to accelerate in the electric field of a thundercloud . Further accelerating electrons are produced over distances of hundreds of metres from energetic knock-on electrons. This leads to the build up of a lightning discharge.

the electric field of a thunder cloud can, over distances of hundreds of metres, produces other electrons which are capable of being accelerated. And so the discharge builds up into a lighting bolt. This process is illustrated in figure 3 which shows a Monte Carlo simulation of an energetic electron in an electric field building up into several further energetic electrons which are each accelerated.

Acknowledgements

I thank my colleagues Anatoly Erlykin and Arnold Wolfendale for their support and stimulating discussion. I also thank Joe Dwyer for explaining the origin of lightning to me. I am grateful to the Dr John C. Taylor Charitable Foundation for financial support.

References

1. E.P. Ney, *Nature* **183**, 451 (1951)
2. H. Svensmark and E. Friis-Christensen, *J. Atmos. Sol.-Terr. Phys.* **59**, 1225 (1997).
3. N.D. Marsh and H. Svensmark *Phys. Rev. Letts.* **85** 5004 (2000).
4. F. Yu, *J. Geo. Phys. Res.* **107** A7,10.1029/2001JA000248 (2002).
5. J. Kirkby et al., *Nature* **476** 429 (2011).
6. M.B. Enghoff et al., *Geo. Phys. Letts.* **38** L09805 (2011).
7. K. G. McCracken and J. Beer, *J. Geo. Res.* **112** A10101 (2007) doi:10.1029/2006JA01117.
8. A.Erlykin, T.Sloan and A.W.Wolfendale, *J. Atmos. and Sol.-Terr. Phys.* **72** 151-156 (2010) and references therein.

A PROPOSAL FOR THE ATLAS COSMIC MUON AND EXOTICS DETECTOR (ACME) AT THE LHC*

JAMES PINFOLD†, RICHARD SOLUK

Centre for Particle Physics, University of Alberta, Edmonton, Alberta, Canada

HELIO TAKAI

Physics Department, Brookhaven National Laboratory, Upton, NY, USA

The ACME group proposes to install a surface array of scintillation detectors at LHC point 1 above the ATLAS detector. This surface array in combination with the muon system of ATLAS allows two complimentary independent measurements of the electromagnetic and hadronic components of cosmic ray showers with particular sensitivity to the knee region of the cosmic ray energy spectrum. ACME will be a sensitive instrument for the study of primary composition, exotics such as centauro and anti-centauro events and the excess of high multiplicity muon bundles observed by such experiments as CosmoALEPH. The surface array also presents an outreach opportunity with students being involved in the construction and testing of detectors.

1. Introduction

Atmospheric cosmic ray showers are generally studied using either deep under-ground detectors which are sensitive to high energy muons, neutrinos and exotics, or with large surface arrays triggering on the electromagnetic or low energy muon components of the shower. The individual particle energies that these two types of detectors are sensitive to is quite different, surface detectors usually have a low threshold, on the order of a GeV, while deep underground detectors require muons with energies of hundreds of GeV to penetrate the overburden above them. As a result these two types of experiments probe different points in the shower development, with the majority of low energy surface particles being generated quite late in the shower while the high energy muons are produced by interactions at a much higher altitude and energy scale. Since the interpretation of cosmic ray air showers depends heavily on the

---

* Official approval of ACME by the ATLAS collaboration is still pending.

† Spokesman of the ACME collaboration. Email: jpinfold@ualberta.ca

interaction models used to simulate them, which generally involve forward physics extrapolated well beyond the level where it has been directly tested, it is highly desirable to have many different measurements, with different energies, to cross-check the simulation results.

With an overburden of 70m the energy threshold for muons reaching ATLAS is on the order of 40GeV and the electromagnetic component of the shower is blocked. This gives it an energy threshold that compliments both the low threshold of surface arrays and the high threshold of deep underground detectors. But the key feature of ACME is that by combining the ATLAS result with the surface array it is possible to have a simultaneous but independent measurement of both the electromagnetic surface shower and the relatively high energy muon component of the shower core. Having these two separate measurements for each shower aids in determining particle composition and yields a much better test of the shower models.

2. LEP/LHC Cosmic Ray Experiments

Several of the LEP detectors started cosmic ray programs. Of special interest were L3+C, which placed scintillators on top of L3 and operated a small surface array and CosmoALEPH which added scintillators at points in the LEP ring near the ALEPH detector. Additional work was done by the CosmoLEP group [1]

Unfortunately these cosmic ray programs started late in the lifetime of the LEP ring which limited the amount of data they were able to acquire. For instance the surface array of L3+C only ran with L3 for a few months. It is therefore highly desirable to start these projects early in the life of the detectors and in fact one experiment, ACORDE, has started at the LHC by placing long narrow scintillators on top of the ALICE detector.

The experiment which is currently most similar to ACME is the Baksan Underground Scintillation Telescope (BUST) which operates both a surface array and an underground detector. However it lacks the fine granularity and precision of the ATLAS muon system, has about 1/3 the detection area and is at a much greater depth, with a 200GeV muon cutoff energy.

3. Physics Goals

3.1. *Composition of primary cosmic rays*

Cosmic ray primaries have an energy spectrum that has been well established at all but the highest energies. Above 10^{10}eV the primary flux falls as $E^{-2.7}$ until the so called 'knee' region around 10^{15}eV where the slope changes to roughly $E^{-3.1}$ (Figure 1 [2]). Unlike the energy spectrum the composition of primaries above 10^{14}eV, the maximum energy that direct balloon and space based detectors have been able to probe, is still not established with their being a great deal of disagreement between experiments (Figure 2 [3]).

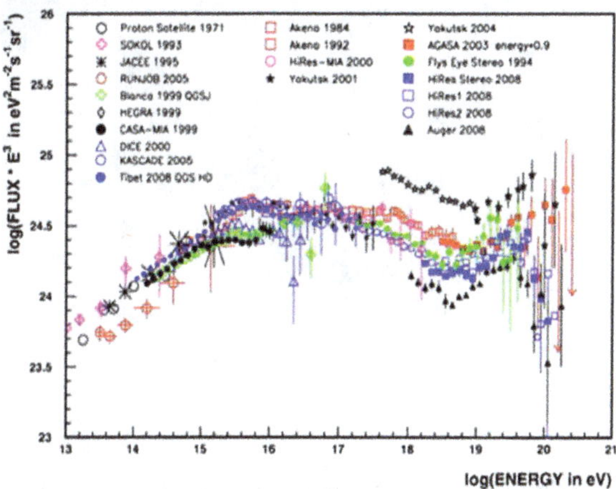

Figure 1. The cosmic ray energy spectrum as measured by various experiments.

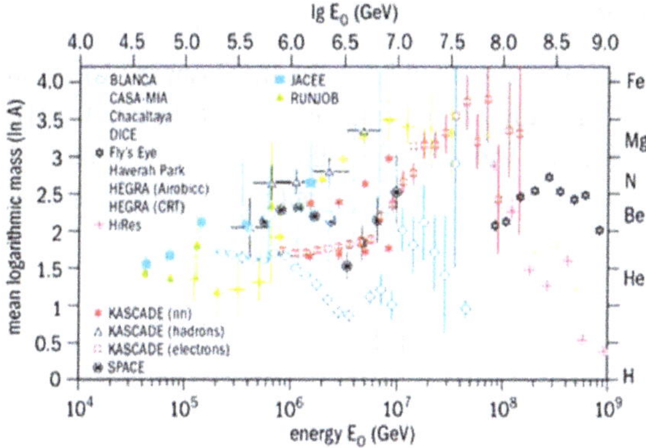

Figure 2: Atomic composition of cosmic ray primaries reported by various experiments.

The size of the ATLAS muon detection system and the energy threshold for muons which can penetrate the overburden above ATLAS make it well suited to investigating the knee region of the cosmic ray spectrum. A key feature of ACME is that in addition to the standard techniques for measuring energy and composition having a surface array allows a comparison to be made between the electromagnetic particle density at the surface and the muon density in ATLAS in a manner similar to figure 3 [4] (which was generated for a slightly greater depth then ATLAS). Primaries of different atomic number form diagonal bands as indicated providing an additional tool for determining composition.

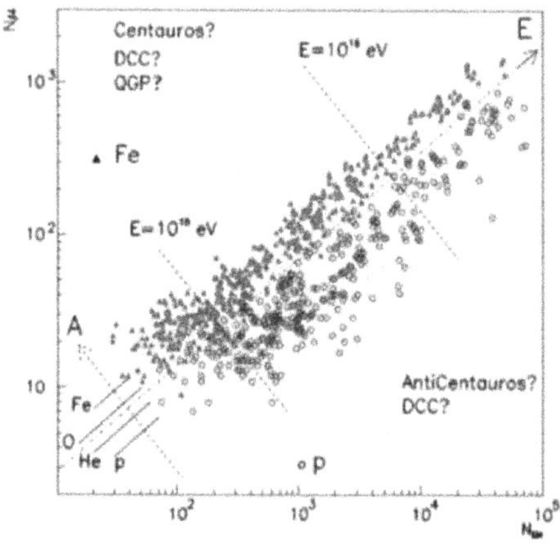

Figure 3. Surface particle density (x-axis) vs underground muon density for muons with energy greater than 70GeV. For iron (triangle) or proton (circle) primaries between $5{\cdot}10^{15}$ and $5{\cdot}10^{16}$eV.

3.2. *High multiplicity muon bundles*

Using its high precision tracking system CosmoALEPH observed multi-muon bundles originating from cosmic rays (Figure 4 [4]). CORSIKA simulations provide a good match to the data for low multiplicity bundles, assuming either proton or iron primaries, but fail to match the data for

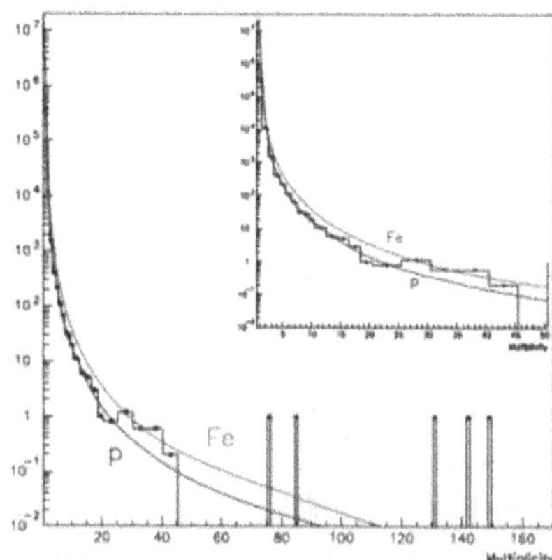

Figure 4. Muon multiplicity as seen by ALEPH along with simulation results for p or Fe primaries.

high multiplicity events some of which had over 100 muon tracks which had penetrated the overburden and reached ALEPH. Other experiments such as BUST[5] and Kolar Gold Fields[6] have reported similar high multiplicity events. The core density and lateral extent of these bundles can be used to estimate the energy and compositions or primaries.

3.3. *Exotics and other physics*

Data from ACME can be compared to that from nearby detectors, such as ACORDE, to look for wide area coincidences which have been suggested may be caused by photo disintegration of heavy primaries near the sun or certain exotic particles.

Some experiments have reported cosmic ray events with an anomalous ratio of charged to neutral particles within limited solid angle regions which could be a signal of the existence of disoriented chiral condensate (DCC) states created in high energy interactions in the shower. Referred to as centauro (or anticentauro) events these would appear in figure 3 as events off the preferred diagonal line.

4. Components of ACME

4.1. *The ATLAS muon system*

The ATLAS muon system consists of a number of different types of muon detectors and covers a cylinder of approximately 22m diameter and 28m length, giving it a cross section for cosmic muons of over 600m^2. The total area of all the detection planes is roughly 12,000m^2 with about one million channels of data readout (Figure 5).

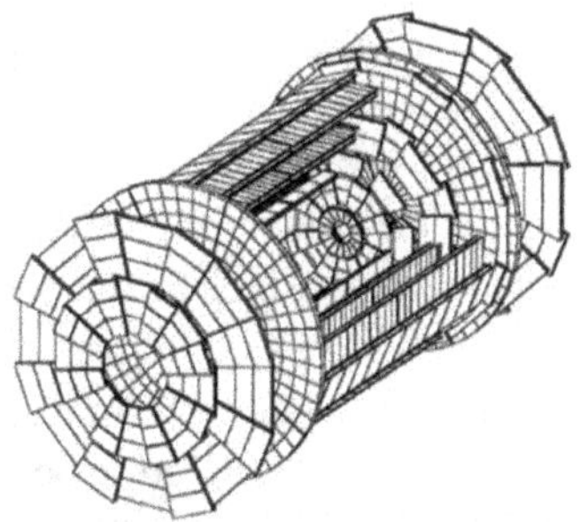

Figure 5. The ATLAS muon detection system.

The momentum resolution is a few percent up to 1 TeV muons. This far exceeds the complexity that could ever be expected for a dedicated cosmic ray experiment. For the purposes of ACME the main elements are 30mm diameter straw tube chambers arranged in overlapping layers, which give ATLAS more than sufficient granularity to track and count high multiplicity muon bundles.

4.2. *The surface array*

The planned surface array consists of insulated containers of liquid scintillator read out by wave length shifting fiber and silicon photomultipliers (Figure 6). A conventional PMT can be used as a fallback solution if signal yields are too low from the wavelength shifter/SiPMT combination. The liquid scintillator is PPO dissolved in linear alkylbenzene.

The detectors will have an area of $0.5m^2$ and be placed in a grid pattern with 10m spacing. Deployment will be in phases with the first step being a proof of concept test using plastic scintillator with the goal of working out triggering issues. Phase 0 consists of roughly 50 detectors on the rooftop above ATLAS (Figure 6), phase 1 doubles the array to the area west of the main buildings while phase 2 then extends the array over the rest of the CERN controlled land at LHC point 1.

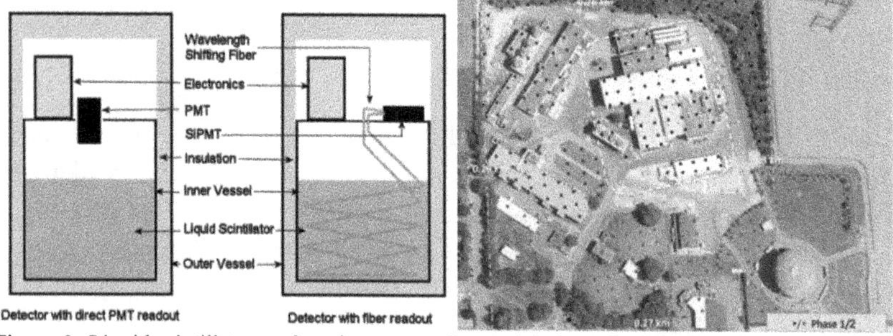

Figure 6. Liquid scintillator surface detectors and an aerial view of point 1 showing the planned detector layout, the white dotted line denotes the initial phase 0 detector deployment.

A signal from the surface array will be sent to the ATLAS level 1 central trigger processor (CTP) to allow the array to trigger ATLAS. The latency in the ATLAS trigger allows approximately 1.8μs for the trigger to travel from the surface to the CTP. Detectors too far away to trigger ATLAS directly will have their data recorded and time stamped using GPS timing. Since most GPS errors are common mode errors the relative timing error between two GPS antennas located within a few hundred meters of each other is only a few ns [7] allowing the surface data to be synchronized with the ATLAS trigger GPS time. Note that a second cosmic muon trigger operating in ATLAS independent of the surface array will be required to ensure that the ATLAS data is recorded.

5. Outreach

Construction and deployment of the surface array represents an outreach opportunity. Colleges that don't normally have research programs and institutions that lack the resources to become involved directly in large projects like ATLAS can easily participate in the ACME experiment. In fact the prototyping of the surface array is being carried out by students from just such a teaching college which doesn't have a strong research program but that has staff and students interested in becoming actively involved in particle physics.

6. Conclusion

The large area, high granularity, precise tracking and shallow depth of the ATLAS detector make it an attractive tool for cosmic ray studies. Combined with a surface array to form ACME, with the surface array generating level 1 ATLAS triggers, yields a detector that has an energy range complementary to existing experiments as well as independent measures of the electromagnetic and hadronic portions of each shower. ACME can be used to study the elemental composition of primary cosmic rays and shower structure in the knee region of the cosmic ray spectrum as well as investigate high multiplicity muon bundles and other exotic processes. Elements of the surface array are also simple enough that their construction and testing represent an excellent outreach opportunity.

Acknowledgments

This work was supported in part by the Teaching and Learning Enhancement Fund of the University of Alberta.

References

1. C. Taylor, *et al. "CosmoLEP, an underground cosmic ray muon experiment in the LEP ring".* CERN/LEPC 99-5.
2. M. Nagano, *New J. Phys.* **11**, (2009).
3. *"Cosmic-ray observations need accelerator data"*, CERN Courier, June 26, 2002.
4. V. Avati, *et al. "CORAL: A cosmic ray experiment in and above the LHC tunnel"*. CERN/2001-003.
5. V.N. Bakatanov, *et al.*, Physics of Atomic Nuclei, **Vol. 69 No. 9**, 1507 (1998).
6. Phys. Lett. **B267**, 138 (1991).
7. W. Brouwer, *et al.* Nuclear Instr. and Methods A, **Vol. 493**, 79 (2002).

THE TRIGGER SYSTEM OF THE TOTEM EXPERIMENT AT LHC

NICOLA TURINI
nicola.turini@cern.ch

Dipartimento di Fisica, Università degli Studi di Siena, Via Roma 55
Siena, 53100, Italy

M. BAGLIESI[1], M. BERRETTI[1], E. BOSSINI[1], R. CHECCHI[1], V. GRECO[1], J. KOPAL[3], E. OLIVERI[1], E. PEDRESCHI[2], F. SPINELLA[2].
[1]Dipartimento di Fisica, Università degli Studi di Siena, Via Roma 55
Siena, 53100, Italy

[2]INFN sezione di Pisa, Largo Pontecorvo 3
Pisa, 56100, Italy

[3]CERN
Geneve, Switzerland

The Totem experiment, dedicated to the measurement of proton proton total cross section and diffractive physics, has developed a trigger system that put together the information from the two diffractive detectors, T2 and T1, installed in the higher eta regions inside the CMS experiment in IP5, and the Roman Pots silicon detectors displaced at 220 m and 147 m from the interaction point. We will describe the algorithms used to perform the online data selection.

1. Trigger overview

1.1. *The hardware*

The trigger of the Totem experiment[1] follow the standard tree structure to produce the level one accept bit (Figure 1). When the level one accept bit is asserted the full DAQ chain of any subdetector start pulling data from the front-end chips buffers into the DAQ computers for data recording.

The Totem philosophy for the readout has been the standardization of the front-end electronics, all the three subdetectors are read by the same front-end chip named VFAT[2].

The VFAT chip has the possibility to produce up to 8 fast signals from 128 inputs that can be used for trigger purposes. The bits are or-ing the input digitized signals into an adaptable structure, for Roman Pots 16 strips are put in or to produce super strip trigger bit, for T2 a phi sector is divided into 8 η super pads, for T1 each sextant produce N trigger bits .

For the Roman Pots and the T2 the trigger bits are feeding local coincidence chips that make majority coincidences into trigger road bits. T1 is instead sending to the global trigger system the raw bits from the VFAT due to a more complex geometry but less multiplicity.

The coincidence chip can choose between N over the total planes of the telescope hits to create the trigger road information. For The Roman Pots the number of planes are five, while for T2 are ten.

The trigger data are sent to the counting room, sitting aside from the CMS cavern, through an optical link into VME receiver boards. We have six such boards, two for each subdetector. Each board will perform online cuts and prepare patterns sent to the final decision board called LONEG.

The synchronization signals and fast commands to the front end electronics are handled by a CMS specific hardware called LTC and local TTCci's sitting on each DAQ crate. In particular the LTC broadcast the CMS clock and orbit signals together with the level one accept. The trigger throttling system is handled by the LTC that blocks triggers when local DAQ buffers fill up and the number of blocked trigger are taken into account for the data analysis.

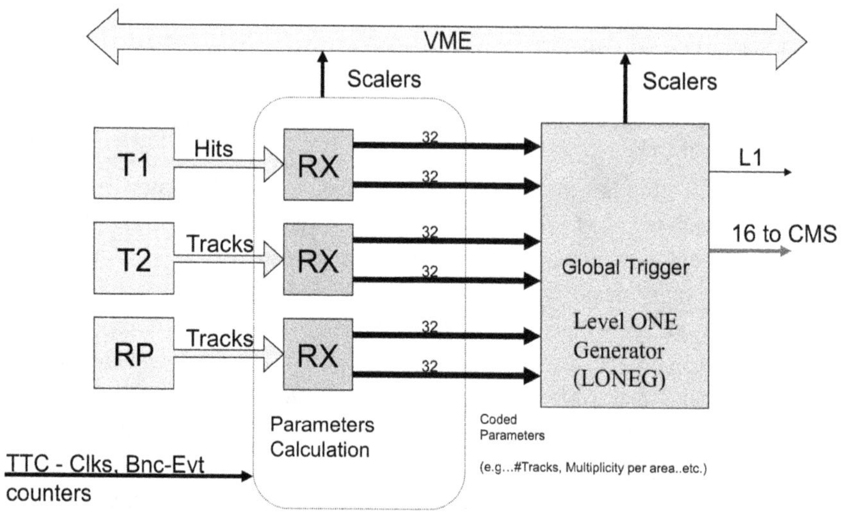

Figure 1. The Totem trigger structure.

1.2. *Roman Pots trigger selection algorithm*

Each Roman pot station has three silicon telescopes, two placed vertically respect to the beam axis and one horizontally.

The vertical pots can measure elastic and low ξ protons, while horizontal pots are sensitive to large ξ protons. Two such stations are placed about 6 meters away to allow the measurement of the protons angle. On each station 5 edgeless silicon[3] detector telescope track the charged particles with the strips at 45° from x beam coordinate (U view), while another 5 planes telescope, with the strips orthogonal with the previous one (V view), allows a 3D reconstruction for single track events. Each detector plane has 512 strips, organized into 32 trigger superstrips. A local coincidence chip select trigger roads when for each telescope we have more than 2-5 out of 5 superstrips collinear with the beam axis. The number of majority coincidence is programmable. We are using 3 out of 5 superstrips in coincidence. The U and V 2D roads are sent separately to the trigger receiver boards.

Most of the background recorded is due to multiple track events, coming mainly from minimum bias collisions, while protons are tagged as single track events. On the receiver boards we first count the number of tracks and we perform, depending on the run requests, a cut on the maximum number of roads for each pot. On high β runs we don't perform this cut and we prefer not to bias the data. The roads of each telescope are put in or and the U and V projection bits are then put in and. We have than one bit for each vertical (top and bottom) pot and one for horizontal. Counting that we have a total of 24 pots we end up with 24 bits for the RP detector. These bits are fed into the LONEG board as two independent trigger patterns, one for the 220m pots and one for the 147m pots.

1.3. *The T2 trigger algorithms*

The T2 detector is made by four telescopes covering each from 5.3 to 6.5 in pseudorapidity units and a little more than 180° in ϕ. There are two telescopes for each side respect to the interaction point 5.

Each telescope is half a cylinder made by ten GEM[4] detector planes. The trigger is based upon 13 ϕ sectors each one covered by 5x24 for a total of 120 pads.

The VFAT reading each sector organize the pad into 3x5 trigger superpad, for a total of 8 of them. In total each T2 GEM detector has 104 trigger superpads.

A Coincidence Chip is looking to ten consecutive sectors from each plane and generates the equivalent trigger road. It performs a majority coincidence N out of 10 superpads where N is programmable. To reduce the noise to negligible counts we typically adopt the cut at 5 out of 10 configuration.

Each quarter is treated as a separate entity and a final grouping is performed into the LONEG.

1.4. *T1 trigger algorithms*

T1 trigger is different due to a more complicate geometry. It is not possible to use Coincidence Chips, instead the hit information is much reduced respect to the other two detectors[5].

The hit information is then sent with the same type of hardware of the other subdetectors to the main trigger. The algorithm is based up on counting the minimum number of hits for each sextant plane, and then making a majority coincidence with the other planes of the same sextant.

For each sextant a single bit is sent to the LONEG for the final trigger decision.

2. The global trigger generation.

2.1. *The Level ONE Generator*

The Level ONE Generator (LONEG) is a board that put together all the subdetectors information and generates the trigger.

The trigger algorithms are implemented into an Altera StratixII FPGA, the minimal latency for the final decision is 4 clock cycles (100 ns).

The idea is to put in time all the detectors with the farthest one, the 220m Roman Pots, and generate up to 16 separate triggers. Each bit is correlated with a specific physics trigger or calibration triggers. The triggers are generated through a look up table and each trigger can be prescaled singularly.

Since Totem needs for his specific kind of measurements to select the LHC bunches to be used for the data acquisition, each trigger is accepted only if coincides with the requested ones. In fact most of the time not all the bunches injected in the machine have the same number of protons, but also sometimes exists some bunch with large emittancy that should be excluded from the list.

The list of good bunches is written in a ring buffer that recall each orbit the accept strobe (fork) in synch with all the requested collisions. In this way we can enable the trigger on the requested bunch structure.

Each trigger is put in coincidence with the fork singularly and then prescaled. The final trigger is simply the or of all the enabled triggers.

2.2. *The trigger DAQ frame*

The trigger system is generating a lot of information any time a level one accept is asserted.

The trigger pattern tells the DAQ which is the source of that event, but also the bunch number is necessary to understand which bunch originates the collision, in the bunch list are added bunches that are not colliding in IP5 to study the beam gas background that require mandatory the knowledge of such information. The orbit number is necessary to follow the time structure of the data taking.

The trigger throttling system of Totem is based on the busy signals that are generated by the DAQ buffers. Any time a buffer gets almost full the Busy is asserted, The LTC start blocking the incoming triggers waiting that the system gets idle again. A counter of the generated triggers is added to count the number of blocked triggers by subtracting the event number counter.

All this information is packed into a frame that is read by the DAQ as any other subdetector and is added to the event record.

3. Conclusions

The Totem trigger system is evolving with the knowledge of the background topology but also has the capability to be reprogrammed any time the machine delivers different collision schemes and special optics. Upgrades are foreseen to allow more flexibility in case of LHC filling schemes that have bunches with different populations, upon request each physics channel can require different bunch lists.

References

1. The TOTEM Experiment at the CERN Large Hadron Collider, The TOTEM Collaboration, G Anelli et al 2008 JINST 3 S08007.
2. P.Aspell, G.Anelli , P.Chalmet , J.Kaplon , K.Kloukinas , H.Mugnier , W.Snoeys. "VFAT2 : A front-end system on chip providing fast trigger information, digitized data storage and formatting for the charge sensitive readout of multi-channel silicon and gas particle detectors" Topical Workshop on Electronics for Particle Physics, Prague, Czech Republic, 03 - 07 Sep 2007, pp.292-296.
3. G. Ruggiero et al., Planar edgeless silicon detectors for the TOTEM experiment, IEEE Trans. Nucl. Sci. 52 (2005) 1899..
4. S.Lami, G. Latino, E. Oliveri, L. Ropelewsky, N. Turini "A triple-GEM telescope for the TOTEM experiment" Nuclear Physics B - Proceedings Supplements, Volume 172, October 2007, Pages 231-233.
5. S. Minutoli, TOTEM T1 electronics system, Presented at the TOTEM T1 CSC engineering design review, 7 March 2006, CERN,Switzerland, http://indico.cern.ch/conferenceDisplay.py?confId=a061338.

LIST OF PARTICIPANTS AND CONFERENCE ORGANIZERS

ABBANEO	Duccio	CERN	Switzerland
AHN	Sang Un	Universitè Blaise Pascal and Konkuk University	France
ALESSANDRO	Bruno	INFN sez. Torino	Italy
ANTONELLI	Matias	Sincrotrone Trieste	Italy
ANTONELLI	Vito	Physics Department Milano University and I.N.F.N. Milano	Italy
ARAMO	Carla	INFN	Italy
AWES	Terry	Oak Ridge National Laboratory	USA
BACCARO	Stefania	ENEA	Italy
BALDINI	Luca	INFN-Sezione di Pisa	Italy
BARBERIS	Emanuela	Northeastern University	USA
BARDEEN	Marjorie	Fermilab	USA
BARILE	Francesco	Università degli studi di Bari and INFN Bari	Italy
BARILLARI	Teresa	Max-Planck-Inst. fuer Physik Muenchen	Germany
BARNEY	David	CERN	Switzerland
BASTIA	Paolo	Thales Alenia Space	Italy
BATES	Richard	The University of Glasgow	UK
BATTILANA	Carlo	CIEMAT	Switzerland
BELLERIVE	Alain	Carleton University	Canada
BERDERMANN	Jens	DESY	Germany
BERRETTI	Mirko	Pisa INFN	Italy
BERTAINA	Mario Edoardo	INFN, Torino	Italy
BERTUCCIO	Giuseppe	Politecnico di Milano	Italy
BINKO	Pavel	ISDC — University of Geneva	Switzerland
BOBIK	Pavol	Institute of Experimental Physics, SAS	Slovakia
BODINE	Laura	University of Washington	USA
BONGI	Massimo	Istituto Nazionale di Fisica Nucleare INFN — Sezione di Firenze	Italy
BOSCHINI	Matteo	INFN Milano Bicocca	Italy
BOSCH-RAMON	Valenti	Dublin Institute for Advanced Studies	Ireland
BOTTACINI	Eugenio	Stanford University	USA

BOULAY	Mark	Queen's University	Canada
BRETZ	Thomas	EPF Lausanne	Switzerland
BRONNER	Christophe	Laboratoire Leprince Ringuet — Ecole Polytechnique	France
BROWN	Ethan	University of Muenster	Germany
BUCCI	Francesca	NFN Firenze	Italy
BURGER	William	Università di Perugia	Italy
CACCIA	Massimo	Università dell'Insubria/INFN Milano	Italy
CARRILLO	Camilo	INFN Naples	Italy
CARVALHO	Joao	LIP	Portugal
CASTELLINA	Antonella	INFN-Sezione di Torino	Italy
CECIRE	Kenneth	University of Notre Dame	USA
CEMMI	Alessia	ENEA	Italy
CERDENO	David G.	Instituto de Fisica Teorica/ Universidad Autonoma de Madrid	Spain
CERVERA	Anselmo	IFIC — CSIC and Universidad de Valencia	Spain
CILMO	Marco	INFN and Università di Napoli Federico II	Italy
CIRELLI	Marco	CERN and CNRS CEA Saclay	Switzerland
CITTERIO	Mauro	INFN — Milano	Italy
COLAFRANCESCHI	Stefano	CERN	Switzerland
CONIAVITIS	Elias	University of Oxford	Switzerland
CONSOLANDI	Cristina	INFN Milano Bicocca	Italy
CONTIN	Andrea	University of Bologna	Italy
COSTANZA	Susanna	INFN — sezione di Pavia	Italy
CRISTOFARI	Pierre	APC CNRS	France
CUOCO	Alessandro	Stockholm University — Oskar Klein Center	Sweden
DAMMER	Jiri	Institute of Experimental and Applied Physics, Czech Technical University in Prague	Czech Republic
DE BARBARO	Pawel	University of Rochester	France
DE SIMONE	Nicola	University and INFN of Rome Tor Vergata	Italy
DECERPRIT	Guillaume	DESY	Germany
DELAQUIS	Sèbastien Claude	LHEP University of Bern	Switzerland
DELLA TORRE	Stefano	INFN Milano-Bicocca	Italy
DELL'ANNA	Massimiliano	Università degli Studi di Genova	Italy
DONATO	Fiorenza	University and INFN Torino	Italy
D'URSO	Domenico	Department of Physics, University of Perugia and INFN	Italy
EBER	Robert	Karlsruhe Institute of Technology	Germany
ERNENWEIN	Jean-Pierre	Aix-Marseille University	France
FAVA	Angela	I.N.F.N. — Padova	Italy
FAVATA	Fabio	ESA	France

FERELLA	Alfredo Davide	University of Zurich UZH	Italy
FERREIRA	Stefan	North-West University	South Africa
FICHTNER	Horst	Ruhr-University Bochum	Germany
FORTY	Roger	CERN	Switzerland
GABICI	Stefano	Laboratoire APC	France
GAGGERO	Daniele	University of Pisa and INFN Pisa	Italy
GARAI	Baishali	Indian Institute of Science	India
GAROFOLI	Justin	Syracuse University	France
GASTALDI	Ugo	INFN-LNL	Italy
GEBAUER	Iris	Karlsruhe Institute for Technology	Germany
GERVASI	Massimo	INFN and University of Milano Bicocca	Italy
GHEZZI	Alessio	Universita' degli Studi di Milano Bicocca e INFN Milano Bicocca	Italy
GIANI	Simone	CERN	Switzerland
GIANOTTI	Paola	INFN — LNF	Italy
GIGLIETTO	Nicola	INFN-bari	Italy
GIORDANO	Raffaele	Universitaà degli Studi di Bari and INFN Sez. bari	Italy
GIORDANO	Francesco	INFN Sezione di Napoli	Italy
GOETT	Johnny	INFN Gran Sasso	Italy
GOLDFARB	Steven	University of Michigan	Switzerland
GRANDI	Davide	INFN — Milano Bicocca	Italy
GRONDIN	Marie-Helene	Max-Planck-Institut für Kernphysik and Landessternwarte	Germany
GROSSE-KNETTER	Joern	II. Physikalisches Institut, Uni Goettingen	Germany
GULMEZ	Erhan	Bogazici University	Turkey
GUSEV	Konstantin	Joint Institute for Nuclear Research, Dubna, Russia/Technische Universitt Mnchen, Munich, Germany/National Research Centre 'Kurchatov Institute', Moscow, Russia	Russian Federation
GUZIK	T. Gregory	Louisiana State University	USA
HAAS	Daniel	SRON Netherlands Institute for Space Research	Netherlands
HAUF	Steffen	TU Darmstadt	Germany
HEEGER	Karsten	University of Wisconsin	USA
HEIDRICH	Nadine	University of Hamburg	Germany
HEINDL	Stefan	Institut für Experimentelle Kernphysik, Karlsruhe Institute of Technology	Germany
HEKTOR	Andi	CERN	Switzerland
HENRIQUES	Ana	CERN	Switzerland
HSU	Ching-Cheng	NIKHEF	The Netherlands
ILIA	Britvitch	Paul Scherrer Institut	Switzerland
INCE	Tayfun	University of Bonn	Switzerland

IVANOV	Marian	GSI Darmstadt	Germany
IVANOV	Mikhail	Central Institute of Aviation Motors	Russian Federation
JAEGER	Andreas	Physikalisches Institut, University of Heidelberg	Germany
JANSEN	Frank	DLR Institute of Space Systems	Germany
JEZ	Pavel	Niels Bohr Institute	Denmark
KANNIKE	Kristjan	NICPB	Estonia
KAUSSEN	Gordon	Institut fuer Experimentalphysik, Hamburg University	Germany
KELLERMANN	Hanna	Max-Planck Institute for Physics	Germany
KHLOPOV	Maxim	APC Laboratory	France
KISLAT	Fabian	Deutsches Elektronensynchrotron	Germany
KNOTIG	Max	Max-Planck Institute for Physics	Germany
KORPAR	Samo	University of Maribor and J. Stefan Institute	Slovenia
KOTERA	Kumiko	Caltech	USA
KOTYNIA	Anna	GSI	Germany
KOZLOV	Valentin	Karslruhe Institute of Technology	Germany
KRESLO	Igor	LHEP, Uni-Bern	Switzerland
KUDELA	Karel	IEP SAS, Kosice, Slovakia	Slovakia
KUKHTIN	Victor	Joint Institute for Nuclear Research	Russian Federation
LACUESTA	Vicente	IFIC	Spain
LANG	Karol	The University of Texas at Austin	USA
LEFLAT	Alexander	CERN	Switzerland
LEMOINE-GOUMARD	Marianne	CENBG — Université Bordeaux I	France
LEROY	Claude	University of Montreal	Canada
LIDVANSKY	Aleksandr	Institute for Nuclear Research, Russian Academy of Sciences	Russian Federation
LONGO	Renata	University of Trieste	Italy
MANEIRA	Jose	LIP-Lisboa	Portugal
MARAWAR	Ravi	National Instruments	USA
MARCOWITH	Alexandre	Laboratoire Univers et Particules de Montpellier-LUPM-	France
MARKI	Raphael	EPFL Ecole Polytechnique Fédérale de Lausanne	Switzerland
MARSELLA	Giovanni	Università del Salento and INFN	Italy
MARTISIKOVA	Maria	German Cancer Research Center	Germany
MASERA	Massimo	Dipartimento di Fisica Sperimentale dell'Universita' di Torino e INFN	Italy
MASIK	Jiri	University of Manchester	Switzerland
MATTANA	Fabio	APC — Centre Franois Arago	France
MATTONE	Cristina	C.A.E.N. SPA	Italia
MAVROMATOS	Nikolaos	Dept. of Physics, King's College London, Strand, London WC2R 2LS, UK	UK
MAZZAFERRO	Luca	INFN Sezione Roma Tor Vergata	Italy
McCAULEY	Thomas	Fermilab	Switzerland

MENNELLA	Aniello	Università degli Studi di Milano — Dipartimento di Fisica	Italy
MEREGAGLIA	Anselmo	IPHC	France
MERLO	Jean-Pierre	Unervisity of Iowa	Suisse
MERSI	Stefano	CERN	Switzerland
MESSINEO	Alberto	Università di Pisa and INFN sez. di Pisa	Italy
MILIANO	Simone	INFN Milano-Bicocca	Italy
MINANO	Mercedes	IFIC — Instituto de Física Corpuscular	Spain
MINI	Giuliano	C.A.E.N. SPA	Italy
MITSOU	Vasiliki	Instituto de Fisica Corpuscular, CSIC and Univ Valencia	Spain
MOCCHIUTTI	Emiliano	INFN	Italy
MOLNAR	Levente	CERN	Switzerland
MONTAGNA	Paolo	Dip. Fisica Nucleare e Teorica Univ. Pavia — INFN Sezione di Pavia	Italy
MONTE	Claudia	INFN Bari	Italy
MORICI	Andrea	ENEA	Italy
NAGEL	Martin	Max-Planck-Institute for Physics	Germany
NAPPI	Eugenio	INFN	Italy
NGUYEN	Nelly	Institute for Experimental Physics, University of Hamburg	Germany
NOVA	Federico	University of Texas at Austin	USA
NOVENTA	Francesco	INFN Milano-Bicocca	Italy
NOVGORODOVA	Olga	DESY	Germany
NOVOSELTSEV	Yury	Institute for Nuclear Research of RAS	Russian Federation
OLIVEIRA	Carlos	i3N — University of Aveiro	Portugal
ORR	Robert	University of Toronto	Canada
PAOLI	Nicola	CAEN SpA	Italy
PAPA	Francesca	Università degli Studi di Milano Bicocca	Italy
PAPOTTI	Giulia	CERN	Switzerland
PARAMATTI	Riccardo	INFN — Roma1	Italy
PARESCHI	Giovanni	INAF-Osservatorio Astronomico di Brera	Italy
PARTRIDGE	Richard	SLAC National Accelerator Center	USA
PEARCE	Mark	KTH	Sweden
PEREVALOV	Denis	Fermilab	USA
PESARESI	Mark	Imperial College, London	UK
PIA	Maria Grazia	INFN Sezione di Genova	Italy
PIANDANI	Roberto	University of Perugia and INFN	Italy
PIEMONTESE	Livio	INFN Sez. Ferrara Send the invoice to INFN Sez. Milano 1	Italy
PINFOLD	James	University of ALberta	Canada
PIZZICHEMI	Marco	University of Milano-Bicocca	Italy
POSPISIL	Stanislav	Institute of Experimental and Applied Physics, Czech Technical University in Prague	Czech Republic

POTGIETER	Marthinus	North-West University, Potchefstroom	South Africa
POZZOBON	Nicola	Dipartimento di Fisica, Università degli Studi di Padova e INFN — Sezione di Padova	Italy
PRALL	Dr. Matthias	Institut fuer Kernphysik/ University of Muenster	Germany
PRICE	Lawrence	Argonne National Laboratory	USA
PROCZ	Szymon	Freiburger Materialforschungszentrum FMF	Germany
PUGLISI	Donatella	Politecnico di Milano	Italy
RAINO'	Silvia	INFN — Sezione di Bari	Italy
RANCOITA	Pier Giorgio	INFN sez. di Milano-Bicocca	Italy
RANDO	Riccardo	University and INFN — Padova	Italy
RANDRIAN-ARIVONY	Kolonia	Carleton University	Canada
RAY	Heather	University of Florida	USA
RAZETO	Alessandro	INFN — Laboratori Nazionali del Gran Sasso	Italy
RIVA	Marco	INFN Milano	Italy
ROMANO	Angela	University of Birmingham	Switzerland
ROZZA	Davide	University of Milano — Bicocca	Italy
RUCHTI	Randy	University of Notre Dame	USA
SALAMON	Andrea	INFN Sezione Roma Tor Vergata	Italy
SAOUTER	Pierre	University of Geneva	France
SARRA	Ivano	laboratori nazionali di frascati INFN	Italy
SCHERER	Klaus	Institut für Theoretische Physik Lehrstuhl IV: Weltraum — und Astrophysik Ruhr-Universität Bochum	Germany
SCHILLER	Stephan	Heinrich-Heine-University Düsseldorf	Germany
SCHOO-RLEMMER	Harm	IMAPP, Radboud University Nijmegen	Netherlands
SEGUINOT	Jacques	College de France	Switzerland
SEO	Seon-Hee	Stockholm University	Sweden
SEQUEIROS	Juan	UNIVERSIDAD DE ALCALA	Spain
SHOEMAKER	David	Massachusetts Institute of Technology	USA
SIMON	Manfred	University of Siegen	Germany
SINITSYNA	V. Georgievna	P.N. Lebedev Physical Institute RAS	Russian Federation
SINITSYNA	Vera Yurievna	P.N. Lebedev Physical Institute RAS	Russian Federation
SLOAN	Terry	Lancaster University	UK
SOLODKOV	Alexander	Institute for High Energy Physics	Russian Federation
SOLUK	Richard	University of Alberta	Canada
SONNEN-SCHEIN	Andrew	Fermilab	USA

SPARKES	Ailsa	University of Edinburgh	UK
SPATARO	Stefano	Università di Torino — INFN	Italy
STORACI	Barbara	Nikhef	France
STRAUSS	Du Toit	Centre for Space Research	South Africa
STRIZENEC	Pavol	CERN	Switzerland
TACCONI	Mauro	INFN Milano Bicocca	Italy
TAKAI	Helio	Brookhaven National Laboratory	USA
TAKEUCHI	Yasuo	Dept. of Physics, Grad. School of Science, Kobe University	Japan
TARONI	Silvia	Università e INFN Perugia	Italy
TESSAROTTO	Fulvio	INFN — Trieste	Italy
TEYSSIER	Cecile	University of Montreal	France
THOMPSON	Joshua	Cornell University	France
THOMSON	Gordon	Department of Physics and Astronomy, University of Utah	USA
TIBOLLA	Omar	ITPA, Würzburg University	Germany
TOBIN	Mark	Physik-Institut der Universität Zürich	Switzerland
TOGO	Vincent	INFN — SEZIONE DI BOLOGNA	Italy
TOMASETTI	Nicola	Università degli Studi di Perugia and INFN	Italy
TOMS	Konstantin	University of New Mexico	France
TREBERSPURG	Wolfgang	Institute of High Energy Physics Vienna	Austria
TREVES	Aldo	Cv	Italy
TRICOMI	Alessia	Università di Catania and INFN Catania	Italy
TSISKARIDZE	Vakhtang	Albert-Ludwigs Universität Freiburg	Germany
TURINI	Nicola	Universita' di Siena e INFN Pisa	Italy
URAS	Antonio	Institut de Physique Nucléaire de Lyon	France
VENTURA	Guglielmo	INFN, Section of Florence	Italy
VIVALDI	Franco	C.A.E.N. SPA	Italy
VOGEL	Adrian	Physikalisches Institut, University of Bonn	Germany
WEBER	Manuel	LHEP University of Bern	Switzerland
WEYDA	Frantisek	Faculty of Science, University of South Bohemia	Czech Republic
WICHOSKI	Ubi	Laurentian University	Canada
WINTZ	Peter	FZ Juelich	Germany
WRIGHT	Alex	Princeton University	USA
YOUNG	Charles	SLAC National Accelerator Laboratory	USA
ZELLER	Marcel	University of Bern LHEP	Switzerland
ZEUNER	Wolfram	CERN	Switzerland
ZIBELL	Andre	LMU Munich	Germany
ZIENTEK	Michal	CERN	Switzerland

www.ingramcontent.com/pod-product-compliance
Lightning Source LLC
Chambersburg PA
CBHW050533190326
41458CB00007B/1770

* 9 7 8 9 8 1 4 4 0 5 0 6 5 *